RESERVOIR DEVELOPMENT

Sustainable Oil and Gas Development Series

RESERVOIR DEVELOPMENT

M. RAFIQUL ISLAM

Gulf Professional Publishing
An imprint of Elsevier

Gulf Professional Publishing is an imprint of Elsevier
50 Hampshire Street, 5th Floor, Cambridge, MA 02139, United States
The Boulevard, Langford Lane, Kidlington, Oxford, OX5 1GB, United Kingdom

Copyright © 2022 Elsevier Inc. All rights reserved.

No part of this publication may be reproduced or transmitted in any form or by any means, electronic or mechanical, including photocopying, recording, or any information storage and retrieval system, without permission in writing from the publisher. Details on how to seek permission, further information about the Publisher's permissions policies and our arrangements with organizations such as the Copyright Clearance Center and the Copyright Licensing Agency, can be found at our website: www.elsevier.com/permissions.

This book and the individual contributions contained in it are protected under copyright by the Publisher (other than as may be noted herein).

Notices

Knowledge and best practice in this field are constantly changing. As new research and experience broaden our understanding, changes in research methods, professional practices, or medical treatment may become necessary.

Practitioners and researchers must always rely on their own experience and knowledge in evaluating and using any information, methods, compounds, or experiments described herein. In using such information or methods they should be mindful of their own safety and the safety of others, including parties for whom they have a professional responsibility.

To the fullest extent of the law, neither the Publisher nor the authors, contributors, or editors, assume any liability for any injury and/or damage to persons or property as a matter of products liability, negligence or otherwise, or from any use or operation of any methods, products, instructions, or ideas contained in the material herein.

Library of Congress Cataloging-in-Publication Data
A catalog record for this book is available from the Library of Congress

British Library Cataloguing-in-Publication Data
A catalogue record for this book is available from the British Library

ISBN: 978-0-12-820053-7

For information on all Gulf Professional publications
visit our website at https://www.elsevier.com/books-and-journals

Publisher: Charlotte Cockle
Senior Acquisitions Editor: Katie Hammon
Editorial Project Manager: Leticia Lima
Production Project Manager: Manju Thirumalaivasan
Cover designer: Christian Bilbow

Typeset by STRAIVE, India

Contents

Foreword vii

1. Introduction 1
1.1 Opening statement 1
1.2 World energy 2
1.3 Role of oil and gas 14
1.4 Key events and future outlook of oil and gas 27
1.5 Sustainability status of current technologies 38
1.6 Summary of various chapters 50

2. Reservoir rock and fluid characterization 53
2.1 Introduction 53
2.2 Unique features 61
2.3 Sustainability criteria 67
2.4 Fluid characterization based on origin 72
2.5 Abiogenic petroleum origin theory 79
2.6 Scientific ranking of petroleum 106
2.7 Characterization of reservoirs 117
2.8 Reservoir heterogeneity 178

3. Complex reservoirs 185
3.1 Introduction 185
3.2 Complex reservoirs 186
3.3 Fracture mechanics in geological scale 190
3.4 Core analysis 214
3.5 Modeling unstable flow 221
3.6 Essence of reservoir simulation 232
3.7 Material balance equation 233
3.8 Representative elemental volume, REV 236
3.9 Thermal stress 238
3.10 Reservoir geochemistry 251
3.11 Fluid and rock properties 264
3.12 Avoiding spurious solutions 265

4. Unconventional reservoirs 267
4.1 Introduction 267
4.2 Unconventional oil and gas production 271
4.3 Current potentials 275
4.4 Sustainable development of unconventional reservoirs 337
4.5 Improving oil and gas recovery from unconventional reservoirs 339
4.6 Reserve growth potential 437
4.7 Quantitative measures of well production variability 461
4.8 Coupled fluid flow and geomechanical stress model 465
4.9 Sustainability pathways 492
4.10 Zero-waste operations 520
4.11 Greening of hydraulic fracturing 523

5. Basement reservoirs 533
5.1 Introduction 533
5.2 World reserve 534
5.3 Reservoir characterization 548
5.4 Organic source of hydrocarbon 576
5.5 Nonconventional sources of petroleum fluids 590
5.6 Scientific ranking of petroleum 597

6. Reserves prediction and deliverability 609
6.1 Introduction 609
6.2 Conventional material balance 615
6.3 Analytical solutions 622
6.4 Inclusion of fluid memory 637
6.5 Anomalous diffusion: A memory application 649
6.6 Results and discussion 662
6.7 The compositional simulator using engineering approach 673

7. Field guidelines 737

7.1 Introduction 738
7.2 Scaling guidelines 738
7.3 Planning with reservoir simulators 770
7.4 Uncertainty analysis 794
7.5 The prediction uncertainty 800
7.6 Recent advances in reservoir simulation 800
7.7 Real-time monitoring 814
7.8 Sustainability analysis of a zero-waste design 817
7.9 Global efficiency calculations 831

8. Conclusions 845

8.1 Facts and fictions 845
8.2 Conclusions 846
8.3 Recommendations 848

References 851

Index 899

Foreword

Petroleum fluids are natural and have been used in harmony with nature from the dawn of civilization. Sustainable petroleum development is inherently logical, yet for decades, the phrase "sustainable petroleum" has become an oxymoron. Scientists are talking about "carbon-free energy"—a scientific oxymoron as if it is a viable alternative to today's technological disaster. In the name of "science," there has been a growing trend of dogmatic solutions forced on the world by the ruling elite. This is not new. Nearly a century ago, a prime scholarly organization, *Nature*, wrote,

> UNLIKE most problems concerning origins, which have but a philosophic, or academic interest, that of the genesis of petroleum has a distinctly practical significance, for if solved, prospectors for mineral oil would be provided with important data and chemists might learn how to produce artificially valuable substances similar to, if not identical with, natural petroleum. Man's fertile imagination has spun not only an embarrassing number of speculations and hypotheses concerning the nature of the raw material or materials from which petroleum has been derived, but also innumerable explanations of the modus operandi of its formation.

Note how this 100-year-old article spins one out of the research trajectory by defining the only "real" intention behind any research, and then leads one toward "settled science." And, what is "settled science"? It is whatever process that is most accepted by scientists, each of whom can be bought. Among the vast majority of the "scientific" world, there is a natural tendency to mock anyone advancing any argument against the so-called "settled science," irrespective of the logicality of the argument. This is then followed by anything the "scientist" would say about anything as "fact," no matter how egregious. Be it manufacturing cow-free burgers and milk or dimming the sun with toxic chemicals would pass for "science," while anyone advancing an "alternate" explanation would be ridiculed. This is not a scholarly forum where real science can survive.[a] As such, this book series on sustainable petroleum development is a remarkably courageous undertaking. It is no surprise that this book starts challenging the sustainability criterion that in itself has seen a yo-yo motion with dozens of definitions floating around. This book is founded on logical discourse—something that has been missing from new science. The readership must be reminded that it is new science that has made the following transition in the past and is poised to continue along the same path.

---

[a] Kraychik, R., 2019, Greenpeace Founder: Global Warming Hoax Pushed by Corrupt Scientists 'Hooked on Government Grants', *Breitbart*. March 7

This book rises above new science rhetoric and prophecies of the profiteers and brings back real science to show how petroleum reservoir engineering can become sustainable. Anyone familiar with petroleum operations will appreciate the breadth and depth of this book.

G.V. Chilingarian
University of Southern California

CHAPTER 1

Introduction

1.1 Opening statement

The evolution of human civilization is synonymous with how it meets its energy needs. Some may argue that the human race has become progressively more civilized with time. Yet, for the first time in human history, an energy crisis has seized the entire globe and the very sustainability of this civilization itself has suddenly come into question. If humanity has actually progressed as a species, it must exhibit, as part of its basis, some evidence that overall efficiency in energy consumption has improved. In terms of energy consumption, this would mean that less energy is required per capita to sustain life today than, say, 50 years earlier. However, exactly the opposite has happened. The oil price has been tumbled into negative territory. For the oil business, which is used to roller coaster rides in prices, this was a new low and highlights the need for a paradigm shift in managing this valuable resource, which is the driver of modern civilization. The scenario has become more complex by invoking 'climate change hysteria', which is not based on science (Islam and Khan, 2019). With the increasing politicization of fossil fuels and the ensuing global 'climate emergency' agenda, the original thrust of sustainability is all but abandoned, as if petroleum resources cannot be developed sustainably. This book on Sustainable Petroleum development is all about introducing technologies that would make petroleum operations sustainable. Islam and Hossain, 2020 presented such technologies for sustainable drilling, whereas this volume is dedicated to presenting technologies for sustainable reservoir development.

A true paradigm shift can be invoked by introducing zero-waste engineering into all aspects of petroleum resource development. The zero-waste mode assures that the proposed recovery mode is sustainable under all scenarios of oil prices. In recent years, Islam and his research group (Islam et al., 2010; Islam and Khan, 2019) have demonstrated that current practices of oil and gas production operations are not sustainable. The principal impediment to sustainability is the introduction of synthetic chemicals, which are introduced at various levels of oil and gas production. Previously, it has been demonstrated that conventional 'renewable' technologies are less sustainable than conventional oil recovery and processing schemes (Islam et al., 2018). In reservoir evaluation and reserve assessment, century-old technologies are used. Fundamental mathematical formulas and scientific descriptions of oil and

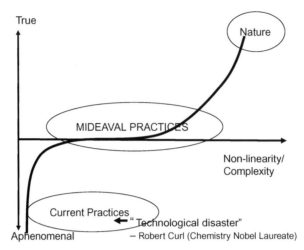

FIG. 1.1 Schematic showing the position of current technological practices related to natural practices.

gas reservoirs have not been updated despite the advent of improved mathematical tools, reservoirs and more accurate scientific models to describe fluid flow through porous media.

Fig. 1.1 shows that current technological practices are focused on short term, linearized solutions that are also inherently unsustainable. As a result, technological disaster prevailed in practically every aspect of the post-renaissance era. Petroleum practices are considered to be the driver of today's society. Here, modern development is essentially dependent on artificial products and processes. We have reviewed the post renaissance transition, calling it the honey-sugar-saccharine-aspartame (HSSA) degradation (Khan and Islam, 2016). In this allegorical transition, honey (with a real source and process) has been systematically replaced by Aspartame, which has both a source and a pathway that are highly artificial. This sets in motion the technology development mode that Nobel Laureate in Chemistry-Robert Curl called "technological disaster."

Sustainable petroleum operations development requires a sustainable supply of clean and affordable energy resources that do not cause negative environmental, economic, and social consequences. In addition, it should consider a holistic approach where the whole system will be considered instead of just one sector at a time (Islam et al., 2010). In 2007, our research group developed an innovative criterion for achieving true sustainability in technological development (described in Islam, 2020). New technology should have the potential to be efficient and functional far into the future in order to ensure true sustainability. Sustainable development is seen as having four elements: economic, social, environmental, and technological.

1.2 World energy

Human civilization is synonymous with carbon-based fuel. The use of fossil fuels for energy began at the onset of the Industrial Revolution. In the beginning, coal was the fuel of choice. Shortly before the introduction of oil and gas as the fossil fuel of convenience, the scarcity of coal in the coming decades was drummed up (Zatzman, 2012). In this, coal offered a peculiar distinction. As pointed out by Clark and Jacks (2007), despite enormous increases in output, the coal industry was credited with little of the national productivity advance either directly, or indirectly through linkages to steam power, metallurgy, or railways. The 'cliometric' account of coal in the Industrial Revolution is represented in Fig. 1.2. The horizontal axis shows cumulative output since the beginning of extraction in the northeast coal field, and the vertical axis shows the hypothetical real cost of extraction per ton, which rises slowly as total extraction increases. However, real extraction costs are only moderately higher at the cumulative output of the 1860s than at the cumulative output of the 1700s. In this portrayal, the supply of coal is elastic. When demand increased, so did output, with little increase in price at the pithead. But the same expansion of output could have occurred earlier or later had demand conditions been appropriate. The movement outward in the rate of extraction was caused by the growth in the population and incomes and incomes, and by improvements in transport and reductions in taxes which reduced the wedge between pithead prices and prices to the final consumers.

Although the alarm that the world would run out of coal was sounded over 100 years ago, this has not been the case. It is true all around the world, but particularly meaningful in the USA, which achieved energy independence only in recent years.

In 1975, the U.S. Geological Survey (USGS) published the most comprehensive national assessment of U.S. coal resources, which indicated that, as of January 1, 1974, coal resources in the United States totaled 4 trillion short tons. Although the USGS has conducted more recent regional assessments of U.S. coal resources, a new national-level assessment of U.S. coal resources has not been conducted. His best estimates are published by the U.S. Energy Information Administration (EIA), which publishes three measures of how much coal is left in the United States. The measures are based on various degrees of geologic certainty and on the economic feasibility of mining coal.

The EIA's estimates of the amount of coal reserves as of January 1, 2020, by type of reserve

- The Demonstrated Reserve Base (DRB) is the sum of coal in both measured and indicated resource categories of reliability. The DRB represents 100% of the in-place coal that could

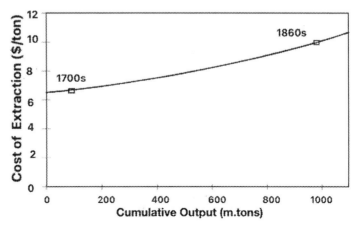

FIG. 1.2 The cliometric account of the coal industry in the Industrial Revolution. *From Clark, G., Jacks, D., 2007. Coal and the industrial revolution, 1700–1869. Eur. Rev. Econ. Hist. 11, 39–72.*

be mined commercially at a given time. The EIA estimates the DRB at about 473 billion short tons, of which about 69% is underground mineable coal.
- Estimated recoverable reserves include only the coal that can be mined with today's mining technology after considering accessibility constraints and recovery factors. The EIA estimates U.S. recoverable coal reserves at about 252 billion short tons, of which about 58% is underground mineable coal.
- Recoverable reserves at producing mines are the amount of recoverable reserves that coal mining companies report to the EIA for their U.S. coal mines that produce more than 25,000 short tons of coal in a year. The EIA estimates these reserves at about 14 billion short tons of recoverable reserves, of which 60% is surface mineable coal.

Fig. 1.3 shows US coal reserves in 2019. Based on U.S. coal production in 2019, of about 0.706 billion short tons, the recoverable coal reserves would last about 357 years, and recoverable reserves at producing mines would last about 20 years. The actual number of years that those reserves will last depends on changes in production and reserve estimates.

Six states had 77% of the *demonstrated reserve base* (DRB) of coal as of January 1, 2020:

- Montana—25%
- Illinois—22%
- Wyoming—12%
- West Virginia—6%
- Kentucky—6%
- Pennsylvania—5%

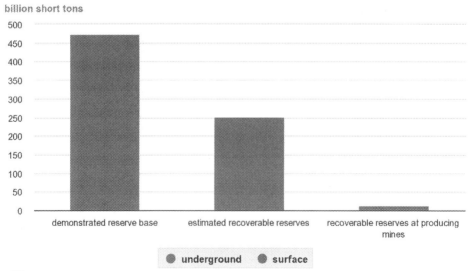

FIG. 1.3 US coal reserves. *From EIA, 2020.*

Twenty five other states had the remaining 23% of the DRB.

U.S. Coal Resource Regions

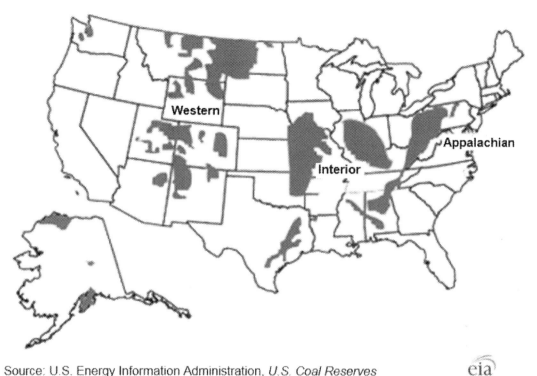

Source: U.S. Energy Information Administration, *U.S. Coal Reserves*

In terms of world reserves, as of December 31, 2016, EIA estimates of total world proved recoverable reserves of coal were about 1144 billion short tons (or about 1.14 trillion short tons), and five countries had about 75% of the world's proved coal reserves. *The top five countries and their share of world proved coal reserves as of 12/31/2016*

- United States—22%
- Russia—15%
- Australia—14%
- China—13%
- India—10%

Table 1.1 shows fuel consumption in the world for the year 2019. This trend shows the impact of the Trump era in the USA.

Growth in energy markets slowed in 2019 in line with weaker economic growth and a partial unwinding of some of the one-off factors that boosted energy demand in 2018. This slowdown was particularly evident in the US, Russia and India, each of which exhibited unusually strong growth in 2018. China was the exception, with its energy consumption accelerating in

TABLE 1.1 Fuel shares of primary energy and contributions to growth in 2019.

Energy source	Consumption (exajoules)	Annual change (exajoules)	Share of primary energy (%)	Percentage point change in share from 2018
Oil	193.0	1.6	33.1	−0.2
Gas	141.5	2.8	24.2	0.2
Coal	157.9	−0.9	27.0	−0.5
Renewables[a]	29.0	3.2	5.0	0.5
Hydro	37.6	0.3	6.4	−0.0
Nuclear	24.9	0.8	4.3	0.1
Total	583.9	7.7		

[a] Renewable power (excluding hydro) plus biofuels.
From BP, 2020. BP Annual Report. https://www.bp.com/content/dam/bp/business-sites/en/global/corporate/pdfs/investors/bp-annual-report-and-form-20f-2020.pdf.

2019. As a result, China dominated the expansion of global energy markets—contributing the largest increment to demand for each individual source of energy other than natural gas, where it was only narrowly surpassed by the US. Despite the support from China, all fuels (other than nuclear) grew at a slower rate than their 10-year averages, with coal consumption declining for the fourth time in 6 years. Nevertheless, renewables still grew by a record increment and provided the largest contribution (41%) to growth in primary energy, with the level of renewable power generation exceeding nuclear power for the first time. The slowdown in energy demand growth, combined with a shift in the fuel mix away from coal and toward natural gas and renewables, led to a significant slowing in the growth of carbon emissions, although only partially unwinding the unusually strong increase seen in 2018. Energy prices fell on the whole, particularly for coal and gas where growth in production outpaced consumption leading to a build up of inventories. Oil prices were a little lower.

Ever since the oil embargo of 1972, the world has been gripped by the fear of an 'energy crisis'. U.S. President Jimmy Carter, in 1978, told the world in a televised speech that the world was in fact running out of oil at a rapid pace—a popular Peak Oil theory of the time—and that the US had to wean itself off of the commodity. Since the day of that speech, worldwide oil output has actually increased by more than 30%, and known available reserves are higher than they were at that time. This hysteria has survived the era of Reaganomics, President Clinton's cold war dividend, President G.W. Bush's post-9-11 era of 'fearing everything but petroleum' and today even the most ardent supporters of the petroleum industry have been convinced that there is an energy crisis looming and that it is only a matter of time before we will be forced to switch to no-petroleum energy sources. During President Obama's time, there had been a marked shift toward so-called renewable energy and the background of 'only a carbon tax can fix the climate change debacle' mantra was firmly established. President Trump has strived to undo much of those biases away from petroleum resources, but the scientific community remains unconvinced. In this chapter, we deconstruct some of the hysteria and unscientific bias that have gripped the scientific community as well as the left leaning segment of the general public.

The general public is being prepared to face an energy crisis that is perceived to be forthcoming. Since the demand for oil is unlikely to decline it inevitably means that the price will increase, probably quite dramatically. This crisis attributed to peak oil theory is proposed to be remedied with (1) austerity measures in order to decrease dependence on energy, possibly decreasing per capita energy consumption, and (2) alternatives to fossil fuels. None of these measures seem appealing because any austerity measure can induce an imbalance in the economic system that is dependent on the spending habits of the population and any alternative energy source may prove to be more expensive than fossil fuel. These concerns create panic, which is beneficial to certain energy industries, including biofuel, nuclear, wind, and others. Add to this problem is the recent hysteria created based on the premise that oil consumption is the reason behind global warming. This in itself has created opportunities for many sectors engaged in carbon sequestration.

In general, there has been a perception that solar, wind and other forms of 'renewable' energy are more sustainable or less harmful to the environment than their petroleum counterparts. It is stated that renewable energy is energy that is collected from renewable resources that are naturally replenished on a human timescale, such as sunlight, wind, rain, tides, waves, and geothermal heat. Chhetri and Islam (2008) have demonstrated that the claim of harmlessness and absolute sustainability is not only exaggerated, it is not supported by science. However, irrespective of scientific research, this positive perception translated into global public support. One such survey was conducted by Ipsos Global in 2011 that found a very favorable rating for non-fossil fuel energy sources (Fig. 1.4). Perception does have economic implications attached to it. The Ipsos study found 75% agreed with the slogan "scientific research makes a direct contribution to economic growth in the UK". However, in the workshops, although participants agreed with this, they did not always understand the mechanisms through which science affects economic growth. There is strong support for the public funding of scientific research, with three-quarters (76%) agreeing that "even if it brings no immediate benefits, research which advances knowledge should be funded by the Government." Very few (15%) think that "Government funding for science should be cut because the money can be better spent elsewhere". This is inspite of public support for cutting Government spending overall. It is not any different in the USA, where perception

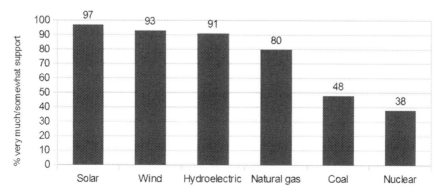

FIG. 1.4 Public perception toward energy sources (Ipsos, 2011).

translates directly into pressure on the legislative body, resulting in improved subsidies for certain activities.

The Energy Outlook considers a range of alternative scenarios to explore different aspects of the energy transition (Fig. 1.4). The scenarios have some common features, such as a significant increase in energy demand and a shift toward a lower carbon fuel mix, but differ in terms of particular policy or technology assumptions. In Fig. 4.2, the Evolving Transition (ET) scenario is a direct function of public perception that dictates government policies, technology and social preferences. Some scenarios focus on particular policies that affect specific fuels or technologies, e.g., a ban on sales of internal combustion engine (ICE) cars, a greater policy push toward renewable energy, or weaker policy support for a switch from coal to gas considered, e.g., faster and even faster transitions (Fig. 1.5).

In the mean time, it is predicted that the so-called decarbonization scheme is in full swing in favor of energy sources other than fossil fuels.[a] The aim is to reduce greenhouse gas (GHG) emissions dramatically, ignoring the fact that non-petroleum energy sources are no less toxic than petroleum emissions (Islam and Khan, 2019). BP (2018) predicts that electric vehicles will play a major role in lowering emissions from transport and boasts about providing a network of 6500 charging points across the UK, and plans to roll out ultra-fast charging on our forecourt network. The assumption in all these is that somehow electric cars are environmentally friendly. This is contrary to the scientific analysis conducted over a decade ago by Chhetri and Islam (2008), who showed that electric vehicles are far more toxic to the environment and far less efficient than regular vehicles, run on internal combustion engines. BP further predicts that by 2040, half of Europe's cars and one third of the world's vehicles will avoid having

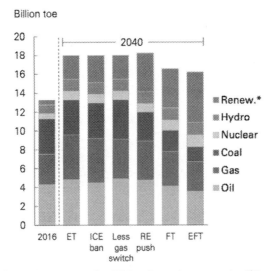

FIG. 1.5 Energy outlook for 2040 as compared to 2016 under various scenarios (*Renewables include wind, solar, geothermal, biomass, and biofuels, from BP Report, 2018).

---

[a] Even electric cars are considered to be a product of decarbonization irrespective of the source of electrical energy.

internal combustion engines (ICE). This process of 'decarbonization' is further accelerated by introducing electrification using 'renewables', hydrogen, e-fuel[b] and even nuclear energy. Clearly, the world stage is ready to accept even nuclear in favor of 'decarbonization'. For instance, Kann et al. (2019) indicated the nuclear option as the number one priority to cut down on GHG emissions.

Fig. 1.6 shows the growth of various energy sources. Only renewable energy made gains. Meanwhile CO_2 emissions declined. It is tempting to conclude that the decline is due to reduction in fossil fuel consumption. This is not scientific as each renewable technology ends up causing greater CO_2 emissions when the entire life cycle is considered.

In 2020, the Covid-19 pandemic played a significant role in shaping the global energy outlook. Because of some degree of lockdown in every country, the Covid-19 pandemic has caused more disruption to the energy sector than any other event in recent history. The 2020 IEA World Energy Outlook (WEO) report (IEA, 2020a,b,c) examined in detail the effects of the pandemic, and in particular how it affects the prospects for rapid clean energy transitions. As shown in Fig. 1.7, the IEA assessment is that global energy demand is set to drop by 5% in 2020, energy-related CO_2 emissions by 7%, and energy investment by 18%. The impacts vary by fuel. The estimated falls of 8% in oil demand and 7% in coal use stand in sharp contrast to a slight rise in the contribution of renewables. The reduction in natural gas demand is around 3%, while global electricity demand looks set to be down by a relatively modest 2% for the year. The global COVID-19 lockdowns caused fossil carbon dioxide emissions to decline by an estimated 2.4 billion tonnes in 2020 - a record drop according to researchers at Future Earth's Global Carbon Project (EurekAlert, 2020). The fall is considerably larger than previous significant decreases - 0.5 (in 1981 and 2009), 0.7 (1992), and 0.9 (1945) billion tonnes of CO_2 ($GtCO_2$). It means that in 2020 fossil CO_2 emissions are predicted to be approximately 34 $GtCO_2$, 7% lower than in 2019. Ironically, the release of 2020s Global Carbon Budget came just ahead of the fifth anniversary of the adoption of the UN Paris Climate Agreement, which aims to reduce the emission of greenhouse gases to limit global warming. Cuts of around 1 to 2 $GtCO_2$ are needed each year on average between 2020 and 2030 to limit climate change in line with its goals. However, the Trump administration was not on board with the agreement and relaxed many of the greenhouse gas emission regulations while achieving very desirable results.

Not surprisingly, emissions from transport account for the largest share of the global decrease. Those from surface transport, such as car journeys, fell by approximately half at the peak of the COVID-19 lockdowns. By December 2020, emissions from road transport and aviation were still below their 2019 levels, by approximately 10% and 40%, respectively, due to continuing restrictions. Total CO_2 emissions from human activities—from fossil CO_2 and land-use change—are set to be around 39 $GtCO_2$ in 2020.

The emissions decrease is notably more pronounced in the US (-12%) and EU27 countries (-11%), where COVID-19 restrictions accelerated previous reductions in emissions from coal use. It appears least pronounced in China (-1.7%), where the effect of COVID-19 restrictions on emissions occurred on top of rising emissions. In addition, restrictions in China occurred early in the year and were more limited in their duration, giving the economy more time to

---

[b] E-fuel involves synthetic fuel, made out of non-petroleum carbon and is branded as 'carbon-free' (Palmer, 2015).

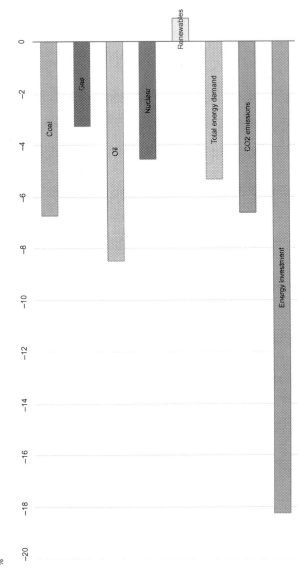

FIG. 1.6 Energy growth of various energy sources.

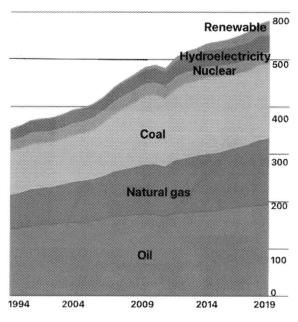

FIG. 1.7 Global energy consumption (BP, 2020), y-axis is energy in exajoules (EJ) (1.1 exajoule = 10^{18} joules).

recover. In the UK, which first introduced lockdown measures in March, emissions are projected to decrease by about 13%. The large decrease in UK emissions is due to the extensive lockdown restrictions and the second wave of the pandemic. In India, where fossil CO_2 emissions are projected to decrease by about 9%, emissions were already lower than normal in late 2019 because of economic turmoil and strong hydropower generation, and the COVID-19 effect is potentially superimposed on this changing trend.

Despite lower emissions in 2020, the level of CO_2 in the atmosphere continues to grow - by about 2.5 ppm (ppm) in 2020 - and is projected to reach 412 ppm average over the year, 48% above pre-industrial levels. Preliminary estimates based on fire emissions in deforestation areas indicate that emissions from deforestation and other land-use changes for 2020 are similar to the previous decade, at around 6 GtCO_2. Approximately 16 GtCO_2 was released, primarily from deforestation, while the uptake of CO_2 from regrowth on managed land, mainly after agricultural abandonment, was just under 11 GtCO_2. Islam and Khan (2019) explained how climate change activists have wrongly targeted fossil fuels as the primary cause of global warming. They identified refining activities, chemical fertilizers, and other modern practices as the primary offenders of environmental integrity. Note that so-called renewable energy operations do not alleviate the problem of the overall CO_2 budget.

Deforestation fires were lower in 2020 compared to 2019 levels, which saw the highest rates of deforestation in the Amazon since 2008. In 2019 deforestation and degradation fires were about 30% above the previous decade, while other tropical emissions, mainly from Indonesia, were twice as large as the previous decade because of unusually dry conditions that promoted peat burning and deforestation. These activities are not conventionally included in global warming analysis.

In the interactive chart, we see global fossil fuel consumption broken down by coal, oil, and gas since 1800. Earlier data, pre-1965, is sourced from Vaclav Smil's work on energy transitions; this has been combined with data published in BP's Statistical Review of World Energy from 1965 onwards.[c]

Fig. 1.7 shows rise in primary energy consumption rose by 1.3% in 2019. This is down from less than half its rate in 2018 (2.8%). Growth was driven by renewables (3.2 EJ) and natural gas (2.8 EJ), which together contributed to three-quarters of the increase. All fuels grew at a slower rate than their 10-year averages, apart from nuclear, with coal consumption falling for the fourth time in 6 years (−0.9 EJ). The pace of nuclear energy is curious. Nuclear energy now provides about 10% of the world's electricity from about 440 power reactors. It is touted to be the world's second-largest source of low-carbon power (29% of the total in 2019). Nuclear power plants are operational in 31 countries worldwide. In fact, through regional transmission grids, many more countries depend in part on nuclear-generated power; Italy and Denmark, for example, get almost 10% of their electricity from imported nuclear power. The latest sustainability criterion that vilifies fossil fuel energy as carbon-intensive has led the way to nuclear energy.

Fig. 1.8 shows various fractions of different energy sources. Oil continues to hold the largest share of the energy mix (33.1%). Coal is the second-largest fuel but lost share in 2019 to account for 27.0%, its lowest level since 2003. The share of both natural gas and renewables rose to record highs of 24.2% and 5.0% respectively. Renewables have now overtaken nuclear

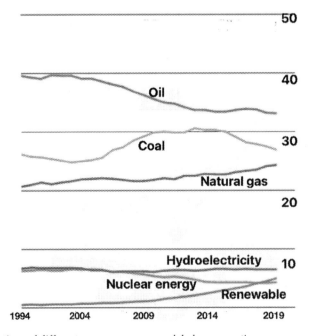

FIG. 1.8 Various fractions of different energy sources on global consumption.

[c]See https://ourworldindata.org/fossil-fuels#note-1.

which makes up only 4.3% of the energy mix. The share of hydroelectricity has been stable at around 6% for several years.

Table 1.2 shows shares of various energy sources. Primary energy consumption rose by 1.3% in 2019, below its 10-year average rate of 1.6% per year, and much weaker than the 2.8% growth seen in 2018. By region, consumption fell in North America, Europe, and CIS and growth was below average in South & Central America. Demand growth in Africa, Middle East, and Asia was roughly in line with historical averages. China was by far the biggest individual driver of primary energy growth, accounting for more than three-quarters of net global growth. India and Indonesia were the next largest contributors, while the US and Germany posted the largest declines in energy terms. There is a shift in terms of energy consumption habits and lifestyles between eastern and western countries.

Looking at energy by fuel, 2019 growth was driven by renewables, followed by natural gas, which together contributed over three-quarters of the net increase. The share of both renewables and natural gas in primary energy increased to record highs. Meanwhile, coal consumption declined, with its share of the energy mix falling to its lowest level since 2003.

BP (2020) summarized coal and other energy consumption as below:
Coal

- Coal consumption declined by 0.6% and its share of primary energy fell to its lowest level in 16 years (27%).
- Increases in coal consumption were driven by the emerging economies, particularly China (1.8 EJ) and Indonesia (0.6 EJ). However, this was outweighed by a sharp fall in OECD demand which fell to its lowest level in our data series (which started in 1965).
- Global coal production rose by 1.5%, with China and Indonesia providing the only significant increases (3.2 EJ and 1.3 EJ respectively). The largest declines came from the US (−1.1 EJ) and Germany (−0.3 EJ).

Renewables, hydro and nuclear energy

- Renewable energy (including biofuels) posted a record increase in consumption in energy terms (3.2 EJ). This was also the largest increment for any source of energy in 2019.

TABLE 1.2 Fuel shares of primary energy and contributions to growth in 2019 (BP, 2020).

Energy source	Consumption (exajoule)	Annual change (exajoule)	Share of primary energy (%)	Percentage point change in share from 2018
Oil	193.0	1.6	33.1	−0.2
Gas	141.5	2.8	24.2	0.2
Coal	157.9	−0.9	27.0	−0.5
Renewables[a]	29.0	3.2	5.0	0.5
Nuclear	24.9	0.8	4.3	0.1
Hydro	37.6	0.3	6.4	−0.0
Total	583.9	7.7		

[a] *Renewable power (excluding hydro) plus biofuels.*

- Wind provided the largest contribution to renewable growth (1.4 EJ) followed closely by solar (1.2 EJ).
- By country, China was the largest contributor to renewable growth (0.8 EJ), followed by the US (0.3 EJ) and Japan (0.2 EJ).
- Hydroelectric consumption rose by a below-average 0.8%, with growth led by China (0.6 EJ), Turkey (0.3 EJ), and India (0.2 EJ).
- Nuclear consumption rose by 3.2% (0.8 EJ), its fastest growth since 2004. China (0.5 EJ) and Japan (0.1 EJ) provided the largest increments.

1.3 Role of oil and gas

Fossil fuel consumption has increased significantly over the past half-century, around eight-fold since 1950, and roughly doubling since 1980. However, the types of fuel we rely on have also shifted, from solely coal toward a combination with oil, and then gas. Today, coal consumption is falling in many parts of the world. Meanwhile, oil and gas are still growing quickly. BP (2020) gives the following update for the year 2019.

Fig. 1.9 shows the world energy balances of the last three decades as reported by the IEA (2020a,b,c). The IEA's World Energy Balances present comprehensive energy balances for all the world's largest energy-producing and consuming countries. It contains detailed data on the supply and consumption of energy for 150 countries and regions, including all OECD countries, over 100 other key countries, as well as world totals.

Energy data is generally collected independently across different commodities. Energy statistics are the simplest format to present all the data together, assembling the individual balances of all products, each expressed in its own physical unit (e.g., TJ for natural gas, kt for coal, etc). These are called commodity balances. However, energy products can be converted into one another through a number of transformation processes. Therefore, it is very useful to also develop a comprehensive national energy balance, to understand how products are

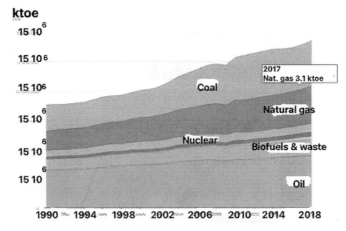

FIG. 1.9 The historical world energy balances. *From IEA, 2020, https://www.iea.org/reports/oil-2020.*

transformed into one another, and to highlight the various relationships between them. By presenting all the data in a common energy unit, the energy balance allows users to see the total amount of energy used and the relative contribution of each different source, for the whole economy and for each individual consumption sector; to compute the different fuel transformation efficiencies; to develop various aggregated indicators and to estimate CO_2 emissions from fuel combustion. The energy balance is a natural starting point to study the evolution of the domestic energy market, forecast energy demand, monitor impacts of energy policies, and assess potential areas for action.

The energy balance takes the form of a matrix, where columns present all the different energy sources (products) categories and rows represent all the different "flows," grouped in three main blocks: energy supply, transformation/energy use, and final consumption (Table 1.3). To develop an energy balance from the set of commodity balances, the two main steps are:

(i) all the data are converted to a common energy unit–also allowing to compute a "total" product; and
(ii) some re-formatting is performed to avoid double counting when summing all products together.

TABLE 1.3 Some useful definitions, related to energy sustainability.

Flow	Short name	Definition
Transformation processes	TOTTRANF	Transformation processes comprise the conversion of primary forms of energy to secondary and further transformation (e.g., coking coal to coke, crude oil to oil products, and fuel oil to electricity). Inputs to transformation processes are shown as negative numbers and output from the process is shown as a positive number. Transformation losses will appear in the "total" column as negative numbers
Main activity producer electricity plants	MAINELEC	Refers to plants that are designed to produce electricity only. If one or more units of the plant is a CHP unit (and the inputs and outputs cannot be distinguished on a unit basis) then the whole plant is designated as a CHP plant. Main activity producers generate electricity for sale to third parties, as their primary activity. They may be privately or publicly owned. Note that the sale need not take place through the public grid
Autoproducer electricity plants	AUTOELEC	Refers to plants that are designed to produce electricity only. If one or more units of the plant is a CHP unit (and the inputs and outputs cannot be distinguished on a unit basis) then the whole plant is designated as a CHP plant. Autoproducer undertakings generate electricity wholly or partly for their own use as an activity that supports their primary activity. They may be privately or publicly owned
Main activity producer plants	MAINCHP	Refers to plants that are designed to produce both heat and electricity (sometimes referred to as co-generation power stations). If possible, fuel inputs and electricity/heat outputs are on a unit basis rather than on a plant basis. However, if data are not available on a unit basis, the convention for defining a CHP plant noted above should be adopted. Main activity producers generate electricity and/or heat for sale to third parties, as their primary activity. They may be privately or publicly owned. Note that the sale need not take place through the public grid

Continued

TABLE 1.3 Some useful definitions, related to energy sustainability—Cont'd

Flow	Short name	Definition
Autoproducer CHP plants	AUTOCHP	Refers to plants that are designed to produce both heat and electricity (sometimes referred to as co-generation power stations). If possible, fuel inputs and electricity/heat outputs are on a unit basis rather than on a plant basis. However, if data are not available on a unit basis, the convention for defining a CHP plant noted above should be adopted. Note that for autoproducer CHP plants, all fuel inputs to electricity production are taken into account, while only the part of fuel inputs to heat sold is shown. Fuel inputs for the production of heat consumed within the autoproducer's establishment are not included here but are included with figures for the final consumption of fuels in the appropriate consuming sector. Autoproducer undertakings generate electricity and/or heat, wholly or partly for their own use as an activity that supports their primary activity. They may be privately or publicly owned
Main activity producer heat plants	MAINHEAT	Refers to plants (including heat pumps and electric boilers) designed to produce heat only and who sell heat to a third party (e.g., residential, commercial or industrial consumers) under the provisions of a contract. Main activity producers generate heat for sale to third parties, as their primary activity. They may be privately or publicly owned. Note that the sale need not take place through the public grid
Autoproducer heat plants	AUTOHEAT	Refers to plants (including heat pumps and electric boilers) designed to produce heat only and who sell heat to a third party (e.g., residential, commercial or industrial consumers) under the provisions of a contract. Autoproducer undertakings generate heat, wholly or partly for their own use as an activity that supports their primary activity. They may be privately or publicly owned
Heat pumps	THEAT	Includes heat produced by heat pumps in transformation. Heat pumps that are operated within the residential sector where the heat is not sold are not considered a transformation process and are not included here—the electricity consumption would appear as residential use
Electric boilers	TBOILER	Includes electric boilers used to produce heat
Chemical heat for electricity production	TELE	Includes heat from chemical processes that is used to generate electricity
Gas works	TGASWKS	Includes the production of recovered gases (e.g., blast furnace gas and oxygen steel furnace gas). The production of pig-iron from iron ore in blast furnaces uses fuels for supporting the blast furnace charge and providing heat and carbon for the reduction of the iron ore. Accounting for the calorific content of the fuels entering the process is a complex matter as transformation (into blast furnace gas) and consumption (heat of combustion) occur simultaneously. Some carbon is also retained in the pig-iron; almost all of this reappears later in the oxygen steel furnace gas (or converter gas) when the pig-iron is converted to steel. In the 1992/1993 annual questionnaires, member countries were asked for the first time to report in transformation processes the quantities of all fuels (e.g., pulverized coal injection [PCI] coal, coke oven coke, natural gas, and oil) entering blast furnaces and the quantity of blast furnace gas and

TABLE 1.3 Some useful definitions, related to energy sustainability—Cont'd

Flow	Short name	Definition
		oxygen steel furnace gas produced. The IEA Secretariat then needed to split these inputs into the transformation and consumption components. The transformation component is shown in the row blast furnaces in the column appropriate for the fuel, and the consumption component is shown in the row iron and steel, in the column appropriate for the fuel. The IEA Secretariat decided to assume a transformation efficiency such that the carbon input into the blast furnaces should equal the carbon output. This is roughly equivalent to assuming an energy transformation efficiency of 40%
Gas works	TGASWKS	Includes the manufacture of town gas. Note: in the summary balances this item also includes other gases blended with natural gas (TBLENDGAS)
Oil refineries	TREFINER	Covers the use transformation of hydrocarbons for the manufacture of finished oil products
Petrochemical plants	TPETCHEM	Covers backflows returned from the petrochemical industry. Note that backflows from oil products that are used for non-energy purposes (i.e., white spirit and lubricants) are not included here but in nonenergy use
Coal liquefaction plants	TCOALLIQ	Includes coal, oil, and tar sands used to produce synthetic oil
Gas-to-liquids (GTL) plants	TGTL	Includes natural gas used as feedstock for the conversion to liquids, e.g., the quantities of fuel entering the methanol production process for transformation into methanol
Blast furnaces	EBLASTFUR	Represents the energy which is used in blast furnaces
Nuclear industry	ENUC	Represents the energy used in the nuclear industry

The world is witnessing several global environmental challenges. As energy production and use are major causes of environmental problems, proper choice of energy technology can have a significant impact on reversing these global problems, such as global warming and climate change. With the incessant campaign against fossil fuel, thus targeting carbon as the 'enemy', every energy source other than fossil fuel has been touted as sustainable. For instance, 'sustainable petroleum technology' is vastly considered to be an oxymoron while others, including nuclear energy is considered to be one of the most efficient and cleanest energy technologies. Particularly, nuclear energy was projected as one of the cheapest energy sources and a reliable alternative to fossil fuel. This was also promoted as an effective solution to reduce CO_2, a precursor to global warming. However, nuclear energy has several environmental, social, and economic issues that have not yet been addressed and remain the most controversial energy source to date (Chhetri and Islam, 2008). Current scientific analyses portray nuclear energy as one of the most efficient technologies. However, these scientific analyses only consider 'local efficiency' as the measure of the true efficiency of a system. If 'global efficiency', which includes long-term environmental sustainability, is considered, the efficiency picture becomes gloomy for nuclear energy. This aspect will be discussed in a later section, but the important here is only fossil fuel uses its own energy throughout the energy cycle. As can be seen in Table 1.3 all energy transformations use petroleum or fossil fuel technology. This is in contrast to other energy sources. For instance, each wind turbine is predominantly made of steel (71%–79% of total turbine mass), fiberglass, resin, or plastic (11%–16%), iron or cast iron (5%–17%), copper (1%), and aluminum (0%–2%). State-of-the-art wind turbine blades are made of carbon fiber, which consists of layers of plastics and plastic resin, both of which are derived from oil and natural gas feedstocks. This involves a very high energy budget supported by fossil fuels. Solar energy, on the other hand, relies on forms of silicon that are used for the construction; namely, single-crystalline, multi-crystalline and amorphous. Other materials used for the construction of photovoltaic cells are polycrystalline thin films such as copper indium diselenide, cadmium telluride, and gallium arsenide. These materials have high energy budget requirements. Wind turbines and solar panels cannot be made solely from other wind turbines and solar panels.

Fossil fuels are required to manufacture wind and solar equipment, transport and construct them, and provide backup electricity when the wind isn't blowing and the sun isn't shining.

Wind and solar facilities currently require massive quantities of steel and concrete, both of which require oil and natural gas in their manufacturing processes. The amount of steel required for wind and solar to replace fossil fuels exceeds the world's capability to produce it for decades. The most significant fossil fuel requirement for wind and solar is to provide the backup electric-generating capacity needed when the wind isn't blowing and the sun isn't shining.

Fast-reacting fossil fuel-based installations—such as highly efficient, combined-cycle, natural gas-fired electric generating plants—are the only viable options to avoid brownouts and blackouts. This practical problem becomes an absurdity when one considers the fact that an effective take over by wind energy of just coal would require 10 billion tons of steel, whereas the total annual global production of steel is only 1.6 billion tons. Conventional sustainability analyses do not consider these factors. Chhetri and Islam (2008) conducted a critical review of energy budget issues, along with environmental, economic, and social aspects. By using a newly developed definition of 'global efficiency', η_g, first developed by Khan and Islam (2012), Chhetri and Islam (2008) showed that solar, wind, and nuclear energy technologies are much less efficient than oil and gas technologies.

Oil

- Oil consumption grew by a below-average 0.9 million barrels per day (b/d), or 0.9%. Demand for all liquid fuels (including biofuels) rose by 1.1 million b/d and topped 100 million b/d for the first time.
- Oil consumption growth was led by China (680,000 b/d) and other emerging economies, while demand fell in the OECD (−290,000 b/d).
- Global oil production fell by 60,000 b/d as strong growth in US output (1.7 million b/d) was more than offset by a decline in OPEC production (−2 million b/d), with sharp declines in Iran (−1.3 million b/d) Venezuela (−560,000 b/d) and Saudi Arabia (−430,000 b/d).
- Refinery utilization fell sharply by 1.2 percentage points as capacity rose by 1.5 million b/d and throughput remained relatively unchanged.

Natural gas

- Natural gas consumption increased by 78 billion cubic meters (bcm), or 2%, well below the exceptional growth seen in 2018 (5.3%). Nevertheless, the share of gas in primary energy rose to a record high of 24.2%. Increases in gas demand were driven by the US (27 bcm) and China (24 bcm), while Russia and Japan saw the largest declines (10 and 8 bcm respectively).
- Gas production grew by 132 bcm (3.4%), with the US accounting for almost two-thirds of this increase (85 bcm). Australia (23 bcm) and China (16 bcm) were also key contributors to growth.
- Inter-regional gas trade expanded at a rate of 4.9%, more than double its 10-year average, driven by a record increase in liquefied natural gas (LNG) of 54 bcm (12.7%).
- LNG supply growth was led by the US (19 bcm) and Russia (14 bcm), with most incremental supplies heading to Europe: European LNG imports (+49 bcm) rose by more than two-thirds.

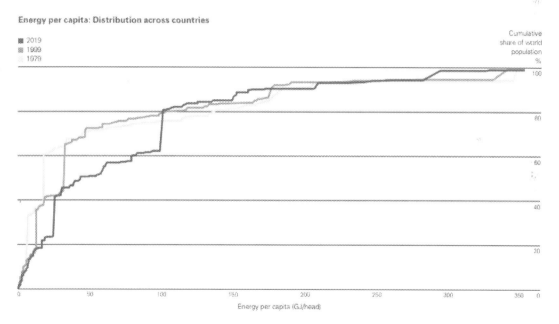

In 2019, 81% of the global population lived in countries where average energy demand per capita was less than 100 GJ/head, two percentage points more than 20 years ago. However, the share of the global population consuming less than 75 GJ/head declined from 76% in 1999 to 57% last year. Average energy demand per capita in China increased from 17 GJ/head in 1979 to 99 GJ/head in 2019.

Ever since the oil price decline in 2014 from a historic high oil price, the world oil market has become vulnerable to uncertainty and economic gloom. It is difficult to separate this oil price issue from the perception of environmental concerns. Even more difficult is to decipher the role of this perception on energy pricing of both renewable and non-renewable resources (Islam et al., 2018).

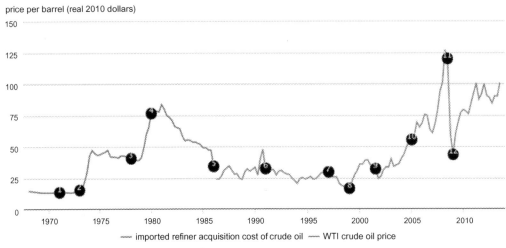

FIG. 1.10 Petroleum is the driver of the world economy and is driven by political events. *Data from Islam, M. R., 2020. Environmentally and Economically Sustainable Enhanced Oil Recovery. Wiley-Scrivener, 786 pp.*

1: US spare capacity exhausted
2: Arab Oil Embargo
3: Iranian Revolution
4: Iran-Irag War
5: Saudis abandon swing producer role
6: Iraq invades Kuwait
7: Asian financial crisis
8: OPEC cuts production targets 1.7 mmbpd
9: 9-11 attacks
10: Low spare capacity
11: Global financial collapse
12: OPEC cuts production targets 4.2 mmbpd

Fig. 1.10 shows long-term oil prices while Fig. 1.11 shows the same for natural gas. As discussed by Islam et al. (2018), the oil price is not governed by supply and demand, it is rather governed by global politics. That politics has become that of contempt of carbon fuel since the installation of the Clinton presidency. With Al Gore's anti-carbon agenda, along with support from IPCC, the energy politics has been governed by what Islam and Khan (2019) called 'climate change hysteria'. Even during the Bush presidency, with the war on terror, the world had little time to reflect on the science of global warming and the likes of Conservatives, such as President G W Bush resorted to the rhetoric of 'oil addiction'. As a consequence, from the mid-1980s to September 2003, the inflation-adjusted price of a barrel of crude oil on NYMEX was generally under US$25/barrel. This would mark oil as the most stable commodity. During 2003, the price rose above $30, reached $60 by 11 August 2005, and peaked at $147.30 in July 2008. This steady rise was first triggered by the war on terror shortly after the 9/11 terror attack in New York. During this time, the USA engaged in costly wars in the Middle East. At the same time, the demand for oil in China soared. Although much of the energy demand of China was offset by coal, the sheer volume of the demand affected the global pricing. It is also true that during the same period the US dollar value dropped (Islam et al., 2018). Added to these is the fact that global petroleum reserve declined (Islam, 2014) and the world became vulnerable to the financial collapse in 2008, which

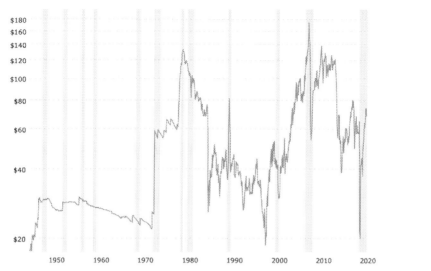

FIG. 1.11 Oil prices in history since Second World War until 2018. *From Islam, M.R., Islam, A.O., Hossain, M.E., 2018a, Hydrocarbons in Basement Formation, Wiley-Scrivener.*

triggered an oil price decline that reverberates until today (Islam et al., 2018). The global financial crisis causes a bubble-bursting sell-off. Prices plummet 78.1% from July to December.

The financial collapse of 2008, along with the recession created a new cycle for oil prices. The recession caused demand for energy to shrink in late 2008, with oil prices collapsing from the July 2008 high of $147 to a December 2008 low of $32. This sudden drop in such a short time remains one of the most important collapses in the history of oil and gas. As the economy recovered, despite losses in bailouts and perpetual wars in the Middle East, oil prices stabilized by August 2009 and generally remained in a broad trading range between $70 and $120 through November 2014, eventually returning to 2003 pre-crisis levels by early 2016. In the global market, during 2011 through 2014, riots and protests from the Arab Spring and the Libyan civil war disrupted the regional output. During 2008 through 2014, unconventional oil and gas, empowered with new fracking technology made a great impact on US energy (Islam, 2014).

The year 2014 is marked by strong production in the United States and Russia that caused oil prices to crash from July to December. This led OPEC to the decision to maintain production further damages the market heading into 2015.

The year 2015 started with the death of King Abdullah—an event many saw as a potential trigger for instability. However, the instability was not as much in a power struggle but more as who assumed power. Initially, Mohamed Bin Salman became the Minister of Defence, whose first act was to orchestrate the Yemen war, which later triggered the most devastating "the worst man-made humanitarian crisis of our time," as recounted by UN officials (Carey, 2018). This political turmoil, however, was overshadowed by the fact that on July 22 of this year, U.S. output reached its highest level in more than 100 years. This is the time, prices hovered near $50 a barrel.

In this context, it is important to look at the past to see the factors that shape oil prices. McGuire (2015) listed the top 25 events that shaped the oil market since the first commercial trading of oil commenced in mid 1800s. These events are:

1. Event (A), 1862–1865: The American Civil War is in full swing, leading to crude oil demand skyrocketing due to its increasing use for lamps and medicinal purposes. This is the period petroleum began to replace heavily taxed and more expensive whale oil, another illuminant with similar qualities. This is the beginning of a culture of introducing a tax to obstruct the natural economy.
2. Event (B), 1865–1890: Prices saw boom and bust over the next 25 years due to fluctuations in U.S. drilling. By 1877, John D. Rockefeller's Standard Oil Company controls more than 95% of all oil refineries in the country. This is the period refining became an integral part of oil production.
3. Event (C), 1890–1892: The United States entered its worst recession to date, causing oil prices to plummet. The period is marked by the excessive financing of railroads, which results in a series of bank failures. Unemployment ranged from 17% to 19%.
4. Event (D), 1891–1894: The Titusville oil fields that gave birth to the U.S. oil industry start to decline. This sets the stage for higher prices in 1895.
5. Event (E), 1894: An international cholera outbreak drew back oil production throughout Europe, contributing to the 1895 spike.
6. Event (F), 1920: The widespread adoption of the automobile drastically raised oil consumption before one of the worst events in world history sent prices to record lows.
7. Event (G), 1931: The onset of the Great Depression reduced demand and sent prices sinking to $0.87 a barrel (roughly $12 a barrel today).
8. Event (H), 1947: Increased spending on advertising after the war lead to a huge boost in nationwide automobile sales. The automotive boom also caused gasoline shortages in many U.S. states.
9. Event (I), 1956–1957: Global prices remained steady due to two equalizing events. The blocked-off Suez Canal from the Suez Crisis took 10% of the world's oil off the market while soaring production outside of the Middle East made up for the absence of a major oil passageway.
10. Event (J), 1972: Total U.S. production peaked near an average of 9 million barrels a day.
11. Event (K), 1973–1974: During the Yom Kippur War, the Organization of Arab Petroleum Exporting Countries (OAPEC), which included Egypt and Syria, imposed an oil embargo against countries supporting Israel. By the end of the embargo in March 1974, oil prices had increased from $3 a barrel ($14 a barrel today) to $12 ($58 today).
12. Event (L), 1978–1979: Iran dramatically cut production and exported during the country's Islamic revolution.
13. Event (M), 1980: The Iran-Iraq War further decreased exports from the Middle Eastern region.
14. Event (N), 1980s: A worldwide supply glut sets in, sending prices from over $35 a barrel (about $100 today) down to $12 (about $28). The former U.S.S.R. and the United States were the top two producers in the world by 1985, respectively producing 11.9 million and 11.2 million barrels per day.

15. Event (O), 1986: Saudi Arabia decides to regain its share of the global oil market by increasing production in the face of crashing prices. The OPEC leader went from 3.8 million barrels a day in 1985 to more than 10 million barrels a day in 1986.
16. Event (P), 1988: The Iran-Iraq War ended in August, allowing both countries to start ramping up production.
17. Event (Q), 1990: Iraq invaded Kuwait after Saddam Hussein accuses Kuwait of stealing Iraq's market share. The conflict involved Iraqi forces setting fire to up to 700 Kuwaiti oil wells. Kuwait cut exports until 1994 as a result.
18. Event (R), 1999: Thailand, Indonesia, and South Korea recovered from the 1997 financial crisis caused by the collapse of Thailand's baht currency. Demand started to soar in the region.
19. Event (S), Early 2000s: Prices started to gain momentum due to growing U.S. and world economies. They headed toward their highest level since 1981.
20. Event (T), 2001–2003: The Sept. 11 attacks and the invasion of Iraq raised concerns about the stability of the Middle East's production.
21. Event (U), Mid-2000s: The combination of declining production and surging Asian demand send prices to record highs.
22. Event (V), 2008: The global financial crisis causes a bubble-bursting sell-off. Prices plummeted 78.1% from July to December.
23. Event (W), 2011: Riots and protests from the Arab Spring wash over the Middle East. The Libyan civil war disrupts the region's output.
24. Event (X), 2014: Strong production in the United States and Russia caused prices to crash from July to December. OPEC's November decision to maintain production further damages the market heading into 2015.
25. Event (Y), 2015: U.S. output reached its highest level in more than 100 years. Prices hover near $50 a barrel as of July 22.

Fig. 1.12 shows the oil price for the period of 2012 through early 2021. Most notably, this covers the Trump era. President Trump's energy policy was radically different from those of previous presidents and were a hallmark of success (Islam et al., 2018). However, like every other policy, Trump received little support from the media or the financial establishment. As early as 2107, it was noticeable that Trump wasn't going to get any support to receive credit for the economic boom, resulting from economic policies (Blackmon, 2017). Within months, however, Trump Administration implemented many policy changes, each of which turned out to be very productive. The most prominent move was Trump's June 1 announcement that the U.S. would withdraw from the Paris climate accord. Environmental Protection Agency (EPA), on the other hand, dismantled President Barack Obama's Clean Power Plan, a signature policy aimed at reducing greenhouse-gas emissions. The science behind the Clean Power Plan assumes that carbon is inherently unsustainable whereas non-carbon methods are sustainable. In June, following a February executive order from Trump, the EPA began the process of rescinding the 2015 Waters of the United States rule, which aimed at protecting smaller bodies of water and streams in the same way that larger ones had been. In December, in the closing weeks of his administration, Obama banned drilling in the Arctic and parts of the Atlantic Ocean; the Trump administration promptly set about undoing that ban, along with restarting the Keystone XL and Dakota Access pipeline projects. Although this ban has been

FIG. 1.12 Oil price in recent years. *From Trading Economics, 2021. https://tradingeconomics.com/commodity/crude-oil.*

halted by one of President Biden's some dozen executive orders of the first day, 4 years of the Trump era enjoyed the deregulation process and brought about historic achievements in both energy and environment. The energy policy, which leaned on oil and gas, along with massive tax cuts laid the foundation of an economy that was parallel to none (Islam et al., 2018). Overall, it involves more oil and gas production, less environmental regulation (Yglesias, 2019). This has created unprecedented energy independence while simultaneously reducing CO_2 emissions. During this period both crude oil price and the gold price rose steadily with a similar slope (Figs. 1.12 and 1.13).

There is, however, a time that US oil prices turned negative for the first time on record shortly after the onset of the Coronavirus pandemic. The price of US crude oil crashed from $18 a barrel to -$38 in a matter of hours, as rising stockpiles of crude threatened to overwhelm storage facilities and forced oil producers to pay buyers to take the barrels they could not store. This market crash was due to the impact of the coronavirus outbreak on oil demand as the global economy slumped due to lockdown measures. However, within a day price rebounded above zero, with the US benchmark West Texas Intermediate for May changing hands at $1.10 a barrel after closing at -$37.63 in New York.

Oil producers have continued to pump near-record levels of crude into the global market even as analysts warned that the impact of the coronavirus outbreak would drive oil demand to its lowest levels since 1995. The emergence of negative oil prices is expected to prompt some oil companies to hasten the shutdown of their rigs and oil wells to avoid plunging deeper into debt or bankruptcy. The price collapse was meant to be a blow to US President Donald Trump who took credit for brokering a historic deal between the OPEC oil cartel and the world's largest oil-producing nations to limit the flood of oil production into the market. The pact to cut between 10 million and 20 million barrels of oil from the market from the

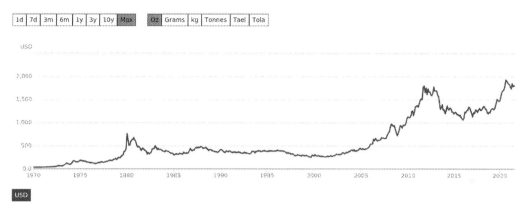

FIG. 1.13   Gold price since 1970. *From Gold.org, 2021, Available at https://www.gold.org/goldhub/data/gold-prices?utm_source=google&utm_medium=cpc&utm_campaign=rwm-goldhub&utm_content=466092663098&utm_term=gold%20price%20history&gclid=CjwKCAiAgJWABhArEiwAmNVTBxJWHZxBdKWr7Oq7x8oLHxOWK-rkOx2JkLgLUgu-5td8taGqMUvTbhoCoLIQAvD_BwE.*

following month was dismissed by many within the market as "too little, too late" to avoid a market crash. However, it turned out President Trump was correct in calling it a "short-term problem". He said the US was filling up its strategic reserves: "If we could buy it for nothing, we're gonna take everything we can get." Brent crude reached highs of almost $69 a barrel in January before plummeting to less than $23 a barrel at the end of March. Many market experts predict the price of Brent will remain below $50 a barrel during 2020. None of these predictions have been true.

1.3.1 Environmental impact

The unprecedented energy boom has been accompanied by environmental gains. Fig. 1.14 shows a steady decline in CO_2 emissions from 1975 through 2019. The trend during the Trump era continued through 2020. While 2020 decline is associated with Covid-19 measures, the decline in fact correlates with the trend that began in 2015. This period is the one when the USA gained energy independence.

Trading economics, 2021, https://tradingeconomics.com/commodity/crude-oil

In 2019, around 5.13 billion metric tons of CO_2 emissions were produced from energy consumption in the United States. In 2018, around 36.6 billion metric tons of carbon dioxide were emitted globally.

FIG. 1.14 Carbon dioxide emissions from energy consumption in the U.S. from 1975 to 2019 (in million metric tons of carbon dioxide). *From Statista, 2021, https://www.statista.com/statistics/1088739/global-oil-discovery-volume/#:~:text=Global%20oil%20discoveries%202011%2D2018&text=The%20total%20volume%20of%20crude,5.45%20billion%20barrels%20of%20oil.*

The year 1997 marked the birth of the Kyoto Protocol. That year, global energy-related CO2 emissions stood at around 24.4 billion metric tons. Despite numerous assurances by policymakers to undertake efforts to reduce pollution, this figure increased to more than 36 billion metric tons of carbon dioxide in 2017.

North America and the Asia Pacific regions are presently the biggest producers of carbon dioxide emissions as a result of a growing thirst for energy derived from fossil fuels such as oil, natural gas, and coal. China is currently the most polluting country in the world, with a 27.5% share of global CO2 emissions in 2018. A comparative analysis between CO2 levels in 1993 and those in 2003 shows that emission levels in China have more than tripled in a span of 10 years. According to a recent forecast, energy-related global CO2 emissions from the consumption of coal, natural gas, and liquid fuels are set to rise to unprecedented levels through 2040, while U.S. CO2 emissions produced by the use of natural gas are set to grow from 1.68 billion metric tons of CO2 equivalent in 2019 to 1.97 billion metric tons in 2050.

In 2014, in Lima, Peru, negotiations were held regarding a post-Kyoto legal framework forcing major polluters, including China, India, and the United States, to pay for CO2 emissions. Although most countries refused to ratify this latter treaty, there has been a worldwide commitment toward so-called "green energy." Global renewable energy consumption soared from 44.4 million metric tons of oil equivalent in 1998 to 561 million metric tons of oil equivalent in 2018. International investments in renewable energy sources increased six-fold, as the world has invested more than 288 billion U.S. dollars in alternative energy sources and technologies as of 2018.

1.4 Key events and future outlook of oil and gas

1.4.1 USA energy outlook

The dramatic increase in oil and gas production in the United States got its boost from the use of fracking in unconventional oil and gas fields (Islam, 2014). The political event that led to the unleashing of this production boom was greatly facilitated by the lifting, in November 2015, of the prohibition to export crude oil, which had been in place since 1973. Such a major political decision came at a time when U.S. domestic refineries are reaching their maximum levels of shale oil processing capacity and oil storage in the U.S. is at a historical high, making crude oil exports a logical move. After the Trump presidency, U.S. crude oil production continued at the fastest rate on record as the increase in prices during the new US regime boosted drilling and completion activities and oil companies employ more horsepower to fracture larger wells (Kemp, 2018). Crude and condensates output hit a record 11.35 million barrels per day in August, up from 10.93 million bpd in July 2018 (EIA, 2018a). Crude output has increased by more than 2 million barrels per day over 2018, an absolute increase that is unparalleled in the history of the U.S. oil industry (Kemp, 2018). Even in percentage terms, output was up in 2018 by nearly 25% over the year, the fastest increase since the 1950s (excluding the recovery from hurricanes). This rate indeed is faster than at the height of the last drilling and fracking boom before prices slumped in the second half of 2014.

Most of the increase is coming from onshore shale fields, where output has risen by more than 1.9 million bpd over the last year, with a smaller contribution from the Gulf of Mexico, where output is up 200,000 bpd. This increase has been for both oil and natural gas. Fig. 1.15 shows how both oil and gas from shale formations have increased for each region. In the first 9 months of 2018, the number of wells drilled in the United States was up by 26% while well completions were up by 24% (EIA, 2018b).

This surge in US domestic output coupled with increased production from Russia, Saudi Arabia, and several other OPEC countries has pushed oil prices lower.

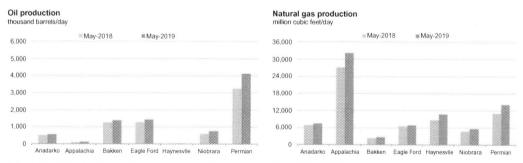

FIG. 1.15 Yearly change in unconventional oil and gas. *From EIA, 2018b.*

U.S. dry shale gas production
billion cubic feet per day

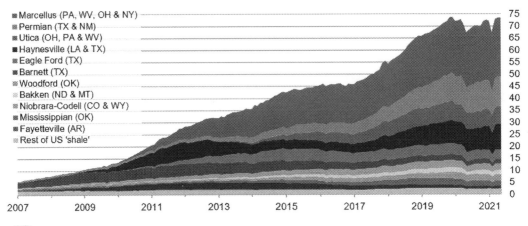

FIG. 1.16 Dry shale gas production since 2006.

Figs. 1.16 and 1.17 show natural gas production and future projections of unconventional gas production. Fig. 1.18 shows the USA has become a natural gas net exporter, starting from 2016.

In the USA, crude oil imports have decreased steadily in recent years as the U.S. crude oil production has increased. After averaging a record high of 10.1 million b/d in 2005, crude oil imports fell by 2.8 million b/d to an average of 7.3 million b/d in 2014. Since then, crude oil imports have increased slightly, averaging 7.7 million b/d in 2018 (Fig. 1.19).

Not all crude oil is of the same quality. Most of the reduction in U.S. imports was of light, sweet crude oil, as those barrels were replaced by domestic production of similar quality. U.S. crude oil exports have also increased as domestic production has risen. U.S. crude oil exports have set annual record highs in each year since 2014, most recently averaging 2.0 million b/d in 2018.

At the same time, U.S. refinery runs have been setting record highs. The increase in refinery output of petroleum products has outpaced the increase in U.S. consumption of petroleum products such as distillate fuel oil, gasoline, and propane, leading to an increase in exports. Considering the overall age of refineries and declining productivity, this is a remarkable feat.

Total U.S. petroleum product exports averaged a record 5.6 million b/d in 2018. Distillate and gasoline exports have increased, particularly to countries in the Western Hemisphere. Propane exports have also increased, mostly to Asian markets.

EIA expects these trends to continue over the next several years. In its March Short-Term Energy Outlook, EIA forecasts that the United States will become a net exporter of crude oil and petroleum products on a monthly basis later this year and on an annual basis in 2020.

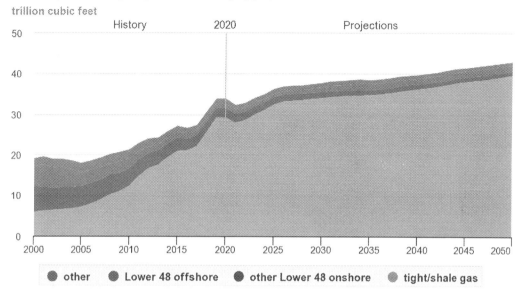

FIG. 1.17   Future projection of dry natural gas production (EIA, 2020b).

1.4.2 China's economic slowdown

Since initiating market reforms in 1978, China has shifted from a centrally-planned to a more market-based economy and has experienced rapid economic and social development. GDP growth has averaged nearly 10% a year—the fastest sustained expansion by a major economy in history (World Bank, 2019). China reached all the Millennium Development Goals (MDGs) by 2015 and made a major contribution to the achievement of the MDGs globally. However, China's GDP growth has gradually slowed since 2012, as needed for a transition to more balanced and sustainable growth. The current lower rate of growth of China's economy can be dismissed as just a transitional phase of a normal economic cycle. That means that after decades of the double-digit growth it is only normal that a period of slower growth ensues. However, the concern is that China's economy may not just be slowing down temporarily but has instead entered a new and prolonged period of weak growth. The GDP has been declining steadily since the record high in 2007. In 2015, China's growth was the weakest of the previous 25 years, accompanied by a collapse of the stock market and a significant devaluation. In January of 2016, some $110 billion left the country, while over $600 billion of capital flight took place during 2015. The most troubling sign, however, is the skyrocketing growth of the national debt, which has tripled since 2007. Nevertheless, China has been the largest single contributor to world growth since the global financial crisis of 2008. The impact of China's level of economic activity on global energy demand (and

30   1. Introduction

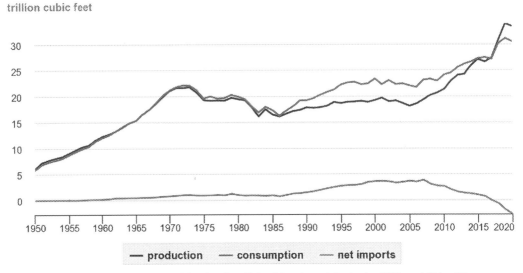

FIG. 1.18 Natural gas production, consumption, and net import (EIA, 2020a,b,c,d,e).

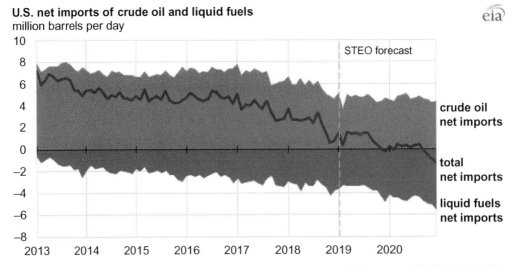

FIG. 1.19 Net oil imports of the USA. *From EIA, 2020a, https://www.eia.gov/todayinenergy/detail.php?id=38672.*

prices) is well known. An economic downturn in China means lower prices for all the commodities that the Asian giant voraciously imports and oil is no exception. In response, the government has stressed its intention to move the economy away from its reliance on exports as a source of growth to the expansion of the domestic market and from massive infrastructure investments and industrial development to the stimulus of a larger and stronger service sector. All these are political decisions that will profoundly change the way China produces and consumes energy.

1.4.3 The Middle East crisis

Bloomberg reported that "about 2.6 million barrels a day are being kept from the market by conflict and sanctions in the region, more than five times the average from 2000 to 2010" (Naím, 2019). The IEA energy outlook for 2016 reports oil production disruptions averaging 3.2 million barrels per day over the last 2 years, mostly due to political instability in Iraq, Libya, South Sudan, and Syria. This significant supply imbalance has been partially compensated by new Iranian exports, which have doubled since last year, reaching 2.1 million barrels per day in May. Such an increase is the result of the lifting of sanctions against Iran by western powers, following the nuclear deal reached in July 2015. However, in April 2019, the USA announced that Iran sanction waiver would end, creating a surge in the oil price (Elliott, 2019). The instability in the Middle East has disrupted the global oil supply and has contributed to fragment and weaken OPEC, leading one of Vladimir Putin's main collaborators, Igor Sechin, to say that "OPEC has practically stopped existing as a united organization." Meanwhile, Libya, Syria Egypt, and the Eastern Mediterranean are all hotspots rife with instability and hydrocarbons. The latest crisis has been the civil war in Libya (Lee, 2019). Fig. 1.20 shows how radical the impact of Libya civil war has been on the oil price. In the Middle East politics far outweighs technology in defining its weight in the world of energy.

1.4.4 Russia's expansionism and sanctions

Russia's 2014 annexation of Crimea triggered economic sanctions by the European Union and the United States. Some of these sanctions directly affect the Russian energy sector and its

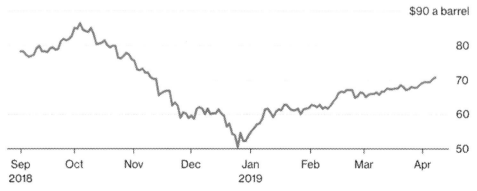

FIG. 1.20   Oil price in recent years. *From Lee, 2019.*

ability to continue to be the foremost supplier of natural gas to Europe. The sanctions include the freezing of exports to Russia of energy-related equipment and technology and the banning of the supply to Russian oil and gas companies of services like drilling, well testing, and completion services. Equally important has been the impact on the natural gas market (discussed in a later section). Many observers predicted that the international coalition that supported the sanctions would quickly fragment, that the sanctions would be watered down or that they would be short-lived and ineffectual. None of these predictions has come to pass. Instead, the Kremlin's decision to annex Crimea and destabilize Ukraine has resulted in major upheavals in Russia's oil and gas industry and an unexpected opening for U.S. gas exporters to European markets.

1.4.5 The implosion of Venezuela and Brazil

Both Venezuela and Brazil had been in the crosshairs of the US. However, after the election of a right-wing president in Brazil, Venezuela has been the sole target of US regime-change politics. The allegations of mismanagement, lack of investment, and massive corruption in Venezuela's state oil company created a crisis that culminated in creating an alternate government, calling President Maduro illegitimate. Venezuela—the country with the largest oil reserve has seen a massive loss in production and exports. In both cases, technology had nothing to do with their downfall. It was all about politics, as summed up by sanctioning of Venezuelan national oil company, PDFSA, and freezing of its asset in the USA.

The most recent political events and their effect on oil prices are shown in Fig. 1.21 This figure shows discounts offered on Dubai crude marker. The effect of events in Saudi Arabia, Iran, Canada, and Venezuela all have effects on the market price.

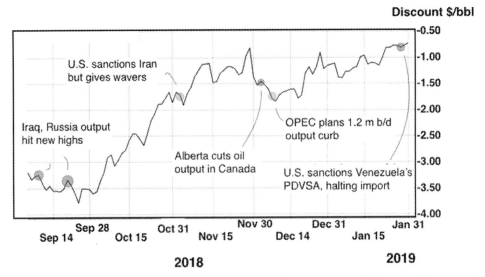

FIG. 1.21 Discounts and correlation with political events. *From Cheong, S., 2015. OPEC Brings Oil Price War Home in Pursuit of Asia's Cash, Bloomberg, October 19.*

In terms of the future, Iraq is scheduled to become a major player. International Energy Agency declared that Iraq would be the third-biggest provider of new oil supplies over the next decade (Smith, 2019). However, the growth rate is slower than that seen earlier this decade as Iraq faces competition for foreign investment and expertise, and struggles to inject enough water to maintain pressure at oil reservoirs. The OPEC member will raise output to almost 6 million barrels a day by 2030, overtaking Canada as the world's fourth-largest producer, as it continues to rehabilitate an oil industry ravaged by decades of conflict and sanctions. Water supplies are one of the industry's most acute needs because relatively low recovery rates mean that Iraqi oil fields rely on the injection of liquids to sustain reservoir pressure. Demand for water in Iraq's oil sector will climb by 60% to more than 8 million barrels a day by 2030, the IEA estimated (Smith, 2019). Fig. 1.22 shows the short-term history and outlook of oil prices.

Brent crude oil spot prices averaged $66 per barrel (b) in March, up to $2/b from February 2019. Brent prices for the first quarter of 2019 averaged $63/b, which is $4/b lower than the same period in 2018. Despite lower crude oil prices than the year before, Brent prices in March were $9/b higher than in December 2018, marking the largest December-to-March price increase from December 2011 to March 2012. EIA forecasts Brent spot prices will average $65/b in 2019 and $62/b in 2020, compared with an average of $71/b in 2018. EIA expects that West Texas Intermediate (WTI) crude oil prices will average $8/b lower than Brent prices in the first half of 2019 before the discount gradually falls to $4/b in late-2019 and through 2020.

EIA (2019) estimates that U.S. crude oil production averaged 12.1 million barrels per day (b/d) in March, up 0.3 million b/d from the February average. EIA forecasts that U.S. crude oil production will average 12.4 million b/d in 2019 and 13.1 million b/d in 2020, with most of the growth coming from the Permian region of Texas and New Mexico.

For the 2019 summer driving season that runs from April through September, EIA forecasts that U.S. regular gasoline retail prices will average $2.76 per gallon (gal), down from an average of $2.85/gal last summer. EIA's forecast is discussed in its Summer Fuels Outlook. The lower forecast gasoline prices primarily reflect EIA's expectation of lower crude oil prices in 2019. For all of 2019, EIA expects U.S. regular gasoline retail prices to average $2.60/gal and

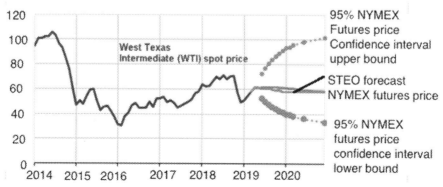

FIG. 1.22   Short-term energy outlook. *From EIA, 2019. Energy Outlook. https://www.eia.gov/outlooks/steo/.*

gasoline retail prices for all grades to average $2.71/gal, which would result in the average U.S. household spending about $100 (4%) less on motor fuel in 2019 compared with 2018.

In many regards, gas prices have been influenced by political factors, but some aspects of natural gas are unique.

Fig. 1.23 shows gas prices over the last few decades. Fig. 1.24 adds the older history of gas prices. This figure shows, stable gas prices were maintained throughout the few decades after the second world war. As can be seen in Fig. 1.23, from 1949 to 1978, wellhead prices averaged $0.21 per thousand cubic feet (mcf). During that period gas prices were regulated. Although phased deregulation began with the passage of the Natural Gas Policy Act of 1978, prices began to rise in the mid-1970s, a period of turmoil in international energy markets that saw a sharp increase in crude oil prices (triggered by the 1973 Arab oil embargo). This rise continued until 1984 at $2.66 per mcf (nominal). Prices subsequently retreated modestly and then

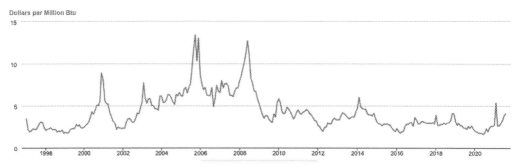

FIG. 1.23 Gas price (in $/million BTU). *From EIA, 2021. EIA report. https://www.eia.gov/dnav/ng/hist/rngwhhdm.htm.*

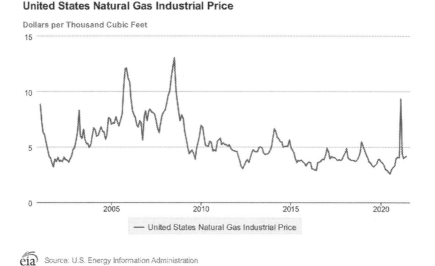

FIG. 1.24 Gas price (in $/1000 Cuft). *From EIA, 2021. EIA report. https://www.eia.gov/dnav/ng/hist/n3035us3m.htm.*

remained fairly stable for several years. From 1986 to 1999, natural gas prices averaged $1.87 per mcf. Following the 9/11 terror attack and the following recession, natural gas prices began to rise, keeping pace with the oil price. By 2004, gas prices in both real and nominal dollars were at record-high levels. The late 2005 rise in gas price was due to severe weather issues related to Hurricane Katrina (in Louisiana) and Hurricane Rita (in the Texas-Louisiana border) that damaged Gulf Coast's production, refining, and distribution facilities. Other events were due to international events triggered by Russia (Islam, 2014). Even before the former USSR broke down, gas export from Russia had been declining. Even though such decline is often correlated with political events and US hegemony, the fact remains that the so-called "gas war" had to break out in order to restore values of natural gas to the level comparable to crude oil or petroleum liquids. Even though Gazprom was privatized in 2005, the Russian government has held a controlling share in Gazprom. The earliest sign of the restoration of natural gas price to an equitable value was in place when on October 2, 2008, the Ukrainian Prime Minister Julia Timoshenko and the then Russian Prime Minister Vladimir Putin had agreed in a memorandum on the Ukraine raising the gas price to world market standards within the next 3 years. Previous to that, Russia had delivered gas to the Ukraine far below world market prices until the end of at the end of 2008, the existing contract between the Russian Gazprom and the Ukrainian gas corporations expired both gas corporations are under state control. A new contract about a new gas price in terms of the October 2008 memorandum and valid from January 1, 2009, was prevented by the Ukraine, although Russia had made an offer to deliver the gas for US$ 250/1000 m^3, which is less than the current world market price. Thus, Gazprom stopped its gas deliveries to the Ukraine on January 1, 2009. And this led to Ukraine unlawfully tapping the transit pipelines running through the Ukraine to other European states. Russia reacted by discontinuing the gas transfer across the Ukraine, completely. What followed after this turmoil is a series of political events, which culminated in the annexation of Crimea by Russia. On 25 November 2015 Gazprom halted its exports of Russian natural gas to Ukraine. According to the Ukrainian government, they had stopped buying from Gazprom because Ukraine could buy natural gas cheaper from other suppliers. According to Gazprom, it had halted deliveries because Ukraine had not paid them for the next delivery. Since then, Ukraine has been able to fulfill its gas supply needs solely from European Union states. In 2018 the Arbitration Institute of the Stockholm Chamber of Commerce ordered that Ukraine's Naftogaz should import 5 billion cubic meters of gas annually from Russia, as required under its 2009 contract with Russia's Gazprom. These events ended up making little impact on the natural gas price.

In recent years (Fig. 1.25), natural gas consumption rose by 96 billion cubic meters (bcm), or 3%, the fastest since 2010 (BP, 2018). Consumption growth was driven by China (31 bcm), the Middle East (28 bcm), and Europe (26 bcm). Consumption in the US fell by 1.2% or 11 bcm. Meanwhile, global natural gas production increased by 131 bcm, or 4%, almost double the 10-year average growth rate. In 2018, Russian growth was the largest at 46 bcm, followed by Iran (21 bcm). With the Iran sanction looming, Russia will likely become the biggest beneficiary of the energy crisis.

In the short term, the Henry Hub natural gas spot price averaged $2.95/million British thermal units (MMBtu) in March, up 26 cents/MMBtu from February. Prices increased as a result of colder-than-normal temperatures across much of the United States, which increased the use of natural gas for space heating. EIA (2019) expects strong growth in U.S.

36   1. Introduction

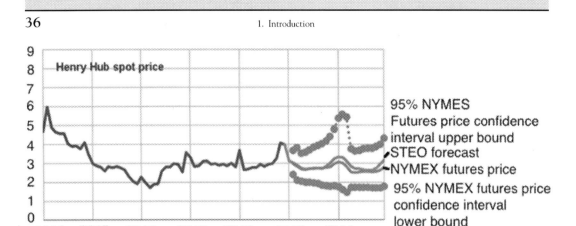

FIG. 1.25 $/million BTU gas price. *From EIA, 2019. Energy Outlook. https://www.eia.gov/outlooks/steo/.*

natural gas production to put downward pressure on prices in 2019 and 2020. EIA expects Henry Hub natural gas spot prices will average $2.82/MMBtu in 2019, down 33 cents/MMBtu from 2018. The forecasted 2020 Henry Hub spot price is $2.77/MMBtu.

EIA (2019) forecasts that dry natural gas production will average 91.0 billion cubic feet per day (Bcf/d) in 2019, up 7.6 Bcf/d from 2018. The EIA expects natural gas production will continue to grow in 2020 to an average of 92.5 Bcf/d.

EIA estimates that natural gas inventories ended March at 1.2 trillion cubic feet (Tcf), which would be 17% lower than levels from a year earlier and 30% lower than the five-year (2014–18) average. EIA forecasts that natural gas storage injections will outpace the previous five-year average during the April-through-October injection season and that inventories will reach 3.7 Tcf at the end of October, which would be 13% higher than October 2018 levels but 1% lower than the five-year average. The impact of political events is rarely included in the EIA forecast, whereas spot prices are primarily dictated by political events.

Fig. 1.26 shows that both oil and gas maintain a steady increase in world demand. Only coal went through a sharper rise during 2003 through 2005, due to excessive growth in

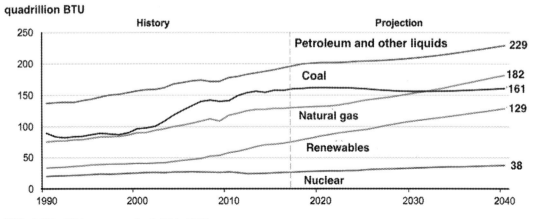

FIG. 1.26   USA energy outlook (EIA, 2018).

Chinese energy consumption. Overall, renewables remained insignificant until many government-funded projects were initiated during the Obama presidency as well as in Europe. As Islam and Khan (2019) pointed out this was more of a policy choice rather than a scientific need. In terms of the project, natural gas shows the highest growth in the 30-year projection. In fact, other than natural gas, only renewable and liquid biofuels show a modest increase, while others drop or remain constant. Natural gas plays an even more intense role when it comes to electrical power usage (Fig. 1.27).

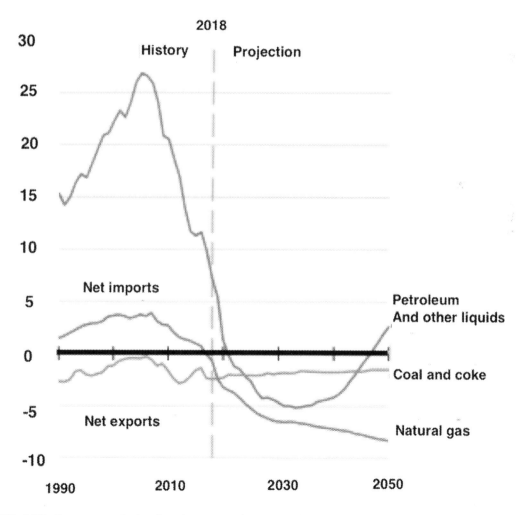

FIG. 1.27 Long term projections based on past performance in the USA (from EIA, 2019), y-axis represents quadrillion British thermal units.

1.5 Sustainability status of current technologies

Not long ago, any sustainable operation in the context of petroleum engineering was considered Until now, there has been no suitable alternative to fossil fuels and all trends indicate continued dominance of the petroleum industry in the foreseeable future (Islam and Khan, 2019). Even though petroleum operations have been based on solid scientific excellence and engineering marvels, only recently it has been discovered that many of the practices are not environmentally sustainable. Practically all activities of hydrocarbon operations are accompanied by undesirable discharges of liquid, solid, and gaseous wastes (Khan and Islam, 2007b), which have enormous impacts on the environment (Islam et al., 2010). Hence, reducing environmental impact is the most pressing issue today and many environmentalist groups are calling for curtailing petroleum operations altogether. Even though there is no appropriate tool or guideline available for achieving sustainability in this sector, there are numerous studies that criticize the petroleum sector and attempt to curtail petroleum activities (Holdway, 2002). There is clearly a need to develop a new management approach to hydrocarbon operations. The new approach should be environmentally acceptable, economically profitable, and socially responsible.

Crude oil is truly a non-toxic, natural, and biodegradable product but the way it is refined is responsible for all the problems created by fossil fuel utilization. The refined oil is hard to biodegrade and is toxic to all living objects. Refining crude oil and processing natural gas use large amounts of toxic chemicals and catalysts including heavy metals. These heavy metals contaminate the end products and are burnt along with the fuels producing various toxic by-products. The pathways of these toxic chemicals and catalysts show that they severely affect the environment and public health. The use of toxic catalysts creates many environmental effects that make irreversible damage to the global ecosystem. Similarly, the use of synthetic chemicals can render a drilling operation as well as enhanced oil recovery operation unsustainable (Islam and Khan, 2019).

Crude oil is a naturally occurring liquid found in formations in the Earth consisting of a complex mixture of hydrocarbons consisting of various lengths. It contains mainly four groups of hydrocarbons among, which saturated hydrocarbon consists of a straight chain of carbon atoms, aromatics consists of ring chains, asphaltenes consists of complex polycyclic hydrocarbons with complicated carbon rings and other compounds mostly are of nitrogen, sulfur, and oxygen. It is believed that crude oil and natural gas are the products of huge overburden pressure and heating of organic materials over millions of years.

Crude oil and natural gases are formed as a result of the compression and heating of ancient organic materials over a long period of time. Oil, gas, and coal are formed from the remains of zooplankton, algae, terrestrial plants, and other organic matter after exposure to heavy pressure and temperature of Earth. These organic materials are chemically changed to kerogen. With more heat and pressure along with bacterial activities, oil and gas are formed. Fig. 1.28. is the pathway of crude oil and gas formation. These processes are all driven by natural forces.

Sustainable petroleum operations development requires a sustainable supply of clean and affordable energy resources that do not cause negative environmental, economic, and social consequences (Dincer and Rosen, 2007, 2011). In addition, it should consider a holistic

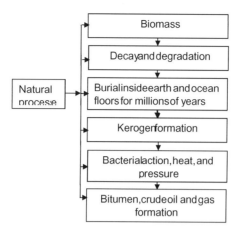

FIG. 1.28 Crude oil formation pathway. *After Chhetri, A.B., Islam, M.R., 2008. Inherently Sustainable Technologies. Nova Science Publishers, 452 pp.*

approach where the whole system will be considered instead of just one sector at a time (Islam et al., 2010). In 2007, Khan and Islam developed an innovative criterion for achieving true sustainability in technological development. New technology should have the potential to be efficient and functional far into the future in order to ensure true sustainability. Sustainable development is seen as having four elements– economic, social, environmental, and technological.

Fig. 1.29 shows the different phases of petroleum operations which are seismic, drilling, production, transportation and processing, and decommissioning, as well as their associated wastes generation and energy consumption. Various types of waste from ships, emission of CO_2, human-related waste, drilling mud, produced water, radioactive materials, oil spills, the release of injected chemicals, toxic release used as corrosion inhibitors, metals, and scraps, flare, etc. are produced during the petroleum operations.

Drilling is a necessary step for petroleum exploration and production. The conventional rotary drilling technique falls short since it is costly and contaminates surrounding rock and water due to the use of toxic components in the drilling fluids. Conventional rotary drilling has been the main technique used for drilling in the oil and gas industry. However, this method has shown its limits regarding the depth of the wells drilled, in addition to the use of toxic components in drilling fluids. The success of a high-risk hydrocarbon exploration and production depends on the use of appropriate technologies. Therefore, to overcome the limitations of conventional rotary drilling techniques, we need to look for other environmentally friendly drilling technologies which may lead to a sustainable drilling operation.

Generally, technology is selected based on criteria such as technical feasibility, cost-effectiveness, regulatory requirements, and environmental impacts. Recently, Khan and Islam (2006) introduced a new approach in technology evaluation based on the novel sustainability criterion. In their study, they not only considered the environmental, economic, and regulatory criteria but investigated the sustainability of technology. 'Sustainability' or 'sustainable technology' has been using in many publications, company brochures, research reports, and government documents which do not necessarily give a clear direction. Sometimes, these conventional

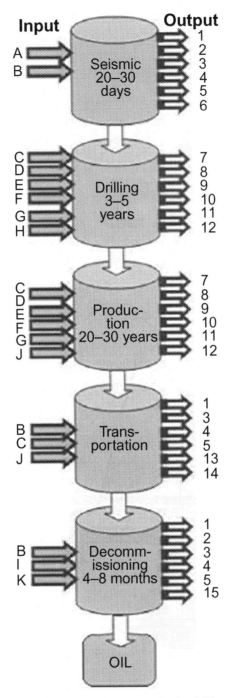

FIG. 1.29 Different phases of petroleum operations are seismic, drilling, production, transportation and processing, decommissioning, as well as their associated wastes generation and energy consumption. Output: 1. Ship source wastes, 2. Dredging effects, 3. Human related wastes, 4. Release of CO_2, 5. Conflicting with fisheries, 6. Sound effects, 7. Drilling muds, 8. Drilling cuttings, 9. Flare, 10. Radioactive materials, 11. Produced water, 12. Release of injected chemicals, 13. Ship source oil spills, 14. Toxic chemicals as corrosion inhibitors, 15. Release of metals and scraps. Input: A. Sound waves, B. Shipping operations, C. Associated inputs related to installation, D. Water-based drilling muds, E. Oil-based drilling muds, F. Synthetic-based drilling mud, G. Well testing fluids, H. Inputting casings, I. Cuttings pieces (Khan and Islam, 2007a).

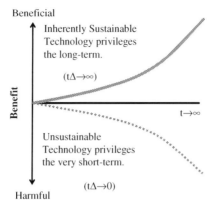

FIG. 1.30 Direction of sustainable and unsustainable technology (Khan and Islam, 2016).

approaches/definitions mislead to achieve true sustainability. Fig. 1.30 shows the directions of true sustainability in technology devolvement. It shows the direction of nature-based, inherently sustainable technology, as contrasted with an unsustainable technology. The path of sustainable technology is its long-term durability and environmentally wholesome impact, while unsustainable technology is marked by Δt approaching 0. Presently, the most commonly used theme in technology development is to select technologies that are good for $t =$ 'right now', or $\Delta t = 0$. In reality, such models are devoid of any real basis (termed "aphenomenal" by Khan and Islam, 2012), and should not be applied in technology development if we seek sustainability for economic, social, and environmental purposes.

In addition to technological details of appropriate drilling technology, the sustainability of this technology is evaluated based on the model proposed by Khan and Islam. Fig. 1.31 shows the detailed steps for its evaluation. The first step of this method is to evaluate a sustainable technology based on time criterion (Fig. 1.31). If the technology passes this stage, it would be evaluated based on criteria such as environmental, economic, and social variants. According to Khan and Islam's method, any technology is considered sustainable if it fulfills the environmental, economic, and social conditions $(C_n + C_e + C_s) \geq$ constant for any time, t, provided that, $dCn_t/dt \geq 0$, $dCe_t/dt \geq 0$, $dCs_t/dt \geq 0$.

To evaluate environmental sustainability, a proposed drilling technique is compared with conventional technology. The current drilling technologies are considered to be the most environmentally concerning activities in the whole petroleum operations. The current practices produce numerous gaseous, liquid, and solid wastes and pollutants, none of which have been completely remedied. Therefore, it is believed that conventional drilling has negative impacts on habitat, wildlife, fisheries, and biodiversity.

For analyzing the environmental consequences of drilling, conventional drilling practices need to be analyzed, which will be continued, chapter by chapter, in this book on sustainability. In conventional drilling, different types of rigs are used. However, the drilling operations are similar. The main tasks of a drill rig are completed by the hosting, circulating, and rotary system, backed up by the pressure-control equipment. A drill bit is attached at the end portion of a drill pipe. Motorized equipment rotates the drill pipe to make it cut into rocks. During

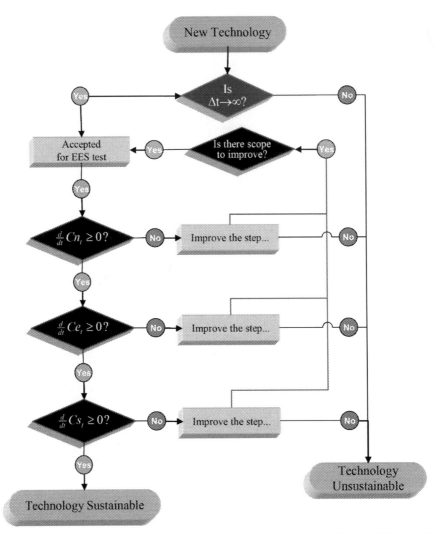

FIG. 1.31 Flowchart of sustainability analysis of a drilling technology (redrawn Islam and Hossain, 2020).

drilling, many pumps and prime movers circulate drilling fluids from tanks through a standpipe into the drill pipe and drill collar to the bit. The muds flow out of the annulus above the blowout preventer over the shale shaker (a screen to remove formation cutting), and back into the mud tanks. Drilling muds are composed of numerous chemicals, some of which are toxic, and which are harmful to the environment and its flora and fauna. These issues will be discussed in the drilling mud chapter. The conventional practice in the oil industry is to use different drilling techniques, where huge capital is involved, and which create huge environmental negative impacts. The technology is also more complicated to handle. Therefore, sustainable petroleum operation is one of the important keys for our future existence on this planet.

1.5 Sustainability status of current technologies

The challenge in seeking sustainability is in taking a fundamentally new approach. It no longer suffices to resort to innovations, these innovations have to be sustainable. In recent years, a unique way has been devised that considers the extent to which oil and gas patents are being leveraged over time by non-petroleum patents. Just as technologies combine in unique ways over time within the petroleum patent universe, technologies developed by petroleum companies are also cited outside of the petroleum universe. Higher synergies between disparate technologies move them into stronger and more central locations within the broader patent universe.

Location in the knowledge network is important because the more centrally located technologies have a greater chance of quickly incorporating innovations from neighboring technology areas. By way of contrast, technologies that are at the fringe of the knowledge network have fewer chances of bridging into totally new knowledge areas and creating genuinely disruptive innovations.

In Fig. 1.32, all patents from 2015 are used to form a map of human knowledge (as embodied by patents). Technologies are circles, and lines appear when technologies draw on each other. In the 2015 example shown below, semiconductor devices, which were among the fastest-growing patent area in the oil and gas industry at the time (Growth and volume by technology), connect widely within the patent universe because of their cross-sectional applications across industries.

Modern civilization is driven by our energy needs. Despite controversies surrounding petroleum operations and their impact on the environment, petroleum resources continue to carry the bulk of energy needs. Sustainable development can alleviate environmental impacts and place petroleum operations in a leadership position even for environmental integrity and long-term sustainability (Islam and Khan, 2019). Drilling is the primary operation that connects us to petroleum resources. As such, it is the most important operation. This technology is a necessary step for both petroleum exploration and production. While Drilling engineering a well-established discipline, the fact that every well is unique makes a drilling operation risky. In the past, risk management due to blow-out concerns and the safety of personnel have been the primary focus of a drilling operation. Over the last few decades, the concerns over environmental integrity and carbon footprint have overwhelmed petroleum operators. The challenge has been to drill faster, with greater precision in more hazardous areas or more technically challenging depths with minimum environmental damage than ever before.

While the background work of planning, involving rock/fluid characterization, environmental impact assessment, and others is a team endeavor, the execution of drilling is the responsibility of the drilling engineer. Drilling petroleum wells continues to be the most daunting task among all engineering undertakings. The most important aspect of preparing the well plan, and subsequent drilling engineering, is determining the expected characteristics and problems to be encountered in the well. A well cannot be planned properly if these environments are unknown. Therefore, the drilling engineer must initially pursue various types of data to gain insight used to develop the projected drilling conditions.

If there had to be the development of sustainable technologies, the role of 'peripheral' technologies becomes clear from Fig. 1.32. The term Energy Innovation Index (EII) was introduced by Deloitte (n.d.) to express the drift in aggregate of innovation within the petroleum industry as a whole by producing a weighted average across all technologies. Technologies with high

FIG. 1.32 The central role of drilling technology. *From Deloitte, n.d. The arc of innovation in the oil and gas industry. https://www2.deloitte.com/us/en/pages/energy-and-resources/articles/tracking-innovation-in-oil-and-gas-patents.html.*

petroleum contributions and a commanding location in the patent network contribute positively toward the O&G aggregate score. When a core technology in the oil and gas industry drifts away from the center of the patent universe or has a decreasing contribution from the petroleum industry, it causes a decline in the EII. When EII is computed, it shows a slight decline over time. This means that the overall petroleum industry seems to be moving toward the edge of the overall patent universe. This should not be interpreted as a slowdown in the pace of innovation within oil and gas, but rather that other industries have likely accelerated their intensity and interconnectedness of innovation faster over the past decade or so—a decade in which advances in IT and communications technologies have become pervasive. It also reflects the fact that many of the meaningful innovations in the oil and gas industry have come from combining existing petroleum technologies rather than "moon shot" innovations combining more distant technologies. In this context, the monitoring technology is worth a mention. New research findings have helped with accurately predicting pore pressure and fracture gradients, making it possible to advance such technologies as underbalanced, managed pressure and air drilling with low rheology drilling muds and cements. Now, if the mud and cement system had to be recalibrated based on sustainable development, one needs to reconstruct the EII, because several aspects of the innovations would have contributions from other fields.

One of the most important technologies developed in the new millennium is the so-called Zipper fracturing. It is a complete methodology commonly executed in many shale developments. It was originally called 'simulfrac' and tested in Barnett shales of the USA. Initially, operators implemented zipper fracs to enhance operational efficiency and to reduce the cycle time between frac stages through the drilling of parallel wells. The objective is to utilize fracture networks in creating greater transmissivity. This technology has become popular ever since its first implementation in 2012 and has proven to enhance production as well as ultimate recovery. EII for this technology shows strong trends because this is the technology that has sustainable technologies embedded in it.

By 2012, lesser-cited categories such as metalworking (patents comprising of new processes, tools, machines, and apparatus made from metal) had emerged as a key linking technology, or a bridge, between earth drilling and other O&G-related technologies. Combined with its high growth rate of 192% from 2006 to 2015, metalworking has become a technology area that has not only grown in volume but seemingly also in its significance within the oil and gas knowledge network.

1.5.1 Challenges in waste management

Drilling and production phases are the most waste-generating phases in petroleum operations. Drilling mud is condensed liquids that may be oil- or synthetic-based wastes and contain a variety of chemical additives and heavy minerals that are circulated through the drilling pipe to perform several functions. These functions include cleaning and conditioning the hole, maintaining hydrostatic pressure in the well, lubrication of the drill bit and counterbalance formation pressure, removal of the drill cuttings, and stabilization of the wall of the drilling hole. Water-based muds (WBMs) are a complex blend of water and bentonite. Oil-based muds (OBMs) are composed of mineral oils, barite, mineral oil, and chemical additives. Typically, a single well may lead to 1000–6000 m^3 of cuttings and muds depending on the nature of cuttings, well depths, and rock types (Khan and Islam, 2007b). A production platform generally consists of 12 wells, which may generate (62×5000 m^3) 60,000 m^3 of wastes (Khan and Islam, 2007b). Fig. 1.29 shows the supply chain of petroleum operations indicating the type of wastes generated.

The current challenge of petroleum operation is how to minimize petroleum wastes and their impact in the long term. Conventional drilling and production methods generate an enormous amount of wastes. Existing management practices are mainly focused to achieve sectoral success and are not coordinated with other operations surrounding the development site. The following are the major wastes generated during drilling and production.

a. Drilling muds
b. Produced water
c. Produced sand

d. Storage displacement water
e. Bilge and ballast water
f. Deck drainage
g. Well treatment fluids
h. Naturally occurring radioactive materials
i. Cooling water
j. Desalination brine
k. Other assorted wastes

The most significant advancement in sustainable petroleum operations has been in the areas of zero-waste engineering (Khan and Islam, 2016). This scheme emerged from petroleum policies from decades ago that required oil and gas companies to investigate no-flare technologies. Bjorndalen et al. (2005) developed a novel approach to avoid flaring during petroleum operations. Petroleum products contain materials in various phases. Solids in the form of fines, liquid hydrocarbon, carbon dioxide, and hydrogen sulfide are among the many substances found in the products. According to Bjorndalen et al. (2005), by separating these components through the following steps, no-flare oil production can be established (Fig. 1.33). Simply by avoiding flaring, over 30% of pollution created by petroleum operations can be reduced. Once the components for no-flaring have been fulfilled, value-added end products can be developed. For example, the solids can be used for minerals, the brine can be purified, and the low-quality gas can be re-injected into the reservoir for EOR.

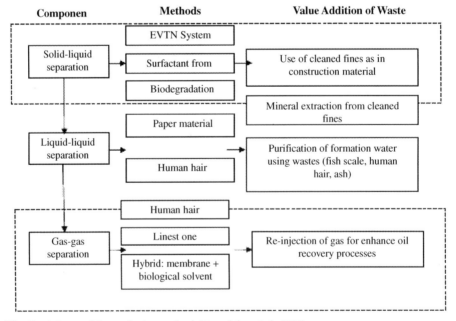

FIG. 1.33 Breakdown of the no-flaring method (Bjorndalen et al., 2005).

1.5.2 A novel desalination technique

Management of produced water during petroleum operations offers a unique challenge. The concentration of this water is very high and cannot be disposed of outside. To bring down the concentration, expensive and energy-intensive techniques are being practiced. Recently, Khan and Islam (2007b) have developed a novel desalination technique that can be characterized as a totally environment-friendly process. This process uses no non-organic chemicals (e.g., membrane, additives). This process relies on the following chemical reactions in four stages:

(1) saline water + CO_2 + NH_3 → (2) precipitates (valuable chemicals) + desalinated water → (3) plant growth in solar aquarium → (4) further desalination.

This process is a significant improvement over an existing US patent. The improvements are in the following areas:

-. CO_2 source is the exhaust of a power plant (negative cost)
-. NH_3 source is sewage water (negative cost + the advantage of organic origin)
-. Addition of plant growth in solar aquarium (emulating the world's first and the biggest solar aquarium in New Brunswick, Canada).

This process works very well for general desalination involving seawater. However, for produced water from petroleum formations, it is common to encounter salt concentration much higher than seawater. For this, water plant growth (Stage 3 above) is not possible because the salt concentration is too high for plant growth. In addition, even Stage 1 does not function properly because chemical reactions slow down at high salt concentrations. This process can be enhanced by adding an additional stage. The new process should function as:

(1) Saline water + ethyl alcohol → (2) saline water + CO_2 + NH_3 → (3) precipitates (valuable chemicals) + desalinated water → (4) plant growth in solar aquarium → (5) further desalination.

Care must be taken, however, to avoid using non-organic ethyl alcohol. Further value addition can be performed if the ethyl alcohol is extracted from fermented waste organic materials.

The process that has the biggest impact on the long-term sustainability of petroleum fluids is refining (or gas processing for gas), in which artificial chemicals are introduced in various forms (Islam, 2014, 2020). To render the process sustainable, there have to be fundamental changes to the refining process. Fig. 1.34 shows the overall picture of conventional refining and how it can be transformed. The economics of this transition is reflected in the fact that the profit made through conventional refining would be directly channeled into a reduced cost of operation.

This figure amounts to the depiction of a paradigm shift. The task of reverting to natural from unnatural has to be performed for each stage involved in the petroleum refining sector. A sustainable refinery can render the process sustainable and create enough incentive to investigate the concept of a Downhole refinery (Islam, 2020).

In our previous work, we have identified the following sources of toxicity in conventional petroleum refining:

- Use of toxic catalyst
- Use of artificial heat (e.g., combustion, electrical, nuclear)

48 1. Introduction

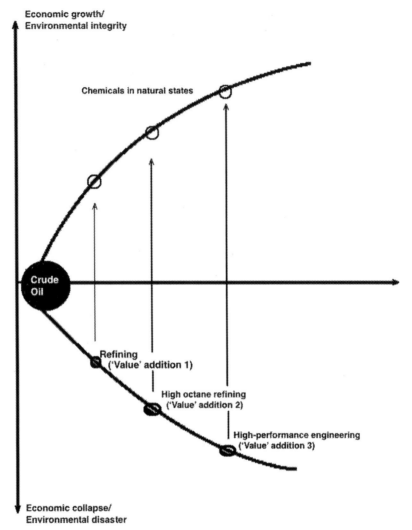

FIG. 1.34 Natural chemicals can turn a sustainable process into a sustainable process while preserving similar efficiency.

The use of toxic catalysts contaminates the pathway irreversibly. These catalysts should be replaced by natural performance enhancers. In this proposed theme, research should be performed in order to introduce catalysts that are available in their natural state. This will make the process environmentally acceptable and will reduce the cost very significantly.

The problem associated with efficiency is often covered up by citing the local efficiency of a single component. When global efficiency is considered, artificial heating proves to be utterly inefficient. Recently, Khan and Islam (2016) have demonstrated that direct heating with solar energy (enhanced by a parabolic collector) can be very effective and environmentally sustainable. They achieved up to 75% of global efficiency as compared to some 15% efficiency when solar energy is used through electricity conversion. They also discovered that the temperature generated by the solar collector can be quite high, even for cold countries. Note that direct solar heating or wind energy doesn't involve the conversion into electricity that would otherwise introduce toxic battery cells and would also make the overall process very low inefficiency.

To introduce a total zero-waste scheme, a green supply chain framework is introduced to achieve sustainability in refining operations. The specific aspects of the model include:

a. Zero emissions (air, soil, water, solid waste, hazardous waste)
b. Zero waste of resources (energy, materials, human)
c. Zero waste in activities (administration, production)
d. Zero use of toxics (processes and products)
e. Zero waste in the product life cycle (transportation, use, end-of-life)

We have seen the scientific analysis that gives an optimistic picture of the reserve development. By properly characterizing reservoirs and using technologies that best suit the broader sustainability picture, one can make the reserve grow continuously (Fig. 1.35). This picture can be further enhanced by using a zero-waste scheme at every stage of EOR operations. This can be further bolstered by using petroleum products in different applications, based on their long-term impact. This concept has been tested by Islam et al., 2018a).

Consider the use of zero-waste engineering, which is the only truly sustainable oil recovery technique. Fig. 1.36 shows how enhanced oil recovery adds to the profitability over conventional primary recovery processes. Enhanced oil recovery schemes are well established and use either added chemicals or energy (e.g., steam) to increase profitability. This profitability

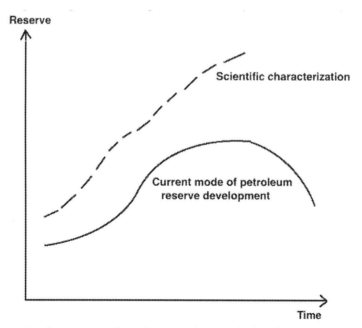

FIG. 1.35 Unconventional reserve growth can be given a boost with scientific characterization.

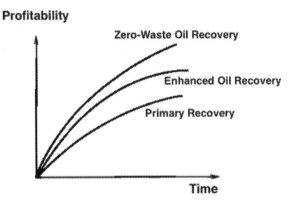

FIG. 1.36 Profitability grows continuously with time when a zero-waste oil recovery scheme is introduced.

goes up tremendously if zero-waste schemes are added. Zero-waste represents the use of waste products from the petroleum operation and other naturally available materials that are abundant in the locality of the petroleum operation site.

In summary, sustainable production ensures exponential economic growth. Such production tactics, when coupled with sustainable economic models offer a true paradigm shift in all aspects of economic development. Based on the discussion presented in this section, the following conclusions can be reached.

1. Total sustainability has eluded petroleum operators due to myopic vision of profit maximization in the short term. A long-term approach involves environmental considerations before profit-making.
2. EOR is an integral part of petroleum operations and a sustainable approach involves considering a zero-waste approach in all aspects, including refining, drilling, and production. Most importantly, however, rendering refinery zero-waste can help sustain the EOR process, due to optimization of the energy and mass cycle.
3. Series of novel zero-waste technologies are introduced in order to demonstrate how steps can be taken to keep the EOR totally sustainable.
4. Sustainable processes show very high global efficiency.
5. By minimizing processing with artificial chemicals, great strides are made toward achieving both environmental and economic sustainability.

1.6 Summary of various chapters

This chapter gives the big picture of energy sustainability in today's world. It deconstructs some of the myths regarding renewable energy and the role of fossil fuel. These myths relate to political positions rather than scientific stands. After demystifying the popular narrative, the challenges of rendering petroleum operations are highlighted.

Chapter 2 presents details of reservoir rock and fluid characterization techniques. The most important and unique features of petroleum reservoirs are discussed. Sustainability criteria

1.6 Summary of various chapters

are presented in order to show the sustainability status of various techniques. In order to introduce a truly sustainable technique, a fluid characterization technique, based on the origin of fluid is introduced. Pathway analysis of crude oil and gas is performed to determine the path of sustainable development. Finally, a sustainable characterization technique, free from some of the spurious assumptions commonly associated with the conventional technique, is presented.

Chapter 3 details important features of complex reservoirs. Reservoir heterogeneity is discussed from a scientific standpoint. Scientifically, the entire history of the rock needs to be considered. Following this analysis, the scaling up to geological scale (both time and space) is performed. The role of thermal stress and geochemistry in shaping reservoir properties is discussed for various types of reservoirs. Finally, changes invoked through drawdown (or fluid injection) are presented and their impact on production is discussed.

Chapter 4 is dedicated to unconventional oil and gas reservoirs. Current potentials are presented in view of extraordinary developments in the areas of fracking over the last 13 years. Various types of unconventional oil and gas plays are reviewed to assess future target areas. Optimization of these resources can increase the total potential significantly. In terms of sustainability, the most important considerations involve the selection of fracking fluids and proppants used in the process. Current practices being unsustainable, an array of sustainable techniques will be presented. Finally, a framework of sustainable process development will be discussed.

Chapter 5 presents the development of basement reservoirs. Only recently these reservoirs have come to prominence. While the potential of these reservoirs in terms of boosting oil production is recognized, conventional techniques of reservoir development do not apply to these reservoirs. This chapter introduces a new approach to reservoir characterization, especially suitable for basement reservoirs. This approach enables one to set correct developmental strategies. As a bonus, the prospect of value addition through waste conversion instead of waste minimization. One of the important features of the newly introduced characterization tool is that it endorses the targeted application of crude oil in its native form, rather than refining them to suit different forms of fuels. This range of applications is described in this chapter.

Chapter 6 presents reserve prediction and deliverability techniques. Conventional material balance considerations are shown along with improvements that have been introduced in recent years. The material balance equations can produce sustainable results only if non-linear solutions are determined. These solutions are available for limited cases and are presented in this chapter. Risk analysis and sustainability considerations are discussed with practical pieces of advice for practicing engineers.

Chapter 7 provides a guideline for practicing engineers. One of the most important aspects of engineering calculations is scaling. This is particularly important for reservoir engineering—a discipline in which the model to prototype ratio is very small and the difference between the two is very different. After establishing a correct set of scaling rules, risk analysis can be performed. This sets the stage for ranking various reservoir development schemes according to their global efficiency. The end result is a comprehensive technique for developing sustainable reservoir development tools.

Chapter 8 presents a comprehensive conclusion.

CHAPTER

2

Reservoir rock and fluid characterization

2.1 Introduction

Until now, 'sustainable petroleum development' is considered to be an oxymoron (Islam et al., 2010). Even today, the way the term 'sustainability' is used, it makes it difficult to discuss petroleum operations and sustainability in the same vein. From the 1973 Arab oil embargo onward, when cheap oil became a thing of the past, there has been a sustained campaign against fossil fuel, in general, and oil and gas, in particular. Then came Al Gore's 'saving the planet from Carbon' awakening. The Greenpeace movement designated Carbon as the existential threat to the current civilization. At the same time, Enron—the most "innovative energy management" company—turned out to be a fraud. Even George, Republican president, chimed in to castigate humanity with the 'oil addiction' line. Carbon is the enemy mantra spread like wildfire, and imposing universal carbon tax reached global pitch. Until now, the world is convinced that petroleum consumption should be minimized, the oil price should be low, and the replacement of petroleum should be subsidized. So, what is left for petroleum engineers to do other than folding shops and hiding behind an alternate fuel 'wall'? If it were not for the opening up of unconventional oil and gas that gave rise to unprecedented surge in oil and gas production in the United States and equally important surge in global reserve in terms of heavy oil and tar sand, no book on sustainable petroleum development would see the light of the day. It is this unprecedented surge in petroleum activities, along with renewed focus on environmental sustainability that a book of current title is necessary and timely. Even when the oil price dropped to the unprecedented negative territory and COVID-19 measures took a great toll on global energy consumption, nothing substantial has happened to global production of oil and gas.

2.1.1 Petroleum

The United States is a hydrocarbon-based culture with petroleum and natural gas being the main sources of hydrocarbons. Unfortunately, the USA is one of the largest importers of

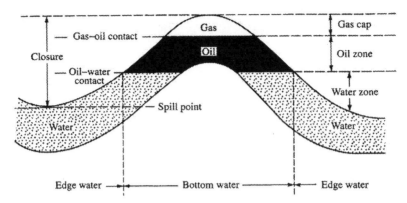

FIG. 2.1 Typical anticlinal petroleum trap.

petroleum and, as the imports of crude oil into the United States continue to rise, it is interesting, perhaps frightening, that the United States now imports approximately 65% of its daily crude oil (and crude oil products) requirements. As per recent events, it seems to be little direction in terms of stability of supply or any measure of self-sufficiency in liquid fuel precursors, other than resorting to military action. This is particularly important for the US refineries, since disruption in supply could cause major shortfalls in feedstock availability. In addition, the crude oils available to the refinery today are quite different in composition and properties to those available some 50 years ago. The current crude oils are somewhat heavier as they have higher proportions of nonvolatile (asphaltic) constituents. Changes in feedstock character, such as this tendency to heavier (higher boiling) materials (heavy oils), require adjustments to refinery operations to handle these heavier crude oils to reduce the amount of coke formed during processing and to balance the overall product slate (Speight and Islam, 2016).

However, petroleum (crude oil) is found in a reservoir, which is a subsurface collection of hydrocarbons contained in porous or fractured rock formation. The hydrocarbons are trapped by impermeable underlying and overlying rock formations (Fig. 2.1). Natural gas also occurs with petroleum as a gas cap (associated natural gas) or it may occur on its own in a gas reservoir (unassociated natural gas).

2.1.2 Natural gas

Natural gas is a gaseous hydrocarbon-based fossil fuel that primarily consists of methane but contains significant quantity of ethane, propane, butane and other hydrocarbons up to octane as well as carbon dioxide, nitrogen, helium, and hydrogen sulfide (Table 2.1). Natural gas is found with petroleum in petroleum reservoirs (associated natural gas) (Fig. 2.1), in natural gas reservoirs (non-associated natural gas), and in coal beds (coalbed methane). Natural gas is often informally referred to as simply gas; before it can be used to produce hydrocarbons, it must undergo extensive processing (refining) to remove almost all materials other than methane (Mokhtab et al., 2006). The by-products of that processing include ethane,

TABLE 2.1 Range of composition (% v/v) of natural gas.

Methane	CH_4	70–90
Ethane	C_2H_6	0–20
Propane	C_3H_8	
Butane	C_4H_{10}	
Pentane and higher boiling hydrocarbons	C_5H_{12+}	0–10
Carbon dioxide	CO_2	0–8
Nitrogen	N_2	0–5
Hydrogen sulfide, carbonyl sulfide	H2S, COS	0–5
Oxygen	O_2	0–0.2
Rare gases: argon, helium, neon, xenon	A, He, Ne, Xe	Trace

propane, butanes, pentanes and higher-molecular-weight hydrocarbons, elemental sulfur, and sometimes helium and nitrogen. Gas processing (gas refining) usually involves several processes to remove: (1) oil; (2) water; (3) elements such as sulfur, helium, and carbon dioxide; and (4) natural gas liquids. In addition, it is often necessary to install scrubbers and heaters at or near the wellhead that serve primarily to remove sand and other large-particle impurities. The heaters ensure that the temperature of the natural gas does not drop too low and form a hydrate with the water vapor content of the gas stream.

In addition to hydrogen sulfide and carbon dioxide, the gas may contain other contaminants, such as mercaptans (also called thiols, R–SH) and carbonyl sulfide (COS). The presence of these impurities may eliminate some of the sweetening processes, since some processes remove large amounts of acid gas but not to a sufficiently low concentration. On the other hand, there are those processes that are not designed to remove (or are incapable of removing) large amounts of acid gases. However, these processes are also capable of removing the acid gas impurities to very low levels when the acid gases are present in low to medium concentrations in the gas. Initially, natural gas receives a degree of cleaning at the wellhead. The extent of the cleaning depends upon the specifications that the gas must meet to enter the pipeline system. For example, natural gas from high-pressure wells is usually passed through field separators at the well to remove hydrocarbon condensate and water. Natural gasoline, butane, and propane are usually present in the gas, and gas-processing plants are required for the recovery of these liquefiable constituents. The production of hydrocarbons (either for fuel use or chemical use) from sources other than petroleum broadly covers liquid fuels that are produced from tar sand (oil sand), bitumen, coal, oil shale, and natural gas. Synthetic liquid fuels have characteristics approaching those of liquid fuels generated from petroleum, but differ because the constituents of synthetic liquid fuels do not occur naturally in the source material used for their production. Thus, the creation of hydrocarbons to be used as fuel from sources other than natural crude petroleum broadly defines synthetic liquid fuels. For much of the twentieth century, the synthetic fuel emphasis was on liquid products derived from coal upgrading or by extraction or hydrogenation of organic matter in coke liquids, coal tars, tar sands, or bitumen deposits.

The potential of natural gas, which typically has 85%–95% methane, has been recognized as a plentiful and clean alternative feedstock to crude oil. Currently, the rate of discovery of proven natural gas reserves is increasing faster than the rate of natural gas production. Many of the large natural gas deposits are located in areas where abundant crude oil resources lie, such as in the Middle East. However, huge reserves of natural gas are also found in many other regions of the world, providing oil-deficient countries with access to a plentiful energy source. The gas is frequently located in remote areas far from centers of consumption, and pipeline costs can account for as much as one-third of the total natural gas cost. Thus, tremendous strategic and economic incentives exist for gas conversion to liquids, especially if this can be accomplished on site or at a point close to the wellhead, where shipping costs become a minor issue.

2.1.3 Natural gas hydrates

Natural gas hydrates (gas hydrates) are crystalline solids in which a hydrocarbon, usually methane, is trapped in a lattice of ice. They occur in the pore spaces of sediments, and may form cement, nodes, or layers. Gas hydrates are found in naturally occurring deposits under ocean sediments or within continental sedimentary rock formations. The worldwide amount of carbon bound in gas hydrates is conservatively estimated at a total of twice the amount of carbon to be found in all known fossil fuels on Earth. Gas hydrates occur abundantly in nature, both in the Arctic regions and in marine sediments. A gas hydrate is a crystalline solid consisting of gas molecules, usually methane, each surrounded by a cage of water molecules. It looks very much like ice. Methane hydrate is stable in ocean floor sediments at water depths greater than 300 m, and where it occurs, it is known to cement loose sediments into a surface layer several hundred meters thick.

The role of methane trapped in marine sediments as a hydrate represents such an immense hydrocarbon reservoir that it must be considered a dominant factor in estimating unconventional energy resources; the role of methane as a greenhouse gas also must be carefully assessed. Hydrates have major implications for energy resources and the climate, but the natural controls on hydrates and their impacts on the environment are very poorly understood. The extraction of methane from hydrates could provide an enormous energy and petroleum feedstock resource. In addition, conventional gas resources appear to be trapped beneath methane hydrate layers in ocean sediments. The immense volumes of gas and the richness of the deposits may make methane hydrates a strong candidate for development as an energy resource.

2.1.4 Tar sand bitumen

Tar sand bitumen is another source of hydrocarbon fuel that is distinctly separated from conventional petroleum. Tar sand, also called oil sand (in Canada), or the more geologically correct term, bituminous sand, is commonly used to describe a sandstone reservoir that is impregnated with a heavy, viscous bituminous material. Tar sand is actually a mixture of sand, water, and bitumen, but many of the tar sand deposits in countries other than Canada lack the water layer that is believed to facilitate the hot water recovery process. The heavy bituminous material has a high viscosity under reservoir conditions and cannot be retrieved from a well by conventional production techniques.

Geologically, the term tar sand is commonly used to describe a sandstone reservoir that is impregnated with bitumen, a naturally occurring material that is solid or near solid and is substantially immobile under reservoir conditions. The bitumen cannot be retrieved from a well by conventional production techniques, including currently used enhanced recovery techniques. In fact, tar sand is defined (FE-76-4) in the United States as (U.S. Congress, 2005):

> The several rock types that contain an extremely viscous hydrocarbon which is not recoverable in its natural state by conventional oil well production methods including currently used enhanced recovery techniques. The hydrocarbon-bearing rocks are variously known as bitumen-rocks oil, impregnated rocks, tar sands, and rock asphalt.

In addition to this definition, several tests must be carried out to determine whether or not in the first instance, a resource is a tar sand deposit (Speight, 2008). A core taken from a tar sand deposit, and the bitumen isolated therefrom, are certainly not identifiable by the preliminary inspections (sight and touch) alone. The relevant position of tar sand bitumen in nature is best illustrated by comparing its position relevant to petroleum and heavy oil. Thus, petroleum is referred to generally as a fossil energy resource (Fig. 2.2) and is further classified as a hydrocarbon resource and, for illustrative (or comparative) purposes in this course. Coal and oil shale kerogen are included in this classification. However, the inclusion of coal and oil shale under the broad classification of hydrocarbon resources has required (incorrectly) the term hydrocarbon which can be expanded to include these resources. It is essential to recognize that resources such as coal, oil shale kerogen, and tar sand bitumen contain large proportions of heteroatomic species. Heteroatomic species are those organic constituents that contain atoms other than carbon and hydrogen, e.g., nitrogen, oxygen, sulfur, and metals (nickel and vanadium).

Coal and oil shale kerogen are included in the category of hydrocarbon resources because these two natural resources (coal and oil shale kerogen) will produce hydrocarbons by thermal decomposition (high-temperature processing). Therefore, if either coal and/or oil shale

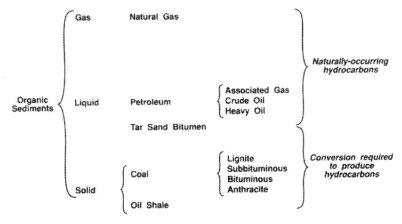

FIG. 2.2 Informal classification of organic sediments by their ability to produce hydrocarbons.

kerogen are to be included in the term hydrocarbon resources, it is more appropriate. They can be classed as hydrocarbon producing resources under the general classification of organic sediments (Fig. 2.2). Thus, tar and bitumen stand apart from petroleum and heavy oil not only from the method of recovery but also from the means by which hydrocarbons are produced. It is incorrect to refer to tar sand bitumen as tar or pitch. In many parts of the world, bitumen is used as the name for road asphalt. Although the word tar is somewhat descriptive of the black bituminous material, it is best to avoid its use with respect to natural materials. More correctly, the name tar is usually applied to the heavy product remaining after the destructive distillation of coal or other organic matter. Pitch is the distillation residue of various types of tar.

2.1.5 Coal

Coal is a fossil fuel formed in swamp ecosystems where plant remains were saved by water and mud from oxidation and biodegradation. Coal is a combustible black or brownish-black organic rock and is composed primarily of carbon along with assorted other elements, including hydrogen and oxygen. It is extracted from the ground by coal mining—either underground mining or open-pit mining (surface mining). Coal included the following classifications: (1) lignite—also referred to as brown coal and is the lowest rank of coal, used almost exclusively as fuel for steam-electric power generation; (2) sub-bituminous coal—the properties of which range from those of lignite to those of bituminous coal and are used primarily as fuel for steam-electric power generation; (3) bituminous coal—a dense coal, usually black, sometimes dark brown, often with well-defined bands of bright and dull material, used primarily as fuel in steam-electric power generation, with substantial quantities also used for heat and power applications in manufacturing and to make coke; and (4) anthracite—the highest rank coal which is a hard, glossy, black coal used primarily for residential and commercial space heating. Coal can be converted into liquid fuels by indirect liquefaction. It involves the gasification of coal to mixtures of carbon monoxide and hydrogen (synthesis gas) followed by the application of the Fischer–Tropsch process in which the synthesis gas is converted to hydrocarbons under catalytic conditions of temperature and pressure.

2.1.6 Oil shale

Oil shale is a fine-grained sedimentary rock containing relatively large amounts of organic matter (called kerogen) from which a significant amount of shale oil and combustible gas can be extracted by destructive distillation. Oil shale, or the kerogen contained therein, does not have a definite geological definition or specific chemical formulas. Different types of oil shales vary by the chemical content, type of kerogen, age, and depositional history, including the organisms from which they were derived. Based upon the environment of deposition, different oil shales can be divided into three groups: terrestrial origin, lacustrine (lake) origin, and marine origin. The term oil shale is a misnomer. The shale does not contain oil, nor is it commonly shale. The organic material is mainly kerogen, and the shale is usually a relatively hard rock, called marl. Properly processed, kerogen can be converted into a substance somewhat similar to petroleum. However, the kerogen in oil shale has not gone through the oil window through which petroleum is produced. Kerogen needs to be converted into a liquid hydrocarbon product which must be heated to a high temperature. This process converts the organic material into a liquid which must be further processed to produce oil. The organic content of oil shale is much higher than that of normal and ordinary rocks. The typical oil shale range starts from 1% to 5% by mass (lean shale) to 15%–20% by mass (rich shale). This natural resource is widely scattered throughout the entire world, and occurrences are scientifically closely linked to the history and geological evolution of the earth. Its utilization has a long history due to its abundance and wider distribution throughout the world. It is also obvious that these shale must have been relatively easy sources for domestic energy requirements for the ancient world. Solid fuels were more convenient in ancient human history due to the ease of handling and transportation. Such examples are plentiful, including wood and coal.

To extract hydrocarbons, oil shale is typically subjected to a thermal treatment and is scientifically categorized as destructive distillation. A collective scientific term for hydrocarbons in oil shale is called kerogen, an ill-defined macromolecule which, when heated, undergoes both physical and chemical changes. Physical changes involve phase changes, softening, expansion, and oozing through pores, while chemical changes typically involve bond cleavages mostly in carbon–carbon bonds that result in smaller and simpler molecules. The chemical change is often termed pyrolysis or thermal decomposition. The pyrolysis reaction is endothermic in nature, requires heat, and produces lighter molecules, thereby increasing the pressure.

2.1.7 Wax

Naturally occurring wax is often referred to as mineral wax. It occurs as a yellow to dark brown solid substance that is composed largely of paraffins. Fusion points vary from 60°C (140°F) to as high as 95°C (203°F). They are usually found associated with considerable mineral matter, as a filling in veins and fissures or as an interstitial material in porous rocks. The similarity in character of these native products is substantiated by the fact that, with minor exceptions where local names have prevailed. The original term ozokerite (ozocerite) has served without notable ambiguity for mineral wax deposits. The word Ozokerite (ozocerite) initiated from the Greek meaning odoriferous wax. It is a naturally occurring hydrocarbon material composed of mainly solid paraffin and cycloparaffin. Ozocerite usually occurs as stringers and veins that fill rock fractures in tectonically disturbed areas. It is predominantly paraffinic material (containing up to 90% non-aromatic hydrocarbons) with a high content (40%–50%) of normal or slightly branched paraffin as well as cyclic paraffin derivatives. Ozocerite contains approximately 85% carbon, 14% hydrogen, and 0.3% sulfur and nitrogen. Therefore, wax is predominantly a mixture of pure hydrocarbons. Any non-hydrocarbon constituents are in the minority. Crude ozocerite is black. However, after refining, its colour varies from yellow to white. It hardens on aging and the hardness varies according to its source and refinement. Ceresin is a white to yellow waxy mixture of paraffin hydrocarbons obtained by purification of ozocerite. The specific gravity of ozocerite ranges from 0.85 to 0.96, and the melting point falls in the range 60–95°C (140–200°F). The flash point is high, of the order of 205°C (400°F). Ceresin (ceresine, cerasin), a chemical relative of ozocerite, has lower melting point at 55–72°C (130–160°F). Both waxes are non-toxic and non-hazardous, thus permitting use in personal-care applications.

2.1.8 Biomass

Biomass refers to (a) energy crops grown specifically to be used as fuel, such as fast-growing trees or switch grass; (b) agricultural residues and byproducts, such as straw, sugarcane fiber, and rice hulls; and (c) residues from forestry, construction, and other wood-processing industries. Biomass is material that is derived from plants (Wright et al., 2010) and there are many types of biomass resources currently used and potentially available. Biomass is a term used to describe any material of recent biological origin, including plant materials such as trees, grasses, agricultural crops, and even animal manure. Other biomass components, which are generally present in minor amounts, include triglycerides, sterols, alkaloids, resins, terpenes, terpenoids, and waxes. This includes everything from primary sources of crops and residues harvested/collected directly from the land, to secondary sources such as sawmill residuals, to tertiary sources of post-consumer residuals that often end up in landfills. A fourth source, although not usually categorized as such, includes the gases that result from anaerobic digestion of animal manures or organic materials in landfills (Wright et al., 2010).

The production of hydrocarbons from renewable plant-based feedstocks utilizing state-of-the-art conversion technologies presents an opportunity to maintain competitive advantage and contribute to the attainment of national environmental targets. Bioprocessing routes have a number of compelling advantages over conventional petrochemical production. However, it is only in the last decade that rapid progress in biotechnology has facilitated the commercialization of a number of plant-based chemical processes. It is widely recognized that further significant production of plant-based chemicals will only be economically viable in highly integrated and efficient production complexes producing a diverse range of chemical products. This bio-refinery concept is analogous to conventional oil refineries and petrochemical complexes that have evolved over many years to maximize process synergies, energy integration, and feedstock utilization to drive down production costs.

Plants offer a unique and diverse feedstock for hydrocarbons. Plant biomass can be gasified to produce synthesis gas, a basic chemical feedstock and also a source of hydrogen for a future hydrogen economy. In addition, the specific components of plants such as carbohydrates, vegetable oils, plant fiber, and complex organic molecules known as primary and secondary metabolites can be utilized to produce a range of valuable monomers, chemical intermediates, pharmaceuticals and materials. They are summarized as follows:

(1) *Carbohydrates* (starch, cellulose, sugars): starch is readily obtained from wheat and potatoes, whilst cellulose is obtained from wood pulp. The structures of these polysaccharides can be readily manipulated to produce a range of biodegradable polymers with properties similar to those of conventional plastics such as polystyrene foams and polyethylene film. In addition, these polysaccharides can be hydrolyzed, catalytically or enzymatically, to produce sugars, a valuable fermentation feedstock for the production of ethanol, citric acid, lactic acid, and dibasic acids such as succinic acid.
(2) *Vegetable oils*: vegetable oils are obtained from seed oil plants such as palm, sunflower, and soya. The predominant source of vegetable oil in many countries is rapeseed oil. Vegetable oils are a major feedstock for the oleo-chemical industry (surfactants, dispersants, and personal care products) and are now successfully entering new markets

such as diesel fuel, lubricants, polyurethane monomers, functional polymer additives, and solvents.
(3) *Plant fibers*: lignocellulosic fibers extracted from plants such as hemp and flax can replace cotton and polyester fibers in textile materials and glass fibers in insulation products.
(4) *Specialties*: plants can synthesize highly complex bioactive molecules, often beyond the power of laboratories. A wide range of chemicals are currently extracted from plants for a wide range of markets. They come from crude herbal remedies through very high-valued pharmaceutical intermediates.

2.2 Unique features

We call crude oil and petroleum fossil fuels because they are mixtures of hydrocarbons that formed from the remains of animals and plants (diatoms) that lived millions of years ago in a marine environment before the existence of dinosaurs (Website 5). Over millions of years, the remains of these animals and plants were covered by layers of sand, silt, and rock. Heat and pressure from these layers turn the remains into what is called crude oil or petroleum. The word petroleum The word petroleum (literally "rock oil"from the Latin petra, "rock"or "stone," and oleum, "oil") was first used in 1556 in a treatise published by the German mineralogist Georg Bauer, known as Georgius Agricola. The reference to biogenic origin (Figs. 2.3 and 2.4) is not universally accepted (Islam et al., 2018).

The origin of oil and gas has been a long-debated theoretical issue. There are two opposing points of view: (1) the organic origin theory and (2) the inorganic origin theory. Organic origin theory considers oil and gas to come from biological processes. Inorganic origin theory explains the origin of oil and gas through inorganic synthesis and mantle degassing. The earliest

Petroleum and natural gas formation

Tiny marine plants and animals died and were buried on the ocean floor. Over time the marine plants and animals were covered by layers of silt and sand.

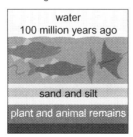
Over millions of years, the remains were buried deeper and deeper. The enormous heat and pressure turned the remains into oil and natural gas

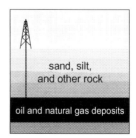
Today, we drill down through layers of sand, silt, and rock to reach the rock formations that contain oil and natural gas deposits.

Source: Adapted from National Energy Education Development Project (public domain)

FIG. 2.3 Petroleum and natural gas formation. *Source: U.S. Energy Information Administration, Website 5.*

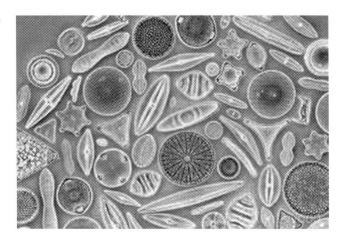

FIG. 2.4 Diatoms are considered to be the source of biogenic petroleum fluids.

organic origin theory was proposed by Lomonosov in 1763 (Dott, 1969). He hypothesized that fertile substances underground, such as oil shale, carbon, asphalt, petroleum and amber, originated in plants.

The hydrocarbon formation theory of kerogen thermal degradation proposed by Tissot and Welte (1984) are the representatives of the organic hydrocarbon generation theory. The hydrocarbon formation theory of kerogen thermal degradation is based on the diagenesis of organic matter resulting from biopolymers into geopolymers, then kerogen. Kerogen is the main precursor material for oil compounds during the process of hydrocarbon generation, when thermal degradation plays a major role. For sufficient hydrocarbon production and commercial oil gathering, sedimentary rocks must experience the hydrocarbon generation and temperature threshold. Mass hydrocarbons are formed at temperatures from 60 to 150°C by heated organic matter (Hunt, 1995). According to this theoretical model, sedimentary organic matter maturity, especially for kerogen, becomes the key factor for evaluating hydrocarbon potential. When the threshold burial depth is reached, kerogen will be changed from immature to mature. The theory has been accepted gradually by the majority of petroleum geologists and plays a major role in oil and gas exploration. By contrast, the non-organic origin of petroleum is considered to be controversial. However, a significant number of reservoir fluids cannot be explained by the conventional organic origin concept.

Over the last 30 years, geochemical research has demonstrated that abiotic methane (CH4), formed by chemical reactions which do not directly involve organic matter, occurs on Earth in several specific geologic environments. Methane as well as higher chain molecules of hydrocarbons can be produced by either high-temperature magmatic processes in volcanic and geothermal areas, or via low-temperature (100°C) gas-water-rock reactions in continental settings, even at shallow depths. The isotopic composition of C and H is a first step in distinguishing abiotic from biotic (including either microbial or thermogenic) CH4 (Etiope and Lollar, 2013). Etiope and Lollar (2013) reviewed the major sources of abiotic CH4 and the primary approaches for differentiating abiotic from biotic CH4, including novel potential tools such as clumped isotope geochemistry. They also proposed a diagnostic approach for differentiation between the two sources.

Curiously, there has been little consideration of the origin of crude oil and gas for the processing into fuel (both refining and gas processing). After crude oil is removed from the ground, it is sent to a refinery where different parts of the crude oil are separated into useable petroleum products. These petroleum products include gasoline, distillates such as diesel fuel and heating oil, jet fuel, petrochemical feedstocks, waxes, lubricating oils, and asphalt. Learn more about refining crude oil—inputs and outputs. A U.S. 42-gal barrel of crude oil yields about 45 gal of petroleum products in U.S. refineries because of refinery processing gains (Fig. 2.5). This increase in volume is due to linearization of long-chain molecules as well as numerous chemicals added to enhance the upgrading process. The amount of individual products produced varies from month-to-month and year-to-year as refineries adjust production to meet market demand and to maximize profitability. The sustainability of the refining process has been discussed in detail by Islam and Khan (2019). One key point of sustainability is custom design based on the origin of oil or gas. Conventionally, this has not been in the petroleum industry.

The most unique feature of petroleum is its origin and the amount of time spent in nature during the natural cycle of maturation. Pang et al. (2015) characterized deep petroliferous

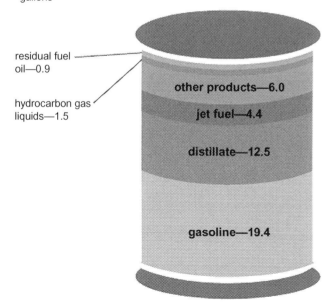

Petroleum products made from a barrel of crude oil, 2019
gallons

FIG. 2.5 Crude oil typically has a volume gain during the refining process.

Note: A 42-gallon (U.S.) barrel of crude oil yields about 45 gallons of petroleum products because of refinery processing gain. The sum of the product amounts in the image may not equal 45 because of independent rounding.

Source: U.S. Energy Information Administration, *Petroleum Supply Monthly*, April 2020, preliminary data

basins based on geological features of oil–gas reservoirs across the world. They identified 10 major geological features:

(1) While oil–gas reservoirs have been discovered in many different types of deep petroliferous basins, most have been discovered in low heat flux deep basins.
(2) Many types of petroliferous traps have been developed in deep basins, and tight oil–gas reservoirs in deep basin traps are arousing increasing attention.
(3) Deep petroleum normally has more natural gas than liquid oil, and the natural gas ratio increases with the burial depth.
(4) The residual organic matter in deep source rocks reduces but the hydrocarbon expulsion rate and efficiency increase with the burial depth.
(5) There are many types of rocks in deep hydrocarbon reservoirs, and most are clastic rocks and carbonates.
(6) The age of deep hydrocarbon reservoirs is widely different, but those recently discovered are predominantly Paleogene and Upper Paleozoic.
(7) The porosity and permeability of deep hydrocarbon reservoirs differ widely, but they vary in a regular way with lithology and burial depth.
(8) The temperatures of deep oil–gas reservoirs are widely different, but they typically vary with the burial depth and basin geothermal gradient.
(9) The pressures of deep oil–gas reservoirs differ significantly, but they typically vary with burial depth, genesis, and evolution period.
(10) Deep oil–gas reservoirs may exist with or without a cap, and those without a cap are typically of unconventional genesis. These reservoirs are discussed in Chapter 4.

Pang et al. (2015) also listed six major steps in the understanding of deep hydrocarbon reservoir formation.

(1) Deep petroleum in petroliferous basins has multiple sources and many different genetic mechanisms.
(2) There are high-porosity, high-permeability reservoirs in deep basins, the formation of which is associated with tectonic events and subsurface fluid movement.
(3) Capillary pressure differences inside and outside the target reservoir are the principal driving force of hydrocarbon enrichment in deep basins.
(4) There are three dynamic boundaries for deep oil–gas reservoirs; a buoyancy-controlled threshold, hydrocarbon accumulation limits, and the upper limit of hydrocarbon generation.
(5) The formation and distribution of deep hydrocarbon reservoirs are controlled by free, limited, and bound fluid dynamic fields.
(6) tight conventional, tight deep, tight superimposed, and related reconstructed hydrocarbon reservoirs formed in deep-limited fluid dynamic fields have great resource potential and vast scope for exploration.

Pang et al. (2015) recommended that further study should pay more attention to the following four aspects:

(1) identification of deep petroleum sources and evaluation of their relative contributions;
(2) preservation conditions and genetic mechanisms of deep high-quality reservoirs with high permeability and high porosity;

(3) facies feature and transformation of deep petroleum and their potential distribution; and.
(4) economic feasibility evaluation of deep tight petroleum exploration and development.

The unique feature of reservoir rocks extends to the realm of extremely diverse rock properties. Because the sample size commonly derived through core extraction is so small (way smaller than the Representative Elemental Volume, REV), inadequate characterization of petroleum reservoir rocks is common. Yet, the knowledge of the in situ stress state within petroleum reservoirs is vital for drilling and hydrocarbon production, as it can influence wellbore stability and hydraulic fracturing stimulation design (Pham et al., 2020). The characterization of in situ stress fields is often complicated by the presence of geological discontinuities such as faults, fractures, and changes in lithology, each of which can render the characterization process impractical.. In areas close to active faults and natural fractures, the stress field rotates as a result of stress perturbation, as evidenced by the rotation of borehole stress indicators such as wellbore breakouts (Lin et al., 2010). Both major faults and relatively small-scale natural fractures can induce significant local stress perturbations that can either rotate the orientation of borehole stress indicators from the prevailing stress orientation or perturb the stress magnitude as a result of compliant shear behavior (Pham et al., 2020).

Variations in rock mechanical properties within different stratigraphic formations can also affect the stress field, as the contrasting elastic properties of different rock units can cause discontinuous stress states in terms of stress magnitude. This has profound implications for any recovery scheme, particularly when waterflood or enhanced oil recovery is considered for reservoir development (Islam, 2020). This means that reservoirs with highly variable mechanical properties, as is often the case for carbonate-hosted reservoirs, have stress states that vary significantly. Consequently, constructing simplified stress models for these highly variable reservoirs is very difficult, hampering the resolution of issues related to wellbore stability and borehole orientation selection for hydraulic stimulation (Pham et al., 2020). Some of the field examples will be discussed in Chapter 8.

The most important feature of petroleum is the fact that it is naturally processed with an inherent quality to be used for multiple purposes. The petroleum industry was first encountered in the archaeological record near Hit in what is now Iraq. Hit is on the banks of the Euphrates river and is the site of an oil seep known locally as The Fountains Of Pitch. There, asphalt was quarried for use as mortar between building stones as early as 6000 years ago. Asphalt was also used as a waterproofing agent for baths, pottery and boats (Purdy, 1995). Such use was also common among the Babylonians, who caulked their ships with asphalt. In Mesopotamia, around 4000 BCE, bitumen-a tarry crude-was used as caulking for ships, as a setting for jewels and mosaics, and as an adhesive to secure weapon handles. The Egyptians used it for embalming, and the walls of Babylon and the famed pyramids were held together with it.

Natural deposits of asphalt occur in pits or lakes as residue from crude petroleum that has seeped up through fissures in the earth. In antiquity, bitumen was the Roman name for asphalt used as cement and mortar. About 2000 years ago, the Chinese used oil and natural gas for heat and light. Bamboo pipes carried gas into the home (Purdy, 1995). This was perhaps the first use of petroleum for generating direct energy.

Ancient Persians, 10th century Sumatrans and pre-Columbian Indians all believed that crude oil had medicinal benefits. Marco Polo found it used in the Caspian Sea region to treat camels for mange, and the first oil exported from Venezuela (in 1539) was intended as a gout treatment for the Holy Roman Emperor Charles V.

The American Indians collected oil for medicine. This was the original use of petroleum in line with other civilizations. The American settlers found its presence in the water supplies a contamination, but they learned to collect it to use as fuel in their lamps.

Native Americans near Sacramento used asphaltum to waterproof their baskets and to glue fibers of a soap-root brush to form a handle with twine. Also, hard asphaltum was used to make blades for knives and arrowheads. Their counterparts on the coast in the Mattole Valley in Northern California also harvested the sticky, dark material that made baskets airtight, secured arrowheads to wooden shafts and for some was said to have served as medicine for colds, coughs, burns and cuts (Rintoul, 1990).

Native Americans traded crude oil that they obtained from oil seeps in upstate New York among other places. The Seneca tribe traded oil for so long that all crude oil was referred to as 'Seneca Oil'. Seneca Oil was supposed to have great medicinal value. It sold for $20 per quart at a time when skilled workers were paid pennies per hour. A petroleum technology text published in 1901 states that: "It is an undisputed fact at present time that petroleum is an excellent remedy for diphtheria" (Purdy, 1995).

It is well known that petrochemicals are the main source of today's pharmaceutical drugs (Anderson, 2005). Hydrocarbons are organic molecules. Most pharmaceutical drugs are made through chemical reactions that involve the use of organic molecules. Petroleum is a plentiful source of organic molecules that feed into the drug synthesis process. Some sources cite as much as 99% of pharmaceutical feedstocks and reagents as coming from petrochemical sources. Even drugs that come from natural sources like plants are still often purified using petrochemicals, resulting in a more efficient and less costly manufacturing process. Others still, like antibiotics derived from natural fungi and microbes—namely, penicillin—often use phenol and cumene as preparatory agents.

The fact is, without petrochemicals it would in many cases be extremely difficult to make and mass produce pharmaceutical drugs, particularly at the scale needed to meet global demand.

Finally, pill capsules and coatings are also most frequently polymer based. In fact, time-release drugs depend on a tartaric acid-based polymer that slowly dissolves, administering just the right dose of medication.

Vaseline, petroleum jelly, or petrolatum, was one of the first petroleum-based "medicines." Even though today, many of its medicinal properties as a topical ointment have been discounted, it is still used to help prevent skin chapping, treat rashes and alleviate nosebleeds. It is recognized by the U.S. Food and Drug Administration as an approved over-the-counter skin protectant.

Many topical medicines, such as those to treat psoriasis, also have their basis in petrochemistry. For instance, salicylic acid gets rid of scales that show up on affected skin. It comes in lotions, creams, ointments and other treatments. In an early (1966) biosynthetic process, researchers at Kerr-McGee Oil Industries prepared salicylic acid via the microbial degradation of naphthalene. It is now commercially biosynthesized from phenylalanine. Numerous other creams and salves used to treat everything from fungus to eczema also find their basis in petrochemistry.

Other medicines, such as laxatives, have their basis in mineral oils, another petrochemical. Mineral oils are a mixture of liquid hydrocarbons produced from the distillation of petroleum and then refined to be suitable for commercial use. When used as a laxative, mineral oil works to alleviate constipation by retaining water in stool and the intestines.

Recently, the research focus has been on replacing petrochemicals with renewable biomass. For instance, Cao et al. (2013) discussed the use of fermented sucrose as an alternative to petrochemicals. They report that the succinate-producing process using microbial fermentation has been made commercially available by the joint efforts of researchers in different fields. In this review, recent attempts and experiences devoted to reducing the production cost of biobased succinate are summarized, including strain improvement, fermentation engineering, and downstream processing. Cheetri and Islam (2008) argued that petroleum resources are the longest-processed materials and, as such, should be valuable for direct applications. In this matter, heavy drugs that use heavy metals are excellent candidates for using crude oil as a raw material. This aspect will be discussed further in the latter section.

2.3 Sustainability criteria

Until now, there has been no suitable alternative to fossil fuels and all trends indicate continued dominance of the petroleum industry in the foreseeable future (Islam and Khan, 2019). Even though petroleum operations have been based on solid scientific excellence and engineering marvels, only recently has it been discovered that many of the practices are not environmentally sustainable. Practically all activities of hydrocarbon operations are accompanied by undesirable discharges of liquid, solid, and gaseous wastes (Khan and Islam, 2007b), which have enormous impacts on the environment (Islam et al., 2010). Hence, reducing environmental impact is the most pressing issue today and many environmentalist groups are calling for curtailing petroleum operations altogether. Even though there is no appropriate tool or guideline available for achieving sustainability in this sector, there are numerous studies that criticize the petroleum sector and attempt to curtail petroleum activities (Holdway, 2002). There is clearly a need to develop a new management approach to hydrocarbon operations. The new approach should be environmentally acceptable, economically profitable, and socially responsible.

The crude oil is truly a non-toxic, natural, and biodegradable product, but the way it is refined is responsible for all the problems created by fossil fuel utilization. The refined oil is hard to biodegrade and is toxic to all living objects. Refining crude oil and processing natural gas use large amounts of toxic chemicals and catalysts, including heavy metals. These heavy metals contaminate the end products and are burnt along with the fuels, producing various toxic by-products. The pathways of these toxic chemicals and catalysts show that they severely affect the environment and public health. The use of toxic catalysts creates many environmental effects that cause irreversible damage to the global ecosystem. Similarly, the use of synthetic chemicals can render a drilling operation as well as enhanced oil recovery operations unsustainable (Islam, 2020).

Crude oil is a naturally occurring liquid found in formations on the Earth consisting of a complex mixture of hydrocarbons consisting of various lengths. It contains mainly four groups of hydrocarbons, among which saturated hydrocarbons consist of straight chains of carbon atoms, aromatics consist of ring chains, asphaltenes consist of complex polycyclic hydrocarbons with complicated carbon rings, and other compounds are mostly nitrogen, sulfur, and oxygen. It is believed that crude oil and natural gas are the products of huge overburden pressure and heating of organic materials over millions of years.

Crude oil and natural gases are formed as a result of the compression and heating of ancient organic materials over a long period of time. Oil, gas and coal are formed from the remains of zooplankton, algae, terrestrial plants and other organic matter after exposure to the heavy pressure and temperature of the Earth. These organic materials are chemically changed into kerogen. With more heat and pressure, along with bacterial activities, oil and gas are formed. Fig. 2.6. is the pathway of crude oil and gas formation. These processes are all driven by natural forces.

Sustainable petroleum operations development requires a sustainable supply of clean and affordable energy resources that do not cause negative environmental, economic, and social consequences (Dincer and Rosen, 2005). In addition, it should consider a holistic approach where the whole system will be considered instead of just one sector at a time (Islam et al., 2010). In 2007, Khan and Islam developed an innovative criterion for achieving true sustainability in technological development. New technology should have the potential to be efficient and functional far into the future to ensure true sustainability. Sustainable development is seen as having four elements– economic, social, environmental, and technological.

Generally, a technology is selected based on criteria such as technical feasibility, cost effectiveness, regulatory requirements, and environmental impacts. Recently, Khan and Islam (2007b) introduced a new approach to technology evaluation based on a novel sustainability criterion. In their study, they not only considered the environmental, economic, and regulatory criteria, but investigated the sustainability of a technology. 'Sustainability' or 'sustainable technology' has been used in many publications, company brochures, research reports, and government documents which do not necessarily give a clear direction. Sometimes, these conventional approaches/definitions mislead us to achieve true sustainability. Fig. 2.7 shows the directions of true sustainability in technology devolvement. It shows the direction of nature-based, inherently sustainable technology, as contrasted with unsustainable technology. The path of sustainable technology is its long-term durability and environmentally wholesome impact, while unsustainable technology is marked by Δt

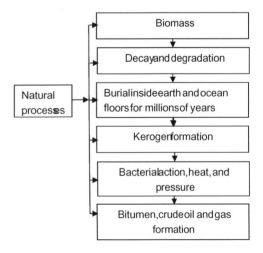

FIG. 2.6 Crude oil formation pathway. *Based on Chhetri, A.B., Islam, M.R., 2008. Inherently Sustainable Technologies. Nova Science Publishers, 452 pp.*

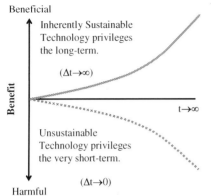

FIG. 2.7 Direction of sustainable and unsustainable technology (Khan and Islam, 2016).

approaching 0. Presently, the most commonly used theme in technology development is to select technologies that are good for $t=$'right now', or $\Delta t = 0$. In reality, such models are devoid of any real basis (termed "aphenomenal" by Khan and Islam, 2012), and should not be applied in technology development if we seek sustainability for economic, social, and environmental purposes.

In addition to the technological details of an appropriate drilling technology, the sustainability of this technology is evaluated based on the model proposed by Khan and Islam. Fig. 2.8 shows the detailed steps for its evaluation. The first step of this method is to evaluate a sustainable technology based on a time criterion (Fig. 2.9). If the technology passes this stage, it will be evaluated based on criteria such as environmental, economic, and social variants. According to Khan and Islam's method, any technology is considered sustainable if it fulfills the environmental, economic, and social conditions $(C_n + C_e + C_s) \geq$ constant for any time, t, provided that, $dCn_t/dt \geq 0$, $dCe_t/dt \geq 0$, $dCs_t/dt \geq 0$.

To evaluate the environmental sustainability, a proposed drilling technique is compared with conventional technology. The current drilling technologies are considered to be the most environmentally concerning activities in the whole petroleum operations. Current practices produce numerous gaseous, liquid, and solid wastes and pollutants, none of which have been completely remedied. Therefore, it is believed that conventional drilling has a negative impact on habitat, wildlife, fisheries, and biodiversity.

For analyzing the environmental consequences of drilling, conventional drilling practices need to be analyzed, which will be continued, chapter by chapter, in this book on sustainability. In conventional drilling, different types of rigs are used. However, the drilling operations are similar. The main tasks of a drill rig are completed by the hosting, circulating, and rotary systems, backed up by the pressure-control equipment. A drill bit is attached to the end portion of a drill pipe. Motorized equipment rotates the drill pipe to make it cut into rocks. During drilling, many pumps and prime movers circulate drilling fluids from tanks through a standpipe into the drill pipe and drill collar to the bit. The mud flows out of the annulus above the blowout preventer over the shale shaker (a screen to remove formation cutting), and back into the mud tanks. Drilling mud is composed of numerous chemicals, some of which are toxic, and which are harmful to the environment and its flora and fauna. These issues will

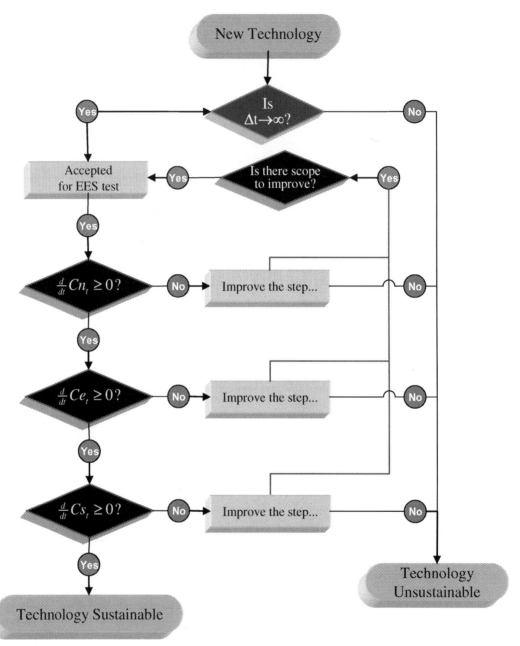

FIG. 2.8 Flowchart of sustainability analysis of a drilling technology. *Redrawn from Islam, M.R., Hossain, M.E., 2020. Drilling Engineering: Towards Achieving Total Sustainability. Elsevier, 800 pp.*

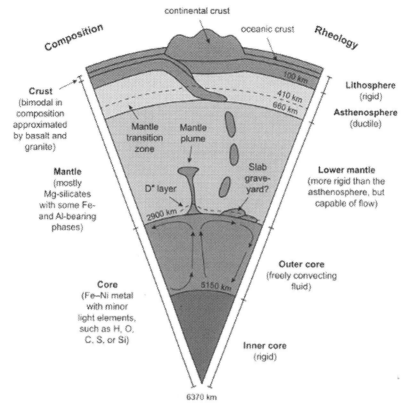

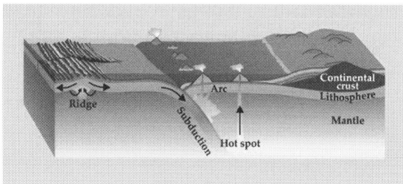

FIG. 2.9 The science of Earth's core is riddled with many unanswered questions. *From Hazen, R.M., Hemley, R.J., Mangum, A.J., 2012. Carbon in Earth's interior: storage, cycling, and life. EoS Trans. Amer. Geophys. Union 93, 17–28.*

be discussed in the drilling mud chapter. The conventional practice in the oil industry is to use different drilling techniques, where huge capital is involved, and which create huge environmental negative impacts. Technology is also more complicated to handle. Therefore, sustainable petroleum operations are one of the important keys to our future existence on this planet.

2.4 Fluid characterization based on origin

2.4.1 The carbon cycle and hydrocarbon

Carbon plays an unparalleled role in human life. After oxygen, carbon is the most important element required for sustaining human life (Islam, 2014). Carbon-based fuel is the most sustainable form of energy (Speight and Islam, 2016). It is the element of life, providing the chemical backbone for all essential biomolecules. Yet, in spite of carbon's importance, scientists remain largely ignorant of the physical, chemical, and biological behavior of carbon-bearing systems more than a few hundred meters beneath the Earth's surface. Little is known about how much carbon is stored on the Earth as a whole, how deep reservoirs form and evolve, or how carbon moves from one deep repository to another. While extraordinary claims have been made regarding the origin of hydrocarbons and how all hydrocarbons are related to living objects, little is known about the interior of the Earth, where 90% of carbon resides (Hazen et al., 2012).

Earth materials from crust to core are in constant motion: Magma generated in the upper mantle brings deep-Earth carbon to the surface in the form of CO_2, diamond, graphite, carbides, and perhaps hydrocarbons, while carbon-bearing seafloor sediments move by subduction from near-surface to deep-Earth reservoirs (Seto et al., 2008). The dynamic interface between the deep and surface carbon cycles thus includes subduction zones, volcanoes, deep hydrocarbon reservoirs, and deep life.

Carbon cycling to and from Earth's deep interior is dependent on speciation in fluids rich in carbon, oxygen, hydrogen, nitrogen, and sulfur. Recent studies reveal significant shifts in the ratio of CO_2 to methane in mantle fluids as a function of depth (Litasov et al., 2011). However, the rheology of carbon compounds, particularly the rates of carbon transfer under mantle conditions has eluded scientists. Hazen et al. (2012) listed possible deep carbon reservoirs (Table 2.2), along with Fig. 2.9 while pointing out that this is an area of great unknowns.

The conventional theory states that primitive chondritic meteorites, thought to reflect the basic composition of the early solar system, contain up to several weight percent carbon, yet estimates of Earth's total carbon inventory, though poorly constrained, are much lower, ranging from 0.07 to 1.5 wt%. This is more than a 20-fold uncertainty. Earth appears to be significantly depleted in highly volatile elements com-pared to chondrites, but uncertainty remains because of large potential reservoirs of car- carbon in Earth's mantle and core. Polymorphs of native carbon, including graphite and diamond, exemplify science's relative ignorance of the deep carbon cycle. These phases, though volumetrically minor, reveal a dynamic cycling of carbon from mantle to crust (Shirey and Richardson, 2011). Preservation of unaltered diamond crystals during eruptions of kimberlite and other ultramafic intrusions suggests transport of magma from mantle depths at velocities exceeding 100 km per hour. However, these are quite speculative in absence of a consistent theory. The conventional theory of origin of earth doesn't explain natural phenomena without resorting to dogmatic assertions. In latter chapters, details of a consistent theory, originally put forward by Islam (2014) will be presented.

Carbon cycling to and from the Earth's deep interior is dependent on speciation in fluids rich in carbon, oxygen, hydrogen, nitrogen, and sulfur. For example, recent studies reveal significant shifts in the ratio of CO2 to methane in mantle fluids as a function of depth (Litasov

TABLE 2.2 Possible deep carbon reservoirs.

Reservoir	Composition	Structure	Atom (% C)	Depth (km)	Abundance (%)
Diamond	C	Diamond	100	>150	≪1
Graphite	C	Graphite	100	<150	≪1
Carbides	SiC, FeC, Fe_3C	Moissanite, cohenite	50	?	?
Carbonates	$(Ca,Mg,Fe)CO_3$	Unknown	20	0 to?	?
Metal	Fe, Ni	Kamecite/awaurite	Minor?	?	?
Silicates	Mg-Si-O	Various	Trace?	?	?
Oxides	Mg-Fe-O	Various	Trace?	?	?
Sulfides	Fe-S	Various	Trace?	?	?
Silicate melts	Mg-Si-O	–	Trace?	?	?
CHON fluids	C-H-O-N	–	Variable	?	?
Methane	CH_4	–	20	?	?
Methane clathrate	$[H_2O + CH_4]$	Clathrate	Variable	?	?
Hydrocarbons	C_nH_{2n+2}	–	Variable	?	?
Organic species	C-H-O-N	–	Variable	?	?
Deep life	C-H-O-N-P-S	–	Variable	<15	?

Question mark denotes unknown.
From Hazen, R.M., Hemley, R.J., Mangum, A.J., 2012. Carbon in Earth's interior: storage, cycling, and life. EoS Trans. Amer. Geophys. Union 93, 17–28.

et al., 2011). Only recently, has it been experimentally determined that CO2 and other carbon-bearing oxides can reduce ferrous iron (Fe2+) to methane and other hydrocarbon products. This discovery makes room for hydrocarbon generation without the involvement of organic matter. Living organisms hold only a small fraction of the Earth's carbon, yet the biological cycling of carbon is relatively rapid. It is of relevance to consider how this active pool of biological carbon links to the slower deep cycle, and is biologically processed carbon represented in deep-Earth reservoirs. Only recently, have surprising discoveries of deep microbial life in terrestrial and oceanic environments pointed to a rich subsurface biota that may rival or eclipse all surface life in total biomass (Gold, 1999). It is conceivable that more surprises will be encountered once we become more knowledgeable about the deep environment and how life can be sustained in extreme environments.

Through a series of chemical reactions and tectonic activity, carbon takes between 100 and 200 million years to move between rocks, soil, ocean, and atmosphere in the carbon cycle. On average, 10^{13} to 10^{14} g (10–100 million metric tons) of carbon move through the carbon cycle every year. In comparison, human emissions of carbon to the atmosphere are on the order of 10^{15} g, whereas the fast carbon cycle moves 10^{16} to 10^{17} g of carbon per year. Only a trace amount of this carbon cycle is affected by human activities. That is not to say human activities cannot alter the natural course of the ecosystem (see Islam et al., 2012 for details).

In nature, the movement of carbon from the atmosphere to the lithosphere begins with rain. Globally, rain is the main source of fresh water for plants and animals rainfall is essential for life across Earth's landscapes. In addition to moving tremendous amounts of water through Earth's atmosphere, rain clouds also move tremendous amounts of energy. When water evaporates from the surface and rises as vapor into the atmosphere, it carries heat from the sun-warmed surface with it. Later, when the water vapor condenses to form cloud droplets and rain, the heat is released into the atmosphere. This heating is a major part of Earth's energy budget and climate. The average annual precipitation of the entire surface of our planet is estimated to be about 1050 mm per year or approximately 88 mm per month, amounting to a standardized value of 5.36×10^{14} m^3 (Pidwirny, 2008).[a] Because the Earth's average annual rainfall is about 100 cm (39 in.), the average time that the water spends in the atmosphere, between its evaporation from the surface and its return as precipitation, is about 1/40 of a year, or about nine days (Encyclopaedia Britannica, 2008).

Of course, water transport can take on many forms such as rain, snow, or hail although the overwhelming majority occurs as rain. Precipitate forms as moist, warm air rises and cools. During the cooling process, the water vapor collects on condensation nuclei like dust particles which results in the formation of clouds. When the water droplets fuse to create large drops too heavy to be sustained in the air by air currents, they begin to fall as precipitation in a process known as coalescence.

Global precipitation is unevenly distributed due to a variety of factors including the pattern of global winds, the changing latitude of the location, and the presence of mountains. The pattern of global winds partially explains why the equatorial belt consistently experiences more precipitation — the trade winds from both hemispheres contribute to a larger push of air upwards. Latitude has an inverse relationship with precipitation. Areas near the poles tend to experience less precipitation because the cold air cannot support the same amount of moisture that warm air can. Equatorial areas, however, experience more solar heating, which encourages the process of convection and, subsequently, precipitation. Mountainous regions are often classified as either windward (wet) or leeward (dry) with respect to the mountains. In this regard, the locations of mountains are strategic for the overall water budget and sustenance of the lifeform.

During the rain, the atmospheric carbon combines with water to form a weak acid—carbonic acid—that falls to the surface in rain. The acid dissolves rocks, releasing calcium, magnesium, potassium, or sodium ions. Various rivers carry the ions to the ocean. Just in one pathway, rivers carry calcium ions into the ocean, where they react with carbonate dissolved in the water. The product of that reaction, calcium carbonate, is then deposited onto the ocean floor, where it becomes limestone.

In the ocean, the calcium ions combine with bicarbonate ions to form calcium carbonate, the active ingredient in antacids and the chalky white substance that dries on your faucet if you live in an area with hard water. Most of the calcium carbonate is made by shell-building (calcifying) organisms (such as corals) and plankton (like coccolithophores and foraminifera). After the organisms die, they sink to the seafloor. Over time, layers of shells and sediment are cemented together and turn to rock, storing the carbon in stone—limestone and its derivatives.

---

[a] The volume of global precipitation is calculated by taking the product of the Earth's surface area and its average annual rainfall.

Only 80% of carbon-containing rock is currently made this way. This carbon has the potential of forming petroleum resources without going through organic transformation. These carbon sources can increase the hydrocarbon bearing capacity of basins through volcanic activities, both in terms of source rock maturity hydrocarbon trapping. Volcanic rocks act as important basin filling material in different types of basins, for instance, rift basins, epicontinental basins, basins in a trench-arc system, back-arc foreland basins, etc. Only recently, volcanic accumulation of oil and gas has been recognized as legitimate exploration sites of hydrocarbon exploration and has been proved in more than 300 basins in 20 countries and regions (Farooqui et al., 2009). It is now known that hydrocarbon in volcanic accumulation have organic as well as inorganic sources (Liu et al., 2012). Volcanic rocks could act as a reservoir or cover within hydrocarbon traps, whose thermal effects could accelerate the maturity of source rocks or destroy preserved hydrocarbon. Lithofacies, including deposits and rocks formed by explosive, effusive, extrusive and subvolcanic processes, could bear hydrocarbon, and the facies combination close to a volcanic conduit shows better porosity and permeability due to an increased number of fractures and reservoir spaces, or an elevated volume of coarse-grained fragmented rocks. Reservoir spaces within volcanic rocks are composed of primary pores, secondary pores and fissures with significant heterogeneity.

Hydrocarbon can be generated through inorganic so-called synthetic reactions. In its simplest forms, methane can be generated from carbon dioxide. Such process has gained recognition in recent years. For instance, Drab et al. (2013) present a two-step process for hydrocarbon synthesis. In the first-step, CO_2 and H_2 are reacted over an iron-based catalyst to produce light olefins. The mechanism of CO_2 hydrogenation has been proposed to occur in two steps. The reverse water-gas shift (Eq. 2.1 below) is endothermic and produces carbon monoxide (CO). This CO is then carried forward in an exothermic synthesis step (Eq. 2.2), producing predominantly monounsaturated hydrocarbons (Eq. 2.3). Carbon dioxide is also hydrogenated directly to methane, in a widely cited thermodynamically favorable and highly competitive side reaction (Eq. 2.4).

$$nCO_2 + nH_2 \rightleftarrows nCO + nH_2O \tag{2.1}$$

$$nCO + 2nH_2 \rightarrow (CH_2)_n + nH_2O \tag{2.2}$$

$$nCO_2 + 3nH_2 \rightarrow (CH_2)_n + 2nH_2O \tag{2.3}$$

$$CO_2 + 4H_2 \rightarrow CH_4 + 2H_2O \tag{2.4}$$

In a natural setting, far more complex reactions can take place that are faster and more efficient due to the presence of naturally occurring catalysts. Similarly, studies have shown that some volcanic minerals undergo catalysis and hydrogenation which can produce more oil and gas source rocks at lower temperature and pressure.

Similarly, producing fuel from inorganic sources is not uncommon in modern day engineering. For example, "water gas" is a well-known gaseous fuel produced by heating carbon and water at relatively high temperatures whereupon a mixture of carbon monoxide and hydrogen is formed as follows:

$$C + H_2O \rightarrow CO + H_2 \tag{2.5}$$

Acetylene has also been produced by the addition of water to calcium carbide at ordinary temperatures as follows:

$$CaC_2 + 2H_2O \rightarrow Ca(OH)_2 + C_2H_{2+} \tag{2.6}$$

As early as in 1931, limestone has been employed as a raw material in the production of calcium carbide and carbon monoxide as described in French Pat. No. 694,459 (1931) as follows:

$$CaCO_3 + 4C \rightarrow CaC_2 + 3CO \tag{2.7}$$

However, none of the above-described methods of producing a gaseous fuel yield a fuel which is characterized by a high heating value and useful as an industrial fuel in place of more expensive petroleum-derived fuels.

In a recently developed process, limestone ($CaCO_3$) is heated at a temperature of about 850C (Chen et al., 1999). to decompose the calcium carbonate into calcium oxide (CaO) and carbon dioxide (CO_2). Appropriate amounts of carbon and water are then added to the furnace, with the temperature being increased up to about 1000°C. to form the gaseous fuel mixture of carbon monoxide (CO), hydrogen (H_2), low molecular weight aliphatic hydrocarbons (e.g., methane), and calcium carbide.

The reactions which result in the formation of the gaseous fuel proceeds as follows:
Step-wise reactions:

$$CaCO_3 \rightarrow CO_2 + CaO \tag{2.8}$$

$$CO_2 + 5C + 3H_2O \rightarrow 5CO + H_2 + CH_4 \tag{2.9}$$

$$CaO + 7C + 3H_2O \rightarrow 4CO + CH_4 + CaC_2 \tag{2.10}$$

The overall reaction is:

$$CaCO_3 + 12C + 6H_2O \rightarrow 9CO + 2H_2 + 2CH_4 + CaC_2 \tag{2.11}$$

In a volcanic setting, the initial temperature is much higher and the presence of catalysts is plentiful. In addition, limestone continues to be decomposed at lower temperature all the way down to 600C. Also, the presence of water either within the magma or from oceanic sources will enhance the effectiveness of the chemical reactions.

Decades ago, authors speculated the nature of hydrocarbon present in the mantle. For instance, Sugisaki and Mimura (1994) analyzed 227 rocks from fifty localities throughout the world showed that mantle derived rocks such as tectonized peridotites in ophiolite sequences (tectonites) arid peridotite xenoliths in alkali basalts contain heavier hydrocarbons (n-alkanes), whereas igneous rocks produced by magmas such as gabbro arid granite lack them. These hydrocarbons compounds found in the mantle-derived rocks are called here "mantle hydrocarbons". Possible origins for the mantle hydrocarbons are as follows:

(1) They were in organically synthesized by Fischer-Tropsch type reaction in the mantle.
(2) They were delivered by meteorites and comets to the early Earth.
(3) They were recycled by subduction.

2.4 Fluid characterization based on origin

The mantle hydrocarbons in the cases of (1) and (2) are abiogenic and those in (3) are mainly biogenic. It appears that hydrocarbons may survive high pressures and temperatures in the mantle, but they are decomposed into lighter hydrocarbon gases such as CH_4 at lower pressures when magmas intrude into the crust; consequently, peridotite cumulates do not contain heavier hydrocarbons but possess hydrocarbon gases up to C_4H_{10}.

The Fischer–Tropsch process involves a series of chemical reactions that produce a variety of hydrocarbons, namely, alkanes, having the formula (C_nH_{2n+2}), as expressed in the following reaction:

$$(2n+1)H_2 + nCO \rightarrow C_nH_{2n+2} + nH_2O \quad (2.12)$$

Typically, n is 10–20. In a laboratory setting, the synthesis of hydrocarbon chains involve a repeated sequence in which hydrogen atoms are added to carbon and oxygen, the C—O bond is split and a new C—C bond is formed. For one –CH$_2$– group produced by $CO + 2H_2 \rightarrow (CH_2) + H_2O$, several reactions are necessary:

- Associative adsorption of CO
- Splitting of the C—O bond
- Dissociative adsorption of 2 H_2
- Transfer of 2H to the oxygen to yield H_2O
- Desorption of H_2O
- Transfer of 2H to the carbon to yield CH_2

The process can be enhanced in presence of catalysts. Under laboratory conditions, most of the alkanes produced tend to be straight-chain. In a natural setting, such straight chain molecules are rarely formed. Instead, complex hydrocarbon chains are formed, along with oxygenated hydrocarbons. In presence of high temperature and pressure, as prevalent in magma, similar reactions take place in presence of water and any carbon source, including limestone. Studies have shown that some volcanic minerals undergo catalysis and hydrogenation which can produce more oil and gas source rocks at lower temperature and pressure.

In terms of Carbon source, it can be graphite, carbon, or limestone, etc. that comes in contact with the lava and reactions take place during the cooling process, the digenesis continuing further.

The remaining 20% contain carbon from living things (organic carbon) that have been embedded in layers of mud. Heat and pressure compress the mud and carbon over millions of years, forming sedimentary rock such as shale. In special cases, when dead organic matter builds up faster than it can decay, akin to a pyrolysis process, layers of organic carbon become oil, coal, or natural gas instead of sedimentary rock like shale.

The slow cycle returns carbon to the atmosphere through volcanoes. Earth's land and ocean surfaces sit on several moving crustal plates. When the plates collide, one sinks beneath the other, and the rock it carries melts under the extreme heat and pressure. The heated rock recombines into silicate minerals, releasing carbon dioxide. The resulting carbon dioxide then synthesizes into carbohydrate and hydrocarbons.

$$\text{Surface rock} \rightarrow \text{Silicate} + CO_2$$

$$CO_2 \rightarrow \text{synthesis to carbohydrate} \rightarrow \text{hydrocarbon}$$

As first pointed out by Lucido (1983), there is no reason for this process to be discontinuous or limited to sedimentary rocks. Islam (2014) takes it further and argues that mass transfer is continuous at all levels of the ecosystem, involving magma and the solar system.

As stated earlier, any transformation of organic matter (carbohydrate) to hydrocarbon should involve pyrolysis-type transformation. During the transformation from organic matter to hydrocarbon, the role of volcanic is mainly to supply a catalyst and thermal energy. Volcanogenic zeolite, olivine, and many transition metals, such as Ni, Co, Cu, Mn, Zn, Ti, V, etc. that are present in a hydrothermal liquid will act as a catalyst in turning organic matter into hydrocarbon through a series of degradation reactions.

Volcanic minerals and the prevalent temperature and pressure conditions of magma have all the necessary features required for first producing hydrogen, then hydrogenation of carbon compounds in the presence of some desired catalysts. One such catalyst is zeolite, which has recently been rediscovered for many applications (Yilmaz and Müller, 2009). In the presence of olivine, the hydrogen production rate is increased with marked improvements in the presence of zeolites (Liu et al., 2012). The following reaction can take place, occurring and reacting with water to produce hydrogen in the organic matter into the hydrocarbon conversion process.

$$6(Mg_{1.5}Fe_{0.5})SiO_4 + 13H_2O \rightarrow 3Mg_3SiO_2O_5(OH)_4 + Fe_3O_4 + 7H_2 \tag{2.13}$$

Similar reactions are likely to take place with Pyrite, which is related to kerogene products.

Liu et al. (2014) point to many reservoirs that contain alkane gas and non-hydrocarbon gas with inorganic origin, such as CH_4 and CO_2. Carbon isotope of carbon dioxide ($\delta^{13}C_{CO2}$) is an important indicator to identify the carbon dioxide origin. Using that technique, Dai et al. (2004) pointed out that the $\delta^{13}C_{CO2}$ value ranges from +7‰ to -39‰ in China, in which the organic origin $\delta^{13}C_{CO2}$ value is from -10‰ to -39.14‰, with the main frequency scope of $-12‰ \sim -17‰$; the inorganic origin $\delta^{13}C_{CO2}$ value is from +10‰ to $-8‰$ with the main frequency scope of $-3‰ \sim -8‰$. This is shown in Fig. 2.10 The source of CO_2 can be further divided into mantle-derived, carbonate pyrolysis, magma degassing and others. The $\delta^{13}C$ value of mantle-derived CO_2 is around -6‰, of which of carbonate pyrolysis is from +3‰ to $-3‰$. CO_2 volume fraction of Well FS9 in the Xujiaweizi area, Songliao Basin, is 89.73%, and the $\delta^{13}C$ value is from $-4.06‰$ to $-5.46‰$. The $\delta^{13}C_{CO2}$ value is $-6.61‰$ of Well FS6, which confirms that CO_2 of the Xujiaweizi region belongs to the mantle-derived category.

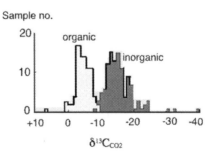

FIG. 2.10 Organic and inorganic origin $\delta^{13}C_{CO2}$ frequency.

Although the principle is well known, only a few mainstream researchers have used this technique to trace the origin of basement oil.

2.4.2 Role of magma during the formation of hydrocarbon from organic sources

Magma provides both mass and energy support for the evolution of organic materials into hydrocarbons. It is well known that magma is rich in chemicals that serve as a catalyst for the thermal degradation of organic matter. As stated earlier, volcanogenic zeolite and olivine can be a catalyst in turning organic matter into hydrocarbons. The magma fluid is rich in metals, such as Ni, Co, Cu, Mn, Zn, Ti, V, each of which can become a catalyst for the transformation in question.

In addition to catalytic support, volcanic activities provide the system with energy to facilitate thermal degradation of organic matter. Thermal energy is useful for both maturation of immature source rocks as well as pyrolytic conversion into hydrocarbons. This same process can lead to the destruction of an oil and gas reservoir that is exposed to over maturation and volatilization of the residual hydrocarbons.

Studies at the Illinois Basin by Schimmelmann et al. (2009) have shown that R_o values increased from 0.62% to 5.03% within 5 m at the coal contact to large intrusion, while R_o values increased from 0.63% to 3.71% within 1 m at the coal contact to small intrusion.

George (1992) found that intrusion made Ro rise from 0.55% to 6.55% when he investigated the maturity of the Scottish Midland Valley oil shale. Raymond and Murchison (1988) found that vitrinite reflectance was significantly higher around bedrock in the carboniferous strata, Midland Valley, Scotland. Overall, the size of the igneous rock affected the maturity of the organic matter. This is explained in terms of energy availability in addition to the catalytic effect.

2.5 Abiogenic petroleum origin theory

Even though the abiogenic origin of hydrocarbons has been there for many decades, recently Kutcherov (2013) formalized it. His particular version considers the generation of hydrocarbons in the outermost layer of the earth where atmospheric CO2, oxygen, and photosynthesis products exist is the sole source of petroleum that seeps downward and migrate through the crust of the Earth. Fig. 2.11 shows the configuration of this theory.

According to the theory of the abyssal abiogenic origin of hydrocarbons, the following conditions are necessary for the synthesis of hydrocarbons:

- adequately high pressure and temperature;
- donors/sources of carbon and hydrogen;
- a thermodynamically favorable reaction environment.

Kutcherov (2013) offered experimental evidence in support of the abiogenic theory. Some examples are:

- polymerization of hydrocarbons takes place in the temperature range 600–1500 degrees C and at pressures range of 20–70 kbar (Kenney et al., 2009);
- these conditions prevail deep in the Earth at depths of 70–250 km.

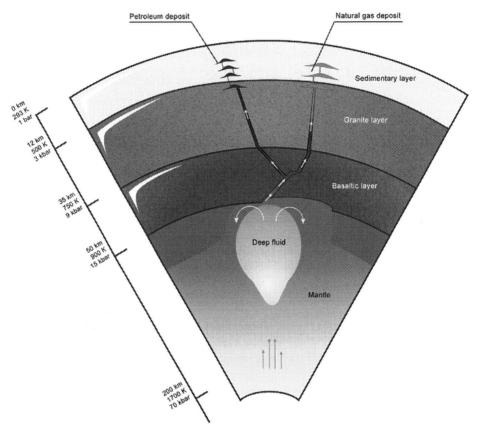

FIG. 2.11 A scheme of genesis of hydrocarbons and petroleum deposits formation. *From Kutcherov, V.G., 2013. Abiogenic deep origin of hydrocarbons and oil and gas deposits formation. In: Kutcherov, V. (Ed.), Hydrocarbon. https://doi.org/10.5772/51549.*

Mao et al. (2011) showed that the addition of minor amounts of iron can stabilize dolomite carbonate in a series of polymorphs that are stable in the pressure and temperature conditions of subducting slabs, thereby providing a mechanism to carry carbonate into the deep mantle. They studied the high pressure/temperature behavior of Fe-dolomite, $Ca_{0.988}Mg_{0.918}Fe_{0.078}-Mn_{0.016}(CO_3)_2$, using synchrotron X-ray diffraction. The Fe-dolomite phase undergoes a pressure-induced phase transition to orthorhombic Fe-dolomite-II at 17 GPa and 300 K and is stable up to 1500 K. A different high-pressure phase was identified that involved Fe-dolomite-III, stable between 36 and 83 GPa up to 1500 K in a monoclinic structure.

2.5 Abiogenic petroleum origin theory

As discussed earlier, Hazen et al. (2012) suggested that the deep interior may contain more than 90% of Earth's carbon. Possible sources of carbon in the crust are shown in Fig. 2.11.

Recent data of Green et al. (2010) show that water-storage capacity in the uppermost mantle is dominated by pargasite and has a maximum of about 0.6 wt% H_2O (30% pargasite) at about 1.5 GPa, decreasing to about 0.2 wt% H2O (10% pargasite) at 2.5 GPa. In addition to this water, another possible source of hydrogen is the hydroxyl group in some minerals (biotite, muscovite). The Earth has distinctive convective behavior, described by the plate tectonics model, in which lateral motion of the oceanic lithosphere of basaltic crust and peridotitic uppermost mantle is decoupled from the underlying mechanically weaker upper mantle (asthenosphere).

We have also seen that water has an important effect on mantle rheology, either by weakening the crystal structure of olivine and pyroxenes by dilute solid solution or by causing low-temperature partial melting. Hazen et al. (2012) show that on the depth of 100 km temperature is about 1250 K and the pressure is 3 GPa. On the depth of 150 km temperature is about 1500–1700 K and the pressure is 5 GPa. Both donors of carbon (carbon itself, carbonates, CO_2) and hydrogen (water, hydroxyl group of minerals) are present in the asphenosphere in sufficient amounts. A thermodynamically favorable reaction environment (reducing conditions) could be created by the presence of FeO. The presence of several present of FeO in basic and ultra-basic rocks of the asthenosphere is documented.

Overall, abiogenic synthesis of hydrocarbons can take place in the basic and ultra-basic rocks of the asthenosphere in the presence of FeO, donors/sources of carbon and hydrogen (Kutcherov, 2013). The possible reaction of synthesis in this case could be presented as follows:

- reduced mantle substance + mantle gases → oxidized mantle substance + hydrocarbons or
- combination of chemical radicals (methylene (CH_2), methyl (CH_3). Different combinations of these radicals define all scale of oil-and-gas hydrocarbons, and also cause close properties and genetic similarity of oils from different deposits of the world.

He reports the occurrence of following reactions under conditions that emulate the mantle environment.

$$nCaCO_3 + (9n+3)FeO + (2n+1)H_2O \rightarrow nCa(OH)_2 + (3n+1)Fe_3O_4 + C_nH_{2n}$$
$$+ 2nCaCO_3 + (9n+3)FeO + (2n+1)H_2O \rightarrow nCa(OH)_2 + (3n+1)Fe_3O_4 + CnH_{2n+2} \quad (2.14)$$

where $n \leq 11$.

Furthermore, the reaction scheme proposed by Kutcherov et al. (2002) indicate the following:

$$CaCO_3 + 12FeO + H_2O \rightarrow CaO + 4Fe_3O_4 + CH_4 CaCO_3$$
$$+ 12FeO + H_2O \rightarrow CaO + 4Fe_3O_4 + CH_4 \quad (2.15)$$

The behavior of pure methane at pressure range 1–14 GPa at the temperature interval 900–2500 K was studied by Kolesnikov et al. (2009). The experiments have shown that methane is stable at temperature below 900 K in the above-mentioned pressure range. In the temperature interval 900–1500 K at 2–5 GPa the formation of heavier alkanes (ethane, propane, and butane) from methane was observed:

$$CH_4 \rightarrow C_2H_6 + C_3H_8 + C_4H_{10} + H_2 \quad (2.16)$$

At temperature higher that 1500 K methane dissociates to carbon (graphite) and molecular hydrogen:

$$CH_4 \rightarrow C + H_2 \quad (2.17)$$

Previously, Kolesnikov et al. (2009) modeled the redox conditions of the mantle by introducing into a system of magnetite (Fe_3O_4), which was partially transformed to iron (0) forming a redox buffer.

The introduction of Fe_3O_4 did not affect the thermobaric conditions of methane transformation. Pure iron and water were detected in the reaction products. This gives us the possibility to suggest the following pathway of reactions:

$$CH_4 + Fe_3O_4 \rightarrow C_2H_6 + Fe + H_2O \quad (2.18)$$

$$CH_4 + Fe_3O_4 \rightarrow C + Fe + H_2O \quad (2.19)$$

The experimental results confirming the possibility of synthesis of natural gas from inorganic compounds under the upper mantle conditions were published by Kutcherov et al. (2010). The experimental results are shown in Table 2.3.

If the carbon donor was $CaCO_3$ (Experiments 1 and 2), the methane concentration in the produced mixture was rich in hydrocarbons. If the carbon donor was individual carbon (experiments 3 and 4), the hydrocarbon composition corresponded to "dry" (methane-rich) natural gas. Rapid cooling (quenching) fixes CH_4 and C_2H_6/C_2H_4 in the reaction products. After cooling for 4 h (Experiment 4), the amount of CH_4 and C_2H_6/C_2H_4 in the reaction products increases by a factor of tens, and heavier hydrocarbons up to C_4H_{10}/C_4H_8 emerge. Thus, the time of cooling of the fluid forming at high pressure (e.g., in the course of its jet migration to the surface) has a significant effect on the final composition of the fluid. A decrease in the rate of cooling of the initial fluid results in synthesis of heavier saturated hydrocarbons in the mixture. Then, at the first stage, in case of quenching (Experiment 3), the reaction is the following:

$$Fe + H_2O \rightarrow FeO + H_2 \quad (2.20)$$

and in case of slow cooling (experiment 4), the reaction is

$$3Fe + 4H_2O \rightarrow Fe_3O_4 + 4H_2 \quad (2.21)$$

The most convincing evidence of the above-mentioned mechanism of oil and gas deposit formations is the existence of such giant gas fields as Deep Basin (from Hazen, 2013) of Milk River and San Juan (the Alberta Province of Canada and the Colorado State, U.S.A.). The formation of these giant gas fields questions the existence of any lateral migration of oil and gas during the oil and gas accumulation process. Those giant gas fields occur in

TABLE 2.3 The experimental results received in CONAC high-pressure chamber and in a split-sphere high-pressure device.

Experiment, reagents (mg), cooling rate	Concentration, mol.%									CH_4, µmol
	CO_2	N_2	CO	CH_4	C_2H_4 C_2H_6	C_3H_6 C_3H_8	C_4H_8 C_4H_{10}	C_5H_{10} C_5H_{12}	
Toroidal high-pressure apparatus (CONAC)										
Experiment A quenching									
$CaCO_3$(104.5) + Fe(174.6) + H_2O(45.3)	0.0	tr.	0,0	71.4	25.8	2.5	0.25	0	6.28
Experiment B quenching									
$CaCO_3$(104.7) + Fe(174.6) + D_2O(42.6)	0.0	tr.	0.0	71.1	25.3	3.2	tr.	tr.	4.83
Split-sphere high-pressure device (BARS)										
Experiment C quenching									
C(24.5) + Fe(60.2) + H_2O (10.1)	0	0	0	96.1	3.84	0	0	0	0.23
Experiment D 4 h. cooling									
C(21.3) + Fe(98.6) + H_2O (15.1)	0	0	0	93.2	6.21	0.42	0.16	0	5.4
Severo-Stavropolskoe gas field	0.23	–	–	98.9	0.29	0.16	0.05	–	
Vuctyl'skoe gas field	0.1	–	–	73.80	8.70	3.9	1.8	6.40	

From Kutcherov, V.G., 2013. Abiogenic deep origin of hydrocarbons and oil and gas deposits formation. In: Kutcherov, V. (Ed.), Hydrocarbon. https://doi.org/10.5772/51549.

synclines where gas must be generated but not be accumulated, according to the hypothesis of biotic petroleum origin and hydrodynamically controlled migration. The giant gas volumes (12.5 10^{12} cu m in Deep Basin, 935 10^9-cu m in San Juan, 255 10^9 cu m in Milk River) are concentrated in the very fine-grained, tight, impermeable argillites, clays, shales and in tight sandstones and siltstones (Masters, 1979). The original gas in place contained within tight formations in the entire Mesozoic section was originally estimated in 1750 TCF for the Deep Basin (Masters, 1979). These rocks are usually accepted as source rocks cap rocks/seals rocks in petroleum geology but by no means of universally recognized reservoir rocks of oil and natural gas. Such accumulation of gas cannot be explained in such tight rocks with no visible tectonic, lithological and stratigraphic barriers to prevent updip gas migration. It's because the buoyancy forces prevailing in gas cannot overcome the capillary forces of these tight formations.

Petroleum of abyssal abiogenic origin and their implacement into the crust of the Earth can take place in the recent sea-floor spreading centers in the oceans. Igneous rocks occupy 99% of the total length (55,000 km) of them while the thickness of sedimentary cover over the spreading centers does not exceed 450–500 m (Rona, 1988). Fig. 2.12 (from Rona, 1988) shows a schematic of various components within a hydrothermal convection system involving downwelling of cold, dense, and alkaline seawater through permeable oceanic crust. Note how heating by the flow in proximity to magmatic heat sources, upwelling of hot, thermally expanded seawater and reaction with minerals under ambient conditions evolve into acid, metal-rich hydrothermal solutions that interact with volcanic rocks along flow paths and discharge into the ocean.

In addition, sub-bottom convectional hydrothermal systems discharge hot (170–430C) water through the "smokers" of the seabed. Up to now more than 100 hydrothermal systems of this kind have been identified and studied by in scientific expeditions using submarines such as «ALWIN», «Mir», «Nautile», and «Nautilus» in the Atlantic, Pacific and Indian oceans respectively (Kutcherov, 2013). Kutcherove (2013) summarized their observations:

- The bottom "smokers" of deepwater rift valleys vent hot water, methane, some other gases and petroleum fluids. Active "plumes" with the heights of 800–1000 m venting methane have been discovered in every 20–40 km between 12°N and 37°N along the Mid-Atlantic Ridge (MAR) over a distance of 1200 km. MAR's sites—TAG (26°N), Snake Pit (23°N),

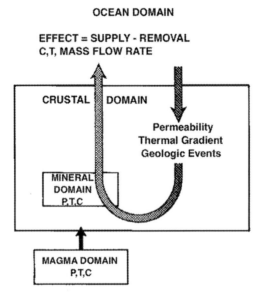

FIG. 2.12 Hydrothermal convection from magma. *Redrawn from Rona, P.A., 1988. Hydrothermal mineralization at oceanic ridges. Can. Mineral. 26 (3), 431–465.*

Logatchev (14°45′N), Broken Spur (29°N), Rainbow (37°17′N), and Menez Gwen (37°50′N) are the most interesting.
- At the Rainbow site, where the bottom outcrops are represented by ultramafic rocks of mantle origin the presence of the following substances were demonstrated (by chromatography/mass-spectrometry): CH_4; C_2H_6; C_3H_8; CO; CO_2; H_2; H_2S, N_2 as well as petroleum consisted of n-C_{16} – n-C_{29} alkanes together with branched alkanes, diaromatics (Charlou et al., 1993, 2002). Contemporary science does not yet know any microbe which really generates n-C_{11} – n-C_{22} alkanes, phytan, pristan, and aromatic hydrocarbons.
- At the TAG site no bottom sediments, sedimentary rocks (Simoneit and Lonsdale, 1982, 1988; Thompson et al., 1988), buried organic matter or any source rocks.
The hydrothermal fluid is very hot (290/321 degree C) for any microbes. There are the Beggiatoa mats there but they were only found at some distances from "smokers."
- Active submarine hydrothermal systems produce the sulfide-metal ore deposits along the whole length of the East Pacific Rise (EPR). At 13oN the axis of EPR is free of any sediment but here aliphatic hydrocarbons are present in hot hydrothermal fluids of black "smokers". In the sulfide-metal ores, here the methane and alkanes higher than n-C25 with the prevalence of the odd number of carbon atoms have been identified (Simoneit, 1988).
- Oil accumulations have been studied by the «ALVIN» submarine and by the deep marine drilling in the Gulf of California (the Guaymas Basin) and in the Escanaba Trough, Gorda Ridge (Gieskes et al., 1988; Koski et al., 1988; Kvenvolden et al., 1987; Simoneit and Lonsdale, 1982) of the Pacific Ocean. These sites are covered by sediments. However, petroleum fluids identified there are of hydrothermal origin according to Simoneit and Lonsdale (1982) and no source rocks were yet identified there.
- As for other sites over the Globe, scientific investigations with the submarines have established that the methane "plumes" occur over the sea bottom "smokers" or other hydrothermal systems in Red Sea, near Galapagos Isles, in Mariana and Tonga deepwater trenches, Gulf of California, etc. (Baker and Massoth, 1987). Non-biogenic methane (105–106 cu m/year) released from a submarine rift off Jamaica (Brooks, 1979) has been also known.

Rona (1988) prepared the schematic, shown in Fig. 2.13.

This schematic cross-sections perpendicular to a spreading axis shows relations between developmental stages of seafloor spreading and hydrothermal activity. Geologic conditions conducive to high-intensity hydrothermal activities are inferred to occur at an early tectonic stage when the rate of extensions exceeds magma supply, which creates a favorable distribution of permeability and heat. In addition, favorable conditions also prevail at a late-magmatic stage when magma supply exceeds extension. "Smokers" (discrete) and "seeps" (diffuse) components of hydrothermal discharge are shown in this figure. Quantitatively speaking the sea-floor spreading centers may vent $1.3 \, 10^9$ cu m of hydrogen and $16 \, 10^7$ cu m methane annually (Welhan and Craig, 1979).

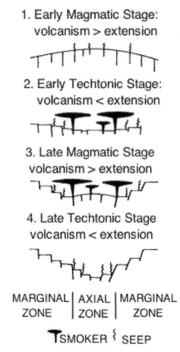

FIG. 2.13 Hydrothermal activities and seafloor spreading.

In summary, Kutcherov (2013) observed that.

- Source rocks accounting for the volume of the petroleum venting described are not available;
- The natural gas and petroleum fluids in the recent sea-floor spreading centers can be explained as a result of the vertical migration of the Mantle fluids.

2.5.1 Diamond as a source of hydrocarbons

In support of the abiogenic theory of petroleum formation, the presence of bitumen and hydrocarbons in native diamonds, carbonado and kimberlites is cited. Studying the native diamonds, carbonado and kimberlites under the microscope many scientists from several countries have found numerous primary fluid inclusions which have been opened due to the specific methods. Fluid contents of primary fluid inclusions have been recovered without any contamination and studied by mass-spectrometry/gas-chromatography. Results of such investigations carried out on the samples from Africa, Asia, Europe, North and South America have been summarized by Kutcherov (2013).

The well-known diamond-producing mines such as the Dan Carl, Finsh, Kimberley, and Roberts Victor are located in the kimberlite pipes of the South Africa. There the

African shield is characterized by the remarkable disjunctive dislocations and non-orogenic magmatism which has produced a great number of the carbonatite and kimberlite intrusions and explosion pipes in the area around Lake Tanganyika, Lake Malawi and Lake Victoria between 70 Ma and 3000 Ma ago (Irvine, 1989). These lakes are in the Great East-African Rift Valley. The Valley's margins and disjunctive edges consist of the African shield crystalline rocks. 258 samples of diamonds from this area have been investigated under the microscope (Deines et al., 1989). The investigation has shown the presence of primary fluid inclusions in all samples investigated. These samples have been disintegrated into the small particles in a vacuum of about 1.3 10–6 Pa and 200C. The gas mixture from each sample was received.

The same hydrocarbons and gases mixtures were detected in natural diamonds from Congo, Brazil (Melton et al., 1974) and Zaire (Giardini et al., 1982). The existence of a vapor phase in the diamond-forming region of the upper mantle is indicated by the ubiquitous presence of entrapped gas in diamonds from all geographic locations that have been examined so far (Melton and Giardini, 1974, 1975). This conclusion is further supported by the qualitative correlation reported by Melton and Giardini (1975) between the volume of occluded gas released by crushing and the volume of exposed solid-inclusion-free internal cavities in some Arkansas diamonds. Past evidence of the existence of a partially-molten state in the diamond-forming region of the mantle at the time of diamond formation is found in the presence of olivine and pyroxene inclusions in the form of both euhedral crystals and octahedral casts. The discovery of silicate glass totally enclosed in diamond is additional evidence of a molten phase (Giardini et al., 1982). The discovery of occluded solid hydrocarbon in diamond, and the quantity of petroleum-precursors occluded in the host rock, kimberlite, supports a value of $>10^{16}$ tons rather than 10^{15} tons as the amount of petroleum and petroleum-forming constituents that have outgassed from the mantle during the past 3 billion years. This larger quantity is sufficient to account directly for petroleum fields as young as Cenozoic. The derived estimate of about 10^{15} tons as the amount of juvenile petroleum remaining to be outgassed from the Earth indicates that crustal accumulation will continue at about today's rate into the foreseeable future.

These authors also determined the $\delta^{13}C$ of CO_2 in an 8.65 ct. diamond from Africa to be -35.2 ppt which is in the range specified for natural petroleum (more negative than -18 ppt). On this basis, they concluded that some carbonaceous material which is considered to be biogenic in origin may, in fact, be abiogenic (Giardini and Melton, 1981). Based on the average amount of petroleum-type compounds in 3.1 Ga old diamonds from Arkansas (33_g g-1), these authors then estimated the amounts of such compounds in the uppermost 400 km of the mantle to be 2×1015 t (Giardini and Melton, 1981). To calculate the transport of hydrocarbons from the mantle to the atmosphere, Giardini and Melton (1982) used hydrogen as a proxy for the hydrocarbons on the basis that all hydrogen can, in principle, be converted into hydrocarbons by Fischer-Tropsch-type reactions. Taking the oldest known petroleum deposits to be 700 Ma and assuming that the transport constant of hydrocarbons from the mantle to the atmosphere is the same as for N_2 (6.5×10^{10} yr^{-1}), Giardini and Melton (1983) estimated that 18.9×10^{12} t of juvenile petroleum has migrated from the mantle to the Earth's surface over the last 700 Ma. This amount is two orders of magnitude more than the present global reserves of petroleum

(Islam, 2014). However, this latter figure is a serious underestimate of the total global oil inventory since biodegraded oils occurring in heavy oil and tar sands make up 50% of the world's oil inventory and the Venezuela and Athabasca Tar Sands alone contain 2.8×10^{11} t of oil. Assuming that this juvenile petroleum was transported to the Earth's surface by lithospheric faults with a total length of 240,000 km and an average width of 0.01 km, the amount of juvenile petroleum transported along these faults was then calculated to be 93×10^6 t km^{-2} Ma^{-1}. This compares with an average accumulation rate of petroleum in 78 giant oil fields of 0.15×10^6 t km^{-2} Ma^{-1}. On this basis, Giardini and Melton (1982) concluded that formation of these giant oil fields required on average only 0.2% of the average outflow of juvenile petroleum precursors. This leads one to conclude that the juvenile petroleum model easily accounts for all known petroleum accumulations. However, these are clearly only order of magnitude calculations and may be substantially in error, particularly in the extrapolation of the concentrations of hydrocarbons in fluid inclusions in a few diamonds to the entire upper mantle and to a gross underestimation in the total global oil inventory.

In Brazil, a set of heavy hydrocarbons was reported in carbonado primary fluid inclusions comprise. Makeev and Ivanukh (2004) reported 9–27 forms of metallic films within the crystal faces of diamonds from Brazil and from the Middle Timan, Ural, and Vishera diamonds in the European part of Russia. These films consist of aluminum, cadmium, calcium, chrome, cerium, copper, gold, iron, lanthanum, lead, magnesium, neodymium, nickel, palladium, silver, tin, titanium, ytterbium, yttrium, zinc, zirconium, and precious metals including even Au_2Pd_3. The thickness of these films is from fractions of a micrometer to several micrometers. These films are the evidence for the growth of diamonds from carbon dissolved in the melt of gold and palladium. The coarseness of the diamond crystals in kimberlite and lamprophyre pipes depends on the sizes of precious metal droplets in the respective zone, in the Earth's upper, transitional, and lower mantle (Table 2.4).

Deines et al. (1989) found no correlation between $\delta^{13}C$ and nitrogen content or aggregation state. However, within the eclogitic and peridotitic paragenesis, groups of diamonds of very similar nitrogen concentration, $\delta^{13}C$, and mineral inclusion composition was recognized. One of these may be common to both kimberlites. Some of the diamonds in the two kimberlites may have had a common origin but subsequent thermal histories, which differed for the sample suites from the two diatremes. The nitrogen content of the diamonds is a function of local nitrogen concentration, temperature and oxygen fugacity. The dependence on the last of these can cause complex relationships between $\delta^{15}N$, $\delta^{13}C$ and nitrogen content of diamonds. $\delta^{13}C$ for 213 diamonds from the different pipes was analyzed. $\delta^{13}C$ is ranged from −1.88 to −16‰. The chemical and isotope peculiarities of natural diamonds reflect the different Mantle media and environments. Diamonds with $\delta^{13}C$ from −15 to −16‰ come from the region at a lower depth than the natural diamonds with $\delta^{13}C$ from −5 to −6‰.

In summary, there is preponderance of evidence that diamonds, carbonado and kimberlites are formed at great depths. The presence of primary hydrocarbon inclusions in diamonds, carbonado and kimberlites show that hydrocarbon Mantle fluids were a material for synthesis of these minerals in the Mantle.

2.5 Abiogenic petroleum origin theory

TABLE 2.4 Results of investigation of gas mixtures from native diamonds, carbonado and kumberlites.

Region	Gas mixture concentration, vol%
Africa diamonds	CH_4, C_2H_4, C_3H_6, solid hydrocarbons, C_2H_5OH, Ar, CO, CO_2, H_2, O_2, H_2O, and N_2
Congo, Brazil, Zaire diamonds	5.8 of CH_4; 0.4 of C_2H_4; 2.0 of C_3H_6; traces of C_4H_8; C_4H_{10} and solid hydrocarbons
Arkansas, U.S.A. diamonds	0.9–5.8 of CH_4; 0.0–5.2 of CH_3OH; 0.0–3.2 of C_2H_5OH; 1.2–9.4 of CO; 5.3–29.6 of CO_2; 1.5–38.9 of H_2; 2.9–76.9 of H_2O; 0.0–87.1 of N_2; 0.0–0.2 of Ar
Brazil carbonado	The homologies of naphthalene ($C_{10}H_8$), phenanthrene ($C_{14}H_{10}$), and pyrene ($C_{16}H_{10}$). Total concentration varies from 20 to 38.75 g/t
East Siberia, Russia diamonds and kumberlites	C_6H_6; $C_{12}H_{10}$; $C_{20}H_{12}$; $C_{16}H_{10}$, and other polynuclear aromatic hydrocarbons. Total concentration is about 0.136 g/t

From Islam, J.S., et al., 2018. Economics of Sustainable Energy. Scrivener-Wiley 628 pp.

2.5.2 Oil and gas deposits in the Precambrian crystalline basement

Fig. 2.14 shows a breakdown of lithologies in which hydrocarbon deposits have been described from around the world, based on the compilation provided by Schutter (2003) and presented by Petford and McCaffrey (2003). The highest reported occurrences are in basalts, followed by andesite and rhyolite tufts and lavas. Although volcanic rocks in this survey constitute close to three-quarters of all hydrocarbon-bearing lithotypes, the majority of production and global reserves appears to be confined predominantly to fractured and weathered granitic rocks. These data reveal that volcanic rocks (basalts, andesites and rhyolites) appear most closely associated with hydrocarbons, despite the fact that most large scale production is currently from granitic and associated plutonic rocks. Unfortunately, there are still

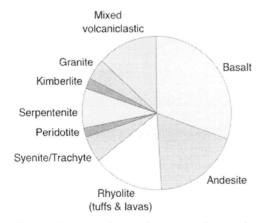

FIG. 2.14 The distribution of hydrocarbons in and around igneous rocks according to lithology. *From Petford, N., McCaffrey, K., 2003. Hydrocarbons in crystalline rocks: an introduction. Geol. Soc. Spec. Publ. 214, 1–5.*

insufficient data to be able to conclude whether hydrocarbons occur in some igneous rocks simply because of post-emplacement migration, or if there is something inherent in magma composition that results in preferential accumulation.

Crystalline crust of the Earth is the basement of 60 sedimentary basins with commercial oil and gas deposits in 29 countries of the world (Islam et al., 2018). In addition, there are 496 oil and gas fields in which commercial reserves occur partly or entirely in the crystalline rocks of that basement. Fifty five of them are classified as giant fields (>500 Mbbls) with 16 non-associated gas, 9 gas-oil and 30 undersaturated oil fields among them. Table 2.5 lists these fields.

They contain 9432 10^9 m^3 of natural gas and 32,837 10^6 tons of crude oil, i.e., 18% of the total world proven reserves of oil and about 5.4% of the total world proven reserves of natural gas.

TABLE 2.5 Giant and supergiant petroleum deposits in the Precambrian crystalline basement.

Deposit	Country	Proven reserves
Achak (gas field)	Turkmenistan	155·10^9 cu m
Gugurtli (gas field)	Turkmenistan	86·10^9 cu m
Brown-Bassett (gas field)	UK	73·10^9 cu m
Bunsville (gas field)	U.S.A.	85·10^9 cu m
Gomez (gas field)	U.S.A.	283·10^9 cu m
Lockridge (gas field)	U.S.A.	103·10^9 cu m
Chayandinskoye (gas field)	Russia	1240·10^9 cu m
Luginetskoye (gas field)	Russia	86·10^9 cu m
Myldzhinskoye (gas field)	Russia	99·10^9 cu m
Durian Mabok (gas field)	Indonesia	68.5·10^9 cu m
Suban (gas field)	Indonesia	71·10^9 cu m
Gidgealpa (gas field)	Australia	153·10^9 cu m
Moomba (gas field)	Australia	153·10^9 cu m
Hateiba (gas field)	Libya	411·10^9 cu m
Bach Ho (oil and gas field)	Vietnam	600·10^6 t of oil and 37·10^9 cu m of gas
Bombay High (oil and gas field)	India	1640·10^6 t of oil and 177·10^9 cu m of gas
Bovanenkovskoye (oil and gas)	Russia	55·10^6 t of oil and 2400·10^9 cu m of gas
Tokhomskoye (oil and gas field)	Russia	1200·10^6 t of oil and 100·10^9 cu m of gas
Coyanosa (oil and gas field)	U.S.A.	6·10^6 t of oil and 37·10^9 cu m of gas
Hugoton-Panhandle (oil and gas)	U.S.A.	223·10^6 t of oil and 2000·10^9 cu m of gas

TABLE 2.5 Giant and supergiant petroleum deposits in the Precambrian crystalline basement—cont'd

Deposit	Country	Proven reserves
Peace River (oil and gas field)	U.S.A.	$19,000·10^6$ t of oil and $147·10^9$ cu m of gas
Puckett (oil and gas field)	U.S.A.	$87.5·10^6$ t of oil and $93·10^9$ cu m of gas
La Vela (oil and gas field)	Venezuela	$54·10^6$ t of oil and $42·10^9·10^9$ cu m of gas
Amal (oil field)	Libya	$583·10^6$ t
Augila-Nafoora (oil field)	Libya	$178·10^6$
Bu Attifel (oil field)	Libya	$90·10^6$
Dahra (oil field)	Libya	$97·10^6$
Defa (oil field)	Libya	$85·10^6$
Gialo (oil field)	Libya	$569·10^6$
Intisar "A" (oil field)	Libya	$227·10^6$
Intisar "D" (oil field)	Libya	$182·10^6$
Nasser (oil field)	Libya	$575·10^6$
Raguba (oil field)	Libya	$144·10^6$
Sarir (oil field)	Libya	$1150·10^6$
Waha (oil field)	Libya	$128·10^6$
Claire (oil field)	UK	$635·10^6$
Dai Hung (oil field)	Vietnam	$60-80·10^6$
Su Tu Den (oil field)	Vietnam	$65·10^6$
Elk Basin (oil field)	U.S.A.	$70·10^6$
Kern River (oil field)	U.S.A.	$200.6·10^6$
Long Beach (oil field)	U.S.A.	$121·10^6$
Wilmington (oil field)	U.S.A.	$326·10^6$
Karmopolis (oil field)	Brazil	$159·10^6$
La Brea-Pariñas-Talara (oil field)	Peru	$137·10^6$
La Paz (oil field)	Venezuela	$224·10^6$
Mara (oil field)	Venezuela	$104.5·10^6$
Mangala (oil field)	India	$137·10^6$
Renqu-Huabei (oil field)	China	$160·10^6$
Shengli (oil field)	China	$3230·10^6$
Severo-Varieganskoye (oil field)	Russia	$70·10^6$
Sovietsko-Sosninsko-Medvedovskoye (oil field)	Russia	$228·10^6$

In the crystalline basement the depths of the productive intervals varies of 900–5985 m. The flow rates of the wells is between 1 and 2 m^3/d to 2400 m^3/d of oil and 1000–2000 m^3/d to 2.3 10^6 m^3/d of gas. The pay thickness in a crystalline basement is highly variable. It is 320 m in the Gomez and Puckett fields, the U.S.A.; 680 m in Xinglontai, China; 760 m in the DDB's northern flank. The petroleum saturated intervals are not necessary right on the top of the crystalline basement. Thus, oil was discovered at distance of 18–20 m below the top of crystalline basement in La Paz and Mara fields (Western Venezuela), 140 m below the basement's top in the Kazakhstan's Oimasha oil field. In the Baltic Shield, Sweden the 1 Gravberg well produced 15 m^3 of oil from the Precambrian igneous rocks of the Siljan Ring impact crater at the depth of 6800 m. In the Kola segment of the Baltic Shield several oil-saturated layers of the Precambrian igneous rocks were penetrated by the Kola ultra-deep well at the depth range of 7004–8004 m.

One of the most success stories of the practical application of the theory of the abyssal abiogenic origin of petroleum in the exploration is the exploration in the Dnieper-Donetsk Basin (DDB), Ukraine (Krayushkin, 2002).

It is a cratonic rift basin running in a NW-SE direction between 30.6°E-40.5°E. Its northern and southern borders are traced from 50.0°N-51.8°N and 47.8°N-50.0°N, respectively. In the DDB's northern, monoclinal flank the sedimentary sequence does not contain any salt-bearing beds, salt domes, nor stratovolcanoes and no sourse rocks. Also this flank is characterized by a dense network of the numerous *syn*-thethic and anti-thethic faults. These faults create the mosaic fault-block structure of crystalline basement and its sedimentary cover, a large number of the fault traps (the faulted anticlines) for oil and natural gas, an alternation uplifts (horsts) and troughs (grabens). The structure of the DDB's northern flank excludes any lateral petroleum migration across it from either the Donets Foldbelt or the DDB's Dnieper Graben. The U.S. Geological Survey (USGS) estimated technically recoverable, conventional, undiscovered oil and gas resources of the Dnieper–Donets Basin Province and Pripyat Basin Province in Russia, Ukraine and Belarus as part of a program to estimate these resources for priority basins around the world. The Dnieper–Donets Basin Province encompasses about 99,000 km^2 and the Pripyat Basin Province, 35,000 km^2 (Fig. 2.15).

Today there are 50 commercial gas and oil fields known in the DDB's northern flank. Data obtained from drilling in many of these areas shows that the crystalline basement of northern flank consists of amphibolites, charnockites, diorites, gneisses, granites, granodiorites, granito-gneisses, migmatites, peridotites, and schists. Thirty two of commercial fields have oil and/or gas accumulations in sandstones of the Middle and Lower Carboniferous age. Sixteen other fields contain reservoirs in the same sandstones but separately from them - in amphibolites, granites and granodiorites of crystalline basement as well. Two fields contain oil pools in the crystalline basement only (Kutcherov, 2013).

Abiogenic petroleum has been discovered in China as well: the giant Xinjiang gas field contains about 400 10^{12} m^3 of abiogenic natural gas (Zhang, 1990). Chinese petroleum geologists estimated this quantity in volcanic island arcs, trans-arc zones of mud volcanism, trans-arc rift basins, trans-arc epicontinental basins, deep fault zones and continental rift basins.

Dai et al. (2004) showed that partial reverse distribution of carbon isotopes is a relatively common phenomenon for biogenic gases in sedimentary basins of China. The causes resulting in a partial carbon isotopic reverse distribution for biogenic gases may include 1)

2.5 Abiogenic petroleum origin theory

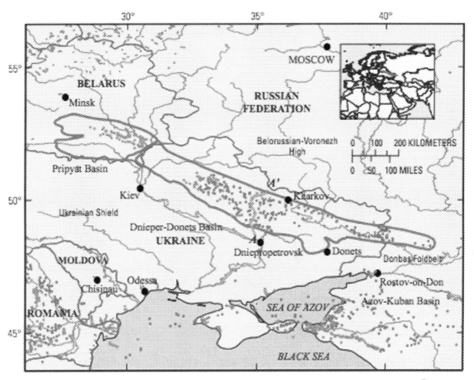

FIG. 2.15 The Generalized map showing the boundaries of the Pripyat Basin and Dnieper Donets Basin geologic provinces (*red (dark gray in the printed version) lines*), centerpoints of oil and gas fields (*green (light gray in the printed version) and red (dark gray in the printed version) circles*, respectively), and the location of geologic cross section A-A′ shown in Fig. 2.16 (*green line (light gray in the printed version)*). Country boundaries are represented by *blue (gray in the printed version) lines* (from USGS website 3).

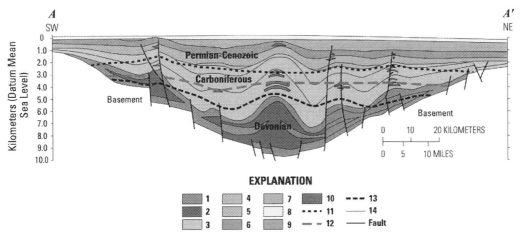

FIG. 2.16 Geologic cross section for the Dnieper-Donets Basin. 2, Devonian evaporites; 3, Carboniferous; 4, Permian; 5, Triassic; 6, Jurassic; 7, Cretaceous; 8, Cenozoic; 9, oil accumulation; 10, gas accumulation; 11, top of overpressure; 12, 100°C isotherm; 13, 0.9% vitrinite reflectance isochore; 14, stratigraphic boundary.

mixing of biogenic and abiogenic alkane gases; 2) mixing of sapropelic and humic sourced gases; 3) mixing of gases from "the same source but with different maturities"; and 4) the selective microbial oxidation of a certain component or some components.

In summary, one can see that by conventional petroleum origin hypothesis can lead to falsely concluding that reservoirs are in petroleum producers. Numerous large reservoirs have abiogenic origin and it is likely that many others remain untapped.

2.5.3 Supergiant oil and gas accumulations

Recently, it has been discovered that abiogenic source theory of petroleum can explain the occurrence of petroleum in a number of giant oil/gas fields, whereas biogenic source theory doesn't. Kutcherov (2013) in his seminal work listed a number of those giant petroleum fields.

Middle East. In the Middle East proven recoverable reserves are equal to 104×10^9 t of oil and 82×10^{12} m^3 of gas respectively as of the end 2012 (Islam, 2014). Nearly one third of that oil and one tenth of the gas are in Saudi Arabia. As can be seen from Table 2.6, most of these reserves are located in 10 supergiant gas and oilfields. Current hydrocarbon production in Saudi Arabia is from reservoirs of Cretaceous and Jurassic age. Geochemical studies of the sediments

TABLE 2.6 Giant fields with Volcanic and Volcano-sedimentary rocks.

Deposit	Country	Proven reserves
Hassi R'Mel (gas field)	Algeria	1522×10^9 m^3
Indefatigable (gas field)	UK	226×10^9 m^3
Leman Bank (gas field)	UK	340×10^9 m^3
Viking (gas field)	UK	130×10^9 m^3
Kenai (gas field)	United States	150×10^9 m^3
Monroe (gas field)	United States	266×10^9 m^3
San Juan (gas field)	United States	935×10^9 m^3
Niigata (gas field)	Japan	70×10^9 m^3
Daqing/Quigshen (oil and gas field)	China	1860×10^6 t of oil and 99×10^9 m^3 of gas
Gagliano (oil and gas field)	Italy	25×10^6 t of oil and 120×10^9 m^3 of gas
Poza Rica (oil and gas field)	Mexico	373×10^6 t of oil and 120×10^9 m^3 of gas
Saraji (oil and gas field)	Iran	160×10^6 t of oil and 142×10^9 m^3 of gas
Ankleshwar (oil field)	India	150×10^6 t
Bombay High (oil field)	India	1640×10^6 t
Ghandar (oil field)	India	200×10^6 t
Beckasap (oil field)	Indonesia	75×10^6 t
Duri (oil field)	Indonesia	258×10^6 t

TABLE 2.6 Giant fields with Volcanic and Volcano-sedimentary rocks—cont'd

Deposit	Country	Proven reserves
Jatibarang (oil field)	Indonesia	90×10^6 t
Minas (oil field)	Indonesia	953×10^6 t
Dagang/Daqang (oil field)	China	102×10^6 t
Liaohe (oil field)	China	120×10^6 t
Buzatchi North (oil field)	Kazakhstan	500×10^6 t
Karazhanbas (oil field)	Kazakhstan	500×10^6 t
Ebano-Panuco (oil field)	Mexico	157×10^6 t
Naranjos Cerro Azul (oil field)	Mexico	192×10^6 t
Forthies (oil field)	UK	348×10^6 t
Piper (oil field)	UK	246×10^6 t
Gela (oil field)	Italy	176×10^6 t
Raguza (oil field)	Italy	290×10^6 t
Hassi Messaoud (oil field)	Algeria	1491×10^6 t
Koturdepe (oil field)	Turkmenistan	230×10^6 t
Mendoza (oil field)	Argentina	100×10^6 t
Sacha (oil field)	Ecuador	70×10^6 t
Verkhnetchon (oil field)	Russia	260×10^6 t

From Kutcherov, V.G., et al., 2010. Synthesis of complex hydrocarbon systems at temperatures and pressures corresponding to the Earth's upper mantle conditions. Dokl. Phys. Chem. 433 (1), 132. https://doi.org/10.1134/S0012501610070079.

and oils suggest that the hydrocarbons were derived from two separate source-rock provinces. Oil production from the large fields in the southern part of the area is from Jurassic carbonate reservoirs. Most of these oils were derived from thermally mature, thinly laminated, organic-rich carbonate rocks of Jurassic age (Callovian-Oxfordian). These source rocks were deposited in an intrashelf basin which is limited to the southern part of the main producing areas. Extensive vertical migration of oils originating in these sediments is prevented by superjacent evaporite seals deposited during the Late Jurassic.

These giant oil fields give oil production from the Jurassic-Cretaceous granular carbonates. All these crude oils have very similar composition referring to a common source. Such source is the Jurassic-Cretaceous thermally mature, thin-bedded organic rich carbonate sequence (3–5 mass %). Organic material is concentrated in dark, 0.5–3.0 mm thin beds alternating with the lightly colored, similarly thin beds poor in organics. This volume relates to an absurd size of the petroleum production sources, as demonstrated by Ayres et al. (2013). More importantly, the calculations account for less than 6% of Saudi Arabia's proven reserve. Where did 94% of Saudi Arabia's recoverable oil come from? This question is not a rhetorical one because any other source of beds of petroleum is absent in the subsurface of Saudi Arabia

TABLE 2.7 Supergiant oil and gas deposits in Saudi Arabia.

Deposit	Estimated reserves
Ghawar	$10.2 \cdot 10^9$–$11.3 \cdot 10^9$ t of oil $1.5 \cdot 10^{12}$ m^3 of gas
Safaniyah	$4.1 \cdot 10^9$ t of oil
Shaybah	$2.0 \cdot 10^9$ t of oil
Abqaiq	$1.6 \cdot 10^9$ t of oil
Berri	$1.6 \cdot 10^9$ t of oil
Manifa	$1.5 \cdot 10^9$ t of oil
Marjan	$1.4 \cdot 10^9$ t of oil
Qatif	$1.3 \cdot 10^9$ t of oil
Abu Safah	$0.9 \cdot 10^9$ t of oil
Dammam	$0.7 \cdot 10^9$ t of oil

as well as of all countries mentioned above, according to Ayres et al. (2013), and Baker et al. (1984). Bahrain, Iran, Iraq, Kuwait, Oman, Qatar, Saudi Arabia, Syria, United Arab Emirates and Yemen occur in the same sedimentary basin—the Arabian-Iranian Basin, where Dunnington (1967) established the genetic relationship, i.e. the single common source of all crude oils (Table 2.7).

The next important oil field is in Canada. There is the unique oil/bitumen belt extended as the arc-like strip of 960 km length from Peace River through Athabasca (Alberta Province) to Lloydminster (Saskatchewan Province). This belt includes such supergiant petroleum fields as Athabasca (125 km width, 250 km length), Cold Lake (50 km, 125 km), Peace River (145 km, 180 km) and Wabaska (60 km, 125 km). Here the heavy (946.5–1.029 kg/m^3) and viscous (several hundred-several million cP) oil saturates the Lower Cretaceous sands and sandstones. These fields contain the "in place" oil reserves equal to 92 10^9–187 10^9 m^3 in Athabasca, 32 10^9–75 10^9 m^3 in Cold Lake, 15 10^9–19 10^9 m^3 in Peace River, 4.5 10^9–50 10^9 m^3 in Wabaska, and 2 10^9–5 10^9 m^3 of oil/bitumen in Lloydminster, totally 170,109–388,109 m^3 (Vigrass, 1968; Wennekers, 1981).

The conventional understanding is, that the oil of Athabasca, Cold Lake, Lloydminster, Peace River and Wabasca generated from dispersed organic matter buried in the argillaceous shales of the Lower Cretaceous Mannville Group only. It is underlain by the Pre-Cretaceous regional unconformity and its thickness varies from 100 to 300 m. Its total volume is about 190 10^3 cubic km with a 65% shale content. Having the data of the total organic carbon concentration (TOC), the hydrocarbon index (HI), constant of transformation (K), and all other values from the accepted geochemical model of the oil generation from the buried organic matter dispersed in clays-argillites, it was concluded that the Mannville Group could only give 71.5 10^9 m^3 of oil. It is in several times less than the quantity of oil (see above) which was totally estimated before 1985 in Athabasca, Cold Lake, Lloydminster, Peace River and Wabasca oil sand deposits (Moshier and Waples, 1985).

If we accept other estimations of the volume of oil/bitumen "in place" in Athabasca, Wabasca, Cold Lake and Peace River area ($\sim$ 122,800 sq. km) conducted by Alberta Energy

TABLE 2.8 Estimations of the volume of oil/bitumen in place in west Canada.

Source of information	Estimated volume (m³)
Lower Cretaceous sands and sandstones [Vigrass, 1968; Wennekers, 1981; Seifert and Lennox, 1985]	170×10^9 to 388×10^9
Lower Cretaceous sands and sandstones, Alberta Energy Utilities Board	270×10^9
Lower Cretaceous sands and sandstones, National Energy Board	397×10^9
Lower Cretaceous sands and sandstones and Upper Devonian carbonates [Wennekers, 1981; Seifert and Lennox, 1985; Hoffmann and Strausz, 1986]	370×10^9 to 603×10^9
Total amount that could be received from biotic source (Mannville clays and shales)	71.5×10^9

From Kutcherov, V.G., et al., 2010. Synthesis of complex hydrocarbon systems at temperatures and pressures corresponding to the Earth's upper mantle conditions. Dokl. Phys. Chem. 433 (1), 132. https://doi.org/10.1134/S0012501610070079.

Utilities Board (AEUB) and the National Energy Board (NEB), Canada the gap between the booked and organically generated quantities is even wider. AEUB estimated 270 10^9 m³ of bitumen "in place", while NEB—397 10^9 m³ (Table 2.8).

In the above-mentioned area there additionally are 200 10^9–215 10^9 m³ of heavy (986–1030 kg/cu m) and viscous (10^6 cP under 16 degrees C) oil at the depth range of 75–400 m in the Upper Devonian carbonates (Grosmont Formation). They occur in the area of 70 10^3 km² beneath the Athabasca/Cold Lake/Lloydminster/Peace River/Wabasca oil sand deposits (Seifert et al., 1985; Hoffmann et al., 1986).

The total estimated reserves of bitumen "in place" in the above-mentioned area are between 370 10^9 and 603 10^9 m³. If besides the Mannville clays and shales which could give 71.5 10^9 m³ of oil only there is no any other petroleum source rock, where is a biotic source for the rest of 82–88% of oil in this area?

Venezuela offers similar observations in the Bolivar Coastal oil field. Bockmeulen et al. (1983) reported that the source rock of petroleum here is the La Luna limestone of the Cretaceous age. The estimated oil reserves are equal to 4.8 10^9 m³ [*The List*, 2006] with an oil density of 820–1000 kg/m³. The same kind of calculations that were done for Saudi Arabia above gives us the following result. One m³ of the oil-generating rock contains 2.5 10^{-2} m³ of kerogen which can generate 2.5 10^{-3} m3 bitumen giving 1.25 10^{-4} m³ of oil within the accepted geochemical model of biotic petroleum origin. Having this oil-generating potential and the 4.8 10^9 m³ of estimated oil reserves of the Bolivar Coastal field as a starting point, the necessary volume of oil source rock would be equal to 3.84 10^{13} m³. This is consistent with the oil-generating basin area of 110 km across *if* the oil source rock is 1000 m thick. The average thickness of La Luna limestone is measured with only 91 m. The diameter of the oil-generating basin would be therefore equal to 370 km and area of this basin is equal approximately 50% of the territory of Venezuela what is geologically highly un-probable.

2.5.4 Gas hydrates—The greatest source of abiogenic petroleum

Gas hydrates are unlike any other form of unconventional gas reservoirs. The process of hydrate formation is unique as it involves the trapping of methane molecules directly within

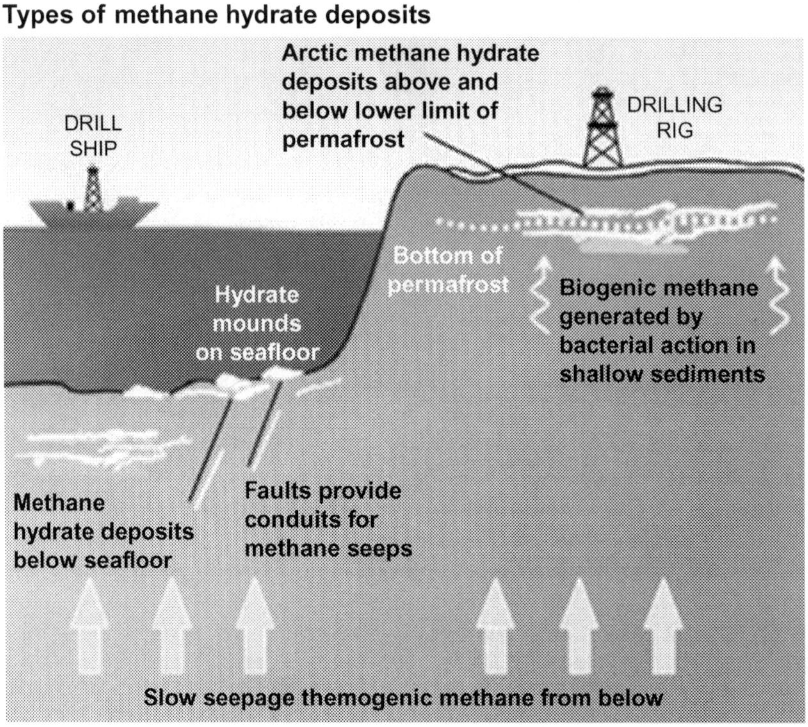

98   2. Reservoir rock and fluid characterization

FIG. 2.17 Hydrate formation in natural environment. *From Islam, M.R., 2014, Unconventional Gas Reservoirs, Elsevier.*

water body. Irrespective of its root (biogeneic or abiogenic), methane gas is purified during the process of hydrate formation.

The gas hydrates represent a huge unconventional resource base: it may amount from 1.5 10^{16} m^3 to 3 10^{18} m^3 of methane (Islam, 2014).

Fig. 2.17 shows how methane seeps through a marine environment and forms hydrates. In both marine and arctic environment, hydrates can form. Before forming solid hydrates, the system is purged of salinity as well as contaminants. The hydrate may mix with solid depositions, but does not lose its potential as an energy retainer. This is evidenced in the phase diagram in Fig. 2.18.

Fig. 2.18 shows the phase diagram of free methane gas and methane hydrate for a pure water and pure methane system. The addition of NaCl to water shifts the curve to the left. Adding CO_2, H_2S, C_2H_6, C_3H_8 to methane shifts the boundary to the right, thereby increasing the area of the hydrate stability field. It means, natural purification of methane before hydrate is formed.

Methane hydrate is a cage-like lattice of ice inside of which are trapped molecules of methane, the chief constituent of natural gas. If methane hydrate is either warmed or depressurized, it will produce natural gas that is pure and ready to burn. When brought to the earth's surface, one cubic meter of gas hydrate releases 164 cubic meters of natural

2.5 Abiogenic petroleum origin theory

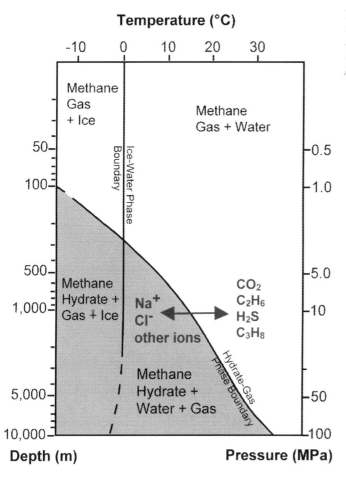

FIG. 2.18 Phase diagram of brine and methane. *From Kvenvolden, K.A., 1993. A primer on gas hydrates. In: The Future of Energy Gases. U.S. Geological Survey Professional Paper 1570, pp. 279–1008.*

gas. Hydrate deposits may be several hundred meters thick and generally occur in two types of settings: under Arctic permafrost, and beneath the ocean floor. Methane that forms hydrate can be both biogenic, created by biological activity in sediments, and thermogenic, created by geological processes deeper within the earth. Hydrates contain the most naturally purified methane available on earth. The processing involves water, thereby making the final product most suitable for human energy consumption.

During gas hydrate formation, water molecule crystallize into cubic lattice structures (Fig. 2.19).

Fig. 2.19 shows the structure of gas hydrates. The rigid cages in the figure are composed of hydrogen-bonded water molecules and each cage contains a methane molecule. During the formation process, hydrate crystals expel salt ions from the crystal structure. Measurements of chlorine in gas hydrate samples indicate far less concentration of salt than sea water (Kvenvolden and Kastner, 1990). The salt concentration of around 2% represents the salinity level of normal drinking water in contrast to some 20% concentration in sea water.

FIG. 2.19 Molecular structure of hydrate.

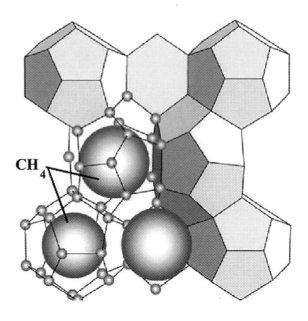

Production Feasibility: Methane hydrates occur in large quantities beneath the permafrost and offshore, on and below the seafloor. DOE R&D is focused on determining the potential and environmental implications of production of natural gas from hydrates. Fig. 2.20 shows the phase diagram for the two types of hydrates.

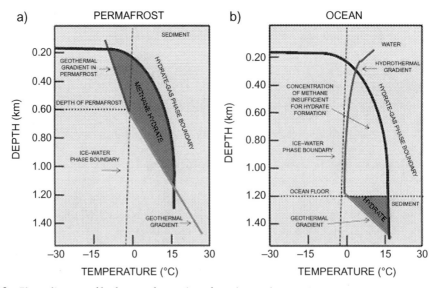

FIG. 2.20 Phase diagram of hydrate under arctic and marine environments.

Top of the supergiant gas hydrate/free natural gas accumulations occur at a depth of 0.4–2.2 m below the sea bottom in the Recent sediments of the world ocean. The bottom of these accumulations is sub-parallel to the sea-bottom surface and intersects beds with anticlinal, synclinal, and tilted forms. This geometry, the geographical distribution of hydrates in the world ocean, their Recent to Pleistocene age and fresh water nature of "combustible ice" could not be explained by terms (source rocks, diagenesis and katagenesis/metagenesis of any buried, dispersed organic matter, lateral migration of natural gas) used in the traditional biotic petroleum origin hypothesis.

According to the theory of the abyssal abiogenic origin of petroleum all gas hydrate/free natural gas accumulations were formed due to the "one worldwide act" i.e. an upward vertical migration of abyssal abiogenic mantle fluid through all the faults, fractures, and pores of rocks and sea-bottom sediments. In that time, not more than 200 thousand years ago those faults, fractures, and pores were transformed by a supercritical geo-fluid (mixture of supercritical water and methane) into a conducting/accumulating/intercommunicating media. Acting as the natural "hydrofracturing" the abyssal geo-fluid has opened up the cavities of cleavage and interstices of bedding in the rocks and sediments as well. According to Dillon et al. (1993) the vertical migration of natural gas still takes place today on the Atlantic continental margin of the United States. Along many faults there the natural gas continues to migrate upwards through the "combustible ice" as through "sieve" that is distinctly seen as the torch-shape vertical strips in the blanking of seismographic records.

On April 10, 2012 Japan Oil, Gas and Metals National Corporation and ConocoPhillips have announced a successful test of technology dealing with safely extract of natural gas from methane hydrates. A mixture of CO_2 and nitrogen was injected into the formation on the North Slope of Alaska. The test demonstrated that this mixture could promote the production of natural gas. This was the first field trial of a methane hydrate production methodology whereby CO_2 was exchanged in situ with the methane molecules within a methane hydrate structure.

2.5.5 Fluid characterization

Conventionally, the following properties are measured for characterizing any crude oil.

Saturation Pressure
Reservoir Temperature, °C
Density at Pb
Separator Test Bo at Pb
GOR at Pb
density, gm/cc 0.779 0.775 STO
API gravity.

These data are used to determine the coefficients of an equation-of-state (EOS), which is then used to form a multi-fluid model for an oilfield. The fluid characterization is then used to estimate reservoir performance with a blackoil reservoir. More data are required if a compositional simulator had to be used. The main task here becomes that of performing crude oil analysis into its hydrocarbon components. Handling of large number of components being cumbersome, the concept of pseudocomponents is introduced.

Typically, a volatile oil sample is collected and conventional PVT experiments—constant composition expansion (CCE), differential liberation expansion (DLE), and multi-stage separator test—are conducted in the laboratory. To describe the reservoir fluid behavior, a multicomponent equation-of-state (EOS) fluid model is developed. The most commonly used EOS is the 1979 Peng-Robinson (PR) equation of state. The EOS fluid model has the standard set of C6- light components (no H_2S), and single carbon number (SCN) fractions from C_7 to C_{29}, and C_{30+} residue. The Søreide specific gravity–molecular weight correlation1is often used for C_7 and heavier components. It is expressed as (Sánchez-Lemus et al., 2016):

$$\gamma_i = a + F_c(M_i - b)^c.$$

$$\gamma_i = a + F_c(M_i - b)^c \qquad (2.22)$$

where γ is the specific gravity, M is the molecular weight and a, b, c, and F_c are constants. The Søreide correlation is tuned by varying F_c while keeping a, b, and c same as default values.

To improve heavy oil characterizations for modeling crude oils in refinery processes, Sánchez-Lemus et al. (2016) devised an in-house Deep Vacuum Fractionation Apparatus (DVFA) that provided reproducible extended distillation data for heavy oils. In addition, an interconversion method was developed to obtain the Normal Boiling Points (NBP) of the distillation fractions. In the current contribution, the physical properties of the distillation cuts collected using the DVFA (the development dataset) are measured and used to evaluate existing correlations and develop new correlations to predict normal boiling point (NBP), specific gravity (SG), and molecular weight (MW) of heavy distillation cuts. The new correlations were also tested on an independent test dataset obtained from the literature. A modified version of Soreide's correlation was proposed for either NBP or MW. The correlation decreased the average absolute relative deviations (AARD) of the NBP from 3.0% to 2.0%, and of the MW from 7.8% to 5.3% for the development dataset and from 2.5% to 2.3% for the test dataset. Similar improvements were obtained for the predicted MW. A new correlation was proposed to predict SG from the H/C ratio and MW, which decreased the AARD from 1.7% to 0.8% for the development dataset and from 3.1% to 1.4% for the test dataset. Of greater significance, the new correlations improved the trend in the predicted properties toward higher boiling cuts and therefore, are expected to provide more accurate estimates of extrapolated properties for heavy oils.

Thiéry et al. (2002) proposed a new method to characterize individual oil-bearing fluid inclusions. It used both the homogenisation temperatures measured by microthermometry, and the degree of gas bubble filling measured by confocal laser scanning microscopy, in conjunction with thermodynamic modeling for describing liquid–gas phase transitions and the volumetric behavior of hydrocarbon mixtures. It is associated with a two-parameter (α, β) compositional model that describes the wide range of compositions of petroleums. We show that this method can give (1) useful estimations of the compositions and pressure–temperature entrapment conditions of oils in fluid inclusions, and (2) insights into the various processes that have affected these fluids either before entrapment (liquid–gas unmixing, gas leaching, mixing, etc.) or after (leakage, etc.) (Fig. 2.21).

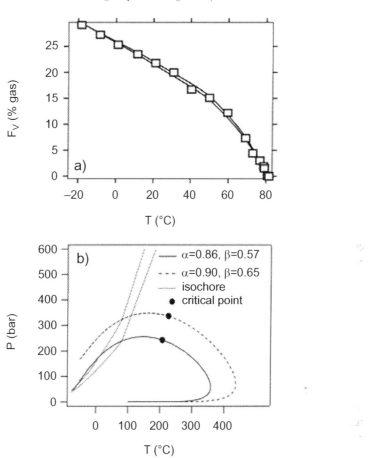

FIG. 2.21 (A) The variation of the degree of gas bubble filling (F_v) as a function of temperature for an oil fluid inclusion from the Alwyn field homogenising at 81.6°C. *Squares* are data measured by confocal laser scanning microscopy. The superimposed curves are calculated with ($\alpha = 0.86, \beta = 0.57$) and ($\alpha = 0.90, \beta = 0.651$), respectively. (B) P–T phase diagram, calculated for the two latter (α–β) sets.

Rhenium-osmium (Re—Os) data from migrated hydrocarbons establish the timing of petroleum emplacement for the giant oil sand deposits of Alberta, Canada, at 112 ± 5.3 million years ago. This date does not support models that invoke oil generation and migration for these deposits in the Late Cretaceous. Most Re—Os data from a variety of deposits within the giant hydrocarbon system show similar characteristics, supporting the notion of a single source for these hydrocarbons. The Re—Os data disqualify Cretaceous rocks as the primary hydrocarbon source but suggest an origin from older source rocks. This approach should be applicable to dating oil deposits worldwide.

Selby and Creaser (2005) introduced a novel dating system to determine age of oil. They stated that the giant oil sand deposits in Alberta formed 112 million years ago - not 60 million years ago, as previously thought. They used the isotopes of two elements found in trace amounts in oil - rhenium and osmium - to accurately pinpoint when oil formed in the western Canada sedimentary basin, which contains much of the world's oil sands. This was the first time that anyone has ever directly determined an age from oil. This uncoupled the migration process from the dating. Previously, the time at which oil was produced from a rock and migrated as a fluid could be deduced from looking at the overall history of that sedimentary basin. However, this is the first time there has actually been a direct determination using any isotopic method to determine the age. Rhenium-osmium (Re—Os) data from migrated hydrocarbons establish the timing of petroleum emplacement for the giant oil sand deposits of Alberta, Canada, at 112 ± 5.3 million years ago. This date does not support models that invoke oil generation and migration for these deposits in the Late Cretaceous. Most Re—Os data from a variety of deposits within the giant hydrocarbon system show similar characteristics, supporting the notion of a single source for these hydrocarbons. The Re—Os data disqualify Cretaceous rocks as the primary hydrocarbon source but suggest an origin from older source rocks. This approach should be applicable to dating oil deposits worldwide.

They collected samples from seven regional deposits within the larger giant oil sands deposit. They sampled and analyzed the same core interval or used an aliquant of material previously collected for organic geochemistry (Fig. 2.20). The distinctive organic geochemistries of the samples have been interpreted to indicate the same or similar sources for all hydrocarbons, possibly the Paleozoic Exshaw Formation. Provost samples 1418 and 1431 also contain minor amounts of the so-called Q compounds [unusual polycyclic alkane and aromatic compounds], known only from the Early Cretaceous Ostracode Zone source rocks, which suggests that the Provost deposit contains minor Ostracode Zone oil in addition to oil from the dominant source.

Our samples contain 3 to 50 parts per billion (ppb) Re and 25 to 290 parts per thousand (ppt) Os (table S1), which are abundances similar to typical organic-rich shale sources of crude oil [crustal abundances of silicate rocks are ~1 to 2 ppb Re and ~50 ppt Os]. These findings lead us to infer that these elements are inherited from their organic-rich source rocks during oil generation and do not originate from crustal units during migration. In shale source rocks, Re and Os are known to be associated with organic matter, but the specific binding compounds are not known. Hydrocarbons analyzed for Re and Os in our samples were isolated by means of organic solvents and filtered to remove any mineral impurities. Thus, we also infer that Re and Os in oil are organically bound, possibly in the asphaltene fraction of oil in which many metals reside.

Li and Yin (2019) used the above model in their research on USGS black shale reference material SBC-1. This reference material was analyzed for Re—Os isotopic composition by three digestion protocols—inverse *aqua regia*, CrO_3-H_2SO_4 and H_2O_2-HNO_3. The results for SBC-1 obtained by inverse *aqua regia* digestion yielded similar Re mass fractions but slightly (~5%) higher Os mass fractions and lower $^{187}Os/^{188}Os$ values than the CrO_3-H_2SO_4 and H_2O_2-HNO_3 digestions. The data set of inverse *aqua regia* digestion exhibited strong correlations in plots of $^{187}Os/^{188}Os$ vs. $1/^{192}Os$ and $^{187}Os/^{188}Os$ vs. $^{187}Re/^{188}Os$, which may signify the incorporation of detrital Re and Os into organic matter in the Re—Os system. Similar correlations were also observed for the CrO_3-H_2SO_4 digestion data set, but not for that of H_2O_2-HNO_3. The data indicate that there is an amount of non-hydrogenous Os in SBC-1 and that CrO_3-H_2SO_4 and H_2O_2-HNO_3 digestions would minimize liberation of the non-hydrogenous Os component. They proposed that SBC-1 may be a more suitable reference material to monitor the influence of detrital Re and Os on Re—Os isochron age data, especially for samples with less organic matter and more siliceous detritus.

They recognized that previous application of the ^{187}Re—^{187}Os system to the dating of organic-rich sedimentary rocks has achieved imprecise Re—Os results because of deficiencies in chemical preparation and mass spectrometry protocols. In addition, different matrix–detritus proportion is also a key factor in the liberated contents of Re and Os. Thus, matrix-matched reference materials (RMs) of similar lithology to the material being analyzed are important for quality control, validation of analytical methods and inter-laboratory comparison. Yin et al. (2018) reported Re—Os data for SGR-1b using three different digestion methods (involving inverse *aqua regia*, CrO_3-H_2SO_4 and H_2O_2-HNO_3) but the results obtained by these methods did not show significant differences. SGR-1b is representative of petroleum- and carbonate-rich shales from the Mahogany zone of the Green River Formation (USA), which possess extremely high contents of organic matter. The RM contains very high Re and Os mass fractions, and the contributions of Re and Os from non-hydrogenous components on the Re—Os system cannot be distinguished. However, for shales with low contents of organic matter and low Re and Os mass fractions, contributions from non-hydrogenous components cannot be ignored. To evaluate the contributions of Re and Os from non-hydrogenous components to the Re—Os system, Li and Yin (2019) present the Re—Os composition for the organic-rich sedimentary rock RM SBC-1 using three different digestion methods: (a) inverse *aqua regia*, (b) CrO_3-H_2SO_4 and (c) H_2O_2-HNO_3. SBC-1 is a marine shale from the Lower Conemaugh Group, upper Pennsylvanian Glenshaw Formation (USA). Compared with SGR-1b, SBC-1 contains much lower organic carbon (C_{org}), Re and Os mass fractions, and higher Al_2O_3 and SiO_2 (Table 2.9). The mineralogy of SBC-1 is dominated by muscovite, quartz, kaolinite and chlorite. Therefore, SBC-1 may be a more suitable reference material to monitor the influence of Re—Os isotopic compositions by detrital material, especially for samples with less organic matter and more siliceous detritus.

Isotopes - versions of elements with different atomic masses - can be used to determine the age of substances. The isotopic method used by Selby and Creaser, for instance, is comparable to carbon-dating, in which the rate of decay of a carbon isotope is used to determine the age of organic matter. Using a mass spectrometer, which analyzes the molecular composition of a

TABLE 2.9 Reference data (USGS) of major element mass fractions (in % m/m) for reference materials SGR-1b and SBC-1.

Analyte	SBC-1	SGR-1b
C_{tot}	2.08	28.0
C_{org}	1.23	24.8
SiO_2	47.6	28.2
Al_2O_3	21.0	6.52
CaO	2.95	8.38
MgO	2.60	4.44
K_2O	3.45	1.66
Na_2O	< 0.15	2.99
Fe_2O_{3T}	9.71	3.03

sample, the researchers spent nearly an entire year examining rhenium and osmium isotopes in large volumes of oil - a meticulous process, says Creaser.

2.6 Scientific ranking of petroleum

From characteristic time analysis, age of hydrocarbon should be considered as the first factor in defining ranking. Table 2.10 shows composition age of a Russian basement reservoir (Ivanov et al., 2014).

The data presented in Table 2.12 are for the Yangiyugan area (the northwestern part of West Siberia). Ivanov et al. (2014) reported that plagiogneisses were formed over the substrate of leucocratic plagiogranites (trondhjemites) under the conditions of the amphibolites facies of metamorphism. The SHRIMP II U–Pb dating of zircons showed that the igneous intrusion of plagiogranites proceeded during the Late Vendian (566 ± 3 Ma). Their metamorphism

TABLE 2.10 Chemical composition (wt%) of minerals from plagiogneis.

Component	1	2c	2e	3	4	5	6	7	8	9
	\multicolumn{10}{c}{Analyses nos.}									
	\multicolumn{10}{c}{Number of analyses}									
	5	4	4	5	7	1	4	6	5	6
P_2O_5	–	–	–	–	–	–	–	–	–	42.33
SiO_2	65.14	37.26	37.66	65.10	37.90	40.96	41.18	46.70	26.10	0.02
TiO_2	0.02	0.08	0.07	0.01	0.11	0.33	0.26	0.45	0.07	–
Al_2O_3	21.23	20.20	20.77	17.71	25.36	15.56	14.77	32.37	19.82	–
Cr_2O_3	–	0.04	0.02	–	0.05	0.01	–	0.04	0.06	–
Fe_2O_3	–	–	–	–	10.63	–	–	–	–	–
FeO	0.16	28.65	29.01	0.47	–	20.77	20.15	3.54	23.30	0.06
MnO	–	2.63	0.50	0.04	0.05	0.62	0.36	0.06	0.38	0.08
MgO	–	2.07	2.83	0.01	0.09	6.34	6.64	1.31	15.73	–
CaO	3.03	8.95	9.59	–	23.31	10.27	10.31	–	0.05	55.39
Na_2O	10.14	–	0.03	0.36	0.03	1.76	1.91	1.05	0.01	0.01
K_2O	0.13	–	–	15.68	–	0.46	0.41	9.33	0.04	–
F	–	–	–	–	0.10	0.22	0.25	0.11	0.18	3.56
Cl	–	–	–	–	–	0.01	0.02	–	–	0.01
sum	99.85	99.88	100.48	99.38	97.63	97.29	96.26	94.96	85.73	101.46

The analyses were carried out using a Cameca SX 100 microanalyzer (the Institute of Geology and Geochemistry, analyst V.V. Khiller); c is the grain center and e is the grain edge; 1—oligoclase; 2—almandine; 3—microcline; 4—epidote; 5—ferropargasite; 6—ferroedenite; 7—muscovite; 8—clinochlore; 9—fluoroapatite.
From Ivanov et al., 2014.

with the formation of plagiogneisses took place in the Early Ordovician (486 ± 4 Ma). The shows of powerful fluid–metasomatic processes of the transformation of rocks during the Carboniferous time are revealed.

Turek and Robinson (1984) listed published data on ages of various basement rocks. This is shown in Table 2.11. In several instances, the age reaches close to billion years. Turek and Robinson (1984) reported similar ages (some exceeding billion years) for a Canadian field (Fig. 2.22). Even though some of these data points were removed at the time, we know now that the age of the earth is estimated to be much older, making those numbers realistic.

Fig. 2.22 shows how different types of oils take up different natural processing time. Of course, in this process, biofuels come the last because their processing time is miniscule in a geologic timeframe. This figure indicates that basement hydrocarbons are naturally processed for the longest time. This is significant considering the fact that if 'natural' breeds sustainability, basement oils should have the most likelihood to be sustainable. However, one must note that there are other factors to consider. For instance, diamond, while naturally processed the longest time, doesn't make a fuel. As we transit to different characteristic time, the usefulness of natural products changes in applications. Another aspect is, when fresh fuel (e.g biofuel) is oxidized, it produces CO_2 that are readily absorbable by the greeneries in the process of photosynthesis. So, even though crude oil has been processed longer than say vegetable oil, the value gained by crude oil is not in the quality of CO_2 that is generated. What is then the gain in crude oil due to longer term natural processing? It is in its applicability to a longer-term application. This aspect will be discussed in the follow-up section. Fig. 2.23 also indicates that in terms of ranking due to natural processing time, the following rule applies:

(1) Basement oil
(2) Unconventional gas/oil

TABLE 2.11 Published isotopic mineral ages for Precambrian basement in southwestern Ontario, Michigan, and Ohio.

Location	Type	Age (Ma)	Mineral/rock
Burford Tp. Brant Co.	K–Ar	920	Biotite/granite gneiss
Romney Tp. Kent Co.	K–Ar	895	Biotite/migmatite
Washtenaw Co.	Rb–Sr	840	Biotite/gneiss
Washtenaw Co.	Rb–Sr	920	Biotite/gneiss
St. Clair Co.	Rb–Sr	900	Biotite/gneiss
	K–Ar	970	Biotite/gneiss
Huron Co.	Rb–Sr	900	Biotite/gneiss–schist
	K–Ar	935	Biotite/gneiss–schist
Sandusky Co.	Rb–Sr	890	Biotite/gneiss–schist
	K–Ar	935	Biotite/gneiss–schist
Wood Co.	Rb–Sr	900	Biotite/biotite gneiss
	K–Ar	960	Biotite/biotite gneiss

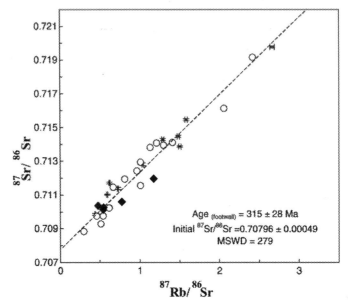

FIG. 2.22 Whole rock Rb—Sr isochron diagram, basement samples.

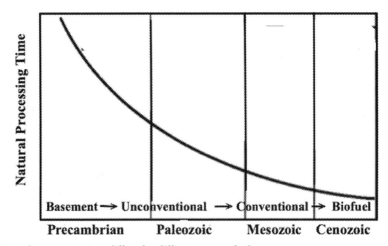

FIG. 2.23 Natural processing time differs for different types of oils.

(3) Conventional gas/oil
(4) Biofuel (including vegetable oil)

The same rule applies to essential oils. These natural plant extracts have been used for centuries in almost every culture and have been recognized for their beneifts outside of being food and fuel. Not only does the long and rich history of essential oils usage validate their efficacy specifically—it also highlights the fact that pristine essential oils are a powerfully

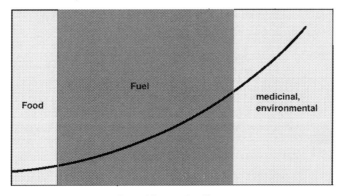

FIG. 2.24 Natural processing enhances intrinsic values of natural products.

effective resource with many uses and benefits—from beauty care, to health and wellness care, fast effective pain relief, and emotional healing, and others. In another word, the benefits of oils are diverse and it is the overall benefit that increases with longer processing time.

Fig. 2.24 is based on the analysis originally performed by Chhetri and Islam (2008) that stated that natural processing is beneficial in both efficiency and environmental impact. However, as Khan and Islam (2012, 2016) have pointed out, such efficiency cannot be reflected in the short-term calculations that focus on the shortest possible duration. The efficiency must be global, meaning it must include long-term effects, in which case natural processing stands out clearly. Fig. 2.24 shows how longer processing time makes a natural product more suitable for diverse applications. This graph is valid even for smaller scales. For instance, when milk is processed as yogurt, its applicability broadens, so does it 'shelf life'. Similarly when alcohol is preserved in its natural state, its value increases. Of course, the most useful example is that of honey that becomes a medicinal marvel after centuries of storage in natural settings (Islam et al., 2015). In terms of fuel, biomass, which represents minimally processed natural fuel, produces readily absorbable CO_2 that can be recycled within days to the state of food products. However, the efficiency of using biomass to generate energy is not comparable to that can be obtained with crude oil or natural gas. Overall, energy per mass is much higher in naturally processed fuel.

Heavy mentals, such as, lithium for manic depression, are used in some of the pharmaceutical drugs. Although several studies have appeared, reporting on heavy metals in drugs, they are limited to heavy metals as a contaminant (Nessa et al., 2016). Process steps involving transition metal catalysts are now commonplace in Active Pharmaceutical Ingredient (API) manufacturing, presenting the real possibility for traces of these metals to remain in the API after purification. Common metals used in this way include chromium, copper, nickel, palladium, platinum, rhodium and platinum, although there are others. Historically, the most likely elements for trace metals in drug products were arsenic, cadmium, lead and mercury, all of which are much more likely to enter the manufacturing chain from natural sources. Islam (2015) pointed out that heavy metals are necessary, albeit in very low functioning of vital organs. While some of these components (e.g. lithium) are used an active ingredient, invariably these drugs are made synthetically. This opens up an opportunity to use crude oil directly as a source of these heavy metals for therapeutic purposes. In general, the following anemia are well known:

Iron: common simple anemia (iron deficiency), results in the loss of functional heme proteins (hemoglobin, myoglobin, etc.), which are responsible for oxygen transport or utilization of oxygen. Pernicious anemia comes from a lack of vitamin B-12 (which contains a cobalt complex called cobalamin), which then in turn interferes with the function of red blood cells.

Zinc: Zinc anemia is mostly due to diet can result in growth retardation.

Copper: Copper anemia in infants results from infants with a poor diet and can cause heart disease (Carver, 2019).

It is well known that physiological functions of humans need metal ions. Less known is the fact that those metals include heavy metals as well. They each have essential roles in human physiology. The recognized essential metals for humans are: Sodium, Potassium, Magnesium, Copper, Vanadium, Chromium, Manganese, Iron, Cobalt, Nickel, Zinc, Molybdenum, and Cadmium. Anemia symptoms are caused by lack of a certain essential metal. Anemia can be associated with malnourishment or faulty metabolic processes, usually caused by a genetic defect (Carver, 2019). In the Preface to the book, Carver (2019) points out that many metals play a significant role in medicine, including not only those that are now recognized as essential (indispensable for human life) and for which a maintenance of metal ion homeostasis is required, but also a large group of elements that are used in diagnosis and therapy of diseases. Islam et al. (2018) made case in favor of natural chemicals as beneficial and essential while the artificial ones are the opposite. However, for natural chemicals to be useful, the concentration cannot exceed the balanced value, beyond which toxicity sets in (e.g. water toxicity). This conclusion is a result of the premise that nature is self healing and is the standard of sustainability (Khan and Islam, 2016). In their discussion Carver (2019) subdivided into aspects of (a) metal-related diseases, (b) metals as medicines, and (c) metal ion toxicity, indicating the broad scope of the subject. In this book, the chapter, titled "Developing vanadium as an antidiabetic or anticancer drug: A clinical and historical perspective", presents the role of chromium in medicine. This summarizes the limited present knowledge on the biological activity and often questioned essentiality of this metal, which is highly toxic in its highest oxidation state [Cr(VI) in chromates] but is now widely used in complexes of its also common oxidation state Cr(III), e.g. as a trace element additive to vitamin pills for humans and in animal nutrition. By contrast, manganese is again an essential element as summarized in the chapter by K. M. Erikson and M. Aschner: "Manganese: Its role in disease and health" (of the same book), which shows the abundance of this element in many domains of the human body and its role as a coenzyme for man biological processes requiring a regular supply. Although cobalt is also known to be essential, its role in clinical applications is limited and the element has been relatively ignored by the pharmaceutical industry. Only few families of complexes have been thoroughly studied regarding the role of the coordinating ligands employed in the formulations of manganese supplementations. For the chapter with the title "Cobalt-Schiff-Base complexes: Preclinical research and potential therapeutic uses" by E. A. Bajema, K. F. Roberts and T. J. Meade, a specific class of complexes has been chosen, for which the results of preclinical studies suggest antimicrobial, antiviral, anticancer, and amyloid-ß inhibitor activities.

Similarly, Copper has been well known to be an essential metal for several decades and its bioinorganic chemistry has been studied for many years to unravel its large variety of basic biological functions. One of the main research areas is currently the significance of copper homeostasis in Alzheimer's disease, since post-mortem analyses of amyloid plaques have indicated an excessive accumulation of copper (less so for Zn and Fe) in the human brain, due to a "mis-metabolism of metal ions". A large number of blood-brain-barrier permeable hydroxyquinoline-derived chelators have been applied as copper scavengers, and some of these were found to be copper-specific (over Zn and Fe). One of the further chapters (by J. Lopez, D. Ramchandany and L. Vahdat) of the same book is dedicated to "Copper depletion as a therapeutic strategy in cancer." Despite promising preclinical data, the clinical experiences (phase II) with various chelating Cu-selective agents have not yet supported this approach regarding inhibition of the evolution of cancer and metastatic spread. A large selection of organic and inorganic complexants were employed, such as prodrugs which are metabolized to give thiocarbamates as chelators for Cu(II), but no satisfactory overall performance has as yet been reached. The well-known therapeutic effect of tetrathiomolybdates may also be partly due to its assistance in copper depletion. Similar conclusions are drawn Fe, V, Cr, Co, Cu.

Metal ions are often used for diagnostic medical imaging. Metal complexes can be used either for radioisotope imaging (from their emitted radiation) or as contrast agents, for example, in magnetic resonance imaging (MRI). Such imaging can be enhanced by manipulation of the ligands in a complex to create specificity so that the complex will be taken up by a certain cell or organ type (Nash, 2005).

For sometime, the following metals have been identified as having therapeutic value. In recent years, the followings are summarized by Lippard (1994).

Platinum: Platinum based compounds have been shown to specifically effect head and neck tumors. These coordination complexes are thought to act to cross-link DNA in tumor cells.

Gold: Gold salt complexes have been used to treat rheumatoid arthritis. The gold salts are believed to interact with albumin and eventually be taken up by immune cells, triggering anti-mitochondrial effects and eventually cell apoptosis. This is an indirect treatment of arthritis, mitigating the immune response.

Lithium: Li_2CO_3 can be used to treat prophylaxis of manic depressive disorder.

Zinc: Zinc can be used topically to heal wounds. Zn^{2+} can be used to treat the herpes virus.

Silver: Silver has been used to prevent infection at the burn site for burn wound patients.

Platinum, Titanium, Vanadium, Iron: cis DDP (cis-diaminedichoroplatinum), titanium, vanadium, and iron have been shown to react with DNA specifically in tumor cells to treat patients with cancer.

Gold, Silver, Copper: Phosphine ligand compounds containing gold, silver, and copper have anti-cancer properties.

Lanthanum: Lanthanum Carbonate often used under the trade-name Fosrenol is used as a phosphate binder in patients suffering from Chronic Kidney disease.

Bismuth: Bismuth subsalicylate is used as an antacid.

Heavy metals are toxic, but their oxides are usually not. Further more, they are actually beneficial if derived from natural sources. Islam et al. (2018) argued that aged crude oil is the best source of these heavy metals. As of now, Food and Drug Administration has approved arsenic trioxide to be used in Acute Promyelocytic Leukemia (APL) (Antman, 2001). There are some reports published on the harmful effects of ayurvedic *Bhasmas* of Indian system of medicine. Actually the *Bhasmas* can be toxic or harmful to humans only if they are not prepared in the correct manner (Thorat and Dahanukar, 1991). The preparations are then prescribed with certain *Anupanas* (accompaniments), e.g., ginger or cumin water, tulsi extract, etc. that have been shown to protect against unwanted toxicity due to varied reasons (Sharma et al., 2002, 2009) including high proportions of trace elements and synergistic or protective effects due to buffering between various constituents. As per Ayurveda, the bioavailability and toxicity of the metals depend on their chemical forms, especially of mercury, although some authors could not ascertain it experimentally (Gochfeld, 2003). An example of nontoxicity of ayurvedically processed (as suggested in *Shastras*) so-called toxic herbs are given as: crude aconite at 2.5 mg/mouse produces 100% mortality. Ayurvedically processed aconite (compound A) the root of the plant was boiled with two parts of cow's urine for 7 h per day for two consecutive days. The root was then thoroughly washed with water and boiled with two parts of cow's milk for the same duration. Processed aconite (compound B) processed only in cow's urine for 7 h per day for 2 consecutive days. Aconite processed only in cow's milk for the same duration (compound C) was also considered safe at 20 mgs. The study exhibited that compound A was totally non toxic followed by compound B and C, respectively, which were also reported to be safer than crude aconite (Thorat and Dahanukar, 1991).

Mercurous mercury, also called calomel, was used as diuretic, antiseptic, skin ointment, vitiligo, and laxative for centuries. Calomel was also used in traditional medicines, but now these uses have largely been replaced by safer therapies. Other preparations containing mercury are still used as antibacterials. *Rasa shastra* experts claim that these medicines, if properly prepared and administered, are safe and therapeutic. Navbal Rasayan (NR) a metal based ayurvedic formulation is used for the treatment of multiple sclerosis; study with NR in animals does not show any toxic effect. However, decrease or attenuation of agonistic activities of histamine, acetylcholine and serotonin needs further exploration (Chandra and Mandal, 2000). Two gold preparations, ayurvedic *Swarna Bhasma* and unani *Kushta Tila Kalan* are claimed to possess general tonic, hepatotonic, nervine tonic, cardiostimulant, aphrodisiac, detoxicant, antiinfective and antiaging properties (Chopra et al., 2006). In modern medicine, gold compounds (e.g., gold disodium thiomalate and auranofin) have been used in the treatment of rheumatoid arthritis for more than 60 years with well documented effects on immune function (Bloom et al., 1988). Marked analgesic (elicited through opioidergic mechanisms) and immunostimulant effects of these preparations with a wide margin of safety have been reported (Baja and Vohora, 1998). Anticataleptic, antianxiety, and antidepressant properties are also observed (Bajaj and Vohora, 2000).

Tamra Bhasma, a metallic ayurvedic preparation, is a time-tested medicine in ayurveda and is in clinical use for various ailments specifically the free radical mediated diseases. Studies show that *Tamra Bhasma* inhibits lipid peroxidation (LPO), prevents the rate of aerial oxidation of reduced glutathione (GSH) content and induces the activity of superoxide dismutase (SOD) in rat liver homogenate in the biphasic manner (Pattanaik, 2003). Table 2.12 shows absorption of heavy metals.

Deficiency of copper in the body causes weight loss, bone disorders, microcytic hypochromic anemia, hypopigmentation, graying of hair and demyelination of nerves etc. (Pandey, 1983). It is reported that *Tamra Bhasma* potentiates the antioxidant activity of animals, when given orally treated animals showed less degree of lipid peroxidation. Results clearly indicated that *Tamra Bhasma* does have antioxidant property in low doses, without any side effect,

TABLE 2.12 Heavy metal absorbing capability of various natural agents.

Category of natural agents	Scientific name	Heavy metals absorbed
Sea weeds	Chlorella emersonii	Cd
	Sargassum muticum	Cd
	Ascophyllum sargassum	Pb,Cd
	Ulva reticulate	Cu(II)
	brown sea weeds	Cr
	Ecklonia species	Cu(II)
Fungal species	Phanerochaete chrysosporium	Ni(II),Pb(II)
	Aspergillus niger	Cd
	Aspergillus fumigatus	Ur(VI)
	Aspergillus terreus	Cu
	Penicillium chrysogenum	Au
Yeast species	Saccharomyces cerevisiae	Uranium
	S. cerevisiae, Kluyveromyces fragilis	Cd
	S. cerevisiae	Methyl mercury and Hg(II)

From Singh, R. et al., 2011. Heavy metals and living systems: an overview. Indian J. Pharmacol. 43(3), 246–253.

even up to 90 days of treatment in the dose of 5 mg/kg body weight. However in higher doses, when given for a longer period, it induced lipid peroxidation, without any effect on the rate of survival but these tested doses are much higher than the human therapeutic doses. Table 2.13 illustrates the effect of *Tamra Bhasma* on the survival of albino rats up to 30 days.

Heavy metals may exert their acute and chronic effects on the human skin through stress signals. Findings suggest that heavy metals reduced the phosphorylation level of small heat shock protein 27 (HSP27), and that the ratio of p-HSP27 and HSP27 may be a sensitive marker or additional endpoint for the hazard assessment of potential skin irritation caused by chemicals and their products (Zhang et al., 2010).

There are some contradictory claims about the effect of heavy metals. Herbal and natural products are safer than the synthetic or modern medicines but even some indigenous herbal products contain heavy metals as essential ingredients. Thus the expanded use of herbal medicine has led to concerns relating to its safety, quality, and effectiveness especially for *Bhasmas* as these are usually made of heavy metals like arsenic, mercury, copper, zinc, gold, and silver. Therefore, contamination of herbal drugs with heavy metals is of prime concern. Prolonged exposure to heavy metals such as cadmium, copper, lead, nickel, and zinc can cause deleterious health effects in humans (Reilly, 1991). Although many of traditional remedies are used safely, there have recently been an increasing number of case reports being published of heavy metal poisoning after the use of traditional remedies, in particular, Indian ayurvedic remedies (Lynch and Braithwaite, 2005). These were started extensively after the study

TABLE 2.13 Effect of *Tamra Bhasma* on the survival of albino rats up to 30 days.

Dose of Bhasma (per kg body weight)	% of survival
Drug vector	100
5 mg	100
10 mg	100
20 mg	100
200 mg	100
600 mg	100
1200 mg	100
2400 mg	100
4800 mg	100

The % has been calculated on the basis of the mean value of six separate experiments. Drug vector is 20% gum acasia in distilled water.
Adopted from Pattanaik, N., 2003. Toxicology and free radicals scavenging property of Tamra Bhasma. Indian J. Clin. Biochem. 18, 18–89.

showed high levels of lead, mercury and arsenic found in ayurvedic products sold in US, and this lead to a strong evidence for further quality and safety issues. The Indian population who frequently purchase ayurvedic herbal supplements, *Bhasmas* and *Rasa*, may not have understood that the traditional formulation contained heavy metals requiring special care and supervision. Inhalation of mercury vapor produces acute corrosive bronchitis and interstitial pneumonitis and, if not fatal, may be associated with central nervous system effects such as tremor or increased excitability. Inhalation of large amounts of mercury vapor can be fatal. With chronic exposure to mercury vapor, the major effects are on the central nervous system. The triad of tremors, gingivitis and erethism (memory loss, increased excitability, insomnia, depression, and shyness) has been recognized historically as the major manifestation of mercury poisoning from inhalation of mercury vapor. Sporadic instances of proteinuria and even nephrotic syndrome may occur in persons with exposure to mercury vapor, particularly with chronic occupational exposure Methyl mercury crosses the placenta and reaches the fetus, and is concentrated in the fetal brain at least 5 to 7 times that of maternal blood. The adverse effects of gold salts particularly on prolonged use (nephrotoxic, bone marrow depression, cutaneous reactions, and blood dyscriasis etc.) are well documented. The preparations under study are not gold salts but calcined preparations of gold used in Ayurveda (SB) and Unani-Tibb (KTK) and involve incorporation of herbal juices (*Aloe vera, Dolichos uniflorus, Rosa damascena*), minerals (mercury, sulfur) and animal origin ingredients (whey, cow's urine) during the ashing process. They constitute unidentified complexes of the metal which may not have properties and biological effects akin to gold salts. *Kushta Tila Kalan* (KTK) and *Swarna Bhasma* (SB) reported to produce immunostimulant, rather than immunosuppressant actions and analgesic actions, without describle untoward effects at the doses used.

2.6 Scientific ranking of petroleum 115

Let's review examples from both oil and natural gas. Historically it has been perpetrated that conventional gas and oil is miniscule compared to unconventional. With it comes the notion that it is more challenging to produce unconventional petroleum resources. In addition, at least for petroleum oil the notion that unconventional resource is more challenging to process is prevalent. This notion is false. With the renewed awareness of the environmental sustainability it is becoming clear unconventional resources offer more opportunities to produce environment-friendly products that conventional resources. Fig. 2.25 shows the pyramid of both oil and gas resources.

On the oil side, the quality of oil is considered to be declining as the API gravity declines. This correlation is related to the processing required for crude oil to be ready for conversion into usable energy, which is related to heating value. Heating value is typically increased by refining crude oil upon addition of artificial chemicals are principally responsible for global warming (Chhetri and Islam, 2008; Islam et al., 2010). In addition, the process is inefficient and resulting products harmful to the environment. Fig. 2.26 shows the trend in efficiency, environmental benefit and real value with the production cost of refined crude. This figure shows clearly there is great advantage to using petroleum products in their natural state. This is the case for unconventional oil. For instance, shale oil burns naturally. The colour of flames (left image of Picture 2.1) indicates that crude oil produced from shale oil doesn't need further

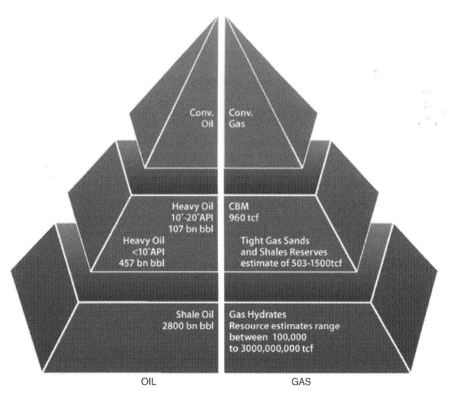

FIG. 2.25 The volume of petroleum resources increases as one moves from conventional to unconventional.

116　　　　　　　　　　　　　　　　　　　2. Reservoir rock and fluid characterization

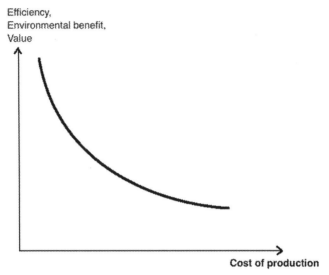

FIG. 2.26　Cost of production increases as efficiency, environmental benefits and real value of crude oil declines. *Modified from Islam, M.R., Chhetri, A.B., Khan, M.M., 2010, Greening of Petroleum Operations, Wiley-Scrivener.*

PICTURE 2.1　Images of burning crude oil from shale oil (left) and refined oil (right).

processing. The right image of Picture 2.1 emerges from burning gasoline and has similar colors to those of the left.

In addition, crude oil from shale oil is 'cleaner' than other forms of crude oil because of the fact that it is relatively low in tar content as well sand particles. Another crucial aspect is the fact that sulfur content or other toxic elements of crude oil have no correlation with unconventional or conventional sources. Also, heavier oils do not have more of these toxic elements and are not in need of refinement to be usable.

Lighter crudes are considered to be easier and less expensive to produce only because modern engineering uses refined version of the crude oil and all refining technologies are specially designed to handle light crude oil. If sustainable refining techniques are used, lighter or conventional oil offers no particular advantage over unconventional one and yet the volume and ease of production of unconventional are greater in unconventional resources. These results suggest strong impacts of API gravity and HP yield on overall refinery efficiency. In this, basement reservoirs offer an interesting solution. They produce petroleum products that are old, yet high API/low HP. Technically, these hydrocarbons should mediocre, but scientifically they should be the best choice. The reason for this discrepancy is the fact by refining we tend to harmonize each type of crude oil. By forcing them to conform to a common standard, we reduce natural synergy these natural resources have.

2.7 Characterization of reservoirs

Most productive reservoirs are heterogeneous with some form of fractured network. All reservoir characterization tools, as well as processing techniques, are uniquely designed to conventional reservoirs and as such are not suitable for fractured formations. Applying these tools and techniques to most prolific reservoirs has the risk of conclusions that are irrelevant and random at best. The main mechanism of fluid flow in these reservoirs is through fractures, natural or induced, whereas existing techniques do not integrate this aspect to reservoir characterization. This section presents a comprehensive reservoir characterization technique, first presented by Islam (2015).

Naturally fractured reservoirs have been classified according to the relative contribution of the matrix and fractures to the total fluid production (Nelson, 2001). Accordingly, the fractured reservoirs are classed as:

Type 1—fractures provide essential porosity and permeability;
Type 2—fractures provide essential permeability; and
Type 3—fractures provide permeability assistance.

Understanding of the in situ stress field is essential for successful and safe production of unconventional reservoirs, including basement formations that are dominated by the flow through fractures (Bickle et al., 2012). Knowledge of the in situ stress orientation is important for understanding borehole stability, fluid flow in naturally fractured reservoirs, and hydraulic fracture stimulation, and most importantly, to develop proper understanding of the fluid flow in the reservoir (Fuchs and Müller, 2001). Artificial fractures are mostly induced during hydraulic fracturing in deviated or horizontal wells. Even in absence of hydraulic fracturing data on hydraulic fracturing can be useful to construct the fracture network of the natural reservoir system.

The World Stress Map (WSM) (Zoback et al., 1985; Heidbach et al., 2004, 2008) was initiated to produce 1:1 million continental-scale stress maps to provide information on the present-day crustal stress field. To accomplish this, the WSM produced guidelines which have become de facto standards for the identification of breakouts from dual-caliper logs (Reinecker et al., 2003) and borehole image logs (Tingey et al., 2008). Borehole breakouts comprise approximately 19% of the data in the WSM database, and are used alongside other data such as focal plane mechanisms to map the distribution of the global stress field (Tingay et al., 2008). The method for characterizing borehole breakouts employed by the WSM concerns the identification of specific breakout intervals defined by a series of guidelines designed to minimize ambiguity. The WSM quality ranking scheme then enables collation with other datasets such as earthquake focal mechanisms to investigate basement stress orientations (Sperner et al., 2003), for example research on deep research boreholes such as Germany's KTB borehole (Emmermann and Lauterjung, 1997).

Maximizing economic recovery from naturally fractured reservoirs is a complex process. It requires a thorough understanding of matrix flow characteristics, fracture network connectivity and fracture-matrix interaction. It involves knowing the geological history. Therefore, key to successful reservoir characterization is in connecting with geologists that can construct an overall picture of the reservoir history. Construction of this history is pivotal. This construction must be scientific, following objective systematic abstraction. The abstraction process has to be bottom up. It involves collecting data in its raw form. These data have to be collected in proper time sequence and at each step, verified from multiple sources. Medieval scholar, Al-Kindus famously said, "Multi-source information is a treasure". Reservoir description should rely on information from many sources. For petroleum reservoir applications, abstraction starts with collection of geological data. In the first phase of abstraction, depiction of the subsurface strata is made. To create this picture, geologists must collect data from any available source, such as outcrop, regional subsurface maps, etc. Based on this geological map, the decision to conduct geophysical survey is made. During the geophysical survey, decisions to implement a certain grid, type of survey, etc., have to be made. The process of geophysical survey offers an example of how multi-source data must be integrated. During this process, geological data are used as a baseline, whereas geophysical data are later used to refine geological data. At the end, the decision to drill is made only after several cycles of abstraction and refinement.

The idea behind reservoir characterization is to know the past. That means, knowing

- Origin of fluid.
- Origin of the reservoir.
- Origin of the fractures.
- Process of erosion, transport, and deposition.
- Process of faulting, fracturing.
- Process of secondary activities (leaching, cementation, etc.)

The next task in reservoir characterization is to know the present. It involves collection of mud data, drill cutting, oil and gas 'shows', pressure data, Temperature data, and core data. These dynamic data are then assembled with static data to refine data on

- stratigraphy
- lithology

TABLE 2.14 Various stages of fracture data collection.

Stage	Well locations	Fracture description	Abstraction/refinement
Exploration	No well present	Based on faults	Geology/geophysics
Delineation	Wells used to refine data	Based on faults + nearby wells	Static data/dynamic data from nearby wells
Drilling	Wells used to correlate	Drilling data	Static data/dynamic data
Logging and coring	Direct evidence	Images and visual inspection	Static data refinement
Primary recovery	Optimization can lead to infill drilling	Based on dynamic data.	Static/dynamic data
Enhanced oil/gas recovery	Injected fluid can act as a tracer	Pathway selected by injected fluid	Static/dynamic data

- fracture density
- fracture orientation
- nature of fractures
- shale breaks
- "sweet spots"

Initial static data, collected during exploration process is the most valuable because it serves as the foundation. As soon as the first well is drilled, data must be collected even during the drilling process. Data on mud loss, rate of penetration, cuttings, azimuth, and others offer valuable insight for refining the subsurface picture.

The core data offer the first set of direct evidence of fractures. These data should be compared with the fault network predicted in the geological map that was refined after the geological survey. Cores also give one an opportunity to determine fracture frequency and the nature of fractures (e.g., plugged or open), and leads toward developing the Rose diagram. The development of the Rose diagram is of utmost importance because that would dictate the direction of flow.

The Drill stem test (DST) data are the first set of dynamic data available. Any discrepancy between core permeability and DST permeability would indicate the existence of fractures. As a well is put on production, it continues to generate dynamic data that are invaluable for the abstraction and refinement process, outlined above.

During the production cycle of a petroleum reservoir, it continues to provide one with valuable data on fracture characterization. Table 2.14 shows how fracture characteristics can be predicted with the data available at each stage of petroleum operations.

2.7.1 Natural and artificial fractures

Natural fractures and drilling-induced wellbore failures provide critical constraints on the state of in situ stress and the direct applicability to problems of reservoir production, hydrocarbon migration, and wellbore stability. Acoustic, electrical, and optical wellbore images provide the means to detect and characterize natural fracture systems.

Fractures are 4D features—a fact that is often neglected in the early stages of reservoir development. While fractures observed in the wellbore will be analyzed to determine aperture and probable production rates. Little effort is made to develop a detailed model of fracture distribution. Even less effort is spent on determining the history of fractures. If properly considered, natural fractures and artificial ones can be shown to be diametrically opposite to each other, including the origin, orientation and propagation. As such, basement reservoirs, just like unconventional ones, offer a unique challenge.

Bell and Gough (1979) noted that stress concentrations around vertical boreholes can cause caving, also known as a borehole breakout. Borehole breakouts and drilling-induced fractures (DIFs) are important indicators of horizontal stress orientation, particularly in a seismic regions and at intermediate depths (>5 km). Approximately one-fifth of the stress orientation indicators in the World Stress Map (WSM) database have been determined from borehole breakouts and DIFs. Furthermore, borehole breakouts and DIFs provide the majority of stress orientation indicators in petroleum and geothermal systems, all of which help characterize a reservoir. Fractures are of utmost importance for basement formations, for which the majority of fluid flows through fractures.

Plumb and Hickman (1985) were able to show that the orientation of the elongations, or breakouts which result in a compressive failure of the well take place in the orientation of S_{hmin}, orthogonal to S_{Hmax} in vertical boreholes.

Upon drilling of a well, borehole breakout occurs when the stresses around the borehole exceed that required to cause compressive failure of the borehole wall (Zoback et al., 1985;). The enlargement of the wellbore is caused by the development of intersecting conjugate shear planes that cause pieces of the borehole wall to spall off (Fig. 2.27). The stress concentration around a vertical borehole is greatest in the direction of the minimum horizontal stress (SH). Hence, the long axes of borehole breakouts are oriented approximately perpendicular to the maximum horizontal compressive stress orientation (Plumb and Hickman, 1985).

DIFs are created when the stresses concentrated around a borehole exceed the limit that required to cause the tensile failure of the wellbore wall. DIFs typically develop as narrow sharply defined features are sub-parallel or slightly inclined to the borehole axis in vertical wells. They are generally not associated with significant borehole enlargement in the fracture direction (note that DIFs and breakouts can form at the same depth in orthogonal directions). The stress concentration around a vertical borehole is at a minimum in the SH direction. Hence, DIFs develop approximately parallel to the SH orientation. When a wellbore is drilled, the material removed from the subsurface is no longer supporting the surrounding rock. As a result, the stresses become concentrated in the surrounding rock (i.e., the wellbore wall). Borehole breakout occurs when the stresses around the borehole exceed that required to cause compressive failure of the borehole wall (Zoback et al., 1985). The enlargement of the wellbore is caused by the development of intersecting conjugate shear planes that cause pieces of the borehole wall to spall off (Fig. 2.27). The stress concentration around a vertical borehole is greatest in the direction of the minimum horizontal stress (SH). Hence, the long axes of borehole breakouts are oriented approximately perpendicular to the maximum horizontal compressive stress orientation (Plumb and Hickman, 1985).

A critical factor in hydraulic fracturing operations is the orientation of the in situ principal stresses. Hydraulic fracturing will propagate along the path of least resistance and create width in a direction that requires the least force. Therefore, hydraulic tensile fractures

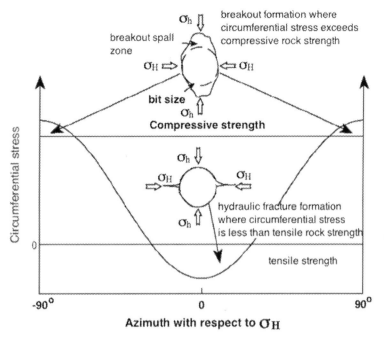

FIG. 2.27 Schematic cross-sections of borehole breakout and drilling-induced fracture (Hillis and Reynolds, 2000).

propagate parallel to the maximum horizontal stress (σ_{Hmax}) in the vertical plane. Consequently, to maximize recovery with minimal energy input it is necessary to drill horizontal wells parallel to the minimum horizontal stress (σ_{hmin}) direction. As a result hydraulic tensile fractures will propagate parallel to the maximum horizontal stress (σ_{Hmax}) in the vertical plane.

Understanding the orientation of the in situ stress is therefore imperative prior to drilling to ensure that wells are deviated favorably with respect to the in situ stress. To construct such a picture, the use of dual-caliper logs is vastly inadequate. The use of borehole image logs to characterize the orientation of σ_{Hmax}. Fig. 2.28 shows the distribution of borehole imaging data across the UK, which by fortunate coincidence corresponds very closely to the area of the UK that is sub-cropped by the potentially economic Bowland–Hodder Shale. This also shows the dual-caliper data distribution across the UK.

Dual-caliper logging, usually undertaken in conjunction with dipmeter tools typically measure four points on the borehole circumference with a vertical resolution of between 25 and 154 mm. Guidelines for breakout identification from caliper logs is detailed in Reinecker et al. (2003). The tool rotates as it is pulled up the wall however when it encounters a zone of borehole elongation the rotation will cease. The tool locks into an elongation zone which, with the aid of the other tool outputs, can be interpreted as a breakout (Reinecker et al., 2003).

Borehole imaging tools provide high resolution borehole images based on (generally) either ultrasonic velocity or resistivity. These tools and their origins are described in Paillet et al.

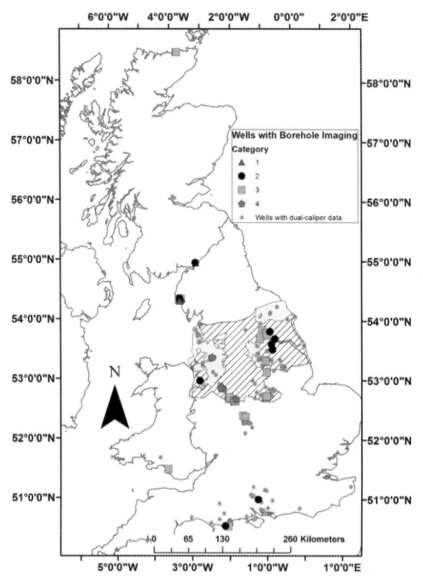

FIG. 2.28 Onshore map of distribution of wells logged with borehole imaging data by category across the UK compared with distribution of dual-caliper logs.

(1990) and Lovell et al. (1999). Borehole imaging tools provide wall coverage of between 20% and 95%, depending on the tool specification and borehole diameter. These imaging tools will be discussed further in latter sections of this chapter.

Drilling induced tensile fractures (DIF) are equivalent to artificial fractures, not different from the ones created by hydraulic fracturing. DIFs result from tensile failure directly induced by the drilling process. These fractures form parallel to the orientation of the greatest far field horizontal stress (σ_{HMAX}) (Moos and Zoback, 1990). These features occur when the sum circumferential stress concentration and the tensile strength are exceeded by the pressure in the well (Moos and Zoback, 1990). As DIFs have widths of only a few mm they can only be identified from high-resolution borehole image logs as they are not associated with any borehole enlargement (Tingay et al., 2008). They are generally narrow well-defined features, which are slightly inclined or sub-parallel to the borehole axis and form perpendicular to breakout orientation (Tingay et al., 2008). Fig. 2.29 shows a section of the Melbourne 1 well in Yorkshire with both breakouts and tensile fractures.

- Review and correction of metadata to ensure that all are properly located and orientated.
- Addition of borehole construction metadata, to include casing intervals and downhole bit size, allowing for accurate section-by-section review of borehole data.
- Where possible, inclinometry surveys from borehole image logs should be reloaded from original media to maximize availability and auditability.
- Review of all available data to ensure that any previously unidentified image logs are included and processed.
- Loading of the complete available digital archive of both the radioactive waste disposal program (whenever available) and also the oil and gas industry which includes outputs from a variety of borehole imaging tools.

2.7.2 Interpretation of borehole images to identify breakouts

The dual caliper analyses have a completely circumferential scatter, on which a slight northwest-southeast trend is visible producing an σ_{Hmax} direction of 149.87° with circular standard deviation of 66.9° calculated according to Mardia (1972). This dataset shows a very clear trend with very limited associated scatter but contains all of the identified breakouts within the data interpreted within this study. Therefore, the primary output of this study is the recognition that the exclusive use of borehole imaging tools has significantly reduced the uncertainty in the orientation of σ_{Hmax}. The interpretation of the borehole imaging tools produces a markedly more precise orientation of σ_{Hmax} with reduced scatter.

The first step in reservoir characterization is the fracture data analysis. The analysis consists of the determination of types of fractures or fracture parameters. Borehole images and production data are used to identify a set of variables such as dip, azimuth, aperture, or density that control hydrocarbon flow. Fracture indicators such as production rates are combined with borehole images to flag the flow contributing fracture zones. This technique has been used successfully in fractured basement reservoirs.

The fracture sets are defined based on fractures dip, length, and azimuth.

2.7.3 Origin of fractures

Fractured systems have been studied by many authors, and the relationship between fracture orientation and folds has been underlined. Numerous parameters affect fractures, including fracture density, size, and shape of the fractures. Until now, no simple general model has been commonly accepted, mainly due to the complexity of geological systems. Reservoir description should rely on information from many sources including static data (well logs, cores, petrophysics, geology, and seismic), and ultimately on dynamic data (formation evaluation well tests, long-term pressure transient tests, tracer tests and longer-term reservoir

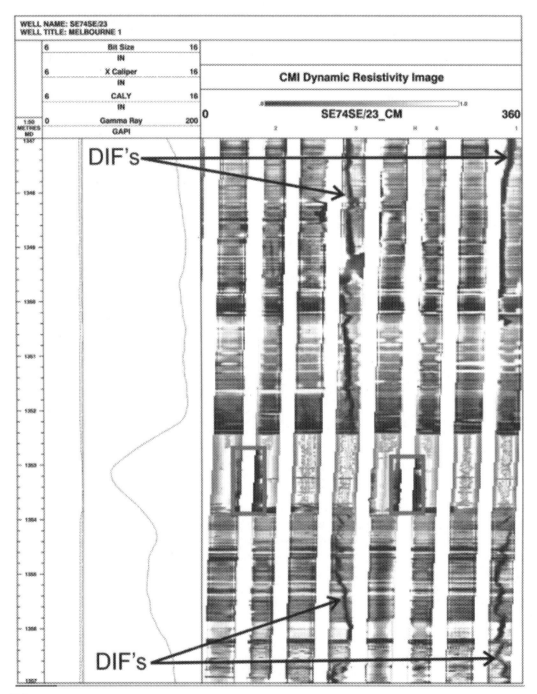

FIG. 2.29 Comparison of resistivity images visualizing Drilling Induced tensile Fractures (DIFs) from PCM Measures in the Melbourne 1 well, Yorkshire (10 m vertical borehole section). Left-hand panel: conventional logs including perpendicular dual-caliper and gamma-ray log. Right: Unwrapped circumferential resistivity borehole imaging (CMI) (clockwise from north), breakouts highlighted by *green (gray in the printed version) boxes*, DIFs terminate across coal horizon (lower gamma-ray) at 1352.4 m which shows clear breakout.

performance). Until now, no simple general model has been commonly accepted, mainly due to the complexity of geological systems. Islam (2014) proposed that each fracture system be studied in view of its origin and history, all in geologic time. This delinearized history analysis can help construct a reservoir characterization scheme.

Often, outcrop data give out useful clues as to the nature of fractures and by knowing the geology of the area one can reconstruct the tectonic history in what can be called 'geological history matching'. To reconstruct the fracture network system, the history of uplift, erosion, and overburden should be studied. If history is reconstructed, what conventionally appears to be an unreliable outcrop can lead to obtaining valuable history matching data that can be fed into a geological model. This figure shows how Figure Uplift and erosion often result in tensional breaking of brittle beds due to deformation along ductile bedding planes. That in itself would be seen as independent of the older fractures that often exist in a basement reservoir. As rock layers return to the surface, stress release allows new fractures to develop.

Picture 2.2 shows some of the outcrops. These fractures can be characterized according to their age or as per the f, g, and h functions of geological time.

The common fracture orientations found in Middle Eastern anticlinal reservoirs are shown in Fig. 2.30. Changes in orientations can be caused by later fault movement associated with variations in tectonic stress through time. In the carbonate reservoirs of Turkey and Iran, the orientation of karstic fractures associated with erosional unconformities is much more variable. The important features are low-angle, stress-relief, and exfoliation fractures which occur as sub-parallel to the unconformity surface in these reservoirs.

Zou et al. (2013) characterized the Junggar basin of China. They characterized the fractures from the perspective of contributing factor into following:

- diagenetic fractures;
- tectonic fractures;
- weathered fractures; and
- dissolving fractures.

PICTURE 2.2 Surface fractures.

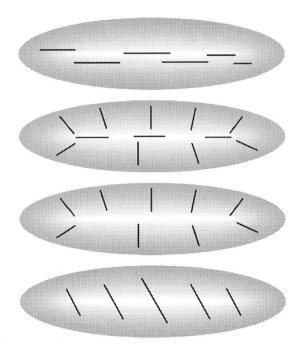

FIG. 2.30   The fracture orientations commonly found in the Middle East.

Diagenetic fractures are further classified into

- condensation shrinkage fractures;
- intergravel fractures; and
- intercrystal fractures.

Condensation shrinkage fracture is the rock volume contracted fracture which is formed by the cooling of the magma. These are the original fractures formed and are of the most random orientation. The fracture width is less than 0.1 mm, its shape is very irregular. It is mainly observed under the thin slice, and it is mainly developed in the andesite and diabase. Intergravel fracture is the fracture that is formed within gravel grains. These fractures are also irregular in shapes. They are important in generating reservoir space and connecting the hole of the volcanic breccias. Intercrystal fracture is developed in aperture of the crystal fragment particles or inner of crystal fragment particles, and it usually forms along the cleavage crack or partition line of twin, of which the shape is irregular and size small (Fig. 2.31A).

Fig. 2.31B shows tectonic micro-fractures with the characteristic width of orders of magnitude smaller than intercyrstal fractures. The surface area of these fractures is quite large making them amenable to high permeability. However, if the same fractures are plugged due to secondary cementation, the permeability would be reduced multifold from the intrinsic permeability of the rock. The next type, weathered fracture, is formed by erosion of volcanic rocks in the presence of surface water and air. Weathered fractures are extremely irregular.

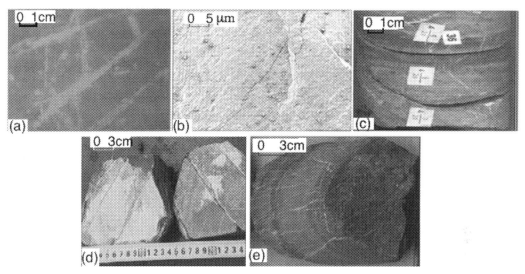

FIG. 2.31 Different types of fractures. (A) intercrystal fractures; (B) unfilled tectonic micro-fractures; (C) dissolving fractures; (D) oblique fractures; (E) mesh fractures. *From Zou, C., et al., 2013. Formation mechanism, geological characteristics and development strategy of nonmarine shale oil in China. Pet. Explor. Dev. 40 (1), 15–27. https://doi.org/10.1016/S1876-3804(13)60002-6.*

Other features dependent on prevailing conditions of the surface exposed to the atmosphere, both chemical changes and tectonic activities. Developed in the top of the volcanic rock, its existence is conducive to post-development.

Dissolving fracture is secondary fracture that is formed by an increase of the width of slots with surface water or groundwater seepage. These fractures have irregular edges. These are the fractures that are quite vulnerable to secondary cementation through deposits of zeolites and calcites and therefore can become much less permeable than the rest of the rock.

In terms of fracture orientation, drilling cores, FMI, and others provide useful information. Fractures can be characterized as.

- high-angle fractures;
- oblique fractures;
- low-angle fractures; and
- mesh fractures.

High-angle fractures are often a group of nearly parallel fractures or two sets of oblique cutting fractures, and the inclination of it is 60°–90°. In general, larger fractures are more susceptible to secondary cementing whereas smaller ones (width less than 1 mm) are more likely to be open. Oblique fractures are those with an angle of inclination of 10°–60°. These fractures tend to be of larger width and therefore more susceptible to secondary cementation with calcites. Low-angle fractures which are similar to the stratification are locally concentrated developed, and the inclination is 0°–10°. Low-angle fractures are mostly narrow hair style cracks, extending short distances.

There are only a few methodologies allowing to consider (and model) the variation of fracture orientation according to the orientation of the geological structures (dip and strike of the geological formations). Borghi et al. (2015) described a methodology that can be used to generate a stochastic discrete fracture network (DFN), in which the fracture orientations are consistent with the orientation (deformation) of the geological formations. Their conceptual model suggests that six main fracture sets occur depending on the position within the fold. Fig. 2.46 show such fracture families. The model includes six main families of fractures:

- Conjugate system C1: This system is roughly perpendicular to the fold axis: 2 conjugate high-angle fracture systems (C1a and C1b) (situations A and D in Fig. 2.32).
- Conjugate system C2: This system is roughly parallel to the fold axis: 2 subvertical conjugate fracture systems (C2a and C2b) (situation B in Fig. 2.32).
- Conjugate system C3: This system is parallel to the fold axis: 2 high-angle conjugate fracture systems (C3a and C3b) (situation E in Fig. 2.32).
- Conjugate systemC4: This system is parallel to the fold axis: 2 low-angle conjugate fracture systems (C4a and C4b) (situation C in Fig. 2.32).
- Fractures with ac orientation: Vertical fractures that have their strike perpendicular to the fold axis (situation A in Fig. 2.32).

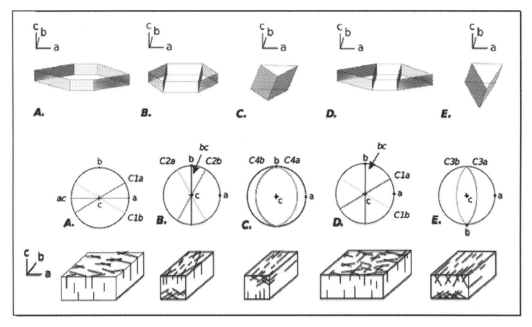

FIG. 2.32 Illustration of the fracture sets in a reference environment. *From Boghi, A., et al., 2015. Stochastic fracture generation accounting for the stratification orientation in a folded environment based on an implicit geological model. Eng. Geol. 187, 135–142.*

Borghi et al. (2015) took the following steps to generate the fractures:

- definition of the number and length of the fractures
- construction of the unrotated fractures
- positioning of the fractures into the geological model and rotation according to the local orientation of the structure.

Liu and Valkó (2019) investigated the types, evolution of fractures, and their relationship with oil and gas migration of the Permian Changxing Formation of the Yuanba gas field of China. Four genetic types of fractures with different occurrences were distinguished:

- intercrystal fractures;
- pressure solution fractures (stylolites);
- structural fractures; and
- overpressure fractures.

The main difficulty in sandstone is that fracture network is not well developed due to lack of proper method that includes fracture properties. For instance, not a single core analysis technique can shed light on fracture properties of a rock. In fact, the presence of fractures or even fissures disqualify core analysis plugs from being considered for further analysis. While several logging tools have emerged that can identify fractures, there is no systematic process to integrate that information into a reservoir characterization tool.

Different features exist in shale-gas plays in that the shale formations are both the source rocks and the reservoir rocks. There is no migration of gas as the very low permeability of the rock causes the rock to trap the gas and it forms its own seal. The gas can be held in natural fractures or pore space, or can be absorbed onto the organic material. Apart from the permeability, total organic content (TOC) and thermal maturity are the key properties of gas potential shale. In general, it can be stated that the higher the TOC, the better the potential for hydrocarbon generation. In addition to these characteristics, thickness, gas-in-place, mineralogy, brittleness, pore space and the depth of the shale gas formation are other characteristics that need to be considered for a shale gas reservoir to become a successful shale gas play. The organic content in these shales, which are measured by their TOC ratings, influence the compressional and shear velocities as well as the density and anisotropy in these formations. Consequently, it should be possible to detect changes in TOC from the surface seismic response.

Gas hydrates, another unconventional natural gas source of abiogenic origin, makes up completely different set of properties. Gas hydrates can be found on the seabed, in ocean sediments, in deep lake sediments, as well as in the permafrost regions. The amount of methane potentially trapped in natural methane hydrate deposits may be significant, which makes them of major interest as a potential energy resource. This methane is also of high quality.

2.7.4 Seismic fracture characterization

As discussed in the previous chapter, any rock deformation is the result of tectonic events that are a unique function of time. With time, events such as magma movement, faulting, earth quake, and fracturing occur in a cyclical form. It is recognized that the state of stress changes with time, affecting rock deformation directly. The stress-strain relationship is

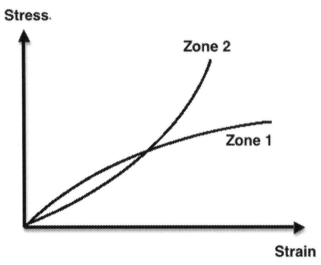

FIG. 2.33 Schematic of the two zones on the Earth's crustal region.

different in different zones, depending on the formation, its conents, and temperatures. Two zones are identified broadly, the shallower zone (Zone 1) where any stress translates into active reaction and changes in strain, and the deeper zone (Zone 2) that is more resilient and the strain deformation is narrow. Fig. 2.33 shows a schematic of this relationship.

In Zone 1, deformation causes brittle failure and rock strength is limited by frictional strength of preexisting faults or fractures, whereas in Zone 2, the prevalent temperature makes it more resilient and faults and fractures can endure greater stress. In this zone, the temperature helps make the flow more ductile with rock strength declining exponentially with increasing temperature. Few studies have investigated the deformation of rock as a function of stress and temperature. However, it is important to understand the nature of deformation as a function of these variables to properly characterize fractures in a reservoir.

Natural fractures develop at lower temperatures with minimal stress. This is how all outcrops of consolidated rocks exhibit natural fracture networks. The existence of fracture eventually leads to the development of a fault when lateral displacement is large enough to invoke such movement across more than one sedimentary bed. This onset of fracture immediately follows the development of fissures with orientation orthogonal to the direction of fault. Depending on the nature of stress, fissures develop into fractures. These fractures become the main vehicle for hydrocarbon transport. The same fractures can onset thermal convection of water. Such flow has tremendous implication for eventual hydrocarbon generation and transport.

Two phenomena add to the complexity of fracture flow. They are: the occurrence of shale breaks and secondary cementation of fractures. Shale breaks decrease overall vertical permeability whereas cementation reduces overall permeability of the reservoir. For the latter, fracture orientation becomes the most important feature of fluid flow. Shale breaks, on the other hand, serve as a barrier to vertical flow and have the capacity to become a storage site for so-called shale gas and oil. While most of the shale gas and oil reservoirs are believed to contain source rock, the shale breaks as well as caprocks also contain significant amount of gas, albeit being trapped within very low-permeability shales.

Because deeper fractures are mainly oriented normal to the direction of minimum in situ compressive stress, the presence of seismic anisotropy can signal a certain pattern of fractures. While this requires true understanding of the relationships between the response of seismic anisotropy and fracture properties, the existence of seismic anisotropy offers one with the base line for fracture characterization.

Several theoretical studies of fracture-induced anisotropy have been reported in the literature. How the presence of the fracture sets affect the elastic modulus of fractured rocks has been discussed in the literature. Based on the simplifying assumption of linear and elastic behavior, it is the background elastic modulus and fracture parameters (fracture density, aspect ratio and saturating fluid) that determine the behavior of seismic waves which propagate through, and are reflected from, the reservoirs.

Fracture models are based on major physical features such as the azimuthal P- and S-wave velocity variations with fracture parameters. The azimuthal variations in P-wave velocity and reflectivity in homogeneously fractured reservoirs as a function of anisotropic parameters have been described by many researchers. These anisotropic parameters are related to fracture parameters through the elastic stiffness tensors. In principle, azimuthal Amplitude vs offset (AVO) responses can be used to detect fractures. Successful use of azimuthal variation of P-wave AVO signatures to determine the principal fracture orientation and density have been reported in the literature.

In principle, the distribution of fractures in reservoir zones can be treated as homogeneous. In practice, the number and variability of small-scale fractures are so large that recovering significant information from P-wave seismic data requires that the distribution characteristics be averaged over the reservoir zone, which leads to a statistical representation. This is equivalent to the notion of assigning pseudo-homogeneity. Heterogeneity due to spatial variations of fracture density could result in spatial variations of velocity anisotropy. It is important to identify the coherent features in reflected seismic data. These features are often used as major seismic characteristics in exploration geophysics. The small incoherent arrivals which occur between the major reflections also contain information about the media. They are currently treated as 'noise', but contain valuable information that can alter the reservoir description if filtered correctly. Currently used techniques are not capable of analyzing these signals with scientific accuracy. The 3-D finite difference modeling technique is effective in studying the azimuthal AVO response and scattering characteristics in heterogeneously fractured media. However, they allow no room for including 'noises'. Although grid memory requirements make computations very expensive and grid dispersion effects limit finite-difference models to small regions, this approach has been successfully used to model energy diffracted at highly irregular interfaces, to study fluid-filled bore-hole wave propagation problems in anisotropic formations, and to study the scattering in isotropic media. In addition, unlike boundary integral techniques, lateral velocity variations can be easily incorporated.

Azimuthal AVO variations have been used in fracture detection and density estimation. However, the sensitivity of reflected P-waves to the discontinuity of elastic properties at a reflected boundary and to the spatial resolution makes it difficult to interpret this attribute unambiguously. The motivation of this thesis is to explore the efficiency, benefits and limitations of using P-waves to characterize fractured reservoirs, theoretically and practically.

Shen (1998) studied the possibility of using P-waves to investigate properties of fractured reservoirs and the diagnostic ability of the P-wave seismic data in fracture detection. This study also considered rheological behavior of rocks at a crustal scale, based on observation and modeling of continental deformation, in particular deformation of the Tibetan plateau. The Tibetan plateau is an ideal location that features continental topography, resulting from the north-south convergence between the Indian and Eurasian plates.

2.7.5 Effects of fractures on Normal Moveout (NMO) velocities and P-wave azimuthal AVO response

Shen (1998) investigated the effects of fracture parameters on anisotropic parameter properties and P-wave NMO velocities, based on developed effective medium models and crack models. Anisotropic parameters of the pseudo transversely isotropic medium model, $S(v)$ and $E(v)$, have different characteristics in gas- and water-saturated, fractured sandstones. When fractures are gas-saturated, $\delta^{(v)}$ and $\varepsilon^{(v)}$ vary with the fracture density alone. In water-saturated, fractured sandstones, both $\delta^{(v)}$ and $\varepsilon^{(v)}$ depend on fracture density and crack aspect ratio. $\delta^{(v)}$ is related to the Vp/Vs of background rocks and $\varepsilon^{(v)}$ is a function of the Vp of background rocks. Studies show that the shear wave splitting parameter, $\gamma^{(v)}$, is most sensitive to crack density and insensitive to saturated fluid content and crack aspect ratio. Properties of P-wave NMO velocities in a horizontally layered medium are the function of $\delta^{(v)}$. The effects of fracture parameters on P-wave NMO velocities are comparable with the influences of $\delta^{(v)}$.

P-wave azimuthal AVO variations are not necessarily correlated with the magnitude of fracture density. Shen (1998) showed that the elastic properties of background rocks have an important effect on P-wave azimuthal AVO responses. Results from 3-D finite difference modeling show that azimuthal AVO variations at the top of gas-saturated, fractured reservoirs which contain the same fracture density are significant in the reservoir model with small Poisson's ratio contrast. Analytical solutions indicate that azimuthal AVO variations are detectable when fracture-induced reflection coefficients can generate a noticeable perturbation in the overall reflection coefficients. Varying fracture density and saturated fluid content can lead to variations in AVO gradients in off fracture strike directions. Shen's numerical results also show that AVO gradients may be significantly distorted in the presence of overburden anisotropy caused by VTI media, which suggests that the inversion of fracture parameters based on an individual AVO curve would be biased without correcting this influence. He recommended that azimuthal AVO variations could be effective for detecting fractures, model analysis studies and combination of P-wave NMO velocities are more beneficial than using reflection amplitude data alone.

2.7.6 Effects of fracture parameters on properties of anisotropic parameters and P-wave NMO velocities

In exploration geophysics, normal moveout (NMO) describes the effect that the distance between a seismic source and a receiver (the offset) has on the arrival time of a reflection in the form of an increase of time with offset. The relationship between arrival time and offset is

2.7 Characterization of reservoirs

hyperbolic, typically described by a wave equation. This relationship is the principal criterion that a geophysicist uses to decide whether an event is a reflection or not. The normal moveout depends on a complex combination of factors including the velocity above the reflector, offset, dip of the reflector and the source receiver azimuth in relation to the dip of the reflector. Of concern is the role of fractures. To understand the effect of fracture parameters on NMO velocities, one needs to understand effects of fracture parameters on anisotropic parameter properties.

Shen (1998) studied various elastic parameters of five sandstones, whose characteristics are summarized in Tables 2.15 and 2.16. He studied for two aspect ratios, i.e., 0.01 and 0.05. Fig. 2.34 shows $\nu^{(v)}$ as a function of fracture density for a gas-saturated sandstone with fracture aspect ratio of 0.01. Results are shown for an aspect ratio of 0.05. These results show that for gas-saturated sandstones, $\nu^{(v)}$ is insensitive to aspect ratio. For the water-saturated case, however, absolute values of $\nu^{(v)}$ increase with fracture density and crack aspect ratio. This latter case (Fig. 2.34) also shows a range of values for different samples, as compared to the gas-filled case that shows little dependence on sample types. It is also noted that $\nu^{(v)}$ is dependent on Vp/Vs of isotropic, unfractured sandstones. The smaller the Vp/Vs, the larger the absolute value of $\nu^{(v)}$ obtained.

$\nu^{(v)}$ shows similar characteristics to $\nu^{(v)}$. The difference is that $\nu^{(v)}$ is the function of the νp of the background medium. In gas-saturated, fractured sandstones, $\nu^{(v)}$ is sensitive to fracture density alone. In water-saturated, fractured sandstone the absolute value of $\nu^{(v)}$ increases with both fracture density and aspect ratio. The smaller the νp, the smaller the absolute value $\nu^{(v)}$ obtained. $\nu^{(v)}$ as a function of crack density and aspect ratio in gas- and water-saturated, fractured sandstones is shown in Fig. 2.34.

TABLE 2.15 Elastic parameters used by Shen (1998).

No. sandstones	Vp (m/s)	Vs (m/s)	ρ (g/cm³)	Vp/Vs	References
No. 1	3368	1829	2.50	1.84	Thomsen (1986)
No. 2	4405	2542	2.51	1.73	Thomsen (1986)
No. 3	4539	2706	2.48	1.68	Thomsen (1986)
No. 4	4476	2814	2.50	1.59	Thomsen (1986)
No. 5	4860	3210	2.32	1.51	Teng and Mavko (1996)

TABLE 2.16 Elastic parameters and fracture parameters of model 1 and model 2.

Model	Vp (m/s)	Vs (m/s)	ρ (g/cm³)	Fracture density (%) and aspect ratio	Poisson's ratio	Type of rocks
Model 1	4358	3048	2.81		0.021	Mesav-erde Shale
	3368	1829	2.50	10 0.01	0.291	Taylor sandstone
Model 2	4561	2988	2.67		0.124	Shale
	4860	3210	2.32	10 0.01	0.113	Sandstone

FIG. 2.34 Variation in anisotropic parameter as a function of fracture density for gas-saturated sandstone and fracture aspect ratio of 0.01. *Redrawn from Shen, F., 1998. Seismic Characterization of Fractured Reservoirs (Part I) Crustal Deformation in Tibetan Plateau (Part II) (Ph.D. dissertation). MIT.*

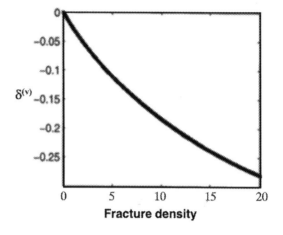

FIG. 2.35 Range of variation in anisotropic parameter as a function of fracture density for water-saturated sandstone and fracture aspect ratio of 0.01. *Redrawm from Shen, F., 1998. Seismic Characterization of Fractured Reservoirs (Part I) Crustal Deformation in Tibetan Plateau (Part II) (Ph.D. dissertation). MIT.*

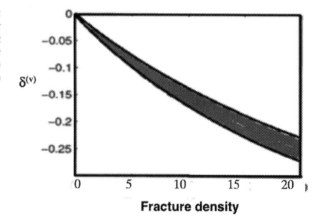

Parameter $\gamma^{(v)}$, is different from both $\nu^{(v)}$ and $\nu^{(v)}$ in that it measures the degree of shear wave splitting at vertical incidence. Fig. 2.35 show that $\nu^{(v)}$ has little dependence on fluid bulk modulus and crack aspect ratio and is the parameter most directly related to fracture density. Therefore, for parallel, penny-shaped cracks, the shear wave splitting parameter, $\nu^{(v)}$, can provide direct information about fracture density with least ambiguity.

These findings show that the variations of parameters $\delta^{(v)}$ and $\varepsilon^{(v)}$ are sensitive to fluid content. For gas-saturated fractures, $\delta^{(v)}$ and $\varepsilon^{(v)}$ vary with the fracture density alone, making them an effective indicator of fractures. On the other hand, when the fractures are filled with water, the magnitudes of $\delta^{(v)}$ and $\varepsilon^{(v)}$ depend on fracture density, crack aspect ratio, and elastic properties of background rocks. On the other hand, the shear wave splitting parameter, $\gamma^{(v)}$, is insensitive to fluid content and crack aspect ratio. It is the parameter most related to crack density (Figs. 2.36–2.39). P-wave NMO velocity is controlled by the vertical P-wave velocity, angle between crack normal and survey line, and the parameter $\delta^{(v)}$.

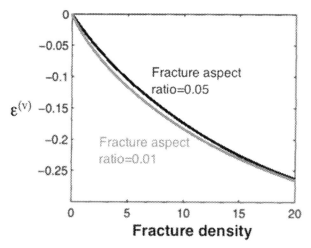

FIG. 2.36 Variation in anisotropic parameter as a function of fracture density for gas-saturated sandstone and fracture aspect ratio of 0.01 and 0.05. *Redrawn from Shen, F., 1998. Seismic Characterization of Fractured Reservoirs (Part I) Crustal Deformation in Tibetan Plateau (Part II) (Ph.D. dissertation). MIT.*

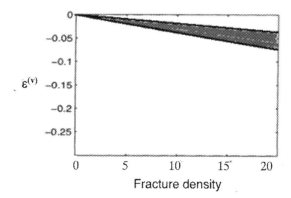

FIG. 2.37 Range of variation in anisotropic parameter as a function of fracture density for water-saturated sandstone and fracture aspect ratio of 0.01. *Redrawn from Shen, F., 1998. Seismic Characterization of Fractured Reservoirs (Part I) Crustal Deformation in Tibetan Plateau (Part II) (Ph.D. dissertation). MIT.*

2.7.7 Reservoir characterization during drilling

Cost-effective drilling techniques and well-completion strategies constitute the most successful technological development in the petroleum industry. At present, 90% of the new wells drilled are horizontal. These wells are likely to intersect natural fractures that are predominantly vertical in most reservoirs. Drilling in this type of formations is often optimized with so-called managed pressure drilling (MPD). MPD uses tools at the surface such as a choke to control the drilling fluid flow rate and bottomhole pressure. The various drilling operations that MPD is comprised of provide economical solutions to many drilling problems such as: managing gas kicks, lost circulation and other well control issues, improving rate of

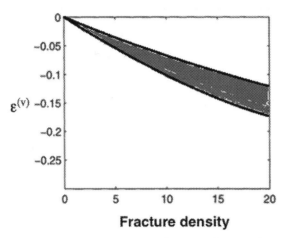

FIG. 2.38 Range of variation in anisotropic parameter as a function of fracture density for water-saturated sandstone and fracture aspect ratio of 0.05. *Redrawn from Shen, F., 1998. Seismic Characterization of Fractured Reservoirs (Part I) Crustal Deformation in Tibetan Plateau (Part II) (Ph.D. dissertation). MIT.*

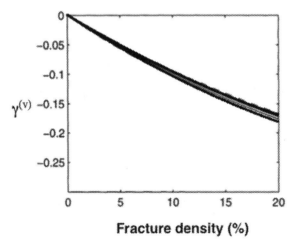

FIG. 2.39 Range of variation in anisotropic parameter as a function of fracture density for water-saturated sandstone and fracture aspect ratio of 0.05. *Redrawn from Shen, F., 1998. Seismic Characterization of Fractured Reservoirs (Part I) Crustal Deformation in Tibetan Plateau (Part II) (Ph.D. dissertation). MIT.*

penetration (ROP), minimizing formation damage, and enabling dynamic reservoir characterization from real-time mud log data. Underbalanced drilling (UBD) is a form of MPD that is particularly useful when drilling horizontal wells in tight gas formations. It also generates data that can turn it into a dynamic reservoir characterization tool.

As shown in Table 2.16, the first set of direct data are produced during drilling. As soon as drilling is commenced, the drilling log becomes available. The drilling log contains information about the progress of the well such as measured depth (MD), true vertical depth (TVD), inclination, weight on bit (WOB), rate of penetration (ROP), and gamma ray. It also provides information about the drilling mud circulation system such as mud pit volume, pump pressure, and mud flow rate. Each of these data is valuable for description of lithology as well as fracture system of a formation.

Any drilling process also accompanies the mud log. This log is generally created by the on-site geologist as the well is drilled. Mud logs contain valuable information regarding formation geology and hydrocarbon in place. As the drill bit penetrates the formation the rock is crushed, and these cuttings are flushed from the well and carried to the surface by the circulating drilling mud. A geologist routinely examines the cuttings and describes the lithology of the formation being penetrated. This information is recorded in the mud log on a depth basis as the well is drilled to create a geologic profile of the entire well. At the same time, total gas measurements are also made and recorded. Total gas measurements indicate the relative concentration of hydrocarbons (methane, ethane, propane, etc.) present in the circulating drilling mud at any given time.

During conventional drilling operations, the drilling mud density is maintained above the reservoir pore pressure for wellbore stability issues. While during this overbalanced drilling useful data are generated that can indicate the presence of fractures and sweet spots (e.g., mud loss, fluid loss, high ROP), information regarding fluid in place is limited. In such system, produced fluids can only occur when an unexpected overpressured zone is encountered or when a transient mud pressure reduction occurs as the drillstring is raised (swabbing). Ever since the advent of underbalanced drilling that uses mud pressure lower than pore pressure, the possibility of extracting in situ fluid to characterize the reservoir has been increased drastically. During underbalanced drilling, hydrocarbon production will occur consistently during drilling, whenever a sweet spot is penetrated. Such 'sweet spots' can be the result of natural fractures or otherwise high-permeability zones. Produced fluid as a result of the underbalanced pressure condition is a major focus of this investigation. Recycled fluids may occur if the gas contained within the circulating mud is not entirely released at the surface. In this instance, the remaining gas will be recirculated through the system and will be detected again on the next pass. Finally, contamination will always occur due to various unavoidable causes. Reasons for contamination of the total gas readings include petroleum products intentionally added to the drilling mud, chemical reactions and degradation of organic mud additives, and even emissions from construction equipment on-site. It is very important to understand all of these processes so that one can effectively interpret the mud log analysis.

The drilling fluid circulation system is essentially a closed loop system in which the mud is pumped down the well through the center of the drillstring, flows through the drill bit nozzle, and is then forced back up to the surface through the annular section. This process ensures that the drill bit stays cool, creates a hydrostatic pressure that is exerted against the wellbore wall for well stability issues, and flushes the cuttings to the surface. At the surface, the mud is transported through a shaker table to remove the cuttings and then released into a mud pit to complete the cycle. To obtain the total gas measurements a gas trap is installed at the mud pit that is able to capture a sample of gas from the mud. An impeller agitates the mud releasing gas into the air. A mixture of this gas-air sample is then sent to the mud logging unit for analysis. The total gas reading is a measure of the relative concentration of all hydrocarbons combined. This concentration is recorded along depth of the drill bit. A correction may be necessary to account for the mud travel time.

While 'gas shows' are routinely used to delineate productive zones as well as plan completion strategies, they can also serve as a tool for formation characterization. This is particularly important for basement petroleum reservoirs.

Most basement and carbonate reservoirs are known to contain a high density of natural fractures and production is in fact dominated by fracture flow. Typically, it is presumed that vertical natural fractures exist in situ. Therefore, to produce economically from these types of reservoirs it is most efficient to drill horizontal wellbores. The lateral sections of these wellbores are often drilled underbalanced. During underbalanced drilling operations, gas is expected to flow into the wellbore consistently throughout the drilling process. The combination of low permeability matrix, high permeability natural fractures, and the underbalanced pressure condition leads to the result of highly complicated mud logs. However, a thorough analysis of the mud log data can reveal critical information about the natural fracture system near the wellbore.

The fundamental premise of this analysis is that for basement and carbonate reservoirs, the bulk of the fluid flow takes place through fractures. Such a premise is justified based on Darcy's law:

$$\vec{v} = -\frac{k}{\mu}\nabla P \qquad (2.23)$$

Here $\vec{v}$ is the velocity, k is the permeability, μ is the viscosity, and P is the pressure. Permeability has the dimension of L^2, which means it is exponentially higher in any fracture than the matrix. For a system with very low permeability, fracture flow accounts for 99% of the flow whereas in terms of volume fractures account for 1% of the volume of the void (or total porosity). This is significant, because in classic petroleum engineering, governing equations are always applied without distinction between storage site (where porosity resides) and flow domain (where permeability is conducive to flow). In fractured reservoirs, fluid flow equations apply to the fracture network whereas storage volume applies to the matrix that has little permeability. In fact, permeability values are so low in the matrix, typical Darcy's law doesn't apply to this domain. It is recommended that Forchheimer equation be used to describe gas flow. This equation is given by:

$$-\nabla P = \frac{\mu}{k}v + \rho\beta v^2 \qquad (2.24)$$

In the above equation, β is an additional proportionality constant that depends on rock properties. For a system with predominantly fracture flow, β would depend on the fracture density and aspect ratio.

In case, fracture network is insignificant and the reservoir matrix permeability is very low, flow in such a system is best described with Brinkman equation, described as:

$$-\frac{\partial P}{\partial x} = u\frac{\mu}{k} - \mu\frac{\partial^2 u}{\partial x^2}. \qquad (2.25)$$

To date the most commonly used model is that proposed by Warren and Root (1963). This so-called dual-porosity model (Fig. 2.40) assumes that two types of porosity are present in the

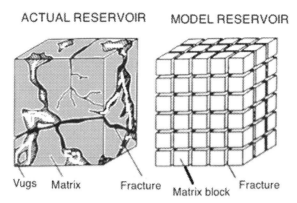

FIG. 2.40 Depiction of Warren and Root model.

formation, one arising from vugs and fracture system whereas the other from matrix. For fractured reservoirs, the matrix permeability is negligible compared to fracture permeability (hence depicted with shades). Warren and Root invoked similar assumptions even for a matrix with relatively high permeability. The approach operates on the concept that fractures have large permeability but low porosity as a fraction of the total pore volume. The matrix rock has the opposite properties: low permeability but relatively high porosity. This approach describes the observation that fluid flow will only occur through the fracture system on a global scale. Locally, fluid may flow between matrix and fractures through interporosity flow, driven by the pressure gradient between matrix and fractures.

Fracture flow is described by Snow's equation (1963), given below:

$$\frac{Q}{\Delta P} = Cw^3 \qquad (2.26)$$

Here w is the fracture aperture and C is a proportionality constant that depends on the flow regime that prevails in the formation. Snow's equation emerges from a simple synthesis of parallel plate flow (Poiseuille law) that assumes permeability to be $b^2/12$, where b is the fracture width.

Fracture geometries are often idealized to simplify modeling efforts. In most cases the width is assumed to be constant, and the fracture is usually considered either a perfect rectangle or a perfect circle. In reality, fracture geometries are very complex (Fig. 2.40), and many different factors could affect the behavior of fluid flow. In Fig. 2.40 that was originally published by Warren and Root (1963), vugs are shown prominently. It is no surprise that they introduced the concept of dual porosity. Indeed, porosity in vugs and in matrix are comparable. For fractured reservoirs, however, the vugs are nonexistent and most fractures have very little storage capacity, making their porosity negligible to that of the matrix. In determining sweet spots within a fractured reservoir, the consideration of very high fracture to matrix permeability, k_f/k_m is of importance. For application in dynamic reservoir characterization using real-time mud log data, the term "sweet spot" is used when the drill bit intersects a transverse natural fracture. Such process is equivalent to numerous passes of history match in the context of reservoir simulation.

2.7.8 Overbalanced drilling

Overbalanced drilling approaches for fracture characterization mostly consist of methods that take advantage of mud-loss data and the rheological properties of the circulating drilling fluid. Drilling mud is usually a non-Newtonian fluid that exhibits shear-thinning behavior. Shear-thinning implies that the fluid viscosity decreases with an increase in the shear rate. During drilling, the drilling mud is constantly circulated through a closed-loop system. If the circulation is stopped, the drilling mud will develop into a thick gel. The mud will remain in this state until a pressure exceeding the mud's yield stress is applied, at which point it will return to its "fluid" state. This non-ideal fluid behavior has allowed engineers to develop methods to characterize fracture permeability using mud-loss data. In a dynamic setting, mud rheology can be calibrated against mud circulation loss, which is directly related to fracture in a tight reservoir.

During overbalanced drilling, when a sweet spot is encountered the mud pressure is greater than the fluid pressure contained in the fracture, resulting in a flood of drilling mud into the fracture. When the drill bit intersects the fracture drilling mud will flow into the fracture. Because the matrix permeability is low, leak off into the matrix is minimal and the mud loss is entirely due to fluid flow through fractures (Fig. 2.41).

This mud flow is reflected in the rheology of the returning mud, thereby, creating a correlation between surface-observed rheology and fracture density as well as fracture geometry.

Liétard (1999) provide type curves describing mud loss volume versus time that can be used to determine the hydraulic width of fractures through a curve matching approach. The type curves are based on an analysis of the local pressure drop in the fracture (Liétard et al., 1999):

$$\frac{dP}{dr} = \frac{12\mu_p v_m}{w^2} + \frac{3\tau_y}{w} \qquad (2.27)$$

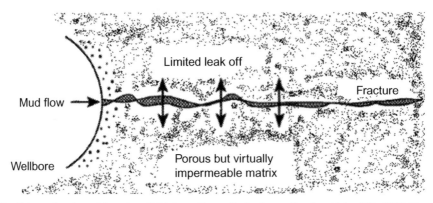

FIG. 2.41 Schematic of mud flow in a tight formation with fractures. *Based on Dyke, C.G., 1996. How sensitive is natural fracture permeability at depth to variation in effective stress? In: Fractured and Jointed Rock Masses, Proceedings of the International ISRM Symposium on Fractured and Jointed Rock Masses. A. A. Balkema, Rotterdam.*

where, v_m is the local velocity of the mud in the fracture, μ_p is the plastic viscosity of the mud, w is the fracture aperture, and τ_y is the yield stress of the mud. This equation was improved by Huang et al. (2011) that suggested the following equation:

$$\left(\frac{\Delta P_{OB}}{\tau_y}\right)^2 w^3 + 6R_w\left(\frac{\Delta P_{OB}}{\tau_y}\right)w^2 - \frac{9}{\pi}(V_m)_{max} = 0 \quad (2.28)$$

In the above equation, R_w is the well radius and V_m is the maximum mud-loss volume. This equation is easier to use than the previously used type curve. However, it is recommended that such correlation be developed for each reservoir.

2.7.9 Underbalanced drilling (UBD)

During UBD operations, a low-density drilling fluid is used to maintain a wellbore pressure profile that is lower than the pore pressure of the formation at all locations along the borehole. One major advantage of UBD over conventional drilling is that formation damage is reduced because a filter-cake is not allowed to form near the wellbore. Wells completed with UBD have been shown to perform three to four times better than their conventional counterparts in the same formation. Among others, lost circulation is minimized with UBD. Overall, the ROP is increased significantly with UBD. Fig. 2.42 shows how switching from overbalanced drilling to UBD can drastically increase ROP. This is especially true for tight gas formations or any formation with harder than normal. This phenomenon is not well understood, but it is thought that the increased ROP can be attributed to the lower confining pressure on the formation rock under UBD conditions and the fact that cuttings are more easily flushed from the bottom of the wellbore reducing the resistance on the drill bit.

The most useful aspect of UBD is in the insight gained during UBD. The deliberate underbalanced pressure difference between the drilling fluid and the formation pressure

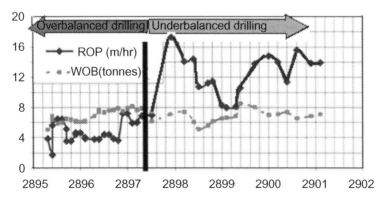

FIG. 2.42 Data from a well drilled overbalanced until a certain depth and then switching to underbalanced operations. Immediately as UBD begins, the ROP greatly increases. *Redrawn from Woodrow, C.K., et al., 2008. One company's first exploration UBD well for characterizing low permeability reservoirs. In: Paper Presented at the SPE North Africa Technical Conference & Exhibition, Marrakech, Morocco, March. Paper Number: SPE-112907-MS. https://doi.org/10.2118/112907-MS.*

causes an inflow of formation fluid into the wellbore along the entire drilled section. This can act as a tracer for dynamic reservoir characterization.

The most important piece of data is the rate of fluid flow from the formation into the wellbore. Very rarely are flow rates actually measured at bottomhole. In almost all cases, the formation fluid flow rate is estimated from surface measurements of inflow and outflow of the drilling fluid. The difference between the mud injected into the wellbore and the outflow of mud from the annulus is often estimated as the formation fluid flow rate. Methods to account for the expansion of gas due to changes in temperature and pressure must be taken into account to obtain accurate data. Other data that models tend to utilize include: bottomhole pressure, rate of penetration, formation porosity, wellbore diameter, and wellbore length. Models generally provide profiles of formation permeability and pore pressure versus depth. Norbeck (2010) presented field data, showing correlation between UBD data and fractures. He extracted field data originally reported by Myal and Frohne (1992). The report investigates the effectiveness of directional drilling in a tight gas formation located in the Piceance Basin of Western Colorado. The formation is known to be highly naturally fractured, and consequently the decision was made to drill a large section of the well underbalanced to reduce formation damage and lost circulation. As can be seen from Fig. 2.43, at least ten major gas shows were detected during drilling. These gas shows were attributed to the presence of natural fractures intersected by the wellbore. An increase in mud density led to suppression of the gas shows but also prevented extraction of gas show data and their correlation with fracture distribution. This type of correlation can lead to depiction of the formation fracture network.

Norbeck (2012) proposed two criteria that can be combined to develop a correlation between mud log data and reservoir properties. They are:

Criterion 1: The first criterion is the use of total gas concentration measurements from mud logs. Using a gas chromatograph, the mud logging unit is able to determine the concentration of gas present in the drilling fluid at any given time. As natural fractures are intersected by the drilling bit, gas flows into the wellbore and the amount of gas is proportional to the average fracture permeability. These concentration values show up as spikes on the total gas concentration of mud log.

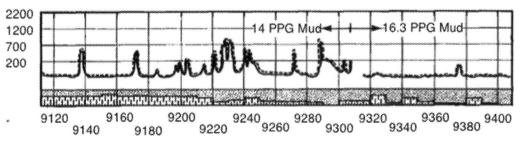

FIG. 2.43 Mud log data from a portion of a well drilled underbalanced in the Piceance Basin. The large gas peaks were attributed to natural fractures that were intersected by the wellbore. *Redrawn from Myal, F.R., Frohne, K.-H., 1992. Drilling and early testing of a sidetrack to the slant-hole completion test well: a case study of gas recovery research in Colorado's Piceance Basin. In: Paper SPE 24382 Presented at the SPE Rocky Mountain Regional Meeting, Casper, Wyoming, USA, 18–21 May. https://doi.org/10.2118/24382-MS.*

Criterion 2: The second criterion is based on observations of the mud pit volume. Observations of the mud pit volume at the surface also show potential as a fracture identification criterion. It is widely accepted that decreases in mud pit volume (mud losses) correspond to encounters with natural fractures while drilling overbalanced. It is logical to assume that the reverse is also true during underbalanced drilling. It means that as a natural fracture is encountered with the drill bit, the formation fluid influx will cause a displacement of drilling fluid in the mud pit. This response is observable at the surface.

Both criteria are related to open fractures that contribute to flow directly. Furthermore, fractures are uniquely correlated if the formation is tight with negligible permeability. To estimate natural fracture permeability, several assumptions have to be made:

1. All natural fractures that have been intersected by the wellbore are transverse to the wellbore and have circular geometry with finite extent (as depicted in Fig. 2.44).
2. Natural fractures have constant aperture. At least, it must be assumed that an equivalent aperture is a reasonable and practical approximation. Tortuosity or other 'eccentricity' factors can be introduced; however, simple geometry is a reasonable assumption.
3. Gas contained within natural fractures is composed 100% of methane. Methane density and viscosity remain constant while flowing through fractures. This approximation avoids the analysis of compositional effect on gas chromatography.
4. Fluid flow through fractures follows the cubic law relationship. This is typical of all existing fracture flow models. The cubic law relationship assumes steady-state, laminar flow between two parallel plates. The cubic law can be derived from a force balance between the forces due to the pressure gradient and the shear resistance on the boundaries, as opposed to the diffusivity equation, which is derived using the principles of conservation of mass. As such, no compressibility term is present in the cubic law relationship. However, the high compressibility of gas will most likely have significant effects on the flow rate through the fracture. Nonetheless, it is assumed that at the high

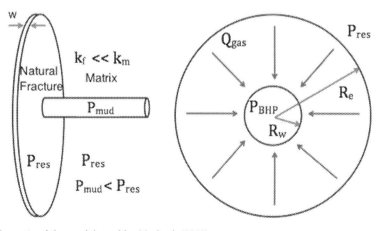

FIG. 2.44 Schematic of the model used by Norbeck (2012).

pressure conditions present in the reservoir, compressibility effects will be negligible over the relatively low magnitude pressure drop between reservoir and bottomhole pressures.
5. Matrix permeability is much lower than fracture permeability.
6. No charging of the fractures occurs during the time spans considered. Considering low permeability of the matrix, this is a reasonable assumption.
7. The gas influx volume is equal to the mud pit volume increase. For most shallow reservoirs, this is a reasonable approximation. Because no direct measurements of gas flow rate at bottomhole are recorded on today's drilling rigs, it is assumed that the observed mud pit volume increase is equal to the volume of gas that entered the wellbore from the fracture. However, compressibility effects could be significant and add to the uncertainty of this analysis. Also, it is well known that methane is highly soluble in oil-based drilling mud, and it is has been reported that observations of mud pit volume increase as a response to a gas kick will be reduced because of solubility effects.

The following estimates can be obtained from the drilling and mud log data:

- Gas flow rate
- Pressure drop (underbalance)
- Methane viscosity
- Wellbore radius

If an assumption about the radial extent of the fracture can be made, then fracture aperture can be determined as follows:

$$w = \left(\frac{Q}{C\Delta P}\right)^{1/3} \tag{2.29}$$

Then, fracture permeability can be estimated by the following equation that is derived from Poiseuille law, as applied in parallel plate flow.

$$k = w^2/12 \tag{2.30}$$

Norbeck (2012) demonstrated through study of six wells that this technique is both practical and accurate. Data from six wells that were drilled underbalanced were collected. Conductive natural fracture zones are determined for each well, thereby estimating fracture permeabilities. The wells were from two tight gas shale formations, one located in the U.S. and the other in Canada. The lateral sections of these horizontal wells range from 3000 to 6000 ft. The lateral sections of all wells were drilled using oil-based mud.

To verify accuracy of the analysis, a special technique had to be applied because no borehole image logs were available. The main validation technique used in that study took advantage of a commonly used horizontal drilling technique in which wells are drilled in parallel. Six wells selected for the study constituted three sets of parallel wells drilled to similar elevations. The spacing of these wells was between 500 and 800 ft. The parallel well sets should penetrate similar natural fracture systems. The results of the natural fracture identification analysis for each parallel well set are compared to determine if any patterns exist that may be indicative of the orientation of natural fracture planes.

In the first field case study, first two wells (Well A-1 and Well A-2) were chosen from the same field. The horizontal spacing between these two wells is roughly 800 ft. Well A-1 runs in a South-North orientation and Well A-2 runs in the opposite direction. Well A-1 was drilled toe-down and Well A-2 was drilled toe-up. The geometric properties and the average reservoir pressure for each well, obtained from DFIT testing, are listed in Table 2.17. The targeted pay zone is roughly 175 ft thick.

The analysis of Well A-1 indicates that ten conductive natural fractures were intersected during the drilling process (see Table 2.18). Two of these natural fractures are within very close proximity to each other and are considered a single conductive natural fracture zone. In total, nine natural fracture zones are present along the lateral of this well. As can be seen in Fig. 2.45, the stretch of lateral between 14,000 and 15,500 ft. MD contains no conductive natural fracture zones. This is a primary example of the insight that can be gained from this type of analysis. This zone can be selected for creating sweet spots through hydraulic fracturing. The fracture apertures range from 13 to 53 μm. The cross-plot indicates a general positive correlation between mud pit volume peak and gas peak for these conductive natural fracture zones (see Fig. 2.45).

TABLE 2.17 Length of lateral sections, average true vertical depth of lateral sections, and average reservoir pore pressures for corresponding true vertical depth for Wells A-1 and A2.

Well name	Length of lateral (ft)	Average TVD of lateral (ft	Average reservoir pressure (psi)
A-1	4280	12,381.8	11,550
A-2	3131	12,389.8	11,575

From Norbeck, J.H., 2010. Identification and Characterization of Natural Fractures while Drilling Underbalanced (MS thesis). Stanford University.

TABLE 2.18 Results of fracture identification in Well A-1.

		Well A-1			
Candidate fracture location (ft)	Fracture aperture (μm)	Fracture permeability (md)	Mud pit volume peak (bbls)	Gas peak (units)	Underbalance (psi)
12,533.0	30	7.500E+04	4.000	988	1093
12,826.0	35	1.021E+05	3.000	171	1081
12,835.0	53	2.341E+05	3.200	156	1081
12,950.0	44	1.613E+05	1.600	286	1080
13,235.0	42	1.470E+05	1.700	182	1143
13,529.0	37	1.141E+05	1.700	69	1138
13,884.0	36	1.080E+05	1.600	74	1135
13,960.0	42	1.470E+05	2.300	150	1133
15,613.0	35	1.021E+05	1.900	103	1178
16,156.0	13	1.408E+04	1.600	644	1234

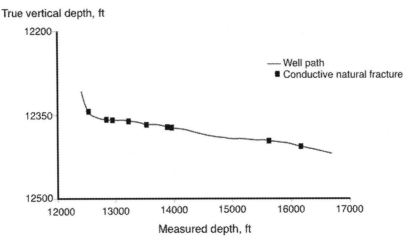

FIG. 2.45 Locations of conductive natural fractures along the lateral of Well A-1. *Redrawn from Norbeck, J.H., 2010. Identification and Characterization of Natural Fractures while Drilling Underbalanced (MS thesis). Stanford University.*

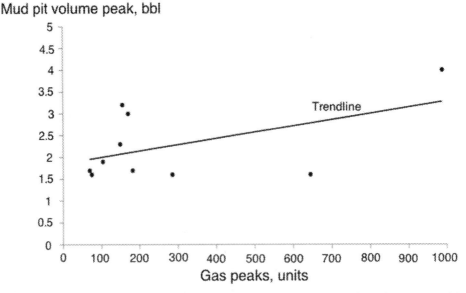

FIG. 2.46 Cross-plot of mud pit volume peak vs. gas peak corresponding to each conductive natural fracture location identified for Well A-1. *Redrawn from Norbeck, J.H., 2010. Identification and Characterization of Natural Fractures while Drilling Underbalanced (MS thesis). Stanford University.*

Similarly, a total of nine conductive natural fracture zones were identified for Well A-2 (Fig. 2.46). The relevant results are listed in Table 2.19. The zones are relatively evenly spaced along the lateral. The fracture apertures are generally smaller than for Well A-1, ranging from 15 to 38 μm. These estimates could be largely due to the higher level of underbalance maintained while drilling A-2. In addition, the observed rise in mud pit volume due to the

2.7 Characterization of reservoirs

TABLE 2.19 Results of fracture identification in Well A-2.

		Well A-2			
Candidate fracture location (ft)	Fracture aperture (μm)	Fracture permeability (md)	Mud pit volume peak (bbls)	Gas peak (units)	Underbalance (psi)
13,467.0	27	6.075E+04	1.177	88	1262
13,867.0	29	7.008E+04	0.811	53	1334
13,943.0	32	8.533E+04	1.000	75	1335
14,192.0	38	1.203E+05	1.176	67	1339
14,598.0	16	2.133E+04	0.774	682	1339
14,788.0	15	1.875E+04	0.790	328	1336
14,990.0	32	8.533E+04	1.170	924	1334
14,991.0	35	1.021E+05	1.460	715	1334
15,296.0	25	5.208E+04	0.768	70	1338
15,750.0	25	5.208E+04	0.930	88	1421

presence of these fractures is relatively low. The cross-plot does not show a strong trend for the relationship between mud pit volume peak and gas peak for these fractures, however, it is a positive correlation (see Fig. 2.47).

Similar results were obtained for other fields as well. On average, the computational tool identifies between nine and 10 conductive natural fracture zones for each well. For each conductive natural fracture zone, the fracture aperture and fracture permeability was estimated.

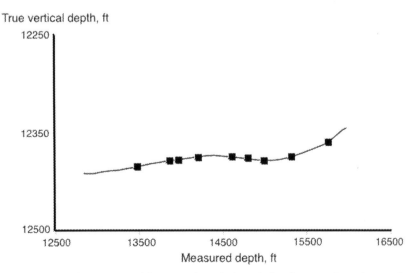

FIG. 2.47 Locations of conductive natural fractures along the lateral of Well A-2. *Redrawn from Norbeck, J.H., 2010. Identification and Characterization of Natural Fractures while Drilling Underbalanced (MS thesis). Stanford University.*

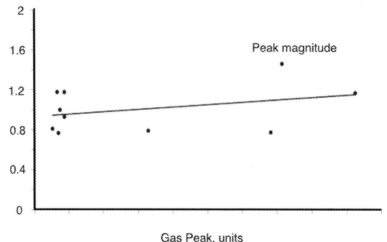

FIG. 2.48 Cross-plot of mud pit volume peak vs. gas peak corresponding to each conductive natural fracture location identified for Well A-2. *Redrawn from Norbeck, J.H., 2010. Identification and Characterization of Natural Fractures while Drilling Underbalanced (MS thesis). Stanford University.*

The estimated fracture apertures all lie within the expected range of values (i.e., 10 to 1000 μm). Overall, although the estimates of fracture aperture may not be entirely accurate, they should be considered as a lower-bound for the true fracture apertures (Fig. 2.48).

In absence of other means of validation, patterns in the locations of conductive natural fracture zones between wells were used. Existence of such a pattern would confirm tectonic continuities that are essential for history matching of the digenesis involved to validate the results because each pair of wells should penetrate similar geologic conditions. Once validated the existence of fractures (or sweet spots), orientation of fractures could be used to refine reservoir characterization. For the case in question, two out of the three parallel well pairs exhibit strong patterns. From visual inspection of the results obtained for Field A, two dominant patterns can be observed, as can be seen in Fig. 2.49. A total of seven pairs of natural fractures are aligned at an orientation of roughly N65°E (see Fig. 2.50). Only one identified natural fracture from Well A-2 does not have a corresponding feature in Well A-1. For each of the seven natural fracture planes identified, the estimated apertures of the corresponding pair of natural fractures compare well, with the exception of Pairs 4 and 5 (see Table 2.20).

2.7.10 Reservoir characterization with image log and core analysis

As presented in Table 2.21, image logging and core analysis present an important stage of reservoir characterization. Techniques for directly assessing near wellbore fracture density, fracture aperture, and fracture orientation are presently available to the industry by means of image log testing. Examples of image log techniques include borehole video camera, acoustic formation image technology (AFIT), and resistivity image logs. These three methods are

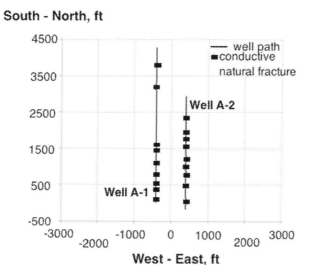

FIG. 2.49 Plan view of Field A. Wells A-1 and A-2 are parallel wells drilled in the South—North direction. The lateral spacing between these wells is roughly 800 ft. *Redrawn from Norbeck, J.H., 2010. Identification and Characterization of Natural Fractures while Drilling Underbalanced (MS thesis). Stanford University.*

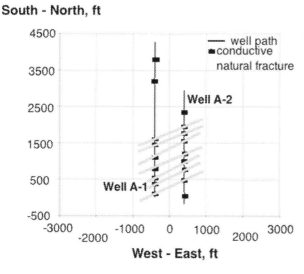

FIG. 2.50 Natural Fracture System Orientation #1 for Field A. A dominant pattern exists that seems to indicate the presence of a natural fracture system oriented at N65°E.

based on different fundamental principles, and each has its own set of advantages and disadvantages. A common drawback is that the image resolution quality is generally too poor to be able to identify conductive features that are believed to be on the order of 100 μm wide. Circumferential Borehole Imaging Log (CBIL), when utilized for potential fractured layers already tagged by other techniques (as acoustic Wave Forms), has been proved as very

TABLE 2.20 Comparison of estimated fracture aperture between pairs of conductive natural fractures for Wells A-1 and A-2 (pairs numbered from North to South).

Pair #	(µm)	(µm)
1	42	29
2	36	32
3	37	38
4	42	16
5	44	15
6	44	34
7	30	25
Average	39	27

TABLE 2.21 List of borehole imaging tools from which BGS holds digital data, details of tool specification, horizontal resolution and wall coverage.

Category	Company	Tool type	Tool names	Specifications	Approx. vertical resolution	Approx. coverage (8.5″ well)
1	Schlumberger	Resistivity & Acoustic	FMI & UBI	Tools run in combination	2.5 mm	100%
2	Schlumberger	Resistivity	FMI	192 button electrodes (24 per pad/flap × 4)	2.5 mm	80%
2	Schlumberger	Acoustic	UBI/BHTV/ATS	Rotating Sensor	5 mm	100%
2	Weatherford	Resistivity	CMI	176 button electrodes (20/24 per pad × 8)	2.5 mm	80%
2	Baker Hughes	Resistivity	STAR	144 button electrodes (12 × 6 pads)	5 mm	80%
3	Schlumberger	Resistivity	4-Pad FMS	64 button electrodes (16 × 4 pads)	2.5 mm	40%
4	Schlumberger	Resistivity	2-Pad FMS	58 button electrodes (29 × 2 pads)	2.5 mm	20%
5	Multiple	Dipmeter	Dual-Caliper	4 button electrodes (1 × 4 pads)	40 mm	>5%

effective and detailed. Each of the following logs also give information that can lead to refinement of reservoir characterization.

- Spontaneous Potential (SP)
- Gamma Ray log (GR)
- Density Log (FDL)
- Neutron Log (CNL)
- Dual-Induction Log (DIL)
- Sonic Log (SON)

Following new array of logs have recently been introduced.

- Array Induction Log (AIT)
- Array Sonic Log (AST)
- Electromagnetic Propagation Log (EPT)
- Nuclear Magnetic Resonance (NMR)

One should highlight here that the representative elemental volume (REV) for fractured reservoirs is greater than the core size as well as the depth of resolution of most imaging tools. As can be seen from Fig. 2.51, below REV, fluctuations occur. Any correlation that is apparent must, therefore, be corroborated/refined with previously available data, starting from data acquired during geological survey.

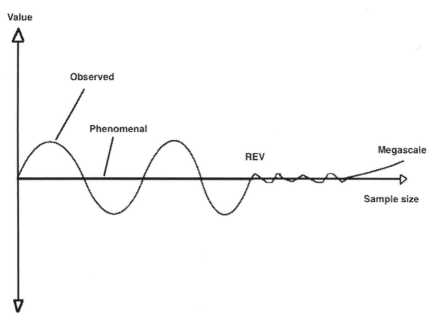

FIG. 2.51 REV in fractured reservoirs is greater than core size. *Redrawn from Islam, M.R., Mousavizadeghan, H., Mustafiz, S., Abou-kassem, J.H., 2010a. Reservoir Simulation: Advanced Approach. Scrivener-Wiley, 468 pp.*

2.7.11 Geophysical logs

In general, following geophysical logs are routinely available for reservoir characterization.

- Gamma Ray (GR) Spectralog: This one is based on collecting Gamma-ray signals from natural rocks in the reservoir. This log can be performed in open as well as cased holes and allows a detailed stratigraphic reconstruction for the entire depth of the well, even in case of cuttings absence due to Total Loss of Circulation (TLC).
- Densilog & Acoustilog—contribute to the stratigraphic-structural reconstruction of the well and are essential for the bulk density and seismic wave velocity determination to give calibration elements for the interpretation of surface gravimetric and seismic surveys. Furthermore these logs are fundamental to computing the formational elastic parameters and their variations in case of presence of fractures.
- Multi-arm Caliper—is very useful not only for the imaging of the hole geometry, but also for structural reconstruction by means of break-out analyses.
- Borehole Imaging Log—allows the 360° mapping of the walls of the hole by analyzing the formational variation of both velocity and resistivity. This is the only specific tool for the direct fracture analyses in terms of nature and geometric parameters.

Usually, during the field recording phase it is possible to make a preliminary individuation of levels which can be potentially fractured. These are very often associated with:

- sharp decrease of bulk density and P wave velocity (VP);
- strong attenuation of the wave form (WF);
- intense and very thin cavings in the walls of the hole;
- peaks of GR in case of mineralized fractures.

Borehole imaging tools provide an image of the borehole wall that is typically based on physical property contrasts. There are currently a wide variety of imaging tools available, though these predominately fall into two categories: resistivity and acoustic imaging tools.

Tingay et al. (2008) investigated borehole breakouts using image logging. They characterized borehole breakouts from resistivity image logs as parallel poorly resolved conductive zones that appear 180° apart on opposite sides of the borehole wall. However, the resolution of borehole breakouts is dependent on the width of the pad compared to the width of the breakout (Tingay et al., 2008). In a limited number of cases, resistivity images are accompanied by ultrasonic borehole images (e.g. Schlumberger's UBI™ tool) which circumferentially record both the amplitude and travel time of the returning wave form. These tools have lower vertical and angular resolution than the resistivity tools.

However, the travel time waveform (TTWF) images from acoustic logs are useful as they are more sensitive to changes in the borehole radius. In TTWF images breakouts appear as broad zones of increased borehole radius observed at 180° from one another (Tingay et al., 2008). Fig. 2.52 shows an example of resistivity (FMI) image and an acoustic (UBI) travel time image from the Sellafield 13A in Cumbria, highlighting the differences between the way these two types of tool show borehole breakout. In the absence of acoustic images, breakouts are therefore identified by darker (less conductive) patches on opposite sides of borehole images and a simultaneous disturbance to the spacing between the imaging pads.

2.7 Characterization of reservoirs 153

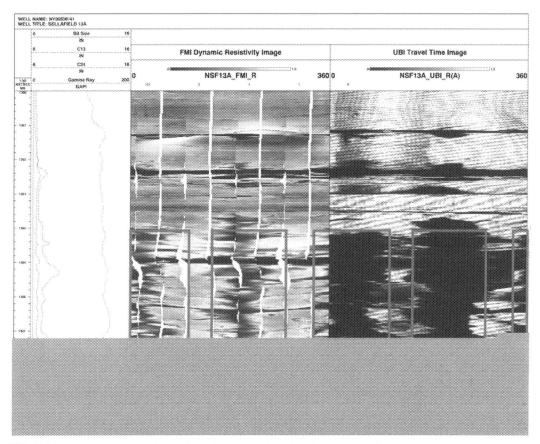

FIG. 2.52 Comparison of methods of visualizing a 4 m long borehole breakout from St Bees Shale Formation, from borehole Sellafield 13A in Cumbria (10 m vertical borehole section). Left-hand panel: conventional logs including perpendicular dual-caliper and gamma-ray. Center panel: Unwrapped circumferential resistivity borehole imaging (FMI) (clockwise from north) with breakout highlighted by the *green (gray in the printed version) boxes*. Right panel: Unwrapped circumferential acoustic borehole amplitude imaging (UBI) (clockwise from north) with breakout highlighted by *green (gray in the printed version) boxes*.

Acoustic tools, emit high-frequency sonar waves. The acoustic imaging tool then records the amplitude of the return echo as well as the total travel time of the sonic pulse. The acoustic wave travel time and reflected amplitude is measured at numerous azimuths inside the wellbore for any given depth. This data is then processed into images of the borehole wall reflectance (based on return echo amplitude) and borehole radius (based on pulse travel time). There are a wide variety of acoustic imaging tools available, some of the more common tools are the Borehole Televiwer (BHTV, from Schlumberger), Ultrasonic Borehole Imager (UBI; from Schlumberger), Circumferential Borehole Imaging Log (CBIL; from Baker Atlas), Simultaneous Acoustic and Resistivity tool (STAR; from Baker Atlas), Circumferential Acoustic Scanning Tool-Visualization (CAST-V; from Halliburton) and the LWD/MWD AcoustiCaliper tool (ACAL; from Halliburton).

The use of AFIT has been implemented to characterize the permeability of feed zones in oil and gas and geothermal wells. A description of the AFIT tool is given by McLean and McNamara (2011):

As the AFIT tool is lowered and raised in the well an acoustic transducer emits a sonic pulse. This pulse is reflected from a rotating, concave mirror in the tool head, focusing the pulse and sending it out into the borehole. The sonic pulse travels through the borehole fluid until it encounters the borehole wall. There the sonic pulse is attenuated and some of the energy of the pulse is reflected back toward the tool. This is reflected off the mirror back to the receiver and the travel time and amplitude of the returning sonic pulse is recorded. Through the use of the rotating mirror (≤ 5 rev/sec) 360° coverage of the inside of the borehole wall can be obtained.

In practice, the interpretation of AFIT data is quite sophisticated. Planar natural fractures appear as sinusoids in the imaged data set, as shown in Fig. 2.53.

Data processing software allows for characterization of geologic features including strike and dip, fracture aperture, and fracture density. The signal amplitude can be used to distinguish between open and closed fractures. Low amplitude signals are seen as a dark features on the acoustic image and often interpreted as open. High amplitude signals are seen as light features on the acoustic image and are thought to be attributed to mineral fill. These high amplitude features are usually considered closed fractures that do not contribute to flow. While McLean and McNamara (2011) report a good level of correlation between measured feedzone fluid velocity and fracture aperture determined from AFIT, the fracture apertures reported range from several centimeters to greater than 50 cm. This implies that the resolution of AFIT can at best distinguish fractures of roughly one to two centimeters. This is nowhere near the level of resolution quality necessary for fracture characterization in highly fractured tight gas reservoirs.

Resistivity image logs, also called formation micro-image (FMI) logs, have been documented as an improved technique to characterize geologic features along the wellbore. These techniques make use of a tool that places an electrode at constant electrical potential against the borehole wall and measuring the current. shows the device. The tool is a small-diameter imaging tool that can be deployed with or without a wireline. The high sampling density of these tools (e.g., 120 samples per foot) provides extremely high resolution in

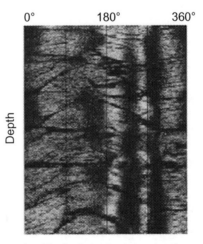

FIG. 2.53 Example of an AFIT image log. The horizontal axis is azimuth around the wellbore. The sinusoids are interpreted as planar geologic features. *From McLean, K., McNamara, D.D., 2011. Fractures interpreted from acoustic formation imaging technology: correlation to permeability. In: Thirty-Sixth Conf. Geothermal Reservoir Eng., Stanford, California.*

the image quality. Microresistivity imaging tools have the ability to visualize features down to 2 mm in width. Careful interpretation of resistivity image logs can provide helpful information about the geologic conditions near wellbore, including dip analysis, structural boundary interpretation, fracture characterization, fracture description, and fracture distribution.

Data obtained from all three of these image log techniques can be analyzed to gain useful insight about the natural fracture system and state of stress system near wellbore. It has been well documented that drilling induced tensile fractures will form in the azimuth of the maximum horizontal principal stress. If the three in situ principal stresses can be determined and the formation fluid pressure is known, then a Coulomb failure analysis can be applied to the natural fractures identified in the image logs. Shear and effective normal stresses acting on each fracture plane can be determined from knowledge of the orientation of the fracture plane with respect to the orientations of the principal stresses. The Mohr-Coulomb failure envelope for fractures is determined from laboratory measurements on prefractured rock. The failure line is constructed assuming no cohesion and using the friction angle of the prefractured rock. Poles to fracture planes are then displayed on the Mohr diagram. Critically stressed fracture planes lie above the Mohr-Coulomb failure envelope (Fig. 2.54). These findings indicate that only a small percentage of the total number of fractures are likely to contribute to flow.

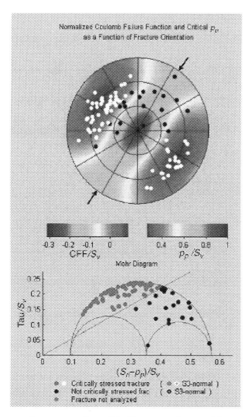

FIG. 2.54 These figures illustrate the concept that critically stressed natural fractures predominantly contribute to fluid flow through reservoirs.

2.7.12 Circumferential borehole imaging log (CBIL)

Recently, the Circumferential Borehole Imaging Log (CBIL), based on the digital acoustic imaging technology has gained popularity among geophysical tools for characterizing fractured formations. All the processing steps are mainly aimed at pointing out all those variations of the rock physic characteristics that can be related to the presence of fracture systems.

The first processing phase involve the Densilog and Acoustilog to compute the Acoustic Impedance, the Reflection Coefficient and the Synthetic Seismogram. The last one is particularly useful for a comparison with surface and well seismic profiles data because seismic reflections have been proved to be very often a signature of fractured horizons.

The wave form analysis, recorded by means of advanced digital acoustic tool, allows to map the image of the instantaneous amplitude. This shows the wave form energy distribution and content evidencing very clearly wave form attenuation due to fractures. Furthermore the S wave velocity (V_S) and of the V_P/V_S ratio are also computed from the wave form analyses. These parameters are combined with the density values and many elastic properties can be computed (see Fig. 2.55). Among these elastic parameters the Fracture Toughness Modulus is particularly sensitive to the presence of fractured levels.

The second processing phase (Fig. 2.56) is aimed at the fracture characterization of both the nature and the structural pattern using data from Multi-arms Caliper and CBIL (orientation-corrected in case of deviated wells). Rough structural information comes from the break-out analysis of the Multi-arms oriented Caliper that allows the definition of the minimum horizontal stress direction ($\sigma 3$) which is orthogonal to the fracture planes considering a vertical direction of the maximum stress ($\sigma 1$).

CBIL data allow detailed structural reconstruction. In the CBIL tool an acoustic transducer, continuously spinning on the 360° of the walls of the hole, emits an acoustic pulse directed into the formation and records both the amplitude and the travel-time of the returning wave. The acoustic amplitude is mainly a function of the acoustic impedance of the formation, so that fractures and their nature (open, mineralized, foliation etc.) can be clearly evidenced. Even though the depth of penetration is not impressive with CBIL, detection of fractures even at the hole surface is useful and practical for reservoir characterization.

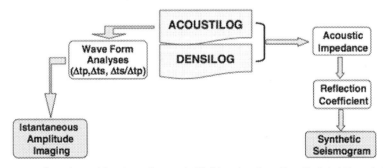

FIG. 2.55 Processing flow chart of density and acoustic Well logging data. *From Batini, A.F., et al., 2002. Geophysical well logging—a contribution to the fractures characterization. In: Proceedings, Twenty-Seventh Workshop on Geothermal Reservoir Engineering Stanford University, Stanford, California, January 28–30. SGP-TR-171.*

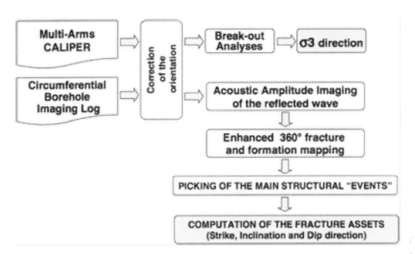

FIG. 2.56 Processing flow chart for fracture analyses from Well logging.

Advanced CBIL processing techniques provide enhanced 360° acoustic amplitude images of the reflected wave. On these images, it is possible to distinguish different types of fractures as a function both of the acoustic impedance variation degree and of their shape and size. These "structural events" can be then picked and all the geometric parameters (i.e., strike, inclination and dip direction) computed. Correlations can also be made with other data (e.g., downhole camera, gamma-ray, geological data) to refine lithological data.

The most effective way to refine CBIL information is to use it in combination with well test data that including thermal gradient and injectivity. The utilization of temperature and pressure log can be a useful tool for the identification of each productive zone in the well and for the direct measurement of the injectivity. As stated earlier in this chapter, underbalanced drilling creates a dynamic database. In absence of underbalanced drilling, similar information can be extracted during an injection test. These test results are affected by the existence of different fractures inside the well, thereby generating data on fracture aspect ratio, density and others. The temperature profile during an injection test will exhibit a change of slope of the thermal gradient where there is a change in the flow rate, i.e., where there is an adsorbing zone: the thermal gradient is proportional to the fluid which passes in the formation. For a gas well, the thermal change is more intense due to the augmented Joule Thomson effect.

The permeability distribution of the reservoir must provide a hydraulic connection throughout all the system; a pressure change in a part of the reservoir (due to exploitation or injection) is propagated in all the system. The propagation velocity of the pressure wave depends on the so-called "hydraulical diffusivity". The well testing is the way for measuring the most important reservoir parameters, as well as the characteristics of the fluid motion. During the drawdown/injection tests the pressure gauge is placed close to the productive zone, and the pressure change is recorded while the well is operated at constant production/injection rate. From the shape of the curve it is possible to identify the reservoir's unique characteristics: the trasmissivity (the permeability-reservoir height product), the skin factor (the well-reservoir coupling factor), the deviation from the ideal radial flow (storage effects, closed or constant pressure boundaries, linear motion of the fluid along preferential paths).

During an interference test the pressure change a given well is recorded, while a drawdown/injection test of another one is performed. This is a very important way for measuring the average characteristics of the reservoir in the volume between the two wells, or for establishing a higher limit of the permeability in the case of negative response.

Batini et al. (2002) presented a comprehensive field case study that utilizes CBIL along with well tests and other geophysical logging. They reconstructed the stratigraphy to determine main rock physical properties for each geological formation of interest. Table 2.22 gives an example of geological characterization performed by means of the GR Spectralog, which gives a value of total GR and of its spectral components: Potassium (K), Thorium (TH) and Uranium (U).

Table 2.23 shows some of the physical characteristics of the rock. This is necessary for further analysis of data. They also conducted core analysis on two core samplings within the interval in question. These data can be utilized to refine static data gathered independent of core sampling. The bulk density of 2.6 g/cm^3 can be compared with the previous indirect measurement from geophysical logs of 2.77 g/cm^3 for the micaschists reservoir rock. Table 2.24 lists these data.

The case study involves a geothermal well that was drilled during May 8, 2000, through September 12, 2000. The formation is cased and the open hole started at a depth of 2202 m. The first important fractured zone has been highlighted at 2600 m; after acidification and hydraulic stimulation an injection test measured a low injectivity: 1.6 m^3/h./bar. Subsequently, a T&P log has been recorded during another stimulation (with 80 kg/s for 2 1/2 h), followed by another medium-duration injection test (with 8 kg/s). Three adsorbing zones have been identified, but, due to the low overall injectivity, it was decided to deepen the well, until the final depth of 4002 m was reached. The following tests were performed:

- Build up immediately after drilling;
- A 17 days production test (the well production could be estimated as 4 kg/s at 1.6 MPa well-head pressure);
- Two T&P logs during the production test, with an indication of six productive fractured zones;
- An interference test, showing a linear motion connecting the two wells;
- Final build up after production test.

TABLE 2.22 Geological characterization from GR Spectralog.

Lithology	Depth interval (m)	GR (GAPI)	K (%)	TH (ppm)	U (ppm)
Neogene Sediments	0–280	44.0 ± 3.8	1.1 ± 0.1	3.9 ± 0.6	2.0 ± 0.5
Flysch	280–550	63.0 ± 5.3	1.9 ± 0.1	2.8 ± 0.77	2.8 ± 0.7
Tectonic Wedges	550–1900)	48.0 ± 7.6	1.92 ± 0.1	6.7 ± 1.2	2.5 ± 0.7
Phyllites	1900–2220	93.5 ± 12.5	2.38 ± 0.6	11.7 ± 1.9	3.2 ± 1.1
Micaschists	2220–3800	109.8 ± 35.5	2.40 ± 0.9	12.7 ± 4.6	4.4 ± 1.6
Gneiss	3800–4000	N/A	N/A	N/A	N/A

TABLE 2.23 Physical characteristics of the reservoir rock (Batini et al., 2002).

Parameter	Value
V_P	4.87 ± 0.32 (km/s)
V_S	2.81 ± 0.20 (km/s)
V_P/V_S	1.7 ± 0.1
Density	2.77 ± 0.07 (g/cm^3)
Acoust. Imp.	12.7 ± 1.6 (kmsec^{-1} gcm^{-3})
Young Mod.	53.25 ± 10.2 (GPa)
Poisson Coef.	0.2 ± 0.06
Fract. Toughn.	0.006 (GPa^{-1})

TABLE 2.24 Core analysis results (Batini et al., 2002).

Core sample	Depth interval (m)	Grain density (g/cm^3)	Bulk density (g/cm^3)	Porosity (%)	Heat capacity (J/g°C)
Micaschists	3085–3088	3.0	2.6	1.3	0.67
Gneiss	3830–3833	2.9	2.6	1.6	0.67

Fig. 2.57 shows the temperature profile. Unfortunately, the drawdown analysis does not give a clear indication of the reservoir characteristics, due to the superposition effects of each production zone. The final build-up shows a slight tendency toward a radial motion, with a stabilized flow rate of 2.2 kg/s at 1.6 MPa. Assuming 1000 m of reservoir height, the formation permeability can be estimated as 0.7 mD and a negative skin factor of −4.2. Table 2.25 shows results of the well test.

The CBIL results gave a means of most definitive confirmation of fractures, as well as quantitative information. The geophysical log processing confirmed that the six depth intervals preliminary identified for the CBIL investigation were particularly affected by signatures related to the presence of fractures (Fig. 2.58).

The CBIL analysis allowed the identification of different kinds of fractures and their geometrical parameters (Fig. 2.58). These last were processed and mapped for each interval as "pole density of all the fracture planes", using the Wulf's lower hemisphere.

stereo-graphical projection. It should be noted that visual inspection of cores is an integral part of the analysis (last column of Fig. 2.59).

For each interval the pole density distribution, for fractures and faults only (foliations excluded), is shown in Fig. 2.60 together with the most representative cycle-graphical traces. These are characterized by a prevalent E-W azimuth direction, the dip direction is almost variable, but the inclination shows a tight variation between 65° and 80°.

A comparison with core fracture analysis is possible only for cores extracted from the same metamorphic formation in the vertical well Sesta 6 bis. They are not oriented, so that the only reliable value is an average slope of about 70° measured on few samples of continuous joints.

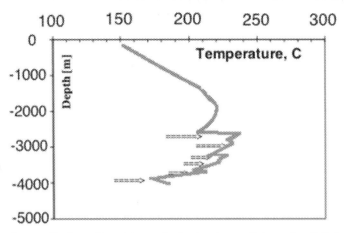

FIG. 2.57 Temperature profile shows the existence of a fractured zone. *Redrawn from Batini, A.F., et al., 2002. Geophysical well logging—a contribution to the fractures characterization. In: Proceedings, Twenty-Seventh Workshop on Geothermal Reservoir Engineering Stanford University, Stanford, California, January 28–30. SGP-TR-171.*

TABLE 2.25 Well test results.

Depth (m)	First T&P Flow rate (kg/s)	Second T&P Flow rate (kg/s)
2640	2.08	1.77
2910	1.11	0.14
3240	N/A	0.33
3400	N/A	0.14
3660	N/A	0.33
3880	2.31	1.44
Total	5.50	4.15

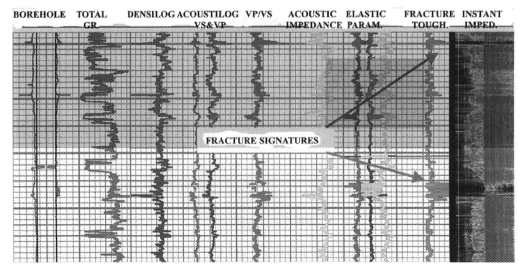

FIG. 2.58 Fracture signatures from geophysical logs. *From Batini, A.F., et al., 2002. Geophysical well logging—a contribution to the fractures characterization. In: Proceedings, Twenty-Seventh Workshop on Geothermal Reservoir Engineering Stanford University, Stanford, California, January 28–30. SGP-TR-171.*

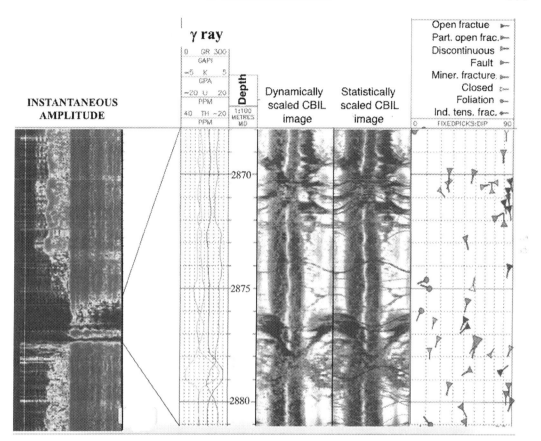

FIG. 2.59 Fracture Analysis from CBIL. *From Batini, A.F., et al., 2002. Geophysical well logging—a contribution to the fractures characterization. In: Proceedings, Twenty-Seventh Workshop on Geothermal Reservoir Engineering Stanford University, Stanford, California, January 28–30. SGP-TR-171.*

A comparison between the fractures detected by geophysical logs and well testings is given in Table 2.26 together with a tentative correlation between fracture asset and productivity.

There is quite a correspondence with fractures detected by well testing in four out of six intervals characterized by geophysical fracture signatures. Excluding the deepest productive zone at 3880 m, not investigated by CBIL, the levels with higher productivity (1.77 and 0.33 kg/s) are associated with sub-vertical fractures (inclination of 70–87°) with an E-W strike direction and Northward dip direction.

2.7.12.1 *Petrophysical data analysis using nuclear magnetic resonance (NMR)*

Recent years have seen a surge in the use of NMR-based logging tools. It is commonly perceived that NMR alone or in combination with conventional logs as well as SCAL data can lead to better determination of petrophysical properties of heterogeneous tight gas sand reservoirs (Hamada, 2009). Hamada reported determination of following parameters of a tight gas formation:

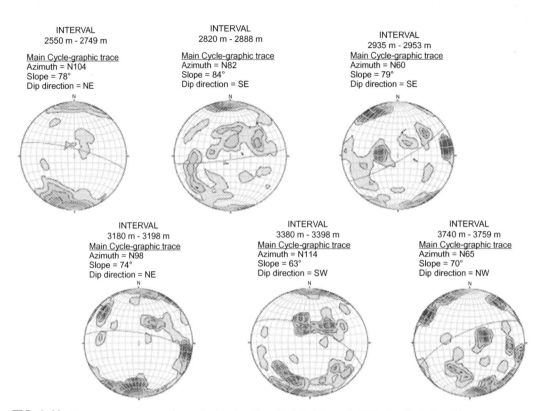

FIG. 2.60 Fracture asset mapped as pole density. *From Batini, A.F., et al., 2002. Geophysical well logging—a contribution to the fractures characterization. In: Proceedings, Twenty-Seventh Workshop on Geothermal Reservoir Engineering Stanford University, Stanford, California, January 28–30. SGP-TR-171.*

TABLE 2.26 Comparison between geophysical logs and well testing.

Fractured levels from CBIL				Fractures from Well Testing	
Depth (m)	Strike direction	Slope and dip direction	Number of samples	Depth (m)	Production flow rate (kg/s)
2550–2750	E-W	87° N	242	2640	1.77
2820–2890	E-W	84° SE	72	Not detected	
	NNW-SSE	46° E	22		
	N-S	50° W	22		
2915–2975	N-S	27° E	36	2910	0.14
3180–3210	E-W	70° N	18	3240	0.33
3380–3410	WSW-ENE	24° SSE	30	3400	0.14
				3660	0.33
3730–3780	Not definable		few	Not detected	
Bottom Log					
				3880	1.44

(1) detailed NMR porosity in combination with density porosity, ϕ_{DMR};
(2) NMR permeability, KBGMR, which is based on the dynamic concept of gas movement and bulk gas volume in the invaded zone; and
(3) Capillary pressure derived from relaxation time T2 distribution, with further possibility of its use in measuring formation saturations, particularly in the transition zone.

Hamada (2009) presented an interesting case study that is used in this section as a template. The case study involves a gas condensate field that produces from a Lower-Mesozoic reservoir. The reservoir is classified as a tight heterogeneous gas shaly sands reservoir. Complex heterogeneity occurs both laterally and vertically due to diagenesis involving kaolinite and illite. As shown in Fig. 2.61, the permeability ranges from 0.01 to 100 mD with a narrow band of porosity ranging from 8 to 10%. The petrophysical analysis indicates narrow 8%–12% porosity range while wide permeability ranges from 0.01 to 100 mD. This cross plot is not useful in its original form as there is no discernible trend. Because fractures are not accounted for in the core analysis, any extension of core data to field scale would be severely skewed. Fig. 2.61 shows that the cloud points were subdivided into six subunits, ranging from high productivity (green) to very low productivity (black). The integration of NMR analysis as well as SCAL can lead to the establishment of facies-independent porosity and permeability models, thereby avoiding the use of lithology-independent T2 cut-off. Incidentally, lithology-independent cut off is a standard for conventional reservoir characterization.

Hamada (2009) presented 3-step procedure for integrating NMR data with SCAL and conventional core data. They are:

(1) The application of Density Magnetic Resonance Porosity (DMR) technique for porosity calculation;

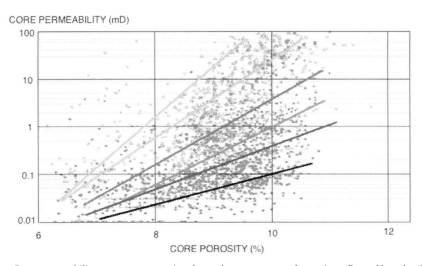

FIG. 2.61 Core permeability vs. core porosity for a heterogeneous formation. *From Hamada, G.M., 2009. Petrophysical properties evaluation of tight gas sand reservoirs using NMR and conventional openhole logs. Open Renew. Energ. J. 2, 6–18.*

(2) Bulk gas Magnetic Resonance Permeability (KBGMR), new technique for permeability calculation beyond the limits of OBM filtrate;
(3) Quantify the effect of OBM filtrate on NMR data and then calibration for approximated capillary pressure from NMR.

Freedman et al. (1998) proposed a combination of density porosity and NMR porosity (ϕ_{DMR}) to determine gas corrected porosity formation and flushed zone water saturation (S_{x0}). Density/NMR crossplot is superior to density/neutron crossplot for detecting and evaluating gas shaly sands. This superiority is due to the effect of thermal neutron absorbers in shaly sands on neutron porosities, which cause neutron porosity readings too high. As a result neutron/density logs can miss gas zones in shaly sands. On the other hand, NMR porosities are not affected by shale or rock mineralogy, and therefore density/ NMR (DMR) technique is more reliable to indicate and evaluate gas shaly sands. Freedman et al. (1998) expressed true porosity as:

$$\phi = \left(\frac{\alpha}{\beta+\alpha}*\phi_D + \frac{\beta}{\beta+\alpha}*\phi_{NMR}\right) \quad (2.31)$$

where ϕ_D is the apparent density porosity, ϕ_{NMR} is porosity determined by NMR, and α and β are given as:

$$\alpha = (1 - HI_g P_g)$$

$$\beta = \frac{\rho_L - \rho_g}{\rho_m - \rho_L}$$

with HI_g indicating hydrogen index for gas.

Eq. (2.31) can be rearranged as follows:

$$\frac{\phi_{Core}}{\phi_{NMR}} = A*\frac{\phi_D}{\phi_{NMR}} + B \quad (2.32)$$

where A and B can be extracted by cross plotting core porosity with NMR porosity, thereby, creating a filter that can be used throughout the reservoir. Such cross plot is shown in Fig. 2.62.

Note that at $S_{gx0} = 0$, the pores are completely filled with liquid (mud filtrate and irreducible water), so the NMR porosity reading and density-porosity should be correct and both should equal to core porosity. As a result, the trend line should intersect at control point, where $\phi_{CORE}/\phi_{NMR} = \phi_D/\phi_{NMR} = 1$. Fluid density for apparent ϕ_D estimation is best fitting at 0.9 g/cc (in this particular case), which is a combination between formation water density and mud filtrate density.

For the above case, A and B are best fitted with the values 0.65 and 0.35, respectively, resulting in the following filter:

$$\phi_{DMR} = 0.65\,\phi_D + 0.35\,\phi_{NMR} \quad (2.33)$$

The results of ϕ_{DMR} transform applications in the three well A, B and C showed very good match between ϕ_{DMR} and core porosities as shown in Figs. 4.60–4.62. As a result, it is considered as an independent facies porosity model. These corrected porosities can be used to estimate permeability in gas bearing formations.

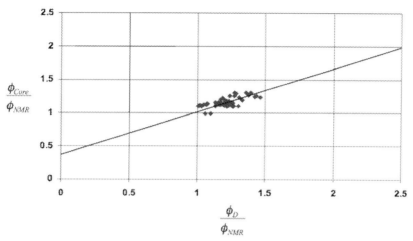

FIG. 2.62 Developing filter out of NMR data.

Figs. 2.63–2.65 present well logs, showing ϕ_D and ϕ_{DMR}. Gamma ray and Caliper curves are shown in the first track (GR&CALI), second track shows depth in meters, the third one is resistivity, the fourth one is neutron-density logs, the fifth track shows comparison between core, density and NMR porosities, sixth track shows comparison between ϕ_{DMR} and core porosity, seventh track shows saturations of gas (green shadow) and water (blue shadow), and the last track shows core permeability in mD.

The next step involves correlation with permeability. Bulk Gas Magnetic Resonance Permeability (k_{BGMR}) is a new technique for permeability estimation in gas reservoirs. It is a dynamic concept of gas movement behind mud cake as a result of permeability formation, gas mobility, capillarity and gravity forces. Because gravity forces are constant, capillarity depends mainly on permeability and mobility depends on permeability and fluid viscosity which is constant for gas; the gas reentry volume directly function in permeability.

The technique involves:

(1) calculation of gas volume in the flushed zone by using Differential spectrum (ΔT_w) Multi acquisition using different waiting times (Tw);
(2) Solve diffusivity equation to solve for the volume of gas. Friedman et al. (1998) gave the following expression:

$$V_{g,xo} = \frac{DPHI - \dfrac{T_{NMR}}{(HI)_f}}{\left[1 - \dfrac{(HI)_g * P_g}{(HI)_f}\right] + \lambda}$$

$V_{g,\,xo}$ = volume in the flushed zone
DPHI = formation porosity from density using filtrate fluid density

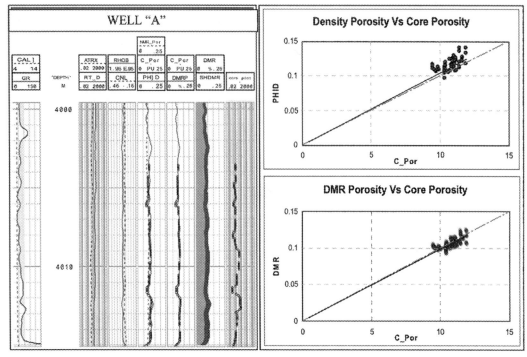

FIG. 2.63 Filter for Well "A". *From Hamada, G.M., 2009. Petrophysical properties evaluation of tight gas sand reservoirs using NMR and conventional openhole logs. Open Renew. Energ. J. 2, 6–18.*

T_{NMR} = total NMR porosity
$(HI)_f$ = Fluid hydrogen index
$(HI)_g$ = Gas hydrogen index
P_g = gas polarization function = $1 - \exp(-W/T_{1,g})$, where W is the wait time and $T_{1,g}$ is the longitudinal relaxation time for gas.

$$\lambda = \frac{\rho_f - \rho_g}{\rho_m - \rho_f}$$

(3) Estimate invasion gas saturation, S_{gx0} with

$$S_{gx0} = \frac{(\phi_D - DMR)*(\rho_m - \rho_L)}{DMR*(\rho_L - \rho_g)}$$

(4) Calculate gas volume approximately by ignoring the gas response in the NMR measurements especially in short TW, and then the gas saturation in the invaded zone as:

$$\text{Bulk gas volume (BG)} = \phi_{DMR} - \phi_{NMR}$$

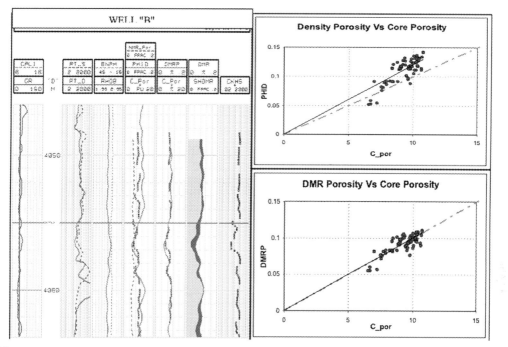

FIG. 2.64 Filter for "Well B" (Hamada, 2009).

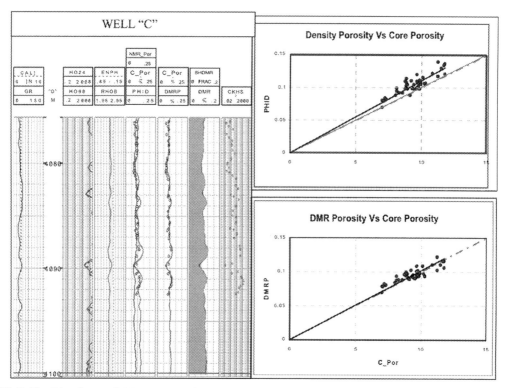

FIG. 2.65 Filter for "Well C". *From Hamada, G.M., 2009. Petrophysical properties evaluation of tight gas sand reservoirs using NMR and conventional openhole logs. Open Renew. Energ. J. 2, 6–18.*

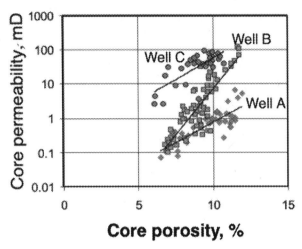

FIG. 2.66 Correlation between core permeability and core porosity. *From Hamada, G.M., 2009. Petrophysical properties evaluation of tight gas sand reservoirs using NMR and conventional openhole logs. Open Renew. Energ. J. 2, 6–18.*

Fig. 2.66 shows core permeability versus core porosity. It reflects how the permeability varies between facies to other within same porosity range. The same method is applied for the three wells A, B and C, and then BG is plotted versus formation permeability, as shown in Fig. 2.67. The correlation is normalized by dividing the gas volume by the total porosity of DMRP to be equal to S_{gx0}, as shown in Fig. 2.68.

$$S_{gx0} = \frac{DMRP - \phi_{NMR}}{DMRP}$$

The correlation between S_{gx0} and permeability shown in Fig. 2.68 has resulted in following permeability transform.

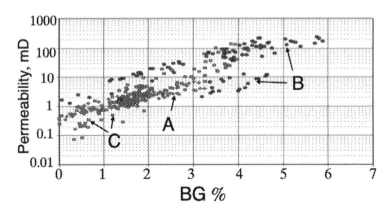

FIG. 2.67 Correlation between pereambility and BG. *From Hamada, G.M., 2009. Petrophysical properties evaluation of tight gas sand reservoirs using NMR and conventional openhole logs. Open Renew. Energ. J. 2, 6–18.*

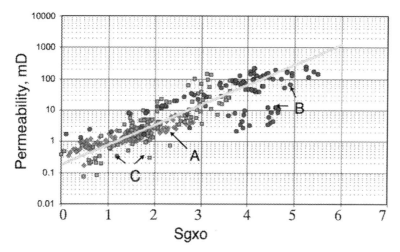

FIG. 2.68 Correlation of permeability vs. S_{gxo}. *From Hamada, G.M., 2009. Petrophysical properties evaluation of tight gas sand reservoirs using NMR and conventional openhole logs. Open Renew. Energ. J. 2, 6–18.*

$$K_{BGMR} = 0.18 * 10^{(6.4 * Sgx0)}$$

Permeability derived using Eq. (2.33) in three wells A, B and C is shown in Figs. 2.69 and 2.70. All three wells A, B and C show a good match between KBGMR permeability with core permeability (Fig. 2.71).

Similarly, correlation was achieved for capillary pressure by Hamada (2009). Fig. 2.72 shows such correlation for Well "B", reported above.

2.7.13 Core analysis

The most direct method for characterizing the natural fractures present in the reservoir is to perform core sample analyses. Fracture density and fracture spacing can be determined through visual analysis of the core. Plugs are typically taken for laboratory experiments to measure permeability and other flow properties. These experiments can only measure an "effective" permeability of the core sample. In general, it is not possible to determine the natural fracture contribution to flow unless the fracture geometry is well known. The key issue, however, is that experience has shown that using laboratory measurements to represent reservoir-scale properties can be vastly misleading. Nonetheless, engineers are faced with the challenge of interpreting the few available direct measurements of the reservoir rock and translating them to field-scale properties. The idea is to use core data to refine data already collected. In addition, several features, including fracture density, mineralization, fracture opening, shale breaks, etc., can be quantified only with visual inspection.

Of relevance is also the fact that total porosity can only be determined through core analysis. For certain gas/oil perspective, this is of great consequence because as much as 50% recoverable gas can go unaccounted without definitive knowledge of porosity. Conventional assessment of porosity through GR analysis can be vastly misleading in fractured reservoirs.

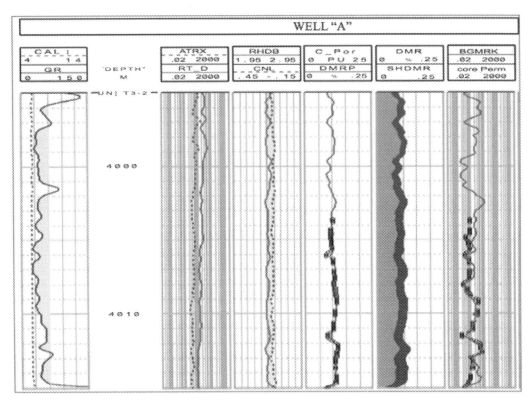

FIG. 2.69 Permeability distribution (track 6) for Well "A". *From Hamada, G.M., 2009. Petrophysical properties evaluation of tight gas sand reservoirs using NMR and conventional openhole logs. Open Renew. Energ. J. 2, 6–18.*

Practically all aspects of core analysis are different in fractured reservoirs from conventional reservoirs. They are highlighted as follows:

1. Low-permeability (matrix and fractures alike) structure itself;
2. Low effective porosity but high total porosity;
3. Response to overburden stress;
4. Impact of the low-permeability structure on effective permeability relationships under conditions of multiphase saturation;
5. Capillary pressure data as well;
6. The role of fractures and fracture/matrix interactions.

The most prominent of the above is the one related to relative permeability graph. In a conventional reservoir, it is clear that there is relative permeability in excess of 2% to one or both fluid phases across a wide range of water saturation. In traditional reservoirs, critical water saturation and irreducible water saturation occur at similar water saturation values. Under these conditions, the absence of common water production usually implies that a reservoir system is at, or near, irreducible water saturation, low-permeability reservoirs; however, one can find that over a wide range of water saturation, there is less than 2% relative

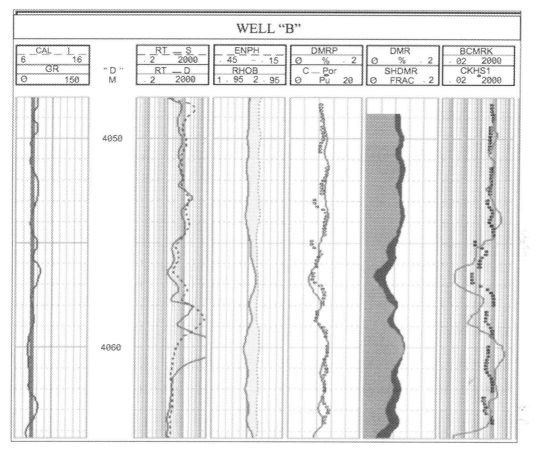

FIG. 2.70 Permeability distribution (track 6) for Well "B". *From Hamada, G.M., 2009. Petrophysical properties evaluation of tight gas sand reservoirs using NMR and conventional openhole logs. Open Renew. Energ. J. 2, 6–18.*

permeability to either fluid phase, and critical water saturation and irreducible water saturation occur at very different water saturation values. In these reservoirs, the lack of water production cannot be used to infer irreducible water saturation. In some very low-permeability reservoirs, there is virtually no mobile water phase even at very high water saturations. The term "permeability jail" coined by Shanely and Byrnes (2004) describes the saturation region across which there is negligible effective permeability to either water or gas.

Fig. 2.73 shows little change in porosity occurs with net increase in stress.

2.7.14 Major forces of oil and gas reservoirs

The theoretical aspect of stable and unstable displacement is discussed in Chapter 3. Many enhanced oil recovery schemes involve the displacement of oil by a miscible fluid. If a scheme is not stable, miscibility cannot occur and all design criteria fail. Whether a displacement is stable or unstable has a profound effect on how efficiently a solvent displaces

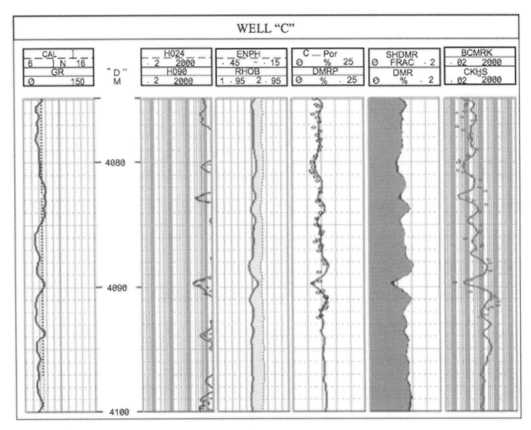

FIG. 2.71 Permeability distribution (track 6) for Well "C". *From Hamada, G.M., 2009. Petrophysical properties evaluation of tight gas sand reservoirs using NMR and conventional openhole logs. Open Renew. Energ. J. 2, 6–18.*

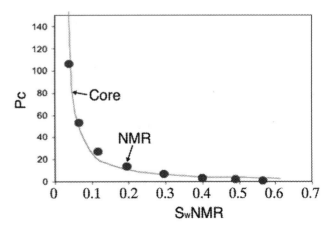

FIG. 2.72 Correlation between core Pc (*blue (light gray in the printed version) dots*) and NMR Pc (*pink (dark gray in the printed version) line*).

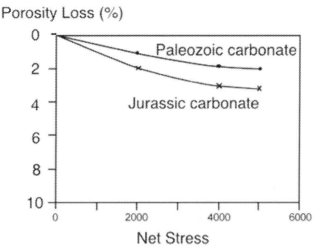

FIG. 2.73 Porosity is only slightly affected by net stress for carbonate formations. *Data from Lucia, F.J., 2007. Carbonate Reservoir Characterization. Springer-Verlag, Berlin Heidelberg, 336 pp; Hariri, M.M., Lisenbee, A.L., Paterson, C.J., 1995. Fracture control on the tertiary epithermal-mesothermal gold deposits northern Black Hills, South Dakota. Explor. Mining Geol. 3 (4), 205–214.*

oil within a reservoir. That is, if viscous fingers are present, the displacement efficiency and, hence, the economic return of the recovery scheme is seriously impaired because of macroscopic bypassing of the oil. As a consequence, it is of interest to be able to predict the boundary which separates stable displacements from those which are unstable.

Coskuner and Bentsen reported a series of stability numbers that deal with miscible displacement. They used linear perturbation to obtain the scaling group. The new scaling group differed from those obtained in previous studies because it had taken into account a variable unperturbed concentration profile, both transverse dimensions of the porous medium, and both the longitudinal and the transverse dispersion coefficient.

It has been shown that stability criteria derived in the literature are special cases of the general condition given here. The stability criterion is verified by comparing it with miscible displacement experiments carried out in a Hele-Shaw cell. Moreover, a comparison of the theory with some porous medium experiments from the literature also supports the validity of the theory. The stability criterion is given below.

$$\frac{U\frac{d\mu}{dC} - kg\frac{d\rho}{dC}\sin\gamma}{\bar{\mu}D}\frac{\partial \bar{C}}{\partial x}\frac{L^2}{\Omega} \cdot \left[\left(\frac{1}{\Omega} + \frac{D'}{D}\right)\left(\frac{1}{\Omega} + 1\right)\right]^{-1} > \pi^2.$$

with $\Omega = \frac{L^2(B^2 + H^2)}{B^2 H^2}$ (for a two-dimensional system in which $H = 0, \Omega = \frac{L^2}{B^2}$) and where

U = displacement velocity
μ = viscosity of mixture
ρ = density of mixture

C = injectant concentration
k = permeability
g = gravitational acceleration
γ = dip angle
ϕ = porosity

Picture 2.3 shows general shapes of viscous fingering in a porous medium. Note that all fingers have similar shape at the onset. As a finger starts to propagate, a dominant one moves faster than the rest of the fingers, leading to bypassing of oils and early breakthrough. This feature is not included in conventional analysis.

When it comes to miscible flood, a mixed scenario of immiscible and miscible displacement processes emerges. It is because every reservoir contains water and an immiscible process is inherently present. This fact is best utilized in designing WAG processes that minimize the use of expensive chemicals. One such example is CO_2 miscible injection.

Note how the displacement front between oil and miscible gas is likely to be unstable because of the unfavorable mobility ratio whereas the displacement front between water and miscible gas (CO_2 in this case) is likely to be stable. This is one of the greatest advantages of WAG that is often overlooked. In addition, WAG makes it possible to access different segments of trapped oil as the wettability natures of gas and water are different. The combination of two different types of flooding has other advantages as well.

The Capillary number expression shows lowering the interfacial tension and or increasing the contact angle will increase the capillary number. This is the basis of chemical injection as well as gas injection (along with various types of gas/water injection). Through the injection of a chemically active fluid, the interfacial tension of the displacement front is decreased, leading to the recovery of residual oil. On the other hand, if rock wettability is changed similar impact of lesser intensity can occur.

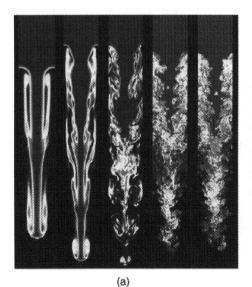
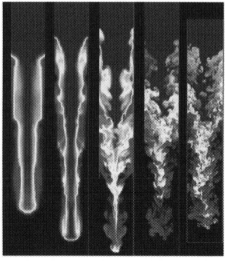

(a)   (b)

PICTURE 2.3 Viscous fingering in a miscible displacement process.

The property of a fluid is directly related to the viscosity of crude oil within the reservoir. These properties are determined by standardized laboratory procedures. Unfortunately the test results do not represent a general characteristic of the reservoir, because the samples are taken from different sites within the reservoir. The prediction of the reservoir fluid properties becomes even more complex when the prevailing conditions within the reservoir change as a result of undergoing processes, leading to unexpected reaction during injection and production operations.

Variations in rock-fluid interaction with changing conditions in a reservoir result in wettability variations, which in turn affect flow parameters such as capillary pressure and relative permeability, affecting both dynamic recovery as well as ultimate recovery from such reservoirs.

Fig. 2.74 shows how end-point permeability values affect residual oil saturation. This figure shows how the use of different gas as well as WAG (water alternated with gas) can affect the ultimate recovery.

For the dynamic part of the displacement process, the most profound impact is through alteration of permeability graphs. Fig. 2.75 shows some examples of how lowering of IFT alters effect permeabilities of both water and oil. The permeability values corresponding to lowest interfacial tension form straight lines. While this is true in a coreflood test, it rarely occurs in the field and discrepancy between laboratory data and field results emerges. In reservoir simulation studies, it has been demonstrated that the oil recovery with straight line permeability curves do not show a oil recovery curve markedly better than others. That happens even after one overcomes the stability problem that occurs when such straight lines are assigned as relative permeability values of an immiscible system (e.g., oil and gas).

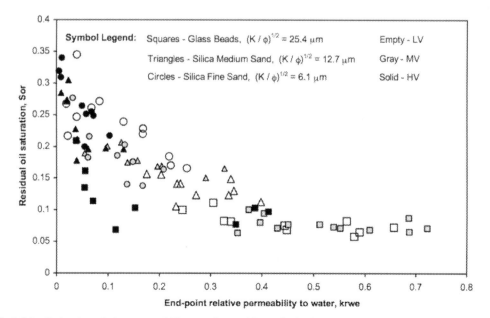

FIG. 2.74 End-point relative permeability correlates with residual oil saturation.

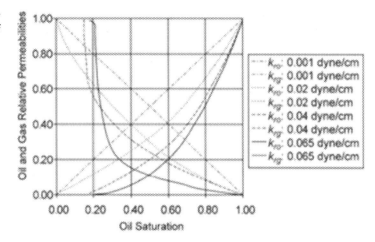

FIG. 2.75 Relative permeability curves are altered by lowering of interfacial tension.

The most frequently encountered saturation endpoints are:

- Residual oil saturation
- Irreducible water saturation
- Trapped-oil and -gas saturations
- Critical gas and condensate saturations

Residual oil, irreducible water, and trapped-gas and trapped-oil saturations all refer to the remaining saturation of those phases after extensive displacement by other phases. Critical saturation, whether gas or condensate, refers to the minimum saturation at which a phase becomes mobile.

The endpoint saturation of a phase for a specific displacement process depends on:

- The structure of the porous material
- The wettabilities with respect to the various phases
- The previous saturation history of the phases
- The extent of the displacement process (the number of pore volumes injected)

The endpoint saturation also can depend on IFTs when they are very low, and on the rate of displacement when it is very high. For certain petroleum reservoirs, lowering of interfacial tension, any chemical treatment, any thermal alteration or onset of fractures, leads to the alteration of relative permeability graphs. Fig. 2.76 shows how such movement can translate into gas production.

Results reported by Chatzis et al. (1983) give general insight on the combined effects of wettability and porous structure on residual saturations. In tests with an unconsolidated sand of nonuniform grain size, the wetting phase (oil) was displaced by a nonwetting phase (air) from an initial saturation of 100% to a residual value. In general, it is known that

- Residual saturation of a wetting phase is less than the residual saturation of a nonwetting phase
- Residual saturation of a nonwetting phase is much more sensitive to heterogeneities in the porous structure

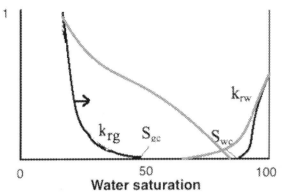

FIG. 2.76 Permeability jail can be removed with thermal or chemical alteration in an unconventional reservoir. *From Islam, M.R., 2014, Unconventional Gas Reservoirs, Elsevier.*

General conclusions on the effects of wettability are useful, but the diverse array of wetting alternatives suggests caution, especially in oil/water reservoir systems. This wide range of wetting possibilities is an obstacle to interpreting or predicting the effect of wettability on endpoint saturations. Indeed, conflicting results for different porous media are likely. For example, Jadhunandan and Morrow (1995) report that residual oil saturation displays a minimum value for mixed-wet media as wettability shifts from water-wet to oil-wet—counter to the results of Bethel and Calhoun (1953) that reported a maximum for media of uniform wettability.

For gas reservoirs, such analysis is important because of the critical gas saturation. The critical gas saturation is that saturation at which gas first becomes mobile during a gasflood in a porous material that is initially saturated with oil and/or water. If, for example, the critical gas saturation is 5%, then gas does not flow until its saturation exceeds 5%. Values of S_{gc} range from zero to 20%. For gas condensate reservoirs, this is of utmost importance. Interest in the mobility of condensates in retrograde gas reservoirs developed in the 1990s, as it was observed that condensates could hamper gas production severely in some reservoirs, particularly those with low permeability. The trend of increasing critical condensate saturations with decreasing permeability, as summarized by Barnum (1995) is reproduced in Fig. 2.77.

This information is of importance for designing of WAG processes. The WAG ratio can also be defined as the ratio of the volume of water injected within the reservoir compared to the volume of injected gas. It plays an important role in obtaining the optimum value of the recovery factor corresponding to an optimal value of the WAG ratio. This optimal WAG ratio is reservoir dependent because the performance of any WAG scheme depends strongly on the distribution of permeability as well as factors that determine the impact of gravity segregation (fluid densities, viscosities, and reservoir flow rates). Studies made by Roger and Grigg (2000), showed that the WAG ratio strongly depends on the reservoir's wettability and availability of the gas to be injected. When the WAG ratio is high, it may cause oil trapping by water blocking or at best may not allow sufficient solvent-oil contact, causing the production performance to behave like a water flood. On the other hand, if the WAG ratio is very small, the gas may channel and the production performance would tend to behave as a gas flood, the pressure declines rapidly, which would lead to early gas breakthrough and high declination on production rate. To find the optimal WAG ratio is necessary to perform sensitivity analysis, proposing different relations of WAG ratio to study the effect on oil recovery.

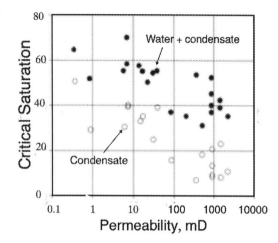

FIG. 2.77 Critical gas saturation for various permeability values of a gas condensate reservoir.

Even though enhanced gas recovery has not been applied in the field at a commercial level, the fundamental mechanisms show that an enhanced gas recovery scheme is more useful and more likely to be successful than pressure maintenance or waterflood. It is because injected gas is less prone to creating channels or fingers in a gas reservoir than an oil reservoir. In addition, miscibility can be achieved in a gas reservoir under easily achievable conditions. Finally, the use of waste gas or gas that would be otherwise flared makes the economics of the project quite appealing.

2.8 Reservoir heterogeneity

The degree of interconnection between the pores of an oil reservoir, are rarely evenly distributed due to non-uniformity of pore size, which gives rise to disordered and complex reservoir fluid flow behavior. Geologically speaking, this is known as the phenomenon of heterogeneous permeability that can manifest different individual layers, forming different homogeneous layers within the oil reservoir with different permeability values.

Most consolidated formations are highly fractured under reservoir conditions. While the presence of fractures makes them ideal for oil and gas recovery because of very high permeability, if the fractures are sealed with secondary cementing activities. Minor porosity of secondary origin occurs locally in sandstones and resulted from the removal of authigenic mineral cements and, to a lesser extent, detrital framework grains. In carbonate-cemented samples, evidence of dissolution includes corrosive contacts between successive carbonate phases and relict cement in pores. Carbonate dissolution features also are observed along the margins of some fractures. Collectively, the dissolution features in sandstones indicate that carbonate cements were previously more widespread before they were partially to extensively dissolved.

Most consolidated formations have fractures in both macroscopic and microscopic scales. Vast majority of these fractures are open without secondary mineralization. The aperture widths commonly exceed 30 μm. One such example is shown in.

An important characteristic of these fractures is that they typically form a dense network that is highly visible on wetted, slabbed rock surfaces if the host sandstones and siltstones have high residual oil saturations. Such fractures are generally absent in rocks that have little or no residual oil. Often, fractures are V-shaped and occluded with pyrite and fine- to coarse-crystalline calcite cement. These fractures, which tend to be small, resemble fluid-escape structures. Such cementation with quartz and cancite can lead to lowering of permeability below the matrix permeability. These types of formations exhibit productivity (or injectivity) lower than one would expect from core and log analysis.

The presence of secondary cementing activities make the porosity-permeability correlation skewed, away from matrix permeability. It is, however, the matrix permeability that is routinely reported in core tests. For a fractured formation, matrix permeability has little to do with permeability of the reservoir. The permeability having a dimension of L^2, a correlation between permeability and porosity turns out to be linear, as long as the porous medium is homogeneous. This linearity no longer holds if the formation is fractured, as shown in Fig. 2.78. If the fractures are open, the medium will have markedly higher permeability than the porosity correlation would indicate. This is of consequence because log analysis usually is performed to determine point porosity of a core. This is later transformed into permeability values using one of numerous correlations available in the literature. Once the transformation is performed, little attention is paid to the origin of these permeability values. Similarly core tests as well as specialized core tests are performed on the homogenous section of the core, leading to the determination of petrophysical properties that have little relevance to the reservoir. Fig. 2.79 shows a typical correlation between porosity and log (permeability) of a

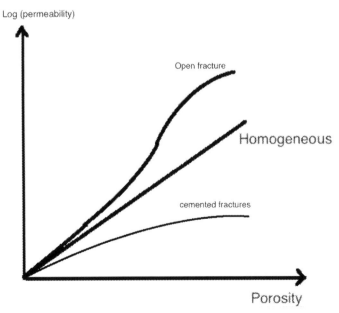

FIG. 2.78 Permeability vs. porosity correlation depends largely on the nature of heterogeneity.

FIG. 2.79 Correlation of porosity vs. permeability for various types of formation.

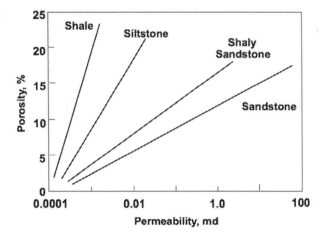

homogeneous formation as compared to the ones that have fractures. An open fracture increases the reservoir permeability drastically, particularly for median porosity range. On the other hand, this trend is reversed if the fractures are closed due to secondary cementing phenomena.

In terms of anisotropy, it can be invoked due to the presence of fractures. Particular orientation of fractures would skew the directional permeability and the magnitude would depend on the aperture size, secondary cementation, and aperture length. Vertical permeability is affected by the extent of vertical fractures and if they are open or cemented.

Fig. 2.80 shows how porosity is correlated with permeability for various types of formations. Fig. 4.91 adds correction factors as a function of fracture frequency. This graph applies to open fractures.

FIG. 2.80 Improvement factor due to open fractures.

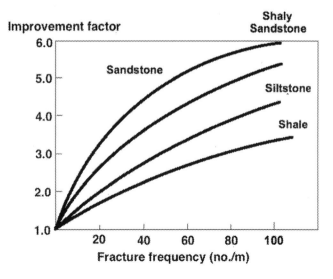

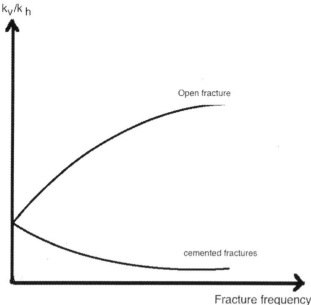

FIG. 2.81 The effect of fractures on k_v/k_h.

Fig. 2.81 shows the trend of vertical permeability (k_v) over horizontal permeability (k_h) with fracture frequency. While fracture frequency is an important factor in determining the anisotropy, the nature of fractures is crucial for determining magnitude of the permeability ratio. When hundreds of wells and kilometer squares of area are considered to estimate flow from reservoirs containing billions of barrels, the accuracy of k_v/k_h analysis can mean a difference of millions of dollars every day throughout the production history.

In terms of EOR, heterogeneity has tremendous impact. To begin with the location of trapped oil and the critical capillary number for mobilization of trapped oil would be different for different pore distribution, which is controlled by fracture characteristics of a reservoir.

To include the influence of fracture and fracture distribution, one must characterize fractures properly. It involves determining frequency and orientation of prominent fractures as identified through examination of cores, micrologs, etc. Commercial software can be used to analyze typical fractures observed in FMS (electrical formation microscanner).

To quantify the role of fracture, a rose diagram should be plotted for open fractures. If there is a trend for closed or cemented fracture, it should be included in the analysis. It is ideal to develop correlation that is specific to a field.

2.8.1 Filtering permeability data

One of the most difficult problems in reservoir engineering is the fact that it is practically impossible to extract a sample that represents the reservoir. It is because the representative elemental volume (REV) for petroleum reservoirs is much larger than the core size commonly encountered in a reservoir. Fig. 2.28 highlights this difficulty. If the core size is below REV, all experiments conducted in it would have little relevance to the reservoir. This problem is further compounded in presence of fractures. In addition, coring in fractured formations is performed by avoiding the fractures as much as possible. This is because it is practically impossible to conduct fluid flow tests in a fractured core. For EOR design, this is of particular concern as the nature of fluid flow in a fractured formation is entirely different from that in a homogenous system. Furthermore, flow that follows a certain regime (e.g., stable and stabilized) is likely to be different in presence of fractures. Because currently used instability number formulations do not use the presence of fractures (open or closed), it becomes even more difficult to predict the onset of fingering during an EOR process. However, fingering can lead to catastrophic failure of an EOR scheme.

- Maps
- Production data
- Well completion history
- Waterflood-injection history
- Any other data that may be available

One of the following averages should be used to construct the permeability (k) x net pay (h) correlation.

For parallel flow, arithmetic average should be used:

$$k_{arith} = \frac{\sum_{i=1}^{n} k_i h_i}{\sum_{i=1}^{n} h_i}$$

For series flow, harmonic average should be used:

$$k_{harm} = \frac{\sum_{i=1}^{n} h_i}{\sum_{i=1}^{n} h_i/k_i}$$

For flow without a particular pattern (random flow), geometric average should be used:

$$k_{geom} = \sqrt[n]{k_1 k_2 k_3 \ldots k_n}$$

At this point, well test results should be gathered and history of permeability variation with time should be collected. Quite often well test permeability becomes a function of time. Such behavior gives out useful clues as to the nature of fluid flow in the reservoir. The knowledge of the operators, combined with the construction of geological history and fluid flow behavior all contribute to determining the true nature of the reservoir. For instance, consistent decline in reservoir permeability, as evidenced from well test data can occur because of fracture closure, asphaltene deposition, or formation of gas pockets in the neighborhood of the well. It is important to conduct laboratory tests to determine natural reaction to cores to stress.

To characterize a fractured formation, one has to know:

- The original of formation
- The overall direction of flow
- Fracture distribution and frequency

2.8.2 Total volume estimate

Total volume estimate comes from the following sequential estimates.

1. Calculation based on geologic and seismic data;
2. Confirmation of hydrocarbon through well test (initial acidization and fracturing may be necessary);
3. The concept of net thickness doesn't apply;
4. Saturation cannot be estimated with logs in most cases;
5. Saturations must be confirmed with special core analysis;
6. End points are the only relevant points in a relative permeability curve;
7. Capillary pressure data is the basis for fine tuning total thickness, which is the determining factor for initial gas in place.

2.8.3 Estimates of fracture properties

Total organic content (TOC) changes in shale formations influence VP, VS, density and anisotropy and thus should be detected on the seismic response. To detect it, different workflows have been discussed by Chopra et al. (2012). Rickman et al. (2008) showed that brittleness of a rock formation can be estimated from the computed Poisson's ratio and Young's modulus well log curves. This suggests a workflow for estimating brittleness from 3D seismic data, by way of simultaneous pre-stack inversion that yields IP, IS, VP/VS, Poisson's ratio, and in some cases meaningful estimates of density. Zones with high Young's modulus and low Poisson's ratio are those that would be brittle as well as have better reservoir quality (higher TOC, higher porosity).

Natural fracture distribution map should be an indicator of how induced fracture will behave. Therefore, the data collected on natural fractures should be processed for designing an induced fracture.

Stress data collected during drilling as well as cuttings and well site logs should be all combined to develop enhanced understanding of reservoir fracture properties.

2.8.4 Special considerations for shale

Some of the considerations of reservoir characterization must be altered for shale due to unique features of shale gas. As force increases in relatively unfractured rock, fractures propagate perpendicular to the direction of maximum principal stress. When they encounter a rock layer boundary (which is naturally weaker), the energy forcing the fracture dissipates laterally, making it harder for a fracture to continue across boundaries. In more fractured rock, pre-existing fractures can cause energy forcing a fracture to dissipate in other directions.

Passey et al. (1990) proposed a technique for measuring total organic content (TOC) in shale gas formations. This technique is based on the porosity-resistivity overlay to locate hydrocarbon bearing shale pockets. Usually, the sonic log is used as the porosity indicator. In this technique, the transit time curve and the resistivity curves are scaled in such a way that the sonic curve lies on top of the resistivity curve over a large depth range, except for organic-rich intervals where they would show crossover between themselves.

This compilation begins with the generation of different attributes from the well-log curves. Then, using the cross-plots of these attributes one can identify the hydrocarbon bearing shale zones. Once this analysis is done at the well locations, seismic data analysis is picked up for computing appropriate attributes. Seismically, pre-stack data is essentially the starting point. After generating angle gathers from the conditioned offset gathers, Fatti's equation (Fatti et al., 1994) can be used to compute P- reflectivity, S-reflectivity, and density which depends on the quality of input data as well as the presence of long offsets. Due to the band-limited nature of acquired seismic data, any attribute extracted from it will also be band-limited, and so will have a limited resolution. While shale formations may be thick, some high TOC shale units may be thin. So, it is desirable to enhance the resolution of the seismic data. An appropriate way of doing it is the thin-bed reflectivity inversion (Chopra et al., 2006). Following this process, the wavelet effect is removed from the data and the output of the inversion process can be viewed as spectrally broadened seismic data, retrieved in the form of broadband reflectivity data that can be filtered back to any bandwidth. This usually represents useful information for interpretation purposes. Thin-bed reflectivity serves to provide the reflection character that can be studied, by convolving the reflectivity with a wavelet of a known frequency band-pass. This not only provides an opportunity to study reflection character associated with features of interest, but also serves to confirm its close match with the original data. Further, the output of thin-bed inversion is considered as input for the model based inversion to compute P-impedance, S-impedance and density. Once impedances are obtained, one can compute other relevant attributes, such as the $\lambda\rho$, $\mu\rho$ and VP/VS. These are used to measure the pore space properties and get information about the rock skeleton. Young's modulus can be treated as brittleness indicators and Poisson's ratio as TOC indicator.

CHAPTER 3

Complex reservoirs

3.1 Introduction

Every petroleum formation is complex. However, each of them is treated as a "model," which is amenable to linear equations that are a hallmark of modern engineering practices. In the modern era, every discipline treats its prototype in a simple form, in conformance with Newtonian mechanics and its derivatives. While this modus operandi made it easy to design an engineering operation, it also created a chasm between the reality of nature, which is always complex and nonlinear, and the modeling, which is implemented at the beginning of an engineering project. In petroleum reservoir simulation, the principle of garbage in and garbage out (GIGO) is well known (Rose, 2000). This principle implies that the input data have to be accurate for the simulation results to be acceptable. The petroleum industry has established itself as the pioneer of subsurface data collection (Islam et al., 2000). Historically, no other discipline has taken so much care in making sure input data are as accurate as the latest technology would allow. The recent superflux of technologies dealing with subsurface mapping, real-time monitoring, and high-speed data transfer is the evidence of the fact that input data in reservoir simulation are not the weak link of reservoir modeling. However, this great feat in modern engineering has not been transmitted into an accurate description of the reservoir because all models used to describe the reservoir are linear. Even when they are rendered "nonlinear" by adding a nonlinear term, the connection between the complex prototype and the simplified model is established.

Fig. 3.1 shows the decision-making process. Data, which are essentially objective and incontrovertible, cannot be automatically transformed into information unless the correct model is used to process data. The process is further convoluted by the fact that disinformation ensues if a correct model is not used. Consequently, every decision based on these models is skewed. In this process linear models are just as misleading as the so-called nonlinear version of linear equations, the nonlinearity being created by adding a term to the original linear equation. In this chapter, complex reservoirs are presented in their original form so the discussion of applying a petroleum development operation begins. Fig. 3.1 shows how every phenomenon has to be modeled with only with nonlinear model.

FIG. 3.1 Natural objects cannot be modeled with linear models.

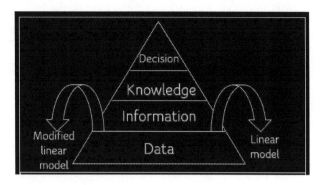

3.2 Complex reservoirs

Every petroleum reservoir is heterogeneous to some degree. Conventionally, it is recognized that when the level of heterogeneity surpasses a certain threshold, the reservoir is called "complex." This complexity can arise either from so-called "structural complexity" or complexity due to oil/gas property changes. Structurally complex reservoirs typically form a distinct class of reservoirs, in which profuse faults and fracture networks, which govern the nature of petroleum trapping and production behavior. This complexity comes from geological features. Historically, these reservoirs have offered a great challenge in extracting oil and gas. These reservoirs offer increasing technical challenges in both discovery and development. Significant progress has been made in recent years in the areas of (i) structural complexity and fault geometry (Jolley et al., 2007):

(i) the detection and prediction of faults and fractures;
(ii) the compartmentalizing effects of fault systems and complex siliciclastic reservoirs; and
(iii) the critical controls that affect fractured reservoirs.

In general, the deeper a reservoir the greater is the level of complexity. As the world demand of petroleum increases, deep petroleum exploration and development have become trendy. As it is nearly impractical to expect any major breakthrough in middle or shallow basins (Tuo, 2002), petroleum exploration turning toward deep basins has become inevitable. After half a century's exploitation in major oilfields across the world, shallow petroleum discoveries tend to be falling sharply (Simmons, 2002). One method is to characterize these formations according to the formation depth and designate them as deep basins depending on the depth. However, the criteria for classifying deep basins also differ somewhat. Representative criteria include 4000 m (Rodrenvskaya, 2001), 4500 m (Barker and Takach, 1992), 5000 m (Samvelov, 1997; Melienvski, 2001), and 5500 m (Manhadieph, 2001; Bluokeny, 2001). Another method to characterize is according to the formation age, i.e., for a given basin, formations older in age and deeper are called deep formations (Sugisaki, 1981). Table 3.1 summarizes the criteria used by different researchers.

A slightly different characterization is proposed by Chinese scholars. Table 3.2 summarizes the criteria used for deep basins. In addition to depth and age, they add formation characteristics to the criteria.

TABLE 3.1 Criteria for deep basins (Pang et al., 2015).

Basis for deep basins	Criteria for deep basins	Targeted area	Researcher and year
Formation depth	>4000 m	Former Soviet Union Caspian Basin	Rodrenvskaya (2001)
	>4500 m		
	>4500 m	Gulf of Mexico, USA	Barker and Takach (1992)
	>5000 m		Samvelov (1997)
		West Siberia Basin, East Siberia Basin	Melienvski (2001)
	>5500 m	South Caspian Basin	Manhadieph (2001)
		Timan-Pechora Basin	Bluokeny (2001)
Formation age	Stratigraphically old formations with largely buried depths	In the USA	Sugisaki (1981)

TABLE 3.2 Criteria of deep basins proposed by Chinese scholars (Pang et al., 2015).

Basis for deep basins	Criteria for deep basins	Targeted area or parameter features
Formation depth	>2500 m	Bohai Bay Basin
	>2800 m	Songliao Basin
	>3500 m	Liaohe Basin
		Bohai Bay Basin
		Bohai Bay Basin
		Yinggehai Basin
	>3500 m	East China basins
	>4500 m	Junggar Basin
		Tarim, Junggar, Sichuan basins
		Sichuan Basin
	>4500 m	West China basins
Formation age & depth	Stratigraphically old with largely buried depths	Varies from basin to basin
Formation characteristics	Formation thermal evolution level	$R_o \geq 1.35\%$
	Formation thermal evolution level or formation pressure	$R_o \geq 1.35\%$ or formation depth overpressure
	Formation thermal evolution level and tightness level	$R_o \geq 1.35\%$ or sandstone formation $\Phi \leq 12\%$, $K \leq 1$ mD, $Y \leq 2$ μm

Pang et al. (2015) point out that deep petroleum formations show 10 major geological features.

(1) While oil-gas reservoirs have been discovered in many different types of deep petroliferous basins, most have been discovered in low heat flux deep basins.
(2) Many types of petroliferous traps are developed in deep basins, and tight oil-gas reservoirs in deep basin traps are arousing increasing attention.
(3) Deep petroleum normally has more natural gas than liquid oil and the natural gas ratio increases with the burial depth.
(4) The residual organic matter in deep source rocks reduces but the hydrocarbon expulsion rate and efficiency increase with the burial depth.
(5) There are many types of rocks in deep hydrocarbon reservoirs, and most are clastic rocks and carbonates.
(6) The age of deep hydrocarbon reservoirs is widely different, but those recently discovered are predominantly Paleogene and Upper Paleozoic.
(7) The porosity and permeability of deep hydrocarbon reservoirs differ widely, but they vary in a regular way with lithology and burial depth.
(8) The temperatures of deep oil-gas reservoirs are widely different, but they typically vary with the burial depth and basin geothermal gradient.
(9) The pressures of deep oil-gas reservoirs differ significantly, but they typically vary with burial depth, genesis, and evolution period.
(10) Deep oil-gas reservoirs may exist with or without a cap, and those without a cap are typically of unconventional genesis.

Complexity in these reservoirs arises due to the fact that the diagenesis process itself is complex. Following phenomena are routinely observed yet none of them can be explained with a linear rock mechanics model.

Over the past decade, six major steps have been made in the understanding of deep hydrocarbon reservoir formation.

(1) Deep petroleum in petroliferous basins has multiple sources and many different genetic mechanisms.
(2) There are high-porosity, high-permeability reservoirs in deep basins, the formation of which is associated with tectonic events and subsurface fluid movement.
(3) Capillary pressure differences inside and outside the target reservoir are the principal driving force of hydrocarbon enrichment in deep basins.
(4) There are three dynamic boundaries for deep oil-gas reservoirs; a buoyancy-controlled threshold, hydrocarbon accumulation limits, and the upper limit of hydrocarbon generation.
(5) The formation and distribution of deep hydrocarbon reservoirs are controlled by free, limited, and bound fluid dynamic fields.
(6) tight conventional, tight deep, tight superimposed, and related reconstructed hydrocarbon reservoirs formed in deep-limited fluid dynamic fields have great resource potential and vast scope for exploration.

Compared with middle-shallow strata, the petroleum geology and accumulation in deep basins are more complex, which overlap the feature of basin evolution in different stages.

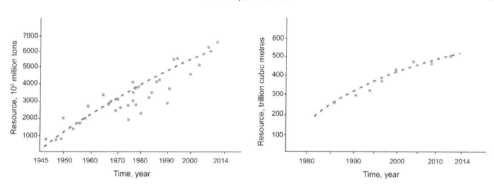

FIG. 3.2 World petroleum resource growth history.

Islam (2014) worked out the science behind unconventional reservoirs, most of which are complex for reasons of rock or fluid property heterogeneities. In 2015, Pang et al. recommended the following four research tasks in order to understand complex reservoirs:

(1) identification of deep petroleum sources and evaluation of their relative contributions;
(2) preservation conditions and genetic mechanisms of deep high-quality reservoirs with high permeability and high porosity;
(3) facies feature and transformation of deep petroleum and their potential distribution; and
(4) economic feasibility evaluation of deep tight petroleum exploration and development.

According to USGS and World Petroleum Investment Environment Database, from 1945 to 2014, the world's normal petroleum resource has increased from 96 billion tons in 1945 to 630 billion tons in 2014, the annual average increase being as high as 8.06% (Fig. 3.2A), and the natural gas resource has also increased from 260 trillion m^3 in 1986 to 460 trillion m^3 in 2013, the annual average increase being as high as 2.85% (Fig. 3.2B). According to data provided by Kutcherov et al. (2008), more than 1000 hydrocarbon fields have been developed at depths of 4500–8103 m, the original recoverable oil reserve of which contributes 7% of the world's total amount and the natural gas reserve makes up 25%. According to IHS data, as of 2010, for the 4500–6000 m deep hydrocarbon fields in the world, the proved recoverable residual oil reserve is 83.8 billion tons or 35.5% of the total recoverable oil reserve, and the natural gas is 65.9 billion ton oil equivalent or 44.4% of the total productive natural gas reserve; for the 6000 m or deeper hydrocarbon fields in the world, the proved recoverable residual oil reserve is 10.5 billion ton or 4.45% of the total productive oil reserve, and the natural gas is 7 billion ton oil equivalent or 4.7% of the total productive natural gas reserve (Fig. 3.3).

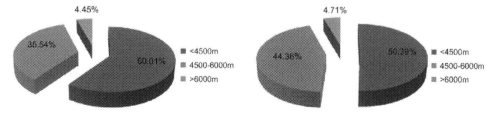

FIG. 3.3 World recovery of oil and gas various depths.

3.3 Fracture mechanics in geological scale

Currently, three major approaches are used to simulate fluid flow through naturally fractured reservoirs, which include continuum, dual porosity/dual permeability, and flow through discrete fracture approaches. In the single continuum approach, the fractured medium is divided into several representative volumes, and bulk macroscopic values of reservoir properties varying from point to point are averaged over the volume, which is often known as blocked-based permeability tensors (Lough et al., 1998; Sarkar et al., 2002; Teimoori et al., 2005). While Lough et al. (1998) calculated the effective permeability tensors by considering flow through matrix and fractures with regular fracture pattern, Sarkar et al. (2002) and Teimoori et al. (2005) on the other hand present comprehensive methodologies for estimating permeability tensor for arbitrarily oriented and interconnected fracture systems.

In the dual continuum approach, the reservoir is divided into two major parts: fractures and matrix. The Warren and Root (1963) model conceptualizes fractured reservoirs with stacked sugar cubes. In this approach, the fractures provide the main flow paths, while the matrix acts as a source of fluid. The fluid transfer between the fractures and the matrix is defined based on transfer functions. Duguid and Lee (1977), Pruess (1985), Gilman (1986), and Bourbiaux et al. (1999) introduced a range of different matrix/fracture transfer functions to simulate the fluid flow in large scales. Following this, a significant number of studies have been carried out both in analytical and numerical frameworks using the dual-porosity approach (Choi et al., 1997; Pride and Berryman, 2003; Gong et al., 2008). Landereau et al. (2001) proposed a new technique for simulating flow in fractured porous media like rock masses. The proposed technique has been created to separate the contributions from the rock matrix and from the fracture systems (which are also treated as an equivalent continua), the so-called dual (double) porosity models, dual (double) permeability models, and dual (double) continuum models, to simplify the complexity in the fracture-matrix interaction behavior and to partially consider the size effects caused mainly by the fractures. Fig. 3.4 shows the advective mass exchange between fractures and porous matrix, as solutes migrate within matrix pores (driven by a regional gradient) and encounter flow in transverse fractures (Fig. 3.5).

It has been shown that the advective transport component in more permeable matrix blocks (e.g., porous sandstones) results in additional solute transfer between fractures and matrix, which is not typically taken into account by dual-porosity concepts (Cortis and Birkholzer, 2008). Therefore, Noetinger and Estebenet (2000) proposed a new technique called the continuous time random walk (CTRW) method. This method is an alternative to the classical dual-porosity models for modeling and upscaling flow and/or solute transport in fractured rock masses. It was originally developed to describe electron hopping in heterogeneous physical systems (Scher and Lax, 1973).

The network of fractures becomes more complicated when there are vugs present in addition to fractures. Fractures and vugs can have profound effects on the porosity exponent (m) and calculated water saturation (S_w) of carbonate rocks. Proper prediction of m in reservoirs avoids the overestimation of S_w commonly caused by the presence of fractures and also avoids the underestimation of S_w commonly caused by vuggy or oomoldic porosity.

3.3 Fracture mechanics in geological scale

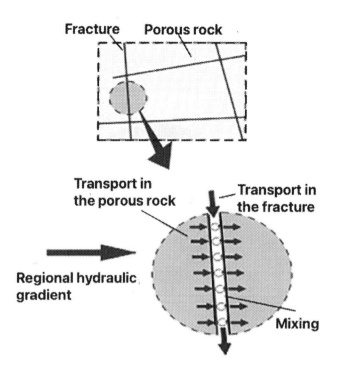

FIG. 3.4 Schematic showing the mixing between fracture and matrix flow as a result of convective transport in the matrix.

The original dual-porosity equation, developed by Aguilera (1974, 1976) and corrected in Aguilera and Aguilera (2003), is as follows:

$$m = \frac{\log\left(\phi_f + \dfrac{1-\phi_f}{\phi_b^{-m_b}}\right)}{\log \phi} \tag{3.1}$$

where ϕ is the total porosity, ϕ_b is the porosity of the bulk rock, m_b is the porosity exponent of the bulk rock, and ϕ_f is the fracture porosity in relation to the total volume. The value of fracture m (m_f) is not used and is implicitly assumed to be 1.0. It means that the fractures are assumed to contribute in parallel to the whole-rock conductivity. The use of parallel resistance implies that the fractures themselves are parallel to the current direction, which is rarely the case. Since fractures inclined to current would give straight line conductance paths as opposed to the tortuous paths in the bulk porosity, m_f should usually be low but not necessarily 1.0. The following equation emerges:

$$m = \frac{\log\left(\phi_f^{m_f} + \dfrac{1-\phi_f^{m_f}}{\phi_b'^{-m_b}}\right)}{\log \phi} \tag{3.2}$$

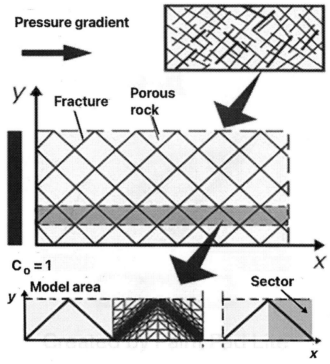

FIG. 3.5 Discrete representation in the idealized fracture network. Flow and transport are from left to right, with a concentration boundary condition on the left side. The total domain is 100 sectors long.

where ϕ_b', according to Aguilera, is the "matrix block porosity affected by m_f" and is defined by the equations

$$\phi_b' = \frac{\phi - \phi_f^f}{1 - \phi_f^f} \tag{3.3}$$

and

$$m = \frac{\log\left[\phi_{nc} + (1 - \phi_{nc}) \cdot \phi_b^{-m_b}\right]}{-\log \phi} \tag{3.4}$$

Note that all of the relationships earlier have been simplified from the original equations by replacing the product of partitioning coefficient and porosity ($v_f\phi$) with the fracture porosity (ϕ_f). For a vuggy reservoir, the following relationship holds:

3.3 Fracture mechanics in geological scale

$$m = \frac{\log\left[\phi_{nc} + (1 - \phi_{nc}) \cdot \phi_b^{-m_b}\right]}{-\log \phi} \tag{3.5}$$

where ϕ_{nc} is the volume fraction of nonconnected vugs relative to the whole rock. Eq. (3.5) has an embedded assumption that nonconnected vugs and bulk rock respond to the current flow as resistors in series. As in the fracture equations, the product of partitioning coefficient and porosity ($\nu_{nc}\phi$) has been replaced by the nonconnected vug porosity (ϕ_{nc}). Note that there is no porosity exponent for the nonconnected vugs.

Archie's law is written as

$$R_{0b} = \frac{R_w}{\phi^{m_b}} \tag{3.6}$$

where R_{0b} is the bulk resistivity and R_w is the water resistivity. Now that the bulk rock has been defined, the enclosing fracture system must be defined. To define m_f other than 1.0, we need a relationship that contains m and that can have nonzero matrix conductivity. Archie's law cannot be used, but effective-medium theory provides just such a relationship. The following is the HB resistivity equation:

$$\phi = \left(\frac{R_w}{R_0}\right)^{\frac{1}{m}} \cdot \left(\frac{R_0 - R_r}{R_w - R_r}\right) \tag{3.7}$$

where R_0 is the whole-rock resistivity and R_r is the matrix resistivity. Eq. (3.7) can be used to define the bulk rock-fracture system as follows:

$$\phi_f = \left(\frac{R_w}{R_0}\right)^{\frac{1}{m_f}} \cdot \left(\frac{R_0 - R_{0b}}{R_w - R_{0b}}\right) \tag{3.8}$$

where R_{0b} is the resistivity for the bulk rock. This derivation assumes that an expression originally derived for granular material (Eq. 3.8) can be used to describe fractures, which is typical of a dual-porosity formulation.

To calculate composite m of the whole rock, we can use Archie's law again:

$$\phi^m = \frac{R_w}{R_0} \tag{3.9}$$

When Eqs. (3.6), (3.8), (3.9) are combined and simplified, we get the following equation:

$$\phi_f = \left[\frac{\phi^{\frac{m}{m_f} - m} \cdot \phi_b^{m_b} - \phi^{\frac{m}{m_f}}}{\phi_b^{m_b} - 1}\right] \tag{3.10}$$

The following is the relationship for calculating m_f:

$$m_f = \frac{\log\left(\phi_f \cdot \sin^2 \theta\right)}{\log \phi_f} \tag{3.11}$$

where θ is the angle between the direction of current flow and the normal to the fracture plane. For multiple fracture directions, Eq. (3.11) can be extended to

$$m_f = \frac{\log\left(\phi_f \cdot \sum_{i=1}^{n} V_i \sin^2\theta_i\right)}{\log \phi_f} \quad (3.12)$$

where V_i is the volume fractions relative to ϕ_f of each set of fractures and θ_i are the respective angles that the normal to each fracture set makes to the current direction. Eq. (3.12) does not take into account what happens at fracture intersections, but it is accurate for ϕ_f at or below 0.1—an extremely large value for fracture porosity (see Appendix A for details).

Vugs and oomoldic porosity present just such a case if the particles, in this case, are the water-filled vugs and the surrounding bulk rock is the enclosing medium. The following is an adaptation of Eq. (3.7) to represent vuggy porosity:

$$1 - \phi_v = \left(\frac{R_{0b}}{R_0}\right)^{\frac{1}{m_v}} \cdot \frac{R_0 - R_w}{R_{0b} - R_w} \quad (3.13)$$

where ϕ_v is the vug porosity with respect to the whole rock and m_v is its exponent. Substitution of R_{0b} in Eq. (3.13) by Eq. (3.6) (Archie's law) yields the effective medium dual-porosity equation for vugs:

$$1 - \phi_v = \left(\frac{\phi^m}{\phi_b^{m_b}}\right)^{\frac{1}{m_v}} \cdot \frac{\phi^{-m} - 1}{\phi_b^{-m_b} - 1} \quad (3.14)$$

As in the case of the derivation for fractures, the resistivities drop out, leaving a relationship independent of resistivity. In addition, as with fractures, Eq. (3.14) cannot be solved directly for m. Accordingly, calculation considerations for this relationship are similar to the considerations discussed for the effective-medium fracture relationship (Eq. 3.10).

More recently, the concept of triple porosity has been developed, one for the matrix, one for the fracture, and one for the vugs:

$$1 - \phi'_v = \left(\frac{\phi_{bv}^{m_{bv}}}{\phi_b^{m_b}}\right)^{\frac{1}{m_v}} \cdot \frac{\phi_{bv}^{-m_{bv}} - 1}{\phi_b^{-m_b} - 1} \quad (3.15)$$

where $\phi'_v = \phi_v/(1 - \phi_f)$, $\phi_{bv} = \phi'_v + \phi_b(1 - \phi'_v)$ and m_{bv} is the composite porosity exponent. The following modified Eq. (3.10) was then used on the results:

$$\phi_f = \frac{\phi^{\frac{m}{m_f} - m} \cdot \phi_{bv}^{m_{bv}} - \phi^{\frac{m}{m_f}}}{\phi_{bv}^{m_{bv}} - 1} \quad (3.16)$$

When doing the calculations, the following equation from Aguilera and Aguilera (2003, their equation A-11) is useful:

$$\phi = \phi_b\left(1 - \phi_f - \phi_v\right) + \phi_f + \phi_v \quad (3.17)$$

3.3.1 Meaningful modeling

For a modeling process to be meaningful, it must fulfill two criteria, namely, the source has to be true (or real), and the subsequent processing has to be true (Islam et al., 2000). For conventional or noncomplex reservoirs, having meaningful data is not a concern. However, it is not the case for unconventional or complex reservoirs (Islam et al., 2018). For instance, most commonly used well log data do not register useful information for unconventional reservoirs that do not allow invasion of sonic signals, mud filtrate, etc., invalidating some of the well logging techniques (Islam, 2014). Processing data are also of concern. Complex reservoirs require special filtering tools that are not readily available.

For reservoir simulation to yield meaningful results, the following logical steps have to be taken:

(1) Collection of data with constant improvement of the data acquisition technique. The data set to be collected is dictated by the objective function, which is an integral part of the decision-making process. Decision-making, however, should not take place without the abstraction process. The connection between the objective function and data needs constant refinement. This area of research is one of the biggest strengths of the petroleum industry, particularly in the information age. Each of the data acquisition tools must be re-evaluated for applicability in unconventional reservoirs.
(2) The gathered data should be transformed into information so that they become useful. With today's technology, the amount of raw data is so huge; the need for a filter is more important than ever before. However, it is important to select a filter that doesn't skew the data set toward a certain decision. Mathematically, these filters have to be nonlinearized (Islam et al., 2010). While the concept of nonlinear filtering is not new, the existence of nonlinearized models is only beginning to be recognized.
(3) Information should be further processed into "knowledge" that is free from preconceived ideas or a "preferred decision." Scientifically, this process must be free from information lobbying, environmental activism, and other forms of bias. Most current models include these factors as an integral part of the decision-making process (Eisenack et al., 2007), whereas a scientific knowledge model must be free from those interferences as they distort the abstraction process and inherently prejudice the decision-making. Knowledge gathering essentially puts information into the big picture. For this picture to be distortion-free, it must be free from nonscientific maneuvering.
(4) Final decision-making is knowledge-based, only if the abstraction from (1) to (4) has been followed without interference. The final decision is a matter of yes or no (or true or false or 1 or 0), and this decision will be either knowledge-based or prejudice-based. Fig. 3.6 shows the essence of knowledge-based decision-making.

The process of a phenomenal- or prejudice-based decision-making is illustrated by the inverted triangle, proceeding from the top down (Fig. 3.7). The inverted representation stresses the inherent instability and unsustainability of the model. The source data from which a decision eventually emerges already incorporate their own justifications, which are then massaged by layers of opacity and disinformation.

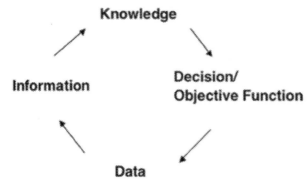

FIG. 3.6 The knowledge model.

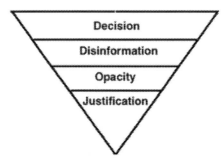

FIG. 3.7 The phenomenal model. *Based on Islam, M.R., Hossain, M.E., Mustafiz, S., Mousavizadegan, S.H., Abou-Kassem, J.H., 2016. Advanced Reservoir Simulation: Toward Developing Reservoir Emulators. John Wiley & Sons, Inc./Scrivener Publishing LLC, Hoboken, NJ/Salem, MA.*

3.3.2 Original oil in place

Typically, the original oil in place is calculated from data on initial saturation distribution and rock properties. Although these data are subjective, petroleum engineers are not involved in qualifying these data. However, these data are the first ones that are manipulated/adjusted during the history matching. It is important to understand that engineers must be involved in characterizing reservoirs before embarking on a modeling project.

Historically, the majority of additions to oil and gas reserves are attributed to the growth of existing fields and reservoirs. In fact, from 1978 to 1990, the growth of known fields in the United States accounted for more than 85% of known additions to proven reserves (Root and Attanasi, 1993; McCabe, 1998). Thus, field growth and reserve growth are essentially synonymous with discussions of domestic resources. The exception occurred when unconventional resources became exploitable. Only recently, it has come to light that conventional reserve growth estimate models are flawed and not applicable to many scenarios, including unconventional formations (Islam, 2014). While evaluating the nature of growth in fields requires an understanding of both geologic and nongeologic factors that affect growth estimations, geology is the pivotal factor in determining the accumulation, quality, and producibility of petroleum resources. Fields may grow when

(1) additional geologic data on existing reservoirs become available and are used to identify new reservoirs or to guide infill drilling;
(2) there are annual updates of reserves data;
(3) field boundaries are extended;
(4) recovery technology is improved;
(5) nongeologic factors such as economics, reporting policies, or politics favor expanded production and development.

Past reserve estimates and characterization both used mathematics and not science as the basis. Attanasi et al. (1999) wrote:

> ...the modeling approach used by the USGS (U.S. Geological Survey) to characterize this phenomenon is statistical rather than geologic in nature.

Islam et al. (2000) discussed in detail the shortcomings of these statistical models and characterized them as unscientific. Yet, volumetric estimates of reserve growth are calculated by using these mathematical approaches and large databases that record field reserves through time. Crovelli and Schmoker (2001) present details of various methods used to estimate reserve growth. The application of these models to unconventional gas is of particular concern. It's because one of the most important bases for statistical models is the past history, which is practically nonexistent for unconventional reserves.

Geologists tend to think in terms of entire reservoirs, in some cases down to facies level, whereas reservoir engineers deal with measurements at the wellbore. To fully understand the geologic factors that affect growth in reserves, this gap in investigative approaches must be bridged. It has become increasingly important to integrate different scales and different observational techniques as secondary and tertiary recovery methods are applied more frequently in mature petroleum provinces.

Fluid flow pathways, governed predominantly by rock porosity and permeability, are a reflection of heterogeneities of varying scales within a reservoir. Because these reservoir heterogeneities are fundamentally geologic in nature (Dyman et al., 2000), an adequate understanding of the reservoir architecture, obtained through evaluation of geologic, engineering, or production data, or a combination of these data sets, can provide the basis for scientific characterization of a reservoir.

Previously, extensive exploration has kept the growth part active. In the United States, however, the main contributor to reserve growth has been infill drilling and enhanced oil recovery. In the infill drilling scheme, one must acknowledge that 90% of the new wells are horizontal. This means the role of horizontal well technology has been embedded in this reserve growth. However, despite the best efforts, the reserve will decline, particularly for the "homogeneous" formations.

For such reservoirs, a vast amount of unconventional or otherwise nonrecognized reserves are associated and are accessible readily. This has been the case for shale gas and tight gas, as has been evidenced in the recent gas boom in the United States. However, the use of horizontal wells and hydraulic fracturing as the sole mode of reservoir development will lead to similar stagnation as in conventional reservoirs (Fig. 3.8). This can be given a boost by accessing reserve that was previously considered to be part of conventional reservoirs and was excluded through the use of "cutoff" points. The dotted line in Fig. 3.9 shows how immediate adjustment can be invoked. This can be followed by subsequent use of enhanced oil/gas recovery schemes.

FIG. 3.8 Three phases of conventional reserves.

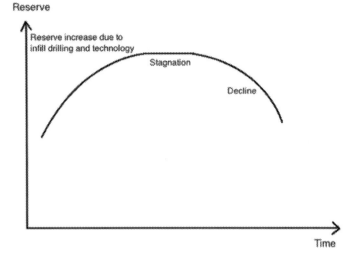

FIG. 3.9 Unconventional reserve growth can be given a boost with scientific characterization.

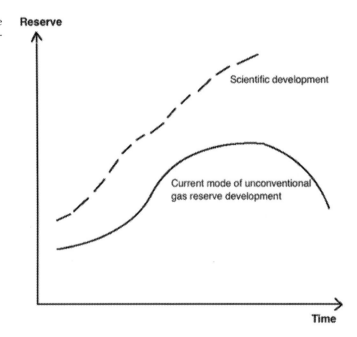

3.3.3 Total volume estimate

Total volume estimate comes from the following sequential estimates:

1. Calculation based on geologic and seismic data.
2. Confirmation of hydrocarbon through well test (initial acidization and fracturing may be necessary).

3. The concept of net thickness doesn't apply.
4. Saturation cannot be estimated with logs in most cases.
5. Saturations must be confirmed with special core analysis.
6. End points are the only relevant points in a relative permeability curve.
7. Capillary pressure data are the basis for fine-tuning total thickness, which is the determining factor for initial gas in place.

3.3.4 Quantitative measures of well production variability

Production history offers a powerful tool for the scientific characterization of a formation. It is not because it offers a refinement of statistical tools or in history matching but because any history is the evidence that can be used to refine a scientific model. Fishman et al. (2008) produced a detailed analysis of production history. This analysis was aimed at identifying the origin of both fluid and rock systems. This report (Fishman et al., 2008) compared historical well production data of the five formations by use of proprietary information. In addition, it considered data from two specific reservoir categories in the Ellenburger Group, which are based on gross geologic differences, to evaluate the possible intraformational variability in production within that formation. Because in most wells, production declines exponentially or hyperbolically as a function of time, cumulative production from older wells (those for which current monthly production is <10% of initial monthly production) asymptotically begins to approximate ultimate recovery. This is typical of conventional reserve analysis. In such wells, variations in cumulative production reflect variations in the volume of reservoir rocks accessed by the wellbore. The slopes of the probability distributions for cumulative production (Fig. 3.10) are direct indicators of the variability as shown by the data set. For example, steeper slopes reflect greater production heterogeneity, whereas a horizontal line represents uniform production characteristics. A dimensionless parameter that is proportional to the slopes of the four probability distributions would provide a quantitative numerical representation of production heterogeneity. Such a parameter, referred to here as a variation coefficient (VC), can be calculated by using a measure of the dispersion (range) of the data set divided by a measure of central tendency such as the mean or the median (Fishman et al., 2008).

The different parts of the distribution behave differently—that is, a single straight line fit does not adequately describe the behavior of the entire distribution of production data. Of interest from a conventional perspective is the central part of the distribution because it represents production from the vast majority of wells. Extreme production behavior, categorized by wells in the upper 5% and lower 20% of the production distribution, makes excellent candidates for unconventional recovery. They either include old wells that would invariably have vast gas content that is deemed uneconomical with conventional analysis, or they include low-permeability formations, suitable for unconventional gas development.

Fig. 3.10 lists wells that were sorted by production from lowest to highest and subdivided into two size classes: a central class representing a productive range of 20%–60% along with the distribution and an upper class representing a productive range of 80%–95%.

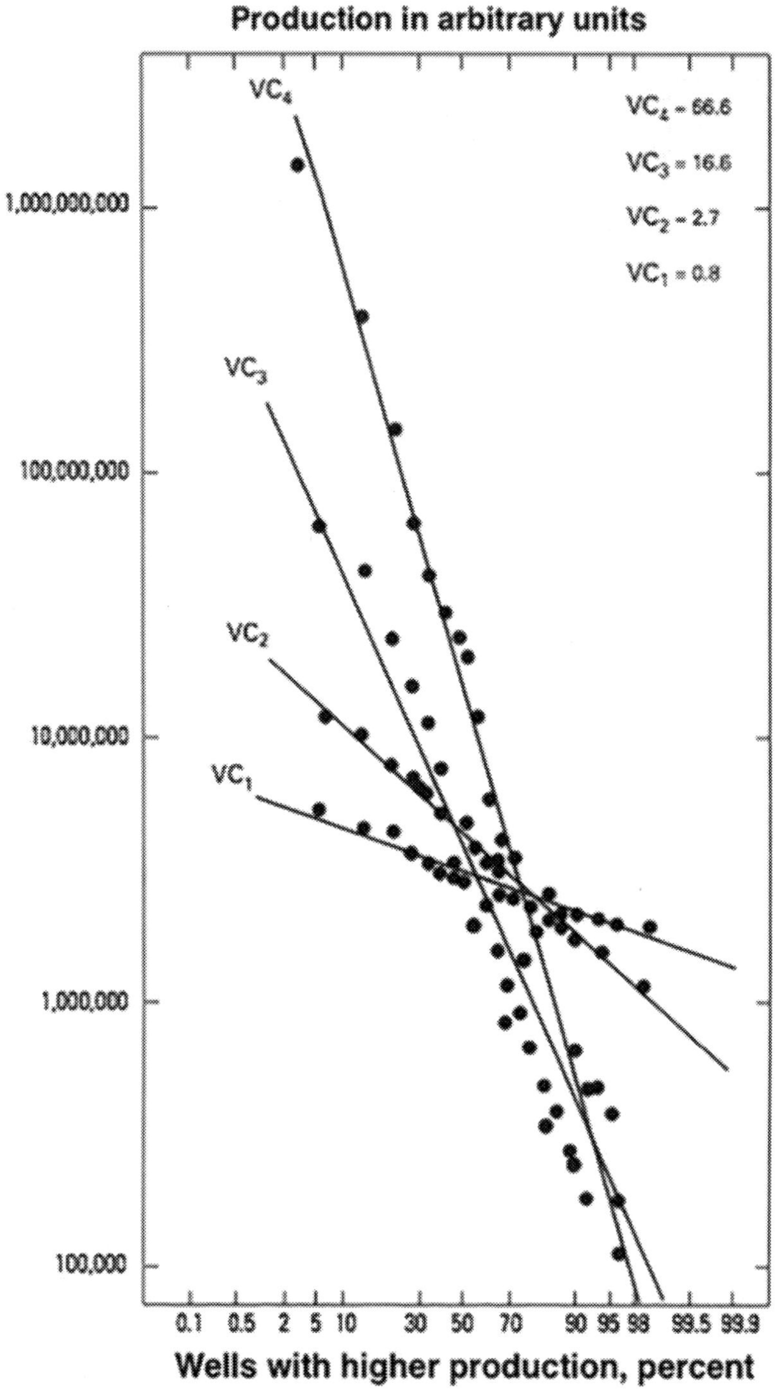

FIG. 3.10 Probability distributions for production from wells of an oil or gas field (distributions based on hypothetical data—peak monthly production, peak yearly production, or cumulative production). Each point represents a well, and four fields (VC1–VC4) are depicted. In this type of plot, lognormal distributions plot as straight lines, and steeper slopes of lines correspond with a greater range of products and thereby greater production variability. The variation coefficient VC = (F5 − F95)/F50 provides a dimensionless numerical value for the variability of each data set and its value increases as the slope increases. *From Fishman, N.S., Turner, C.E., Peterson, F., Dyman, T.S., Cook, T., 2008. Geologic controls on the growth of petroleum reserves. In: U.S. Geological Survey Bulletin 2172–I, 53 pp.*

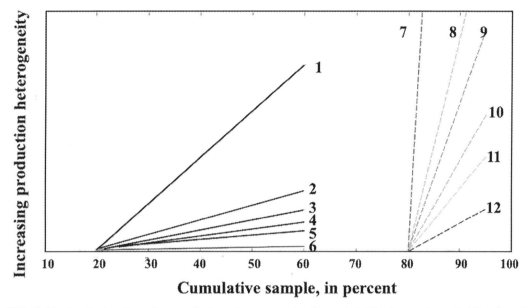

FIG. 3.11 Production data of gas wells in various categories (Legend: 1: Ellenburger, karst; 2: Minnelusa; 3: Wasatch; 4: Ellenburger, platform; 5: Morrow; 6: Frio; 7: Ellenburger, karst; 8: Minnelusa; 9: Ellenburger, platform; 10: Morrow; 11: Frio).

Fig. 3.11 shows fields producing oil from the reservoirs representing the (1) fluvial category of the Frio Formation; (2) incised valley-fill category of the Morrow Formation; (3) Green River, source category of the Wasatch Formation; (4) Minnelusa category of the Minnelusa Formation, and both the (5) platform and karst categories of the Ellenburger Group.

3.3.5 Reservoir heterogeneity

The degree of interconnection between the pores of an oil reservoir is rarely evenly distributed due to nonuniformity of pore size, which gives rise to disordered and complex reservoir fluid flow behavior. Geologically speaking, this is known as the phenomenon of heterogeneous permeability that can manifest different individual layers, forming different homogeneous layers within the oil reservoir with different permeability values.

Most consolidated formations are highly fractured under reservoir conditions. While the presence of fractures makes them ideal for oil and gas recovery because of their very high permeability, the fractures are sealed with secondary cementing activities. Minor porosity of secondary origin occurs locally in sandstones and resulted from the removal of authigenic mineral cements and, to a lesser extent, detrital framework grains. In carbonate-cemented samples, evidence of dissolution includes corrosive contacts between successive carbonate phases and relict cement in pores. Carbonate dissolution features are also observed along the margins of some fractures. Collectively, the dissolution features in sandstones indicate that carbonate cements were previously more widespread before they were partially or extensively dissolved.

Most consolidated formations have fractures in both macroscopic and microscopic scales. The vast majority of these fractures are open without secondary mineralization. The aperture widths commonly exceed 30 μm.

An important characteristic of these fractures is that they typically form a dense network that is highly visible on wetted, slabbed rock surfaces if the host sandstones and siltstones have high residual oil saturations. Such fractures are generally absent in rocks that have little or no residual oil. Often, fractures are V-shaped and occluded with pyrite and fine- to coarse-crystalline calcite cement. These fractures, which tend to be small, resemble fluid-escape structures. Such cementation with quartz and calcite can lead to lowering of permeability below the matrix permeability. These types of formations exhibit productivity (or injectivity) lower than one would expect from the core and log analysis.

The presence of secondary cementing activities makes the porosity-permeability correlation skewed, away from matrix permeability. It is, however, the matrix permeability that is routinely reported in core tests. For a fractured formation, matrix permeability has little to do with the permeability of the reservoir. The permeability having a dimension of L^2, a correlation between permeability, and porosity turns out to be linear, as long as the porous medium is homogeneous. This linearity no longer holds if the formation is fractured, as shown in Fig. 3.12. If the fractures are open, the medium will have markedly higher permeability than the porosity correlation would indicate. This is of consequence because log analysis usually is performed to determine the point porosity of a core. This is later transformed into permeability values using one of the numerous correlations available in the literature. Once the transformation is performed, little attention is paid to the origin of these permeability values. Similarly, core tests and specialized core tests are performed on the homogeneous section of the core, leading to the determination of petrophysical properties that have little relevance to the reservoir. Fig. 3.13 shows a typical correlation between porosity and log (permeability) of a homogeneous formation as compared with the ones that have

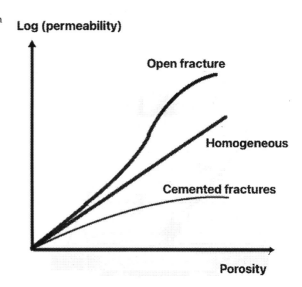

FIG. 3.12 Permeability vs. porosity correlation depends largely on the nature of heterogeneity.

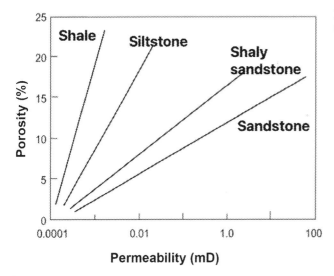

FIG. 3.13 Correlation of porosity vs. permeability for various types of formation.

fractures. An open fracture increases the reservoir permeability drastically, particularly for the median porosity range. On the other hand, this trend is reversed if the fractures are closed due to secondary cementing phenomena.

In terms of anisotropy, it can be invoked due to the presence of fractures. The particular orientation of fractures would skew the directional permeability, and the magnitude would depend on the aperture size, secondary cementation, and aperture length. Vertical permeability is affected by the extent of vertical fractures and if they are open or cemented.

Fig. 3.14 shows how porosity is correlated with permeability for various types of formations. Fig. 3.15 adds correction factors as a function of fracture frequency. This graph applies to open fractures.

Fig. 3.13 shows the trend of vertical permeability (k_v) over horizontal permeability (k_h) with fracture frequency. While fracture frequency is an important factor in determining the anisotropy, the nature of fractures is crucial for determining the magnitude of the permeability ratio. When hundreds of wells and kilometer squares of area are considered to estimate flow from reservoirs containing billions of barrels, the accuracy of k_v/k_h analysis can mean a difference of millions of dollars every day throughout the production history.

The effects of stratification and heterogeneity can be distinct in different reservoirs, affecting various parameters such as capillary pressure, relative permeability, and mobility ratios. The presence of anisotropy and heterogeneity in a reservoir affects the displacement of the native fluids by the injected fluid. Channeling of the solvent through high-permeability regions reduces the storage and displacement efficiency of the displacing solvent. In addition, it can offset viscous fingering through perturbation in case the mobility ratio is not favorable or gravity stabilization is not a dominant mechanism. In case WAG is being considered, heterogeneity and anisotropy have become the most dominant factor that affects oil recovery. It is so because they control the injection and sweep patterns, as well as vertical and areal sweep, viscous fingering, gravity stabilization, and dispersive forces. In addition, horizontal wells

FIG. 3.14 Improvement factor due to open fractures.

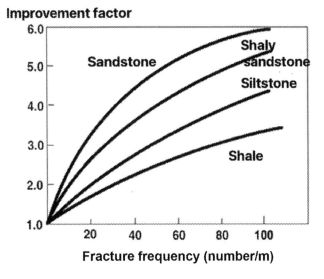

FIG. 3.15 The effect of fractures on k_v/k_h.

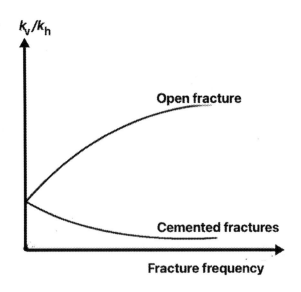

play a role different from vertical wells, and all calculations pertaining to vertical wells become irrelevant.

In the case of water or gas injection or any other EOR scheme, heterogeneity has a tremendous impact. To begin with, the location of trapped oil and the critical capillary number for mobilization of trapped oil would be different for different pore distribution, which is controlled by fracture characteristics of a reservoir. Fig. 3.16 shows how different residual saturations would emerge for different pore distributions.

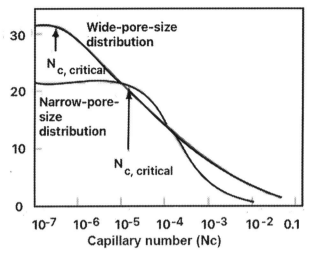

FIG. 3.16   Pore size can be affected by fracture distribution and thereby impact residual oil mobilization.

To include the influence of fracture and fracture distribution, one must characterize fractures properly. It involves determining the frequency and orientation of prominent fractures as identified through examination of cores, micrologs, etc. Commercial software can be used to analyze typical fractures observed in electrical formation microscanner (EMS).

To quantify the role of fracture, a rose diagram should be plotted for open fractures. If there is a trend for closed or cemented fracture, it should be included in the analysis. It is ideal to develop a correlation that is specific to a field.

An example of the Rose diagram is shown in Fig. 3.17. It is a plot with each grouping of data being a petal of the rose. One starts the plot from a center point and draws a line outward (compass direction using a protractor) a distance that matches the number of recorded joints. A Rose diagram includes both frequency and orientation with the assumption that all fractures have the same dimension. The Rose diagram gives one the dominant orientation of fractures. This is of utmost importance in designing injection production strategies. It is useful to catalog core data, along with microlog information for each well. This information then can create isofrequency maps for the reservoir. It is useful for reservoir simulation.

The idea is to transit from macropore structure to a scalable model of the reservoir. Such a model is essential to conduct useful reservoir modeling studies. This was done for the Weyburn CO_2 project—the project that is noted as the largest CO_2 sequestration project in history. Fig. 3.18 shows an example from the Weyburn project.

3.3.6   Nonlinear filtering permeability data

One of the most difficult problems in reservoir engineering is the fact that it is practically impossible to extract a sample that represents the reservoir. It is because the representative elemental volume (REV) for petroleum reservoirs is much larger than the core size commonly

206 3. Complex reservoirs

FIG. 3.17 Rose diagram helps quantify the role of fractures.

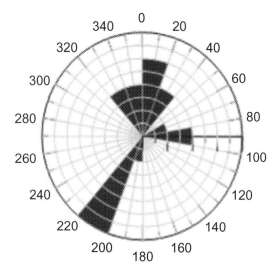

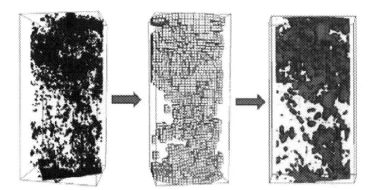

FIG. 3.18 Transiting from macropore scale to an initial reservoir model, as experienced in Weyburn project of Canada.

encountered in a reservoir. Fig. 3.18 highlights this difficulty. If the core size is below REV, all experiments conducted in it would have little relevance to the reservoir. This problem is further compounded in the presence of fractures. The presence of fracture increases the size of REV and places a cored sample squarely in the oscillating region of Fig. 3.19. In addition, coring in fractured formations is performed by avoiding the fractures as much as possible. This is because it is practically impossible to conduct fluid flow tests in a fractured core. For EOR design, this is of particular concern as the nature of fluid flow in a fractured formation is entirely different from that in a homogeneous system. Furthermore, the flow that follows a certain regime (e.g., stable and stabilized) is likely to be different in the presence of fractures. Because currently used instability number formulations do not use the presence of

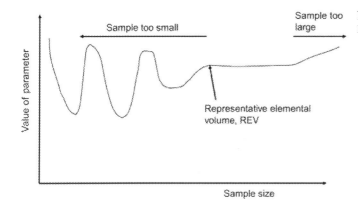

FIG. 3.19 REV for a reservoir is much larger than the core samples collected.

fractures (open or closed), it becomes even more difficult to predict the onset of fingering during an EOR process. However, fingering can lead to the catastrophic failure of an EOR scheme.

To avoid this problem, filtering should be performed. Any such filter has to be custom-designed for a specific field, if not specific well. For that, the following data should be collected:

- Logs.
- Core samples.
- Production data.
- Well completion history.
- Waterflood-injection history.
- Any other data that may be available.

At first, the full 3D schematic of the reservoir should be constructed. The idea is to transit from microscopic scale to reservoir scale, which involves developing the scaling laws.

This schematic is not quantitatively accurate but represents the overall trend. To decide on the schematic and the type of flow that is expected in the reservoir, geophysicists, geologists, and engineers must work together.

Based on that, one of the following averages should be used to construct the permeability (k) × net pay (h) correlation.

For parallel flow, the arithmetic average should be used:

$$k_{arith} = \frac{\sum_{i=1}^{n} k_i h_i}{\sum_{i=1}^{n} h_i} \qquad (3.18)$$

For series flow, the harmonic average should be used:

$$k_{harm} = \frac{\sum_{i=1}^{n} h_i}{\sum_{i=1}^{n} h_i/k_i} \qquad (3.19)$$

For flow without a particular pattern (random flow), the geometric average should be used:

$$k_{geom} = \sqrt[n]{k_1 k_2 k_3 \ldots k_n} \qquad (3.20)$$

At this point, well test results should be gathered, and a history of permeability variation with time should be collected. Quite often, well test permeability becomes a function of time. Such behavior gives out useful clues as to the nature of fluid flow in the reservoir. The knowledge of the operators, combined with the construction of geological history and fluid flow behavior all contribute to determining the true nature of the reservoir. For instance, a consistent decline in reservoir permeability, as evidenced from well test data, can occur because of fracture closure, asphaltene deposition, or the formation of gas pockets in the neighborhood of the well. It is important to conduct laboratory tests to determine the natural reaction to cores to stress. Fig. 3.20 shows an example of such a test. In this laboratory simulation with artificially consolidated cores, the original stress corresponds to the initial conditions of a reservoir, which is subject to little stress because of the overall balance of forces. This is expected because, over geological time, any reservoir reaches a state that can be termed a steady state in conventional terminology. This stress is increased over time as more and more fluid is extracted from the reservoir. During this time, original permeability can be reduced to <50% of the initial permeability. Many field histories support this observation. For instance, in several of the giant oil fields of Hassi Messaoud, Algeria, a similar decline in well test permeability has been observed. The decline is steeper for more heterogeneous zones, the ones that have profuse fracturing.

As can be seen in Fig. 3.20, the core is crushed if the axial stress is very high (exceeding 100 MPa), resulting in a sudden increase in permeability. Even then, however, original permeability is not restored. In theory, such an occurrence of partial restoration of permeability can occur in the field. Several fields report such behavior. Almost all of these have fractures with an intense network of secondary cementation.

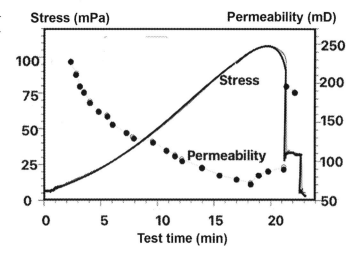

FIG. 3.20 Laboratory test results under an overburden pressure of 50 MPa.

To characterize a fractured formation, one has to know the following:

- The origin of the formation
- The overall direction of flow
- Fracture distribution and frequency

One of the most useful tools for designing local filters is the compilation and plotting of hk from well tests and hk from cores. In the case of relatively homogeneous formation, the correlation between the two will follow the trend of a 45-degree straight line. Fig. 3.21 demonstrates this point. On the other hand, if there are open fractures with significant aperture and length in the reservoir, the data points will fall over the straight line (meaning HKR is greater than unity, where HKR = ratio of $hk_{welltest}/hk_{core}$), the highest points representing the maximum departure from the median line and highest frequency of open fractures. On the other hand, when points are located under the median straight line (meaning HKR less than unity), it signals the existence of secondary cementation that caused fractures to be plugged (Fig. 3.21). For this case, the fraction of closed fractures over open fractures for various locations must be determined before a useful filter can be constructed.

This information is crucial to developing the filter. The filter uses the following data:

- **HKR**
- Frequency of open fractures
- Frequency of closed fractures
- Overall orientation of fractures
- Overall orientation of sedimentation

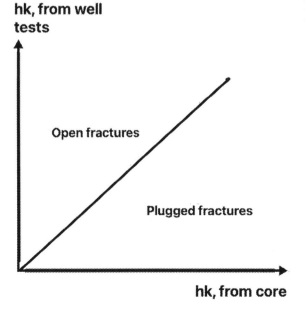

FIG. 3.21 Determination of the nature of fractures from hk data.

3.3.7 Estimation of fracture properties

Total organic content (TOC) changes in shale formations influence VP, VS, density, and anisotropy and thus should be detected on the seismic response. To detect it, different workflows have been discussed by Chopra et al. (2012). Rickman et al. (2008) showed that the brittleness of a rock formation can be estimated from the computed Poisson's ratio and Young's modulus well log curves. This suggests a workflow for estimating brittleness from 3D seismic data, by way of simultaneous prestack inversion that yields IP, IS, VP/VS, Poisson's ratio, and in some cases meaningful estimates of density. Zones with high Young's modulus and low Poisson's ratio are those that would be brittle and have better reservoir quality (higher TOC, higher porosity). Such a workflow works well for good quality data and is shown in the flowchart in the succeeding text (Fig. 3.22).

Stress data collected during drilling and cuttings and well site logs should be all combined to develop an enhanced understanding of reservoir fracture properties.

3.3.8 Special considerations for shale

Some of the considerations of reservoir characterization must be altered for shale due to the unique features of shale gas.

Passey et al. (1990) proposed a technique for measuring TOC in shale gas formations. This technique is based on the porosity-resistivity overlay to locate hydrocarbon-bearing shale pockets. Usually, the sonic log is used as the porosity indicator. In this technique, the transit time curve and the resistivity curves are scaled in such a way that the sonic curve lies on top of the resistivity curve over a large depth range, except for organic-rich intervals where they would show crossover between themselves.

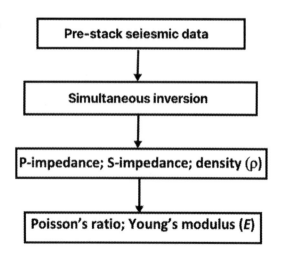

FIG. 3.22 Workflow for stress evaluation.

3.3 Fracture mechanics in geological scale   211

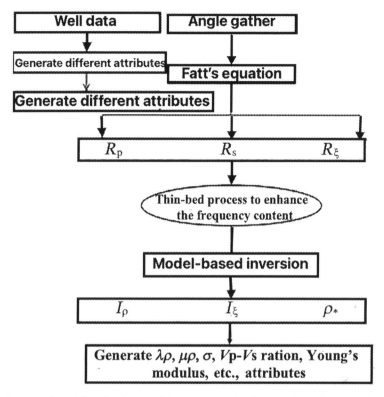

FIG. 3.23 An integrated workflow for characterizing unconventional reservoirs with conventional tools. *Modified from Sharma, R.K., Chopra, S., 2016. Identification of sweet spots in shale reservoir formations. First Break 34, 39–47.*

An integrated workflow in which well data and seismic data are used to characterize the hydrocarbon-bearing shale can be developed as shown in the following workflow (Fig. 3.23).

This compilation begins with the generation of different attributes from the well log curves. Then, using the cross plots of these attributes, one can identify the hydrocarbon-bearing shale zones. Once this analysis is done at the well locations, seismic data analysis is picked up for computing appropriate attributes. Seismically, prestack data are essentially the starting point. After generating angle gathers from the conditioned offset gathers, Fatti's equation (Fatti et al., 1994) can be used to compute P-reflectivity, S-reflectivity, and density, which depends on the quality of input data and the presence of long offsets. Due to the band-limited nature of acquired seismic data, any attribute extracted from it will also be band-limited, and so will have a limited resolution. While shale formations may be thick, some high TOC shale units may be thin. So, it is desirable to enhance the resolution of the seismic data. An appropriate way of doing it is the thin-bed reflectivity inversion (Chopra et al., 2006; Puryear and

Castagna, 2008). Following this process, the wavelet effect is removed from the data, and the output of the inversion process can be viewed as spectrally broadened seismic data, retrieved in the form of broadband reflectivity data that can be filtered back to any bandwidth. This usually represents useful information for interpretation purposes. Thin-bed reflectivity serves to provide the reflection character that can be studied, by convolving the reflectivity with a wavelet of a known frequency band pass. This not only provides an opportunity to study reflection character associated with features of interest but also serves to confirm its close match with the original data. Further, the output of thin-bed inversion is considered as input for the model-based inversion to compute P-impedance, S-impedance, and density. Once impedances are obtained, one can compute other relevant attributes, such as the $\lambda\rho$, $\mu\rho$, and VP/VS. These are used to measure the pore space properties and get information about the rock skeleton. Young's modulus can be treated as brittleness indicators and Poisson's ratio as a TOC indicator.

Fig. 3.24 shows the optimization scheme. It shows various steps to optimization of the subsurface map.

Fig. 3.25 shows pressure change and derivatives for various flow regimes, as determined by Abdelazim and Rahman (2016). This plot was generated based on the hybrid approach, as outlined in Abdelazim et al. (2014). Note that the fractures were generated stochastically using Gaussian stochastic simulation, in which each fracture feature is generated based on the random realization and it continues until the total fracture intensity and fractal dimension of the studied area are met. The three distinct flow regimes identified are (Fig. 3.26)

- the early time
- the mid time
- the late time flow regimes

As expected (Brons and Marting, 1961; Odeh and Babu, 1990), the early time flow regime corresponds to spherical flow and is marked by a negative slope of the pressure derivative curve. This signals partial penetration of the reservoir. By contrast, the mid-time flow regime is a short radial flow marked in the pressure derivative curve as a flat trend. The late time flow regime is a linear flow as recognized by a positive half slope of the derivative curve, which is caused by the fluid flow in discrete fractures. This is the first regime that shows the effects of fractures that invoke numerous segments of linear flow within a broad radial or spherical flow regime. This is the regime that allows one to determine the permeability of the formation in the direction of the flow vectors and the flow area normal to the flow vectors. In the case of the fractured reservoir, the slope of the straight line fitting the data on the derivative plot can be used to determine the fracture length (Bourdet, 2002).

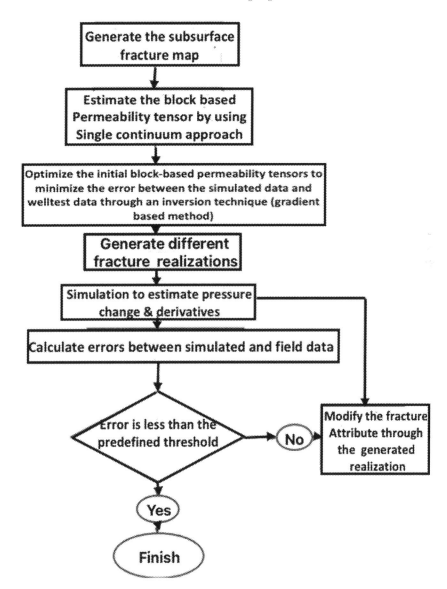

FIG. 3.24 The different steps used in optimizing the subsurface fracture map. *From Abdelazim, R., Rahman, S.S., 2016. Estimation of permeability of naturally fractured reservoirs by pressure transient analysis: an innovative reservoir characterization and flow simulation. J. Petrol. Sci. Eng. 145, 404–422.*

FIG. 3.25 Plot of fracture intensity vs. mean square permeability. *From Islam, M.R., 2014. Unconventional Gas Reservoirs. Elsevier, Netherlands, 624 pp.*

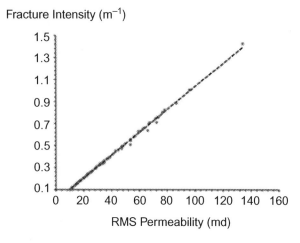

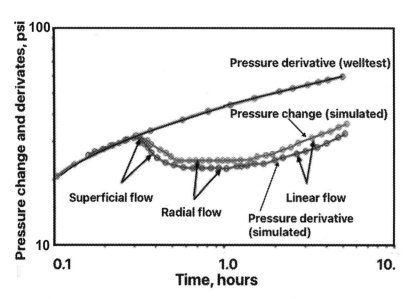

FIG. 3.26 Pressure change and pressure derivatives after inversion at wellbore location using the optimized subsurface fracture map presented in Fig. 3.25.

3.4 Core analysis

The most direct method for characterizing a complex reservoir is physical observation and direct analysis of real cores. Fracture density and fracture spacing can be determined through visual analysis of the core (e.g., Gale et al., 2007). Plugs are typically taken for laboratory experiments to measure permeability and other flow properties. These experiments can only measure an "effective" permeability of the core sample. In general, it is not possible to determine the natural fracture contribution to flow unless the fracture geometry is well known. The key issue, however, is that experience has shown that using laboratory measurements to represent reservoir-scale properties can be vastly misleading. Nonetheless, engineers are faced with the challenge of interpreting the few available direct measurements of the reservoir rock and translating them to field-scale properties. The idea is to use core data to refine data already collected. In addition, several features, including fracture density, mineralization, fracture opening, and shale breaks, can be quantified only with a visual inspection.

Of relevance is also the fact that total porosity can only be determined through core analysis. For complex reservoirs, such as unconventional basement, this is of great consequence because as much as 50% of recoverable gas or light oil can go unaccounted without definitive knowledge of porosity (Islam, 2014). Conventional assessment of porosity through GR analysis can be vastly misleading in basement reservoirs.

Practically, all aspects of core analysis are different in basement reservoirs from conventional reservoirs. They are highlighted as follows:

1. Low-permeability (matrix and fractures alike) structure itself
2. Low effective porosity but high total porosity
3. Response to overburden stress
4. Impact of the low-permeability structure on effective permeability relationships under conditions of multiphase saturation
5. Capillary pressure data as well
6. The role of fractures and fracture/matrix interactions

The most prominent of the aforementioned is the one related to the relative permeability graph. In a conventional reservoir, it is clear that there is relative permeability in excess of 2% to one or both fluid phases across a wide range of water saturation. In traditional reservoirs, critical water saturation and irreducible water saturation occur at similar water saturation values. Under these conditions, the absence of common water production usually implies that a reservoir system is at, or near, irreducible water saturation and low-permeability reservoirs; however, one can find that over a wide range of water saturation, there is <2% relative permeability to either fluid phase, and critical water saturation and irreducible water saturation occur at very different water saturation values. In these reservoirs, the lack of water production cannot be used to infer irreducible water saturation. In some very low-permeability reservoirs, there is virtually no mobile water phase even at very high water saturation. The term "permeability jail" coined by Shanley et al. (2004) describes the saturation region across which there is negligible effective permeability to either water or gas.

Low-permeability reservoirs are usually characterized by high to very high capillary pressures at relatively moderate wetting-phase saturations (Fig. 3.27). In many cases, wetting-phase saturations of 50% (close to Sgc) are associated with capillary pressures in excess of 1000 psia, suggesting that a large number of pore throats are <0.1 μm in diameter and are of the micro- to the nanoscale. In many low-permeability sandstone reservoirs, wetting-phase saturation continues to decrease with increasing capillary pressure. (See Fig. 3.28.)

The relationship between relative permeability, capillary pressure, and position within a trap in a tight formation is shown in Fig. 3.27 (Shanley et al., 2004). It shows a reservoir body that thins and pinches out in a structurally updip direction. In this case, significant water

FIG. 3.27 Typical relative permeability and capillary pressure curve for an unconventional gas reservoir. *From Islam, M.R., 2014. Unconventional Gas Reservoirs. Elsevier, Netherlands, 624 pp.*

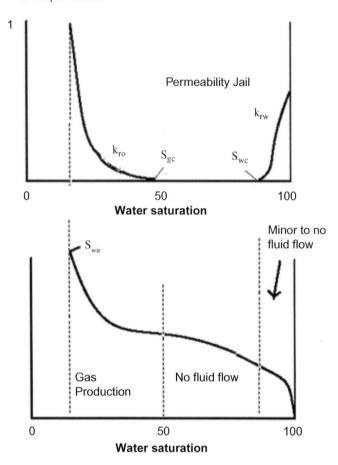

production is restricted to very low structural positions near the free water. In many cases, the effective permeability to water is so low that there is little to no fluid flow at or below the free water level. Above the free water level, a wide region of little to no fluid flow exists. Farther updip, water-free gas production is found (Fig. 3.29).

In terms of porosity, overburden pressure plays a significant role in complex reservoirs. This is unlike carbonate reservoirs or conventional sandstone reservoirs. Fig. 3.29 shows little change in porosity occurs with a net increase in stress.

For unconventional reservoirs, such as Shaley formation and tight sand, porosity is affected by overburden stress. Mechanical compaction is thus an inevitable consequence of burial and basin evolution. Often, this effect is modeled as a function of depth that is most affected by the overburden. However, depth is only a position coordinate that specifies the present-day location. Consequently, depth is a poor measure of the processes that have acted upon a sedimentary section through time, which is the primary function that governs all features of a formation. In that sense, everything, including the diagenetic process, remains

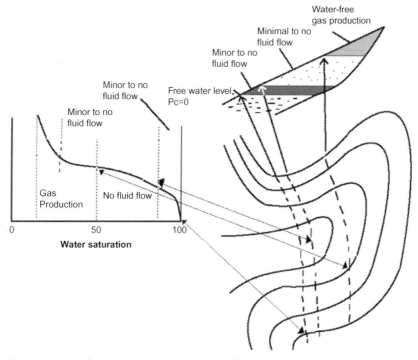

FIG. 3.28 Representation of the relationships between capillary pressure and position within a trap of a tight reservoir.

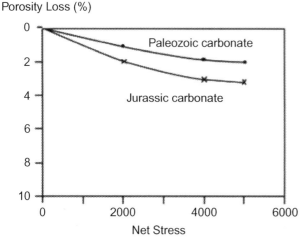

FIG. 3.29 Porosity is only slightly affected by net stress for carbonate formations. *Data from Lucia, 2007. Carbonate Reservoir Characterization: An Integrated Approach. Springer, 332 pp. Hariri, M.M., Lisenbee, A.L., Paterson, C.J., 1995. Fracture control on the Tertiary Epithermal-Mesothermal Gold-Silver Deposits, Northern Black Hills, South Dakota. Explor. Min. Geol. 4(3), 205–214.*

dynamic. Conventional models of shale compaction relate porosity to effective stress using the empirical relationship between void ratio and effective stress established in soil mechanics (Burland, 1990; Yang and Aplin, 2004). Okiongbo (2011) explored the effect of petroleum generation by evaluating the variation in porosity and effective stress in the Kimmeridge Clay Formation (KCF) above and within the oil window.

Fig. 3.30 shows porosity and effective stress relationship for above and below oil window. A common interpretation of the effective stress-porosity relationship above and within the oil window is that the loss of porosity is faster in the pregeneration zone than within the oil window. In addition, cementation and deeper burial create a relatively stiffer matrix for the KCF sediments within the oil window making it more difficult to compact at high effective stresses. In the presence of organic matter, porosity may be impacted by biogenic products. The presence of fractures and fissures alters the dynamics of porosity and pathways for the escape of organic gaseous products. The exact nature of the origin of porosity is not known, but there are factors that are not well understood, particularly in the context of basement reservoirs. Fig. 3.30 also shows a correlation developed by Yang and Aplin (2004) and Skempton (1970) and Burland (1990). These models express porosity as a unique function of clay and its natural, fine-grained clastic sediments. Yang and Aplin's (2004) model is based on well data from the North Sea and data derived from the studies of Skempton (1970) and Burland (1990). Although the clay content is assumed to be the same (~65%) in all the data sets, a remarkable difference exists in terms of their organic carbon content. Skempton (1970) and Burland (1990)

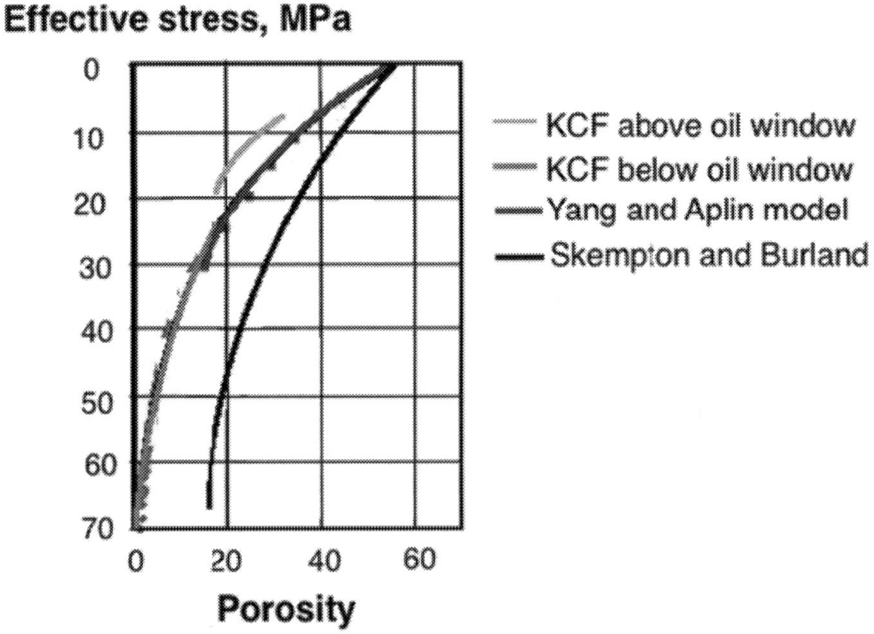

FIG. 3.30 Porosity variation with effective stress. 1: KCF above oil window; 2: KCF below oil window; 3: Yang and Aplin model; 4: Skemton and Burland model. *After Okiongbo, K.S., 2011. Effective stress-porosity relationship above and within the oil window in the North Sea Basin. Res. J. Appl. Sci. Eng. Technol. 3(1), 32–38.*

data sets are low TOC samples (TOC < 1 wt%), the North Sea data set used by Yang and Aplin (2004) has TOC ranging between 1% and 5%, while Okiongbo (2011) data that are portrayed with red and orange lines have TOC ranging between 5% and 10%. In addition, Yang and Aplin's (2004) data set exclude chemically compacted sediments. Above the oil window, the figure indicates similarity between the trend of Yang and Aplin (2004) and that from this study on extrapolation to low effective stresses (<5 MPa). Significant variation only occurs at effective stresses >5.

Within the oil window, the figure shows a close agreement between Yang and Aplin's data and Okiongbo data. Overall, the following factors play a role:

- TOC
- Mechanical compaction during digenesis
- Digenesis
- Thermal effects

While the first three components have been investigated with some details, thermal effects have not received much attention. Ehrenberg et al. (2009) related the geological age of rocks with their digenesis to assess the impact of temperature on porosity. Fig. 3.31 shows the effect of geological age on porosity. The impacts of age and lithology are captured in this figure. For petroleum reservoirs, however, the presence of hydrocarbon adds complications. The release of gas increases pore stress, and any escape and migration by hydrocarbon fluids add complications to the nature of porosity. This dependence is different in carbonate formations from silicate formations (including clay and shales). While depth adds to the overburden stress adding to the compaction of pores, it also adds temperature that affects hydrocarbon-bearing pores in a manner different from water-bearing pores. Adding to this is the chemical compaction and cementation that are directly affected by thermal conditions and overburden stress.

Fig. 3.32 shows that overburden stress alone would decrease the porosity. One must caution, however, basement reservoirs cannot be modeled uniquely with overburden stress. The degrees of "stress sensitivity" will be a function of the lithology and the pore throat size distributions. Rocks with "slot pores" will be more stress sensitive than rocks with more round pore throats. Slot pores are created by quartz overgrowths during digenesis. Whenever fractures (natural or induced) are present, the presence of fractures can alter the flow behavior very differently from conventional reservoirs. This has a tremendous impact on permeability. Fig. 3.32 shows gas permeability at NOB pressure as a function of gas permeability at ambient pressure (both Klinkenberg corrected). For high-permeability (10–100 mD) core plugs, the permeability under the original overburden pressure is slightly less than the value of unstressed permeability for that same core plug. However, as the permeability of the core plugs decreases, the effect of net overburden pressure on the core plug increases substantially. For the core plugs that had values of unstressed permeability of around 0.01 mD, the values of permeability under net overburden stress were about an order of magnitude lower or 0.001 mD. The lower permeability rocks are the most stress-sensitive because the lower permeability core samples have smaller pore throat diameters than the higher-permeability rocks. As overburden stress increases, the diameter of the pore throat decreases. Because the permeability of a rock is roughly proportional to the square of the diameter of the pore throat, the permeability

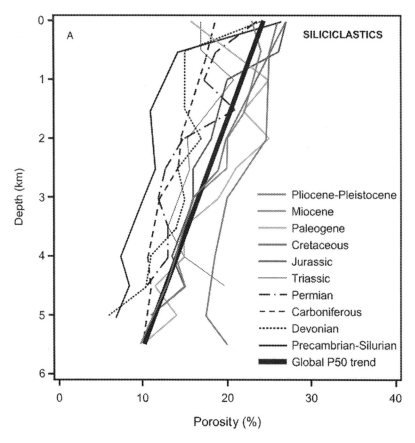

FIG. 3.31 Effect of geological age on porosity. *From Ehrenberg, S.N., Nadeau, P.H., Steen, Ø., 2009. Petroleum reservoir porosity versus depth: influence of geological age. AAPG Bull. 93, 1263–1279.*

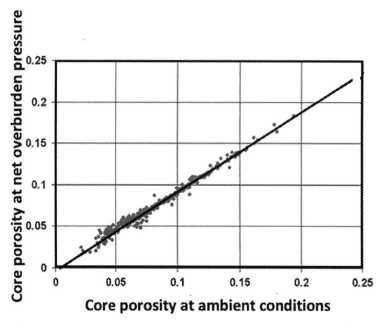

FIG. 3.32 Porosity variation under net overburden conditions. *From Petrowiki.spe.org.*

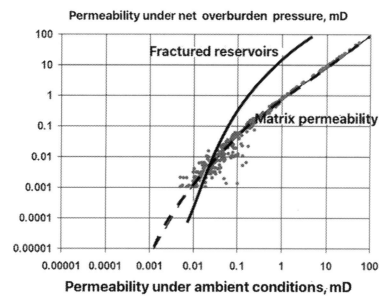

FIG. 3.33   Effect of overburden stress on matrix and fracture permeability.

reduction in low-permeability rocks is much more dramatic than in high-permeability rocks. This behavior would explain the trapping mechanism presented earlier in this section. It also explains why core permeabilities have little relevance when it comes to fractured formations. Fig. 3.32 also shows that a fractured formation would show a much steeper dependence on overburden stress. This behavior indicates that flow rates would be increased under low-stress conditions (Fig. 3.33).

3.5 Modeling unstable flow

An unstable displacement front emerges if there is viscous fingering during a water or gas injection. While it is known that conventional simulation doesn't apply for such a case, rarely anyone suggests an alternative. Typically, modelers are frustrated by the fact that predicted water or gas breakthrough doesn't match the observed breakthrough, which typically is much sooner during unstable flow than unstable flow. In this section, a procedure is outlined to properly predict unstable displacement in a field scenario.

There are three main forces prevalent in any petroleum reservoir. They are capillary forces, gravity forces, and mobility forces. It is often useful to study these forces in terms of dimensionless numbers. Three main dimensionless numbers have been identified for many decades. They are capillary number, mobility ratio, and gravity numbers. The interplay of these three numbers is the essence of oil and gas recovery. In the 1980s, another number, which is a modified form of "mobility ratio," was introduced by Peters and Flock (1981) and Bentsen (1985).

Capillary number, N_c is defined as

$$N_c = \frac{v\mu}{\sigma} \tag{3.21}$$

The following formula is also valid:

$$N_c = \frac{k\left(\frac{\Delta p}{l}\right)}{\sigma} \tag{3.22}$$

where
 μ = Displacing fluid viscosity.
 v = Darcy's velocity.
 σ = Interfacial tension (IFT) between the displaced and the displacing fluids.
 k = Effective permeability to the displaced fluid.
 $\Delta p/L$ = Pressure gradient.

Of the denominator would contain the term cos θ, where θ is the contact angle, which is a function of rock wettability to certain fluids. The capillary number shows what can be done in a reservoir, the idea being to maximize its value so that more irreducible oil can be recovered. Fig. 3.34 shows how the capillary number is linked to residual hydrocarbon saturation. Note that x-axis is in the log scale. This trend means that N_c has to be increased several orders of magnitude before any impact on residual oil saturation is invoked. However, for basement reservoirs, residual saturation bears a different meaning. These reservoirs can be exploited continuously by increasing N_c. Fig. 3.35 shows some of the correlations reported in the literature. This graph is also valid for gas reservoirs and is of significance in basement reservoirs, where the trapping of gas is the main mechanism of gas saturation.

The earlier equation shows that in theory, the displacing phase viscosity alone will increase N_c. If the increase is manifold, this increase will be sufficient to decrease the residual oil saturation. For instance, by using polymer, the displacing phase viscosity can be increased some 1000-fold. This would explain why on a laboratory scale, polymer injection has yielded high recovery efficiency. However, a large increase in the aqueous phase leads to lowering of

FIG. 3.34 General trend of N_c vs. residual saturation.

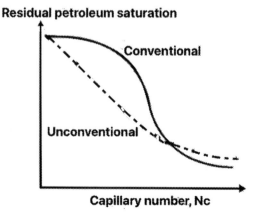

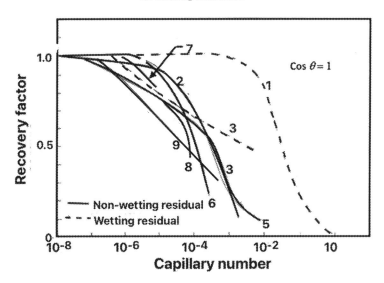

FIG. 3.35 Several correlations between capillary number and residual oil saturation. Dimensionless recovery vs. capillary number (N_c) (1: Bombrowski and Bromwell; 2: Foster; 3: Du Prey; 4: Du Prey; 5: Gupta; 6: Taber; 7: Wagner and Leach; 8: Abrams; 9: Moore and Slobod). Recovery factor = $(S_{or}/S_{or,wf})$ × (Oil saturation after EOR/Oil saturation after waterflood), $S_{or,wf}$ indicates oil saturation after waterflood.

injectivity. This can be severe for reservoirs that need EOR, hence creating a dilemma for operators.

The next component that can be manipulated is the Darcy velocity. However, the extent of the increase of this velocity is limited due to the fact that flow instability in porous media can be triggered with high displacement rates, causing early breakthroughs. In fact, Bentsen's work from the 1980s shows that the following relationship exists between instability number (which includes Darcy velocity) and breakthrough recovery. Fig. 3.36 shows such a

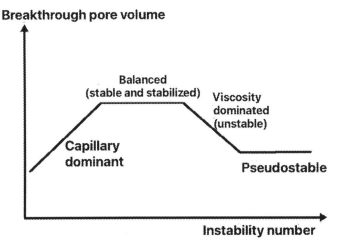

FIG. 3.36 General trend of breakthrough recovery and instability number.

correlation. Note that at very low velocity, capillary forces dominate, and very little oil is recovered because of the travel of injected water through water films in the porous media. This velocity is unrealistic in field applications and has been demonstrated only in laboratory models. Fig. 3.36 shows how a balance between viscous and capillary forces leads to the formation of a stable and stabilized front. This segment has the highest breakthrough oil recovery. This segment corresponds to the classic Buckley-Leverett profile. Practically, all laboratory models are operated in this regime. For instance, unsteady-state relative permeability measurement is performed using flow rates that would correspond to this flow regime. This fact is heavily consequential as none of these relative permeability graphs applies to a flow regime other than stable and stabilized. In case the flow regime is unstable, viscous forces dominate the process, and early breakthrough takes place. As the instability number is increased, the breakthrough recovery declines. For very high instability number values, however, the decline is arrested, and breakthrough recovery becomes insensitive to instability number.

For gas injection, laboratory models and the theoretical interpretation thereof show that a different trend emerges. This is shown in Fig. 3.37 Note that the stable region is very small and the unstable region is extended to very large instability numbers. The pseudostable region is missing within practical limitations of instability numbers.

In general, it is understood that increasing the velocity of displacement is not an option for reducing residual oil saturation or increasing recovery through EOR. However, the support for this conclusion came much later than the original conclusion. Until now, the science of instability is little understood. The science of scaling up laboratory results of such cases is in its infancy.

The next factor considered is IFT. Reduction of the IFT is at the core of all chemical flooding techniques. For perfect miscibility, the IFT is reduced to zero, signaling total recovery. However, such is never the case in the reservoir. Even when the operating conditions (e.g., injection pressure greater than minimum miscibility pressure) are met, natural media doesn't offer conditions conducive to instant or perfect miscibility. There is always a transition zone for which the concentration of each component goes through a gradient.

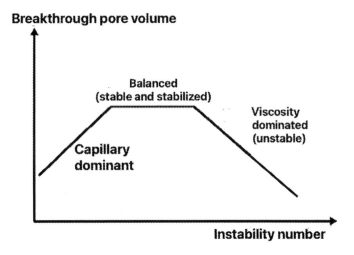

FIG. 3.37 Instability number vs. breakthrough recovery for immiscible gas injection.

It is conventionally known that if miscibility conditions prevail, the recovery is very high, even if perfect miscibility is not achieved. The most important condition, however, is the stability of the displacement front. Stability was historically connected to mobility ratio, given by the following equation:

$$M = \lambda_{ing}/\lambda_{ed} \quad (3.23)$$

where

$$\lambda_{ing} = Mobility of the displacing fluid, \left(\frac{k_{ing}}{\mu_{ing}}\right) \lambda_{ed} = Mobility of the displaced fluid, \left(\frac{k_{ed}}{\mu_{ed}}\right) \quad (3.24)$$

Several authors studied the relation between a capillary number and residual oil saturation that is presented in Fig. 3.38. Thus, the significant increase in oil/gas recovery was due to lowering the IFT between the fluids and pressure gradient promoting a reduction in the capillary number.

Since the mobility ratio is less than one ($M < 1$), the displacement is stable, a fairly sharp "shock front" separates the mobile oil and water phases, and the permeability to water stabilizes fairly quickly. In other case, where the mobility ratio is slightly greater than one ($M > 1$) is considered unfavorable, because it indicates that the displacing fluid flows more readily than the displaced fluid (oil), and it can cause channeling of the displacing fluid and, as a result, by-passing of some of the residual oil. Under such conditions and in the absence of viscous instabilities, more displacing fluid is needed to obtain a given residual oil saturation.

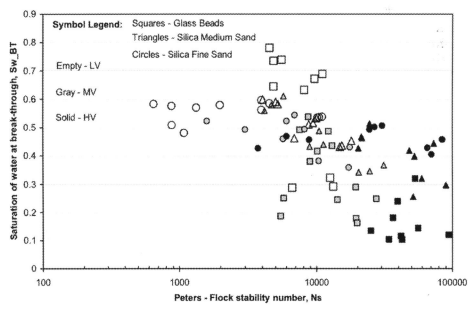

FIG. 3.38 Correlation between breakthrough recovery and Peters-Flock stability number. *HV*, high velocity; *LV*, low velocity; *MV*, medium velocity.

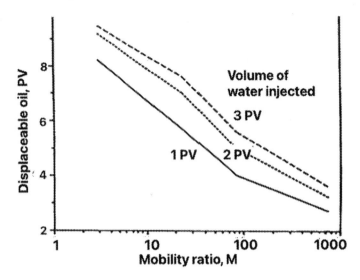

FIG. 3.39 Correlation of mobility ratio with oil recovery for waterflood (y axis shows recoverable oil).

However, if the value is close enough or equal to unity, the displacement is nearly piston-like and denoted a favorable mobility ratio. Mobility ratio influences the microscopic (pore level) and macroscopic (areal and vertical sweep) displacement efficiencies.

For waterflood, it's M that is the unique function. This comes from Buckley-Leverett equation (Fig. 3.39).

The mathematical relationship between microscopic and macroscopic recoveries efficiencies are represented using the oil recovery factor (R_f), and it can be given by the following formula:

$$R_f = E_v \times E_h \times E_m$$

where

$$E_m = E_I(S_{ol} - S_{or}) \tag{3.25}$$

The macroscopic sweep efficiency is defined by the horizontal and the vertical sweep efficiencies. The horizontal sweep efficiency is related to the mobility ratio, and the vertical sweep efficiency depends on the viscous-to-gravity force ratio. The vertical sweep efficiency is related to the difference in density of the injected fluid and in situ fluid. From early days on, there have been efforts to relate recovery efficiency to mobility ratio. However, until the 1980s, it only considered mobility ratio as a unique function of recovery efficiency. While this is a simplistic representation, this was the only correlation available for decades. Many operators still use this correlation to design both waterflood and chemical flood projects. This equation is a deduction from Buckley-Leverett equation that shows the following relationships to hold. Here, f_w and f_o are fractional flows of water and oil, respectively.

$$f_o = \frac{1}{1+M}, \quad f_w = \frac{1}{1+\frac{1}{M}} \tag{3.26}$$

The idea is to manipulate the injection fluid viscosity or mobility to maximize oil recovery potential. This correlation does not include the effect of viscous fingering. Such fingering is a possibility when the mobility ratio is greater than one unless there is a gravity advantage to the injected fluid. For instance, if gas is injected from structurally higher locations, it would stabilize the displacement front. This phenomenon was long recognized but not included in a dimensionless number until the early works of Flock and Bentsen. Even after their pioneering work, the petroleum industry continued to use an older version that included N_c concept or its variation to delineate the onset of fingering. For instance, see the following equation outlined by Lake (1989):

$$N_L = \left(\frac{\phi}{K}\right)^{1/2} \frac{\mu_w v L}{k_{rwe} \sigma \cos\theta} \tag{3.27}$$

The aforementioned criterion was applied by several researchers, and all determined that this definition of instability is grossly insufficient. A far better set of results were obtained using Peters and Flock (1981) criteria. Peters and Flock (1981) worked with velocity potential to come up with a stability criterion for a cylindrical system. It was for immiscible fluid, and for the first time, they introduced a variable that contains the dimension of the reservoir (or model thereof). It meant the same system that would show no fingering on the lab scale will show fingering on the field scale. While in chemical engineering, the dimension is always incorporated in determining the stability of flow in an open duct; for petroleum reservoir applications, it was new. For petroleum reservoirs, the characteristic dimension has always been considered to be pore diameter and not the reservoir dimensions.

Peters and Flock performed a stability analysis of the equations of two-phase flow to determine under what conditions viscous fingers tend to grow and propagate through a porous medium during immiscible flow. Bartley and Ruth (2001) conducted a series of experiments and demonstrated that Peters and Flock's number correlate broadly with their experimental observation of water breakthrough and none supported the criterion offered by Lake.

No such correlation existed in terms of capillary number or the modified capillary number as proposed by Lake. See Fig. 3.40.

Note that the work of Peters and Flock is based on the earlier theoretical work of Chuoke et al. (1959). Values of the Peters and Flock number N_s that are <13.56 indicate stable displacements, and values of $N_s > 13.56$ indicate unstable displacements, meaning viscous fingering occurs. The Peters and Flock stability number depends on the mobility ratio M, a wettability number N_w, and a calculated value for a characteristic velocity v_c. In the context of Peters' and Flock's work, "stable" means that extensive viscous fingers do not form during a waterflood or they become damped out; "unstable" means that viscous fingers continue to grow and propagate through the porous medium. They do not include gravity effects. It turns out this is a significant omission and would form the basis for modification of the theory and the approach by Bentsen (1985).

In this process, fractures play an intermediary role. Older theories didn't include the role of macroscopic dimensions and considered pore geometry only. Later theories considered macroscopic dimensions but didn't include the role of fractures. Saghir and Islam (1999) showed through a series of theoretical works that fractures are important because they trigger

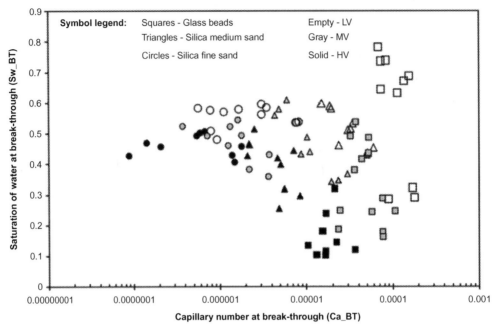

FIG. 3.40 There is no correlation between the capillary number and water breakthrough.

instability. Because conventional theories do not distinguish between fracture dimension and pore geometry and interplay thereof, the role of fracture is ignored. This aspect will be discussed in later sections.

Bentsen (1985) developed a different instability criterion based on force potential as applied on a rectangular system. He introduced the concept of pseudocapillary pressure. Both these criteria have gravity numbers in them. The following is the expression developed by Bentsen (1985):

$$I_{sr} = \frac{\mu_w v (M - 1 - N_g)}{k_{wr}\sigma_e} \times \frac{M^{5/3} + 1}{(M+1)(M^{1/3}+1)^2} \frac{4L_x^2 L_y^2}{L_x^2 + L_y^2} \quad (3.28)$$

where N_g is the gravity number defined as

$$N_g = \frac{\Delta \rho g k_{or} \cos \alpha}{\mu_o v} \quad (3.29)$$

Note that for a vertical injection, N_g assumes the largest value possible. In case N_g is larger than the expression $(M - 1)$, the displacement is unconditionally stable. This one shows the value of a gravity-stabilized displacement process that occurs when gas is injected from the top or water is injected from the bottom. This is the case for both miscible and immiscible displacement processes.

In the earlier expression, σ_e is the pseudointerfacial tension, which is

$$\sigma_e = \frac{C_1 \sigma \cos\theta}{\phi} = \frac{2-v^-}{1-v} dA_c \qquad (3.30)$$

A_c is the area under the capillary pressure curve, and v is a parameter dictated by the curvature of the capillary pressure.

For the dynamic part of the displacement process, the most profound impact is through alteration of permeability graphs. Fig. 3.41 shows some examples of how the lowering of IFT alters the effect permeabilities of both water and oil. The permeability values corresponding to the lowest IFT form straight lines. While this is true in a coreflood test, it rarely occurs in the field, and the discrepancy between laboratory data and field results emerges. In reservoir simulation studies, it has been demonstrated that the oil recovery with straight-line permeability curves does not show an oil recovery curve markedly better than others. That happens even after one overcomes the stability problem that occurs when such straight lines are assigned as relative permeability values of an immiscible system (e.g., oil and gas).

The most frequently encountered saturation endpoints are as follows:

- Residual oil saturation
- Irreducible water saturation
- Trapped-oil and trapped-gas saturations
- Critical gas and condensate saturations

Residual oil, irreducible water, and trapped-gas and trapped-oil saturations all refer to the remaining saturation of those phases after extensive displacement by other phases. Critical

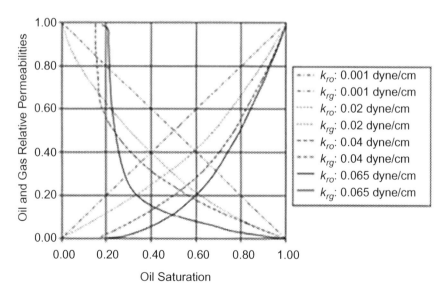

FIG. 3.41 Relative permeability curves are altered by lowering of interfacial tension.

saturation, whether gas or condensate, refers to the minimum saturation at which a phase becomes mobile.

The end point saturation of a phase for a specific displacement process depends on the following:

- The structure of the porous material
- The wettabilities with respect to the various phases
- The previous saturation history of the phases
- The extent of the displacement process (the number of pore volumes injected)

The end point saturation also can depend on IFTs when they are very low and on the rate of displacement when it is very high. For basement petroleum reservoirs, lowering of IFT, any chemical treatment, any thermal alteration, or onset of fractures leads to the alteration of relative permeability graphs. Fig. 3.42 shows how such movement can translate into gas production.

Results reported by Chatzis et al. (1983) give general insight into the combined effects of wettability and porous structure on residual saturations. In tests with unconsolidated sand of nonuniform grain size, the wetting phase (oil) was displaced by a nonwetting phase (air) from an initial saturation of 100% to a residual value. In general, it is known that

- residual saturation of a wetting phase is less than the residual saturation of a nonwetting phase and
- residual saturation of a nonwetting phase is much more sensitive to heterogeneities in the porous structure.

General conclusions on the effects of wettability are useful, but the diverse array of wetting alternatives suggests caution, especially in oil/water reservoir systems. This wide range of wetting possibilities is an obstacle to interpreting or predicting the effect of wettability on end point saturations. Indeed, conflicting results for different porous media are likely. For example, Jadhunandan and Morrow (1995) report that residual oil saturation displays a minimum value for mixed-wet media as wettability shifts from water-wet to oil-wet—counter to the results of Bethel and Calhoun (1953) that reported a maximum for media of uniform wettability.

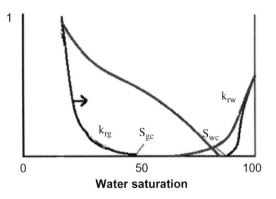

FIG. 3.42 Permeability jail can be removed with thermal or chemical alteration in an unconventional reservoir. *From Islam, M.R., 2014. Unconventional Gas Reservoirs. Elsevier, Netherlands, 624 pp.*

For gas reservoirs, such analysis is important because of the critical gas saturation. The critical gas saturation is that saturation at which gas first becomes mobile during a gasflood in a porous material that is initially saturated with oil and/or water. If, for example, the critical gas saturation is 5%, then gas does not flow until its saturation exceeds 5%. Values of S_{gc} range from 0% to 20%. For gas condensate reservoirs, this is of utmost importance. Interest in the mobility of condensates in retrograde gas reservoirs developed in the 1990s, as it was observed that condensates could hamper gas production severely in some reservoirs, particularly those with low permeability. The trend of increasing critical condensate saturations with decreasing permeability, as summarized by Barnum et al. (1995) is reproduced in Fig. 3.43.

This information is of importance for designing of WAG processes. The WAG ratio can also be defined as the ratio of the volume of water injected within the reservoir compared with the volume of injected gas. It plays an important role in obtaining the optimum value of the recovery factor corresponding to an optimal value of the WAG ratio. This optimal WAG ratio is reservoir-dependent because the performance of any WAG scheme depends strongly on the distribution of permeability and factors that determine the impact of gravity segregation (fluid densities, viscosities, and reservoir flow rates). Studies made by Rogers and Grigg (2000) showed that the WAG ratio strongly depends on the reservoir's wettability and availability of the gas to be injected. When the WAG ratio is high, it may cause oil trapping by water blocking or at best may not allow sufficient solvent-oil contact, causing the production performance to behave like a waterflood. On the other hand, if the WAG ratio is very small, the gas may channel, and the production performance would tend to behave as a gasflood; the pressure declines rapidly, which would lead to early gas breakthrough and high declination in production rate. To find the optimal WAG ratio is necessary to perform sensitivity analysis, proposing different relations of WAG ratio to study the effect on oil recovery.

Even though enhanced gas recovery has not been applied in the field at a commercial level, the fundamental mechanisms show that an enhanced gas recovery scheme is more useful and

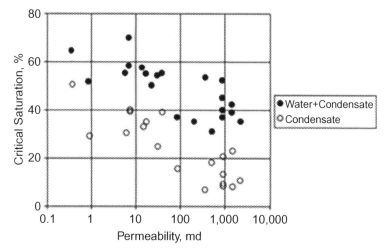

FIG. 3.43 Critical gas saturation for various permeability values of a gas condensate reservoir.

more likely to be successful than pressure maintenance or waterflood. It is because injected gas is less prone to creating channels or fingers in a gas reservoir than an oil reservoir. In addition, miscibility can be achieved in a gas reservoir under easily achievable conditions. Finally, the use of waste gas or gas that would be otherwise flared makes the economics of the project quite appealing.

3.6 Essence of reservoir simulation

Today, practically all aspects of reservoir engineering problems are solved with reservoir simulators, ranging from well testing to the prediction of enhanced oil recovery. For every application, however, there is a custom-designed simulator. Even though, quite often, "comprehensive," "all purpose," and other denominations are used to describe a company simulator, every simulation study is a unique process, starting from the reservoir description to the final analysis of results. Simulation is the art of combining physics, mathematics, reservoir engineering, and computer programming to develop a tool for predicting hydrocarbon reservoir performance under various operating strategies. The first step in the development of a reservoir simulator is expressing the physical situation in terms of mathematical description. This is called the *formulation* step. The *formulation* step outlines the basic assumptions inherent to the simulator, states these assumptions in precise mathematical terms, and applies them to a control volume in the reservoir.

It is possible to bypass the step of formulation in the form of PDEs and directly express the fluid flow equation in the form of a nonlinear algebraic equation as pointed out in Abou-Kassem et al. (2006). In fact, by setting up the algebraic equations directly, one can make the process simple and yet maintain accuracy. This approach is termed the "engineering approach" because it is closer to the engineer's thinking and to the physical meaning of the terms in the flow equations. Both the engineering and mathematical approaches treat boundary conditions with the same accuracy if the mathematical approach uses second-order approximations. The engineering approach is simple and yet general and rigorous.

3.6.1 Assumptions behind various modeling approaches

Reservoir performance is traditionally predicted using three methods, namely, (1) analogical, (2) experimental, and (3) mathematical. The analogical method consists of using mature reservoir properties that are similar to the target reservoir to predict the behavior of the reservoir. This method is especially useful when there is limited available data. The data from the reservoir in the same geologic basin or province may be applied to predict the performance of the target reservoir. Experimental methods measure the reservoir characteristics in the laboratory models and scale these results to the entire hydrocarbon accumulation. The mathematical method applied basic conservation laws and constitutive equations to formulate the behavior of the flow inside the reservoir and the other characteristics in mathematical notations and formulations. The two basic equations are the material balance equation or continuity equation and the equation of motion or momentum equation. These two equations are expressed for different phases of the flow in the reservoir and combine to obtain single

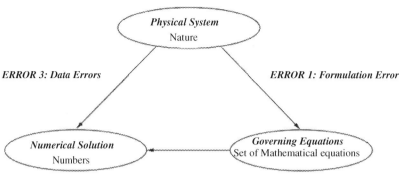

FIG. 3.44 Sources of errors in modeling petroleum reservoirs.

equations for each phase of the flow. However, it is necessary to apply other equations or laws for enhanced oil recovery. As an example, the energy balance equation is necessary to analyze the reservoir behavior for the steam injection or in situ combustion reservoirs.

Fig. 3.44 shows schematically how different errors are committed systematically during reservoir simulation. The first error is the error in the formulation. Sources of this error include the missing or incorrect physical model. Such models do not include an important process (e.g., forces due to surface tension). In addition, constitutive relationships are not a good approximation (e.g., friction law for pipes and channels not as applicable to the open ocean). The next set of numerical errors includes problems in the following:

- Algorithm
- Discretization
- Boundary condition specification and domain selection

The final set of errors comes from the following:

- Data analysis techniques
- Invalid assumptions and approximations
- Inadequate interpretation

The mathematical model traditionally includes material balance equation, decline curve, statistical approaches, and also analytical methods. Darcy's law is almost used in all the available reservoir simulators to model fluid motion. The numerical computations of the derived mathematical model are mostly based on the finite difference method. All these models and approaches are based on several assumptions and approximations that may produce erroneous results and predictions.

3.7 Material balance equation

The material balance equation is known to be the classical mathematical representation of the reservoir. According to this principle, the amount of material remaining in the reservoir after a production time interval is equal to the amount of material originally present in the reservoir minus the amount of material removed from the reservoir due to production plus the amount of material added to the reservoir due to injection. This equation describes the fundamental physics of the production scheme of the reservoir. There are several assumptions in the material balance equation:

- Rock and fluid properties do not change in space.
- Hydrodynamics of the fluid flow in the porous media is adequately described by Darcy's law.
- Fluid segregation is spontaneous and complete.
- Geometrical configuration of the reservoir is known and exact.
- PVT data obtained in the laboratory with the same gas-liberation process (flash vs. differential) are valid in the field.
- Sensitive to inaccuracies in measured reservoir pressure. The model breaks down when no appreciable decline occurs in reservoir pressure, as in pressure maintenance operations.

The material balance equation is directly linked to the original gas in place. To consider the production of an unconventional gas reservoir, a sufficient amount of gas in place within the reservoir is a concern. For shale, this usually means it is a hydrocarbon source that generated large volumes of either thermal or biogenic gas. To have generated such large quantities of gas, shale needs to be rich in organic matter, relatively thick, and to have been exposed to the source of heat in excess of usual global geothermal gradients. The presence of adsorbed gas, trapped gas, and free gas in fractures contributes to the complexity of the problem.

A general equation for calculating gas in place is.

$$G = A \times h \times \phi \times S_g \tag{3.31}$$

where
G = gas in place.
A = area of the entire trap.
h = average thickness.
S_g = gas saturation.
ϕ = porosity.

Note that conventionally, thickness is considered to be net thickness, from which shale breaks are deducted. For analysis of unconventional gas, the total thickness must be considered. Similarly, there cannot be any cutoff point for porosity. For unconventional reservoirs, most porosity occurs within very tight, Shaley formations, and removing those components would seriously falsify the total reserve estimate and subsequent prediction.

Total thickness can be calculated from a geological estimate of the trap and geophysical information. This can later be refined with capillary pressure data. The best estimate of gas saturation is through core analysis.

Organic carbon concentration in the shales is important in deciding its productive potential. Among all shale basins studied here, it appears that organic carbon concentration in productive shales ranges anywhere between 1% and 10% or more. In some unconventional reservoirs, intervals with high carbon concentration exhibit higher gas in place and, generally, the highest matrix porosity and the lowest clay content. It is difficult to come up with a deterministic value of TOC (%) as it often differs from basin to basin.

The thermal maturity of the shales is an important parameter in deciding the oil window and gas window. Prospective shale must be within the thermal maturity window for shale gas production. It is said that shales act as semipermeable membranes and allow only smaller molecules to pass through the sieves, and larger molecules choke pore throat and can't pass through. So, it is important to locate the transition from gas to oil window as wells in the oil window are subjected to poor performance if developed as gas wells. Thermal maturity is generally represented by vitrinite reflectance (R_0%). Vitrinite reflectance is measured in the core analysis. Vitrinite reflectance is one of the organic geochemical indicators of petroleum maturation. The principal maceral groups in coals provide the basic Van Krevelen diagram, which depicts the path of their evolution during carbonization. Paths progressively approach origin depicting 100% carbon. The macerals are distinguished by their plan precursors. Vitrinite is one of the macerals that includes both telinite, in which woody structures are present, and collinite, essentially structureless matrix, cement, and cavity infilling. Vitrinite is not fluorescent. It is primarily humic organic material.

Such analysis also applies to other unconventional reservoirs for which the reservoir rock is the same as the source rock. This includes several shale plays within the sandstone and even carbonate formations, volcanic reservoirs, and gas hydrates.

Water saturation is very difficult to measure in shale gas reservoirs. The reason is that all water saturation equations developed to date are designed for nonshale lithology and the concept of net pay based on non-Shaley zones in a reservoir. So, water saturation could be measured from the core analysis more reliably in shale reservoirs.

In the material balance process, the gas storage mechanism is part of the history of the reservoir. It is important to know how the gas is stored before producing it. For shale gas, there are three possibilities:

1. Most of the gas (>50%) is adsorbed on the shale matrix, and the remaining is stored in the matrix.
2. Most of the gas is stored in the matrix and fractures, and adsorption is not an important phenomenon.
3. Most of the gas is stored in fractures. Matrix storage is not possible due to the absolute absence of porosity and permeability.

Out of these, Item 3 is rarely seen, but it is perceived that it is a possibility. When the gas is adsorbed on the matrix, shale gas reservoirs can be treated as a special case of CBM reservoirs. Dewatering of the shales is required before actual gas production. So, wells are treated at low operating pressures. In other Devonian shales, where both phenomena are present, that is, adsorption and matrix storage, both exist, the production mechanism is decided on a well-to-well basis. Reservoirs where matrix storage is major, it is seen that these wells have high initial decline but produce for a longer time, so payout period for these wells is long, but wells that produce for 30–50 years as matrix gas diffuse into fractures slowly.

3.8 Representative elemental volume, REV

If we direct our attention to the fluid contained in porous space, we also encounter a lot of difficulties to describe the phenomena associated with the fluid itself, such as motion and mass transport. If we consider the fluid itself, it consists of many molecules. To describe the motion of the fluid, a large number of equations should be provided to solve the problem at the molecular level. Since the final goal is to describe the fluid motion in porous media, it is necessary to have a higher level and treat the fluid as continua. The concept of a particle is essential to the treatment of fluids as continua. A particle is an ensemble of many molecules contained in a small volume. The size of a particle should be larger than the mean free path of a single molecule and, however, be sufficiently small as compared with the considered fluid domain that by averaging fluid and flow properties over the molecules included in it, meaningful values, that is, values relevant to the description of bulk fluid properties, will be obtained. The mean values are then related to some centroid of the particle, and then, at every point in the domain occupied by a fluid, we have a particle possessing definite dynamic and kinematic properties.

The concept of the fluid continuum and the definition of density as an example of the fluid properties are illustrated in Fig. 3.45. It shows that the particle size (elementary volume) should be of the order of magnitude of the average distance λ between the molecules (mean free path of molecules). However, to capture the fluid inhomogeneity, the size of the particle should be less than an upper limit. This may be characterized by the length, L (or L_x, L_y, and L_z in the direction of three coordinates). The length L may be considered as characteristic length

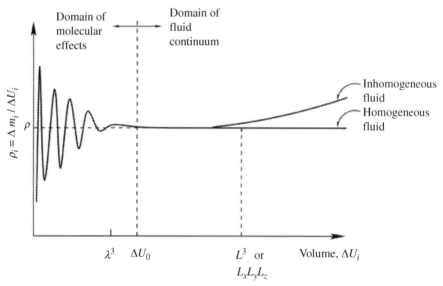

FIG. 3.45 The fluid continuum domain and the variation of fluid density. *Modified from Bear, J., 1975. Dynamics of Flow in Porous Media. American Elsevier Publishing Co., New York.*

for the macroscopic changes in fluid properties such as density. The volume L^3 (or $L_xL_yL_z$) may be used as the upper limit for the particle volume, ΔU_i.

If we turn our attention to the solid medium confining the fluid, the size of a representative elementary porous medium volume around a point P should be determined. The size of REV should be much smaller than the size of the entire fluid flow domain to associate the resulting averages with a point P. On the other hand, it must be sufficiently larger than the size of a single pore that includes a sufficient number of pores to permit the meaningful statistical average required in the continuum concept (Bear, 1975). The definition of REV and representative elementary property (REP) is illustrated in Fig. 3.46 The property of the medium and more importantly permeability which is a property of medium and fluid both should be averaging around a volume ΔU_0 as REV. The REV may be defined as the volume ΔU_0 at which the fluctuation in the REP is not significant beyond it.

The REV may be defined as a critical averaging volume beyond which there is no significant fluctuation in the representative property as the addition of extra voids or solids has a minor effect on the averaged property. Fig. 3.47 shows the result of a thought experiment. It is impossible to carry out actual measurements over enough scales to confirm the behavior shown in this plot. However, it indicates that the variability of a specified property of the rock matrix is varied erratically at a small scale, less than the REV, and has a zero (or small) variability at an intermediate scale, in the range of REV. The variation of the specified property such as porosity will be increased with increasing the scale larger than REV. These results are consistent with the general observation. The variability of the average porosity in a set of 1″ diameter core plugs is not zero, and this variability is certainly smaller than the variability of porosity on the scale of several μm (say 100 μm). The statistical investigation using the variance of the mean of the porosity by Lake and Srinivasan (2004) shows the same trend for small scale. However, they do not find any region of stable porosity at intermediate even though the parameters were chosen to emphasize such stability (Fig. 3.48). This shows that

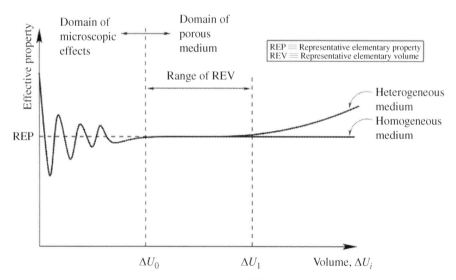

FIG. 3.46 Definition of representative elemental volume (REV) and representative elementary property (REP).

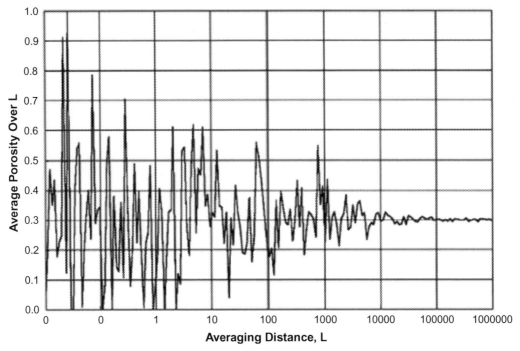

FIG. 3.47 The variation of average porosity as a function of the one-dimensional proxy, L, of the support (measurement) volume. *From Lake, L.W., Srinivasan, S., 2004. Statistical scale-up of reservoir properties: concepts and applications. J. Petrol. Sci. Eng. 44(1–2), 27–39.*

the only way to obtain a stable average is for the averaging volume to exceed the largest scale of heterogeneity. This is not a satisfactory outcome. It says that the REV is the largest scale of heterogeneity in the field. Since any cell-by-cell computation or any measurement is below the REV scale, there are some uncertainties in the modeled rock properties.

3.9 Thermal stress

The most commonly used methods to describe the fluid flow in oil reservoirs employ constant rock properties. As stated earlier, this is done because any transient effect leads to a set of nonlinear governing equations that are not amenable to solutions with the most commonly used simulators. Unfortunately, all reservoirs undergo changes in the rock properties due to variation in pore pressure and are often affected by the Joule Thompson effect during the production of the crude oil. Irrespective of the actual magnitude of this change, it is important to consider, a comprehensive formulation must be made. Common characteristics of fractured reservoirs are sensitivity of permeability and porosity to effective stress. The in-situ stress, in itself, can be of mechanical or thermal origin. The thermal stress is more pronounced in thermally enhanced oil recovery schemes. However, thermal stress also occurs if a cold fluid is injected into a hot reservoir. Such occurrence is common in high API oil reservoirs as well as during hydraulic fracturing (Chaalal et al., 2017). Interestingly, literature reviews reveal that the research in this area has been focused mainly on the thermal recovery of heavy oil. Few investigations, however, have been done on the onset and propagation of fractures under thermal stress or mechanical stress (Chaalal et al., 2017; Islam, 2020). The study of thermal stress in a homogeneous, isotropic medium started many decades ago (Timoshenko and Woinowsky-Krieger, 1959). However, typically, only linearized versions were solved either with analytical methods or with the perturbation approach (Das and Navaratna, 1962; Stavsky, 1963). However, if a body has anisotropic properties, it is more complicated to predict the temperature field and the stresses; the numerical method appears to be the method of suitable choice (Chen, 1988). Similarly, Wu (1997) proposed three different methods to calculate thermal stress. The first is an analytical method that involves an exact solution of linearized equations, similar to those encountered in elastic theory. The second method uses the technique of approximations such as the perturbation procedure. Finally, the third method uses numerical techniques to solve the governing partial differential equations. However, as Islam et al. (2000) pointed out, all numerical methods rely on linearization prior to final solutions, thus making the process incapable of detecting correct solutions. Correct solutions require solutions of nonlinear equations with the possibility of multiple solutions. Such an approach has been taken by Mustafiz et al. (2008), which is limited to one-dimensional problem.

Even though various forms of semianalytical methods have been proposed to alleviate the difficulty of a large number of grid blocks, numerical methods continue to be common in determining the stress field of a body subjected to thermal stresses. In these cases, the emphasis has been focused on the thermal stress coupled with mechanical stress that originates from pore pressure as well as for overburden stress. Gai (2004) summarized some of the salient aspects of Geomechanics modeling and highlighted the need for modeling thermal stress.

Ghassemi (2012) showed that rock mechanics research and improved technologies can impact areas related to in situ stress characterization, initiation and propagation of artificial and natural fractures, and the effects of coupled hydro-thermochemo-mechanical processes on fracture permeability and induced seismicity. Their work was geared toward geothermal well design. Typically, one or more wells are drilled into the fracture network and cold water is injected into one part of the well system and hot water/steam is recovered from the other (Fig. 3.48).

The knowledge of the stress state in geothermal systems can help delineate fractured zones and zones of possible fluid accumulation and flow. The reservoir stress state is influenced by rock discontinuities, and rheology, heterogeneities, as well as poroelastic, and thermal stresses caused by injection/production operations. As stated earlier, this becomes a factor even during primary production if there is significant thermal and/or mechanical change. During the last few decades, several techniques have been developed for developed to determine the in situ stress, including mini-frac tests, analysis of breakouts and drilling-induced cracks, hydraulic testing of preexisting fractures (HTPF) (Baumgartner and Rummel, 1989; Cornet and Valette, 1984), and focal mechanism inversion. A detailed description of these techniques can be found in Bell (2003), Evans et al. (1999), and Amadei and Stephansson (1997). Usually, the vertical stress (σ_v) component is determined using a density log. For determination of the minimum horizontal stress (σ_h) a leak-off test can be used. Also, hydraulic

FIG. 3.48 Schematic of a geothermal system.

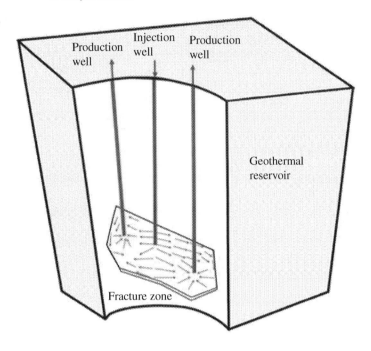

fracturing (micro-frac or mini-frac) to find the closure pressure is often employed. The injection rates for the leak-off test are in the range of 0.04–0.016 m^3/min for about 1 m^3 total. These values are 0.0038–0.038 m^3/min for micro-frac (0.008–0.38 m^3) tests and 0.795–1.590 m^3/min for mini-frac (1.59–159 m^3) tests (De Bree and Walters, 1989). The leak-off test originally was designed to test casing-shoe integrity to ensure safe drilling of the next wellbore section. However, the test has been used for stress measurement by continuing it until the rate of pressure increase declines, and interpreting the departure from linearity of the injection pressure vs. pumped volume as the fracture initiation pressure (Addis et al., 1998). Other interpretations of the test data use Kirsch's solution to calculate σ_h from the leak-off pressure or consider the instantaneous shut-in pressure as the value of σ_h. In light of the uncertainties in the leak-off test data, an extended leak-off test has been proposed which involves 3–4 more pressurization cycles over a period of 1 h (Addis et al., 1998). Methods such as leak-off tests extended leak-off tests and mini-fracs have been used in some geothermal reservoirs to measure stress. Temperature differences between reservoir rocks and fluids and injected fluids can complicate the interpretation of results. Curiously, under this scenario that stress considerations are most crucial.

Even where these techniques can be used to determine the minimum in-situ stress, there are no direct ways to accurately determine the magnitude of the maximum horizontal stress (σ_H). The breakdown pressure (used in calculating σ_H) is rate-dependent, size-dependent, and fluid-dependent, and many techniques for its interpretations have been suggested (Guo et al., 1993; Ito and Hayashi, 1991).

Another method for estimating the maximum horizontal stress (σ_H) relies on the tendency of a deep wellbore wall to fail in compression, at which point the tangential stress reaches a maximum and overcomes the rock's compressive stress. Such compressive failures around the wellbore are called stress-induced wellbore breakouts (Gough and Bell, 1981; Plumb and Hickman, 1985; Zoback et al., 1985). Because the possibility of borehole failure depends on the in situ rock stress and strength, and its location is governed by the in situ stress and borehole orientation, it is possible to use breakout analysis as a tool to constrain the maximum horizontal in situ stress magnitude (Bell and Gough, 1979; Zoback et al., 1985; Zoback and Healy, 1992; Brudy et al., 1997) and the in-situ rock strength (Peska and Zoback, 1995).

In a vertical well, the zone of compressive failure is centered along the azimuth of the minimum horizontal compression. Hence, one can directly deduce the orientation of all principal stresses. However, breakouts may rotate with depth as the wellbore azimuth changes, or as a consequence of a change in the petrophysical and structural characteristics of the reservoir. Interpretation of these cases is more elaborate. Qian and Pedersen (1991) proposed a numerical inversion method for estimating the in-situ stress state according to breakout data for inclined wells. In addition, Djurhuus and Aadnoy (2003) developed an analytical method to determine the in-situ stress orientations from borehole image logs; however, their method requires knowledge of the magnitude of the in situ stress.

Most breakout inversion techniques use an isotropic elastic stress analysis which does not consider the progressive breakage of rock. In addition, the effects of coupled thermal and poro-mechanical processes on the stability of boreholes in geothermal reservoirs are not considered. When rocks are heated or cooled, the bulk solid and the pore fluid undergo a volume change. A volumetric expansion can result in significant pressurization of the pore fluid, depending on the degree of containment and the thermal and hydraulic properties of the fluid as well as the solid. When heated, water trapped in the pores may undergo pressure increases on the order of 1.5 MPa/°C for conditions typical of the earth's upper crust (Williams and McBirney, 1979). The net effect is a coupling of thermal and poro-mechanical processes, which occur on various time scales, and the significance of their interaction depends on the problem of interest. For example, when drilling wells in high-temperature rocks, strong coupling between thermal and poro-mechanical effects might develop that can significantly impact the stress/pore pressure distribution around a wellbore and thus borehole failure and fracture initiation. This is caused by the contrast in thermal and hydraulic diffusivities of rock.

The thermo-poroelastic effects on wellbore stability and its use for constraining in situ stress and rock strength have been considered by Li et al. (1998) and Tao and Ghassemi (2010), respectively using the assumption of rock isotropy and homogeneity. The results showed that poro-thermo-mechanical effects influence both failure potential and mode; cooling tends to prevent compressive failure and radial spalling, whereas heating tends to enhance failure in compression and can cause tensile failure by an excessive increase of pore pressure. The inhomogeneous nature of rocks can lead to qualitatively different borehole failure with elongations that are parallel to the maximum horizontal principal stress orientation. These are suggested to be the result of a pervasive, cooling-induced, tensile micro-cracking process prior to macroscopic failure localization (Berard and Cornet, 2003). This phenomenon can also influence the validity of using regular breakouts, as thermal stress can change near-wellbore rock properties. Temperature, anisotropy, and mismatch in grain thermal expansion coefficient, initial porosity, and grain size contribute to thermal cracking (Fredrich and Wong, 1986) in the wellbore region so that Kirsch's solution may not be applicable. The impact of rock strength anisotropy on breakouts was considered by Vernik and Zoback (1990) and was found to be significant. It can be expected that poro-thermoelastic anisotropy can radically change the pattern of pore pressure and stress distributions around the wellbore and thus breakout orientation and size in other situations.

In petroleum reservoir development, interpretation of a mini-frac test to extract the closure pressure is considered an effective method for measuring the minimum horizontal stress, and many techniques have been developed for this purpose (Guo et al., 1993). Therefore, it would seem that the integration of mini-frac data with other available methods can provide a reasonable estimate of the complete stress state magnitude. Such an approach has indeed been used in, e.g., Coso geothermal field (Sheridan and Hickman, 2004; Nygren and Ghassemi, 2004) and the Desert Peak EGS experiment (Hickman and Davatzes, 2010).

An alternative approach is the inversion of well-constrained earthquake data from seismic stations. The inversion techniques provide the orientation of the three principal stress axes and the relative magnitude of the intermediate principal stress with respect to the maximum and minimum principal stress (Michael, 1987; Gephart and Forsyth, 1984). The orientations are determined by minimizing the average difference between the slip vector and the orientation of the maximum shear stress on the inverted faults. However, this approach cannot provide the stress magnitudes, and its effectiveness suffers from uncertainties in earthquake data and its interpretations. Furthermore, the stress state varies with pore pressure and temperature changes accompanying injection/extraction operation. The interactions between the original 3D stress state and rock discontinuities and heterogeneities play an important role both in near-wellbore areas and the reservoir at large.

Early numerical modeling of stimulation in geothermal systems considered hydrothermal effects while generally neglecting rock mechanical aspects (Hopkirk et al., 1981; Kohl et al., 1995). The advances in computers and computational techniques have led to the development of several numerical models for the analysis of more complex forms of reservoir stimulation (Sesetty and Ghassemi, 2012; Weng et al., 2011; Zhang and Jeffrey, 2006; Koshelev and Ghassemi, 2003); however, these elastic models neglect the details of fracture propagation and interaction. Other approaches have used complex and real variable boundary element methods (Olson, 2008; Dobroskok et al., 2005; Bobet and Einstein, 1998) to model fracture coalescence. Poroelastic and thermoelastic displacement discontinuity methods (Zhou and Ghassemi, 2011; Ghassemi and Zhou, 2011; Ghassemi and Roegiers, 1996; Carter et al., 2000) or the finite element method (FEM) (Boone et al., 1991), extended finite element method (XFEM) (Yazid et al., 2009) have also been developed. These approaches have been useful for studying near the wellbore, and planar fracture propagation and help to better understand aspects of fracture intersection, but none can handle the complex problem of multiple fracture initiation and propagation.

Moreover, experimental analysis (Finnie et al., 1979) and analysis of cooling by injection (Perkins and Gonzalez, 1985; Ghassemi and Zhang, 2006) show that high-stress zones develop in the vicinity of the main fracture, indicating the potential for multiple initiations and propagation events. Cooling-induced stresses cause a complete rotation of the stress field such that stress parallel to the secondary cracks becomes the in-plane major principal stress (higher than the component in the normal direction) and may exceed the in situ stresses of the geothermal reservoir (Perkins and Gonzalez, 1985). The cracks can propagate into the rock matrix perpendicular to the main fracture and increase the permeability of the reservoir (Fig. 3.49). Such secondary cracks can be particularly important in reservoir development

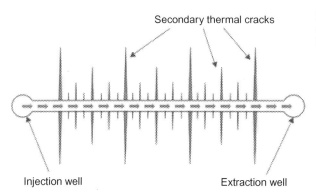

FIG. 3.49 Formation of secondary thermal cracks perpendicular to the main fracture cooled by injected water.

in view of their potential role in enhancing the heat exchange area or increasing fluid loss. To increase heat extraction, the secondary thermal fractures should be sufficiently long and open to allow the fluid to flow deep inside the reservoir matrix where the heat is stored.

The formation and propagation of thermal fractures in response to cooling have been treated theoretically (Bažant and Ohtsubo, 1977; Bažant et al., 1979; Nemat-Nasser et al., 1978). These stability analyses predict that many small cracks appear shortly after cooling of the surface; however, some of them will be arrested upon further cooling. In these studies, it was concluded that because the growth of one crack will suppress the propagation of its nearest neighbors, only every second crack will grow further until the next bifurcation point is reached. However, these analyses did not include the high compressive in situ stresses which are typical for geothermal applications. Barr and Cleary (1983) numerically studied the effect of thermal crack penetration into a geothermal reservoir by assuming parallel fracture geometry without considering the propagation in time of many thermal cracks at unequal rates. As pointed out by Nemat-Nasser (1983), a more complete analysis would consider the possibility of unequal crack growth. Recently, Tarasovs and Ghassemi (2010) developed a complex variable boundary element numerical method for investigating the growth behavior of many cracks under the influence of a nonstationary thermal field resulting from cold water injection (Fig. 3.23) and a compressive stress field. The model has been used to study the influence of the main physical parameters of the system on the length and spacing of thermally driven fractures. Several simulations for various combinations of relevant parameters were performed. Fig. 3.50 shows the crack length vs. cooling

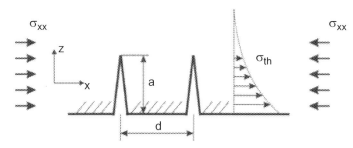

FIG. 3.50 An array of edge cracks loaded by thermally induced stress σ_{th} and a far-field compressive stress σ_∞.

depth for three different minimum compressive in situ stress values (20, 35, and 50 MPa). The results for a random array of cracks in a half-space under uniform cooling show the cracks' length to be approximately proportional to the cooling depth L, or proportional to the square root of time. For a given cooling depth, a larger in situ stress results in smaller fracture lengths. The process of crack pattern formation is self-similar, i.e., the crack pattern repeats itself on different time and length scales, and depends on the parameter ξ. The characteristic length ξ, is the ratio of the energy required to create a new crack surface and the energy that is generated in the solid by the thermal shock in the presence of in situ stress (Fig. 3.51).

More recently, Ghassemi and Tarasovs (2015) investigated the development and propagation of a system of crack from a cooled wellbore, and secondary thermal cracks in a geothermal reservoir due to cold water injection is studied. The extent of the thermal fractures and their spacing is estimated using a combination of the real boundary integral equation method for the temperature solution, and the complex variable boundary integral equation for fracture propagation solution. The results showed that the influence of thermal stress on fractures and their propagation is considered for the problem of a sudden cooling of a rock half-space, injection/extraction process in fractures, and cooling of a wellbore.

In a review article, Obembe showed that the subject of heat transfer in oil reservoirs has gained huge attention, due to its diverse range of applications in petroleum reservoir management and thermal recovery for enhanced oil recovery.

Note that temperature changes in the rock induce thermoelastic stresses Hojka et al. These stresses cause thermal strains that can eventually lead to rupture or shear. The temperature dependency of the thermal stresses can cause enormous variations in the magnitude of the thermo-elastic stresses over reasonable temperature ranges. Thermal stress is complex and cannot be handled easily, and in general, a multidimensional search problem must be solved, which can be difficult in practice. However, theoretical .estimates of the stresses can be calculated from equations of fluid-saturated, pore-elastic solids by incorporating thermo-elastic rock volume changes. The calculation is straightforward if the effects due to the nonlinear deformation module and the elastoplastic processes are neglected.

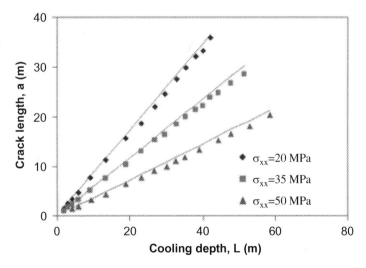

FIG. 3.51 Crack length as a function of cooling depth L. *Dots* represent results of the simulation, *lines*—approximation by an analytical expression for a cooled infinite space (Tarasovs and Ghassemi, 2010). Note that σ_{xx} is the far-filed stress component perpendicular to the crack.

It is clear from the preceding that the calculation based on both nonlinear deformation and elastoplastic processes may be a demanding problem from the computational point of view. Biot proposed equations for pore-elastic theory, and Schiffman studied the stresses and included the thermal effects into Biot's theory. Notice again that the inclusion of the thermal conduction to the original theory was considered to be a difficult task and was only accomplished in the mid 80s.

Furthermore, Kurashige reported the first coupling of conduction and convection in the context of thermal stress. Kurashige provided the numerical solution for the governing equations. Hojka et al. pointed out that Kurashige work presented inappropriate formulation. They postulated that Kurashige allowed the thermo-elastic stresses to reach the hot fluid front and such action makes the model unrealistic.

Further evidence of the weakness of such a model arises from the fact that Kurashige imposed steady-state pressure distribution, leading to poroelastic responses that are not time-dependent.

An elaborated technique was performed by Vafai and Sozen to solve the problem. The technique consists of solving the energy balance twice using the approach originally outlined by Chan and Banerjee to solve the energy balance equation with a two-equation model.

This essentially involves solving the energy balance equation with the thermal properties of the rock and fluid separately. A heat transfer coefficient was incorporated to account for the transfer of energy between the fictitiously separated rock and fluid body. The study demonstrated that there is a lag between fluid and solid temperatures because heat propagation in the fluid is due to convection and conduction whereas the propagation of heat in the solid phase is governed only by conduction.

Hojka et al. derived the transient temperature and stress fields in a two-dimensional domain during constant rate, constant temperature fluid injection. The energy equation was considered to be different for the fluid and the rock. However, the complete conduction equation was not solved for the solid. The procedure consisted of calculating the pore pressure first then the results permitted to evaluate the solid temperature and volume change potentials. Subsequently, rock responses due to thermo-elastic and poroelastic changes were calculated. They re-affirmed the previous observation that the pressure front propagates faster than the thermal front. One of their major findings was that small changes of pressure do not induce large poroelastic stresses; leading them to believe that these effects can be neglected. They observed that the thermoelastic stress changes are limited to the heated or cooled zone in the region between the borehole wall and the injected fluid front. Finally, they demonstrated that shear failure or hydraulic fractures can be initiated on the wall depending mainly on the temperature of the injected fluid.

Dusseault outlined stress changes during a thermal operation. He identified the following effects that make the thermal loading a nonlinear problem:

– Rotation and alteration of principal stress fields;
– Changes in volume and stress (in magnitude as well as direction);
– Yield and dilation of the reservoir, leading to changes in absolute and relative permeability's;

- Change in the overburden and under burden properties due to tensile or compressive cracking or shear yield;
- Reservoir shearing or overburden cracking due to high pore pressure, thermal stresses, and weakening from yield.

The existence of so many governing parameters and their interactions make the problem intractable. Dusseault suggested that conceptual models be used to aid understanding of the problem before coupling with a reservoir model. He reported a series of configurations for the Geomechanics model but did not make any effort to couple the model with a fluid flow simulator.

Hojka et al. reported the coupling of convection and conduction for the plane strain borehole. They limited their investigation to single-phase flow. In addition, the focus of this investigation was the fluid flow and energy balance equation. No effort was made to couple fluid with Geomechanics modeling.

Stress changes due to temperature: a simple Model Dusseault presented the following simple model of stress changes as a function of temperature changes assuming the following:

- a permeable reservoir at initial stress conditions $\sigma_{v1}\sigma_{hi}$, σ_{h2}, pore pressure u.
- temperature to principal stresses corresponds to vertical and horizontal directions.
- linear elastic behavior and Hook's Law applies:

$$\varepsilon_j = \left(\frac{1}{E}\right)\left[\Delta\sigma_i^1 - v\left(\Delta\sigma_k^1\right)\right] + \beta\Delta T \quad (i, j, k = x, y, z\, cyclic) \tag{3.32}$$

E_i, v, and β are Young's modulus, Poisson's ratio, and the coefficient of linear thermoplastic expansion. The stress-strain law is written in terms of effective stress changes Terzaghi's effective stress law is assumed to apply for incremental principal stress changes ($\Delta\sigma_i$):

$$\sigma_{ij}^1 = \sigma_{ij} = \alpha\rho\delta_{ij}\, or\, \Delta\sigma_i^1 = \Delta\sigma_i - \Delta\mu \tag{3.33}$$

where α is Biot's parameter; $\alpha = 1.0$ in this analysis. The reservoir is laterally extensive, thus the total vertical stress is approximately unchanged during uniform pressure or temperature changes; $\Delta\sigma_v = 0$. Assuming a uniform pressure change $\Delta T = 0$), one may show:
when a lateral strain, $\varepsilon_x = \varepsilon_y = 0$, Eq. (1) is reduced to

$$\Delta\sigma_x^1 = \Delta\sigma_x^1 = \frac{v}{1-v}\Delta\sigma_x^1 \tag{3.34}$$

Note that stress changes because of injection or production can lead to shearing or tensile yield, depending on initial conditions, Poisson's ratio, and the process.

In the more general case, Δu and $\Delta T \neq 0$, where u is the internal energy. Using the same boundary conditions and the equations above, one may show:

$$\Delta\sigma_k^1 = \Delta\sigma_x^1 = \Delta\sigma_y^1 = \frac{v}{1-v}\Delta\sigma_x^1 + \frac{\beta E}{1-v}\Delta T \tag{3.35}$$

h is a dummy variable.

Note that a temperature change causes a stress change proportional to the stiffness (E), whereas a pore pressure change does not.

This model was adopted by S. Goodarzi et al., who simulated CO_2 sequestration and storage.

Poroelastic model.

diffusion/diffusion processes are characterized by the theory of poroelasticity introduced by Biot (1941). Rice and Cleary (1976) have recast Biot's theory in terms of physical concepts. The equations governing the responses of fluid-infiltrated porous solids are expressed as:

$$2G\varepsilon ij2G\zeta = \sigma ij - v1 + v\sigma kk\delta ij + \alpha(1-2v)1 + vp\delta ij = \alpha(1-2v)1 + v(\sigma kk + 3Bp)2G\varepsilon ij$$
$$= \sigma ij - v1 + v\sigma kk\delta ij + \alpha(1-2v)1 + vp\delta ij2G\zeta = \alpha(1-2v)1 + v(\sigma kk + 3Bp)$$

$$\begin{aligned} 2G\varepsilon_{ij} &= \sigma_{ij} - \frac{v}{1+v}\sigma_{kk}\delta_{ij} + \frac{\alpha(1-2v)}{1+v}p\delta_{ij} \\ 2G\zeta &= \frac{\alpha(1-2v)}{1+v}\left(\sigma_{kk} + \frac{3}{B}p\right) \end{aligned} \tag{3.36}$$

where the indices take the values 1, 2, and 3 and repeated indices imply summation. The constitutive equations are expressed in terms of the total stress σ_{ij}, the pore pressure p, and their respective conjugate quantities, the solid strain $\varepsilon_{ij,}$ and variation of fluid volume per unit reference pore volume ζ. The basic material constants are the shear modulus G, the drained and undrained Poisson's ratios v and v_u, and the Biot's effective stress coefficient α. B is Skempton's pore pressure coefficient:

$$B = \frac{3(v_u - v)}{\alpha(1-2v)(1+v_u)} \tag{3.37}$$

Linear poroelastic processes are described by the constitutive equations, Darcy's law, the equilibrium equations, and the continuity equations. A set of five material constants, G, ν, ν_u, α, and κ, are needed to fully characterize a linear isotropic poroelastic system. These equations are combined into field equations in terms of u_i and p which consist of an elasticity equation with a fluid coupling term,

$$G\nabla^2 u_i + \frac{G}{1-2v}u_{k,ki} - \alpha p_{,i} = -F_i \tag{3.38}$$

and a diffusion equation with a solid coupling term,

$$\frac{\partial p}{\partial t} - \kappa M \nabla^2 p = -\alpha M \frac{\partial \varepsilon_{kk}}{\partial t} + M(\varphi - \kappa f_{i,i}) \tag{3.39}$$

where κ is the permeability coefficient, which is equal to k/μ, k is the intrinsic permeability, and μ is the fluid dynamic viscosity, φ is the source density (the rate of injected fluid volume per unit volume of the porous solid), $f_i = \rho_f g_i$ is the body force per unit volume of fluid, F_i is the body force per unit volume of the bulk material, and M is Biot's Modulus:

$$M = \frac{2G(v_u - v)}{\alpha^2(1 - 2v_u)(1 - 2v)} \tag{3.40}$$

The diffusion of pore pressure is coupled with the rate of change of the volumetric strain.

The response of a pressurized fracture can be obtained by superposition of two transient solutions corresponding to two nonzero boundary conditions on the fracture surface (Carter and Booker, 1982). These two fundamental loading modes are Mode 1

$$\sigma_n(x, y, z, t)p(x, y, z, t) = -H(t), = 0; \quad \sigma_n(x, y, z, t) = -H(t), p(x, y, z, t) = 0 \tag{3.41}$$

Mode 2

$$\sigma_n(x, y, z, t)p(x, y, z, t) = 0, = H(t) \quad \sigma_n(x, y, z, t) = 0, p(x, y, z, t) = H(t). \tag{3.42}$$

where x, y, z correspond to the coordinates of the surface of the pressurized fracture, $H(t)$ denotes the heaviside step function. The initial conditions for both modes are stress-free and zero pore pressure everywhere. Fig. 3.52 illustrates the decomposed boundary conditions.

Load decomposition for a pressurized fracture in a poroelastic rock: mode 1 (stress loading) is represented by a unit normal stress, σ_n, applied on the fracture surface; mode 2 (pore pressure loading) is represented by a unit pore pressure, p, (equal to σ_n) applied on the fracture surface.

The responses of the model such as stress distribution, pore pressure distribution, and aperture opening can be obtained in terms of response functions F_1 and F_2 for modes 1 and 2, respectively (Carter and Booker, 1982). Considering the existence of far-field stress S_0 normal to the fracture surface and pore pressure p_0 (Fig. 3.53), the response due to applied constant hydraulic pressure p_f can be found by superposition of the responses of mode 1 and mode 2.

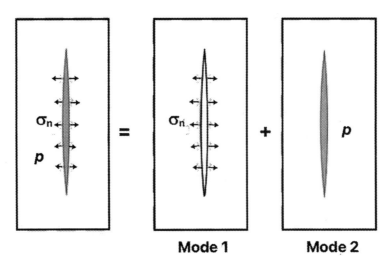

FIG. 3.52   Different modes of fracture.

3.9 Thermal stress

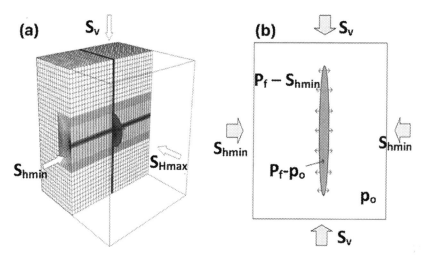

FIG. 3.53 Directions of stress and the resulting fracture.

A 3D mesh for the numerical simulation domain: (A) side view of the domain interior showing the circular fracture in *red*; (B) boundary conditions for the pressurized fracture in (A) showing a vertical section in the *yz*-plane.

Fig. 3.54 shows the mode 1 (stress loading) transient fracture opening profiles. Dimensionless time $t^* = ct/R^2$ is used for transient evolution of the fracture profile. For comparison, an elastic FEM simulation using a drained Poisson's ratio is also included. As illustrated in the figure, the FEM poroelastic results approach these asymptotic limits (short- and long-term responses). The long-term poroelastic results overlap with the elastic solution using drained Poisson's ratio (Fig. 3.55).

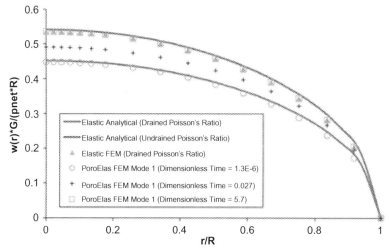

FIG. 3.54 The mode 1 (stress loading) transient fracture opening profiles.

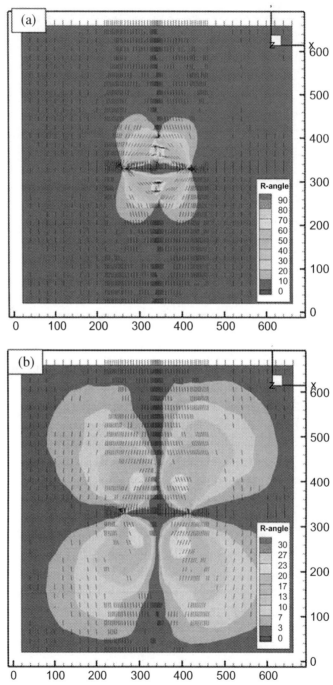

FIG. 3.55 Unsymmetrical distributions of reorientation angle (*R*-angle) of the minimum principal stress for the heterogeneous medium from a top view slice cutting through the center of the fracture: (A) $t = 7$ min; (B) $t = 24$ h.

3.10 Reservoir geochemistry

Petroleum geochemistry has played an important role in many areas of exploration and production of fossil fuels. This is even more meaningful in the case of complex reservoirs. It is commonly identified that five types of reservoir fluids exist according to the differences in composition. These also affect the sizes and shapes of each fluid's phase diagram (McCain, 1993). The five types of fluids are black oils, volatile oils, retrograde gas-condensates, wet gases, and dry gases. Each type is produced by different engineering techniques. The type of fluid is critical to production decisions and, therefore, must be determined early in the life of a reservoir. The reservoir fluid type determines the: Method of fluid sampling. Laboratory tests were used in analyzing the samples. Surface equipment types and sizes. Procedures for determining oil and gas in place. Techniques for predicting oil and gas reserves. Processes for predicting future production rates. Plan of depletion. Selection of secondary or enhanced recovery methods.

Many of the more recent developments can be seen to have developed in parallel with developments in analytical chemistry such as gas chromatography and gas chromatography-mass spectrometry. For the past few decades, such analytical techniques have been used to search for trace amounts of compounds known as biomarkers present in oils and source rock extracts which can be used to provide valuable information on the origin and history of the oil (Carls et al., 2016). These biomarkers are molecular fossils derived from biochemicals in previously living organisms. Carls et al. (2016) introduced a new method, the principal advantage of which is that it provides sample-specific identification, whereas the Nordtest approach is based on multisample statistics. In this method, biomarkers were conserved relative to other constituents, and thus concentrations (per g oil) in initial beach samples were greater than those in fresh oil because they were lost more slowly than more labile oil constituents such as straight-chain alkanes and aromatic hydrocarbons. The purpose of the present study was to investigate whether ANSCO can be definitively identified by biomarker content, determine whether the biomarkers were weathering, and distinguish oil sources in time-series samples. They examined four classes of biomarkers, isoprenoids (acyclic terpenoids), triterpanes (mostly tricyclic), hopanes (pentacyclic triterpanes), and steranes (tetracyclic terpenoids; Table 3.3). The applied novel pattern-matching procedures to compare samples with ANSCO to verify its presence in specific samples (or not). Alternative hydrocarbon sources were similarly compared with oil from sample beaches to determine whether these sources were or were not explanatory. The results were confirmed with Nordtest 11 plots.

Over the past few decades, tremendous efforts have been expended to develop and utilize techniques as an aid to solving reservoir and production problems. In this paper, it is proposed to provide an overview of major developments that have occurred in several areas of geochemistry in recent years. This includes developments in reservoir geochemistry, such as the use of high-resolution gas chromatography for reservoir continuity studies and high-temperature gas chromatography for characterization of wax deposits.

The concept of organic geochemistry has been with us for many decades, if not centuries, in one form or another. For example, the presence of asphalt blocks in the Dead Sea many centuries ago was indicative of petroleum accumulations in the area, although the people

TABLE 3.3 Biomarkers and their abbreviations[a]

Biomarker	Abbreviation		Target ions
Isoprenoids			
Norpristane	norprist		57
2,6,10,14-Tetramethylpentadecane (pristane)	prist		57
2,6,10,14-Tetramethylhexadecane (phytane)	phyt		57
Triterpanes			
C23 tricyclic terpane	TR23	*	191
C24 tricyclic terpane	TR24	*	191
C25 tricyclic terpane (a)	TR25a	*	191
C25 tricyclic terpane (b)	TR25b	*	191
C24 tetracyclic terpane	TET24	*	191
C26 tricyclic terpane (a)	TR26a	*,b	191
C26 tricyclic terpane (b)	TR26b		191
C28 tricyclic terpane (a)	TR28a	*	191
C28 tricyclic terpane (b)	TR28b	*	191
C29 tricyclic terpane (a)	TR29a	*	191
C29 tricyclic terpane (b)	TR29b	*	191
Hopanes			
18α(H),21β(H)-22,29,30-trisnorhopane	Ts	*	191
17α(H),21β(H)-22,29,30-trisnorhopane	Tm	*	191
17α(H),18α(H),21β(H)-28,30-bisnorhopane	H28	*	191
17α(H),21β(H)-25-norhopane	NOR25H		191
17α(H),21β(H)-30-norhopane	H29	*	191
18α(H),21β(H)-30-norneohopane	C29Ts	*	191
17α(H),21β(H)-30-norhopane (normoretane)	M29	*	191
18α(H) and 18β(H)-oleanane	OL		191
17α(H),21β(H)-hopane	H30	*	191
17α(H)-30-nor-29-homohopane	NOR30H	*	191
17β(H),21α(H)-hopane (moretane)	M30	*	191
22S-17α(H),21β(H)-30-homohopane	H31S	*	191
22R-17α(H),21β(H)-30-homohopane	H31R	*	191
Gammacerane	GAM	*	191

TABLE 3.3 Biomarkers and their abbreviations[a]—cont'd

Biomarker	Abbreviation		Target ions
22S-17α(H),21β(H)-30,31-bishomohopane	H32S	*	191
22R-17α(H),21β(H)-30,31-bishomohopane	H32R	*	191
22S-17α(H),21β(H)-30,31,32-trishomohopane	H33S	*	191
22R-17α(H),21β(H)-30,31,32-trishomohopane	H33R	*	191
22S-17α(H),21β(H)-30,31,32,33-tetrakishomohopane	H34S	*	191
22R-17α(H),21β(H)-30,31,32,33-tetrakishomohopane	H34R	*	191
22S-17α(H),21β(H)-30,31,32,33,34-pentakishomohopane	H35S	*	191
22R-17α(H),21β(H)-30,31,32,33,34-pentakishomohopane	H35R	*	191
Steranes			
C_{22} 5α(H),14β(H),17β(H)-sterane	S22	*	217 218
C_{27} 20S-13β(H),17α(H)-diasterane	DIA27S	*	217 218
C_{27} 20R-13β(H),17α(H)-diasterane	DIA27R	*	217 218
C_{27} 20S-5α(H),14α(H),17α(H)-cholestane	C27S	*	217 218
C_{27} 20R-5α(H),14β(H),17β(H)-cholestane	C27bbR	*	217 218
C_{27} 20S-5α(H),14β(H),17β(H)-cholestane	C27bbS	*	217 218
C_{27} 20R-5α(H),14α(H),17α(H)-cholestane	C27R	*	217 218
C_{28} 20S-5α(H),14α(H),17α(H)-ergostane	C28S	*	217 218
C_{28} 20R-5α(H),14β(H),17β(H)-ergostane	C28bbR	*	217 218
C_{28} 20S-5α(H),14β(H),17β(H)-ergostane	C28bbS	*	217 218
C_{28} 20R-5α(H),14α(H),17α(H)-ergostane	C28R	*	217 218
C_{29} 20S-5α(H),14α(H),17α(H)-stigmastane	C29S	*	217,218
C_{29} 20R-5α(H),14β(H),17β(H)-stigmastane	C29bbR	*	217 218
C_{29} 20S-5α(H),14β(H),17β(H)-stigmastane	C29bbS	*	217 218
C_{29} 20R-5α(H),14α(H),17α(H)-stigmastane	C29R	*	217 218

[a] Asterisks mark analytes are used for pattern matching; the number of triterpanes, hopanes, and steranes used for modeling is 10, 20, and 15, respectively.
[b] TR26a and TR26b cannot be resolved with current column settings at our laboratory, and thus were combined for modeling.

resident in the area at that time had no idea of the significance of these asphalt blocks, in terms of our current petroleum-based economy. However, in the early part of the 20th century, a significant change occurred. The generally acknowledged "father" of organic geochemistry is Alfred E. Treibs (1899–1983), who discovered and described, in 1936, porphyrin pigments in shale, coal, and crude oil, and traced the source of these molecules to their biological precursors (Kvenvolden, 2006). Treibs most notably identified the presence of porphyrins in crude oils and realized the structural relationship between these compounds and naturally occurring chlorophyll. This relationship was a very important step in establishing the theory of a biogenic origin for crude oils. From the classical work of Treibs in the 1920s and 1930s, it was a quantum leap to the 1970s and developments in analytical work directly associated with the lunar project and Viking missions to Mars. The search for extraterrestrial life was a major driving force behind the development of the hyphenated analytical technique of gas chromatography-mass spectrometry which, for the first time, permitted both separation and identification of individual organic compounds in very complex mixtures, exactly the impetus geochemistry required to move forward into the decades which followed (Phillip and Lewis, 1987). A parallel development was the concept of biomarkers, biological markers, or chemical fossils, as proposed by Calvin and Eglinton in their classic *Scientific American* paper (Eglinton and Calvin, 1967). In brief, it was proposed that there are many compounds in the geological record, generally hydrocarbons, that can be structurally related to their functionalized precursors occurring in living organisms or plants. This relationship, once established, therefore permits one to look for the presence of certain hydrocarbons in the geological record and make inferences concerning the type of source material responsible for the original sediments and possibly the nature of the depositional environment. More recent efforts have also demonstrated that certain compounds can be associated with specific organisms evolving at a certain time during the geological record, and hence the presence of these compounds can also be used to constrain the age of the source rock. Organic geochemistry is now a widely recognized geoscience in which organic chemistry has contributed significantly not only to geology (i.e., petroleum geochemistry, molecular stratigraphy) and biology (i.e., biogeochemistry), but also to other disciplines, such as chemical oceanography, environmental science, hydrology, biochemical ecology, archaeology, and cosmochemistry.

Islam et al. (2018) discussed this discovery but added that petroleum also has a nonorganic source. The two origins combined make the petroleum resource the most productive energy source.

The advances that were made in the 1970s came, in a large part, from a great deal of work looking at organic compounds in recent sediments where it was relatively easy to associate specific organic compounds with specific sources of organic matter. However, in the mid-1970s Wolfgang Seifert, a student of Treibs with experience of Chevron Oil in California, realized the significance and potential importance of biomarkers and their application to petroleum-related problems. Other oil companies may claim responsibility for applying geochemistry and biomarkers to such problems, but a look at the published record will show that it was the work of Seifert and his collaborators that really took hold of the concept and demonstrated that it would work in a way that is now being used routinely by oil companies on a world-wide basis. Analytical techniques continue to improve in terms of sensitivity and resolution and carbon number range. Much of the early work was done by simply looking at compounds containing up to about 40 carbon atoms. With the advent of high-temperature gas chromatography, this range has now been extended to 120 carbon atoms, potentially opening up a whole new arena of geochemical opportunities.

In recent decades, there has been a marked tendency to apply geochemical concepts to areas concerned with the exploitation and production of fossil fuels. Such areas were previously considered the domain of petroleum and reservoir engineers but the development of reservoir geochemistry has lead to its use in the investigation of many production and reservoir characterization problems. These applications are exceedingly important when it is remembered that in general two-thirds of the oil in a reservoir remains in place, even following secondary and tertiary recovery techniques. Processes responsible for poor production include reservoir heterogeneity; the formation of petroleum-derived barriers, such as tar mats, asphaltenes, bitumens, or waxes; interactions between the oil and the mineral matrix; low viscosity of waxy oils; and petroleum biodegradation. An understanding of the processes responsible for the emplacement of petroleum fluids, the nature of the inorganic/organic interactions in the reservoir, and knowledge of variations in petroleum compositions with time during initial production and secondary and tertiary recovery processes will lead to improved methods for the recovery of residual oils, over and above those currently available. In addition, more effective reservoir management will also result from a better understanding of reservoir geochemical problems.

Petroleum exploration also requires the support of predictive techniques to provide a clear picture of the mechanisms of oil generation occurring in a given sedimentary basin. The prediction of the quantity and quality of hydrocarbons generated by organic matter as well as the timing of generation are of primary importance to reduce the risks involved in petroleum exploration. The empirical approach to natural maturation can only provide partial information since naturally occurring series describing the whole process of maturation from diagenesis to metagenesis are very rare. Moreover, no complete mass balance can be obtained from these data since the migration of hydrocarbons and gases can lead to an underestimation of the amount of oil generated. This creates an obvious problem in evaluating marginal as well as unconventional reservoirs (Islam, 2014). This necessitates laboratory pyrolysis studies that have become an important tool to simulate maturation processes of organic matter and provide information on the timing, quantity, and nature of hydrocarbons and gases generated as well as on the behavior of the residual kerogen. Philp and Mansuy (1997) presented a comprehensive review of reservoir geochemistry and some developments in petroleum geochemistry, specifical aspects of biomarker geochemistry and the utilization of pyrolysis techniques for simulation of maturation.

Kvenvolden (2006) reviewed the previous 70 years of history of organic geochemistry. He identified the following phases:

1936–1959 discovery of the porphyrin pigments in shale, coal, and crude oil, and tracing the source of these molecules to their biological precursors. Thus, the year 1936 marks the beginning of organic geochemistry.

1959–1980s: Formal organization of organic geochemistry when the Organic Geochemistry Division (OGD) of The Geochemical Society was founded in the United States, followed 22 years later (1981) by the establishment of the European Association of Organic Geochemists (EAOG). Organic geochemistry (1) has its own journal, Organic Geochemistry (beginning in 1979) which, since 1988, is the official journal of the EAOG, (2) convenes two major conferences [International Meeting on Organic Geochemistry (IMOG), since 1962, and Gordon Research Conferences on Organic Geochemistry (GRC), since 1968] in alternate years, and (3) is the subject matter of several textbooks.

1990-current: Organic geochemistry is now a widely recognized geoscience in which organic chemistry has contributed significantly not only to geology (i.e., petroleum geochemistry, molecular stratigraphy) and biology (i.e., biogeochemistry), but also to other disciplines, such as chemical oceanography, environmental science, hydrology, biochemical ecology, archaeology, and cosmochemistry.

Geochemistry also plays an important role in predicting what types of oils may be associated with wax problems and the timing of the onset of wax deposition. Wax deposition generally results from changes to the supercritical character of petroleum fluids during accumulation and production. The temperatures in deeper reservoirs may exceed the critical temperature of the low molecular weight petroleum constituents, e.g., methane and ethane, and these compounds may act as supercritical solvents for the high molecular weight hydrocarbons. Loss of reservoir pressure during production will result in a reduction of the carrying capacity of the supercritical solvent system and will lead to wax precipitation. The waxes which form typically contain a relatively high concentration of high molecular weight hydrocarbons above C_{30}, and possibly up to C_{100}, which may not be observable in the oils collected at the well-head. The study of higher molecular weight (C_{40}) hydrocarbons had been largely overlooked prior to the development of high-temperature gas chromatography (HTGC) columns and supercritical fluid chromatography which now permits characterization of hydrocarbons extending to the C_{120}.

Fig. 3.56 shows a timeline of events in the development of the history of organic geochemistry. In 70 years, organic geochemistry, an amalgamation of aspects of organic chemistry and

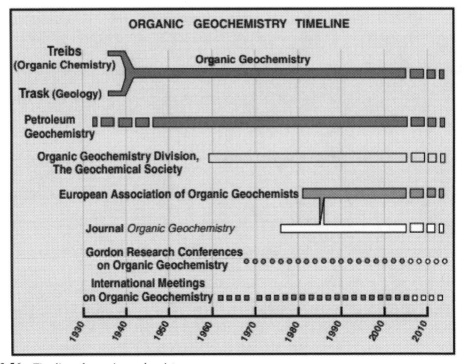

FIG. 3.56 Timeline of organic geochemistry.

geology, has become a mature and widely recognized geoscience. The beginnings can be traced back to the early part of the 20th century when Alfred E. Treibs turned his attention to geological materials and discovered the presence of significant organic molecules in shale, coal, and crude oil. Not only did he find complex organic molecules but, perhaps more important, he traced the source of these molecules to their biological precursors. Formal recognition of organic geochemistry came about in 1959 when an organization was established to promote organic geochemical research, and conferences focusing on organic geochemical topics have been convened and have met regularly since 1962. Because of the proliferation of scientific papers on organic geochemistry, a need was identified for a journal, and Organic Geochemistry was founded in 1977. Outstanding contributions to the subject have been acknowledged with the presentation of various awards starting two years after the journal was founded. Organic geochemistry is a highly interdisciplinary field, comprising fundamental and applied research, and it is now an important component of many studies in geology and other fields such as oceanography, hydrology, archaeology, and cosmochemistry.

3.10.1 Exploration geochemistry

The term exploration geochemistry covers issues related to the generation of oil and its migration from the source rock to the reservoir. This applies to both conventional (Islam et al., 2018) and unconventional reservoirs (Islam, 2014). However, one topic that has been extremely important in exploration geochemistry is the use of biomarkers for source determination, maturity, characterization of depositional environments, determination of the extent of biodegradation, and correlation of oils with their suspected source rocks. This is mainly because the vast majority of researchers in the west rely on the biogenic origin of crude oil. All of this information forms an important part of any exploration and development program, and when combined with geological and geophysical information can produce a comprehensive picture of the basin being evaluated. In this section, a brief overview of the information obtainable from various classes of biomarkers is discussed, as reported by Philp (2007).

1. *n*-Alkanes. The *n*-alkanes are the most extensively studied group of biomarkers since they can be readily analyzed by GC alone. Alkanes are derived from a variety of sources, and their carbon number distributions have been widely used to differentiate marine vs. terrestrial source materials. The odd/even carbon number preference of these distributions will change systematically with maturity, and biodegradation will also change the distribution by initially removing the lower number of carbon compounds preferentially over the longer chain compounds. A notable characteristic of oils derived from terrigenous source materials as being their waxy nature resulting from the high wax content of their higher plant source materials. The discovery of cutans in various plants has introduced another possible source of *n*-alkanes upon the thermal breakdown of this material. *n*-Alkane distributions in many crude oils typically maximize in the C_{20}–C_{40} range but with the advent of high-temperature GC columns, the analysis of oils, bitumens, and waxes from a variety of source materials has shown the existence of compounds extending to at least C_{70} and beyond.

2. Isoprenoids. The ratio of the isoprenoids, pristane, and phytane, has long been associated with the nature of depositional environments. Many oils often contain high concentrations of long-chain isoprenoids ranging up to at least C_{40} or C_{45} as determined by GCMS and single-ion monitoring of the ion at m/z 183. The majority of these compounds are regular isoprenoids or head-to-head isoprenoids commonly associated with archaebacteria. While the presence of these compounds may reflect the enhanced levels of microbial activity in the original oxic depositional environments, the possibility that the regular isoprenoids are also derived from polyprenols is known to be associated with higher plant waxes cannot be excluded.
3. Sesquiterpanes. Abundant concentrations of sesquiterpanes were unequivocally identified for the first time in several Australian crude oils known to be sourced from the terrestrial source material. Early work of Alexander et al. (1983) identified the major sesquiterpanes previously observed by Philp et al. (1981) as having drimane- and eudesmane-based structures, with more recent studies showing that these compounds extend to at least C_{24}. In view of the apparent absence of eudesmanes in samples older than the Devonian Period, it was concluded initially that the eudesmanes were associated with an input from higher plants. The abundance of drimanes in the older sediments led to the conclusion that they were associated with a microbial input, although more recently drimane precursors have also been found in higher plants. The possibility of using these compounds as maturity indicators has not been investigated in detail but is something that may be potentially useful. Other sesquiterpanes such as the resin-derived cadalanes and cedrene and cuparene derivatives in Chinese oils known to contain a predominance of higher plant source material have also been identified.
4. Diterpanes. Tri- and tetracyclic terpanes in the C_{19} and C_{20} range have long been known to be associated with resins from higher plants. The presence of their hydrocarbon analogues in oils and source rocks has been used to establish a higher plant input to the source rocks. Diterpanes are abundant in many oils known to contain higher plant materials from countries such as Australia, New Zealand, Taiwan, Papua New Guinea, and Indonesia in the Southern Hemisphere. Diterpenoids based on the labdane, abietane, primarane, beyerane, kaurane, and phyllocladane structures typically occur in gymnosperms. Kauranoid-type diterpenoids are probably a ubiquitous component of all angiosperms. In contemporary resins, primarane-type diterpenoids occur in conifers of the Pinaceae and Cupressaceae families and the southern conifer families of Podocarpaceae and Araucariaceae. Precursors of phyllocladane occur widely in the Podocarpaceae family and those of kauranes in the Podocarpaceae, Araucariaceae, and Taxodiaceae families. Diterpanes have also found limited use as maturity and depositional environmental indicators.
5. Sesterterpanes. Significant concentrations of the 17,21-C_{24}-secohopane were found in oils derived from terrigenous source materials, particularly Gippsland Basin oils from Australia, although this may be related to a relatively oxic depositional environment rather than a direct source indicator. More recent work has shown the presence of a variety of degraded compounds in the oils derived from terrigenous source materials, including those from New Zealand, Nigeria, Indonesia, and Taiwan, and including the C_{24} des-A-ring analogues of the oleananes, lupanes, and ursanes which are typically very abundant in many of these oils (Fig. 3.57). While it may be proposed that these des-A-ring compounds

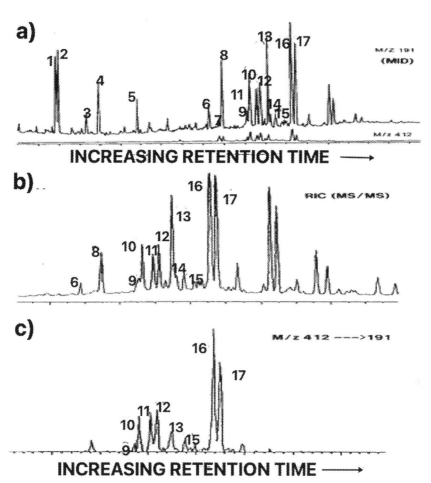

FIG. 3.57 Crude oils of Tertiary age derived from predominantly terrestrial source materials contain relatively high concentrations of des-A-sesterterpanes, which are thought to be derived from the C_{30} triterpanes associated with the angiosperm input to the source materials. In this figure, the des-A-sesterterpanes are the peaks labeled 1–4. The C_{30} triterpanes associated with the angiosperms are peaks 9, 11, 12, 14, and 16.

form via some method of A-ring degradation, it should be noted that several ring-A fissioned derivatives of pentacyclic triterpen-3-ols have been reported.

6. Pentacyclic terpanes. Hopanes. The distribution of hopanes in oils is typically very simple although there are a large number of variations that can be related to the depositional environment since virtually all the hopanes are derived from a bacterial source. In most oils and source rock extracts, the distributions are dominated by regular hopanes, maximizing at C_{30} with the concentration of the extended hopanes above C_{31} decreasing exponentially, characteristic of the distribution associated with more oxic-type environments. Methylhopanes and the various series of norhopanes identified in sample sets from other

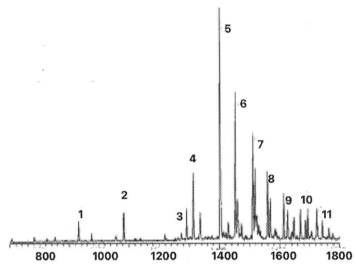

FIG. 3.58 m/z 191 chromatogram for an oil derived from a carbonate source rock. Characteristic terpane features of this type of oil include the following: relatively high concentrations of the C23 tricyclic terpane (1) and the C_{24} 17,21-secohopane (2); the C29 hopane (5). than the C30 (6) hopane; and the C35 extended hopane (11) > than the C34 extended hopane (10). Ts (3) and Tm (4) are the 18R(H)- and 17R(H)-C27-tris(norhopanes), respectively; C28 is the 29,30-bis(norhopane), and the peaks labeled 7–9 are the extended hopanes from C_{31} to C_{33}.

environments and source materials are generally absent or are in very low concentrations in oils derived from terrigenous sources although Haven et al. (1988) did report the occurrence of methylhopanes in coal samples. Oils derived from carbonate source rocks have very characteristic distributions with the C_{29} hopane, often being present in greater abundance than the C30 component since the C_{29} compound can be derived from both the regular hopane series and the 30-norhopane series (Fig. 3.58). Hence if both series of hopanes are present, as they often are, in carbonate-derived oils, then the relative contribution of the C29 component will be significantly higher. In addition, variations in the distribution of the extended hopanes, beyond C_{31}, often show a significant increase in the relative concentrations of the C35 components in oils from carbonate environments as a result of the highly reducing conditions during deposition.

7. Nonhopanoid terpanes. Nonhopanoid terpanes derived from higher plant sources are far more abundant than the hopanoids in many oils from terrigenous source materials. Higher plants contain a wide variety of triterpenoids and many of their hydrocarbon derivatives can be found in the corresponding oils and source rocks. These compounds are based on pentacyclic structures such as oleananes, lupanes, and ursanes, along with their demethylated analogues (Fig. 3.59). These compounds are abundant in the oils of New Zealand, Taiwan, Beaufort-Mackenzie Basin, Canada, and Nigeria. Oleananes may be derived from a number of naturally occurring precursors, including taraxer-14-en-3â-ol and olean-12-en-3â-ol, which through a series of complex reduction and isomerization

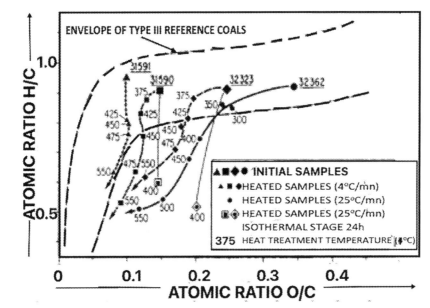

FIG. 3.59 Comparison of natural and artificial evolution paths (open pyrolysis) in a van Krevelen diagram.

reactions can produce the 18R(H)- and 18â(H)-isomers of oleanane. Cadalane-type triterpanes are derived from precursors present in dammar resins and particularly abundant in oils from Indonesia. Although the predominant members of the series are in the C30 range, it has been shown that analogues of these compounds are present in the higher molecular weight region with components up to C60 and C75 and possibly higher. The presence of the saturated hydrocarbons is often accompanied by 1,6-dimethylnaphthalene and cadalene in the aromatic fraction.

8. Steranes. Steranes in oils and source rock extracts are derived from naturally occurring sterols. Conversion pathways are complex and affected by both diagenetic and thermal reactions. However, in general, the carbon number distributions reflect the distributions of the sterols in the original source materials. Sterane distributions in oils derived predominantly from terrigenous source materials are very simple and are dominated by the C_{29} steranes which are in turn derived from the C_{29} sterols present in higher plants. However, in the situation with a mixed input of higher plant material and algal material, the problem is more complex since, in addition to the C_{27} and C_{28} steranes, derived from algae, some of the C_{29} steranes may also be derived from the algal material. Hence in many cases, the sterane distributions are more commonly used for correlation purposes rather than trying to provide an absolute determination of the nature of the source material. Another characteristic feature in oils and source rock extracts is the presence of diasteranes. It was proposed initially that the concentration of diasteranes relative to the steranes reflected the presence of clay minerals and their ability to catalyze sterene rearrangement reactions. More recently it has become apparent that the oxicity of the depositional environment provides an additional clue as to the fate of the sterols since in a highly anoxic,

or reducing, environment the sterenes formed from the sterols will be reduced, thus reducing the amount of sterenes available for the rearrangement reaction. In a more oxic-type environment less of the sterenes will be reduced and available for rearrangement to diasterenes.

3.10.2 Thermal characterization of organic matter and artificial maturation

In the preceding sections, the importance of geochemistry in various aspects of reservoir and exploration problems has been noted. Prospecting for oil and gas accumulations requires the evaluation of the generation potential of petroleum source rocks as well as their maturity levels and kerogen type since the abundance of petroleum in a sedimentary basin will be determined to a large extent by these parameters. Such information helps in the basic understanding of oil generation and in locating source rocks in a petroleum basin when associated with geological settings. Predictive techniques such as artificial maturation are also an important aspect of petroleum exploration. Understanding the mechanisms of maturation of source rock and quantifying oil generation provide information and parameters particularly useful in the elaboration and calibration of kinetic models. To assist in these processes, various pyrolysis techniques have been used for the characterization of organic matter and to simulate maturation processes.

3.10.2.1 Source rock evaluation

To evaluate the potential of a source rock to generate oil or gas, the sample is pyrolyzed with a very rapid heating rate in a flow of carrier gas (Ar, He, or N_2) which transfers the pyrolysis products directly to a detector (thermal conductivity detector and flame ionization detector) or to the front end of the chromatographic column. The most widely used technique for the evaluation of source rock potential to generate oil or gas is based on the Rock Eval pyrolysis system. The method consists of pyrolyzing a small sample of rock (about 100 mg) in an inert atmosphere (helium) at 250°C for 3–4 min and then from 250 to 600°C at 25°C/min. The pyrolysis products are swept to a thermal conductivity detector for the quantification of oxygen-containing compounds and to a flame ionization detector for the analysis of hydrocarbons. Three quantities are measured:

(1) the free and adsorbed hydrocarbons already present in the sample and vaporized at 300°C (peak S1);
(2) the hydrocarbons generated by thermal cracking of the kerogen between 300°C and 600°C (peak S2)
(3) the CO_2 issuing from kerogen cracking is trapped in the 300–390°C temperature range and detected at the end of the pyrolysis program (peak S3). Rock Eval pyrolysis systems can also be equipped with a "Carbon module" which permits determination of the total organic carbon (TOC) content by an oxidation phase at 600°C for 7 min at the end of the preceding cycle. Several ratios can be calculated from the generated data. The ratio S1/[S1 + S2] is also known as the production index (PI) and represents the amount of gas and petroleum that can be generated from the organic matter during maturation in the absence of migration. The petroleum potential of the kerogen can be expressed as the hydrogen index, HI, (S2/TOC). The maximum pyrolysis temperature, T_{max}, recorded at the

maximum of hydrocarbon generation during pyrolysis (at the top of the peak S2), is a maturity index. T_{max} increases with the increasing maturity of the samples. The ratio S3/TOC is defined as the oxygen index (OI). Two diagrams can be used to classify organic matter. The HI vs. OI diagram is analogous to the H/C vs. O/C van Krevelen diagram used to define kerogen type, and the HI vs. T_{max} diagram is used to define both kerogen type and maturity of the kerogen. Two major factors can affect the Rock Eval pyrolysis results of whole rock samples. The "mineral matrix effect" which reflects the activity of different mineral phases will lower the HI values. The presence of heavy bitumens in whole rock samples can affect the S2 peak and T_{max}. The heavy ends of bitumens are not vaporized at 250°C and can contribute to the S2 peak, thus inducing an overestimation of the petroleum potential of source rock. Solvent extraction of the rock prior to Rock Eval pyrolysis provides a more accurate estimate of the S2 peak in this situation but means that the sample has to be analyzed twice in order to get both S1 and S2 data. Pyrolysis-gas chromatography and pyrolysis-gas chromatography-mass spectrometry system (Py-GC and Py-GC-MS) have several similarities to Rock Eval pyrolysis with the major difference being that the pyrolysis products are actually resolved by gas chromatography and identified by mass spectrometry. Py-GC and py-GCMS are commonly used to define the kerogen types and to provide a rapid semiquantitative chemical characterization of the solid at the molecular level.

3.10.2.2 Maturation studies

Open pyrolysis was widely used in the 1980s to simulate the maturation of organic matter. The pyrolysis is performed in a similar manner to that described above but with much lower heating rates (4–25°C/min with a 24 h isothermal stage at the end). The pyrolysis temperatures range between 250°C and 600°C. The pyrolysis products obtained in this way contain olefinic species rarely present in crude oil and the evolution of the pyrolysis products with increasing maturity is far different from that expected under geologic conditions. There are also discrepancies in product yields and their evolution with maturation, with product yields being highly dependent on heating rates (decrease of the oil yield with decreasing heating rate). Analyses of the residual kerogens following open pyrolysis also showed differences manifested in the results of elemental analyses, petroleum potential, T_{max} (as determined by Rock Eval pyrolysis), and ratios determined by IR which often show evolutionary pathways that deviate from the natural trends (Fig. 3.59).

To improve the quality of the simulation, low heating rates are required, but this, in turn, leads to lower product yields. Poor results obtained in terms of simulation of natural processes can be attributed to the nature of the reacting medium developed in such a pyrolysis system. Since the pyrolysis products are swept out of the pyrolysis device as soon as they are generated, the open medium prevents any intimate contact between the hydrocarbon species. In this way, the hydrogen transfer from the source to radical species is very limited.

Closed pyrolysis systems can be divided into those which use additional water and those which are operated under supposed anhydrous conditions. (i) Hydrous pyrolysis was developed in the late 1960s when organic geochemists realized the ubiquity of water in sedimentary basins and its eventual importance in the process of organic matter maturation. The experiments generally are performed in stainless steel reactors in the presence of added water at temperatures below the supercritical temperature of the water. The ability of this pyrolysis

FIG. 3.60 Comparison of natural and artificial evolution paths (b anhydrous and O hydrous pyrolysis) in a van Krevelen diagram.

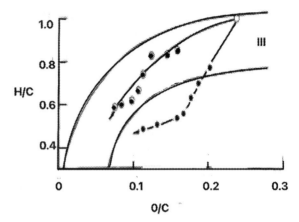

system to reproduce the main processes of organic matter maturation is well-established. The nature and the distribution of pyrolysis products resemble those of natural crude oil, and elemental analyses of the residual solid after pyrolysis show that the experimental trend is very close to the natural maturation pathway in a van Krevelen diagram (Fig. 3.60). One can argue its ability to simulate hydrocarbon expulsion (as proposed by Lewan et al.) since the pyrolysis-inducing high temperatures and excess water can strongly modify the parameters of expulsion mechanisms as defined under geological conditions. If the results are good in terms of quality, some quantitative differences appear when results are compared with those of other pyrolysis or natural systems.

This simulation of the geothermal process offers a formidable challenge from a scaling perspective. It will be discussed in a later chapter.

3.11 Fluid and rock properties

We discussed the approach to defining the fluid and rock properties. The fluid is considered to be a continuum media, and the rock properties are average REV in spite of the fact the existence of the REV is tenuous because it has never been identified in real media. The size of the REV is related to how locally correlate the property on the pore (microscopic) scale. It should be large enough for statistically meaningful averaging. If we follow the traditional definition of REV as given in Fig. 3.61, it should be small enough to avoid heterogeneity. As a very rough number, the typical REV size is somewhere around 100–1000 grain diameters. It should be emphasized that despite all ambiguities, the notion of REV is essential to allow us to use continuous mathematics. Our main focus is on all properties needed in the flow computations. These properties formed a central part of the inputs to a numerical simulator.

3.11.1 Fluid properties

Petroleum deposits vary widely in properties. The bulk of the chemical compounds present are hydrocarbons that are composed of hydrogen and carbon. A typical crude oil

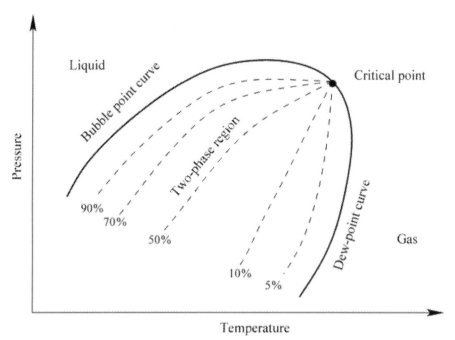

FIG. 3.61   A typical pressure-temperature diagram of crude oil.

contains hundreds of different chemical compounds and normally is separated into crude fractions according to the range of boiling points of the compound included in each fraction. Hydrocarbons may be gaseous, liquid, or solid at normal pressure and temperature depending on the number and arrangement of the carbon atoms in the molecules. Hydrocarbons are moved from a gaseous state to a solid state with increasing the number of carbon atoms. The compound with up to four carbons is gaseous, with 4–20 are in the liquid state, and those with more than 20 carbon atoms are solid. Liquid mixtures, such as crude oils, may contain gaseous or solid compounds or both in solution. Several nonhydrocarbons may occur in crude oils and gases such as sulfur (S), nitrogen (N), and oxygen (O).

The petroleum reservoirs may be classified according to the state of the fluid compound and divided into two broad categories of oil and gas reservoirs. These may be subdivided into different groups according to the hydrocarbon compounds, the initial temperature and pressure of the reservoir, and so on. The phase diagrams can be applied to express the behavior of the reservoir fluid in a graphical form. The pressure-temperature diagram is one of them that can be applied to illustrate the reservoir fluid behavior and to classify reservoirs. A typical phase diagram of a multicomponent compound is given in Fig. 3.35. In this diagram, the critical point is the point at which all properties of the liquid and gaseous state are identical.

3.12 Avoiding spurious solutions

There are numerous reasons a computer modeling run can fail. These include software errors. Most modelers are aware of cases, where cautions must be exercised. Although an experienced modeler can avoid these occurrences of computer "glitches," few documented their experiences. As such, this section is written to aid an inexperienced modeler. Consider the following advice:

1. Making Jacobian diagonally dominant at times of computational trouble

Unexpectedly during simulation, the system of equations during iteration can become ill-conditioned, leading to failure to converge. To avoid this situation, *always* add a small number (10^{-6}) to all entries in the main diagonal of the Jacobian matrix. This way, the diagonal value will never become zero, which would cause the system to break down.

The addition of this small number will not affect the solution of the finite difference equations.

2. Creating Jacobian matrix

Jacobian matrix is useful for inversion of the solution matrix. The Jacobian matrix consists of the first-order derivative information of a vector-valued function with respect to some parameters of a mathematical problem. This sensitivity information not only can be used in optimization but also help to validate the implemented model. Large sensitivities in the simulation output indicate that small perturbations of model parameters might lead to uncontrollable behavior of the model. To avoid generating spurious solutions or computer crashes through the division of zero, a common technique is to use numerical partial differentiation of a function. This approach is used by the Computer Modeling Group (CMG) by numerical partial differentiation.

Another approach is *y*-analytical partial differentiation of F_i with respect to all unknown variables. Estimation of numerical differentiation of a property function (e.g., $B, \mu, K_r,$ and ρ) is carried out by locating the segment that encloses the specific value of the unknown (dependent variable) in the function table and then taking the slope of that segment.

3. Handling phase appearance and disappearance

Whenever a phase disappears, meaning ceases to be part of the flow equation, the flow equation becomes constrained with zero saturations and/or irrelevant parameters within the flow equation. One of the major difficulties arising in fully implicit black oil or thermal simulation occurs during phase appearance or disappearance. Usually, variable or equation substitution is used to circumvent this problem. Another approach, which uses pseudo *K* values, originally was suggested by Crookston et al. Essentially, this method prevents phase disappearance by altering the *K* values so that a small amount of the phase in question remains in the system. This approach also can be used in black-oil systems, where a small amount of free gas (pseudogas) is always present. The pseudogas method is easier to code than the variable substitution method. However, the small amount of pseudogas may lead to an inaccurate solution or less efficiency than variable substitution.

4. Handling phase saturation overshoot (negative values).

This problem can occur during the implicit solution of the flow equations. If a phase (*o, w,* or *g*) overshot occurs during iterations, then reassign only that phase saturation as This problem can occur during the implicit solution of the flow equations. If a phase (o, w, or g) overshot occurs during iterations, then reassign only that phase saturation and continue iterations.

CHAPTER 4

Unconventional reservoirs

4.1 Introduction

Unconventional reserves are typically undervalued. The potential of these resources is far greater than what is set out to be the norm, with multilateral and fracking. The true potential lies within the development of custom-designed schemes based on the nature of resources. The first question to ask is if there is natural productivity. Of course, if the answer is yes, then the reservoir falls under the realm of conventional technology. However, if there is no natural productivity or if the productivity is too low to sustain economic developments, the next question to be asked is if there is a fracture network in place. Even though unconventional gas reservoirs are mostly fractured, whenever they don't have clear fracture networks, injectivity is a major issue. The case becomes akin to tar sand that has very low injectivity. For this set of reservoirs, SAGD equivalent schemes are recommended.

In 2013, the IEA accepted the notion that as technologies and economic constraints changes, definitions for unconventional and conventional oils also change. IEA (2013) announced

> Conventional oil is a category that includes crude oil - and natural gas and its condensates. Crude oil production in 2011 stood at approximately 70 million barrels per day. Unconventional oil consists of a wider variety of liquid sources including oil sands, extra heavy oil, gas to liquids and other liquids. In general conventional oil is easier and cheaper to produce than unconventional oil. However, the categories "conventional" and "unconventional" do not remain fixed, and over time, as economic and technological conditions evolve, resources hitherto considered unconventional can migrate into the conventional category.

However, the above definition is not universal. As Gordon (2012) points out, the US Department of Energy (DOE) recognized that "unconventional oils have yet to be strictly defined." The US DOE divides unconventional oil into four types: heavy oil, extra heavy oil, bitumen, and oil shale. Some analysts also include gas-to-liquids (GTL) processes for converting natural gas to oil and coal-to-liquids (CTL) processes for converting coal to oil in the unconventional oil category. These unconventional oil-processing techniques broaden the feedstock of unconventional oils to include unconventional natural gas, such as tight gas, shale gas, coal-bed methane, and methane hydrates. GTL processing entails converting natural gas and other simple gaseous hydrocarbons into more complex petroleum products. Methane-rich gases are converted into liquid synthetic fuels through direct conversion or syngas as an intermediate using the Fischer Tropsch or Mobil processes. CTL processing entails the liquefaction of solid coal. This can be done directly by dissolving coal in a solvent at high temperature and pressure and then refining these liquids to yield high-grade fuel characteristics. Indirect liquefaction gasifies the coal into a mixture of hydrogen and carbon monoxide (syngas), condensing this over a catalyst and using the GTL processes to produce liquid petroleum products. In unconventional petroleum resources, the terms "shale oil" and "tight oil" are often used interchangeably in public discourse, shale formations are only a subset of all low permeability tight formations, which include sandstones and carbonates, as well as shales, as sources of tight oil production (EIA, 2013). Within the United States, the oil and natural gas industry typically refers to tight oil production rather than shale oil production because it is a more encompassing and accurate term with respect to the geologic formations producing oil at any particular well. EIA has adopted this convention and develops estimates of tight oil production and resources in the United States that include, but are not limited to, production from shale formations.

Islam (2014) took a scientific approach and included even gas hydrates in the repertoire of unconventional oil and gas. Islam et al. (2018) included basement reservoirs in that category. They also argued that economic considerations should not be used to define the category of unconventionality as with sustainable development techniques, costs are not correlated with the overall efficiency or sustainability. The component is the introduction of zero-waste concept.

Recently, it has become evident that unconventional petroleum reservoirs offer the best hope of radical increase global petroleum reserve. Since the introduction of the extensive development of the US shale gas and tight oil in 2008 onward, great changes in the global energy industry have evolved around unconventional reservoirs. Islam (2020) recently demonstrated that the potential of unconventional petroleum reserves is far greater than any enhanced oil recovery scheme designed to date. However, when the two are combined and added to that is the scientific basis for the revised assessment of world gas reserve, the energy outlook becomes very bright. The unconventional energy resource was largely overlooked until the new millennium. Before 2000 only a few small oil and gas companies were pursuing such plays led by independent, Mitchell Energy and Development Corporation of The Woodlands, TX. Mitchell's successful commercial development of the Mississippian Barnett Shale of the Ft. Worth Basin, Texas spawned an energy renaissance in North America. This chapter explores the science behind unconventional oil and gas and provides one with a basis that the realistic estimate of total available oil and gas can easily exceed all estimates of the past. With newfound oil and gas reserves and technologies that harness them with environmental integrity, the world could easily see another 100+ years of crisis-free energy outlook.

Fig. 4.1 shows IEA's vision of a sustainable development scenario. IEA vision is premised on considering petroleum resources as inherently unsustainable. As such, any technology, which is not fossil fuel-based, is considered to be sustainable. This sustainability consideration is stoutly flawed. This chapter shows how true sustainability can be reached.

The geology and reservoir features of unconventional reservoirs are markedly different from the conventional ones considered in the past. This requires nothing short of revolution

4.1 Introduction

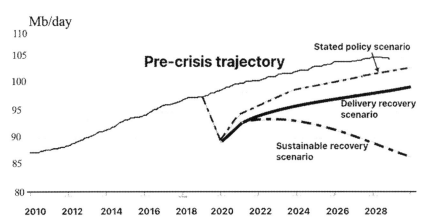

FIG. 4.1 Global oil demand by scenario. *From IEA, 2020a, https://www.iea.org/reports/oil-2020.*

on the classical petroleum geological theories and related engineering (Jian, 2017). Jia (2017) summarized the recent progress in global unconventional oil and gas exploration and development and pointed out that the unconventional oil and gas revolution not only has a significant economic significance of oil and gas resource increment but also brings great innovation to the theory of petroleum geology, thus having important scientific significances. Islam (2014) critically reviewed existing theories and laws of both Newtonian mechanics and quantum mechanics origins and concluded that current scientific technologies are not compatible with the reality of unconventional petroleum reservoirs. In Chapter 5, some of the geological aspects of unconventional reservoirs in the context of basement reservoirs. This chapter summarizes the core contents of various aspects of hydrocarbon generation, reservoir, distribution, and development in both classical petroleum engineering sense and sustainable zero-waste engineering sense.

The rapid development of unconventional oil and gas in the past decade has given rise to all-around changes in the supply-demand structure, recovery method, and technical innovation in the petroleum industry. This surge in unconventional reservoir development was triggered by events in the United States. In particular, unconventional shale resource systems have provided North America with abundant energy supplies and reserves for the present and future decades. Javie (2014) reviewed the history of such developments in the United States. In the last decade, the development of unconventional shale resource systems has been phenomenal in the United States with abundant new supplies of natural gas and oil. These resource systems are all associated with petroleum source rocks, either within the source rock itself or in juxtaposed, no source rock intervals. Three key characteristics of these rocks, apart from being associated with organic-rich intervals, are (Javie, 2014):

- their low porosity (less than 15%);
- ultra-low permeability (<0.1 mD); and
- brittle or nonductile.

These characteristics play a role in the storage, retention, and requirement of high-energy stimulation to obtain petroleum flow. Understanding each of these features is important for developing such reservoirs. The organic richness and hydrogen content certainly play a role in petroleum generation, but they also play a role in the retention and expulsion fractionation of generated petroleum. Often a substantial portion of the porosity evolves from the decomposition of organic carbon that creates organoporosity in addition to any matrix porosity. Open fracture-related porosity is seldom important. Hybrid systems, i.e., organic-lean intervals overlying, interbedded, or underlying the petroleum source rock, are the best-producing shale resource systems, particularly for oil due to their limited adsorptive affinities and the retention of polar constituents of petroleum in the source rock. Javie (2014) points to the paradox that the best shale gas systems are those where the bulk of the retained oil in the source rock has been converted to gas by the process, called *cracking*. Such conversion cracks the retained polar constituents of petroleum as well as saturated and aromatic hydrocarbons to condensate-wet gas or dry gas at high thermal maturity. Such high-level conversion also creates the maximum organoporosity, while enhancing pore pressure. Bitumen (petroleum)-free total organic carbon (TOC) is comprised of two components, a generative and a nongenerative portion. The generative organic carbon (GOC) represents the portion of organic carbon that can be converted to petroleum, whereas the nongenerative portion does not yield any commercial amounts of petroleum due to its low hydrogen content. Organoporosity is created by the decomposition of the GOC as recorded in volume percent. In the oil window, this organoporosity is filled with petroleum (bitumen, oil, and gas) and is difficult to identify, whereas in the gas window any retained petroleum has been converted to gas and pyrobitumen making such organoporosity visible under high magnification microscopy. This complex nature of the formation is reflected in the production decline analysis that shows the variable production potential of many North American shale gas and oil systems and their high decline rates. The most productive North American shale gas systems are shown to be the Marcellus and Haynesville shales, whereas the best shale oil systems are the hybrid Bakken and Eagle Ford systems.

Unconventional hydrocarbon resources have a wide distribution, are not substantially impacted by the hydrodynamic effect (also called "continuous depositional mine"), and consist of coalbed methane, tight gas, shale gas, gas hydrate, native bitumen, and oil shale, and so on (Jian et al., 2015). Unlike conventional oil and gas resources, unconventional petroleum fluids are distributed continuously mainly in slopes or syncline areas of depositional basins. Meanwhile, source rocks are integrated with or close to reservoirs, with high accumulating efficiency. This naturally leads to poor reservoir properties of reservoirs, in which hydrocarbons accumulate mainly in nano-sized micro-pores of source rocks, in tight reservoirs interbedded with or close to source rocks. In the geological theory of unconventional hydrocarbons, the lower limits of physical properties of target reservoirs are continuously extended for the hydrocarbon exploration, from good reservoirs with micrometer-millimeter pore throats to source rocks or near source rocks with nanometer-level pore throats. To accommodate for this extraordinary characteristic of unconventional reservoirs, Islam (2014) suggested overhauling the reservoir characterization tool. This reservoir characterization tool should be adapted to capture the special migration and filtration mechanisms of unconventional hydrocarbons to develop sustainable production strategies.

4.2 Unconventional oil and gas production

The United States continues to produce historically high levels of crude oil and natural gas and continues to increase exports of crude oil, petroleum products, and liquefied natural gas. EIA expects crude oil production to increase to 14 million barrels per day by 2022—an increase of nearly 7.6 million barrels per day in a decade—but to then level off, increasing by less than 400,000 barrels per day over the next decade as the agency expects operators to move to less productive plays and well productivity to decline. Oil production is expected to begin a slow decline in the mid-2030s, falling another 500,000 barrels per day over the next decade and declining below 12 million barrels per day by 2050.

Onshore tight oil development in the lower 48 continues to be the main driver of crude oil production, accounting for about 70% of cumulative domestic production in the forecast. Production in the Lower 48 is expected to peak at 13.84 million barrels per day in 2032, accounting for 96% of total production that year. Fig. 4.2 shows the projected outlook for the coming decades. Although renewable energy sources are projected to grow, they do not come close to fossil fuel contributions.

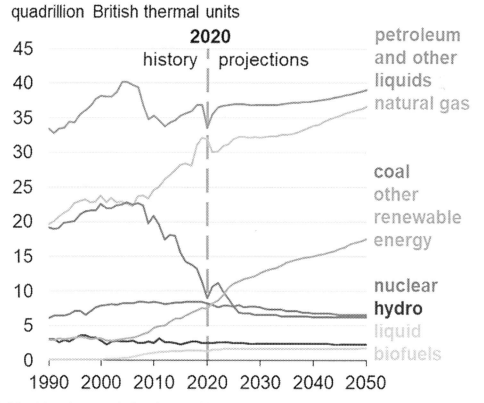

FIG. 4.2 Oil production outlook in the United States. *From EIA, 2020c, World Energy Outlook, https://www.eia.gov/outlooks/steo/report/global_oil.php.*

The continued development of tight oil and shale gas resources supports growth in natural gas plant liquids (NGPL) production, which reaches 6.6 million barrels per day by 2028. Natural gas plant liquids are light hydrocarbons predominantly found in natural gas wells and are diverted from the natural gas stream by natural gas processing plants. These hydrocarbons include ethane, propane, normal butane, isobutane, and natural gasoline. EIA expects production to grow by 26% during the projection period as a result of demand increases by the global petrochemical industry. Most NGPL production growth is projected to occur before 2025 as producers focus on natural gas plant liquids-rich plays, where NGPL-to-gas ratios are highest and increased demand spurs greater ethane recovery.

EIA (2020b) expects natural gas dry production to grow 1.9% per year from 2020 to 2025, which is slower than the 5.1%-per-year average growth rate from 2015 to 2020. Natural gas production grows at a faster rate than consumption after 2020, leading to an increase in US exports of natural gas. Natural gas consumption is expected to remain relatively flat through 2030 because of slower industrial sector growth and a decline in electric power consumption. After 2030, consumption growth is expected to increase almost 1% per year as natural gas use in the electric power and industrial sectors is expected to increase.

To satisfy the growing demand for natural gas, US natural gas production expands into less prolific and more expensive-to-produce areas, putting upward pressure on production costs. Natural gas prices are expected to remain lower than $4 per million British thermal units through 2050 because of an abundance of lower-cost resources, primarily in tight oil plays in the Permian Basin. These lower-cost resources allow higher production levels at lower prices during the projection period.

EIA expects natural gas production from shale gas and tight oil plays to continue to grow, both as a share of total US natural gas production and in absolute volume. This growth is a result of the size of the associated resources, which extend over nearly 500,000 mile^2, and improvements in technology that allow the development of these resources at lower costs. Onshore production of natural gas from sources other than tight oil and shale gas, like coalbed methane, is generally expected to continue to decline through 2050 because of unfavorable economic conditions for producing these resources. Offshore natural gas production remains relatively flat during the projection period, driven by production from new discoveries that generally offset declines in legacy fields (Table 4.1).

The unconventional supply revolution that has redrawn the global oil map will likely expand beyond North America before the end of the decade, the International Energy Agency (IEA) says in its annual 5-year oil market outlook released today. The report also sees global oil demand growth slowing, OPEC capacity growth facing headwinds, and growing regional imbalances in gasoline and diesel markets.

The IEA's *Medium-Term Oil Market Report 2014* says that while no single country outside the United States offers the unique mix of above- and below-ground attributes that made the shale and light, tight oil (LTO) boom possible, several countries will seek to replicate the US success story. The report projects that by 2019, a tight oil supply outside the United States could reach 650,000 barrels per day (650 kb/d), including 390 kb/d from Canada,

TABLE 4.1 Drilling rig activities in USA in 2021.

Region	New well oil production per rig (barrels/day) January 2021	February 2021	Change	New well gas production per rig (thousand cubic feet/day) January 2021	February 2021	Change
Anadarko	1015	989	(26)	5730	5587	(143)
Appalachia	177	179	2	27,373	27,646	273
Bakken	2358	2318	(40)	3147	3093	(54)
Eagle Ford	2227	2205	(22)	7493	7418	(75)
Haynesville	19	19	—	11,406	11,408	2
Niobrara	2019	1938	(81)	6423	6166	(257)
Permian	1173	1138	(35)	2259	2191	(68)
Rig-weighted average	1058	1041	(17)	7057	6906	(151)

Note: The *Drilling Productivity Report* (DPR) rig productivity metric *new-well oil/natural gas production per rig* can become unstable during periods of rapid decreases or increases in the number of active rigs and well completions. The metric uses a fixed ratio of estimated total production from new wells divided by the region's monthly rig count, lagged by 2 months. The metric does not represent new-well oil/natural gas production per newly completed well.
The DPR metric *legacy oil/natural gas production change* can become unstable during periods of rapid decreases or increases in the volume of well production curtailments or shut-ins. This effect has been observed during winter weather freeze-offs, extreme flooding events, and the 2020 global oil demand contraction. The DPR methodology involves applying smoothing techniques to most of the data series because of inherent noise in the data.
January 2021 Supplement: Base production in North Dakota has fully recovered after a significant reduction.
September 2020 Supplement: With low rig counts, the inventory of drilled but uncompleted (DUC) wells provides a short-term reserve for completions of new wells.
August 2020 Supplement: Rig counts fall but new-well production per rig rises as new-well production persists.
March 2020 Supplement: Base production accounts for a material share of total US tight oil production.

100 kb/d from Russia, and 90 kb/d from Argentina. US LTO output is forecast to roughly double from 2013 levels to 5.0 million barrels per day (mb/d) by 2019.

"We are continuing to see unprecedented production growth from North America, and the United States in particular. By the end of the decade, North America will have the capacity to become a net exporter of oil liquids," IEA Executive Director Maria van der Hoeven said as she launched the report in Paris. "At the same time, while OPEC remains a vital supplier to the market, it faces significant headwinds in expanding capacity."

Aging fields are an issue for almost all OPEC producers, but above-ground woes have escalated: security concerns are a growing issue in several producers and investment risks have deterred some international oil companies. The report notes that as much as three-fifths of OPEC's expected growth in capacity by 2019 is set to come from Iraq. The projected addition of 1.28 mb/d to Iraqi production by 2019, a conservative forecast made before the launch last week of a military campaign by insurgents that subsequently claimed several key cities in northern and central Iraq, faces considerable downside risk.

The annual report, which gives a detailed analysis and 5-year projections of oil demand, supply, crude trade, refining capacity, and product supply, sees global demand rising by 1.3% per year to 99.1 mb/d in 2019. Yet the report also expects the market to hit an "inflection point," after which demand growth may start to decelerate due to high oil prices, environmental concerns, and cheaper fuel alternatives. This will lead to fuel-switching away from oil, as well as overall fuel savings. In short, while "peak demand" for oil—other than in mature economies—may still be years away, and while there are regional differences, peak oil demand growth for the market as a whole is already in sight.

Unconventional oil and gas have been playing an increasingly important role in global petroleum production, which has been demonstrated by the commercial production of oil sands, tight gas, and coalbed methane (CBM) and the rapid increase of shale gas and tight oil production caused by the US shale gas revolution (Islam, 2014). Unconventional oil resources include oil shale, oil sands, heavy oil, and tight oil (including shale oil).

Conventional oil production of 69 mb/d represents by far the largest share of global oil output of 85 mb/d. In addition, 6.5 mb/d comes from liquids production from the US shale plays, and the rest is made up of other natural gas liquids and unconventional oil sources such as oil sands and heavy oil.

With global demand expected to grow by 1.2 mb/d a year in the next 5 years, the IEA has repeatedly warned that an extended period of sharply lower oil investment could lead to a tightening in supplies. Exploration spending is expected to fall again in 2017 for the third year in a row to less than half 2014 levels, resulting in another year of low discoveries. The level of new sanctioned projects so far in 2017 remains depressed.

"Every new piece of evidence points to a two-speed oil market, with new activity at a historic low on the conventional side contrasted by remarkable growth in US shale production," said Dr. Fatih Birol, the IEA's executive director. "The key question for the future of the oil market is for how long can a surge in US shale supplies make up for the slow pace of growth elsewhere in the oil sector."

The US shale industry has lowered its costs to such an extent that in many cases it is now more competitive than conventional projects. The average break-even price in the Permian basin in Texas, for example, is now at USD 40–45/bbl. Liquids production from US shale plays is expected to expand by 2.3 mb/d by 2022 at current prices, and expand even more if prices rise further.

The offshore sector, which accounts for almost a third of crude oil production and is a crucial component of future global supplies, has been particularly hard hit by the industry's slowdown. In 2016, only 13% of all conventional resources sanctioned were offshore, compared with more than 40% on average between 2000 and 2015.

In the North Sea, for instance, oil investments fell to less than USD 25 billion in 2016, about half the level of 2014. Coincidentally, this is now approaching the level of spending in offshore wind projects in the North Sea, which has doubled to about USD 20 billion in the same period.

Production of unconventional oil currently amounts to 1.55 mbd and is projected to rise to 3.05 mbd by 2020 and 3.75 mbd by 2030. In 2011, unconventional oil contributed 2% to global oil demand and this is projected to rise to only 3% by 2030.

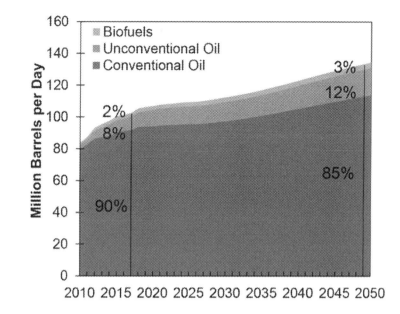

FIG. 4.3 Evolution in world production of liquid petroleum and biofuel.

Globally, fossil fuels supply 81% of primary energy. In 2019, 80% of US primary energy consumption came from fossil fuels. Conventional and unconventional fossil fuels differ in their geologic locations and accessibility; conventional fuels are often found in discrete, easily accessible reservoirs, while unconventional fuels are found in pore spaces throughout a wide geologic formation, requiring advanced extraction techniques. If unconventional oil resources (oil shale, oil sands, extraheavy oil, and natural bitumen) are accounted for, the global oil reserves quadruple current conventional reserves. The price of crude oil peaked in 2008 at $145.31 per barrel, making unconventional fossil fuels more cost-competitive. However, in 2020, the price of crude oil temporarily fell below zero. Partially due to sustained low oil prices, over 200 oil and gas producers have filed for bankruptcy since 2015. The Energy Policy Act of 2005 includes provisions to promote US oil sands, oil shale, and unconventional natural gas development (Fig. 4.3).

4.3 Current potentials

Global oil discoveries fell to a record low in 2016 as companies continued to cut spending and conventional oil projects sanctioned were at the lowest level in more than 70 years, according to the International Energy Agency, which warned that both trends could continue this year.

Oil discoveries declined to 2.4 billion barrels in 2016, compared with an average of 9 billion barrels per year over the past 15 years. Meanwhile, the volume of conventional resources sanctioned for development last year fell to 4.7 billion barrels, 30% lower than the previous year as the number of projects that received a final investment decision dropped to the lowest level since the 1940s.

This sharp slowdown in activity in the conventional oil sector was the result of reduced investment spending driven by low oil prices. It brings an additional cause of concern for global energy security at a time of heightened geopolitical risks in some major producer countries, like Venezuela.

The slump in the conventional oil sector contrasts with the resilience of the US shale industry. There, investment rebounded sharply and output rose, on the back of production costs being reduced by 50% since 2014. This growth in the US shale production has become a fundamental factor in balancing low activity in the conventional oil industry.

Fig. 4.4 shows recent discoveries globally. During the second term of Obama's presidency, the number of oil discoveries continued to decline. This trend reversed during Trump's presidency, starting in 2017, in sync with reduced regulations and increased incentives for the

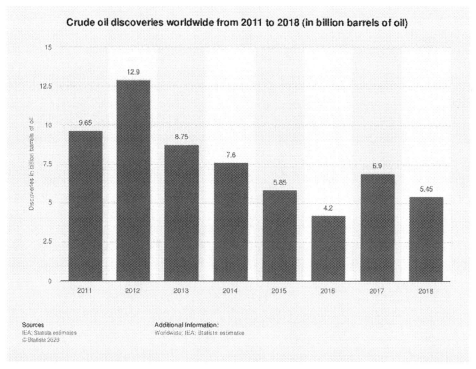

FIG. 4.4 Oil discoveries during 2011–18. *From Statista, 2021a, https://www.statista.com/statistics/1088739/global-oil-discovery-volume/#:~:text=Global%20oil%20discoveries%202011%2D2018&text=The%20total%20volume%20of%20crude,5.45%20billion%20barrels%20of%20oil.*

fossil fuel production. This led to achieving energy solvency in 2019, the first time in 70 years of the US history. The total volume of crude oil discoveries made worldwide in 2018 amounted to approximately 5.45 billion barrels of oil. The volume of oil discovered from 2015 onwards is notably smaller than the discoveries made in 2014 and previous years due to decreased oil exploration as a result of the fall in oil prices that occurred.

Fig. 4.5 shows the global oil demand of recent history. Global oil demand will grow by 5.7 mb/d over the 2019–25 period at an average annual rate of 950 kb/d. This is a sharp reduction on the 1.5 mb/d annual paces seen in the past 10-year period. Following a difficult start in 2020 (−90 kb/d) due to the coronavirus, growth rebound\ed to 2.1 mb/d in 2021 and decelerates to 800 kb/d by 2025 as transport fuels demand growth stagnates. During the same period, oil demand growth slows because demand for diesel and gasoline nears a plateau as new efficiency standards are applied to internal combustion engine vehicles and electric vehicles hit the market. Petrochemical feedstocks LPG/ethane and naphtha will drive around half of all oil products demand growth, helped by continued rising plastics demand and cheap natural gas liquids in North America.

Following a record increase of more than 2.2 mb/d in 2018, the pace of the US expansion slowed to 1.6 mb/d last year as independent producers cut spending and scaled back drilling activity. Further spending cuts are expected for 2020, with capital discipline remaining a priority.

In our base case, which assumes $60/bbl Brent, growth is expected to grind to a halt in the early 2020s and production will plateau around 20–2.5 mb/d higher than in 2019.

Due to its fast ramp-up and rapid decline, the US LTO production is more responsive to a change in the oil price than conventional sources of supply. Recent price volatility could have a major impact on the US production. A price of $40/bbl would cause LTO output to decline from 2021, and fall by 1.1 mb/d to 2025, compared with a growth of 2.2 mb/d in our base case (Fig. 4.6).

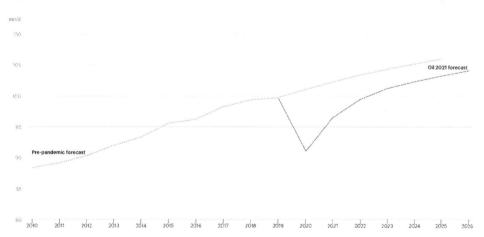

FIG. 4.5 History and oil demand globally. *From IEA, 2021. Oil Demand Forecast, 2010-2026. Pre-Pandemic and in Oil. IEA, Paris. https://www.iea.org/data-and-statistics/charts/oil-demand-forecast-2010-2026-pre-pandemic-and-in-oil-2021.*

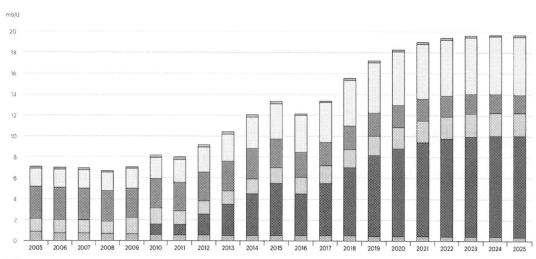

FIG. 4.6 Total supply in the United States.

Global oil supply looks comfortable through the forecast period. The world's oil production capacity is expected to rise by 5.9 mb/d by 2025, which more than covers growth in demand.

The United States leads the way as the largest source of new supply. Brazil, Guyana, Iraq and the UAE also deliver impressive gains. Colombia, the United Kingdom, Russia, Egypt, Nigeria and Angola post the biggest declines. Total non-OPEC oil supply rises by 4.5 mb/d to reach 69.5 mb/d by 2025. As for OPEC, even though sanctions and economic distress have wiped out 2.5 mb/d of production from Iran and Venezuela since 2017, effective crude oil capacity rises by 1.2–34.1 mb/d.

Gains in supply are heavily front-loaded, however, and robust non-OPEC growth through 2021 suggests that there is likely to be a role for OPEC+ market management during the first part of the period. From 2022, the United States loses steam allowing OPEC producers from the Middle East to turn up the taps to help keep the oil market in balance.

The most important source of unconventional oil and gas source is shale formation. EIA (2020a) provides an initial assessment of world shale oil and shale gas resources. The first edition was released in 2011 and updates are released on an ongoing basis. Four countries were added in 2014: Chad, Kazakhstan, Oman, and the United Arab Emirates (UAE) and are available as supplemental chapters to 2013.

4.3.1 Shale gas and tight oil reserves

In Table 4.2, the discrepancy between unproven recoverable and technical recoverability is high owing to the fact that today's technologies are all derived from the arsenal of conventional reservoirs and there is little in common between these two types of formations (Fig. 4.7).

TABLE 4.2　Global reserve in shale gas and tight oil.

Region	Country	Wet shale gas (trillion cubic feet)	Tight oil (billion barrels)	Date updated
North America				
	Canada	572.9	8.8	5/17/13
	Mexico	545.2	13.1	5/17/13
	United States[a]	622.5	78.2	4/14/15
Australia				
	Australia	429.3	15.6	5/17/13
South America				
	Argentina	801.5	27.0	5/17/13
	Bolivia	36.4	0.6	5/17/13
	Brazil	244.9	5.3	5/17/13
	Chile	48.5	2.3	5/17/13
	Colombia	54.7	6.8	5/17/13
	Paraguay	75.3	3.7	5/17/13
	Uruguay	4.6	0.6	5/17/13
	Venezuela	167.3	13.4	5/17/13
Eastern Europe				
	Bulgaria	16.6	0.2	5/17/13
	Lithuania/Kaliningrad	2.4	1.4	5/17/13
	Poland	145.8	1.8	5/17/13
	Romania	50.7	0.3	5/17/13
	Russia	284.5	74.6	5/17/13
	Turkey	23.6	4.7	5/17/13
	Ukraine	127.9	1.1	5/17/13
Western Europe				
	Denmark	31.7	0.0	5/17/13
	France	136.7	4.7	5/17/13

Continued

TABLE 4.2 Global reserve in shale gas and tight oil—cont'd

Region	Country	Wet shale gas (trillion cubic feet)	Tight oil (billion barrels)	Date updated
	Germany	17.0	0.7	5/17/13
	Netherlands	25.9	2.9	5/17/13
	Norway	0.0	0.0	5/17/13
	Spain	8.4	0.1	5/17/13
	Sweden	9.8	0.0	5/17/13
	United Kingdom	25.8	0.7	5/17/13
North Africa				
	Algeria	706.9	5.7	5/17/13
	Egypt	100.0	4.6	5/17/13
	Libya	121.6	26.1	5/17/13
	Mauritania	0.0	0.0	5/17/13
	Morocco	11.9	0.0	5/17/13
	Tunisia	22.7	1.5	5/17/13
	West Sahara	8.6	0.2	5/17/13
Sub-Saharan Africa				
	Chad	44.4	16.2	12/29/14
	South Africa	389.7	0.0	5/17/13
Asia				
	China	1115.2	32.2	5/17/13
	India	96.4	3.8	5/17/13
	Indonesia	46.4	7.9	5/17/13
	Mongolia	4.4	3.4	5/17/13
	Pakistan	105.2	9.1	5/17/13
	Thailand	5.4	0.0	5/17/13
Caspian				
	Kazakhstan	27.5	10.6	12/29/14
Middle East				
	Jordan	6.8	0.1	5/17/13
	Oman	48.3	6.2	12/29/14
	United Arab Emirates	205.3	22.6	12/29/14
46 Countries' total		7576.6	418.9	

[a] Includes data from U.S. Geological Survey, Assessment of Potential Oil and Gas Resources in Source Rocks of the Alaska North Slope, Fact Sheet 2012–3013, February 2012. U.S. Energy Information Administration, Annual Energy Outlook 2015 Assumptions Report.
bbl, barrels; *Tcf*, trillion cubic feet.
From EIA, 2020a. EIA Report. Available from: https://www.eia.gov/energyexplained/coal/how-much-coal-is-left.php.

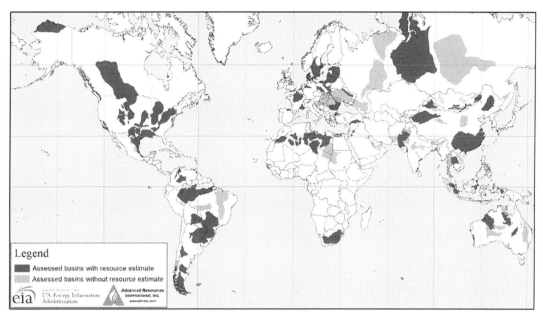

FIG. 4.7   Global map for shale oil and gas basins.

To assess the in-place and recoverable shale gas and shale oil resources, EIA/ARI "World Shale Gas and Shale Oil Resource Assessment" developed this methodology. It relies on geological information and reservoir properties assembled from the technical literature and data from publicly available company reports and presentations. The methodology for conducting the basin- and formation-level assessments of shale gas and shale oil resources includes the following five topics:

(1) Conducting preliminary geologic and reservoir characterization of shale basins and formation(s).
(2) Establishing the areal extent of the major shale gas and shale oil formations.
(3) Defining the prospective area for each shale gas and shale oil formation.
(4) Estimating the risked shale gas and shale oil in-place.
(5) Calculating the technically recoverable shale gas and shale oil resource.

The shale gas and shale oil resource assessment for Argentina's Neuquen Basin was used to illustrate certain of these resource assessment steps.

(1) Conducting Preliminary Geologic and Reservoir Characterization of Shale Basins and Formation(s): The resource assessment begins with the compilation of data from multiple public and private proprietary sources to define the shale gas and shale oil basins and to select the major shale gas and shale oil formations to be assessed. The stratigraphic columns and well logs, showing the geologic age, the source rocks and other data, are used to select the major shale formations for the further study. Preliminary geological and

reservoir data are assembled for each major shale basin and formation, including the following key items:
- Depositional environment of shale (marine vs nonmarine)
- Depth (to top and base of shale interval)
- Structure, including major faults
- Gross shale interval
- Organically rich gross and net shale thickness
- Total organic content (TOC, by wt.)
- Thermal maturity (R_o)

(2) Establishing the Areal Extent of Major Shale Gas and Shale Oil Formations: Having identified the major shale gas and shale oil formations, the next step is to undertake more intensive study to define the areal extent for each of these formations. For this, detailed, regional and local crosssections, identifying the shale oil and gas formations of interest, are collected. The regional cross-sections are used to define the lateral extent of the shale formation in the basin and/or to identify the regional depth and gross interval of the shale formation. This is illustrated for the Vaca Muerta and Los Molles shale gas and shale oil formations in the Neuquen Basin in Fig. 4.8.

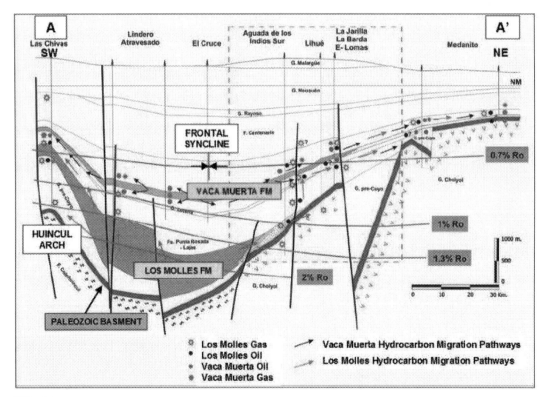

FIG. 4.8 Neuquen Basin SW-NE Cross Section (Structural settings for the two shale gas and shale oil formations, Vaca Muerta and Los Molles).

(3) Defining the Prospective Area for Each Shale Gas and Shale Oil Formation: The criteria used for establishing the prospective area include:
- Depositional Environment. The depositional environment of the shale, particularly whether it is marine or nonmarine. Marine-deposited shales tend to have lower clay content and tend to be high in brittle minerals such as quartz, feldspar and carbonates. Brittle shales respond favorably to hydraulic stimulation. Shales deposited in nonmarine settings (lacustrine and fluvial) tend to be higher in clay, more ductile and less responsive to hydraulic stimulation. Fig. 4.9 provides an illustrative ternary diagram useful for classifying the mineral content of the shale for the Marcellus Shale in Lincoln Co., West Virginia.
- Depth. The depth criterion for the prospective area is greater than 1000 m but less than 5000 m (3300–16,500 ft). Areas shallower than 1000 m have lower reservoir pressure and thus lower driving forces for oil and gas recovery. In addition, shallow shale formations have risks of higher water content in their natural fracture systems. Areas deeper than 5000 m have risks of reduced permeability and much higher drilling and development costs.
- Total Organic Content (TOC). In general, the average TOC of the prospective area needs to be greater than 2%. Fig. 4.10 provides an example of using a gamma ray log to identify the TOC content for the Marcellus Shale in the New York (Chenango Co.) portion of the Appalachian Basin. Organic materials such as microorganism fossils and plant matter provide the requisite carbon, oxygen and hydrogen atoms needed to create natural gas and oil. As such TOC and carbon type (Types I and II) are important measures of the oil generation potential of a shale formation.

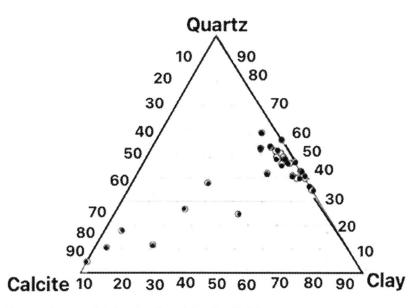

FIG. 4.9 Ternary diagram of shale mineralogy (Marcellus Shale).

Top Marcellus ~1,300'

Beaver Meadow #1 Well, Chenango County, New York

- High TOC in Marcellus concentrates potassium–40 isotope, visible as high radioactivity (100 to 300 units) on gamma ray log.
- Gamma ray count correlates reasonably with TOC.
- The Beaver Meadow #1 well has approximately 150 feet of organically rich (TOC >3% by wt.) shale.

Organically Rich Marcellus ~200'

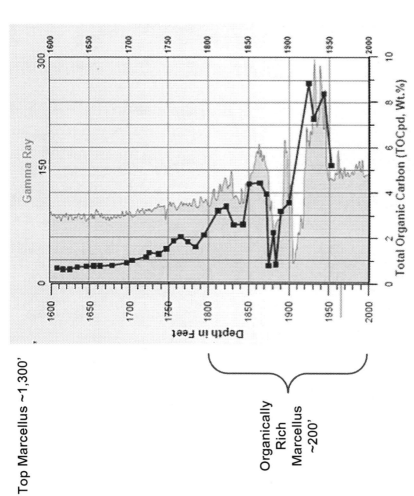

FIG. 4.10 Relationship of gamma ray and total organic carbon.

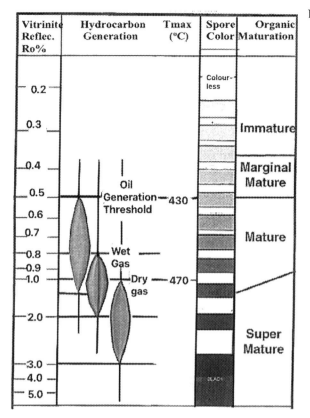

FIG. 4.11 Thermal maturation scale.

- Thermal Maturity: Thermal maturity measures the degree to which a formation has been exposed to high heat needed to break down organic matter into hydrocarbons. The reflectance of certain types of minerals ($R_o\%$) is used as an indication of Thermal Maturity, Fig. 4.11. The thermal maturity of the oil prone prospective area has a R_o greater than 0.7% but less than 1.0%. The wet gas and condensate prospective area has a R_o between 1.0% and 1.3%. Dry gas areas typically have an R_o greater than 1.3%. In this graph, these three hydrocarbon "windows" are identified.
- Geographic Location: The prospective area is limited to the onshore portion of the shale gas and shale oil basin. The prospective area, in general, covers less than half of the overall basin area. The prospective area contains a series of higher quality shale gas and shale oil areas, including a geologically favorable, high resource concentration "core area" and a series of lower quality and lower resource concentration extension areas.

Finally, shale gas and shale oil basins and formations that have very high clay content and/or have very high geologic complexity (e.g., thrusted and high stress) are assigned a high prospective area risk factor or are excluded from the resource assessment. Subsequent, more intensive and smaller-scale (rather than regional scale) resource assessments may identify the more favorable areas of a basin, enabling portions of the

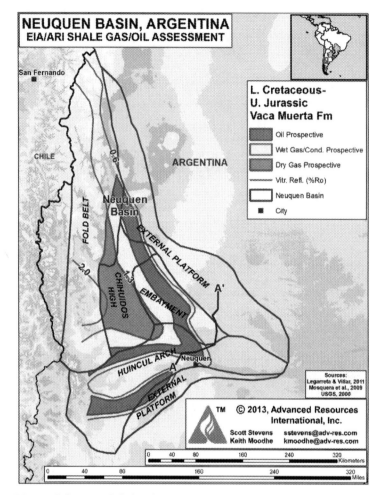

FIG. 4.12   Vaca Muerta shale gas and shale oil prospective areas, Neuquen Basin.

basin currently deemed nonprospective to be added to the shale gas and shale oil resource assessment. Similarly, advances in well completion practices may enable more of the very high clay content shale formations to be efficiently stimulated, also enabling these basins and formations to be added in future years to the resource assessment. The Neuquen Basin's Vaca Muerta Shale illustrates the presence of three prospective areas—oil, wet gas/condensate and dry gas, Fig. 4.12.

(4) Estimating the Risked Shale Gas and Shale Oil In-Place (OIP/GIP): Detailed geologic and reservoir data are assembled to establish the oil and gas in-place (OIP/GIP) for the prospective area.

   (a) Oil In-Place. The calculation of oil in-place for a given areal extent (acre, square mile) is governed by two key characteristics of the shale formation, namely, net organically

4.3 Current potentials

rich shale thickness and oil-filled porosity. In addition, pressure and temperature govern the volume of gas in solution with the reservoir oil, defined by the formation volume factor of the reservoir.

- Net Organically Rich Shale Thickness. The overall geologic interval that contains the organically rich shale is obtained from prior stratigraphic studies of the formations in the basin being appraised. The gross organically rich thickness of the shale interval is established from log data and cross-sections, where available. A net to gross ratio is used to account for the organically barren rock within the gross organically rich shale interval and to estimate the net organically rich thickness of the shale.
- Oil- and Gas-Filled Porosity. The study assembles porosity data from core and/or log analyses available in the public literature. When porosity data are not available, emphasis is placed on identifying the mineralogy of the shale and its maturity for estimating porosity values from analogous US shale basins. Unless other evidence is available, the study assumes the pores are filled with oil, including solution gas, free gas and residual water.
- Pressure. The study methodology places particular emphasis on identifying overpressured areas. Overpressured conditions enable a higher portion of the oil to be produced before the reservoir reaches its "bubble point" where the gas dissolved in the oil begins to be released. A conservative hydrostatic gradient of 0.433 psi per foot of depth is used when actual pressure data is unavailable because water salinity data are usually not available.
- Temperature. The study assembles data on the temperature of the shale formation. A standard temperature gradient of 1.25°F per 100 ft of depth and a surface temperature of 60°F are used when actual temperature data are unavailable. The above data are combined using established reservoir engineering equations and conversion factors to calculate OIP per square mile.

$$\text{OIP} = \frac{7758(A*h)*\varphi*(S_o)}{B_{oi}} \quad (4.1)$$

where

A is area, in acres (with the conversion factors of 7758 barrels per acre foot).
h is net organically rich shale thickness, in feet.
φ is porosity, a dimensionless fraction (the values for porosity are obtained from log or core information published in the technical literature or assigned by analogy from the US shale oil basins; the thermal maturity of the shale and its depth of burial can influence the porosity value used for the shale).
(S_o) is the fraction of the porosity filled by oil (S_o) instead of water (S_w) or gas (S_g), a dimensionless fraction, the established value for porosity (φ) is multiplied by the term (S_o) to establish oil-filled porosity; the value S_w defines the fraction of the pore space that is filled with water, often the residual or irreducible reservoir water saturation in the natural fracture and matrix porosity of the shale; shales may also contain free gas (S_g) in the pore space, further reducing oil-filled porosity.

B_{oi} is the oil formation gas volume factor that is used to adjust the oil volume in the reservoirs, typically swollen with gas in solution, to oil volume in stocktank barrels; reservoir pressure, temperature and thermal maturity (R_o) values are used to estimate the B_{oi} value. The procedures for calculating B_{oi} are provided in a standard reservoir engineering text. In addition, B_{oi} can be estimated from correlations.

In general, the shale oil in the reservoir contains solution or associated gas. A series of engineering calculations, involving reservoir pressure, temperature and analog data from the US shale oil formations are used to estimate the volume of associated GIP and produced along with the shale oil. As the pressure in the shale oil reservoir drops below the bubble point, a portion of the solution gas separates from the oil creating a free gas phase in the reservoir. At this point, both oil (with remaining gas in solution) and free gas are produced.

(b) Free GIP. The calculation of free GIP for a given areal extent (acre, square mile) is governed, to a large extent, by four characteristics of the shale formation, namely, pressure, temperature, gas-filled porosity and net organically rich shale thickness.

- Pressure. The study methodology places particular emphasis on identifying areas with overpressure, which enables a higher concentration of gas to be contained within a fixed reservoir volume. A conservative hydrostatic gradient of 0.433 psi per foot of depth is used when actual pressure data is unavailable.
- Temperature. The study assembles data on the temperature of the shale formation, giving particular emphasis on identifying areas with higher than average temperature gradients and surface temperatures. A temperature gradient of 1.25°F per 100 ft of depth plus a surface temperature of 60°F are used when actual temperature data is unavailable.
- Gas-Filled Porosity. The study assembles the porosity data from core or log analyses available in the public literature. When porosity data are not available, emphasis is placed on identifying the mineralogy of the shale and its maturity for estimating porosity values from analogous US shale basins. Unless other evidence is available, the study assumes the pores are filled with gas and residual water.

The above data are combined using established PVT reservoir engineering equations and conversion factors to calculate free GIP per acre. The calculation of free GIP uses the following standard reservoir engineering equation:

$$GIP = \frac{43{,}560 * A\, h\varphi (S_g)}{B_g} \qquad (4.2)$$

$$\text{Where}: B_g = \frac{0.02829 zT}{P}$$

where

A is area, in acres (with the conversion factors of 43,560 square feet per acre and 640 acres per square mile).
h is net organically rich shale thickness, in feet.
φ is porosity, a dimensionless fraction (the values for porosity are obtained from log or core information published in the technical literature or assigned by analogy from US shale gas

basins; the thermal maturity of the shale and its depth of burial can influence the porosity value used for the shale).

(S_g) is the fraction of the porosity filled by gas (S_g) instead of water (S_w) or oil (S_o), a dimensionless fraction, the established value for porosity (φ) is multiplied by the term (S_g) to establish gas-filled porosity; the value S_w defines the fraction of the pore space that is filled with water, often the residual or irreducible reservoir water saturation in the natural fracture and matrix porosity of the shale; liquids-rich shales may also contain condensate and/or oil (S_o) in the pore space, further reducing gas-filled porosity.

P is pressure, in psi (pressure data is obtained from well test information published in the literature, inferred from mud weights used to drill through the shale sequence, or assigned by analog from US shale gas basins; basins with normal reservoir pressure are assigned a conservative hydrostatic gradient of 0.433 psi per foot of depth; basins with indicated overpressure are assigned pressure gradients of 0.5–0.6 psi per foot of depth; basins with indicated underpressure are assigned pressure gradients of 0.35–0.4 psi per foot of depth).

T is temperature, in degrees Rankin (temperature data is obtained from well test information published in the literature or from regional temperature versus depth gradients; the factor 460°F is added to the reservoir temperature (in °F) to provide the input value for the gas volume factor (B_g) equation).

B_g is the gas volume factor, in cubic feet per standard cubic feet and includes the gas deviation factor (z), a dimensionless fraction. (The gas deviation factor (z) adjusts the ideal compressibility (PVT) factor to account for nonideal PVT behavior of the gas; gas deviation factors, complex functions of pressure, temperature and gas composition, are published in standard reservoir engineering text.)

(c) Adsorbed GIP. In addition to free gas, shales can hold significant quantities of gas adsorbed on the surface of the organics (and clays) in the shale formation. A Langmuir isotherm is established for the prospective area of the basin using available data on TOC and on thermal maturity to establish the Langmuir volume (VL) and the Langmuir pressure (PL). Adsorbed GIP is then calculated using the formula below (where P is original reservoir pressure).

$$GC = (VL*P)/(PL+P) \tag{4.3}$$

The above gas content (GC) (typically measured as cubic feet of gas per ton of net shale) is converted to gas concentration (adsorbed GIP per square mile) using actual or typical values for shale density. (Density values for shale are typically in the range of 2.65 g/cc and depend on the mineralogy and organic content of the shale.) The estimates of the Langmuir value (VL) and pressure (PL) for adsorbed GIP calculations are based on either publically available data in the technical literature or internal (proprietary) data developed by Advanced Resources from prior work on various US and international shale basins. In general, the Langmuir volume (VL) is a function of the organic richness and thermal maturity of the shale, as illustrated in Fig. 4.13. The Langmuir pressure (PL) is a function of how readily the adsorbed gas on the organics in the shale matrix is released as a function of a finite decrease in pressure. The free GIP and adsorbed GIP are combined to estimate the resource concentration (Bcf/mi^2) for the prospective area of the shale gas basin. Fig. 4.14 illustrates the relative contributions of free (porosity) gas and adsorbed (sorbed) gas to total GIP, as a function of pressure.

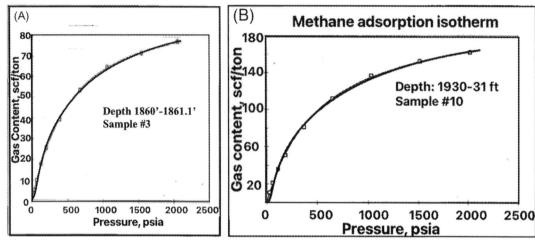

FIG. 4.13 Marcellus shale adsorbed gas content.

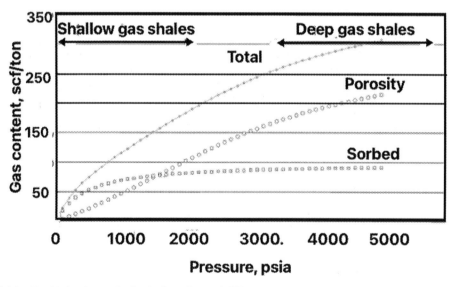

FIG. 4.14 Combining free and adsorbed gas for total GIP.

(5) Establishing the Success/Risk Factors: Two judgmentally established success/risk factors are used to estimate risked OIP and GIP within the prospective area of the shale oil and gas formation. These two factors are as follows:

- Play Success Probability Factor. The shale gas and shale oil play success probability factor captures the likelihood that at least some significant portion of the shale formation will provide oil and/or gas at attractive flow rates and become developed. Certain shale oil formations, such as the Duvernay Shale in Alberta, Canada, are already under development and thus would have a play probability factor of 100%. More speculative shale oil formations with limited geologic and reservoir data may only have a play success probability factor of 30% to 40%. As exploration wells are drilled, tested and produced and information on the viability of the shale gas and shale oil play is established, the play success probability factor will change.

- Prospective Area Success (Risk) Factor: The prospective area success (risk) factor combines a series of concerns that could relegate a portion of the prospective area to be unsuccessful or unproductive for shale gas and shale oil production. These concerns include areas with high structural complexity (e.g., deep faults, upthrust fault blocks); areas with lower thermal maturity (R_o between 0.7% and 0.8%); the outer edge areas of the prospective area with lower net organic thickness; and other information appropriate to include in the success (risk) factor. The prospective area success (risk) factor also captures the amount of available geologic/reservoir data and the extent of exploration that has occurred in the prospective area of the basin to determine what portion of the prospective area has been sufficiently "de-risked." As exploration and delineation proceed, providing a more rigorous definition of the prospective area, the prospective area success (risk) factor will change. These two success/risk factors are combined to derive a single composite success factor with which to risk the OIP and GIP for the prospective area.

The history of shale gas and shale oil exploration has shown that with time the success/risk factors improve, particularly the prospective area success factor. As exploration wells are drilled and the favorable shale oil reservoir settings and prospective areas are more fully established, it is likely that the assessments of the size of the shale gas and shale oil in-place will change.

(6) Estimating the Technically Recoverable Resource. The technically recoverable resource is established by multiplying the risked OIP and GIP by a shale oil and gas recovery efficiency factor, which incorporates a number of geological inputs and analogs appropriate to each shale gas and shale oil basin and formation. The recovery efficiency factor uses information on the mineralogy of the shale to determine its favorability for applying hydraulic fracturing to "shatter" the shale matrix and also considers other information that would impact shale well productivity, such as:
- presence of favorable micro-scale natural fractures;
- the absence of unfavorable deep cutting faults;
- the state of stress (compressibility) for shale formations in the prospective area; and
- the extent of reservoir overpressure as well as the pressure differential between the reservoir original rock pressure and the reservoir bubble point pressure.

Three basic shale oil recovery efficiency factors, incorporating shale mineralogy, reservoir properties and geologic complexity, are used in the resource assessment.
* Favorable Oil Recovery. A 6% recovery efficiency factor of the oil in-place is used for shale oil basins and formations that have low clay content, low to moderate geologic complexity and favorable reservoir properties such as an overpressured shale formation and high oil-filled porosity.

- Average Oil Recovery. A 4%–5% recovery efficiency factor of the oil in-place is used for shale gas basins and formations that have a medium clay content, moderate geologic complexity and average reservoir pressure and other properties.
- Less Favorable Gas Recovery. A 3% recovery efficiency factor of the oil in-place is used for shale gas basins and formations that have medium to high clay content, moderate to high geologic complexity and below average reservoir pressure and other properties.

A recovery efficiency factor of up to 8% may be applied in a few exceptional cases for shale areas with reservoir properties or established high rates of well performance.

A recovery efficiency factor of 2% is applied in cases of severe underpressure and reservoir complexity. Table 4.3 provides information on oil recovery efficiency factors assembled for a series of US shale oil basins that provide input for the oil recovery factors presented above (Table 4.4).

Three basic shale gas recovery efficiency factors, incorporating shale mineralogy, reservoir properties and geologic complexity, are used in the resource assessment.

TABLE 4.3 Tight oil data base used for establishing oil recovery efficiency factors.

Basin	Formation/play	Age	Reservoir pressure	Thermal maturity (% R_o)	Formation volume factor (B_{oi})
Williston	Bakken ND Core	Mississippian-Devonian	Overpressured	0.80%	1.35
	Bakken ND Ext	Mississippian-Devonian	Overpressured	0.80%	1.58
	Bakken MT	Mississippian-Devonian	Overpressured	0.75%	1.26
	Three Forks ND	Devonian	Overpressured	0.85%	1.47
	Three Forks MT	Devonian	Overpressured	0.85%	1.27
Maverick	Eagle Ford Play #3A	Late Cretaceous	Overpressured	0.90%	1.75
	Eagle Ford Play #3B	Late Cretaceous	Overpressured	0.85%	2.01
	Eagle Ford Play #4A	Late Cretaceous	Overpressured	0.75%	1.57
	Eagle Ford Play #4B	Late Cretaceous	Overpressured	0.70%	1.33
Ft. Worth	Barnett Combo—Core	Mississippian	Slightly overpressured	0.90%	1.53
	Barnett Combo—Ext	Mississippian	Slightly overpressured	0.80%	1.41

TABLE 4.3 Tight oil data base used for establishing oil recovery efficiency factor—cont'd

Basin	Formation/play	Age	Reservoir pressure	Thermal maturity (% R_o)	Formation volume factor (B_{oi})
Permian	Del. Avalon/BS (NM)	Permian	Slightly overpressured	0.90%	1.70
	Del. Avalon/BS (TX)	Permian	Slightly overpressured	0.90%	1.74
	Del. Wolfcamp (TX Core)	Permian-Pennsylvanian	Slightly overpressured	0.92%	1.96
	Del. Wolfcamp (TX Ext.)	Permian-Pennsylvanian	Slightly overpressured	0.92%	1.79
	Del. Wolfcamp (NM Ext.)	Permian-Pennsylvanian	Slightly overpressured	0.92%	1.85
	Midi. Wolfcamp Core	Permian-Pennsylvanian	Overpressured	0.90%	1.67
	Midi. Wolfcamp Ext.	Permian-Pennsylvanian	Overpressured	0.90%	1.66
	Midi. Cline Shale	Pennsylvanian	Overpressured	0.90%	1.82
Anadarko	Cana Woodford Oil	Upper Devonian	Overpressured	0.80%	1.76
	Miss. Lime Central OK Core	Mississippian	Normal	0.90%	1.29
	Miss. Lime—Eastern OK Ext.	Mississippian	Normal	0.90%	1.20
	Miss. Lime—KS Ext.	Mississippian	Normal	0.90%	1.29
Appalachian	Utica Shale—Oil	Ordovician	Slightly overpressured	0.80%	1.46
D-J	D-J Niobrara Core	Late Cretaceous	Normal	1.00%	1.57
	D-J Niobrara East Ext	Late Cretaceous	Normal	0.70%	1.26
	D-J Niobrara North Ext. W1	Late Cretaceous	Normal	0.70%	1.37
	D-J Niobrara North Ext. #2	Late Cretaceous	Normal	0.65%	1.28

TABLE 4.4 Oil recovery efficiency for 28 US tight oil plays (black oil, volatile oil and condensates).

Basin	Formation/play	Age	Oil in-place (MBbls/mi^2)	Oil recovery (MBbls/mi^2)	Oil recovery efficiency (%)
Williston	Bakken ND Core	Mississippian-Devonian	12,245	1025	8.4%
	Bakken ND Ext.	Mississippian-Devonian	9599	736	7.7%
	Bakken MT	Mississippian-Devonian	10,958	422	3.9%
	Three Forks ND	Devonian	9859	810	8.2%
	Three Forks MT	Devonian	10,415	376	3.6%
Maverick	Eagle Ford Play #3A	Late Cretaceous	22,455	1827	8.1%
	Eagle Ford Play #3B	Late Cretaceous	25,738	2328	9.0%
	Eagle Ford Play #4A	Late Cretaceous	45,350	1895	4.2%
	Eagle Ford Play #4B	Late Cretaceous	34,505	2007	5.8%
Ft. Worth	Barnett Combo—Core	Mississippian	25,262	377	1.5%
	Barnett Combo—Ext.	Mississippian	13,750	251	1.8%
Permian	Del. Avalon/BS (NM)	Permian	34,976	648	1.9%
	Del. Avalon/BS (TX)	Permian	27,354	580	2.1%
	Del. Wolfeamp (TX Core)	Permian-Pennsylvanian	35,390	1193	3.4%
	Del. Wolfeamp (TX Ext.)	Permian-Pennsylvanian	27,683	372	1.3%
	Del. Wolfeamp (NM Ext.)	Permian-Pennsylvanian	21,485	506	2.4%
	Midi. Wolfeamp Core	Permian-Pennsylvanian	53,304	1012	1.9%
	Midi. Wolfeamp Ext.	Permian-Pennsylvanian	46,767	756	1.6%
	Midi. Cline Shale	Pennsylvanian	32,148	892	2.8%
Anadarko	Cana Woodford—OH	Upper Devonian	11,413	964	8.4%

TABLE 4.4 Oil recovery efficiency for 28 US tight oil plays (black oil, volatile oil and condensates)—cont'd

Basin	Formation/play	Age	Oil in-place (MBbls/mi^2)	Oil recovery (MBbls/mi^2)	Oil recovery efficiency (%)
	Miss. Lime—Central OK Core	Mississippian	28,364	885	3.1%
	Miss. Lime—Eastern OK Ext.	Mississippian	30,441	189	0.6%
	Miss. Lime—KS Ext	Mississippian	21,881	294	1.3%
Appalachian	Utica Shale—Oil	Ordovician	42,408	906	2.1%
D-J	D-J Niobrara Core	Late Cretaceous	33,061	703	2.1%
	D-J Niobrara East Ext.	Late Cretaceous	30,676	363	1.2%
	D-J Niobrara North Ext. #1	Late Cretaceous	28,722	1326	4.6%
	D-J Niobrara North Ext. #2	Late Cretaceous	16,469	143	0.9%

- Favorable Gas Recovery. A 25% recovery efficiency factor of the GIP is used for shale gas basins and formations that have low clay content, low to moderate geologic complexity and favorable reservoir properties such as an overpressured shale formation and high gas-filled porosity.
- Average Gas Recovery. A 20% recovery efficiency factor of the GIP is used for shale gas basins and formations that have a medium clay content, moderate geologic complexity and average reservoir pressure and properties.
- Less Favorable Gas Recovery. A 15% recovery efficiency factor of the GIP is used for shale gas basins and formations that have medium to high clay content, moderate to high geologic complexity and below average reservoir properties.

A recovery efficiency factor of 30% may be applied in exceptional cases for shale areas with exceptional reservoir performance or established rates of well performance. A recovery efficiency factor of 10% is applied in cases of severe underpressure and reservoir complexity. The recovery efficiency factors for associated (solution) gas are scaled to the oil recovery factors, discussed above.

(a) Two Key Oil Recovery Technologies. Because the native permeability of the shale gas reservoir is extremely low, on the order of a few hundred nano-darcies (0.0001 md) to a few milli-darcies (0.001 md), efficient recovery of the oil held in the shale matrix requires two key well drilling and completion techniques, as illustrated by Fig. 4.15:
- Long Horizontal Wells. Long horizontal wells (laterals) are designed to place the oil production well in contact with as much of the shale matrix as technically and economically feasible.

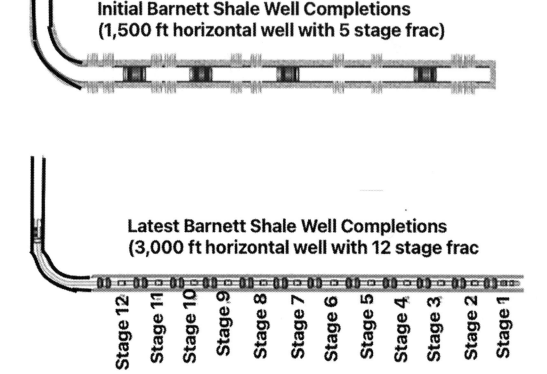

FIG. 4.15 Lower damage, more effective horizontal well completions provide higher reserves per well.

- Intensive Well Stimulation. Large volume hydraulic stimulations, conducted in multiple, closely spaced stages (up to 20), are used to "shatter" the shale matrix and create a permeable reservoir. This intensive set of induced and propped hydraulic fractures provides the critical flow paths from the shale matrix to the horizontal well. Existing, small scale natural fractures (micro-fractures) will, if open, contribute additional flow paths from the shale matrix to the wellbore.
(b) Importance of Mineralogy on Recoverable Resources. The mineralogy of the shale, particularly its relative quartz, carbonate and clay content, significantly determines how efficiently the induced hydraulic fracture will stimulate the shale, as illustrated by Fig. 4.16:
  - Shales with a high percentage of quartz and carbonate tend to be brittle and will "shatter," leading to a vast array of small-scale induced fractures providing numerous flow paths from the matrix to the wellbore, when hydraulic pressure and energy are injected into the shale matrix, Fig. 4.16A.

A. Quartz-Rich (brittle) — Quartz-rich — Barnett Shale

B. Clay-Rich (Ductile) — Clay-rich — Cretaceous Shale

FIG. 4.16 The properties of the reservoir rock greatly influence the effectiveness of hydraulic stimulations.

- Shales with a high clay content tend to be ductile and to deform instead of shattering, leading to relatively few induced fractures (providing only limited flow paths from the matrix to the well) when hydraulic pressure and energy are injected into the shale matrix, Fig. 4.16B.

(c) Significance of Geologic Complexity. A variety of complex geologic features can reduce the shale gas and shale oil recovery efficiency from a shale basin and formation:
  - Extensive Fault Systems. Areas with extensive faults can hinder recovery by limiting the productive length of the horizontal well, as illustrated by Fig. 4.17.
  - Deep Seated Fault System. Vertically extensive faults that cut through organically rich shale intervals can introduce water into the shale matrix, reducing relative permeability and flow capacity.
  - Thrust Faults and Other High Stress Geological Features. Compressional tectonic features, such as thrust faults and up-thrusted fault blocks, are an indication of basin areas with high lateral reservoir stress, reducing the permeability of the shale matrix and its flow capacity.

Tables 4.5 and 4.6, for the Neuquen Basin and its Vaca Muerta Shale formation, provide a summary of the resource assessment conducted for one basin and one shale formation in Argentina including the risked, technically recoverable shale gas and shale oil, as follows:
  - 308 trillion cubic feet (Tcf) of risked, technically recoverable shale gas resource, including 194 Tcf of dry gas, 91 Tcf of wet gas and 23 Tcf of associated gas, Table 4.2.
  - 16.2 billion barrels of technically recoverable shale oil resource, including 2.6 billion barrels of condensate and 13.6 billion barrels of volatile/black oil, Table 4.3.

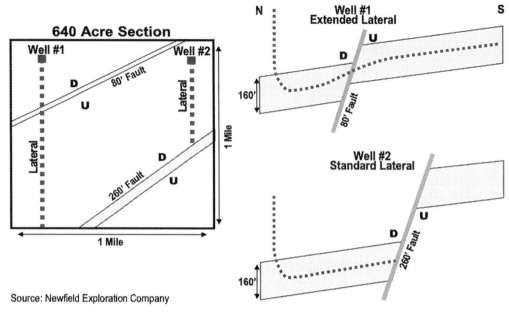

FIG. 4.17 3D seismic helps design extended vs limited length lateral wells.

TABLE 4.5 Shale gas reservoir properties and resources of Argentina.

<table>
<tr><td rowspan="4">Basic Data</td><td colspan="2">Basin/Gross Area</td><td colspan="3">Neuquen
(66,900 mi²)</td></tr>
<tr><td colspan="2">Shale Formation</td><td colspan="3">Vaca Muerta</td></tr>
<tr><td colspan="2">Geologic Age</td><td colspan="3">U. Jurassic - L. Cretaceous</td></tr>
<tr><td colspan="2">Depositional Environment</td><td colspan="3">Marine</td></tr>
<tr><td rowspan="5">Physical Extent</td><td colspan="2">Prospective Area (mi²)</td><td>4,840</td><td>3,270</td><td>3,550</td></tr>
<tr><td rowspan="2">Thickness (ft)</td><td>Organically Rich</td><td>500</td><td>500</td><td>500</td></tr>
<tr><td>Net</td><td>325</td><td>325</td><td>325</td></tr>
<tr><td rowspan="2">Depth (ft)</td><td>Interval</td><td>3,000 -9,000</td><td>4,500 -9,000</td><td>5,500 -10,000</td></tr>
<tr><td>Average</td><td>5,000</td><td>6,500</td><td>8,000</td></tr>
<tr><td rowspan="4">Reservoir Properties</td><td colspan="2">Reservoir Pressure</td><td>Highly Overpress.</td><td>Highly Overpress.</td><td>Highly Overpress.</td></tr>
<tr><td colspan="2">Average TOC (wt. %)</td><td>5.0%</td><td>5.0%</td><td>5.0%</td></tr>
<tr><td colspan="2">Thermal Maturity (% Ro)</td><td>0.85%</td><td>1.15%</td><td>1.50%</td></tr>
<tr><td colspan="2">Clay Content</td><td>Low/Medium</td><td>Low/Medium</td><td>Low/Medium</td></tr>
<tr><td rowspan="4">Resource</td><td colspan="2">Gas Phase</td><td>Assoc. Gas</td><td>Wet Gas</td><td>Dry Gas</td></tr>
<tr><td colspan="2">GIP Concentration (Bcf/mi²)</td><td>66.1</td><td>185.9</td><td>302.9</td></tr>
<tr><td colspan="2">Risked GIP (Tcf)</td><td>192.0</td><td>364.8</td><td>645.1</td></tr>
<tr><td colspan="2">Risked Recoverable (Tcf)</td><td>23.0</td><td>91.2</td><td>193.5</td></tr>
</table>

TABLE 4.6 Shale oil reservoir properties and resources of Argentina.

<table>
<tr><th colspan="3"></th><th colspan="2">Neuquen</th></tr>
<tr><td rowspan="4">Basic Data</td><td colspan="2">Basin/Gross Area</td><td colspan="2">(66,900 mi^2)</td></tr>
<tr><td colspan="2">Shale Formation</td><td colspan="2">Vaca Muerta</td></tr>
<tr><td colspan="2">Geologic Age</td><td colspan="2">U. Jurassic - L. Cretaceous</td></tr>
<tr><td colspan="2">Depositional Environment</td><td colspan="2">Marine</td></tr>
<tr><td rowspan="5">Physical Extent</td><td colspan="2">Prospective Area (mi^2)</td><td>4,840</td><td>3,270</td></tr>
<tr><td rowspan="2">Thickness (ft)</td><td>Organically Rich</td><td>500</td><td>500</td></tr>
<tr><td>Net</td><td>325</td><td>325</td></tr>
<tr><td rowspan="2">Depth (ft)</td><td>Interval</td><td>3,000 - 9,000</td><td>4,500 - 9,000</td></tr>
<tr><td>Average</td><td>5,000</td><td>6,500</td></tr>
<tr><td rowspan="4">Reservoir Properties</td><td colspan="2">Reservoir Pressure</td><td>Highly Overpress.</td><td>Highly Overpress.</td></tr>
<tr><td colspan="2">Average TOC (wt, %)</td><td>5.0%</td><td>5.0%</td></tr>
<tr><td colspan="2">Thermal Maturity (% Ro)</td><td>0.85%</td><td>1.15%</td></tr>
<tr><td colspan="2">Clay Content</td><td>Low/Medium</td><td>Low/Medium</td></tr>
<tr><td rowspan="4">Resource</td><td colspan="2">Oil Phase</td><td>Oil</td><td>Condensale</td></tr>
<tr><td colspan="2">OIP Concentration (MMbbl/mi^2)</td><td>77.9</td><td>22.5</td></tr>
<tr><td colspan="2">Risked OIP (B bbl)</td><td>226.2</td><td>44.2</td></tr>
<tr><td colspan="2">Risked Recoverable (B bbl)</td><td>13.57</td><td>2.65</td></tr>
</table>

4.3.2 Coal bed methane (CBM)

Coal mine degasification started in United States in 1970s. The vast amount of gas recovered was discharged into the atmosphere in the beginning. This was liable to make global warming worse because methane is 23 times more effective in trapping infra-red radiations (radiative forcing) than CO_2. Hence, the gas was processed to meet the gas pipeline specifications and was marketed for additional profit. Shortly after that CBM production from deeper coal seams (that were not mined) started in western United States. The CBM production peaked at 1.8 Tcf, about 10% of the total gas production in the United States in 2008 (Thakur et al., 2020).

Coalbed methane (CBM) is an important source of natural gas. Coalbed methane is generally formed due to thermal maturation of kerogen and organic matter. However, coal seams with regular groundwater recharge see methane generated by microbial communities living in situ. Coalbed methane (CBM; also known as coal-bed methane, coalbed natural gas, coal seam gas) is a type of unconventional natural gas generated and stored in coal beds. Sorbed gas is released and produced from coal following the reduction of hydrostatic pressure with the removal of water from coal cleats and other fractures during drilling. Coal mine methane (CMM), on the other hand, is gas produced in association with underground coal mining operations.

Research and developments on CBM were started in the early 1980s. Production of coalbed methane was first started in 1971 in the underground coal mines for the reduction of explosive gas hazards. In recent times, it has gained prominence in countries such as United States, Australia, China, Russia and Canada. In 2006, World Coal Organization has estimated the global CBM resources total 143 TCM, of which 1 TCM was actually recovered from reserves (USGS, 2006). China has the richest CBM reserve of 30–35 TCM. According to Wang et al. (2016), China has the world's third-largest ready reserves of CBM with an estimated reserve of 36.8 TCM among which 10 TCM can be exploited. At the end of the year 1990, coalbed methane reserve in the United States was measured about more than 700 Tcf among which 100 Tcf gasin-place was economically recoverable (US Geological Survey, 2000). Among the recoverable CBM reserve of the United States, 57 Tcf CBM was estimated in Alaska. According to the Energy Information Agency's (EIA, 2013) Annual Energy Outlook published report total US recoverable natural gas resources were estimated to be 2327 Tcf, up from 1259 Tcf in 2000. Gas-in-place was estimated in Russia was ranging from 600 to 2825 Tcf. Southeast British Columbia has a proven reserve of CBM resource for more than 12 Tcf and 8 Tcf for northwest British Columbia. A resource of about 60 Tcf was found in North East British Columbia and 0.3–1.6 Tcf reserve was found in Vancouver Island (Ryan and Dawson, 1993). The United States started exploratory CBM well in the year 1984. At the end of the year 2000, the United States explored a total number of 13,973 coalbed methane production wells (EPA, 2001). At the end of 2000, coalbed methane production was about 7% of the total United States dry gas production and 9% of proven dry gas reserves (EIA, 2001). The United States is the global leader as the CBM producing country in the world which has produced over 1.91 Tcf of gas sold in the year 2009 (Pashin et al., 2011). China government has setup a target production of 30 BCM in the year 2015 and 50 BCM by 2020, whereas in the year 2010 their target production was 10 BCM (Honglin et al., 2007). According to world coal organization, 2012 report China has produced 402 billion cubic feet (Bcf) CBM in 2011 whereas the production has increased to 542 Bcf in the year 2012 (Pearson et al., 2012). Australian Petroleum Production and Exploration Association Limited has provided an annual CBM production report of Australia in the year 2008 which was estimated amount of 4 BCM at the increase rate of 39% per year from the year 1990 (APPEA, 2009). In 2011 Australia's gross CBM production was 252 Bcf. The prognosticated CBM resources in India are about 92 Tcf. Around 16,613 km and 70% of the total reserves have been allocated for CBM explorations in total four rounds of biddings (Chattaraj et al., 2016). Commercial CBM production of India started in the year 2007 (Great Eastern Energy Corporation Ltd., accessed March 10, 2009) and in 2011 it was 10.5 Bcf (Pearson et al., 2012) while it was just over 2 Bcf in the year 2009. As per DGH report current CBM production is around 0.77 MMSCMD.

CBM production involves using water or other fluids to create a "crack" through which methane can flow easily into a well. Even though coal-bed methane as a major source of natural gas has been recognized for many decades, recently it has become an important source of energy in United States, Canada, Australia, and other countries. This timeline is in sync with the surge in shale gas and tight oil production. CBM is one of the "purest" forms of natural gas. It is usually free from hydrogen sulfide. It is explained by the fact that coal matrix is an excellent site for adsorption of sulfide molecules (Islam, 2014). The methane is in a near-liquid state, lining the inside of pores within the coal matrix. The open fractures in the coal (called the cleats) can also contain free gas and is saturated with water. Any CBM would thus invariably have a high water-gas ratio. *Coalbed methane*, which is methane obtained from coal seams, or *beds*, is a source of methane that is added to the US natural gas supply. In 2019, US coalbed methane production was equal to about 3% of total US dry natural gas production (Fig. 4.18).

Unlike much natural gas from conventional reservoirs, coalbed methane contains very little heavier hydrocarbons such as propane or butane, and no natural-gas condensate. However, it often contains a few percent of CO_2. This is not harmful but can limit transportation ability due to lowering of heating value and risk of corrosion. Table 4.7 shows the heating efficiency of various fuel gases. Note that coal gas, in its natural state, has relatively high heating value.

Despite the focus on shale gas exploration and exploitation in recent years, CBM resources remain an important source of non-conventional gas (Mastalerz, 2014), with a total of 250×10^{15} m^3 estimated worldwide (Murray, 1996). Meanwhile, production of CBM in the United States continued to decline in 2017 while CBM proved reserves increased by 12%. CBM is still an important resource globally. Research on CBM remains active, however, as indicated by >80 technical papers published in 2018. Fig. 4.19 shows a map showing world CBM as well as coal resources. Related production, and exploration activities as summarized in Table 4.8, revised with data from Kelafant (2016). Table 4.9 shows production data from around the world.

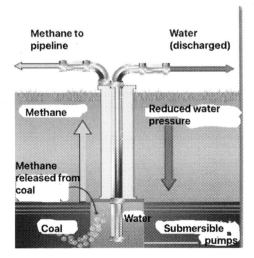

FIG. 4.18 CBM production involves using water or other fluids to create a "crack" through which methane can flow easily into a well.

TABLE 4.7 Heating value fuel gases gross and net heating values (Engineering ToolBox, n.d.).

Gas	Gross heating value (Btu/ft^3)	Gross heating value (Btu/lb)	Net heating value (Btu/ft^3)	Net heating value (Btu/lb)
Acetylene (ethyne)—C$_2$H$_2$	1498	21,569	1447	20,837
Benzene	3741	18,150	3590	17,418
Blast furnace gas	92	1178	92	1178
Blue water gas		6550		
Butane—C$_4$H$_{10}$	3225	21,640	2977	19,976
Butylene (Butene)	3077	20,780	2876	19,420
Carbon monoxide—CO	323	4368	323	4368
Carbon to CO		3960		3960
Carbon to CO$_2$		14,150		14,150
Carburetted water gas	550	11,440	508	10,566
Coal gas	149	16,500		
Coke oven gas	574	17,048	514	15,266
Digester gas (sewage or biogas)	690	11,316	621	10,184
Ethane—C$_2$H$_6$	1783	22,198	1630	20,295
Ethyl alcohol saturated with water	1548	12,804		
Ethylene	1631	21,884	1530	20,525
Hexane	4667	20,526	4315	18,976
Hydrogen (H$_2$)	325	61,084	275	51,628
Hydrogen sulfide	672	7479		
Landfill gas	476			
Methane—CH$_4$	1011	23,811	910	21,433
Methyl alcohol saturated with water	818	9603		
Naphthalene	5859	17,298		
Natural gas (typical)	950–1150	19,500–22,500	850–1050	17,500–22,000
Octane saturated with water	6239	20,542	3170	10,444
Pentane	3981	20,908	3679	19,322
Producer gas		2470		
Propane—C$_3$H$_8$	2572	21,564	2371	19,834
Propene (propylene)—C$_3$H$_6$	2332	20,990	2181	19,630
Propylene	2336	21,042	2185	19,683
Sasol	500	14,550	443	13,016
Sulfur		3940		
Toluene	4408	18,291	4206	17,301
Water gas (bituminous)	261	4881	239	4469
Xylene	5155	18,410		

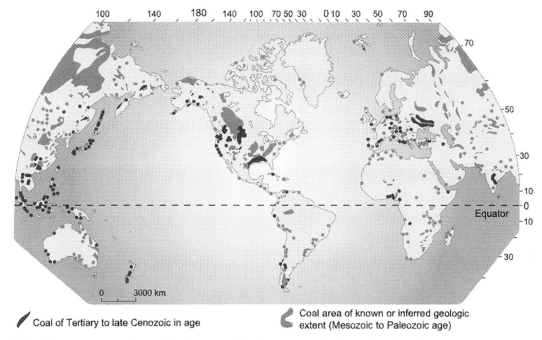

FIG. 4.19 The present-day worldwide geographic distribution of coal related to the Paleozoic-Mesozoic age and Tertiary age. *From Flores, Romeo M., 2013. Coal and Coalbed Gas: Fueling the Future. Elsevier Science & Technology, Saint Louis.*

TABLE 4.8 CBM resources as estimated in 2014 and 2018.

Country	Mastalerz (2014) 2010 resources, Tcf	Kelafant (2016) Tcf
Russia	2824	
China	1100	1300
Alaska	1037	
United States (minus Alaska)	700	
Australia	500	203
Canada	500	801
Indonesia	435	
Poland	424	
France	368	
Germany	100	
United Kingdom	100	
India	70	120
Ukraine	60	
Zimbabwe	40	
Kazakhstan	25	27
Southern Africa		110

TABLE 4.9 Annual production rate from various countries.

Annual CBM production by country (2010 data) (from Mastalerz, 2014)	
Country	Production, Bcf
United States (minus Alaska)	1886
Canada	320
Australia	190
China	50
Alaska	1
Russia	0.5
India	0.4
Kazakhstan	0.4

The EIA (2009a) shows a map of US lower 48 states CBM fields (as of April 2009). US annual CBM production peaked at 1.966 trillion cubic feet (Tcf) in 2008 (EIA, 2009b, 2010, 2018e). CBM production declined to 980 billion cubic feet (Bcf) in 2017 (EIA, 2018c), the lowest level since 1996, representing 3.6% of the US total natural gas production of 27.3 Tcf (EIA, 2018d). Production of coalbed methane gas generally continues to decline through 2050 because of unfavorable economic conditions for producing that resource. Fig. 4.20 shows the data.

Increased US natural gas production is the result of continued development of shale gas and tight oil plays.

According to EIA (2018c,e), the top 7 CBM-producing US states during 2017 (production in billion cubic feet, Bcf) were Colorado (338), New Mexico (234), Wyoming (135), Virginia (99), Alabama (62), Oklahoma (36), and Utah (36). Annual CBM production decreased for each state over the previous year except Alabama (EIA 2018c; Fig. 4.21). Cumulative US CBM production from 1989 through 2017 was 36.7 Tcf. According to EIA (2018c), annual peak CBM

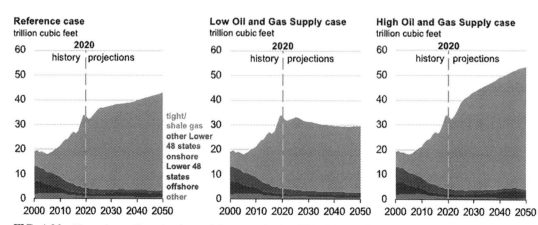

FIG. 4.20 Natural gas oil production and future projection (EIA, 2021, AEO, 2021).

production in the top seven CBM producing US states during 2017 occurred in the following years: Colorado (2010), New Mexico (1997), Wyoming (2008), Virginia (2009), Alabama (1998), Oklahoma (2007), and Utah (2002) (Figs. 4.21 and 4.22).

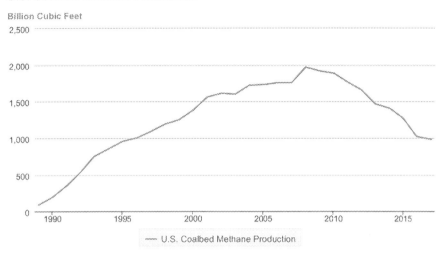

FIG. 4.21 United States CBM production (1989–2017). *Compiled from EIA, 2018c. U.S. Crude Oil and Natural Gas Proved Reserves, Year-End 2016. U.S. Energy Information Administration, 46 pp. http://www.eia.gov/naturalgas/crudeoilreserves/index.cfm.*

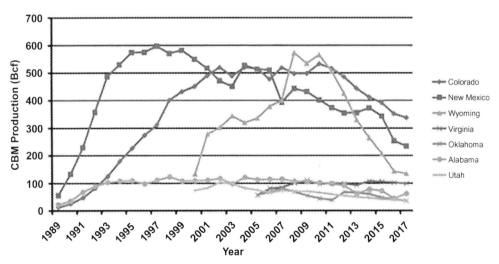

FIG. 4.22 Annual CBM production of the top seven US states during 2017 (1989–2017). *Compiled from EIA, 2018c. U.S. Crude Oil and Natural Gas Proved Reserves, Year-End 2016. U.S. Energy Information Administration, 46 p. http://www.eia.gov/naturalgas/crudeoilreserves/index.cfm; EIA, 2018e. U.S. CBM Production. U.S. Energy Information Administration. http://www.eia.gov/dnav/ng/ng_prod_coalbed_s1_a.htm.*

US annual CBM proved reserves peaked at 21.87 Tcf in 2007 (EIA, 2009b, 2010, 2018f) and declined to 11.878 Tcf in 2017 (EIA, 2018c) representing 2.7% of the US total natural gas reserves of 438 Tcf (EIA, 2018c,g) (Fig. 4.23). Annual CBM proved reserves by US state (through 2017) are available at EIA (2018f).

The International Energy Agency (https://www.iea.org/ugforum/ugd/cbm/) has CBM production data by country for Australia (2000–14), Canada (2000–14), China (2006–14), Czech Republic (2000–14), France (2000–14), Germany (2000–14), India (2007–14), Poland (2000–14), Russia (2010–14), Ukraine (2012–14), United Kingdom (2000–14) (Figs. 4.24 and 4.25).

Table 4.10 shows the estimated CBM reserve for selected countries.

Application of horizontal drilling technology with hydrofracturing can easily double the US CBM production. For working mines, the benefits of CBM production are many as listed below (Thakur et al., 2020):

(1) Safety in mines and prevention of disasters.
(2) Increased productivity for coal resulting in reduced cost of mining.
(3) Additional revenue from gas sales.

There are currently 12 coal basins in the United States producing CBM. Table 4.11 lists the main basins with their location, CBM production, and reserves.

Hydrocarbon generation from source rock is mainly controlled by temperature, pressure, geologic time as well as by quantity, type and structure of the kerogens. Although, hydrocarbon can

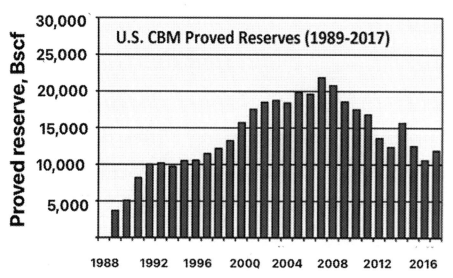

FIG. 4.23 United States CBM proved reserves (1989–2017). *Compiled from EIA, 2009b. U.S. Crude Oil, Natural Gas, and Natural Gas Liquids Reserves, 2007 Annual Report. U.S. Energy Information Administration, DOE/EIA-0216(2007), 145 pp; EIA, 2010. Summary: U.S. Crude Oil, Natural Gas, and Natural Gas Liquids Proved Reserves 2009. U.S. Energy Information Administration, 28 pp; EIA, 2018c. U.S. Crude Oil and Natural Gas Proved Reserves, Year-End 2016. U.S. Energy Information Administration, 46 pp.*

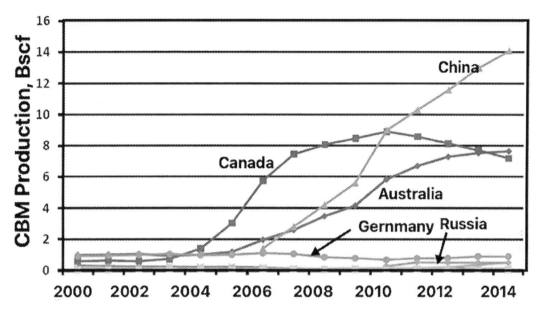

FIG. 4.24 World CBM production from International Energy Agency (2000–14; https://www.iea.org/ugforum/ugd/cbm/).

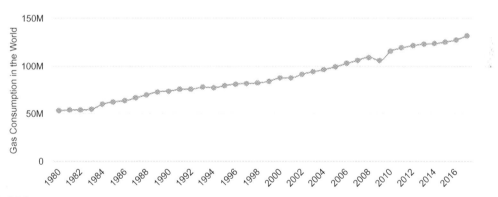

FIG. 4.25 World natural gas consumption.

be of either organic or nonorganic origin, it is mostly recognized that hydrocarbon generation may be caused by any one or both of the following processes (Chattaraj et al., 2016):

(1) Biogenic, due to microbial activity and
(2) Thermogenic, through cracking of organic matter due to temperature and pressure.

Methane generally formed and stored in the coal reservoirs. Most of the gases is sorbed rather than found in free stage (McLennan et al., 1995). In coalbed methane, sorption

TABLE 4.10 Estimates of world coalbed methane reserve.

Country	Estimated coal reserve (10^9 tons)	1992 estimated Tcf	1987 estimated Tcf
United States	3000	388	30–41
Russia	5000	700–5860	118–790
China	4000	700–875	31
Canada	300	212–2682	92
Australia	200	282–494	NA
Germany	300	106	2.83
India	200	35	0.7
South Africa	100	35	NA
Poland	100	106	0.4–1.5
Other countries	200	177–353	NA
Total gas in place		30,958–33,853	275–11,296

From Thakur, P., et al., 2020. Coal Bed Methane: Theory and Applications, second ed. Elsevier, 426 pp.

TABLE 4.11 US coal basins and coalbed methane (CBM) reserves.

Coal depth	Basins	CBM production (Bcf/Year)	CBM (Tcf) reserves
I. Shallow (100–1500 ft)	(a) Powder River Basin	280	100[a]
	(b) Cherokee Basin	5	20[a]
	(c) Illinois Basin	1	21
	(d) Northern Appalachian Basin	10	61
II. Medium depth (1500–3000 ft)	(a) Central Appalachian Basin	94	21
	(b) Warrior Basin	52	21–22
	(c) Raton Basin	105	11
	(d) Arkoma Basin	100	3
III. Deep (+3000 ft)	(a) San Juan Basin	650	84
	(b) Uinta Basin	40	42
	(c) Piceance Basin	5	84
	(d) Green River Basin	20	83
Total		1362	571[b]

[a] Estimated from coal tonnage in the basin.
[b] Does not include Alaska.
From Thakur, P., et al., 2020. Coal Bed Methane: Theory and Applications, second ed. Elsevier, 426 pp.

phenomenon is caused by weak molecular attractions and the gas is ultimately held in the coal hydrostatically. CBM is generally not free-flowing to the well bore; so at the time of extraction, the reservoir pressure is reduced by dewatering the coal seams. Rice (1993) has described another type of CBM generation of a later biogenic stage which is produced by bacterial activity after reaching the thermal maturity (Fig. 4.26).

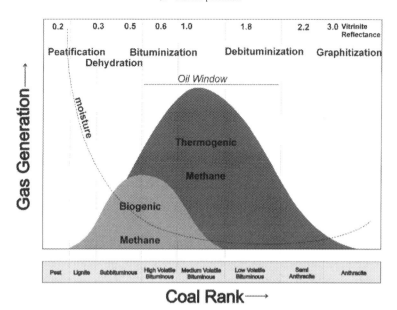

FIG. 4.26 Generation of methane with rank enhancement of coal (Moore, 2012).

Most of the gases are stored in adsorbed form and behaves differently from the conventional gas reservoirs. Parts of the gas also gets dissolved in water within the coal matrix that can be extracted through depressurization. Sorption phenomenon is caused by weak molecular attractions known as Van der Waal forces. However, the gas storage capacity of coal is very high and many times that of conventional reservoirs at low pressures (Mavor and Nelson, 1997). Commercial well completion for the exploration of coalbed methane is generally done by hydro-fracturing. As the reservoir rock is dewatered, most of the sorbed gas diffuses to fractures where it is released and migrates to the well-bore (McLennan et al., 1995; Mavor and Nelson, 1997). Permeability and porosity are the main determining factor for the flow characteristics of the coal reservoir gases. As it generally contains limited natural fractures, extraction of gas is not possible without hydraulic fracturing. The amount of methane in the coal bed usually increases with an increase in surface area of organic matter and/or clays. Higher free-gas content in CBM wells generally results in higher initial rates of production. This is because the free gas resides in fractures and pores and is easily released than the adsorbed gas. The high, initial flow rates decline rapidly to a low and steady rate within about a year, as adsorbed gas is slowly released from the coalbed. Coal gas is generally stored in the reservoir in three forms (Mohanty et al., 2013):

(1) Adsorbed gas, i.e., the gas attached to organic matter or to clays
(2) Free gas, i.e., the gas held within the tiny spaces in the rock (pores, porosity or microporosity) or in spaces created by the rock cracking (fractures or micro-fractures)
(3) Solution gas, i.e., the gas held within other liquids, such as bitumen and oil

Coal is a porous medium with dual porosity. Most of the gas is stored in the form of an adsorbed layer on a micropore rather than free gas in the pore space (Karacan and Okandan,

2000). Therefore the volume stored in compressed form is relatively small compared to conventional reservoirs. During mining, the stress pattern alters and it develops stress around the face causing weakening and fracturing of the strata. Permeability also increases at the distress zone (Harpalani et al., 1985). This enhances the path for the flow of gas toward the face and mine workings. Hence transportation of gas occurs in two stages:

(1) Gas flows through the cleat system in a laminar pattern which is described by Darcy's law of fluid flow (Fig. 4.27).
(2) Diffusion occurs within the coal matrix due to the difference in concentration gradient which follows Fick's law of diffusion.

Gas diffusion in coals is significantly influenced by coal type, rank, microstructure and mineralization, and so on (Mohanty, 2011). Molecular or bulk diffusion happens in the large pores size and high-pressure zone due to intermolecular collisions. This type of diffusion follows Fick's first law of diffusion in the vector form.

Knudsen diffusion occurs where the pores are very small and mean free path (distance between molecular collisions) is greater than the pore diameter. The gaseous molecules collide with the pore walls and rarely with other molecules. This normally occurs at very low pressures and channels of small size, usually of the order of 10–100 nm. Under such conditions, the molecules bounce from wall to wall, rather than colliding with themselves by the concentration difference. Surface diffusion occurs when the gas molecules transform along the pore wall surface and it happens for the micropores diffusion where the pore diameter is <1 nm.

Gas transport in CBM reservoirs Coalbed is a heterogeneous reservoir with dual porosity. The transportation mechanism of methane in the reservoir initiates with the desorption of methane from coal. Desorbed gas gathers in the micropore spaces in the coal matrix. With the increase of the gas concentration in the pore spaces, the gas diffuses through the matrix to the cleat/fracture network and then flows to the wellbore through the cleat/fracture (macropore) network (Harpalani and Schraufnagel, 1990). In summary, the gas transport

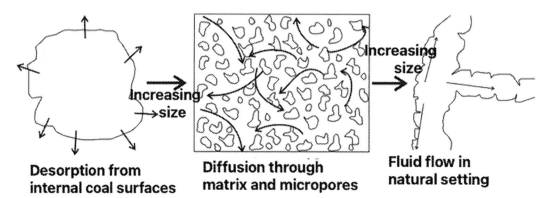

FIG. 4.27 Transport of methane through coalbed (Harpalani and Schraufnagel, 1990).

in coal beds includes three steps viz. desorption from the coal surface, diffusion through the coal matrix, and Darcy's flow through the cleat/fracture network. This flow of gas through the cleats or the macropores is assumed to be laminar (Shi and Durucan, 2003). Three types of diffusion mechanisms have been recognized for adsorbing gas in the matrix and these are as follows (Saulsberry et al., 1996; Shi and Durucan, 2003): (1) Molecular/bulk diffusion (due to its intermolecular collisions), (2) Knudsen diffusion (collisions between molecule and wall) and (3) Surface diffusion (transport through physically adsorbed layer).

The following countries have significant coal bed methane resources.

4.3.2.1 Canada

In Canada, British Columbia is estimated to have approximately 90 trillion cubic feet (2.5 trillion cubic meters) of coalbed gas. Alberta, in 2013, was the only province with commercial coalbed methane wells and is estimated to have approximately 170 trillion cubic feet (4.8 trillion cubic meters) of economically recoverable coalbed methane, with overall reserves totaling up to 500 trillion cubic feet (14 trillion cubic meters).

Coalbed methane is considered a nonrenewable resource, although the Alberta Research Council, Alberta Geological Survey, and others have argued coalbed methane is a renewable resource because the bacterial action that formed the methane is ongoing. This is subject to debate since it has also been shown that the dewatering that accompanies CBM production destroys the conditions needed for the bacteria to produce methane and the rate of formation of additional methane is undetermined. This debate is currently causing a right of ownership issue in the Canadian province of Alberta, as only nonrenewable resources can legally be owned by the province.

4.3.2.2 United Kingdom

Although gas in place in Britain's coal fields has been estimated to be 2900 billion cubic meters, it may be that as little as 1% might be economically recoverable. Britain's CBM potential is largely untested. Some methane is extracted by coal mine venting operations and burned to generate electricity. Assessment by private industry of coalbed methane wells independent of mining began in 2008, when 55 onshore exploration licenses were issued, covering 7000 km^2 of potential coalbed methane areas. IGas Energy became the first in the United Kingdom to commercially extract coalbed methane separate from mine venting; as of 2012, the Igas coalbed methane wells at Doe Green, extracting gas for electrical generation, were the only commercial CBM wells in the United Kingdom.

4.3.2.3 United States

The United States coalbed methane production in 2017 was 1.76 trillion cubic feet (Tcf), 3.6% of all the US dry gas production that year. The 2017 production was down from the peak of 1.97 Tcf in 2008. Most CBM production came from the Rocky Mountain states of Colorado, Wyoming, and New Mexico.

CBM proved reserves and production have grown nearly every year since 1989. Today it accounts for 9% of total domestic natural gas production and nearly 8% of US proved reserves. The US Energy Information Administration believes CBM will continue to provide an important share of domestic energy between now and 2035 (Fig. 4.28).

FIG. 4.28 US natural gas production, 1990–2035.

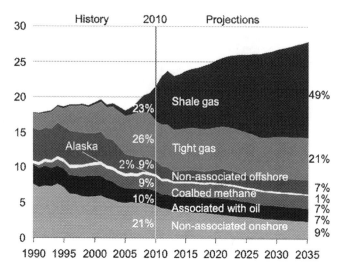

The US Geological Survey (USGS) reports more than 700 trillion cubic feet (Tcf) of CBM in-place resources, over 100 Tcf of which is economically recoverable. CBM can be recovered from underground coal before, during, or after mining operations. Significant quantities of CBM can also be extracted from "unminable" coal seams that are relatively deep, thin, or of poor or inconsistent quality. Vertical and horizontal wells are used to develop CBM resources.

The United States remains the world leader in CBM exploration, booked reserves, and production. Currently, there is commercial coalbed gas production or exploration in approximately 12 US basins. However, activity has slowed substantially in response to low gas prices. The major producing areas are the Powder River, San Juan, Black Warrior, Central Appalachian, Raton, and Uinta basins. Other United States areas with significant exploration or production are the Cherokee, Arkoma, Illinois, Hanna, Gulf Coast, and Greater Green River basins. Development continues in all major US basins, and the principal environmental issue confronting the industry is water disposal. Production operations are maturing, and the US DOE has sponsored a series of studies on produced water management and CO_2-enhanced coalbed methane recovery. Of major interest is a new pilot program that is being led by Virginia Tech in the Appalachian Basin of Virginia, which is scheduled to begin injection of up to 20,000 short tons (18,144 metric tonnes) of CO_2 into multiple coal seams to determine the viability of enhanced recovery and geologic storage. The US Energy Information Administration (EIA) has released CBM production and reserve numbers through the end of 2010. CBM production in 2010 was 1886 billion cubic feet (Bcf) (53.4 billion m^3), decreasing by 1.5% from 2009. Booked reserves decreased from 18,578 Bcf (526 billion m^3) in 2009 to 17,508 Bcf (496 billion m^3) in 2010 representing a decrease of 1070 Bcf (30 billion m^3) (5.7%). CBM represented 8.5% of 2010 dry gas production and 5.7% of proved dry gas reserves in the US Interestingly, CBM production is declining only slightly as a proportion of US gas production but is declining significantly in terms of proved dry gas reserves. This decline is related to the booking of major shale gas reserves, which is significantly changing US gas markets (Fig. 4.29). Most

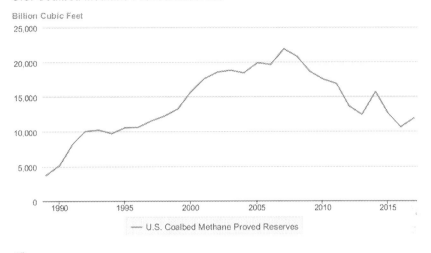

FIG. 4.29 Historic US CBM production 1990–2010. 1 Billion cubic feet (Bcf) = 28.3 million m^3.

CBM activity in the eastern US is focused on the Appalachian Basin of southwestern Virginia and the Black Warrior Basin of Alabama, with several companies actively developing joint CBM and coalmine methane (CMM) projects. In southwestern Virginia, production has decreased substantially from 111 Bcf (3.1 billion m^3) in 2009 to 97 Bcf (2.7 billion m^3) in 2010. West Virginia production declined from 31 to 17 Bcf (0.9–0.5 billion m^3) over the same period. Pennsylvania production decreased from 16 to 3 Bcf. In Alabama, production decline was less pronounced, with 105 Bcf (2.97 billion m^3) being produced in 2009 and 102 Bcf (2.89 billion m^3) being produced in 2010. The Midcontinent region consists of the Cherokee, Forest City, Arkoma, and Illinois Basins. Horizontal drilling has been an effective development strategy, although major increases of production in recent years are now being offset by slowed development. Kansas production decreased modestly from 43 Bcf (1.21 billion m^3) in 2009 to 41 Bcf (1.16 billion m^3) in 2010 (Fig. 4.30 and Table 4.12).

Most CBM activity in the eastern US is focused on the Appalachian Basin of southwestern Virginia and the Black Warrior Basin of Alabama, with several companies actively developing joint CBM and coal-mine methane (CMM) projects. In southwestern Virginia, production has decreased substantially from 111 Bcf (3.1 billion m^3) in 2009 to 97 Bcf (2.7 billion m^3) in 2010. West Virginia production declined from 31 to 17 Bcf (0.9–0.5 billion m^3) over the same period. Pennsylvania production decreased from 16 to 3 Bcf. In Alabama, production decline was less pronounced, with 105 Bcf (2.97 billion m^3) being produced in 2009 and 102 Bcf (2.89 billion m^3) being produced in 2010.

The Midcontinent region consists of the Cherokee, Forest City, Arkoma, and Illinois Basins. Horizontal drilling has been an effective development strategy, although major increases of production in recent years are now being offset by slowed development. Kansas production

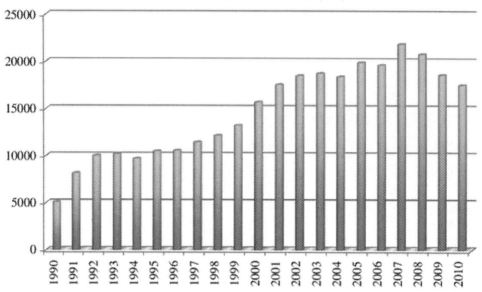

FIG. 4.30 Historic US CBM reserve trends from 1990 to 2010. 1 Billion cubic feet (Bcf) = 28.3 million m^3.

decreased modestly from 43 Bcf (1.21 billion m^3) in 2009 to 41 Bcf (1.16 billion m^3) in 2010, whereas Oklahoma decreased from 55 Bcf (1.56 billion m^3) in 2009 to 45 Bcf (1.27 billion m^3) in 2010, continuing a steep decline trend that began in 2007. The principal issue affecting CBM development in the eastern and midcontinental US is competition with shale gas, which has introduced significant price pressure. Although production operations persist, few wells are being drilled, and reserves are not being replaced.

Infill drilling of Fruitland CBM wells in the San Juan Basin (Colorado and New Mexico) decreased markedly in 2009 due to recession, but activity is starting to accelerate. Colorado and New Mexico continue to dominate CBM production and reserves. Cumulative production for Colorado and New Mexico represents 50% of total US CBM production. In 2010, CBM production in Colorado increased from 498 to 533 Bcf (14.1–15.1 billion m^3), and production in New Mexico declined slightly from 432 to 402 Bcf (12.2–11.4 billion m^3). In addition, activity is rebounding in the Powder River Basin of Wyoming, and production increased in 2010 from 535 to 566 Bcf (15.1–16 billion m^3), accounting for 30% of US CBM production.

International activity has been on the rise, and operations in the Qinshui Basin of China remain active, thus proving the CBM potential of intensely fractured semianthracite and anthracite. As in the United States, depressed natural gas prices are slowing Canadian development. Development is intensifying in the Bowen, Surat, and Sydney Basins of Australia, as well as the Karoo Basin of South Africa. CBM in eastern Australia is being produced from high-permeability coal seams that can contain large quantities of oil-prone organic matter, and the produced gas is being considered for export into Asian liquefied natural gas

TABLE 4.12 Historic US CBM production by year (1989–2010; Billion cubic feet, Bcf).

Year	US	AL	CO	NM	OK	UT	VA	WV	WY	Others
1990	196	36	26	133						1
1991	348	68	48	229						3
1992	539	89	82	358						10
1993	752	103	125	486						38
1994	851	108	179	530						34
1995	956	109	226	574						47
1996	1003	98	274	575						56
1997	1090	111	312	597						70
1998	1194	123	401	571						99
1999	1252	108	432	582						130
2000	1379	109	451	550		74			133	62
2001	1562	111	490	517		83			278	83
2002	1614	117	520	471		103			302	101
2003	1600	98	488	451		97			344	122
2004	1720	121	520	528		82			320	149
2005	1732	113	515	514	58	75	56	30	336	35
2006	1758	114	477	510	68	66	81	18	378	46
2007	1753	114	519	394	82	73	85	25	401	60
2008	1966	107	497	443	69	71	101	28	573	77
2009	1914	105	498	432	55	71	111	31	535	76
2010	1886	102	533	402	45	66	97	17	566	58

(LNG) markets. Several LNG plants (up to five or six) are being considered in Australia. Hence, companies are striving to book reserves to support the expenditure for LNG plant development as quickly as possible. Significant potential exists in the Gondwanan coal basins of India, and some fields have been developed.

Potential also exists in the coal basins of Europe and the Russian platform, and development in these areas is focusing mainly on CMM. Exploration programs have been initiated in recent years to explore CBM in the structurally complex European coal basins of western Europe, including Germany. Russia continues to promote CBM exploration and development but defining a market for the gas and predicting gas prices are problematic for future development. However, the coal basins in Russia may contain the largest CBM resources in the world. Once a market for this gas is identified, then CBM exploration in Russia should increase significantly.

On March 12, 2013, the Japanese Ministry of Economy, Trade and Industry announced the commencement of the first offshore gas hydrate production test. The test was conducted in the Nankai Area off the coasts of Atsumi and Shima peninsulas in water depths of approximately 1000 m (3280.8 ft).

4.3.3 Gas hydrate

Production was initiated through depressurization of hydrate-bearing turbidite sands located 300 m (984.3 ft) beneath the seafloor. Sustained natural gas production was established with a drillstem test at a rate of 0.7 million cubic feet per day (MMcfd; 19.8 thousand m^3/day). The test continued until March 18, 2013, at which point there was both a malfunction of the pump used for depressurization and a simultaneous increase in sand production. A total of 4 million cubic feet (113 thousand m^3) of gas was recovered in total, an amount higher than had been predicted. Initial analysis of the test indicates that the dissociation front reached the monitoring wells located 20 m (65.6 ft) from the test well.

Abandonment of the site will be completed by August 31, 2013. The brief test was not designed to yield commercial production rates; however, the results will be used to implement the next phase of the MH21 program, which will include commercial development. That phase is scheduled for fiscal years 2016–18.

The Nankai test was conducted with the deep sea drillship "Chikyu." The produced gas was either vented or flared, depending on flow rates and weather conditions. In preparation for the production test, a part of the production well (AT1-P) and two temperature-monitoring boreholes (AT1-MC/MT1) were drilled in February and March 2012. During drilling operations, intensive geophysical logging was conducted. In addition, a dedicated borehole in the same area was drilled to recover pressure cores. This was undertaken to obtain detailed data regarding the geology, geomechanics, geochemistry, microbiology, and petrophysics of the hydrate-bearing sediments.

4.3.3.1 United States gas hydrate program

The US DOE's Methane Hydrate Program continues to pursue several important areas of gas hydrate research and characterization despite severe budget constraints. The selected projects are designed to increase the understanding of gas hydrates in the context of future energy supply and changing climates. US Geological Survey (USGS) personnel continue their involvement in resource evaluation, including consultation with assessment programs being conducted outside of the United States.

4.3.3.2 Ignik Sikumi gas hydrate exchange trial

The results of the Ignik Sikumi Gas Hydrate Exchange Trial were released in late 2012, including a presentation at the Arctic Technology Conference. The test was carried out from February 15 to April 10, 2012, in Prudhoe Bay Field, Alaska. The project team injected a mixture of carbon dioxide (CO_2) and nitrogen into hydrate-bearing sand and demonstrated that this mixture could promote the production of natural gas. This test was the first ever field trial of a methane hydrate production methodology whereby CO_2 was exchanged in situ with the methane molecules resulting in methane gas and CO_2 hydrate.

After measurement and compositional analysis, gas from the Ignik Sikumi test was flared. During the test, 210 thousand cubic feet (5.9 thousand m^3) of a N_2/CO_2 gas mixture was pumped into the methane hydrate-bearing formation. The injected gas was 23% N_2 and 77% CO_2. The recovered gas was 2% CO_2, 16% N_2, and 82% CH_4, demonstrating that significant exchange had taken place within the reservoir of carbon dioxide for methane. Although not a technology for near-term production of methane from hydrate deposits, the test provides a path for carbon sequestration in the future. All data developed in the test were released in March 2013. The Prudhoe Bay production test, delayed by a number of issues, remains under review by the partners. The Gulf of Mexico Joint Industry Project (JIP) has concluded. The pressure core system and laboratory equipment developed for the JIP were used for the Japanese program.

4.3.3.3 Gas hydrate in India

A planned logging-while-drilling (LWD) drilling program is under review for offshore India in 2014, with site selection finalized in April 2013. This program is focused on reservoir delineation and resource assessment and is targeting hydrate-bearing sands. Two legs are planned, with the first dedicated to LWD logging. The previous gas hydrate field program (2006) targeted seismic BSRs (bottom simulating reflectors) and recovered significant amounts of gas hydrate, but in fine-grained sediments having low permeability.

4.3.3.4 Gas hydrate in China

China commenced exploratory gas hydrate drilling and coring in Spring, 2013. Results and other details of the program have not yet been released.

4.3.3.5 United States

From 2011 to 2015, the US Gas Hydrate program at the DOE had been functioning at a low level compared, with the period 2001–10, as the DOE has moved away from fossil energy. That approach has changed under Energy Secretary Ernest Moniz, and fossil energy is now being included in the federal "All of the Above" philosophy of energy.

Areas of gas hydrate focus are a continuation of the characterization of gas hydrate in the Gulf of Mexico and a production test in Alaska. A solicitation is being released from DOE regarding the Alaska test, and the State of Alaska is working with DOE to identify a site for the test on state lands west of Prudhoe Bay Field. Industry participation will be necessary.

On October 22, 2014, the DOE announced the selection of a multiyear, field-based research project designed to gain further insight into the nature, formation, occurrence, and physical properties of methane hydrate-bearing sediments in the Gulf of Mexico for gas hydrate resource appraisal. Under this program, the University of Texas at Austin, along with The Ohio State University, Columbia University-Lamont Doherty Earth Observatory, the Consortium for Ocean Leadership, and the US Geological Survey, will characterize and prioritize known and prospective drilling locations with a high probability of encountering concentrated methane hydrates in sand-rich reservoirs. The 4-year program includes $41,270,609 of DOE funding and a cost-share of $17,030,884.

The US program has a two-pronged approach, focusing on both the North Slope of Alaska and the Gulf of Mexico. The next US field test is being developed under the leadership of the University of Texas at Austin, in partnership with The Ohio State University, Columbia

University-Lamont-Doherty Earth Observatory, the Consortium for Ocean Leadership, and the US Geological Survey. Their 4-year exploratory program will characterize prospective drilling locations in the Gulf of Mexico, then in 2018 drill and collect pressure cores and well logs as well as conducting short-duration pressure drawdown tests.

The Gulf of Mexico project includes a focused drilling program that will acquire conventional cores, pressure cores, and downhole logs; measure in situ properties, and measure reservoir response to short-duration pressure perturbations. The field program will also serve to deploy and test several coring and hydrate characterization tools developed through previous DOE-supported research efforts. Postcruise analyses will determine the in situ concentrations, the physical properties, the lithology, and the thermodynamic state of methane hydrate-bearing sand reservoirs. The goal of the project is to improve the ability to estimate the occurrence and distribution of marine gas hydrates and lay the groundwork needed to simulate production behavior from sand-rich reservoirs.

On the North Slope, the program is pursuing two options with Japan for a long-term test. The program is evaluating unleased acreage the State of Alaska has set aside temporarily for this test and is also exploring the possibility of testing within one of the producing units. Prior DOE programs in Alaska have explored gas hydrate reservoir potential and alternative production strategies, and additional testing programs are in development.

A new Methane Hydrate Advisory Committee has been established and met in Galveston March 27–28, 2014. The Committee advises the Secretary of Energy on potential applications of methane hydrate, assists in developing recommendations and priorities for the methane hydrate research and development program, and submits to Congress one or more reports on an assessment of the research program and an assessment of the DOE 5-year research plan. The Committee's charter stipulates that up to 15 members can be appointed by the Secretary of Energy, representing institutions of higher education, industrial enterprises and oceanographic institutions, and state agencies.

4.3.3.6 Japan

Following the successful drillstem test in 2013, Japan's gas hydrate program continues to evolve. In a move to become a commercial rather than governmental project, a new industrial joint venture corporation was formed on October 1, 2014. The "Japan Methane Hydrate Operating Company" (JMH) was formed with the agreement and capital participation of 11 companies engaging in oil and natural gas development and in plant engineering. Commercial production of natural gas from hydrate is expected to commence from the Nankai area on Japan's Pacific margin by 2018.

Japan has expanded its gas hydrate program beyond the Nankai area and is funding programs to explore the Sea of Japan and conduct an assessment. The areas being explored in the Sea of Japan are off the Joetsu region and off of the Akita and Yamagata regions (Fig. 4.31). Based on the initial results of a June 2014 drilling program, the assessment is focused on "shallow" hydrates, the term used by Japan's Agency for Natural Resources and Energy (ANRE) for chimney/fracture-fill occurrences. This represents a new development since throughout the world, all commercial considerations for production of gas from gas hydrate have involved hydrate-bearing sands. It is unclear whether a viable production technology for fracture-fill hydrate is being investigated.

Basins drilled by the Japanese ANRE expedition in 2014.

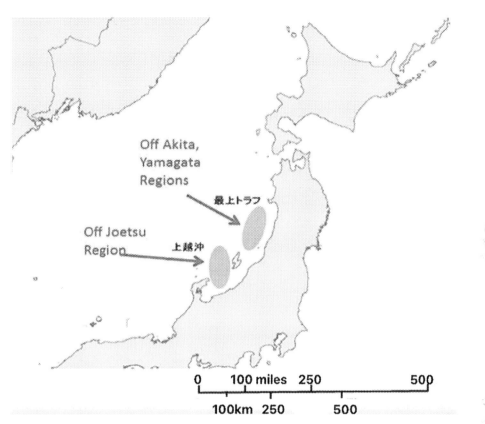

FIG. 4.31 Basins drilled by the Japanese Agency for Natural Resources and Energy (ANRE) expedition in 2014.

4.3.3.7 *India*

The second leg of India's gas hydrate drilling program commenced in February 2015, utilizing the drillship D/V Chikyu. Roughly 170 days of operations are planned, with 20 deep LWD holes and 10 core sites in the northern Bay of Bengal. The program is targeting hydrate-bearing sands with a focus on reservoir delineation and resource assessment. The goal is to identify an optimal site for a future production test.

Results of India's first gas hydrate drilling program were published in December 2014 in Marine and Petroleum Geology. That program included a 113.5-day voyage from April 28 to August 19, 2006, during which the expedition cored or drilled 39 holes at 21 sites (one site in the Kerala-Konkan Basin, 15 sites in the Krishna-Godavari Basin, four sites in the Mahanadi Basin, and one site in the Andaman deep offshore areas). The drilled holes penetrated a total of more than 9250 m (30,348 ft) of sedimentary section and recovered nearly 2850 m (9350 ft) of the core. Twelve holes were logged with LWD tools and an additional 13 holes were wireline logged.

4.3.3.8 South Korea

As a part of the Korean National Gas Hydrate Program, a production test in the Ulleung Basin had been planned for 2015 but its current status is uncertain. The targets are the gas hydrate-bearing sand reservoirs that were found during the Second Ulleung Basin Gas Hydrate Drilling Expedition (UBGH2) in 2010.

4.3.3.9 European Union

The European Cooperation for Science and Technology (COST) has initiated MIGRATE (Marine Gas Hydrates: An Indigenous Resource of Natural Gas for Europe), a program to integrate the expertise of a large number of European research groups and industrial players to promote the development of multidisciplinary knowledge on the potential of gas hydrates as an economically feasible and environmentally sound energy resource. In particular, MIGRATE aims to determine the European potential inventory of exploitable gas hydrates, to assess current technologies for their production, and to evaluate the associated risks. National efforts will be coordinated through Working Groups focusing on (1) resource assessment, (2) exploration, production, and monitoring technologies, (3) environmental challenges, (4) integration, public perception, and dissemination. Study areas will span the European continental margins, including the Black Sea, the Nordic Seas, the Mediterranean Sea, and the Atlantic Ocean.

MIGRATE will examine the potential of gas hydrates as an economically feasible and environmentally sound energy resource. Stefan Bünz, associate professor at Centre for Arctic Gas Hydrate, Environment and Climate (CAGE) at The Arctic University of Norway, was elected the Vice-Chair of the action. Three working groups have been established in MIGRATE: resource assessment; exploration, production, and monitoring technologies; and environmental and geohazard challenges.

4.3.3.10 Turkey

After many years of planning, Turkey has begun an extensive evaluation of the nation's gas hydrate potential in the Black Sea. The program is being led by Dokuz Eylül University in conjunction with the Turkish National Oil Company (TPAO). This comprehensive program includes depositional modeling that integrates onshore and offshore studies, hydroacoustic and geophysical surveys (multibeam, sonar, chirp, high-resolution seismic acquisition, and bathymetry surveys), water column sampling, sediment sampling, laboratory studies, and computer modeling. The initial cruise began in March and data collection will continue for more than 1 year.

After 3 years, a second phase is planned that will include 3D seismic and electromagnetic data acquisition, along with evaluation and development of production technology.

4.3.3.11 New Zealand

As part of a larger project focused on understanding the dynamic interaction of gas hydrates and slow-moving active sediment mass flows, a joint New Zealand-German research team mapped a large area of hydrate-bearing sediment off New Zealand's eastern coast in April and May 2014 (Fig. 4.32). The project utilized 3D and 2D seismic data and found evidence of gas hydrate along with 99 gas plumes venting from the seafloor. The plumes formed columns

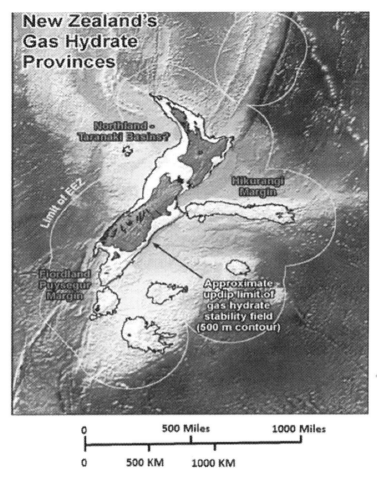

FIG. 4.32 New Zealand's gas hydrate provinces.

extending up to 250 m (820 ft) into the water column. The venting and the presence of gas hydrate have significant implications for slope failure along New Zealand's coastal margin.

New Zealand also has an active research program investigating the resource potential of New Zealand's gas hydrate deposits. The program is led by GNS Science, in collaboration with the National Institute of Water and Atmospheric Research (NIWA), the University of Otago, and the University of Auckland; with funding from the Ministry of Business, Innovation, and Employment. The current program builds on a 2010–12 pilot program funded by the Foundation for Research, Science, and Technology.

The key objectives for the resource assessment program are to study the regional distribution of gas hydrate and to characterize individual gas hydrate reservoirs. The initial area of investigation is a zone outside of the Hikurangi Margin. This characterization effort is utilizing analysis of seismic data to improve the understanding of gas hydrate reservoir rocks and investigation of gas hydrate formation mechanisms. Initial production modeling has been completed as well as the first assessment of seafloor communities that may be affected by gas hydrate production. The overarching goal within the current program is to identify targets for scientific exploration drilling.

4.3.4 Bitumen and heavy oil

This commodity commonly consists of bitumen and heavy oil principally in unlithified sand. Heavy oil reservoirs can also include porous sandstone and carbonates. Oil sands petroleum includes those hydrocarbons in the spectrum from viscous heavy oil to near-solid bitumen, although these accumulations also can contain some lighter hydrocarbons and even gas. These hydrocarbons are denser than conventional crude oil and considerably more viscous (Fig. 4.33), making them more difficult to recover, transport, and refine. Heavy oil is just slightly less dense than water, with specific gravity in the 1.000–0.920 g/cc range, equivalent to API gravity of 10°–22.3°. Bitumen and extraheavy oil are denser than water, with API gravity less than 10°. Extraheavy oil is generally mobile in the reservoir, whereas bitumen is not. At ambient reservoir conditions, heavy and extraheavy oils have viscosities greater than 100 centipoise (cP), the consistency of maple syrup. Bitumen has a gas-free viscosity greater than 10,000 cP (Cornelius, 1987), equivalent to molasses. Many bitumens and extraheavy oils have in-reservoir viscosities many orders of magnitude large. There are a variety of factors that govern the viscosity of these high-density hydrocarbons, such as their organic chemistry,

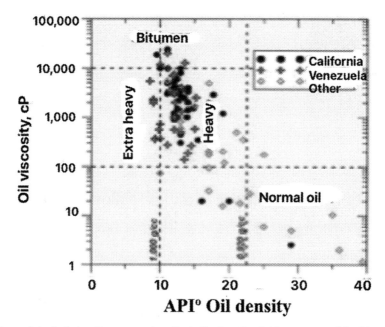

FIG. 4.33 Cross-plot of oil density versus viscosity indicating the fields represented by bitumen, heavy and extra-heavy oils. Actual properties are plotted for a variety of oils from producing oil sand accumulations. *From Koottungal, L., 2012. 2012 worldwide EOR survey. Oil Gas J. 110 (2). Available at www.ogj.com/articles/print/vol-110/issue4/general-interest/special-report-eor-heavy-oil-survey/2012-worldwide-eor-survey.html.*

the presence of dissolved natural gas, and the reservoir temperature and pressure. The viscosity of heavy oil or bitumen is only approximated by its density.

Some heavy oils are the direct product of immature (early) oil generation. However, bitumen and most heavy oils are the products of in-reservoir alteration of conventional oils by water washing, evaporation (selective fractionation) or, at reservoir temperatures below 80°C (176°F), biodegradation (Blanc and Connan, 1994), all of which reduce the fraction of lower molecular weight components of the oil. These light-end distillates are what add commercial value to crude oil. Thus, in addition to being more difficult and costly to recover and transport than conventional oil, heavy oil and bitumen have lower economic value. Upgrading to a marketable syncrude requires the addition of hydrogen to the crude to increase the H/C ratio to values near those of conventional crudes. Heavy oil and bitumen normally contain high concentrations of NSO compounds (nitrogen, sulfur, and oxygen) and heavy metals, the removal of which during upgrading and refining further discounts the value of the resource. Heavy and some extraheavy oils can be extracted in situ by injection of steam or superhot water, CO_2, or viscosity-reducing solvents such as naphtha. Bitumen normally is recovered by surface mining and processing with hot water or solvents.

The International Energy Agency (IEA, 2014) estimates the total world crude oil resources are between 9 and 13 trillion barrels (1.4–2.1 trillion m^3), of which just 30% is conventional crude oil. The remaining 70% is unconventional crude, which is divided into 30% oil sands and bitumen, 25% extraheavy oil, and 15% heavy oil. Heavy oil and bitumen deposits occur in more than 70 countries across the world. Meyer et al. (2007) observed that heavy oils are found in 192 sedimentary basins and bitumen accumulations occur in 89 basins. However, these unconventional oils are not uniformly distributed. The global in-place resources of bitumen and heavy oil are estimated to be 5.9 trillion barrels (938 billion m^3), with more than 80% of these resources found in Canada, Venezuela, and the United States. The largest oil sand deposits in the world, having a combined in-place resource of 3.05 trillion barrels (484.9 billion m^3), are along the shallow up-dip margins of the Western Canada sedimentary basin and the Orinoco foreland basin, eastern Venezuela. Western Canada has several separate accumulations of bitumen and heavy oil that together comprise 1.85 trillion barrels (294.1 billion m^3). The Orinoco Heavy Oil Belt is a single extensive deposit containing 1.2 trillion barrels (191 billion m^3) of extraheavy oil. Both basins have extensive world-class source rocks and host substantial conventional oil pools in addition to the considerably larger resources within shallow oil sands.

Table 4.11 Estimated Global In-Place Heavy Oil and Bitumen Resources, Technically Recoverable Reserves, and Percentage of Global Reserves per Region.

Globally, there are just over 1 trillion barrels (159 billion m^3) of technically recoverable unconventional oils, 434.3 billion barrels (69.1 billion m^3) of heavy oil, including extraheavy crude, and 650.7 billion barrels (103.5 billion m^3) of bitumen. South America, principally Venezuela, has 61.2% of the heavy oil reserves and North America, mainly western Canada, has 81.6% of the bitumen reserves.

Heavy oil, in general, is more easily produced, transported, and marketed than bitumen. Consequently, it tends to be in a more advanced stage of development than bitumen deposits. Countries with very large reserves of conventional crude oil, particularly Saudi Arabia and Kuwait, have been slow to develop their heavy oil resource, whereas countries with small or dwindling conventional oil reserves are exploiting heavy oil to a greater degree.

4.3.4.1 Canada

Nearly all of the bitumen being commercially produced in North America is from Alberta, Canada. Canada is an important strategic source of bitumen and of the synthetic crude oil (SCO) obtained by upgrading bitumen. Bitumen and heavy oil are also characterized by high concentrations of nitrogen, oxygen, sulfur, and heavy metals, which results in increased costs for extraction, transportation, refining, and marketing compared to conventional oil (Attanasi and Meyer, 2010). Research and planning are ongoing for transportation alternatives for heavy crude, bitumen, and upgraded bitumen using the new and existing infrastructure of pipelines and railways. Such integration has been called a virtual "pipeline on rails" to get the raw and upgraded bitumen to US markets. SCO from bitumen and (or) partially upgraded bitumen is being evaluated for potential long-distance transport to refineries in the Midwest and Gulf states of the United States and to existing or proposed terminals on the west coast of North America. Associated concerns include effects on the price of crude oil, and the environmental impacts that are associated with land disturbance, surface reclamation, habitat disturbance, and oil spills or leaks with associated potential pollution of surface and ground waters.

Excellent sources of information on Alberta oil sands and carbonate-hosted bitumen deposits are the resource assessments and regulatory information by the Alberta Energy Regulator (AER, 2015). Estimated in-place resources for the Alberta oil sands are 1845 billion barrels (293.4 billion m^3) (AER 2015, p. 3). The estimated remaining established reserves of in situ and mineable crude bitumen are 166 billion barrels (26.4 billion m^3). Only 5.9% of the initial established crude bitumen has been produced since commercial production began in 1967 (AER, 2015). Cumulative bitumen production for Alberta in 2014 was 10.4 billion barrels (1.65 billion m^3). The bitumen that was produced by surface mining was upgraded; in situ bitumen production was marketed as nonupgraded crude bitumen. Alberta bitumen production has more than doubled in the last decade and is expected to increase to greater than 4 million barrels per day (0.6 million m^3 per day) by 2024. Over the last 10 years, the contribution of bitumen to Alberta's total primary energy production has increased and in 2014 represented over half of Alberta's primary energy production. A breakdown of the production of energy in Alberta from all sources, including renewable sources, is given in Table 4.13.

Crude bitumen is heavy and extraheavy oil that at reservoir conditions has a very high viscosity such that it will not naturally flow to a well bore. Administratively, in Alberta, the geologic formations (whether clastic or carbonate) and the geographic areas containing the bitumen are designated as the Athabasca, Cold Lake or Peace River oil sands areas. Most of the in-place bitumen is hosted within unlithified sands of the Lower Cretaceous Wabiskaw-McMurray deposit in the in situ development area, followed by the Grosmont carbonate-bitumen deposit, and the Wabiskaw-McMurray deposit in the surface mineable area (Table 4.14).

A number of factors (including economic, environmental, and technological criteria) are applied to the initial in-place volumes of crude bitumen to attain the established reserves. In Alberta, there are two types of reserves for crude bitumen—those that are anticipated to be recovered by surface mining techniques (generally in areas with <65 m (<213 ft) of overburden in the Athabasca area), and those to be recovered by underground in situ and largely

TABLE 4.13 Summary of Alberta's energy reserves, resources, and production at the end of 2014 (Alberta Energy Regulator, 2015).

Region	Heavy oil (BBO) Resources	Heavy oil (BBO) Reserves	%	Bitumen (BBO) Resources	Bitumen (BBO) Reserves	%
N. America	185.8	35.3	8.1	1659.1	530.9	81.6
S. America	2043.8	265.7	61.2	1.1	0.1	0.0
Europe	32.7	4.9	1.1	1.4	0.2	0.0
Russia	103.1	13.4	3.1	259.2	33.7	5.2
Middle East	651.7	78.2	18.0	0.0	0.0	0.0
Asia	211.4	29.6	6.8	267.5	42.8	6.6
Africa	40.0	7.2	1.7	430.0	43.0	6.6
Western Hemisphere	2315.4	301.0	69.3	1659.4	531.0	81.6
Eastern Hemisphere	1025.4	133.3	30.7	920.8	119.7	18.4
World total	3340.8	434.3		2580.1	650.7	

The heavy oil category includes extraheavy oil. BBO, billion barrels of oil. 1 barrel = 0.159 m^3.
BBO, billion barrels of oil.

thermal technologies in areas with >65 m (>213 ft) of overburden. The principal technology of choice for Athabasca is Steam-Assisted Gravity Drainage (SAGD), for Cold Lake, it is Cyclic Steam Stimulation (CSS), and for Peace River, it is thermal and primary recovery.

In situ oil sands production continues to be the largest growth area. Compared to surface mining, in situ operations, like SAGD, involve lower capital costs, a smaller "footprint" and reduced environmental impacts. A modest increase in both conventional and tight-formation development is expected, largely due to improvements in multistage hydraulic fracturing from horizontal wells that are targeting these previously uneconomic, but potentially large, resources. In 2012, in situ recovery overtook mining as the favored means of bitumen recovery, according to the Alberta Energy Regulator.

Crude oil prices began to fall in June 2014 and continued to decline for the rest of the year. Despite falling oil prices, upgraded and nonupgraded bitumen production from oil sands continued to grow at a steady pace in 2014. With the decrease in oil prices, economic returns of crude bitumen projects are expected to be affected. Even with this low price environment, oil sands projects under construction continue to move ahead while producers continue to evaluate the viability of new oil sands projects.

4.3.4.2 *Venezuela*

The Faja Petrolifera del Orinoco (Orinoco Heavy Oil Belt) in eastern Venezuela is the world's single largest oil accumulation. The total estimated oil-in-place is 1.2 trillion barrels (191 billion m^3) of which 310 billion barrels (49.3 billion m^3) are considered technically recoverable (Villarroel et al., 2013). The Faja is 55,314 km^2 (21,357 mi^2) in size and extends 600 km (373 mi) in an east-west arcuate band that is up to 90 km (56 mi) wide (Fig. 4.34). The deposit

TABLE 4.14 Initial in-place volumes of crude bitumen as of December 31, 2014 (Alberta Energy Regulator 2015).

Oil sands area oil sands deposit	Initial volume in-place (10^6 m^3)	Area (10^3 ha)	Average pay thickness (m)	Average reservoir parameters Mass (%)	Pore volume oil (%)	Average porosity (%)
Athabasca						
Upper Grand Rapids	5817	359	8.5	9.2	58	33
Middle Grand Rapids	2171	183	6.8	8.4	55	32
Lower Grand Rapids	1286	134	5.6	8.3	52	33
Wabiskaw-McMurray (mineable)	20,823	375	25.9	10.1	76	28
Wabiskaw-McMurray (in situ)	131,609	4694	13.1	10.2	73	29
Nisku	16,232	819	14.4	5.7	68	20
Grosmont	64,537	1766	23.8	6.6	79	20
Subtotal	242,475					
Cold Lake						
Upper Grand Rapids	5377	612	4.8	9.0	65	28
Lower Grand Rapids	10,004	658	7.8	9.2	65	30
Clearwater	9422	433	11.8	8.9	59	31
Wabiskaw-McMurray	4287	485	5.1	8.1	62	28
Subtotal	29,090					
Peace River						
Bluesky-Gething	10,968	1016	6.1	8.1	68	26
Belloy	282	26	8.0	78	64	27
Debolt	7800	258	25.3	5.1	66	18
Shunda	2510	143	14.0	5.3	52	23
Subtotal	21,560					
Total	293,125					

1 m^3 = 35.3 ft^3; 1 hectare (ha) = 2.47 acres; 1 m = 3.28 ft.
From: Unconventional Energy Resources: 2015 Review.

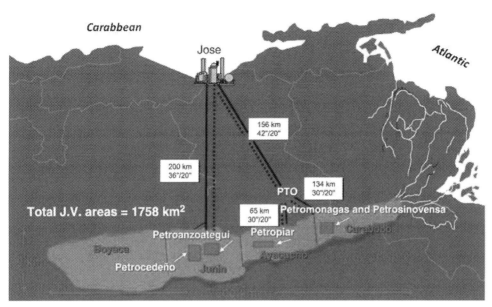

FIG. 4.34 The Faja Petrolifera del Orinoco in eastern Venezuela *(light green)* showing the four production units *(red text)*, four current production projects *(white text)*, and pipelines connecting the projects to the Jose upgrading facility on the coast. 1 km = 0.6 mi; 1 km² = 0.386 mi². *From Villarroel, T., Mambrano, A., Garcia, R., 2013. New progress and technological challenges in the integral development of the Faja Petrolifera del Orinoco, Venezuela. In: Hein, F.J., Leckie, D., Larter, S., Suter, J.R. (Eds.), Heavy-Oil and Oil-Sand Petroleum Systems in Alberta and Beyond. AAPG Studies in Geology, vol. 64. AAPG, Tulsa, OK, pp. 669–688.*

lies immediately north of the Orinoco and Arauca Rivers in the southern portions of the states of Guarico, Anzoategui, and Monagas. The Faja follows the extreme up-dip edge of the foreland basin of the young Serrania del Interior thrust belt, the source of the oil, where Neogene-age sediments overlie the crystalline basement of the Guayana Shield. To the north, in the foothills of the Serrania del Interior, there are numerous conventional oilfields, the majority in structural traps within the thrust belt.

Extra-heavy oil having an average API gravity of 8.5° is in stratigraphic traps within the highly porous and permeable sands of the lower and middle Miocene Oficina Formation. These sands were carried off the Guayana Shield by river systems flowing north and northeastward to be deposited in fluvio-deltaic and estuarine complexes on the south rim of the foreland basin (Martinius et al., 2013). Upper Miocene marine shales of the Freites Formation form the top seal to the Faja oil accumulation. The net thickness of oil-impregnated sands is highest within the paleo-deltas, giving rise to a highly irregular distribution of resource richness within the Faja.

At present, there are four active heavy oil recovery projects operating in the Faja (Fig. 4.34), each begun in successive years between 1998 and 2001. Petroleos de Venezuela (PDVSA) is the sole owner/operator of Petroanzoategui and is the senior joint venture partner in the other three projects with a partner as the operator: BP in Petromonagas, Chevron in Petropiar, and Total with Statoil in Petrocedeño. In what is referred to as the "first stage" of development, the four projects are now producing collectively about 640,000 bopd (101,760 m^3/d) using cold production methods (Villarroel et al., 2013). These methods are possible due to the highly porous and permeable properties of the reservoir sands and the gas-charged and foamy character of the extraheavy oil. The dissolution of dissolved natural gas in the oil during production aids in propelling the oil from the sand and toward the wellbore. The foaming of the oil and reservoir temperatures of about 50°C (122°F) helps overcome its viscosity, which is on the order of thousands of centipoises. The oil is extracted from horizontal wells as long as 1.5 km (0.9 mi) with the aid of downhole progressive cavity pumps and multiphase pumps at the wellhead. A major challenge is the optimal placement of the long horizontal wells in these complex heterogeneous fluvial-deltaic sands (Martinius et al., 2013).

To enhance production, a 50° API naphtha diluent is commonly injected into the horizontal wells to further decrease viscosity. The recovery factor for this cold production is about 10%. The naphtha-charged oil is transported by pipeline about 200 km (124 mi) to the Jose upgrading facility on the Caribbean coast (Fig. 4.34). Here the naphtha is separated from the oil and returned to the projects via dedicated diluent pipelines (Fig. 4.34). The oil is upgraded in one of four delayed coking units to a 32° API syncrude that is exported as "Zuata Sweet." As the projects prepare for the next phase of development, a variety of established enhanced oil recovery (EOR) technologies are being tested in pilots, including thermal methods (SAGD, CSS) and reservoir flooding using polymer-viscosified water.

To increase the rate of extraheavy oil production by expanding operating areas, PDVSA has entered into joint venture agreements with various national or quasinational oil companies. However, at present, more heavy crude is being produced than can be processed in the Jose upgraders, which are more than a decade old. The lack of investment funds has prevented PDVSA from adequately maintaining and expanding the pipelines and upgrading facilities. This situation has been exacerbated by the recent drop in global oil price.

4.3.4.3 *United States*

The goal of the United States to move toward greater energy independence could include production from existing US oil sands deposits using surface mining or in situ extraction. Current US bitumen production is mainly for local use on roads and similar surfaces. This is due mainly to the different characters and scale of the deposits compared to Canada and Venezuela, but in part it is because, outside of heavy oilfields in California and Alaska, the United States has not developed the infrastructure required to produce oil sands as a commercially viable fuel source. Schenk et al. (2006) compiled total measured, plus speculative, estimates of bitumen in-place of about 54 billion barrels (8.6 billion m^3) for 29 major oil sand accumulations in Alabama, Alaska, California, Kentucky, New Mexico, Oklahoma, Texas, Utah, and Wyoming (Table 4.15). However, these older estimates of total oil sand resources provide only limited guidance for commercial, environmentally responsible development of the oil sand deposits. In addition, the estimates do not factor in commercially viable heavy oil resources. The resources in each of the states have distinct characteristics that influence current and future exploitation.

California has the second largest heavy oil proved reserves in the world, second only to Venezuela. California's oilfields, of which 52 each have reserves greater than 100 million barrels (15.9 million m^3), are located in the central and southern parts of the state (Fig. 4.35). As of the beginning of 2014, California's proved reserves were 2878 million barrels

TABLE 4.15 Previous estimates of bitumen-heavy oil resource-in-place, measured and total, including speculative, in the United States.

State	No. deposits	°API range	Measured (MMB)	Total (MMB)
Utah	10	−2.9 to 10.4	11,850	18,680
Alaska	1	7.1–11.5	15,000	15,000
Alabama	2	NA	1760	6360
Texas	3	−2.0 to 7.0	3870	4880
California	6	0.0–17.0	1910	4470
Kentucky	4	10	1720	3410
New Mexico	1	12	130	350
Wyoming	2	NA	120	145

1 MMB = 158,987 m^3.
MMB, million barrels.
Data from Schenk, C.J., Pollastro, R.M., Hill, R.J., 2006. Natural Bitumen Resources of the United AAPG EMD Bitumen and Heavy Oil Committee Commodity Report – May 2019. U. S. Geological Survey Fact Sheet 2006-3133, 2 pp. http://pubs.usgs.gov/fs/2006/3133/pdf/FS2006-3133_508.pdf.

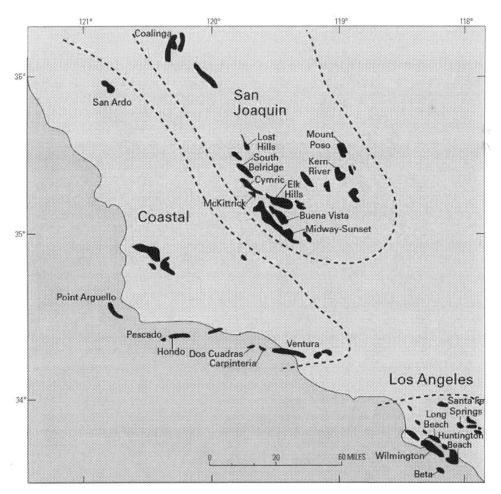

FIG. 4.35 Principal oilfields of California. 1 mile = 1.6 km. *From Tennyson, M.E., 2005. Growth history of oil reserves in major California oil fields during the twentieth century. In: Dyman, T.S., Schmoker, J.W., Verma, M. (Eds.), Geologic, En-*

(457.6 million m^3) (EIA, 2014a). The dominantly heavy oilfields of the southern San Joaquin basin have 2014 proved reserves of 1813 million barrels (288.3 million m^3). Most of the fields were discovered and put into primary production in the period 1890–1930. However, with the introduction of water flooding, thermal recovery, and other EOR technologies starting in the 1950s and 1960s, oil recoveries improved dramatically and the proved reserves increased several fold (Tennyson, 2005).

Nearly all of the oil is sourced from organic-rich intervals within the thick Miocene Monterey diatomite, diatomaceous mudstone and carbonate. Due to a combination of Type IIS kerogen, modest burial and thermal heating, and generally shallow depths of oil pools, the oil tends to be heavy and relatively viscous. These are thermally immature, partially biodegraded oils. Roughly 40% of the oil is produced by steam flooding, cyclic steam stimulation, or other thermal recovery methods. Thermally produced oil comes mainly from fields in the San Joaquin basin (Fig. 4.35). In general, the reservoirs are poorly or unconsolidated sandstones intercalated within or overlying the Monterey Formation. In addition to the heavy oil accumulations that are being produced, California has numerous shallow bitumen deposits and seeps that are not currently exploited. The total resource is estimated to be as large as 4.7 billion barrels (747.3 million m^3) (Kuuskraa et al., 1987). Five of the six largest oil sand deposits are in the onshore Santa Maria basin (central Coastal zone in Fig. 4.35), covering a total area of over 60 mile^2 (155 km^2).

The California heavy oils are exceptional in that they sell with little or no discount compared to the WTI benchmark. From 2011 through mid-2014, the price of benchmark Midway-Sunset 13° API crude had remained near \$100/barrel (\$100/0.159 m^3) (EIA, 2015). The oil price dropped to a low of \$42.93 in January 2015, but then gradually increased to \$57.48 in early May 2015. In the existing heavy oilfields of California, where natural gas burned to generate steam is the principal operational cost factor, a dramatic drop in both oil and gas price may reduce new capital expenditures, but not ongoing oil production. During the past decade, oil production in California has steadily declined (EIA, 2014a). The rate of decline is being slowed, and in some fields reversed, through the application of fully integrated reservoir characterization and improved recovery technologies that are resulting in higher recovery factors (Dusseault, 2013; Beeson et al., 2014), up to 70%–80% in some fields.

Alaska's heavy oil and bitumen deposits on the North Slope are very large (24–33 billion barrels, or 3.8–5.2 billion m^3) and they hold promise for sustained commercially successful development. Since early 1980s (Werner, 1987), two very large, shallow heavy oil-impregnated sands have been known to overlie the Kuparuk River field and underlie a 1800-ft (549 m)-thick permafrost (Fig. 4.36). These are the Ugnu Sands (8°–12° API) at depths of 2000–5000 ft (610–1524 m) and the West Sak Formation (16°–22° API) at 2300–5500 ft (701–1676 m). The size of the deposits is well defined with the numerous wells tapping the underlying conventional oilfields. For the Lower Ugnu Sands and West Sak Formation, the resources are 12–18 billion barrels (1.9–2.9 billion m^3) and 12 billion barrels (1.9 billion m^3), respectively. The reservoirs are fluvial-deltaic sands deposited during the Late Cretaceous-earliest Paleocene across the north and northeast prograding Brooks Range coastal plain (Hulm et al., 2013).

Production of viscous (50–5000 cP) oil from the West Sak pools began in the early 1990s, reaching the current level of 4000–5000 barrels (636–795 m^3) of oil per day in 2004. To date, over 100 million barrels (15.9 million m^3) have been recovered from the formation using a

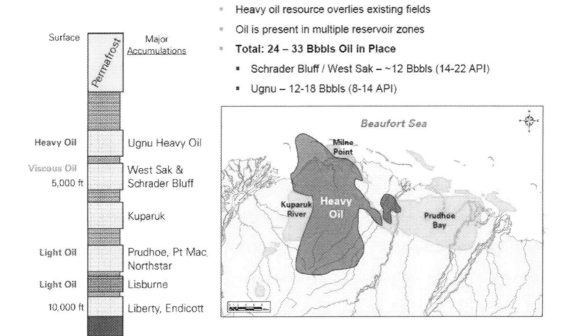

FIG. 4.36 Location of shallow, heavy oil accumulations on the North Slope of Alaska. Heavy oil deposits overlie the Kuparuk field and parts of the Prudhoe and Milne Point fields and occur in sands within the Ugnu, West Sak, and Schrader Bluff formations. 1 ft = 0.3 m; 1 barrel (oil) = 0.159 m^3. *From Pospisil, G., 2011. BP Exploration (Alaska) Inc. Presentation on January 6. Available at www.aoga.org/wp-content/uploads/2011/01/8-Pospisil-Heavy-Viscous-Oil.pdf.*

combination of vertical wells and water flood. The heavy oil in the Ugnu Sands presents a much greater technical challenge due to its higher viscosity (5000 to >20,000 cP) and the friability of the reservoir sand (Chmielowski, 2013). A pilot project at Milne Point using the CHOPS ("cold heavy oil production with sand") recovery process (Young et al., 2010), although technically successful, has been suspended, at least for the present.

Utah's bitumen and heavy oil deposits are found throughout the eastern half of the state (Schamel, 2013a,b). In northeast Utah, the largest accumulations are located along the southern margin of the Uinta Basin underlying vast portions of the gently north-dipping East and West Tavaputs Plateaus. This highland surface above the Book and Roan Cliffs on either side of the Green River (Desolation) Canyon is supported by sandstone and limestones of the Green River Formation (lower Eocene). Here the resource-in-place is at least 10 billion barrels (1.59 billion m^3), nearly all of it reservoired in fluvial-deltaic sandstone bodies within the lower member of the Green River Formation. On the northern margin of the Uinta Basin, heavy oil occurs in a variety of Mesozoic and Tertiary reservoirs on the hanging wall of the Uinta Basin Boundary Fault. The proven resource is <2 billion barrels (<0.32 billion m^3), but the potential for additional undiscovered heavy oil and bitumen is great. In both areas, the source of the heavy oil is organic-rich lacustrine calcareous mudstone in the Green River Formation. These naphthenic oils have API gravities in the 5.5°–17.3° range, are only weakly biodegraded in the subsurface, and are sulfur-poor (0.19–0.76 wt%). The known oil sand reservoirs are lithified and oil-wet.

New resource-in-place estimates for the major deposits are determined from the average volume of bitumen/heavy oil measured in cores distributed across the deposit, as delineated by wells and surface exposures (Table 4.16). The deposits on the south flank of the basin are extensive and large, but the actual concentrations (richness) of resource are small. For the vast P.R. Spring-Hill Creek deposit, the average richness is just 25.9 thousand barrels per acre (4.1 thousand m^3 per 0.4 ha); it is only slightly higher for the entire Sunnyside accumulation west of the Green River. However, a small portion of the Sunnyside deposit having unusually thick reservoir sands within a monoclinal structure trap has measured average richness as large as 638.3 thousand barrels (101.5 thousand m^3) per acre. The two principal deposits on the north flank of the basin, Asphalt Ridge and Whiterocks, are relatively small, but they contain high concentrations of heavy oil (Table 4.16).

In the southeast quadrant of Utah, there are numerous shallow bitumen accumulations on the northwest and west margins of the Pennsylvanian-Permian Paradox Basin. The deposits are hosted in rocks of late Paleozoic and early Mesozoic age. With the exception of the Tar Sand Triangle and Circle Cliffs deposits, most accumulations are small and/or very lean. Normally, the oils are heavier than 10° API and highly biodegraded. In contrast to the Uinta Basin deposits, this bitumen is derived from a marine source rock and is aromatic with high sulfur content (1.6–6.3 wt%), but low nitrogen (0.3–0.9 wt%).

TABLE 4.16 Estimated resource and richness of principal bitumen deposits in Utah.

Bitumen-heavy oil deposit	Resource estimate MMB	Areal extent square miles	Richness, average (MB/acre)	°API gravity	Reservoir unit
P.R. Spring-Hill Creek	7790	470	25.9	5.9–13.8	Lower Green River ss
Sunnyside	3500–4000	122	45–51	7.1–10.1	Lower Green River ss
Sunnyside "core"	1160	2.7	638.3		Lower Green River ss
Asphalt Ridge	1360	16	132.9	10.0–14.4	Mesaverde ss (U Cret.)
Whiterocks	98	0.45	338	11.4–13.5	Navajo Ss (Tr.-Jr.)
Tar Sand Triangle	4250–5150	198	33.5–40.6	−3.6 to 9.6	White Rim Ss. (L. Perm)
TST "core"	1300–2460	30–52	67.7–73.9		White Rim Ss. (L. Perm)

1 MMB = 158,987 m^3; 1 MB = 159 m^3; 1 mile^2 = 259 ha. 1 acre = 0.4049 ha.
MMB, million barrels; *MB*, thousand barrels.
From Schamel 2013a,b.

The Uinta Basin heavy oils and bitumens are highly viscous; the Tar Sand Triangle bitumen is only slightly less viscous. Both groups of oils have viscosities that are orders of magnitude greater than that of the 13° API heavy oil produced by steam flood in the southern San Joaquin Basin, California. So far, the Utah oil sands have resisted attempts at commercial development. However, two projects are scheduled to be operational before the end of 2015. The Calgary-based US Oil Sands PR Spring commercial demonstration project at Seep Ridge is designed to produce liquids from surface-mined oil sand using a closed-loop solvent extraction process. The mine and processing site was prepared in 2014, and the process extraction equipment modules are scheduled to be delivered and assembled mid-year 2015. The mine and extractor could be operating by October 2015. Toronto-based MCW Energy Group Limited has announced plans to build two 2500 bopd (400 m^3 per day) closed-loop solvent extractors at the existing Temple Mountain mine at the south end of Asphalt Ridge.

4.3.4.4 Russia

Heavy oil constitutes roughly 13.1% of the total Russian oil reserves, which official estimates place at 22.5 billion m^3 (141.8 billion barrels). Recoverable heavy oil occurs in three principal petroleum provinces: (1) the southern, up-dip portion of the West Siberian Basin, (2) the Volga-Ural Basin, and (3) Timan-Pechora Basin, on the southwest and northwest foreland (Ural Mountains), respectively. Resource and reserve summaries of the deposits within the three principal petroleum provinces are described herein.

The heavy oil and bitumen accumulations of the Volga-Ural province, Russia's second largest oil producing region, are within Carboniferous-Permian age reservoirs on or flanking the enormous Tatar dome. There are 194 known heavy oil-bitumen fields, most of which are reservoired within shallow Permian rocks in the central and northern parts of the province. Tatarstan holds Russia's largest natural bitumen resources; there are 450 deposits in Upper Permian sandstones with 1.163 billion m^3 (7.3 billion barrels) of resource-in-place. The heavy oil and bitumen of this province have high sulfur content (up to 4.5%) and contain transition metals (Ni, Mo). A very large portion of the total oil reserves is heavy oil.

In the Timan-Pechora basin, the heavy oil and bitumen resources occur in shallow pools on the Timan arch forming the southwest margin of the basin and in shallow anticlinal traps within the basin center. Several known accumulations are in production. The Yaregskoye oilfield is located in East-Pechora Swell and the Usinskoye oilfield is located in the Kolvino Swell. The Yaregskoye field in Komi Republic, containing about 375 million m^3 (2358 million barrels) of heavy oil proved recoverable reserves in Devonian formation, is the largest field in the Timan-Pechora petroleum province. The second largest is the Usinskoye oilfield which contains OOIP around 67.9 million m^3 (425 million barrels) of heavy oil in Permian-Carboniferous age reservoirs.

4.3.5 Underground refining

The growing size or "footprint" of the surface mines and their tailings ponds is an environmental problem needing to be addressed. To this end, Syncrude Canada Limited is preparing to build a C$1.9 billion centrifuge plant at its Mildred Lake mine, which when operational in 2015 will reduce the waste slurry from the separators to a less-hazardous, near-dry sediment requiring far less surface storage (Curran, 2014).

Oil sand developers in Canada largely have been successful in reaching the goal of reducing CO_2 emissions by 45% per barrel (45% per 0.159 m^3), as compared to 1990 levels. Also in Canada, developers are legislated to restore oil sand mining sites to at least the equivalent of their previous biological productivity. For example, at development sites near Fort McMurray, Alberta, the First Nation aboriginal community, as part of the Athabasca Tribal Council, and industry have worked together to reclaim disturbed land (Boucher, 2012) and industry has reclaimed much of the previous tailings pond areas into grasslands that are now supporting a modest bison herd (about 500–700 head).

- Unconventional natural gas (UG) comes primarily from three sources: shale gas found in low permeability shale formations; tight gas found in low-permeability sandstone and carbonate reservoirs; and coalbed methane (CBM) found in coal seams.
- Although several countries have begun producing UG, many global resources have yet to be assessed. According to current estimates, China has the largest technically recoverable shale gas resource with 1115 trillion cubic feet (Tcf), followed by Argentina (802 Tcf) and Algeria (707 Tcf). Global tight gas resources are estimated at 2684 Tcf, with the largest in Asia/Pacific and Latin America. Resources of CBM are estimated at 1660 Tcf, with more than 75% in Eastern Europe/Eurasia and Asia/Pacific.
- Recoverable US resources are estimated at 1611 Tcf from shale and tight gas and 105 Tcf from CBM.
- UG, particularly shale and tight gas, is most commonly extracted through hydraulic fracturing, or "fracking." A mixture of fluid (usually water) and sand is pumped underground at extreme pressures to create cracks in the geologic formation, allowing gas to flow out. When the pressure is released, a portion of the fluid returns as "flowback," and the sand remains as a "proppant," keeping the fractures open.
- UG accounted for 87% of total US natural gas production in 2019 and is expected to account for 92% of production by 2050 (Fig. 4.37).
- Tight oil, or shale oil, is found in impermeable rocks such as shale or limestone and is extracted through fracking and is often extracted concurrently with natural gas.
- Over the past decade, tight oil production has expanded significantly. In 2019, 63% (7.7 million barrels per day) of crude oil production in the United States came from tight oil. Top tight oil producing states include Texas, North Dakota, New Mexico, Colorado, and Oklahoma.
- It is estimated that the United States has 174 Bbbl of technically recoverable tight oil.
- Negative health effects in newborns from in utero exposure to fracking sites have been found (Fig. 4.38).
- Oil sands, i.e., "tar sands" or "natural bitumen," are a combination of sand (83%), bitumen (10%), water (4%), and clay (3%). Bitumen is a semisolid, tar-like mixture of hydrocarbons.
- Known oil sands deposits exist in 23 countries. Canada has 73% of global estimated oil sands, approximately 2.4 trillion barrels (bbls) of oil. The United States has 1.6% of global oil sands resources; however, 56% of US crude oil imports came from Canada in 2019, and 64% of Canadian production comes from oil sands.

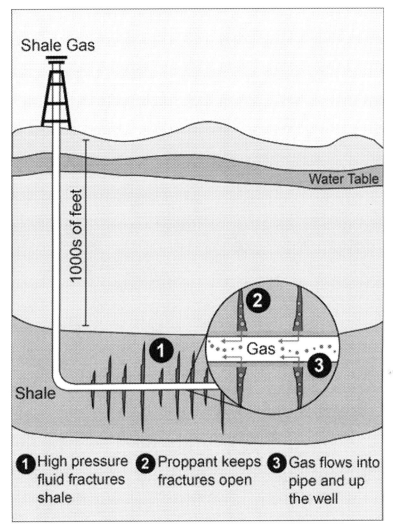

FIG. 4.37 Proppant movement in hydraulic fracturing.

- Deposits less than 250 ft below the surface are mined and processed to separate the bitumen. Deeper deposits employ in situ (underground) methods, including steam or solvent injection to liquify the bitumen so that it can be extracted from the ground. Bitumen must be upgraded to synthetic crude oil (SCO) before it is refined into petroleum products.
- Two tons of oil sands produce one barrel of SCO (Fig. 4.39).
- Oil shale is a sedimentary rock with deposits of organic compounds called kerogen, which has not undergone enough geologic pressure, heat, and time to become conventional oil. Oil shale can be heated to generate petroleum-like liquids.

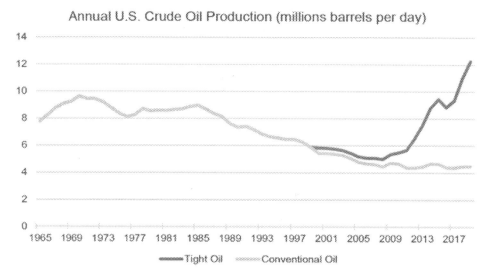

FIG. 4.38 Tight oil production history in USA.

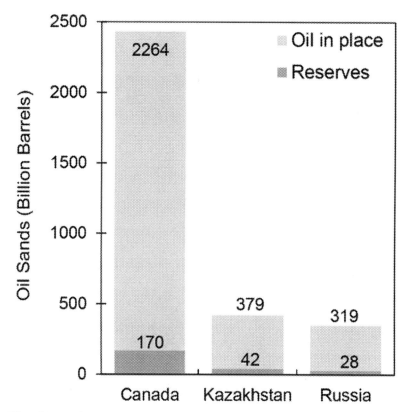

FIG. 4.39 Oil sand in top three countries.

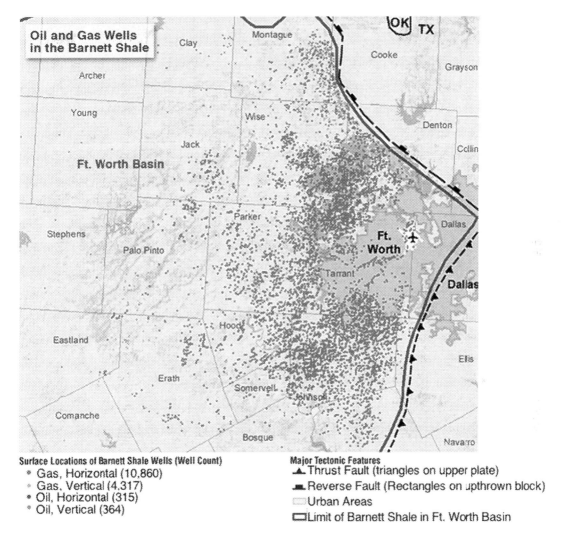

FIG. 4.40 Location map for Barnett shale. *From EIA. 2011. North American Shale Plays. Available at http://www.eia.gov/oil_gas/rpd/northamer_gas.pdf.*

- Oil shale deposits exist in 33 countries. The United States has the largest oil shale resource in the world, approximately 6 trillion bbls of oil in-place, however, oil shale is far from commercial development (Figs. 4.40 and 4.41).

4.4 Sustainable development of unconventional reservoirs

Sustainability of the unconventional technology must be custom designed its geological nature and the economics that can support its development and growth. It shows

4. Unconventional reservoirs

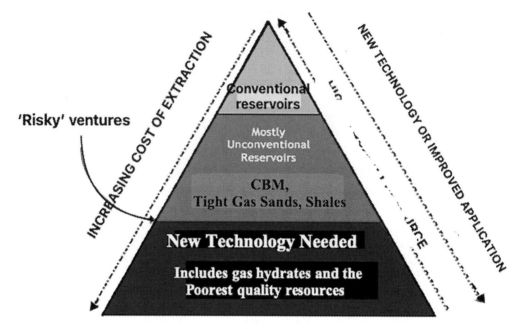

FIG. 4.41 Conventional perception of resource triangle.

conventional engineering analysis is not the only one that doesn't apply to unconventional reservoirs, conventional accounting theories as well as energy pricing models don't apply also. It shows with details why essential features of unconventional gas are such that none of the conventional techniques and principles should apply. This ranges from geology to chemistry to engineering for coal bed methane, shale gas, tight gas as well as methane gas hydrate.

Recent years have brought a rapid expansion of activities around unconventional gas. The successes in unconventional gas exploitation have been mainly attributed to advances in engineering. In particular, hydraulic fracturing and horizontal wells are considered to be the game changer. To date, geophysics and geology have yet to assume a major role in the development of this important resource.

Islam et al. (2012) recently argued any "game changer" must accompany real change, which means change from the first premise to the process itself. If previous chapters established, unconventional gas reservoirs are indeed uniquely different from conventional ones, why should we assume the reservoir characterization tools and procedure of conventional reservoirs would apply to unconventional ones?

Previous techniques included reservoir characterization based on production or data of the present. Even in exploration, unconventional gas reservoirs are first identified and "proved," followed by the search of additional resources that are marked as "emerging" and "speculative." This defies the logic that all unconventional gas emerges from conventional sources. The new technique reversed that and introduced reservoir characterization based on geology first. Based on this technique, all the US oil and gas fields are characterized based on origin, then migration of the fluid. This results in a truly scientific characterization of unconventional resources. Consider the significance when applied on worldwide reserve.

4.5 Improving oil and gas recovery from unconventional reservoirs

4.5.1 Existing projects

This section is an overview of existing enhanced oil recovery schemes. Even though the focus is conventional oil and gas, the discussion helps the readership identify technologies that are proper for oil and gas recovery from unconventional sources.

Primary recovery is the oil recovery by natural drive mechanisms. Such natural drives may be through solution gas expansion, water drive, gas-cap drive, or simply gravity drainage. Secondary recovery is known to be the oil recovery technique in which gas or water is injected to maintain the reservoir pressure. Tertiary recovery is any oil recovery scheme, conventionally applied after secondary recovery. However, for over two decades, there is a tendency to use the term *enhanced oil recovery* to define a wide range of recovery processes. The enhanced oil recovery (EOR) is an oil recovery scheme that uses the injection of fluids, not normally present in the reservoir. For instance, chemical injection, steam injection, in situ combustion, or even microbial enhanced recovery will be considered as EOR. This definition, while encompasses many recovery schemes beyond the scope of *tertiary recovery* (recovery scheme that follows secondary recovery), does not include techniques, such as electromagnetic heating even when this could be an effective technique for increasing oil production from a reservoir. However, recent publications acknowledge electromagnetic heating as an EOR process. A proper definition of EOR should, therefore, be any oil recovery technique that improves oil recovery from a reservoir beyond primary recovery. While fluid injection may be required for some techniques, energy dissipation may be sufficient in some cases. This definition may seem to be too broad because it does not exclude waterflood or pressure maintenance gas injection from the definition of EOR. Many waterflood and gas injection schemes are indeed displacement-type recovery processes and should be called an EOR scheme. A purely pressure maintenance scheme is usually well defined and no confusion as to its distinction from EOR schemes exists.

The word "EOR" fell out of grace shortly after tax incentive for EOR schemes were repealed in 80s. This saw the sudden drop of EOR projects in the United States that peaked in 1986.

More recently, environmental concerns have become part and parcel of EOR activities. In part this is fueled by renewed awareness of the environmental impact of unsustainable engineering practices, even though it is perceived as a greenhouse gas emission problem.

Energy production and use are considered major causes of greenhouse gas emissions. The emission of greenhouse gases, particularly CO_2, is of great concern today. Even though CO_2 is considered one of the major greenhouse gases, production of natural CO_2 is essential for maintaining life on Earth. Note that all CO_2 is not the same and plants do not accept all types of CO_2 for photosynthesis. There is a clear difference between old CO_2 from fossil fuels and new CO_2 produced from renewable biofuels (Islam et al., 2010b). The CO_2 generated from burning fossil fuel is an old and contaminated CO_2. Because various toxic chemicals and catalysts are used for oil and natural gas refining/processing, the danger of generating CO_2 with higher isotopes cannot be ignored. Hence, it is clear that CO_2 itself is not a culprit for global warming, but the industrial CO_2 that is contaminated with catalysts and chemicals likely becomes heavier with higher isotopes and plants cannot accept this CO_2. Plants always accept the lighter portion of CO_2 from the atmosphere. Thus, CO_2 has to be distinguished between natural and industrial CO_2, based on the source from which it is emitted and the pathway that the fuel follows from the source to combustion. Islam et al. (2010) showed that even though the total CO_2 is increasing in the atmosphere, the natural CO_2 has been decreasing since the industrial revolution. They further argued that industrial CO_2 is responsible for global warming. Thus, generalizing CO_2 as a precursor for global warming is an absurd concept and is not valid. See discussion below for more details.

In absence of solidly founded scientific investigations, world governments have at least agreed that the greenhouse emission is a concern and have been under pressure to ratify Kyoto Agreement. Even though the Kyoto protocol as well as Copenhagen agreement (or lack of it) have not been implemented, it managed to extract more funding for greenhouse mitigation projects than EOR projects. For instance, in 2011, the petroleum sector in the United States invested $71.1 billion, where as $73.7 billion came from the private sector other than the petroleum industry. The rest (23%) that amounts to $43.4 billion came from the government that tagged the amount to greenhouse gas mitigation projects. It turns out the 39% from the "other industry" also invested in these "environmental" projects.

Most recent DoE/EIA report (DoE/EIA, 2013) shows how climate change concerns and specifically greenhouse gas emissions have been instrumental in determining future energy outlook and regulatory policies. Regulators and investment companies been pushing energy companies to invest in technologies that are less GHG-intensive. Federal government grants are often linked to environmental aspects of petroleum engineering. Even within the petroleum industry, the focus has shifted toward mitigation of greenhouse gases. This provides a unique opportunity for coupling EOR projects that can bring in double dividends, namely environmental integrity and financial boon.

The word "EOR" fell out of grace shortly after tax incentives for EOR schemes were repealed in 1980s. This saw the sudden drop of EOR projects in the United States that peaked in 1986. Fig. 4.42 shows the number of EOR projects during 1971–2010. The biggest "victim" EOR tax break withdrawal was the chemical methods. The number of chemical methods dropped to practically zero. Chemical injection projects were also among the first ones introduced. There were numerous patents that showed, based on laboratory results, that chemical methods would be grand success, often reporting as much as 99% of the oil recovery. It turned out, all projects went into applications before properly scaling up the laboratory results. In 1990, Islam and Ali showed that the scaling laws are such that practically all chemical processes should have no chance of success in field scale although they showed very promising

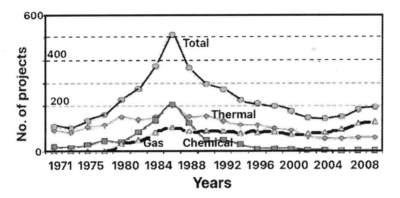

FIG. 4.42 Number of EOR projects during 1971–2006. *From Alvarado, V., Manrique, E., 2010. Enhanced oil recovery: an update review. Energies 3 (9), 1529–1575. https://doi.org/10.3390/en3091529.*

results during the laboratory tests. These field projects were costly and they produced practically no result in terms of additional recovery. However, some oil recovery was tagged to chemical recovery, most likely due to tax credits. The chemical project died off in 1998, but even before that they were producing no additional oil. This demise of chemical injection projects, curiously, did not impact total recovery due to EOR practices, even though the number of projects declined sharply. This showed that projects were being tagged as "EOR" in the past to gain tax credits. During 1990s, thermal recovery mainly involved steam flood whereas a gas injection relied heavily on CO_2 injection. The rise in CO_2 and other miscible injection projects maintained an overall status quo in terms of number of projects and increasing oil recovery. This is a trend that continues to-date.

In the 2000s, there has been somewhat of a resurrection of chemical methods. From several of the inventions of Chemical companies, there has been concerted effort to use new line of polymers to control mobility of waterflood schemes. These numbers are shown in Fig. 4.43. In this figure, both CO_2 and other gas miscible injections projects are lumped together. As can be seen from this figure, only gas injection projects show steady overall growth.

Prior to 2014, The Oil and Gas Journal (OGJ) used to publish a bi-annual update on EOR projects, most of which were from the United States. Traditionally, the United States has been the pioneer of EOR and other technologies. Ever since 2014, the IEA has been conducting its extensive reviews of the global status of EOR projects. These reviews are lumped together with OGJ reviews to show global trend since 1971, when the first EOR project review was conducted (Fig. 4.44). The number of global projects has been increasing continuously since 1996, when chemical injection projects were shut down in the United States. It turns out that chemical projects were kept alive outside of the United States and they grew in number. Some of these projects relate to North Sea operations that used new line of chemicals, such as polymers and surfactants. Success of these projects has been mixed and will be discussed in latter sections. Among all EOR applications, CO_2 miscible injection has been the most consistently used one. In recent years, the appeal of CO_2 sequestration has added to the growing number of projects involving CO_2 injection.

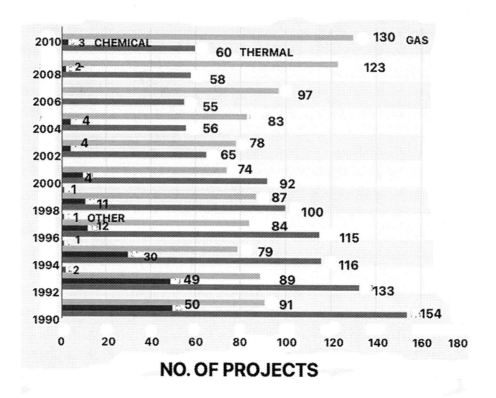

FIG. 4.43 EOR projects in post Cold war era. *From IEA, 2018a. World Energy Outlook 2017, China. https://www.iea.org/weo/china/.*

IEA (2018a) estimates that there are currently around 375 EOR projects operating globally, producing just over 2 million barrels per day in 2018 (Fig. 4.44). While this is a 0.7 mb/d increase from the last assessment carried out in 2014 by the OGJ, this amounts to a stable 2% of the global oil production (Fig. 4.45).

As stated earlier, historically EOR production has been uniquely from North America. However, in recent years, other countries have joined in. Malaysia has started offshore EOR production in 2016. EOR is expected to increase the oil recovery factor at Malaysia's producing fields by 5%–10% from around 33%–37% on average before EOR. Initially, around 10 EOR projects are anticipated to start during 2016–26. The process used in Malaysia is WAG (water alternating gas). Ever since, the United Arab Emirates, Kuwait, Saudi Arabia, India, Colombia and Ecuador have all started pilot EOR projects. Meanwhile Oman has also registered a major increase in EOR production. As a result, while in 2013 three quarters of all EOR projects (providing 0.8 mb/d) were located in North America, today this proportion has fallen to 40%. There has also been a wave of efforts to apply EOR technologies in offshore fields; today there are around 15 offshore projects, which mostly inject natural gas.

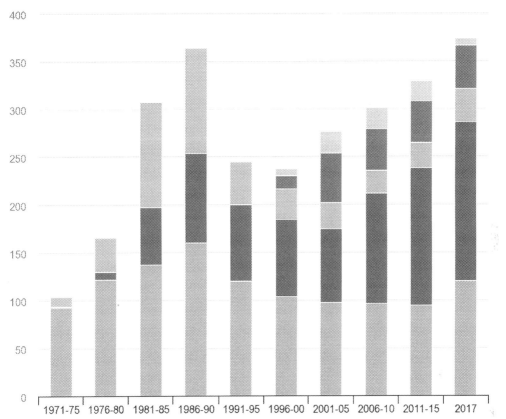

FIG. 4.44 Global EOR projects, in the figure, bottom stack represents thermal, then CO_2, then chemical, then other gas, and finally other EOR. *From IEA, 2018a. World Energy Outlook 2017, China. https://www.iea.org/weo/china/.*

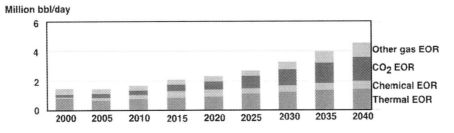

FIG. 4.45 Global oil production due to EOR activities.

Outside of the United States, Oman is the first country to employ EOR. It is also first project that used a set of sustainable technologies. Since 2007, Oman has steadily increased its oil production to near record levels through steam injection and other advanced EOR solutions. According to the National Centre of Statistics and Information (NCSI), gas used at Oman's oilfields account for more than 20% of the country's total gas use, with fuel for EOR

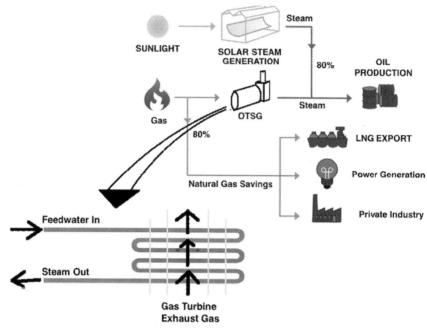

FIG. 4.46 Solar EOR of Oman.

representing a significant portion of that. This will continue to increase as EOR expands to contribute a third of the country's crude oil production by 2020. At the same time, more gas is needed for power generation, desalination and industrial development. Due to the use of solar EOR pilot project, built in 2011, 50 tonnes of steam per day is produced, proving the effectiveness and cost efficiency of GlassPoint's technology (Islam, 2020). Fig. 4.46 shows the schematic of the EOR technology.

The vast majority of the materials needed to manufacture the steam generators were sourced locally within Oman itself. As pointed out by Khan and Islam (2012, 2016), the direct solar heating has much higher efficiency than photo voltaic generation of heat. The process is further improved by using the Through Heat Recovery Steam Generators (OTSGs). The process essentially maximizes usage of energy to a level of close to zero-waste. When looked at it within bigger picture, the project brings in savings from avoiding use of gas, utilizing local resources, and from long-term environmental sustainability. By using solar to generate steam, Oman can save up to 80% of the gas currently used for EOR. The saved natural gas then brings in added revenue by exporting as LNG.

Historically EOR has been seen as unsustainable without government incentive. While Oman's EOR project is no exception (it was heavily funded through government and private agencies), it became an example of how EOR can be sustainable. Other countries still rely on some form of support or strategic choice, sponsored by the government. Today over 80% of global EOR production benefits from some sort of government incentive or is prioritized by national oil companies as part of their efforts to maximize the return from national resources. This model is derived from the United States, which started since the inception of

EOR projects tax credits that were initially related to research but later became full-blown subsidies. In the 1980s, as domestic production began to decline, the Crude Oil Windfall Profit Tax 1980 kick-started the US EOR industry by significantly reducing its tax burden. Although that incentive was revoked, new incentives involving global warming and CO_2 sequestration have been in place. Most recently, the US section 45Q tax credit has been amended to provide a tax reduction of \$35/tCO_2 for 12 years for CO_2 stored in EOR operations. These incentives have, however, failed to generate growth in EOR applications. The reason for this failure is that the current society is not really convinced that EOR can be performed as an economically viable process. Such perception is further supported by the newfound perception that EOR methods are inherently toxic to the environment and would further increase long-term liability due to environmental impact. Another feature has been added. As oil companies practically shut down fundamental research, dating back to the Reagan era, the new business model has created a niche for EOR among contractors and service companies. Five midsize oil and gas companies currently operate the majority of CO_2-EOR projects in the United States.

Costs for EOR have come down since 2014, but the costs of other projects—including shale and offshore developments—have come down more quickly. For the moment at least, EOR technologies struggle to compete with other investment opportunities. This is the result of short-term thinking, which was first triggered three decades ago when companies' planning was being based on quarterly returns. With this mindset, Enron underwent a spectacular collapse but the underlying lessons remain elusive.

Fig. 4.47 shows the global energy need under two different scenarios. Fig. 4.47A incorporates existing energy policies as well as an assessment of the results likely to stem from the implementation of announced policy intentions. Note how under this scenario, the biggest suppression is in fossil fuels, such as coal, oil, and natural gas. The total consumption is brought down significantly under the Sustainable Development Scenario (SDS), which is a solution adopted from the Paris Accord climate change policies (Fig. 4.47B). This scenario also claims to be optimal for air quality and universal access to modern energy. In this scenario, the steepest decline is with coal, followed by natural gas and oil. While nuclear, hydro, and bioenergy remain similar to the option in Fig. 4.47A, "other renewables" are projected to be posting the highest growth.

In 2013, IEA estimated that by increasing recovery rates in conventional reservoirs, enhanced oil recovery (EOR) technologies of that time would have the potential to unlock another 300 billion barrels on top of the current existing resources. This was an amount comparable to the resource additions from light oil. The fluid of choice was CO_2 into reservoirs, thus gaining double dividend from the environmental sustainability side. This number was later revised to project the real scenario with only a fraction of the previously cited amount being produced. Indeed, growth in EOR production is modest to the mid-2020s. Over this period, the rise of the US shale alongside contributions from Brazil and Canada leaves little room for EOR to grow (IEA, 2018b). Now it is projected that, by the mid-2020s, EOR will start coming through in larger volumes. By then, a greater number of regions and countries would have also become mature production provinces, and so are more inclined to pursue efforts to maintain production or slow declines by supporting new EOR developments. Between 2025 and 2040, total EOR production is expected to grow from 2.7 mb/d to more than 4.5 mb/d, and it accounts for around 4% of global production in 2040 (Fig. 4.48).

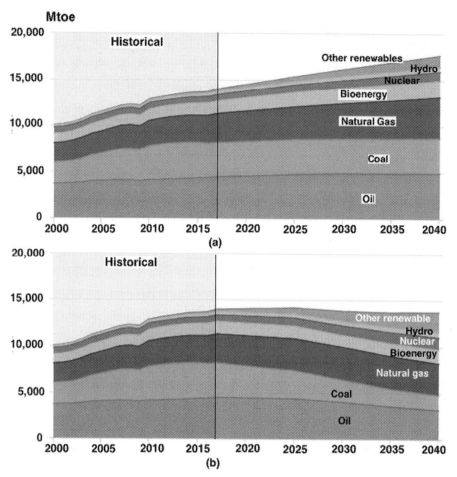

FIG. 4.47 (A) Status quo; (B) sustainable. *From IEA, 2018b, https://www.iea.org/weo/.*

For even this modest fraction of world production, repeatedly, government programs and tax subsidies for industry projects, including CO_2 credits in the case of CO_2-EOR, are considered to be essential to any realistic growth. While other factors, such as a concerted effort to screen fields and determine EOR potential in resource-rich areas; the timely piloting of EOR-projects in countries where it has not previously been used; better understanding of the subsurface and technological advances such as decreasing the volume of chemicals that need to be injected, little discussion is carried out as to environmental and economic sustainability.

In the Sustainable Development Scenario, total EOR production grows to around 4 mb/d in 2040 (see Fig. 4.48). This is smaller than the New Policies Scenario since oil demand and prices are lower. However, there is much larger production from CO_2-EOR given additional policy support for efforts to advance carbon capture utilization and storage (CCUS). In this scenario, climate imperatives emerge as the main rationale for pushing forward EOR technologies. Once again, the sole emphasis for the coming decade is the environmental appeal of the

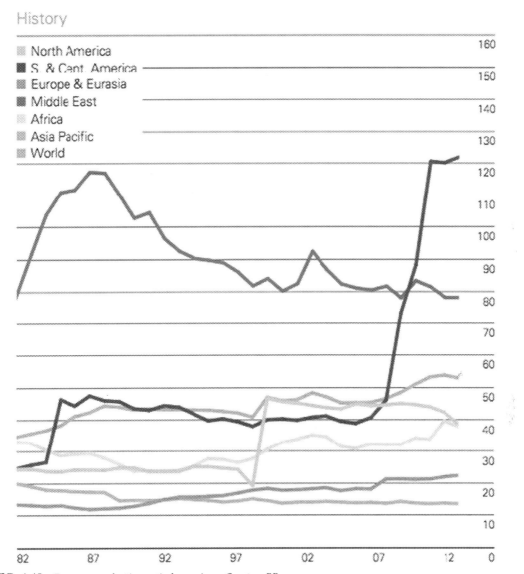

FIG. 4.48 Reserve production ratio by regions. *Courtesy BP.*

CO$_2$ injection option, which itself depends heavily on being subsidized through Climate Change programs, Carbon tax, and others.

As indicated earlier, the reserve to production ratio of oil is declining around the world, some hitting critical needs for EOR. The exception is the middle eastern region that continues to produce under par, as evident from Fig. 4.48. World proved oil reserves at the end of 2012 reached 1668.9 billion barrels. This is sufficient to meet 60 years of global production, without tapping into additional sources. Note that additional sources include heavier or nonconventional resources and new discoveries. Global proved reserves have increased by 26%, or nearly 350 billion barrels, over the past decade. This trend is likely to continue.

Of significance is the fact that there is much more nonconventional petroleum reserve than the convention "proven" reserve. This point is made in Fig. 4.49. Even though it is generally assumed that more abundant resources are "dirtier," hence in need of processing that can render the resource economically unattractive, sustainable recovery techniques can be developed that are more efficient for these resources and also economically attractive and environmentally appealing (Islam et al., 2010). In addition, natural gas quality is little affected by the environment. For instance, gas hydrate that represents the most abundant source of natural gas is actually far cleaner than less abundant resources.

It has been shown by Islam et al. (2018a) that the need for higher price and/or increased technological challenge is fictitious and is erased if scientific energy pricing along with sustainable technology is used. The current investment strategy has fueled this misconception.

In terms of oil industry, the main focus has been on nonconventional petroleum extraction. For instance, Fig. 4.50 shows major investments in oil sands in Canada. In 2013, 7299 trillion cubic feet of shale gas 345 billion barrels of shale/tight oil has been added. In the United States, the focus has been on unconventional oil and gas. A government report published in the

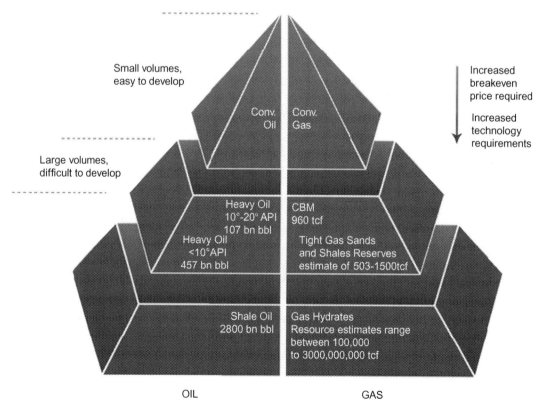

FIG. 4.49 Three is a lot more oil and gas reserve than the "proven" reserve.

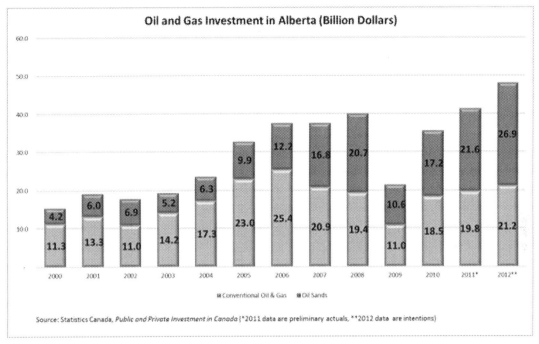

FIG. 4.50 Major investment in oil sands in Canada.

summer of 2013 revealed that US domestic crude-oil production exceeded imports last week for the first time in 16 years (Bloomberg, 2013). Output was 32,000 barrels a day higher than imports in the 7 days ended May 31, according to weekly data today from the Energy Information Administration, the Energy Department's statistical arm. Production had been lower than international purchases since January 1997. This surge in oil is attributed to the influx of horizontal drilling and hydraulic fracturing (popularly known as "fracking"). For over 20 years, horizontal drilling has been the most common drilling technique in the United States. However, the unlocking of tight formations, including shale, has become the most important reason for the surge. Large schemes of fracking have been implemented in the states of North Dakota, Oklahoma, and Texas. According to the EIA data, the surge in oil and gas production helped the US meet 88% of its own energy needs in February, the highest monthly rate since April 1986. Crude inventories climbed to the highest level in 82 years in the week ended May 24, 2013.

This has been accentuated with an increasing efficiency in refining. Fig. 4.51 shows how refining capacity has grown despite the declining number of refineries.

4.5.2 Need for EOR

There are primarily three reasons given for increasing oil recovery. They are:

(1) Primary recovery technique leave behind more than half of the original oil in-place. This is a tremendous reserve to forego.

350   4. Unconventional reservoirs

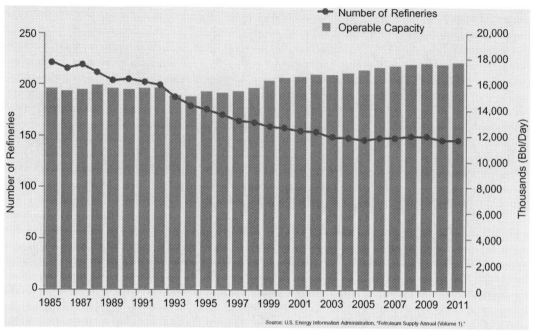

FIG. 4.51   Last few decades have observed an increase in efficiency of refineries.

(2) Increased drilling activities do not increase new discoveries of petroleum reserve. While this has been replaced with "new" technological opportunities (e.g., fracking technology creating oil and gas reserve in unconventional reserve), the argument is made to justify EOR.

(3) Environmental concern of CO_2 emission. Ever since signing of Kyoto Agreement, the US government has led the movement of CO_2 sequestration, thereby increasing oil recovery.

From the beginning of oil recovery, scientists have been puzzled by the huge amount of oil left over following primary recovery. Naturally occurring drive mechanisms recover anything from 0% to 70% of the oil in-place. In most cases, recovery declines rapidly as viscosity of oil increases. For instance, primary recovery is less than 5% when oil viscosity exceeds 100,000 cP. This is not to say that heavy oil recovery was the primary incentive for EOR, even though most EOR projects in United States, Canada, and Venezuela involve heavy oil recovery. The primary incentive for EOR is the fact that a typical light oil reservoir would have more than 50% of the original oil in-place left over while a small investment can recover over 70% of the oil in-place. For heavy oil, the room for improvement is much higher. Even though theoretically there is much more recovery potential of heavier energy sources all the way up to biomass (Fig. 4.52), the current recovery techniques are geared toward light oil. This figure shows that natural gas is the most efficient with the most environmental integrity. This argument has been sharpened in Chapter 3 that breaks down natural gas further into various forms of unconventional reservoirs. Within petroleum itself, the "proven reserve" is miniscule compared to the overall potential, as depicted in Fig. 4.7.

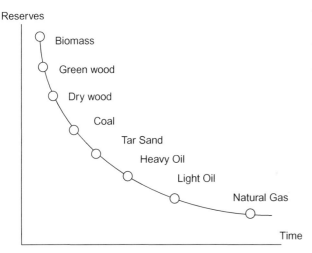

FIG. 4.52 As natural processing time increases so does reserve of natural resources. *From Chhetri, A.B., Islam, M.R., 2008. Inherently Sustainable Technologies. Nova Science Publishers, 452 pp.*

Of course, if one includes solar energy that would be the highest reserve possible. Because all energy source utilization techniques are equipped with processing light oil as a reference, the primary focus of EOR has been light oil. In early 2000, the US tertiary recover was estimated to be 12% (Fig. 4.53). This number has held steady until the huge surge in unconventional recovery of oil and gas that increased the oil and gas production by 40% in 2013. It is difficult to characterize unconventional recovery under a known category of oil and gas production. In any event, the knowledge of EOR is invaluable for developing any form of petroleum production scheme.

One of the most important criteria for selection of an EOR scheme is the reserve production ratio. This ratio is low for the United States. Consequently, the United States has been the leader in implementing EOR techniques. Of the total recoverable oil in the United States, 12% lends itself to EOR (Fig. 4.54). Further clarification must be made about the term "recoverable" oil. Some of the OPEC countries calculate this number by multiplying total petroleum in place with the recovery factor, which has little scientific merit. Other countries back calculate initial oil in-place by dividing "recoverable reserve" by the recovery factor, which is often low and without scientific justification. A recent survey shows little is known about how numbers such as "recovery factor," "recoverable reserve," and so on come to exist. However, it is commonly accepted that the "recoverable reserve" that is published worldwide and is used as the basis for determining OPEC quota for production is the most accurate starting point. This itself creates confusion in the scientific community that is vastly unfamiliar with how those "recoverable reserve" numbers are calculated. As for the pie chart in Fig. 4.54, it concerns the contribution of various recovery techniques in the United States. It turns out that the recovery factor in the United States is much higher. For instance, many miscible floods as well as steamfloods have recovery factors in the vicinity of 85% and 70%, respectively. To avoid confusion as to what this pie chart should represent in countries that have yet to start an EOR scheme, it is best to determine primary recovery potential and expect 30% of that oil during EOR schemes for light oil reservoirs. For heavy oil reservoirs, this percentage of recovery is much higher, mainly because the primary oil recovery factor is very low. In addition, many heavy oil formations do not "see"

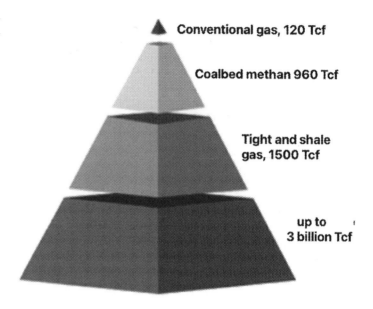

FIG. 4.53 "Proven" reserve is miniscule compared to total potential of oil.

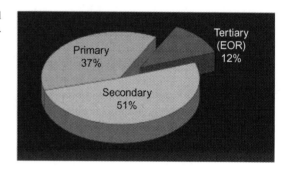

FIG. 4.54 Until 2012, tertiary recovery in the United States is estimated to be 12% of recoverable oil in-place.

primary recovery as is the case for most heavy oil reservoirs in the United States. Some Canadian heavy oil reservoirs do produce in primary mode but with an extremely low recovery factor (less than 10% of initial oil in place, IOIP).

There are three primary techniques of EOR: gas injection, thermal injection, and chemical injection. Gas injection, itself can be miscible or immiscible, although vast majority of the applications are planned to be miscible. Gas injected can be natural gas, nitrogen, or carbon dioxide (CO_2). As seen earlier, gas injection projects account for nearly 60% of EOR production in the United States. Thermal injection, which involves the introduction of heat, accounts for 40% of EOR production in the United States, with most of it occurring in California, whereas any other methods are insignificant. Chemical injection, which involves altering fluid/rock interface characteristics, is considered to be an improved over waterflooding (thus often called "Improved oil recovery"). Chemical recovery has been responsible for less than 1% of total EOR production over last few decades.

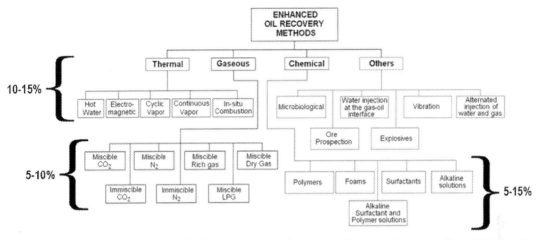

FIG. 4.55 Various available EOR methods, with their typical percentage incremental recovery. *From Adil, M., et al., 2018. Experimental study on electromagnetic-assisted ZnO nanofluid flooding for enhanced oil recovery (EOR). PLoS ONE 13 (2), e0193518. https://doi.org/10.1371/journal.pone.0193518.*

Fig. 4.55 shows various EOR methods, along with their potential recovery fractions. These numbers, however, are theoretical and do not represent the reality of oil fields. In extending laboratory-tested techniques to the field, scaling criteria have to be accurate and that is not the case in petroleum engineering. Although efforts have been made decades ago to have the scientific scaling criteria in place for both chemical (Islam and Farouq Ali, 1990) and in situ combustion (Islam and Farouq Ali, 1991), it has been proven practically impossible to accurately model field phenomena, involving EOR, with any acceptable features. For instance, for modeling a chemically active process, the experimental model has to be tested in equal time frame as the field—an absurdity. In addition, there are problems involving chemical (i.e., surfactant flooding) and gas (i.e., CO_2) EOR methods, for which the change of properties of injection fluids under the extreme condition is one of the major challenges. Most importantly, chemical processes are often constrained by the high cost of chemicals, possible formation damages, losses of chemicals, and long-lasting environmental damage. Similar problems occur with thermal methods for which the depth of the reservoir can render them inefficient due to the high energy cost, as well as heat loss from generation source to undesired reservoir levels.

4.5.3 Gas injection

Gas injection or miscible flooding is presently the most-commonly used approach in enhanced oil recovery. During early periods, it was the case because gas injection offered simple pressure maintenance, without the burden of paying for injection fluids. However, miscibility was not easily achieved and the tradition of purifying the injected gas became common. In addition, liquid nitrogen injection took hold in many projects. A miscible displacement process maintains reservoir pressure and improves oil displacement because the interfacial tension between oil and water is reduced. This refers to removing the interface between the two interacting fluids. Theoretically, the use of a miscible fluid is equivalent to having total displacement efficiency. With recent climate change "hysteria," carbon dioxide injection has become the EOR fluid of choice as CO_2 injection offers the double dividend of "greenhouse gas sequestration." As such, today's most commonly used gases are CO_2, natural gas or nitrogen. The fluid most commonly used for miscible displacement is carbon dioxide because it reduces the oil viscosity and is less expensive than liquefied petroleum gas. Oil displacement by carbon dioxide injection relies on the phase behavior of the mixtures of that gas and the crude, which are strongly dependent on reservoir temperature, pressure and crude oil composition. In addition, swelling of the liquid phase increases mobility in presence of CO_2. Most predictions of miscible flood recovery consider first contact miscibility. In the field, however, it is invariably mutlicontact and for heterogeneous formations, often miscibility doesn't occur for a broad transition zone.

4.5.4 Thermal injection

Viscosity of any petroleum fluid decreases exponentially with the rise in temperature. Any decrease in viscosity translates into linear increase in flow rate. As such, thermal injection schemes are popular with heavy oil reservoirs. Heat is transmitted through steam of through starting fire in situ. Other methods such as electrical heating, electromagnetic heating, hot water injection, and others have been proposed as well, but steam injection remains the most popular among all thermal EOR schemes. Many reactions take place under intense heat, including thermal cracking, visbreaking, upgrading, and so on. The increased heat reduces the surface tension and increases the mobility of the oil. The heated oil may also vaporize and then condense forming improved oil. Methods include cyclic steam injection, steam flooding and combustion. These methods improve the sweep efficiency and the displacement efficiency.

Steamflood is the most commonly used thermal injection technique. Steam injection has been used commercially since the 1960s in California fields. Steam flooding is one means of introducing heat to the reservoir by pumping steam into the well with a pattern similar to that of water injection. Eventually the steam condenses to hot water; in the steam zone the oil evaporates, and in the hot water zone the oil expands. As a result, the oil expands, the viscosity drops, and the permeability increases. To ensure success the process has to be cyclical. This is the principal enhanced oil recovery program in use today.

Solar EOR is a form of steam flooding that uses solar arrays to concentrate the sun's energy to heat water and generate steam. Solar EOR is proving to be a viable alternative to gas-fired steam production for the oil industry. In 2015 solar thermal enhanced oil recovery projects were planned in California and Oman, this method is similar to thermal EOR (TEOR) but uses a solar array to produce the steam. Solar TEOR setups have significant environmental the economic advantages over gas-based setups. On the environmental side, several studies have demonstrated the advantages of solar-based EOR setups. Specifically, TEOR can potentially reduce carbon emissions of EOR from 23.8 g CO_2/MJ with a gas setup to 0.1 g CO_2/MJ with a solar setup (Sandler et al., 2014).

Economically, solar-based setups also offer several advantages. Solar-based setups, after initial investment, do not have the high marginal costs of production as do gas-based setups, and are not subject to market fluctuations in natural gas prices. Chhetri and Islam (2008) showed that while solar setups are subject to significantly higher initial investment costs, the lowered marginal costs make solar TEOR economically viable at a much lower price point in both the near- and long-term time horizons. This can be farther facilitated by using direct thermal heating and using local products. This aspect will be discussed in a latter section. The environmental benefit is clear if one considers that direct solar heating will not involve the use of toxic solar panels.

In November 2017, GlassPoint and Petroleum Development Oman (PDO) completed construction on the first block of the Miraah solar plant safely on schedule and on budget, and successfully delivered steam to the Amal West oilfield. This would be the largest in its class once fully operational. The same month (November 2017), GlassPoint and Aera Energy announced a joint project to create California's largest solar EOR field at the South Belridge Oil Field, near Bakersfield, California. The facility is projected to produce approximately 12 million barrels of steam per year through a 850 MW thermal solar steam generator. It is slated to cut carbon emissions from the facility by 376,000 metric tons per year.

Fire flood or in situ combustion is one of the first known thermal recovery technique. From early on, several screening criteria for selection of a candidate field focused on high porosity and high viscosity. As early as in 1981, DoE came up with the following screening criteria after observing numerous oil field data and critically analyzing selection criteria proposed at the time (Islam and Khan, 2019) (Table 4.17).

During in situ combustion, the combustion generates the heat within the reservoir itself. Continuous injection of air or other gas mixture with high oxygen content will maintain the flame front, which can be control by controlling the air injection rate. As the fire burns,

TABLE 4.17 Selection criteria for in situ combustion.

Depth (ft)	Not critical (<4500 favorable)
Net thickness (ft)	>10 (10–60 favorable)
Porosity (%)	>20
Permeability (md)	>100 (450–1300 favorable)
API gravity (deg)	12–25
Viscosity (cP)	60–1000
S_o (%)	>50
S_g (%)	<10
ϕS_o (fraction)	>0.13 (1000 B/AF)
Gas cap	None
Lithology	Noncarbonate

From Kujawa and Lechtenber, 1981.

it moves through the reservoir toward production wells. Heat from the fire reduces oil viscosity and helps vaporize reservoir water to steam. The steam, hot water, combustion gas and a bank of distilled solvent all act to drive oil in front of the fire toward production wells. In theory, in situ combustion offers the most sustainable way of heating the reservoir as the combustion as well as the drive is self sustained.

There are three methods of combustion: Dry forward, reverse and wet combustion. Dry forward uses an igniter to set fire to the oil. As the fire progresses, the oil is pushed away from the fire toward the producing well. An analogy of this is cigarette smoking as the smoker inhales. The reverse injection process is just the opposite and similar to blowing on a burning cigarette as the front moves toward the smoker. In an oil field, it the ignition occurs from opposite directions. In wet combustion water is injected just behind the front and turned into steam by the hot rock. This quenches the fire and spreads the heat more evenly. Theoretically wet combustion is the most effective one as it avoids heating up one part of the formation in favor of homogenous heating of the reservoir.

For heavy oil and tar sand, steam assisted gravity drainage (SAGD) and more recently Vapor extraction (VAPEX) have gained popularity, although the latter one is found to be too expensive. In the 80s and 90s electromagnetic heating made some debut in Alberta, spearheaded by a company in Calgary. Although this technology had some theoretical merit, it never gained traction in the oil field (Wadadar and Islam, 1994). In 1992, Islam and Chakma introduced a combined electromagnetic and gas injection process. In laboratory scale, it showed very high recovery for both heavy oil and tar sand. Such a process was later followed up by Bansal and Islam (1994), who used CO_2 and other gases along with a horizontal well configuration. They should that CO_2 immiscible can recover as much as miscible when a scaled model is used.

4.5.5 Chemical injection

This EOR process involves injection of artificial chemicals that alters the capillary number and/or the mobility ratio to increase production in the short term or reduce residual saturation to recover more oil in the long run. Most often, the chemicals used are in concentration and are designed to be function at low concentration. Even at low concentration, their costs are prohibitive for continuous use. So, they are used in small slugs followed by a chaser fluid. Often sacrificial agents are used to minimize adsorption of the chemicals. In most laboratory studies, chemical injections, involving surfactant and polymers show great performance. However, practically none of them produce good results in the field. Historically, chemical floods have produced little other than giving tax benefits.

4.5.6 Microbial injection

Microbial injection involves injecting certain strains of bacteria along with their nutrients to sustain degradation of oil in situ, thus generating surfactants and other chemicals that can act as an automated chemical plant within the reservoir. After the injected nutrients are consumed, the microbes go into near-shutdown mode, their exteriors become hydrophilic, and they migrate to the oil-water interface area, where they cause oil droplets to form from the larger oil mass, making the droplets more likely to migrate to the wellhead. Bacteria have also been applied to seal off unwanted sections of the reservoir or to consolidate unconsolidated sands of an aquifer (Jack et al., 1991). Microbes are useful for generating carbonate precipitates that can consolidate porous media with beneficial effects (Gollapudi et al., 1995). Microbial treatment has also been used in remedying wax formation. Bacteria decompose these waxes and free up fluid flow. However, this particular application of MEOR is not usually considered to be a recovery technique rather than a production stimulation technique.

In principle, microbial enhanced oil recovery (MEOR) is the most promising technique in terms of environmental and economic sustainability. As early as 1980s, there have been reports of successful MEOR applications, particularly involving stripper wells, where the economics is marginal. Historically, the potential of microorganisms to degrade heavy crude oil to reduce viscosity was the primary impetus for considering MEOR. Mostly sulfate reducing bacteria (SRB) are involved. Earlier studies of MEOR (1950s) were based on three broad areas: injection, dispersion, and propagation of microorganisms in petroleum reservoirs; selective degradation of oil components to improve flow characteristics; and production of metabolites by microorganisms and their effects (Shibulal et al., 2014). Another appeal is the fact that thermophilic spore-forming bacteria can thrive in very extreme conditions in oil reservoirs, thus MEOR can be combined with hot water injection or at the front of the steam front, which contains hot water instead of stem (Al-Maghrabi et al., 1999). Recently, Shibulal et al. (2014) presented a detailed review of MEOR with thermophilic spore-forming bacteria. According to them, up to 50% of the residual oil can be extracted by this exceptionally low operating cost MEOR technology. They point out that MEOR can overcome the main hindrances of efficient oil recovery such as low reservoir permeability, high viscosity of the crude oil, and high oil-water interfacial tensions, which in turn result in high capillary forces retaining the oil within the reservoir rock.

4.5.7 Other techniques

Often Carbon dioxide (CO_2) injection in liquid form is considered to be a unique EOR system, separate from miscible gas injection. This is the case because CO_2 injection holds an additional appeal in terms of environmental sustainability. It can function in formations deeper than 2000 ft, where CO_2 will be in a supercritical state. CO_2 does not have to be miscible, although most applications in light oil formations are designed to maintain miscibility. In high pressure applications with lighter oils, CO_2 is miscible with the oil, with resultant swelling of the oil, and reduction in viscosity, and possibly also with a reduction in the surface tension with the reservoir rock. In the case of low pressure reservoirs or heavy oils, CO_2 will form an immiscible fluid, or will only partially mix with the oil. Some oil swelling may occur, and oil viscosity can still be significantly reduced. Immiscible CO_2 injection has not been explored in light oil reservoirs but holds great promises, particularly in the form of flue gas, which is far less expensive than purified CO_2, which itself is cheaper than other miscible fluids, such as propane and butane.

Water-alternating-gas (WAG) and other variations of it such as Simultaneous WAG (SWAG) are considered to be more effective than waterflood and gas injection alone. Instead of fresh water, saline water is often used to avoid clay swelling or precipitation within carbonate formations (while using CO_2 as a slug). Water and carbon dioxide are injected into the oil well for larger recovery, as they typically have low miscibility with oil. The use of both water and carbon dioxide also lowers the mobility of carbon dioxide, thus preventing fingering due to adverse mobility ratio. Making the gas more effective at displacing the oil in the well. In the context of CO_2 sequestration and simultaneous EOR, Islam and Chakma (1992) showed the benefit of using WAG. In this topic, Kovscek et al. (2005) showed that using small slugs of both carbon dioxide and water allows for faster recovery of the oil. In 2014, Dang et al. showed that using water with a lower salinity allows for greater oil removal, and greater geochemical interactions.

The use of water with gas arises from mobility control concerns. To improve on the mobility control issues, suggestions have been made to inject chemicals simultaneously. This so-called. Chemically enhanced water alternating gas injection (CEWAG) is getting significance importance in EOR after WAG due to its ability to improve both the displacement and sweep efficiency (Kumar and Mandal, 2017). The unique feature of the process includes use of alkaline, surfactant, and polymer as a chemical slug during the CEWAG process to reduce the interfacial tension (IFT) with simultaneous improvement of mobility ratio.

A new technology, Plasma-Pulse technology, has been introduced recently (Patel et al., 2018). This technology patented technology (Ageev and Molchanov, 2015) originated from Russia and has some connection to Soviet era research. The Plasma-Pulse Oil Well EOR uses low energy emissions to create the same effect that many other technologies, although with localized effects. Plasma pulse technology (PPT) treatment is administered with an electric wireline conveyed plasma pulse generator tool that is run in the well and positioned alongside the perforations. Using energy stored in the generator's capacitors, a plasma arc is created that emits a tremendous amount of heat and pressure for a fraction of a second. This in turn creates a broad band of hydraulic impulse acoustic waves that are powerful enough to clean perforations and near wellbore damage. It may be perceived as a well stimulation tool, but the wavers generated through the treatment continue to resonate deep into the reservoir, exciting the fluid molecules and increasing the reservoirs natural resonance to the degree that it can break larger hydrocarbon molecules to smaller one and simultaneously reducing adhesion tension which results in increased mobility of hydrocarbons (Patel et al., 2018). The plasma pulse technology has been successfully used on production as well as injection wells. Current clients and users of the new technology include ConocoPhillips, ONGC, Gazprom, Rosneft and Lukoil, and others.

Among others, sonic as well as ultrasonic irradiation have been cited as useful for enhanced oil recovery. However, few field reports have demonstrated success of the sonic method, whereas ultrasonic one is largely a well stimulation tool, which is effective for well bore cleanup as well as asphaltene removal (Bjorndahlen and Islam, 2004).

Fig. 4.56 shows the full picture of the EOR options. Note that the categories are not exclusive as various options can fall under different categories simultaneously.

4.5.8 Pivotal criterion for selection of EOR projects

In considering EOR, the reservoir/production ratio (R/P) is the most important criterion.

Table 4.18 shows total oil reserve as well as reserve/production ratio of top oil producing countries. Each country is marked for its need for EOR. Note that the need does not imply suitability nor does it mean that other countries would not benefit from an EOR scheme.

4.5 Improving oil and gas recovery from unconventional reservoirs 359

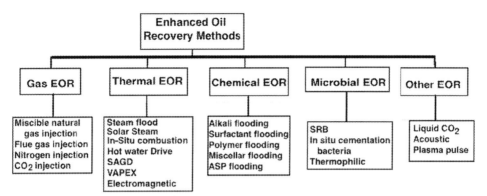

FIG. 4.56 Various EOR techniques with subcategories.

TABLE 4.18 Summary of proven reserve data as of 2018.

	Country	Reserve, bbl
1	Venezuela	302,300,000,000
2	Saudi Arabia	266,200,000,000
3	Canada	170,500,000,000
4	Iran	157,200,000,000
5	Iraq	148,800,000,000
6	Kuwait	101,500,000,000
7	United Arab Emirates	97,800,000,000
8	Russia	80,000,000,000
9	Libya	48,360,000,000
10	Nigeria	37,450,000,000
11	Kazakhstan	30,000,000,000
12	China	25,630,000,000
13	Qatar	25,240,000,000
14	Brazil	12,630,000,000
15	Algeria	12,200,000,000
16	Angola	9,523,000,000

Rank	Country	(Million bbl)	Reserves-to-Production (R/P) Ratios
1	Venezuela	302.300	393.6
2	Saudi Arabia	266.2	61
3	Canada	170.5	95.8
4	Iran	157.2	86.5
5	Iraq	148.8	90.2

Continued

TABLE 4.18 Summary of proven reserve data as of 2018—cont'd

Rank	Country	(Million bbl)	Reserves-to-Production (R/P) Ratios
6	Kuwait	101.5	91.9
7	United Arab Emirates	97.8	68.1
8	Russia	80	25.8
9	United States	50	10.5
10	Libya	48.36	153.3
11	Nigeria	37.45	51.6
12	Kazakhstan	30	44.8
13	China	25.63	18.3
14	Qatar	25.24	36.1
15	Brazil	12.63	12.6
16	Algeria	12.2	21.7
17	Angola	9.523	15.6
18	Ecuador		8.3
19	Azerbaijan		24.1
20	Mexico		8.9
21	Norway		11

Data from CIA Factbook, 2020. https://www.cia.gov/the-world-factbook/. (Accessed December 2020); BP, 2019. Statistical Review of World Energy, 68th ed. https://www.bp.com/content/dam/bp/business-sites/en/global/corporate/pdfs/energy-economics/statistical-review/bp-stats-review-2019-full-report.pdf.

Fig. 4.57 shows R/P ratios for top 20 petroleum reserve holding countries. Note that there is an overall trend of higher R/P with higher reserve values. The straight line represents the standard, which is drawn from Venezuela and the origin. The countries that fall under this line represent excessive production compared to their reserve. Among them, Saudi Arabia represents the fastest depleting oil fields. Curiously, Canada, Iraq and Iran have been maintaining modest R/P ratios despite having midsize reserve. Western hemisphere countries in general have low R/P. United Kingdom (not listed above) has an R/P value of 6.3 with a reserve of 2.3 billion (BP, 2018). This is the lowest among "oil rich countries." The United States represents the lowest R/P for countries possessing a reserve 50 billion or more. At par with the United States, Russia has also maintained low R/P (of 25), signaling the fact its reserve is depleted fast.

Islam (2014) estimated EOR recovery potentials for various countries. When recovery potential is estimated, the actual recovery with EOR can be calculated. These numbers are listed in Table 4.19. The sustainable EOR is the most economic and environmentally appealing option. Unless such scheme is implemented, infill drilling is the most economic scheme. This is especially true for reservoirs with high reserve/production ratios.

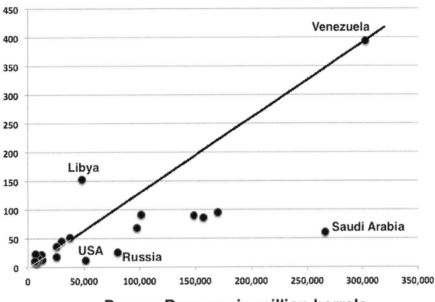

FIG. 4.57 R/P ratio vs proven reserve for top oil producing countries.

There has been some skepticism regarding "reserve" numbers. It is difficult to comment on the validity of the "reserve" numbers. Islam (2014) cited a survey that shows that there is no standard for this reserve or the definition of "proven" reserve is so varied and subjective that there is a need or a comprehensive study of this subject. It seems clear that politics plays a significant role in claiming "proven reserve." While the developing countries in general and OPEC countries in particular are cited for politicizing the reserve numbers, history tells us there is a systematic lack of transparency in both numbers and the process involved in determining these numbers.

As can be seen in Table 4.20, recovery alone cannot be an evidence of declining reserve because the recovery to reserve ratio varies largely among different countries.

Fig. 4.58 also shows how countries with the exception of Venezuela have added no new reserve in the last decade. Despite this, there have been claims that major OPEC countries have inflated their reserves to gain more share in the competitive world market. This scenario is a pessimistic one because other countries do not actively look for or necessarily declare new reserves or reserves that have become "recoverable" because of technological improvements. The most remarkable case here is Saudi Arabia (Table 4.21).

It turns out that the R/P ratios are very similar for gas reserves and far less than coal reserve. As pointed out by Zatzman (2012), before oil, the most hysterical line of the energy industry was that we would run out of coal. Table 4.22 shows global values for oil, gas and coal.

TABLE 4.19 Reserve recovery ratios for different countries.

Rank	Country	Reserves 10^9 bbl (2018)	Reserve/production ratio Years (2017)	EOR reserve 10^9 bbl	EOR suitability with existing technology	EOR suitability with sustainable technology
1	Venezuela	302.3	393.6	45.4	Low	High
2	Saudi Arabia	266.2	61	39.9	Medium	High
3	Canada	170.5	95.8	25.57	Low	Medium
4	Iran	157.2	86.5	23.58	Low	Medium
5	Iraq	148.8	90.2	22.32	Low	Medium
6	Kuwait	101.5	91.9	15.2	Low	Medium
7	United Arab Emirates	97.8	68.1	14.7	Low	Medium
8	Russia	80	25.8	12	High	High
9	Kazakhstan	50	44.8	7.5	High	High
10	Libya	48.36	153.3	7.25	Medium	High
11	Nigeria	37.45	51.6	5.61	High	Medium
12	Qatar	30	36.1	4.5	Medium	Medium
13	China	25.63	18.3	3.84	High	High
14	United States	25.24	10.5	3.79	High	High
15	Angola	12.63	15.6	1.89	High	High
16	Algeria	12.2	21.7	1.83	High	High
17	Brazil	9.523	12.8	1.43	High	High

Data from BP, 2017, 2018.

Table 4.22 shows the variation in R/P ratio for certain countries of interest. These numbers are only approximations. Uncertainty in reserve calculations comes from the fact the technology is evolving, both in recovery techniques and delineation of reservoirs. For instance, different estimates may or may not include oil shale, mined oil sands or natural gas liquids. Yet others would not include basement reservoirs in the calculation. In addition, proven reserves include oil recoverable under current economic conditions, which are variable depending on the overall state of economy and other factors of a country. The case in point is Canada's proven reserve that increased suddenly in 2003 when the oil sands of Alberta were seen to be economically viable. Similarly, Venezuela's proven reserves jumped in the late 2000s when the heavy oil of the Orinoco was judged economic. When the United States made great advances in recovering unconventional oil and gas in 2008, the US reserve increased

TABLE 4.20 Variation in reserves for top oil producing countries.

Rank	Country	Reserves 10^9 bbl (2012)	Reserves 10^9 bbl (2016)	Reserves 10^9 bbl (2017)	Reserves 10^9 bbl (2018)
1	Venezuela	296.5	300.9	300.9	302.3
2	Saudi Arabia	265.4	266.5	266.5	266.2
3	Canada	175	171.5	169.7	170.5
4	Iran	151.2	158.4	158.4	157.2
5	Iraq	143.1	153	142.5	148.8
6	Kuwait	101.5	101.5	101.5	101.5
7	United Arab Emirates	136.7	97.8	97.8	97.8
8	Russia	80	109.5	80	80
9	Kazakhstan	49	30	48.3	50
10	Libya	47	48.4	37.06	48.36
11	Nigeria	37	37.1	36.52	37.45
12	Qatar	25.41	25.2	30.00	30
13	China	20.35	25.7	25.62	25.63
14	United States	26.8	48	25.24	25.24
15	Angola	13.5	11.6	12.70	12.63
16	Algeria	13.42	12.2	12.20	12.2
17	Brazil	13.2	12.6	8.273	9.523

Data from BP and CIA Factbook.

significantly. Environmental concerns add to those uncertainties, particularly because those concerns are also a part of the political decisions.

Twenty-five percent of the world's recoverable reserve is in Saudi Arabia and until now the entire recovery process is through primary. Because the recovery over reserve ratio is still fairly high, EOR suitability of Saudi fields is a question mark. While it is considered to low risk to develop Saudi reservoirs for secondary recovery because of the low-cost of implementation of waterflood schemes, the benefit of implementing suitable EOR schemes directly after primary remain very high, at least in theory. In addition, it is of importance to note that Saudi Arabia has significant amount of tar and other heavy oil deposits that are ignored in their reserve estimates. However, considering latest technological breakthroughs in tar sand and heavy oils, due to mega projects in Canada, Saudi heavy oil reserves can every well become very prominent in the world scale. Developments in the next most important case is that of Venezuela. Venezuela has the highest reserve to production ratio in the world. With EOR implementations, it has the capacity to double the daily output or total recoverable reserve.

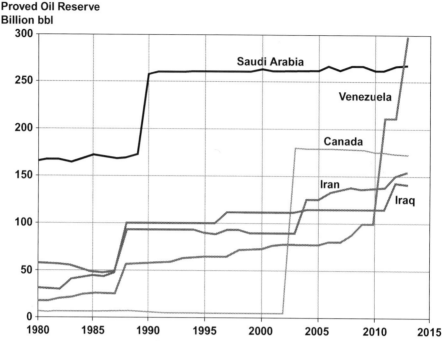

FIG. 4.58 Declared reserve for various countries. *Updated from Islam, J.S., et al., 2018. Economics of Sustainable Energy. Scrivener-Wiley, 628 pp.*

Fig. 4.59 shows how the global shares have evolved during 2015–17. During this period, the global reserve has dropped slightly. The shares of Africa remain constant at 8.14%. In the meantime, shares of Middle Eastern reserve rise slightly from 51.8% to 52%. The shares of North America oil drop, albeit in small magnitude due to adjustment in Canadian reserve. European reserve fluctuates slightly while South and Central American reserve increases very slightly. Figs. 4.60 and 4.61 show gradual evolution in global reserve and fluctuations in shares of different regions.

Table 4.23 shows total oil reserve as well as reserve/production ratio of top oil producing countries. Each country is marked for its need for EOR. Note that the EOR need does not imply suitability nor does it mean that other countries would not benefit from an EOR scheme (Fig. 4.62).

4.5.9 Role of the US leadership

In the post World War II era, the United States has been the undisputed leader of technology development. Other countries do not even come close both research and development and application of new technologies. A recent report (Science News, 2018) shows that the United States leads in providing business, financial and information services, accounting for 31% of the global share, followed by the entire European Union (EU) at 21%. China is the third largest producer of these services (17% global share) and continues to grow at a

TABLE 4.21 Variations in reserve/production ratios for various countries.

Countries	Reserve/production ratio years (2012)	Reserve/production ratio years (2016)	Reserve/production ratio years (2017)	Comments
Venezuela	387	341.1	393.6	
Saudi Arabia	81	59	61	
Canada	178	105	95.8	
Iran	101	94.1	86.5	
Iraq	163	93.5	90.2	
Kuwait	121	88	91.9	
United Arab Emirates	156	65.5	68.1	
Russia	22	26.5	25.8	
Kazakhstan	55	49	44.8	
Libya	76	310	153.3	
Nigeria	41	49.3	51.6	
Qatar	63	36.3	36.1	
China	14	17.5	18.3	
United States	10	10.5	10.5	
Angola	19	17.5	15.6	
Algeria	22	21.1	21.7	
Brazil	17	13.3	12.8	

Data from various BP reports and EIA reports.

TABLE 4.22 Global RPR of oil, natural gas and coal (BP, 2018).

Fuel	Unit of measure	Reserves	Annual production	RPR (years)	RPR 2017
Oil	Billions of tons	240	5	51	50.2
Coal	Billions of tons	890	8	114	134
Natural gas	Trillions of cubic meters	190	4	53	52.6

far faster rate (19% annual growth) than the United States. However, considering that China's population is almost three times that of the United States, China's per capita output doesn't come close to the United States. While production of aircraft and spacecraft, semiconductors, computers, pharmaceuticals, and measuring and control instruments are the most talked about items, the United States's dominance of technology is the clearest in petroleum technology development (Islam et al., 2018a). Likewise, EOR technology is dominated by the United States.

In early 2000, the US tertiary recovery was estimated to be 12% (Fig. 4.63). This number has held steady until the huge surge in unconventional recovery of oil and gas that increased the oil and gas production by 40% in 2013. It is difficult to characterize recovery under "secondary" and "tertiary" because rarely an oilfield undergoes such sequential development. Often, "secondary" or "tertiary" begins soon after the oilfield is put on production, due to unfavorable fluid or rock characteristics. For instance, for many heavy and tar sand, steam injection starts from the.

Conventionally, perhaps the most important criteria for selection of an EOR scheme is the reserve/production ratio. This ratio is low for the United States. Consequently, the United States has been the leader in implementing EOR techniques. Of the total recoverable oil in the United States, 12% lends itself to EOR. It turns out that the recovery factor in the United States is much higher. For instance, many miscible floods as well as steamfloods have recovery factors in the vicinity of 85% and 70%, respectively. To avoid confusion as to what this pie chart should represent in countries that have yet to start an EOR scheme, it is best to determine primary recovery potential and expect 30% of that oil during EOR schemes for light oil reservoirs. For heavy oil reservoirs, this percentage of recovery is much higher, mainly because the primary oil recovery factor is very low. In addition, many heavy oil formations do not "see"

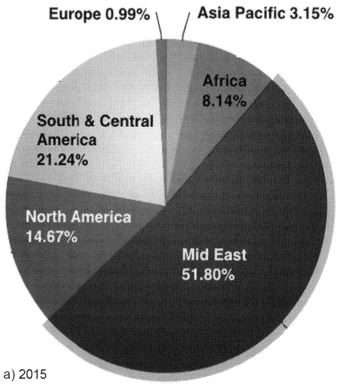

FIG. 4.59 Changes in global reserve shares. *From BP, 2018. BP Annual Report. https://www.bp.com/content/dam/bp/business-sites/en/global/corporate/pdfs/investors/bp-annual-report-and-form-20f-2018.pdf.*

(Continued)

4.5 Improving oil and gas recovery from unconventional reservoirs 367

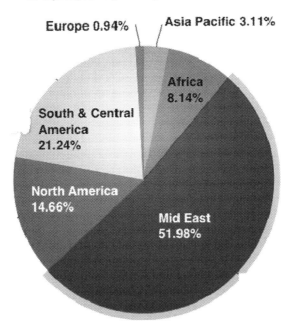

b) 2016

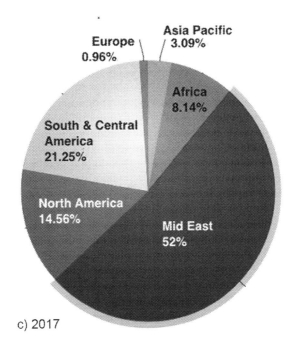

c) 2017

FIG. 4.59, CONT'D

368　　　　　　　　　　　　4. Unconventional reservoirs

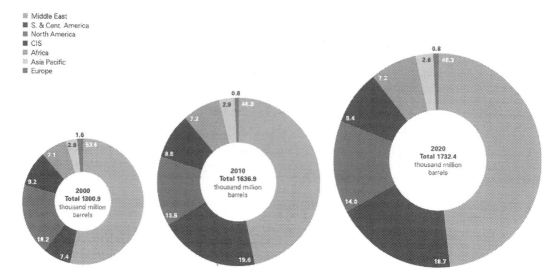

FIG. 4.60 Distribution of proved reserve for various regions. *Af*, Africa; *AP*, Asia Pacific; *EuE*, Europe and Eurosia; *SCA*, South and Central America. *From BP, 2021. BP Annual Report. Available at https://www.bp.com/content/dam/bp/business-sites/en/global/corporate/pdfs/energy-economics/statistical-review/bp-stats-review-2021-full-report.pdf; BP, 2017. BP Annual Report. Available at https://www.bp.com/content/dam/bp/business-sites/en/global/corporate/pdfs/investors/bp-annual-report-and-form-20f-2017.pdf.*

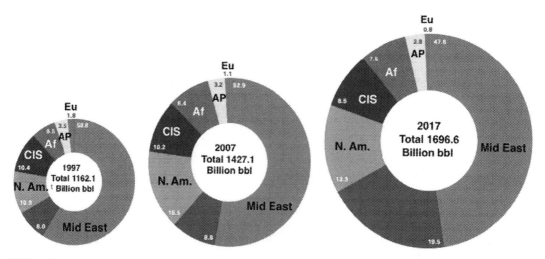

FIG. 4.61 Distribution of proved reserve for various regions. *Af*, Africa; *AP*, Asia Pacific; *Eu*, Europe. *From BP, 2018. BP Annual Report. https://www.bp.com/content/dam/bp/business-sites/en/global/corporate/pdfs/investors/bp-annual-report-and-form-20f-2018.pdf.*

TABLE 4.23 Reserve recovery ratio for different countries.

Rank	Country	Reserves 10⁹ bbl	Reserve/production ratio Years	EOR reserve 10⁹ bbl	EOR suitability with existing technology	EOR suitability with sustainable technology
1	Venezuela	296.5	387	44.5	Low	High
2	Saudi Arabia	265.4	81	39.8	Medium	High
3	Canada	175	178	26.25	Low	Medium
4	Iran	151.2	101	22.7	Low	Medium
5	Iraq	143.1	163	21.5	Low	Medium
6	Kuwait	101.5	121	15.2	Low	Medium
7	United Arab Emirates	136.7	156	20.5	Low	Medium
8	Russia	80	22	12	High	High
9	Kazakhstan	49	55	7.35	High	High
10	Libya	47	76	7.0	Medium	High
11	Nigeria	37	41	5.5	High	Medium
12	Qatar	25.41	63	3.8	Medium	Medium
13	China	20.35	14	3.1	High	High
14	United States	26.8	10	4.0	High	High
15	Angola	13.5	19	2.0	High	High
16	Algeria	13.42	22	2.0	High	High
17	Brazil	13.2	17	2.0	High	High

primary recovery as is the case for most heavy oil reservoirs in the United States. Some Canadian heavy oil reservoirs do produce in primary mode but with an extremely low recovery factor (less than 10% of IOIP). With regards to EOR technology development and implementation, the United States has been the world leader in implementing EOR. Fig. 4.64 shows that the United States is ahead both in time of implementation and recovery fraction with EOR in the world scale. This trend continues today despite the new found domestic resources in unconventional oil and gas and shale oil megaproject of Canada. Such leadership emerges from the US superiority in related areas of new drilling and well technologies, intelligent reservoir management and control, advanced reservoir monitoring techniques, and the application of different enhancements of primary and secondary recovery processes.

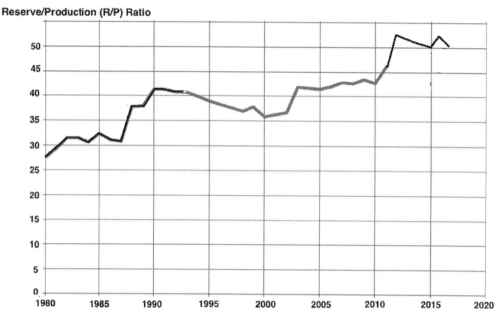

FIG. 4.62 Global R/P ratios during 1980–2017. *Data from BP reports.*

FIG. 4.63 The US oil production under different categories in 2000. *Data from Moritis, G., 2008. Worldwide EOR survey. Oil Gas J. 106, 41–42, 44–59.*

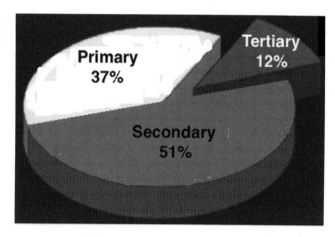

It is well known that EOR projects have been strongly influenced by economics and crude oil prices. The initiation of EOR projects depends on the preparedness and willingness of investors to manage EOR risk and economic exposure and the availability of more attractive investment options. For an economic scheme to be successful in the long term, the technology cannot be expensive beyond the rate of return. In addition, the incremental recovery has to be substantially more than recovery with status quo. In this regard, the recovery/reserve ratio is important and is the most important criterion for implementation of EOR (Fig. 4.65).

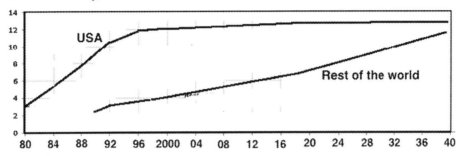

FIG. 4.64 Incremental recovery owing to EOR. *Data from IEA, 2017. World Energy Outlook: Poverty and Prosperity, https://www.iea.org/publications/freepublications/publication/WEO2017SpecialReport_EnergyAccessOutlook.pdf.*

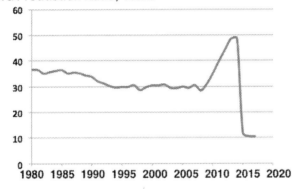

FIG. 4.65 The US reserve/production (R/P) ratio variation over the years. *Data from various EIA reports, BP report.*

Fig. 4.66 shows R/P ratio variation for the United States. Until 2009, there was a steady decline in the ratio, although there are fluctuations. Fig. 4.66 shows this point with a magnified shot of the previous graph. Before 2008 financial market collapse, the US production policy was kept at its natural state. Afterward, this moved into a different mode, during which period the R/P ratio rose sharply. This culminated into over 50% increase in R/P ratio while at the same time accompanied by a 40% hike in domestic oil and gas recovery in the year 2013. This is the period a large amount of tight gas and oil and other unconventional oil and gas reserve was added to the US repertoire as the Fracking technology along with horizontal well technology made a huge difference in both recoverable reserve and production rate. Such drastic increase in R/P ratio was followed by sharp decline as the US oil and gas production reached a record high. As a consequence, the R/P ratio dropped below the pace of pre-2009 era. This despite the fact that proved reserves of crude oil in the United States increased 19.5% (6.4 billion barrels) to 39.2 billion barrels at Year-End 2017, setting a new US record for crude oil proved reserves. The previous record was 39.0 billion barrels set in 1970 (EIA, 2018c). Similarly, proved reserves of natural gas increased by 123.2 trillion cubic feet (Tcf) (36.1%) to 464.3 Tcf at year-end 2017. This too was a new US record for total natural gas proved reserves.

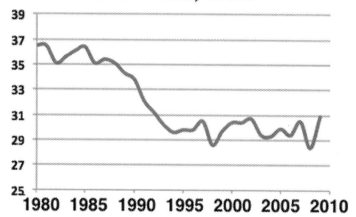

FIG. 4.66 The US reserve/production (R/P) ratio variation over the years. *Data from various EIA reports.*

The previous US record was 388.8 Tcf, set in 2014, and that record was due to expansion of unconventional gas reserves. This latest increase in reserve is accompanied with a hike in the US production of total natural gas by 4% from 2016 to 2017, reaching a new record level. Overall, the United States has exceeded the global pace of oil and gas production (Fig. 4.67).

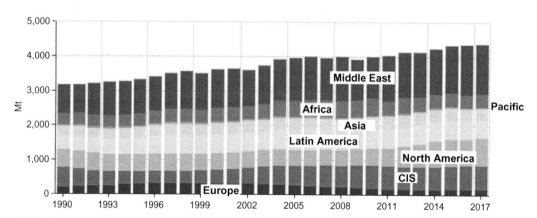

FIG. 4.67 Crude oil production continues to rise overall (Enerdata, 2018). Similar expansion has occurred in the natural gas sector as well. Proven wet natural gas reserves increased in each of the five largest natural gas producing states (Texas, Wyoming, Louisiana, Oklahoma, and Pennsylvania) in 2011. Pennsylvania's proven natural gas reserves, which more than doubled in 2010, rose an additional 90% in 2011, contributing 41% of the overall US increase. Combined, Texas and Pennsylvania added 73% of the net increase in US proved wet natural gas reserves. Expanding shale gas developments in these and other areas, particularly the Pennsylvania and West Virginia portions of the Marcellus formation in the Appalachian Basin, drove overall increases.

Fig. 4.68 shows the history of the US crude oil proved reserve. Prior to 2008, there was overall decline in the US reserve. Beyond 2008, upon expansion in the unconventional oil and gas formations, there has been a steady rise in the reserve. Of course, the proved reserve is linked to oil price, in the sense that a low oil price can render certain reserve untenable with the current technology cost. For instance, in 1980, proved reserves in the United States were 36.5 billion barrels. At the 1980 rate of US production, that was enough oil for just over 10 years of production. Of course, that did not happen. In reality, between 1980 and the end of 2014, the United States produced 111 billion barrels of oil. Despite the 111 billion barrels that were produced, US crude oil reserves at the end of 2014 had grown to 48 billion barrels. On the other hand, the sharp increase since 2008 is due to both oil price and technological advancement. Oil at $100/bbl enabled the shale oil boom by making it economical to combine hydraulic fracturing ("fracking") and horizontal drilling in previously uneconomical formations. This pushed a lot of oil from the resource category into the proved reserves category. Similarly, the decline in oil reserve in 2014 is due to oil price collapse. Contrary to the US reserve, the global reserve has maintained a consistent pattern. For instance, in 1980, the global oil reserve was 683 billion barrels, which has burgeoned to 1.7 trillion barrels in 2014.

Natural gas has maintained similar trends in the United States. Fig. 4.69 shows the history of natural gas reserve in various locations of the United States. Both federal offshore and Alaska show steady decline in the reserve, whereas reserve in the lower 48 onshore has been fluctuating (due to the points discussed earlier), leading to total reserve showing fluctuations.

Fig. 4.70 shows gas production history of the United States. Gas production has been steadily rising with anomalies showing during the oil embargo of 1973 and during 1990s and early 2000s. After 2008, however, the reserve has increased at a greater pace than before. This is the time, the so-called gas war in Europe took place that saw great hikes in gas price.

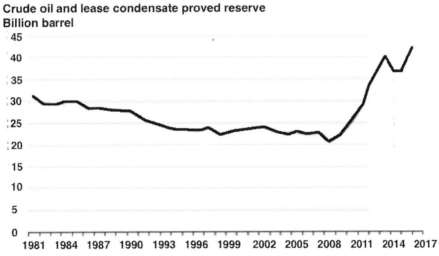

FIG. 4.68 History of the US crude oil and lease condensate proved reserve. *Data from EIA reports; BP, 2018. BP Annual Report. https://www.bp.com/content/dam/bp/business-sites/en/global/corporate/pdfs/investors/bp-annual-report-and-form-20f-2018.pdf.*

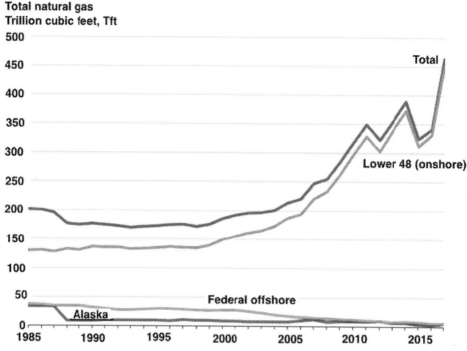

FIG. 4.69 The United States reserve variation in recent history. *From EIA, 2018. International Energy Outlook. Available at: https://www.eia.gov/outlooks/ieo/executive_summary.php.*

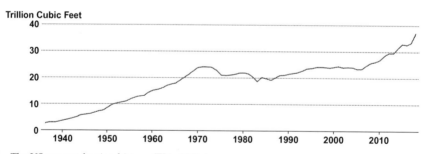

FIG. 4.70 The US gas production history (EIA, 2018c).

Fig. 4.71 shows the gas reserve production (R/P) ratio of the United States. The numbers are similar to the oil R/P values. However, the trends are not similar because gas prices have been governed by different set of rules from those of oil prices (Zatzman, 2012).

The need for EOR comes from the declining nature of world-wide reserve. Practically all countries have reached peak recovery rate as evidenced from Fig. 4.72. This is the case despite the fact that new discoveries continue to grow. This is not to subscribe to the theory of peak oil because this graph doesn't consider increase in reserve due to the improvement of recovery technologies and the addition of unconventional oil and gas. This unconventional reserve is

4.5 Improving oil and gas recovery from unconventional reservoirs

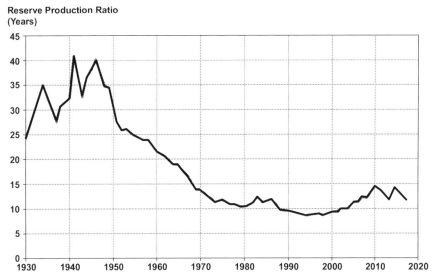

FIG. 4.71 The US gas reserve-production history. *Data from EIA, 2018. International Energy Outlook. Available at: https://www.eia.gov/outlooks/ieo/executive_summary.php.*

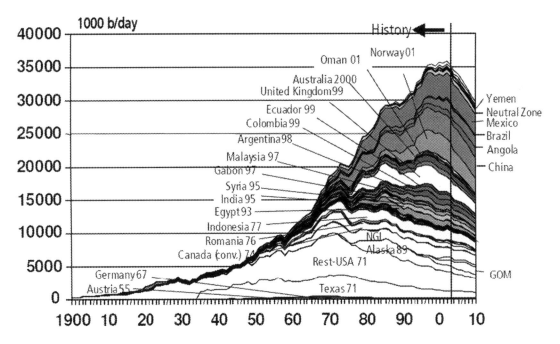

FIG. 4.72 Recovery rates decline around the world.

the principal reason the United States has seen some 40% increase in domestic oil and gas recovery in Year 2013. Fig. 4.72 shows how the US reserve in oil and gas have progressively declined since 1970 only to pick up in 2011 onward. Despite the perception otherwise, petroleum reservoirs are discovered in almost a linear fashion. The addition of offshore oilfields follows similar pattern. This argument has been made in the past to justify status quo, which in the supporters' view only need infill drilling. This view counters the argument that says that the need for EOR comes from the declining nature of worldwide reserve. From the production perspective, however, practically all countries have reached peak recovery rate as evidenced from Fig. 4.72. This is the case despite the fact that new discoveries continue to grow. This is not to subscribe to the theory of peak oil because this graph does not consider increase in reserve due to the improvement of recovery technologies and the addition of unconventional oil and gas (Speight and Islam, 2016).

There is a scientific group that believes that the above graph is misleading. As can be seen in Table 4.23, recovery alone cannot be an evidence of declining reserve because the recovery to reserve ratio varies largely among different countries. Fig. 4.73 lends credibility to this statement.

Fig. 4.73 also shows how countries with the exception of Venezuela have added no new reserve in the last decade. Despite this, there have been claims that major OPEC countries have inflated their reserves to gain more share in the competitive world market. This scenario is a pessimistic one because other countries do not actively look for or necessarily declare new reserves or reserves that have become "recoverable" because of technological improvements. The most remarkable case here is Saudi Arabia. World's 25% of recoverable reserve is in Saudi

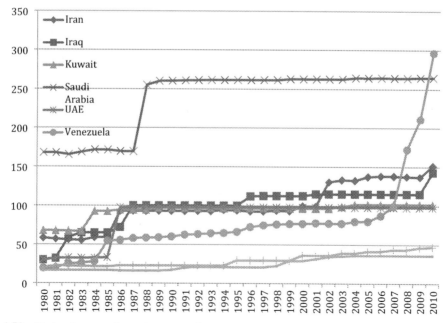

FIG. 4.73   Declared reserve for various countries.

Arabia and until now the entire recovery process is through primary. Because the recovery over reserve ratio is still fairly high, EOR suitability of Saudi fields is a question mark. While it is considered to low risk to develop Saudi reservoirs for secondary recovery because of the low-cost of implementation of waterflood schemes, the benefit of implementing suitable EOR schemes directly after primary remain very high, at least in theory. In addition, it is of importance to note that Saudi Arabia has significant amount of tar and other heavy oil deposits that are ignored in their reserve estimates. However, considering latest technological breakthroughs in tar sand and heavy oils, due to mega projects in Canada, Saudi heavy oil reserves can every well become very prominent in the world scale. Developments in The next most important case is that of Venezuela. Venezuela has the highest reserve to production ratio in the world. With EOR implementations, it has the capacity to double the daily output or total recoverable reserve.

In United States, total oil reserve has been declining since 1970 only to peak up in 2011. Wet natural gas reserve, on the other hand has been declining since 1981 but started to increase shortly after 1999. Fig. 4.74 shows the US reserve as a function of time. The decline in reserve was reversed shortly after the 2008 financial crisis. Such rise in reserve is attributed to horizontal drilling and hydraulic fracturing in shale and other "tight" (very low permeability) formations. Because the oil price increased sharply during that period, more drilling was performed. This added to the proven reserve of the United States. US proven reserves of natural gas began growing sharply in the mid-2000s as operators adopted expanded horizontal drilling programs and applied new hydraulic fracturing techniques in shale formations. Starting with 2009, similar horizontal drilling programs were applied in several of the nation's tight oil formations—reserves additions from tight oil plays have reversed the long-term trend of generally declining proved US oil and lease condensate reserves. Proved reserves of crude oil and lease condensate increased in each of the five largest crude oil and lease condensate areas (Texas, the Gulf of Mexico federal offshore, Alaska, California, and North Dakota) in 2011. Of these, Texas had the largest increase by a large margin, about 1.8 billion barrels (46% of the net increase), resulting mostly from ongoing development in the Permian and Western Gulf Basins in the western and south-central portions of the state. North Dakota reported the second largest increase, 771 million barrels (20% of the net increase), driven by development activity in the Williston Basin. Collectively, North Dakota and Texas accounted for two-thirds of the net increase in total United States proved oil reserves in 2011.

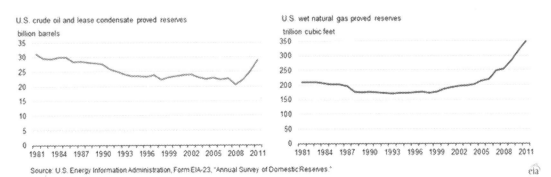

FIG. 4.74 The United States reserve variation in recent history.

Proven wet natural gas reserves increased in each of the five largest natural gas producing states (Texas, Wyoming, Louisiana, Oklahoma, and Pennsylvania) in 2011. Pennsylvania's proven natural gas reserves, which more than doubled in 2010, rose an additional 90% in 2011, contributing 41% of the overall US increase. Combined, Texas and Pennsylvania added 73% of the net increase in United States proved wet natural gas reserves. Expanding shale gas developments in these and other areas, particularly the Pennsylvania and West Virginia portions of the Marcellus formation in the Appalachian Basin, drove overall increases.

In terms of technical recoverability, both oil and gas reserves changed over the last decade. Fig. 4.75 shows how technical recoverability has changed for both oil and gas reserves in the United States. Even with reduced aggressive research, technological developments in various aspects of petroleum engineering made it possible to upgrade the reserve estimates. This decline has been accompanied with increasing sulfur content of the US crude.

Figs. 4.76 and 4.77 demonstrates the need for EOR. EOR involves making up for the loss of natural production cycle to meet the growing need of petroleum. However, for reservoirs with high reserve/production ratio, it is most cost effective to infill drill. By carefully selecting infill drilling sites, the recovery factor can be increased even with primary production mode. For reservoirs that have observed significant rise in watercut during primary production, one should consider local improvement of mobility ratio by adding chemicals, such as polymer. However, increasing EOR performance with polymer is not recommended because polymer slugs do not travel in the reservoir beyond a few meters. The economics of EOR changes drastically if waste gas, produced gas or locally available gas is used for EOR injection. Fig. 4.78 presents a qualitative comparison among various modes of EOR. This figure that is modified

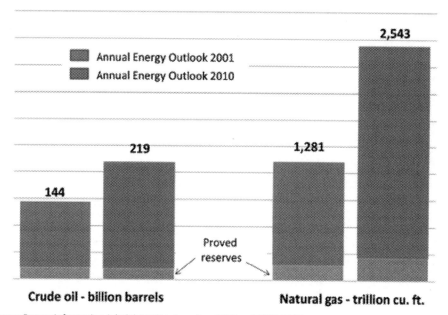

FIG. 4.75 Technically recoverable oil and gas reserve in the United States.

4.5 Improving oil and gas recovery from unconventional reservoirs 379

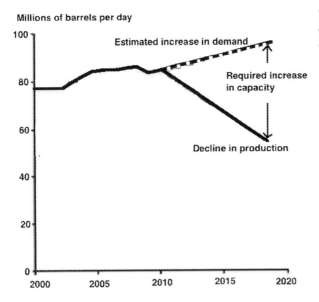

FIG. 4.76 The need for EOR is evident in production and oil quality decline. *From Islam, M.R., 2014. Unconventional Gas Reservoirs. Elsevier.*

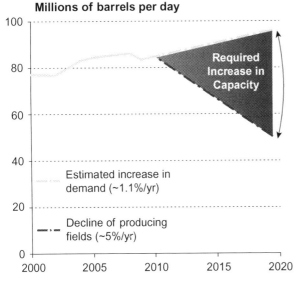

FIG. 4.77 The need for EOR is evident in production and oil quality decline.

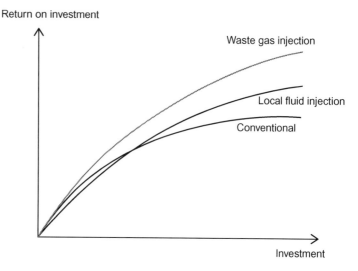

FIG. 4.78 For the same investment, return is much different depending on type of fluid injected.

from Zatzman and Islam (2007) shows how local fluid injection gives higher return in investment than conventional turn key projects, even though the return is lower at early stages of EOR. Using local fluid requires more investment in infrastructure than turn key projects but the investment pays off quickly and much higher return is posted at later stages. Local fluids may be produced hydrocarbon gas, locally available CO_2, or other gas/fluid available in and around the reservoir. Waste gas, on the other hand, shows higher return throughout the duration of the project. Waste gas may include produced hydrocarbon gas that is normally flared, flue gas, sour gas, or any others that are considered to be liability to the producer.

4.5.10 EOR for unconventional formations

Most EOR techniques are based on oil viscosity reduction and/or improvement of mobility ratio by increasing the displacement phase viscosity or by reducing oil viscosity and/or the interfacial tension between injected fluid and oil. They can be categorized into two broad types. They are thermal and chemical recovery processes. This follows the natural cleaning technique of hot water wash with soap. It turns out the most potent cleaning agents of petroleum (second most abundant fluid) in nature are (1) Water (most abundant fluid); (2) Clayey material (most abundant solid material on earth); (3) Wood ash (solid products of oxidation of organic products); (4) Carbon dioxide (gaseous product of oxidation of organic products). It turns out that organic products are the second most abundant solid on earth and oxygen is the most abundant gas. Hot water naturally offers the most effective cleaning product as long as the heating is done with the most abundant energy source, viz. solar energy. Finally, one should note heating through combustion of organic energy source (e.g., petroleum and wood) is the second most efficient heating mechanism.

Fig. 4.79 shows unconventional oil production in the United States along with projection under two different scenarios. This figure shows that the Lower 48 onshore tight oil development continues to be the main driver of total US crude oil production, accounting for about

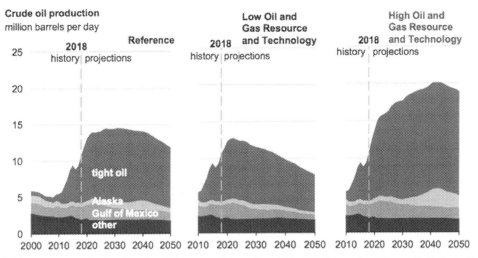

FIG. 4.79 Projection of tight oil under different conditions. *From EIA, 2019. International Energy Outlook. Available at https://www.eia.gov/outlooks/ieo/pdf/ieo2019.pdf.*

68% of cumulative domestic production in the Reference case during the projection period. US crude oil production is expected to level off at about 14 million barrels per day (b/d) through 2040 in the Reference case as tight oil development moves into less productive areas and well productivity declines. In the Reference case, oil and natural gas resource discoveries in deepwater in the Gulf of Mexico lead Lower 48 states offshore production to reach a record 2.4 million b/d in 2022. Many of these discoveries resulted from exploration when oil prices were higher than $100 per barrel before the oil price collapse in 2015 and are being developed as oil prices rise. Offshore production then declines through 2035 before flattening through 2050 as a result of new discoveries offsetting declines in legacy fields. Alaska crude oil production increases through 2030, driven primarily by the development of fields in the.

National Petroleum Reserve-Alaska (NPR-A), and after 2030, the development of fields in the 1002 Section of the Arctic National Wildlife Refuge (ANWR). Exploration and development of fields in ANWR is not economical in the Low Oil Price case. In addition, this area remains protected despite the efforts of the Trump administration to open to oil and gas exploration and exploitation.

Even though thermal recovery would include several methods in addition to steam flooding, steam flooding remains by far the most widely successful thermal EOR technique. Recently, it has been realized that mobility control can be realized by using surfactant with steam when steam/foam is generated. Ever since this realization, most emerging technologies in steam flooding involve some kind of surfactant application. EOR in light oil reservoirs have mainly focused on surfactant and/or polymer injection. Hundreds of patents have been issued on different forms of surfactant and polymer injection (in form of surfactant-water flood, micellar flood, surfactant/polymer enhanced waterflooding, etc.). Even though the chemical EOR has been recently marked as too expensive ever since the drop in oil prices in 1982, surfactants continue to play an important role in virtually all forms of successful EOR, be it in form of foam (mobility control in gas injection), steam/foam (mobility control in steam flooding), micellar, alkaline/polymer flooding or others.

The largest field to undergo EOR is the Cantarell field of Mexico. Discovered in 1976, by 1981 the Cantarell field was producing 1.16 million barrels per day. However, the production rate dropped to 1 million barrels per day in 1995. At this point nitrogen gas injection was used as an EOR fluid. The nitrogen injection project, including the largest nitrogen plant in the world, installed onshore at Atasta Campeche, started operating in 2000, and it increased the production rate to 1.6 million barrels per day to 1.9 million barrels per day in 2002 and to 2.1 million barrels per day of output in 2003, which ranked Cantarell the second fastest producing oil field in the world behind Ghawar Field in Saudi Arabia (Swart et al., 2016). This is despite having a much smaller size of oil fiend than Ghawar. Such rapid extraction of oil led to premature breakthrough of the Nitrogen front, thus lowering the heating value of the gas. As such, the nitrogen had to be removed to maintain the quality of the produced gas.

With regard to offshore, slightly more than 10 years after its discovery, Lula Field of Brazil has the world's largest oil production in ultra-deep waters. This field employed EOR scheme from the beginning of its operations. This impressive goal was achieved through a robust reservoir-oriented strategy to minimize risks and maximize value during the fast-track development of this giant field (Rosa et al., 2018). The methodology applied included four steps. The first step consisted of performing an adequate data acquisition during exploratory and early development phases. The acquired data guided the second step, which is the implementation of pilot projects to gather important dynamic information and to anticipate profits. The third step was the deployment of definitive production systems based on robust drainage strategies, built under different reservoir scenarios. The fourth step was production management through the application of innovative technologies in ultra-deep offshore environment, such as water-alternating gas (WAG) wells, 4D seismic monitoring, and massive use of intelligent completion. Seismic data and special processing enabled the identification and characterization of the reservoirs located under a thick saline formation. The drilling of reservoir data acquisition wells along with extended well tests was important to appraise critical reservoir regions and to verify communication between different reservoir zones. Extensive fluid sampling and advanced thermodynamic modeling allowed the understanding of Lula Fields complex fluids. Two pilot projects were positioned over 20 km apart from each other on this extensive reservoir, allowing interference tests between the pilots and new drilled wells. Afterward, eight following production systems were conceived using previously acquired data and flexibilities to accomplish modifications into the project based on new information acquired during the drilling campaign. The reservoir characterization was rendered dynamic, with continual adjustment of operating parameters. Such dynamic reservoir management of injection and production parameters produced promising results. It is expected that Lula Field will reach a peak of production of around 1 million bpd (Rosa et al., 2018). Such production levels were never achieved before in a single ultra-deep-water field.

Safaniya oil field, located 200 km north of Dhahran in the Persian Gulf, Saudi Arabia, is currently the largest offshore oil field in the world. This oilfield offers a unique insight into the world of oil and gas development. Discovered in 1954, the Safaniya oil field began producing 50,000 barrels of oil a day from 18 wells in April 1957. The daily output of the field was increased by seven times in 1962 with 25 producing wells. Today, with an estimated oil reserve of 50 billion barrel, this field currently produces 1.2 million bbl/day. Even though this field has been producing for over 60 years, it still produces in primary recovery mode. At present, this mature offshore field is being upgraded in two phases as part of Saudi Aramco's Safaniya Master Development Plan to maintain the field's production level of around 1.2 million barrels per day. Phase one of the development plan envisages the upgrade of crude-gathering facilities, improving the site's crude oil transport capacity and providing adequate power supply for central and north Safaniya. Phase Two involves the upgrade of existing well heads and the installation of artificial lift infrastructure such as electric submersible pumps (ESPs).

In the United States, the focus has been on unconventional oil and gas. The recent record-breaking production of oil and gas in the United States is attributed to the influx of horizontal drilling and hydraulic fracturing (popularly known as "fracking"). For over 20 years, horizontal drilling has been the most common drilling technique in the United States. However, the unlocking of tight formations, including shale, has become the most important reason for the surge. Large schemes of fracking have been implemented in the states of North Dakota, Oklahoma and Texas. In 2018, The US Energy Information Administration (EIA) has added new play production data to its shale gas and tight oil reports. Last December, the US shale and tight plays produced approximately 65 Bcf/D of natural gas and 7 million b/d of crude oil, accounting for 70% and 60% of the US production in those areas, respectively. These totals represent a significant jump in the last 10 years: shale gas and tight oil accounted for 16% of total US gas production and approximately 12% of the US total crude oil production.

EIA updated its production volume estimates to include seven additional shale gas and tight oil plays, increasing the share of shale gas by 9% and tight oil by 8% compared with previously estimated shale production volumes. The change captures increasing production from new, emerging plays as well as from older plays that had previously been in decline, but are now rebounding because of advancements in horizontal drilling and hydraulic fracturing.

These plays include the Mississippian formation, located mainly within the Anadarko Basin in Oklahoma. While the play has produced liquids and natural gas for some time, newer completion techniques have driven recent production gains (Fig. 4.80).

In the US Energy Information Administration's International Energy Outlook 2016 (EIA, 2016) and Annual Energy Outlook 2016, natural gas production worldwide is projected to increase from 342 billion cubic feet per day (Bcf/d) in 2015 to 554 Bcf/d by 2040. The largest component of this growth is natural gas production from shale resources, which grows from

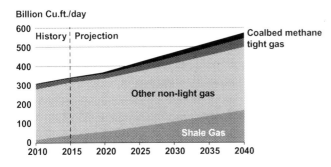

FIG. 4.80 Future prospect of unconventional gas (EIA, 2019).

42 Bcf/d in 2015 to 168 Bcf/d by 2040. Shale gas is expected to account for 30% of world natural gas production by the end of the forecast period.

Although currently only four countries—the United States, Canada, China, and Argentina—have commercial shale gas production, technological improvements over the forecast period are expected to encourage development of shale resources in other countries, primarily in Mexico and Algeria. Together, these six countries are projected to account for 70% of global shale production by 2040.

In the United States, shale gas production accounted for more than half of US natural gas production in 2015 and is projected to more than double from 37 Bcf/d in 2015 to 79 Bcf/d by 2040, which is 70% of total US natural gas production in the AEO2016 Reference case by 2040.

Several AEO2016 side cases illustrate the effect of technological improvements on cost and productivity. Shale gas production in 2040 is projected to be 50% higher under the High Oil and Gas Resources and Technology case, reaching 112 Bcf/d, while in the Low Oil and Gas Resources and Technology case, production is projected to be 50% lower than the Reference case, reaching 41 Bcf/d.

Canada has been producing shale gas since 2008, reaching 4.1 Bcf/d in 2015. Shale gas production in Canada is projected to continue increasing and to account for almost 30% of Canada's total natural gas production by 2040.

China has been among the first countries outside of North America to develop shale resources. In the past 5 years, China has drilled more than 600 shale gas wells and produced 0.5 Bcf/d of shale gas as of 2015. Shale gas is projected to account for more than 40% of the country's total natural gas production by 2040, which would make China the second-largest shale gas producer in the world after the United States.

Argentina's commercial shale gas production was just 0.07 Bcf/d at the end of 2015, but foreign investment in shale gas production is increasing. Pipeline infrastructure in Argentina is adequate to support current levels of shale gas production, but it will need to be expanded as production grows. Current shortages of specialized rigs and fracturing equipment are expected to be resolved, and shale production is projected to account for almost 75% of Argentina's total natural gas production by 2040.

Algeria's production of both oil and natural gas has declined over the past decade, which prompted the government to begin revising investment laws that stipulate preferential treatment for national oil companies in favor of collaboration with international companies to develop shale resources. Algeria has begun a pilot shale gas well project and developed a 20-year investment plan to produce shale gas commercially by 2020. Algerian shale production is projected to account for one-third of the country's total natural gas production by 2040.

Mexico is expected to gradually develop its shale resource basins after the recent opening of the upstream sector to foreign investors. At present, Mexico is expanding its pipeline capacity to import low-priced natural gas from the United States. Mexico is expected to begin producing shale gas commercially after 2030, with shale volumes contributing more than 75% of total natural gas production by 2040.

Fig. 4.81 shows projection of both shale gas and other gas production for these countries.

Fig. 4.82 shows the distribution of shale oil and gas reserve of the world. Note that assessment for certain countries, such as Saudi Arabia is without resource estimate and as such the actual numbers are likely to be much higher.

4.5 Improving oil and gas recovery from unconventional reservoirs

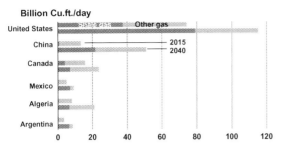

FIG. 4.81 Future prospect of unconventional oil and gas in various countries (EIA, 2016).

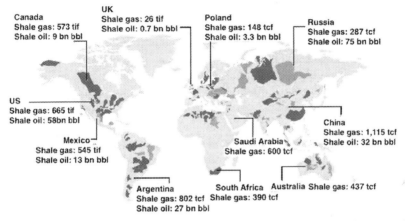

FIG. 4.82 Global unconventional shale oil and gas (*dark spots*: with resource estimates; *light spots*: without resource estimate).

USGS (2014) lists major factors that would dictate suitability of an unconventional oil and gas. These factors are classified under two broad categories.

(A) Market attractiveness:
  (i) Size of potential resources: involves initial oil/gas in place, irrespective of reservoir characteristics.
  (ii) Enabling fiscal regime: involves investment promotion regime, tax exemptions, government subsidies
(B) Ease of implementation:
  (i) Geological considerations
  (ii) Land access and operability
  (iii) Unconventional services sector
  (iv) Oil and gas distribution network
  (v) Regional connectivity
  (vi) Skilled workforce

Interestingly Acceture's (n.d.) analysis does not include environmental considerations. It is no surprise, therefore, that such expansion plan is invariably countered with legitimate concerns over environmental impact of oil and gas development. The most important feature of any sustainable development is, the environmental considerations are inherently imbedded. True sustainability also implies conformity with economic constraints.

Of significance is the fact that there is much more nonconventional petroleum reserve than the convention "proven" reserve. This point is made in Fig. 4.83. Even though it is generally assumed that more abundant resources are "dirtier," hence in need of processing that can render the resource economically unattractive, sustainable recovery techniques can be developed that are more efficient for these resources and also economically attractive and environmentally appealing (Islam et al., 2010). In addition, natural gas quality is little affected by the environment. For instance, gas hydrate that represents the most abundant source of natural gas is actually far cleaner than less abundant resources. Finally, Islam et al. (2018) showed how unconventional oils have more diverse applications, some of which are lucrative. As such, these factors add appeal to the development of unconventional oil and gas.

It has been shown in Chapter 3 that the need for higher price and/or increased technological challenge is fictitious and is erased if scientific energy pricing along with sustainable technology are used. Current investment strategy has fueled this misconception.

In terms of oil industry, the main focus has been in nonconventional petroleum extraction. For instance, Fig. 4.84 shows major investments in oil sands in Canada. In 2013, 7299 trillion cubic feet of shale gas 345 billion barrel of shale/tight oil has been added. In Canada, the theme of "climate change" features prominently in every aspect of energy management. As such, Canada is a pioneer in ratifying Paris Agreement. Fig. 4.85 shows projected CO_2 emissions in 2030.

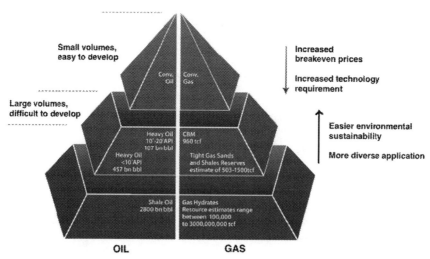

FIG. 4.83 Three is a lot more oil and gas reserve than the "proven" reserve. *From Islam, M.R., 2014. Unconventional Gas Reservoirs. Elsevier.*

4.5 Improving oil and gas recovery from unconventional reservoirs

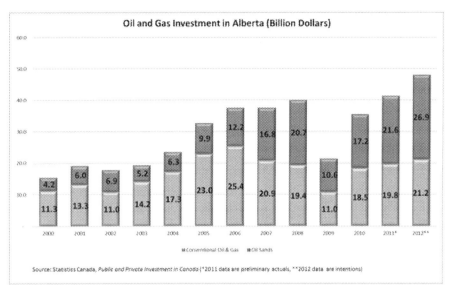

FIG. 4.84 Major investment in oil sands in Canada. *From Islam, J.S. et al., 2018. Economics of Sustainable Energy. Scrivener-Wiley, 628 pp.*

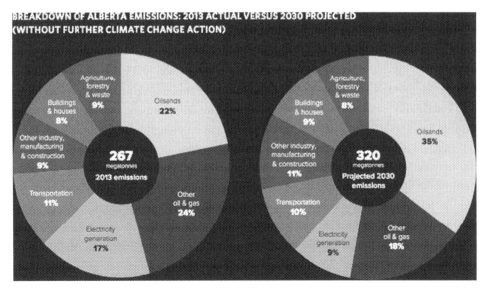

FIG. 4.85 Past emissions and projected emissions of Alberta, Canada.

388   4. Unconventional reservoirs

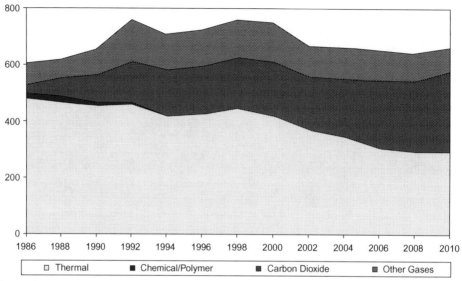

Source: Oil and Gas Journal, Biennial Enhanced Oil Recovery Project Surveys

FIG. 4.86 Contribution of various EOR projects.

Recent reports show that the contribution of EOR in total oil production has increased steadily throughout last two decades, despite fluctuating oil prices. In addition, the number of EOR projects has steadily declined since the peak in 1986. This shows more efficiency of the EOR projects, indicating a trend that efficiency is the focus of the future EOR schemes. This trend is likely to continue even if the oil price continues to rise. Oil industries seem to be convinced that they can no longer afford to experiment with EOR schemes that do not show immediate improvement in oil production (Fig. 4.86).

In using the United States as a reference, one must be cautious about the context in the United States. In the United States, tax benefits for EOR projects were repealed in mid-80s during Reagan era. This followed the sudden decline of the number of projects that were declared as "EOR." In absence of tax breaks, there was no benefit of declaring a project as EOR. Of particular consequence was the "chemical/polymer" processes. There has been nil contribution of these techniques over the last two decades. With the exception of China, no other country developed any commercial EOR project using chemical methods. There have been several pilot tests involving chemical methods in the North Sea, but the results have been dismal. During the slump period, the production continued to rise, albeit with a small slope. The rise subsided after 2000. Note that this is the year oil price was unusually low. The original prediction was $200 per barrel in 2000. Instead, it was 10 times that number. In that decade, oil price was so low it was hard to justify investments in aggressive EOR techniques. The notion of sustainable EOR that gives one multiple dividend and is environmentally appealing was only a theoretical concept at that time. During that period recoveries with gas, thermal and carbon dioxide schemes started to decline keeping pace with decreasing number of EOR projects. The number of thermal projects decreased because main fields in California began to reach maturity and the use of expensive mobility control agents with thermal EOR didn't bear fruit. While this is theoretically demonstratable, the petroleum industry had to experiment with it before shutting down many thermal projects. For the same reason chemical EOR practically disappeared. By contrast, carbon dioxide projects continued to be in operation and after 2004, the number actually increased. Even though the number of projects with "other gases" was also increased, CO_2 projects showed marked increase in oil recovery. CO_2 projects continued to increase with the new incentive related to green house gas emission, as discussed in earlier sections. Since 2002 EOR gas injection projects outnumber thermal projects for the first time in the last three decades. However, thermal projects have shown a slightly increase since 2004 due to the increase of High Pressure Air Injection (HPAI) projects in light oil reservoirs. This technique originally perfected with heavy oil and tar sand (through in situ combustion projects) has the potential of increasing oil recovery from light oil reservoirs to a great extent. The technique is simple and cost effective.

Chemical EOR methods that were highly successful in laboratory tests have failed miserably in field trials. Once again, this was theoretically expected but the industry couldn't anticipate in absence of scaled model studies or even scaling laws that capture chemical flooding effectively. Only two projects in chemical injection were reported in 2008. However, there is an consorted efforts in the United States as well as the rest of the world to promote chemical flooding, particularly those involving mobility control agents. The focus now is not to create miscible fronts with micelle, etc. Instead, new genre of chemical flooding schemes involve the introduction of new polymers and surface active agents. This reattachment to chemical flooding is reminiscent of 70s and 80s research in which hundreds of patents "proved" that chemical flooding would be effective in the field but not a single project became cost effective or even technically successful, despite enjoying tax benefits in the United States for such projects. These techniques are not likely to produce positive results, as will be discussed in the case studies section.

4.5.11 Carbon dioxide injection

The most significant development in terms of EOR has been in CO_2 projects. Fig. 4.87 shows various US basins have shown increased recovery throughout the last decade. It is this time that there has been a global effort to link CO_2 to global warming. The use of CO_2 provides one with double dividends. Based on this principle, numerous CO_2 projects have surfaced. While theoretically, any CO_2 project is both effective and environment friendly, a CO_2 project cannot be sustainable unless proper process is followed. This aspect will be considered in a later section.

The second most important considerations in CO_2 floods is the fact that it is considered to be inexpensive, at least in the United States (\$US 1–2/Mscf). In addition, the United States has existing network of pipelines that can be readily used for distribution of CO_2. There is one significant case study in Weyburn field of Canada, for which an entire pipeline was created to dispatch CO_2 from the United States to Canada. This CO_2 was deemed most cost effective than Canadian CO_2 that would have to be extracted from local coal fired power plants. The project received \$1 billion in government grants and more in tax rebates and flagged as the most important CO_2 sequestration project of the time. This project was "profitable" only because of the government grant and some 10-fold increase in oil price. This will be discussed in latter sections.

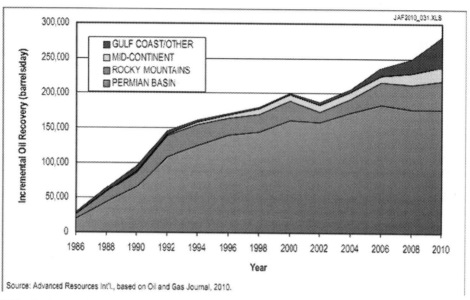

FIG. 4.87 Evolution of EOR projects in the United States. *From Oil & Gas Journal EOR Surveys 1976–2010.*

It is also important to note that the CO_2 pipeline system in the United States was built in a 30 years (1975–2005) time span when oil prices and tax incentives were sufficiently attractive to ensure security of supply as main drivers. These are viable only because of government interference in name of climate change funding and investment. Fig. 4.88 shows evolution of CO_2 projects in the United States and average crude oil prices for the last 30 years. This figure is extracted from Alvarado and Manrique (2010). They used oil prices of the refiner average domestic crude oil acquisition cost reported by the Energy Information Administration (EIA). For reference purposes, crude oil price used in Fig. 4.89 was arbitrarily selected for every month of June except for year 2010 (oil price as of March 2010). These CO_2 projects led to significant recovery (Fig. 4.90).

Although it can be concluded that CO_2-EOR ("from natural sources") is a proven technology with oil prices > $20/bbl, this EOR method represents a specific opportunity in the United States and not necessarily can be extrapolated to all producing basins in the world. This conclusion is based on the selection criteria listed in Table 4.24. This cannot be generalized to other countries, where different economic, environmental, and technical conditions prevail. From sustainability point of view, there must be questions that should be asked in proper sequence. For instance, if the technological feasibility question is asked before availability of the carbon dioxide, the answer would be irrelevant at best. Similarly, the presence of an existing infrastructure for both CO_2 purification and distribution can alter the decision tree. Finally, recent findings indicate that CO_2 is quite effective in recovering heavy oil. With the new incentive of CO_2 sequestration, heavy oil reservoirs offer the greatest potential for CO_2 injection.

Fig. 4.90 shows the strategy developed by the government of Alberta. This program shows equal importance to conventional and heavy oil formations. Scaled model studies show that

4.5 Improving oil and gas recovery from unconventional reservoirs 391

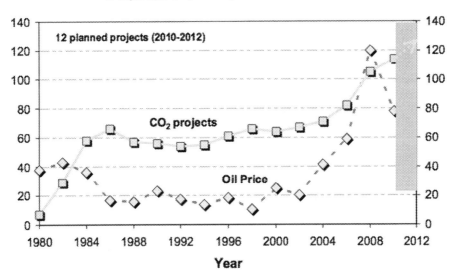

FIG. 4.88 Evolution of CO_2 projects and oil prices in the United States. *From Oil & Gas Journal EOR Surveys 1980–2010 and U.S.; EIA, 2010. Annual Energy Outlook. Available at: https://www.nrc.gov/docs/ML1111/ML111170385.pdf*

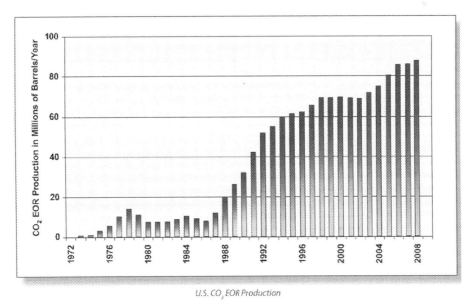

FIG. 4.89 CO_2 EOR recovery in the United States throughout history.

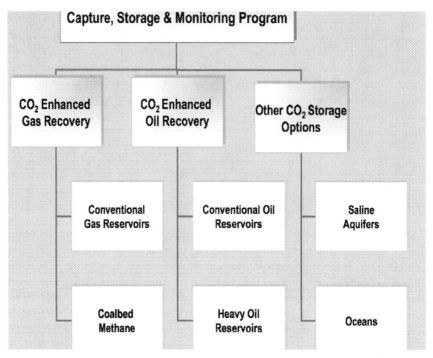

FIG. 4.90   Alberta government strategy.

TABLE 4.24   Screening criteria for CO_2 projects as used in the United States.

Depth, ft	<9800 and >2000
Temperature, °F	<250, but not critical
Pressure, psia	>1200–1500
Permeability, md	>1–5
Oil gravity, °API	>27–30
Viscosity, cP	≤10–12
Residual oil saturation after waterflood, fraction of pore space	>0.25–0.30

heavy oil recovery with CO_2 can lead to 70% of the oil in-place. This is tremendous considering the fact that primary recovery of heavy oil is less than 5% and similar recovery factor with steam flooding would require significant cost increase while having bigger footprint on the environment.

Fig. 4.91 shows the importance given to CO_2-enhanced gas recovery. The use of CO_2 injection in enhanced oil recovery is a mature well practice technology. Enhancing gas recovery through the injection of CO_2 however is yet to be tested in the field. Numerous simulation

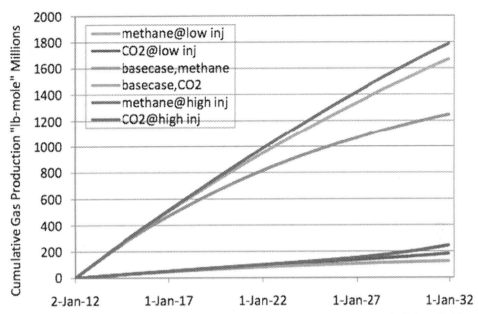

FIG. 4.91 Natural gas production with CO_2 injection schemes. *From Islam, M.R., Mousavizadeghan, H., Mustafiz, S., Abou-kassem, J.H., 2010a. Reservoir Simulation: Advanced Approach. Scrivener-Wiley, 468 pp.*

studies ever since the early work of Islam and Chakma (1993) appeared to support high recovery of gas and heavier components from a gas reservoir along with high capacity of CO_2 sequestration. Although there are some published simulation studies that have been carried out to comprehend by which process CO_2 sequestration in a depleted gas reservoir could lead to enhance gas recovery, none of these studies have ever attempted to manifest the effect of mixing (CO_2-CH_4) on the recovery process prior to depleted reservoir. These studies were mainly aimed to reduce greenhouse gas emission in the atmosphere and sequestrating in a depleted gas reservoir or in an aquifer. In the year 2005, a project by Gas de France Production Netherland was in progress to assess the feasibility CO_2 injection prior to depletion of the gas reservoir (K12-B) for EGR and storage. However, since then no follow up results have been published on the final gain in reserve recovery.

Generally, high natural gas recovery factors along with concerns with degrading of the natural gas resource through mixing of the natural gas and CO_2 have led to very little interest shown in CO_2-EGR. In terms of sequestration, natural gas reservoirs can be a perfect place for carbon dioxide storage by direct carbon dioxide injection. This is because of the ability of such reservoirs to permeate gas during production and their proven integrity to seal the gas against future escape (Oldenburg et al., 2001).

However, displacement of natural gas by injection of CO_2 at supper critical state has not been studied extensively and not well understood. Despite of the fact that CO_2 and natural gas are mixable, their physical properties such as viscosity, density and solubility are potentially favorable for reservoir re-pressurization without extensive mixing. This phenomenon of gas-gas mixing can be controlled by controlling the operating parameters.

The injected CO_2 in geological formations undergo geochemical interactions, such as structural, stratigraphic and hydrodynamic trapping. The injected CO_2 is trapped either in the form of physical trapping as a separate phase or as a chemical trapping where it reacts with other minerals present in the geological formation (International Energy Agency, 2010). As time passes, CO_2 becomes immobilized in the geological formation as a function of given long time scales. This is known as geological sequestration. Oldenburg et al. (2003) simulated CO_2 as a storage gas. The results suggested that CO_2 injection as a supercritical fluid allows more CO_2 storage as the pressure increases due to its high compressibility factor. Thus, an expansion of the compressed is expected due to changes in pressure and temperature. As a result, there will be a point when gas production no longer is economically feasible.

In terms of economics, not unsurprisingly, Gaspar et al. (2005) claimed the major obstacle for applying CO_2-EGR is the high costs involved in the process of CO_2 capture and storage. The experience from oil recovery schemes indicate that the economics look quite different when purity in injected CO_2 is not sought. It turns out that the purity doesn't need to be high and naturally available CO_2 or even flue gas would accomplish the same outcome. It is in line with pressure maintenance schemes in oil reservoirs. This option that would make CO_2 injection appealing without tax incentive as claimed by IEA (2010).

Islam et al. (2010) conducted economic feasibility study of carbon dioxide into a natural gas reservoir and found the scheme economically attractive because of EGR. Fig. 4.91 shows results of CO_2 injection at high and low injection rates. Natural gas production is the highest for CO_2 injection at high rate because the mixing is the greatest under high injection rates. However, one should note that this study used a stable displacement front. This is a reasonable assumption because CO_2 is more viscous and denser than natural gas. Such results are not expected in oil reservoirs. In terms of overall gas injection for EGR, there are 50 projects in North America that employs sour gas injection for treatment of natural gas and produced CO_2 has been injected in Dutch sector of North Sea for years (K-12B gas reservoir).

Based on the CO_2 capture, utilization and sequestration strategy, government of Alberta has drafted a comprehensive scheme as shown in Fig. 4.92. "CO_2 Backbone" is a network or manifold of pipelines that can be used for transporting CO_2 from emission hubs as well as taking CO_2 to customer sites. The idea is to create an infrastructure based on the "CO_2 culture." Because CO_2 is ultimately a valuable commodity, it is suggested that industrial complexes, including pharmaceutical industry, be developed along the backbone. This is a powerful template for developing a comprehensive carbon dioxide-based EOR technique.

Fig. 4.93 shows locations for various CO_2 sequestration projects around the world. These projects are in support of greenhouse gas mitigation.

In general, it is accepted that the oil production from individual field is declining. This notion comes from the fact that natural depletion occurs in every petroleum well. However, both country-wide and global oil production rates have been rising, apart from interruptions due to political reason. Fig. 4.94 shows oil production history of some of the top oil producing countries. Iran is not included in this figure. However, Iran follows similar trend and after dropping below 1.5 million bbl/day after the Iranian revolution, the production rate has increased gradually and hovers around 4 million bbl/day today.

4.5 Improving oil and gas recovery from unconventional reservoirs

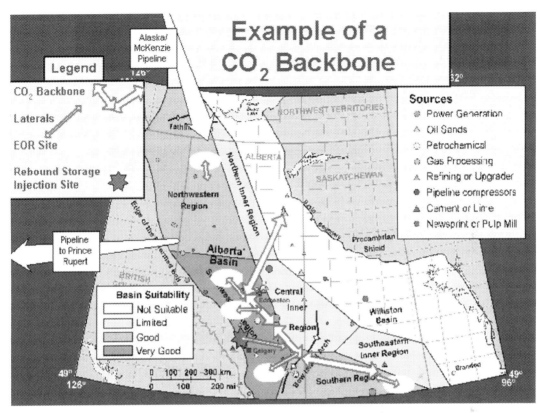

FIG. 4.92 Alberta's plan to implement comprehensive carbon management scheme.

FIG. 4.93 CO_2 sequestration demonstration projects around the world.

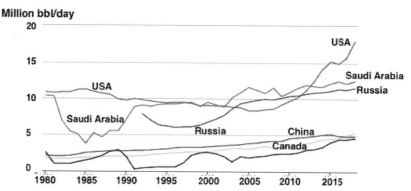

FIG. 4.94 Oil production rate history for top oil producers. *From EIA, 2019. International Energy Outlook. Available at https://www.eia.gov/outlooks/ieo/pdf/ieo2019.pdf.*

There are primarily three reasons given for increasing oil recovery:

(1) Primary recovery techniques leave behind more than half of the original oil in-place. This is a tremendous reserve to forego.
(2) Increased drilling activities do not increase new discoveries of petroleum reserve. While this has been replaced with new technological opportunities (e.g., fracking technology creating oil and gas reserves in unconventional reserve), the argument is made to justify EOR.
(3) Environmental concern of CO_2 emission. Ever since the signing of Kyoto Agreement, the US government has led the movement of CO_2 sequestration, thereby increasing oil recovery.

From the beginning of oil recovery, scientists have been puzzled by the huge amount of oil leftover following primary recovery. Naturally occurring drive mechanisms recover anything from 0% to 70% of the oil in-place. In most cases, recovery declines rapidly as viscosity of oil increases.

For instance, primary recovery is less than 5% when oil viscosity exceeds 100,000 cP. This is not to say that heavy oil recovery was the primary incentive for EOR, even though most EOR projects in the United States, Canada, and Venezuela involve heavy oil recovery. The primary incentive for EOR is the fact that a typical light oil reservoir would have more than 50% of the original oil in-place leftover, while a small investment can recover over 70% of the oil in-place. For heavy oil, the room for improvement is much higher. Even though theoretically there is much more recovery potential of heavier energy sources all the way up to biomass (Fig. 4.95), the current recovery techniques are geared toward light oil. This figure shows that natural gas is the most efficient with the most environmental integrity. The argument that is made in this figure is if natural gas, light oil, or any other energy source is burnt without adding artificial chemicals in the stream (e.g., during refining), the entire combustion output is fully sustainable and the CO_2 that is produced is 100% recyclable. Each molecule of produced CO_2 would end up contributing to the formation of greeneries. Greeneries then end up as biomass, which contributes to enriching the ecosystem. As such, the energy resource is infinity as long as sustainability is maintained.

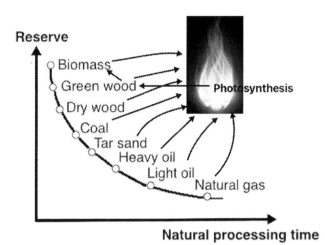

FIG. 4.95 Key to sustainability in energy management.

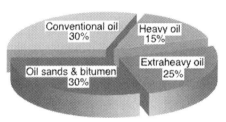

FIG. 4.96 Distribution of the World's proven reserve. *From Alboudwarej, H., et al., 2006. Highlighting heavy oil. Oilfield Rev. 18 (2), 34–53.*

Within petroleum itself, the "proven reserve"[a] is nearly 1.7 Trillion Barrel (BP, 2018). Out of this reserve, conventional light oil is only 30% (Fig. 4.96). It means that devising a thermal EOR technique is paramount. Any thermal EOR technique involves adding heat, which increases the mobility of the oil exponentially. Fig. 4.96 shows one example of such exponential decrease in viscosity, which correlates directly with flow rate. The task at hand becomes the delivery of sustainable heat to the formation (Fig. 4.97).

The "easy oil," which is the target of "oil wars" involves only miniscule compared to the overall potential, as depicted in Fig. 4.98. Because all energy source utilization techniques are equipped with processing light oil as a reference, the primary focus of EOR has been light oil. A much larger portion of the global oil reserve involves heavy oil, tar sand, shale, and other reservoirs, which require some form of EOR to produce. This reserve can be doubled by using sustainable technology, which increases the overall efficiency and can be utilized in otherwise marginal oil reservoirs. The use of such technology can double the current reserve, even when

---

[a] Proven oil reserves are reserves that are known to exist and that are recoverable under current technological and economic conditions.

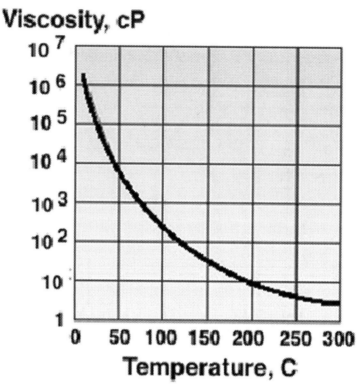

FIG. 4.97 Viscosity change is invoked by temperature. *From Alboudwarej, H., et al., 2006. Highlighting heavy oil. Oilfield Rev. 18 (2), 34–53.*

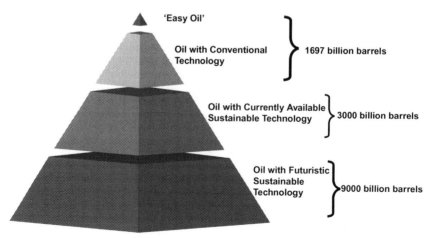

FIG. 4.98 Much more oil can be recovered with a double dividend of environmental benefit with sustainable technologies.

no new technology is implemented. When the potential of novel technologies is included a much bigger oil reserve becomes accessible. Most importantly, the exploitation of oil with sustainable technology produces only environment-friendly gases that are readily assimilated with the ecosystem. With it comes the double dividend of economic benefit because all truly sustainable technologies are also the least expensive.

Fig. 4.99 shows drilling activities in the United States over the last decade. This represents an enhanced level of drilling throughout to match with the production boost during the same period. Note that the reserve/recovery ratio in the United States is quite low. Such intense drilling activities would represent far greater output in high ratio reservoirs. As stated earlier, infill drilling can increase both production rate and recoverable reserve for cases for which reserve/recovery ratio is low, as is in most OPEC countries (Fig. 4.100).

The number of drilled but uncompleted wells in seven key oil and natural gas production regions in the United States has increased over the last 2 years, reaching a high of 8504 wells in February 2019, according to well counts in EIA's Drilling Productivity Report (DPR). The most recent count, at 8500 wells in March 2019, was 26% higher than the previous March.

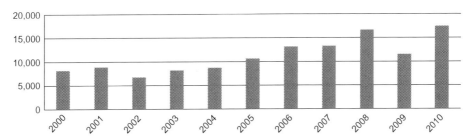

FIG. 4.99 Drilling activities in the United States for various years (EIA, 2014b).

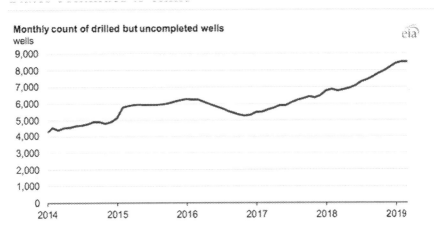

FIG. 4.100 Uncompleted drilling activities in the United States. *From EIA, 2019. International Energy Outlook. Available at https://www.eia.gov/outlooks/ieo/pdf/ieo2019.pdf.*

Drilled but uncompleted wells, also known as DUCs, are oil and natural gas wells that have been drilled but have not yet undergone well completion activities to start producing hydrocarbons. The well completion process involves casing, cementing, perforating, hydraulic fracturing, and other procedures required to produce crude oil or natural gas.

The number of DUCs has generally increased since the end of 2016. A high inventory of DUCs may be attributable to economic factors or resource constraints. For example, a low oil and natural gas price environment may postpone well completion activities in areas where the wellhead break-even price is too high relative to the current market price. Another example may be the lack of available well completion crews to perform hydraulic fracture activities in areas of high demand. Takeaway capacity, or the ability to transport hydrocarbons through pipelines away from the resource, may also place additional constraints when pipeline networks are insufficient to accommodate supply (Fig. 4.101).

Most of the recent increase in the DUC count has been in regions dominated by oil production, especially the Permian region that spans western Texas and eastern New Mexico. As of March 2019, nearly half of the total DUCs included in the DPR were in the Permian region. The Permian Basin experienced takeaway constraints in the second half of 2018, but recent pipeline capacity additions in the region have reduced some of the takeaway constraints. Other pipeline projects are planned or currently under construction.

In contrast to oil-directed regions, the number of DUCs in natural gas-dominated DPR regions such as the Appalachian and Haynesville regions has decreased by nearly half over the past 3 years, from 1230 wells in March 2016 to 713 wells in March 2019. New pipelines in these regions have increased the ability to transport natural gas to demand centers in the Northeast and Midwest.

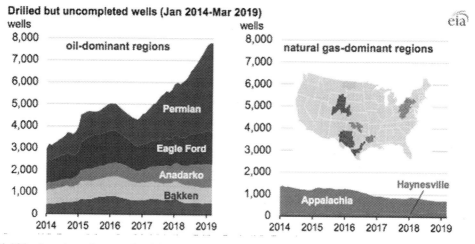

FIG. 4.101 Locations of uncompleted drilled wells. *From EIA, 2019. International Energy Outlook. Available at https://www.eia.gov/outlooks/ieo/pdf/ieo2019.pdf.*

4.5.12 Thermal methods

By far the most important EOR scheme in United States and the world has been the thermal EOR. It has been in operation in heavy oil formations for over five decades. Obviously, the advantage gained by exponential decrease in oil viscosity due to linear increase in temperature has been the focal point of all thermal EOR schemes. Among thermal methods, steam injection has been the most dominant EOR scheme. Simplicity of the scheme and the unique latent heat properties of water are the major reasons why oil industry has been active in steam flooding. Besides, primary oil recovery being practically impossible, huge heavy oil reserves are left as target of the steam injection scheme.

Just as water is the best cleaner, thereby, making waterdrive the most common drive for oil production, steam injection is the most common technique for heavy oil reservoirs. Table 4.25 lists the contrasting features of water and oil. This list makes it clear why the use of water is the most effective technique for oil recovery.

TABLE 4.25 Contrasting features of water and petroleum.

Water	Petroleum
Source of all organic matter	End product of all organic matter
Most abundant fluid on earth	Second most abundant fluid on earth
Oxygen—85.84; Sulfur—0.091 Hydrogen—10.82; Calcium—0.04 Chloride—1.94; Potassium—0.04 Sodium—1.08; Bromine—0.0067 Magnesium—0.1292; Carbon—0.0028	Carbon—83%–87% Hydrogen—10%–14% Nitrogen—0.1%–2% Oxygen—0.05%–1.5% Sulfur—0.05%–6.0% Metals—<0.1%
Mostly homogeneous	Hydrocarbon (15%–60%), naphthenes (30%–60%), aromatics (3%–30%), with asphaltics making up the remainder.
Reactivity of water toward metals. Alkali metals react with water readily. Contact of cesium metal with water causes immediate explosion, and the reactions become slower for potassium, sodium and lithium. Reaction with barium, strontium, calcium are less well known, but they do react readily.	Nonreactive toward metal.
Nonmetals like Cl_2 and Si react with water $Cl_{2(g)} + H_2O_{(l)} \rightarrow HCl_{(aq)} + HOCl_{(aq)}$ $Si_{(s)} + 2H_2O_{(g)} \rightarrow SiO_{2(s)} + 2H_{2(g)}$ Some nonmetallic oxides react with water to form acids. These oxides are referred to as acid anhydrides.	Reaction with nonmetals is faster
High cohesion	Low cohesion
Unusually high surface tension; susceptible to thin film	Unusually low surface tension
Adhesive to inorganic	Adhesive to organic
Unusually high specific heat	Unusually low specific heat

Continued

TABLE 4.25 Contrasting features of water and petroleum—cont'd

Water	Petroleum
Unusually high heat of vaporization	Unusually low heat of vaporization
Has a parabolic relationship between temperature and density	Has monotonous relationship between temperature and density
Unusually high latent heat of vaporization and freezing	Unusually low latent heat of vaporization and freezing
Versatile solvent	Very poor solvent
Unusually high dielectric constants	Unusually low dielectric constants
Has the ability to form colloidal sols	Destabilizes colloids
Can form hydrogen bridges with other molecules, giving it the ability to transport minerals, carbon dioxide and oxygen	Poor ability to transport oxygen and carbon dioxide
Unusually high melting point and boiling point	Unusually low melting point and boiling point
Unusually poor conductor of heat	Unusually good conductor of heat
Unusually high osmotic pressure	Unusually low osmotic pressure
Nonlinear viscosity pressure and temperature relationship (extreme nonlinearity at nano-scale, Khan and Islam, 2016)	Mild nonlinearity in viscosity pressure and temperature relationship
Enables carbon dioxide to attach to carbonate	Absorbs carbon dioxide from carbonate
Allows unusually high sound travel	Allows unusually slow sound travel
Large bandwidth microwave signals propagating in dispersive media can result in pulses decaying according to a nonexponential law (Peraccini et al., 2009)	Faster than usual movement of microwave
Unusually high confinement of X-ray movement (Davis et al., 1995)	Unusually high facilitation of X-ray movement.

From Hutchinson, G.E., 1957. A Treatise on Limnology, Geography, Physics and Chemistry. John Wiley and Sons, New York; CRC Handbook of Chemistry and Physics, 1983. 62nd ed. (Handbook of Chemistry and Physics) Hardcover – Jan. 1, ISBN-10: 0849304628.

Steam injection process involves conversion of scale-free water to high-quality steam of about 232°C temperature and at a pressure higher than the corresponding saturation pressure. In general, using direct fired heaters the water is converted to steam. Using insulated distribution lines, steam is transported to various injection wells. Steam injection can be done by two different methods. They are: steam stimulation and steam displacement.

In the stimulation method a predetermined volume of steam is injection into well and the well is shut in to allow to stimulate the wellbore area. After a few days of shut in the well starts to production. If necessary the stimulation process repeat again. In the steam displacement process, continuous injection of steam, usually apply at lower rates. The steam is injected in place as to distance and direction form production wells. Steam injection is highly sophisticated process and it requires extensive engineering and analytical inputs.

It is estimated that there are 85–110 billion barrels of heavy oil reserve in the United States. Since the 1960s, steam has become the predominant enhanced oil recovery method for these high viscosity, heavy oil reservoirs world-wide. However, factors such as steam-channeling, gravity segregation, and reservoir heterogeneity often result in poor contact of the heavy oil formation by the injected steam, leading to low recoveries. One method of conformance control that has received considerable attention is the use of surfactant foams that reduce steam mobility. Numerous laboratory and technically successful field studies have been reported.

Fig. 4.102 shows the production of Syncrude and bitumen in Alberta. Syncrude and bitumen represent the two extremes of the viscosity spectrum. Syncrude is synthesized from natural gas, whereas bitumen represents the heaviest (and most viscous) components of petroleum. Note that both products grew exponentially in the last few decades, ever since implementation in 80s. While the economics of these products have been reported to be attractive, often the government contribution in building the infrastructure has been overlooked or not included in the analysis. Without significant government involvement, these projects would not be implemented, particularly during the time when oil price was in the range of $10/bbl. With the increase in oil price, these schemes have become attractive and mega projects are being implemented in bitumen extraction and processing. The future of Syncrude has a somewhat conflicted scenarios. Gas price has increased making it more comparable with oil than previous years. In addition, Alberta has suffered from lack of enough natural gas to meet local needs. This is due to the fact that the population of the province has increased manifold in a country that has seen almost zero growth in population over the same period.

For a long time, steam has been used as the driving fluid in heavy oil reservoirs. The steam injection scheme has been very popular because of its simplicity. However, steam injection leads to an unfavorable mobility ratio in most applications. Besides, gravity lay over is a problem with most reservoirs with little or no dip. Injected steam because of its low density, rises to the top of the reservoir and tends to form a channel beneath the cap rock to the

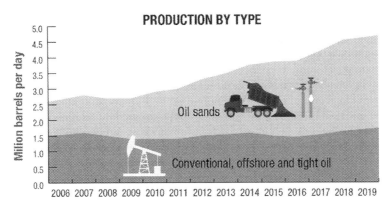

FIG. 4.102 Synthetic crude and bitumen production from Canada's tar sand. *Source: NRCan, 2021. Report. Available at https://www.nrcan.gc.ca/science-and-data/data-and-analysis/energy-data-and-analysis/energy-facts/crude-oil-facts/20064.*

production well. Early steam breakthrough can occur at producing wells owing to override, channeling, unfavorable and viscous fingering mobility ratio, resulting in low oil recovery efficiency. Because of high steam mobility, there is little pressure differential between injector and producer once steam breakthrough occurs. The majority of subsequently injected steam follows this established path of least resistance and the process efficiency is impaired. Injecting surfactants to generate foam in situ can reduce steam mobility and improve the volumetric sweep efficiency in oil reservoirs. There have been many examples of increased oil production in California heavy oil reservoirs when steam in foam was used.

In the last three decades, there have been many attempts to improve steam injection efficiency by the use of additives. Among many additives tried, the aqueous surfactant solution appears to be the most promising. The objective of such surfactant injection is either to increase the pressure gradient across the region of interest by generation of foam or to use the surface active properties of the surfactant to reduce the oil-water interfacial tension and to alter the relative permeability curve. Following is a list of research areas in this topic:

(1) *Surfactant Selection Criteria for Steam Flooding.* In selecting surfactants for application in thermal recovery, two criteria are set, namely, the resistance of surfactants to hydrolytic degradation and to thermal degradation. It is a common practice to study surfactants at elevated temperatures exterior to porous media. The foam tube test is the most commonly applied technique for determining foam stability exterior to porous media (de Vries, 1958). Some studies found foam stability outside of porous media to be an important indicator of potential mobility reduction within porous media (Doscher and Hammershaimb, 1981). In other studies, however, no such correlation was found (Dilgren et al., 1982). The tube test may represent foam behavior in very large pore throats and may not represent foam stability in a confined case as in a real porous medium. This observation has been further confirmed by Zhong et al. (1999).

Handy et al. (1982) indicated that thermal stability is a critical factor in the choice of a foaming agent for thermal EOR processes. It has been demonstrated through many studies that foam can be used for flow diversion in a steam flood process. Recently, Djabbarah et al. (1990) reported thermal stability of several surfactants. Despite many disjointed efforts, a comprehensive selection criterion applicable to steam flooding has not been developed yet (Zhong et al., 1999).

(2) *Microscopic Behavior of Surfactant-Steam Flooding.* It is important to understand microscopic behavior of a system before a field application can be recommended. In steam flooding research, little effort has been spent in studying microscopic behavior and extending that observation to the scaled-up version. Several theories have been proposed to try to explain surface phenomena for a surfactant-steam system (Falls et al., 1988). However, very little agreement among researchers exist sand fundamental questions, such as the role of gas rate on foam apparent viscosity, mechanism of bubble generation, effect of surfactant concentration, or the effect of temperature on foam flow cannot be answered without some degree of ambiguity.

(3) *Role of Residual Oil on Foam.* This fundamental aspect of the steam/foam process has not been addressed properly. Most papers on the topic claim to offer different solutions. One possible way to address this process is to conduct research on the microphysical aspect of the process (George and Islam, 1998).

When heat is combined with water, producing steam becomes an effective displacement tool for additional heavy oil recovery. The decrease in heavy oil viscosity being log with increase in temperature, any heating unlocks tremendous amount of oil from the porous medium. Fig. 4.103 shows general trend in viscosity vs temperature. Note that the temperature scale is linear whereas the viscosity scale is logarithmic. It translates into a sharp decline in viscosity for moderate increase in temperature. Darcy's law being linear, such decrease in viscosity leads to immediate flow rate increase. In addition, the larger change in viscosity takes place in the lower temperature region and the sharpest decline is higher viscosity oils.

Also affected by temperature is the interfacial tension. This alteration in interfacial difference comes from the fact that surface tensions of water and various petroleum fluids are affected differently, even though each of them varies linearly. Fig. 4.104 shows how surface tension varies for various liquids.

Fig. 4.105 provides a summary of the measured endpoint residual oil saturations to water flood as a function of temperature for certain Canadian bitumen samples.

Fig. 4.105 shows two different regimes exist as a function of temperature. At the lower temperature range (below 100°C), there is a rapid decline in oil saturation. In the range of 120–200°C, the decrease rate is subsided. However, at higher range of temperature (beyond 200°C), the saturation declines rapidly once more.

It is well documented that residual oil saturation tends to reduce at constant temperature by steamflooding in comparison to conventional waterflooding at the same temperature condition. This is believed to be due to turbulence effects associated with the vaporization of pellicular films of water underlying trapped bitumen as well as possible changes in interfacial

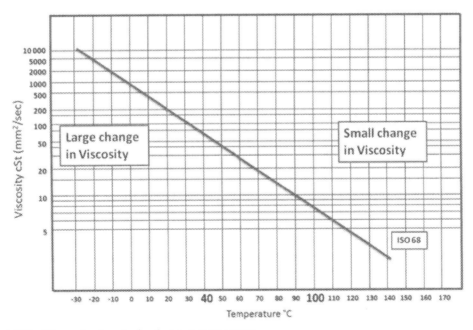

FIG. 4.103 Change in viscosity for change in temperature.

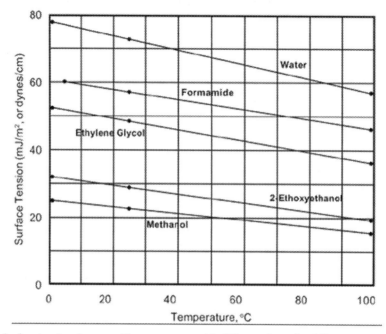

FIG. 4.104 Surface tension changes with temperature with different slopes for different chemicals.

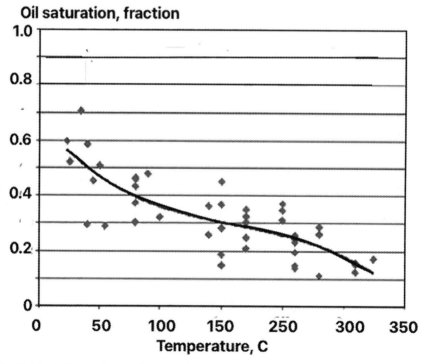

FIG. 4.105 Residual oil saturation as a function of temperature. *From Bennion, D.B., et al., 2006. A correlation of the low and high temperature water-oil relative permeability characteristics of typical western canadian unconsolidated bitumen producing formations. In: Paper Presented at the Canadian International Petroleum Conference, Calgary, Alberta, June. Paper Number: PETSOC-2006-136.*

tension (IFT) and wettability during the steam displacement process. Also active is the steam distillation factor that can improve the efficiency of oil recovery with steam flooding. Overall, steam flooding represents optimum cleaning of oil.

Fig. 4.106 illustrates the trend of pre and post steamflood residual oil saturation as a function of steamflood temperature. This figure demonstrates the superiority of steam over hot water injection at the same temperature.

Cyclic steam injection (Huff and Puff), steamflooding and Steam-Assisted Gravity Drainage (SAGD) have been the most widely used recovery methods of heavy and extra-heavy oil production in sandstone reservoirs during last decades. Thermal EOR projects have been concentrated mostly in Canada, Former Soviet Union, United States and Venezuela, and Brazil. Recently, China has made good progress in thermal EOR. Steam injection began approximately five decades ago. Mene Grande and Tia Juana field in Venezuela, and Yorba Linda and Kern River fields in California are good examples of steam injection projects over four decades. They are considered to be some of the most successful EOR projects of all time. The lessons learned have been immense. However, little of that knowledge has been transferred to conventional light oil recovery processes. There is a one-way disconnection between EOR in heavy oil and EOR in light oil because a great deal of the knowledge from light oil recovery has been transferred to recently developed heavy oil processes, such as steamfloods in the Crude E Field in Trinidad, Schoonebeek oil field in Netherlands and Alto do Rodrigues in Brazil. In addition, heavy oil recovery processes such as VAPEX has used light oil solvent flooding technologies, developed in the 60s and 70s. Ironically, "mistakes" of light oil recovery, particularly when it applies to much discredited chemical flooding, have filtered through

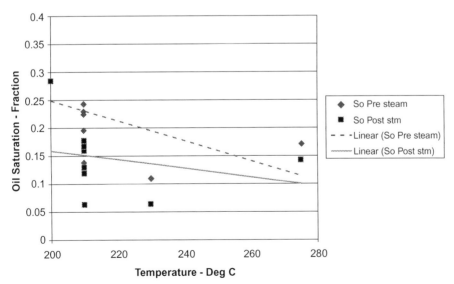

FIG. 4.106 Residual oil reduction with temperature for pre- and poststream flood. *From Bennion, D.B., et al., 2006. A correlation of the low and high temperature water-oil relative permeability characteristics of typical western Canadian unconsolidated bitumen producing formations. In: Paper Presented at the Canadian International Petroleum Conference, Calgary, Alberta, June. Paper Number: PETSOC-2006-136.*

heavy oil recovery schemes. If it wasn't for the subsidy of the government and the tax credits offered to stimulate heavy oil and tar sand recovery, these projects would not be viable.

Capitalizing on the success of steam floods, numerous "improvements" have been suggested for steam-related recovery techniques. They include the use of solvents, gases, chemical additives, and foam in an attempt to control the mobility of the displacement front. Laboratory results shows great recovery potentials of these "novel" techniques. However, similar to chemical flood schemes, field experimentation with this mobility control chemicals have failed to produce satisfactory results. This failure is mainly due to the fact that (1) any use of solvent is deemed uneconomical; (2) it is impossible to control mobility with chemicals beyond a few feet from the wellbore; (3) original steam flood of cyclic steam injection produce significant amount of heavy oil, leaving behind little room for improvement. One example is the LASER (for Liquid Addition to Steam for Enhancing Recovery) process, which consists of the injection of C5+ liquids as a steam additive in cyclic steam injection processes. Although the LASER process was tested at pilot scale in Cold Lake, Canada, the process has not been expanded at a commercial scale. As stated in the previous paragraph, commercial viability of these projects is nil because of inherent issues.

Steam injection has also been tested in medium and light oil reservoirs being crude oil distillation and thermal expansion the main recovery mechanisms in these types of reservoirs. Because light oil reservoirs are often fractured that pose a scenario different from conventionally homogenous formations of heavy oil, considerations must be made in designing steam flood in light oil reservoirs. To be remembered also that light oil reservoirs are already hot and the temperature range from which the maximum decrease in viscosity occurs does not apply to light oil reservoirs. Any heat in the formation will expand the rock/fluid system in such a way that the displacement front is altered. Steam in light oil reservoirs will distillate the crude oil, creating in situ refining. The precipitation of heavier component and ensuing adsorption on the rock surface can change the rock wettability that may favor the oil production. Steam injection in light oil formations does hold promises but has rarely been investigated with rigor.

On the other extreme of the oil viscosity spectrum, Steam Assisted Gravity Drainage (SAGD) has been employed in recovering tar sand in Canada. This process is particularly suitable for unconsolidated reservoirs with high vertical permeability and has become standard in many fields of Canada. Even though SAGD pilot tests have been reported in China, United States, and Venezuela, commercial applications of this EOR process have been reported in Canada only and more specifically those implemented in McMurray Formation, Athabasca (e.g., Hanginstone, Foster Creek, Christina Lake and Firebag, among others). These projects were all subsided by the government of Alberta that spent practically all extra revenues of additional income due to oil boom in the province on these and similar landmark projects. From technological perspective, these projects have been successful. However, their economics have been good only because of the new surge in oil prices. Some argue that they were attractive even when oil price was $12/bbl. These calculations do not account for government subsidies and the tax breaks. A more realistic estimate is the oil price has to be at least $20/bbl for these projects to be economically viable.

Fig. 4.107 shows reservoir depths, average horizontal permeability and formation of several SAGD (pilot and large scale) projects, as documented in the literature. Among these projects, only those developed in McMurray Formation (blue bars of Fig. 4.107) operate

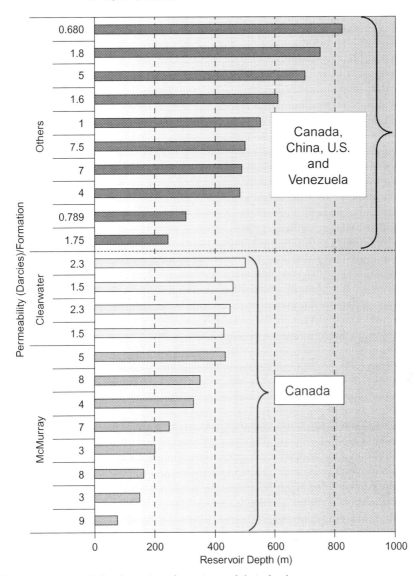

FIG. 4.107 Average permeability for various formation and their depth.

commercially. SAGD projects tested in Clearwater formation in Cold Lake, Canada (yellow bars of Fig. 4.107) have proven to be uneconomic.

Commercial SAGD projects in McMurray formation validate the importance of the geology and reservoir characteristics for this EOR method, findings that have been reported by Rottenfusser and Ranger (2004) among others. For any formation beyond 400 m depth, the nature of vertical permeability is such that the horizontal extent of the steam flood becomes more dominant leading to loss of steam in nonextractable zones. With such loss, economics of the system cannot be attractive. From technical perspective, there is a need to study the lateral extents of SAGD wells so that the fact that vertical permeability is lower than horizontal permeability can be used to the benefit of the project.

With the current level oil prices, it is anticipated that the SAGD processes will continue to expand, mainly in Athabasca's McMurray formation. More research and pilot projects should be done for implementation of SAGD to formations that are deeper than 400 m or have low vertical permeability.

Alternatives to SAGD have been proposed. As stated earlier, most alternatives involve "improvements" with chemicals that are meant to reduce mobility of the displacement phase and/or increase extraction of the oil through mixing with solvents (e.g., VAPEX, SW-SAGD, ES-SAGD). In addition, the well configuration or number of wells is also changed for some applications. As examples, one can cite X-SAGD, Fast SAGD and single well SAGD or SW-SAGD. Well configuration should be designed based on individual formations and typically one should not adhere to a rigid set of well configurations. The use of chemicals, on the other hand, is unlikely to yield positive results because of inherent technical flaws. In addition, they are not sustainable from both economic and environmental aspects.

It is recognized that in-situ combustion (ISC) is the second most important thermal recovery method. Even though ISC has been applied in tar sand and extra heavy oil formations, evidence has surfaced that tells us that it is most applicable to medium heavy or even light oil formations. This new evidence explains why most of the ISC pilot projects have yielded inconclusive or failed pilot results. Heavy oil and tar sands are wrong candidates for ISC. In the last decade, ongoing ISC projects in heavy oil reservoirs such as Battrum Field in Canada, Suplacu de Barcu, Romania, Balol, Bechraji, Lanwa and Santhal in India, and Bellevue in the United States demonstrate that a much better candidate for ISC is medium heavy oil formation.

It is worth noting also that hot air injection is the first EOR scheme known to the modern petroleum industry. It is not well publicized because it was not implemented by design. The injection of air leads to in situ combustion and every oil reservoir is a potential candidate of this application. It turns out that recently popularized High Pressure Air Injection or (HPAI) is only an offshoot of the original hot air injection concept. The successful application of air injection projects in light oil reservoirs like West Hackberry in the US demonstrate that this recovery process is a viable EOR strategy for high dipping angle reservoirs combined with Double Displacement (DDP) strategies. Since 2000, the number of ISC projects has been steady with 10 projects in sandstone formations, whereas the number of HPAI projects in US light oil reservoirs has shown an important increase during the same period (Fig. 4.108). These HPAI projects have been implemented in carbonate formations exclusively.

In-situ combustion in light oil reservoirs doesn't need to be at high pressure. Simple air injection can lead to the onset of the in situ combustion, making the drive turn into an effective recovery technique. There have been several reports on such applications, such as the one reported by Duiveman et al. (2005) and Yu et al. (2008) on air injection projects in Handil Field, Indonesia and Hu 12 Block, Zhong Yuan Field in China, respectively. Although Handil Field HPAI pilot (0.5–1 cP oil) reported injectivity problems due to lack of reservoir communication in the pilot area the results were reported as encouraging. Injectivity problem in this field is most likely due to reasons other than oil viscosity. During injection of air, low temperature oxidation may occur leading to precipitation of plugging agents that wouldn't generally

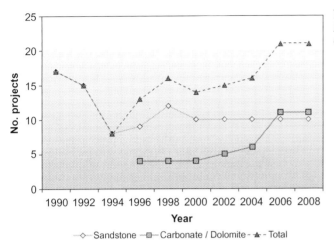

FIG. 4.108 Trends in ISC and HPAI. *From Moritis, G., 2008, Worldwide EOR survey. Oil Gas J. 106, 41–42, 44–59.*

occur under original field conditions. In addition, air injection that doesn't necessarily have a fire front at the leading edge can lead to intense viscous fingering making the process extremely inefficient. This may result in low recovery. However, such instability problem cannot be alleviated with the use of chemicals, mainly because of chemicals do not travel beyond a few meters in the formation. In addition, the use of chemicals is inherently uneconomic and can render the process environmentally unsustainable.

In an attempt to improve air injection, foam assisted water alternating air was used in a pilot project in China (Yu et al., 2008). This reservoir has an oil viscosity of 3.9 cP. The results were reported to be encouraging but it is difficult to determine how much of the result can be assigned to improvements with foam. It is likely that the presence of foam didn't alter the mechanism of water injection alternated with air. An improvement is expected when one combines these two techniques. Other examples can be given from Rio Preto West onshore Brazil reported by Moritis (2008) and studies reported by Hughes and Sarma (2006), Sarma and Das (2009) and Teramoto et al. (2005), and Onishi et al. (2007) evaluating technical feasibilities and potential of HPAI in Australia and Japan, respectively. All these suggest both technical feasibility and future potential of HPAI in light oil formations.

Other alternatives to ISC has been proposed as well. One alternative involves "Toe-to-Heel Air Injection" or "THAI." It is an integrated reservoir-horizontal wells process, which uses air injection to propagate a combustion front from the toe-position to the heel of the horizontal producer. Fig. 4.109 is a schematic representation of the basic features of the process. This process is meant to minimize gravity override.

The stability of the THAI process depends on two key factors: (1) a high temperature burning zone, which is more advanced in the top part of the oil layer, exhibiting controlled (stable) gas override behavior, and (2) deposition of coke, or heavy residue, inside the horizontal producer. The coke which is deposited inside the horizontal producer acts as a gas seal.

THAI is a new, more advanced variant of the conventional in-situ combustion (ISC) process, which operates as a short-distance, as opposed to long-distance displacement process. This is equivalent to SAGD version of steam flooding. Due to the well arrangement used in

FIG. 4.109 Schematic of THAI process. *From Greaves, M., Xia, T., 2004. Downhole upgrading of Wolf Lake oil using THAI/CAPRI processes-tracer tests. Am. Chem. Soc. Div. Pet. Chem. Prepr. 49 (1), 69–72.*

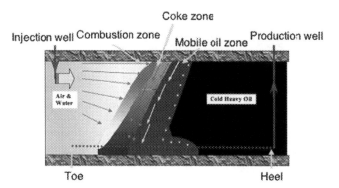

THAI, the mobilized oil ahead of the combustion front only travels a short distance (down) to the exposed section of the horizontal producer. Since THAI operates at much higher temperatures than SAGD, it can achieve significant in-situ upgrading, and thereby maximizing oil recovery. THAI is currently the subject of a pilot development at Christina Lake, Canada.

CAPRI (controlled atmospheric pressure resin infusion) is the catalytic extension of the THAI process, incorporating an annular layer of catalyst, emplaced on the outside of the perforated horizontal producer well, along its whole length. The reaction conditions created ahead of the combustion front, prior to reactants passing down through the mobile oil zone to contact the catalyst, are established by the THAI process. Further upgrading of the produced oil is achieved by catalytic conversion, as the mobilized oil passes through the catalyst layer. This process is the first step toward downhole refining. While this system works well in laboratory, field application of refining technique is economically unattractive. This would not be the case if (1) expensive catalysts are replaced with natural, yet effective catalysts; (2) the produced fluid is considered to be upgraded, thereby being assigned a higher grade at the refinery; (3) custom-designed well placement for each application, depending on formation and fluid characteristics.

Fig. 4.110 shows some of the test results using THAI and CAPRI. The test was combination of dry and wet THAI/CAPRI test, in which water was injected together with the injected air (CAPRI), as tracer during the second wet combustion period. A stable, high temperature combustion front (500–600°C) was propagated along the horizontal producer, during the dry and wet combustion periods. The excellent sweep of the combustion front, in a "Toe-to-Heel" manner, achieved a high oil recovery, at 87% OOIP. The figure shows the variation of the API gravity and viscosity for samples of the produced oil collected during the experiment. Before the combustion front reached the catalyst layer, the degree of thermal upgrading of the produced oil was only about 2 API points. This is very low, compared to a normal THAI test on Wolf Lake heavy oil. This is because no clay was added into the sandpack for this particular trial. The effect of catalyst on the produced oil is clearly evident in Fig. 4.110. The API gravity of the produced oil jumped from an API value of 14, up to 24, during the period of 240–400 min. As the combustion front approached the catalyst section along the horizontal producer, mobilized oil, already partially upgraded in the THAI process, underwent further catalytic conversion reactions. During the second wet combustion mode, with water injection,

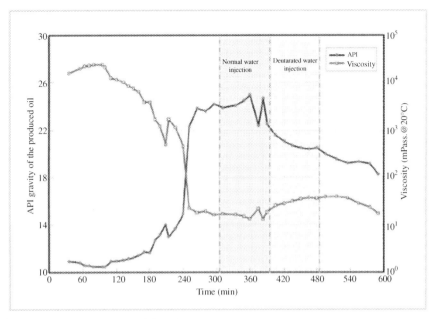

FIG. 4.110   Upgrading with THAI and CAPRI.

the upgrading trend was reduced slightly from 22°API to 20°API. The viscosity of the produced oil achieved by CAPRI was 10–40 mPas, down from the original 24,400 mPas for Wolf Lake Crude Oil.

To seek increased efficiency, other forms of thermal EOR have been proposed. For instance, downhole steam generation (Eson, 1982; Donaldson, 1997), electric heating (Sierra et al., 2001) or electromagnetic heating (Islam and Wadadar, 1991; Das, 2008), and microwave (Hascakir et al., 2008) technologies. Some of these technologies were touted decades ago in the form of electrical heating. Later on, great promises were made with electromagnetic heating. A company, called EOR (Electromagnetic Oil Recovery) was formed in 80s in Calgary. This company implemented a number of field applications of the electromagnetic heating technology. However, none produced satisfactory result and the company went bankrupt. This technique promised to eliminate one of the bigger technical problems with conventional steam injection in regions where permafrost exists. Several options have been tried for the use of electricity to heat oil reservoirs. These methods can be classified according to the mechanism of thermal dissipation that dominates the recovery process (Pizarro and Trevisan, 1990). They range from dielectric heating with high frequency range to radio frequency in the microwave range. Wadadar and Islam (1994) had investigated the possibility of using electromagnetic heating with horizontal wells. Islam and Chilingar (1995) reported the results of a series of numerical simulation tests based on the method originally proposed by Islam and Chakma (1992). They showed that by coupling electromagnetic heating with other EOR schemes, one can increase the recovery of heavy oil or tar sand significantly. To-date, however, not a single field project has been reported to be successful on the use of this technology.

4.5.13 Chemical methods

Chemical EOR methods lived their best times in the 1980s, most of them in sandstone reservoirs. These methods typically promise alteration of rock and/or fluid properties so that irreducible oil saturation is decreased. The total of active projects using chemical peaked in 1986 with polymer flooding as the most important chemical EOR method. However, since 1990s, oil production from chemical EOR methods has been negligible around the world except for China (Han, et al., 1999; Delamaide et al., 1994; Wang et al., 2002). Nevertheless, chemical flooding has been shown to be sensitive to volatility of oil markets despite recent advances (e.g., low surfactant concentrations) and lower costs of chemical additives. Technically polymer alone doesn't increase oil recovery efficiency. In addition, polymer doesn't propagate within the formation beyond a few meters, mainly because polymer adsorption rate is very high and if polymer concentration is increased, one runs into serious injectivity problem. Polymer seems to be effective when combined with surfactants and sacrificial chemicals. However, cost of such process remains prohibitively high.

Polymer flooding has been tried for a long time and is considered to be a mature technology and still the most important EOR chemical method in sandstone reservoirs based on the review of full-field case histories. Even though all forms of chemical methods have been practically abandoned in the United States, According to the EOR survey presented by Moritis in 2008, there are ongoing pilots or large-scale polymer floods in Argentina (El Tordillo Field), Canada (Pelican Lake), China with approximately 20 projects (e.g., Daqing, Gudao, Gudong and Karamay fields, among others), and India (Jhalora Field). It is important to mention that a commercial polymer flood was developed in North Burbank during the 1980s, demonstrating that this EOR method may still have potential to increase oil recovery in mature basins (i.e., mature floods with movable and/or by passed oil). North Burbank reinitiated polymer flooding on a 19-well pattern in December. Other reported polymer flooding projects include Brazilian Carmópolis, Buracica and Canto do Amaro fields. India also reports a polymer flood in Sanand Field. Oman documented a polymer flood pilot developed in Marmul Field and almost 20 years later a large-scale application is under way (Moritis, 2008). In addition, Argentina (El Tordillo Field), Brazil (Voador offshore Field), Canada (Horsefly Lake Field) and Germany (Bochstedt Field) announced plans to implement polymer flood projects.

Colloidal Dispersion Gels (CDG's) and BrightWater also represent novel polymer-based technologies that are currently under evaluation at field scale. These chemicals are meant for mobility control of a displacement drive. By injecting these mobility control agents, reservoir heterogeneities are homogenized, thereby avoiding channeling or fluid loss in unproductive areas. Documented CDG's projects include Daqing Field in China, El Tordillo and Loma Alta Sur fields in Argentina and in multiple US oilfields. Regarding BrightWater, at the present time Milne Point in Alaska is the only field application discussed or documented in the public domain. While it is expected that the number of CDG's and BrightWater field applications will increase in the near future based on recent field and laboratory studies underway, none of them is expected to give positive results. If the oil price continues to be high, producers would be satisfied with the investment, irrespective of the actual benefit of the recovery scheme.

While polymer flooding has been the most applied EOR chemical method in sandstone reservoirs, the injection of alkali, surfactant, alkali-polymer (AP), surfactant-polymer (SP) and

Alkali-Surfactant-Polymer (ASP) have been tested in a limited number of fields. In this application, alkali plays the role of a sacrificial agent. New genre of surfactants have been developed that reduce the dynamic interfacial tension to a very low number (Islam and Farouq Ali, 1990; Taylor et al., 1990). These surfactants are highly unstable extremely toxic, and exuberantly costly.

As mentioned earlier, micellar polymer flooding had been the second most used EOR chemical method in light and medium crude oil reservoirs until the early 1990s. Although this recovery method was considered a promising EOR process since the 1970s, the high concentrations and cost of surfactants and co-surfactants, combined with the low oil prices during mid-1980s limited its use. The development of the ASP technology since mid-1980s and the development of the surfactant chemistry have brought up a renewed attention for chemical floods in recent years, especially to boost oil production in mature and waterflooded fields.

Several EOR chemical methods, other than polymer flood, have been extensively documented in the literature during the last two decades. However, at the present time Daqing Field represents one of the largest, if not the largest, ASP flood implemented as of today. ASP flooding has been studied and tested in Daqing for more than 15 years though several pilots of different scales. According to the EOR survey presented by Moritis in 2008, there are ongoing ASP pilots in Delaware Childers Field (Oklahoma) and planned ASP floods in Lawrence Field (Illinois) and Nowata Field (Oklahoma), and SP floods in Midland Farm Unit, Texas (Grayburg Carbonate Fm.) and in Minas Field, Indonesia. The surging number of the chemical projects, however, doesn't tell the full story. There are many other unreported cases that are either in the plan or have been implemented but have not been reported because of lack of success. Fundamentally, ASP or any other chemical EOR technique is economically unattractive and environmentally disastrous. This can be changed only by resorting to other nonconventional sources of chemicals that are either naturally available in local areas or are liability of an operation site because it is a waste of by product of other activities.

4.5.14 Gas injection

Even though CO_2 injection has been discussed in a previous section, this section is introduced to familiarize readers with the basic of gas injection projects. EOR gas flooding has been the most widely used recovery methods of light, condensate and volatile oil reservoirs. Both miscible and immiscible gas injection schemes hold tremendous potential for future applications in the mainly untapped reservoirs of the world, especially in regions where the reserve/production rate ratios are high. However, a gas injection process can be severely flawed if the displacement front is not stabilized. It is essential to take advantage of the gravitational forces. Recently, horizontal wells have been proposed to enhance gravity segregation while maintaining stable displacement fronts.

Numerous field reports in Canada (Shell and Husky Oil) show that horizontal wells can be used to successfully conduct stable miscible displacement of both light and heavy oil. The key to success in miscible or immiscible gas injection appears to be the accurate prediction of the frontal stability, which is very sensitive to reservoir heterogeneity.

Another method of reducing viscous fingering during gas injection is the use of foam. The foam increases the viscosity of the displacing gas phase to the extent that an otherwise unstable front (with gas only) can become stable. In addition, foam has a homogenizing effect when the displacement front encounters a heterogeneous spot in the reservoir. Even though no study has been reported on the topic of frontal stability in a heterogeneous medium, foam is likely to eliminate some of the problem associated with frontal instability due to heterogeneity.

Launched in 2000, the Weyburn-Midale CO_2 Project in Saskatchewan, Canada, is the world's largest full-scale, in-field study of CO_2 injection and storage in depleted oil fields. When completed, the 11-year International Energy Agency project (funded in part by DOE) will permanently store 40 million metric tons of CO_2 while increasing oil production by 18,000 barrels per day.

Although nitrogen (N_2) injection has been proposed to increase oil recoveries under miscible conditions favoring the vaporization of light fractions of light oils and condensates, today few N_2 floods are ongoing in sandstone reservoirs. Immiscible N_2 floods are reported in Hawkins Field (Texas) and Elk Hills (California) based on the Moritis EOR survey in 2008. No new N_2 floods in sandstone reservoirs have been documented in the literature during the last few years. High pressure air injection schemes and their success tell us that N_2 injection is not an effective option and certainly not economically and environmentally sustainable.

Hydrocarbon gas injection projects in onshore sandstone reservoirs have been employed for many decades. Initially, they were used as a means of pressure maintenance to arrest the pressure decline that occurs in any naturally depleting reservoir (Fig. 4.111). These projects made a relatively marginal contribution in terms of total oil recovered in Canada and the United States other than on the North Slope of Alaska, where large natural gas resources are available for use that do not have a transportation system to markets. Conventionally, the term "EOR" gas methods include mainly hydrocarbon gases such as Water-Alternating-Gas (WAG) injection schemes, enriched gases or solvents and its combinations. Even though pressure maintenance is not considered to be an EOR technique, the process remains the same and therefore one must investigate the possibility of additional oil and gas recovery with pressure-maintenance. Most of immiscible and miscible EOR hydrocarbon gas floods in the United States are on the North Slope of Alaska while in Canada a miscible gas flood is reported in Brassey Field. The situation of hydrocarbon gas injection projects is different in offshore sandstone reservoirs. This aspect will be discussed in a latter section. Conventionally, it is said that if there is no other way to monetize natural gas, then a more practical use of

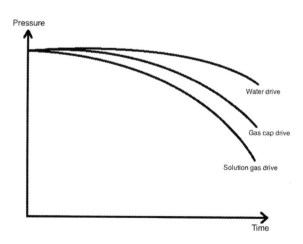

FIG. 4.111 Pressure maintenance program involves artificially boosting pressure-time curve.

natural gas would be to use it in pressure maintenance projects or in WAG processes. However and if available, the substitution of hydrocarbon gases by nonhydrocarbon gases (N_2, CO_2, acid gas, air) oil recovery will make more natural gas available for domestic use or export while still maintaining reservoir pressure and increasing oil recoveries. This recommendation applies to both EOR and EGR.

On the other hand, CO_2 flooding has been the most widely used EOR recovery method for medium and light oil production in sandstone reservoirs during last decades, especially in the United States due to the availability of cheap and readily available CO_2 from natural sources. There has been an increasing trend in the number of CO_2 field projects in the United States during the last decade in both, sandstone and carbonate reservoirs.

The number of CO_2 floods is expected to continue to grow in US sandstone reservoirs. Some examples of planned CO_2-EOR projects in the United States include Cranfield Field, Heidelberg West (from anthropogenic sources) and Lazy Creek Field in Mississippi and Sussex Field in Wyoming. Number of CO_2 floods in Wyoming sandstone reservoirs are also expected to increase based on a recent evaluation presented by Moritis (2008). This particular project depends on the CO_2 availability. In all these projects, availability of CO_2 is considered as the determining factor of EOR application.

Additionally, Holtz (2008) reported an overview of sandstone gulf coast and Louisiana CO_2-EOR projects to estimate EOR reserve growth potential in the area including sandstone reservoirs in the Gulf of Mexico. Table 4.26 summarizes major features of certain CO_2 EOR projects. Moritis (2008) EOR survey reports up to nine (9) active immiscible CO_2 floods operating since mid-1970s. The experience of various countries represent different lessons that can be learned. For instance, Midale project in Canada uses trucked CO_2 from another Canadian field, whereas Weyburn projects imports CO_2 that is specifically generated for this project and pipelined (over 200 km) from the United States. Yet, Weyburn is the only project

TABLE 4.26 Selected projects involving CO_2 injection.

Country	Type of formation	Field name	CO_2 source	Incentive
Brazil	Sandstone	Buracica and Rio Pojuca	Anthropogenic, ammonia plant	EOR and storage
Canada	Sandstone	Pembina and Joffre	Anthropogenic	EOR
	Limestone	Weyburn	CO_2 from coal burning	EOR and storage
	Limestone	Midale	Transported truck	EOR
				PTAC EOR EGR storage multipurpose
Croatia	Sandstone	Ivanić Field	Transported truck	EOR
Hungary	Sandstone	Budafa and Lovvaszi	Anthropogenic	EOR
	Sandstone	Szank	Sweetening plant	EOR and storage
Trinidad	Sandstone		Ammonia plant	EOR

worldwide that has the classification of being an EOR and Sequestration project simultaneously. The experience of Trinidad is noteworthy. This project uses waste gas from a nearby Ammonia plant. As discussed earlier, waste gas from a chemical plant represents very high economic boon as well as environmental sustainability. Yet, this project is not considered to be a model for greenhouse gas sequestration. Canada's PTAC project represents the most comprehensive application of CO_2 projects. In terms of CO_2-EOR, Alberta projects 3.6 billion barrels additional oil recovery over the next two decades. At the same time, a significant amount of greenhouse gas would be sequestered. In addition, new industries that make use of CO_2 as a commodity would be developed.

Due to the lack of rigorous scientific investigation, the designs of EOR projects involving CO_2 or other greenhouse gases have been flawed from technical, environmental, and economical perspective. For instance, on the technical side, most flow rates selected, even with horizontal wells, are high enough to induce viscous fingering, with the only exception being the projects involving injection of gas from the top of an anticline or highly dipped formation. The problem is further accentuated for a heterogeneous formation or a formation previously flooded with water or with high watercut. Flow instability in these cases diminish or eliminate the possibility of maintaining a stable front. All laboratory experiments, however, are conducted under stable conditions, thereby representing optimistic conditions that will not prevail in field. Any field or pilot design based on these experiments is likely to yield disappointing results (Islam et al., 2010, 2012).

Based on laboratory tests, it is often proposed that pure CO_2 be used for maintaining miscibility. Because of flawed definition of environmental integrity, it is suggested that the use of purified CO_2 and sequestration of it would be beneficial to the environment. A truly scientific criterion would indicate that processing CO_2 makes it lose its natural properties that make it a principal player of the photosynthesis process (Khan and Islam, 2007b; Chhetri and Islam, 2008). As a consequence, "purified" CO_2 is toxic to the environment at the same time being costly. Ironically, the need for pure CO_2 arises from the premise that miscibility prevails in the reservoir—a premise that doesn't apply to most CO_2 applications. In presence of immiscible flow, there is no need for maintaining high concentration of CO_2 in the injected gas. In addition, if miscibility is not achieved, there is little advantage to having pure CO_2, and even waste gas, produced gas, and so on would suffice. Pure CO_2 in high watercut zones would only result in a loss of valuable CO_2. On the other hand, the injection of greenhouse gas or produced gas would increase accessibility of the untapped oil.

The economics of any EOR project is unacceptable with expensive chemicals (e.g., pure CO_2, surfactant, polymer, and alkali). While pure CO_2 is technically capable of recovering additional oil, it has been demonstrated through numerous field trials, conventional chemical floods do not yield acceptable results. This is because, these chemicals do not travel more than a few meters within the reservoir. This is the reason, chemical flooding has failed to recover any additional oil and the chemical techniques have been discontinued for several decades. Today, only China applies chemical flooding techniques and only a handful of operations have been tested in the North Sea. This is mainly because these operators produce their own chemicals.

In terms of emerging technologies, High Pressure Air Injection (HPAI) is the method with greatest potentials. This method combines the positive effects of CO_2 injection (through oxidation of in situ oil) as well as thermal methods. In case, CO_2 or other gases are not locally available, HPAI should be considered.

The following sequential screening is recommended.

(1) Screen the type of fluid (gas or water) available
(2) Consider stability with both water and gas
(3) Consider miscible injection only with stable cases (in presence of natural dip for which CO_2 or other gases can be injected from a structurally higher position)
(4) Consider reinjection of produced gas (including sour gas in original concentration), flue gas, and finally air, depending of availability.
(5) Laboratory tests should be performed using the above fluids and not using idealized fluid.
(6) Numerical simulation should be considered only for well placement, injection protocols and similar strategic issues. One must note that reservoir simulators are incapable of modeling unstable flow.

For scaling an EOR process or to determine stability of the displacement front, the following steps are required:

(1) Determine the end-point permeabilities.
(2) Determine the capillary pressure curve.
(3) Determine interfacial tensions.
(4) Estimate flood pattern dimensions.
(5) Estimate frontal velocity from the injection well.
(6) Calculate the gravity number.
(7) Calculate the instability number.
(8) If the instability number is less than π^2, follow the conventional approach (velocity matched with that of the field as per the scaling requirement).
(9) If the instability number is greater than π^2, calculate the laboratory velocity such that the instability number in the laboratory matches with that of the field.

Khan and Islam (2007b) give full details of these steps, including the definition of the instability number, capillary number, mobility ratio, and so on.

It is important to note that, for gas injection, the instability number continues to affect the recovery, meaning the higher the flow rate the less recovery there will be. This is because the pseudo-stable regime is never reached with gas. Gravity plays an important role during gas injection because the value of the gravity number can stabilize a process.

During miscible displacement, the displacement front develops a transition zone that can vary in length significantly. If the crude oil in question is not light, the transition zone can be much wider. The problem with a wide transition zone is that the miscibility can be lost altogether. The lack of miscibility or the extension of the transition under any displacement situation would translate into an inadequate sweep of the reservoir, resulting in low oil recovery. The effect of the transition zone length has not been studied in the past. Inherently related to this problem is the storage or mitigation aspect of CO_2 displacement. Unless efforts are made to define the miscible/immiscible system, the performance prediction is bound to be inaccurate. In addition, it is important to predict the sustenance of a miscible front. The lack of miscibility may in turn lead to the onset of viscous fingering.

4.5.15 Carbon sequestration enhanced gas recovery (CSEGR)

It was discussed earlier that EGR has great potentials, despite being ignored as a commercial project. It is also presented that Alberta's CO_2-backbone template includes EGR as part of overall CO_2 sequestration and oil and gas recovery. This section presents further discussion on EGR.

It has been recognized for decades that depleted natural gas reservoirs are promising targets for carbon dioxide sequestration. However, the same reservoirs are not devoid of methane. In addition, the possibility of using CO_2 sequestration to enhanced gas recovery should be investigated.

The CSEGR process consists of collecting CO_2, for example by scrubbing CO_2 from flue gases at fossil-fueled power plants or collecting by-product CO_2 from refineries, pressurizing the CO_2 to supercritical conditions for transport in a pipeline, transporting the CO_2 to a depleted natural gas reservoir, injecting the CO_2 into the reservoir, and enhancing the production of CH_4 from the reservoir. After some period of enhanced CH_4 recovery, the production wells would be sealed and the reservoir would be filled with CO_2 up to initial reservoir pressure. The injected CO_2 would then be sequestered in the gas reservoir just as CH_4 was stored over geologic time prior to its production as an energy resource. A schematic illustrating the CSEGR process for a gas-fired power plant is shown in Fig. 4.112.

Even though depleted natural gas reservoir is cited above, the technique equally applies to active gas reservoirs. The use of a power plant is a quick utilization natural gas. Depending on the size of the reservoir or the overall flow rate, the utilization scheme can be any other

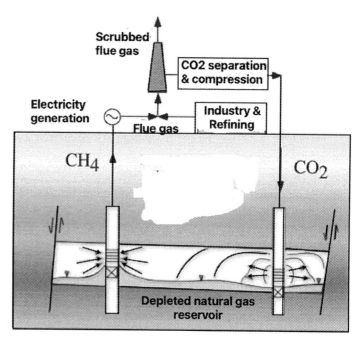

FIG. 4.112 Various components of CSEGR.

potential use, such as fertilizer, cement factory, and so on. The idea is to maximize the value addition of the natural gas.

The flue gas typically is processed to capture high quality CO_2. Once again, purification of CO_2 doesn't need to be carried out as most reservoirs are amenable to reinjection of low-quality CO_2 that will offer similar benefit as pure CO_2 but without having to spend more on CO_2 purification. As discussed earlier, purification with expensive and toxic solvents actually increases the footprint of the process, thereby defeating the purpose of CSEGR.

During the injection process, following factors must be considered:

(1) injectivity of CO_2 in a gas reservoir;
(2) effects of CO_2 injection pressure on injectivity and flow;
(3) cooling around the injection well due to phase change and Joule-Thomson effects;
(4) flow of CO_2 within the reservoir;
(5) mixing of CO_2 and CH_4 in the reservoir; and
(6) repressurization and production of CH_4.

Conventionally, CO_2 is injected under supercritical conditions. This has implication on the design of compressor. The supercritical condition is also subject to strong cooling at injection port due to (1) flashing of supercritical liquid-like CO_2 to gas, and (2) Joule-Thomson cooling as the CO_2 gas expands in the low pressure reservoir. In addition, the injected CO_2 dries the formation, another potential heat consuming process. Given that the formation and residual gas and liquid are at somewhat elevated temperature ($T > 40°C$), heat will be available for the expanding gas. Eventually however, the temperature around the well may become quite low leading to the possibility of hydrate formation and associated decreases in injectivity. Pure carbon dioxide hydrate can form at approximately 0°C at 20 bars pressure.

Assuming there is sufficient permeability, the injected CO_2 should flow in the reservoir due to pressure gradient and gravitational effects. If there is liquid CO_2 immediately around the wellbore, it will flow strongly downward through the gas reservoir due to its large density. Such gravity segregation must be accounted for during considerations of the well placement in a structurally inclined formation. It is preferable that high-pressure CO_2 be injected through a structurally lower well so that gravity stabilization of the front takes place. Once flashed to gas, CO_2 is also notably denser than CH_4 at all relevant pressures (see Fig. 4.113) and will tend to flow downwards, displacing the native CH_4 gas and repressurizing the reservoir. Because CO_2 gas is more viscous than CH_4, the displacement will be stable.

The reservoir processes of CO_2 injection and enhanced CH_4 production are shown schematically in Fig. 4.114. As observed in the figure, CO_2 injection can deflect the water table, giving rise to repressurization at a large distance from the injection well. CO_2 can be used effectively to minimize water coning. Selection of injection well should be made in such a way that water coning is minimized while gas mobilization and displacement are maximized. Heterogeneity in the formation may lead to preferential flow paths for the injected CO_2. This phenomenon may be favorable for injectivity and carbon sequestration in that it allows greater amounts of CO_2 to be injected. However, preferential flow may lead to early breakthrough and is therefore detrimental to enhanced gas recovery. Furthermore, the development of larger gas composition gradients and subsequent mixing by molecular diffusion is enhanced by preferential flow.

4. Unconventional reservoirs

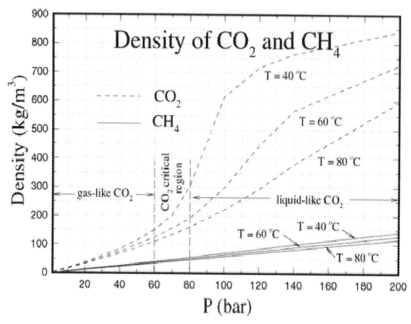

FIG. 4.113 Density of CO_2 and CH_4 as a function of pressure for various temperatures.

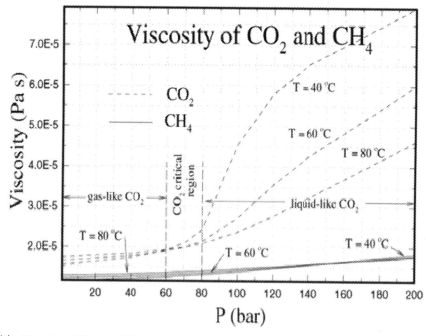

FIG. 4.114 Viscosity of CO_2 and CH_4 as a function of pressure for various temperatures.

Heterogeneity in the gas reservoir plays a dual role during CO_2 injection. CO_2 gains access to high permeability zones, thereby increasing injectivity. On the other hand, heterogeneity may lead to channeling or fingering, which would lead to early breakthrough. This case can be alleviated by reducing the injection pressure or by using horizontal well for gas injection. In case there is a dip, CO_2 should be injected through the lower section of the reservoir to take advantage of gravity.

Prior reservoir simulation studies show favorable results for CSEGR in terms of the feasibility of reservoir re-pressurization and enhanced production of CH_4. While simulations demonstrate that reservoir heterogeneity promotes early breakthrough, standard techniques for controlling preferential flow may be used in the field where required. Our simulations for a particular system, combined with the inherently favorable attributes of CO_2 in general, such as a comparatively high density and viscosity, suggest that CSEGR is feasible. Further progress in evaluating the feasibility of CSEGR requires field testing. We propose a Phase I field test to investigate the reservoir processes involved in injecting high- pressure CO_2 into depleted natural gas reservoirs. Key processes will be the injection of CO_2 and associated cooling of the formation, flow and transport of CO_2 gas through the reservoir, re-pressurization of the reservoir, and the enhanced production of CH_4. The Rio Vista Gas Field in California is a promising target for the field test, however, we will consider other locations if they have better potential to satisfy the objectives of the pilot study.

Heterogeneity in the formation may lead to preferential flow paths for the injected CO_2. This phenomenon may be favorable for injectivity and carbon sequestration in that it allows greater amounts of CO_2 to be injected. However, preferential flow may lead to early breakthrough and is therefore detrimental to enhanced gas recovery. Furthermore, the development of larger gas composition gradients and subsequent mixing by molecular diffusion is enhanced by preferential flow.

In case, flow instability occurs, adheres to high pressure injection should be avoided. Instead of trying to achieve miscibility at high pressure, it is recommended that immiscible injection at lower pressure conditions be planned. Numerical simulation results shows that such scenario will eventually produce much of the residual natural gas as long as well placement is designed with flow instability in mind. The use of chemicals to viscosity CO_2 or to create gels in situ should be avoided. Field trials of these schemes under the auspices of oil production have been dismal.

One aspect of CO_2 sequestration is rarely considered. If CO_2 is injected in natural porous media (limestone or sandstone), its properties are altered after it is stored. In case the injected CO_2 has contaminants, such as heavy metals or toxic additives (e.g., glycol, DEA, TEA), the porous medium is likely to adsorb the contaminants, purifying CO_2. The resulting CO_2 is more amenable to photosynthesis than CO_2 that is produced during combustion of refined crude oil (Fig. 4.115).

There has been just one published field test of EGR. This test was conducted during 1986–94 in the Hungarian field Budafa Szinfelleti, a weak water drive sandstone reservoir of 5–40 mD permeability (Papay, 2003). For that pilot study, EGR started when the natural gas recovery was 67% OGIP and the injected gas was an impure CO_2 stream, consisting of 80% CO_2 and 20% CH_4 from an adjacent natural CO_2 pool.

The incremental gas recovery was 11.6% OGIP, or 35% of the gas in place at the initiation of CO_2 injection. CO_2 breakthrough occurred 1.5 years after the start of injection; the distance

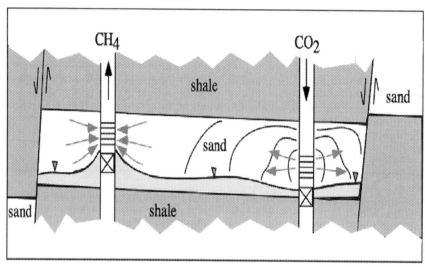

FIG. 4.115 Cross section of the CSEGR site.

between injection and production wells was 500 m. This recovery is in sharp contrast to 70% recovery expected from core flood tests. The discrepancy can be explained through the existence of viscous instability and/or loss in sweep efficiency due to heterogeneity. Because the density of CO_2 is 3–9 times higher than that of CH_4, instability would occur if CO_2 is injected from the top of a formation.

Turta et al. (2007) reported laboratory test results of sweep efficiency of CO_2 and flue gas in an EGR scheme. These results are shown in Table 4.27. This table confirms previous postulation that flue gas can significantly improve the economics of the EGR process due to its relative abundance and lower cost as compared to pure CO_2. In addition, the delay in the production of CO_2 when flue gas was used (as the displacing agent) means that the equipment is less vulnerable to corrosion. The problem will arise in the fact that the flue gas contains nitrogen

TABLE 4.27 Comparison between flue gas and CO_2 efficiency.

Test #	Injection gas	S_{wi} %	Methane recovery at 1% contamination % OGIP	Methane recovery with 10% contamination % OGIP	Methane recovery at 20% N_2 contamination % OGIP
11	CO_2	18	61	71	—
12[a]	14% CO_2 in N_2	18	66	76	84
13	14% CO_2 in N_2	20	75	82	87
18[a]	CO_2	19	62	70	—

[a] Velocity reduced.

concentration higher than the allowable limit in pipelining. This is not a technical issue but an issue of policy, which needs to be revised for unconventional gas applications.

Everything about unconventional gas is challenging, mainly because conventional tools of characterization do not apply to unconventional gas. From origin to digenesis, that is the overall history of most unconventional reservoirs is different from conventional ones. Only tight gas and tight oil formations are equivalent to low-permeability sandstone. As seen in the previous chapter, tight gas and oil are low-permeability version of sandstone. However, origin, depositional setting, stratigraphy, structure, geochemistry, geomechanics, seismic character, and petrophysical properties are different for other unconventional reservoirs. Some of the biggest challenges are:

- understanding the origin of the gas, i.e., how and where these rocks are charged with gas;
- determine a pattern of the highly-productive "sweetspots";
- determine what factors influence the large variation in both drainage area and permeability; and
- determine of a pattern of fracture distribution and orientation

Even though it is commonly assumed that shale gas is also the source rock, there are numerous scenarios that have shale gas interbedded with sandstone or even limestone. Determining the origin of this gas is not a trivial task. The age of this gas can be determined by correlating gas properties with time. Arising from the premise that natural processing increases the quality of petroleum fluids, a scenario depicted in Fig. 4.116 can be developed. The actual numbers in this figure will arise from database of the region. In absence of such data, geological age of the rock can be determined and gas property data be evaluated to assess the origin of the gas.

Because the processes involving the formation of unconventional gas are different, the quality at the beginning of the gas generation is different. Even if it is assumed that the original gas is biogenic for each of these cases, gas hydrate methane is the purest form of methane form the beginning because the original gas is purified with water and only a pure form of methane gas can form hydrate structures. This process is diametrically opposite to the

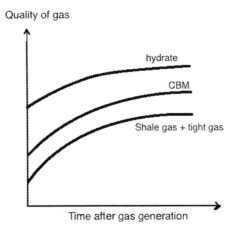

FIG. 4.116   Variation of gas qualities as a function of time.

formation of diamond, albeit in a different time scale. This discussion of diamond formation is relevant for both explaining purity of gas hydrate methane as well as the formation of coal bed methane.

Coal is considered to be the fossilized form of large volumes of biomass. One theory suggests that coal is formed in presence of water and with a low thermal gradient. Any gas that forms has the ability to escape or remain absorbed in the aqueous phase as well as adsorbed in the coal mass. Diamond, on the other hand, is formed when the cooling rate is much faster. Crystals are formed with a very high bonding energy without water molecules. As a result coal oxidizes at a relatively low temperature (similar to biomass oxidation temperature). On the other hand, diamond only oxidizes at a very high temperature, making it impractical to be an energy source for human consumption.

Volcanic rocks can be yet another form of unconventional oil/gas source. Vegetation entrained in ash flows may contain enough water to protect it from heat of emplacement. Subareal volcanism may create lakes and swamps with kerogen-rich sediments, and the volcanically warmed water in these basins can trigger nutrient growth, further enhancing the production of organic material. While there has been little consideration of exploring the potential of volcanic oil and gas reservoirs, it is well known that many conventional reservoirs have volcanic bedrocks that enhance oil and gas recovery from conventional reservoirs. Once hydrocarbons are found in igneous reservoirs, assessing hydrocarbon volumes and productivity presents several challenges. Volcanic reservoirs are extremely hard. Seismic and logging fail to characterize these formations. The difficulty arises from the fact that volcanic rocks act as a shield for transmission of practically all signals that are used in logging. Log interpretation in igneous reservoirs often requires adapting techniques designed for other environments. For instance, resistivity logs that are based on conventional reservoirs will be less suitable for highly resistive matrix, low porosity, shale-free clean formations. Similar anomalies will occur with density log, neutron logs. Also, magnetic elements are more due to igneous origin. So NMR tool is ineffective in such formations. While sonic logs have better chance of being functional, volcanic rocks are so highly reflective, they wouldn't allow these signals to travel through.

Furthermore, because mineralogy varies greatly in these formations, methods that work in one volcanic province may fail in another. Usually, a combination of methods is required. In terms of porosity, direct measurement of volcanic rock porosity is the most preferred.

The formation of volcanic rocks and entrapment of fluid within it can be explained by following the history of volcanic rocks. Fig. 4.117 depicts formation of volcanic rocks. Emplacement of igneous rocks. Plutonic rocks, formed by cooling of magma within the Earth, display well-developed crystals with little porosity. Plutons and laccoliths—bulging igneous injections into sedimentary layers—are examples of plutonic rock. Volcanic rocks, formed when magma extrudes onto the surface and cools rapidly, show very fine crystalline or even glassy textures. While during the cooling process, cooling of volcanic fluids either underground or at the surface, and agglomeration of fragments ejected during explosive eruptions will develop different tendency of fracturing. This explains why volcanic rocks do not show well defined patterns in terms of fractures. The onset of fractures itself is distinctly different in volcanic rocks from others, such as sedimentary rocks. When magma contains large amounts of water and gases, buildup of excessive pressure will cause sudden eruption, leading to the development of fissures and eventual fractures. During cooling of volcanic rocks, tremendous

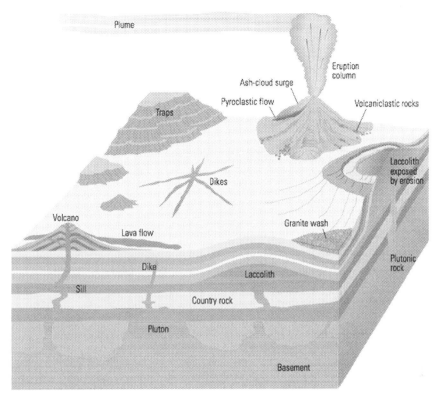

FIG. 4.117 Formation of volcanic rocks. *Courtesy Schlumberger. 2009. Oilfield review. Spring 21 (1).*

restoring force is developed internally because rock is a poor thermal conductor and there is differential stress that develops along the temperature variation line. This restoring force or stress depends on the nature of cooling. Nature and tendency of internal force guides tendency or direction of fracturing.

During the cooling process the neighboring rocks are also affected. Volcanic eruption near sedimentary basins will intrude into sediments extending over several kilometers and during cooling will create fractures and subsequently faults. This process essentially creates the basis for cyclic nature of fault, fracture, fissures and faults. Fig. 4.118 shows the cyclic nature of the process. In the subsurface, volcanic events trigger earth quakes that can create faults within consolidated rocks. Any fault is accompanied with fissures and cracks that eventually can grow into fractures. These fractures themselves make a rock system vulnerable to faults that lead the way to volcanic fluid invasion and earth quake.

This is in conformance with what has been known as burial-thermal diagenesis. The diagenetic evolution of the middle member of the Bakken Formation is depicted in conjunction with the reconstructed burial-thermal history of the unit in the deep part of the Williston Basin (Fig. 4.119). The burial-thermal model takes into account differences in deposition and erosion, as well as variations in thermal regime during the basin's history. Thermal data constrain the

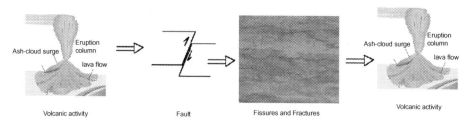

FIG. 4.118 Volcanic activities and their impact on sedimentary rock create cycles of tectonic events.

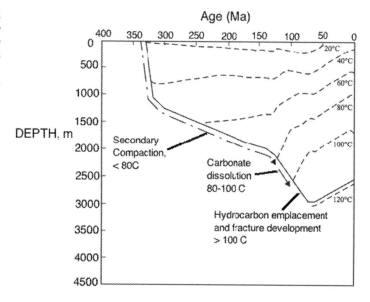

FIG. 4.119 Reconstructed burial and thermal history curves of the Bakken Formation showing relative timing of major diagenetic events in the deep part of the Williston Basin. *Redrawn from Schlumberger. 2009. Oilfield review. Spring 21 (1).*

maximum amount of erosion during the late Tertiary to about 500 m, which agrees closely with previously reported erosional estimates.

Fig. 4.120 highlights the permeability-porosity characteristics of volcanic and conventional rocks. Volcanic rocks have much lower porosity of the matrix and the permeability of this rock is much higher than permeability encountered in conventional reservoirs. In contrast, other unconventional gas reservoirs, such as tight gas and shale will have higher porosity but lower permeability. Of course, this is due to the fact that shale and tighter formations have high surface area but little effective porosity. It is, therefore, difficult to penetrate shale gas and tight sand altogether and similarly it is difficult to penetrate the matrix of a volcanic reservoir. In terms of fluid production, shale gas or tight sand form a typical "permeability" as shown in Fig. 4.120.

While Fig. 4.120 highlights the difficulties associated with EGR, it also unlocks potential techniques that would work in such situations. It is known that fracturing as well as drilling horizontal wells recover significant amount of oil and gas from shale and tight sand. Both these processes increase gas saturation within fractures and the wellbore, thereby increasing

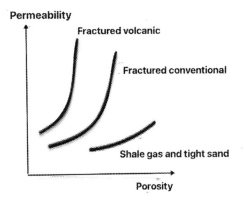

FIG. 4.120 Fractured volcanic reservoir properties compared with fractured conventional.

the relative permeability to gas. The same effect can be invoked by using in situ combustion (or equivalent to High Pressure Air Injection HPAI) or selected gas injection (e.g., with CO_2, or flue gas).

For very tight but fractured formations, the possibility of using cyclic thermal fluid injection should be considered. Because diffusive heat and mass transfer is the most prominent mechanism prevalent in these formations, hot gas and/or steam injection can be beneficial. Heat injection can also trigger thermal cracking.

Fig. 4.121 also shows how injection of gas can move the actual placement of a reservoir toward the left side, rendering the gas phase mobile. The presence of fissures and fractures can facilitate this process, as gas injectivity is moderate under such conditions.

For CBM, the most effective recovery technique is production of water that contain almost the entire amount of methane in absorbed state. The possibility of enhanced gas recovery for CBM is not well explored. However, it is well known that the production can be greatly enhanced with multilaterals, which have been widely successful in recent years. It is also well known that the addition of heat can increase the sweep efficiency of water tremendously. Steam injection in such cases is not economically feasible. However, in situ combustion holds great promises. Coal burning is slow and conforms to the slow mode of steam generation

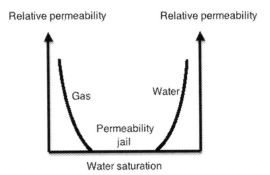

FIG. 4.121 Existence of "permeability jail" in tight gas and shale formation.

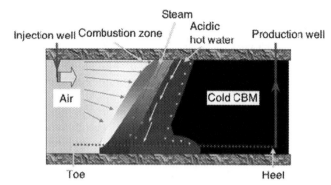

FIG. 4.122   In situ combustion in coal bed methane.

within the reservoir. Combustion of coal will lead to the formation of CO_2 that itself will enhance the displacement of gas as well as penetration of fractures. Fig. 4.122 shows how such scheme would work.

Similar scheme is also applicable to oil-bearing volcanic reservoirs. However, for such reservoirs, the fuel is the hydrocarbon and not the matrix itself as is the case for CBM.

For gas hydrate formations, typically thermal, depressurization, and chemical stimulation have been considered for enhancing gas recovery. Among these, chemical stimulation is the least desirable. Chemical stimulations are economically impractical and the efficiency of the process is poor. Depressurization is effective if in situ pressure is high. Otherwise, depressurization doesn't yield reasonable gas production. Fig. 4.123 shows various types of gas traps within a gas hydrate reservoir. Two of the three types of traps are structural whereas the third one is strategraphic. In case, there is a gas trap, it become easier to initiate depressurization by putting the trap gas in production. This production destabilizes the hydrate zone and stream of gas is released making the process commercially viable. In case the volume of the trapped gas is not sufficient to create instability within the hydrate body, an innovative technique

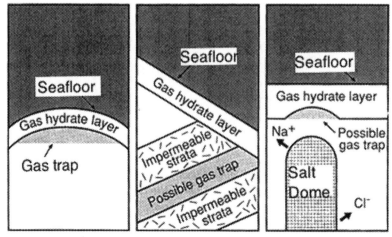

FIG. 4.123   Different types of gas traps. *From Kvenvolden, K.A., 1993. Gas hydrates—geological perspective and global change. Rev. Geophys. 31 (2), 173–187.*

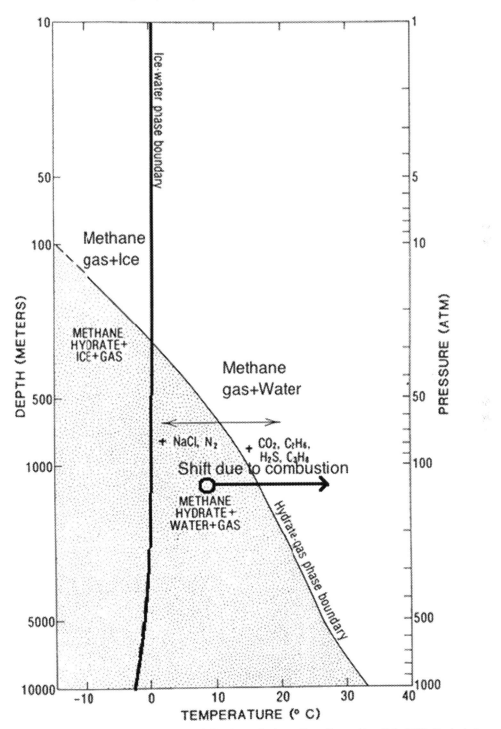

FIG. 4.124 In situ combustion can trigger instability in gas hydrate. *From Kvenvolden, K.A., 1993. Gas hydrates—geological perspective and global change. Rev. Geophys. 31 (2), 173–187.*

would be to initiate in situ combustion by injecting air into the trap. This would trigger immediate instability due to sudden thermal change (Fig. 4.124). This initial instability is followed by generation of CO_2 that acts as yet another stimulus for further hydrate instability (Fig. 4.124). At this point, production of methane can be resumed because all three modes of hydrate dissociation have been activated. Even though the use of geothermal energy has been discussed in the literature as a means of stimulating hydrate production, air injection to trigger in situ heat generation is the most effective technique.

At present, three ways of transport of gas from hydrates are known. They are: (1) by conventional pipeline; (2) by converting the gas hydrates to liquid middle distillates via the newly-improved Fischer-Tropsch process and loading it onto a conventional tanker or barge; or (3) by reconverting the gas into solid hydrate and shipping it ashore in a close-to-conventional ship or barge. The latter option was proposed in 1995 by a research team at the Norwegian Institute of Technology, which determined that the use of natural gas hydrate for the transportation and storage of natural gas was a serious alternative to gas liquefaction since the upfront capital costs are 25% lower. Yet another positive factor is that it is far safer to create, handle, transport, store, and regasify natural gas hydrate than liquefied natural gas.

Because natural gas hydrates are often found in shallow reservoirs, mining them is also an option. More importantly, natural gas hydrates offer the best candidates for downhole electricity generation. Efficiency of such process would be much higher than conventional electricity generation with gas turbines (Figs. 4.125 and 4.126).

As an example, the case of fracture-free hydrate reservoirs can be cited. Fig. 4.127 shows how natural gas hydrates can invade permeable and porous beds that are surrounded by fractured formations. It is difficult to penetrate these formations with an injected fluid due to poor injectivity. However, access can be gained through fractures to trigger dissociation of the natural gas hydrate. Once set in motion, such dissociation process snowballs and the reservoir can produce high volume of in situ gas.

In a fracture-free formation, the next most suitable scheme is called "thermal cracking equivalent." If for whatever reason the formation in question is not accessible, in situ combustion should be considered to trigger production. Unlike conventional in situ combustion that needs special completion to deal with high temperature, air injection is recommended. Air injection can trigger combustion that would lead to steam formation, CO_2, generation and creation of thermal cracks, all of which will increase gas recovery or render the gas recovery process economically attractive. Fig. 4.128 shows this particular feature of unconventional gas production.

The lowest priority option of fracture-free unconventional gas reservoir is displacement type scheme. Because of the low injectivity of these reservoirs, a displacement-type recovery scheme can be achieved only with chemically active water or gas. For instance, alkaline water or carbon dioxide can both become effective displacement agents. In case of CO_2 injection, high solubility of CO_2 leads to the formation of carbonic acid that itself causes leaching effects. Because of the tightness of the formation, the injection rate has to be slow. However, diffusive forces will eventually drive the injected fluid to affect a large area that "softened," thereby being able to produce gas, either from a different well or through huff and puff mode with the same well. This is depicted in Fig. 4.129.

While all the above mentioned options are applicable to unconventional reservoirs without a fracture, network, the presence of fracture networks calls for a new line of strategy. In this case, if the fractures are open, it is recommended that intensive fracking be used. Note that

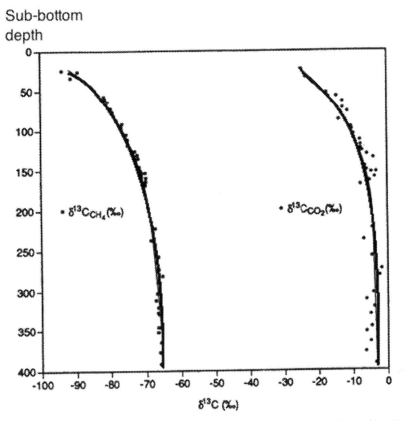

FIG. 4.125 Typical distribution of carbon isotopic compositions with depth. *From Kvenvolden, K.A., 1993. Gas hydrates—geological perspective and global change. Rev. Geophys. 31 (2), 173–187.*

open fractures and low productivity is typical of shallow unconventional reservoirs. Fracturing doesn't work well in deep formations, for which fracturing is both prohibitively expensive and least effective. Hydraulic fractures would be formed in a direction orthogonal to the principal direction of natural fractures. However, prior to creating new fractures, the injected fluid would open existing fractures. In terms of volume, these newly opened fractures can account for very significant increase in gas production. Once hydraulic fractures are created and proppants placed, the overall productivity is sustained at a high value. The placement of hydraulic fractures is more challenging with a horizontal well (Fig. 4.130). However, this challenge is easily overcome when the formation is shallow. For deeper formations, hydraulic fractures are more difficult to create and to sustain.

For reservoirs with open fractures, the question to be asked is if multilaterals are useful. Most of the cases, multilaterals translate into increased productivity. For a multilateral to be effective, following conditions have to be fulfilled:

(1) horizontal well should intersect most of the fractures;
(2) $k_h/k_v > 10$, in the event of lower vertical permeability, hydraulic fracturing is desirable in case they are feasible (e.g., when natural fractures are predominantly horizontal) and liner cementing can be avoided;

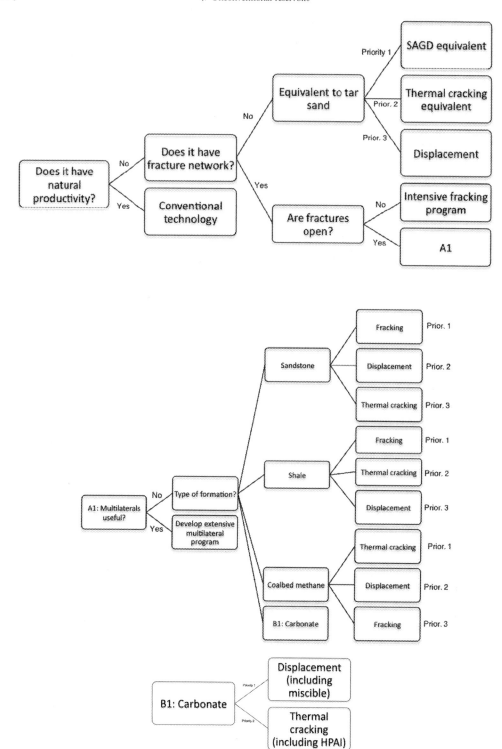

FIG. 4.126 Flow chart for determining appropriate unconventional gas production schemes.

4.5 Improving oil and gas recovery from unconventional reservoirs 435

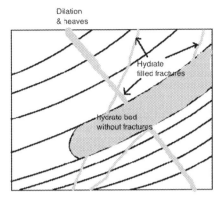

FIG. 4.127   Example of SAGD-like recovery scheme for natural gas hydrate (*yellow* (*gray* in print version) marks hydrate).

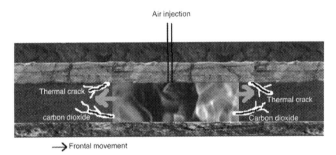

FIG. 4.128   Air injection can trigger several enabling features that will render a recovery process economically attractive.

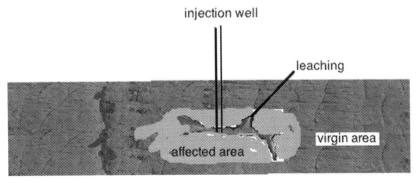

FIG. 4.129   Chemically active water or gas can affect a large area in a tight reservoir.

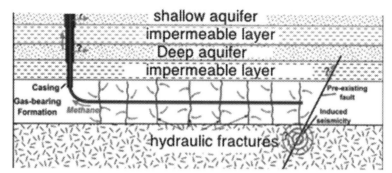

FIG. 4.130 Hydraulic fracturing is most effective in shallow unconventional formations.

(3) Flow near wellbore is Darcian (as non-Darcy effects can reduce flow effectiveness);
(4) Pressure drop in the horizontal wellbore is not significant, in which case telescopic liner can be used to complete the horizontal well.

Fig. 4.131 shows how the productivity improvement factor (PIF) is generally impacted for various cases of reservoirs. Naturally fractured formations offer the best probability of success over vertical wells, even for the unconventional gas reservoirs as long as open fractures are present.

For cases for which multilaters fail to yield satisfactory results, different types of schemes are recommended for different types of formations. Table 4.28 summarizes the final recommendation for this category.

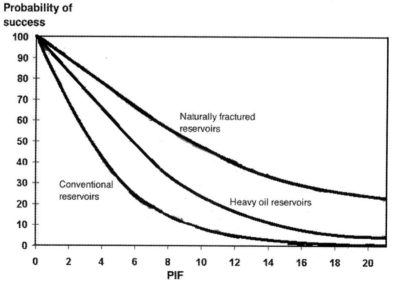

FIG. 4.131 Probability of success with horizontal wells for various types of reservoirs.

TABLE 4.28 Various types of unconventional reservoirs with different priorities.

Type of formation	Priority 1	Priority 2	Priority 3
Sandstone	Fracking	Displacement	Thermal cracking
Shale	Fracking	Thermal cracking	Displacement
CBM	Thermal cracking	Displacement	Fracking
Carbonate	Displacement	HPAI	

4.6 Reserve growth potential

Historically, the majority of additions to oil and gas reserves are attributed to growth of existing fields and reservoirs. From 1978 to 1990, growth of known fields in the United States accounted for more than 85% of known additions to proven reserves (Root and Attanasi, 1993. Thus, field growth and reserve growth are essentially synonymous for discussions of domestic resources. The exception occurred when unconventional resources became exploitable. However, it would be illogical to apply the principle of conventional reserve growth to unconventional reserves that have little in common with the conventional reserves in terms of technology (Islam, 2014). Even though it is promoted that unconventional gas and oil are of lesser quality and are in need of more expensive technology for future development, previous sections demonstrated that notion to be false. The quality of gas improves as the quantity of the resource increases, the case in point being hydrate. For basement reservoirs, it is also true, vast amount of petroleum can be exploited with little or no exploration. In addition, many existing conventional reservoirs have links to unconventional resources, including basement reservoirs and even the abandoned fields can be accessed for tapping the nonconventional resources.

While evaluating the nature of growth in fields requires understanding of both geologic and nongeologic factors that affect growth estimations, geology is the pivotal factor in determining accumulation, quality, and producibility of petroleum resource. Fields may grow when

(1) additional geologic data on existing reservoirs become available and are used to identify new reservoirs or to guide infill drilling;
(2) there are annual updates of reserves data;
(3) field boundaries are extended;
(4) recovery technology is improved; and/or
(5) nongeologic factors such as economics, reporting policies, or politics favor expanded production and development.

Past reserve estimates and characterization both used mathematics and not science as the basis. Attanasi et al. (1999) wrote: "…the modeling approach used by the USGS (U.S. Geological Survey) to characterize this phenomenon is statistical rather than geologic in nature."

Islam et al. (2016) discussed in detail the shortcomings of these statistical models and characterized them as unscientific. Yet, volumetric estimates of reserve growth are calculated by using these mathematical approaches and large data bases that record field reserves through time. Crovelli and Schmoker (2001), present details of various methods used to estimate reserve growth. Application of these models to unconventional gas is of particular concern. It's because one of the most important bases for statistical models is past history, which is practically nonexistent for unconventional reserves.

Geologists tend to think in terms of entire reservoirs, in some cases down to facies level, whereas reservoir engineers deal with measurements at the well bore. To fully understand the geologic factors that affect growth in reserves, this gap in investigative approaches must be bridged. It has become increasingly important to integrate different scales and different observational techniques as secondary and tertiary recovery methods are applied more frequently in mature petroleum provinces.

Fluid-flow pathways, governed predominantly by rock porosity and permeability, are a reflection of heterogeneities of varying scales within a reservoir. Because these reservoir heterogeneities are fundamentally geologic in nature, an adequate understanding of the reservoir architecture, obtained through evaluation of geologic, engineering, or production data, or a combination of these data sets, can provide the basis for scientific characterization of unconventional gas reservoirs.

Proper estimates and growth potential of unconventional gas depends on geological classification and linking of conventional resources with unconventional ones through scientific characterization. This is done in the following sections.

4.6.1 Reservoir categories in the United States

Fishman et al. (2008) evaluated the geologic factors that affect reserve growth in both siliciclastic (largely sandstone) and carbonate (lime-stone and dolomite) reservoirs. This study included 10 formations in the United States (one of which extends into southern Canada) that represent various depositional environments in both siliciclastic and carbonate settings. This is shown in Table 4.29.

TABLE 4.29 Depositional environments and rock units selected for study of reserve growth, and geologic age and general location of units.

Depositional environment and formation studied	Age	General location
Eolian sandstone		
Norphlet Formation	Upper Jurassic	Gulf of Mexico Basin
Minnelusa Formation	Pennsylvanian-Permian	Powder River Basin
Fluvial or deltaic-shallow marine		
Frio Formation	Tertiary (Oligocene)	Gulf of Mexico Basin
Morrow Formation	Pennsylvanian (Morrowan)	Anadarko and Denver Basin
Marine shale		
Barnett Shale	Mississippian (Chesterian)	Fort Worth Basin
Bakken Formation	Devonian-Mississippian	Williston Basin

TABLE 4.29 Depositional environments and rock units selected for study of reserve growth, and geologic age and general location of units—cont'd

Depositional environment and formation studied	Age	General location
Marine carbonates		
Ellenburger Group	Ordovician (Early Ordovician)	Permian Basin
Smackover Formation	Upper Jurassic (late Oxfordian)	
Submarine sands		
Spraberry Formation	Permian (Leonardian)	Gulf of Mexico Basin
		Permian Basin
Nonmarine fluvial-deltaic		
Wasatch Formation	Tertiary (Paleocene-Eocene)	Unita-Piceance Basin

Reservoirs were then categorized based on geological criteria, such as source rock, depositional setting, and postdepositional alteration of the reservoirs. The following environments were studied:

(1) Eolian environments—Norphlet Formation of the Gulf of Mexico Basin (Fig. 4.132) and Minnelusa Formation of the Powder River Basin (Fig. 4.133);
(2) Interconnected fluvial, deltaic, and shallow marine environments—Frio Formation of the Gulf of Mexico Basin (Fig. 4.134) and Morrow Formation of the Anadarko and Denver Basins (Fig. 4.134);
(3) Deeper marine environments—Barnett Shale of the Fort Worth Basin (Fig. 4.134) and Bakken Formation of the Williston Basin (Fig. 4.134);
(4) Marine carbonate environments—Ellenburger Group of the Permian Basin (Fig. 4.134) and Smackover Formation of the Gulf of Mexico Basin (Fig. 4.134);
(5) Submarine fan environment—Spraberry Formation of the Midland Basin (Fig. 4.134); and
(6) Fluvial environment—Wasatch Formation of the Uinta-Piceance Basin (Fig. 4.134).

4.6.1.1 Eolian reservoirs

4.6.1.1.1 Norphlet formation

The Middle to Upper Jurassic Norphlet Formation of the Gulf of Mexico Basin consists largely of eolian sandstones, with minor black shale, conglomerate, and red beds; thicknesses are as much as 100 ft. The Norphlet produces oil and gas largely in Alabama, offshore in Mobile Bay, and in Mississippi (Fig. 4.132). Principal reservoirs in the Norphlet are eolian sandstones (Table 4.30), which are known to have excellent porosity (as much as 20%) and permeability (as much as 500 mD).

As can be seen from Table 4.30, broad similarities in reservoir characteristics throughout the area of production suggest that only a single reservoir category is warranted. This reservoir was considered to be a single one with homogeneous rock and fluid properties. While

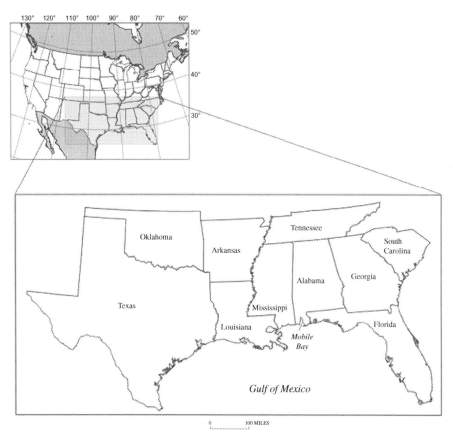

FIG. 4.132 Gulf of Mexico Basin region, the petroleum-producing region of the Norphlet and Smackover Formations. Both formations produce in both onshore and offshore locations; the Norphlet produces from Mobile Bay. *From Fishman, N.S., Turner, C.E., Peterson, F., Dyman, T.S., Cook, T., 2008. Geologic Controls on the Growth of Petroleum Reserves. U.S. Geological Survey Bulletin 2172-I, 53 pp.*

such a reservoir is not considered to be a source of unconventional gas, this reservoir offers the least expensive access to a vast resource that is conventionally not included in all analyses. For instance, any cutoff point of shale in shale breaks or caprocks eliminates as much as 10% of the reserve that lies within low-permeability, low effective-porosity formations. As conventional reserves become insensitive to infill drilling or other means (including enhanced oil/gas recovery), the potential of unconventional gas increases. Fig. 4.135 shows general trend in reserve for conventional reservoirs. Previously, extensive exploration has kept the growth part active. In the United States, however, the main contributor to reserve growth has been infill drilling and enhanced oil recovery. In the infill drilling scheme, one must acknowledge 90% of the new wells are horizontal. This means the role of horizontal well technology has been embedded in this reserve growth. However, despite the best of efforts the reserve will decline, particularly for the "homogeneous" formations.

4.6 Reserve growth potential

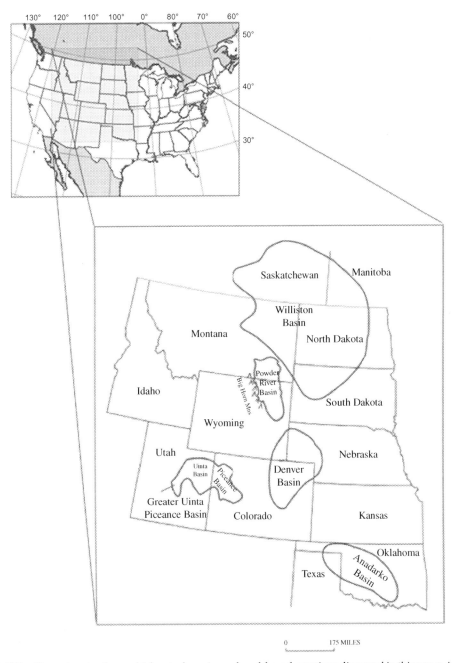

FIG. 4.133 General region from which petroleum is produced from formations discussed in this paper, including the Minnelusa (Powder River Basin), Morrow (Anadarko and Denver Basins), Bakken (Williston Basin), and Wasatch (Uinta and Piceance Basins) Formations. *From Fishman, N.S., Turner, C.E., Peterson, F., Dyman, T.S., Cook, T., 2008. Geologic Controls on the Growth of Petroleum Reserves. U.S. Geological Survey Bulletin 2172-I, 53 pp.*

442   4. Unconventional reservoirs

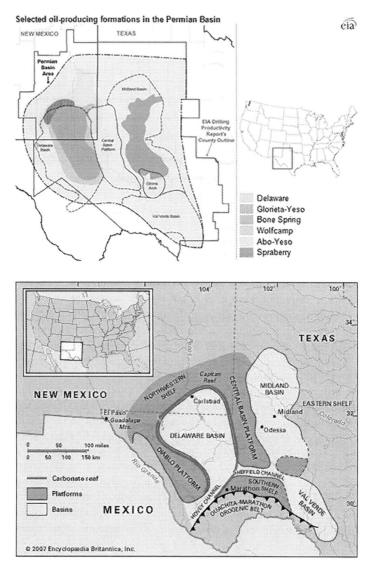

FIG. 4.134 Area from which petroleum is produced from the Frio Formation, Barnett Shale, Ellenburger Group, and Spraberry Formation. Extent of depositional environments in the Frio (such as the Norias delta complex or the Buna barrier-strandplain) from Galloway et al. (1982). For the Barnett, the locations of the Llano uplift and Ouachita thrust belt mark the southern and eastern limits of the Fort Worth Basin, respectively. Horseshoe Atoll is a Pennsylvanian structure that effectively separates productive rocks of the Spraberry Formation (to the south) from nonproductive rocks (to the north). *From Fishman, N.S., Turner, C.E., Peterson, F., Dyman, T.S., Cook, T., 2008. Geologic Controls on the Growth of Petroleum Reserves. U.S. Geological Survey Bulletin 2172-I, 53 pp.*

TABLE 4.30 Norphlet Formation, Gulf of Mexico Basin—Summary of geological characteristics and reserve growth potential of reservoirs.

	Depositional characteristics			Reservoir characteristics				
					Porosity (bulk rock)			
Reservoir category	Environment	Reservoir facies	Non reservoir facies	Lithology	Principal pore space	Diagenetic enhancement	Diagenetic occlusion	Porosity
Norphlet	Sand sea	Eolian sands	Overlying and interbedded marine shale and interdune sediments	Sandstone	Primary intergranular and secondary intergranular and m oldie	Dissolution of early authigenic cements and authigenic chlorite	Local quartz, anhydrite, halite, illite. Intense quartz cementation may seal some accumulations	As much as 20% in onshore reservoirs and 12% in deeper offshore reservoirs

Reservoir characteristics			Stratigraphic controls		Structural controls		
Permeability	Fractures	Source rock	Reservoir location	Traps or seals	Reservoir location	Traps or seals	Oil or gas
Generally high; as much as 500 mD	May be complexly faulted	Overlying marine shale of Smack-over Formation; interbedded or interfingering organic-rich shale in Norphlet Formation	Updip pinchout against basement complex	Overlying shale and interbedded interdune, sabkha, or playa units	Reservoir rocks thicken in basement-controlled grabens and are absent or thin over basement-controlled highs	Anticlines, faulted anticlines, faults associated with basement structures and halokinesis of Louann Salt	Dominantly nonassociated gas (cracked) and minor oil

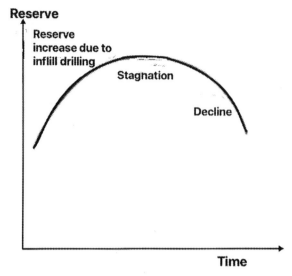

FIG. 4.135 Three phases of conventional reserve.

For such reservoirs, a vast amount of unconventional reservoirs are associated and are accessible readily. This has been the case for shale gas as well as tight gas, as has been evidenced in the recent gas boom in the United States. However, the use of horizontal wells and hydraulic fracturing as the sole mode of reservoir development will lead to similar stagnation as in conventional reservoirs (Fig. 4.136). This can be given a boost by accessing reserve that was previously considered to be part of conventional reservoirs and was excluded through the use of "cutoff" points. The dotted line in Fig. 4.136 shows how immediate boost can be invoked. This can be followed by subsequent use of enhanced gas recovery schemes, as outlined in

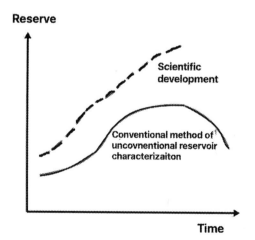

FIG. 4.136 Unconventional reserve growth can be given a boost with scientific characterization.

TABLE 4.31 Minnelusa Formation, Powder River Basin—Summary of geological characteristics and reserve growth potential of reservoirs.

Reservoir category	Depositional characteristics			Reservoir characteristics			
	Environment	Reservoir facies	Non reservoir facies	Lithology	Principal pore space	Porosity (bulk rock)		
						Diagenetic enhancement	Diagenetic occlusion	Porosity
Minnelusa	Coastal sand sea	Eolian dunes	Overlying shallow marine shales, anhydrite, and carbonate rocks	Sandstone, quartz arenite, sublitharenite	Primary and secondary intergranular; moldic	Dissolution of early authigenic cements and of some unstable detrital grains	Quartz, carbonates minerals, and anhydrite/gypsum where not dissolved. Cemented zones may act as seals	Averages 12–24%; but may be as high as 47%
Leo	Coastal dunes	Eolian dunes	Overlying shallow marine shales, anhydrite, and carbonate rocks	Sandstone, quartz arenite, sublitharenite	Primary and secondary intergranular; moldic	Dissolution of early authigenic cements and of some unstable detrital grains	Quartz, carbonates minerals, and anhydrite/gypsum where not dissolved. Cemented zones may act as seals	Averages 12%–24%

previous chapters of this book. Ironically, the greatest potential for most economic recovery of unconventional gas lies within high productivity, "homogeneous" formations as exemplified in this category of Norphlet formation.

4.6.1.1.2 Minnelusa formation

The Pennsylvanian to Early Permian Minnelusa Formation of the Powder River Basin, northeastern Wyoming, consists largely of eolian sandstones, with minor shale and carbonate; thicknesses are as much as 1200 ft. Most production is in the north-central and northeastern parts of the basin; lesser production is in the southerly and southeastern parts (Fig. 4.133). Principal reservoirs are the eolian sandstones (Table 4.31), which can have excellent porosity (as much as 47%) and permeability (as much as 830 mD).

Reservoirs in the Minnelusa Formation are placed into two categories, Minnelusa and Leo (Table 4.31). This twofold division was necessary because of differences in stratigraphic position, depositional environment, and geographic distribution of producing wells; in addition, reservoirs in the two categories may have different source rocks. Reservoir rocks of the Leo category have been variously referred to by previous workers as the "Leo sandstone," "Leo section," "Leo Formation," or the "Leo sandstone of the Minnelusa Formation" (Fishman et al., 2008).

4.6.1.2 Interconnected fluvial, deltaic, and shallow marine reservoirs

4.6.1.2.1 Frio formation

The Oligocene Frio Formation of the Gulf of Mexico Basin consists largely of sandstone and shale deposited in various environments; it is as much as 15,000 ft thick. The Frio produces largely from onshore and offshore locations in Texas. Principal reservoirs in the Frio are sandstones (Table 4.32), which are known to have good to excellent porosity (as much as 35%) and variable permeability (as much as 3500 mD).

Reservoir categories defined in the Frio Formation are fluvial, deltaic, strandplain-barrier, and shelf sandstones (Table 4.32). These four categories were selected principally because reservoirs within them differ in terms of their broad depositional and geographic settings, structural setting, proximity to structures and potential source rocks, and reservoir characteristics.

These reservoirs contain thick shale deposits that are always excluded in conventional reserve calculations. However, these shales are the source of unconventional gas and are accessible through current development schemes. As these conventional resources are subject enhanced oil recovery, unconventional resources should be considered because often such resources would contain higher saturation of oil and gas than the conventional reserve, particularly for matured reservoirs.

4.6.1.2.2 Morrow formation

The Lower Pennsylvanian Morrow Formation of the Anadarko and Denver Basins consists largely of sandstone and shale; it is as much as 1500 ft thick. The Morrow produces oil and gas in Oklahoma, Texas, Kansas, and Colorado (Fig. 4.133). Principal reservoirs in the Morrow are sandstones (Table 4.33), which are known to have good porosity (as much as 22%) and permeability (as much as several darcies).

Petroleum reservoirs in the Morrow Formation were placed into three categories—incised valley-fill, deltaic, and shallow marine (Table 4.33). These categories were selected because

TABLE 4.32 Frio Formation, Gulf of Mexico Basin—Summary of geological characteristics and reserve growth potential of reservoirs.

Reservoir category	Depositional characteristics				Reservoir characteristics				
						Porosity (bulk rock)			
	Environment	Reservoir facies	Nonreservoir facies	Lithology	Principal pore space	Diagenetic enhancement	Diagenetic occlusion	Porosity
Fluvial, chiefly the Gueydan and Chita/Corrigan fluvial systems	Chiefly fluvial with associated channel fill, point bar, crevasse splay, and floodplain sediments	Channel sands, point bars, and crevasse splay sands	Floodplain and lacustrine muds	Feldspathic litharenite, litharenite, and sublitharenite sandstone	Intergranular and moldic	Dissolution of unstable detrital grains and earlier formed cements, resulting in secondary pore space	Quartz, calcite, and clay cements; mechanical compaction	15%–35%
Deltaic; chiefly the Norias and Houston delta complexes	Delta-plain, delta-front, and delta-flank environments of a prograding continental margin in the Gulf Basin. Norias contains more sediment and more sand, and was less influenced by marine processes than Houston	Distributary channel, delta-from and deltaflank, and channel-mouth bar sands	Prodelta and shelf shales	Feldspathic litharenite, litharenite, and sublitharenite sandstone	Intergranular and moldic	Dissolution of unstable detrital grains and earlier formed cements, resulting in secondary pore space	Quartz, calcite, and clay cements; mechanical compaction	10%–35%
Strandplain-barrier, chiefly the Buna and	Shoreface, beach, barrier, and lagoonal deposits	Shoreface, beach, and barrier sands	Marsh and lagoonal muds	Feldspathic litharenite, litharenite, and	Intergranular and moldic	Dissolution of unstable detrital grains and	Quartz, calcite, and clay cements;	20%–35%

Continued

TABLE 4.32 Frio Formation, Gulf of Mexico Basin—Summary of geological characteristics and reserve growth potential of reservoirs—cont'd

	Depositional characteristics			Reservoir characteristics				
						Porosity (bulk rock)		
Reservoir category	Environment	Reservoir facies	Nonreservoir facies	Lithology	Principal pore space	Diagenetic enhancement	Diagenetic occlusion	Porosity
Greta/Carancahua barrier strandplains	adjacent to deltaic depocenters			sublitharenite sandstones		earlier formed cements, resulting in secondary pore space	mechanical compaction	
Shelf; offshore Gulf Coast Basin	Shelf, slope, and perhaps submarine fan environments in deeper parts of the Gulf Coast Basin	Shelf, slope, and possibly fan sandstones	Marine shales and siltstones	Feldspathic litharenite, litharenite, and sublitharenite sandstones	Intergranular and moldic	Dissolution of unstable detrital grains and earlier formed cements, resulting in secondary pore space	Quartz, calcite, and clay cements; mechanical compaction	As much as 30%

Reservoir characteristics		Stratigraphic controls		Structural controls		
Permeability	Source rock	Fractures	Reservoir location	Traps or seals	Reservoir location	Traps or seals	Oil or gas
20–1500 mD	Shales that underlie reservoirs	Important in hydrocarbon migration from source to reservoir	Gueydan system largely a single drainage; leads to stacked channels and lateral amalgamation of channels. Chita Corrigan largely multiple channels with somewhat less stacking of sands	Stratigraphic component of trap is the interval where facies change to mud-rich floodplain rocks; mud-rich rocks arc seals	Production best where fluvial and splay sands cross anticlines, faulted anticlines, or growth-fault trends, and faults served as conduits for upward petroleum migration	Rollover anticlines, particularly on downdip side of Vicksburg growth fault	Oil and gas

TABLE 4.32 Frio Formation, Gulf of Mexico Basin—Summary of geological characteristics and reserve growth potential of reservoirs—cont'd

10–2400 mD	Important in hydrocarbon migration from source to reservoir; also juxtapose reservoirs and seals	Shales that underlie or arc basin ward facies of reservoirs	Abundant sediment supply and single fluvial system input lead to vertically stacked sandy deltaic lobes (Norias), whereas Houston delta fed by several smaller fluvial systems that led to numerous small dispersed lobes with less continuous sands	Stratigraphic component of trap is at abrupt facies changes from reservoir to fine-grained rocks; mud-rich rocks arc seals	Syndepositional movement on growth faults and salt diapirs but no thickening of deltaic sediments, including reservoir rocks	Anticlines and faulted anticlines, some of which are associated with growth faults (Noria and Houston) or salt diapirism (Houston): also growth faults juxtapose reservoirs with seals or compartmentalize reservoirs	Associated gas and oil from more proximal pans, and nonassociated gas from more distal parts
8–3500 mD	Important in hydrocarbon migration from source to reservoir; also juxtaposes reservoirs and seals	Shales that underlie or arc basin ward facies of reservoirs	Greater marine influence on Houston delta led to greater redistribution of sands into strandplain systems than on sands that originated in Norias delta	Stratigraphic component of trap is the interval where facies change to mud-rich floodplain rocks; mud-rich rocks arc seals	Vertical stacking of sands and strike-parallel orientation of sands greatly influenced by orientation and movement of growth faults	Anticlines, rollover anticlines, and faulted anticlines	Associated gas and oil
As much as 1500 mD	Important in hydrocarbon migration from source to reservoir; also juxtaposes reservoirs and seals	Shales that interbed with or underlie reservoir rocks	Stratigraphic controls on reservoir location unclear	Stratigraphic component of trap is at abrupt change from reservoir to fine-grained rocks; fine-grained rocks serve as seals	Sediment accumulation in submarine canyons or intraslope basins that formed from active faulting or salt diapirs (or both)	Faulted anticlines and salt-related structures. Seals formed by fault-related juxtaposition of reservoirs with impermeable rocks	Largely gas

TABLE 4.33 Morrow Formation, Anadarko and Denver Basins—Summary of geological characteristics and reserve growth potential of reservoirs.

	Depositional characteristics				Reservoir characteristics				
						Porosity (bulk rock)			
Reservoir category	Environment	Reservoir facies	Nonreservoir facies	Lithology	Principal pore space	Diagenetic enhancement	Diagenetic occlusion	Porosity	
Incised valley fill	Braided streams that grade upward into meandering and estuarine environments	Dominantly coarser grained fluvial sands that fill incised valleys	Floodplain, estuarine, and marine mudstone	Sandstone; varies from quartz arenite to litharenite or arkosic	Intergranular: variable volume of moldic porosity due to dissolution of detrital grains	Secondary pore space from dissolution of early formed authigenic cements and some unstable detrital grains	Extensive cement in lower parts of channel sands with calcite or iron carbonate minerals, or both	12%–21%	
Deltaic	Lower delta plain	Point bar, meander channel, stream-mouth bar, and distributary channel sands	Overbank, backswamp marsh, prodelta, and marine mudstone	Sandstone; varies from quartz arenite to litharenite or arkosic		Secondary pore space from dissolution of early formed authigenic cements and some unstable detrital grains	Late-stage calcite or iron carbonate minerals, or both	12%–22%	
Shallow marine	Near-shore and marginal marine	Beach, barrier island, and shoreline parallel sand bar sands	Marine shale and siltstone	Sandstone; varies from quartz arenite to litharenite or arkosic: locally fossiliferous		Secondary pore space from dissolution of early formed authigenic cements and some unstable detrital grains	Late-stage calcite or iron carbonate minerals, or both, mechanical compaction	4%–20%	

TABLE 4.33 Morrow Formation, Anadarko and Denver Basins—Summary of geological characteristics and reserve growth potential of reservoirs—cont'd

Reservoir characteristics			Stratigraphic controls		Structural controls		Oil or gas
Permeability	Fractures	Source rock	Reservoir location	Traps or seal	Reservoir location	Traps or seals	
As much as several darcies	Could have helped hydrocarbons to migrate from any overlying or underlying sources	Possibly marine muds of the Morrow Formation, where mature in Anadarko Basin: other organic-bearing formations outside the Morrow	Downcutting and formation of paleovalleys localized fluvial channel-reservoirs, dominantly in upper pan of Morrow	Underlying marine limestone or shale and overlying floodplain muds	Paleostructures and perhaps subsidence from dissolution of underlying evaporates may have localized areas of downcutting and incision	Anticlines may influence but arc secondary to stratigraphic controls	Associated gas and oil
1–100 mD	Could have helped hydrocarbons to migrate from any overlying or underlying sources	Possibly marine muds of the Morrow Formation, where mature in Anadarko Basin, other organic-bearing formations outside the Morrow	Unclear	Lateral pinch out of sands into fine-grained marine muds	Unclear	Anticlines may influence but arc secondary to stratigraphic controls	Dominantly gas
<1–200 mD	Could have helped hydrocarbons to migrate from any overlying or underlying sources	Possibly marine muds of the Morrow Formation, where mature in Anadarko Basin; other organic-bearing formations outside the Morrow	Location of sands in part a function of longshore currents, dominantly in lower part of Morrow	Lateral pinch out of sands into fine-grained marine muds	Unclear	Anticlines may influence but arc secondary to stratigraphic controls	Dominantly non associated gas

reservoirs within them differ in terms of their broad geographic and depositional setting. The differing depositional settings of the reservoir categories have led to differing reservoir-rock characteristics, such as porosity and permeability, which bear directly on the reservoir properties and contained resources. The shallow marine category offers the greatest potential for unconventional gas reserves. However, caprock of high-porosity reservoirs also contain large volume of natural gas.

4.6.1.3 Deeper marine shales
4.6.1.3.1 Barnett shale

The Middle to Late Mississippian Barnett Shale of the Fort Worth Basin, Texas, consists largely of black marine shales with some limestone; it is as much as 650 ft thick. Most production is of nonassociated gas, principally in the northeastern part of the basin (Fig. 4.134). Reservoirs of this self-sourced unit are marine shales in the Barnett (Table 4.34), which have very low porosity (less than 6%) and extremely low permeability (a few nanodarcies).

Reservoirs in the Barnett Shale are grouped in a single category termed the shale (unconventional) category (Table 4.34). Until recently, the lower shale member has been the more productive, although considerable production is now being realized from the upper shale member as well. Both members characteristically have a high content of organic material, which is largely Type-II. In general, the current average content of organic material in both members is 4%–5%, although in places the Barnett is thought to have contained as much as 20% total organic carbon when it was deposited. The organic material serves as the source of the gas, thereby defining these reservoirs as self-sourced and unconventional.

This formation is characteristically categorized as unconventional gas reserve. However, this formation makes up for only a fraction of total unconventional reserve that would be evident through scientific characterization.

4.6.1.3.2 Bakken formation

The Late Devonian to Early Mississippian Bakken Formation (of the Williston Basin of North Dakota, Montana, and the Canadian provinces of Saskatchewan and Manitoba) consists largely of marine shale with minor sandstone; it is as much as 140 ft thick. The Bakken produces mostly oil, principally in North Dakota and Montana and lesser amounts in Saskatchewan and Manitoba. Reservoirs in the Bakken are principally marine shales, although smaller reservoirs are found in interbedded near-shore to shoreface sandstones (Table 4.35). Porosity of the shales is very low (typically less than 5%) as is their permeability (<0.01–60 mD). Porosity of sandstone reservoirs is higher (as much as 10%) as is permeability (<0.01–109 mD).

Two categories of reservoirs were defined in the Bakken Formation—shale (unconventional) and siltstone-sandstone (unconventional) (Table 4.35). These two categories were selected because they have different characteristics, stratigraphic positions, and geographic distributions. In each, however, the petroleum is thought to be generated within the Bakken, so both categories are considered to be unconventional, similar to those in the Barnett Shale. Recent success in producing from these unconventional sources with conventional technology (e.g., horizontal well and fracturing) point out that there is tremendous potential for expanding this resource base.

TABLE 4.34 Barnett Shale, Fort Worth Basin—Summary of geological characteristics and reserve-growth potential of reservoirs.

	Depositional characteristics			Reservoir characteristics			
					Porosity (bulk rock)		
Reservoir category	Environment	Reservoir facies	Nonreservoir facies	Principal pore space	Diagenetic enhancement	Diagenetic occlusion	Porosity
Shale (unconventional)	Offshore marine	Marine shale	Dense limestone	Matrix, but very low	Uncertain	Calcite along fractures	Very low, typically <6%

Reservoir characteristics		Stratigraphic controls			Structural controls		
Permeability	Fractures	Source rock	Lithology	Traps or seals	Reservoir location	Traps or seals	Oil or gas
Very low, typically in the range of nanodarcies	Naturally fractured in deeper parts of basin and over structures; fractures reduce productivity	Organic-rich shale in the Barnett that also serves as reservoir rock	Organic-rich shale	Gas trapped by fine-grained nature of shale reservoir	Best production away from fractured areas	Open faults tended to leak gas out of formation, whereas calcite-filled faults prevented gas migration	Nonassociated gas

TABLE 4.35 Bakken Formation, Williston Basin—Summary of geological characteristics and reserve-growth potential of reservoirs.

Depositional characteristics

Reservoir category	Environment	Reservoir facies	Source rock	Nonreservoir facies
Shale (unconventional)	Deep marine, below wave base	Black, organic-rich mudstone	Black, organic-rich mudstone; is also the reservoir rock	Overlying shallow marine carbonates and shales
Siltstone-sandstone (unconventional)	Near-shore and shoreface	Siltstone and very fine to medium-grained sandstone	Organic-rich mud in Bakken, interbedded with or perhaps downdip from reservoirs	Enclosing black mudstone

Reservoir characteristics

Reservoir category	Lithology	Principal pore space	Porosity (bulk rock) Diagenetic enhancement	Diagenetic occlusion	Porosity
Shale (unconventional)	Black mudstone	Fracture	Little or none	Little or none	Very low, typically <5%
Siltstone-sandstone (unconventional)	Dolomitic siltstone and sandstone	Fracture	Dissolution of carbonate cement	Carbonate cement	Can be >10% but typically 3%–10%

Reservoir characteristics

Permeability	Fractures	Stratigraphic controls Reservoir location	Traps or seals	Structural controls Reservoir location	Traps or seals	Oil or gas
<0.01–60 mD	Critical for production	Apparently not important	Apparently not important	Fracture zones overlying anticlinal or monoclinal folds and solution fronts in underlying salts	Minimal; reservoirs unconventional	Oil
<0.01–109 mD	Critical for production	Local thickening owing to subsidence associated with dissolution of underlying salts	Overlying shales of the Bakken	Fracture zones overlying anticlinal or monoclinal folds and solution fronts in underlying salts	Updip against enclosing mudstone strata	Oil

4.6.1.4 Marine carbonate reservoirs

4.6.1.4.1 Ellenburger group

The Early Ordovician Ellenburger Group of the Permian Basin consists largely of marine carbonate rocks; the group is as much as 1500 ft thick. Units in the Ellenburger produce oil and gas chiefly in Texas. Principal reservoirs in the Ellenburger are in karstified parts of a carbonate platform and in dolomitized carbonate muds (Table 4.36). Reservoirs in karstified rocks have low but variable porosity (2%–7%) and moderate but variable permeability (2–750 mD). Reservoirs in dolomitized muds have higher porosity (2%–14%) but lower permeability (1–44 mD) than karstified reservoirs.

Reservoirs in the Ellenburger Group are placed into three categories (Table 4.36)—karstified, platform, and tectonically fractured—based primarily on differences in the nature and volume of porosity and permeability, geographic distribution, produced petroleum, and the degree to which structure influenced reservoir development. This threefold division is similar to that presented by others. This is the category rarely connected to unconventional gas. However, significant amount of oil and gas exists within such reservoirs that fit the description of unconventional oil and gas, including in low-permeability patches, caprock, and volcanic rocks. Some of these resources are in high pressure and temperature conditions. They form a special candidate for reverse thermal recovery. This involves injection of cold water to induce thermal fracturing owing to large temperature gradient. In most reservoirs, this is easy to accomplish. These formations typically are not suitable for hydraulic fracturing for reasons described in previous chapters.

4.6.1.4.2 Smackover formation

The Upper Jurassic Smackover Formation in onshore parts of Texas, Arkansas, Louisiana, Mississippi, Alabama, and Florida, as well as offshore in the Gulf of Mexico Basin, consists largely of carbonate rocks with minor black shale and siltstones; it is as much as 1000 ft thick. Most oil and gas is produced from onshore locations in the above-listed states. Principal reservoirs in the Smackover are in carbonate rocks deposited in a ramp setting (Table 4.36) that have good to excellent porosity (as much as 35%) and variable permeability (<1–4100 mD). Reservoir categories in the Smackover Formation are salt structure, basement structure, graben, stratigraphic, and updip fault (Table 4.36). These categories, which were defined or later refined through regional studies by other workers were selected because of differences in their geographic extent and in the role that structures played in both source-rock deposition and petroleum trapping.

This type of formations are not typically considered to generate unconventional gas. However, significant amount of gas is present in these formations in caprcoks, salt domes and other locations. Salt domes have not been included in this book as a source of potential unconventional gas. However, they do contain natural gas in many instances and each case should be investigated to access unconventional gas with minimal cost.

4.6.1.5 Submarine fan reservoir

4.6.1.5.1 Spraberry formation

The Early Permian Spraberry Formation of the Midland Basin consists largely of turbiditic sandstones, with minor black shales, silty dolostones, and argillaceous siltstones; it is as much as 1000 ft thick. Most production of oil is in west-central Texas, in the Midland Basin. Principal reservoirs in the Spraberry are the turbiditic sandstones (Table 4.37), which have good

TABLE 4.36 Ellenburger Group, Permian Basin—Summary of geological characteristics and reserve-growth potential of reservoirs.

	Depositional characteristics				Reservoir characteristics			
						Porosity (bulk rock)		
Reservoir category	Environment	Reservoir facies	Nonreservoir facies	Lithology	Principal pore space	Diagenetic enhancement	Diagenetic occlusion	Porosity
Karstified, principally in Central Basin platform and Midland Basin	Shallow aggrading marine carbonate platform	Inner platform	Reef, forereef, supratidal	Dolomitized mudstone	Inter breccia fragment and within fractures	Dissolution of lime mud leading to karstification and brecciation, intercrystal-line intercrystalline porosity owing to dolomitization of muds	Late-stage saddle dolomite	Average, 3% Range, 2%–7%
Platform, dominantly in southern and eastern parts of Midland Basin	Shallow aggrading marine carbonate platform	Middle to outer platform	Reef, forereef, supratidal	Dolomitized packstone and mudstone	Intercrystalline	Intercrystalline porosity owing to dolomitization	Late-stage saddle dolomite	Average, 14% Range, 2%–14%
Tectonically fractured, dominantly in the eastern Delaware Basin	Shallow aggrading marine carbonate platform	Inner platform	Reef, forereef, supratidal	Dolomitized mudstone	Fracture (tectonic)	Dissolution of lime mud leading to karstification and brecciation, intercrystal-line intercrystalline porosity owing to dolomitization of muds	Late-stage saddle dolomite	Average, 4% Range, 1%–8%

TABLE 4.36 Ellenburger Group, Permian Basin—Summary of geological characteristics and reserve-growth potential of reservoirs—cont'd

Reservoir characteristics		Stratigraphic controls			Structural controls		
Permeability	Fractures	Source rock	Reservoir location	Traps or seals	Reservoir location	Traps or seals	Oil or gas
Mean, 32 mD Range, 2–750 mD	Channeled pore fluids that allowed vertical infiltration of dissolving waters into various stratigraphic horizons to promote karstification	Overlying Ordovician Simpson Group	Lime muds remaining after early dolomitizatio, which became horizons subject to dissolution leading to karstification	Traps and seals include overlying Simpson Group and unkarsted Ellenburger dolomite. Scats also include impermeable cave-fill sediments and collapse zone adjacent to reservoirs	Anticlines, faulted anticlines, and fault-bounded anticlines	Uncertain	Principally oil with some associated gas and gas condensate
Average, 12 mD Range, <1–44 mD	Focused early dolomitizing fluids, which resulted in intercrystal-line porosity and permeability	Overlying Devonian Woodford Shale?	Lime muds that were dolomitized	Traps and seals include overlying Simpson Group	Anticlines, faulted anticlines	Uncertain	Largely oil
Average, 4 mD Range, 1–100 mD	Early fracturing promoted karsrification, whereas later fracturing improved porosity and permeability of the reservoir	Overlying Ordovician Simpson Group	Lime muds that were dolomitized	Traps and seals include overlying Simpson Group	Fractured anticlines and faults critical	Uncertain	Nonassociated gas

TABLE 4.37 Mackover Formation, Gulf Coast region—Summary of geological characteristics and reserve-growth potential of reservoirs.

	Depositional characteristics				Reservoir characteristics		
						Porosity (bulk rock)	
Reservoir category	Environment	Reservoir facies	Nonreservoir facies	Lithology	Principal pore space	Diagenetic enhancement	Diagenetic occlusion	Porosity
Salt structure, dominantly in southern and eastern Texas, southern Arkansas, southern and central Mississippi, southwestern Alabama, and northern Louisiana	Slow regressive to stillstand marine carbonate ramp	Ramp, higher energy shoaling facies	Subtidal mudstone, wackestone, supratidal units, and outer ramp dolostones	Largely dolomitic oolitic grainstones and packstones	Dominantly intercrystalline where dolomitized, oomoldic in updip regions, intergranular in basinal regions	Intercrystalline owing to dolomitization; ooid dissolution: late calcite dissolution; diagenesis most pronounced on structural highs	Late-stage saddle dolomite, anhydrite, and calcite	2%–35%
Basement structure, primarily in eastern Texas, central Mississippi, southern Arkansas, and southwestern Alabama	Slow-regressive to stillstand marine carbonate ramp	Ramp, higher energy shoaling facies	Subtidal mudstone, wackestone, supratidal units, and outer ramp dolostones	Largely dolomitic oolitic grainstones and packstones	Principally oomoldic; minor primary interparticle and intercrystalline where dolomitized	Principally oomoldic; minor intercrystalline owing to minor dolomitization; diagenesis pronounced on structural highs	Late-stage calcite and dolomite	As much as 20%
Graben, principally along Arkansas-Louisiana border	Slow-regressive to stillstand marine carbonate ramp	Ramp, higher energy shoaling facies	Subtidal mudstone, wackestone, supratidal units, and outer ramp dolostones	Oolitic limestone, locally dolomitic	Considerable interparticle pore space preserved: also oomoldic	Some interparticle and intercrystalline owing to dolomitization; some oomoldic	Partial cementation by calcite	4%–19%

TABLE 4.37 Mackover Formation, Gulf Coast region—Summary of geological characteristics and reserve-growth potential of reservoirs—cont'd

| Stratigraphic, principally in southern Arkansas | Slow, regressive to stillstand marine carbonate ramp | Ramp, higher energy shoaling facies | Subtidal mudstone, peloid packstone, wackestone, supratidal units, and outer ramp dolostones | Oolitic, oncolitic, or skeletal grainstone limestone minimally dolomitized | Considerable interparticle: some oomoldic and intercrystalline where dolomitized | Some interparticle and intercrystalline owing to dolomitization; considerable early- and late-stage dissolution of particles and late-stage cement | Cements such as early and late stage calcite and anhydrite; some compaction | 3%–30% |
| Updip fault, principally in eastern Texas, southern Arkansas, central Mississippi, southwestern Alabama, and Florida Panhandle | Slow-regressive to Stillstand marine carbonate ramp | Ramp, higher energy shoaling facies | Subtidal mudstone, wackestone, supratidal units, and outer ramp dolostones | Oolitic limestone, locally dolomitic | Principally oomoldic | Ooid dissolution common: some dolomitization | Early calcite cement | 10%–20% |

Reservoir characteristics

			Stratigraphic controls		Structural controls		
Permeability	Fractures	Source rock	Reservoir location	Traps or seals	Reservoir location	Traps or seals	Oil or gas
<1–4100 mD	Large-scale open fractures not now widespread: however, fractures probably served as conduits for hydrocarbon migration	Organic-rich units in lower part of Smackover Formation	Shoaling sequences best developed on positive features formed by salt diapirism during deposition	Fine-grained beds in overlying Buckner Formation acted as seals	Salt anticlines, faulted salt anticlines, faulted salt-pierced anticlines	Faults seal some reservoirs	Dominantly oil and associated gas with minor condensate
60–350 mD	Faults now act as seals owing to impermeability of fault zones but earlier probably served as conduits for	Organic-rich units in lower part of Smackover Formation	Facies changes up on basement highs; shoaling on positive basement highs during deposition.	Stratigraphic and structural trap with overlying Buckner Formation: pinchouts on	Regional fault zones, anticlines, faulted anticlines	Downdip fault zone served as reservoir seal	Dominantly oil in updip areas: associated gas or gas

Continued

TABLE 4.37 Mackover Formation, Gulf Coast region—Summary of geological characteristics and reserve-growth potential of reservoirs—cont'd

Reservoir characteristics			Stratigraphic controls		Structural controls		
Permeability	Fractures	Source rock	Reservoir location	Traps or seals	Reservoir location	Traps or seals	Oil or gas
<1–1000 mD	Faults now act as seals owing to impermeability of fault zones but earlier probably served as conduits for hydrocarbon migration	Organic-rich units in lower part of Smackover Formation	Little evidence of halokinesis Shoaling sequences best developed on horst blocks adjacent to grabens	basemen: highs serve as seals Structural and stratigraphic trap; overlying Buckner Formation serves as seal	Fault zones and faulted anticlines	Faults seal some reservoirs	condensate in basinal areas Dominantly oil
1–250 mD	Faults probably served as conduits for hydrocarbon migration	Organic-rich units in lower pan of Smackover Formation	Facies changes and regressive units overlying reservoirs	Structural and stratigraphic trap; overlying Buckner Formation serves as seal	Likely; structures limited deposition of reservoir rocks or facilitated pinchouts	Faults seal some reservoirs	Dominantly oil: some associated gas
3–280 mD	Faults now act as seals owing to impermeability of fault zones but earlier probably served as conduits for hydrocarbon migration	Organic-rich units in lower part of Smackover Formation	Near updip limit of Smackover deposition	Dominantly structural trap: fault systems serve as seals	Uplift on faults juxtaposed reservoirs and impermeable beds	Fault zones	Dominantly oil: some gas or gas condensate

porosity (as much as 18%) but relatively low permeability (maximum, 10 mD). A single reservoir category, submarine sand, was defined for the Spraberry Formation.

Even though not explicitly recognized, these reservoirs form an excellent candidate for unconventional gas. These are thick formations that can have very high reserve once the cutoff point for porosity is removed.

4.6.1.6 *Fluvial reservoir*
4.6.1.6.1 Wasatch formation

The Paleocene-Eocene Wasatch Formation of the Uinta-Piceance Basin of Utah and Colorado consists largely of overbank and lacustrine mudstones with some fluvial and fluvial-dominated deltaic sandstones; it is as much as 5000 ft thick. The Wasatch produces oil and associated gas mostly in the Uinta Basin of northeastern Utah, although minor gas is also produced in the Piceance Basin of Colorado. Principal reservoirs in the Wasatch are the fluvial sandstones (Table 4.37), which are known to have good porosity (maximum, 15%) but low permeability (maximum, 40 mD).

Reservoirs in the Wasatch Formation are categorized as Green River source and Mesaverde source (Table 4.38). The two categories are distinguished by (1) the source of the petroleum produced from each, (2) the nature of the petroleum produced from each, and (3) the geographic distribution of production. This division is important because it recognizes that petroleum produced from the Wasatch comes from two different source rocks; hence, two petroleum systems generated economic amounts of petroleum within the greater Uinta-Piceance Basin.

In addition to unconfirmed patches of unconventional plays, these formations contain three types of continuous-type unconventional sources. They are:

- Oil in fractured Upper Cretaceous marine shale;
- gas in tight sandstone;
- coal bed methane

Other unconventional gas may be present in heavy oil and tar sand formations (Table 4.39).

4.7 Quantitative measures of well production variability

Production history offers a powerful tool for scientific characterization of a formation. It is not because it offers refinement of statistical tools, but because any history is an evidence that can be used to refine a scientific model. Fishman et al. (2008) produced a detailed analysis of production history. This analysis was aimed at identifying origin of both fluid and rock systems. This report (USGS, 2008) compared historical well production data of the five formations by use of proprietary information. In addition, it considered data from two specific reservoir categories in the Ellenburger Group (karst and platform), which are based on gross geologic differences, to evaluate the possible intraformational variability in production within that formation. Because in most wells production declines exponentially or hyperbolically as a function of time, cumulative production from older wells (those for which current monthly production is less than 10% of initial monthly production) asymptotically begins to approximate ultimate recovery. This is typical of conventional reserve analysis. In such wells,

TABLE 4.38 Spraberry Formation, Midland Basin—Summary of geological characteristics and reserve-growth potential of reservoirs.

Depositional characteristics

Reservoir category	Environment	Reservoir facies	Nonreservoir facies
Submarine sand	Deep-water submarine basin and fan	Submarine fan and turbidite sandstones	Silty dolostone, organic-rich shale, and argillaceous sandstone

Reservoir characteristics

Lithology	Porosity (bulk rock)			Porosity
	Principal pore space	Diagenetic enhancement	Diagenetic occlusion	
Sandstone	Largely intergranular but some minor moldic	Dissolution of preexisting authigenic cements and unstable detrital grains	Mechanical compaction and authigenic cements such as illite, chlorite, quartz, and dolomite	Matrix porosity usually 5%–15% but may be as high as 18%

Stratigraphic controls

Source rock	Reservoir location	Traps or seals
Interbedded organic-rich shales	Most reservoirs downdip from the ancient Horseshoe Atoll at mouth of submarine canyons or where facies change from channel to interchannel deposits	Pinchouts of reservoir rocks updip and downdip into fine-grained rocks serve as traps. Shales seal reservoirs

Reservoir characteristics

Fractures	Permeability
Very common; multiple orientations observed; fractures cemented to various degrees	Average matrix permeability low, <1 mD, but may be as high as 10 mD

Structural controls

Reservoir location	Traps or seals	Oil or gas
Uncertain	Mostly stratigraphic traps; one small field on an anticline	Largely oil

TABLE 4.39 Reservoir characteristics of certain formations.

Depositional characteristics

Reservoir category	Environment	Reservoir facies	Nonreservoir facies	Source rock
Green River source	Fluvial, deltaic, and lacustrine	Fluvial, channel sandstone, and sands deposited in lacustrine deltas	Overlying and interbedded overbank, floodplain, delta plain, and lacustrine mudstone and claystone	Organic-rich lacustrine mudstones of Green River Formation, which largely interfingers with the Wasatch
Mesaverde source	Fluvial, deltaic, and lacustrine	Fluvial, channel sandstone, and sands deposited in lacustrine deltas	Overlying and interbedded overbank, floodplain, delta plain, and lacustrine mudstone and claystone	Coals and organic-rich shale of the Mesaverde Group, which underlies the Wasatch

Reservoir characteristics

Porosity (bulk rock)

Lithology	Principal pore space	Diagenetic enhancement	Diagenetic occlusion	Porosity
Sandstones, lithic arkoses, or feldspathic litharenites	Intergranular, principally secondary; some minor moldic	Dissolution of early authigenic cements and unstable detrital grains	Some quartz and carbonate cements and authigenic clays	Ranges up to 15% at shallow depths (<4000 ft) but <10% at greater depths (>8500 ft)
Sandstones, lithic arkoses, or feldspathic litharenites	Intergranular, principally secondary; some minor moldic	Dissolution of early authigenic cements and unstable detrital grains	Some quartz and carbonate cements and authigenic clays	Ranges up to 15% at shallow depths (<4000 ft) but <10% at greater depths (>8500 ft)

Reservoir characteristics

Permeability	Fractures
Generally low; as much as 40 mD but commonly <0.1 mD	Reservoirs may be complexly faulted: faults allow production
Generally low; as much as 40 mD but commonly <0.1 mD	Reservoirs may be complexly faulted: faults allow production; migration along fractures

Stratigraphic controls

Reservoir location	Traps or seals
Reservoir rocks deposited adjacent to and in deltas within ancient Lake Uinta	Overlying and interbedded shales, mudstones, and claystones trap and seal reservoirs
Reservoir rocks deposited adjacent to and in deltas within ancient Lake Uinta	Overlying and interbedded shales, mudstones, and claystones trap and seal reservoirs

Structural controls

Reservoir location	Traps or seal	Oil or gas
Uncertain	Secondary to stratigraphic traps or seals	Dominantly oil; some associated gas
In areas where gas could migrate up fractures that cut from source to reservoir rocks	Secondary to stratigraphic traps or seals	Nonassociated gas

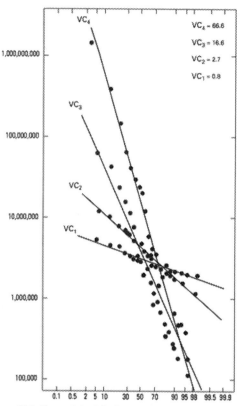

FIG. 4.137 Probability distributions for production from wells of an oil or gas field (distributions based on hypothetical data—peak monthly production, peak yearly production, or cumulative production). Each point represents a well, and four fields (VC1–VC4) are depicted. In this type of plot log normal distributions plot as straight lines, and steeper slopes of lines correspond with a greater range of production and thereby greater production variability. The variation coefficient VC = (F5–F95)/F50 provides a dimensionless numerical value for the variability of each data set, and its value increases as slope increases. *From Fishman, N.S., Turner, C.E., Peterson, F., Dyman, T.S., Cook, T., 2008. Geologic Controls on the Growth of Petroleum Reserves. U.S. Geological Survey Bulletin 2172-I, 53 pp.*

variations in cumulative production reflect variations in the volume of reservoir rocks accessed by the well bore. The slopes of the probability distributions for cumulative production (Fig. 4.137) are direct indicators of the variability as shown by the data set. For example, steeper slopes reflect greater production heterogeneity, whereas a horizontal line represents uniform production characteristics. A dimensionless parameter that is proportional to the slopes of the four probability distributions of Fig. 4.137 would provide a quantitative numerical representation of production heterogeneity. Such a parameter, referred to here as a variation coefficient (VC), can be calculated by using a measure of the dispersion (range) of the data set divided by a measure of central tendency such as the mean or the median.

Different parts of the distribution behave differently—that is, a single straight-line fit does not adequately describe the behavior of the entire distribution of production data. Of interest from a conventional perspective is the central part of the distribution because it represents production from the vast majority of wells. Extreme production behavior, categorized by wells in the upper 5% and lower 20% of the production distribution, forms excellent candidates for unconventional recovery. They either include old wells that would invariably have vast gas content that is deemed uneconomical with conventional analysis or they include low-permeability formation, suitable for unconventional gas development.

Fig. 4.137 lists wells that were sorted by production from lowest to highest and subdivided into two size classes: a central class representing a productive range of 20%–60% along the distribution and an upper class representing a productive range of 80%–95%.

Fig. 4.133 shows fields producing oil from the reservoirs representing the (1) fluvial category of the Frio Formation, (2) incised valley-fill category of the Morrow Formation, (3) Green River-source category of the Wasatch Formation, (4) Minnelusa category of the Minnelusa Formation, and both the (5) platform and karst categories of the Ellenburger Group. Table 4.40 contains the basic data used in calculating production variability for each reservoir category. A minimum of 35 producing wells were used to describe the production behavior for each category and to calculate upper class and central class rates of recovery and slope ratios for each. We also identified a well productive life of at least 10 years on the basis of data in the IHS Energy Group production file. For example, 6301 wells were selected from IHS data as Frio Formation producers in all or parts of Starr, Hidalgo, Brooks, Jim Hills, and Kleburg Counties, Texas. A computer program then calculated upper and central class rates of recovery and slope ratios on the basis of a subset of these wells that met our selection criteria. The six reservoirs analyzed have produced more than 2 billion barrels of oil and 12 trillion cubic feet of gas from nearly 13,000 producing wells. The results are plotted in Fig. 4.138. Of interest is the role of depositional environment, diagenesis, and lithology on reservoir productivity. Comparing the slope ratios and variation coefficients of reservoirs with different geologic characteristics may provide insight into productivity analysis and ultimately into estimating field growth through time. An opposite trend will be followed by unconventional oil and gas. Whenever conventional resources hit stagnation, unconventional resource potentials increase.

4.8 Coupled fluid flow and geomechanical stress model

Coupling different flow equations has always been a challenge in reservoir simulators. In this context, Pedrosa and Aziz (1986) introduced the framework of hybrid grid modeling. Even though this work was related to coupling cylindrical and Cartesian grid blocks, it was used as a basis for coupling various fluid flow models (Islam and Chakma, 1990). Coupling flow equations to describe fluid flow in a setting, for which both pipe flow and porous media flow prevail continues to be a challenge (Islam et al., 2010). For unconventional gas, horizontal wells are of great importance. Even though it is commonly perceived that pressure drop in the horizontal well within a gas reservoir is negligible, this is not the case in many scenarios, particularly the ones involving multiphase flow. As example, one can cite coal bed methane, condensate, and even some tight gas formations with oil flow.

TABLE 4.40 Location of, number of fields and wells in, cumulative production of, and largest fields in each reservoir category analyzed in this study.

Reservoir category	Location	No. fields	No. wells	Cum. oil (MMBO)	Cum. gas (Bcf)	Largest oil fields	Cum. oil (MMBO)	Cum. gas (Bcf)
Frio	Texas[a]	272	6301	534.4	11,193.0	Seeligson	238.1	2306.0
						Tijerina-Canales-Blucher	87.0	753.8
						Stratton	41.8	1841.5
Morrow	Colorado[b]	38	386	74.7	116.8	Arapahoe	23.2	35.7
						Mt. Pearl	13.6	41.2
						Sorrento	12.6	8.5
Wasatch	Utah[c]	24	436	89.9	139.2	Altamont	48.6	74.5
						Bluebell	34.9	48.6
						Cedar Rim	4.4	6.6
Ellenburger (karst)	Texas[d]	141	2784	1155.8	1042.5	Andector	178.4	70.2
						TXL	129.3	29.8
						Pegasus	96.3	361.2
Ellenburger (ramp)	Texas[e]	134	928	65.0	29.0	Barnhart	16.7	11.9
						Swenson-Barron	5.8	1.2
						Swenson-Garza	4.2	0.7
Minnelusa	Wyoming[f]	315	1936	586.8	14.9	Raven Creek	44.2	0.03
						Timber Creek	16.2	1.2
						Dillinger Ranch	16.2	1.6

[a] *All or parts of Stair, Hidalgo, Brooks, Jim Hills, and Kleburg Counties, Texas.*
[b] *Morrow Formation producing wells in Colorado.*
[c] *Wasatch Formation producing wells in Utah.*
[d] *All or parts of Andrews, Winkler, Ector, Midland, Upton, and Crane Counties, Texas.*
[e] *All or parts of Borden, Garza, Scurry, Coke, Mitchell, Irion, Reagan, and Crockett Counties, Texas.*
[f] *All of Campbell, Crook, and Johnson Counties, Wyoming.*

Geomechanical stresses are very important in production schemes. However, due to strong seepage flow, disintegration of formation occurs and sand is carried toward the well opening. The most common practice to prevent accumulation as followed by the industry is take filter measures, such as liners and gravel packs. In general, such measures are very expensive to use and often, due to plugging of the liners, the cost increases to maintain the same

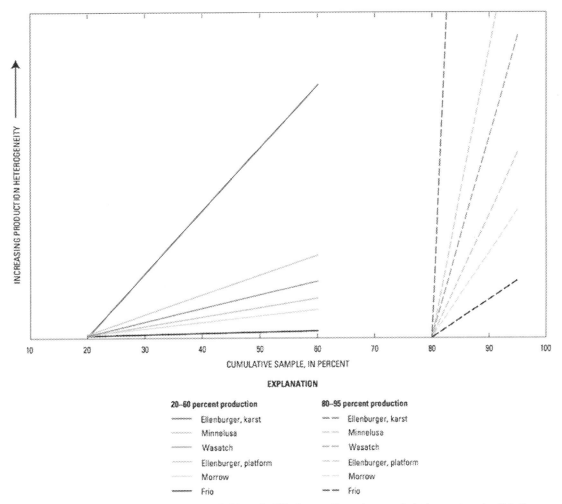

FIG. 4.138 Production data of gas wells in fields in the Ellenburger Group karst and platform categories, Frio Formation fluvial category, Morrow Formation incised-valley category, Minnelusa Formation Minnelusa category, and Wasatch Formation Green River-source category. *From Fishman, N.S., Turner, C.E., Peterson, F., Dyman, T.S., Cook, T., 2008. Geologic Controls on the Growth of Petroleum Reserves. U.S. Geological Survey Bulletin 2172-I, 53 pp.*

level of production. In recent years, there have been studies in various categories of well completion including modeling of coupled fluid flow and mechanical deformation of medium (Vaziri et al., 2001). Vaziri et al. (2001) used a finite element analysis developing a modified form of the Mohr–Coulomb failure envelope to simulate both tensile and shear-induced failure around deep wellbores in oil and gas reservoirs. The coupled model was useful in predicting the onset and quantity of sanding. Nouri et al. (2006) highlighted the experimental part of it in addition to a numerical analysis and measured the severity of sanding in terms of rate and duration. It should be noted that these studies (Nouri et al., 2002, 2006; Vaziri et al., 2001) took into account the elasto-plastic stress-strain relationship with strain softening to capture sand production in a more realistic manner. Although, at present these studies lack validation with field data, they offer significant insight into the mechanism of sanding and have potential in smart-designing of well-completions and operational conditions.

Settari et al. (2008) applied numerical techniques to calculate subsidence induced by gas production in the North Adriatic. Due to the complexity of the reservoir and compaction mechanisms, Settari (2009) took a combined approach of reservoir and geomechanical simulators in modeling subsidence. As well, an extensive validation of the modeling techniques was undertaken, including the level of coupling between the fluid flow and geomechanical solution. The researchers found that a fully coupled solution had an impact only on the aquifer area, and an explicitly coupled technique was good enough to give accurate results. On grid issues, the preferred approach was to use compatible grids in the reservoir domain and to extend that mesh to geomechanical modeling. However, it was also noted that the grids generated for reservoir simulation are often not suitable for coupled models and require modification.

In fields, on several instances, subsidence delay has been noticed and related to over consolidation, which is also termed as the threshold effect (Merle et al., 1976; Hettema et al., 2002). Settari et al. (2008) used the numerical modeling techniques to explore the effects of small levels of over-consolidation in one of their studied fields on the onset of subsidence and the areal extent of the resulting subsidence bowl. The same framework that Settari et al. (2008) used can be introduced in coupling the multiphase, compositional simulator and the geomechanical simulator in future.

4.8.1 Fluid flow modeling under thermal stress

The temperature changes in the rock can induce thermo-elastic stresses (Hojka et al., 1993), which can either create new fractures or alter the shapes of existing fractures, changing the nature of primary mode of production. The thermal stress occurs as a result of the difference in temperature between injected fluids and reservoir fluids or due to the Joule Thompson effect. However, in the study with unconsolidated sand, the thermal stresses are reported to be negligible in comparison to the mechanical stresses (Chalaturnyk and Scot, 1995). Similar trend is noticeable in the work by Chen et al. (1995), which also ignored the effect of thermal stresses, even though a simultaneous modeling of fluid flow and geomechanics is proposed.

Most of the past research has been focused only on thermal recovery of heavy oil. Modeling subsidence under thermal recovery technique (Tortike and Farouq Ali, 1987) was one of the early attempts that considered both thermal and mechanical stresses in their formulation. There are only few investigations that attempted to capture the onset and propagation of fractures under thermal stress. Zekri et al. (2006) investigated the effects of thermal shock on fractured core permeability of carbonate formations of UAE reservoirs by conducting a series of experiments. In addition, the stress-strain relationship due to thermal shocks was noted. Apart from experimental observations, there is also the scope to perform numerical simulations to determine the impact of thermal stress in various categories, such as water injection, gas injection/production, etc.

4.8.2 Challenges of modeling unconventional gas reservoirs

Most modeling approaches in unconventional gas reservoirs involve flow equations that are not valid for unconventional cases. In addition, complexities arise from complex geometry that prevails in unconventional reservoirs. For instance, the traditional approach of modeling fractured shale-gas reservoirs with regular Cartesian grids could be limited in that it cannot efficiently represent complex geologies, including nonplanar and nonorthogonal fractures, and cannot adequately capture the elliptical flow geometries expected around the fracture tips in such fractured reservoirs. It also suffers from the fact that the number of grids can easily grow into millions because of the inability to change the orientation and shape of the grids away from the fracture tips, thus requiring extremely fine discretization in an attempt to describe all possible configurations. This problem is also aggravated by the large number (up to 60) of hydraulic fractures in a clustered fracture system. The problem is more intense for natural fractures.

Moridis et al. (2010) identified the distinct fracture systems present in producing shale-gas and tight-gas reservoirs. Fig. 4.139 shows a graphical illustration of the following four fracture systems.

- Primary or hydraulic fractures: These are fractures that are typically created by injecting hydrofracturing fluids (with or without proppants) into the formation. Proppants provide high-permeability flow paths that allow gas to flow more easily from the formation matrix into the well. Artificial fractures are predominantly orthogonal to natural fractures.
- Secondary fractures: These fractures are termed "secondary" because they are induced as a result of changes in the geomechanical status of a rock when the primary fractures are being created. Microseismic fracture mappings suggest that they generally intersect the primary fractures, either orthogonally or at an angle. Most prior studies assumed ideal

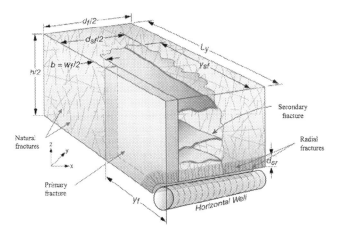

FIG. 4.139 Identification of the four fractured systems. *From Moridis, G.J., Blasingame, T.A., Freeman, C.M., 2010. Analysis of mechanisms of flow in fractured tight-gas and shale-gas reservoirs. In: Paper presented at the SPE Latin American and Caribbean Petroleum Engineering Conference, Lima, Peru, December. Paper Number: SPE-139250-MS. https://doi.org/10.2118/139250-MS.*

configurations with orthogonal and planar fracture intersections so as to simplify the gridding; but in this work, we have developed an unstructured mesh-maker that facilitates the gridding of nonorthogonal, nonplanar and other nonideal fracture geometries.
- Natural fractures: As the name implies, these fractures are native to the formation in the original state, prior to any well completion or fracturing process. These fractures are result of faults that are part of broader tectonic of the reservoir.
- Radial fractures: These are fractures that are created as a result of stress releases in the immediate neighborhood of the horizontal well. This is integral part of the invasion due to drilling.

Efforts have been made to simplify the above geometry in a tractable form. It is recognized the importance of explicitly gridding secondary fractures so as to quantify the interaction between primary and secondary networks as distinct systems, using either a regular orthogonal pattern or a more random and complex system. In this work, we have identified two classes of possible fracture geometries/orientations:
- Regular or ideal fractures: These are idealized fracture geometries, which are usually planar and orthogonal. A perfectly planar (or orthogonal fracture) is the idealized geometry used in numerical studies using Cartesian grids. Fig. 4.140A gives an illustration of this fracture geometry.
- Irregular or nonideal fractures: These are the kinds of fracture geometries we are likely to encounter in real life. They could be nonorthogonal, meaning that the fractures intersect either the well (for primary fractures) or primary fractures (for secondary fractures) at angles other than 90°, and they could be complex, meaning that the fractures are not restricted to a flat, nonundulating plane. Fig. 4.140B and C give a diagrammatic illustration of two such scenarios.

Olorode (2011) reported modeling using nonorthogonal fractures (Fig. 4.141). Fig. 4.141 shows an almost identical rate profile for the nonplanar and nonorthogonal fracture systems at an angle of 60° to the horizontal. The fracture interference in these two cases becomes evident at the same time with the planar case that has a fracture half-length equal to the apparent length, l_a. The nonorthogonal fracture case with $\theta = 30°$ exhibits fracture interference earlier than the other cases, and this can be attributed to the fact that the apparent fracture half-length, l_a, is smaller in comparison to the other nonideal cases. The cumulative production plot shows that the nonideal fractured systems have lower production than the planar cases with the same fracture half-length. This implies that to the extent possible, fractures should be designed such that their angle of inclination with the horizontal well should be as close to 90° as possible. This can be achieved by ensuring that the horizontal well is drilled in the direction of the minimum principal stress, as fractures usually propagate perpendicularly to the direction of the minimum principal stress.

Other challenges in modeling unconventional reservoirs lie within describing fracture flow and considering interactions between matrix and fracture. The original fracture flow model was developed in 1963 by Warren and Root. It assumed that the matrix has no permeability and the porosity is distributed in fractures as well as the matrix. Kazemi et al. (1976) used a slab matrix model with horizontal fractures and unsteady state matrix-fracture flow to represent single-phase flow in the fractured reservoir. This was the first extension of the Warren Root model. The assumptions included homogeneous behavior and isotropic matrix and

4.8 Coupled fluid flow and geomechanical stress model

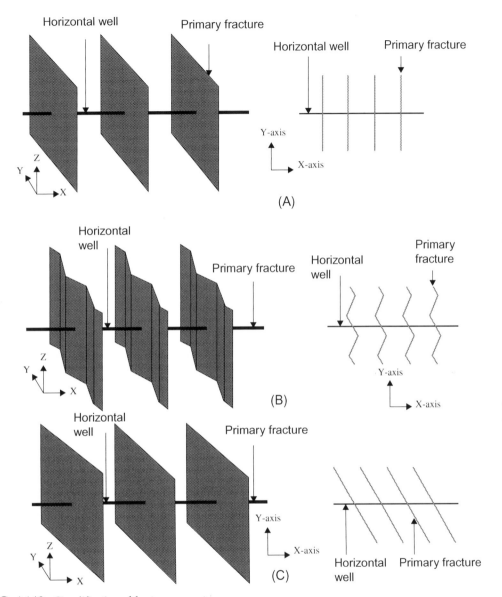

FIG. 4.140 Simplification of fracture geometry.

fracture properties. He further assumed that the well is centrally located in a bounded radial reservoir. A numerical reservoir simulator was used. It was concluded that the results were similar to the Warren and Root model when applied to a drawdown test in which the boundaries have not been detected. Two parallel straight lines were obtained on a semilog plot. The first straight line may be obscured by wellbore storage effects and the second straight line may lead to overestimating ω when boundary effects have been detected.

472 4. Unconventional reservoirs

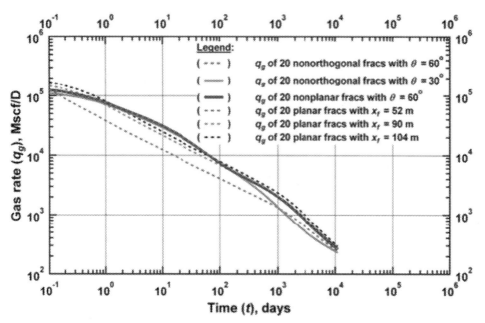

FIG. 4.141 Flow profiles for nonplanar and nonorthogonal fractures. *From Olorode, O.M., 2011. Numerical Modeling of Fractured Shale-Gas and Tight-Gas Reservoirs Using Unstructured Grids (MSc thesis). Texas A&M.*

De Swaan (2013) presented a model which approximates the matrix blocks by regular solids (slab and spheres) and utilizes heat flow theory to describe the pressure distribution. It was assumed that the pressure in the fractures around the matrix blocks is variable and the source term is described through a convolution term. Approximate line-source solutions for early and late time are presented. The late time solutions are similar to those for early time except that modified hydraulic diffusivity terms dependent on fracture and matrix properties are included. The results are two parallel lines representing the early and late time approximations. The late time solution matches Kazemi et al. (1976) for the slab case. De Swaan's model does not properly represent the transition period.

Chen et al. (1985) presented methods for analyzing drawdown and buildup data for a constant rate producing well centrally located in a closed radial reservoir. The slab model similar to De Swaan (2013) and Kazemi et al. (1976) is used. Five flow regimes are presented. Flow regimes 1, 2 and 3 are associated with an infinite reservoir and Flow regime 1 occurs when there is a transient only in the fracture system. Flow regime 2 occurs when the transient occurs in the matrix and fractures. Flow regime 3 is a combination of transient flow in the fractures and "pseudosteady state" in the matrix. Pseudosteady state in the matrix occurs when the no-flow boundary represented by the symmetry center line in the matrix affects the response. Two new flow regimes associated with a bounded reservoir are also presented. Flow regime 4 reflects unsteady linear flow in the matrix system and pseudosteady state in the fractures. Flow Regime 5 occurs when the response is affected by all the boundaries (pseudosteady-state).

Streltsova (1983) applied a "gradient model" (transient matrix-fracture transfer flow) with slab-shaped matrix blocks to an infinite reservoir. The model predicted results which differ from the Warren and Root model in early time but converge to similar values in late time. The model also predicted a linear transitional response on a semilog plot between the early and late time pressure responses which has a slope equal to half that of the early and late time lines. This linear transitional response was also shown to differ from the S-shaped inflection predicted by the Warren and Root model.

Cinco-Ley and Fernando Samaniego (1982) utilized models similar to De Swaan and Najurieta and presented solutions for slab and sphere matrix cases. They utilize new dimensionless variables—dimensionless matrix hydraulic diffusivity, and dimensionless fracture area. They describe three flow regimes observed on a semilog plot—fracture storage dominated flow, "matrix transient linear" dominated flow and a matrix pseudosteady state flow. The "matrix transient linear" dominated flow period is observed as a line with one-half the slopes of the other two lines. It should be noted that the "matrix transient linear" period yields a straight line on a semilog plot indicating radial flow and might be a misnomer. The fracture storage dominated flow is due to fluid expansion in the fractures. The "matrix transient linear" period is due to fluid expansion in the matrix. The matrix pseudosteady state period occurs when the matrix is under pseudosteady state flow and the reservoir pressure is dominated by the total storativity of the system (matrix + fractures). It was concluded that matrix geometry might be identified with their methods provided the pressure data is smooth.

Lai et al. (1983) utilize a one-sixth of a cube matrix geometry transient model to develop well test equations for finite and infinite cases including wellbore storage and skin. Their model was verified with a numerical simulator employing the Multiple Interacting Continua (MINC) method.

Ozkan et al. (1987) presented analysis of flow regimes associated with flow of a well at constant pressure in a closed radial reservoir. The rectangular slab model similar to De Swaan and Kazemi is used. Five flow regimes are presented. Two new regimes are presented-Flow regime 4 reflects unsteady linear flow in the matrix system and occurs when the outer boundary influences the well response and the matrix boundary has no influence. Flow Regime 5 occurs when the response is affected by all the boundaries.

Stewart and Ascharsobbi (1988) presented an equation for interporosity skin which can be introduced into the pseudosteady state and transient models. It should be noted that all the transient models previously described were developed for the radial reservoir cases (infinite or bounded).

El-Banbi (1998) was the first to present transient dual porosity solutions for the linear reservoir case. New solutions were presented for a naturally fractured reservoir using a dual porosity, linear reservoir model. Solutions are presented for a combination of different inner boundary (constant pressure, constant rate, with or without skin and wellbore storage) and outer boundary conditions (infinite, closed, constant pressure).

Kucuk and Sawyer (1988) presented a model for transient matrix-fracture transfer for the gas case. Previous work had been concerned mainly with modeling slightly compressible (liquid) flow. They considered cylindrical and spherical matrix blocks cases. They also incorporate the pseudopressure definitions for gases. Techniques for analyzing buildup data are also presented for shale gas reservoirs. Their model results plotted on a dimensionless basis matched Warren and Root (1963) and Kazemi (1968) for very large matrix blocks at early time but differ at later times. They also conclude from their tests that naturally fractured reservoirs do not always exhibit the Warren and Root behavior (two parallel lines).

Carlson and Mercer (1991) coupled Fick's law for diffusion within the matrix and desorption in their transient radial reservoir model for shale gas. This is the closest to Brinkman formulation that any fluid flow model came in terms of unconventional reservoir modeling. Modifications included use of the pressure-squared forms valid for gas at low pressures to linearize the diffusivity equation. They provide a Laplace space equation for the gas cumulative production from their model and use it to history match a sample well. They also show that semiinfinite behavior (portions of the matrix remain at initial pressure and is unaffected by production from the fractures) occurs in shale gas reservoirs regardless of matrix geometry. They present an equation for predicting the end of this semiinfinite behavior.

Gatens et al. (1987) analyzed production data from about 898 Devonian shale wells in four areas. They present three methods of analyzing production data—type curves, analytical model and empirical equations. The empirical equation correlates cumulative production data at a certain time with cumulative production at other times. This avoids the need to determine reservoir properties. Reasonable matches with actual data were presented. The analytical model is used along with an automatic history matching algorithm and a model selection procedure to determine statistically the best fit with actual data.

Watson (1989) present a procedure that involves selection of the most appropriate production model from a list of models including the dual porosity model using statistics. The analytical slab matrix model is utilized. Reservoir parameters are estimated through a history matching procedure that involves minimizing an objective function comparing measured and estimated cumulative production. They incorporate the use of a normalized time in the analytical model to account for changing gas properties with pressure. Reasonable history matches were obtained with sample field cases but forecast was slightly underestimated.

Spivey and Semmelbeck (1995) presented an iterative method for predicting production from dewatered coal and fractured gas shale reservoirs. The model used is a well producing at constant bottomhole pressure centered in a closed radial reservoir. A slab matrix is incorporated into these solutions. These solutions are extended to the gas case by using an adjusted time and adjusted pressure. Their method also uses a total compressibility term accounting for desorption.

None of these formulations can account for smooth transition between flow regimes. For gas hydrate as well as CBM, additional difficulties arise from the inability of conventional models to simulate phase transition and/or desorption. This is the only model that would be able to model different scenarios with the same model, as depicted in Fig. 4.142.

A dynamic reservoir characterization tool is needed to introduce real-time monitoring. This tool can use the inversion technique to determine permeability data. At present, cuttings need to be collected before preparing petrophysical logs. The numerical inversion requires the solution of a set of nonlinear partial differential equations. Conventional numerical methods require these equations to be linearized prior to solution (discussed early in the paper). In this process, many of the routes to final solutions may be suppressed (Mustafiz et al., 2008b) while it is to be noted that a set of nonlinear equations should lead to the

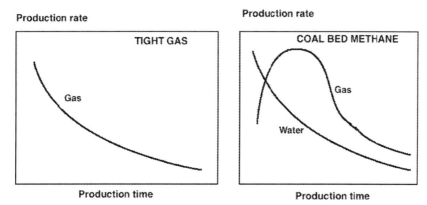

FIG. 4.142 Conventional simulation techniques cannot simulate even fundamental features of various unconventional gas flow.

emergence of multiple solutions. Therefore, it is important that a nonlinear problem is investigated for multiple-value solutions of physical significance.

4.8.3 Comprehensive modeling

This section introduces the governing partial differential equations that describe fluid flow in reservoir, where the matrix/fracture scheme controls the flow behavior. To represent flow behavior in porous media, suitable boundary and initial conditions must be considered and selected, and appropriate equations derived. Such governing equations describing the flow mechanism are the cornerstones of numerical modeling of an unconventional reservoir.

4.8.3.1 Governing equations

In this section, the diffusivity equation of Forchheimer, the modified Brinkman Model and the Comprehensive Model are discussed. Although modeled previously, Darcy's diffusivity equation is included because of its wide use throughout the oil industry. Here it will be used as a reference to compare its predictions with alternative models proposed herein. Forchheimer's model, popular in gas reservoirs' description, is used to derive a new diffusivity equation. What may be called Modified Brinkman's Model (MBM)—MBM's diffusivity equation incorporating both viscous and viscous/frictional terms from the original equation, plus the inertial term from Forchheimer's basic equation—is also developed. The viscous terms of Brinkman's original equation are Darcy's viscous term and another viscous/frictional term representing the forces generated by friction between layers of the fluid and its medium. The Comprehensive is the ultimate proposed model; it has been called "comprehensive" because it includes all terms that might influence fluid flow in porous media in both Darcian and non-Darcian domains. It includes Darcy's viscous term, Forchheimer's inertial term, Brinkman's viscous/frictional term and a Navier-Stokes convective term.

Although the situation of hydraulic fracturing, with the assumption of single horizontal fracture acting in the middle of a matrix system, can be modeled as a separate case, it is more appropriate to describe such a case using the models proposed in this section. Such models are valid for any hydraulic fracture aperture and length, any matrix size and extent, and also taking into account inconsistent and heterogeneous matrix systems, are highly preferable. Another unique model presented here deals with the coupled fluid flow/stress, and it should substantially enhance predictions of depletion while a reservoir is still in production. This model couples fluid flow conceptualization from this study with another stress model that updates petrophysical properties (porosity and permeability) according to effective stress variations arising from depletion, or other injection/flooding recovery techniques. Because, the model uses different flow equations in different sections of the reservoir, there is no need to make further assumptions.

4.8.3.1.1 Darcy's model

Darcy's original equation (Eq. 4.4) has been used to derive a diffusivity equation. This is the most commonly used equation in petroleum engineering analysis. This equation has been used to predict reservoir pressure at any point of space and time within the reservoir. For single phase linear flow, the pressure gradient in the x-coordinate is predicted by this equation:

$$\frac{\partial^2 P}{\partial x^2} = \frac{\phi \mu c}{k} \frac{\partial P}{\partial t} \qquad (4.4)$$

where c, the total compressibility, is the sum of c_f (fluid compressibility) and c_r (rock compressibility).

In two dimensions, the diffusivity equation becomes:

$$\frac{\partial^2 P}{\partial x^2} + \frac{\partial^2 P}{\partial y^2} = \frac{\phi \mu c}{k} \frac{\partial P}{\partial t} \qquad (4.5)$$

If the single phase considered flowing in the porous media is liquid, porosity (ϕ), permeability (k), viscosity (μ) and compressibility (c) are assumed to be constant. This assumption rests on the general and credible belief that liquids flowing in porous media are slightly compressible and the change in rock compressibility is negligible. Therefore, slight changes in these properties result in only negligible changes during flow. However, as will be seen in later section, this is not always the case. Many experimental investigations and field evidence show significant changes in rock properties during processes of depletion.

4.8.3.2 Forchheimer's model

To develop a diffusivity equation similar to that of Darcy (Eqs. 4.4, 4.5) for flow behavior replacing Darcy's linear equation by Forchheimer's nonlinear equation, consider an element of the reservoir through which a single phase (water, for example) is flowing in the x-direction and apply the conservation of mass balance equation (Fig. 4.143):

Mass rate in − Mass rate out = Mass rate of accumulation.

In symbols:

$$(v_x \rho_x \Delta y \Delta z) - (v_{x+\Delta x} \rho_{x+\Delta x} \Delta y \Delta z) = (\Delta x \Delta y \Delta z)\phi \frac{(\rho_{t+\Delta t} - \rho_t)}{\Delta t} \qquad (4.6)$$

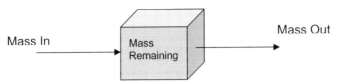

FIG. 4.143 Schematic of the porous system considered for derivation of the diffusivity equation.

Dividing Eq. (4.74) by $\Delta x \Delta y \Delta z$:

$$-\frac{(v_{x+\Delta x}\rho_{x+\Delta x}) - (v_x \rho_x)}{\Delta x} = \frac{\phi(\rho_{t+\Delta t} - \rho_t)}{\Delta t} \quad (4.7)$$

Take the limit as Δx and Δt go to zero simultaneously:

$$\frac{\partial(v\rho)}{\partial(x)} = -\phi \frac{\partial \rho}{\partial t} \quad (4.8)$$

This is the continuity equation in a linear system. We may therefore write this equation in the y and z coordinates:

$$\frac{\partial(v\rho)}{\partial(y)} = -\phi \frac{\partial \rho}{\partial t} \quad (4.9)$$

$$\frac{\partial(v\rho)}{\partial(z)} = -\phi \frac{\partial \rho}{\partial t} \quad (4.10)$$

Taking the derivative of the left-hand side of Eq. (4.6) in two dimensions, we obtain:

$$\rho \frac{\partial v}{\partial x} + v \frac{\partial \rho}{\partial x} + \rho \frac{\partial v}{\partial y} + v \frac{\partial \rho}{\partial y} = -\phi \frac{\partial \rho}{\partial t} \quad (4.11)$$

Now, Forchheimer's equation in the x-direction states:

$$-\frac{\partial P}{\partial x} = \frac{\mu}{k} v + \beta \rho v^2 \quad (4.12)$$

and in the y-direction:

$$-\frac{\partial P}{\partial y} = \frac{\mu}{k} v + \beta \rho v^2 \quad (4.13)$$

β is called the non-Darcy flow coefficient, Forchheimer's coefficient or turbulence coefficient, (1/ft). It is attributed to non-Darcy flow behavior. Several formulae are available for evaluating β, for example:

$$\beta = \frac{5.5 \times 10^9}{k^{5/4} \phi^{3/4}} \quad (4.14)$$

Taking the derivative with respect to x of both sides of Eq. (4.10) and with respect to y of both sides of Eq. (4.11):

$$-\frac{\partial^2 P}{\partial x^2} = \frac{\mu}{k}\frac{\partial v}{\partial x} + 2\beta\rho v\frac{\partial v}{\partial x} \qquad (4.15)$$

$$-\frac{\partial^2 P}{\partial y^2} = \frac{\mu}{k}\frac{\partial v}{\partial y} + 2\beta\rho v\frac{\partial v}{\partial y} \qquad (4.16)$$

Rearranging Eqs. (4.23), (4.24):

$$-\frac{\partial^2 P}{\partial x^2} = \rho\left(\frac{\mu}{k\rho} + 2\beta v\right)\frac{\partial v}{\partial x} \qquad (4.17)$$

$$-\frac{\partial^2 P}{\partial y^2} = \rho\left(\frac{\mu}{k\rho} + 2\beta v\right)\frac{\partial v}{\partial y} \qquad (4.18)$$

From Eq. (4.19):

$$\rho\left(\frac{\partial v}{\partial x}\right) = -\phi\frac{\partial \rho}{\partial t} - v\frac{\partial \rho}{\partial x} \qquad (4.19)$$

Substituting the value of $\rho\left(\frac{\partial v}{\partial x}\right)$ in Eq. (4.15):

$$-\frac{\partial^2 P}{\partial x^2} = \left(\frac{\mu}{k\rho} + 2\beta v\right)\left(-\phi\frac{\partial \rho}{\partial t} - v\frac{\partial \rho}{\partial x}\right) \qquad (4.20)$$

Dividing both sides of Eq. (4.18) by (-1):

$$\frac{\partial^2 P}{\partial x^2} = \left(\frac{\mu}{k\rho} + 2\beta v\right)\left(\phi\frac{\partial \rho}{\partial t} + v\frac{\partial \rho}{\partial x}\right) \qquad (4.21)$$

By chain rule: $\frac{\partial \rho}{\partial x} = \frac{\partial \rho}{\partial P}\frac{\partial P}{\partial x}$ and $\frac{\partial \rho}{\partial t} = \frac{\partial \rho}{\partial P}\frac{\partial P}{\partial t}$ substituting in Eq. (4.19) we get:

$$\frac{\partial^2 P}{\partial x^2} = \left(\frac{\mu}{k\rho} + 2\beta v\right)\left(\phi\frac{\partial \rho}{\partial P}\frac{\partial P}{\partial t} + v\frac{\partial \rho}{\partial P}\frac{\partial P}{\partial x}\right) \qquad (4.22)$$

Rearrange Eq. (4.20):

$$\frac{\partial^2 P}{\partial x^2} = \left(\frac{\mu}{k\rho} + 2\beta v\right)\left(\phi\frac{\partial P}{\partial t} + v\frac{\partial P}{\partial x}\right)\frac{\partial \rho}{\partial P} \qquad (4.23)$$

$\frac{\partial \rho}{\partial P} = c\rho$, substituting in Eq. (4.21):

$$\frac{\partial^2 P}{\partial x^2} = \left(\frac{\mu}{k\rho} + 2\beta v\right)\left(\phi\frac{\partial P}{\partial t} + v\frac{\partial P}{\partial x}\right)c\rho \qquad (4.24)$$

By rearrangement of Eq. (4.22):

$$\frac{\partial^2 P}{\partial x^2} = c\phi\left(\frac{\mu}{k} + 2\rho\beta v\right)\frac{\partial P}{\partial t} + cv\left(\frac{\mu}{k} + 2\rho\beta v\right)\frac{\partial P}{\partial x} \qquad (4.25)$$

Eq. (4.23) is the nonlinear diffusivity equation in one dimension for the Darcy/non-Darcy (Forchheimer Model). It can be expressed in two dimensions:

$$\frac{\partial^2 P}{\partial x^2} + \frac{\partial^2 P}{\partial y^2} = c\left(\frac{\mu}{k} + 2\beta\rho v\right)\left(\phi\frac{\partial P}{\partial t} + v\left[\frac{\partial P}{\partial x} + \frac{\partial P}{\partial y}\right]\right) \qquad (4.26)$$

4.8.3.3 Modified Brinkman's model

The common belief is that non-Darcian behavior can be caused by a number of factors and not the inertial term suggested by Forchheimer alone. Three aspects of non-Darcian behavior have been addressed in this section. The model presented herein is based on the two terms included in the previously derived Forchheimer's model (Darcy's viscous and Forchheimer's inertial) and the viscous/frictional term introduced in Brinkman's basic equation.

Having derived a diffusivity equation based on Forchheimer's fundamental equation, a diffusivity equation based on a combination of Darcy's, Forchheimer's and Brinkman's terms may be derived as following:

Consider the flow in the x-direction for now. The pressure drop in the x-direction would be affected by Darcy's, Forchheimer's and Brinkman's terms. The fundamental equation inferred from this set-up takes the following form:

$$\frac{\partial P}{\partial X} = -\frac{\mu V}{K} - \rho\beta V^2 + \mu'\frac{\partial^2 V}{\partial X^2} \qquad (4.27)$$

Re-arrange:

$$\frac{\partial^2 V}{\partial X^2} = \frac{1}{\mu'}\left(\frac{\partial P}{\partial X} + \frac{\mu}{K}V + \rho\beta V^2\right) \qquad (4.28)$$

From the same Eq. (4.6), we can obtain:

$$\rho\frac{\partial V}{\partial X} + V\frac{\partial \rho}{\partial X} = -\phi\frac{\partial \rho}{\partial t} \qquad (4.29)$$

Take the derivative of both sides of Eq. (4.29) with respect to X:

$$\rho\frac{\partial^2 V}{\partial X^2} + \frac{\partial V}{\partial X}\frac{\partial \rho}{\partial X} + V\frac{\partial^2 \rho}{\partial X^2} + \frac{\partial \rho}{\partial X}\frac{\partial V}{\partial X} = -\phi\frac{\partial^2 \rho}{\partial t \partial X} \qquad (4.30)$$

Substitute Eq. (4.28) into Eq. (4.30):

$$\frac{\rho}{\mu'}\left(\frac{\partial P}{\partial X} + \frac{\mu}{K}V + \rho\beta V^2\right) + \frac{\partial V}{\partial X}\frac{\partial \rho}{\partial X} + V\frac{\partial^2 \rho}{\partial X^2} + \frac{\partial V}{\partial X}\frac{\partial \rho}{\partial X} = -\phi\frac{\partial^2 \rho}{\partial t \partial X} \qquad (4.31)$$

Re-arrange:

$$\frac{\rho}{\mu'}\frac{\partial P}{\partial X} + \frac{\rho\mu}{\mu' K}V + \frac{\rho^2\beta}{\mu'}V^2 + 2\frac{\partial V}{\partial X}\frac{\partial \rho}{\partial X} + V\frac{\partial^2 \rho}{\partial X^2} = -\phi\frac{\partial^2 \rho}{\partial t \partial X} \qquad (4.32)$$

Using chain rule for: $\frac{\partial^2 \rho}{\partial X^2}$ we obtain:

$$\frac{\partial^2 \rho}{\partial X^2} = \frac{\partial}{\partial X}\left(\frac{\partial \rho}{\partial X}\right) = \frac{\partial}{\partial X}\left(\frac{\partial \rho}{\partial P}\frac{\partial P}{\partial X}\right) = \frac{\partial}{\partial X}\left(\rho C\frac{\partial P}{\partial X}\right) = \rho C\frac{\partial^2 P}{\partial X^2} \qquad (4.33)$$

Same way we have: for: $\frac{\partial^2 \rho}{\partial t \partial X}$ we obtain:

$$\frac{\partial^2 \rho}{\partial t \partial X} = \frac{\partial}{\partial t}\left(\frac{\partial \rho}{\partial X}\right) = \frac{\partial}{\partial t}\left(\frac{\partial \rho}{\partial P}\frac{\partial P}{\partial X}\right) = \frac{\partial}{\partial t}\left(\rho C \frac{\partial P}{\partial X}\right) = \rho C \frac{\partial^2 P}{\partial t \partial X} \quad (4.34)$$

And for: $\frac{\partial \rho}{\partial X}$ we obtain:

$$\frac{\partial \rho}{\partial X} = \frac{\partial \rho}{\partial P}\frac{\partial P}{\partial X} = \rho C \frac{\partial P}{\partial X} \quad (4.35)$$

$$C = \frac{1}{\rho}\frac{\partial \rho}{\partial P} \quad (4.36)$$

Substitute for $\frac{\partial^2 \rho}{\partial X^2}$, $\frac{\partial^2 \rho}{\partial t \partial X}$ and $\frac{\partial \rho}{\partial X}$ from Eqs. (4.34), (4.35), (4.36) respectively into Eq. (4.32):

$$\frac{\rho}{\mu'}\frac{\partial P}{\partial X} + \frac{\rho \mu}{\mu' K}V + \frac{\rho^2 \beta}{\mu'}V^2 + 2\rho C \frac{\partial V}{\partial X}\frac{\partial P}{\partial X} + V\rho C \frac{\partial^2 P}{\partial X^2} = -\phi \rho C \frac{\partial^2 P}{\partial t \partial X} \quad (4.37)$$

Effective viscosity "μ'" is defined as:
$\mu' = \frac{\mu}{\varphi}$, substitute into Eq. (4.37):

$$\frac{\rho \phi}{\mu}\frac{\partial P}{\partial X} + \frac{\rho \phi}{K}V + \frac{\rho^2 \beta \phi}{\mu}V^2 + 2\rho C \frac{\partial V}{\partial X}\frac{\partial P}{\partial X} + V\rho C \frac{\partial^2 P}{\partial X^2} = -\phi \rho C \frac{\partial^2 P}{\partial t \partial X} \quad (4.38)$$

Divide both sides of Eq. (4.38) by $-\phi \rho C$:

$$-\frac{1}{C\mu}\frac{\partial P}{\partial X} - \frac{1}{CK}V - \frac{\rho \beta}{C\mu}V^2 - \frac{2}{\phi}\frac{\partial V}{\partial X}\frac{\partial P}{\partial X} - \frac{V}{\phi}\frac{\partial^2 P}{\partial X^2} = \frac{\partial^2 P}{\partial t \partial X} \quad (4.39)$$

Again, using chain rule for $\frac{\partial V}{\partial X}$ we obtain:

$$\frac{\partial V}{\partial X} = \frac{\partial V}{\partial P}\frac{\partial P}{\partial X} \quad (4.40)$$

Substitute in Eq. (4.39):

$$-\frac{1}{C\mu}\frac{\partial P}{\partial X} - \frac{1}{CK}V - \frac{\rho \beta}{C\mu}V^2 - \frac{2}{\phi}\frac{\partial V}{\partial P}\left(\frac{\partial P}{\partial X}\right)^2 - \frac{V}{\phi}\frac{\partial^2 P}{\partial X^2} = \frac{\partial^2 P}{\partial t \partial X} \quad (4.41)$$

$\frac{\partial P}{\partial X}$ is a small quantity and $(\frac{\partial P}{\partial X})2$ is even smaller, therefore: $(\frac{\partial P}{\partial X})2 \approx 0$, substitute in Eq. (4.41):

$$-\frac{1}{C\mu}\frac{\partial P}{\partial X} - \frac{1}{CK}V - \frac{\rho \beta}{C\mu}V^2 - \frac{V}{\phi}\frac{\partial^2 P}{\partial X^2} = \frac{\partial^2 P}{\partial t \partial X} \quad (4.42)$$

Eq. (4.42) is the diffusivity equation in the x-direction, for the y-direction we obtain:

$$-\frac{1}{C\mu}\frac{\partial P}{\partial Y} - \frac{1}{CK}V - \frac{\rho \beta}{C\mu}V^2 - \frac{V}{\phi}\frac{\partial^2 P}{\partial Y^2} = \frac{\partial^2 P}{\partial t \partial Y} \quad (4.43)$$

The summation of Eqs. (4.42), (4.43) gives the diffusivity equation in two dimensions:

$$-\frac{1}{C\mu}\left(\frac{\partial P}{\partial X} + \frac{\partial P}{\partial Y}\right) - \frac{2}{CK}V - \frac{2\rho \beta}{C\mu}V^2 - \frac{V}{\phi}\left(\frac{\partial^2 P}{\partial X^2} + \frac{\partial^2 P}{\partial Y^2}\right) = \frac{\partial}{\partial t}\left(\frac{\partial P}{\partial X} + \frac{\partial P}{\partial Y}\right) \quad (4.44)$$

4.8.3.4 The comprehensive model

In this section, the principle idea underlying the Comprehensive Model is introduced. This is translated into a mathematical form that can be easily used to describe fluid flow in porous media.

The idea is very simple: the Navier-Stokes equations, representing fluid flow between two parallel plates, are applied. Fig. 4.144 schematically describes such a system.

Fluid flow between two parallel plates or in a pipe as described by the Navier-Stokes equations is well documented. The representative equations give highly accurate results. They have been used in many applications and modified to suit many other circumstances. The pressure gradient in the x-direction, as described by the basic Navier-Stokes equations, takes this form:

$$-\frac{\partial P}{\partial x} = \rho\left(\frac{\partial U}{\partial t} + U\frac{\partial U}{\partial x} + V\frac{\partial U}{\partial y}\right) \tag{4.45}$$

In the y-direction, the pressure gradient takes the following form:

$$-\frac{\partial P}{\partial y} = \rho\left(\frac{\partial V}{\partial t} + U\frac{\partial V}{\partial x} + V\frac{\partial V}{\partial y}\right) \tag{4.46}$$

The summation of Eqs. (4.45), (4.46) is the total pressure gradient in two dimensions as described by the Navier-Stokes equations:

$$-\left(\frac{\partial P}{\partial x} + \frac{\partial P}{\partial y}\right) = \rho\left(\frac{\partial U}{\partial t} + \frac{\partial V}{\partial t} + U\frac{\partial U}{\partial x} + V\frac{\partial U}{\partial y} + U\frac{\partial V}{\partial x} + V\frac{\partial V}{\partial y}\right) \tag{4.47}$$

Introducing grain particles in the space between the two plates (Fig. 4.145), would hinder the flow and cause extra resistance forces against Navier-Stokes prescribed fluid flow. Hence, a modification is required to account for this resistance.

The full boundary layer equations were analyzed to investigate forced convection from a horizontal flat porous medium filling the space between the two plates described in Fig. 4.145. The porous material is assumed to be 100% saturated with the flowing fluid. The Navier-Stokes equations are adapted from the case described in Fig. 4.145 where the space between

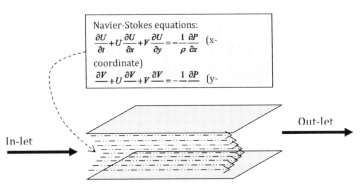

FIG. 4.144 Schematic of fluid flow between two parallel plates described by the Navier-Stokes equations.

FIG. 4.145 Schematic of fluid flow between two parallel plates with pile of grain particles in between causing extra resistance to flow.

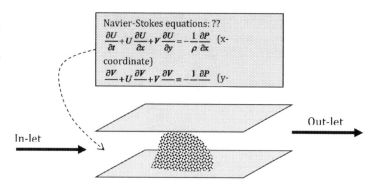

the two plates is occupied by both the volume of grains and the space left between the grain particles through which the flowing fluid runs. Obviously, the fluid will be flowing only through that space (pore space) between grain particles opposed by convective forces in addition to the other viscous, frictional and inertial forces usually associated with fluid flow in a porous system. Therefore, the term "porosity" denoted by "ϕ," become helpful in expressing the ratio of total pore space volume (space between grain particles) to the total volume of the system (total volume bounded by the two plates consists of both grain particles and pores in between):

$$\phi = \frac{\text{(Pore Space Volume)}}{\text{(Total Volume)}} \tag{4.48}$$

The pressure gradient mentioned in both Eqs. (4.45), (4.46) describes the situation where flow is subjected to the "total volume" between the plates. In the case considered in Fig. 4.145, the convective forces are causing resistance to flow and contributing to a pressure gradient only within "pores space" through which the flow is actually taking place. We can define the pores space volume according to Eq. (4.48) then:

$$\text{Pore Space Volume} = \phi \times \text{Total Volume} \tag{4.49}$$

Hence, to express the effect of convective forces on the pressure gradient in the scheme represented by Fig. 4.145, we should multiply by "ϕ" to determine how much of the bulk volume between the plates is exposed to the flow. So Eqs. (4.45), (4.46) become:

$$-\phi\left(\frac{\partial P}{\partial x} + \rho\frac{\partial U}{\partial t}\right) = \rho\left(U\frac{\partial U}{\partial x} + V\frac{\partial U}{\partial y}\right) \tag{4.50}$$

$$-\phi\left(\frac{\partial P}{\partial y} + \rho\frac{\partial V}{\partial t}\right) = \rho\left(U\frac{\partial V}{\partial x} + V\frac{\partial V}{\partial y}\right) \tag{4.51}$$

However, the pressure gradient is affected not only by the convection term from Navier-Stokes. It is also affected by viscous forces (Darcy's term), inertial forces (Forchheimer's term) and the viscous frictional drag forces (Brinkman's term). Considering the contribution of these terms, the above equations would take the following form, in which the x-direction component becomes:

$$-\frac{\partial P}{\partial x} = \rho \left(\frac{\partial U}{\partial t} + \frac{1}{\phi} \left[U \frac{\partial U}{\partial x} + V \frac{\partial U}{\partial y} \right] + \frac{\mu}{K} U + \rho \beta U^2 - \frac{\mu}{\phi} \frac{\partial^2 U}{\partial y^2} \right) \quad (4.52)$$

The y-direction component becomes:

$$-\frac{\partial P}{\partial y} = \rho \left(\frac{\partial V}{\partial t} + \frac{1}{\phi} \left[U \frac{\partial V}{\partial x} + V \frac{\partial V}{\partial y} \right] + \frac{\mu}{K} V + \rho \beta V^2 - \frac{\mu}{\phi} \frac{\partial^2 V}{\partial x^2} \right) \quad (4.53)$$

Rearrange Eqs. (4.52), (4.53) to evaluate velocity (U and V) at any time:

$$\frac{\partial U}{\partial t} = \frac{\mu}{\rho} \frac{\partial^2 U}{\partial y^2} - \frac{\mu \phi}{\rho K} U - \beta \phi U^2 - \frac{1}{\phi} \left(U \frac{\partial U}{\partial x} + V \frac{\partial U}{\partial y} \right) - \frac{\phi}{\rho} \frac{\partial P}{\partial x} \quad (4.54)$$

$$\frac{\partial V}{\partial t} = \frac{\mu}{\rho} \frac{\partial^2 V}{\partial y^2} - \frac{\mu \phi}{\rho K} V - \beta \phi U V^2 - \frac{1}{\phi} \left(U \frac{\partial V}{\partial x} + V \frac{\partial V}{\partial y} \right) - \frac{\phi}{\rho} \frac{\partial P}{\partial y} \quad (4.55)$$

Eqs. (4.54), (4.55) represent the final form of the Comprehensive Model that can be applied to porous media flow. It is capable of addressing a matrix/fracture scheme, including Darcian and non-Darcian domains (Fig. 4.146).

4.8.3.5 Toward solving nonlinear equations

Petroleum reservoir engineering problems are known to be inherently nonlinear. Consequently, solutions to the complete multiphase flow equations have been principally attempted with numerical methods. However, simplified forms of the problem have been

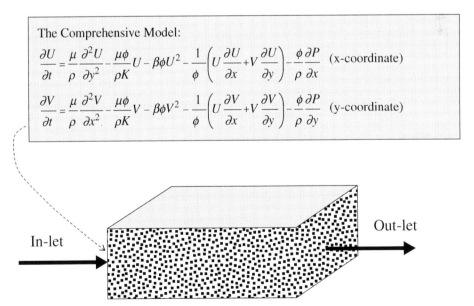

FIG. 4.146 Schematic of fluid flow between two parallel plates filled with grain particles in between representing porous media flow.

solved some 60 years ago, when the Buckley-Leverett formulation was introduced. Ever since that pioneer work that neglected the capillary term, this formulation has been widely accepted in the petroleum industry. By using the method of characteristic, the multiphase one-dimensional fluid flow was solved. However, the resulting solution was a triple-valued one for a significant region. For decades, the existence of multiple solutions was considered to be the result of nonlinearity. Buckley and Leverett introduced shock utilizing the concept of material balance and two decades later, when numerical solutions were possible, it was discovered that the triple-value problem disappeared if the complete flow equation, including the capillary pressure form, is solved. Numerical methods, however, are not free from linearization. Every numerical solution imposes linearization at some point of the solution scheme. Therefore, a numerical technique cannot be used to definitively state the origin of multiple solutions. In this section, a semianalytical technique, the Adomian decomposition method (ADM), capable of solving nonlinear partial differential equations without any linearizing assumptions, is used to unravel the true nature of the one-dimensional, two-phase flow. Results show that the Buckley-Leverett shock is neither necessary nor accurate portrayal of the displacement process. By using the ADM, the solution profile observed through, numerous experimental studies was rediscovered. This paper opens up an opportunity to seek approximate but close to exact solutions to the multiphase flow problems in porous media.

4.8.3.6 *Adomian domain decomposition method*

Since Adomian (1986) proposed his decomposition technique, the Adomian decomposition method (ADM) has gained significant interests among researchers, particularly in the fields of physics and mathematics. They applied this method to many deterministic and stochastic problems (Adomian, 1994). In this method, the governing equation is transformed into a recursive relationship and the solution appears in the form of a power series. The ADM has emerged as an alternative method to solve various mathematical models including algebraic, differential, integral, integro-differential, partial differential equations (PDEs) and systems, higher-order ordinary differential equations, and others.

Guellal et al. (1997) utilized the method to solve an elliptical boundary value problem with an auxiliary condition. Laffez and Abbaoui (1996) used it in modeling thermal exchanges associated to drilling wells. The application of this method is also noticeable in medical research; in which the decomposition technique is used to solve differential system of equations in modeling of HIV immune dynamics (Adjedj, 1999). Wazwaz and El-Sayed (2001) reported that the ADM could be useful in solving problems without considering linearization, perturbation, or unjustified assumptions that may alter the nature of the problem under investigated. In addition, there have been suggestions that this method can be more advantageous over numerical methods by providing analytic, verifiable, rapidly convergent approximations, which add insight into the character and the behavior of the solution as obtainable in the closed form solution (El-Sayed and Abdel-Aziz, 2003). In particular, they noticed the strength of the ADM in handling nonlinear problems in terms of rapid convergence. Biazar and Ebrahimi (2011) also expressed similar notion about rapid convergence and further added the advantage of the technique in terms of considerable savings in computation time when they attempted to solve hyperbolic equations.

Even though petroleum problems are some of the most intrigue candidates for the ADM, there have been sparse attempts to utilize the method in the petroleum problems. Recently, Mustafiz et al. (2005) reported the decomposition of the non-Darcy flow equations in porous media. In a different paper, Mustafiz and Islam (2005) transformed the diffusivity equations in well-test applications into canonical forms. There is some success with Adomian method in modeling drilling activities, such as perforation by drilling (Rahman et al.,2007) specified drilling operations (Biazar et al., 2005).

The Buckley and Leverett (1942) analysis is considered to be the first pioneer work in the study of linear displacement of a fluid by another fluid. The solution of their displacement study on two-phase fluid excluded the effect of capillary and gave multiple results for saturation at a given position. Realizing the fact that such co-existence of multiple saturation values is physically unrealistic, Buckley and Leverett (1942) used the fundamental concept of material-balance to explain the shock. However, the lack of theoretical justification of shock was a major constraint in understanding the displacement phenomenon more vividly than before, and would not be re-evaluated until the next significant work published by Holmgren and Morse (1951). They utilized the Buckley-Leverett theory to calculate the average water saturation at breakthrough and explained dispersion as a consequence of capillary effects. Welge (1952) continued the effort in finding the average saturation at water breakthrough in an oil reservoir. He also found that the method of *tangent construction* by Terwilliger et al. (1951) was equivalent to the shock proposed by Buckley and Leverett.

To solve the displacement equation including capillary as well as gravity, Fayers and Sheldon (1959) attempted a Lagrangian approach. They, however, did not succeed to determine the time required to obtain a particular saturation, which later was explained by Bentsen (1978) revealing the fact that the distance traveled by zero saturation is governed by a separate equation. Bentsen also noted that at slower injection rates, the input boundary condition of constant normalized saturation that Fayers and Sheldon used was incorrect in formulation. In addition, there have been numerical investigations in the past to solve the displacement equation. Hovanessian and Fayers (1961) were able to avoid profiles of multiple valued saturations by considering the capillary pressure.

The Adomian decomposition method is applied to solve the nonlinear Buckley-Leverett equation. The solution for the water saturation is expressed in a series form. The base element of the series solution is obtained using the solution of the linear part of the Buckley-Leverett equation without the effect of the capillary pressure using the characteristic method. The other elements of the series solution are obtained recursively using the Adomian polynomial. The modification of the ADM in selection of the base element of the series solution makes the ADM feasible in solution of the nonlinear Buckley-Leverett equation. The computation is carried out for a reservoir of certain properties and initial conditions.

4.8.4 Governing equations

The Buckley-Leverett equation is given by

$$\frac{\partial S_w}{\partial t} + \frac{q}{A\phi}\frac{\partial f_w}{\partial S_w}\frac{\partial S_w}{\partial x} = 0 \qquad (4.56)$$

where f_w is expressed in the form

$$f_w = \left(\frac{1}{1+\frac{k_{ro}\mu_w}{k_{rw}\mu_o}}\right)\left\{1+\frac{Akk_{ro}}{q\mu_o}\left[\frac{\partial P_c}{\partial x}-(\rho_w-\rho_o)g\sin\alpha\right]\right\}. \qquad (4.57)$$

This equation indicates that the fractional flow rate of water depends on reservoir characteristics, water injection rate, viscosity and direction of flow. The effect of capillary pressure, P_c, which appears in the fractional flow equation, on saturation profiles is important, since these profiles affect the ultimate economic oil recovery (Bentsen, 1978). The ratio of effective permeability to viscosity is defined as the mobility which is shown for water and oil, respectively

$$\lambda_w = \frac{kk_{rw}}{\mu_w} \text{ and } \lambda_o = \frac{kk_{ro}}{\mu_o}. \qquad (4.58)$$

The mobility ratio of oil to water is defined by

$$M = \frac{\lambda_o}{\lambda_w} = \frac{k_{ro}\mu_w}{k_{rw}\mu_o}. \qquad (4.59)$$

The assumptions associated with the Buckley-Leverett equation are:
- oil and water phases are assumed incompressible;
- the porous medium is assumed incompressible, which implies that porosity is constant;
- injection and production are taken care of by means of boundary conditions, which indicate that there are no external sink or source in the porous medium;
- the cross-sectional area that is open to flow is constant;
- the saturation-constraint equation for two-phase flow is valid; and
- the fractional flow of water is dependent on water saturation only.

The expression of the fractional flow rate of water in Eq. (4.57) suggests that Eq. (4.58) is a nonlinear differential equation and it is due to the effect of capillary pressure. In the simplest case of horizontal flow and neglecting the effects of capillary pressure variation along the reservoir, the expression for f_w in Eq. (4.57) is simplified to a linear differential equation, which is given by

$$f_w = \frac{1}{1+M}. \qquad (4.60)$$

The water saturation distribution along the reservoir can be found at different time steps by knowing the injection flow rate, the initial water saturation distribution, and the variation of fractional water flow rate. The water is normally the wetting fluid in the water-oil two phase flow systems. The relative permeability of the water and oil for a specific reservoir are obtained with the drainage and imbibitions process on the core in the laboratory. However, in this paper, the variation of relative permeability to water and oil as a function of water saturation are obtained using from following empirical relationships respectively

4.8 Coupled fluid flow and geomechanical stress model

$$k_{rw} = \alpha_1 S_{wn}{}^{n1} \qquad (4.61)$$

$$k_{ro} = \alpha_2 (1 - S_{wn})^{n2} \qquad (4.62)$$

where the normalized water saturation, S_{wn}, is defined as

$$S_{wn} = \frac{S_w - S_{wi}}{1 - S_{wi} - S_{or}} \qquad (4.63)$$

If the effects of capillary pressure are included in a horizontal reservoir, Eq. (4.57) takes the form

$$f_w = \frac{1}{1 + \frac{k_{ro}\mu_w}{k_{rw}\mu_o}} \left[1 + \frac{A k k_{ro}}{q \mu_o} \frac{\partial P_c}{\partial x} \right]. \qquad (4.64)$$

Capillary pressure at any point is directly related to the mean curvature of the interface, which, in turn, is a function of saturation (Leverett, 1941). Therefore, it can be safely assumed that capillary pressure is only a function of water saturation. By applying the chain rule,

$$P_c = f(S_w) \rightarrow \frac{\partial P_c}{\partial x} = \frac{\partial P_c}{\partial S_w} \frac{\partial S_w}{\partial x} \qquad (4.65)$$

and by incorporating Eqs. (4.63), (4.64) in Eq. (4.65), the following partial differential equation is obtained

$$\frac{\partial S_w}{\partial t} + \frac{q}{A\phi} \frac{\partial}{\partial S_w}\left(\frac{1}{1 + \frac{k_{ro}\mu_w}{k_{rw}\mu_o}}\right) \frac{\partial S_w}{\partial x} + \frac{k k_{ro}}{\mu_o \phi} \frac{\partial}{\partial S_w}\left(\frac{1}{1 + \frac{k_{ro}\mu_w}{k_{rw}\mu_o}}\right) \frac{\partial P_c}{\partial S_w}\left(\frac{\partial S_w}{\partial x}\right)^2$$
$$+ \frac{k}{\mu_o \phi} \left(\frac{1}{1 + \frac{k_{ro}\mu_w}{k_{rw}\mu_o}}\right) \frac{\partial k_{ro}}{\partial S_w} \frac{\partial P_c}{\partial S_w}\left(\frac{\partial S_w}{\partial x}\right)^2 + \frac{k k_{ro}}{\mu_o \phi} \left(\frac{1}{1 + \frac{k_{ro}\mu_w}{k_{rw}\mu_o}}\right) \frac{\partial^2 P_c}{\partial S_w^2}\left(\frac{\partial S_w}{\partial x}\right)^2 = 0. \qquad (4.66)$$

Eq. (4.66) is a nonlinear partial differential equation and the nonlinearity arises because of the inclusion of capillary pressure in it.

4.8.4.1 Adomian decomposition of Buckley-Leverett equation

In the Adomian decomposition method, the solution of a given problem is considered as a series solution. Therefore, the water saturation is expressed as

$$S_w(x,t) = \sum_{n=0}^{\infty} S_{wn}. \qquad (4.67)$$

If a functional equation is considered for water saturation, it can be written in the form

$$\sum_{n=0}^{\infty} S_{wn} = f(x,t) + \sum_{n=0}^{\infty} A_n. \tag{4.68}$$

where $f(x,t)$ is a given function and A_n (S_{w0}, S_{w1}, S_{w2}, ..., S_{wn}) or, simply, A_n's are called Adomian polynomials. The Adomian polynomials are expressed as

$$A_n = \frac{1}{n!} \frac{d^n}{d\lambda^n} \left[N\left(\sum_{i=0}^{\infty} \lambda^i S_{wn}(x,t)_i \right) \right]_{\lambda=0} \tag{4.69}$$

where N is an operator and λ is a parameter. The elements of the series solution for $S_{wn}(x,t)$ are obtained recursively by

$$S_{w0}(x,t) = f(x,t), \quad S_{w1}(x,t) = A_0, \quad \cdots \quad S_{wk}(x,t) = A_{k-1} \quad \cdots. \tag{4.70}$$

By this arrangement, the linear and nonlinear part of the functional (Eq. 4.68) is replaced by a known function using the recursive (Eq. 4.70). By integrating (Eq. 4.66) with respect to t

$$S_w = S_w(x,0) - \int_0^t \left[\frac{q}{A\phi} \frac{\partial}{\partial S_w} \left(\frac{1}{1 + \frac{k_{ro}\mu_w}{k_{rw}\mu_o}} \right) \frac{\partial S_w}{\partial x} \right] dt$$

$$- \int_0^t \left[\frac{kk_{ro}}{\mu_o \phi} \frac{\partial}{\partial S_w} \left(\frac{1}{1 + \frac{k_{ro}\mu_w}{k_{rw}\mu_o}} \right) \frac{\partial P_c}{\partial S_w} \left(\frac{\partial S_w}{\partial x} \right)^2 \right.$$

$$+ \frac{k}{\mu_o \phi} \left(\frac{1}{1 + \frac{k_{ro}\mu_w}{k_{rw}\mu_o}} \right) \frac{\partial k_{ro}}{\partial S_w} \frac{\partial P_c}{\partial S_w} \left(\frac{\partial S_w}{\partial x} \right)^2$$

$$\left. + \frac{kk_{ro}}{\mu_o \phi} \left(\frac{1}{1 + \frac{k_{ro}\mu_w}{k_{rw}\mu_o}} \right) \frac{\partial^2 P_c}{\partial S_w^2} \left(\frac{\partial S_w}{\partial x} \right)^2 \right] dt. \tag{4.71}$$

4.8 Coupled fluid flow and geomechanical stress model

Comparing Eq. (4.71) with Eq. (4.68) and taking into account of Eq. (4.70), the elements of the water saturation series are obtained

$$S_{w0}(x,t) = S_w(0,x) - \int_0^t \left[\frac{q}{A\phi} \frac{\partial}{\partial S_{w0}} \left(\frac{1}{1 + \frac{k_{ro}\mu_w}{k_{rw}\mu_o}} \right) \frac{\partial S_{w0}}{\partial x} \right] dt$$

$$S_{w1}(x,t) = -\int_0^t \left[\frac{kk_{ro}}{\mu_o\phi} \frac{\partial}{\partial S_{w0}} \left(\frac{1}{1 + \frac{k_{ro}\mu_w}{k_{rw}\mu_o}} \right) \frac{\partial P_c}{\partial S_{w0}} + \frac{k}{\mu_o\phi} \left(\frac{1}{1 + \frac{k_{ro}\mu_w}{k_{rw}\mu_o}} \right) \frac{\partial k_{ro}}{\partial S_{w0}} \frac{\partial P_c}{\partial S_{w0}} \right.$$

$$\left. + \frac{kk_{ro}}{\mu_o\phi} \left(\frac{1}{1 + \frac{k_{ro}\mu_w}{k_{rw}\mu_o}} \right) \frac{\partial^2 P_c}{\partial S_{w0}^2} \right] \left(\frac{\partial S_{w0}}{\partial x} \right)^2 dt$$

$$S_{w2}(x,t) = -\frac{d}{d\lambda} \int_0^t \left[\frac{kk_{ro}}{\mu_o\phi} \frac{\partial}{\partial (S_{w0}+S_{w1})} \left(\frac{1}{1 + \frac{k_{ro}\mu_w}{k_{rw}\mu_o}} \right) \frac{\partial P_c}{\partial (S_{w0}+S_{w1})} \right.$$

$$+ \frac{k}{\mu_o\phi} \left(\frac{1}{1 + \frac{k_{ro}\mu_w}{k_{rw}\mu_o}} \right) \frac{\partial k_{ro}}{\partial (S_{w0}+S_{w1})} \frac{\partial P_c}{\partial (S_{w0}+S_{w1})} \quad (4.72)$$

$$\left. + \frac{kk_{ro}}{\mu_o\phi} \left(\frac{1}{1 + \frac{k_{ro}\mu_w}{k_{rw}\mu_o}} \right) \frac{\partial^2 P_c}{\partial (S_{w0}+S_{w1})^2} \right] \left(\frac{\partial S_{w0} + \lambda S_{w1}}{\partial x} \right)^2 dt$$

$$S_{w3}(x,t) = -\frac{1}{2!}\frac{d^2}{d\lambda^2} \int_0^t \left[\frac{kk_{ro}}{\mu_o\phi} \frac{\partial}{\partial (S_{w0}+S_{w1}+S_{w2})} \left(\frac{1}{1 + \frac{k_{ro}\mu_w}{k_{rw}\mu_o}} \right) \frac{\partial P_c}{\partial (S_{w0}+S_{w1}+S_{w2})} \right.$$

$$+ \frac{k}{\mu_o\phi} \left(\frac{1}{1 + \frac{k_{ro}\mu_w}{k_{rw}\mu_o}} \right) \frac{\partial k_{ro}}{\partial (S_{w0}+S_{w1}+S_{w2})} \frac{\partial P_c}{\partial (S_{w0}+S_{w1}+S_{w2})}$$

$$\left. + \frac{kk_{ro}}{\mu_o\phi} \left(\frac{1}{1 + \frac{k_{ro}\mu_w}{k_{rw}\mu_o}} \right) \frac{\partial^2 P_c}{\partial (S_{w0}+S_{w1}+S_{w2})^2} \right] \left(\frac{\partial S_{w0} + \lambda S_{w1} + \lambda^2 S_{w2}}{\partial x} \right)^2 dt$$

$$\ldots \ldots \ldots \ldots \ldots \ldots \ldots \ldots$$

The method is applied to a given reservoir with the initial water saturation of 0.18. This value is corresponding to the irreducible water saturation of the reservoir. Water is injected into the reservoir with a linear flow rate of 1 ft/day. The oil and water viscosities are 1.73 cP and 0.52 cP, respectively. The flow of the displaced phase (oil) ceases at $S_{oc} = 0.1$. The porosity of the medium is 0.25 with an absolute permeability of $k = 10$ md.

The normalized water saturation and the water and oil relative permeability are obtained using

$$S_{wn} = \frac{S_w - 0.18}{1 - 0.1 - 0.18}, \quad k_{rw} = 0.59439\, S_{wn}^4, \quad k_{ro} = (1 - S_{wn})^2 \qquad (4.73)$$

A typical plot of variation of relative permeability to water, k_{rw}, relative permeability to oil, k_{ro}, fractional flow curve, f_w and its derivative, df_w/dS_w are shown in Fig. 4.147. The capillary pressure data are also known as shown in Table 4.41.

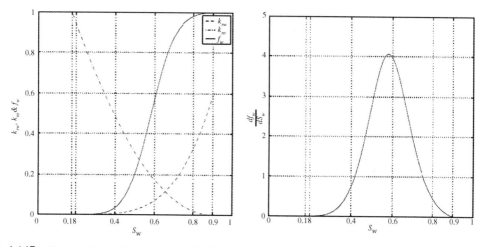

FIG. 4.147 The variation of the water and oil relative permeability, the fractional water flow rate and the differentiation of fractional water flow rate with the water saturation.

TABLE 4.41 Capillary pressure data.

S_{wn}	P_{cr} [atm]	S_{wn}	P_{cr} [atm]	S_{wn}	P_{cr} [atm]	S_{wn}	P_{cr} [atm]
0.00	3.9921	0.11	1.2036	0.30	0.3600	0.65	0.0550
0.01	3.5853	0.15	0.8745	0.36	0.2699	0.72	0.0350
0.02	3.1987	0.18	0.7010	0.42	0.1980	0.87	0.0100
0.05	2.2577	0.21	0.5709	0.48	0.1450	0.95	0.0027
0.08	1.6209	0.25	0.4592	0.56	0.0920	1.00	0.0000

The solution for the first base element of the series solution of the water saturation, S_{w0}, is obtained through the solution of the first equation in Eq. (4.72). It can be written by integrating from the both sides of the first equation in Eq. (4.72) and using Eq. (4.59) that

$$\frac{\partial S_{w0}}{\partial t} + \frac{q}{A\phi} \frac{\partial}{\partial S_{w0}} \left(\frac{1}{1+M}\right) \frac{\partial S_{w0}}{\partial x} = 0. \tag{4.74}$$

This equation suggests that S_{w0} is constant along a direction that is called characteristic direction. The characteristic direction can be obtained by

$$\left(\frac{dx}{dt}\right)_{S_{w0}} = \frac{q}{A\phi} \frac{\partial}{\partial S_{w0}} \left(\frac{1}{1+M}\right)_t \tag{4.75}$$

that is the Buckley-Leverett frontal advance equation. Integrating respect to time, the distribution of S_{w0} is found in the form of

$$x(t, S_{w0}) = \frac{qt}{A\phi} \frac{\partial}{\partial S_{w0}} \left(\frac{1}{1+M}\right)_t. \tag{4.76}$$

The solution of Eq. (4.66) gives the variation of the S_{w0} along the reservoir at certain time. The distribution of S_{w0} along the reservoir is obtained based on the definition of the mobility ratio (Eq. 4.59) and the relations for the relative permeability of water and oil in Eq. (4.73). It corresponds to the variation of the water saturation when the effect of the capillary pressure is ignored. The computations are carried out at different time of $t = 0.5, 1, 2, 5$ and 10 [days]. The solution for the water saturation without the capillary pressure effect shows the unrealistic physical situation that Buckley-Leverett mentioned in their pioneer paper with multiple-saturations at each distance (x-position) as given partly in Fig. 4.148. To avoid multiple saturation values at a particular distance, a saturation discontinuity at a distance, x_f is generally created in such a way that the areas ahead of the front and below the curve are equal to each other.

The other elements of the series solution for the water are obtained recursively using Eq. (4.72) and the solution for S_{w0}. The solution for S_{w1} is obtained using the solution of S_{w0} at a certain point for a given time. To induce the nonlinear dependence of capillary on saturation, during decomposition, the capillary pressure and its derivatives are approximated by using the cubic spline, as shown in Fig. 4.149. The interpolating splines are preferred over interpolating polynomials as they do not suffer from oscillations between the knots and are smoother and more realistic than linear splines. Such technique is also used to observe the effects of linearization in pressure in reservoir flow equations (Mustafiz et al., 2006). The solutions for S_{w2} and S_{w3} are obtained recursively using Eq. (4.72). The computations shows that the series solution converges very fast and it is not necessary to go further than S_{w3}.

The distribution of the water saturation along the reservoir is given in Fig. 4.146. The results of the water saturation without the effect of the capillary pressure are also depicted in Fig. 4.146 for the sake of comparison. This figure shows that by considering the effects of capillary pressure, it is possible to avoid unrealistic multiple saturation values. Moreover, the decomposition approach shows notable prediction of saturation profiles along the length. The gradual and mild changes observed in the saturation profile here, are perhaps less severe than the conventional prediction of shock-fronts.

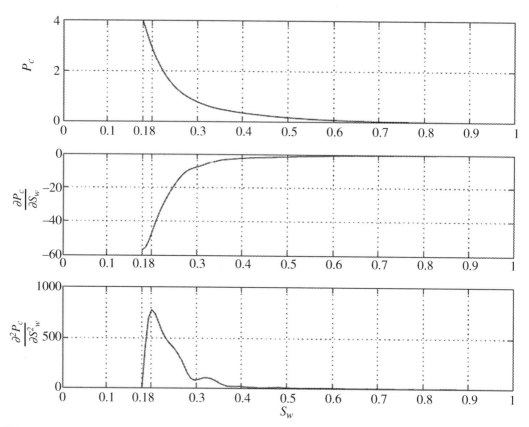

FIG. 4.148 The capillary pressure variation and its first and second derivatives as a function of the water saturation.

4.9 Sustainability pathways

The crude oil is truly a nontoxic, natural, and biodegradable product but the way it is refined is responsible for all the problems created by fossil fuel utilization. The refined oil is hard to biodegrade and is toxic to all living objects. Refining crude oil and processing natural gas use large amount of toxic chemicals and catalysts including heavy metals. These heavy metals contaminate the end products and are burnt along with the fuels producing various toxic by-products. The pathways of these toxic chemicals and catalysts show that they severely affect the environment and public health. The use of toxic catalysts creates many environmental effects that make irreversible damage to the global ecosystem. A detailed pathway analysis of formation of crude oil and the pathway of refined oil and gas clearly shows that the problem of oil and gas operation lies during synthesis or their refining.

4.9.1 Pathways of crude oil formation

Crude oil is a naturally occurring liquid found in formations in the Earth consisting of a complex mixture of hydrocarbons consisting of various lengths. It contains mainly four

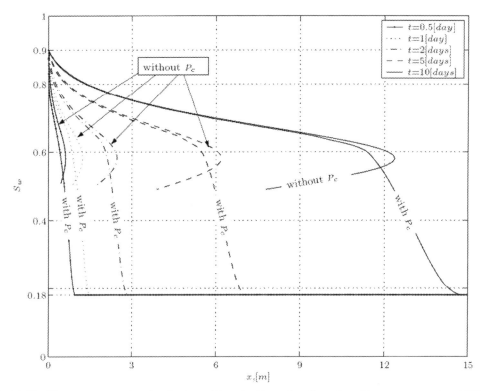

FIG. 4.149 The water saturation distribution with and without the effect of capillary pressure using ADM.

groups of hydrocarbons among, which saturated hydrocarbon consists of straight chain of carbon atoms, aromatics consists of ring chains, asphaltenes consists of complex polycyclic hydrocarbons with complicated carbon rings and other compounds mostly are of nitrogen, sulfur and oxygen. It is believed that crude oil and natural gas are the products of huge overburden pressure and heating of organic materials over millions of years.

Crude oil and natural gases are formed as a result of the compression and heating of ancient organic materials over a long period of time. Oil, gas and coal are formed from the remains of zooplankton, algae, terrestrial plants and other organic matters after exposure to heavy pressure and temperature of Earth. These organic materials are chemically changed to kerogen. With more heat and pressure along with bacterial activities, oil and gas are formed. Fig. 4.150 is the pathway of crude oil and gas formation. These processes are all driven by natural forces.

4.9.2 Pathways of oil refining

Fossil fuels derived from the petroleum reservoirs are refined to suit the various application purposes from car fuels to aeroplane and space fuels. It is a complex mixture of hydrocarbons varying in composition depending on its source. Depending on the number of carbon atoms the molecules contain and their arrangement, the hydrocarbons in the crude oil have

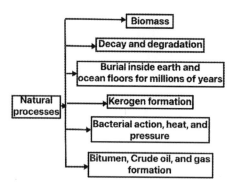

FIG. 4.150 Crude oil formation pathway. *After Chhetri, A.B., Islam, M.R., 2008. Inherently Sustainable Technologies. Nova Science Publishers, 452 pp.*

different boiling points. To take the advantage of the difference in boiling point of different components in the mixture, fractional distillation is used to separate the hydrocarbons from the crude oil. Fig. 4.151 shows general activities involved in oil refining.

Petroleum refining begins with the distillation, or fractionation of crude oils into separate hydrocarbon groups. The resultant products of petroleum are directly related to the properties of the crude processed. Most of the distillation products are further processed into more conventionally usable products changing the size and structure of the carbon chain through several processes by cracking, reforming and other conversion processes. To remove the impurities in the products and improve the quality, extraction, hydrotreating and sweetening are applied. Hence, an integrated refinery consists of fractionation, conversion, treatment and blending including petrochemicals processing units.

Oil refining involves the use of different types of acid catalysts along with high heat and pressure (Fig. 4.152). The process of employing the breaking of hydrocarbon molecules is the thermal cracking. During alkylation, sulfuric acids, hydrogen fluorides, aluminum chlorides and platinum are used as catalysts. Platinum, nickel, tungsten, palladium and other catalysts are used during hydro processing. In distillation, high heat and pressure are used as catalysts.

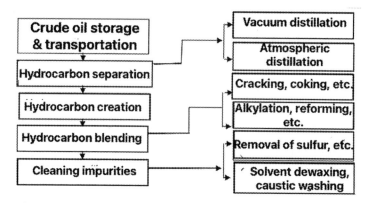

FIG. 4.151 General activities in oil refining (Chhetri and Islam, 2008).

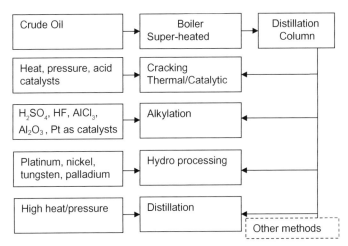

FIG. 4.152 Pathway of oil refining process. *After Chhetri, A.B., Islam, M.R., 2008. Inherently Sustainable Technologies. Nova Science Publishers, 452 pp.*

The use of these highly toxic chemicals and catalysts creates several environmental problems. Their use will contaminate the air, water and land in different ways. Use of such chemicals is not a sustainable option. The pathway analysis shows that current oil refining process is inherently unsustainable.

Refining petroleum products emits several hazardous air toxins and particulate materials. They are produced while transferring and storage of materials and during hydrocarbon separations. Table 4.42 shows the emission released during the hydrocarbon separation process and handling.

Table 4.43 shows the primary waste generated from an oil refinery. In all processes, air toxics and hazardous solid materials, including volatile organic compounds are present.

There are various sources of emissions in the petroleum refining and petrochemical industries, and the following are the major categories of emission sources.

Crude oil is always refined to create value added products. Refining translates directly into value addition. However, the refining process also involves cost-intensive usage of catalysts.

TABLE 4.42 Emission from a refinery (Islam et al., 2010b).

Activities	Emission
Material transfer and storage	– Air release: Volatile organic compounds – Hazardous solid wastes: anthracene, benzene, 1,3-butadiene, curnene, cyclohexane, ethylbenzene, ethylene, methanol, naphthalene, phenol, PAHs, propylene, toluene, 1,2,4-trimethylbenzene, xylene
Separating hydrocarbons	– Air release: Carbon monoxide, nitrogen oxides, particulate matters, sulfur dioxide, VOCs – Hazardous solid waste: ammonia, anthracene, benzene, 1,3-butadiene, curnene, cyclohexane, ethylbenzene, ethylene, methanol, naphthalene, phenol, PAHs, propylene, toluene, 1,2,4-trimethylbenzene, xylene

TABLE 4.43 Primary wastes from oil refinery (Islam et al., 2010b).

Cracking/coking	Alkylation and reforming	Sulfur removal
Air releases: carbon monoxide, nitrogen, oxides, particulate matter, sulfur, dioxide, VOCs	Air releases: carbon monoxide, nitrogen oxides, particulate matter, sulfur dioxide, VOCs	Air releases: carbon monoxide, nitrogen oxides, particulate, matter, sulfur dioxide, VOCs
Hazardous/solid wastes, wastewater, ammonia, anthracene, benzene, 1,3-butadiene, copper, cumene, cyclohexane, ethylbenzene, ethylene, methanol, naphthalene, nickel, phenol, PAHs, propylene, toluene, 1,2,4-trimethylbenzene, vanadium (fumes and dust), xylene	Hazardous/solid wastes: ammonia, benzene, phenol, propylene, sulfuric acid aerosols or hydrofluoric acid, toluene, xylene Wastewater	Hazardous/solid wastes: ammonia, diethanolamine, phenol, metals Wastewater

Catalysts act as denaturing agent. Such denaturing creates products that are also unnatural and for them to be useful special provisions have to be made. For instance, vehicle engines are designed to run with gasoline, aircraft engines with kerosene, diesel engines with diesel, and so on. In the modern era, there have been few attempts to use crude oil in its natural state, and the main innovations have been in the topic of enhancing performance with denatured fluids. Consequently, any economic calculations presume that these are the only means of technology development and makes any possibility of alternate design invariably unsuitable for economic considerations.

Catalysis started to play a major role in every aspect of chemical engineering beginning with the 20th century, in sync with plastic revolution. Today, more than 95% of chemicals produced commercially are processed with at least one catalytic step. These chemicals include the food industry. Fig. 4.153 shows the introduction of major industrial catalytic processes as a function of time. Even though it appears that catalysis is a mature technology, new catalysts continue to be developed. The focus now has become in developing catalysts that are more efficient and muffle the toxicity. World catalysis sales accounted for $7.4 billion in 1997 and today it is estimated to be over $20 billion in 2018.

The processing and refining industry depend exclusively on the use of catalysts that themselves are extracted from natural minerals through a series of unsustainable processing, each step involving rendering a material more toxic while creating profit for the manufacturer. The following operations, mostly involving hydroprocessing applications, use numerous catalysts:

- tail gas treating;
- alkylation pretreatment;
- paraffin isomerization;
- xylene isomerization;
- naphtha reforming (fixed and moving bed);
- gasoline desulfurization;
- naphtha hydrotreating;
- distillate hydrotreating;
- fluidized catalytic cracking pretreatment;
- hydrocracking pretreatment;

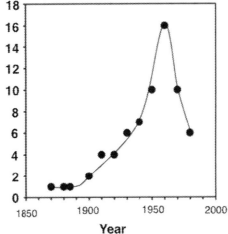

FIG. 4.153 Summary of the historical development of the major industrial catalytic processes per decade in the 20th century. *From Farnetti, E., Di Monte, R., Kašpar, J., 2000. Inorganic and Bio-Inorganic Chemistry – vol. II – Homogeneous and Heterogeneous Catalysis. Encyclopedia of Life Support Systems (EOLSS).*

- hydrocracking;
- lubricant production (hydrocracking, hydrofinishing and dewaxing);
- fixed and ebullated-bed residue hydrotreating; and
- catalyst-bed grading products.
- Process description

Each of these processes involve selection of proprietary reactor internals that are part of conventional design optimization. The entire optimization takes place after fixing the chemicals to be used during the refining process. Some examples are:

- Catalytic naphtha reforming;
- Dimerization, Isomerization (C_4);
- Isomerization (C_5 and C_6);
- Isomerization (xylenes);
- Fluid catalytic cracking (FCC);
- Hydrocracking, Mild hydrocracking;
- Hydrotreating/hydrogenation/saturation;
- Hydrorefining;
- Polymerization;
- Sulfur (elemental) recovery
- Steam hydrocarbon reforming;
- Sweetening;
- Claus unit tail gas treatment;
- Oxygenates;
- Combustion promoters (FCC);
- Sulfur oxides reduction (FCC).

Yet, each step of the refining process is remarkably simple and can work effectively without the addition of natural material in their natural state (without extraction of toxic chemicals). The distillation of crude oil into various fractions will give naphtha as a fraction which ranges from C_5 to 160 degrees (initial to final boiling point). This fraction is further treated to remove sulfur, nitrogen and oxygen which is commonly known as "hydrotreating" and rearranged for improving octane number which can be done by "continuous catalytic reforming (for heavy naphtha which starts from C_7)" or "isomerization (for light naphtha which contains only C_6 and C_7 molecules)" and after that blended for desired spec (BS-III, BS-IV or euro IV, euro V, etc.) and sold in market as gasoline through gas stations.

Some of the processes are given below.

4.9.3 Catalytic cracking

Cracking is the name given to breaking up large hydrocarbon molecules into smaller and more useful bits. This is achieved by using high pressures and temperatures without a catalyst, or lower temperatures and pressures in the presence of a catalyst.

The source of the large hydrocarbon molecules is often the naphtha fraction or the gas oil fraction from the fractional distillation of crude oil (petroleum). These fractions are obtained from the distillation process as liquids but are re-vaporized before cracking.

The hydrocarbons are mixed with a very fine catalyst powder. The efficiency being inversely proportional to the grain size such powder forms are deemed necessary. For this stage, zeolites (natural aluminosilicates) that are more efficient than the older mixtures of aluminum oxide and silicon dioxide can render the process move toward sustainability. For decades, it has been known that zeolites can be effective catalysts (Turkevich and Ono, 1969).

The whole mixture is blown rather like a liquid through a reaction chamber at a temperature of about 500°C. Because the mixture behaves like a liquid, this is known as fluid catalytic cracking (or fluidized catalytic cracking). Although the mixture of gas and fine solid behaves as a liquid, this is nevertheless an example of heterogeneous catalysis—the catalyst is in a different phase from the reactants.

The catalyst is recovered afterward, and the cracked mixture is separated by cooling and further fractional distillation. There is not a single unique reaction taking place in the cracker. The hydrocarbon molecules are broken up random way to produce mixtures of smaller hydrocarbons, some of which have carbon-carbon double bonds. One possible reaction involving the hydrocarbon $C_{15}H_{32}$ might be:

$$\underset{\text{Catalyst}}{C_{15}H_{32}} \rightarrow 2\,\underset{\text{ethane}}{C_2H_4} + \underset{\text{propane}}{C_3H_6} + \underset{\text{ocatane}}{C_8H_{18}}$$

This is only one way in which this particular molecule might break up. The ethene and propene are important materials for making plastics or producing other organic chemicals. The octane is one of the molecules found in gasoline. A high-octane gasoline fetches height value in the retail market.

4.9.4 Isomerization

Hydrocarbons used in petrol (gasoline) are given an octane rating which relates to how effectively they perform in the engine. A hydrocarbon with a high octane rating burns more smoothly than one with a low octane rating.

Molecules with "straight chains" have a tendency to preignition. When the fuel/air mixture is compressed it tends to explode, and then explode a second time when the spark is passed through them. This double explosion produces knocking in the engine.

Octane ratings are based on a scale on which heptane is given a rating of 0, and 2,2,4-trimethylpentane (an isomer of octane) a rating of 100. To raise the octane rating of the molecules found in gasoline to enhance the combustion efficiency in an engine, the chemical branch of oil industry rearranges straight chain molecules into their isomers with branched chains.

One process uses a platinum catalyst on a zeolite base at a temperature of about 250°C and a pressure of 13–30 atm. It is used particularly to change straight chains containing 5 or 6 carbon atoms into their branched isomers. The problem here, of course, is that platinum is highly toxic to the environment in its pure form (after mineral processing). The same result can be achieved by using platinum ore and making adjustment to the volume of the reactor. Because platinum ore is natural, it would be free from the toxicity of pure platinum. It is also possible that there are other alternatives to platinum ore—a subject that has to be researched.

4.9.5 Reforming

Reforming is another process used to improve the octane rating of hydrocarbons to be used in gasoline. It is also a useful source of aromatic compounds for the chemical industry. Aromatic compounds are ones based on a benzene ring.

Once again, reforming uses a platinum catalyst suspended on aluminum oxide together with various promoters to make the catalyst more efficient. The original molecules are passed as vapors over the solid catalyst at a temperature of about 500°C. This process has two levels of toxic addition that has to be corrected. The first one is platinum and the second one is aluminum oxide and its related promoters. We have already seen how zeolite contains aluminum silicate that can replace aluminum oxide. In addition, other natural materials are available that can replace aluminum oxide.

Isomerization reactions occur but, in addition, chain molecules get converted into rings with the loss of hydrogen. Hexane, for example, gets converted into benzene, and heptane into methylbenzene.

The overall picture of conventional refining and how it can be transformed is given in Fig. 4.154. The economics of this transition is reflected in the fact that the profit made through conventional refining would be directly channeled into reduced cost of operation.

This figure amounts to the depiction of a paradigm shift. The task of reverting to natural from unnatural has to be performed for each stage involved in the petroleum refining sector. Table 4.44 shows various processes involved and the different derivatives produced. Note that each of the products later becomes a seed for further use in all aspects of our lifestyle. As a consequence, any fundamental shift from unsustainable to sustainable would reverberate globally.

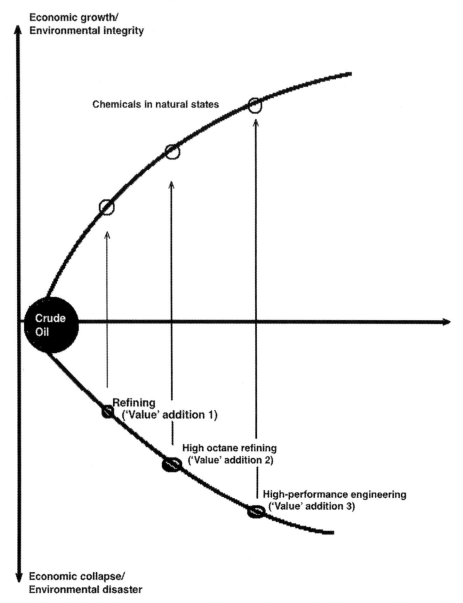

FIG. 4.154 Natural chemicals can turn an sustainable process into a sustainable process while preserving similar efficiency.

4.9.6 Pathways of gas processing

Natural gas is a mixture of methane, ethane, propane, butane and other hydrocarbons, water vapor, oil and condensates, hydrogen sulfides, carbon dioxide, nitrogen, some other gases and solid particles. The free water and water vapors are corrosive to the transportation

4.9 Sustainability pathways

TABLE 4.44 Overview of petroleum refining processes (Islam et al., 2010b).

Process name	Action	Method	Purpose	Feedstock(s)	Product(s)
Fractionation processes					
Atmospheric distillation	Separation	Thermal	Separate fractions	Desalted crude oil	Gas, gas oil, distillate, residual
Vacuum distillation	Separation	Thermal	Separate w/o cracking	Atmospheric tower residual	Gas oil, lube stock, residual
Conversion processes—Decomposition					
Catalytic cracking	Alteration	Catalytic	Upgrade gasoline	Gas oil, coke distillate	Gasoline, petrochemical feedstock
Coking	Polymerize	Thermal	Convert vacuum residuals	Gas oil, coke distillate	Gasoline, petrochemical feedstock
Hydro-cracking	Hydrogenate	Catalytic	Convert to lighter HC's	Gas oil, cracked oil, residual	Lighter, higher-quality products
[a]Hydrogen steam reforming	Decompose	Thermal/catalytic	Produce hydrogen	Desulfurized gas, O_2, steam	Hydrogen, CO, CO_2
[a]Steam cracking	Decompose	Thermal	Crack large molecules	Atm tower hvy fuel/distillate	Cracked naphtha, coke, residual
Visbreaking	Decompose	Thermal	Reduce viscosity	Atmospheric tower residual	Distillate, tar
Conversion processes—Unification					
Alkylation	Combining	Catalytic	Unite olefins and isoparaffins	Tower isobutane/cracker olefin	Iso-octane (alkylate)
Grease compounding	Combining	Thermal	Combine soaps and oils	Lube oil, fatty acid, alky metal	Lubricating grease
Polymerizing	Polymerize	Catalytic	Unite 2 or more olefins	Cracker olefins	High-octane naphtha, petrochemical stocks
Conversion processes—Alteration or rearrangement					
Catalytic reforming	Alteration/dehydration	Catalytic	Upgrade low-octane naphtha	Coker/hydro-cracker naphtha	High oct. Reformate/aromatic
Isomerization	Rearrange	Catalytic	Convert straight chain to branch	Butane, pentane, hexane	Isobutane/pentane/hexane

Continued

TABLE 4.44 Overview of petroleum refining processes (Islam et al., 2010b)—cont'd

Process name	Action	Method	Purpose	Feedstock(s)	Product(s)
Treatment processes					
Amine treating	Treatment	Absorption	Remove acidic contaminants	Sour gas, HCs w/CO_2 and H_2S	Acid free gases and liquid HCs
Deslating	Dehydration	Absorption	Remove contaminants	Crude oil	Desalted crude oil
Drying and sweetening	Treatment	Abspt/therm	Remove H_2O and sulfur cmpds	Liq Hcs, LPG, alky feedstk	Sweet and dry hydrocarbons
[a]Furfural extraction	Solvent extr.	Absorption	Upgrade mid distillate and lubes	Cycle oils and lube feedstocks	High quality diesel and lube oil
Hydrodesulfurization	Treatment	Catalytic	Remove sulfur, contaminants	High-sulfur residual/gas oil	Desulfurized olefins
Hydrotreating	Hydrogenation	Catalytic	Remove impurities, saturate HC's	Residuals, cracked HC's	Cracker feed, distillate, lube
Phenol extraction	Solvent extr.	Abspt/therm.	Improve visc. index, color	Lube oil base stocks	High quality lube oils
Solvent deasphalting	Treatment	Absorption	Remove asphalt	Vac. tower residual, propane	Heavy lube oil, asphalt
Solvent dewaxing	Treatment	Cool/filter	Remove wax from lube stocks	Vac. tower lube oils	Dewaxed lube basestock
Solvent extraction	Solvent extr.	Abspt/precip.	Separate unsat. oils	Gas oil, reformate, distillate	High-octane gasoline
Sweetening	Treatment	Catalytic	Remove H_2S, convert mercaptan	Untreated distillate/gasoline	High-quality distillate/gasoline

[a] *Note: These processes are not depicted in the refinery process flow chart.*

equipment. Hydrates can plug the gas accessories creating several flow problems. Other gas mixtures such as hydrogen sulfide and carbon dioxide are known to lower the heating value of natural gas by reducing its overall fuel efficiency. There are certain restrictions imposed on major transportation pipelines on the make-up of the natural gas that is allowed into the pipeline called pipe "line quality" gas. This makes mandatory that natural gas be purified before it is sent to transportation pipelines. The gas processing is aimed at preventing corrosion, environmental and safety hazards associated with transport of natural gas.

The presence of water in natural gas creates several problems. Liquid water and natural gas can form solid ice-like hydrates that can plug valves and fittings in the pipeline. Natural gas containing liquid water is corrosive, especially if it contains carbon dioxide and hydrogen sulfide. Water vapor in natural gas transport systems may condense causing a sluggish flow. Hence, the removal of free water, water vapors, and condensates is a very important step during gas processing. Other impurities of natural gas, such as carbon dioxide and hydrogen sulfide generally called as acid gases must be removed from the natural gas prior to its transportation. Hydrogen sulfide is a toxic and corrosive gas which is rapidly oxidized to form sulfur dioxide in the atmosphere. Oxides of nitrogen found in traces in the natural gas may cause ozone layer depletion and global warming.

Fig. 4.155 illustrates the pathway of natural gas processing from reservoir to end uses. This figure also shows various emissions from natural gas processing from different steps. After the exploration and production, natural gas stream is sent through the processing systems.

Fig. 4.156 is the schematic of general gas processing system. Glycol dehydration is used for water removal from the natural gas stream. Similarly, methanolamines (MEA) and Diethanolamine (DEA) are used for removing H_2S and CO_2 from the gas streams (Fig. 4.156). Since these chemicals are used for gas processing, it is impossible to completely free the gas from these chemicals. Glycols and amines are very toxic chemicals. Burning of ethylene glycols produces carbon monoxide and when the natural gas is burned in the stoves, it is possible that the emission produces carbon monoxide. Carbon monoxide is a poisonous gas and very harmful for the health and environment. Similarly, amines are also toxic chemicals and burning the gas contaminated by amines produces toxic emissions. Despite the prevalent notion that natural gas burning is clean, the emission is not free from environmental problems. It is reported that one of the highly toxic compounds released in natural gas stoves burning (LPG in stoves) is isobutene, which causes hypoxia in the human body (Sugie, 2004).

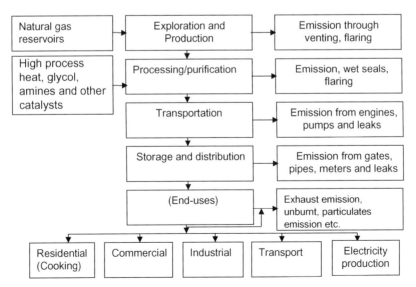

FIG. 4.155 Natural gas "well to wheel" pathway.

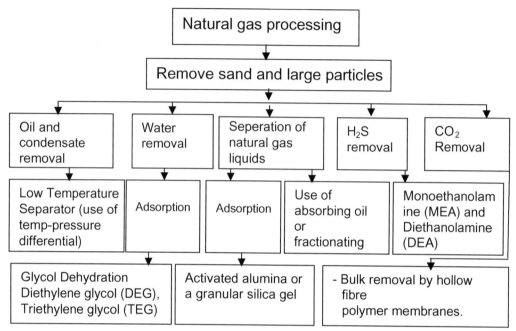

FIG. 4.156 Natural gas processing methods. *Redrawn from Islam et al., 2010.*

4.9.7 Pathways of glycol and amines

Conventional natural gas processing process consists of applications of various types of chemicals and polymeric membranes. These are all synthetic products that are derived from petroleum sources but after a series of denaturing. The common chemicals used to remove water, CO_2 and H_2S are Diethylene glycol (DEG) and Triethylene glycol (TEG) and Monoethanolamines (MEA), Diethanolamines (DEA) and Triethanolamine (TEA). These are synthetic chemicals and have various health and environmental impacts. Synthetic polymers used as membrane during gas processing are highly toxic and their production involves using highly toxic catalysts, chemicals, excessive heat and pressures (Islam et al., 2010). Hull et al. (2002) reported combustion toxicity of ethylene-vinyl acetate copolymer (EVA) reported higher yield of CO and several volatile compounds along with CO_2. Islam et al. (2010) reported that the oxidation of polymers produces more than 4000 toxic chemicals, 80 of which are known carcinogens.

A study on electro oxidation of methanol and glycol and found that electro-oxidation of ethylene glycol at 400 mV forms glycolate, oxalate and formate (Fig. 4.157). The glycolate was obtained by three-electron oxidation of ethylene glycol, and was an electrochemically active product even at 400 mV, which led to the further oxidation of glycolate. Oxalate was found stable, no further oxidation was seen and was termed as nonpoisoning path. The other product of glycol oxidation is called formate which is termed as poisoning path or CO poisoning path. The glycolate formation decreased from 40% to 18% and formate increased from 15% to 20% between 400 and 500 mV. Thus, ethylene glycol oxidation produced CO instead of

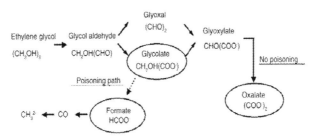

FIG. 4.157 Ethylene glycol oxidation pathway in alkaline solution.

CO_2 and follows the poisoning path over 500 mV. The glycol oxidation produces glycol aldehyde as intermediate products. Hence, the use of these products in refining will have several impacts in the end uses, and are not sustainable at all.

Glycol ethers are known to produce toxic metabolites such as the teratogenic methoxyacetic acid during biodegradation, the biological treatment of glycol ethers can be hazardous (Fischer and Hahn, 2005). Abiotic degradation experiments with ethylene glycol showed that the by-products are monoethylether (EGME) and toxic aldehydes, e.g., methoxy acetaldehyde (MALD). Glycol passes into body by inhalation, ingestion or skin. Toxicity of ethylene glycol causes depression and kidney damage (MSDS, 2005). High concentration levels can interfere with the ability of the blood to carry oxygen causing headache and a blue color to the skin and lips (*methemoglobinemia*), collapse and even death. High exposure may affect the nervous system and may damage the red blood cells leading to anemia (low blood count). During a study of carcinogenetic and toxicity of propylene glycol on animals, the skin tumor incidence was observed. Glycol may form toxic alcohol inside human body if ingested as fermentation may take place.

Amines are considered to be toxic chemicals. It was reported that occupational asthma was found in people handling of a cutting fluid containing diethanolamine (Piipari et al., 1998). Toninello et al. (2006) reported that the oxidation products of some biogenic amines appear to be also carcinogenic. DEA also reversibly inhibits phosphatidylcholine synthesis by blocking choline uptake (Lehman-McKeeman and Gamsky, 1999). Systemic toxicity occurs in many tissue types including the nervous system, liver, kidney, and blood system that may cause increased blood pressure, diuresis, salivation, and pupillary dilation. Diethanolamine causes mild skin irritation to the rabbit at concentrations above 5%, and severe ocular irritation at concentrations above 50% (Beyer et al., 1983). Ingestion of diethylamine causes severe gastrointestinal pain, vomiting, and diarrhea, and may result in perforation of the stomach possibly due to the oxidation products and fermentation products (Table 4.45).

Bjorndalen et al. (2005) developed a novel approach to avoid flaring during petroleum operations. Petroleum products contain materials in various phases. Solids in the form of fines, liquid hydrocarbon, carbon dioxide, and hydrogen sulfide are among the many substances found in the products. According to Bjorndalen et al. (2005), by separating these components through the following steps, no-flare oil production can be established (Fig. 4.158). Simply by avoiding flaring, over 30% of pollution created by petroleum operation can be reduced. Once the components for no-flaring have been fulfilled, value added end products can be

TABLE 4.45 The HSSA pathway in energy management schemes.

Natural state	First stage of intervention	Second stage of intervention	Third stage of intervention
Honey	Sugar	Saccharin	Aspartame
Crude oil	Refined oil	High-octane refining	Chemical additives for combating bacteria, thermal degradation, weather conditions, etc.
Solar	Photovoltaics	Storage in batteries	Re-use in artificial light forms
Organic vegetable oil	Chemical fertilizer, pesticides	Refining, thermal extractions	Genetically modified crops\s
Organic saturated fat	Hormones, antibiotics	Artificial fat (transfat)	No-transfat artificial fat
Wind	Conversion into electricity	Storage in batteries	Re-usage in artificial energy forms
Water and hydro-energy	Conversion into electricity	Dissociation utilizing toxic processes	Recombination through fuel cells
Uranium ore	Enrichment	Conversion into electrical energy	Re-usage in artificial energy forms

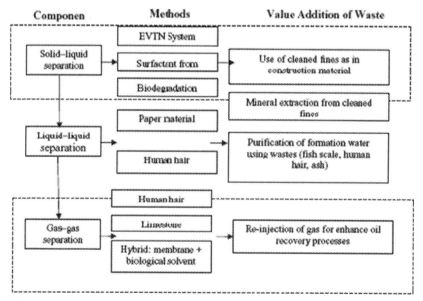

FIG. 4.158 Breakdown of the no-flaring method (Bjorndalen et al., 2005).

developed. For example, the solids can be used for minerals, the brine can be purified, and the low-quality gas can be re-injected into the reservoir for EOR.

4.9.8 Direct use of solar energy

The petroleum industry rarely considered the use of solar energy as any petroleum site has plenty of fossil fuel in hand. However, as we have observed in previous chapters, solar-assisted steam injection projects have been initiated in Oman and California. With renewed focus on environmental impacts, companies are turning to solar energy, especially in places, where the sunlight is plentiful.

For any solar high-temperature system, a solar contractor is necessary. Such collector will have much higher efficiency than conversion to electricity if used directly. Fig. 4.159 shows a widely used steam power plant, which transforms heat energy to electrical energy.

The solar collector efficiency indicates the fraction of solar energy that can be transferred to the thermal fluid in the receiver. The parabolic solar collector efficiency varies much on the fluid temperature. Markus Eck and Steinmann (2005) reported that the collector efficiency shows higher at low temperature ranges (Fig. 4.160).

At low fluid temperatures, the thermal loss is minimal, as shown in Fig. 4.161. From the figure, it is found that at a fluid temperature of 100°C (78°C above ambient temperature), the efficiency of solar collector is 75%.

The solar transmission efficiency is dependent on the heat transfer loss from the thermal fluid to the fluid in the generator and the bubble pump. An efficient system will have more than 90% efficiency of transmission.

If the efficiency of the solar system is calculated from the solar energy on the parabolic surface to the heat transfers to the heating fluid, the overall efficiency will be:

$$\text{Overall energy transfer efficiency} = \text{Collector efficiency } (75\%) \\ \times \text{Transmission efficiency } (90\%).$$

$$\text{Overall energy transfer efficiency} = 67.5\% \tag{4.77}$$

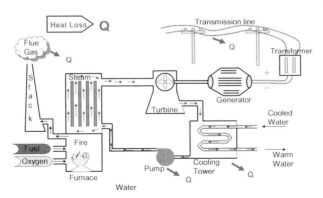

FIG. 4.159 Typical steam power plant.

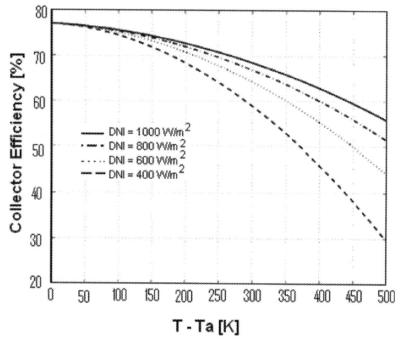

FIG. 4.160 Collector efficiency at different direct normal irradiance (DNI) as a function of fluid temperatures above the ambient temperatures. *Redrawn from Markus Eck, M., Steinmann, W.D., 2005. Modelling and design of direct solar steam generating collector fields. J. Solar Energy Eng. 127 (3), 371–380.*

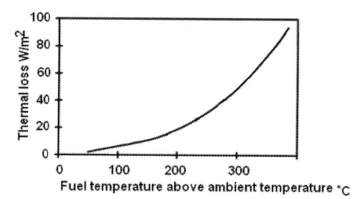

FIG. 4.161 The thermal loss of the collector with respect to fluid temperature above the ambient temperature. *Redrawn from Odeh, S., Morrison, G.L., 2006. Optimization of parabolic solar collector system. Int. J. Energy Res. 30 (4), 259–271. doi:10.1002/er.1153.*

It can be speculated that the extraction process of energy from different processes do not differ much. So the consideration of a solar system is beneficial as it has other benefits as discussed earlier.

There are some existing efficient methods to concentrate the dispersed solar energy and transfer to the desired places. The most common method is the use of a parabolic trough (Fig. 4.162) for the concentration of solar energy to obtain high temperatures without any serious degradations in the collector's efficiency. The parabolic trough collector consists of large curved mirror, which can concentrate the sunlight by a factor of 80 or more to a focal line depending upon the surface area of the trough. In the focal line of these is a metal absorber tube, which is usually embedded into an evacuated glass tube that reduces heat losses (Fig. 4.162). A special high-temperature, resistive selective coating additionally reduces radiation heat losses (Fig. 4.163).

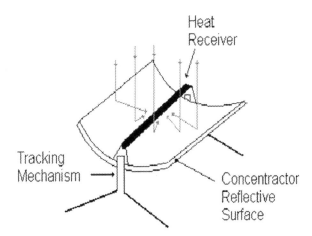

FIG. 4.162 Parabolic trough.

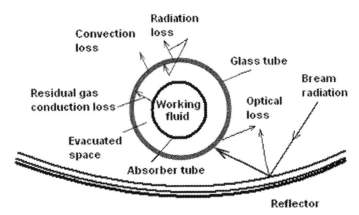

FIG. 4.163 Cross section of collector assembly. *Redrawn from Odeh, S., Morrison, G.L., 2006. Optimization of parabolic solar collector system. Int. J. Energy Res. 30 (4), 259–271. doi:10.1002/er.1153.*

FIG. 4.164 Constructed parabolic trough.

California power plants, known as solar electric generating systems have a total installed capacity of 354 MW (Kalogirou, 1997). These system use thermo-oil as a heat transfer fluid, which can reach up to 400°C (Herrmann et al., 2004). The parabolic collector effectively produces heat at a temperature between 50°C and 400°C (Kalogirou, 2004).

Khan and Islam (2016) reported the use of a parabolic trough has been constructed that is adjustable and moves along the direction of the sun so that maximum solar energy can be achieved anytime of the day (Fig. 4.164). Each parabolic trough has a surface area of 4 m^2 (2.25 m × 1.8 m) that can radiate almost 1.6–4 kW to the absorber, depending on the direct normal irradiance, which is again dependent on the geographical area. Taking 600 w/m^2 as DNI (direct normal irradiance) and considering the energy transfer efficiency (Eq. 4.77) from solar surface to the heating point, it is found that one surface (4 m^2) can supply 1.62 kW. A heating load of 10.47 kW will require seven such parabolic collectors, which can supply necessary energy to run a refrigerator or an air cooler having a 1 ton cooling load. The number of collectors will vary from place to place, depending on the DNI of any place and the climate of that place. The experimental data show that the parabolic collector can absorb 0.80 kW during early summer in a cold country when the environmental temperature is nearly 21°C. The thermal fluid used by Khan and Islam (2016) was vegetable oil. It is circulated by the solar pump and that is why no electricity is needed. The choice of waste vegetable oil itself is another step toward achieving true sustainability (Fig. 4.165).

4.9.9 Effective separation of solid from liquid

Organic waste products such as, cattle manure, slaughter house waste, vegetable waste, fruit peels and pits, dried leaves and natural fibers, wood ash, natural rocks (limestone, zeolite, siltstone, etc.) are all viable options for the separation of fines and oil. In 1999, a patent was issued to Titanium Corporation for separation of oil from sand tailings. However, this technique uses chemical systems, such as NaOH and H_2O_2, which are both expensive and environmentally hostile. Drilling wastes have been found to be beneficial in highway

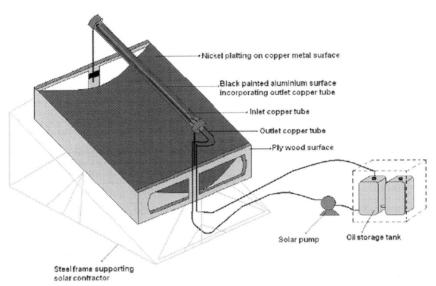

FIG. 4.165 Experimental solar trough. *From Khan, M.M., Islam, M.R., 2016. Zero Waste Engineering: A New Era of Sustainable Technology Development, second ed. Scrivener Wiley.*

construction (Wasiuddin et al., 2002). Studies have shown that the tailings from oil sands are high in various mineral contents. To extract the minerals, usually the chemical treatment is used to modify the surface of minerals. Treatment with a solution derived from natural material has a great potential. Also, microwave heating has the potential to enhance selective floatability of different particles. This aspect has been studied by Gunal and Islam (2000). Temperature can be a major factor in the reaction kinetics of a biological solvent with mineral surfaces. Various metals respond in a different manner under microwave condition, which can make significant change in floatability. The recovery process can be completed through transferring the microwave-treated fines to a flotation chamber (Henda et al., 2005).

Application of bio membranes to separate solid from liquid has also given considerable attention recently. Even though synthetic membranes are being used for some application, they are highly energy intensive to produce, toxic and costly.

4.9.10 Effective separation of liquid from liquid

The current practice involves separation of oil and water and the water is disposed as long as it has a hydrocarbon concentration below an allowable limit. However, it is expected that such practice cannot be sustained and further purification of water is necessary. Consequently, this task involves on both separation of oil and water and heavy metals removal from the water. The oil-water emulsion has been produced and it was found that the selected paper fiber material gives 98%–99% recovery of oil without producing any water (Khan and Islam, 2012). An emulsion made up of varying ratios of oil and water was utilized in different sets of experiments. Almost all the oil-water ratios gave the same separation efficiency with the material used. The mentioned paper material, which is made of long fibrous wood pulp treated with water proofing material as filtering medium. Water proofing agent for paper used in these experiments is "Rosin Soap" (rosin solution treated with caustic soda). This soap is then treated with alum to keep the pH of the solution within a range of 4–5. Cellulose present in the paper reacts reversibly with rosin soap in presence of alum and forms chemical coating around the fibrous structure and acts as coating to prevent water to seep through it. This coating allows long chain oil molecules to pass making the paper a good conductor for oil stream. Because of the reversible reaction, it was also observed the performance of filter medium increases with the increased acidity of the emulsion and vice versa. As well, the filter medium is durable to make its continuous use for a long time keeping the cost of replacement and production cut-down. The material used as filtering medium is environmental friendly and useful for down-hole conditions. Further, other paper-equivalent alternate materials, which are inexpensive and down-hole environment appealing, can be used for oil-water separation. Human hair has been proven to be effective in separation of oil and water as well as heavy metal removal.

Similarly, natural zeolites can also effectively function to separate liquid-liquid from different solutions. Such zeolites can adsorb some liquid leaving others to separate depending on their molecular weight.

4.9.11 Effective separation of gas from gas

Most of the hydrocarbons found in natural gas wells are complex mixtures of hundreds of different compounds. A typical natural gas stream is a mixture of methane, ethane, propane, butane and other hydrocarbons, water vapor, oil and condensates, hydrogen sulfides, carbon dioxide, nitrogen, some other gases and solid particles. The free water and water vapors are corrosive to the transportation equipment. Hydrates can plug the gas accessories creating several flow problems. Other gas mixtures such as hydrogen sulfide and carbon dioxide are known to lower the heating value of natural gas by reducing its overall fuel efficiency. There are certain restrictions imposed on major transportation pipelines on the make-up of the natural gas that is allowed into the pipeline called pipe "line quality" gas. This makes mandatory that natural gas is purified before it is sent to transportation pipelines. The gas processing is aimed at preventing corrosion, environmental and safety hazards associated with transport of natural gas.

The presence of water in natural gas creates several problems. Liquid water and natural gas can form solid ice-like hydrates that can plug valves and fittings in the pipeline. Natural gas containing liquid water is corrosive, especially if it contains carbon dioxide and hydrogen sulfide. Water vapor in natural gas transport systems may condense causing a sluggish flow. Hence, the removal of free water, water vapors, and condensates is a very important step during gas processing. Other impurities of natural gas, such as carbon dioxide and hydrogen sulfide generally called as acid gases must be removed from the natural gas prior to its transportation. Hydrogen sulfide is a toxic and corrosive gas which is rapidly oxidized to form sulfur dioxide in the atmosphere. Oxides of nitrogen found in traces in the natural gas may cause ozone layer depletion and global warming. Hence, an environment-friendly gas processing is essential in order for greening the petroleum operations.

4.9.12 Natural substitutes for gas processing chemicals (glycol and amines)

Glycol is one of the most important chemicals used during the dehydration of natural gas. In search of the cheap and abundantly material, clay has been considered as one of the bets substitute of toxic glycol. Clay is a porous material containing various minerals such as silica, alumina, and several others. The dry clay as a plaster has water absorption coefficient of 0.067–0.075 ($kg/m^2 S^{1/2}$) where weight of water absorbed is in kg, surface area in square meter and time in second. The preliminary experimental results have indicated that clay can absorb considerable amount of water vapor and can be efficiently used in dehydration of natural gas (Fig. 4.166). Moreover, glycol can be obtained from some natural source, which is not toxic as synthetic glycol. Glycol can be extracted from Tricholoma Matsutake (mushroom) which is an edible fungus (Ahn and Lee, 1986). Ethylene glycol is also found as a metabolite of ethylene which regulates the natural growth of the plant (Blomstrom and Beyer, 1980). Orange peel oils can replace this synthetic glycol. These natural glycols derived without using nonorganic chemicals can replace the synthetic glycols. Recent work of Miralai et al. (2007) demonstrated that such considerations are vital.

Amines are used in natural gas processing to remove H_2S and CO_2. Monoethanolamine (MEA), DEA and TEA are the members of alkanolamine compound. These are synthetic chemicals the toxicity of which has been discussed earlier. If these chemicals are extracted from natural sources, such toxicity is not expected. Monoethanolamine is found in the hemp oil which is extracted from the seeds of hemp (Cannabis Sativa) plant. 100 g of hemp oil contain 0.55 mg of monoethanolamine (Chhetri and Islam, 2008). Moreover, an experimental study showed that olive oil and waste vegetable oil can absorb sulfur dioxide. Fig. 4.167 indicates the decrease in pH of de-ionized water with time. This could be a good model to remove sulfur compounds from the natural gas streams. Calcium hydroxides can also be utilized to remove CO_2 from the natural gas.

4.9.13 Membranes and absorbents

Various types of synthetic membranes are in use for the gas separation. Some of them are liquid membranes and some are polymeric. Liquid membranes operate by immobilizing a liquid solvent in a microporous filter or between polymer layers. A high degree of solute

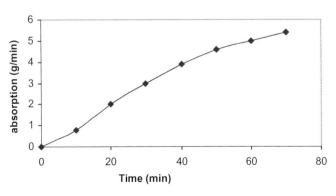

FIG. 4.166 Water vapor absorption by Nova Scotia clay (Chhetri and Islam, 2008).

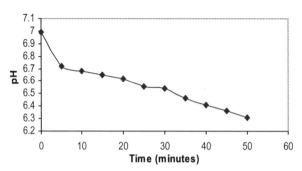

FIG. 4.167 Decrease of pH with time due to sulfur absorption in de-ionized water (Chhetri and Islam, 2008).

removal can be obtained when using chemical solvents. When the gas or solute reacts with the liquid solvent in the membrane, the result is an increased liquid phase diffusivity. This leads to an increase in the overall flux of the solute. Furthermore, solvents can be chosen to selectively remove a single solute from a gas stream to improve selectivity. Saha and Chakma (1992) suggested the attachment of a liquid membrane in a microporous polymeric membrane. They immobilized mixtures of various amines such as monoethanolamine (MEA), diethanolamine (DEA), amino-methyl-propanol (AMP), and polyethylene glycol (PEG) in a microporous polypropylene film and placed it in a permeator. They tested the mechanism for the separation of carbon dioxide from some hydrocarbon gases and obtained separation factors as high as 145.

Polymeric membranes have been developed for a variety of industrial applications, including gas separation. For gas separation, the selectivity and permeability of the membrane material determines the efficiency of the gas separation process. Based on flux density and selectivity, a membrane can be classified broadly into two classes, porous and nonporous. A porous membrane is a rigid, highly voided structure with randomly distributed interconnected pores. The separation of materials by a porous membrane is mainly a function of the permeate character and membrane properties, such as the molecular size of the membrane polymer, pore size, and pore-size distribution. A porous membrane is similar in its structure and function to the conventional filter. In general, only those molecules that differ considerably in size can be separated effectively by microporous membranes. Porous membranes for gas separation do exhibit high levels of flux but inherit low selectivity values. However, synthetic membranes are not as environment-friendly as biodegradable bio membranes.

The efficiency of polymeric membranes decreases with time due to fouling, compaction, chemical degradation, and thermal instability. Because of this limited thermal stability and susceptibility to abrasion and chemical attack, polymeric membranes have found application in separation processes where hot reactive gases are encountered. This has resulted in a shift of interest toward inorganic membranes.

Inorganic membranes are increasingly being explored to separate gas mixtures. Besides having appreciable thermal and chemical stability, inorganic membranes have much higher gas fluxes when compared to polymeric membranes. There are basically two types of inorganic membranes, dense (nonporous) and porous. Examples of commercial porous inorganic membranes are ceramic membranes, such as alumina, silica, titanium, glass, and porous metals, such as stainless steel and silver. These membranes are characterized by high permeabilities and low selectivities. Dense inorganic membranes are specific in their separation behaviors, for example, Pd-metal-based membranes are hydrogen specific and metal oxide membranes are oxygen specific. Palladium and its alloys have been studied extensively as potential membrane materials. Air Products and Chemical Inc. developed the Selective Surface Flow (SSF) membrane. It consists of a thin layer (2–3 nm) of nano-porous carbon supported on a macroporous alumina tube.

A variety of bio-membranes are also in use today. These membranes such as human hair can be used instead of synthetic membranes for gas-gas separation. The use of human hair as biomembrane has been illustrated by Khan and Islam (2012). Initial results indicated that human hairs have characteristics similar to hollow fiber cylinders, but are even more effective because of the flexible nature and a texture that can allow the use of a hybrid system through solvent absorption along with mechanical separation. Natural absorbents such as silica gels can also be used for absorbing various contaminants from the natural gas stream. Khan and Islam (2012) showed that synthetic membranes can be replaced by simple paper membranes for oil water separation. Moreover, limestone has the potential to separate sulfur dioxide from natural gas. When caustic soda is combined with wood ash, it was found to be an alternative to zeolite. Since caustic soda is a chemical, waste materials such as okra extract can be a good substitute. The same technique can be used with any type of exhaust, large (power plants) or small (cars). Once the gas is separated, low quality gas can then be injected into the reservoir for enhanced oil recovery technique. This will enhance the system efficiency. Moreover, low quality can be converted into power by a turbine. Bjorndalen et al. (2005) developed a comprehensive scheme for the separation of petroleum products in different form using novel materials with value addition of the by-products.

4.9.14 A novel desalination technique

Management of produced water during petroleum operations offers a unique challenge. The concentration of this water is very high and cannot be disposed of outside. to bring down the concentration, expensive and energy-intensive techniques are being practiced. Recently, Khan and Islam (2012, 2016) have developed a novel desalination technique that can be characterized as totally environment-friendly process. This process uses no nonorganic chemical (e.g., membrane, additives). This process relies on the following chemical reactions in four stages:

(1) saline water + CO_2 + NH_3 ➔ (2) precipitates (valuable chemicals)

+ desalinated water ➔ (3) plant growth in solar aquarium ➔ (4) further desalination

This process is a significant improvement over an existing US patent. The improvements are in the following areas:

- CO_2 source is exhaust of a power plant (negative cost)
- NH_3 source is sewage water (negative cost + the advantage of organic origin)
- Addition of plant growth in solar aquarium (emulating the world's first and the biggest solar aquarium in New Brunswick, Canada).

This process works very well for general desalination involving sea water. However, for produced water from petroleum formations, it is common to encounter salt concentration much higher than sea water. For this, water plant growth (Stage 3 above) is not possible because the salt concentration is too high for plant growth. In addition, even Stage 1 does not function properly because chemical reactions slow down at high salt concentrations. This process can be enhanced by adding an additional stage. The new process should function as:

(1) Saline water + ethyl alcohol → (2) saline water + CO_2

+ NH_3 → (3) precipitates (valuable chemicals)

+ desalinated water → (4) plant growth in solar aquarium → (5) further desalination

Care must be taken, however, to avoid using nonorganic ethyl alcohol. Further value addition can be performed if the ethyl alcohol is extracted from fermented waste organic materials.

4.9.15 A novel refining technique

As discussed in Chapter 7, a sustainable refinery can render the process sustainable and create enough incentive to investigate the concept of Downhole refinery.

Khan and Islam (2007b) identified the following sources of toxicity in conventional petroleum refining:

- Use of toxic catalyst
- Use of artificial heat (e.g., combustion, electrical, and nuclear).

The use of toxic catalysts contaminates the pathway irreversibly. These catalysts should be replaced by natural performance enhancers. Such practices have been proposed by Chhetri and Islam (2008) in the context of biodiesel. In this proposed project, research will be performed to introduce catalysts that are available in their natural state. This will make the process environmentally acceptable and will reduce to the cost very significantly.

The problem associated with efficiency is often covered up by citing local efficiency of a single component (Islam et al., 2010). When global efficiency is considered, artificial heating proves to be utterly inefficient (Khan and Islam 2012; Chhetri and Islam, 2008). Recently, Khan and Islam (2016) demonstrated that direct heating with solar energy (enhanced by a parabolic collector) can be very effective and environmentally sustainable. They achieved up to 75% of global efficiency as compared to some 15% efficiency when solar energy is used through electricity conversion. They also discovered that the temperature generated by the solar collector can be quite high, even for cold countries. In hot climates, the temperature can exceed 300°C, making it suitable for thermal cracking of crude oil. In this project, the design of a direct heating refinery with natural catalysts will be completed. Note that the direct solar heating or wind energy doesn't involve the conversion into electricity that would otherwise introduce toxic battery cells and would also make the overall process very low in efficiency.

4.9.16 Use of solid acid catalyst for alkylation

Refiners typically use either hydrofluoric acid (HF), which can be deadly if spilled, or sulfuric acid, which is also toxic and increasingly costly to recycle. Refineries can use a solid acid catalyst, unsupported and supported forms of heteropolyacids and their cation exchanged salts, which has recently proved effective in refinery alkylation. A solid acid catalyst for alkylation is less widely dispersed into the environment compared to HF. Changing to a solid acid catalyst for alkylation would also promote more safety at a refinery. Solid acid catalysts are an environment-friendly replacement for liquid acids, used in many significant reactions, including alkylation of light hydrocarbon gases to form isooctane (alkylate) used in reformulated gasoline. Use of organic acids and enzymes for various reactions is to be promoted.

The catalysts that are in use today are very toxic and wasted after a series of use. This will create pollution to the environment, so using catalysts with fewer toxic materials significantly reduces pollution. The use of Nature-based catalysts such as zeolites, alumina, and silica should be promoted. Various biocatalyst and enzymes, which are nontoxic and from renewable origin, are to be considered for future use.

4.9.17 Use of bacteria to breakdown heavier hydrocarbons to lighter ones

Since the formation of crude oil is the decomposition of biomass by bacteria at high temperature and pressure, there must be some bacteria that can effectively break down the crude oil into lighter products. A series of investigations are necessary to observe the effect of bacteria on the crude oil.

4.9.18 Use of cleaner crude oil

Crude oil itself is comparatively cleaner than distillates as it contains less sulfur and toxic metals. The use of crude oil for various applications is to be promoted. This will not only help to maintain the environment because of its less toxic nature but also be less costly as it avoids expensive catalytic refining processes. Recently, the direct use of crude oil is of great interest.

Several studies have been conducted to investigate the electricity generation from saw dust (see Islam et al., 2010 for details). Fig. 4.168 shows the schematic of a scaled model developed by our research group in collaboration with Veridity Environmental Technologies (Halifax, Nova Scotia). A raw sawdust silo is equipped with a powered auger sawdust feeder. The saw dust is inserted inside another feeding chamber that is equipped with a powered grinder that pulverizes sawdust into wood flour. The chamber is attached to a heat exchanger that dries the saw dust before it enters into the grinder. The wood flour is fed into the combustion chamber with a powered auger wood flour feeder. The pulverization of sawdust increases the surface area of the particles very significantly. The additional energy required to run the feeder and the grinder is provided by the electricity generated by the generator itself, requiring no additional energy investment. In addition, the pulverization chamber is also used to dry the saw dust. The removal of moisture increases flammability of the feedstock. The combustion chamber itself is equipped with a start-up fuel injector that uses biofuel. Note that initial temperature required to startup the combustion chamber is quite high and cannot be achieved

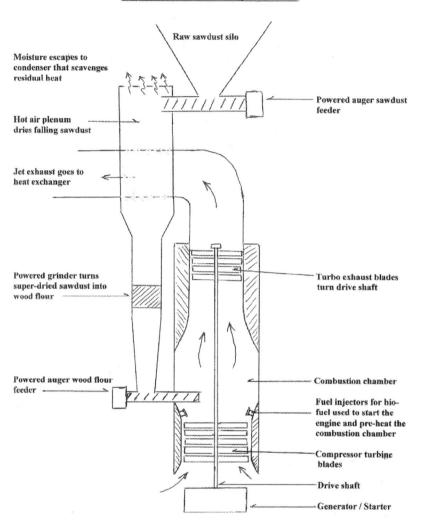

FIG. 4.168   Schematic of sawdust fueled electricity generator.

without a liquid fuel. The exhaust of the combustion chamber is circulated through a heat exchanger to dry sawdust prior to pulverization. As the combustion fluids escape the combustion chamber, they turn the drive shaft blades rotate to turn the drive shaft, which in turn, turns the compressor turbine blades. The power generator is placed directly under the main drive shaft.

Fernandes and Brooks (2003) compared black carbon (BC) derived from various sources. An interesting feature of this study was that they studied the impact of different sources on the composition, extractability and bioavailability of resulting BC. By using molecular fingerprints, the concluded that fossil BC may be more refractory than plant derived BC. This is an important finding as recently there has been some advocacy that BC from fossil fuel may have cooling effect, nullifying the contention that fossil fuel burning is the biggest contributor to global warming. BC from fossil fuel may have higher refractory ability, however, there is no study available to date to quantify the cooling effect and to determine the overall effect of BC from fossil fuel. As for the other effects, BC from fossil appear to be on the harmful side as compared to BC from organic matters. For instance, vegetarian fire residues, straw ash and wood charcoals had only residual concentrations of n-alkanes (<9 μg/g) and polyclyclic aromatic (PAHs) of less than 0.2 μg/g. These concentrations compared with Diesel soot, urban dust and chimney soot PAH concentrations of greater than 8 μg/g and n-alkanes greater than 20 μg/g.

This design shows that even the solid fuels can be used to produce electricity at high efficiencies. Burning of saw dust produces fresh carbon dioxide compared burning of fossil fuels which produces older carbon dioxide (Islam et al., 2010). The use of crude oil (which is liquid) in such a system will be even more efficient than solid fuel. If the crude oil is used directly similar to saw dust electricity generator, the environmental problems associated with fossil fuel use and refining can be minimized increasing the economic efficiency of the fossil fuel use. The CO_2 produced from the direct use of crude oil will be acceptable to plants as it is fresh CO_2 and no catalysts and chemicals are used for processing. (Chhetri and Islam, 2008; Islam et al., 2010) argued that the CO_2 produced from refined petroleum products may not be acceptable to plants as this CO_2 is contaminated with toxic catalysts and chemicals. Hence, the use of direct crude oil will solve the major environmental problems the humanity is facing today.

4.9.19 Use of gravity separation systems

Heavier fractions can be settled out through the density difference method. Various settling tanks in different stages can be designed that allow sufficient time to settle the fractions based on their density. Even though it will not solve all the problems, some of the environmental problems can be reduced by this method. This will be less costly compared to other processes but more time-consuming.

4.9.20 A novel separation technique

Chaalal and Islam (2001) developed a fully sustainable water purification technique that can be used to purify water from heavy metals, including radioactive elements. The setup is shown in Fig. 4.169. The technique involves treating the effluent in an algae-packed column. The permeability of the packed column is high enough to allow continuous flow with only hydrostatic pressure of the line connected to the source trough. The selection of algae depends on the type of contaminant to be removed. For instance, for the case of Strontium, *C. vulgaris* was chosen by Chaalal and Islam (2001). The packed column is connected to the air-curtain driven fluidized bed/membrane system. This reactor is a very successful air curtain driven fluidized bed reactor, coupled with a membrane system (see Fig. 4.169). Compressed air is

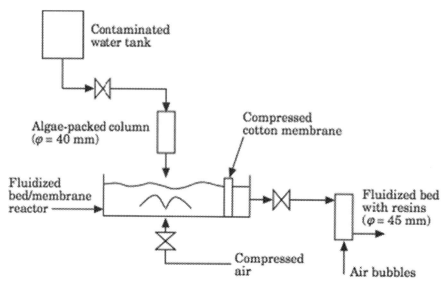

FIG. 4.169 The schematic of the separation unit. *From Chaalal and Islam, 2001.*

injected into the reactor through a series of perforations in a transverse tube to create fluid circulation with an air curtain. The effluent then moves through to fluidized bed, packed with resins. For the process of this study, a flow rate of 600 mL/h was used. This volume was injected through 16 orifices of the perforated tubing.

4.10 Zero-waste operations

Khan and Islam (2012) developed a green supply chain framework to achieve sustainability in refining operations. The framework proposed supply chain model is developed based on the work of Lakhal et al. (2007). It analyses the structure of the supply chain from production, transportation, and distribution to end users. The specific aspects of the model include:

Five zeros of waste or emissions (corresponding to the five circles in the Olympic flag):
(a) Zero emissions (air, soil, water, solid waste, and hazardous waste)
(b) Zero waste of resources (energy, materials, and human)
(c) Zero waste in activities (administration, production)
(d) Zero use of toxics (processes and products)
(e) Zero waste in product life cycle (transportation, use, and end of life)

Green inputs and outputs: The zero-waste approach is defended by the Zero-waste Organization using a visionary goal of zero waste to represent the endpoint of "closing-the-loop," so that all materials are returned at the end of their life as industrial nutrients, thereby avoiding any degradation of nature. A 100% efficiency of use of all resources—energy, material, and human—is promoted by Zero-waste, working toward a goal of reducing costs, easing demands on scarce resources, and providing greater availability for all. These principles of Zero waste applies to products reduce impact during manufacture, transportation, during use, and at end of life. Such an approach, which is always the norm in Nature, is only beginning to be proposed in the petroleum sector (Bjorndalen et al., 2005) or even in the renewable energy sector (Khan and Islam, 2012).

4.10.1 Zero emissions (air, soil, water, solid waste, hazardous waste)

Primary activities in refinery processes are materials transfer and storage, separating hydrocarbons (e.g., distillation), creating hydrocarbons (e.g., cracking/coking, alkylation, and reforming), blending hydrocarbons, and removing impurities (e.g., sulfur removal) and cooling are the major operations to release the solid, liquid and gaseous emission. These emissions should be completely controlled so as not to release in the atmosphere, land and water bodies.

4.10.2 Zero waste of resources (energy, material, and human)

The most important resource in the refinery process is energy. Unlike the manufacturing industry, labor costs do not constitute a high percentage of expenses in a refinery. In this continuous process, there is no waste of materials in general. The waste of human resources could be measured by ratios of accident (number of work accidents/number of employers) and absenteeism due to illness (number of days lost for illness/number of work days × number of employees). In the "Olympic" refinery, ratios of accident and absenteeism would be near to zero.

The refining process uses a huge amount of energy. Approximately 2% of the energy contained in crude oil is used for distillation. The "advanced" or "progressive" distillation, involving the use of modern chemical engineering techniques, cuts down the energy required to distill crude oil by 30%–65%. This technique requires large-scale rebuilding of distillation units to enable separation of crude oil components. This would avoid the ineptness of conventional distillation processes, like heating oil at high temperatures and separating products as they cool. Advanced or progressive distillation may also include more effective heat exchange, further reducing the energy required for distillation. Considering this potential reduction of energy consumption, it can be estimated that an "Olympic" refinery would use only 1% contained in the crude.

4.10.3 Zero waste in administration activities

Considerable cost savings could be achieved by using resource-efficient products and good environmental practices. Good practices should target energy-efficiency, waste reduction, water conservation, and other resource-efficient practices for the environment. By taking advantage of these practices, refineries can avoid resource waste and save money. An "Olympic" refinery should have a list of good practices and encourage employers to respect them. For example, one workstation (computer and monitor) left running after business hours, causes power plants to emit nearly one ton of CO_2 per year. That emission could be cut by 80% if the workstation is switched off at night and set to "sleep mode" during idle periods in the day. If every computer and monitor in the United States was turned off at night, the nation could shut down eight large power stations and avoid emitting 7 million tons of CO_2 every year.

4.10.4 Zero use of toxics (processes and products)

A number of procedures are used to turn heavier components of crude oil into lighter and more useful hydrocarbons. These processes use catalysts or materials that help chemical reactions without being used up themselves. Refinery catalysts are generally toxic and must be replaced or regenerated after repeated use, turning used catalysts into a waste source. The refining process uses either sulfuric acid or hydrofluoric acid as catalysts to transform propylene, butylenes, and/or isobutane into alkylation products or alkylate. Vast quantities of sulfuric acid are required for this process. Hydrofluoric acid (HF), also known as hydrogen fluoride, is extremely toxic and can be lethal. Using catalysts with fewer toxic materials significantly reduces pollution. Eventually, organic acids and enzymes, instead of catalysts must be considered. Thermal degradation and slow reaction rates are often considered to be the greatest problems of using organic acid and catalysts.

However, recent discoveries have shown that this perception is not justified. There are numerous organic products and enzymes that can withstand high temperatures and many of them induce fast reactions. More importantly, recent developments in biodiesel indicate that the process (Islam et al., 2010) itself can be modified to eliminate the use of toxic substances. The same principle applies to other materials, for example, corrosion inhibitors, bactericides, and so on. Often, toxic chemicals lead to high corrosion vulnerability and even more toxic corrosion inhibitors are required. The whole process spirals down to a very unstable process, which can be eliminated with the new approach (Al-Darbi et al., 2002).

4.10.5 Zero waste in product life cycle (transportation, use, and end of life)

The complex array of pipes, valves, pumps, compressors, and storage tanks at refineries are potential sources of leaks into air, land, and water. If they are not contained, liquids can leak from transfer and storage equipment and contaminate soil, surface water, and groundwater. This explains why, according to industrial data, approximately 85% of monitored refineries have confirmed groundwater contamination as a result of leaks and transfer spills. To prevent the risks associated with transportation of sulfuric acid and on-site accidents associated with the use of hydrofluoric acid, refineries can use a solid acid catalyst that has recently proven effective for refinery alkylation. A solid acid catalyst for alkylation is much less able than HF to disperse into the environment in a short time frame. Changing to this method would promote inherent safety at a refinery, rather than merely improving accident mitigation and response. An "Olympic" green refinery supply chain should have storage tanks and pipes above ground to prevent groundwater contamination. There is room for improving the efficiency of these tanks with natural additives. Frequently, the addition of synthetic materials makes an otherwise sustainable process unstable. Advances in using natural materials for improving material quality have been made by Saeed et al. (2003).

4.10.6 Zero waste in reservoir management

We have observed the scientific analysis that gives an optimistic picture of the reserve development. By properly characterizing reservoirs and using technologies that best suit the broader sustainability picture, one can make the reserve grow continuously (Fig. 4.170). This picture can be further enhanced by using zero-waste scheme at every stage of EOR operations.

Consider the use of zero-waste engineering, which is the only truly sustainable oil recovery technique. Fig. 4.171 shows how enhanced oil recovery adds to the profitability over conventional primary recovery processes. Enhanced oil recovery schemes are well established and use either added chemicals or energy (e.g., steam) to increase profitability. This profitability goes up tremendously if zero-waste schemes are added. Zero-waste in this case represents the use of waste products from the petroleum operation and use other naturally available materials that are abundant in the locality of the petroleum operation site.

In summary, sustainable production ensures exponential economic growth. Such production tactic, when coupled with sustainable economic models offer a true paradigm shift in all aspects of economic development.

4.11 Greening of hydraulic fracturing

Hydraulic fracturing stimulates wells drilled into these formations, making profitable otherwise prohibitively expensive extraction. Ever since 2008, when unconventional formations made a remarkable comeback with the combination of hydraulic fracturing with horizontal drilling. A mixture of water, proppants and chemicals is pumped into the rock or coal formation. There are, however, other ways to fracture wells. Sometimes fractures are created by

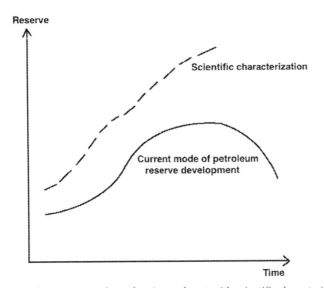

FIG. 4.170 Unconventional reserve growth can be given a boost with scientific characterization.

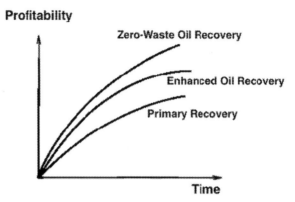

FIG. 4.171 Profitability grows continuously with time when zero-waste oil recovery scheme is introduced.

injecting gases such as propane or nitrogen, and sometimes acidizing occurs simultaneously with fracturing. Acidizing involves pumping acid (usually hydrochloric acid), into the formation to dissolve some of the 34,444 rock material to clean out pores and enable gas and fluid to flows more readily into the well.

A 2013 report (Down Stream Strategies, n.d.) has shown that more than 90% of fracking fluids may remain underground. Used fracturing fluids that return to the surface are often referred to as flowback, and these wastes are typically stored in open pits or tanks at the well site prior to disposal. These materials are both toxic and costly (Kahrilas et al., 2015). They identified the following mechanisms involved with hydraulic fracturing fluids. The pathway is shown in Fig. 4.172

(1) hydrolysis;
(2) direct or indirect photolysis;
(3) aerobic biodegradation in the water or soil;
(4) other chemical reaction with oxygen present;
(5) complexation underground with dissolved inorganic species;
(6) anaerobic biodegradation;
(7) other chemical reaction (e.g., nucleophilic substitution or polymerization) under anoxic conditions, high pressure, and elevated temperature. Artwork is conceptual and not drawn to scale.

Starting from mid 19th century, fracturing rock to increase oil production became an option. During early years, liquid and solidified nitroglycerin was used to induce fractures with an in situ explosion. The concept of (hydraulic) fracturing with pressure instead of explosives grew in the 1930s (Kreipl and Kreipl, 2017). Beginning in 1953, water-based fluids were developed using different types of gelling agents. Gelling agents were necessary to increase proppant carrying ability of the chase fluid. Nowadays, aqueous fluids such as acid, water, brines, and water-based foams are used in most fracturing treatments. The breakdown of the fluids to decrease viscosity is mostly carried out by use of oxidizing agents. Thereby, the technology is facing concerns regarding microseismicity, air emissions, water consumption, and the endangerment of groundwater due to the risk of perforating protective layers and the

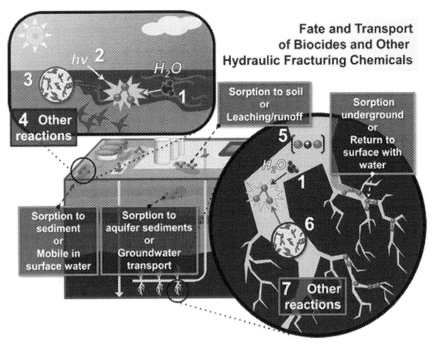

FIG. 4.172 Fate and transport of hydraulic fracturing chemicals (including biocides) in different environments.

ooze of chemicals through the surface (Kreipl and Kreipl, 2017). Furthermore, particularly both cross-linking and breaking agents pose serious risks for humans respectively are environmentally hazardous in terms of eco-toxicity—while the degradation effect of common oxidizing agents is relatively low in cases of high-temperature fracturing treatments. According to Kreipl and Kreipl (2017), the viscosity of both common hydrogels with and without oxidizing agents can be reduced to the same level when heated to 130°C or above. Furthermore, in both cases no non-Newtonian behavior could be observed after the temperature treatment (anymore). Therefore, they developed a hydrogel that allows for optimized cross-linking without toxic linkers and that can be dissolved without environmentally hazardous chemicals.

As the conventional fluids continued to draw criticism with retards to their environmental impact, new technologies such as waterless fracturing is the key to effectively improving unconventional resources recovery, while addressing the issue in reducing water consumption and environmental footprints. Fu and Liu (2019) investigates the development of two major waterless fracturing fluids, foams and liquid N_2/CO_2, including the advantages and challenges faced with waterless fracturing, fracturing mechanisms, and fluid properties such as stability and rheology. Based on the literature review, it is believed that foam has a great potential to be a promising fracturing fluid in improving productivity and long-term production with benefits such as fast cleanup, improved proppant transport, and minimal environmental footprint. Foam properties such as stability and rheology have been continuously improved with technological advances in the stabilizing agents. Foams stabilized by nanoparticles are reported to significantly improved foam stability and rheology under reservoir conditions over conventional surfactants. Other fracturing fluids such as liquid CO_2/N_2 and gas fracturing fluids are designed to clear formation damage near the wellbore or for scenarios where long fractures are not desired, and both are faced with various technical challenges.

Fig. 4.173 shows how both foam viscosity and foam half-life are improved with the use of nanoparticle induced surfactants. The height of foams stabilized by nanoparticles with a concentration of 0.3 wt% or higher hardly changed over 90 h, while the case without nanoparticles degraded completely.

By using the mixture of nanoparticles, surfactant, and polymer, Xue et al. (2016) reported stable and more viscous CO_2 foam. Stable high foam qualities of CO_2 foams were obtained as supercritical CO_2 and nanosilica, LAPB and HPAM flowed through a porous medium, where foams were generated. They found that nanoparticles led to higher foam apparent viscosity for foam qualities up to 0.95 (Table 4.46). Adding polymers increases viscoelasticity of the aqueous phase, which further decreased the lamella drainage rate and inhibited foam bubble coalescence. It is noticed that adding nanoparticles has led to finer foams judging from the reduced diameter of the foam bubbles, and the improved texture has resulted in the increased apparent foam viscosity. Adding polymers also seems to have a similar effect in reducing bubble diameter, but the mechanisms of apparent viscosity improvement are different. Nanoparticles improve the apparent viscosity by improving foam texture, and the structural integrity of the 3D nanoparticle network, while polymer mainly improves the viscosity of the

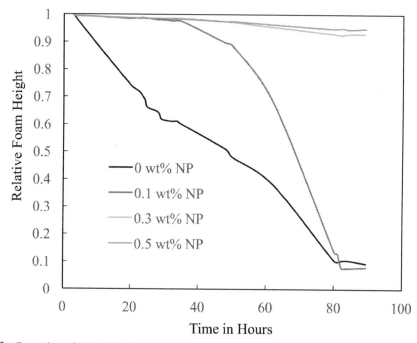

FIG. 4.173 Foam degradation with time.

TABLE 4.46 Apparent viscosity of CO_2/water foams at 3000 psi and room temperature.

Mixtures	CO_2 volume fraction	D_{sm} (µm)	U_{poly}	$\mu_{continuous\ phase}$ (cP)	μ_{foam} (cP)
0.08% LAPB and 1% NP	0.9	72	0.98	1	74
0.1% HPAM, 0.08% LAPB and 1% NP	0.9	49	0.4	5	160
0.88% HPAM, 0.08% LAPB and 1% NP	0.9	23	0.41	80	274
0.88% HPAM, 0.08% LAPB	0.9	46	0.29	80	147
0.88% HPAM, 0.08% LAPB and 1% NP	0.95	36	0.36	80	105
0.88% HPAM, 0.08% LAPB	0.95	74	0.34	80	53

Note: D_{sm} is the Sauter mean diameter of foam bubbles and U_{poly} is the dimensionless polydispersity.

continuous phase. The decreased polydispersity of polymers indicates that the polymers may have aggregated with the nanoparticle and surfactants adding to the structural integrity at the interfaces.

The problem then is reduced to using natural surfactants and natural nanomaterials. Such approach has been discussed by Islam (2020). The production and use of nanoparticles leads to the emission of manufactured or engineered nanoparticles into the environment. Those particles undergo many possible reactions and interactions in the environment they are exposed to. If artificial, these reactions and the resulting behavior and fate of nanoparticles in the environment will lead to chain reactions with irreversible damage to the environment (Mokhatab et al., 2016). Others have studied for decades naturally occurring nanoparticulate (1–100 nm) and colloidal (1–1000 nm) substances. While the knowledge gained from these investigations is nowhere near sufficiently complete to create a detailed model of the behavior and fate of engineered and natural nanoparticles in the environment, a valuable starting point is the fact that their pathways are divergent, although properties are similar.

It is known that processes such as surface reactions, stability, mobility, and dissolution play a major role in controlling their fate and behavior in the aqueous environment (Fig. 4.174). The extent of these processes is regulated, among other things, by surface properties of the particles and environmental conditions such as pH value, ionic strength, and natural organic matter. Nowadays, the release of engineered nanomaterials (ENMs) as a consequence of increased production, use in consumer products, and industrial applications provokes questions regarding appropriate risk assessment strategies for the prediction of their fate and behavior in various environment media. At present, there is no scientific method to assess fundamental difference between natural and artificial material, but the correct sustainability criterion can show the distinction between the two (Khan and Islam, 2007b). Chhetri and Islam (2008) used Khan and Islam's sustainability criterion to establish that natural materials are inherently sustainable whereas artificial materials are inherently implosive.

The processes which affect the fate and behavior of NIOPs are illustrated in Fig. 4.174. NIOPs can dissolve or grow, can aggregate or be deposited on surfaces, can be coated with organic and inorganic constituents of water and/or can be transformed in terms of their crystal structure. The fate and behavior of the iron particles is controlled by the hydrochemical conditions and the properties of the NIOPs. For example, the transport of natural NIOPs is

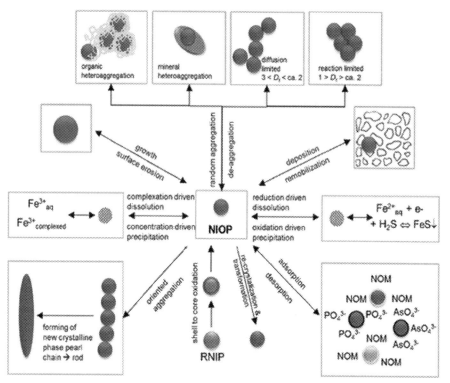

FIG. 4.174 Graphical representation of possible reactions of nanoparticulate materials in natural aquatic media, with a nano iron oxide particle (NIOP) used as an example. *From Wagner, S., et al., 2014. Spot the difference: engineered and natural nanoparticles in the environment—release, behavior, and fate. Angew. Chem.*

regulated by the Ca^{2+} concentration. Engineered NIOPs have a distinct shape, which is very homogeneous between particles compared to naturally occurring NIOPs (Fig. 4.175).

Mobile natural colloids include silicates such as clay minerals, oxides and hydroxides of Fe and Al, colloidal silica, carbonates, organic matter, and "biocolloids" such as viruses and bacteria. These entities may interact with dissolved compounds, thus explaining the major importance of natural colloids for the transport of matter in aquatic environments. Such natural colloids can be mobilized (translocated), immobilized, or generated by changes in the geochemical and hydraulic parameters in aquifers or in surface waters. The NP transport behavior in aquifers is governed by several processes (e.g., deposition, dissolution, filtration, aggregation; Fig. 4.176). These processes are controlled by the properties of the NPs and the chemical conditions of their surroundings.

Conventional oil and gas wells use, on average, 300,000 pounds of proppant, coalbed fracture treatments use anywhere from 75,000 to 320,000 pounds of proppant and shale gas wells can use more than 4 million pounds of proppant per well.

In addition to large volumes of water, a variety of chemicals are used in hydraulic fracturing fluids. The oil and gas industry and trade groups are quick to point out that chemicals typically make up just 0.5% and 2.0% of the total volume of the fracturing fluid. When

FIG. 4.175 Electron micrographs of manufactured iron oxide nanoparticles applicable as contrast agents in magnetic resonance imaging and natural iron oxide particles extracted from a floodplain sediment. *From Wagner, S., et al., 2014. Spot the difference: engineered and natural nanoparticles in the environment—release, behavior, and fate. Angew. Chem.*

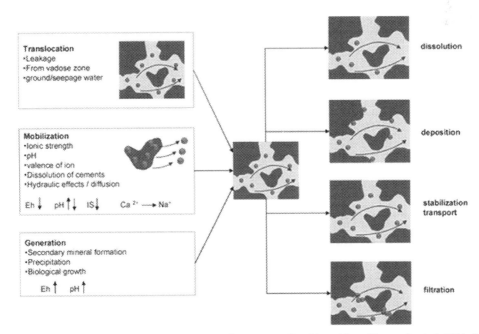

FIG. 4.176 Particle transport processes in saturated porous media. *Adapted from Wagner, S., et al., 2014. Spot the difference: engineered and natural nanoparticles in the environment—release, behavior, and fate. Angew. Chem.*

millions of gallons of water are being used, however, the amount of chemicals per fracking operation is very large. For example, a four million gallon fracturing operation would use from 80 to 330 tons of chemicals.

As part of New York State's Draft Supplemental Generic Environmental Impact Statement (SGEIS) related to Horizontal Drilling and High-Volume Hydraulic Fracturing in the Marcellus Shale, the Department of Environmental Conservation complied a list of chemicals and additives used during hydraulic fracturing. Table 4.47 provides examples of various types of hydraulic fracturing additives proposed for use in New York. Chemicals in brackets [ ] have not been proposed for use in the state, but are known to be used in other states or shale formations.

TABLE 4.47 Various chemicals used during hydraulic fracturing.

Additive type	Description of purpose	Examples of chemicals
Proppant	"Props" open fractures and allows gas/fluids to flow more freely to the well bore.	Sand [Sintered bauxite; zirconium oxide; ceramic beads]
Acid	Cleans up perforation intervals of cement and drilling mud prior to fracturing fluid injection, and provides accessible path to formation.	Hydrochloric acid (HCl, 3%–28%) or muriatic acid
Breaker	Reduces the viscosity of the fluid to release proppant into fractures and enhance the recovery of the fracturing fluid.	Peroxydisulfates
Bactericide/biocide	Inhibits growth of organisms that could produce gases (particularly hydrogen sulfide) that could contaminate methane gas. Also prevents the growth of bacteria which can reduce the ability of the fluid to carry proppant into the fractures.	Gluteraldehyde; 2-Bromo-2-nitro-1,2-propanediol
Buffer/pH adjusting agent	Adjusts and controls the pH of the fluid to maximize the effectiveness of other additives such as crosslinkers.	Sodium or potassium carbonate; acetic acid
Clay stabilizer/control	Prevents swelling and migration of formation clays which could block pore spaces, thereby reducing permeability.	Salts (e.g., tetramethyl ammonium chloride) [potassium chloride]
Corrosion inhibitor	Reduces rust formation on steel tubing, well casings, tools, and tanks (used only in fracturing fluids that contain acid).	Methanol; ammonium bisulfate for oxygen scavengers
Crosslinker	The fluid viscosity is increased using phosphate esters combined with metals. The metals are referred to as crosslinking agents. The increased fracturing fluid viscosity allows the fluid to carry more proppant into the fractures.	Potassium hydroxide; borate salts

TABLE 4.47 Various chemicals used during hydraulic fracturing—cont'd

Additive type	Description of purpose	Examples of chemicals
Friction reducer	Allows fracture fluids to be injected at optimum rates and pressures by minimizing friction.	Sodium acrylate-acrylamide copolymer; polyacrylamide (PAM); petroleum distillates
Gelling agent	Increases fracturing fluid viscosity, allowing the fluid to carry more proppant into the fractures.	Guar gum; petroleum distillate
Iron control	Prevents the precipitation of carbonates and sulfates (calcium carbonate, calcium sulfate, barium sulfate) which could plug off the formation.	Ammonium chloride; ethylene glycol; polyacrylate
Solvent	Additive which is soluble in oil, water and acid-based treatment fluids which is used to control the wettability of contact surfaces or to prevent or break emulsions.	Various aromatic hydrocarbons
Surfactant	Reduces fracturing fluid surface tension, thereby aiding fluid recovery.	Methanol; isopropanol; ethoxylated alcohol

All these chemicals have naturally available substitutes, which are technically equivalent in performance. In any case, the slight benefit potentially received with artificial proppants cannot be justified considering the cost and environmental impact.

Basement reservoirs

5.1 Introduction

The majority of hydrocarbons are found on Earth naturally occur in the form of crude oil or natural gas. After water, petroleum fluids are the most abundant compounds on earth (Islam, 2014). When organic matters are decomposed, naturally they provide for an abundant source of hydrocarbons. This is, however, is one of the two potential sources of petroleum fluids. Islam et al. (2018) assign a significant portion of world reserve to nonorganic sources. That would automatically improve the current world petroleum reserve to a great extent. It is recognized in the literature that there are several specific forms of hydrocarbons:

- **Dry gas**: contains largely methane.
- **Wet gas**: contains ethane propane, butane.
- **Condensates**: hydrocarbon with a molecular weight such that they are gas in the subsurface where temperatures are high, but condense to liquid when reach cooler, surface temperatures.
- **Liquid hydrocarbons**: commonly known as oil, or crude oil before refining.
- **Plastic hydrocarbons**: asphalt.
- **Solid hydrocarbons**: coal and kerogen, matter in sediments that is insoluble in normal petroleum solvents.
- **Gas hydrates**: solids composed of water molecules surrounding gas molecules, usually methane but also H_2S, CO_2, and other less common gases.

More than 100,000 types of hydrocarbons are known. The main reason for this diversity is that carbon atoms can unite in many different ways to form a complex chain or ring frameworks. Different arrangements of atoms yield different molecules.

The energy scenario is remarkably improved when basement reservoirs are considered. The term "basement" in "basement reservoirs" refers to crystalline formations ranging from intrusive and extrusive magmatic bodies (especially granites) to the family of low to medium-grade metamorphic rocks. Hydrocarbons have been under production from these types of rocks around the world for many decades but since around 1990 there has been growing interest and exploration in these formations where matrix porosity is negligible, and storage and production are dominated by the fracture system (a Type 1 reservoir of Nelson 2001). For example, the Arabian shield basement of Yemen and the Tertiary basement granites offshore Vietnam are two classic areas for this type of development, and significant production has been achieved. Other less publicized locations have also been investigated including prospects on the UK continental shelf.

At present, basement reservoirs are not developed fully, just like unconventional reservoirs. The Basement reservoir is a subset of naturally fractured reservoirs from any metamorphic or igneous rock (regardless of age) which is uncomfortably overlain by a sedimentary sequence. The term "basement" refers to crystalline formations ranging from intrusive and extrusive magmatic bodies (especially granites) to the family of low- to medium-grade metamorphic rocks (Bawazer et al., 2018). The term basement is used for a range of intrusive or extrusive igneous and metamorphic rocks beneath an unconformity at the base of a sedimentary sequence. A large portion of the world's oil reserves is found in naturally fractured reservoirs; fractured reservoirs contain more than 20% of the world's remaining oil and gas resources (Islam et al., 2018).

5.2 World reserve

Oil reserves denote the amount of crude oil that can be technically recovered at a cost that is financially feasible at the present price of oil. This classic definition is adopted by the Society of Petroleum Engineers. With this definition, reserves will change with the price, unlike oil resources, which include all oil that can be technically recovered at any price. Reserves may be for a well, a reservoir, a field, a nation, or the world. This poses some challenges in identifying true resources. Islam et al. (2018) deconstructed this denomination of reserve and suggested proper evaluation based on science rather than economic constraints. That would allow one to properly evaluate a reserve and thus recommend sustainable recovery techniques (Islam, 2020).

The total estimated amount of oil in an oil reservoir, including both producible and nonproducible oil, is called oil in place. The term "proven reserve" is coined to standardize oil in place. Proven reserves are those quantities of petroleum which, by analysis of geological and engineering data, can be estimated with reasonable certainty to be commercially recoverable, from a given date forward, from known reservoirs and under current economic conditions, operating methods, and government regulations. Unfortunately, this definition also lacks scientific rigor and explains why different organizations come up with different numbers while describing the reserve in the same country. The ratio of reserves to the total amount of oil in a particular reservoir is called the recovery factor. Determining a recovery factor for a given field depends on several features of the operation, including the method of oil recovery used and technological developments. This adds further uncertainty to the analysis. In addition, many oil-producing nations do not reveal their reservoir engineering field data and instead provide unaudited claims for their oil reserves. The numbers disclosed by some national governments are suspected of being manipulated for political reasons (see Islam et al., 2018a for details) (Tables 5.1 and 5.2).

Yet another list is given by CIA Factbook (Table 5.3).

TABLE 5.1 Reserve and recovery factors for various countries.

Proven reserves (millions of barrels)	U.S. EIA (start of 2020)		OPEC (end of 2017)		BP (end of 2015)		Reserves-to-production ratio	
Country	Rank	Reserves	Rank	Reserves	Rank	Reserves	Production (million bbl/year, 2016)	Years
Venezuela	1	302,809	1	302,809	1	300,900	831.1	362
Saudi Arabia	2	267,026	2	266,260	2	266,000	3818.1	78
Canada	3	167,896	22	4,421	3	172,200	1336.8	126
Iran	4	155,600	3	208,600	4	155,600	1452.9	109
Iraq	5	145,019	4	147,223	5	143,100	1624.8	88
Kuwait	6	104,000	6	104,000	6	104,000	1067.2	95
UAE	7	98,630	7	98,630	8	98,630	1133.7	86
Russia	8	80,000	7	80,000	6	102,400	3851.3	21
Libya	9	48,363	8	74,363	9	78,400	366.1	131
United States	10	47,053	10	32,773	10	55,000	3239.7	10
Nigeria	11	36,972	9	37,453	11	37,100	730.0	51
Kazakhstan	12	30,000	11	30,000	12	30,000	582.2	52
China	13	25,620	12	25,627	14	18,500	1452.9	17
Qatar	14	25,244	13	25,244	13	25,244	555.9	45
Brazil	15	12,999	14	12,634	15	13,000	918.1	14
Algeria	16	12,200	15	12,200	17	12,200	492.1	15
Angola	17/18	8273	16	8,384	16	12,700	645.9	22
Ecuador	17/18	8273	17	8,273	19/20	8000	200.2	
Mexico	20	7300	19	6537	18	10,800	798.2	9
Azerbaijan	21	7000	18	7000	21	7000	304.2	23
Norway	22	6611	20	6376	19/20	8000	601.5	11
Oman	23	5373	21	5373	23	5300	367.5	15
India	24	4600	23	4495	22	5680	267	17
Egypt	25/26	4400	24/25	4400	29	3500	180.4	25
Vietnam	25/26	4400	24/25	4400	25	4000	110.2	40
Indonesia	27/28	3600	29	3310	27/28	3600	304.2	11

Continued

TABLE 5.1 Reserve and recovery factors for various countries—Cont'd

Proven reserves (millions of barrels) Country	U.S. EIA (start of 2020) Rank	Reserves	OPEC (end of 2017) Rank	Reserves	BP (end of 2015) Rank	Reserves	Reserves-to-production ratio Production (million bbl/year, 2016)	Years
Malaysia	27/28	3600	28	3600	27/28	3600	241.3	15
Yemen	29	3000			30	3000		
United Kingdom	30	2564	32	2069	31	2800	343.1	7.5
Syria	31/32	2500	30	2500	32	2500		
Uganda	31/32	2500						
Argentina	33	2185	31	2162	33	2400		
Colombia	34	2002	34	1665	34	2300		
Gabon	35	2000	33	2000	35	2000		
Australia	36	1821	26	3985	26	4000		
Congo, Republic of the (Brazzaville)	37	1600			36	1600		
Chad	38	1500			37	1500		
Brunei	39/40	1100	35	1100	39/40	1100		
Equatorial Guinea	39/40	1,100			39/40	1100		
Kenya	41	750				750		
Ghana	41	660						
Romania	42/43	600			42/46	600		
Turkmenistan	42/43	600	36	600	42/46	600		
Uzbekistan	44	594	37	594	42/46	600		
Italy	45	557			42/46	600		
Denmark	46	491	38	439	42/46	600		
Peru	47	473			38	1,400		
Tunisia	48	425			47/48	400		
Thailand	49	396			47/48	400		
Ukraine	50	395	39	395				
Pakistan	51	350						
World total		1,779,685		1,535,773		1,750,600	29427	59

5.2 World reserve

TABLE 5.2 Shows updated numbers from BP (2020).

	At end 1999	At end 2009	At end 2018	At end 2019			
	Thousand million barrels	Thousand million barrels	Thousand million barrels	Thousand million barrels	Thousand million tonnes	Share of total	R/P ratio
Canada	181.6	175.0	170.8	**169.7**	27.3	9.8%	82.3
Mexico	21.5	11.9	5.8	**5.8**	0.8	0.3%	8.3
US	29.7	30.9	68.9	**68.9**	8.2	4.0%	11.1
Total North America	232.8	217.8	245.5	**244.4**	36.3	14.1%	27.2
Argentina	3.1	2.5	2.4	**2.4**	0.3	0.1%	10.5
Brazil	8.2	12.9	13.4	**12.7**	1.8	0.7%	12.1
Colombia	2.3	1.4	1.8	**2.0**	0.3	0.1%	6.1
Ecuador	2.6	2.7	1.6	**1.6**	0.2	0.1%	8.4
Peru	0.9	1.1	0.9	**0.9**	0.1	b	16.5
Trinidad & Tobago	0.8	0.8	0.2	**0.2**	a	b	8.1
Venezuela	76.8	211.2	303.8	**303.8**	48.0	17.5%	c
Other S. & Cent. America	1.3	0.8	0.5	**0.5**	0.1	b	12.7
Total S. & Cent. America	95.9	233.3	324.7	**324.1**	50.9	18.7%	143.8
Denmark	0.9	0.9	0.4	**0.4**	0.1	b	11.7
Italy	0.6	0.5	0.6	**0.6**	0.1	b	17.0
Norway	10.9	7.1	8.6	**8.5**	1.1	0.5%	13.5
Romania	1.2	0.6	0.6	**0.6**	0.1	b	22.0
United Kingdom	5.0	2.8	2.7	**2.7**	0.4	0.2%	6.6
Other Europe	2.0	2.0	1.6	**1.6**	0.2	0.1%	15.0
Total Europe	20.7	14.0	14.6	**14.4**	1.9	0.8%	11.6
Azerbaijan	1.2	7.0	7.0	**7.0**	1.0	0.4%	24.6
Kazakhstan	5.4	30.0	30.0	**30.0**	3.9	1.7%	42.6
Russian Federation	112.1	105.6	107.2	**102.7**	14.7	6.2%	25.5
Turkmenistan	0.5	0.6	0.6	**0.6**	0.1	b	6.2
Uzbekistan	0.6	0.6	0.6	**0.6**	0.1	b	26.3
Other CIS	0.3	0.3	0.3	**0.3**	a	b	17.6

Continued

TABLE 5.2 Shows updated numbers from BP (2020)—Cont'd

	At end 1999	At end 2009	At end 2018	At end 2019			
	Thousand million barrels	Thousand million barrels	Thousand million barrels	Thousand million barrels	Thousand million tonnes	Share of total	R/P ratio
Total CIS	120.1	144.0	145.7	**145.7**	19.8	8.4%	27.3
Iran	93.1	137.0	155.6	**155.6**	21.4	9.0%	120.6
Iraq	112.5	115.0	145.0	**145.0**	19.6	8.4%	120.6
Kuwait	96.5	101.5	101.5	**101.5**	14.0	5.9%	92.8
Oman	5.7	5.5	5.4	**5.4**	0.7	0.3%	15.2
Qatar	13.1	25.9	25.2	**25.2**	2.6	1.5%	36.7
Saudi Arabia	262.8	264.6	297.7	**297.6**	40.9	17.2%	68.9
Syria	2.3	2.5	2.5	**2.5**	0.3	0.1%	291.2
United Arab Emirates	97.8	97.8	97.8	**97.8**	13.0	5.6%	67.0
Yemen	1.9	3.0	3.0	**3.0**	0.4	0.2%	84.2
Other Middle East	0.2	0.3	0.2	**0.2**	[a]	[b]	2.6
Total Middle East	685.8	753.1	833.9	**833.8**	112.9	48.1%	75.3
Algeria	11.3	12.2	12.2	**12.2**	1.5	0.7%	22.5
Angola	5.1	9.5	8.2	**8.2**	1.1	0.5%	15.8
Chad	–	1.5	1.5	**1.5**	0.2	0.1%	32.4
Republic of Congo	1.7	2.0	3.0	**3.0**	0.4	0.2%	24.1
Egypt	3.8	4.4	3.1	**3.1**	0.4	0.2%	12.3
Equatorial Guinea	0.6	1.7	1.1	**1.1**	0.1	0.1%	16.7
Gabon	2.6	2.0	2.0	**2.0**	0.3	0.1%	25.1
Libya	29.5	46.4	48.4	**48.4**	6.3	2.8%	107.9
Nigeria	29.5	37.2	37.0	**37.0**	5.0	2.1%	48.0
South Sudan	n/a	n/a	3.5	**3.5**	0.5	0.2%	69.1
Sudan	0.3	5.0	1.5	**1.5**	0.2	0.1%	40.2
Tunisia	0.3	0.4	0.4	**0.4**	0.1	[b]	23.2
Other Africa	0.7	0.6	3.9	**3.9**	0.5	0.2%	33.8
Total Africa	84.7	123.0	125.7	**125.7**	16.6	7.2%	41.0
Australia	4.7	4.1	2.4	**2.4**	0.3	0.1%	13.4

TABLE 5.2   Shows updated numbers from BP (2020)—Cont'd

	At end 1999	At end 2009	At end 2018	At end 2019		
	Thousand million barrels	Thousand million barrels	Thousand million barrels	Thousand million barrels	Thousand million tonnes	Share of total	R/P ratio
Brunei	1.3	1.1	1.1	1.1	0.1	0.1%	24.8
China	15.1	21.6	26.2	26.2	3.6	1.5%	18.7
India	5.0	5.8	4.5	4.7	0.6	0.3%	15.5
Indonesia	5.2	4.3	3.2	2.5	0.3	0.1%	8.7
Malaysia	2.1	3.6	2.8	2.8	0.4	0.2%	11.9
Thailand	0.4	0.4	0.3	0.3	a	b	1.7
Vietnam	1.8	4.5	4.4	4.4	0.6	0.3%	51.0
Other Asia Pacific	1.4	1.1	1.2	1.4	0.2	0.1%	16.3
Total Asia Pacific	37.0	46.6	46.0	45.7	6.1	2.6%	16.4
Total World	**1277.1**	**1531.8**	**1735.9**	**1733.9**	**244.6**	**100.0%**	**49.9**
of which: OECD	256.4	234.7	261.3	260.1	38.3	15.0%	25.1
Non-OECD	1020.7	1297.1	1474.6	1473.7	206.3	85.0%	60.4
OPEC	821.8	1040.8	1214.8	1214.7	171.8	70.1%	93.6
Non-OPEC	455.3	491.0	521.1	519.2	72.8	29.9%	23.9
European Union	8.8	6.0	5.1	5.0	0.7	0.3%	9.0
Canadian oil sands: Total	175.2	169.8	163.5	162.4	26.4	9.4%	
of which: Under active development	11.9	26.5	21.2	20.1	3.3	1.2%	
Venezuela: Orinoco Belt	–	133.4	261.8	261.8	42.0	15.1%	

[a] Less than 0.05.
[b] Less than 0.05%.
[c] More than 500 years.
n/a not available.
Notes: Total proved reserves of oil—generally taken to be those quantities that geological and engineering information indicates with reasonable certainty can be recovered in the future from known reservoirs under existing economic and operating conditions. The data series for total proved oil reserves does not necessarily meet the definitions, guidelines, and practices used for determining proved reserves at the company level, for instance as published by the US Securities and Exchange Commission, nor does it necessarily represent bp's view of proved reserves by country. Reserves-to-production (R/P) ratio—if the reserves remaining at the end of any year are divided by the production in that year, the result is the length of time that those remaining reserves would last if production were to continue at that rate. Source of data—the estimates in this table have been compiled using a combination of primary official sources, third-party data from the OPEC Secretariat, World Oil, Oil & Gas Journal, and Chinese reserves based on official data and information in the public domain. Canadian oil sands "under active development" are an official estimate. Venezuelan Orinoco Belt reserves are based on the OPEC Secretariat and government announcements. Reserves and R/P ratio for Canada includes Canadian oil sands. Reserves and R/P ratio for Venezuela includes the Orinoco Belt. Saudi Arabia's oil reserves include NGLs from 2017. Reserves include gas condensate and natural gas liquids (NGLs) as well as crude oil. Shares of total and R/P ratios are calculated using thousand million barrels figures.

TABLE 5.3 Reserve according to CIA factbook.

Rank	Country	bbl	Date of information
1	Venezuela	302,300,000,000	1 January 2018 est.
2	Saudi Arabia	266,200,000,000	1 January 2018 est.
3	Canada	170,500,000,000	1 January 2018 est.
4	Iran	157,200,000,000	1 January 2018 est.
5	Iraq	148,800,000,000	1 January 2018 est.
6	Kuwait	101,500,000,000	1 January 2018 est.
7	United Arab Emirates	97,800,000,000	1 January 2018 est.
8	Russia	80,000,000,000	1 January 2018 est.
9	Libya	48,360,000,000	1 January 2018 est.
10	Nigeria	37,450,000,000	1 January 2018 est.
11	Kazakhstan	30,000,000,000	1 January 2018 est.
12	China	25,630,000,000	1 January 2018 est.
13	Qatar	25,240,000,000	1 January 2018 est.
14	Brazil	12,630,000,000	1 January 2018 est.
15	Algeria	12,200,000,000	1 January 2018 est.
16	Angola	9,523,000,000	1 January 2018 est.
17	Ecuador	8,273,000,000	1 January 2018 est.
18	Azerbaijan	7,000,000,000	1 January 2018 est.
19	Mexico	6,630,000,000	1 January 2018 est.
20	Norway	6,376,000,000	1 January 2018
21	Oman	5,373,000,000	1 January 2018 est.
22	Sudan	5,000,000,000	1 January 2018 est.
23	India	4,495,000,000	1 January 2018 est.
24	Egypt	4,400,000,000	1 January 2018 est.
25	Vietnam	4,400,000,000	1 January 2018 est.
26	South Sudan	3,750,000,000	1 January 2017 est.
27	Malaysia	3,600,000,000	1 January 2018 est.
28	Indonesia	3,310,000,000	1 January 2018 est.
29	Yemen	3,000,000,000	1 January 2018 est.
30	Syria	2,500,000,000	1 January 2018 est.
31	Uganda	2,500,000,000	1 January 2018 est.
32	Argentina	2,162,000,000	1 January 2018 est.
33	United Kingdom	2,069,000,000	1 January 2018 est.

TABLE 5.3 Reserve according to CIA factbook—Cont'd

Rank	Country	bbl	Date of information
34	Gabon	2,000,000,000	1 January 2018 est.
35	Australia	1,821,000,000	1 January 2018 est.
36	Colombia	1,665,000,000	1 January 2018 est.
37	Congo, Republic of the	1,600,000,000	1 January 2018 est.
38	Chad	1,500,000,000	1 January 2018 est.
39	Brunei	1,100,000,000	1 January 2018 est.
40	Equatorial Guinea	1,100,000,000	1 January 2018 est.
41	Ghana	660,000,000	1 January 2018 est.
42	Romania	600,000,000	1 January 2018 est.
43	Turkmenistan	600,000,000	1 January 2018 est.
44	Uzbekistan	594,000,000	1 January 2018 est.
45	Italy	487,800,000	1 January 2018 est.
46	Denmark	439,000,000	1 January 2018 est.
47	Peru	434,900,000	1 January 2018 est.
48	Tunisia	425,000,000	1 January 2018 est.
49	Ukraine	395,000,000	1 January 2018 est.
50	Thailand	349,400,000	1 January 2018 est.
51	Turkey	341,600,000	1 January 2018 est.
52	Pakistan	332,200,000	1 January 2018 est.
53	Trinidad and Tobago	243,000,000	1 January 2018 est.
54	Bolivia	211,500,000	1 January 2018 est.
55	Cameroon	200,000,000	1 January 2018 est.
56	Belarus	198,000,000	1 January 2018 est.
57	Papua New Guinea	183,800,000	1 January 2018 est.
58	Congo, Democratic Republic of the	180,000,000	1 January 2018 est.
59	Albania	168,300,000	1 January 2018 est.
60	Chile	150,000,000	1 January 2018 est.
61	Niger	150,000,000	1 January 2018 est.
62	Spain	150,000,000	1 January 2018 est.
63	Burma	139,000,000	1 January 2018 est.
64	Philippines	138,500,000	1 January 2018 est.
65	Germany	129,600,000	1 January 2018 est.
66	Poland	126,000,000	1 January 2018

Continued

TABLE 5.3 Reserve according to CIA factbook—Cont'd

Rank	Country	bbl	Date of information
67	Bahrain	124,600,000	1 January 2018 est.
68	Cuba	124,000,000	1 January 2018 est.
69	Cote d'Ivoire	100,000,000	1 January 2018 est.
70	Suriname	84,200,000	1 January 2018 est.
71	Guatemala	83,070,000	1 January 2018 est.
72	Netherlands	81,130,000	1 January 2018 est.
73	Serbia	77,500,000	1 January 2018 est.
74	Croatia	71,000,000	1 January 2018 est.
75	France	65,970,000	1 January 2018 est.
76	New Zealand	51,800,000	1 January 2018 est.
77	Japan	44,120,000	1 January 2018 est.
78	Austria	41,200,000	1 January 2018 est.
79	Kyrgyzstan	40,000,000	1 January 2018 est.
80	Georgia	35,000,000	1 January 2018 est.
81	Bangladesh	28,000,000	1 January 2018 est.
82	Hungary	24,000,000	1 January 2018 est.
83	Mauritania	20,000,000	1 January 2018 est.
84	Bulgaria	15,000,000	1 January 2018 est.
85	Czechia	15,000,000	1 January 2018 est.
86	South Africa	15,000,000	1 January 2018 est.
87	Israel	12,730,000	1 January 2018 est.
88	Lithuania	12,000,000	1 January 2018 est.
89	Tajikistan	12,000,000	1 January 2018 est.
90	Greece	10,000,000	1 January 2018 est.
91	Slovakia	9,000,000	1 January 2018 est.
92	Benin	8,000,000	1 January 2018 est.
93	Belize	6,700,000	1 January 2018 est.
94	Barbados	2,534,000	1 January 2018 est.
95	Taiwan	2,380,000	1 January 2018 est.
96	Jordan	1,000,000	1 January 2018 est.
97	Morocco	684,000	1 January 2018 est.
98	Ethiopia	428,000	1 January 2018 est.

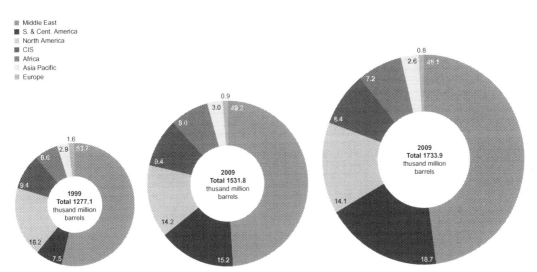

FIG. 5.1 Overall reserve growth in the world. *From BP, 2020. BP Annual Report. https://www.bp.com/content/dam/bp/business-sites/en/global/corporate/pdfs/investors/bp-annual-report-and-form-20f-2020.pdf.*

Of importance is the fact that the highest proven oil reserves including nonconventional oil deposits are in Venezuela, Saudi Arabia, Canada, and Iran. However, the USA has made the biggest stride in achieving net energy exporter status for, first time in 75 years (Blas, 2018). The shift to net exports is the dramatic result of an unprecedented boom in American oil production, with thousands of wells pumping from the Permian region of Texas and New Mexico to the Bakken in North Dakota to the Marcellus in Pennsylvania. This aspect is discussed in a different chapter, the point here being conventional petroleum reserves are becoming irrelevant in view of the fact that unconventional resources are becoming profitable to develop.

Fig. 5.1 shows how the overall world reserve has increased over the last few decades. This growth keeps pace with increasing demand. Note that the R/P ratios for most productive countries are quite high and this number is likely to remain steady considering the fact that the proven reserve continues to increase.

These numbers are only approximations. Uncertainty in reserve calculations comes from the fact the technology is evolving, both in recovery techniques and delineation of reservoirs. For instance, different estimates may or may not include oil shale, mined oil sands, or natural gas liquids. Yet others wouldn't include basement reservoirs in the calculation. In addition, proven reserves include oil recoverable under current economic conditions, which are variable depending on the overall state of the economy and other factors of a country. The case in point is Canada's proven reserve that increased suddenly in 2003 when the oil sands of Alberta were seen to be economically viable. Similarly, Venezuela's proven reserves jumped in the late 2000s when the heavy oil of the Orinoco was judged economic. In the meantime, crude oil production continues to increase overall (Fig. 5.2). When the United States made great advances in recovering unconventional oil and gas in 2008, the U.S. reserve increased significantly. Environmental concerns add to those uncertainties, particularly because those concerns are also a part of political decisions.

544   5. Basement reservoirs

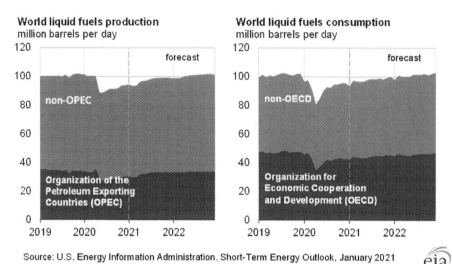

FIG. 5.2 World energy outlook and forecast. *From EIA, 2020b. https://www.eia.gov/energyexplained/natural-gas/where-our-natural-gas-comes-from.php.*

Proven wet natural gas reserves increased in each of the five largest natural gas-producing states (Texas, Wyoming, Louisiana, Oklahoma, and Pennsylvania) in 2011. Pennsylvania's proven natural gas reserves, which more than doubled in 2010, rose an additional 90% in 2011, contributing 41% of the overall U.S. increase. Combined, Texas and Pennsylvania added 73% of the net increase in the U.S. proved wet natural gas reserves. Expanding shale gas developments in these and other areas, particularly the Pennsylvania and West Virginia portions of the Marcellus formation in the Appalachian Basin, drove overall increases. In more recent years, even after the Covid-19 related dip, the trend continues. See Fig. 5.2. This figure shows that EIA estimates that for 2020 as a whole, non-OPEC production declined by 2.3 million b/d from 2019 levels. This was due to Covid-19 restrictions. More than 90% of this decline came from the three largest non-OPEC producers: the United States, Russia, and Canada. Non-OPEC production was its lowest for the year during the second quarter, but production began rising in the third quarter as global oil demand increased. The EIA expects production of non-OPEC petroleum and other liquid fuels to increase by 1.2 million b/d in 2021. In 2022, EIA expects non-OPEC production to rise by 2.3 million b/d, surpassing 2019 production levels. Canada and Brazil lead forecast non-OPEC production growth in 2021 and Russia and the United States will lead growth in 2022.

The EIA (2020b) expects that Canada's total liquid fuels production fell by 0.2 million b/d in 2020. This decrease is the result of both 2019 government-ordered production cuts in Alberta that continued into 2020 and economics-driven shut-ins because of the effect of low oil prices and falling demand for oil exports. In late October, the Alberta government announced it would stop setting monthly oil production limits. Although the government will extend its regulatory authority to curtail oil production through December 2021, pausing production cuts will allow producers to use available export pipeline capacity. As of the end of

2020, the EIA estimates that most shut-in production as a result of responses to COVID-19 has been restored, faster than previously estimated. In 2021, the EIA expects Canada's production to increase by 0.4 million b/d and surpass the first quarter of 2020 production, driven by the removal of government-ordered curtailments and expansions of previously deferred oil sands projects. The EIA does not expect any new upstream projects to come online in Canada during the forecast period. Any additional crude oil production will come from expansions or debottlenecking of existing projects. Forecast production in Canada grows by 0.1 million b/d in 2022.

After the United States, Russia is the second-largest producer of liquid fuels among non-OPEC countries. EIA expects production in Russia to grow in 2021 and 2022 after declining sharply in 2020 because of the OPEC+ agreement, in which Russia participates, limited crude oil production. Russia experienced the largest liquid fuels production decline in 2020 among OPEC+ producers: a decline of 1.0 million b/d from 2019 production. EIA expects Russia's liquid fuels production to increase by 0.1 million b/d in 2021 and by 0.9 million b/d in 2022. After the OPEC+ agreement ends in early 2022, EIA expects Russia's production to return to 11.5 million b/d by April 2022, almost the same level as in the first quarter of 2020.

The EIA also expects production growth in Norway during 2021 and 2022. Norway's Ministry of Petroleum and Energy enacted unilateral production limits on the Norwegian continental shelf from June to December 2020. The limits applied to production at existing fields and delayed the start of new fields and kept growth in total liquids production in 2020 to less than 0.3 million b/d. After production limits expire, the EIA forecasts production growth of 0.2 million b/d in 2021 and 0.1 million b/d in 2022 as existing fields increase production and new fields come online, including the much-delayed Martin Linge field. The ramp-up in new fields during 2021 will contribute to the year-over-year growth in both 2021 and 2022. The Johan Sverdrup field, which was the main driver of growth in Norway's production in 2020, will also contribute to growth in 2021, 2022, and beyond. The EIA forecasts Phase 1 of the Johan Sverdrup field to return to its pre-COVID-19 peak production of 470,000 b/d in early 2021 and surpass that before the end of 2021. In addition, Phase 2 of the Johan Sverdrup field is scheduled to come online in the fourth quarter of 2022 and add more than 0.2 million b/d productions at full capacity (Fig. 5.3).

In terms of technical recoverability, both oil and gas reserves changed over the last decade. Overall, even with reduced aggressive research, technological developments in various aspects of petroleum engineering made it possible to upgrade the reserve estimates. This decline has been accompanied by the increasing sulfur content of the U.S. crude. Fig. 5.4 shows general trends in the sulfur content of crude oil in the United States.

Together, Figs. 5.4 and 5.5 show that the overall quality of the U.S. crude declined from 1985 through 2010. After that period, there has been a steady improvement in the USA crude quality. This is owing to the production from unconventional reservoirs. While these reservoirs are difficult to develop and exploit. Fig. 5.6 shows both API gravity and sulfur content of crude oil from around the world. Light and sweet crude oil is the most desirable. However, any change in the quality of the crude implies both economic and technological drain on the crude oil. Light sweet grades are desirable because they can be processed with far less sophisticated and energy-intensive processes/refineries. The figure shows select crude types from around the world with their corresponding sulfur content and density characteristics. One

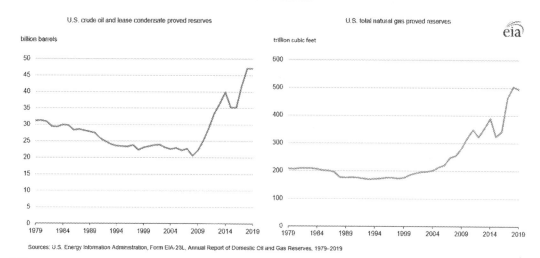

FIG. 5.3 U.S. reserve variation in recent history.

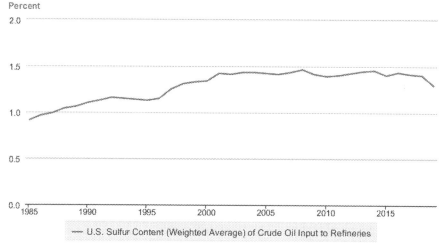

FIG. 5.4 The sulfur content of the U.S. crude over the last few decades.

particular advantage of certain EOR techniques is in situ upgrading of in situ oil. While no data is available on the quality of oil recovered with enhanced oil recovery (EOR) as compared to the same without EOR, it is reasonable to assume that in situ upgrading would improve the quality of produced oil.

The selected crude oils in Fig. 5.6 show the "sweetness" of various crude oils from around the world. These grades were selected for the recurrent and recently updated EIA report,

5.2 World reserve

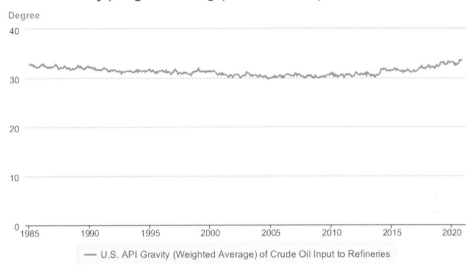

FIG. 5.5   API variation in the U.S. crude.

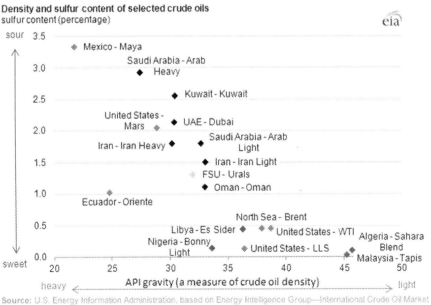

FIG. 5.6   Worldwide crude oil quality.

"The Availability and Price of Petroleum and Petroleum Products Produced in Countries Other Than Iran."

5.3 Reservoir characterization

Fractured basement reservoirs actually had been known since decades ago but were often abandoned because of the common perception that these reservoirs are not economically attractive. However, as early as the 1940s, it was known that a significant portion of oil production is often associated with basement reservoirs. For instance, Eggleston (1948) estimated that California's oil production due to fractured basement reservoirs is in the range of 6%. Note that these are sand reservoirs, often with unconsolidated sands. Eggleston described basement reservoirs with the following statement:

> The main sources of oil production in California are sand reservoirs. In recent years, however, oil production from fractured rock reservoirs has entered the limelight and has gained a certain mount of prominence. Most of the interest has been caused by the occurrence of oil in fractured basement rocks, to some extent a geological anomaly, and not because of oil production from fractured cherts. Production from fractured cherts is not new or in any way unique. Some of the older fields in California have produced most, if not all of their oil, from fractured rock reservoirs. Such fields as Lompoc, Orcutt, and Casmalia discovered in the years 1903 and 1904, have produced oil for many years from fractured cherts in the Monterey formation of Miocene age.
>
> Fractured rock reservoirs can in general be divided into two groups or classes. This division can be made both as to time of discovery and as to geologic age. The early discoveries made in 1903 and 1904 were in cherts of Miocene age. Recent discoveries made in 1903 and 1904 were in cherts of Miocene age. Recent discoveries have been made fractured basement rocks probably Jurassic age. (p. 1352)

Basement reservoirs can be found all over the world, but following countries have significant proven reserves in the basement reservoirs (Fig. 5.7). Fractured basement reservoirs

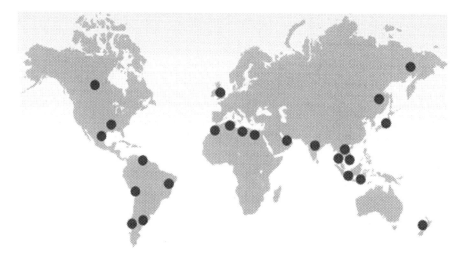

FIG. 5.7 Basement reservoirs are distributed around the world.

occur in more than 25 basins in over 30 countries. The most famous is the White Tiger offshore oil field in the Cuu Long Basin in Vietnam. The reservoir was discovered in 1986 and has cumulative production of 180 MBBL. Approximately 85% of hydrocarbon production in Vietnam is from fractured basement rocks (Gutmanis, 2010). The Wilmington Field in the United States was discovered in 1945. The Wilmington Oil Field is the third-largest field in the contiguous United States with an ultimate recovery estimated at three billion barrels of oil. The field is located on the 13 miles long and 3 miles wide Wilmington Anticline that extends from onshore San Pedro to offshore Seal Beach and is divided vertically by faults creating separate producing entities called Fault Blocks. Oil is produced from five major sand intervals ranging in depths from 2000 feet to 11,000 feet where over two and one-half billion barrels of oil have been recovered. Oil and Gas are recovered through primary production, secondary water flooding, and stream flooding. A total of 6150 wells have been drilled to date (Website 1).

Website 1, http://www.longbeach.gov/energyresources/about-us/oil/history/#:~:text=The%20Wilmington%20Oil%20Field%20is, three%20billion%20barrels%20of%20oil, accessed January 29, 2021.

Overall, the following countries are known to have basement reservoirs that contribute significantly.

Algeria, Argentina, Brazil, Canada, Chile, China, Egypt, Former USSR, India, Indonesia, Japan, Korea, Libya, Mexico, Morocco, New Zealand, Peru, Thailand, UK, USA, Venezuela, Yemen, Malaysia, and Vietnam.

A global summary of different types of igneous rocks, describing hydrocarbon deposits, was prepared by Petford and McCaffrey (2003) based on the review by Schutter (2003) (Fig. 5.2). The distribution of hydrocarbons in igneous rocks shows that basalts, andesites, and rhyolites constitute 75% of hydrocarbon-bearing igneous rocks. Although it is not common that hydrocarbons are retrieved from crystalline basement rocks, naturally fractured basement reservoirs have been known and exploited by the hydrocarbon industry since 1948 (Eggleston, 1948) (Fig. 5.8).

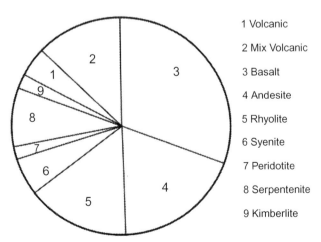

FIG. 5.8 Distribution of hydrocarbons in igneous rocks after Petford and McCaffrey (2003).

Basement rock can be very old, going up to 2.5 billion years of age. It is largely hard rock such as granite. The basement rocks are often highly metamorphosed and complex, with many different types of rock—volcanic, intrusive igneous, and metamorphic. They are considered to be older than the time frame of the formation of hydrocarbons in petroleum reservoirs. There are several theories on how hydrocarbons ended up in the basement rocks. Each theory has its shortcomings that will be discussed in this chapter.

New igneous rock may freshly intrude into the crust from underneath or may form underplating, where the new igneous rock forms a layer on the underside of the crust. It is said that the majority of continental crust on the planet is around 1–3 billion years old, and it is theorized that there was at least one period of rapid expansion and accretion to the continents during the Precambrian. The natural position in geological time for the oil-producing layer would be above the basement, but this conclusion has the assumption that basement and nonbasement reservoirs have the same source of hydrocarbons. Over millions of years, movement caused by tectonic forces can cause disruption in the layers of rock. The basement can be forced up by as much as a kilometer. This process itself can cause an extensive fracture network. Consequently, it can also move the oil-producing layer being at a lower level than the basement. Often, this is followed by the trapping of the oil by the structural trap that itself can have a seal of muds and clays above. As the oil-producing rock forces out hydrocarbons, they move up the flank and into the basement through the fracture network.

Islam (2014) outlined how mixed convection within crystalline basement rocks affects the final lithology of the rock as well as oil composition. Often a high-temperature bubble can be formed due to the presence of thermal insulation, similar to the lithological traps. The energy for such a convective system can come from the shallow mantle, bodies of hot rock in the upper crust, and/or from radiogenic heat produced by elements in the rock.

More than 125 hydrocarbon fields with fractured basement reservoirs have been recognized worldwide (e.g., P'an, 1982; Koning, 2003a, b; Cuong & Warren, 2009; Gutmanis, 2009; Trice, 2014), but their geology and the processes that lead to the accumulation of significant volumes of hydrocarbons are poorly understood (Holdsworth et al., 2020. As a result, the true potential of basement reservoirs or proper development of them is a daunting task. In most basement plays, oil is thought to migrate from an organic-rich mudstone source rock into a trap formed from a palaeo-high, termed a "buried hill" trap (Biddle & Wielchowsky, 1994). The seal is typically provided by a blanketing sequence of clay-rich mudstone and the reservoir formed from naturally fractured crystalline basement rocks.

5.3.1 Fractures and anomalies

As early as the 1940s, it was recognized that oil-producing basement reservoirs are anomalous. Established diagenesis theories don't seem to apply to them. Most basement hydrocarbons are hosted in structural highs of varying but generally moderate to large elevation (hundreds or even 1000m +). The highs are formed by fault-controlled blocks, often in rift settings, or by palaeo-hills buried below sedimentary cover. The second type of setting comprises intrusive igneous bodies (plutons) within sedimentary sequences (Bonter & Trice, 2019). Additionally, some of the overlying, or fault-juxtaposed, sedimentary units may also be charged so that a composite hydrocarbon play is developed. For example, sandstone "washes" above top granite surfaces or carbonate formations developed on or adjacent to the fault blocks. Most basement charging is considered to have taken place by lateral or up-dip migration from source areas in nearby structural lows. The migration route is through higher permeability sedimentary units, active faults, and perhaps also at the basement/cover interface, thence into the basement itself.

Given the very low matrix permeability of most basement rocks, oil and other associated fluids are typically transported and stored via well-connected, open fracture systems. The geological characteristics and formation mechanisms of these fracture systems are reservoir-specific and are challenging to characterize as the fracture network (or hydrodynamic fracture network associated with fluid transportation) cannot be imaged directly using seismic reflection data. Borehole images and core samples can give insights into the fracture network, particularly when integrated with dynamic data such as pressure transient analysis, interference, and production logging data.

Bonter and Trice (2019) provided a case study from the basement fields West of Shetland on the UK continental shelf. Other examples are found in the rift basins offshore Vietnam (CuuLong Basin) and onshore in Yemen (eg Sab'atayn Rift). Up-dip and also lateral charging along fault structures is likely to occur in periods of tectonic activity when active faults are believed to be open conduits for the migration of fluids by the episodic and repeated mechanisms of "fault-valving" and "seismic pumping," which are intimately associated with the earthquake process (Sibson, 2002). Wherever there is a basement component it can mean these faults behave in a different way from what is normally expected. It is known that unusual earthquakes that happen on different cycles can have an impact on the basement reservoirs. Reversely, the knowledge of basement structures could also help us forecast and manage the effects of geohazards like earthquakes, volcanic eruptions, or landslides. For example, there are places where old faults have been put under stress by present-day tectonic forces.

The basement formation is nearly impermeable. However, deep-seated dissolution and cavity formation, as well as structural processes, create pore space in carbonate rocks, and cavities and cavity-filling minerals may both be deep burial phenomena. Dissolution along with fractures, for example, is widely reported (Nelson, 1985). If deep-seated dissolution is important in a given reservoir, predictions based on shallow (epigenetic) dissolution and palaeokarst processes may be misleading. The possibility of large-scale (volumetrically significant) dissolution at great depth (>1 km) has been doubted (e.g., Ehrenberg, 2006; Ehrenberg et al., 2012). But workers agree that discerning cavities formed by shallow (epigenetic) from those owing to deep-seated (mesogenetic) dissolution is challenging, as by definition the cavities themselves preserve scant evidence of their origin.

Hydrocarbon reservoirs in carbonate rocks in which pores and fractures have been enlarged by dissolution are common in many petroliferous basins in the world (Loucks, 1999; Ahr, 2011; Garland et al., 2012). These reservoirs typically have low to moderate recovery efficiencies owing to heterogeneity and anisotropy in the size, spatial arrangement, and connectivity of open pores and fractures. The behavior of such reservoirs is notoriously challenging to characterize and predict. In some instances, a palaeokarst model is a useful paradigm for predicting open pores and fractures. In modern karst settings, corrosive surficial conditions and groundwaters from various subsurface and geomorphological (landscape) features, including caves. Some of these features collapse and coalesce during subsequent burial (James and Choquette, 1983). Such palaeokarst features are typically marked by evidence of near-surface conditions at the time of formation, including cave-fill sediments and distributions that match palaeotopography. Palaeokarst porosity is reduced by cavity-fill sediments, compaction (stylolites), and mineralization. Mineral fills might not be diagnostic of porosity formation at shallow depths, as cements may accumulate during subsequent burial or hydrothermal fluid flow. Such reservoirs are widely recognized in China (e.g., Jiu et al., 2020).

5.3.2 Formation evaluation

The optimum approach to formation evaluation is to acquire as much high-quality data as possible, with the objective of reducing interpretation ambiguity through the provision of multiple measurement methods. This approach is particularly important in the analysis and quantification of resources within the fractured basement play where key measurements can be at the lowermost acceptable range of their signal-to-noise ratio. The formation evaluation challenge is further compounded by the paucity of relevant and detailed analytical case studies available in the public domain. Hurricane has struggled to find suitable field analogues for comparison with Lancaster; that stated, however, the data gathered to date from the field does demonstrate that Lancaster shares many similarities with certain characteristics of other fractured basement reservoirs globally, including:

- the trapping mechanism;
- porosity ranges;
- the significance of tectonic faulting to improve the productivity of the fracture network; the positive effect of dissolution and subaerial exposure in enhancing the hydrodynamic fracture network; the value of horizontal wells targeted to seismically mapped targets in order to achieve high production rates.

5.3.3 Reserve growth potential of basement oil/gas

Historically, the majority of additions to oil and gas reserves are attributed to the growth of existing fields and reservoirs. In fact, from 1978 to 1990, the growth of known fields in the United States accounted for more than 85% of known additions to proven reserves (Root and Attanasi, 1993; McCabe, 1998). Thus, field growth and reserve growth are essentially synonymous with discussions of domestic resources. The exception occurred when unconventional resources became exploitable. However, it would be illogical to apply the principle of conventional reserve growth to unconventional reserves that have little in common with conventional reserves in terms of technology (Islam, 2014). Even though it is promoted that unconventional gas and oil are of lesser quality and are in need of more expensive technology for future development, previous sections demonstrated that notion to be false. In fact, the quality of gas improves as the quantity of the resource increases, the case in point being hydrate. For basement reservoirs, it is also true, a vast amount of petroleum can be exploited with little or no exploration. In addition, many existing conventional reservoirs have links to unconventional resources, including basement reservoirs and even the abandoned fields can be accessed for tapping the nonconventional resources.

While evaluating the nature of growth in fields requires an understanding of both geologic and nongeologic factors that affect growth estimations, geology is the pivotal factor in determining the accumulation, quality, and producibility of petroleum resources. Fields may grow when

(1) additional geologic data on existing reservoirs become available and are used to identify new reservoirs or to guide infill drilling;
(2) there are annual updates of reserves data;
(3) field boundaries are extended;
(4) recovery technology is improved; and/or
(5) nongeologic factors such as economics, reporting policies, or politics favor expanded production and development.

Past reserve estimates and characterization both used mathematics and not science as the basis. Attanasi et al. (1999) wrote: "...the modeling approach used by the USGS (U.S. Geological Survey) to characterize this phenomenon is statistical rather than geologic in nature."

Islam et al. (2016) discussed in detail the shortcomings of these statistical models and characterized them as unscientific. Yet, volumetric estimates of reserve growth are calculated by using these mathematical approaches and large databases that record field reserves through time. Crovelli and Schmoker (2001), Verma (2003), and Klett (2003) present details of various methods used to estimate reserve growth. The application of these models to unconventional gas is of particular concern. It's because one of the most important bases for statistical models is past history, which is practically nonexistent for unconventional reserves.

Geologists tend to think in terms of entire reservoirs, in some cases down to facies level, whereas reservoir engineers deal with measurements at the wellbore. To fully understand the geologic factors that affect growth in reserves, this gap in investigative approaches must be bridged. It has become increasingly important to integrate different scales and different observational techniques as secondary and tertiary recovery methods are applied more frequently in mature petroleum provinces.

Fluid-flow pathways, governed predominantly by rock porosity and permeability, are a reflection of heterogeneities of varying scales within a reservoir. Because these reservoir hetereogeneities are fundamentally geologic in nature (Dyman et al., 2000), an adequate understanding of the reservoir architecture, obtained through evaluation of geologic, engineering, or production data, or a combination of these data sets, can provide the basis for scientific characterization of unconventional gas reservoirs.

Proper estimates and growth potential of unconventional gas depend on geological classification and linking of conventional resources with unconventional ones through scientific characterization.

5.3.4 Case studies

Ukar et al. (2020) use observations from outcrops and core to show that fracture-related, deep dissolution, and porous fault rock are widespread in Yijianfang Formation rocks in China. Evidence from the Keping Uplift outcrops, 100 km SW from producing Yijianfang Formation carbonate rocks in the Halahatang oilfield show that multiple stages of fracture-related dissolution are common in exposed Middle Ordovician rocks. Stable isotopic analyses indicate cavity-filling cements precipitated from fluids at depths of c.220–2000 m. Crosscutting relations and abutting show that some fractures and vugs formed and became cement-filled prior to later dissolution events and the formation of subsequent generations of fractures and vugs. Together these relations show that dissolution (probably multiple instances), not only cementation, occurred at depth.

Bonter and Trice (2019) presented a detailed case study on the Lancaster field of the UK. The Lancaster area has good 3D seismic data coverage of reasonable quality allowing faults penetrating the top basement to a depth of at least 150 m to be identified with confidence. These have been mapped out across the top basement surface using automatic picking in the uppermost 10 ms of the seismic volume utilizing coherency analysis, the parameters of which were constrained based on manual interpretation of seismic and well data. All seismically mapped basement faults intersected by Lancaster wells drilled to date have been demonstrably porous and permeable (e.g., Belaidi et al., 2016; Bonter et al., 2018). The associated fracture network appears to form a deeply penetrating and well-connected fissure system (Slightam, 2012; Trice, 2014). Belaidi et al. (2016) subdivided the network into scale-dependent fracture types referred to as "microfractures," "fractures," "joints" and "faults" (Fig. 5.9); they also identified "veins," which were taken to be fractures completely occluded by impermeable mineral fills. In addition, "microfractures" (and veins) were identified and characterized from sidewall cores using petrophysical measurements, 3D scanning techniques for porosity and density, and preliminary thin-section analysis. The existence of microfracture signals an intricate network of fractures of various sizes and a very high potential of increasing recovery by manipulating the production rate. The term "fracture" was used to describe lineaments with observable trace lengths longer than the sampling medium, be it thin section, core, or borehole diameter. Thin-section analysis of the cores indicates that the minimum effective microfracture and fracture apertures are 20 μm (Belaidi et al., 2016). The majority of Lancaster cores are acquired through the process of rotary sidewall coring, all of which exhibit natural fractures. This observation is a testament to the high fracture frequency of the Lewisian basement as the sidewall cores were targeted to minimize the presence of fractures to optimize core recovery.

Oil-bearing fissure networks dominated by sediment and vuggy calcite fills are widely developed in both the basement and well-cemented Jurassic Rona Sandstone cover rocks in well

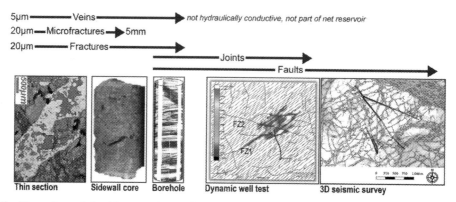

FIG. 5.9 The scales and classification scheme of fractures used for the Lancaster basement by Belaidi et al. (2016) and Bonter et al. (2018) and the range of data types used to image and understand these fractures.

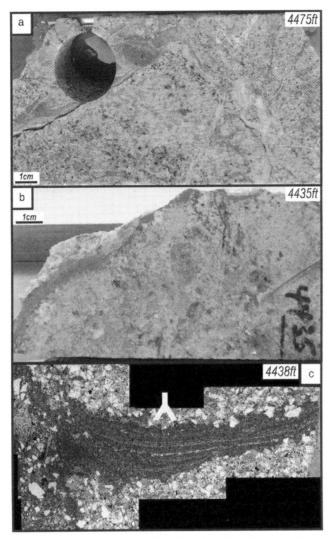

FIG. 5.10 Sediment-filled fractures from legacy well 205/21-1A. (A) Laminated sediment and carbonate mineral fill in fractured or fissured metatonalitic basement; (B) oil-stained sediment-sparry calcite fill in fracture (top) cutting paler well-cemented Rona Sandstone cover sequence breccia; (C) PPL view a thin-section montage of laminated sediment with grading giving way up, filling a triangular fissure in carbonate-cemented low-porosity Rona Sandstone. The total horizontal distance across the bottom of 8c is 20 mm.

205/21-1A (Fig. 5.10A–C). Filled fractures with fine depositional laminations are preserved locally and consistently young up the core (Fig. 5.10C). Basement-rooted fractures are also inferred to cut the Victory Formation sandstones in 205/21a-7 as significant mud losses were encountered in this unit during drilling of 205/21a-6, 205/21a-7, and 205/21a-7Z (C. Slightam. pers. comm., 2019). Similar subterranean cavity fills are seen in basement fractures elsewhere along the Rona Ridge (Holdsworth et al., 2019)

In core, a key isochron age from a fluorite-bearing fracture that postdates several types and stages of mineral-filled vugs (dissolution pores) provide evidence that substantial dissolution had occurred in Middle Ordovician carbonates by the end of the Devonian. We present a geological model based on our data and reported structural, petrological, geochemical, geophysical, petrophysical, radiometric, and kinematic findings to show the evolution of the mechanisms and timing of dissolution that led to the formation of cavities in the Yijianfang Formation. The estimated fault-related porosity volume is at least c.0.5%, which is augmented by fault-related mesogenetic dissolution and fault-transverse extension.

Bonter and Trice (2019) present a case study involving basement reservoir characterization. Hosting up to 3.3 billion barrels of oil in place, the upfaulted Precambrian crystalline rocks of the Lancaster field, offshore west of Shetland, give key insights into how fractured hydrocarbon reservoirs can form in such old rocks. The Neoarchean (c.2700–2740 Ma) charnockitic basement is cut by deeply penetrating oil-, mineral- and sediment-filled fissure systems seen in geophysical and production logs and thin sections of the core. Mineral textures and fluid inclusion geothermometry suggest that a low-temperature (The Lancaster Field is located West of Shetland in blocks 205/21a, 205/22a, and 205/26b, license P1368 Central, and is in the relatively shallow water of c.150 m. The Lancaster prospect was first drilled by Hurricane in 2009. The structure had already been drilled in 1974, by a previous operator, with well 205/21-1A, which was plugged and abandoned after retrieving a total depth (TD) core from the basement and undertaking two drill stem tests (DSTs). Slight oil seepage from the core is noted at the end of the well report and small volumes of oil (c.2%) were recovered from the tests, along with significant brine returns. A temperature anomaly confirmed that there was flow from the fractured basement, although a fluid sample a short distance above this influx point consisted entirely of water and so the end of the well report concluded that it was probable that the small volumes of oil were being produced from the overlying sediments.

Ultimately, this well was abandoned and the structure was left undrilled for 35 years. Hurricane reinterpreted the well data and considered that the evidence of oil within the core was significant enough to demonstrate oil presence at a specific depth. The lack of oil within the fluid sample was interpreted by Hurricane as circumstantial, a product of low oil volume being flowed. Hurricane's interpretation of the recovered fluid was that it was consistent with drilling brine filtrate, and, therefore, there was no evidence of formation water influx into the wellbore. Therefore, the temperature anomaly was most likely due to oil influx into a severely compromised wellbore, with complications arising from the invasion of the drilling fluid into the fracture system—a perennial issue which Hurricane has had to deal with many times since.

The reinterpretation of well 205/21-1A provided confidence that exploring the fractured basement reservoir at this location would prove fruitful. Hurricane's first exploration well at Lancaster, 205/21a-4, was the first UK exploration well drilled specifically to evaluate the basement as an exploration play (Trice, 2014). Hurricane continued this success with a sidetrack (205/21a-4Z) in 2010 to further appraise the basement, and a subsequent horizontal production well (205/21a-6) in 2014 that demonstrated extremely high reservoir productivity. An extensive drilling campaign in 2016–17 included two further wells on Lancaster—another deep inclined well (205/21a-7) designed to investigate the depth of the oil column; and a second horizontal producer (205/21a-7Z), which confirmed the high productivity observed with well 205/21a-6.

Data from the Lancaster wells has been supported by the drilling of neighboring assets in Hurricane's acreage. Whirlwind (205/21a-5) was drilled in 2010, while Lincoln (205/26b-12) and Halifax (205/23-3A) were drilled during the 2016–17 drilling campaign. The drilling and testing results across Hurricane's assets demonstrate the similarity of this reservoir on the Rona Ridge and present an exciting opportunity within the UK Continental Shelf (UKCS). The progress in unlocking this basement play, specifically related to Lancaster, was documented by Hurricane prior to the 2016–17 drilling campaign (Belaidi et al., 2016), and through this drilling campaign, Hurricane moved the field closer to development and further de-risked the play.

The development is designed to be phased with production from the first phase achieved in 2019.

The teamwork of a multidisciplinary team at Hurricane incorporated drilling and mudlogging data, high-resolution gas chromatography, logging while drilling (LWD) and wireline logs, DST, and production logging tool (PLT) data to analyze and model the reservoir. It is the combination of these disparate datasets which is key to Hurricane's analysis and has led to the technical de-risking that has underpinned the final investment decisions leading to the first phase of the Lancaster development.

Lancaster is the first basement reservoir to have been put onto production in the UK, with the first oil achieved in May this year (2019). Current production comes from two horizontal wells that are tied back to the Aoka Mizu, a floating production, storage, and offloading vessel (FPSO). This is termed an early production system (EPS) as it is the first phase of full field development. The EPS is designed to obtain early, commercially sustainable production from the Lancaster Field ahead of full-scale development, while at the same time providing crucial information about reservoir behavior. Lancaster has assigned 2P Reserves of 37 MMstb based on the planned EPS, and a further 486 MMstb of 2C Contingent Resources assigned to the full field. In total, this attributes a Base Case recoverable resource of 523 MMstb to the Lancaster Field (RPS Energy 2017). The acquisition of reservoir data during the EPS will allow these numbers to be revised over time.

The reservoir has been dated to c.2.7 Ga and that extremely long and varied geological history has resulted in a unique and well-connected hydrodynamic fracture network. This well-connected fracture network is reflected in the intense fracturing seen on image logs throughout the reservoir section, as well as the extremely high and consistent productivity index (PI) from two horizontal wells. No matching fracture networks have been located from the literature or through access to commercially available databases. Consequently, the use of analogs is confined to the understanding of general fractured reservoir behavior, the impact of applied reservoir management schemes, and comparisons of formation evaluation techniques. These comparisons have helped Hurricane to develop its formation evaluation program, which is summarized in Table 5.4 in terms of the datasets acquired from the Lancaster wells (Tables 5.5 and 5.6).

There are several challenges when acquiring and interpreting well data in fractured basement reservoirs, many of which arise from the behavior of using overbalanced drilling mud in highly permeable fractures. Hurricane has recently published, in cooperation with Schlumberger, some of the considerations that have been noted during the exploration and appraisal of Lancaster (Bonter et al., 2018).

TABLE 5.4 Hurricane well data acquisition for formation evaluation.

	Well name	205/21a-4	205/21a-4Z	205/21a-5	205/21a-6	205/21a-7	205/21a-7Z	205/26b-12	205/23-3A
	Asset name Year drilled Well type	Lancaster 2009 Inclined	Lancaster 2010 Inclined	Whirlwind 2010 Inclined	Lancaster 2014 Horizontal	Lancaster 2016 Inclined	Lancaster 2016 Horizontal	Lincoln 2016 Inclined	Halifax 2017 Inclined
LWD data	At-bit gamma		✓	✓	✓	✓	✓	✓	✓
	Dual gamma ray		✓	✓	✓	✓	✓	✓	✓
	Gamma ray	✓							
	Compensated thermal neutron		✓	✓	✓	✓	✓	✓	✓
	Electromagnetic wave resistivity		✓	✓	✓	✓	✓	✓	✓
	Compensated wave resistivity	✓							
	Azimuthal deep resistivity		✓	✓					
	Azimuthal litho-density image		✓	✓	✓	✓	✓	✓	✓
	Azimuthally focused resistivity image		✓	✓	✓	✓	✓	✓	✓
	Pressure while drilling	✓	✓	✓	✓	✓			
	Bimodal acoustic tool			✓					
Mud gas	Standard	✓	✓	✓	✓	✓	✓	✓	✓
	Flair	✓					✓		
	GC tracer			✓		✓			
Wireline logs	Micro-imager-/triple combo-/spectroscopy	✓	✓	✓		✓		✓	✓
	Ultrasonic imager	✓	✓			✓		✓	✓
	Nuclear magnetic resonance		✓			✓		✓	✓
	Large-size sidewall core		✓	✓		✓		✓	✓
	Modular formation tester	✓	✓	✓	✓	✓		✓[a]	✓
Well test	Drill stem test	✓	✓	✓	✓	✓	✓		
	Production logging tool		✓		✓	✓			

[a] Modular formation tester (MDT) run aborted for operational reasons.

TABLE 5.5 Summary of Hurricane well tests.

Well name	Asset name	Year	Well type	Mud system	Maximum stable rate ESP	Natural flow
205/21a-4	Lancaster	2009	Inclined	Drilplex (MMO mud)	–	200 bopd
205/21a-4Z	Lancaster	2010	Inclined	NaCl brine (no viscosifier)	–	2399 bopd
205/21a-5	Whirlwind	2010 (drilled) 2011 (tested)	Inclined	CaCl/CaBr brine	–	2–400 bopd
205/21a-6	Lancaster	2014	Horizontal	CaCl brine (viscosified)	9800 bopd	5300 bopd
205/21a-7	Lancaster	2016	Inclined	CaCl brine (viscosified)	10930 bopd	6300 bopd
205/21a-7Z	Lancaster	2016	Horizontal	CaCl brine (viscosified)	15375 bopd	6520 bopd
205/23-3A	Halifax	2017	Inclined	CaCl brine (viscosified)	–	Minor flow

TABLE 5.6 Summary of dual-porosity model porosity and permeability values.

Facies	Fracture set	Porosity (%)	Permeability (mD)
Fault zone	Highly conductive joints	2.3	796
	Supportive fractures	2.9	0.06
Fractured basement	Highly conductive joints	0.7	796
	Supportive fractures	2.9	

Fracture identification from logs, core, and drilling data form the core workstream in evaluating the reservoir. Hurricane refers to all discrete fractures that can be interpreted from image logs as joints, and therefore have an associated depth, dip, azimuth, and classification (Belaidi et al., 2016).

Conventional logs respond to joints, particularly those with wide apertures or in discrete intervals of relatively high frequency, although the resolution is insufficient to fully characterize them. Therefore, higher-resolution image logs are used to pick and characterize joints where they display a response indicative of drilling mud invasion. This "invasion response" is typified by a reduction in resistivity as shown by the LWD AFR (azimuthally focused resistivity) tool and wireline FMI (formation micro-imager) log. It is also shown by a reduction in density on the LWD ALD (azimuthal litho-density) and acoustic amplitude or impedance typical of porosity on the UBI (ultrasonic borehole imaging) wireline log. High-resolution logs are processed such that low resistivity and porous responses produce a dark response on the images. This comparison is particularly useful in guiding joint interpretation in horizontal wells, where only the lower-resolution LWD data are available.

In Chapter 3, a fracture characterization technique is presented. Fracture characterization is a challenge. While some published methods for estimating fracture aperture from electrical logs do have practical application in many instances, the conclusion from work on Lancaster is that the technique is unreliable as the fracture aperture results are influenced by mud resistivity and tool effects. Attempts to use electrical logs to establish fracture aperture have been augmented by manual picking of joint apertures when the apparent aperture is sufficient to do so. This manual technique is recognized to be the most effective (Islam, 2014). The minimum aperture that can be reliably measured manually is 2 cm, so any joint with an aperture in excess of 2 cm is defined as a wide aperture joint. Fig. 5.11 shows three examples of wide aperture joints from well 205/21a-7. All of these wide aperture joints display sufficient expression to pick a top and base, allowing for manual measurement of the apparent aperture. These three examples are also associated with significant porosity responses, demonstrating that they are contributing to the poro-perm system. There are some differences between them, identified by comparing the FMI to the UBI images. In Fig. 5.11A, a typical response is observed from drilling fluid invasion causing an FMI response which is mirrored by the UBI—indicating an open fracture. Fig. 5.11B shows some clastic or mineral infill of a large fracture, as seen on the reduced UBI response, but the continuous high-porosity response indicates some dissolution control and does not indicate a completely closed feature. Fig. 5.11C could represent clastic or cuttings infill within a wide aperture joint; the joint boundaries are sharp and there is only a slight occlusion of the open aperture visible on the UBI. Again, the porosity log responds to this large feature.

These wide aperture joints can be effectively measured and assessed for how open or closed they may be, but Fig. 5.11 also highlights the problem with quantifying fracture porosity from image logs. This is an acknowledged industry challenge that can be seen in this case as being due to the difficulty in estimating how open these wide aperture joints are, as well as calculating the apertures of the many subordinate joints that can be seen in these three images. However, as Lancaster is a Type 1 NFR, the view is taken that all of the porosity is associated with fractures. Consequently, bulk porosity measurements will reflect fracture porosity. This assumption has been corroborated by comparing bulk porosity measurements to core porosity, joint frequency plots, joint aperture estimates from electrical imaging logs, and fracture frequency derived from drilling data (Bonter et al., 2018).

The preferred method of establishing bulk porosity is the Bateman-Konen technique, a density-neutron cross-plot methodology that simultaneously solves for matrix density and porosity, and thereby avoids problems caused by the mineralogical variation in the host rock (Bateman & Konen, 1977). The technique has been compared between wireline and LWD datasets to confirm the validity of relying on LWD porosity alone in horizontal wells. Generally, the LWD and wireline porosity matches well (note that the LWD was not run to TD in this well).

Bateman-Konen porosity was compared with porosity derived from nuclear magnetic resonance (NMR) logs. The NMR has the advantage of being completely independent of lithology and mineralogy and responds to effective porosity. However, within the high-salinity mud environment and relatively low porosity ranges of the Lancaster reservoir, the NMR is at the lower end of its effective operational limit. Despite these operational constraints,

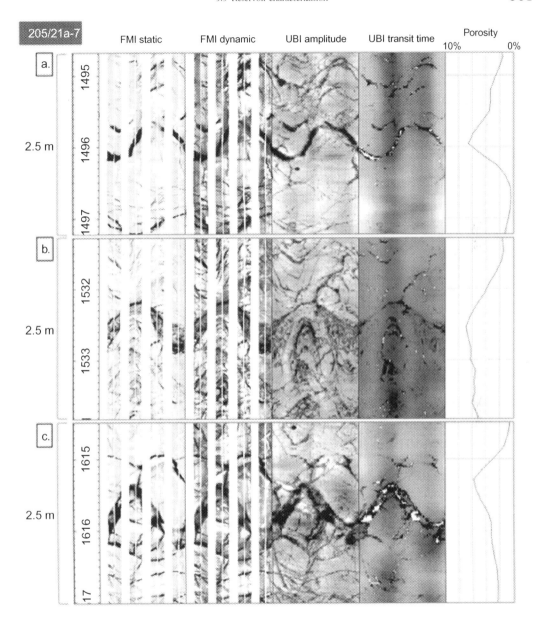

FIG. 5.11 Examples of aperture imaging, well 205/21a-7.

NMR porosity is generally supportive of Bateman-Konen, although tends to read slightly higher, presenting a potential upside porosity case. Both the Bateman-Konen and NMR porosity logs can be observed to respond to the darker (more jointed) portions of the image log in this well; thus demonstrating that porosity peaks are associated with those joints with wider apparent apertures, or with clusters of joints. However, there is also a background porosity calculated from the bulk porosity logs which is due to microfractures contributing to the overall storage system of the reservoir.

To provide lithological data and control on the Bateman-Konen (1977) porosity calculation, Hurricane acquired a large number of rotary sidewall cores. The sidewall cores are purposely targeted away from discrete joints to maximize their recovery and provide grain density data. Despite this intention, numerous cores are associated with microfractures and thereby provide a data source to help quantify the porosity associated with microfractures. These sidewall core porosities provide a constraint on the downside from the bulk porosity logs, but do include some higher values that match the bulk porosity logs quite well [e.g., a sidewall core porosity value of 12% at 1560 m measured depth (MD)]. Some of these surprisingly large porosity values in sidewall cores may be caused by discrete joints that could not be avoided when running the coring tool. However, most of the porosity seen in thin sections from sidewall cores is associated with microfractures. These microfractures either do not cut the entire borehole or are below the resolution of the image logs, but the porosity contribution they provide demonstrates that even the smallest fractures are open features contributing to the bulk porosity of the reservoir.

In terms of characterizing the fractured basement reservoir as a whole, the conceptual model shown in Fig. 5.10 focuses on the interaction of regional and cross joints within fault zones and fractured basement. As seen from the bulk porosity logs and sidewall core analysis, microfractures are also present within the reservoir between these joints, contributing to the overall porosity. The significance of defining "fault zones" is because porosity is enhanced within these damaged volumes of rock surrounding seismic-scale faults.

Faults interpreted on seismic are matched to fault zones interpreted from well logs, with the fault map evolving over time pre- and post-drilling (Slightam, 2012). More recently, automated fault identification has been achieved using Ant Tracking in-house, providing assistance to the manual fault interpretation, particularly in terms of the connectivity of the fault network at top basement level. This technique does not provide a good image of the faults within the basement itself using the seismic vintage Hurricane has available, but a comparison between the Ant Tracking extracted at top basement level and the fault zone interpretation at depth from well logs shows that there is a good match between the two, so therefore the majority of the faults are near-vertical (Fig. 5.12).

Through an extensive formation evaluation program, it is clear that porosity within the Lancaster reservoir is related both to discrete joints and background microfractures, and that porosity is enhanced by seismic-scale faulting. There is also a demonstrable link between the faults interpreted from the seismic data and the fault zones interpreted from well logs, allowing some prediction of porosity enhancement (within fault zones) away from well control.

Understanding the behavior of a Type 1 NFR, such as Lancaster, requires an integrated approach to combine the dynamic data from well test interpretation with the static information from well log analysis. As in the case of formation evaluation from wireline and LWD data, the impact of drilling fluid invasion needs to be carefully considered in well test analysis. Each well test has benefitted from a learning process through analysis of previous wells. For example, the use of a mixed metal oxide (MMO) mud throughout the reservoir section of well 205/21a-4 was selected due to safety concerns related to anticipated high loss rates while drilling. This mud system proved to be excellent at preventing losses but severely compromised the well test due to operational constraints preventing the deployment of acid

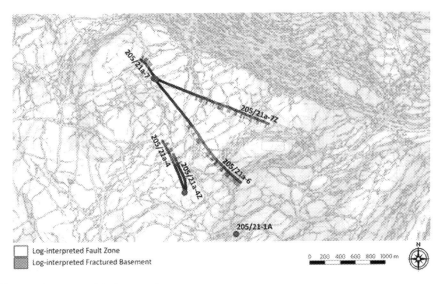

FIG. 5.12   The fracture and fault network.

to break down the mud prior to testing. This well, however, flowed oil to the surface and demonstrated that the reservoir was indeed productive.

Therefore, the following well (205/21a-4Z) was drilled with brine, with careful control on pressure at the bit to keep as close to balance as possible and limit losses. Again, this strategy was successful for drilling but, due to the lack of the lifting capacity of this brine, a large quantity of cuttings were left downhole which induced a significant skin in the near-wellbore environment, limiting the productivity that could be achieved. Later wells employed varying brine compositions including viscosifying agents to enhance the lifting capacity of the mud and avoid this skin issue.

The most productive well tests have been the two horizontal wells on Lancaster, 205/21a-6 and 205/21a-7Z, which had a PI of 160 and 147 stb/day/psi, respectively. The rates of these well tests were constrained by the surface equipment on the rig in both the ESP (electrical submersible pump) and natural flowing cases. A strategy of using brine with improved lifting capacity and a rigorously monitored hole-cleaning program combined with the horizontal well path contributed to minimizing skin in both these wells. Losses while drilling were accepted but carefully monitored and controlled as much as possible to achieve an acceptable balance between safe drilling practice and a successful well test.

Data from the pressure build-ups following the natural flowing periods of these two productive horizontal wells has been used to interpret the behavior of the fracture network away from well control and is shown in Fig. 5.13. In each test, the early time (up to 1 h) is characterized by a complicated storage response. By analyzing higher-resolution data around the shut-in on well 205/21a-7Z, the true wellbore storage effect can be seen occurring within the first 3–4 s. The remaining storage effect is interpreted as being caused by large fractures connecting to the wellbore so well that they appear to increase the volume stored within the wellbore. Therefore, this early time response is seeing a very large volume of well-connected reservoir fluid.

564 5. Basement reservoirs

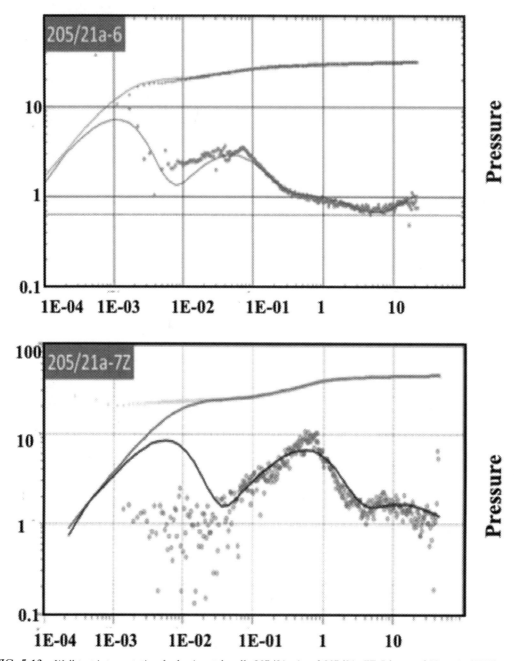

FIG. 5.13 Well test interpretation for horizontal wells 205/21a-6 and 205/21a-7Z, Islam and Hossain (2020).

After the early time storage effects, both horizontal wells display a period of increasing transmissivity away from the wellbore, followed by a dip and recovery (Fig. 5.13) that can be matched with a classic dual-porosity response (Warren and Root, 1963). This is interpreted consistently between the wells as being caused by a highly productive medium, containing approximately one-third of the STOIIP (stock tank oil-initially-in-place), which is supported by a second, less productive, medium containing the remaining two-thirds. In terms of relating this to the static reservoir model, as the reservoir is a Type 1 NFR, both mediums are subdivisions of the connected fracture network. The porosity evaluation shows that there is a continuum of fracture sizes contributing to the reservoir storage system, from microfractures up to wide joints with apparent open apertures of decimeter scale. Although the extremes of this continuum are easy to identify, it is difficult to make a distinction between fractures within the range that contribute to either the productive or supportive medium as the dynamic response may be due to a combination of factors such as aperture, orientation, fracture fill and connectivity with the rest of the network.

PLT logging provides a method to link the well test data to the fracture interpretation, demonstrating which features flow under different conditions during a well test. Well 205/21a-4Z exhibited several interesting features during three PLT runs in 2010. The cumulative flow from these PLT runs shows a constant increase in contribution flowing into the wellbore, indicating that there are productive fractures throughout the entire tested interval. However, there are certain points at which the flow increases more dramatically allowing for the assignment of specific flow zones, associated with particularly wide aperture joints. This is further demonstrated in Fig. 5.13 where significant flow contribution was seen at the base of Fault Zone 1A, caused by a joint with a 45 cm open aperture. This aperture has been estimated from manual methods as described above and has been corrected for dip and borehole inclination. Similarly, the entire flow contribution for well 205/21a-7 [maximum stable rate of 10 930 bopd (barrels of oil per day) using an ESP] was provided by a series of wide aperture joints near the top of the wellbore, each having apertures of several centimeters.

A combination of PLT data and pressure build-up analysis demonstrates that flow in the fractured basement reservoir is dominated by highly productive joints and supported by other elements of the fracture network. This indicates an extremely well-connected hydrodynamic fracture network and is fully consistent with static observations from well logs.

The static and dynamic data that Hurricane has gathered are integrated into a single reservoir model that honors the conceptual model described in Fig. 5.10 and reflects the outcrop analogs that Hurricane has analyzed. Hurricane produces a static model in-house which is designed for simulation, modeling faults vertically, and stair-stepping them laterally so that the grid cells remain orthogonal. Grid cells are built with an areal extent of 10×10 m, and maintaining this geological grid-scale required the use of a high-resolution simulator (Bonter et al., 2018; Bonter and Trice, 2019).

With the facies model constructed, the grid cells can be populated with reservoir properties. These are applied as single values within each facies due to the lack of any guidance away from well control other than the presence or absence of fault zones. Each grid cell within the model represents a block of the basement that contains many joints and many more microfractures, so the bulk porosity methods described previously apply very well to this model. The permeability values are derived from the well test analysis where the primary permeability is provided by the highly conductive joint system. This permeability value is applied to the productive joint set in both facies. Through the Warren & Root dual-porosity interpretation, a secondary permeability value is calculated for a given block size of 10 m (matching the cell geometry) that represents the supportiveness of the secondary medium in the dual-porosity response and can be used for the supportive fracture set in this model. It should be noted that the values applied are bulk values, applicable to a representative elementary volume (REV) that matches the grid cell size, rather than being applicable to individual fractures. It should also be noted that these values are a first pass that are being updated with new data and interpretation, but that by combining the dual-porosity observations from the well test analysis with the porosity calculation, a reasonable dual-porosity model has been constructed that honors both the static and dynamic observations.

Discrete fracture network (DFN) modeling is used to aid in constraining the geocellular grid. To this end, DFN modeling has been applied to model joint sets away from the near wellbore environment and to establish which of these joint sets are associated with the fluid flow recorded during a DST. The simulated joint network established from DFN modeling is validated by modeling pseudo-wells within the DFN modeled space at locations and orientations equivalent to the source well paths. Comparisons of joint populations from well data with the DFN simulated data give confidence that the DFN representation of joint distribution is consistent with measured data. Once calibrated, the DFN can be used to simulate the pressure response from a given well and thereby used to evaluate which joint sets are the primary contributors to flow. The mapped pressure response indicates that fault zones and regional joints provide the primary conduits for fluid flow into the wellbore, with cross joints and microfractures providing secondary pressure support. The DFN model has also been used to construct pseudo-PLT responses that can be used to infer fracture apertures, which can be compared to measurements of wide aperture joints estimated from electrical image logs. Finally, DFN modeling has been used to constrain the size of REVs which provides a mechanism of cross-checking the geological character of the reservoir that the geocellular grid is attempting to model. Further integration of the DFN and geocellular modeling work is ongoing to improve the inclusion of factors such as permeability anisotropy; however, the conclusion of this work requires data from the planned inter-well testing during the EPS, the results from which are not expected to be available until the latter part of 2019.

The full-field simulation model constructed for Lancaster consists of a geocellular grid with approximately 80 million cells. This number of cells is a consequence of the desire to maintain a fine resolution to the model for robust fault-zone modeling, as described above. Building the static model with a view to simulation ensured that an appropriate level of geological complexity could be maintained throughout the process, avoiding common industry problems with upscaling and lack of integration between geological and reservoir engineering workflows. Acceptable compromises were introduced during the static model build, such as maintaining an orthogonal grid and including all faults as vertical, so that there was congruence between the static and dynamic models. In fact, there is no difference between the two and by using a high-resolution simulator the static model can be taken straight through to dynamic simulation with no alterations or upscaling.

Initially, the full-field simulation model was designed to establish base cases from which to evaluate the data that will be acquired from interference and long-term production during the EPS phase of the Lancaster Field development. Further to this, it has allowed several discrete uncertainty cases to be modeled. This began with a study on the three primary uncertainties of fracture porosity, oil-water contact (OWC) depth, and aquifer strength. Those scenarios that did not match the well test data from well 205/21a-6 without requiring significant changes to other assumptions were immediately discounted—this included the cases with the shallowest OWC and a lack of aquifer support. Other discrete cases that could be used to match the well test data were used to produce forecasts for single well production to evaluate reservoir behavior as well as developing end-member predictions for the two-well EPS. As this work was progressing concurrently with the 2016–17 drilling campaign, once these data were incorporated they reinforced the simulation methodology and the concepts that underpin the reservoir model while refining some of the input ranges. The initial results have provided confidence to continue with the simulation modeling approach. New data from the start-up of the Lancaster Field earlier this year will be gradually incorporated into the analysis, though this process takes time. This will include variable rate production and longer shut-in periods than previously acquired during well tests, in addition to long-term production data. The simulation work is, therefore, iterative and refinements will be ongoing as production and data continue to be acquired from the EPS.

The Lancaster EPS consists of a two-well tieback to the Aoka Mizu FPSO. Long-term dynamic data gathering is the primary technical objective of the EPS, to complement the well-test data gathered to date. By producing from two wells, Hurricane will be able to investigate the connectivity of the fracture network with interference testing, enabling estimates of permeability anisotropy which will provide invaluable data to position future wells in the next phases of development. By taking this phased approach to development, Hurricane can ensure that the Lancaster Field is developed optimally to enhance ultimate recovery from the field. It also provides an economically viable solution with lower risk in the early phase of the development.

Real-time data from the FPSO is being streamed back to shore, for analysis and integration into the reservoir model. Hurricane should be able to begin eliminating the least likely cases from the uncertainty analysis and focus on improving the predictions from the simulation modeling. Individual flowlines from the two wells enable accurate attribution of the production, while high-resolution downhole gauges provide reservoir pressure and temperature both during production and during regular shut-in periods, which will be used for reservoir characterization and compared to those pressure build-ups shown in Fig. 5.13.

A successful EPS on Lancaster has a significant impact on the rest of Hurricane's portfolio, as the surrounding assets can be expected to behave in a similar way to Lancaster—unlocking the potential from Lancaster may enable Hurricane to repeat this success on the Lincoln, Warwick, Halifax, and Whirlwind fields. This could have a dramatic impact not just on the West of Shetland region, but the UKCS as a whole. The fractured basement is underexplored in the UKCS and success in proving the play could be highly beneficial for the UK oil sector.

Ukar et al. (2020) reported a case study in China. At >7 km depths in the Tarim Basin, hydrocarbon reservoirs in Ordovician rocks of the Yijianfang Formation contain large cavities (c.10 m or more), vugs, fractures, and porous fault rocks. Although some Yijianfang Formation outcrops contain shallow (formed near the surface) palaeokarst features, cores from the Halahatang oilfield lack penetrative palaeokarst evidence. Outcrop palaeokarst cavities and opening-mode fractures are mostly mineral-filled but some show evidence of secondary dissolution and fault rocks are locally highly (c.30%) porous. Cores contain textural evidence of the repeated formation of dissolution cavities and subsequent filling by cement. Calcite isotopic analyses indicate depths between c.220 and 2000 m. Correlation of core and image logs shows abundant cement-filled vugs associated with decametre-scale fractured zones with open cavities that host hydrocarbons. An Sm-Nd isochron age of 400 ± 37 Ma for fracture-filling fluorite indicates that cavities in the core formed and were partially cemented prior to the Carboniferous, predating Permian oil emplacement. Repeated creation and filling of vugs, timing constraints, and the association of vugs with large cavities suggest dissolution related to fractures and faults. In the current high-strain-rate regime, corroborated by velocity gradient tensor analysis of global positioning system (GPS) data, a rapid horizontal extension could promote the connection of porous and/or solution-enlarged fault rock, fractures, and cavities.

The Halahatang oilfield is part of the Tabei Uplift, one of the most prolific hydrocarbon accumulation zones in the Tarim Basin. The Tabei Uplift has the largest proven oil/gas reserves of the Tarim Basin with over 3 billion tons oil-equivalent (c.21.5 billion barrels oil-equivalent) discovered in Ordovician carbonate rocks currently at 6000–7000 m depth (Zhu et al., 2016). Recently, giant oil accumulations have been discovered in the lower Paleozoic section at depths >8000 m (Zhu et al., 2019).

The Tabei Uplift is an inherited structural high formed on the pre-Sinian metamorphic basement. This area experienced Sinian-Ordovician rifting during the Caledonian orogeny, followed by Silurian-Permian uplift during the Hercynian orogeny (290–250 Ma), transition into a foreland basin (fore bulge) in the Triassic-Jurassic during the Indo-China orogeny, Cretaceous-Neogene extension related to the Yanshan orogeny, and late Neogene-Quaternary rapid subsidence (foreland basin) during the Himalayan orogeny (Zhu et al., 2013, 2017, 2019; Fig. 5.14). The Halahatang oilfield is located in the Halahatang Depression and it is bordered by the Luntai Fault to the north, the Shuntuoguole Low Uplift to the south, the Lunnan Low to the east, and the Yingmaili Low Uplift to the west (Zhu et al., 2012; Chang et al., 2013). Other important oilfields in the Tabei Uplift include the Tahe and Lunnan oilfields to the east (Chang et al., 2013).

During the Cambrian, dolomitic sequences with intercalated gypsum and salt layers were deposited in an array of half-grabens developed as a result of regional extension (Lin et al., 2012; Gao and Fan, 2013) (Fig. 5.14). Throughout the Ordovician, thick reef and shoal deposits accumulated in intra-platform, platform-margin and slope environments in a series of uplifts (i.e., Tazhong-Bachu, Tabei uplifts) with intervening depressions (i.e., Kuqa, Awati, Manjiaer) (Kang, 2003; Gao and Fan, 2013). In the Late Ordovician, the carbonate platform was drowned and overlain by a thick section of the deepwater continental shelf- and slope-facies of dark mudstone and carbonaceous mudstone. The Silurian-Devonian was characterized by widespread marine sandstone deposition in intracratonic contractional basins, giving way to alternate marine-nonmarine deposition in the Carboniferous-Permian. Almost continuous Paleozoic sedimentation was interrupted by sporadic volcanism, including the Permian flood basalt magmatism of the Tarim Large Igneous Province. From the Triassic to the Quaternary, the Tarim Basin has been dominated by terrestrial siliciclastic deposition.

Hydrocarbon reservoirs in the Tabei Uplift mainly occur in Middle Ordovician Yingshan and Yijianfang Formations in the Lunnan Low Uplift and Halahatang Depression areas (Zhu et al., 2016, 2017). Muddy limestones and mudstones of the Tumuxiuke, Lianglitage, and Sangtamu Formations form the caprocks of these reservoirs (Fig. 5.15).

This study builds on analyses of 227 hand samples from 16 cores and visual inspection of c.450 m of full-bore formation microimages (FMI) and log data of the studied cores (Baqués

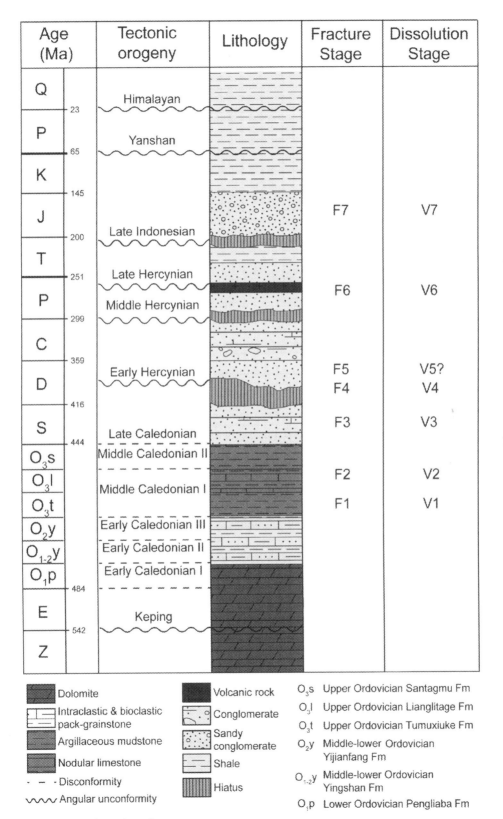

FIG. 5.14 See figure legend on next page.

FIG. 5.14 Stratigraphy and lithology of Paleozoic-Cenozoic sediments in the Tarim Basin (Zhang et al., 1983; Liu and Xiong, 1991). Almost continuous Paleozoic sedimentation was interrupted bysporadic volcanism, including the Permian flood basalt magmatism of the Tarim Large Igneous Province (e.g., Xu et al., 2014). From the Triassic to the Quaternary, the Tarim Basin has been dominated by terrestrial siliciclastic deposition. Hydrocarbon reservoirs in the Tabei Uplift mainly occur in MiddleOrdovician Yingshan and Yijianfang Formations in the Lunnan LowUplift and Halahatang Depression areas (Zhu et al., 2016, 2017) (Fig. 1). Muddy limestones and mudstones of the Tumuxiuke, Lianglitage, and Sangtamu Formations form the caprocks of these reservoirs. Results: This study builds on analyses of 227 hand samples from 16 cores and visual inspection of c.450 m of full-bore formation microimages (FMI) and log data of the studied cores (Baqués et al., 2020). Baques et al. reported an in-depth description of the diagenetic history recorded in these cores by means of detailed petrographic, isotopic, and fluid inclusion analyses of fracture and vug cements. Here we focus on constraining the timing of each diagenetic event in the microstructural sequence and tie them to outcrop observations and the geological history of the area (see Appendix A for methods). Fractures and vugs in core: In the carbonate reservoirs studied, porosity does not correlate with permeability (Zhu et al., 2019). Such rocks constitute Type 1 "fracture type" and "dissolution vug-fracture type" or "fracture-cavity" reservoirs, with very low porosity and permeability where most of the reservoir permeability is controlled by fractures and vugs (Nelson, 1985; Bagrintseva et al., 1989). Based on petrological, cross-cutting, SEM-EDS, cathodoluminescence (CL), isotopic, and fluid inclusion thermometric and compositional analyses, Baqués et al. (2020) identified seven main groups of fractures (F) and associated vugs (V) in Halahatang Middle Ordovician carbonate core samples (group 1oldest, group 7 youngest; Figs. 2 and 3). Fractures and vugs were grouped on the basis of cross-cutting relationships, mineral fill, and isotopic compositions irrespective of their orientations because individual fractures are highly variable in orientation so that the latter is not a good measure of timing (set) and/or type. Fractures show a spectrum of cement-fill degrees that is dependent on size. Consequently, the conventional terms "vein" and "joint" are not helpful. We use the unambiguous descriptive term "fracture" to refer to opening-mode fractures and specify the cement attributes where necessary. The oldest group of fractures (F1) comprises bed-perpendicular and bed-oblique, compacted (distorted) fractures filled with calcite cement that grades from equant to palisade morphology (Fig. 3a). Some F1 fractures contain a small amount of internal yellow carbonate sediment. The second group of fractures (F2) consists of bed-perpendicular, oblique, and subhorizontal fractures filled with blocky calcite (Fig. 3a). F3 fractures are bed-perpendicular and oblique, calcite, and bitumen-filled (Fig. 3b), and may preserve as much as 20% partial open porosity in fractures c.1 mm wide. A fourth group (F4) encompasses bed-perpendicular and oblique fractures filled with calcite ± bitumen ± minor late pyrite cement (Fig. 3c). Some F4 are partially corroded. F5 fractures are infilled by euhedral fluorite, barite, celestite, and calcite cements in variable proportions, and are associated with calcite, celestite, and/or calcite-filled silica host-rock replacement (Fig. 3d). F6 fractures are calcite-filled (Fig. 3e) and show elevated $^{87}Sr/^{86}Sr$ isotopic ratios compared with the host rock. Finally, group 7 comprises bedding-oblique, stylolitic, and sheared fractures (F7) containing insoluble residual material, bitumen, dolomite, and sheared calcite cement (Fig. 3f). Some fractures in this category contain cement bridges and crack-seal textures indicating cementation synchronous with fracture opening (see Lander & Laubach, 2015; Ukar and Laubach, 2016; Baqués et al., 2020) (Fig. 3g). Such fractures may preserve up to 25% porosity. Several dissolution features are evident in the core including bed-parallel (S1) to bed-perpendicular stylolites (S2) (Fig. 3b, g, and i), an early diagenetic alteration of the host rock ("mottle fabrics") (Fig. 3h), and several types of vugs [dissolution pores larger than the average grains (c.20 μm)] up to 2 mm in diameter (Fig. 3b–i) (Baqués et al., 2020). Vugs are usually associated with fractures and an increased abundance of both occurs where the core is highly broken (bagged pieces) and FMI shows decimetre- to decametre-scale fractured zones adjacent to open cavities (Fig. 4). The earliest generation of vugs (V1) are infilled by carbonate sediment having mottled textures (Fig. 3h), whereas subsequently formed V2–V7 (youngest) vugs (Fig. 3b, c, g, and i) are mainly infilled by blocky calcite cement with different isotopic and cathodo luminescence characteristics similar to those of fracture cements (see Baqués et al., 2020). The most widespread and abundant type of vugs (V3) are filled with calcite and bitumen and are usually associated with F3fractures, stylolites, and V1 mottled fabrics, highlighting the connection between fractures and dissolution in these rocks (Fig. 3h and i). We did not observe vugs associated with F5 fractures but based on the fact that this association is common for all other identified fracture and vug groups, V5 vugs are probable. Similar to vugs, stylolites, both bed-parallel and tectonic, are Fig. 2. Stratigraphy and lithology of Paleozoic-Cenozoic sediments in the (3) (PDF). The nature and origins of decameter-scale porosity in Ordovician carbonate rocks, Halahatang oilfield, Tarim Basin, China. Available from: https://www.researchgate.net/publication/341033923_The_nature_and_origins_of_decameter-scale_porosity_in_Ordovician_carbonate_rocks_Halahatang_oilfield_Tarim_Basin_China [accessed Jul 03 2021].

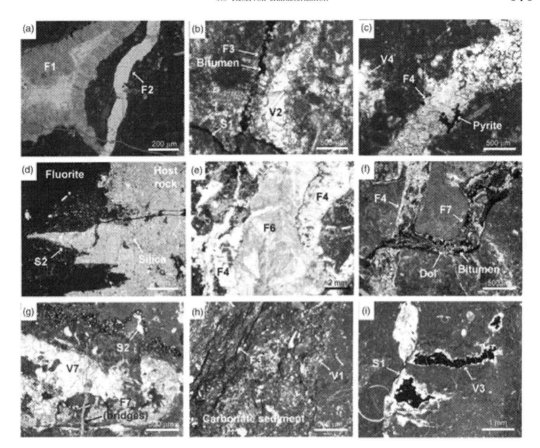

FIG. 5.15 Plane-light optical photomicrographs of fractures and vugs in the core. (A) F1 fracture filled with equant to palisade calcite cement cut by an F2 fracture filled with blocky cement. (B) V2 vug filled with blocky calcite cement cut by an F3 fracture filled with calcite and bitumen. Both terminate at a bed-parallel stylolite (S1) containing bitumen. (C) Pyrite-bearing F4 fracture associated with a V4 vug filled with calcite (+ bitumen + minor late pyrite, outside of the field of view) cement. (D) F5 fracture infilled by euhedral fluorite, barite, celestite, and calcite cement associated with silica host-rock replacement. (E) F6 fracture cutting F4 fractures filled by calcite cement. (F) Bedding-oblique, stylolitic, and sheared F7 fractures containing insoluble material, bitumen, dolomite, and calcite cement cutting and displacing a cemented F4 fracture. (G) V7 vugs cutting an F7 fracture partially filled with synkinematic calcite bridges. Porosity-bearing F7 fractures may have acted as corrosive fluid conduits. Group 7 fractures and vugs terminate against an S2 stylolite. (H) V1 vugs infilled by carbonate sediment (mottle fabrics) cut by F3 fractures. (I) V3 vugs lined by calcite and later filled by bitumen. Vugs terminate against or are cut by bitumen-bearing S1 stylolites.

et al., 2020). Baques et al. reported an in-depth description of the diagenetic history recorded in these cores by means of detailed petrographic, isotopic, and fluid inclusion analyses of fracture and vug cements. Here we focus on constraining the timing of each diagenetic event in the microstructural sequence and tie them to outcrop observations and the geological history of the area (see Appendix A for methods).

In the carbonate reservoirs studied, porosity does not correlate with permeability (Zhu et al., 2019). Such rocks constitute Type 1 "fracture type" and "dissolution vug-fracture type" or "fracture-cavity" reservoirs, with very low porosity and permeability where most of the reservoir permeability is controlled by fractures and vugs (Nelson, 1985; Bagrintseva et al., 1989). Based on petrological, cross-cutting, SEM-EDS, cathodoluminescence (CL), isotopic, and fluid inclusion thermometric and compositional analyses, Baqués et al. (2020) identified seven main groups of fractures (F) and associated vugs (V) in Halahatang Middle Ordovician carbonate core samples (group 1 oldest, group 7 youngest). Fractures and vugs were grouped on the basis of cross-cutting relationships, mineral fill, and isotopic compositions irrespective of their orientations because individual fractures are highly variable in orientation so that the latter is not a good measure of timing (set) and/or type. Fractures show a spectrum of cement-fill degrees that is dependent on size. Consequently, the conventional terms "vein" and "joint" are not helpful. We use the unambiguous descriptive term "fracture" to refer to opening-mode fractures and specify the cement attributes where necessary.

The oldest group of fractures (F1) comprises bed-perpendicular and bed-oblique, compacted (distorted) fractures filled with calcite cement that grades from equant to palisade morphology. Some F1 fractures contain a small amount of internal yellow carbonate sediment. The second group of fractures (F2) consists of bed-perpendicular, oblique, and subhorizontal fractures filled with blocky calcite. F3 fractures are bed-perpendicular and oblique, calcite and bitumen-filled, and may preserve as much as 20% partial open porosity in fractures c.1 mm wide. A fourth group (F4) encompasses bed-perpendicular and oblique fractures filled with calcite ± bitumen ± minor late pyrite cement. Some F4 are partially corroded. F5 fractures are infilled by euhedral fluorite, barite, celestite, and calcite cements in variable proportions, and are associated with calcite, celestite, and/or calcite-filled silica host-rock replacement. F6 fractures are calcite-filled (Fig. 3e) and show elevated $^{87}Sr/^{86}Sr$ isotopic ratios compared with the host rock. Finally, group 7 comprises bedding-oblique, stylolitic, and sheared fractures (F7) containing insoluble residual material, bitumen, dolomite, and sheared calcite cement. Some fractures in this category contain cement bridges and crack-seal textures indicating cementation synchronous with fracture opening (see Lander and Laubach, 2015; Ukar and Laubach, 2016; Baqués et al., 2020). Such fractures may preserve up to 25% porosity.

Several dissolution features are evident in the core including bed-parallel (S1) to bed-perpendicular stylolites (S2), an early diagenetic alteration of the host rock ("mottle fabrics"), and several types of vugs (dissolution pores larger than the average grains (c.20 μm)) up to 2 mm in diameter (Baqués et al., 2020). Vugs are usually associated with fractures and an increased abundance of both occurs where the core is highly broken (bagged pieces) and FMI shows decimetre- to decametre-scale fractured zones adjacent to open cavities. The earliest generation of vugs (V1) are infilled by carbonate sediment having mottled textures, whereas subsequently formed V2–V7 (youngest) vugs are mainly infilled by blocky calcite cement with different isotopic and cathodoluminescence characteristics similar to those of fracture cements (see Baqués et al., 2020). The most widespread and abundant type of vugs (V3) are filled with calcite and bitumen and are usually associated with F3 fractures, stylolites, and V1 mottled fabrics, highlighting the connection between fractures and dissolution in these rocks (Fig. 3h and i). We did not observe vugs associated with F5 fractures but based on the fact that this association is common for all other identified fracture and vug groups, V5 vugs are probable. Similar to vugs, stylolites, both bed-parallel and tectonic, are ubiquitous and formed throughout the geological history of these rocks. All stylolites contain bitumen and some bed-parallel stylolites also contain dolomite and replacement silica.

Highly variable, but generally high (>80°C) homogenization temperatures of fluid inclusions within the fracture and vug cements, as well as high salinities (5%–20% NaCl equiv.) indicate that fluid inclusions probably re-equilibrated with high-temperature burial fluids and do not provide information on the true temperature and depth of precipitation of the cements. Consequently, Baqués et al. (2020) determined fracture and vug cementation depths on the basis of stable isotopic analyses (Craig, 1965) assuming a geothermal gradient of 2.79°C per 100 m for the Tabei Uplift during the Middle Ordovician (Wang et al., 2014). Because cemented fractures and vugs are cross-cut by the next generation of filled fractures and vugs, the calculated depths of cement precipitation based on stable isotopic analyses represent dissolution depths. Host rock diagenesis, including mottled fabrics (V1), developed at <220 m, within the shallow burial diagenetic environment, whereas F1 fractures formed and were cemented near the surface within the marine-phreatic level. Group 2 fractures and vugs (F2, V2) developed within the shallow burial diagenetic environment (c.220 m), group 3 within the intermediate burial diagenetic environment (c.625 m) and group 4 within the deep burial diagenetic environment (c.2000 m). Groups 5–7 are probably the result of reported hydrothermal activity in the area (e.g., Liu et al., 2017a, b) and Mg-rich fluid infiltration, and their depth is undetermined. All fracture and vug cements of group 3 (included) and younger contain primary inclusions of oil indicating the presence of hydrocarbons in the system at the time of their filling by cement (Figs. 5.16–5.18).

5.3.4.1 Sm-Nd fluorite date from core

To decipher the timing of fluorite mineralization in the Halahatang oilfield, we conducted Sm-Nd radiometric dating of fluorite, barite, and calcite aliquots of one F5

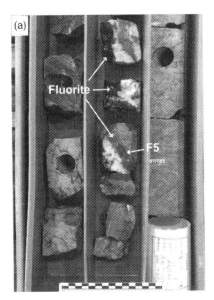
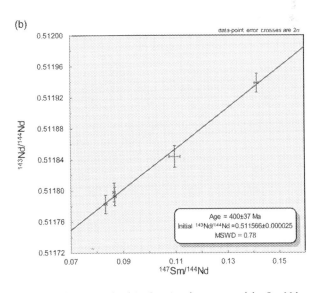

FIG. 5.16 (A) Photograph of RP4 core showing fluorite-, barite- and calcite-bearing fracture used for Sm-Nd radiometric dating. (B) 147Sm/144Nd v. 143Nd/144Nd isochron showing the results of the five analysed samples and the calculated age.

FIG. 5.17 (A) Outcrop photograph showing karstic palaeotopography (karst towers and dolines) and collapse deformation structures, and karstic chaotic breccias present on the topmost section of the Yingshan Formation in the XikeEr outcrop. Silurian red beds unconformably overlie Ordovician carbonates, indicating subaerial exposure. (B) Polymictic breccia bodies exposed c.60 m below the unconformity. (C) Cave-sediment fill showing laminated sediments and geopetal structures evidencing their deposition in the vadose zone. (D) Laminated cave-sediment fillings showing nonluminescent to bright-red-luminescent concentric calcite overgrowths as seen under optical cathodoluminescence (CL). (E) Silica and barite mineralization postdating calcite overgrowths. (F) Redissolution cavities and calcite precipitated in cavities developed at the top surface of the karstic system (karst towers). (G) Palisade calcite crystals, over 30 cm thick, precipitated in the remaining open porosity of the redissolved cavities.

fracture fill from core RP4. Fluorite typically contains elevated REE concentrations that may show sufficient variations in Sm/Nd ratios to allow dating using the isochron approach (Turner et al., 2003). The ^{147}Sm/^{144}Nd ratios of the samples analyzed range from 0.083 (calcite) to 0.14 (fluorite + calcite + barite; whole fracture). All five analyzed samples plot in a linear array and yield an age of 400 ± 37 Ma, with an initial εNd value of 0.511566 ± 0.000025 (2σ, MSWD = 0.78).

Structural diagenetic features in outcrops in the Bachu-Keping Uplift region, western Tarim Basin, share many characteristics with Ordovician rocks in core from producing parts of the basin, although with some important differences. Karstic palaeotopography (karst towers and dolines) and cave-fill sediments (chaotic breccias and laminated cave-sediment fill) are present in the topmost section of the Yingshan Formation in the XikeEr outcrop, indicating

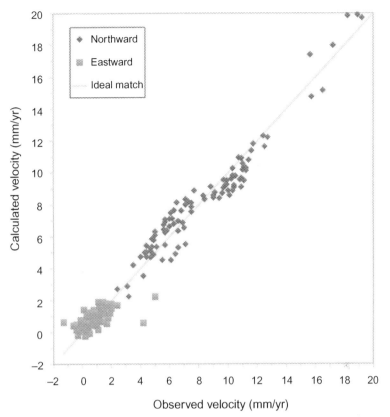

FIG. 5.18 Validation plot comparing observed velocity data and velocity calculated using the new method described in Appendix B.

that subaerial exposure and karstification occurred prior to deposition of unconformable Silurian red beds. Approximately 60 m below the unconformity, large polymictic breccia bodies are exposed. The breccia clasts consist of Ordovician carbonate clasts and the matrix consists of a reddish muddy to sandy sediment. Cave breccias and cave sediment fills show evidence of sedimentation and cementation in the vadose zone such as laminated and geopetal structures. Laminated cave-sediment fillings consist of nonluminescent to bright-red-luminescent subhedral calcite crystals. Silica and barite mineralization postdate calcite overgrowths. These phases probably reflect interaction with silica-rich fluids that percolated through overlying Silurian siliciclastic rocks or high-temperature fluids. The palaeokarst is affected by redissolution that is best developed on the top surface of the karstic system (karst towers), forming vugs and cavities that are partially cemented by palisade calcite crystals that can be over 30 cm thick. Such vugs closely resemble those widespread in Halahatang cores. Oxygen and carbon isotopic analyses, especially the depleted values in $\delta^{18}O$ compared with cave-sediment fill and host-rock calcite overgrowths within cave infills and palisade calcite, indicate precipitation from higher-temperature fluids.

5.4 Organic source of hydrocarbon

The main constituents of petroleum are hydrogen and carbon. Carbon molecules are bound to hydrogen molecules to form hydrocarbons that form the main matrix of all petroleum fluids. The simplest configuration is depicted in Fig. 5.19. This steady-state depiction shows how petroleum products range from simplest (e.g., methane) to most complex (e.g., asphaletene) forms of hydrocarbons. While methane molecules have only one carbon atom, asphaltenes have over hundred carbon atoms bound to equivalent (C:H ratio is approximately 1:1.2) number of hydrogen atoms. In addition, an asphaltene molecule would have three nitrogen atoms, two oxygen atoms, and two sulfur atoms. In the hydrocarbon spectrum of crude oil and gas, natural gas consists mainly of methane with smaller amounts of longer chains up to pentane. Liquid petroleum is referred to as crude oil and consists of a wide range of more complex hydrocarbons and minor quantities of asphaltenes. Tars, on the other hand, are semisolid that has mostly longer-chain hydrocarbons and asphaltenes (Fig. 5.19).

Hydrocarbons of organic sources are formed organically be two pathways:

(1) Through the generation of hydrocarbons directly by organisms. This constitutes perhaps 10%–20% of the hydrocarbons in the crust. They generally contain more than 15 C atoms and are easily recognized structures (biomarkers).
(2) Through the conversion of organic matter (lipids, proteins, and carbohydrates) into kerogen, then to bitumen, and finally to petroleum as it gets buried to higher temperatures.

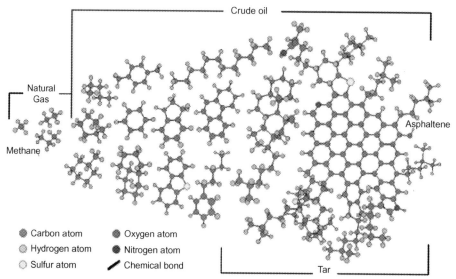

FIG. 5.19 Schematic of crude oil molecular composition. *From USGS Website 1. https://www.usgs.gov/energy-and-minerals/energy-resources-program.*

The thermal alteration of hydrocarbons continues with continued burial depth. The maturation and degradation follow two pathways again:

(1) One where the H/C ratio decreases, i.e., hydrogen stripped from compounds and number of carbon atoms in compounds increase. Ultimately this H/C ratio reaches 0 (i.e., graphite and coal).
(2) One where the H/C ratio increase which ultimately reaches a ratio of 4 (i.e., the compound is methane).

Hydrocarbon generation is the natural result of the maturation of buried organic matter. Organic matter (organic carbon) in sediments underlying the oceans is derived from different sources, including marine phytoplankton, Phytobentos in shallow water with sufficient light, bacteria, and allochtonous (i.e., land derived) material. The organic carbon produced in the water column varies from 0.1% to 5% depending on various factors such as: (i) oxygen depletion in bottom waters or in sediment as a result of high organic input, (ii) adsorption of certain compounds to mineral particles, (iii) preservation of organic compounds as shell constituents, (iv) changes in the rate of deposition of sediment organic matter, (v) high input of terrigenous organic compounds, which are more stable than organic matter, and (vi) dominant input of argillaceous sediments where oxygenation of pore water is restricted.

Organic matters undergo changes in composition with increasing burial depth and temperature. The three steps in the transformation of organic matter to petroleum hydrocarbons are termed diagenesis, catagenesis, and metagenesis. The general scheme of evolution of the organic fraction and the hydrocarbon produced is depicted in Fig. 5.20 (Tissot and Welte, 1984). Petroleum hydrocarbons exist as a gaseous, liquid, and solid phases, depending on temperature, pressure, burial time, and composition of the system.

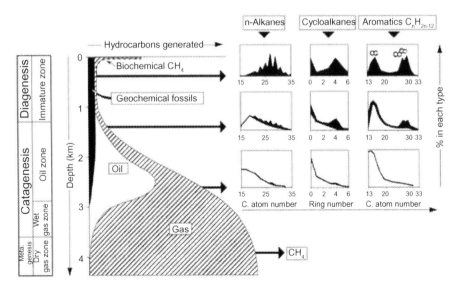

FIG. 5.20 General scheme of the evolution of the organic fraction and the hydrocarbon produced (Tissot and Welte, 1984).

Assessments of global coal, oil, and natural gas occurrences usually focus on conventional hydrocarbon reserves, i.e., those occurrences that can be exploited with current technology and present market conditions. The focus on reserves seriously underestimates long-term global hydrocarbon availability. Greenhouse gas emissions based on these estimates may convey the message that the world is running out of fossil fuels. As a result, emissions would be reduced automatically. If the vast unconventional hydrocarbon occurrences are included in the resource estimates and historically observed rates of technology change are applied to their mobilization, the potential accessibility of fossil sources increases dramatically with long-term production costs. They are not significantly higher than present market prices. Although the geographical hydrocarbon resource distribution varies significantly, a regional breakdown for 11 world regions indicates that neither hydrocarbon resource availability nor costs are likely to become forces.

Natural products are found in nature and usually have a pharmacological or biological activity for use in pharmaceutical drug design. A natural product can be considered as such even if it can be prepared by the total synthesis in the laboratory or in an industrial setting. In the more general sense, fossil fuels are natural products insofar as the precursors to the fossil fuels were originally derived from living organisms. Finally, the natural forces (i.e., including but not limited to temperature, pressure, aerial oxidation bacteria, etc.) caused the starting materials to be converted to fossil fuel.

The predominant theory assumes the development of stagnant water conditions in some of the expanded oceans caused the bottom waters to be depleted in oxygen (anoxic), which allowed portions of decaying plankton (e.g., algae, copepods, bacteria, and archaea) that originally lived in the upper oxygen-bearing (oxic) waters to be preserved as a sediment layer enriched in organic matter. Such a condition is necessary for the anaerobic decomposition of organic matter (Fig. 5.21). In a laboratory setting such a process is called pyrolysis, which is typical thermolysis, and is most commonly observed in organic materials exposed to high temperatures. In a laboratory, such decomposition takes place starting at 200–300°C (390–570°F). In nature, a similar incident occurs when organic matters come into contact with lava in volcanic eruptions. General composition of biomass is presented in Table 5.7. Recently, Luo et al. (2004) determined composition for specific organic matters. This is presented in Tables 5.8 and 5.9.

Table 5.10 shows the effect of temperature on pyrolysis products. With an increase in temperature, the relative amount of char and condensable liquid fraction decreases continuously, whereas the fraction of gas formed increases.

Because different types of reactions get activated at different temperature values, an optimum temperature value emerges. Fig. 5.21 shows product distribution with temperature. These results would be affected under a different set pressure condition, as is often the case for the natural decomposition of organic matters in geologic time. In addition, the secondary reaction of volatile components would also influence the composition of final products. Fig. 5.22 shows how the composition of incondensable gas is influenced by temperature. As the temperature is increased, a higher proportion of CO and CH_4, while a lower proportion of CO_2 emerges. During the pyrolysis process, CO_2 is produced at a relatively low temperature. As higher temperature conditions prevail, secondary thermal cracking of volatile produces CO and CH_4 rather than CO_2.

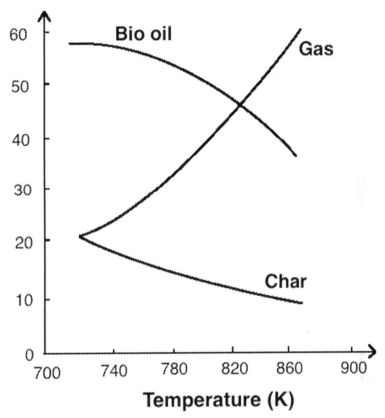

FIG. 5.21 Product distribution with temperature. *Redrawn from Luo, Z., et al., 2004. Research on biomass fast pyrolysis for liquid fuel. Biomass Bioenergy 26, 455–462.*

In a geologic setting, organic-rich sediment layers are buried by deposition of overlying sediments in the subsiding basin, the sediments are compressed and eventually lithified into rocks referred to as black shale, bituminous limestone, or coal. Over millions of years of digenesis, new organisms come in contact with the sedimentary rock and often enrich preexisting deposits. These microorganisms consume portions of the organic matter as a food source and generate methane as a by-product. This methane, which is typically the main hydrocarbon in natural gas, has a distinct neutron deficiency in its carbon nuclei (i.e., carbon isotopes), which allows microbial natural gas to be readily distinguished from methane generated by thermal processes later in a basin's subsidence history. The microbial methane may remain in the organic-rich layer or it may bubble up into the overlying sediment layers and escape into the ocean waters or atmosphere. If impermeable sediment layers hinder the upward migration of microbial gas, the gas may collect in underlying porous sediments that end up being the reservoir (Figs. 5.23 and 5.24).

TABLE 5.7 General composition of biomass.

Gas compounds	Composition (kg/100 kg of dry biomass)
Carbon dioxide	5.42
Carbon monoxide	6.56
Methane	0.035
Ethane	0.142
Hydrogen	0.588
Propane	0.152
Ammonia	0.0121
Bio-oil compounds	
Acetic acid	5.93
Propionic acid	7.31
Methoxyphenol	0.61
Ethylphenol	3.80
Formic acid	3.41
Propyl-benzoate	16.36
Phenol	0.46
Toluene	2.27
Furfural	18.98
Benzene	0.77
Other compounds	
Water	10.80
Char/Ash	16.39

From Islam, M.R., 2014. Unconventional Gas Reservoirs. Elsevier.

Economically significant accumulations of microbial natural gas have been estimated to account for 20 percent of the world's produced natural gas (USGS website 1). This microbial methane often forms the basis for accumulations referred to as coal-bed methane and shale gas. During the course of digenesis, the burial of the organic-rich rock layer may continue in some subsiding basins to depths of 6000–18,000 feet (1830–5490 m). At these depths, the organic-rich rock layer is exposed to temperatures of 150–350°F (66–177°C) for a few million to tens of millions of years. The organic matter within the organic-rich rock layer begins to cook during this period of heating and portions of it thermally decompose into crude oil and natural gas (i.e., thermogenic gas).

Classical views for the deposition of oil-prone and organic-rich sediments in deep environments invoke two principal types of sedimentologic settings. The first is confined basins in which stratified oxygen-depleted waters lead to anoxic preservation of organic matter in

TABLE 5.8 Typical properties of wood-derived crude bio-oil.

Physical property	Typical value
Moisture content (%)	15–30
pH	2.5
Specific gravity	1.20
Elemental analysis	
C (%)	55–58
H (%)	5.5–7
O (%)	35–40
N (%)	0–0.2
Ash (%)	0–0.2
Higher heating value (HHV) (MJ/Kg)	16–19
Viscosity (at 40°C and 25% water) (cp)	40–100

From Liu, Y., et al., 2014. Electromembrane extraction of salivary polyamines followed by capillary zone electrophoresis with capacitively coupled contactless conductivity detection. Talanta 128, 386–392.

TABLE 5.9 Composition of specific organic matters

	Water (%)	Ash (%)	Volatile (%)	Fixed carbon (%)	Heating value (kJ/kg)	C (%)	H (%)	N (%)	S (%)	O (%)
F. mandshurica	1.85	3.58	76.9	17.9	20,200	48.3	5.95	0.18	0.19	40.1
C. lanceolata	3.27	0.74	81.2	14.8	19,200	46.5	6.04	0.17	0.12	43.1
P. indicus	1.96	0.44	79.3	18.4	20,500	49.1	5.98	0.22	0.12	42.3
Rice straw	3.61	12.2	67.8	16.4	16,400	40.3	5.55	1.02	0.15	37.3

Modified from Jahirul, M.I., et al., 2012. Biofuels production through biomass pyrolysis—a technological review. Energies 5 (12), 4952–5001. https://doi.org/10.3390/en5124952.

TABLE 5.10 Effect of temperature on pyrolysis.

Temp. °C	Char % dry feed	Lqd. % dry feed	Gas % dry feed
500	29	53	18
600	24	52	24
700	20	50	30
800	17	47	36
900	15	40	45
1000	14	30	56

From Islam, M.R., 2014. Unconventional Gas Reservoirs. Elsevier.

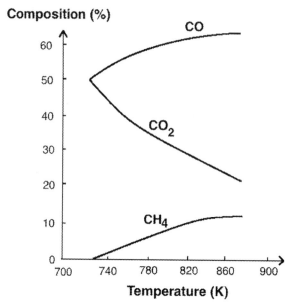

FIG. 5.22 Incondensable gas composition with temperature. *Redrawn from Luo, Z., et al., 2004. Research on biomass fast pyrolysis for liquid fuel. Biomass Bioenergy 26, 455–462.*

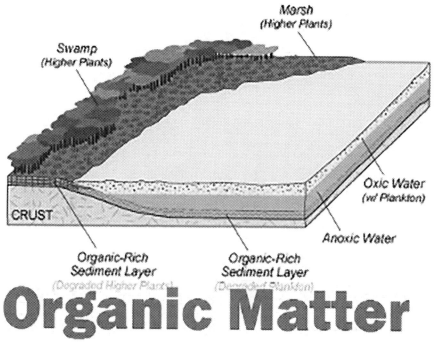

FIG. 5.23 Formation of organic-rich sediment layer. *From USGS Website 1. https://www.usgs.gov/energy-and-minerals/energy-resources-program.*

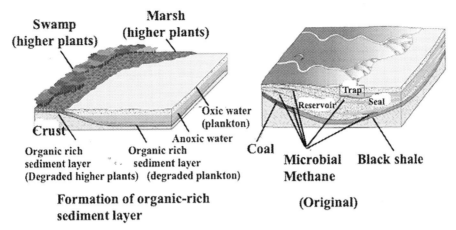

FIG. 5.24 Early burial of sediment layers in basin. *From USGS Website 1. https://www.usgs.gov/energy-and-minerals/energy-resources-program.*

the water column and in underlying sediments. The second is an open-ocean setting where episodic mass transfers due to slope sediment instability lead to the rapid burial of outer-shelf- and upper slope-derived organic matter and its consequent preservation due to limited oxic or anoxic degradation. Other studies have shown, however, that organic matter in modern deep-sea sediments may occur in high amounts where oxygen is not significantly depleted.

Another mechanism may account for the occurrence of organic-rich deposits, namely relative variations in the magnitude of nonorganic burial fluxes. This one forms the basis of the work published by Bertrand and Rangin (2003).

The overall process of hydrocarbon generation through thermal maturation is highly affected by the nature of organic matter. If the original source of the organic matter is mostly higher plants (e.g., trees, shrubs, and grasses), natural gas will be the dominant petroleum generated with lesser amounts of crude oil generation. If the original source of the organic matter is plankton (e.g., algae, copepods, and bacteria), crude oil will be the dominant petroleum generated with lesser amounts of natural gas generation. It turns out that the earth has a much more volume of plankton material than higher plants (Fig. 5.25).

It is well known that phytoplankton is central to the pelagic ecosystem due to the fact that it traps almost all the energy used by the ecosystem. The biomass of phytoplankton is a subject of intense studies dealing with aquatic ecological research. Phytoplankton forms the basis of fundamental fodder of the food chain in nature. Phytoplankton is the main primary producers at the bottom of the marine food chain. Phytoplankton use photosynthesis to convert inorganic carbon into protoplasm. They are then consumed by microscopic animals called zooplankton. Zooplankton comprises the second level in the food chain and includes small crustaceans, such as copepods and krill, and the larva of fish, squid, lobsters, and crabs. In turn, small zooplankton is consumed by both larger predatory zooplankters, such as krill and forage fish which are small, schooling, and filter-feeding fish. This makes up the third level in the food chain.

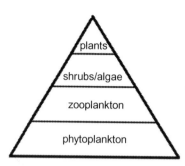

FIG. 5.25 The volume of biomass increases as the size of the living object decreases.

It is known that algal carbon content is extremely difficult to determine directly and is therefore usually estimated from other parameters. Counting and volume assessment of cells, and measurement of pigment concentration, are widely used to estimate algal biomass (Smayda, 1978; Jeffrey et al., 1997). However, both methods have technical limitations. Direct cell counting and species volume measurement are laborious, requiring specialists in taxonomy and the use of preserved samples. Chlorophyll *a* is common to all photosynthetic organisms. Furthermore, it is the most abundant photosynthetic pigment and it is relatively easy and rapid to quantify. Consequently, its concentration is used extensively for estimating phytoplankton biomass. A variety of techniques is at present available offering varying degrees of accuracy. However, the ratio of chlorophyll *a* to cell carbon depends on external and internal factors, such as phytoplankton taxonomic composition, cell physiological conditions, temperature, nutrient concentrations, and light intensity.

The effects of phytoplankton taxonomic composition and external factors on the decoupling between chlorophyll and biovolume were evaluated by Felip and Catalan (2000). These correlations identified the importance of light and temperature that eventually would impact the maturation of any organic matter. Table 5.11 shows the relevant data used and Table 2.6 shows the results. Table 5.12 shows that the low chlorophyll content related to phytoplankton biovolume (residual negative) was significantly related to high light and

TABLE 5.11 Statistical descriptors of some phytoplankton parameters observed during the seasonal cycle studied in Lake Redó.

	Mean	**SD**	**Maximum**	**Minimum**	*n*
Abundance (cell mL^{-1})	2954	2925	16060	31	289
Biovolume (mm^3 m^{-3})	206	174	1486	14	289
Average cell size (μm^3)	98	86	1113	31	289
Chlorophyll *a* (μg L^{-1})	1.54	1.09	5.72	0.14	233
Chlorophyll *a* per biovolume (μg mm^{-3})	9.06	5.03	23.59	0.53	233
Chlorophyll *a* per cell (pg cell^{-1})	0.69	0.64	6.1	0.08	233

From Felip and Catalan (2000).

TABLE 5.12 Chlorophyll content and phytoplankton biovolume for two biomass estimators. Factors were selected from the literature. The percentage of biovolume of each phytoplanktonic group was used as a descriptor of sample taxonomic composition.

Variable	R (n = 232)
% Chlorococcales	0.4400
% Desmidiaceae	0.2760
% Cryptophytes	0.0611
% *Stichogloea doederlinii*	−0.0179
% Volvocales	−0.0235
% Flagellated chrysophytes	−0.0872
Inorganic N	−0.1069
Temperature	−0.4119
Light	−0.4561
% Dinoflagellates	−0.4802

From Felip and Catalan (2000).

temperature values and a predominance of dinoflagellates, reflecting the characteristics of the epilimnion during the summer stratification. In contrast, large amounts of chlorophyll per unit biovolume were related to low light and temperature values, and to the predominance of small cells, such as, some chlorophyte species. This study indicates the nature of carbohydrate generation that would impact the generation of hydrocarbons.

The global volume of the growing stock was estimated at 386 billion cubic meters in 2000. The regions with the largest volume were Europe (including the Russian Federation) with 30 percent (116 billion cubic meters) and South America with 29 percent (111 billion cubic meters) (Table 5.13). Oceania shows the lowest growing stock with 11 billion cubic meters or 3 percent of the global volume.

Globally, terrestrial and oceanic habitats produce a similar amount of new biomass each year (56.4 billion tonnes C terrestrial and 48.5 billion tonnes C oceanic). The amount of biomass is the order of magnitude higher for smaller species (Whitman et al., 1998). Table 5.14 shows data on global biomass production.

Organic-rich rocks that have not been thermally matured are referred to as being thermally immature. Even before they are identified as crude oil, they can become energy sources. It is evidenced from the fact that oil shale economic quantify of oil if they are heated at around 540 C. Oil shale is a sedimentary rock containing up to 50% mostly immature organic matter. Once extracted from the ground, the rock can either be used directly as fuel for a power plant or be processed to produce shale oil and other chemicals and materials. Similar to crude oil and gas, oil shales are widely distributed around the world—some 600 deposits are known, with resources of the associated shale oil totaling almost 500 billion tonnes, or approximately 3.2 trillion barrels (EASAC, 2007). Yet, these deposits are vastly unexploited—the exceptions being Estonia and Brazil.

TABLE 5.13 Forest volume and above-ground biomass by region.

Region	Forest area million ha	Volume By area m³/ha	Volume Total Gm³	Biomass By area t/ha	Biomass Total Gt
Africa	650	72	46	109	71
Asia	548	63	35	82	45
Oceania	198	55	11	64	13
Europe	1039	112	116	59	61
North and Central America	549	123	67	95	52
South America	886	125	111	203	180
Total	3869	100	386	109	422

From FAO, 2012. FAO Report. Available at: http://www.fao.org/3/i3027e/i3027e00.htm.

TABLE 5.14 Global biomass production.

Producer	Biomass productivity (gC/m²/yr)	Total area (million km²)	Total production (billion tons C/yr)
Swamps and Marshes	2500		
Tropical rainforests	2000	8	16
Coral reefs	2000	0.28	0.56
Algal beds	2000		
River estuaries	1800		
Temperate forests	1250	19	24
Cultivated lands	650	17	11
Tundras	140		
Open ocean	125	311	39
Deserts	3	50	0.15

Data from Ricklefs, R.E., Miller, G.L., 2000. Ecology, fourth ed. Macmillan. ISBN 978-0-7167-2829-0.

Because petroleum has a lower density than the water that occupies pores, voids, and cracks in the source rock and the overlying rock and sediment layers, the density difference leads to upward migration of petroleum due to buoyancy forces. Such migration continues until a trap is reached. Often, such migration may not take place due to lack of permeable channels, in which case source rocks remain impregnated with petroleum. The Barnett Shale in the Fort Worth basin of Texas is a good example of this type of accumulation, where the natural gas generated over geological eras is found within the source rock.

On the other extreme, in some basins, petroleum may not encounter a trap and continue migrating upward into the overlying water or atmosphere as petroleum seeps. This

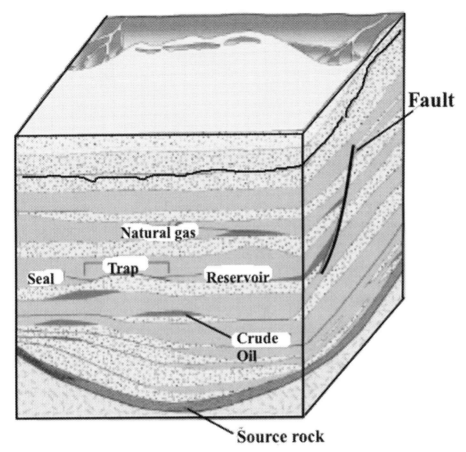

FIG. 5.26 Continued burial of sediment and rock layers in subsiding basin. *From USGS Website 1. https://www.usgs.gov/energy-and-minerals/energy-resources-program.*

petroleum would seep through the surface or ocean floor, leaving behind a residual tar enriched in large complex hydrocarbons and asphaltenes (Fig. 5.26). Tar deposits range in size from small local seeps like the La Brea tar pits of California to regionally extensive occurrences as observed in the Athabasca tar sands of Alberta, where a 50,000 square mile reservoir of heavy crude oil, possibly holds 2 trillion barrels of recoverable oil (Fig. 5.27).

Burial of the source rock may continue to depths greater than 20,000 ft. (6100 m) in some sedimentary basins. At these depths, temperatures greater than 350°F (177°C) and pressures greater than 15,000 psi (103 MPa) begin to transform the remaining organic matter into more natural gas and a residual carbon referred to as char. This char itself is a function of swelling, surface area, porosity, mass release, true density, and chemical compositions and may lead to changes in the rock properties, thus altering the overall characteristics of the formation. Oil trapped in reservoirs that are sometimes buried to these depths also decomposes to natural gas and char. The char, which is also called pyrobitumen, remains in the original reservoir while the generated natural gas may migrate upward to shallower traps within the overlying

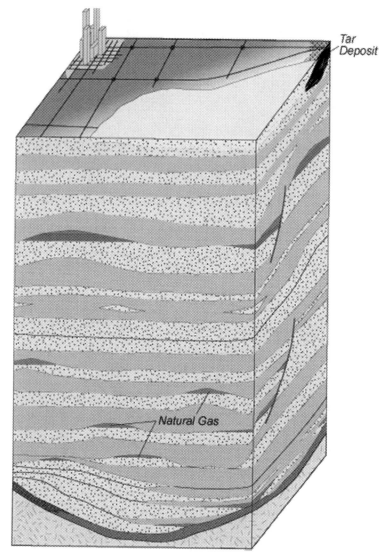

FIG. 5.27 Deeper burial of rock layers in subsiding basin. *From USGS website 2.*

rock layers of the basin. The Gulf Coast basin that extends into the offshore of Louisiana and the Anadarko basin of the US mid-continent are good examples of these deep basins (Fig. 5.28).

Further burial to temperatures and pressures above 600°F (316°C) and 60,000 psi (414 MPa), respectively, represent metamorphic conditions in which the residual char converts to graphite with the emission of molecular hydrogen gas. In this process, it is important to understand the mobility of inorganic elements on the char or biomass surface because if the metals are active sites for exothermic reactions then the temperature at the metal sites might

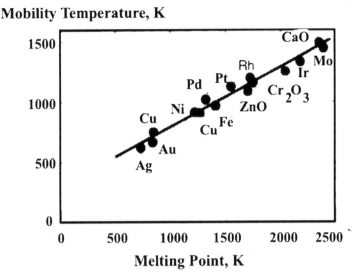

FIG. 5.28 Relationship between the mobility of particles supported on graphite and their bulk melting temperatures. *From Baker, R.T.K., 1982. The relationship between particle motion on a graphite surface and Tammann temperature. J. Catal. 78 (2), 473–476.*

be higher which may increase the mobility of the metals. It also unlocks the key to understanding the role of catalysts in both natural processes as well as artificial refining. In particular, air gasification is exothermic whereas gasification with steam or CO_2 is endothermic, which could result in significant temperature differences at the site where the reaction takes place. The more mobile the particle is, the more it may agglomerate on the surface. Fig. 5.28 shows the relationship between mobility temperature and melting point for various materials.

The resulting metamorphic rocks of this process of maturation are graphitic slate, schist, or marble. The prevailing thermal and pressure conditions dictate that water remaining in these rocks should react with the graphite to form either methane or carbon dioxide depending on the amount of molecular hydrogen present. Similarly, a crystalline metamorphic rock composed primarily of calcium carbonate ($CaCO_3$), is susceptible to further decomposition, especially under thermal stress. It is recognized that high-grade metamorphism typically breaks down and destroys carbonate minerals. Schist rocks are formed in presence of a moderate level of heat and pressure. As schist rocks are formed through a metamorphosis of mudstone/shale, or some types of igneous rock, traces of chars and petroleum residue work as cementing material, thus impacting rock properties.

Currently, the deepest wells in sedimentary basins do not exceed 32,000 ft (9760 m). Therefore, the significance of natural gas generation under these extreme conditions remains uncertain. Sedimentary basins vary considerably in size, shape, and depth all over the Earth's crust (Fig. 5.29).

The extraction of liquid hydrocarbon fuel from several sedimentary basins has been an integral part of modern energy development. Hydrocarbons are mined from tar sands and oil shale. These reserves require distillation and upgrading to produce synthetic crude and

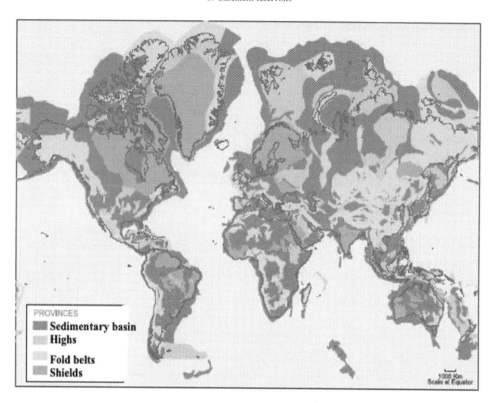

FIG. 5.29 General outline of major sedimentary basins. *From USGS Website 1. https://www.usgs.gov/energy-and-minerals/energy-resources-program.*

petroleum. A future source of methane may be methane hydrates found on ocean floors. There are two possible sources for the hydrocarbons such as inorganic and organic. Inorganic hydrocarbons form from the reduction of primordial carbon or oxidized forms at high temperatures in the earth. Organic accumulation of hydrocarbons produced directly by living organisms as well as the thermal alteration of biologically formed organic matter. It is generally recognized that most hydrocarbons are produced by the organic method. A few hydrocarbons in the crust may be from inorganic sources, however, the majority of them are from organic. We discuss the nonorganic sources of hydrocarbon in the next section.

5.5 Nonconventional sources of petroleum fluids

Even though it has been long recognized that every theory depends on a set of fundamental assumptions and in case those assumptions are not true, the theory has no validity, few theories have challenged mainstream theories. In particular, Geological theories are all based on the assumption that the earth was entirely inhabitable without a life form and all lives evolved from a single cell organism. Such sequence of events comes from the linear thinking that created a flurry of cosmic theories, including the Big Bang that assumes that the initial superflux of the matter was all hydrogen atoms that eventually evolved into other elements. Only recently, such notions have been challenged from multiple fronts.

In 2014, Islam presented the origin of the universe from water, making water ubiquitous in the cosmos. Similarly, the possibility of earth being an entity of its own (rather than being formed of "star dusts") was discussed. In 2009, Biello (2009) argued that even if the hypothesis that the earth was part of the superdense core starts, the fact that the oxygen element is the third most abundant element in the universe despite being highly reactive with all elements in nature explains that the breathable air of today originated from tiny organisms. Organisms such as cyanobacteria or blue-green algae are considered to be responsible for maintaining a high percentage (21%) of elemental oxygen in the atmosphere. Cynobacteria is considered to be the most primitive photosynthetic prokaryotes which are supposed to have appeared on this planet during the Precambrian period (Ash and Jenkins, 2006). Possibly, these are the first photosynthetic microorganisms that persisted over a period of 2 to 3 billion years, performing an important role in the evolution of higher forms. Cyanobacteria are a unique assemblage of organisms that occupy and predominate a vast array of habitats as a result of several general characteristics; some belonging to bacteria and others unique to higher plants.

Cyanobacteria, also known as Cyanophyta, is a phylum of bacteria that obtain their energy through photosynthesis, and are the only photosynthetic prokaryotes able to produce oxygen (Singh et al., 2016). The name "cyanobacteria" comes from the color of the bacteria. Singh et al. (2016) point out several unique features of cyanobacteria such as oxygenic photosynthesis, high biomass yield, growth on nonarable lands and a wide variety of water sources (contaminated and polluted waters), generation of useful by-products and bio-fuels, enhancing the soil fertility and reducing green house gas emissions, have collectively offered these bio-agents as the precious bio-resource for sustainable development. Cyanobacterial biomass is the effective bio-fertilizer source to improve soil physico-chemical characteristics such as water-holding capacity and mineral nutrient status of the degraded lands. The unique characteristics of cyanobacteria include their ubiquity presence, short generation time, and capability to fix the atmospheric nitrogen. While genetically engineered cyanobacteria have been devised with novel genes for the production of several bio-fuels such as bio-diesel, bio-hydrogen, bio-methane, synga, etc., the relevance of these bacteria in this chapter is their ability to produce hydrocarbon directly from nonorganic sources.

The unique features of these microbes are that they conduct photosynthesis, converting sunlight, water, and carbon dioxide into carbohydrates and oxygen. These bacteria perform photosynthesis for all the plants, as part of the continuous cycle, similar to Fig. 5.30.

Sarsekeyeva et al. (2015) identified the role of cyanobacteria that, by virtue of being a part of marine and freshwater phytoplankton, significantly contribute to the fixation of atmospheric carbon via photosynthesis. It is reasonably inferred that ancient cyanobacteria participated in the formation of the earth's oil deposits. While the biomass of modern cyanobacteria may be converted into bio-oil by pyrolysis, in geologic time, hydrocarbons produced by cyanobacteria can also form the basis of so-called fossil fuels. Modern cyanobacteria grow fast; they do not compete for agricultural lands and resources; they efficiently convert excessive amounts of CO_2 into biomass, thus participating in both carbon fixation and organic chemical production.

Cyanobacteria increasingly attract the attention as promising cell factories for the production of renewable biofuels and chemicals from just CO_2 and water at the expense of sunlight.

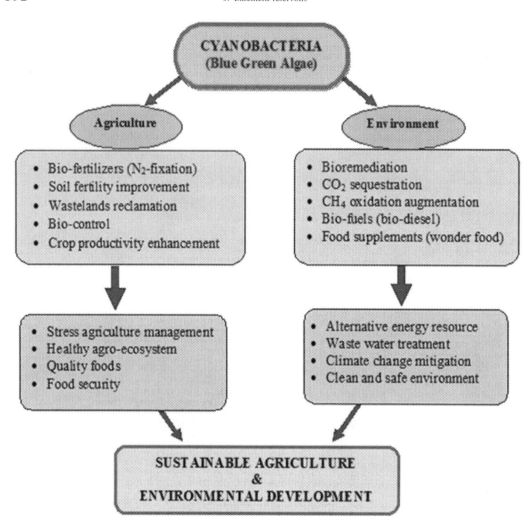

FIG. 5.30 Sustainable tracks in agricultural practices. *From Singh, J.S., et al., 2016. Cyanobacteria: a precious bioresource in agriculture, ecosystem, and environmental sustainability. Front. Microbiol. 7, 529. doi:10.3389/fmicb.2016.00529.*

Recently the genetic alterations in the cyanobacterial metabolic pathways look somewhat amazing. Although not all hopes and calculations ultimately lead to viable and productive phenotypes, some progress has been achieved in cyanobacterial biofuel-related biotechnology. The biochemical pathways that have been modified to produce cyanofuels are depicted in Fig. 5.31. Strategies for metabolic optimization in cyanobacteria can be grouped into four areas: 1. Improvement of CO2 fixation; 2. Optimization of pathway flux; 3. Improvement of tolerance of toxic products; 4. Elimination of competing pathways.

Cyanobacteria directly fix atmospheric or water-dissolved CO_2, and they require only sunlight, water, and a minimum set of inorganic trace elements for growth. Cyanobacteria are capable of handling the excess CO2 (the emissions we are trying to avoid) directly into

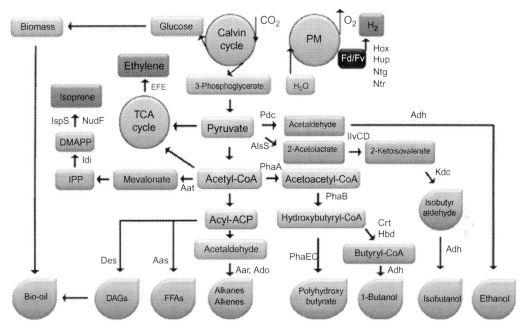

FIG. 5.31 Engineered and natural biochemical pathways of cyanobacteria that are employed for the production of valuable compounds.

hydrocarbons for biofuels. They can be also used to recover the water wastes from organic and inorganic contaminants (Fig. 5.32).

Even though it is assumed that these cyanobacteria are from the Archean eon, note that it's based on the assumption that anaerobic bacteria are more primitive than cyanobacteria. In any event, ancient organisms—and their "extremophile" descendants today—thrived in the absence of oxygen, relying on sulfate for their energy needs. With the same fundamental premise, it then follows that the isotopic ratio of sulfur transformed, indicating that for the first time oxygen was becoming a significant component of Earth's atmosphere (Farquhar et al., 2000). The evolutionary models assume that oxygen levels rose after the "boring billion" to a level amenable to the evolution of animals. Such an assumption, however, doesn't explain how oxygen percentage in the atmosphere became balanced at a value of 21. It is thought that cyanobacteria are capable of living within an anaerobic environment. In addition, these bacteria are also considered to be "extremophiles," meaning they are capable of surviving extreme conditions although they don't particularly thrive under any of the extreme features of the environment. In absence of oxygen, energy needs are met from sulfate compounds.

Climate, volcanism, plate tectonics all played a key role in regulating the oxygen level during various periods. What is missing from the literature is any indication that the oxygen content of the earth was low and what phenomena led to the rise of oxygen level to the value that is ideal for sustaining life or when it reached the balanced value of 21% in the atmosphere.

Cyanobacteria are still active today and operate in a pattern that is conducive to converting magma into the greenery. Historically, volcano materials have turned into a bedrock for

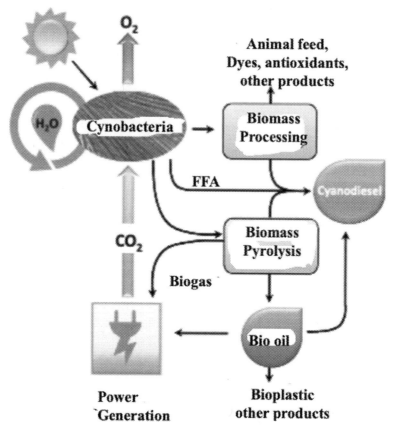

FIG. 5.32 The use of cyanobacteria to produce cyanodiesel and other valuable products. *From Sarsekeyeva, F., et al., 2015. Cyanofuels: biofuels from cyanobacteria. Reality and perspectives. Photosynth. Res. 125 (1–2). doi:10.1007/s11120-015-0103-3.*

greeneries in a short period of time. Only recently, McNamara et al. (2018) reported the birth of an island in the thick of volcanic eruption. Within eight months, The eruption is thought to have had a more lasting impact on the environment, however, by attracting barnacles, corals, algae, and oysters.

5.5.1 Migration of basement oil

By definition, basement rocks cannot overlie younger sedimentary source rocks, and so hydrocarbons will not have migrated directly upwards into them, as occurs in a classic petroleum system. However, in areas where the basement has been uplifted so that it is adjacent to a deep source, a migration pathway can exist either through the basement itself or through a carrier bed that connects the source rock to the basement. An alternative migration route could be from a source rock that is draped over the basement high, allowing hydrocarbons to be expelled directly into the basement underneath (Fig. 5.33).

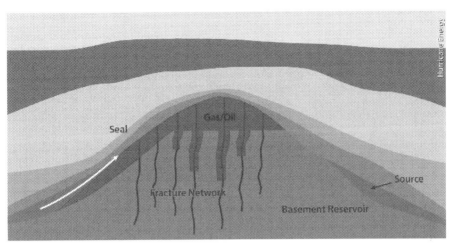

FIG. 5.33 Trap structure in a basement reservoir.

One of the great attractions of fractured basement reservoirs is that they can be considered to be stratigraphic traps, in that oil can be found outside of structural closure, which can lead to extensive flank oil accumulations. This has been demonstrated in basement discoveries in Yemen, and the potential also exists for the same to occur on Hurricane's Typhoon Prospect.

5.5.2 Exploring basement reservoirs

Although basement rocks are complex, the process and workflow involved in exploring hydrocarbons in them are not greatly different from that applied to reservoirs with intergranular porosity, it is important to gather and analyze a wide range of data, in particular, to assess if an effective fracture system is present. By integrating a variety of technical information using off-the-shelf technology and proprietary techniques, the exploration potential of the basement can be comprehensively evaluated, geological risk minimized, and prospective well locations identified. The most important recent technological development to assist in exploring basement reservoirs is probably the advent of 3D seismic data, which has allowed for effective imaging of fault networks. 3D seismic, combined with horizontal drilling and advances in both wireline and LWD imaging technology, has enabled fracture sweet spots to be targeted and effectively quantified by exploration and development wells.

5.5.3 Abiogenic sources

While petroleum (oil and gas) occurrences are commonly thought of as being derived from biological substances, i.e., by the thermal decomposition of organic matter or by microbial processes (Lollar et al., 2002), since the very beginning of the "petroleum age" there has been a substantial amount of skepticism voiced about its exclusively biological origin (e.g., Mendeleev, 1877; Kudryavtsev, 1951; Kenney et al., 2009). According to the proponents of an abiogenic origin of hydrocarbons, CH_4 and oil are also, perhaps even predominantly, formed well below the usual depths of oil and gas deposits. By comparing the carbon and hydrogen stable isotope signatures of CH_4 seeping from hard rock mines of Canadian and Fennoscandian shields, with microbial and thermogenic CH_4, Lollar et al. (2002) confirmed the abiogenic formation of hydrocarbons within the Earth's crust. Abiogenic hydrocarbon formation is attributed to the reduction of CO_2 in hydrothermal water-rock interactions similar to Fischer-Tropsch reactions and serpentinization of ultramafic rock. However, the lack of abiogenic isotope signatures in current conventional gases means these authors doubt that abiogenic formation is a significant hydrocarbon source. In contrast, the "modern Russian-Ukrainian" theory of the origins of hydrocarbons postulates that petroleum is a primordial material of deep origin that is transported at high pressures via "cold" eruptive processes into the crust of the Earth (Kenney et al., 2002). Much of this theory has developed through chemistry and chemical thermodynamics. One conclusion from the theory is that the origin of hydrocarbon molecules (except CH_4) from biogenic ones in the temperature and pressure regimes of the Earth's near-surface crust is in violation of the second law of thermodynamics (Kenney, 1996).

The abiogenic theory has been applied extensively across the former Soviet Union and has revealed 80 oil and gas fields in the Caspian district with production from the crystalline basement rock. Other examples have been found in the western Siberian cratonic-rift sedimentary basin with numerous fields producing from the crystalline basement; on the northern flanks of the Dneiper-Donotz basin; and elsewhere in Azerbaijan, Tatarstan, and Asian Siberia.

The theory also postulates that many of the world's oil and gas fields are being continuously recharged from deeper horizons with abiogenic hydrocarbons, so making the fields "effectively inexhaustible," and therefore oil and gas fields now considered exhausted should be thoroughly investigated to ascertain the quantities of oil and gas that may have accumulated since the fields were shut-in. In view of the above considerations, Kenney et al. (2009) believe that petroleum abundances are limited by little more than the quantities of its constituent elements that were incorporated into the Earth at the time of its formation and the petroleum industry is only now "entering its adolescence." Gold (1999) argues that hydrocarbons are of primordial origin without biological derivation. The presence of hydrocarbons in the planetary system, e.g., the atmospheres of Jupiter, Saturn, or Neptune contain enormous amounts of CH_4 and other hydrocarbons, while Titan, a satellite of Saturn, has a CH_4 and ethane atmosphere that is unlikely the result of biological processes. Rather, carbonaceous chondrite meteorites are thought to have brought carbon to the Earth during its formation. Subjected to the appropriate heat and pressure domains present at great depths, e.g., in the vicinity of magma cooling, the carbonaceous material would produce hydrocarbons, chiefly CH_4, as well as hydrogen and CO_2, which is subsequently outgassed to the upper layers of the Earth's crust. Such an environment is also quite amenable for the formation of heavier hydrocarbon molecules. The discovery of microbial life in ocean vents too deep for photosynthesis is used by Gold (1999) in support of the abiogenic gas theory. Instead of photosynthesis, the large amounts of CH_4 in these hydrothermal vents migrating from deeper levels supply the energy required for life in chemical form. In addition, the existence of anaerobic bacteria in rock structures and CH_4 environments at depths of more than 4 km is another indication of abiogenic CH_4. In turn, bacteria brought along the upward-migrating hydrocarbons explain the existence of certain biomarkers in extracted petroleum commonly associated with the decomposition of organic matter. The abiogenic origin of hydrocarbon debate dates back to the mid-19th century and since then the theory has undergone extensive development, refinement, and application. More than 4000 articles on the theory have been published (Odell, 2004) and numerous scientific conferences have been held to debate and evaluate the theory. The existence of abiogenic formation of CH_4 and hydrocarbon gases within the Earth has been confirmed (Lollar et al., 2002), but their substantial or exclusive contribution to, and replenishment of, current gas and oil deposits making hydrocarbons a quasirenewable source as proposed by the Russian-Ukrainian theory remains a matter of controversy.

5.6 Scientific ranking of petroleum

From characteristic time analysis, the age of hydrocarbon should be considered as the first factor in defining ranking. Basement hydrocarbons are known to be the oldest. Table 5.15 shows the composition age of a Russian basement reservoir (Ivanov et al., 2018).

The data presented in Table 5.16 are for the Yangiyugan area (the northwestern part of West Siberia). Ivanov et al. (2018) reported that plagiogneisses were formed over the substrate

TABLE 5.15 Chemical composition (wt%) of minerals from plagiogneis.

	\multicolumn{10}{c}{Analyses nos.}
	1	2c	2c	3	4	5	6	7	8	9
					\multicolumn{2}{c}{Number of analyses}				
Component	5	4	4	5	7	1	4	6	5	6
P_2O_5	–	–	–	–	–	–	–	–	–	42.33
SiO_2	65.14	37.26	37.66	65.10	37.90	40.96	41.18	46.70	26.10	0.02
TiO_2	0.02	0.08	0.07	0.01	0.11	0.33	0.26	0.45	0.07	–
Al_2O_3	21.23	20.20	20.77	17.71	25.36	15.56	14.77	32.37	19.82	–
Cr_2O_3	–	0.04	0.02	–	0.05	0.01	–	0.04	0.06	–
Fe_2O_3	–	–	–	–	10.63	–	–	–	–	–
FeO	0.16	28.65	29.01	0.47	–	20.77	20.15	3.54	23.30	0.06
MnO	–	2.63	0.50	0.04	0.05	0.62	0.36	0.06	0.38	0.08
MgO	–	2.07	2.83	0.01	0.09	6.34	6.64	1.31	15.73	–
CaO	3.03	8.95	9.59	–	23.31	10.27	10.31	–	0.05	55.39
Na_2O	10.14	–	0.03	0.36	0.03	1.76	1.91	1.05	0.01	0.01
K_2O	0.13	–	–	15.68	–	0.46	0.41	9.33	0.04	–
F	–	–	–	–	0.10	0.22	0.25	0.11	0.18	3.56
Cl	–	–	–	–	–	0.01	0.02	–	–	0.01
sum	99.85	99.88	100.48	99.38	97.63	97.29	96.26	94.96	85.73	101.46

From Ivanov, K.S., Erokhin, Yu.V., Ponomarev, V.S., 2018. Age and composition of granitoids from the basement of Krasnoleninsky oil and gas region (Western Siberia). Izv. UGGU 2 (50), 7–14.

TABLE 5.16 Published isotopic mineral ages for Precambrian basement in southwestern Ontario, Michigan, and Ohio.

Location	Type	(Ma)	Mineral/rock
Ontario			
Burford Tp. Brant Co.	K-Ar	920	Biotite/granite gneiss
Romney Tp. Kent Co.	K-Ar	895	Biotite/migmatite
Michigan			
Washtenaw Co.	Rb-Sr	840	Biotite/gneiss
Washtenaw Co.	Rb-Sr	920	Biotite/gneiss
St. Clair Co.	Rb-Sr	900	Biotite/gneiss
	K-Ar	970	Biotite/gneiss
Ohio			
Huron Co.	Rb-Sr	900	Biotite/gneiss-schist
	K-Ar	935	Biotite/gneiss-schist
Sandusky Co.	Rb-Sr	890	Biotite/gneiss-schist
	K-Ar	935	Biotite/gneiss-schist
Wood Co.	Rb-Sr	900	Biotite/biotite gneiss
	K-Ar	960	Biotite/biotite gneiss

of leucocratic plagiogranites (trondhjemites) under the conditions of the amphibolites facies of metamorphism. The SHRIMP II U-Pb dating of zircons showed that the igneous intrusion of plagiogranites proceeded during the Late Vendian (566 ± 3 Ma). Their metamorphism with the formation of plagiogneisses took place in the Early Ordovician (486 ± 4 Ma). The shows of powerful fluid-metasomatic processes of the transformation of rocks during the Carboniferous time are revealed.

Turek et al. (2008) listed published data on the ages of various basement rocks. This is shown in Table 5.16. In several instances, the age reaches close to a billion years. Turek et al. (2008) reported similar ages (some exceeding billion years) for a Canadian field (Fig. 5.34). Even though some of these data points were removed at the time, we know now that the age of the Earth is estimated to be much older, making those numbers realistic.

Fig. 5.35 shows how different types of oils take up the different natural processing times. Of course, in this process, biofuels come last because their processing time is miniscule in a geologic timeframe. This figure indicates that basement hydrocarbons are naturally processed for the longest time. This is significant considering the fact that if "natural" breeds sustainability, basement oils should have the most likelihood to be sustainable. However, one must note that there are other factors to consider. For instance, diamond, while naturally processed for the

5.6 Scientific ranking of petroleum 599

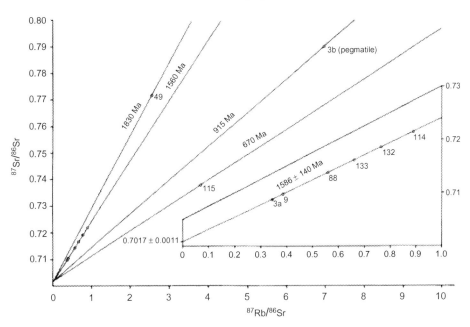

FIG. 5.34 Whole-rock Rb-Sr isochron diagram, basement samples.

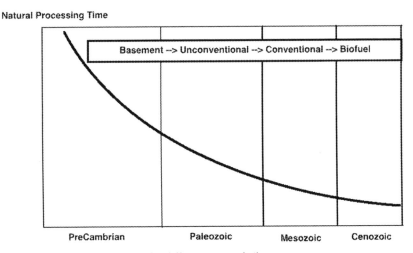

FIG. 5.35 Natural processing time differs for different types of oils.

longest time, doesn't make a fuel. As we transit to different characteristic times, the usefulness of natural products changes in applications. Another aspect is, when fresh fuel (e.g., biofuel) is oxidized, it produces CO_2 that is readily absorbable by the greeneries in the process of photosynthesis. So, even though crude oil has been processed longer than say vegetable oil, the value gained by crude oil is not in the quality of CO_2 that is generated. What is then the gain in crude oil due to longer-term natural processing? It is in its applicability to a longer-term application. This aspect will be discussed in the follow-up section. Fig. 5.35 also indicates that in terms of ranking due to natural processing time, the following rule applies:

(1) Basement oil
(2) Unconventional gas/oil
(3) Conventional gas/oil
(4) Biofuel (including vegetable oil)

The same rule applies to essential oils. These natural plant extracts have been used for centuries in almost every culture and have been recognized for their benefits outside of being food and fuel. Not only does the long and rich history of essential oils usage validate their efficacy specifically, it also highlights the fact that pristine essential oils are a powerfully effective resource with many uses and benefits—from beauty care to health and wellness care, fast effective pain relief, and emotional healing, and others. In another word, the benefits of oils are diverse and it is the overall benefit that increases with longer processing time.

Fig. 5.36 is based on the analysis originally performed by Chhetri and Islam (2008) that stated that natural processing is beneficial in both efficiency and environmental impact. However, as Khan and Islam (2012) have pointed out, such efficiency cannot be reflected in the short-term calculations that focus on the shortest possible duration. The efficiency must be global, meaning it must include long-term effects, in which case natural processing stands out clearly. Fig. 5.36 shows how a longer processing time makes a natural product more suitable for diverse applications. This graph is valid even for smaller scales. For instance, when milk is processed as yogurt, its applicability broadens, so does its "shelf life." Similarly, when alcohol is preserved in its natural state, its value increases. Of course, the most useful example is that of honey, which becomes a medicinal marvel after centuries of storage in natural

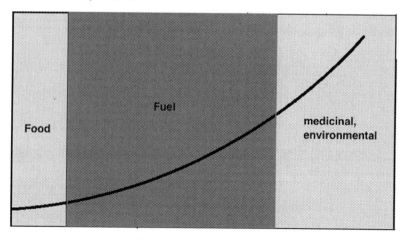

FIG. 5.36 Natural processing enhances the intrinsic values of natural products.

settings (Islam et al., 2015). In terms of fuel, biomass, which represents minimally processed natural fuel, produces readily absorbable CO_2 that can be recycled within days to the state of food products. However, the efficiency of using biomass to generate energy is not comparable to that which can be obtained with crude oil or natural gas. Overall, energy per mass is much higher in naturally processed fuel.

Let's review examples from both oil and natural gas. Historically it has been believed that conventional gas and oil are miniscule compared to unconventional. With it comes the notion that it is more challenging to produce unconventional petroleum resources. In addition, at least for petroleum oil, the notion that unconventional resource is more challenging to process is prevalent. This notion is false. With the renewed awareness of the environmental sustainability, it is becoming clear unconventional resources offer more opportunities to produce environment-friendly products than conventional resources. Fig. 5.37 shows the pyramid of both oil and gas resources.

On the oil side, the quality of the oil is considered to be declining as the API gravity declines. This correlation is related to the processing required for crude oil to be ready for conversion into usable energy, which is related to heating value. Heating value is typically increased by refining crude oil upon the addition of artificial chemicals that are principally responsible for global warming (Chhetri and Islam, 2008; Islam et al., 2010). In addition, the

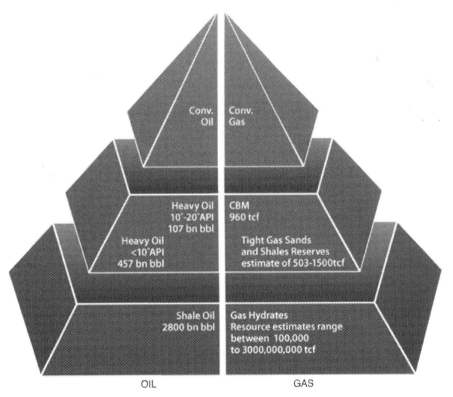

FIG. 5.37 The volume of petroleum resources increases as one moves from conventional to unconventional.

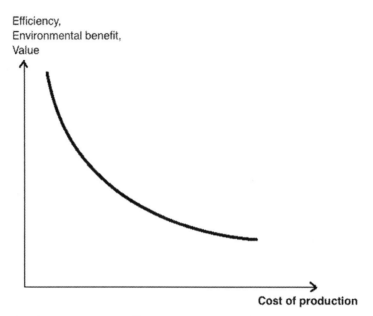

FIG. 5.38 Cost of production increases as efficiency, environmental benefits, and real value of crude oil declines. *Modified from Islam, M.R., Chhetri, A.B., Khan, M.M., 2010. Greening of Petroleum Operations. Wiley-Scrivener.*

process is inefficient and results in products that are harmful to the environment. Fig. 5.38 shows the trend inefficiency, environmental benefit, and real value with the production cost of refined crude. This figure shows clearly there is a great advantage to using petroleum products in their natural state. This is the case for unconventional oil. For instance, shale oil burns naturally. The color of flames indicates that crude oil produced from shale oil doesn't need further processing. The right image of Picture 5.1 emerges from burning gasoline and has similar colors to those of the left.

In addition, crude oil from shale oil is "cleaner" than other forms of crude oil because it is relatively low in tar content as well sand particles. Another crucial aspect is the fact that sulfur content or other toxic elements of crude oil have no correlation with unconventional or conventional sources. Furthermore, heavier oils do not have more of these toxic elements and are not in need of refinement to be usable.

Lighter crudes are considered to be easier and less expensive to produce only because modern engineering uses a refined version of the crude oil and all refining technologies are specially designed to handle light crude oil. If sustainable refining techniques are used, lighter or conventional oil offers no particular advantage over unconventional one and yet the volume and ease of production of unconventional are greater in unconventional resources. Figs. 5.39 and 5.40 illustrate the overall refinery efficiency in each of the three refinery groups, namely Low API, High API/Low HP (heavy product), and High API/High HP.

These results suggest strong impacts of API gravity and HP yield on overall refinery efficiency. In this, basement reservoirs offer an interesting solution. They produce old petroleum products, yet high API/low HP. Technically, these hydrocarbons should be mediocre, but scientifically they should be the best choice. The reason for this discrepancy is the fact that

PICTURE 5.1 Images of burning crude oil from shale oil (left) and refined oil (right).

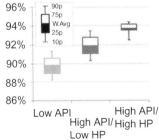

FIG. 5.39 Overall refining efficiency for various crude oils. *Modified from Han, J., et al., 2015. A comparative assessment of resource efficiency in petroleum refining. Fuel 157, 292–298.*

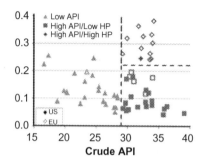

FIG. 5.40 Crude API gravity and heavy product yield of the studied US and EU refineries (The yield of heavy products, such as residual fuel oil, pet coke, asphalt, slurry oil, and reduced crude, is calculated as a share of all energy products by energy value). *From Han, J., et al., 2015. A comparative assessment of resource efficiency in petroleum refining. Fuel 157, 292–298.*

by refining we tend to harmonize each type of crude oil. By forcing them to conform to a common standard, we reduce the natural synergy these natural resources have.

This effect is further visible when one considers heavy product yields for various refineries as a function of Crude oil API gravity. Fig. 4.49 shows an overall trend for the three groups of oils, described above. Note the almost no overlaps in the key parameters between the Low API and High API/High HP group. Among the two High API groups, the Low HP group is clearly more resource-efficient than the High HP group. This conclusion would be different if the heavy products were used properly as a whole rather than extracting them to refine the crude into gasoline. This process makes the refining system inherently inefficient as well as unsustainable from both technical and environmental perspectives.

If one takes natural gas into consideration, resources that are more readily available are considered to be less toxic. For instance, biogas is the least toxic, whereas it is most plentiful. As can be seen in Fig. 5.41, as one transits from conventional gas to coal bed methane (CBM) to tight gas and shale gas all the way to hydrates, one encounters more readily combustible natural resources. In fact, CBM burns so readily that coal mine safety primarily revolves around the combustion of methane gas. Processing of gas doesn't involve making it more combustible, it rather involves the removal of components that do not add to the heating value or create safety concerns (e.g., water, CO_2, H_2S).

Fig. 5.41 shows how the volume of resources goes up as one moves from conventional to unconventional resources. In this process, the quality of gas also increases. For instance, hydrate has the purest form of methane and can be burnt directly with little or no safety concerns. At the same time, the volume of natural gas in the hydrate is very large. The concentration of "sour" gas components also decreases with the abundance of resources. Such a trend can be explained by the processing time of a particular resource. There is a continuity in nature that dictates that the natural processing increases both the value and global efficiency of energy sources (Chhetri and Islam, 2008). Fig. 5.42 depicts the nature of the volume of natural resources as a function of processing time.

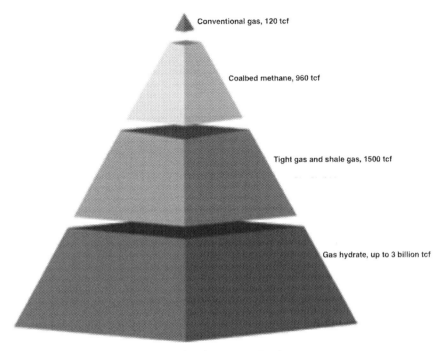

FIG. 5.41   The current estimate of conventional and unconventional gas reserve.

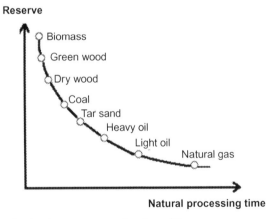

FIG. 5.42   The abundance of natural resources as a function of time.

In this picture, "natural gas" relates to petroleum products in a conventional sense. This figure shows natural gas in general is most suitable for clean energy generation. Within unconventional gas sources, there exists another correlation between reserve volume and processing time.

In general, the processing time for various energy sources is not a well-understood science. Scientists are still grappling with the origin of the Earth or the universe, some discovering only recently that water was and remains the matrix component for all matter. We have seen in Fig. 5.40 how natural evolution on Earth involved a distinctly different departure point not previously recognized. Pearson et al. (2014) observed a "rough diamond" found along a shallow riverbed in Brazil that unlocked the evidence of a vast "wet zone" deep inside the Earth that could hold as much water as all the world's oceans put together. This discovery is important for two reasons. Water and carbon are both essential for living organisms. They also mark the beginning and end of a life cycle. All natural energy sources have carbon or require carbon to transform energy in usable form (e.g., photosynthesis).

Even though theoretically there is much more recovery potential of heavier energy sources all the way up to biomass (Fig. 5.43), the current recovery techniques are geared toward light oil. This figure shows that natural gas is the most efficient with the most environmental integrity. This argument has been sharpened in the previous chapter that breaks down natural gas further into various forms of unconventional reservoirs. Within petroleum itself, the "proven reserve" is miniscule compared to the overall potential, as depicted in Fig. 5.44.

Of course, if one includes solar energy that would be the highest reserve possible. Because all energy source utilization techniques are equipped with processing light oil as a reference, the primary focus of EOR has been light oil. In early 2000, U.S. tertiary recovery was estimated to be 12% (Islam, 2014). This number has held steady until the huge surge in the unconventional recovery of oil and gas that increased oil and gas production by 40% in 2013. It is difficult to characterize unconventional recovery under a known category of oil and gas production. In any event, the knowledge of EOR is invaluable for developing any form of petroleum production scheme.

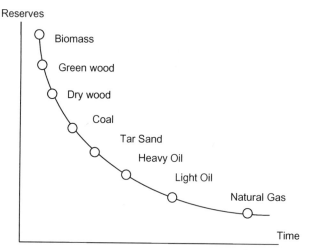

FIG. 5.43 As natural processing time increases so does the reserve of natural resources. *From Chhetri, A.B., Islam, M.R., 2008. Inherently Sustainable Technologies. Nova Science Publishers, 452 pp.*

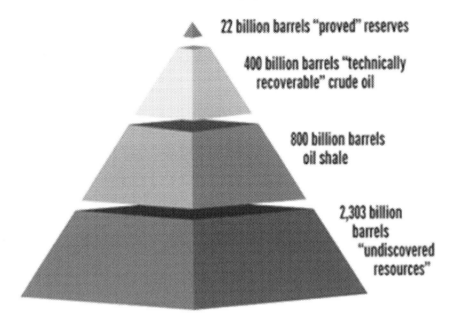

FIG. 5.44 Proven reserve is miniscule compared to the total potential of oil.

CHAPTER 6

Reserves prediction and deliverability

6.1 Introduction

The need for a proper description of reservoir rocks and reservoir fluids for reservoir appraisal and exploration cannot be over-emphasized; as such various domains related to this subject exist in the petroleum industry. This includes reservoir characterization, reservoir geophysics, reservoir rock description, reservoir geo-mechanics, and petroleum geo-statics. However, of equal importance are the changes that the rock properties undergo over time. It is logical to expect that the current behavior of any particle of matter is controlled by its origin of formation and previous history.

Description of the permeability and porosity of petroleum-bearing formations have been attempted by numerous theoretical and empirical means with varying degrees of success. The majority of the reported studies considered static porous media. A limited number of studies have considered porous media undergoing pore evolution by various mechanisms.

The basic law describing the diffusion process of fluids in any porous medium is the well-known Darcy law (Darcy, 1856). In its simplest sense, it states that the flux is proportional to the gradient of pressure as was also established by Fick (1855, 1995). Since its discovery, many researchers have then contributed to the ever-burgeoning literature on fluid flow in porous media by extending the Darcy law either by accounting for slip, inertia, and so on, and have successfully obtained well-known fluid flow equations (Bear, 1972; Brinkman, 1949; Lauriat and Prasad, 1989; Sposito, 1980; Whitaker, 1986). Most authors who have studied diffusion problems in porous media use the classic empirical law of Darcy stating proportionality between the fluid mass flow rate and the gradient of the pore pressure in the same direction.

Darcy's law is based on some simplifying assumptions about the properties of the porous medium and also of the flowing fluid(s). However, if any of the simplifying assumptions—for example, the porous medium is heterogeneous, non-isothermal conditions prevail, and the fluid properties are strong functions of pressure and temperature—does not apply, Darcy's law in its simplest form cannot be used to model fluid flow in such situations.

The majority of the available commercial simulators lack in their capability to predict permeability accurately. Instead, they focus more on providing efficient numerical scheme(s) and procedures for the solution of the governing flow partial differential equations.

These approximate solutions, however, are limited due to their inherent inability to predict permeability properly. Therefore, it is the accurate prediction of permeability that is instrumental to improving the reliability of existing commercial simulators. In this chapter, a mathematical model is developed using the concept of memory to address the evolution of rock (porosity and permeability) and fluid properties in a reservoir through the memory formalism by modifying the Darcy law by introducing the Grunwald-Letnikov approximation of the Riemann-Liouville fractional derivative to capture the variation of permeability and viscosity over time.

6.1.1 Fluid classification

Energy Information Administration (EIA) treats it as crude oil. Many different classifications can be applied to crude oil, depending on the different physical or chemical properties. However, the most common way is to describe oil by its density, often better known as gravity number. The American Petroleum Institute (API) defines the gravity number according to Eq. (6.1) (Dake, 1978).

$$°API = \frac{141.5}{\text{specific gravity}} - 131.5 \qquad (6.1)$$

Specific gravity is often defined as the ratio of densities for crude oil and water at 15.6°C, although slight deviations from this may use other reference points, such as 0°C, 20°C or the maximum volumetric mass of water which is at 3.98°C. API gravity ranges from 0 to 60°, where dense oils have low values and highly viscous oils have high values. Condensates typically have API gravities over 45° and Canadian tar sands from the Athabasca can be found in the range of 6–10° (Peters et al., 2005). Oil with less than 10°API is denser than water and may be called extra-heavy oil or natural bitumen (USGS, 2006), depending on viscosity. Heavy oils have gravities of less than 20°API, but more than 10°API (USGS, 2006). Medium crudes can be found between 20°API and 30°API. Light crudes have more than 30°API (Robelius, 2007). Generalized approximate relationships between API gravity and gas-oil ratio, reservoir depth, percentage sulfur and trace metal content are described by Tissot and Welte (1978, 1984). The API gravity classification is a simple system and worked well, as long as there was one dominating quality type of crude oil in use. As new oil fields were brought into production and new crude oil blends entered the market, the simple gravity classification scheme was insufficient to fully measure the quality of crude oil. Even so, the API system is still in use for certain crude oils and products (Speight and Islam, 2016). An improvement in classification can be performed by including the content of various important pollutants, especially sulfur. The sourness of crude oil refers to the sulfur content. The Society of Petroleum Engineers (SPE) defines sour crude oil as an oil containing free sulfur or other sulfur compounds whose total sulfur content is more than 1% (SPE, 2009). Crude oils with low sulfur content are commonly called—sweet. It is more complicated to refine heavy and sour crude oils, and consequently, they are worth less on the market compared to the light and sweet crude oils. Heavy crude needs more processing to yield high-quality products due to their low API-gravity, high viscosity, high initial boiling point, high carbon residue, and low hydrogen content (Nygren, 2008). The most valuable oil is light and sweet crude oil.

6.1.2 Reserve estimates

All oil and gas fields represent a limited geological structure, and consequently, they have an upper limit of how much hydrocarbons they contain. The size of the trap and reservoir, which can be defined by geological and geophysical methods, gives an estimate of the potential volume of oil in the field before the drilling has begun. As borehole data and production data become available, the reserve estimate will tend towards increasing accuracy (Dake, 1978). The total volume of oil in a field is commonly referred to as either oil initially in place (OIIP) or oil originally in place (OOIP) or sometimes just oil in place (OIP). This is equivalent to the total amount of oil residing in the pores of one or more reservoirs making up a field (Robelius, 2007). It is relatively straightforward to calculate OIIP if the areal extent and thickness of the reservoir are known together with the average porosity and saturation levels (Robelius, 2007). In practice, OIIP estimates get more complicated since both porosity and saturation vary throughout the reservoir. Conventional oil fields only capture a tiny amount of all the oil that is generated from the source rocks in a petroleum system and it should be remembered that there are other types of oil-bearing formations as well. For instance, the Elm Coulee field and the Bakken formation in Northern USA can be described as a—continuous-type II reservoir, which means that the hydrocarbons have not accumulated in a discrete reservoir with a limited areal extent (EIA, 2006). Oil in place for unconventional formations can vary enormously depending on conditions or assumptions made in the estimation process. Far from all the oil in place can be recovered from a given reservoir. The recoverable amount of the oil in place is classified as the reserve (Eq. 6.2). The recovery factor (RF) is a dynamic value, representing the estimated percentage of the total oil in place volume that can be recovered. RF depends on numerous parameters, such as rock and fluid properties, reservoir drive mechanism and production technology, variations in the formation and the development process (Robelius, 2007). In some modern reservoir simulators, it is not necessary to use OIIP or RF at all in order to estimate reserves.

$$\text{Reserve} = \text{Recovery Factor} * \text{Oil in Place} \tag{6.2}$$

The recoverable percentage of the OIIP can vary from less than 10% to more than 80% depending on individual reservoir properties and recovery methods, but the global average is as low as around 20% (Miller, 1995). Meling (2005) estimates the global average recovery factor to 29%, which is expected to be improved to 38% with new technologies. Laherrere (2003) writes that improved recovery factors due to technical progress cannot be justified with available data from individual oil fields or global data sets. However, technology can bring significant increases in recoverable volumes in more unconventional oil formations. Initially, oil is recovered through the energy that is occurring naturally in the reservoir (buoyancy energy, pressure energy, etc.), for instance through gas drive or water drive mechanisms. This can be called the primary recovery method and usually, 10%–30% of the oil in place can be recovered this way (Kjärstad and Johnsson, 2009). Differences from field to field can occur since individual reservoir properties can greatly influence recovery success. Secondary recovery methods utilize injection of water and/or gas to maintain pressure, thus feeding additional energy to the reservoir. About 30%–50% of the oil in place can be recovered by the use of primary and secondary recovery methods (IHS, 2007; Kjärstad and Johnsson, 2009). Today, almost 100% of all oil fields suitable for secondary recovery methods are using it

(AAPG, 1970, 1994). However, as it was discussed in the preceding chapters, none of them used sustainable technologies and there is room for improvements from both theoretical and application perspectives. By custom designing enhanced oil/gas recovery, each petroleum field can be given a second lifeline (Islam, 2020).

In this chapter, the issues related to reserve growth and dynamic optimization of the recovery process are discussed. Also, accounting for the memory of porous rock in fluid flow equations is usually neglected in all commercial reservoir simulation applications. However, we believe it is the most important property to be considered when developing fluid flow equations. The inclusion of memory accounts for the history of a porous rock and how it will behave in the future. Memory formalism accounts for the interaction of matter and molecules within the flow pathway, as opposed to just focusing on the permeability of porous materials encountered during the flow of the fluid. That is memory formalism approach switches the frame of analysis from the external observation of flow. It has been reported that some fluids possess characteristics that suggest there are other properties not accounted for in viscous fluids. The memory of the fluid has been said to describe these special characteristics.

6.1.3 Oil field formation

Oil and gas fields are the basic units in any oil and gas production unit. To find an oil or gas field, a complete petroleum system with all its necessary elements must first have been formed. The essential elements are source rock, reservoir rock, seal rock, and overburdened rock, and the required processes include trap formation and the generation-migration-accumulation of petroleum. All essential elements must be properly placed in time and space such that the processes required for forming a petroleum accumulation can occur. If any of the conditions are unfulfilled, there will not be any petroleum field (AAPG, 1994). A brief introduction to some of the necessary elements will follow. A suitable source rock, containing organic material, must be present somewhere relatively close to the reservoir. Without a source rock, no petroleum can be formed. The source rock must be buried to a suitable depth and once sufficient thermal energy has been passed on to the organic matter to break chemical bonds, the petroleum produced will be expelled and starts its mitigation towards the surface (Walters, 2006). If a source rock is not buried deep enough, the heat will not be enough to cause a chemical transformation of organic matter into petroleum. Oil shale, consisting of sedimentary rocks with significant organic content that yields substantial amounts of petroleum and gas upon destructive distillation (Dyni, 2005), can be seen as an ideal source rock, which never entered the oil window where it could produce petroleum. A field consists of one or several subsurface reservoirs, where hydrocarbons are located. Hydrocarbon reservoirs are not subterranean ponds or pools of oil, just waiting to be extracted as people often believe. In reality, hydrocarbons reside in the microscopic pore space of rocks, which are tiny void areas within the internal structure of the rocks. The situation is somewhat similar to a sponge soaked with water. If no suitable reservoir rock is present, there is no place that hydrocarbons can gather and form a commercially extractable accumulation. The basic properties of reservoirs are described in the next section. A tight and impermeable layer, commonly called seal or cap rock, must be present in order to trap hydrocarbons and prevent further mitigation (Selley, 1998). Otherwise, upward movement will continue, due to buoyancy, until the hydrocarbons reach the surface, where they will be broken down and destroyed by microorganisms. Entrapment is an absolute necessity for any commercially exploitable oil or gas accumulation (Robelius, 2007). If the seal is imperfect, small amounts can mitigate to the surface and form oil seepages (Tiratsoo, 1984). In the early days of petroleum exploration, oil seepages were an important tool for locating reservoirs. Several important oil fields, such as the Mexican giant Cantarell, were found as a result of oil seepages. There are many types of seals capable of generating a wide array of petroleum traps. A trap can be described—as any geometric arrangement of rock, regardless of origin, that permits significant accumulations of oil or gas, or both, in the subsurface‖ according to Biddle and Wielchowsky (1994). Low permeability materials and rocks make ideal seals, due to high capillary entry pressure. Mudrocks are the most common seals, while salt and other evaporates are the most effective ones (Selley, 1998). Structural traps, caused by tectonic processes after the deposition of the rock beds, are the most common seal among the world's largest oil fields (AAPG, 1970). Stratigraphic traps, caused by changes in rock lithology, and combination traps, consisting of both structural and stratigraphic traps, are the other two main types of petroleum traps. The term porosity refers to the percentage of pore volume compared to the total bulk volume of a rock. High porosity means that the rock can contain more oil per volume unit. A simplified picture of this can be seen in Fig. 6.1. Greater burial depth generally leads to a compaction of the sediments, which results in a decreased porosity (Selley, 1998).

Any type of rock can be a reservoir as long as the pore space is large enough to store fluids and connected well enough to effectively allow flows. However, sedimentary rocks such as sandstones and carbonates are the most frequently occurring reservoir types and sedimentary reservoirs dominate the world's known oil fields (Tiratsoo, 1984). Porosities of more than 15% are deemed good or even excellent for oil reservoirs (Hyne, 2001). Oil, gas, and water saturation levels are important factors and refer to the percentage of the pore volume that is occupied by oil or gas. An oil saturation level of 20% means that 20% of the pore volume is occupied by oil, while the rest is gas or water. A closer discussion on this can be found in Dake (1978). If oil is supposed to be able to flow, the oil saturation must be over a certain value, often referred to as the critical oil saturation (Robelius, 2007). A main driving force of secondary mitigation from the source rock to the reservoir is the buoyancy force. Most sedimentary rocks have their pores filled with water to some extent in normal circumstances (Selley, 1998). Oil is less dense than water and the difference in density will cause a buoyancy force, driving the oil upwards in the water. The principle goes back to Ancient Greek and Archimedes' classical treatise On Floating Bodies. The situation is also similar to that of a hot air balloon. As long as the oil droplet is smaller than the narrowest part of the rock pore, commonly called the pore throat, it will continue to move upwards (Selley, 1998). Throughout the development of the reservoir, the pore content might be change due to production or other parameters affecting the reservoir. It should also be noted that rock properties seldom are known in all locations of the reservoir. The first role is as a storage space for oil and other hydrocarbons, the second role is as a transmission network for fluid flows. Consequently, the pores must be connected to allow movement of the hydrocarbons within the reservoir. This was first investigated by French engineer Henry Darcy in the 1850s, who studies fluid flows through a bed of packed sand. He derived a phenomenological expression to describe the behavior and this is today known as Darcy's law (Eq. 6.3). Darcy's law is analogous to Fourier's heat conduction law or Fick's law of diffusion. Alternatively, Darcy's law can be derived from the Navier-Stokes

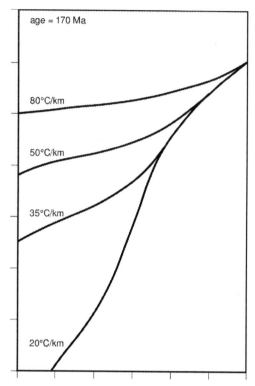

FIG. 6.1 Computed variation of porosity with depth for Jurassic (170 Ma) sandstones for basins having temperature gradients of 80, 50, 35, and 20°C/km. *From Bjørkum, P.A., et al., 1998. Porosity prediction in quartzose sandstones as a function of time, temperature, depth, stylolite frequency, and hydrocarbon saturation. AAPG Bull. 82 (4), 637–647.*

equations (Neuman, 1977). The ability of a rock to permit fluid movement is called permeability, usually denoted k.

$$q = \frac{kA}{\mu} \frac{\partial P}{\partial L} \tag{6.3}$$

where

q = volumetric flow rate,
k = permeability,
A = cross-sectional area,
μ = fluid viscosity, and
$\partial P/\partial L$ = pressure drop over the length of the flow path.

Permeability can differ in different directions, and generally, horizontal permeability is greater than vertical (Selley, 1998). Pumice stone, well-known for floating on water, can have porosities of up to 90% but lacks any permeability at all due to highly isolated pores (Dandekar, 2006). Consequently, it makes pumice a bad reservoir rock. The reverse can also be true, as low-porosity rocks such as micro fractured carbonate can allow unimpeded flows in the fractures. Good reservoirs are dependent on both porosity and permeability. In general, reservoir rocks do not demonstrate any solid theoretical relationship between these properties, making practical relationships and empirical surveys important. Fractures, cracks, and rifts can transmit fluids well, thus partially bypassing permeability problems caused by the pore structure, and this has been known to have a major influence on reservoir flows in certain reservoirs. This is also a property that can be affected with suitable technology. For example, fracturing techniques are used to enhance production in many Danish chalk reservoirs with low permeability. A more complete overview of reservoir rocks and their fluid properties can be found in Dandekar (2006).

6.2 Conventional material balance

Mass and energy balance equations form the core of every type of fluid movement in petroleum reservoirs. The fundamental principle is material balance, in which mass in = mass out + mass left within a confined domain. All mathematical models are based on this material balance principle. The following basic material balance equation holds.

$$N_p[B_o + B_g(R_p - R_s)] + W_p B_W - W_{inj}B_{winj} - G_{inj}B_{ginj}$$
$$= N\left\{[B_o - B_{oi} + B_g(R_{si} - R_s)] + \frac{B_{oi}}{B_{gi}}m(B_g - B_{gi})\right.$$
$$\left. + B_{oi}(1+m)\left(\frac{c_f + c_w S_w}{1 - S_w}\Delta p\right)\right\} + W_e B_w$$

Where, various terms hold the following meanings (Table 6.1)

Gas material balance is a simplified version of the general material balance equation. When the general equation is reduced to its simplest form containing only gas terms, it appears as shown below:

$$G = \frac{G_p B_g}{(B_g - B_{gi})}$$

In this equation, it is assumed that gas expansion is the only driving force causing production. This form is commonly used because the expansion of gas often dominates over the expansion of oil, water, and rock. B_g is the ratio of gas volume at reservoir conditions to gas volume at standard conditions. This is expanded using the real gas law.

$$B_g = \frac{V_{res}}{V_{std}} = \frac{Z_{res}nRT_{res}p_{std}}{Z_{std}nRT_{std}p_{res}}$$

The reservoir temperature is considered to remain constant. The compressibility factor (Z) for standard conditions is assumed to be 1. The number of moles of gas does not change from reservoir to surface. Standard temperature and pressure are known constants. When Bg is replaced and the constants are canceled out, the gas material balance equation then simplifies to:

$$\frac{p}{Z} = -\left(\frac{p_i}{Z_i}\frac{1}{G}\right)G_p + \frac{p_i}{Z_i}$$

TABLE 6.1 Various terms of the general material balance equation.

Terms	Meaning
$F = N_p[B_o + B_g(R_p - R_s)] + W_p B_w - W_{inj} B_{winj} - G_{inj} B_{ginj}$	The volume of withdrawal (production and injection) at reservoir conditions is determined by the oil, water, and gas produced at the surface.
$E_t = E_o + \frac{B_{oi}}{B_{gi}} m E_g + B_{oi}(1+m) E_{fw}$	Total expansion.
$E_o = B_o - B_{oi} + B_g(R_{si} - R_s)$	If the oil column is initially at the bubble point, reducing the pressure will result in the release of gas and the shrinkage of oil. The remaining oil will consist of oil and the remaining gas still dissolved at the reduced pressure.
$E_g = B_g - B_{gi}$	Gas expansion factor. For example, as the reservoir depletes, the gas cap expands into reservoir volume previously occupied by oil.
$E_{fw} = \frac{c_f + c_w S_w}{1 - S_w} \Delta p$	Even though water has low compressibility, the volume of connate water in the system is usually large enough to be significant. The water will expand to fill the emptying pore spaces as the reservoir depletes. As the reservoir is produced, the pressure declines, and the entire reservoir pore volume is reduced due to compaction. The volume change expels an equal volume of fluid as production and is therefore additive in the expansion terms.
$m = \frac{G B_{gi}}{N B_{oi}}$	The ratio of gas cap to the original oil in place. A gas cap also implies that the initial pressure in the oil column must be equal to the bubble point pressure.
$W_e B_w$	If the reservoir is connected to an active aquifer, then once the pressure drop is communicated throughout the reservoir, the water will encroach into the reservoir resulting in a net water influx.

From Fekete, 2021. http://www.fekete.com/ (Accessed 23 February 2021).

When plotted on a graph of p/Z versus cumulative production, the equation can be analyzed as a linear relationship. Several measurements of static pressure and the corresponding cumulative productions can be used to determine the x-intercept of the plot—the original gas-in-place (OGIP), shown as G in the equation.

For a volumetric gas reservoir, gas expansion (the most significant source of energy) dominates depletion behavior; and the general gas material balance equation is a very simple yet powerful tool for interpretation. However, in cases where other sources of energy are significant enough to cause deviation from the linear behavior of a p/Z plot, a more sophisticated tool is required. For this, a more advanced form of the material balance equation has been developed, and the standard p/Z plot is modified to maintain a linear trend with the simplicity of interpretation.

The success and usefulness of the P/Z plot in conventional reservoirs led to its application in unconventional reservoirs such as coal bed methane (CBM) and shale/tight gas reservoirs. In his work on CBM, King (1993) introduced p/Z^* to replace p/Z. By modifying Z, parameters to incorporate the effects of adsorbed gas were incorporated so the total gas-in-place is interpreted rather than just the free gas-in-place; and a straight line analysis technique is still used. This concept has been extended to additional reservoir types with Fekete's p/Z^{**} method (Moghadam et al., 2011).

The reservoir types considered in the advanced material balance equation are overpressured reservoirs, water-drive reservoirs, and connected reservoirs. The total Z^{**} equation is shown below with the modified material balance equation.

$$\frac{p}{Z^{**}} = \frac{p_i}{Z_i^{**}}\left(1 - \frac{G_p}{G}\right)$$

$$Z^{**} = \frac{p}{\left[\frac{1}{S_{gi}}\frac{p}{Z}(S_{gi} - c_{wip} - c_{ep} - c_d) + \frac{p_i}{Z_i}\left(\frac{G}{G_f} - 1\right)\right]\frac{G_f}{G}}$$

Despite its usefulness, the P/Z plot may give inaccurate results when applied directly to unconventional reservoirs such as coal/shale. This is because, in its conventional form, it did not include other sources of gas storage such as connected reservoirs or adsorption which is present in coal/shale reservoirs (Moghadam et al., 2011). This led to the modification of the P/Z plot for it to be suitably applied to unconventional reservoirs, especially for coal/shale gas reservoirs. Unconventional reservoirs such as coal/shale are characterized by gas adsorption, hence incorporating adsorption into the derivation of the P/Z method is necessary for accurate prediction of hydrocarbons in place for such reservoirs. This requires an adsorption model that can correctly represent the adsorption phenomenon within these reservoirs. Langmuir isotherm represented this phenomenon for the traditional P/Z plot used in unconventional gas reservoirs. Despite the limitations of Langmuir isotherm such as adsorption being a function of only pressure, it remains the only model currently incorporated in most P/Z plots to evaluate the production performance of unconventional gas reservoirs using material balance. Several P/Z methodologies have been developed for use in unconventional reservoirs with the use of classical Langmuir isotherm. Table 6.2 summarizes the different methodologies used in material balance calculations for unconventional gas reservoirs.

TABLE 6.2 Summary of methods used in applications of MBE in unconventional gas reservoirs.

Method	Description	Advantages/limitation
King's Method (1993)	Modified P/Z plot for unconventional reservoirs like coal/shale gas reservoirs. Effect of adsorption was introduced into the methodology.	Only suitable for under-pressured coal and not under saturated coal. An iterative solution made calculation tedious. Langmuir Isotherm used in prediction of gas adsorption.
Jensen and Smith Modified Method (1997)	Modified Kings Methodology with a more practical based evaluation of the estimated recovery and remaining reserves. Neglected the effect of water saturation in their solution. Suitable for reservoirs with a high adsorption rate	Use of Langmuir isotherm meant only single temperature could be used for adsorption calculations
Seidle Method – Modified King Method (1999)	Another modification of King's methodology but with the assumption of constant water saturation over time instead of average water saturation	Avoided the iterative solution adopted by king in its calculation. Langmuir isotherm is used to account for gas adsorption.

Continued

TABLE 6.2 Summary of methods used in applications of MBE in unconventional gas reservoirs—cont'd

Method	Description	Advantages/limitation
Ahmed et al. Method (2006)	Expressed the material balance as an equation of a straight line to enable calculations of original gas in place and also predicting average reservoir pressure	The method is applicable to any coal/shale that behaves according to Langmuir isotherm. Limited to only a single temperature for evaluation of adsorption
Moghadam et al. Method (2011)	A similar but more rigorous and advanced form of the material balance proposed by King.	Used to define the total compressibility of the system for analyzing fluid flow in unconventional gas reservoirs. Adsorption was described by Langmuir isotherm
Firanda Method (2011)	Introduced different drive mechanisms in the material balance equation proposed by King. These included water expansion, rock compaction, connate water expansion and moisture expansion.	Offered a similar methodology used by Ahmed and Roux by expressing material balance as a straight line. The adsorption isotherm used is Langmuir isotherm.
New Approach	The similar approach adopted by Ahmed and Roux by expressing material balance as a straight line. Average reservoir pressure, as well as future performance of the reservoir, can be obtained.	Avoids iterative solution of Kings approach. Adopts temperature-dependent gas adsorption models such as *Bi*-Langmuir and exponential model in its methodology to account for gas adsorption at several temperatures.

Temperature plays an important role in gas adsorption. Hence any adsorption model should be capable of expressing the adsorption as a function of both pressure and temperature. The majority of gas in shale gas reservoirs is from adsorbed gas and an ample knowledge of gas adsorption behavior over a wide range of pressure and temperature is needed. Typical sorption capacities for unconventional gas reservoirs requires measurement over an extended range of pressures especially for shales gas reservoirs where pressures need to be greater than 20 MPa and temperatures over 100°C for it to be considered as in situ reservoir conditions (Gasparik et al., 2015). Therefore, in order to accurately describe the adsorption capacities of these reservoirs, multiple adsorption capacities at different temperatures need to be collected. Langmuir's model is limited in providing such adsorption measurements at multiple temperatures.

Since gas adsorption is also a function of temperature, geothermal gradients will contribute substantially to the adsorption capacity of these reservoirs. An example is the black Warrior basin where there is a temperature variation of about 26.8–51.85°C within 0.3–1.8 km depth range (Pashin and McIntyre, 2003). For shale plays with much thinner thickness where the temperature might not play a crucial role, the use of the Bi-Langmuir model could help describe the adsorption phenomenon more accurately (Lu et al., 1995). This is because the Bi-Langmuir model takes into account the heterogeneities of the shale play. For example in Devonian shale, both clay and kerogen content plays a significant role in accounting for adsorption (Lu et al., 1995). A key assumption of Langmuir isotherm is the homogeneity of the adsorbent. However, this may not be suitable or true for coal/shale gas systems since different materials such as clay minerals and kerogen content may contribute to gas adsorption on shale (Lu et al., 1995). Hence, the need to introduce temperature-dependent adsorption models in material balance calculations for unconventional gas reservoirs.

Previous modifications of the material balance equation for unconventional gas reservoirs especially for coal and shale have included the adsorption capability of these resources. The additional term of gas adsorption has been modeled based on Langmuir isotherm. Langmuir isotherm describes an equilibrium relation between the free gas and the adsorbed gas under isothermal conditions. Thus gas adsorption has been expressed only as a function of pressure. Material balance calculations can be further improved when the effects of temperature on adsorption are considered. The role of temperature in adsorption has been highlighted by several researchers (Lu et al., 1995; Gasparik et al., 2015; Pashin and McIntyre, 2003).

Over the years, several researchers have modified the classical Langmuir model to include a temperature term that makes it appropriate to describe the adsorption capacity of shale/coal. Lu et al. (1995) proposed the use of the Bi-Langmuir model to describe the adsorption of gas at several temperatures. One term of the model describes gas adsorption on clay minerals while the other accounts for gas adsorption on kerogen. The model, therefore, is suitable for non-homogeneous adsorbents especially in the Devonian shale where two mineral compositions of clay and kerogen are said to be mainly responsible for gas storage. Ye et al. (2016) also proposed a variation of the classical Langmuir isotherm equation by introducing an exponential relation that expresses the Langmuir volume (VL) as dependent on temperature. Thus, the constant Langmuir volume (VL) as expressed in the Langmuir equation is replaced by a VL that is a function of temperature. Based on applying the models to different sets of shale gas data, it has been established that the modified Langmuir model with temperature dependency can accurately describe shale gas adsorption under various pressures and temperatures (Fianu et al., 2018; Lu et al., 1995; Ye et al., 2016). To include the effect of temperature, temperature-dependent gas adsorption models were incorporated by Fianu et al. (2018) into material balance calculations for prediction of original gas in place as well as determining both average reservoir pressure and future performance in coal/shale gas reservoirs. The material balance equation has been expressed as a straight line with both *Bi*-Langmuir and Exponential models used in the prediction of gas adsorption rather than the Langmuir isotherm. With this methodology, several adsorption capacities can be obtained at multiple temperatures which will allow for better estimation of the original gas in place and future gas production. The results from this works show that temperature-dependent gas adsorption models can be used in place of Langmuir isotherm to account for the effect of temperature variations and a more accurate representation of the adsorption of gas in coal/shale gas reservoirs.

They outlined the following steps.

The steps needed to carry out a prediction of future reservoir performance can be outlined below and also by the algorithm in Fig. 6.2.

Step 1: A future reservoir pressure below the current reservoir pressure needs to be selected.

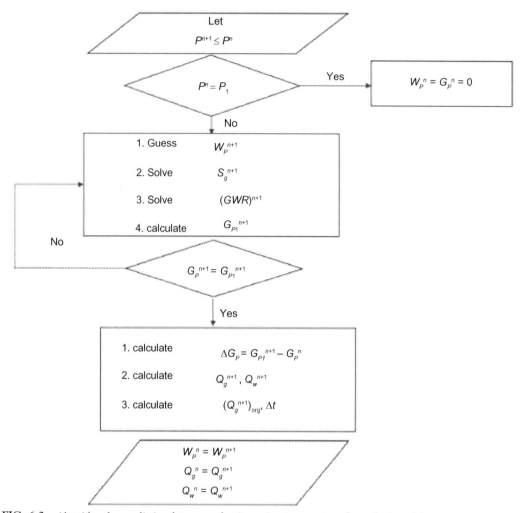

FIG. 6.2 Algorithm for predicting future production using temperature-dependent models.

Step 2: If the current pressure is equal to the initial reservoir, then set current water production and current Gas produced to zero.
Step 3: Guess or estimate the cumulative water production and solve future cumulative gas production, gas saturation, and gas-water ratio.
Step 4: Recalculate cumulative gas production by applying material balance.
Step 5: if the two values of cumulative gas production in Step 3 and 4 agree, then the assumed value of water production in Step 3 is correct. If not, assumed a new value for water production and go through Step 3 to Step 5.
Step 6: next calculate incremental gas production, gas, and water flow rates.

Step 7: replace the calculated old values of water production, cumulative gas production, gas, and water rate with the new values calculated and repeat steps 1–7.

6.2.1 Overpressured reservoir

At typical reservoir conditions, gas compressibility is orders of magnitude greater than that of the formation rock or residual fluids. In reservoirs at high initial pressures, the gas compressibility is much lower, in the same order of magnitude as the formation. A typical example of this would be an overpressured reservoir, which is a reservoir at a higher pressure than the hydrostatic column of water at that depth—in other words, a higher than expected initial pressure given the depth. In this situation, ignoring the formation and residual fluid compressibility will result in over-prediction of the original gas-in-place. The initial depletion will show effects of both depletion and reservoir compaction and the slope of a p/Z plot will be shallower. Once the pressure is much lower than the initial pressure, gas expansion is dominant and a steeper slope is observed on the p/Z plot. When matching on the shallower slope of this bow-shaped trend, all later pressure data will be lower than the analysis line, and the estimated original gas-in-place will be higher than the true original gas-in-place. The plot below shows an overpressured reservoir matched on the initial data and the analysis line of the advanced material balance method.

Based on the definition of compressibility, the following equation represents the total effect of formation and residual fluid compressibility:

$$\Delta V_{ep} = \frac{B_{gi}G}{S_{gi}}\left[\left(1 - e^{-\int_p^{p_i} c_f dp}\right) + S_{wi}\left(e^{\int_p^{p_i} c_w dp} - 1\right) + S_{oi}\left(e^{\int_p^{p_i} c_o dp} - 1\right)\right] \quad (6.4)$$

The approximate form of this equation, found by considering compressibility for oil, water, and the formation as constant; and e^x as $1 + x$, is:

$$\Delta V_{ep} = \frac{B_{gi}G}{S_{gi}}(c_f + S_{wi}c_w + S_{oi}c_o)(p_i - p) \quad (6.5)$$

To use this compressibility in the material balance equation, the change in pore volume is taken relative to the initial pore volume. The rigorous and approximate forms are shown below.

Rigorous form:

$$C_{ep} = \left(1 - e^{-\int_p^{p_i} c_f dp}\right) + S_{wi}\left(e^{\int_p^{p_i} c_w dp} - 1\right) + S_{oi}\left(e^{\int_p^{p_i} c_o dp} - 1\right) \quad (6.6)$$

Approximate form:

$$C_{ep} = (c_f + S_{wi}c_w + S_{oi}c_o)(p_i - p) \quad (6.7)$$

6.2.2 Water-drive reservoir

Some gas reservoirs may be connected to aquifers that provide pressure support to the gas reservoir as it is depleted. In this case, the pressure decrease in the gas reservoir is balanced by water encroaching into the reservoir. As this happens, the pore volume of gas is decreasing and the average reservoir pressure is maintained. Often this reservoir will show a flat pressure trend after some depletion. An example of this behavior on a p/Z plot is shown below.

The change in reservoir volume due to net encroached water can be determined from the following equation:

$$\Delta V_{wip} = 5.615(W_e - W_p B_w) \tag{6.8}$$

To use this in the material balance, the change in pore volume is taken relative to the initial pore volume, shown below.

$$c_{wip} = \frac{5.615(W_e - W_p B_w)}{B_{gi} G / S_{gi}} \tag{6.9}$$

When dealing with this equation, the major unknown value to be determined is water encroachment from the aquifer (W_e). Two aquifer models are provided to determine net encroached water: Schilthuis Steady-State Model and Fetkovich Model.

6.3 Analytical solutions

6.3.1 Production fundamentals

Once a reservoir has been located, the actual extraction of its hydrocarbon content can begin. The extraction process is often referred to as production, although one may think that extraction is a more suitable word since the hydrocarbons are removed from the reservoir. A good introductory description of production methods and various technical components can be found in Robelius (2007).

6.3.1.1 Reservoir flow relations

Within the reservoir, the flow of fluids is the governing factor for the extraction process. In order to be produced, the hydrocarbon fluids must reach the production wells and consequently, the rock properties affecting fluid mobility will have a major influence on the amount that can be extracted and also on how fast it can be extracted. Viscosity, gravity drainage, and capillary effects are the main forces governing the flow (Satter et al., 2008). Viscous forces dominate the behavior of fluids, both produced and injected, in a reservoir. Under viscous conditions, flow rates are laminar and proportional to the pressure gradient that exists in the reservoir (Satter et al., 2008). However, there are examples of tilted reservoirs and dipping formations, where gravity drainage is the prime driving force. Capillary forces are a result of surface tension between the fluid phase and the pore walls, something that can form sealing conditions if the capillary entry pressure is high. Gravity and capillary forces act in opposite directions and can be used to determine the initial distribution and saturation of oil, gas, and water in any hydrocarbon-bearing porous structure (Satter et al., 2008). The movement of fluids in a reservoir depends on the following factors:

- Depletion (leading to a decrease in reservoir pressure)
- Compressibility of the rock/fluid system
- Dissolution of the gas phase into the liquid
- Formation slope Capillary rise through microscopic pores
- Additional energy provided from aquifer or gas cap
- External fluid injection
- Thermal, miscible, or similar manipulation of fluid properties

In most reservoirs, more than one factor is responsible for the flow of fluids and closer discussion on this can be found in Satter et al. (2008). Some parameters can be affected by man-made measures, while others cannot. The slope of the hydrocarbon-bearing formation is an example of a flow parameter that is fixed, while the external fluid injection is dynamic and dependent on installed technology and product strategy. Compressibility determines how much the reservoir can be compacted, which is similar to squeezing a sponge to get more fluid out. It is a function of several parameters, including the type of minerals that make up the rock mass, the degree of sorting, the degree of mineral decomposition or alteration, cementation, and especially porosity (Nagel, 2001). Highly compactable reservoirs usually have reservoir porosity greater than 30% (Nagel, 2001). Compaction has been known to cause a significant increase in the available drive energy for hydrocarbon recovery. In the Norwegian giant field Valhall, compaction has been claimed to make up 50% of the total drive energy (Cook et al., 1997). The Norwegian Ekofisk field is estimated to recover an additional 243 to 280 million barrels as a result of increased reservoir compaction (Sylte et al., 1999). In the Bolivar Coastal oil fields in Venezuela, compaction drive has been estimated to constitute as much as 70% of the total drive energy (Escojido, 1981) and if steam flooding is used, compaction drive contribution can reach 80% of the total energy for the same region (Finol and Sancevic, 1995). Compaction can also lead to subsidence, which is the sinking of the ground level above the reservoir. Wilmington and Ekofisk oil fields are both well-known examples due to the magnitude of the subsidence as well as the cost of remediation (Nagel, 2001). Lake Maracaibo and the nearby Bolivar Coastal Region are other examples of how reservoir depletion has caused severe subsidence and flooding and similar effects are true for the giant Groningen gas field, where subsidence of only a few decimetres poses a significant threat since large portions of the Netherlands are below sea level and protected by dikes (Nagel, 2001). Water injection can solve subsidence problems and has also been shown to be a cost-effective way to control compaction (Pierce, 1970). However, injecting water might take the edge off subsidence issues but this also leads to the loss of compaction drive, significant energy for driving hydrocarbon flows in the reservoir. Altogether, this shows how various flow factors can balance each other. An increase in one of the driving forces can lead to a decrease in another and vice versa. All together, reservoir flows are a complex problem, with many interdependent variables.

6.3.2 Oil production modeling

The production of an oil field tends to pass through several stages. This can be described by an idealized production curve. A version of this curve can be seen in Fig. 6.3. After the

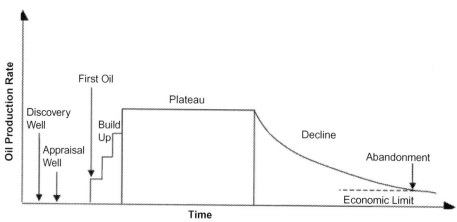

FIG. 6.3   A theoretical production curve, describing the various stages of maturity. *Source: Robelius (2007).*

discovery well, an appraisal well is drilled to determine the development potential of the reservoir. Further development follows and the first oil production marks the beginning of the build-up phase. Later the field enters a plateau phase, where the full installed extraction capacity is used, before finally arriving at the onset of decline, which ends in abandonment once the economical limit is reached. For many fields, especially smaller ones, the plateau phase can be very short and resemble more to a sharp peak, while large fields can stay several decades at the plateau production level. The lifetime of a field and the shape of the production curve are often related to the kind of hydrocarbon that is produced. For instance: condensate flows very easily and can be extracted almost all at once, which results in high decline rates (I). NGL is a by-product of natural gas, hence following the gas production curve intimately (I).

Fluid flows in porous media can be simulated to high levels of complexity or simplicity, largely depending on details in the flow model. These types of flow processes generally lead to complicated behavior and mathematical models must include statistical analysis, fractal and/or stochastic procedures. Many reservoir simulation models are dependent on various numerical models. One example is the ECLIPSE oil and gas simulator from Schlumberger Information Solutions (2009), which uses an implicit three-dimensional finite-difference approach to solve material and energy balance equations in a multiphase fluid system with up to four components in a subsurface reservoir with complex geometry. Traditionally finite difference methods dominate, but finite elements and streamlined numerical models are also used. Recently even more advanced computational techniques, such as neural networks and fuzzy logic (Zellou and Ouenes, 2007) or algebraic multi grids (Stüben et al., 2007), have been utilized to model reservoir flows. Combining the reservoir flow models with drilling and development plans along with economic investment models for the field can result inaccurate descriptions of actual production and how it changes over time. However, precise prediction of fluid flows usually requires detailed data and knowledge of many important reservoir properties and parameters, such as permeability, pressure, and similar. Drilling plans and details around installations and development schemes are also seldom openly available. In practical cases, much of the necessary data for accurate modeling is rarely available for outsiders, since oil companies and producers do not release it. Consequently, simplified models have been developed by various researchers and engineers to mitigate this shortcoming. Some examples of simplified models for production forecasting are depletion rate analysis and the utilization of decline curves, which are applicable to individual fields. Decline rate and depletion rate analysis are the main focus for this work. Decline curve analysis has a long history and has been used for more than 50 years within the oil and gas industry. Similarly, depletion is strongly linked to the fundamental reservoir flow relations and analysis of the depletion rate is, therefore, a sound approach. Peak oil discussions are often connected to the Hubbert curve, which aims to depict the collective behavior of a large number of fields. A significant number of models and methods for creating future outlooks on a global or regional scale are available (Bentley and Boyle, 2007), but they cannot be used for single fields. Due to this constraint, further discussion of these modeling approaches will not be pursued.

6.3.3 Decline rate analysis

The decline rate refers to the decrease in petroleum extraction over time. In many cases the decline rate is calculated on annual basis, yielding the change in produced volume from one year to another. See Eq. (6.10) for a general definition. It should be noted that the decline rate can be positive in some cases, representing an increasing production.

$$\text{Decline rate}_n = \frac{\text{Production}_n - \text{Production}_{n-1}}{\text{Production}_{n-1}} \tag{6.10}$$

The decline might be caused by politics, malfunctions, sabotage, depletion, and other factors. The driving force behind decline can be political or socioeconomic, representing man-made restrictions on the utilization of a reservoir. The decline can also be driven by natural forces, such as depletion of recoverable volumes within a reservoir and the resulting decline in reservoir pressure that diminishes the flow rates. In reality, the decline is often driven by several factors. Politics-driven decline usually disappears once the political tensions have been resolved, and this was clearly seen after the oil crises of the 1970s when the Middle East resumed their oil export to the western countries. Similarly, the economics-driven decline might be seen in fields where lack of payments, service, modernization, and investments has reduced the production flow. Also, in this case, decline usually disappears once more investments have been made or the economic situation returned to normal. The depletion-driven decline occurs when the recoverable resources become exhausted and the production flow is reduced due to the physical limitations of the reservoir. The depletion-driven decline is different from other forms of decline and much harder to compensate for since it can only be alleviated by expanding the recoverable reserves of the reservoir, which will ultimately be limited by the physical extent of the formation, permeability, or other geological parameters. Depletion is a key factor for the fluid flows within the reservoir and its connection to flow fundamentals makes it an important parameter for understanding oil production. In order to conceptually understand how depletion affects fluid flows, a simplified example can be considered. In gas fields, the ideal gas law and related special cases are often useful and pedagogic tools. One should also remember that the behavior of real gases deviates from ideal gases, notably at high pressures and temperatures. However, this can be handled with gas deviation factors (Satter et al., 2008). Boyle's Law, first formulated by Robert Boyle (1627–1691), describes the inverse proportionality of the absolute pressure and volume of a gas if the temperature is kept constant within an isolated system (Eq. 6.11). This is often applicable in gas reservoirs, since they are reasonably isolated and in thermal equilibrium with the surrounding bedrock, resulting in constant temperature.

$$\text{Pressure} * \text{Volume} = \text{Constant} \tag{6.11}$$

The law can also be rewritten into a relationship between pressures and volumes before and after a certain isothermal change

$$p_1 v_1 = p_2 v_2 \tag{6.12}$$

where

p_1, p_2 = pressure of gas at states 1 and 2, respectively.
v_1, v_2 = specific volume of gas at states 1 and 2, respectively.

Gas extraction removes mass without changing the volume of gas in the different states, i.e.. As a result, pressure must fall to maintain balance. From Darcy's law (Eq. 6.3) it follows that decreasing pressure leads to decreased flow rates if all other things are equal. Consequently, extraction of gas from a reservoir will result in declining production with time, in other words, a depletion-driven decline. The situation becomes more complicated with oil extraction or other forms of production strategies, but the general situation is the same as in the simplified gas reservoir case. In fields where the production strategy is to maintain reservoir pressure, for instance by water or gas injection, the extracted volumes of oil and water will remain relatively constant through the life of the field, in agreement with the material balance equation (Satter et al., 2008). However, the oil production will ultimately fall and water production increase as more and more injected water begins to diffuse into the production wells. As the reservoir depletes, the well will eventually produce too much water to be economically viable, despite the fact that reservoir pressure might still be high. The ratio of water compared to the volume of total liquids produced is referred to as water cut. In mature fields, the water cut can reach very high levels, up to 90% and more has been recorded in parts of China. American oil fields in Kern Bluff and Mount Poso areas reports water cut of over 99% (California Department of Conservation, 2007).

6.3.4 Decline curves

Arps (1945) created the foundation of decline curve analysis by proposing simple mathematical curves, i.e., exponential, harmonic or hyperbolic, as a tool for creating a reasonable outlook for the production of an oil well once it has reached the onset of decline. His original approach has later been developed further and is still used as a benchmark for industry for analysis and interpretation of production data due to its simplicity (II, III). It should also be noted that there is a strong connection between the physical models for reservoir flows and empirical simplifications based on decline curves. The exponential curve, introduced by Arps (1945), is actually the analytical long-term solution to the flow equation of a well with constant bottom hole flowing pressure (Hurst, 1934; van Everdingen and Hurst, 1949). The biggest advantage of decline curve analysis is that it is virtually independent of the size and shape of the reservoir or the actual drive mechanism (Doublet et al., 1994), thus avoiding the need for detailed reservoir or production data. The only data required for decline curve analysis and extrapolation is production data, which is relatively easy to obtain for a large number of fields. Decline curves of various forms can be used to create reasonable outlooks for the fluid production of a single well or an entire field. However, it should be emphasized that in many field cases a single curve is not sufficient to obtain a good fit and it may be necessary to use a combination of curves to obtain good agreement (Haavardsson and Huseby, 2007). The importance of individual fields diminishes as the total number of studied fields becomes large and generalized field behavior can be identified (II, III). In such cases, a simple decline curve can successfully be used to forecast total production from a large set of fields, as under- and overestimations for individual fields cancel out each other in the long run. Consequently, decline rate analysis and decline curves can be a convenient tool for identifying long-term trends and projecting reasonable production behavior into the future. The Arps decline curves are simplistic and focused on obtaining expressions with mathematical tractability that could be utilized in a simple and straightforward manner. In the models, it is assumed that the declining production starts at a given time, t_0 with an initial production rate of r_o and the initial cumulative production, Q_o. The production rate at time $t > t_0$ is denoted by $q(t)$ and the corresponding cumulative production at the same time is defined by the integral $Q(t) = \int_{t_0}^{t} q(u)du$.

The simplest decline curves are characterized by three parameters, the initial production rate $r_0 > 0$, the decline rate $\lambda > 0$ and the shape parameter $\beta \in [0,1]$. If the production is allowed to continue without end and the integral $Q(t) = \int_{t_0}^{t} q(u)du$ converges as $t \to \infty$ it is possible to calculate the ultimate cumulative production of the decline phase, which can be summed with Q_0 to give the fields URR. Normally production is stopped when the economic/energetic limit is reached, i.e. when keeping the equipment running requires more money and/or energy than it yields. This cut-off point can be denoted $r_c < r_0$ and is found by solving $q(t) = r_c$ with respect to t, where the solution occurs at t_{cut}. By inserting t_{cut} as the upper limit for $Q(t)$, one can now calculate the technically recoverable volume, denoted by V_{rec}.

The key properties of the Arps exponential and harmonic decline curves can be seen in Table 6.3. The generalized hyperbolic case is described in Table 6.4. Note that the exponential and harmonic curves only are special cases of general hyperbolic decline. Modification of the shape parameter β can alter the shape of the production rate function and be used to determine what kind of decline curve is suitable for fitting against empirical data. The value of the decline parameter λ governs how steep the decrease in production will be. The exponential decline curve is by far the most convenient to work with and still agrees well with actual data. Hyperbolic and harmonic decline curves involve more complicated functions and are, consequently, less practical to utilize. The disadvantage of the exponential decline curve is that it sometimes tends to underestimate production far out in the tail part of the production curve, as decline often flattens out towards a more harmonic and hyperbolic behavior in that region. Production from individual fields normally rises to a peak or plateau, after which it declines as a result of falling pressure and/or the breakthrough of water. While each field will have a unique (and not necessarily smooth) production profile as a result of both its physical characteristics and the manner in which it is developed and managed, the same broad pattern is generally observed. As an example, Fig. 6.4 shows the production history of the Thistle field

TABLE 6.3 Key properties of Arps exponential and harmonic decline curves.

	Exponential	Harmonic
β	$\beta = 0$	$\beta = 1$
$q(t)$	$r_0 \exp(-\lambda(t-t_0))$	$r_0[\exp(-\lambda(t-t_0))]^{-1}$
$Q(t)$	$Q_0 + \frac{r_0}{\lambda}(1 - \exp(-\lambda(t-t_0)))$	$Q_0 + \frac{r_0}{\lambda} \ln(1 + \lambda(t-t_0))$
URR	$Q_0 + \frac{r_0}{\lambda}$	Not defined
t_{cut}	$t_0 + \frac{1}{\lambda} \ln\left(\frac{r_0}{r_c}\right)$	$t_0 + \frac{1}{\lambda}\left[\frac{r_0}{r_c} - 1\right]$
V_{rec}	$Q_0 + \frac{r_0 - r_c}{\lambda}$	$Q_0 + \frac{r_0}{\lambda} \ln\left(\frac{r_0}{r_c}\right)$

TABLE 6.4 Key properties of Arps generalized hyperbolic decline curve.

	Hyperbolic
β	$\beta \in [0,1]$
$q(t)$	$r_0[1 + \lambda\beta(t-t_0)]^{-1/\beta}$
$Q(t)$	$Q_0 + \frac{r_0}{\lambda(1-\beta)}\left[1 - \left(1 + \lambda\beta(t-t_0)^{1-\frac{1}{\beta}}\right)\right]$
URR	$Q_0 + \frac{r_0}{\lambda(1-\beta)}$
t_{cut}	$t_0 + \frac{1}{\lambda\beta}\left[\left(\frac{r_0}{r_c}\right)^\beta - 1\right]$
V_{rec}	$Q_0 + \frac{r_0}{\lambda(\beta-1)}\left[\left(\frac{r_0}{r_c}\right)^{\beta-1} - 1\right]$

in the North Sea (Sorrell et al., 2009). Second, most of the oil in a region tends to be located in a small number of large fields, with the balance being located in a much larger number of small fields. This pattern can be observed at all levels of aggregation from individual basins to the entire world. For example, the IEA (2008) estimates that there are some 70,000 oilfields in production worldwide, but in 2007 approximately half of global production derived from only 110 fields, one quarter from only 20 fields and as much as one-fifth from only 10 fields. Indeed, as much as 7% of production is derived from a single field—Ghawar in Saudi Arabia. Around 500 "giant" fields account for around two-thirds of all the crude oil that have ever been discovered. Third, these large fields tend to be discovered relatively early in the exploration history of a region, in part because they occupy a larger surface area. Subsequent discoveries tend to be progressively smaller and often require more effort to locate. Again, this broad pattern can be observed at all levels, although it is always modified by technical, political, and economic factors, such as restrictions on the areas available for exploration. As an illustration, over half of the world's "giant" fields were discovered.

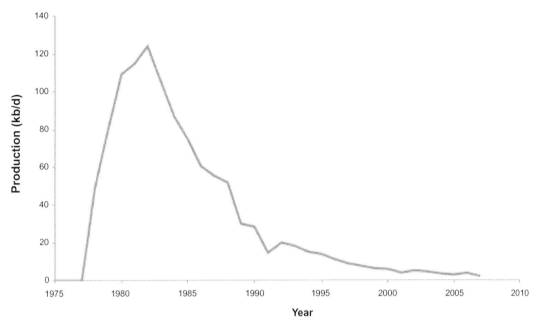

FIG. 6.4 Production cycle of the Thistle field in the North Sea.

Third, these large fields tend to be discovered relatively early in the exploration history of a region, in part because they occupy a larger surface area. Subsequent discoveries tend to be progressively smaller and often require more effort to locate. Again, this broad pattern can be observed at all levels, although it is always modified by technical, political, and economic factors, such as restrictions on the areas available for exploration. As an illustration, over half of the world's "giant" fields were discovered. An assessment of the evidence for a near-term peak in global oil production 7 more than fifty years ago, while less than one-tenth have been discovered since 1990 (Robelius, 2007). The implications of these features of the oil resource can be illustrated with the help of a simple model (Fig. 6.5) (Bentley and Boyle, 2007). Here, each triangle represents the production from a single field, with one field being brought into production each year. It is assumed that fields are developed in declining order of size, with each field being 10% smaller than the previous. The result is that, at some point, the additional production from the small fields that were discovered relatively late becomes insufficient to compensate for the decline in production from the large fields that were discovered relatively early, leading to a regional peak in production. Under these assumptions, the peak occurs when around one-third of the resources in the region have been produced. Since it also occurs when there are large quantities of reserves in the producing fields, the reserve to production (R/P) ratios are relatively stable and new fields are continuing to be discovered, the peak may not necessarily be anticipated.

Assume that the formation is homogeneous, isopachous, and isotropic, the reservoir is filled with slightly compressible fluid with constant compressibility, the flow behavior is

630　　　　　　　　　　　6. Reserves prediction and deliverability

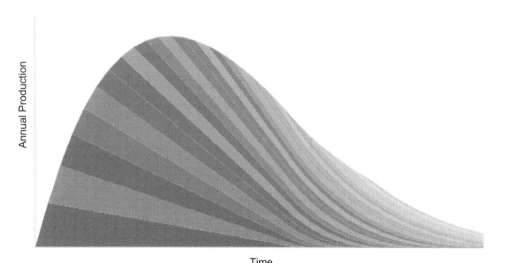

FIG. 6.5　Stylised model of a regional peak in oil production.

isothermal and follows Darcy's law, and that the gravity and capillary force are negligible. Under these fundamental assumptions, the definite solution problem of a single well producing at a constant rate can be described as follows under basic SI.

The governing equation for the linear flow of a single-phase fluid through a pressure-sensitive formation in the Cartesian coordinate system is a partial-differential equation, as shown in Eq. (6.1),

$$\frac{\partial}{\partial x}\left[\frac{k(p)}{\mu(p)B(p)}\frac{\partial p}{\partial x}\right] = \frac{1}{d_1}\frac{\phi(p)c_t(p)}{B(p)}\frac{\partial p}{\partial t}, \quad (6.13)$$

where k is the permeability, μ is the fluid viscosity, B is the fluid formation volume factor (FVF), $d_1 = 2.64 \times 10^{-4}$ is the unit-conversion factor, ϕ is the porosity, $c_t(p) = c_f(p) + c_r$ is the total compressibility, cf. and cr are the fluid compressibility and the rock compressibility, respectively, and the fluid can be gas, oil, or water.

Eq. (6.13) is a nonlinear diffusivity equation that is universally applicable to both single-phase slightly compressible liquid and compressible-gas flow through a pressure-sensitive matrix or fracture. The initial and boundary conditions for Eq. (6.13) are given as

$$p(x, 0) = p_i, \quad p(0, t) = p_{wf}, \quad p(x \to \infty, t) = p_i. \quad (6.14)$$

To lump pressure-dependent parameters in Eq. (6.13), the diffusivity coefficient and transmissibility are defined as

$$\eta = d_1\frac{k(p)}{\phi(p)c_t(p)\mu(p)}, \quad T = d_2\frac{k(p)}{\mu(p)B(p)}, \quad (6.15)$$

where η is the diffusivity coefficient, T is the transmissibility, and d_2 is the unit-conversion factor for transmissibility, which equals to 1.127×10^{-3} and 6.329×10^{-6} for slightly compressible liquid and compressible gas, respectively.

$$\eta \left[\frac{1}{r}\frac{\partial}{\partial r}\left(r\frac{\partial p}{\partial r}\right)\right] = \frac{\partial p}{\partial t}$$
$$\text{where } \eta = \frac{K}{\varphi \mu C_t}.$$
(6.16)

The aforementioned problem during the flow period can be divided into three stages:

Infinite acting radial flow, which refers to the period of time after the flow period when the pressure wave spreads outwards but does not reach the boundary. At this time, the boundary has not yet been touched, and the formation can be regarded as an infinite formation. As to a large reservoir, the infinite acting radial flow period is long, while as to a micro-reservoir, the infinite acting radial flow period is short.

Transition stage, which refers to the period of time after the flow period when the pressure wave spreads outwards and has reached the boundary, while the change in pressure in the vicinity of the outer boundary has not yet been stabilized.

Steady-state or pseudo-steady state flow, which refers to the period of time after the pressure variation all over the formation has been stabilized. As to the state of constant pressure outer boundary, the pressure all over the reservoir does not change, while as to the state of closed outer boundary, the pressure all over the reservoir uniformly drops with time.

6.3.5 Steady-state flow

If a well has been producing at a constant rate for a long time, and if the pressure distribution around the whole wellbore remains constant, the flow state is termed as steady-state flow. The time of commencement of steady-state flow is denoted by t_{ss}. When $t > t_{ss}$, the pressure at any point of the formation does not change with time, that is, $\partial p/\partial t = 0$.

As to the natural water drive reservoir with significant surrounding water or the water flooding reservoir, the oil-water boundary may possibly form a "constant pressure boundary," making the well production stay at a steady flow state or approximately reach a steady flow state. The pressure distribution in the vicinity of such a well is shown in Fig. 6.6. It must be pointed out that as to gas wells, such a steady flow state generally would not occur, and the so-called "constant pressure boundary" would not exist either, because of a large difference of gas and water viscosities, and it is usually characterized by a composite model in which the outer region becomes poor.

6.3.6 Pseudo-steady state flow

The so-called pseudo-steady state flow virtually is a kind of transient flow. After the wells in a closed reservoir have been producing at a stable flow rate for a certain time, the pressure spreads to all the boundaries around. Thereafter, the pressure at all points of the closed reservoir will drop at the same speed and then a pseudo-steady state flow occurs, as shown in Fig. 1.3.

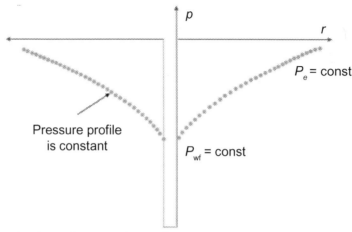

FIG. 6.6 Pressure distribution during steady-state flow.

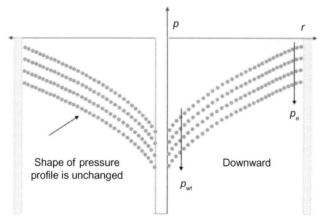

FIG. 6.7 Pressure distribution during pseudo-steady state flow.

If the time of commencement of pseudo-steady state flow is t_{ss}, when $t > t_{pss}$, at any point in the formation, $\partial p/\partial t = $ Const. As observed from Fig. 6.7, the shape of the pressure distribution curve remains unchanged at this time, where the pressure distribution curves of different times parallel to each other and only their heights are different.

In radial coordinate, one obtains

$$\frac{\partial^2 p}{\partial r^2} + \frac{1}{r}\frac{\partial p}{\partial r} = \frac{\phi \mu c_1}{0.000264\, k}\frac{\partial p}{\partial t} \tag{6.17}$$

where

k = permeability, md
r = radial position, ft
p = pressure, psia
c_t = total compressibility, psi^{-1}
t = time, hrs
ϕ = porosity, fraction
μ = viscosity, cp

When the reservoir contains more than one fluid, total compressibility should be computed as

$$c_t = c_o S_o + c_w S_w + c_g S_g + c_f \tag{6.18}$$

where c_o, c_w, and c_g refer to the compressibility of oil, water, and gas, respectively, while S_o, S_w, and S_g refer to the fractional saturation of these fluids. Note that the introduction of c_t into Eq. (6.5) does not make Eq. (6.5) applicable to multiphase flow; the use of ct, as defined by Eq. (6.6), simply accounts for the compressibility of any immobile fluids that may be in the reservoir with the fluid that is flowing.

Liu and Hamid Emami-Meybodi (2021) proposed a unified relationship between pressure and fluid properties by using exponential functions applicable to any type of fluid, μ, leads to

$$\mu = \mu_i \exp(-\varepsilon_\mu p_D), \quad \text{with } \varepsilon_\mu = (a_0 + b_0 p_D)\Delta p, \tag{6.19a}$$

$$c_t = c_{ti} \exp(\varepsilon_c p_D), \quad \text{with } \varepsilon_c = (a_1 + b_1 p_D)\Delta p, \tag{6.19b}$$

$$B = B_i \exp(\varepsilon_B p_D), \quad \text{with } \varepsilon_B = (a_2 + b_2 p_D)\Delta p, \tag{6.19c}$$

where $p_D = (p - p_i)/(p_{wf} - p_i)$ is the dimensionless pressure, $\Delta p = (p_i - p_{wf})$ is the pressure drawdown, ε represents the pressure-dependent exponent for each fluid or rock property, and a_0, a_1, a_2, b_0, b_1, and b_2 are the fitting constants (in psi^{-1}).

The exponential functions are also adopted for the relations between pressure and rock properties,

$$k = k_i \exp(-\varepsilon_k p_D), \quad \text{with } \varepsilon_k = \gamma \Delta p, \tag{6.20a}$$

$$\phi = \phi_i \exp(-\varepsilon_\phi p_D), \quad \text{with } \varepsilon_\phi = c_r \Delta p, \tag{6.20b}$$

where γ is the permeability modulus.

Liu and Emami-Meybobi introduced linearization is done by introducing pseudo functions, which themselves have a setoff assumptions attached to them.

The pseudo pressure is commonly applied to linearize the pressure-dependent transmissibility on the space side of the diffusivity equation, which is defined as

$$\psi = \frac{\mu_i B_i}{k_i} \int_0^p d_2 \frac{k(p)}{\mu(p) B(p)} dp = \frac{1}{T_i} \int_0^p T(p) dp. \tag{6.21}$$

6. Reserves prediction and deliverability

By inserting Eq. (6.21) into Eq. (6.13), the pseudo-pressure-based diffusivity equation and the initial and boundary conditions are

$$\frac{\partial^2 \psi}{\partial x^2} = \frac{1}{\eta(\psi)} \frac{\partial \psi}{\partial t}, \tag{6.22a}$$

$$\psi(x, 0) = \psi_i, \quad \psi(0, t) = \psi_{wf}, \quad \psi(x \to \infty, t) = \psi_i, \tag{6.22b}$$

$$\psi(x, 0) = \psi_i, \quad \psi(0, t) = \psi_{wf}, \quad \psi(x \to \infty, t) = \psi_i.$$

$$\psi_D(x_D, 0) = 0, \quad \psi_D(0, t_D) = 1, \quad \psi_D(x_D \to \infty, t_D) = 0.$$

Table 6.5 provides the definition of pseudovariables and related parameters in both dimensional and nondimensional forms. Based on the defined dimensionless parameters in Table 6.5, Eq. (6.22) in nondimensional form can be written as

$$\frac{\partial^2 \psi_D}{\partial x_D^2} = \frac{1}{\eta_D(\psi_D)} \frac{\partial \psi_D}{\partial t_D}, \tag{6.23a}$$

$$\psi_D(x_D, 0) = 0, \quad \psi_D(0, t_D) = 1, \quad \psi_D(x_D \to \infty, t_D) = 0. \tag{6.23b}$$

TABLE 6.5 Summary of the dimensional and dimensionless variables.

Parameters	Dimensional	Dimensionless
Pressure	p	$p_D = \dfrac{p - p_1}{p_{wf} - p_1}$
Time	t	$t_D = \dfrac{\eta_1 t}{W^2}$
Distance	x	$x_D = \dfrac{x}{W}$
Diffusivity coefficient	$\eta = \dfrac{k}{\phi c_t \mu}$	$\eta_D = \dfrac{\eta}{\eta_1}$
Transmissibility	$T = \dfrac{k}{\mu B}$	$T_D = \dfrac{T}{T_1}$
Pseudopressure	$\psi = \dfrac{1}{T_1} \int_0^p T(p) dp$	$\psi_D = \dfrac{\psi - \psi_1}{\psi_{wf} - \psi_1}$
Pseudotime	$\tau = \dfrac{1}{\eta_1} \int_0^t \eta(\bar{p}) dt$	$\tau_D = \int_0^{t_D} \eta_D(\bar{\psi}_D) dt_D$
Flow rate	q	$q_D = \dfrac{q_{sc}}{H(\psi_1 - \psi_{wf})}$

By inserting Eqs. (6.23a) and (6.23b) into Eq. (6.22a), we can relate the dimensionless diffusivity coefficient and transmissibility to the dimensionless pressure by the exponential functions, which is

$$\eta_D = \exp[-(\alpha_0 p_D + \alpha_1 p_{2D})], \text{with } \alpha_0 = (\gamma - c_r + a_1 - a_0)\Delta p,$$
$$\alpha_1 = (b_1 - b_0)\Delta p_\eta D = \exp[-\alpha 0 pD + \alpha 1 pD2], \text{with } \alpha 0 = \gamma - cr + a1 - a0\Delta p, \quad \alpha 1 = b1 - b0\Delta p \quad (6.24a)$$

$$T_D = \exp[-(\beta_0 p_D + \beta_1 p_{2D})], \text{with } \beta_0 = (\gamma + a_2 - a_0)\Delta p,$$
$$\beta_1 = (b_2 - b_0)\Delta p T_D = \exp[-\beta 0 pD + \beta 1 pD2], \text{with } \beta 0 = \gamma + a2 - a0\Delta p, \quad \beta 1 = b2 - b0\Delta p \quad (6.24b)$$

where $\alpha 0$, $\alpha 1$, $\beta 0$, and $\beta 1$ are exponents for pressure-dependent diffusivity coefficient and transmissibility, respectively. As discussed previously, for a slightly compressible fluid, the fitting constants $a_1 = b_0 = b_1 = b_2 = 0$ and a_0 and a_2 are conceived as $c\mu$ and cl, respectively. Therefore, the pressure-dependent exponents can be expressed as $\alpha 0 = (\gamma - cr - c\mu)\Delta p \alpha 0 = \gamma - cr - c\mu\Delta p$, $\beta 0 = (\gamma + cl - c\mu)\Delta p \beta 0 = \gamma + cl - c\mu\Delta p$, and $\alpha 1 = \beta 1 = 0 \alpha 1 = \beta 1 = 0$.

All constants are summarized and exponents introduced in the developed exponential function in Table 6.6. According to Eq. (6.24a), $\alpha 0$ is positively correlated with permeability modulus and the pressure dependence of fluid compressibility (a_1) and negatively correlated with rock compressibility and the pressure dependence of viscosity (a_0). Therefore, $\alpha 0 > 0 \alpha 0 > 0$ when the sum of γ and the pressure dependence of ct are higher than the sum of cr and the pressure dependence of μ. In such a case, η increases with pressure. In contrast, η decreases as pressure increases when $\alpha 0 < 0 \alpha 0 < 0$. As the absolute value of $\alpha 0$ increases, the difference between the pressure dependence of kk and $ctct$ and the pressure

TABLE 6.6 Summary of pressure-dependent constants used in proposed exponential functions for pressure-dependent fluid and rock properties.

Parameters	Description	Equation
a_0, b_0	Fitting constants for the exponent of pressure-dependent viscosity ε_μ	Eq. (6.19a)
a_1, b_1	Fitting constants for the exponent of pressure-dependent total compressibility ε_c	Eq. (6.19b)
a_2, b_2	Fitting constants for the exponent of pressure-dependent FVF ε_B	Eq. (6.19c)
ε_μ	Exponent for pressure-dependent viscosity μ	Eq. (6.19a)
ε_c	Exponent for pressure-dependent total compressibility c_t	Eq. (6.19b)
ε_B	Exponent for pressure-dependent FVF B	Eq. (6.19c)
ε_k	Constant exponent for pressure-dependent Permeability k	Eq. (6.20a)
ε_ϕ	Constant exponent for pressure-dependent porosity ϕ	Eq. (6.20b)
α_0, α_1	Exponents for pressure-dependent diffusivity coefficient η	Eq. (6.24a)
β_0, β_1	Exponents for pressure-dependent transmissibility T	Eq. (6.24b)

dependence of ϕ and μ increases. The definition in Eq. (6.24a) also indicates that $\alpha 1$ is a higher-order pressure dependence of ηD and affected by the pressure dependence of ct (b_1) and μ (b_0). The increase in the absolute value of $\alpha 1$ indicates that the difference of the pressure dependence of ct and μ increases. Similarly, Eq. (6.24b) reveals that $\beta 0$ is positively correlated with γ and the pressure dependence of B (a_2) and negatively correlated with the pressure dependence of μ (a_0). As defined in Eq. (6.24b), $\beta 1$, representing the higher-order pressure dependence of TD, is affected by the difference of the pressure dependence of B and μ.

We defined the NL as the root mean square of ηD deviation from its initial value, $\eta D = 1$, with respect to dimensionless pseudo pressure ψD,

$$NL = \sqrt{\int_0^1 [\eta_D - 1]^2 d\psi_D}. \tag{6.25}$$

By combining Eqs. (6.25), (6.26), the nonlinearity can be calculated as

$$NL = \sqrt{\frac{\int_0^1 T_D[\eta_D - 1]^2 dp_D}{\int_0^1 T_D dp_D}} = \sqrt{\frac{\int_0^1 \exp\{-[(2\alpha_0 + \beta_0)p_D + (2\alpha_1 + \beta_1)p_D^2]\}dp_D - 2\int_0^1 \exp\{-\}}{\int_0^1 \exp[-(\beta_0 p_D + \beta_1 p_D^2)]dp}}$$

$$\int_0^1 \exp[-(F_0 p_D + F_1 p_D^2)]dp_D = \begin{cases} \dfrac{\sqrt{\pi}}{2\sqrt{F_1}} \exp\left(\dfrac{F_0^2}{4F_1}\right)\left[\mathrm{erf}\left(\dfrac{2F_1 + F_0}{2\sqrt{F_1}}\right) - \right. \\ -\dfrac{1}{F_0}[\exp(-F_0) - 1], F_1 = 0 \\ \dfrac{\sqrt{\pi}}{2\sqrt{-F_1}} \exp\left(\dfrac{F_0^2}{4F_1}\right)\left[\mathrm{erfi}\left(\dfrac{-2F_1 - F_0}{2\sqrt{-F_1}}\right.\right. \end{cases} \tag{6.26}$$

where F_0 represents $2\alpha_0 + \beta_0, \alpha_0 + \beta_0$, or β_0, F_1 represents $2\alpha_1 + \beta_1$

where F_0 represents $2\alpha 0 + \beta 0 2\alpha 0 + \beta 0$, $\alpha 0 + \beta 0 \alpha 0 + \beta 0$, or $\beta 0$, F_1 represents $2\alpha 1 + \beta 1 2\alpha 1 + \beta 1$, $\alpha 1 + \beta 1 \alpha 1 + \beta 1$, or $\beta 1$, erf is the error function, and $erfi$ is the imaginary error function.

For the slightly compressible fluid ($\alpha 1 = \beta 1 = 0 \alpha 1 = \beta 1 = 0$), the nonlinearity equation given by Eq. (6.26) is simplified to

$$N_L = \sqrt{\frac{\beta_0}{1 - e^{\beta_0}}\left(\frac{e^{-2\alpha_0} - e^{\beta_0}}{2\alpha_0 + \beta_0} - \frac{2e^{-\alpha_0} - 2e^{\beta_0}}{\alpha_0 + \beta_0}\right) + 1}. \tag{6.27}$$

Based on Eqs. (6.26), (6.27), the limiting case of $NL = 0$ can be found when $\alpha 0 = \alpha 1 = \beta 1 = 0 \alpha 0 = \alpha 1 = \beta 1 = 0$. In addition, when $\alpha 0$ approaches $+\infty$, NL reaches the limit of $NL = 1$. However, when $\alpha 0$ approaches $-\infty$, NL increases rapidly with $\alpha 0$ without any limit. The sensitivity of NL to the exponents will be explained in more detail in the section

Results and Discussion. The typical range of the fitting constants and exponents for different fluid and rock properties will also be discussed in that section.

The dimensionless diffusivity equation, Eq. (6.28), is still a nonlinear partial-differential equation because of the pseudo-pressure-dependent diffusivity coefficient, ηD. The pseudo-time approach is commonly used to account for the nonlinearity associated with the diffusivity coefficient ηD on the time side of the pseudo-pressure-based diffusivity equation in Eq. (6.27). The pseudotime is normally defined by evaluating the diffusivity coefficient at an average pressure throughout the entire formation at each timestep (Liu and Emami-Meybodi, 2021); that is,

$$\tau_D = \int_0^{t_D} \eta_D(\overline{\psi}_D) dt_D. \tag{6.28}$$

Inserting Eq. (6.27) into Eq. (6.28) gives a linearized dimensionless pseudovariable-based diffusivity equation,

$$\frac{\partial^2 \psi_D}{\partial x_D^2} = \frac{\partial \psi_D}{\partial \tau_D}, \tag{6.29}$$

$$\psi D(xD, 0) = 0, \quad \psi D(0, \tau D) = 1, \quad \psi D(xD \to \infty, \tau D) = 0. \psi D x D, 0 = 0, \quad \psi D 0, \tau D = 1,$$

$$\psi D x D \to \infty, \tau D = 0. \tag{6.30}$$

The solution for Eq. (6.31) is

$$\psi_D = \operatorname{erfc}\left(\frac{x_D}{2\sqrt{\tau_D}}\right). \tag{6.31}$$

Applying Darcy's law, we can calculate the volumetric flux at the inner boundary as

$$q_{sc} = A_c T(p) \frac{\partial p}{\partial x}\Big|_{x=0} = A_c T_i \frac{\psi_i - \psi_{wf}}{\sqrt{\pi \eta_i^\tau}}. \tag{6.32}$$

By rearranging Eq. (6.32), one can obtain the rate-normalized pseudo pressure (RNP) = $(\psi i - \psi wf)/qsc = m\tau - \sqrt{} = (\psi i - \psi wf)/qsc = m\tau$, where $m = \pi \eta i \sqrt{A_c T i} m = \pi \eta i A_c T_i$ is the slope of the straight line in the specialty plot of RNP vs. $\tau - \sqrt{\tau}$. Accordingly, the cross-sectional area can be obtained from
or the product of the cross-sectional area and the square root of the initial permeability,

$$A_c \sqrt{k_i} = \frac{B_i}{m} \sqrt{\frac{\pi}{\phi_i c_{ti}}}. \tag{6.33}$$

6.4 Inclusion of fluid memory

This section explores the memory concept in the context of transport mechanisms in porous media. The author undertakes a thorough literature review of published articles related to the history, its definition, advantages, limitations, solution techniques, and experimental studies based on memory formalism. Results show that a vast amount of numerical and theoretical articles aimed at understanding fluid flow behavior for Newtonian (water) and non-Newtonian fluids (polymers and heavy crude oil) using the memory formalism exist but are not included in a reservoir simulator. Furthermore, limited experimental works have been reported in comparison to analytical/numerical studies. Finally, some future potential research areas are identified to fill the gap of the current state-of-the-art in the area. This chapter contributes to the fundamental understanding and the further development of fluid flow models that accurately describe the local and global phenomena.

Rheology is simply defined as the science of deformation and flow. Likewise, Newtonian fluid mechanics is primarily the investigation of the interplay of inertial and viscous forces and, where applicable, surface tension. As eluded to, many common or industrial fluids display unusual behavior due to their complex microstructure. Such fluids range from suspensions (e.g., bread dough or concrete), foams, granular media (such as sand or coal) to polymeric fluids such as oils, molten plastics, paint, blood, or egg white. These fluids behave differently from Newtonian fluids due to the presence of long-chain molecules, which affect the flow behavior in the way which they align to the motion of the fluid. They are stretched out by the drag forces, and consequently want to retract back to their unstressed configuration in an elastic behavior.

All fluid motion is governed by the law of conservation of mass, as well as the law of transfer of linear and angular momentum; and these laws are sometimes supplemented by the energy balance equation if thermal effects are considered. Assuming an incompressible fluid, the conservation of mass is stated simply by:

$$\nabla . v = 0, \tag{6.34}$$

From the conservation of linear momentum using the Navier-Stokes equation we have:

$$\rho \left(\frac{\partial v}{\partial t} + (v.\nabla)v \right) = -\nabla p + \nabla . T, \tag{6.35}$$

T represents the fluid stress developed in response to the deformation, with the total stress tensor given by:

$$\sigma = -pI + T, \quad \sigma = \sigma^T, \tag{6.36}$$

The conservation of angular momentum imposes that σ is a symmetric tensor.

In most fluid flow problems, the conservation equations are not sufficient to determine the unknowns; hence, constitutive equations or relations need to be introduced to relate the fluid momentum to the stress tensor. A viscous fluid is a fluid that resists forces exerted upon it through internal friction. Most fluids with small molecules such as gases and water obey this model. For a complete review of Newtonian fluids and the definition of the aforementioned terms (coefficients) refer to published texts on fluid mechanics by Kundu et al. (2015).

However, a viscoelastic fluid shows a combination of viscous and elastic characteristics when subjected to deformation. These fluids show some distinguished characteristics such as hysteresis, stress relaxation, and creep.

In addition, the stress tensor for viscoelastic fluids depends on the previous history (memory) of fluid motion and the current motion of the fluid. There are numerous methods and models including mechanical analogues such as Maxwell's one-dimensional linear model

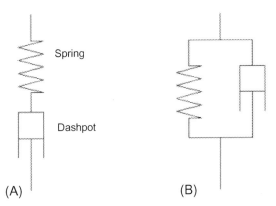

FIG. 6.8 Representations of (A) Maxwell element (Sibley 2010) (B) Kelvin-Voigt element (Sibley 2010).

where the fluid is modeled for both viscous elements and elastic elements. Other models include Kelvin-Voigt linear model (Fig. 6.8). Most of the proposed mathematical models of viscoelasticity are described either in differential or integral form. The differential representations have been preferred to the integral forms in many applications due to their simplicity in implementing numerical models. However, the integral representations predict the time dependence of viscoelastic fluids better.

Naturally occurring fluids show a nonlinear behavior and are usually associated with large deformations, with nonlinear response in the presence of such deformation (Cherizol et al., 2015).

6.4.1 Constitutive equations

There are numerous constitutive relations available in the literature. Here, only the most popular models are listed.

1. Upper-Convected-Maxwell Model.
2. K-BKZ Model.
3. Giesekus-Leonov Model.
4. White-Metzner Model.
5. Phan-Thien-Tanner Model.

Table 6.7 presents a summary of the above models, their advantages, and their limitations in general.

6.4.2 Application of memory in diffusion in porous media

The increase in global energy demand over the past few decades has heralded numerous studies and research devoted to understanding the diffusion process in a porous medium within an oil and gas reservoir. Just as important are water shortages that occur in some regions of the world, which have contributed to the increasing need for research devoted to the

TABLE 6.7 Summary of available viscoelastic models.

Constitutive model	Description	Advantages	Disadvantages	Applications
Upper-Convected-Maxwell Model. (1950)	A differential type model that describes the flow of polymers with linear or quasi-linear viscoelasticity.	Recaptures all of the linear viscoelastic modeling. Reduces to Newtonian/Neo-Hookean behavior for the limiting cases of slow and fast flow respectively. Predicts first normal stress difference and extensional thickening.	Does not predict second normal stress difference. The extensional thickening predicted is too severe. Assumes constant viscosity.	The behavior of a viscoelastic fluid under large deformations. Boger fluids.
K-BKZ Model. (1962)	An integral type model that combines linear viscoelasticity and nonlinear elasticity in terms of a memory integral constitutive law.	Ease of usability for material data selection. Guaranteed good fit with many experimental data.	The integral constitutive equations pose significant Numerical computation difficulty.	Prediction of stresses within fibrous suspensions.
Oldroyd-B Model. (1950)	This differential type model adds an extra Newtonian component to the previous UCM model.	Same as UCM. Simple to implement in simulations.	Unphysical singularity in extensional flow. Not suitable for injection molding process due to the lack of shear-thinning characteristic.	Predicting the rheological characteristics of polymer liquids.
Giesekus-Leonov Model. (1966 &1982)	A differential type model that accounts for quadratic nonlinearity. The model reduces the deviatoric stress component into a solvent contribution and a polymer contribution.	Provides excellent fits for shearing flow. Gives more realistic results compared to the linear models.	Limitations in predicting or describing the extensional flow.	Prediction of tension thickening regions during elongational flow.
White-Metzner Model (1963)	A differential type model, and an extension of the Maxwell model since it allows for the incorporation of experimental data. i.e., Viscosity as a function of shear rate.	Ease of selection of material parameters.	Dependence on experimental data.	Fins wide application in shear-thinning fluids. Suitable for fluids with fast time-dependent motions.

TABLE 6.7 Summary of available viscoelastic models—cont'd

Constitutive model	Description	Advantages	Disadvantages	Applications
Phan-Thien-Tanner Model. (1977 & 1978)	Differential type model that is similar in form to Giesekus Model. The difference stems from the additional nonlinear term.	Fits data reasonably well for a variety of different types of deformation. A robust method for numerical simulations.	Spurious oscillations are observed in the start-up of steady shearing when $\xi \neq 0$. Fails to predict accurately both the shear viscosity as well as transient and extensional properties at higher shear rates.	Finds application in the simulation of polymer solution flows.

flow of water in porous media (Caputo, 1998). Furthermore, the diffusion process is useful in understanding contaminant transport.

The basic law describing the diffusion process of fluids in any porous medium is the well-known Darcy's Law (Darcy, 1856). Many researchers have proposed different extensions to the classic Darcy's law by accounting for slip, inertia, and so on (Bear, 1975; Sposito, 1980; Brinkman, 1949; Whitaker, 1986). Likewise, many studies related to the diffusion of water in porous media have been well reported in the literature with different researchers following different approaches. However, most studies focused on the fluid pressure field as opposed to the flux of the fluid.

It is not uncommon for some fluids to react chemically with the porous medium leading to changes in the pores. Likewise, solid particles embedded with the reservoir fluids under suitable conditions can be deposited or attached along the pore throats, or pore walls. Furthermore, the pore size has been reported to be affected by temperature variations within the porous media. Furthermore, it has been reported in the literature that during steam injecting, alterations in fluid and rock properties do occur (Hossain et al., 2009).

6.4.3 Definition of memory

Memory was defined in simple terms as the effect of past events on the present and future course of developments (Hossain et al., 2007). Accordingly, any memory formulation should be at least a function of time and space. Caputo and Kolari (2001) in their study describe how memory functions capture the past events based on Eq. (6.37). They concluded that the contribution of variable f at any time t, is a weighted mean of its past values.

$$\frac{\partial^n f(t)}{\partial t^n} = \frac{1}{\Gamma(1-n)} \int_0^t (t-\tau)^{-n} (df(\tau)/d\tau) d\tau, \tag{6.37}$$

In the aforementioned definition, it is assumed that $0 < n < 1$.

State-of-the-art Memory-based Models

The amount of scientific and engineering applications described by fractional calculus is huge. The concept of fractional derivatives incorporated into constitutive equations is not a new theme and there are plenty of examples to be found in the literature to model rheological properties of solids, frequency-independent quality factor Fennoscandian uplift, heat diffusion, and other fields of research in Zhang (2003). Likewise, fractional diffusion models have been successfully proposed to model sub/super-diffusive transport in the absence and presence of an external field (Metzler et al., 1999), tumor invasion (Iomin et al., 2004; Iomin 2005, 2006), and studies related to the dynamics of interfaces between nanoparticles and substrate (Chow, 2005).

Hristov (2013) reported on the application of the integral-balance method to diffusion models with memory terms expressed with weakly singular (power-law) kernels. In that study, examples were presented where the fading memory term was represented by Volterra integrals and by a time-fractional, Riemann-Liouville derivative. He focused on problems concerning diffusion in short times where the relaxation takes place. The diffusion equation presented was described mathematically by:

$$\frac{\partial u}{\partial t} = a_E \int_0^t \frac{1}{(t-\tau)^\alpha} \frac{\partial^2 u(x,\tau)}{\partial x^2} d\tau, \tag{6.38}$$

He derived approximate solutions to the aforementioned equation utilizing the Integral-Balance method (Heat-Balance Integral Method (HBIM) and Double-integration method (DIM)) and a semi-empirical method termed Frozen Front Approach (FFA).

Kolomietz (2014) applied the kinetic theory to the nuclear Fermi liquid. He investigated the effects of Fermi-surface distortion, the relaxation parameters, and memory terms on the nuclear dynamics of the Fermi liquid. He observed that the distortions lead to the scattering of particles on the Fermi surface and the relaxation of collective motion (Landau, 1959). Furthermore, he concluded that the memory effects depended on the relaxation time. The proposed non-Markovian Navier-Stokes equation is presented below.

$$m\rho \frac{\partial u_v}{\partial t} + m\rho \nabla_\mu u_v u_\mu + \nabla_v P + \rho \nabla_v \frac{\delta E_{opt}}{\delta \rho} \\ - \nabla_\mu \int_{t_0}^t dt' \exp\left(\frac{t'-t}{\tau}\right) P(r,\tau') \frac{\partial}{\partial t'} \Lambda_{v\mu}(r,t') = 0, \tag{6.39}$$

Interestingly, as $\tau \to 0$, i.e., a short relaxation time limit, Eq. (8.6) reduces to the classical viscous Navier-Stokes equation. Kolomietz concluded that the memory integral in Eq. (8.6) gives rise to both the time irreversible viscosity and the additional, time-dependent conservative force.

Balhoff et al. (2012) developed a pore-scale network model to understand non-Newtonian fluid flow in porous media. They were able to investigate the accuracy and efficiency of other popular equations used for modeling non-Newtonian fluids considering shear-thinning fluids. The effect of fluid yield stress and zero fluid yield stress were also considered. In addition, for yield-stress fluids, they were able to determine the required threshold pressure gradient required to initiate flow, and the qualitative pathway in which the fluid would travel at this threshold. Different approaches were proposed for determining the threshold gradient such as minimum threshold path (MTP) algorithms (Sochi and Blunt, 2008), the invasion percolation with memory algorithm (Kharabaf and Yortsos, 1997) (IPM), and path of minimum pressure (Sochi, 2010) (PMP).

Hristov (2013) proposed a new technique based on the final penetration depth assumption to obtain an approximate solution to the heat conduction equation with fading memory expressed by Jeffrey's kernel. He observed that the infinite speed of propagation of the flux is implicitly assumed for the conventional diffusion conservation equations. However, real-life processes do occur at infinite speed. Therefore, he introduced a damping function to correct the unrealistic nature of the conventional equations. For the heat conduction problem, he introduced a Volterra-type integral which has the damping function as was conceived by Cattaneo (1958). The proposed method consists of the classic Fourier's law, the heat flux, and its time derivative related to its history (Carillo, 2005). The heat flux and the relaxation (damping) function $F(x, t)$ are assumed to satisfy Eq. (8.7):

$$q(x,t) = -\int_0^\infty F(x,t)\nabla t(x,t-\tau)d\tau, \qquad (6.40)$$

Eq. (6.41) presents the final form of the modified energy conservation balance:

$$\frac{\partial T(x,t)}{\partial t} = a_1 \frac{\partial^2 T(x,s)}{\partial x^2} + \frac{a_2}{\tau}\int_0^t e^{\frac{t-s}{\tau}} \frac{\partial^2 T(x,s)}{\partial x^2} ds, \qquad (6.41)$$

Carillo et al. (2014) derived analytical solutions for the integro-differential equations describing a rigid heat conductor with memory. They were also able to prove the uniqueness of their solution. According to the authors, the introduction of memory effects provided an alternative way to account for non linearities in problems where linear models cannot be applied. Two different models were considered to understand the role of memory on some physical properties of the material. They concluded that for heat diffusion problems the heat flux was related to the temperature-gradient history. Likewise, for isothermal viscoelasticity problems, the stress tensor is related to the strain history. Likewise, other studies have been devoted to the properties of free energy functionals for materials with memory. [For my information, interested readers can reference herein (Coleman and Dill, 1973)].

Hristov (2011, 2013, 2014) investigated the start-up problem considering second-grade viscoelastic fluids. He proposed an integral balance solution by assuming a parabolic profile with an unspecified exponent. The stress tensor component relevant to the problem was considered as:

$$T_{xy} = \mu \frac{\partial u}{\partial y} + \alpha_1 D_t^\beta \frac{\partial u}{\partial y}, \qquad (6.42)$$

Hence, in the absence of body force the equation of motion reduces to:

$$\rho \frac{\partial u}{\partial t} = \mu \frac{\partial^2 u}{\partial y^2} + \alpha_1 D_t^\beta \frac{\partial^2 u}{\partial y^2}. \qquad (6.43)$$

where D_t^β is defined as the Riemann-Liouville operator. They concluded that the approximate solution had some hidden terms repressing the effects of the Newtonian viscosity and the visco-elasticity.

Sprouse (2010) proposed a numerical solution based on the short memory approach to solving a class of fractional diffusional heat equations using an explicit finite difference scheme. The short memory approach assumes that prior events up to a certain time can be neglected. The proposed technique was applied to a neural glial network to better describe the mechanism of extracellular ATP diffusion that results in calcium signaling between astrocytes modeled by Eq. (8.11).

$$\frac{du}{dt} = \alpha D^{1-\gamma}\nabla^2 u - \beta u. \tag{6.44}$$

Hossain and Islam (2009) investigated the influence of time and reservoir rock properties on cumulative oil production. They introduced a modified material-balance equation (MBE) by including a stress-strain formulation for both the porous rock and fluid. The authors claim that the proposed MBE can also be applied to fractured formations with dynamic features. Furthermore, an improvement of 5% in oil recovery was observed from the new MBE over the classic MBE. However, the proposed MBE requires accurate rock and fluid compressibility data obtained from laboratory measurements or from reliable correlations.

Hossain and Islam (2009) presented a thorough review of memory applications pertaining to fluid flow porous media problems. The authors showed how different researchers related fluid memory to various fluid properties; i.e., the stress, density, and free energies, and so on.

Hossain and Islam (2009) proposed an extension of the classic flow diffusivity equation by incorporating fluid and rock memory. This was derived by introducing the Caputo fractional derivative to the classic Darcy's Law. The authors believed the introduction was necessary to account for the variation of fluid and formation properties with time. The proposed model is a nonlinear integro-differential equation refer to Eqs. (6.45) and (6.46). Furthermore, they proposed an explicit finite difference scheme to solve the resulting nonlinear equation.

$$\frac{1}{\eta}\frac{\partial \eta}{\partial x}Z + c_f \frac{\partial p}{\partial x}Z + \frac{\partial}{\partial x}Z = \frac{\phi c_t}{\eta}\frac{\partial p}{\partial t}, \tag{6.45}$$

$$Z = \frac{\int_0^t (t-\xi)^{-\alpha} \left[\frac{\partial^2 p}{\partial \xi \partial x}\right] d\xi}{\Gamma(1-\alpha)}. \tag{6.46}$$

Hossain and Islam (2009) proposed a stress-strain relationship applicable to non-Newtonian fluid flow in porous media. They included temperature variations, surface tension, pressure variations, and fluid and rock memory. The authors investigated the effect of memory on the stress-strain curve assuming a homogeneous, isotropic porous media. They concluded that the stress-strain behavior was a strong function of time, distance, and the memory parameter. The proposed stress-strain model equations are represented below:

$$\tau_T = \frac{k\Delta p A_{yz}\Gamma(1-\alpha)}{\mu_0^2 \eta \rho_0 \phi y l} \times \left[\left(\frac{\partial \sigma}{\partial T}\frac{\Delta T}{\alpha_D M_a}\right)e^{\frac{E}{RT}}\right] \times \frac{\partial u_x}{\partial y}, \tag{6.47}$$

$$I = \int_0^t \frac{(t-\xi)^{-\alpha}}{k}\left(\frac{\partial c}{\partial \xi}\frac{\partial p}{\partial \xi} - \frac{c\partial k}{k\partial \xi}\frac{\partial p}{\partial \xi} + c\frac{\partial^2 p}{\partial \xi^2}\right)d\xi. \tag{6.48}$$

El-Shahed (2003) proposed an analytical solution for the equation of motion describing the vibrations produced by semilunar heart valves. He introduced a fractional derivative of order alpha (Caputo definition) into the governing semi-differential equation to model the vibrations of the valves. By applying Laplace transformation principles he presented a closed-form solution in terms of the Mittag-Leffler function. One of the advantages of his work lies in the simplicity of his solution, they are usually used for validating the results of numerical solutions, and also for a better fit of experimental data.

Iaffaldano et al. (2006) carried out experimental investigations to understand the permeability reduction observed during the diffusion of water in sand layers. Based on their results they concluded that the reduction in permeability observed was a result of grain rearrangement and compaction (leading to a reduction in porosity). This phenomenon was shown qualitatively previously by Elias and Hajash (1992). In addition, the authors proposed a modified diffusion model (refer to Eqs. 6.49, 6.50, and 6.51) applicable to porous media by incorporating fluid memory. The proposed diffusion model was able to give a good fit with the volumetric flux observed during their experiments.

$$\gamma q(x,t) = -\left[c + d\frac{\partial^n}{\partial t^n}\right]\nabla p(x,t), \qquad (6.49)$$

$$ap(x,t) = \alpha\rho(x,t), \qquad (6.50)$$

$$\nabla \cdot q(x,t) + \frac{\partial\rho(x,t)}{\partial t} = 0. \qquad (6.51)$$

Caputo (1998, 2000) proposed a modified form of Darcy's Law by introducing the time-fractional derivative to account for the local permeability changes in the porous media (refer to Eqs. 8.19, 8.20). According to him, the time-fractional derivative has the ability to capture all local variations in pressure, however, this modification is only recommended when considering local phenomena. Furthermore, he derived the pressure distribution in a fluid in a half-space under varying boundary conditions. Finally, he suggested a method to determine the two parameters defining his memory diffusion model.

$$A\left(\partial^a/\partial t^a\right)\nabla^2(\sigma_{kk} + 3p/B) = \left(\partial/\partial t\right)(\sigma_{kk} + 3p/B), \qquad (6.52)$$

$$A = \eta\left[\frac{2G(1-\nu_u)}{(1-2\nu)}\right]\left[\frac{B^2(1+\nu_u)^2(1-2\nu)}{9(1-\nu_u)(\nu_u-\nu)}\right]. \qquad (6.53)$$

Caputo and Plastino (2003) proposed a modified constitutive relation to better describe the diffusion process of fluids in porous media. They proposed a space fractional derivative of pressure be introduced to Darcy's Law (refer to Eqs. 8.21, 8.22). In this model, the memory effect is introduced through the space fractional derivative, and its purpose is to capture the effect of the medium previously affected by the fluid. Therefore, as can be deduced from Eq. (8.21) the volumetric flow is proportional to the spatial fractional derivative plus another term as opposed to the classical Darcy equation. Furthermore, they provided a rigorous derivation of the solution of few classic problems and presented the closed-form formulae. They noticed that the Green function acted as a low-pass filter in the frequency domain. Hence, while the time-memory is suitable for accounting for local phenomena, the space memory captures the variations in space.

$$q = \alpha \frac{\partial^{1+n}}{\partial x^{1+n}} p + \beta \frac{\partial}{\partial x} p, \qquad (6.54)$$

Combing with continuity equation to give the following pressure equation

$$-\frac{p_t}{k} = \alpha \frac{\partial^{2+n}}{\partial x^{2+n}} p + \beta \frac{\partial^2}{\partial x^2} p. \qquad (6.55)$$

Other studies by Di Giuseppe et al. (2010) and Caputo and Plastino (2004) were associated with modifications to the classical Darcy equation and one or more constitutive equations. For example, in one case they modified the equation relating the density variation in a fluid to the pressure. The modification there involved introducing the fractional-order derivative to both equations (see Eqs. 8.23, 8.24). As in previous studies, this memory formalism accounted for the permeability changes over time. Furthermore, analytical solutions were derived for fluid pressure in a porous layer with constant pressure maintained at both boundaries.

$$\left(a + b \partial^{m_2} / \partial t^{m_2}\right) p = \left(\alpha + \beta \partial^{m_2} / \partial t^{m_2}\right) \{m(x, t) - m_0\}, \qquad (6.56)$$

$$\left(\gamma + \varepsilon \partial^{n_1} / \partial t^{n_1}\right) q = -\left(c + d \partial^{n_2} / \partial t^{n_2}\right) \nabla p. \qquad (6.57)$$

where

$$0 \leq n_1 < 1, \ 0 \leq n_2 < 1 \text{ and } 0 \leq m_1, 0 \leq m_2 < 1$$

Caputo (2000) investigated memory effects on the volumetric flux along with its spectral properties. He showed using a practical example that the memory formalism implied low-pass filtering of the volumetric flow rate or a band pass centered in the low-frequency spectrum. Furthermore, he provided a summary of how to determine the new parameters introduced by the memory formalism while considering the diffusion in porous layers using the observed pressure or flux at several frequencies.

Caputo and Cametti (2008) investigated the diffusion process in biological structures e.g. membranes. They proposed that a space-dependent diffusion constant be introduced to Fick's diffusion equation to accurately describe the transport process. According to them, solid particles could obstruct some of the pores, leading to a non-constant permeability. Finally, they concluded that their approach was a generalization of Fick's equation to describe the diffusion process in more complex systems.

Table 6.8 presents a summary of some flow equations with memory concepts in a porous medium. More studies devoted to memory applications in fluid flow in porous media can be found in Gatti and Vuk, (2005), Lu and Hanyga (2005), Arenzon et al. (2003), Mifflin and Schowalter (1986), Ciarletta and Scarpetta (1989), Nibbi (1994), Broszeit (1997), Kar et al. (2014), Hayat et al. (2011), Wang and Tan (2008, 2011), Postelnicu (2007), Malashetty and Begum (2011), Rudraiah et al. (1990), and Shivakumara et al. (2011).

6.4.4 Basset force: A history term

When a particle is immersed in a fluid, the forces acting on such particle due to the fluid can be grouped into two; forces due to the virtual mass effect and the Basset force.

TABLE 6.8 Summary of flow equations with memory concept.

Author and year	Description	Assumptions/limitations
Hristov (2013)	Proposed an integral balance approach to solving the heat diffusion model including the effects of memory.	1. Unsteady state 2. Isotropic media 3. The method is based on a final penetration depth.
Hristov (2013)	Applied the integral balance approach to describe the flow of a second-grade fluid resulting from constant surface shear stress.	1. Incompressible fluid. 2. Isotropic and homogeneous porous media. 3. Neglected inertia effects. 4. Memory effects incorporated by the Reimann-Louiville fractional differentiation.
Sprouse (2010)	Developed a numerical scheme to handle a fractional diffusion equation applying the short memory effect technique.	1. The proposed scheme suffers from stability issues. 2. Fractional derivative defined based on GL definition. 3. Constant diffusion coefficient.
Hossain et al. (2009)	Proposed a new fluid flow model for describing crude oil flow inside porous media.	1. Linear isotropic, homogeneous porous media. 2. Slightly compressible and viscous fluid. 3. The proposed scheme suffers from stability issues.
Hossain et al. (2009)	Proposed a comprehensive stress-strain model for the flow of crude oil in a reservoir	1. Linear isotropic, homogeneous porous media. 2. Slightly compressible and viscous fluid. 3. The fluid memory was incorporated through the Caputo fractional derivative.
Iaffaldano et al. (2006)	Proposed a new diffusion model applicable to fluid transport porous media.	1. Linear, isotropic, homogeneous porous medium. 2. Incompressible fluid. 3. The porosity of the media is not considered. 4. Transient flowing conditions.
Caputo and Plastino (2004)	Proposed a modification to Darcy's law and an additional constitutive relation. Incorporated the Caputo fractional derivative to account for the memory of the fluid and rock.	1. Neglects the elastic reaction of the matrix since the equation of diffusion is uncoupled from the equation of elasticity. 2. Isotropic and homogeneous porous media. 3. Neglected inertia effects 4. The permeability is assumed to vary with time.

Continued

TABLE 6.8 Summary of flow equations with memory concept—cont'd

Author and year	Description	Assumptions/limitations
Caputo and Plastino (2003)	Proposed a modification to Darcy law by introducing a space fractional derivative to accurately describe the diffusion process in porous media.	1. Assumed the sand layers to be very thick. 2. Linear, isotropic, homogeneous porous medium. 3. Incompressible and viscous fluid. 4. The porosity of the media is not considered. 5. Permeability is assumed to vary with time.
Caputo (1999)	Proposed a modification to Darcy law through the Caputo fractional derivative to account for permeability reduction with time.	1. Memory formalism to simulate permeability reduction with time in geothermal areas. 2. Linear, isotropic, homogeneous porous medium. 3. Incompressible and viscous fluid. 4. The porosity of the media is not considered.
Slattery (1967)	Proposed an extension of Darcy's law to describe the flow of viscoelastic fluids in porous rocks.	1. Steady-state flow conditions. 2. Isotropic and homogeneous porous rocks were considered. 3. Neglected the effects of inertia. 4. Model parameters depend entirely upon the local thermodynamic state.

Crowe et al. (2011) defined the Basset force (also called history term) as a term accounting for viscous effects as well as the temporal delay in boundary layer development as the relative velocity changes with time. However, most studies usually neglect its effect due to difficulty in its computing, even though sometimes this force term is very large especially when a body is accelerated at a high rate (Blevins, 1984; Johnson, 1998).

The Basset-Boussinesq-Oseen equation (BBO equation), named after Joseph Valentin Boussinesq, Alfred Barnard Basset, and Carl Wilhelm Oseen, was proposed to model the motion of, as well as forces acting, on a small particle in unsteady flow for low Reynolds number. Fan and Zhu (2005) formulated a BBO equation applicable to spherical particles of certain diameter in a fluid of certain density as shown below:

$$BF = \frac{3}{2}d_p^2 \sqrt{(\pi \rho_f \mu)} \int_{t_0}^{t} \frac{1}{\sqrt{t-\tau}} \frac{d}{d\tau}(U_f - U_p) d\tau. \tag{6.58}$$

Through a force balance, they equated the particle's rate of change of momentum to the sum of five force terms acting on the particle. The forces include; stokes drag, pressure gradient, added mass, Basset force, and other forces, i.e., body force.

White and Corfield (2006) studied the acceleration of a flat plate assuming a step change in velocity is applied parallel to the plate-fluid interface plane. He derived an equation to predict the fluid velocity as a function of time and distance from the plate. He investigated the cumulative effect of the accelerations on the shear stress by assuming the plate acceleration comprised of series of step changes in velocity. Accordingly, the shear stress experienced on a flat late due to a step-change in velocity was represented by:

$$\tau = \sqrt{\frac{\rho_c \mu_c}{\pi}} \int_0^t \frac{\frac{\partial u_p}{\partial t'}}{\sqrt{t-t'}} dt', \tag{6.59}$$

Applications of the Basset history also arise in numerical models developed to simulate fluid flow with solid particles. There have been concerns about the significant computer resources needed to calculate the Basset history when a large number of particles are considered. Resource-saving methods for the calculation of Basset force were proposed by Michaelides (1992) and (Bombardelli et al., 2008). Other researchers chose to omit the Basset history in their models: see the following studies: (Schmeeckle and Nelson, 2003; Kholpanov and Ibyatov, 2005; Lee et al., 2006). Fortunately, avoiding the history term does not affect calculations for particles of intermediate or large size. However, for simulations of fine evaporating particles, the history term accounts for approximately 20% of the total force (Liang and Michaelides, 1992). Therefore, it would not be appropriate to neglect the history term in such calculations as it would lead to errors in particle velocities and positions.

Vojir and Michaelides (1994) proposed the use of an integro-differential operator to handle the Basset force term in describing the motion of a sphere. This technique transforms the governing equation of motion to a second-order, ordinary, differential equation (ode), explicit in velocity. The resulting ode was solved numerically to quantify the effect of the history term on the particle velocity and trajectories in unsteady flow conditions. Lukerchenko (2010) carried out numerical investigations to understand the effect of the Basset force during particle motion. He noted that the Basset force could be divided into; force(s) due to particle's motion in a fluid and secondly due to the particle-bed collisions. Furthermore, he stated the conditions for which the Basset force should be neglected during calculations.

6.5 Anomalous diffusion: A memory application

Anomalous diffusion (sub/super diffusion) occurs when a particle plume spreads at a rate that cannot be predicted by classical Brownian motion models. This lead to various articles and research tailored towards the application of fractional space derivatives to describe such phenomena. Generally, when a fractional derivative replaces the second derivative in a diffusion model, it leads to some form of enhanced diffusion (super diffusion). As reported by Podlubny in his many books, the main advantage of fractional calculus (fractional derivatives) is that they provide an excellent tool for the description of memory and other hereditary properties of various materials and processes (Podlubny 1998, 2001).

For many one-dimensional diffusion models based on constant diffusion coefficient, as well as advection–dispersion models, many analytical solutions have been proposed with the help of various methods such as:

- Laplace transforms (Caputo, 2003; Iaffaldano et al., 2006; Antimirov et al., 1993).
- Fourier transforms (Benson et al., 2000; Chaves, 1998);

- Homotopy method (HAM) (Cavlak and Bayram, 2014; Kumar et al., 2015; Liu et al., 2014; Mohebbi et al., 2014; Ray and Gupta, 2014; Ray and Sahoo, 2015; Wei et al., 2014);
- Adomian decomposition method (ADM) (Jafari et al., 2015; Wu and Huang, 2014; Babolian et al., 2014; Hariharan and Kannan, 2014; Duan et al., 2014; Povstenko, 2015; Shi et al., 2014; Baleanu et al., 2014);
- Variational iteration method (Malik et al., 2014);
- Wavelet methods (Hariharan and Kannan 2014); and
- Pseudo-spectral methods (Sweilam et al., 2014) to name a few.

However, practical problems do not allow for such over simplification of constant-coefficient of diffusion, or constant source terms. Hence, the only technique to handle such problems involves the development of efficient numerical schemes that can account for such difficulties (non-linearity, variable coefficients, etc.). In the next section, some of the numerical schemes presented by various researchers to study the anomalous diffusion models are discussed.

6.5.1 Fractional-order transport equations and numerical schemes

According to (Miller and Ross, 1993) and (Samko et al., 1993), the origin of fractional derivatives dates back to the more familiar integer-order derivatives. These fractional differential equations (FDEs) have garnered increasing attention in most fields of science and engineering. They have been successfully used to describe many physical and chemical processes, as well as biological systems, and so on. Fig. 6.9 presents a chart showing different applications of fractional modeling.

The main reason for discussing diffusion equations of fractional order in this article is to describe the phenomena of anomalous diffusion in transport processes, for example, its application in modeling fluid flow in porous media. A plethora of numerical schemes have been

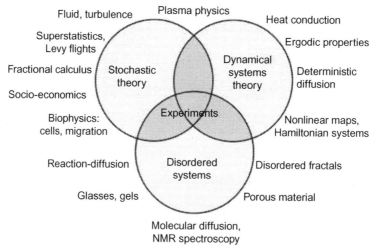

FIG. 6.9 Fractional Modeling, a new Meta-Discipline.

successfully proposed for solving this class of equations (space and/or time FDEs). Some of which are mentioned in the following pages.

Tian et al. (2015) presented a second-order approximation "WSGD" (weighted and shifted Grunwald difference) for the Riemann Liouville fractional derivative to solve a class of space fractional diffusion equation. The proposed scheme was shown to be applicable for both one-dimensional and two-dimensional problems with or without constant diffusion coefficients with the fractional-order derivative between ($1 \leq \alpha \leq 2$). Furthermore, the stability and convergence of the proposed scheme were investigated assuming constant diffusion coefficients only.

Chen and Deng (2014) proposed a fourth-order approximation 'WSLD' (weighted and shifted Lubich difference operators) to the Reimann-Liouville fractional derivative to solve a class of space fractional derivative diffusion. The authors applied their proposed scheme to solve one-dimensional and two-dimensional fractional diffusion equations with variable diffusion coefficients with great success. Finally, they proved that the scheme was unconditionally stable with a global truncation error of $O(k^2, h^4)$. The terms h and k denote the space and time steps respectively.

Lynch et al. (2003) proposed a finite difference method to solve a certain class of fractional order, a partial differential equation that describing an anomalous diffusion problem. The numerical method was centered on an explicit- and a semi-explicit scheme for fractional-order between ($1 \leq \alpha \leq 2$). The authors also derived the accuracy and stability of the presented schemes, the major drawback of their schemes is their stability. The interested reader should refer to the original manuscript for the complete overview.

Meerschaert and Tadjeran (2004) considered a one-dimensional, radial flow, the fractional, advection-dispersion model with variable coefficients. They used the finite difference method to obtain approximate solutions using an Implicit-Euler scheme. The fractional-order derivative was defined based on the modified Grünwald approximation. The authors showed that the fractional derivative model was able to predict the early arrival observed in the field that the classical advection-dispersion equations failed to predict. For their study, the fractional-order was within the range ($1 < \alpha \leq 2$). They insisted that without the modification to the Grünwald approximation for the fractional derivative, the implicit Euler method was also unstable.

Langlands and Henry (2005) presented an implicit finite difference scheme applicable to describing anomalous diffusion equation models. In addition, they investigated the accuracy and stability of their scheme. They showed that the order of their scheme was $O(\Delta x^2)$ in the spatial grid size and $O(\Delta t^{1+\gamma})$ in the fractional time step, where $0 \leq 1 - \gamma < 1$ is the order of the fractional derivative and $\gamma = 1$ is standard diffusion. The developed schemes are unconditionally stable for $0 < \gamma \leq 1$.

Likewise, Tadjeran et al. (2006) developed an unconditionally stable, second-order (in time and space) accurate numerical scheme, to provide approximate solutions to a class of initial boundary-value, fractional, one-dimensional diffusion equation with variable coefficients in a finite domain. Their approach was based on the classical Crank-Nicholson method combined with a Richardson extrapolation scheme in the x-direction. The fractional-order derivative was approximated by the shifted Grünwald formula.

In addition, Yuste and Acedo (2005) considered developing an approximate solution to a one-dimensional fractional diffusion equation. They proposed a forward in time, and centered in space (FTCS) finite difference method, along with the Grünwald-Letnikov discretization of the Riemann-Liouville derivative to obtain an explicit FTCS scheme. The authors were able to show that their analytical stability bounds were in excellent agreement with numerical tests.

Not satisfied with the explicit scheme presented earlier, Yuste (2006) proposed a modification of weighted average methods for ordinary diffusion equations with an accuracy of order $(\Delta x)^2$ and Δt to fractional-order diffusion equations. However, for the fractional version of the Crank-Nicholson method, it is possible to obtain a second-order accuracy with respect to time step, if a second-order approximation to the fractional time-derivative is employed. In addition, they presented accurate stability criteria valid for different discretization schemes of the fractional derivative, considering arbitrary weight factor, and arbitrary order of the fractional derivative.

Murio (2008) proposed an implicit, finite-difference numerical scheme, for a class of fractional-order diffusion equations. In this case, the time-fractional derivative was defined with the Caputo derivative. He showed that his numerical solution was second-order accurate in space and first-order in time while carrying out an extensive stability analysis to prove the stability of his numerical scheme.

However, Ding and Zhang (2011) showed that the stability analysis carried out by Murio (2008) was not accurate; Hence, they carried out a proper stability analysis of the above implicit finite difference scheme. Furthermore, Scherer et al. (2008) presented an explicit and implicit finite-difference numerical scheme applicable to the treatment of some fractional extensions of the energy balance used for predicting the temperature field in oil strata. The fractional-order derivative was approximated based on Grünwald-Letnikov's definition. The other details presented about the numerical scheme were related to stability, convergence and error behavior. The interested reader can refer to the original manuscript.

Chen et al. (2008) proposed a finite difference implicit and explicit scheme that was centered on the relationship between the Riemann-Liouville and Grünwald-Letnikov definitions of fractional derivatives for a class of fractional reaction-sub-diffusion equations. Furthermore, they investigated the stability and the convergence of their schemes by Fourier analysis.

Cui (2009) developed a high-order (fourth-order accurate in space), implicit, finite difference scheme for solving a class of one-dimensional fractional diffusion equations. The discretization consisted of approximating the Riemann-Liouville derivative by the Grünwald-Letnikov approximation. In addition, the local truncation error and stability were analyzed. Interested readers can refer to the original manuscript for more details.

Liu et al. (2009) presented an implicit, finite-difference, numerical scheme to describe a class of fractional-order diffusion equation with a nonlinear source term. Such models can be used to describe processes that become less anomalous as time progresses by the inclusion of a second fractional time derivative acting on the diffusion term. They employed an energy method to derive stability criteria.

Mustapha (2010) proposed a hybrid implicit scheme to describe a special class of the fractional sub-diffusion equation involving the fractional-order parameter in the range $-1 < \alpha < 0$. The hybrid scheme consisted of an implicit, finite-difference, Crank-Nicolson method for the time discretization and a linear finite elements discretization in space. For $0 < \alpha < 1$ (i.e., positive memory term) the numerical solution of the same problem has been extensively investigated. For example, refer to the works herein (McLean et al., 1996; McLean and Mustapha, 2007).

Sweilam et al. (2012) proposed a finite difference scheme that could handle efficiently a class of nonlinear, fractional-order, Riccati differential equation. The fractional derivative was approximated by the Caputo definition. The resulting nonlinear system of algebraic equations obtained from the finite difference discretization was solved using the Newton iteration method.

Sunarto et al. (2014) derived an implicit finite difference numerical scheme to approximate a class of one-dimensional, linear, time-fractional diffusion equations with the fractional-order derivative approximated based on Caputo's definition. The corresponding system of algebraic linear equations was then solved using the Accelerated Over-Relaxation (AOR) iterative technique. In addition, as proof of the effectiveness of the scheme, they compared the performance with that of the Successive Over-Relaxation (SOR) method and Gauss-Seidel (GS) iterative method by carrying out the numerical test.

Iyiola and Zaman (2014) proposed a fractional-order time derivative equation as compared to the classical first-order derivative in time for cancer tumor models. Three different cases of the net killing rate were considered including the case where the net killing rate of the cancer cells was dependent on the concentration of the cells. In their work, a new analytical technique called q-Homotopy Analysis Method (q-HAM) was proposed to handle the resulting time-fractional partial differential equations. The obtained solution was in the form of a convergent series with easily computable components. Based on their numerical analysis, they were able to give some recommendations on the appropriate order (fractional) of derivatives in time to be used in killing cancer tumors.

Recently, Mustapha and AlMutawa (2012) proposed an implicit, finite difference numerical scheme applicable to a class of fractional diffusion equation (refer to Table 6.9). The discretization consists of a Crank-Nicolson method combined with the second central finite difference for the spatial discretization. The approximation leads to an error of order $O(h^2, \max(1, \log k - 1) + k^{2+\alpha})$. The terms h and k denote the maximum space and time steps, respectively. However, they emphasized that employing a non-uniform time does compensate for the singular behavior of the exact solution at $t = 0$. Table 6.9 below presents the model equations proposed by all the aforementioned researchers.

6.5.2 Porous media considerations

Memory was defined in simple terms as the effect of past events on the present and future course of developments (Hossain et al., 2007). Accordingly, Zhang (2003) defined memory as a function of time and space, and forward time events depend on previous time events.

In this literature review, we focus more on rock-dependent memory by considering the evolution of permeability and porosity due to various processes. Reports from laboratory studies have shown that permeability depends on a variety of parameters such as porosity, pore size, and shape, clay content, stress, pore pressure, fluid type, and saturation.

The earliest permeability-porosity models are the empirical straight-line correlations using either a semi-logarithmic relation (Archie, 1950; Timur, 1968) or a fully logarithmic relation (Wyllie and Rose, 1950).

TABLE 6.9 Fractional partial differential equation models.

Author and year	Fractional derivative definition and type of discretization	Model equation & notes
McLean et al. (1996)	Riemann-Liouville derivative/finite difference for time step and finite elements space discretization with $0 < \alpha < 1$.	$\dfrac{\partial u(x,t)}{\partial t} - D_t^{-\alpha}\left(\dfrac{\partial^2 u(x,t)}{\partial x^2}\right) = f(t)$, Subject to boundary conditions. Unconditionally stable scheme.
Lynch et al. (2003)	Riemann-Liouville derivative/explicit finite difference discretization with $1 \leq \alpha \leq 2$.	$\dfrac{\partial n}{\partial t} = F(x,n) + \chi D^\alpha[n(x,t) - n(0,t)]$, Subject to boundary conditions. Conditionally stable.
Meerschaert and Tadjeran (2004)	Riemann-Liouville derivative/Implicit Euler finite difference discretization with ($1 < \alpha \leq 2$).	$\dfrac{\partial c(r,t)}{\partial t} = -v(r)\dfrac{\partial c(r,t)}{\partial r} + d(r)\dfrac{\partial^\alpha c(r,t)}{\partial r^\alpha} = f(r,t)$, Subject to boundary conditions. Unconditionally stable scheme.
Langlands and Henry (2005)	Riemann-Liouville derivative/Implicit finite difference for time steeping and finite elements space discretization.	$\dfrac{\partial y}{\partial t} = \dfrac{\partial^{1-\gamma}}{\partial t^{1-\gamma}}\left(\dfrac{\partial y^2}{\partial x^2}\right)$, Subject to boundary conditions. Unconditionally stable scheme.
Yuste and Acedo (2005)	Riemann-Liouville derivative/explicit finite difference approximation.	$\dfrac{\partial u(x,t)}{\partial t} = K_\gamma D_t^{1-\gamma}\left(\dfrac{\partial^2 u(x,t)}{\partial x^2}\right)$, Subject to boundary conditions. Conditionally stable scheme.
Yuste (2006)	Riemann-Liouville derivative/ weighted average finite difference approximation.	$\dfrac{\partial u(x,t)}{\partial t} = K_\gamma D_t^{1-\gamma}\left(\dfrac{\partial^2 u(x,t)}{\partial x^2}\right)$, Subject to boundary conditions. Unconditionally stable for the implicit scheme and other schemes conditionally stable.
Tadjeran et al. (2006)	Riemann-Liouville derivative/explicit finite difference approximation $1 < \alpha < 2$.	$\dfrac{\partial u(x,t)}{\partial t} = d(x)\dfrac{\partial^\alpha u(x,t)}{\partial x^\alpha} + q(x,t)$, Subject to boundary conditions. Conditionally stable scheme.
Scherer et al. (2007)	Caputo fractional derivative/ Explicit and Implicit finite difference approximation with $0 < \alpha < 1$.	$\dfrac{\partial^\alpha u}{\partial t^\alpha} = c^2\dfrac{\partial^2 u}{\partial z^2}, 0 < r,z,t < \infty$. Subject to boundary condition. $z=0: \dfrac{\partial^\alpha u}{\partial t^\alpha} = \dfrac{\partial^2 u}{\partial r^2} + \dfrac{1-2\gamma}{r}\dfrac{\partial u}{\partial r} + \beta\dfrac{\partial u}{\partial z}, 0 < r,t < \infty$. Conditionally stable scheme. Unconditionally stable for the implicit scheme and other schemes conditionally stable.
Chen et al. (2007)	Riemann-Liouville derivative/Implicit and explicit finite difference approximation $1 < \gamma < 2$.	$\dfrac{\partial u(x,t)}{\partial t} = D_t^{1-\gamma}\left[K_\gamma\left(\dfrac{\partial^2 u(x,t)}{\partial x^2}\right) - \kappa u(x,t)\right] + f(x,t)$, Subject to boundary conditions. Unconditionally stable for the implicit scheme and other schemes conditionally stable.

TABLE 6.9 Fractional partial differential equation models—cont'd

Author and year	Fractional derivative definition and type of discretization	Model equation & notes
Murio (2008)	Caputo derivative/Implicit finite difference approximation.	$\frac{\partial^\alpha u(x,t)}{\partial t^\alpha} = \frac{\partial^2 u(x,t)}{\partial x^2}$, Subject to boundary conditions. Unconditionally stable scheme.
Liu et al. (2009)	Riemann-Liouville derivative/Implicit finite difference discretization with $0 < \alpha < \beta \leq 1$.	$\frac{\partial u(x,t)}{\partial t} = \left(A\frac{\partial^{1-\gamma}}{\partial t^{1-\gamma}} + B\frac{\partial^{1-\beta}}{\partial t^{1-\beta}}\right)\left[\frac{\partial^2 u(x,t)}{\partial x^2} + f(x,t)\right] + g(u,x,t)$, Subject to boundary conditions. Unconditionally stable scheme.
Cui (2010)	Riemann-Liouville derivative/Implicit finite difference scheme.	$\frac{\partial u(x,t)}{\partial t} = D_t^{1-\gamma}\left\{K_\gamma \frac{\partial^2 u(x,t)}{\partial x^2}\right\} + f(x,t)$, Subject to boundary conditions. Unconditionally stable scheme.
Mustapha (2010)	Riemann-Liouville derivative/Implicit finite difference for time step and finite elements space discretization.	$\frac{\partial u(x,t)}{\partial t} - D_t^{-\alpha}\left(\frac{\partial^2 u(x,t)}{\partial x^2}\right) = f(t)$, Subject to boundary conditions. Unconditionally stable scheme.
Sunarto et al. (2012)	Caputo derivative/Implicit finite difference approximation.	$\frac{\partial^\alpha u(x,t)}{\partial t^\alpha} = a(x)\frac{\partial^2 u(x,t)}{\partial x^2} + b(x)\frac{\partial u(x,t)}{\partial x} + c(x)u(x,t)$, Subject to boundary conditions. Unconditionally stable scheme.
Mustapha and Al-Mutawa (2012)	Riemann-Liouville derivative/Implicit finite difference approximation	$\frac{\partial u(x,t)}{\partial t} - \beta_\alpha\left(\frac{\partial^2 u(x,t)}{\partial x^2}\right) = f$, Subject to boundary conditions. Unconditionally stable scheme.
Chen and Deng (2014)	Riemann-Lioville derivative/Crank-Nicholson discretization for the time step and WSLD for the space discretization	$\frac{\partial u(x,y,t)}{\partial t} - f(x,y,t) = d_+(x,y)D_{xL,x}^\alpha(x,y,t)$ $+ d_-(x,y)D_{x,xR}^\alpha(x,y,t)$ $+ e_+(x,y)D_{yL,y}^\beta(x,y,t)$ $+ d_-(x,y)D_{y,yR}^\beta(x,y,t)$ Subject to boundary conditions. Unconditionally stable scheme. Where $1 < \alpha, \beta < 2$.
Tian et al. (2015)	Riemann-Lioville derivative/Crank-Nicholson discretization for the time step and WSGD for the space discretization.	$\frac{\partial u(x,t)}{\partial t} = K_1 D_{a,x}^\alpha u(x,t) + K_2 D_{x,b}^\alpha u(x,t) + f(x,t)$, Subject to boundary conditions Unconditionally stable scheme.

To describe the permeability-porosity relationship of a porous medium, the Kozeny-Carman equation (Kozeny, 1927; Carman, 1937) assumes that a fluid flowing through the porous medium follows a bundle of tortuous capillary paths without interacting with each other. Since the equation is based on conductive hydraulic flow paths, there are some inherent limitations; for example, it cannot handle the nonzero-transport threshold, the cementation effects, and the gate or valve effects to predict zero permeability when pore throats are plugged to create isolated pores with non-zero porosity. Thus, the Kozeny-Carman equation is mostly applicable for high permeability and unconsolidated porous media such as a pack of relatively large grains like sand or glass beads. Many studies have concluded that the Kozeny-Carman equation does not perform well in geological porous formations that are of low permeability and containing a substantial degree of cementation and grain embedment and fusing.

Bhat and Kovscek (1998) proposed a network model to study the effects of mineral dissolution and re-precipitation during the injection of steam into California diatomites. They observed that for moderate increase or reduction in porosity, a power-law expression describing the permeability alteration with a power-law exponent in the range of 8–9. However, it must be emphasized that this power-law relation is only applicable for moderate ranges of deposition and dissolution. Such power-law relation has been presented in the experimental work conducted by Kohl et al. (1996), who measured the permeability of diatomite plugs flushed with hot, saturated silica-laden water. Mavko and Nur (1997) porosity and permeability models were developed to include the percolation threshold in the Kozeny-Carman relation. They concluded that a third-power dependence on porosity could be retained while accurately fitting the observed permeability in certain well-sorted materials.

Civan (1998) presented a model to estimate the porosity and permeability variations in a porous media by considering the effects of solids deposition and compaction. The equations presented could easily be modified to account for the terminal or limit porosity and permeability. Again, (Civan, 2000) proposed a new model for interpretation, correlation, and prediction of the variations of permeability and porosity in petroleum reservoirs, in which the pore topology is evolving by geochemical and geo-mechanical rock and fluid interactions during formation damage and stimulation processes. The newly developed permeability and porosity relationship consider pore evolution by precipitation and dissolution, and stress and temperature variations. The presented equations offer improved correlation applying (Civan, 1996) power-law equation with the parameters correlated by exponential functions, which are well-behaved smooth functions as opposed to the previous polynomial functions.

Finally, Civan (2001) developed a kinetic model for porosity and permeability variation by dissolution and precipitation. Here, the author considers the porous media to consist of interconnected spherical shaped pores with associated empirical shape factors, correcting for deviations from actual irregular pore shapes. He derived analytical solutions for three and two-dimensional porous media considering applications to cores and thin sections, respectively. These solutions were shown to accurately describe a variety of reported experimental data and allowed for the convenient determination of the various model parameters. Table 6.10 summarizes some of the available permeability relationships.

6.5.3 Theoretical development

To determine the pressure distribution over space and time, a momentum equation is considered as the governing equation for the flow of a single-phase, incompressible fluid in a reservoir. The partial differential equations have a familiar form because the system has been averaged over representative elementary volumes.

TABLE 6.10 Some of the available permeability correlations.

Author	Equation	Application/limitations
Kozeny-Carman (1956)	$K = B\frac{\phi^3}{S^2}$	1. Applicable for high permeability and unconsolidated porous media, such as a pack of relatively large grains like sand or glass beads. 2. An apparent failure of the Kozeny-Carman relations is often observed at low porosities where permeability decreases much more rapidly with decreasing porosity. 3. The neglect of the percolation threshold porosity in the Kozeny-Carman relation
Schlumber organization (1962)	$K = 6.25 * 10^{-4} \frac{\phi^6}{S_{iwr}^2}$	1. Based on field observations in oil sands.
Timur (1968)	$K = 0.136 \frac{\phi^{4.4}}{S_{iwr}^2}$	1. Applicable to sandstone reservoirs. 2. Essential that an accurate residual water saturation estimate be provided
Bourbie et al. (1987)	$K = B\phi^n d^2$	1. Where n ranges from the derived value of 3 for large porosities to values of 7–8 at very low porosities. 2. Overcomes the limitation of Kozeny-Carman at low porosities.
Pape et al. (2000)	$K = \frac{\phi r^2}{8\Gamma^2}\left(\frac{2\phi}{M^2(1-\phi)}\right)^{\frac{2}{D-1}}$	1. Applicable to sandstone
Mavko and Nur (1997)	$K = C\frac{(\phi-\phi_c)^3}{(1+\phi_c-\phi)^2}d^2$	1. Applicable in clean, well-sorted rocks. 2. Accounts for percolation threshold porosity in the Kozeny-Carman relation
Civian (2005)	$\sqrt{\frac{K}{\phi}} = \Gamma\left(\frac{\phi}{a-\phi}\right)^n$	1. Based on a bundle of the leaky-tubes model of tortuous preferential flow paths involving cross-flow between them depending on the pore interconnectivity in porous media
Nooruddin and Hossain (2009)	$K = \left(\frac{1}{f_g a^2 S_{Vgr}^2}\right)\frac{\phi^{2m+1}}{(1-\phi)^2}$	1. A total of six data sets from sandstone and carbonate reservoirs Were used to verify and validate the proposed model.

Fluid flow in porous media is calculated from Darcy's law which is based on some simplifying assumptions about the properties of the porous medium and also of the flowing fluid(s). However, if any of the simplifying assumptions, for example, the porous medium is heterogeneous, non-isothermal conditions and the fluid properties a strong function of pressure and temperature, Darcy law in its simplest form cannot be used to model fluid flow in such conditions.

Darcy's law is modified to introduce the notion of fluid memory. For a 1D system the model equation can be written as:

$$u = -\eta D_t^{1-\gamma}\left(\frac{\partial p}{\partial x}\right), \quad (6.60)$$

The variable, η, is defined as the ratio of pseudo-permeability of the medium with memory to fluid viscosity measured in S.I. units of ($m^3 s^{2-\gamma} kg^{-1}$). In addition, the operator, $D_t^{1-\gamma}$, is to be discretized using the Grunwald-Letnikov definition of the Riemann-Liouville fractional time derivative defined below (MacDonald et al., 2015).

$$D_t^{1-\gamma} u(t) = \lim_{\tau \to 0} \tau^{\gamma-1} \sum_{k=0}^{t/\tau} \frac{(-1)^k \Gamma(2-\gamma)}{k! \Gamma(2-\gamma-k)} u(t - m\tau), \quad (6.61)$$

From the aforementioned definition, the Grunwald-Letnikov fractional derivative is nonlocal. That is, determining the value of a real derivative requires knowledge of previous values from $k = 0$ to $k = t/\tau$. This takes into account the history of the system from the time of interest ($k = 0$) all the way back to the beginning point ($k = t/\tau$).

Therefore, the proposed fluid flow model accounts for the variation of fluid and formation properties with time. Interestingly, the proposed model reduces to the conventional Darcy-based model if the term γ equals 1. The two important properties in the new flow model are the terms γ and η.

6.5.4 Mass conservation

Considering a porous medium with a control volume fully saturated with fluid as depicted by Fig. 6.10 Applying the mass conservation principle, the governing differential equation describing the fluid flow in this porous medium is:

$$\frac{\partial}{\partial x}(\rho_o u_x A_x)\Delta x + \rho_o q^{well} = \frac{V_b}{\alpha_c} \frac{\partial(\rho_o \phi)}{\partial t}, \quad (6.62)$$

Using our definition of velocity and expressing in the equation;

$$\frac{\partial}{\partial x}\left(\rho_o \eta A_x \left\{ D_t^{1-\gamma}\left(\frac{\partial p}{\partial x}\right) \right\}\right)\Delta x + \rho_o q^{well} = \frac{V_b \partial(\rho_o \phi)}{\alpha_c \partial t}, \quad (6.63)$$

Expressing the density of the fluid in terms of stock tank conditions;

$$\rho_o = \frac{\rho_{o,sc}}{B_o}, \quad (6.64)$$

Therefore, inserting Eq. (6.64) into Eq. (6.63)

$$\frac{\partial}{\partial x}\left(\frac{A_x \eta}{B_o} D_t^{1-\gamma}\left(\frac{\partial p}{\partial x}\right)\right)\Delta x + q_{sc} = \frac{V_b}{\alpha_c} \frac{\partial}{\partial t}\left(\frac{\phi}{B_o}\right), \quad (6.65)$$

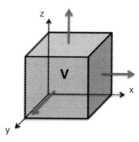

FIG. 6.10 Control volume fully saturated with a fluid.

6.5 Anomalous diffusion: A memory application

$$\text{Where } q_{sc} = \frac{q^{well}}{B_o}, \tag{6.66}$$

Therefore, Eq. (6.65) can be re-written as

$$\frac{\partial}{\partial x}\left(g_x(p)D_t^{1-\gamma}\frac{\partial p}{\partial x}\right)\Delta x + q_{sc} = \frac{V_b}{\alpha_c}\frac{\partial}{\partial t}\left(\frac{\phi}{B_o}\right), \tag{6.67}$$

where coefficient $g_x(p) = \frac{\eta A_x}{B_o}$.

6.5.4.1 Composite variable, η

The most important variable in the newly developed flow model is the term η, which is dependent on the particular rock-fluid system. Hence, the proper definition of the type of reservoir fluid and formation properties is paramount to proper modeling of the newly formulated flow equations.

To state explicitly that there is no universal permeability-porosity relationship valid to all porous media is a well-known fact considering the number of publications on this topic alone. For example, porosity is invariant under a homothetic transformation (e.g., uniform, isotropic stretching) of the pore space whereas permeability is not. Secondly, in a given material, not all pores are equally effective in conducting fluid flow. Two media with the same porosity but different proportions of effective and non-effective pore space must therefore have different permeabilities. However, luckily for researchers, the evolution of permeability and porosity in rocks can be constrained provided that the processes changing the pore space are known (Bernabé et al., 2003).

To solve the aforementioned fractional model, the (Mavko and Nur, 1997) permeability correlation will be used as an example.

The composite variable can be defined by Eq. (6.68) with all terms in field units.

$$\eta = \beta_c \frac{[K(\phi_c, \phi, C, d)]}{\mu(p)} t^{1-\gamma}, \quad \text{where } K(mD) = C\frac{(\phi - \phi_c)^3}{(1+\phi_c-\phi)^2}d^2, \tag{6.68}$$

For analysis, we assume a single-phase slightly compressible fluid with porosity varying as a function of pressure:

$$\phi = \phi_0 \exp[c_s(p-p_i)], \tag{6.69}$$

In addition, in this study, we assume that the oil viscosity above the bubble-point pressure can be correlated by:

$$\mu = \mu_{ob} \exp[c_\mu(p-p_b)], \tag{6.70}$$

where μ_{ob}, is the oil viscosity at the bubble-point pressure obtained from Eq. (6.70).

$$\mu_{ob} = 6.59927 \times 10^5 R_s^{-0.597627} T^{-0.941624} \times \gamma_g^{-0.555208} API^{-1.487449}, \tag{6.71}$$

Likewise, the following equations present the correlations used for determining other reservoir fluid properties:

$$p_b = -620.592 + 6.23087\frac{R_s\gamma_o}{\gamma_g B_o^{1.38559}} + 2.89868\,T, \tag{6.72}$$

$$B_o = B_{ob}\exp[-c_o(p-p_b)], \qquad (6.73)$$

$$B_{ob} = 1.122018 + 1.410 \times 10^{-6}\frac{R_s T}{\gamma_o^2}, \qquad (6.74)$$

$$c_o = \frac{\left(1433 + 5R_s + 17.2T - 1180\gamma_g + 12.61\,API\right)}{(p \times 10^5)}, \qquad (6.75)$$

$$c_w = (A + BT + CT^2) \times 10^6, \qquad (6.76)$$

With the coefficients A, B and C presented below:

$$A = 3.8546 - 1.34 \times 10^{-4} p, \qquad (6.77)$$

$$B = -1.052 \times 10^{-2} + 4.77 \times 10^{-7} 7p, \qquad (6.78)$$

$$C = 3.9267 \times 10^{-5} - 8.8 \times 10^{-10} p, \qquad (6.79)$$

6.5.5 Implicit formulation

In finite-difference methods, the space-time solution's domain is discretized. We shall use the following notation: Δt is the temporal mesh or time step, Δx is the spatial mesh along the reservoir length. The coordinates of the mesh points are $xi = i\Delta x$ and $tn = n\Delta t$, and the values of the solution $p(x,t)$ on these grid points are $p(xi, tn) \equiv pin \approx Pin$, where i denotes the numerical estimate of the exact value of $p(x,t)$ at the point (xi,tn).

The right-hand side of Eq. (6.68) can also be expressed in terms of the unknown pressure as follows:

$$\frac{\partial}{\partial t}\left(\frac{\phi}{B_o}\right) \approx \frac{1}{\Delta t}\left(\left(\frac{\phi}{B_o}\right)^{n+1} - \left(\frac{\phi}{B_o}\right)^n\right), \qquad (6.80)$$

Using Eqs. (6.68)–(6.76), the accumulation term can be expressed in a conservative form as:

$$\frac{\partial}{\partial t}\left(\frac{\phi}{B_o}\right) \approx \left(\frac{c_s\phi_0}{B_o^{n+1}} + \frac{\phi^n c_o}{B_o^0}\right)\frac{(p^{n+1}-p^n)}{\Delta t}, \qquad (6.81)$$

Therefore, from Eq. (6.67)

$$\frac{\partial}{\partial x}\left(g_x(p)D_t^{1-\gamma}\frac{\partial p}{\partial x}\right)\Delta x + q_{sc} = \chi(p)\frac{(p^{n+1}-p^n)}{\Delta t}, \qquad (6.82)$$

where coefficient $\chi(p) = \frac{V_b}{\alpha_c}\left(\frac{c_s\phi_0}{B_o^{n+1}} + \frac{\phi^n c_o}{B_o^0}\right)$.

Likewise, to approximate $D_t^{1-\gamma}$, the operator, the Gruunwald-Letnikov approximation defined in Eq. (6.63) will be adopted. Here, t represents the current moment in time and Δt represents the time step.

6.5 Anomalous diffusion: A memory application

Next, we define a function, $\psi(\gamma, k)$, such that

$$\psi(\gamma, k) = \frac{(-1)^k \Gamma(2-\gamma)}{k! \Gamma(2-\gamma-k)}, \tag{6.83}$$

Noting that

$$\psi(\gamma, k) = -\psi(\gamma, k-1) \frac{(2-\gamma-k)}{k}, \quad \text{for } k > 0, \text{ with } \psi(\gamma, 0) = 1, \tag{6.84}$$

Hence, from Eq. (6.82)

$$\chi(p) \frac{(p^{n+1} - p^n)}{\Delta t} = \lim_{\tau \to 0} \tau^{\gamma-1} \sum_{k=0}^{t/\tau} \psi(\gamma, k) \frac{\partial}{\partial x} \left(g_x(p) \frac{\partial p(t-k\tau)}{\partial x} \right) \Delta x + q_{sc}, \tag{6.85}$$

The next step is to discretize the spatial derivative in Eq. (6.85) to allow for numerical implementation.

$$C_{p,i}^{n+1}\left(p_i^{n+1} - p_i^n\right) = q_{sc,i}^{n+1}$$

$$+ \frac{1}{(\Delta t)^{1-\gamma}} \sum_{k=0}^{t/\Delta t} \psi(\gamma, k) \left[T_{i+\frac{1}{2}}^{n+1-k}\left(p_{i+1}^{n+1-k} - p_i^{n-k}\right) - T_{i-\frac{1}{2}}^{n+1-k}\left(p_i^{n+1-k} - p_{i-1}^{n+1-k}\right) \right], \tag{6.86}$$

where the fluid transmissibility and pseudo-compressibility introduced in Eq. (6.86) are defined below.

$$T = \frac{\beta_c \eta A_x}{B \Delta x (\Delta t)^{1-\gamma}}, \quad \text{and} \quad C_{p,i}^{n+1} = \frac{\chi_i^{n+1}}{\Delta t}$$

Hence, we have

$$C_{p,i}^{n+1}\left(p_i^{n+1} - p_i^n\right) - q_{sc,i}^{n+1}$$

$$= \left[T_{i-\frac{1}{2}}^{n+1} p_{i-1}^{n+1} - \left(T_{i-\frac{1}{2}}^{n+1} + T_{i+\frac{1}{2}}^{n+1}\right) p_i^{n+1} + T_{i+\frac{1}{2}}^{n+1} p_{i+1}^{n+1} \right] \tag{6.87}$$

$$+ \sum_{k=1}^{t/\Delta t} \psi(\gamma, k) \left[\left(T_{i-\frac{1}{2}}^{n+1-k} p_{i-1}^{n+1-k} - \left(T_{i-\frac{1}{2}}^{n+1-k} + T_{i+\frac{1}{2}}^{n+1-k}\right) p_i^{n+1-k} + T_{i+\frac{1}{2}}^{n+1-k} p_{i+1}^{n+1-k} \right) \right],$$

Taking all the like terms together in Eq. (6.87),

$$-T_{i-\frac{1}{2}}^{n+1} p_{i-1}^{n+1} + \left(T_{i-\frac{1}{2}}^{n+1} + T_{i+\frac{1}{2}}^{n+1} + C_{p,i}^{n+1}\right) p_i^{n+1} - T_{i+\frac{1}{2}}^{n+1} p_{i+1}^{n+1}$$

$$= C_{p,i}^{n+1} p_i^n + q_{sc,i}^{n+1} + \sum_{k=1}^{t/\Delta t} \psi(\gamma, k) \delta_i^{n+1-k}, \tag{6.88}$$

where δ_i^{n+1-k} is the finite-difference kernel given by:

$$\delta_i^{n+1-k} = \left(T_{i-\frac{1}{2}}^{n+1-k} p_{i-1}^{n+1-k} - \left(T_{i-\frac{1}{2}}^{n+1-k} + T_{i+\frac{1}{2}}^{n+1-k}\right) p_i^{n+1-k} + T_{i+\frac{1}{2}}^{n+1-k} p_{i+1}^{n+1-k} \right), \tag{6.89}$$

Eq. (6.89) is applicable to the interior control volumes, where i is the grid counter and N_x is the number of control volumes in the x direction. That is, $1 < i < N_x$.

6.5.6 Numerical simulation

To test the finite difference scheme, Islam et al. (2016) consider a reservoir of length 3200 ft, width 300 ft, and height of 100 ft. The initial porosity and permeability are given by 35% and 235 mD respectively. The reservoir is assumed to be completely sealed at the right boundary and a specified flux boundary condition at the left boundary, for which the initial pressure is 7500 psi. The properties of the crude used for this computation have the following ranges of measured properties presented in Table 6.11. The reservoir is deemed to have a fractional-order derivative of order 0.9, i.e., ($\gamma = 0.9$) with $\Delta x = 20$ feet, $\Delta t = 0.05$ days and $t = 20$ days. To solve this flow problem Eq. (6.88) is implemented with MATLAB.

The memory-based flow model is solved using Eq. (6.88). To solve this equation, however, $p(x,0) = p_0$ is applied as an initial condition.

Boundary conditions: Both internal and external boundaries are considered constant-pressure boundaries.

To implement the boundary conditions, Eq. (6.88) has to be modified to incorporate them.

6.6 Results and discussion

Four hypothetical cases have been tested.

Case 1: Model Validation with neither production nor injection wells present in the reservoir with constant pressure maintained at the right boundary.

TABLE 6.11 Other simulation inputs.

Parameter	Value
ϕ_c	0.025
C	0.05
d	250 μm
API	31
T	200 °F
R_s	129 scf/stb
γ_g	0.748
Q_{xo}	350 bbl/d
p_{xL}	7500 psi
c_s	3.5e-6 psi^{-1}
c_μ	8.422e-5 psi^{-1}

It seemed logical to validate the proposed memory model with the classical fluid flow model in order to ascertain the accuracy of the proposed scheme. From Eq. (6.61), when $\gamma = 1$, the Grunwald-Letnikov operator reduces to the identity operator. Therefore, from Eq. (6.60) we have:

$$u = -\frac{\beta_c [K(\phi_c, \phi, C, d)]}{\mu(p)} \left(\frac{\partial p}{\partial x}\right), \tag{6.90}$$

Following this concept, our proposed model was validated with the results presented in Fig. 6.11. We observed that for the total length of simulation time the prediction of pressures by both models was equal (Fig. 6.11).

Case 2: No producer or injector present in the reservoir with constant pressure maintained at the right boundary.

In this case, the pressure variation in the reservoir for different values of γ in the newly proposed flow model was investigated. In this case, it was assumed that the pressure disturbance was due to an influx from the left boundary. In addition, no producer or injector wells were present within the reservoir domain. Fig. 6.12 presents the variation of pressure with distance throughout the reservoir domain based on the new memory-based flow equations for different values of γ. The developed numerical scheme has been found to be unconditionally stable, i.e., regardless of the value of γ consistent and accurate numerical results are always obtained.

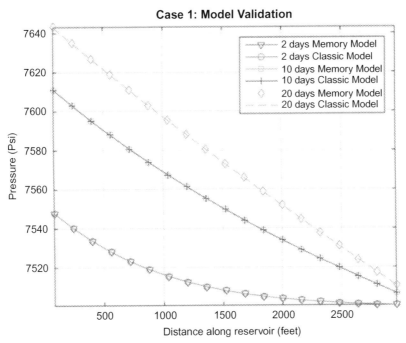

FIG. 6.11 Case 1: Memory model validation with Classic Model.

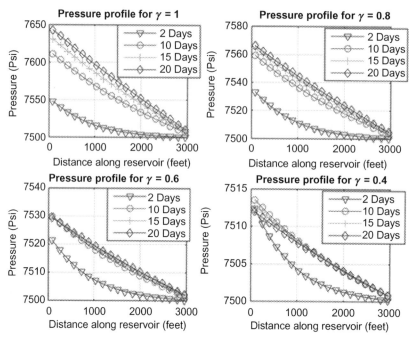

FIG. 6.12   Case 2: Pressure profile across reservoir length for different values of γ.

The term γ has a significant effect on the predicted pressure profile within the reservoir, as shown in Fig. 6.12. Although at first glance we noticed that the predicted pressures were higher as the value of γ increases, likewise the diffusion process seemed a bit slower. Therefore, it seems logical to investigate the influence of this term at different simulation times.

To understand the effect of γ, we carried out a sensitivity study of this term at different time intervals. Due to space restrictions, however, we can only present the results of the pressure along the reservoir after 2, 10, and 20 days for different values of γ as shown in Fig. 6.13.

From Fig. 6.13, we conclude that as the fractional order of derivative increases the pressure profile predicted across the reservoir length increases and vice versa. It should be mentioned that after 2 days the effect of the fractional order beyond 2000 ft is negligible (Fig. 6.13A) since the pressures are almost identical. This might be due to the fact that the effect of the influx on the left boundary has not progressed more than 2000 ft into the reservoir at that time. Finally, we can also conclude based on Fig. 6.13 that as time increases the effect of γ on the predicted pressure becomes significant. Therefore, for the proposed model, accurate or reliable estimates of γ are essential for proper production forecasting.

Case 3: Producers and injectors distributed in the reservoir with constant pressure maintained at the right boundary.

In this section, we introduce a producer and an injector into the reservoir domain. For simplicity, both wells are operated under a constant flow rate for the whole simulation runtime. The location and distribution of the wells within the reservoir are presented in Table 6.12.

The variation of pressure across the reservoir at different time intervals for the two values of γ is presented in Figs. 6.14 and 6.15. For all values of γ, the reservoir pressure profile has

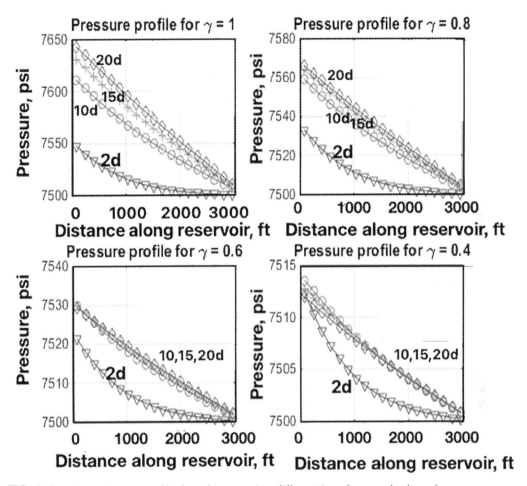

FIG. 6.13 Case 2: Pressure profile along the reservoir at different times for several values of γ.

TABLE 6.12 Distribution of injector and producer in the domain.

Position	Rate (STB/d)
10	−300
50	200

some troughs and peaks at the locations where the wells are situated. For example, the trough at around 200 ft. corresponds to the effect of the production well, hence the observed reduction in reservoir pressure. For the injection well located around 1000 ft. we notice a peak. Furthermore, a similar conclusion can be made about the predicted pressure profiles; that as the value of γ increases the predicted pressure within the hydrocarbon reservoir increases.

6. Reserves prediction and deliverability

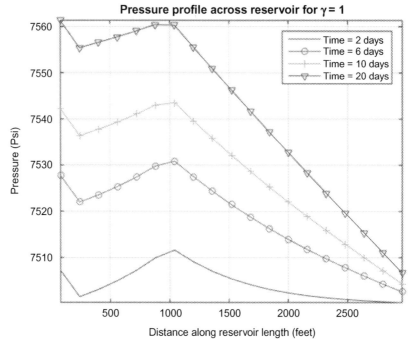

FIG. 6.14 Case 3: Pressure profile across reservoir length with source terms for $\gamma = 1$.

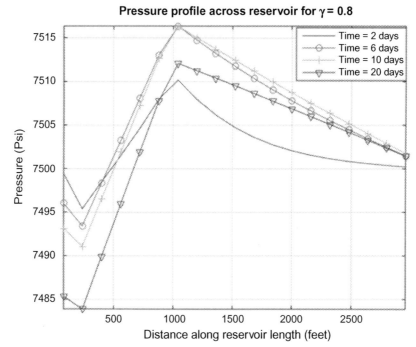

FIG. 6.15 Case 3: Pressure profile across reservoir length with source terms for $\gamma = 0.8$.

6.6 Results and discussion

Typically, the wellbore flowing pressure is very useful information, especially in transient well testing such as drawdown and fall-off analysis. In the next section, we investigate the effect of γ on the measured wellbore pressures, i.e. the injector and the producer.

One can notice in Fig. 6.16B that initially there is a drop in the producer wellbore pressure as a result of oil production (drawdown); however, there is a reversal in this trend due to the influx from the left boundary. That is, the producer's wellbore pressure starts to increase only in the case of $\gamma = 1$. On the other hand, the injector's wellbore pressure (Fig. 6.16A) initially increases in all cases due to the continuous injection of fluid. However, for values of γ less than 1 there is a change, i.e. the wellbore pressure of this well starts to drop due to the sub-diffusion process occurring in the hydrocarbon reservoir. It is observed from this study that the value of γ is of great significance in this proposed model.

Case 4: One producer in the reservoir with no flow boundaries at both sides.

To understand the behavior of the function η, we assume that both boundaries of the reservoir are sealed completely. A well (at Block 1) is producing at a rate of 1500 STB/d.

Variation of pressure with distance: The pressure profile across the reservoir based on the proposed memory model for $\gamma = 0.9$ at different times is shown in Fig. 6.17. As expected due to oil production from the well situated at block 1 the pressure within the reservoir decreases with time. Fig. 6.17 clearly shows a very large pressure gradient is observed very close to the well location.

Variation of wellbore pressure with time. The wellbore pressure is shown in Fig. 6.18 below. As with the previous cases, the wellbore flowing pressure drops with time due to

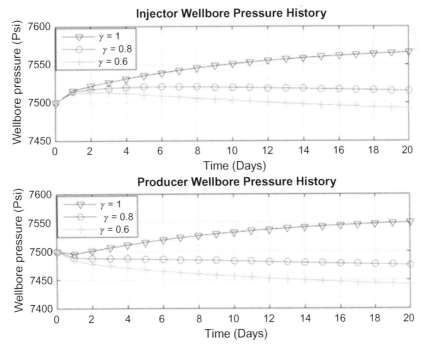

FIG. 6.16 (A) and (B) Injector and Producer wellbore pressure history for different values of γ.

6. Reserves prediction and deliverability

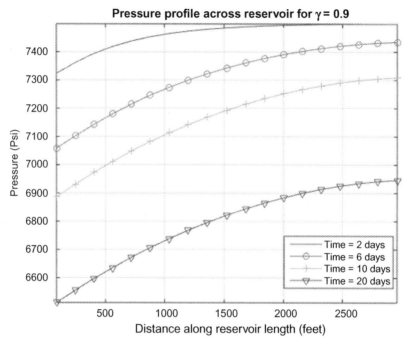

FIG. 6.17 Case 4: Pressure profile across the reservoir for $\gamma = 0.9$.

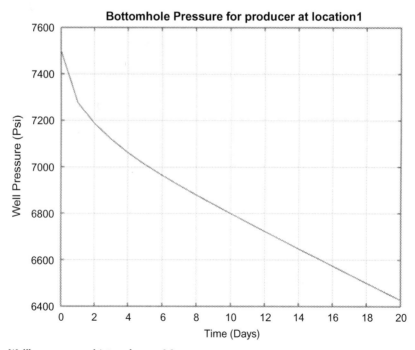

FIG. 6.18 Wellbore pressure history for $\gamma = 0.9$.

the production of oil. (this paragraph and the figure below should be moved from the end of the paper to this location, which is more appropriate. Change all subsequent figure numbers).

Variation of η with distance. The variation of η along the reservoir towards the outer boundary is shown in Fig. 6.19 at different time intervals for $\gamma = 0.9$.

We observe that the variation of η is significant around the wellbore due to the large pressure gradient observed at that location. However, away from the wellbore, this term is almost constant, which is a trend observed at all times. Furthermore, as time proceeds the value of η increases since this term is a function of time too. The proposed memory model is able to capture this variation through the notion of memory.

Next, the magnitude of the proposed memory term η is compared with its analogous form in the classic fluid model. Fig. 6.20 presents the variation of η along the reservoir for the proposed memory for $\eta = 0.9$ with the classic fluid model at 6 and 20 days.

Porosity change with distance. The variation of porosity across the reservoir at different times is shown in Fig. 6.21. Since the rock porosity is somewhat dependent on the fluid pressure, we expect the porosity profile to be similar in trend with the pressure profile; that is very close to the production well, where a large pressure gradient exists, a relatively rapid change in porosity occurs. This is indeed revealed by Fig. 6.21. Also, the decrease in porosity towards the well is attributed to shrinkage of the pore space caused by the increase in effective pressure exerted on the pores. Logically, we also expect that as the fluid pressure increases, the porosity values increase proportionally.

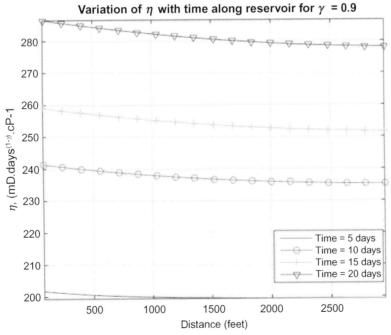

FIG. 6.19 Case 4: Variation of η along the reservoir for $\gamma = 0.9$.

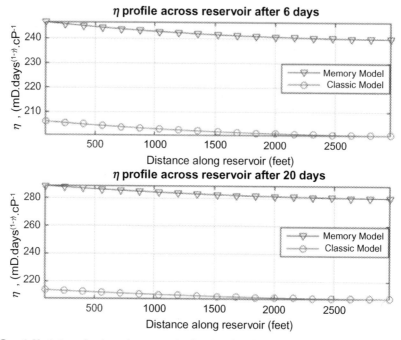

FIG. 6.20 Case 4: Variation of η along the reservoir after 6 and 20 days for the proposed memory model and the classic flow model.

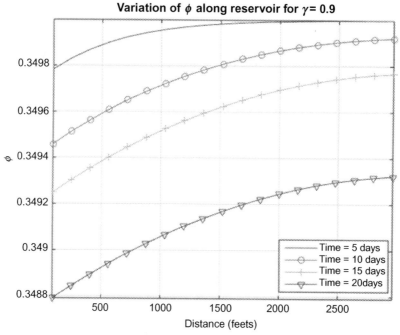

FIG. 6.21 Case 4: Porosity profile at different times for proposed memory model.

6.6 Results and discussion

Again, we compare the porosity distribution across the reservoir using the classic model and the proposed memory model for $\gamma = 0.9$ after 6 and 20 days. We observe from Fig. 6.22A that the porosity values predicted by both models are almost identical. However, as time proceeds the effect of fluid and rock memory becomes significant leading to differences in the porosity values predicted. Fig. 6.22B shows that the Classic (Darcy-based) model predicts higher porosity values near the wellbore region, while further away from the wellbore the proposed memory model predicts higher porosity values.

Porosity change with time. The porosity histories at blocks 4 and 100 of the reservoir model were computed using the proposed memory model with $\gamma = 0.9$. These are shown in Fig. 6.23. The pressure histories observed at both blocks are shown in Fig. 6.24. Due to the large pressure gradient observed very close to the wellbore, i.e., block 4, a rapid reduction in porosity is observed at this block as compared with block 100.

Finally, the wellbore pressure is shown in Fig. 6.25 below. As expected, the wellbore flowing pressure reduces with time due to the production of oil in the reservoir to maintain the flow rate.

In this section, a rigorous model is proposed to describe fluid flow in a porous medium that considers the fluid and medium properties as space and time-dependent. This model can also be applicable to any non-Newtonian fluid flow during an enhanced oil recovery (EOR) process. Furthermore, we have successfully introduced a fully implicit, unconditionally stable numerical scheme to solve the non-linear, time-fractional flow equation when the partial time-fractional derivative is interpreted in the sense of Grunwald-Letnikov. The algorithm

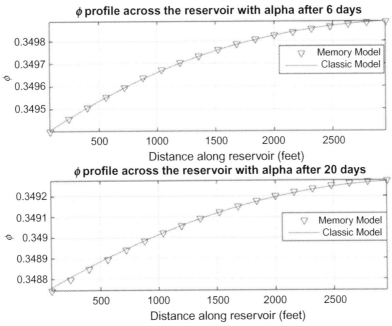

FIG. 6.22 Case 4: Porosity profiles along the reservoir as predicted by the classic model and the memory model.

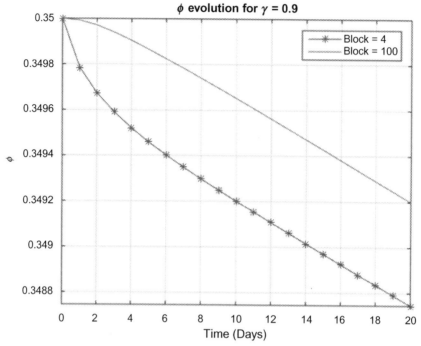

FIG. 6.23 Porosity history at block 4 and 100 for $\gamma = 0.9$.

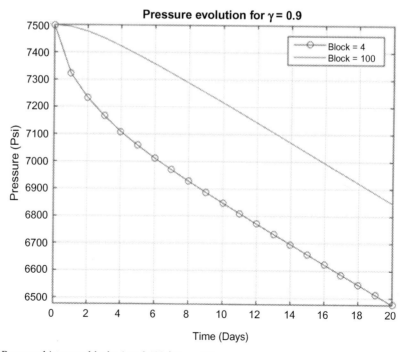

FIG. 6.24 Pressure history at blocks 4 and 100 for $\gamma = 0.9$.

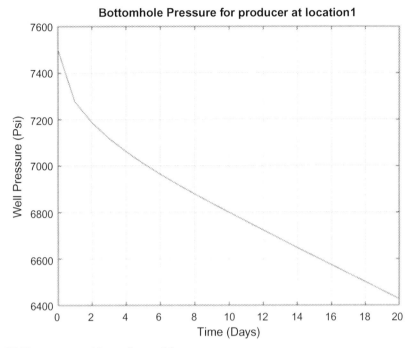

FIG. 6.25 Wellbore pressure history for $\gamma = 0.9$.

is based on a simple approximation to the time-fractional derivative and requires at each time step an iterative solution of a triangular system of nonlinear algebraic equations.

6.7 The compositional simulator using engineering approach

The basic formulations of the compositional model are described for an infinitesimal elementary volume in Chapter 5 of Islam et al. (2016). These are partial differential equations that are nonlinear. Due to the complexity of the reservoir properties and geometry, the nonhomogeneity of the reservoir fluid, and the nonlinear behavior of the partial differential equations governing the flow inside the porous matrix, analytical solutions can be obtained for very simplified cases and geometries, such as the Buckley-Leverett solution for oil-water production without the effect of capillary pressure. However, the analytical solution for the case of oil/water production without the effect of capillary pressure provides multiple solutions that need special attention to check the validity of them and the possibilities of the existence of different solutions for a certain problem.

In general, a numerical solution of the flow equations in porous media is imperative. The mathematical model should be converted into numerical formulations that are solved for a set of discrete points in the reservoir. These numerical formulations are normally a set of algebraic equations that are called discretized equations. The algebraic equations involving the unknown values of dependent variables are derived from the governing partial differential equations. The values of all the dependent variables at these grid points are considered as basic unknowns. Different algorithms may be applied to solve the discretized equations. In general, the method should be fast and provide a set of valid solutions with an acceptable degree of accuracy. The region that is attributed to a certain grid point is called the gridblocks. The number of gridblocks depends on the complexity of the reservoir properties and geometry, applied algorithm, and the capabilities of the computational hardware facilities. It should be selected in a range that provides the solutions in a reasonable time frame.

It is necessary to apply proper numerical methods to handle both heterogeneity and anisotropy. Large discontinuities in medium properties require the construction of numerical schemes with a proper definition of the effective conductivity across cell interfaces. To satisfy local continuity in flux between grid cells with strong discontinuities in permeability, control-volume methods are especially well suited. For time-dependent problems with large solution gradients, it is important that the methods can be combined with a fully implicit time stepping technique.

6.7.1 Finite control volume method

There may be several different ways to derive discretized equations for a given differential equation such as finite-difference; finite control volume; and finite element methods. The control volume methods ensure integral conservation of mass, momentum, and energy over any group of control volumes and over the whole solution domain. This characteristic exists for any number of grid points (and not only for the limiting case of a large number of grid points) and is the most attractive feature from the reservoir engineer's point of view. Therefore, we discuss details of the finite control volume method to solve the partial differential equations derived in Chapter 5.

Integrating the governing partial differential equations over a finite control volume (CV) gives discretized equations in the finite control volume method. The finite control volume may be called a computational cell or a grid block. The first step is the division of the solution domain into a finite number of control volumes or gridblocks. There may be two types of gridblocks, namely, structured and unstructured gridblocks. For the sake of discussing the finite control volume method, we considered a simple, structured grid block. It is important to note that gridblocks should not overlap and each control volume face should be unique to the two gridblocks which lie on either side of it. A typical finite volume gridblocks in 1-, 2-, and 3-dimensional reservoir representation are shown in Fig. 6.26 using a Cartesian coordinate system. Different methods can be employed to select positions of computational nodes and boundaries of computational cells or in other words to generate a structured computational grid. The boundaries of the gridblocks are usually decided as a first step and then a computational node is assigned at the center of each gridblocks.

6.7.2 Reservoir discretization in Cartesian coordinates

Reservoir discretization means that the reservoir is described by a set of blocks (or points) whose properties, dimensions, boundaries, and locations in the reservoir are well defined.

6.7 The compositional simulator using engineering approach 675

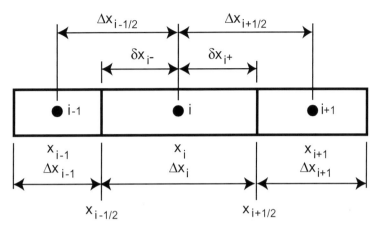

FIG. 6.26 Relationships between block and its neighboring blocks in 1D flow.

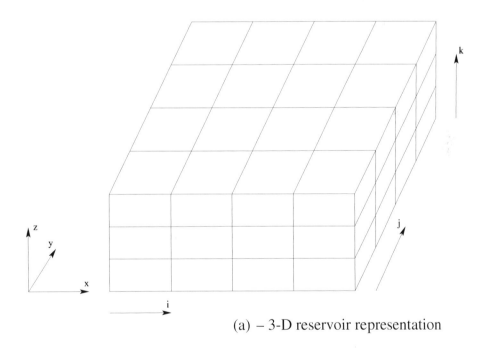

(a) – 3-D reservoir representation

FIG. 6.27 Representation of reservoir in x-, y-, and z-directions.

Fig. 6.27 shows reservoir discretization in the x-direction, using the block-centered grid, as one focuses on the block i. The figure shows how the blocks are related to each other—block i and its neighboring blocks (blocks $i-1$ and $i+1$)—, block dimensions ($\Delta x_i, \Delta x_{i-1}, \Delta x_{i+1}$), block boundaries ($x_{i-1/2}, x_{i+1/2}$), distances between the point that represents the block and block boundaries ($\delta x_{i-}, \delta x_{i+}$), and distances between the points representing the blocks

($\Delta x_{i-1/2}, \Delta x_{i+1/2}$). Reservoir discretization uses similar terminology in the y- and z-directions. Thus, reservoir discretization involves dividing the reservoir into n_x blocks in the x-direction, n_y blocks in the y-direction, and n_z blocks in the z-direction. Any reservoir block in a discretized reservoir is identified as a block (i,j,k), where i, j and k are respectively the orders of the block in x-, y-, and z-directions with $1 \leq i \leq n_x$, $1 \leq j \leq n_y$, $1 \leq k \leq n_z$, see Fig. 6.26. In addition, each block is assigned elevation and rock properties such as porosity and permeability in the x-, y-, and z-directions. The transfer of fluids from one block to the rest of the reservoir takes place through the immediate neighboring blocks. When the whole reservoir is discretized, each block is surrounded by a set (group) of neighboring blocks.

A control volume and the neighboring blocks to it are shown in Fig. 6.28A in 1D flow along the x-axis, Fig. 6.28B in 2D flow in the x-y plane, and Fig. 6.28C in 3D flow in x-y-z space.

Fig. 6.28C shows that block (i,j,k) surrounded by blocks $(i-1,j,k)$ and $(i+1,j,k)$ in the x-direction, by blocks $(i,j-1,k)$ and $(i,j+1,k)$ in the y-direction, and by blocks $(i,j,k-1)$ and $(i,j,k+1)$ in the z-direction. The boundaries between the block (i,j,k) and its neighboring (or surrounding) blocks are termed $(i-1/2,j,k)$ and $(i+1/2,j,k)$ in the x-direction, $(i,j-1/2,k)$ and $(i,j+1/2,k)$ in the y-direction, and $(i,j,k-1/2)$ and $(i,j,k+1/2)$ in the z-direction.

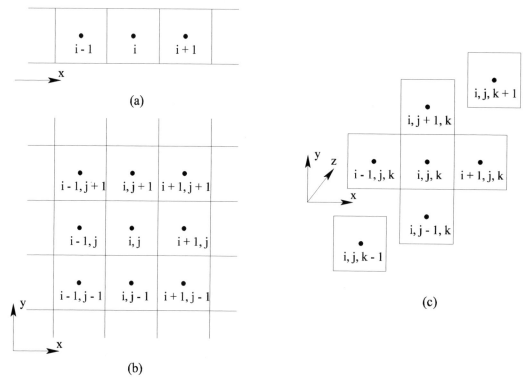

FIG. 6.28 A block and its neighboring blocks: (A) 1D flow, (B) 2D flow; and (C) 3D flow.

6.7.3 Discretization of governing equations

The governing equations are the mass and energy balance equations along with a motion equation such as the Darcy law for a general case of a non-isothermal reservoir. These governing equations should be integrated over each computational cell and over the desired time interval. At first, the mass balance equation for each component of the reservoir fluid is described and then the energy balance equation will be dealt with.

6.7.3.1 Components mass conservation equation

The mass conservation equation for the component i is integrated over the control volume C.V. and the time interval Δt.

$$\int_{t}^{t+\Delta t} \left\{ \int_{C.V} \left[\frac{\partial}{\partial t}\left(\phi \sum_{j} \rho_j y_{i,j} S_j\right) + \nabla \cdot \sum_{j}\left(\rho_j y_{i,j} V_j - \rho_j S_j \underline{D}_{i,j} \nabla y_{i,j}\right) + q_i = 0 \right] dU \right\} dt \quad (6.91)$$

Using the divergence theorem, the integral of the divergent of convective and dispersion part of the mass flux over the control volume can be converted into a surface integral over the boundaries of the control volume S. Therefore, it may be written that:

$$\int_{t}^{t+\Delta t} \left\{ \int_{C.V.} \left[\frac{\partial}{\partial t}\left(\phi \sum_{j} \rho_j y_{i,j} S_j\right) \right] dU + \int_{S}\left[\sum_{j}\left(\rho_j y_{i,j} V_j - \rho_j S_j \underline{D}_{i,j} \nabla y_{i,j}\right)\right] \cdot \mathbf{n}\, dS + \int_{C.V.} q_i\, dU = 0 \right\} dt \quad (6.92)$$

(a) The accumulation term

The first term shows that variation of the mass of component i in a time interval Δt over a control volume say l. It is assumed that each grid block has fixed dimensions and the volume of the grid block l is U_l. It is also assumed that the properties of each fluid component are homogeneous along each control volume. The rock properties are also fixed throughout a given grid block. Thus, the integration of the first term may be given:

$$\int_{t}^{t+\Delta t} \left[\int_{C.V.} \frac{\partial}{\partial t}\left(\phi \sum_{j} \rho_j y_{i,j} S_j\right) dU \right] dt = \int_{C.V.} \left[\int_{t}^{t+\Delta t} \frac{\partial}{\partial t}\left(\phi \sum_{j} \rho_j y_{i,j} S_j\right) dt \right] dU$$

$$= \int_{C.V.} \left(\phi \sum_{j} \rho_j y_{i,j} S_j\right)\bigg|_{t}^{t+\Delta t} dU \quad (6.93)$$

$$= \left[\left(\phi \sum_{j} \rho_j y_{i,j} S_j\right)_{t+\Delta t} - \left(\phi \sum_{j} \rho_j y_{i,j} S_j\right)_{t}\right]_l U_l$$

It is assumed $t = n \cdot \Delta t$, or in the other word the time t corresponds to the time step n and the time $t + \Delta t$ corresponds to time step $n + 1$. Using (6.3), the first term in (6.2) may be given as:

$$\begin{array}{c}\text{The accumulation}\\ \text{of component } i\\ \text{in grid block } l\end{array} = \left[\left(\phi \sum_j \rho_j y_{i,j} S_j\right)_l^{n+1} - \left(\phi \sum_j \rho_j y_{i,j} S_j\right)_l^n\right] U_l \qquad (6.94)$$

(b) The mass flux through the boundaries:

The second bracket in Eq. (6.2) shows the net flux of components i through the boundaries of the control volume l. This surface integral is approximated as the summation of products of the integrand at the center of cell faces and cell faces area.

$$\int_S \left[\sum_j \left(\rho_j y_{i,j} V_j - \rho_j S_j \underline{D}_{i,j} \nabla y_{i,j}\right)\right] \cdot \mathbf{n}\, dS$$
$$= \sum_l \left\{\left[\sum_j \left(\rho_j y_{i,j} V_j - \rho_j S_j \underline{D}_{i,j} \nabla y_{i,j}\right)\right] \cdot \mathbf{n}\right\} S_k \qquad (6.95)$$

The notation S_k is the area of the face k of a given control volume. For a cubic grid block with the faces S_x, S_y and S_z in the direction x, y and z of a Cartesian coordinate system, the convective part of the mass flux of component i may be given according to Eq. (6.5) in the form:

$$\begin{array}{c}\text{The net convective mass}\\ \text{flux of the component } i\end{array} = \sum_l \left(\sum_j \rho_j y_{i,j} V_j\right) \cdot \mathbf{n}\, S_k$$

$$= \left[\left(\sum_j \rho_j y_{i,j} u_j\right)_{x_{i+\frac{1}{2}}} - \left(\sum_j \rho_j y_{i,j} u_j\right)_{x_{i-\frac{1}{2}}}\right] S_x$$
$$+ \left[\left(\sum_j \rho_j y_{i,j} v_j\right)_{y_{i+\frac{1}{2}}} - \left(\sum_j \rho_j y_{i,j} v_j\right)_{y_{i-\frac{1}{2}}}\right] S_y \qquad (6.96)$$
$$+ \left[\left(\sum_j \rho_j y_{i,j} w_j\right)_{z_{i+\frac{1}{2}}} - \left(\sum_j \rho_j y_{i,j} w_j\right)_{z_{i-\frac{1}{2}}}\right] S_z$$

Using Eq. (5.17), the mass flux due to dispersion for the cubic control volume is given as:

6.7 The compositional simulator using engineering approach

$$\text{The net dispersive mass flux of the component } i = -\sum_l \left(\sum_j \rho_j S_j \underline{D}_{i,j} \nabla y_{i,j} \right) \cdot n S_k$$

$$= \begin{bmatrix} \left(\sum_j \rho_j S_j \left(D_{x,x}^{i,j} + D_{y,x}^{i,j} + D_{z,x}^{i,j} \right) \frac{\partial y_{i,j}}{\partial x} \right)_{x_{i-\frac{1}{2}}} \\ -\left(\sum_j \rho_j S_j \left(D_{x,x}^{i,j} + D_{y,x}^{i,j} + D_{z,x}^{i,j} \right) \frac{\partial y_{i,j}}{\partial x} \right)_{x_{i+\frac{1}{2}}} \end{bmatrix} S_x$$

$$+ \begin{bmatrix} \left(\sum_j \rho_j S_j \left(D_{x,y}^{i,j} + D_{y,y}^{i,j} + D_{z,y}^{i,j} \right) \frac{\partial y_{i,j}}{\partial y} \right)_{y_{i-\frac{1}{2}}} \\ -\left(\sum_j \rho_j S_j \left(D_{x,y}^{i,j} + D_{y,y}^{i,j} + D_{z,y}^{i,j} \right) \frac{\partial y_{i,j}}{\partial y} \right)_{y_{i+\frac{1}{2}}} \end{bmatrix} S_y \qquad (6.97)$$

$$+ \begin{bmatrix} \left(\sum_j \rho_j S_j \left(D_{x,z}^{i,j} + D_{y,z}^{i,j} + D_{z,z}^{i,j} \right) \frac{\partial y_{i,j}}{\partial z} \right)_{z_{i-\frac{1}{2}}} \\ -\left(\sum_j \rho_j S_j \left(D_{x,z}^{i,j} + D_{y,z}^{i,j} + D_{z,z}^{i,j} \right) \frac{\partial y_{i,j}}{\partial z} \right)_{z_{i+\frac{1}{2}}} \end{bmatrix} S_z$$

The summation of Eqs. (6.6), (6.8) gives the total flux at the time t.

$$\text{The total net mass flux of the component } i = \sum_l \left\{ \left[\sum_j \left(\rho_j y_{i,j} V_j - \rho_j S_j \underline{D}_{i,j} \nabla y_{i,j} \right) \right] \cdot n \right\} S_k$$

$$= \begin{bmatrix} \left(\sum_j \rho_j \left[y_{i,j} u_j - S_j \left(D_{x,x}^{i,j} + D_{y,x}^{i,j} + D_{z,x}^{i,j} \right) \frac{\partial y_{i,j}}{\partial x} \right] \right)_{x_{i+\frac{1}{2}}} \\ -\left(\sum_j \rho_j \left[y_{i,j} u_j - S_j \left(D_{x,x}^{i,j} + D_{y,x}^{i,j} + D_{z,x}^{i,j} \right) \frac{\partial y_{i,j}}{\partial x} \right] \right)_{x_{i-\frac{1}{2}}} \end{bmatrix} S_x$$

$$+ \begin{bmatrix} \left(\sum_j \rho_j \left[y_{i,j} v_j - S_j \left(D_{x,y}^{i,j} + D_{y,y}^{i,j} + D_{z,y}^{i,j} \right) \frac{\partial y_{i,j}}{\partial y} \right] \right)_{y_{i+\frac{1}{2}}} \\ -\left(\sum_j \rho_j \left[y_{i,j} v_j - S_j \left(D_{x,y}^{i,j} + D_{y,y}^{i,j} + D_{z,y}^{i,j} \right) \frac{\partial y_{i,j}}{\partial y} \right] \right)_{y_{i-\frac{1}{2}}} \end{bmatrix} S_y \qquad (6.98)$$

$$+ \begin{bmatrix} \left(\sum_j \rho_j \left[y_{i,j} w_j - S_j \left(D_{x,z}^{i,j} + D_{y,z}^{i,j} + D_{z,z}^{i,j} \right) \frac{\partial y_{i,j}}{\partial z} \right] \right)_{z_{i+\frac{1}{2}}} \\ -\left(\sum_j \rho_j \left[y_{i,j} w_j - S_j \left(D_{x,z}^{i,j} + D_{y,z}^{i,j} + D_{z,z}^{i,j} \right) \frac{\partial y_{i,j}}{\partial z} \right] \right) \end{bmatrix} S_z$$

Eq. (6.8) gives the integrand of the time integral between the time t and $t + \Delta t$. The integrand is replaced by the function $F(t)$. The integral of $F(t)$ the interval Δt is equal to the area under the curve $F(t)$ between t and $t + \Delta t$. This area is equal to the area of a rectangle as shown in Fig. 6.29. $F(t^m)$ is an average value of $F(t)$ in the interval of Δt. It is set that $F^m = F(t^m)$.

$$\int_{t}^{t+\Delta t} F(t)dt = F^m \times \Delta t \tag{6.99}$$

The value of this integral can be calculated using the aforementioned equation provided that the value of F^m is known. In reality, however, F^m is not known and, therefore, it needs to be approximated. The approximation can be obtained by using different types of numerical integration. Some of them are illustrated in Fig. 6.30.

(a) Fig. 6.30A shows a fully explicit approximation of the integral $\int_{t}^{t+\Delta t} F(t)dt$. The value of $F(t)$ at time step n is considered to approximate the integration.

$$\int_{t}^{t+\Delta t} F(t)dt = F(t^n) \times \Delta t \tag{6.100}$$

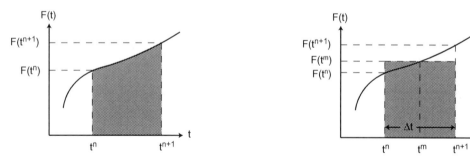

FIG. 6.29 Representation of the time integral of $F(t)$ as the area under the curve (left) and the equivalent of it as the area under a rectangle with a dimension equal to an average value $F(t^m)$.

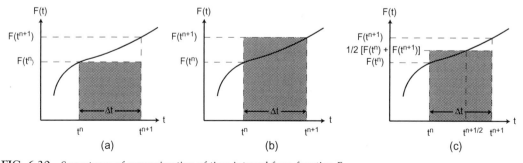

FIG. 6.30 Some types of approximation of time integral for a function F.

6.7 The compositional simulator using engineering approach

(b) It may apply a fully implicit scheme and apply the value of the function F(t) at a time step (n + 1). It is illustrated in Fig. 6.30B.

$$\int_{t}^{t+\Delta t} F(t)dt = F(t^{n+1}) \times \Delta t \quad (6.101)$$

(c) The trapezoidal integration rule may also be applied and approximate the function F(t) with a linear function. This is shown in Fig. 6.30C.

$$\int_{t}^{t+\Delta t} F(t)dt = \frac{1}{2}\left[F(t^{n}) + F(t^{n+1})\right] \times \Delta t \quad (6.102)$$

Using Eq. (6.9), the total net mass flux through the boundaries of the control volume l in a time interval Δt can be given in the following form.

$$\text{The total net mass flux of the component } i = \left\{\sum_{k}\left\{\left[\sum_{j}\left(\rho_j y_{i,j} V_j - \rho_j S_j \underline{D}_{i,j} \nabla y_{i,j}\right)\right] \cdot \mathbf{n}\right\} S_k\right\}^m \times \Delta t \quad (6.103)$$

The total mass flux in an extended form can be given by using Eq. (6.8).

$$\text{The total net mass flux of the component } i = \left\{ \begin{aligned} & \left[\begin{aligned} & \left(\sum_{j} \rho_j \left[y_{i,j} u_j - S_j \left(D_{x,x}^{i,j} + D_{y,x}^{i,j} + D_{z,x}^{i,j} \right) \frac{\partial y_{i,j}}{\partial x} \right] \right)_{x_{i+\frac{1}{2}}} \\ & - \left(\sum_{j} \rho_j \left[y_{i,j} u_j - S_j \left(D_{x,x}^{i,j} + D_{y,x}^{i,j} + D_{z,x}^{i,j} \right) \frac{\partial y_{i,j}}{\partial x} \right] \right)_{x_{i-\frac{1}{2}}} \end{aligned} \right] S_x \\ & + \left[\begin{aligned} & \left(\sum_{j} \rho_j \left[y_{i,j} v_j - S_j \left(D_{x,y}^{i,j} + D_{y,y}^{i,j} + D_{z,y}^{i,j} \right) \frac{\partial y_{i,j}}{\partial y} \right] \right)_{y_{i+\frac{1}{2}}} \\ & - \left(\sum_{j} \rho_j \left[y_{i,j} v_j - S_j \left(D_{x,y}^{i,j} + D_{y,y}^{i,j} + D_{z,y}^{i,j} \right) \frac{\partial y_{i,j}}{\partial y} \right] \right)_{y_{i-\frac{1}{2}}} \end{aligned} \right] S_y \\ & + \left[\begin{aligned} & \left(\sum_{j} \rho_j \left[y_{i,j} w_j - S_j \left(D_{x,z}^{i,j} + D_{y,z}^{i,j} + D_{z,z}^{i,j} \right) \frac{\partial y_{i,j}}{\partial z} \right] \right)_{z_{i+\frac{1}{2}}} \\ & - \left(\sum_{j} \rho_j \left[y_{i,j} w_j - S_j \left(D_{x,z}^{i,j} + D_{y,z}^{i,j} + D_{z,z}^{i,j} \right) \frac{\partial y_{i,j}}{\partial z} \right] \right)_{z_{i-\frac{1}{2}}} \end{aligned} \right] S_z \end{aligned} \right\}^m \times \Delta t \quad (6.104)$$

The average value of the integrand may be obtained by using the methods that have already been described for a function $F(t)$.

(c) The source term:

The source term is associated with the production rate of the component i or the injection rate of it into the reservoir in a time interval Δt. The component i may produce in different phases. It can be associated with the phases volumetric rates from a single well located in the block l as shown in Fig. 6.31.

$$\int_{C.V.} q_i dU = \sum_j y_{ij} \rho_j Q_j \tag{6.105}$$

where Q_j is the volumetric rate of phase j at a given time t. The total production rate between t and $t + \Delta t$ is obtained by integrating over that time interval. Using Eq. (6.9), it may be given by an average value multiply by the time interval Δt.

$$\int_t^{t+\Delta t} \left(\int_{C.V.} q_i dU \right) dt = \left(\sum_j y_{ij} \rho_j Q_j \right)^m \times \Delta t \tag{6.106}$$

Fluid production rates in multiphase flow are dependent on each other through at least the relative permeability of different phases. In other words, the specification of the production rate of any phase implicitly dictates the production rates of the other phases. The concern in single-block wells is to estimate the production rate of phase j from well block l under different well-operating conditions.

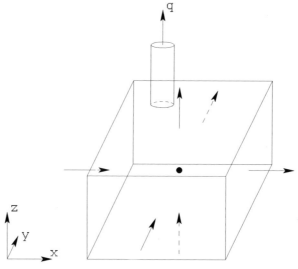

FIG. 6.31 A grid block l with production well with a rate $Q = \sum_j Q_j$.

The mass balance equation in discretized form can be obtained by substituting Eqs. (6.4), (6.13), and (6.19) in Eq. (6.3).

$$\frac{U}{\Delta t}\left[\left(\phi\sum_j \rho_j y_{i,j} S_j\right)^{n+1} - \left(\phi\sum_j \rho_j y_{i,j} S_j\right)^n\right] + \left\{\sum_k \left\{\left[\sum_j \left(\rho_j y_{i,j} V_j - \rho_j S_j \underline{D}_{i,j} \nabla y_{i,j}\right)\right] \cdot \mathbf{n}\right\} S_k\right\}^m$$

$$+ \left(\sum_j y_{ij}\rho_j Q_j\right)^m = 0$$

(6.107)

The Eq. (6.19) is for a grid block l of the discretized reservoir. If there are n_x gridblocks in the x-direction, n_y gridblocks in y-direction, and n_z gridblocks in z-direction, there will be a total of $n_x \times n_y \times n_z$ discretized mass balance equations for the component i. If there exist N_c components in the reservoir fluid, the total number of discretized mass balance equations is $N_c \times (n_x \times n_y \times n_z)$. The solution of these sets of algebraic equations provides the unknown variables.

6.7.3.2 Energy balance equation

The energy balance equation is given in Eq. (5.39) for the case of $T_f = T_r = T$ and in Eqs. (5.40), (5.41) for the case when $T_f \neq T_r$. These equations are integrated over the control volume C.V. and the time interval Δt to obtain the control volume formulations.

Case a: Equal temperature $T_f = T_r = T$

We consider the case of equal temperature for the rock and fluid and the discretized formulation is derived for a grid block l. The partial differential equation for the energy balance equation for the case of $T_f = T_r = T$ is integrated over the finite control volume C.V. and the time interval Δt.

$$\int_t^{t+\Delta t}\left\{\int_{C.V.}\left\{\frac{\partial}{\partial t}\left[\phi\sum_j \rho_j S_j(\hat{u}_j + \mathbf{g}\cdot \mathbf{R}_j)\right] + \frac{\partial}{\partial t}[(1-\phi)\rho_r C_r(T-T_0)] - \nabla\cdot(k\nabla T)\right.\right.$$
$$\left.\left. -\phi\left\{\nabla\cdot\left[\sum_j \rho_j S_j(h_j + \mathbf{g}\cdot \mathbf{R}_j)V_j\right]\right\} + e_s\right\}dU\right\}dt = 0$$

(6.108)

Using the divergence theorem, the integral of the divergent of the energy flux due to the conduction and convection over the control volume can be converted into a surface integral over the boundaries of the control volume S. Therefore, it may be written that:

$$\int_t^{t+\Delta t}\left\{\int_{C.V.}\left\{\frac{\partial}{\partial t}\left[\phi\sum_j \rho_j S_j(\hat{u}_j + \mathbf{g}\cdot \mathbf{R}_j)\right] + \frac{\partial}{\partial t}[(1-\phi)\rho_r C_r(T-T_0)]\right\}dU \right.$$
$$\left. -\int_S \left[(k\nabla T) + \phi\sum_j \rho_j S_j(h_j + \mathbf{g}\cdot \mathbf{R}_j)V_j\right]\cdot\mathbf{n}\,ds + \int_{C.V.} e_s\,dU\right\}dt = 0$$

(6.109)

The energy balance equation in discretized form can be obtained with the same procedure as given for the mass balance equation in 6.1.2.1. The final equation for the case of equal temperature is:

$$\begin{aligned}\frac{U}{\Delta t} &\left\{\left[\phi\sum_j\rho_jS_j(\hat{u}_j+g\cdot R_j)+(1-\phi)\rho_rC_r(T-T_0)\right]^{n+1}\right.\\ &\left.-\left[\phi\sum_j\rho_jS_j(\hat{u}_j+g\cdot R_j)+(1-\phi)\rho_rC_r(T-T_0)\right]^n\right\}\\ &+\left\{\sum_k\left\{\left[(k\nabla T)+\phi\sum_j\rho_jS_j(h_j+g\cdot R_j)V_j\right]\cdot n\right\}S_k\right\}^m+(E_s)^m=0\end{aligned} \quad (6.110)$$

where $E_s = \int_{C.V.} e_s\, dU$.

Case b: Different temperature $T_f \neq T_r$

Eqs. (5.40), (5.41) are the energy balance equations for the case where the rock and fluid temperatures are not equal. These partial differential equations are integrated over the finite control volume C.V. and the time interval Δt.

$$\begin{aligned}&\int_t^{t+\Delta t}\left\{\int_{C.V.}\left\{\frac{\partial}{\partial t}\left[\phi\sum_j\rho_jS_j(\hat{u}_j+g\cdot R_j)\right]-\nabla\cdot(k_f\nabla T_f)\right.\right.\\ &\left.\left.-\phi[\nabla\cdot(e_f V)]+h(T_r-T_f)+e_s\right\}dU\right\}dt=0\\ &\int_t^{t+\Delta t}\left\{\int_{C.V.}\left\{\frac{\partial}{\partial t}[(1-\phi)\rho_rC_r(T_r-T_0)]-\nabla\cdot(k_r\nabla T_r)-h(T_r-T_f)\right\}dU\right\}dt=0\end{aligned} \quad (6.111)$$

The notation $e_f = \sum_j\rho_jS_j(h_j+g\cdot R_j)$ is the total energy of a unit volume of the fluid system. These sets of equations can be written by using the divergence theorem as follows:

$$\begin{aligned}&\int_t^{t+\Delta t}\left\{\int_{C.V.}\left\{\frac{\partial}{\partial t}\left[\phi\sum_j\rho_jS_j(\hat{u}_j+g\cdot R_j)\right]\right.\right.\\ &\left.\left.+h(T_r-T_f)+e_s\right\}dU-\int_S[k_f\nabla T_f+\phi e_f V]\cdot n\, ds\right\}dt=0\\ &\int_t^{t+\Delta t}\left\{\int_{C.V.}\left\{\frac{\partial}{\partial t}[(1-\phi)\rho_rC_r(T_r-T_0)]\right.\right.\\ &\left.\left.-h(T_r-T_f)\right\}dU-\int_S k_r\nabla T_r\cdot n\, ds\right\}dt=0\end{aligned} \quad (6.112)$$

The discretized form of energy balance equation for the general case of $T_f \neq T_r$ is:

$$\begin{cases} \dfrac{U}{\Delta t}\left\{\left[\phi\sum_j\rho_j S_j(\hat{u}_j+g\cdot R_j)\right]^{n+1} \\ -\left[\phi\sum_j\rho_j S_j(\hat{u}_j+g\cdot R_j)\right]^{n}\right\} + \left\{\sum_k\{[k_f\nabla T_f+\phi e_f\, V]\cdot n\}S_k \\ +\int_{C.V.} h(T_r-T_f)dU+E_s\right\}^m = 0 \\[1em] \dfrac{U}{\Delta t}\left\{\begin{array}{l}[(1-\phi)\rho_r C_r(T_r-T_0)]^{n+1}\\ -[(1-\phi)\rho_r C_r(T_r-T_0)]^n\end{array}\right\} + \left\{\sum_k k_r\nabla T_r\cdot nS_k \\ -\int_{C.V.} h(T_r-T_f)dU\right\}^m = 0 \end{cases} \quad (6.113)$$

6.7.4 Discretization of motion equation

The motion equations are described in Chapter 5, Section 5.2.6. They describe the relationship for fluid volumetric velocity. A petroleum reservoir is generally a heterogeneous anisotropic medium. In the most simplified form, the volumetric flux of a component α of the reservoir fluid is approximated by the Darcy law, $V_j = -\dfrac{\underline{K}_j}{\mu_j\nabla\Phi_j}$ as given in Eq. (5.57). Using the notation of relative permeability, the Darcy law can be given in the following form.

$$V_j = -\underline{K}\dfrac{k_{rj}}{\mu_j}\nabla\Phi_j = -\underline{K}\lambda_j\nabla\Phi_j \quad (6.114)$$

The notation λ_j is the mobility of phase j. The notation $\underline{K}$ is the space-dependent permeability tensor. Off-diagonal elements in the permeability tensor may exist if the coordinate directions are not aligned with the principal directions $\underline{K}$.

If a one-dimensional fluid flow is considered in x-direction, the discretized form of flow rate per unit cross-sectional area (the volumetric velocity) for a phase j from a block $i-1$ to block i as shown in Fig. 6.31 according to Darcy's law Eq. (6.26) is:

$$V_{jx}|_{x-1/2} = \dfrac{k_x k_{rj}}{\mu_j}\dfrac{\Phi_{j_{i-1}}-\Phi_{j_i}}{\Delta x_{i-1/2}} \quad (6.115)$$

The potential difference between block $i-1$ and block i is

$$\Phi_{j_{i-1}}-\Phi_{j_i} = p_{j_{i-1}}-p_{j_i}-\rho_{j_{i-1}}g(z_{j_{i-1}}-z_{j_i}) \quad (6.116)$$

for phase j. Substituting Eq. (6.28) in Eq. (6.27) yields

$$V_{jx}|_{x-1/2} = \dfrac{k_x k_{rj}}{\mu_j}\dfrac{(p_{j_{i-1}}-p_{j_i})-\rho_{j_{i-1}}g(z_{j_{i-1}}-z_{j_i})}{\Delta x_{i-1/2}} \quad (6.117)$$

Eq. (6.29) can be rewritten as:

$$V_{jx}|_{x-1/2} = \dfrac{k_x|_{x_{i-1/2}}}{\Delta x_{i-1/2}}\left(\dfrac{k_{rj}}{\mu_j}\right)\bigg|_{x_{i-1/2}}\left[(p_{j_{i-1}}-p_{j_i})-\rho_{j_{i-1}}g(z_{j_{i-1}}-z_{j_i})\right] \quad (6.118)$$

Likewise, the fluid volumetric velocity of phase j from block $i+1$ to block i in Fig. 6.1 is expressed as:

$$V_{jx}\big|_{x+1/2} = \frac{k_x|_{x_{i+1/2}}}{\Delta x_{i+1/2}} \left(\frac{k_{rj}}{\mu_j}\right)\bigg|_{x_{i+1/2}} \left[(p_{j_{i+1}} - p_{j_i}) - \rho_{j_{i-1}} g(z_{j_{i+1}} - z_{j_i})\right] \qquad (6.119)$$

Note that the fluid volumetric velocity of phase j from block i to block $i+1$ is the negative of the value given by Eq. (6.31). It may be written that:

$$V_{jx}\big|_{x\mp 1/2} = \frac{k_x|_{x_{i\mp 1/2}}}{\Delta x_{i\mp 1/2}} \left(\frac{k_{rj}}{\mu_j}\right)\bigg|_{x_{i\mp 1/2}} \left[(p_{j_{i\mp 1}} - p_{j_i}) - \rho_{j_{i-1}} g(z_{j_{i\mp 1}} - z_{j_i})\right] \qquad (6.120)$$

For multidimensional flow in rectangular coordinates, the fluid volumetric velocity of phase j from neighboring blocks to block (i,j,k) in the x, y, and z directions are:

$$\begin{aligned}
V_{jx}\big|_{x\mp 1/2} &= \frac{k_x|_{x_{i\mp 1/2,j,k}}}{\Delta x_{i\mp 1/2,j,k}} \left(\frac{k_{rj}}{\mu_j}\right)\bigg|_{x_{i\mp 1/2,j,k}} \left[\left(p_{j_{i\mp 1,j,k}} - p_{j_{i,j,k}}\right) - \rho_{j_{i-1}} g\left(z_{j_{i\mp 1,j,k}} - z_{j_{i,j,k}}\right)\right] \\
V_{jy}\big|_{y\mp 1/2} &= \frac{k_y|_{y_{i,j\mp 1/2,k}}}{\Delta} \left(\frac{k_{rj}}{\mu_j}\right)\bigg|_{y_{i,j\mp 1/2,k}} \left[\left(p_{j_{i,j\mp 1,k}} - p_{j_{i,j,k}}\right) - \rho_{j_{i-1}} g\left(z_{j_{i,j\mp 1,k}} - z_{j_{i,j,k}}\right)\right] \\
V_{jz}\big|_{z\mp 1/2} &= \frac{k_z|_{z_{i,j,k\mp 1/2}}}{\Delta z_{i,j,k\mp 1/2}} \left(\frac{k_{rj}}{\mu_j}\right)\bigg|_{z_{i,j,k\mp 1/2}} \left[\left(p_{j_{i,j,k\mp 1}} - p_{j_{i,j,k}}\right) - \rho_{j_{i-1}} g\left(z_{j_{i,j,k\mp 1}} - z_{j_{i,j,k}}\right)\right]
\end{aligned} \qquad (6.121)$$

The fluid volumetric velocity is $\mathbf{V}_j = V_{jx}\hat{i} + V_{jy}\hat{j} + V_{jz}\hat{k}$.

6.7.5 Uniform temperature reservoir compositional flow equations in a 1-D domain

The numerical formulation for the component mass balance equation and the energy balance equation with the Darcy law were explained in previous sections. The numerical model is based on the finite control volume method. The formulations should be combined to obtain a set of equations for practical applications. At first, we consider a constant temperature reservoir and developed the compositional formulation for single and multidimensional reservoirs. We will combine the component mass balance equation with the Darcy law to obtain the flow equation. We start with the case of 1-D domain and then developed the formulation for a multidimensional domain.

The mass balance for component i for a block l in 1-D flow over a time step $\Delta t = t^{n+1} - t^n$ is derived in this chapter using the Darcy law as the motion equation which describes the volumetric velocity relationship. The general form of discretized form of mass balance equation for a 3-D flow is given in Eq. (6.122). It is assumed that there exists no mass transfer due to thermal and molecular diffusion. The mass balance equation for 1-D flow x-direction is simplified as, Fig. 6.1:

$$\sum_j \left\{ \frac{(U)_l}{\Delta t}\left[\frac{\left(\phi \rho_j y_{i,j} S_j\right)^{n+1} - }{\left(\phi \rho_j y_{i,j} S_j\right)^n}\right]_l + \left[\frac{\left(\rho_j y_{i,j}\right)_{i+1/2} V_{jx_{i+1/2}} A_{x_{i+1/2}} -}{\left(\rho_j y_{i,j}\right)_{i-1/2} V_{jx_{i-1/2}} A_{x_{i-1/2}}}\right]^m + \left(y_{ij} \rho_j Q_j\right)^m \right\} = 0 \qquad (6.122)$$

6.7 The compositional simulator using engineering approach

Substituting Eq. (6.33) in Eq. (6.35), it can be written that:

$$\sum_j \left\{ \begin{array}{l} \left[(\rho_j y_{i,j})_{i-1/2}^m \left(\frac{A_x k_x}{\Delta x}\right)_{x_{i-1/2}} \left(\frac{k_{rj}}{\mu_j}\right)^m \bigg|_{x_{i-1/2}} \left[(p_{j_{i-1}} - p_{j_i}) - \rho_{j_{i+1}} g(z_{j_{i-1}} - z_{j_i})\right] + \right. \\ \left. (\rho_j y_{i,j})_{i+1/2}^m \left(\frac{A_x k_x}{\Delta x}\right)_{x_{i+1/2}} \left(\frac{k_{rj}}{\mu_j}\right)^m \bigg|_{x_{i+1/2}} \left[(p_{j_{i+1}} - p_{j_i}) - \rho_{j_{i+1}} g(z_{j_{i+1}} - z_{j_i})\right] \right] \\ - \left(y_{ij} \rho_j Q_j\right)^m = \frac{(U\phi)_l}{\Delta t} \left[(\rho_j y_{i,j} S_j)^{n+1} - (\rho_j y_{i,j} S_j)^n \right]_l \end{array} \right\} \quad (6.123)$$

There may exist three phases of gas, oil, and water, ($j = g, o$ & w) in the reservoir fluid. To consider a simplified and practical sense, the oil and water phases are assumed immiscible; i.e., the hydrocarbon oil components do not dissolve in the water phase and the water component does not dissolve in the oil phase. Furthermore, the water phase does not transport any component other than the water component. The water component may evaporate into the gas phase. Therefore, two-phase oil and gas physical equilibrium describes mass transfer between the oil and gas phases, and the Dalton and Raoult laws describe the mass transfer of the water component between the water and gas phases. Let the oil phase contains n_c hydrocarbon components, the water phase consists of the water component only. Therefore, the gas phase consists of $n_c + 1$ components. For hydrocarbon component i, where $i = 1, \cdots, n_c$,

$$\begin{array}{l} (\rho_g y_{i,g})_{l-1/2}^m \left(\frac{A_x k_x}{\Delta x}\right)_{x_{l-1/2}} \left(\frac{k_{rg}}{\mu_g}\right)_{x_{l-1/2}}^m \left[(p_{g_{l-1}} - p_{g_l}) - \rho_{g_{l-1}} g(z_{g_{l-1}} - z_{g_l})\right] \\ + (\rho_g y_{i,g})_{l+1/2}^m \left(\frac{A_x k_x}{\Delta x}\right)_{x_{l+1/2}} \left(\frac{k_{rg}}{\mu_g}\right)_{x_{l+1/2}}^m \left[(p_{g_{l+1}} - p_{g_l}) - \rho_{g_{l+1}} g(z_{g_{l+1}} - z_{g_l})\right] \\ + (\rho_o y_{i,o})_{l-1/2}^m \left(\frac{A_x k_x}{\Delta x}\right)_{x_{l-1/2}} \left(\frac{k_{ro}}{\mu_o}\right)_{x_{l-1/2}}^m \left[(p_{o_{l-1}} - p_{o_l}) - \rho_{o_{l-1}} g(z_{o_{l-1}} - z_{o_l})\right] \\ + (\rho_o y_{i,o})_{l+1/2}^m \left(\frac{A_x k_x}{\Delta x}\right)_{x_{l+1/2}} \left(\frac{k_{ro}}{\mu_o}\right)_{x_{l+1/2}}^m \left[(p_{o_{l+1}} - p_{o_l}) - \rho_{o_{l+1}} g(z_{o_{l+1}} - z_{o_l})\right] \\ - \left(y_{ig} \rho_g Q_g + y_{io} \rho_o Q_o\right)^m = \frac{(U)_l}{\Delta t} \left[\begin{array}{l} (\phi \rho_g y_{i,g} S_g + \phi \rho_o y_{i,o} S_o)^{n+1} \\ -(\phi \rho_g y_{i,g} S_g + \phi \rho_o y_{i,o} S_o)^n \end{array} \right]_l . \end{array} \quad (6.124)$$

For the water component, where $i = n_c + 1$,

$$\begin{array}{l} (\rho_g y_{i,g})_{l-1/2}^m \left(\frac{A_x k_x}{\Delta x}\right)_{x_{l-1/2}} \left(\frac{k_{rg}}{\mu_g}\right)_{x_{l-1/2}}^m \left[(p_{g_{l-1}} - p_{g_l}) - \rho_{g_{l-1}} g(z_{g_{l-1}} - z_{g_l})\right] \\ + (\rho_g y_{i,g})_{l+1/2}^m \left(\frac{A_x k_x}{\Delta x}\right)_{x_{l+1/2}} \left(\frac{k_{rg}}{\mu_g}\right)_{x_{l+1/2}}^m \left[(p_{g_{l+1}} - p_{g_l}) - \rho_{g_{l+1}} g(z_{g_{l+1}} - z_{g_l})\right] \\ + (\rho_w)_{l-1/2}^m \left(\frac{A_x k_x}{\Delta x}\right)_{x_{l-1/2}} \left(\frac{k_{rw}}{\mu_w}\right)_{x_{l-1/2}}^m \left[(p_{w_{l-1}} - p_{w_l}) - \rho_{w_{l-1}} g(z_{w_{l-1}} - z_{w_l})\right] \\ + (\rho_w)_{l+1/2}^m \left(\frac{A_x k_x}{\Delta x}\right)_{x_{l+1/2}} \left(\frac{k_{rw}}{\mu_w}\right)_{x_{l+1/2}}^m \left[(p_{w_{l+1}} - p_l) - \rho_{w_{l+1}} g(z_{w_{l+1}} - z_{w_l})\right] \\ - \left(y_{ig} \rho_g Q_g + \rho_w Q_w\right)^m = \frac{(U)_l}{\Delta t} \left[\begin{array}{l} (\phi \rho_g y_{i,g} S_g + \phi \rho_w S_w)^{n+1} \\ -(\phi \rho_g y_{i,g} S_g + \phi \rho_w S_w)^n \end{array} \right] \end{array} \quad (6.125)$$

The derivation (6.35) and (6.36) are based on the assumption of the validity of Darcy's Law to estimate the phase volumetric velocities between the block i and its neighboring blocks $i-1$ and $i+1$. Such validity is widely accepted by petroleum engineers.

To simplify presenting the compositional mass balance equations in the multi-dimensional domain, it is defined:

$$\left(\mathbb{T}_{i,o}^m\right)_{l\mp 1/2} = \left(\rho_o y_{i,o}\right)_{l\mp 1/2}^m \left(\frac{A_x k_x}{\Delta x}\right)_{x_{l\mp 1/2}} \left(\frac{k_{ro}}{\mu_o}\right)_{x_{l\mp 1/2}}^m = \left(\rho_o y_{i,o}\right)_{l\mp 1/2}^m \left(\mathbb{T}_o^m\right)_{l\mp 1/2} \quad (6.126)$$

where $i = 1, \cdots, n_c$;

$$\left(\mathbb{T}_{i,g}^m\right)_{l\mp 1/2} = \left(\rho_g y_{i,g}\right)_{l\mp 1/2}^m \left(\frac{A_x k_x}{\Delta x}\right)_{x_{l\mp 1/2}} \left(\frac{k_{rg}}{\mu_g}\right)_{x_{l\mp 1/2}}^m = \left(\rho_g y_{i,g}\right)_{l\mp 1/2}^m \left(\mathbb{T}_g^m\right)_{l\mp 1/2} \quad (6.127)$$

where $i = 1, \cdots, n_c + 1$; and

$$\left(\mathbb{T}_{i,w}^m\right)_{l\mp 1/2} = \left(\rho_w\right)_{l\mp 1/2}^m \left(\frac{A_x k_x}{\Delta x}\right)_{x_{l\mp 1/2}} \left(\frac{k_{rw}}{\mu_w}\right)_{x_{l\mp 1/2}}^m = \left(\rho_w\right)_{l\mp 1/2}^m \left(\mathbb{T}_w^m\right)_{l\mp 1/2} \quad (6.128)$$

where $i = n_c + 1$. According to Abou-Kassem (2007b), it may be defined that:

$$\left(\mathbb{T}_{i,j}^m\right)_{l\mp 1,l} = \left(\mathbb{T}_{i,j}^m\right)_{l\mp 1/2} = G_{l\mp 1/2}\left(\frac{k_{rj}}{\mu_j}\right)_{x_{l\mp 1/2}}^m = G_{l\mp 1,l}\left(\frac{k_{rj}}{\mu_j}\right)_{x_{l\mp 1,l}}^m \quad (6.129)$$

where $G_{l\mp 1,l} = G_{l\mp 1,l} = \left(\frac{A_x k_x}{\Delta x}\right)_{x_{l\mp 1/2}}$ are a geometric function. The notation $(\mathbb{T}_{i,j}^m)_{l\mp 1, l}$ is called the transmissibility between the blocks $l \mp 1, l$ for the phase j. Using the definition (6.129), (6.38), (6.39), and (6.40), the mass balance equation for a hydrocarbon component takes a compact form as:

$$\left(\mathbb{T}_{i,g}^m\right)_{l-1,l}\begin{bmatrix}(p_{g_{l-1}} - p_{g_l})-\\ \rho_{g_{l-1}}g(z_{g_{l-1}} - z_{g_l})\end{bmatrix} + \left(\mathbb{T}_{i,g}^m\right)_{l+1,l}\begin{bmatrix}(p_{g_{l+1}} - p_{g_l})-\\ \rho_{g_{l+1}}g(z_{g_{l+1}} - z_{g_l})\end{bmatrix}$$
$$+ \left(\mathbb{T}_{i,o}^m\right)_{l-1,l}\begin{bmatrix}(p_{o_{l-1}} - p_{o_l})-\\ \rho_{o_{l-1}}g(z_{o_{l-1}} - z_{o_l})\end{bmatrix} + \left(\mathbb{T}_{i,o}^m\right)_{l-1,l}\begin{bmatrix}(p_{o_{l+1}} - p_{o_l})-\\ \rho_{o_{l-1}}g(z_{o_{l+1}} - z_{o_l})\end{bmatrix} \quad (6.130)$$
$$- \left(y_{ig}\rho_g Q_g + y_{io}\rho_o Q_o\right)^m = \frac{(U)_l}{\Delta t}\begin{bmatrix}\left(\phi\rho_g y_{i,g}S_g + \phi\rho_o y_{i,o}S_o\right)^{n+1}\\ -\left(\phi\rho_g y_{i,g}S_g + \phi\rho_o y_{i,o}S_o\right)^n\end{bmatrix}_l.$$

For the water component, where $i = n_c + 1$,

$$\left(\mathbb{T}_{i,g}^m\right)_{l-1,l}\begin{bmatrix}(p_{g_{l-1}} - p_{g_l})-\\ \rho_{g_{l-1}}g(z_{g_{l-1}} - z_{g_l})\end{bmatrix} + \left(\mathbb{T}_{i,g}^m\right)_{l+1,l}\begin{bmatrix}(p_{g_{l+1}} - p_{g_l})-\\ \rho_{g_{l+1}}g(z_{g_{l+1}} - z_{g_l})\end{bmatrix}$$
$$+ \left(\mathbb{T}_{i,w}^m\right)_{l-1,l}\begin{bmatrix}(p_{w_{l-1}} - p_{w_l})-\\ \rho_{w_{l-1}}g(z_{w_{l-1}} - z_{w_l})\end{bmatrix} + \left(\mathbb{T}_{i,w}^m\right)_{l+1,l}\begin{bmatrix}(p_{w_{l+1}} - p_l)-\\ \rho_{w_{l+1}}g(z_{w_{l+1}} - z_{w_l})\end{bmatrix} \quad (6.131)$$
$$- \left(y_{ig}\rho_g Q_g + \rho_w Q_w\right)^m = \frac{(U)_l}{\Delta t}\begin{bmatrix}\left(\phi\rho_g y_{i,g}S_g + \phi\rho_w S_w\right)^{n+1}\\ -\left(\phi\rho_g y_{i,g}S_g + \phi\rho_w S_w\right)^n\end{bmatrix}_l.$$

6.7 The compositional simulator using engineering approach

The transmissibility between blocks l, n for a phase, j can be given as:

$$\left(\mathbb{T}_{i,j}^m\right)_{l,n} = G_{l,n}\left(\frac{k_{rj}}{\mu_j}\right)_{l,n}^m = \left(\mathbb{T}_j^m\right)_{l,n}\left(\rho_j y_{i,j}\right)_{l,n}^m \tag{6.132}$$

for $j = g, o \,\&\, w$. The directional geometric factors for the direction x, y, and z according to Table 6.13 are:

Following the works of Abou-Kassem et al. (2007a), the mass balance equation can be written for a block l in a compact form as:

$$\sum_{n=\psi_n}\left\{\left(\mathbb{T}_{i,g}^m\right)_{l,n}\left[\begin{array}{c}(p_{g_n}-p_{g_l})-\\ \rho_{g_n}g(z_{g_n}-z_{g_l})\end{array}\right] + \left(\mathbb{T}_{i,o}^m\right)_{l,n}\left[\begin{array}{c}(p_{o_n}-p_{o_l})-\\ \rho_{o_n}g(z_{o_n}-z_{o_l})\end{array}\right]\right\}$$
$$-\left(y_{ig}\rho_g Q_g + y_{io}\rho_o Q_o\right)^m = \frac{(U)_l}{\Delta t}\left[\begin{array}{c}\left(\phi\rho_g y_{i,g}S_g + \phi\rho_o y_{i,o}S_o\right)^{n+1}\\ -\left(\phi\rho_g y_{i,g}S_g + \phi\rho_o y_{i,o}S_o\right)^n\end{array}\right]_l \tag{6.133}$$

for a hydrocarbon component $i = 1, \cdots, n_c$. For the water component $i = n_c + 1$, the mass balance equation in a compact form is:

$$\sum_{n=\psi_n}\left\{\left(\mathbb{T}_{i,g}^m\right)_{l,n}\left[\begin{array}{c}(p_{g_n}-p_{g_l})-\\ \rho_{g_n}g(z_{g_n}-z_{g_l})\end{array}\right] + \left(\mathbb{T}_{i,w}^m\right)_{l,n}\left[\begin{array}{c}(p_{o_n}-p_{o_l})-\\ \rho_{o_n}g(z_{o_n}-z_{o_l})\end{array}\right]\right\}$$
$$-\left(y_{ig}\rho_g Q_g + \rho_w Q_w\right)^m = \frac{(U)_l}{\Delta t}\left[\begin{array}{c}\left(\phi\rho_g y_{i,g}S_g + \phi\rho_w S_w\right)^{n+1}\\ -\left(\phi\rho_g y_{i,g}S_g + \phi\rho_w S_w\right)^n\end{array}\right]_l. \tag{6.134}$$

where $\psi_n = \{l - 1, l + 1\}$ is the neighboring block to the block l. Eqs. (6.44), (6.45) are for an interior grid block or in other words, are for blocks $l = 2, 3, \cdots, n_x - 1$. To include the boundary

TABLE 6.13 Directional Geometric factors in a rectangular grid.

Direction	Geometric Factor
x	$G_{x_{i\mp 1/2,j,k}} = \dfrac{2}{\Delta x_{i,j,k}/\left(A_{x_{i,j,k}}k_{x_{i,j,k}}\right) + \Delta x_{i\mp 1,j,k}/\left(A_{x_{i\mp 1,j,k}}k_{x_{i\mp 1,j,k}}\right)}$
y	$G_{y_{i,j\mp 1/2,k}} = \dfrac{2}{\Delta y_{i,j,k}/\left(A_{x_{i,j,k}}k_{x_{i,j,k}}\right) + \Delta y_{i,j\mp 1,k}/\left(A_{x_{i,j\mp 1,k}}k_{x_{i,j\mp 1,k}}\right)}$
z	$G_{z_{i,j,k\mp 1/2}} = \dfrac{2}{\Delta z_{i,j,k}/\left(A_{x_{i,j,k}}k_{x_{i,j,k}}\right) + \Delta z_{i,j,k\mp 1}/\left(A_{x_{i,j,k\mp 1}}k_{x_{i,j,k\mp 1}}\right)}$

From Islam, M.R., et al., 2016. Advanced Reservoir Simulation: Towards developing reservoir emulators. Scrivener-Wiley, 520 pp.

blocks, it may be possible to use a fictitious well instead of the boundary conditions and extend a general form of the formulation that can be applied for all gridblocks either interior blocks or boundary blocks. The general form of the formulation is:

$$\sum_{n=\psi_n}\left\{\left(\mathbb{T}_{i,g}^m\right)_{l,n}\left[\begin{array}{c}(p_{g_n}-p_{g_l})-\\ \rho_{g_n}g(z_{g_n}-z_{g_l})\end{array}\right]+\left(\mathbb{T}_{i,o}^m\right)_{l,n}\left[\begin{array}{c}(p_{o_n}-p_{o_l})-\\ \rho_{o_n}g(z_{o_n}-z_{o_l})\end{array}\right]\right\}+\sum_{n=\xi_n}\left(q_i^m\right)_{l,n}$$
$$-\left(y_{ig}\rho_g Q_g+y_{io}\rho_o Q_o\right)_l^m=\frac{(U)_l}{\Delta t}\left[\begin{array}{c}\left(\phi\rho_g y_{i,g} S_g+\phi\rho_o y_{i,o} S_o\right)^{n+1}\\ -\left(\phi\rho_g y_{i,g} S_g+\phi\rho_o y_{i,o} S_o\right)^n\end{array}\right]_l \quad (6.135)$$

for a hydrocarbon component $i = 1, \cdots, n_c$. The notation $(q_i^m)_{l,n}$ is a fictitious well rate to model the boundary. It is equal to:

$$\left(q_i^m\right)_{l,n}=\left(y_{ig}\rho_g Q_g+y_{io}\rho_o Q_o\right)_{l,n}^m \quad (6.136)$$

For the water component $i = n_c + 1$, the mass balance equation in a compact form for all gridblocks is:

$$\sum_{n=\psi_m}\left\{\left(\mathbb{T}_{i,g}^m\right)_{l,n}\left[\begin{array}{c}(p_{g_n}-p_{g_l})-\\ \rho_{g_n}g(z_{g_n}-z_{g_l})\end{array}\right]+\left(\mathbb{T}_{i,w}^m\right)_{l,n}\left[\begin{array}{c}(p_{o_n}-p_{o_l})-\\ \rho_{o_n}g(z_{o_n}-z_{o_l})\end{array}\right]\right\}+\sum_{n=\xi_n}\left(q_i^m\right)_{l,n}$$
$$-\left(y_{ig}\rho_g Q_g+\rho_w Q_w\right)_l^m=\frac{(U)_l}{\Delta t}\left[\begin{array}{c}\left(\phi\rho_g y_{i,g} S_g+\phi\rho_w S_w\right)^{n+1}\\ -\left(\phi\rho_g y_{i,g} S_g+\phi\rho_w S_w\right)^n\end{array}\right]_l \quad (6.137)$$

where:

$$\left(q_i^m\right)_{l,n}=\left(y_{ig}\rho_g Q_g+\rho_w Q_w\right)_{l,n}^m. \quad (6.138)$$

6.7.6 Compositional mass balance equation in a multidimensional domain

Reservoir blocks have a three-dimensional shape whether fluid flow is 1D, 2D, or 3D. The number of existing neighboring blocks and the number of reservoir boundaries shared by a reservoir block add up to six, as is the case in 3D flow. Existing neighboring blocks contribute to flow to or from the block, whereas reservoir boundaries may or may not contribute to flow depending on the dimensionality of flow and the prevailing boundary conditions. The dimensionality of flow implicitly defines those reservoir boundaries that do not contribute to flow at all. In 1D-flow problems, all reservoir blocks have four reservoir boundaries that do not contribute to flow. In 1D flow in the x-direction, the reservoir south (b_s), north (b_n), lower (b_l), and upper (b_u) boundaries do not contribute to flow to any reservoir block, including boundary blocks. These four reservoir boundaries (b_s, b_n, b_l & b_u) in this case are discarded as if they did not exist. As a result, an interior reservoir block has two neighboring blocks (one to the left and another to the right of the block) and no reservoir boundaries, whereas a boundary reservoir block has one neighboring block (Block 2 for $n = 1$ and Block $n_x - 1$ for $n = n_x$) and one reservoir boundary (b_w for $n = 1$ and b_E for $n = n_x$). In 2D flow problems, all reservoir blocks have two reservoir boundaries that do not contribute to flow at all. For example, in 2D flow in the xy plane, the reservoir lower and upper boundaries do not contribute to flow to any reservoir block, including boundary blocks. These two reservoir boundaries (b_l, b_u) are discarded as if they did not exist. As a result, an interior reservoir block has four neighboring blocks and no reservoir boundaries, a reservoir block that falls on one reservoir boundary has three neighboring blocks and one reservoir boundary, and a reservoir block that falls on two reservoir boundaries has two neighboring blocks and two reservoir boundaries. In 3D flow problems, any of the six reservoir boundaries may contribute to flow depending on the specified boundary condition. An interior block has six neighboring blocks. It does not share any of its boundaries with any of the reservoir boundaries. A boundary block may fall on one, two, or three of the reservoir boundaries. Therefore, a boundary block that falls on one, two, or three reservoir boundaries has five, four, or three neighboring blocks, respectively. If the notation ψ_n is a set showing the neighboring blocks and ξ_n denotes a set showing the number of boundaries, the aforementioned discussion leads to a few conclusions related to the number of elements contained in sets ψ_n and ξ_n.

(1) For an interior reservoir block, set ψ_n contains two, four, or six elements for a 1D, 2D, or 3D flow problem, respectively, and set ξ_n contains no elements or, in other words, is empty.
(2) For a boundary reservoir block, set ψ_n contains less than two, four, or six elements for a 1D, 2D, or 3D flow problem, respectively, and set ξ_n is not empty.
(3) The sum of the number of elements in sets ψ_n and ξ_n for any reservoir block is a constant that depends on the dimensionality of flow. This sum is two, four, or six for a 1D, 2D, or 3D flow problem, respectively.

Therefore, Eqs. (6.46), (6.48) are applicable for 1D, 2D, or 3D flow problems. In 1D flow in the x-direction (see Fig. 6.32A), ψ_n for a block $n \equiv i$ has a maximum of two elements of

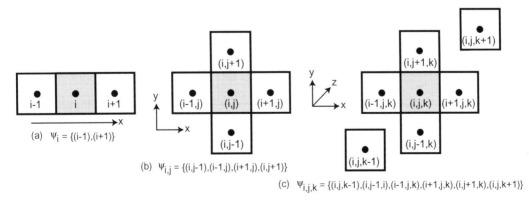

FIG. 6.32 A Block and its neighboring blocks in 1D, 2D, and 3D using engineering notation.

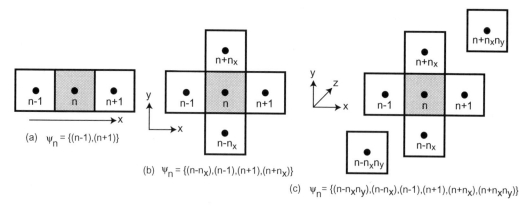

FIG. 6.33 A Block and its neighboring blocks in 1D, 2D, and 3D using natural ordering.

$[(i-1),(i+1)]$, and ξ_n is empty or has a maximum of one element of (b_w, b_E). In 2D flow in the x-y plane (Fig. 6.32B), ψ_n for $n \equiv (i,j)$ has a maximum of four elements of $[(i-1,j), (i,j-1), (i,(j+1), (i+1,j)]$, and ξ_n is empty or has a maximum of two elements of (b_S, b_W, b_E, b_N). In 3D flow in the x-y-z plane (Fig. 6.32C), ψ_n for $n \equiv (i,j,k)$ has a maximum of six elements of $[(i-1,j,k), (i,j-1,k), (i,j,)(k-1), (i+1,j,k), (i,j+1,k), (i,j,k+1)]$, and ξ_n is empty or has a maximum of three elements of $(b_L, b_S, b_W, b_E, b_N, b_U)$. The relations for the fictitious well, Eqs. (6.47), (6.48), are not only valid for the (i,j,k) notation but also valid for any block-ordering scheme. For example, if blocks are ordered using natural ordering with blocks being ordered along the x-axis, followed the y-axis, and finally along the z-axis, then the set of neighboring blocks are defined as shown in Fig. 6.33. That is to say, for block n, $\psi_n = [(n-1), (n+1)]$ for 1D flow along the x-axis (Fig. 6.33A), $\psi_n = [(n-n_x), (n-1), (n+1), (n+n_x)]$ for 2D flow in the x-y plane (Fig. 6.8B), and $\psi_n = [(n-n_xn_y), (n-n_x), (n-1), (n+1), (n+n_x), (n+n_xn_y)]$ for 3D flow in the xyz space (Fig. 6.33C). In summary, $\psi_n =$ a set whose elements are the existing neighboring blocks to block n in the reservoir, $\xi_n =$ a set whose elements are the reservoir boundaries $(b_L, b_S, b_W, b_E, b_N, b_u)$ that are shared by block n. It is either an empty set for interior blocks or a set that contains one element for boundary blocks that fall on one reservoir boundary, two elements for boundary blocks that fall on two reservoir boundaries, or three elements for blocks that fall on three reservoir boundaries. An empty set implies that the block does not fall on any reservoir boundary; i.e., block n is an interior block, and hence $\sum_{l=\xi_n}(q_i^m)_{l,n}$ for $i = 1, 2, \cdots, n_c + 1$. The notation $(q_i^m)_{l,n}$ is component i flow rate of the fictitious well that represents transfer of component i between reservoir boundary l and block n because of a boundary condition.

6.7.7 Implicit formulation of compositional model in multi-dimensional domain

The explicit, implicit, and Crank-Nicolson formulations are derived from Eqs. (6.46), (6.48) by specifying the approximation of time t^m as t^n, $t^{n+1/2}$, or t^{n+1}, which are illustrated in Fig. 6.30. The explicit formulation, however, is not used in the multiphase flow because of time step limitations, and the Crank-Nicolson formulation is not commonly used. Consequently,

6.7 The compositional simulator using engineering approach

we limit our presentation to the implicit formulation. In the following equations, fluid gravity is dated at old-time level n instead of the new time level $n+1$, as this approximation does not introduce any noticeable errors (Coats, George, and Marcum, 1974). The implicit model is presented below for a hydrocarbon component $i = 1, \cdots, n_c$.

$$\sum_{n=\psi_n} \left\{ \left(\mathbb{T}_{i,g}^{n+1}\right)_{l,n} \left[(p_{g_n} - p_{g_l})^{n+1} - \rho_{g_n}^{n+1} g(z_{g_n} - z_{g_l}) \right] + \left(\mathbb{T}_{i,o}^{n+1}\right)_{l,n} \left[(p_{o_n} - p_{o_l})^{n+1} - \rho_{o_n}^{n+1} g(z_{o_n} - z_{o_l}) \right] \right\} + \sum_{n=\xi_n} \left(q_i^{n+1}\right)_{l,n}$$

$$- \left(y_{ig}\rho_g Q_g + y_{io}\rho_o Q_o\right)_l^{n+1} = \frac{(U)_l}{\Delta t} \left[\left(\phi\rho_g y_{i,g} S_g + \phi\rho_o y_{i,o} S_o\right)^{n+1} - \left(\phi\rho_g y_{i,g} S_g + \phi\rho_o y_{i,o} S_o\right)^n \right]_l \quad (6.139)$$

where $(q_i^{n+1})_{l,n}$ is a fictitious well rate to model the boundary.

$$\left(q_i^{n+1}\right)_{l,n} = \left(y_{ig}\rho_g Q_g + y_{io}\rho_o Q_o\right)_{l,n}^{n+1} \quad (6.140)$$

The implicit model for the water component $i = n_c + 1$ is in the form:

$$\sum_{n=\psi_m} \left\{ \left(\mathbb{T}_{i,g}^{n+1}\right)_{l,n} \left[(p_{g_n} - p_{g_l})^{n+1} - \rho_{g_n}^{n+1} g(z_{g_n} - z_{g_l}) \right] + \left(\mathbb{T}_{i,w}^{n+1}\right)_{l,n} \left[(p_{o_n} - p_{o_l})^{n+1} - \rho_{o_n}^{n+1} g(z_{o_n} - z_{o_l}) \right] \right\} + \sum_{n=\xi_n} \left(q_i^{n+1}\right)_{l,n}$$

$$- \left(y_{ig}\rho_g Q_g + \rho_w Q_w\right)_l^{n+1} = \frac{(U)_l}{\Delta t} \left[\left(\phi\rho_g y_{i,g} S_g + \phi\rho_w S_w\right)^{n+1} - \left(\phi\rho_g y_{i,g} S_g + \phi\rho_w S_w\right)^n \right]_l \quad (6.141)$$

where:

$$\left(q_i^{n+1}\right)_{l,n} = \left(y_{ig}\rho_g Q_g + \rho_w Q_w\right)_{l,n}^m \quad (6.142)$$

$$\left(\mathbb{T}_{i,o}^{n+1}\right)_{l,n} = \left(\mathbb{T}_o^{n+1}\right)_{l,n} (y_{io}\rho_o) \quad \text{for } i = 1, 2, \cdots, n_c \quad (6.143)$$

$$\left(\mathbb{T}_{i,g}^{n+1}\right)_{l,n} = \left(\mathbb{T}_g^{n+1}\right)_{l,n} (y_{ig}\rho_g) \quad \text{for } i = 1, 2, \cdots, n_c + 1 \quad (6.144)$$

$$\left(\mathbb{T}_{i,w}^{n+1}\right)_{l,n} = \left(\mathbb{T}_w^{n+1}\right)_{l,n} (y_{iw}\rho_w) \quad \text{for } i = n_c + 1 \quad (6.145)$$

$$\left(\mathbb{T}_j^{n+1}\right)_{l,n} = G_{l,n} \left(\frac{k_{rj}}{\mu_j}\right)_{l,n}^{n+1} \quad \text{for } j = g, o, w \quad (6.146)$$

694

Each block contributes $n_c + 6$ equations:

- $n_c + 1$ component flow equations;
- two capillary pressure relationships, and
- three constraint equations.

There are $n_c + 6$ unknowns in each block:

- n_c-mass fractions in the oil phase $y_{i,\,o}$ for $i = 1, 2, \cdots, n_c$;
- three-phase pressures (p_g, p_o, p_w); and
- three-phase saturations (S_g, S_o, S_w).

The component flow equation for one of the oil components say the lightest oil component ($c = 1$), can be replaced with a pressure equation. The pressure equation is obtained by adding all $n_c + 1$ component flow equations. The mass fraction constraints for oil and gas phases are used when necessary to replace the sum of mass fractions in the oil and gas phases with the value of one. The resulting pressure equation in this case is

$$\sum_{n=\psi_n} \left\{ \begin{array}{l} \left(\mathbb{T}_g^{n+1}\right)_{l,n} \rho_g^{n+1} \left[\begin{array}{l} \left(p_{g_n} - p_{g_l}\right)^{n+1} - \\ \rho_{g_n}^n g(z_{g_n} - z_{g_l}) \end{array} \right] \\ + \left(\mathbb{T}_o^{n+1}\right)_{l,n} \rho_o^{n+1} \left[\begin{array}{l} \left(p_{o_n} - p_{o_l}\right)^{n+1} - \\ \rho_{o_n}^n g(z_{o_n} - z_{o_l}) \end{array} \right] \\ + \left(\mathbb{T}_w^{n+1}\right)_{l,n} \rho_w^{n+1} \left[\begin{array}{l} \left(p_{w_n} - p_{w_l}\right)^{n+1} - \\ \rho_{w_n}^n g(z_{o_n} - z_{o_l}) \end{array} \right] \end{array} \right\} + \sum_{n=\xi_n} \left(\rho_g Q_g + \rho_o Q_o + \rho_w Q_w\right)_{l,n}^{n+1}$$

$$- \left(\rho_g Q_g + \rho_o Q_o + \rho_w Q_w\right)_l^{n+1} = \frac{(U)_l}{\Delta t} \left[\begin{array}{l} \left(\phi \rho_g y_{i,g} S_g + \phi \rho_o y_{i,o} S_o\right)^{n+1} \\ - \left(\phi \rho_g y_{i,g} S_g + \phi \rho_o y_{i,o} S_o\right)^n \end{array} \right]_l \quad (6.147)$$

6.7.8 Reduced equations of an implicit compositional model in multidimensional domain

The compositional model reduced equations are obtained by eliminating the equations, which do not contain flow terms from the set of equations comprising the model. We limit our presentation to the $p_o - S_w - S_g$ formulation, i.e., the formulation that uses p_o, S_w and S_g as the primary unknowns for phase flow in the reservoir. The secondary unknowns in this formulation are p_g, p_w and S_o. The primary unknowns in composition are the n_c mass fractions in the oil phase, ($y_{i,\,o}$ for $i = 1, 2, \cdots, n_c$). Other formulations using $p_o - p_w - p_g$, $p_o - P_{cow} - P_{cgo}$, $p_o - P_{cow} - S_g$ or $p_o - S_w - P_{cgo}$ break down for negligible or zero capillary pressures. To obtain the reduced set of equations of the model for each block, we express the secondary unknowns in the component equations in terms of the primary unknowns and eliminate the secondary unknowns from the flow equations. In the reduced set of equations, each block contributes $n_c + 3$ equations:

- the oil-phase composition constraint;
- the gas-phase composition constraint;
- $n_c + 1$ oil component flow equations;
- the water component flow equation; and
- the pressure equation.

The $n_c + 3$ primary unknowns in each block are:

- n_c mass fractions in the oil phase $y_{i,o}$ for $i = 1, 2, \cdots, n_c$;
- the saturation of the gas and water phases S_g, S_w; and
- the pressure of the oil-phase p_o.

The equations used to eliminate the secondary unknowns (p_w, p_g, S_o) are the capillary pressure relationships:

$$p_w = p_o - P_{cow}(S_w) \quad \text{and} \quad p_g = p_o - P_{cgo}(S_g); \tag{6.148}$$

and the phase saturation constraint equation.

$$S_o = 1 - S_w - S_g \tag{6.149}$$

Once the primary unknowns are solved, the phase saturation constraint and capillary pressure relationships are used to solve for the secondary unknowns for each reservoir block.

The equations and primary unknowns for each reservoir block are aligned to obtain a diagonally dominant matrix and to alleviate the need for pivoting during forward Gaussian elimination. The model-reduced equations have the following order:

- the gas-phase composition constraint equation;

$$\sum_{i=1}^{n_c+1} y_{i,g}^{n+1} = 1 \tag{6.150}$$

- the oil-phase composition constraint equation;

$$\sum_{i=1}^{n_c} y_{i,o}^{n+1} = 1 \tag{6.151}$$

- the second lightest oil component flow equation ($c = 2$);

$$\sum_{n=\psi_n} \left\{ \left(\mathbb{T}_{2,g}^{n+1}\right)_{l,n} \begin{bmatrix} (p_{g_n} - p_{g_l})^{n+1} - \\ \rho_{g_n}^{n+1} g(z_{g_n} - z_{g_l}) \end{bmatrix} + \left(\mathbb{T}_{2,o}^{n+1}\right)_{l,n} \begin{bmatrix} (p_{o_n} - p_{o_l})^{n+1} - \\ \rho_{o_n}^{n+1} g(z_{o_n} - z_{o_l}) \end{bmatrix} \right\} + \sum_{n=\xi_n} \left(q_2^{n+1}\right)_{L,n} - \begin{pmatrix} y_{2,g} \rho_g Q_g \\ + y_{2,o} \rho_o Q_o \end{pmatrix}_l^{n+1}$$

$$= \frac{(U)_l}{\Delta t} \begin{bmatrix} \left(\phi \rho_g y_{2,g} S_g + \phi \rho_o y_{2,o} S_o\right)^{n+1} \\ - \left(\phi \rho_g y_{2,g} S_g + \phi \rho_o y_{2,o} S_o\right)^n \end{bmatrix}_l \tag{6.152}$$

- the third lightest oil component flow equation ($c=3$);

$$\sum_{n=\psi_n} \left\{ \left(\mathbb{T}_{3,g}^{n+1}\right)_{l,n} \begin{bmatrix} (p_{g_n}-p_{g_l})^{n+1}- \\ \rho_{g_n}^{n+1} g(z_{g_n}-z_{g_l}) \end{bmatrix} + \left(\mathbb{T}_{3,o}^{n+1}\right)_{l,n} \begin{bmatrix} (p_{o_n}-p_{o_l})^{n+1}- \\ \rho_{o_n}^{n+1} g(z_{o_n}-z_{o_l}) \end{bmatrix} \right\} + \sum_{n=\xi_n} \left(q_3^{n+1}\right)_{l,n} - \begin{pmatrix} y_{3,g}\rho_g Q_g \\ +y_{3,o}\rho_o Q_o \end{pmatrix}_l^{n+1}$$

$$= \frac{(U)_l}{\Delta t} \left[\left(\phi\rho_g y_{3,g} S_g + \phi\rho_o y_{3,o} S_o\right)^{n+1} - \left(\phi\rho_g y_{3,g} S_g + \phi\rho_o y_{3,o} S_o\right)^n \right]_l . \tag{6.153}$$

- ...;
- the heaviest oil component flow equation ($c=n_c$);

$$\sum_{n=\psi_n} \left\{ \left(\mathbb{T}_{n_c,g}^{n+1}\right)_{l,n} \begin{bmatrix} (p_{g_n}-p_{g_l})^{n+1}- \\ \rho_{g_n}^{n+1} g(z_{g_n}-z_{g_l}) \end{bmatrix} + \left(\mathbb{T}_{n_c,o}^{n+1}\right)_{l,n} \begin{bmatrix} (p_{o_n}-p_{o_l})^{n+1}- \\ \rho_{o_n}^{n+1} g(z_{o_n}-z_{o_l}) \end{bmatrix} \right\} + \sum_{n=\xi_n} \left(q_{n_c}^{n+1}\right)_{l,n} - \begin{pmatrix} y_{n_c,g}\rho_g Q_g \\ +y_{n_c,o}\rho_o Q_o \end{pmatrix}_l^{n+1}$$

$$= \frac{(U)_l}{\Delta t} \left[\left(\phi\rho_g y_{n_c,g} S_g + \phi\rho_o y_{n_c,o} S_o\right)^{n+1} - \left(\phi\rho_g y_{n_c,g} S_g + \phi\rho_o y_{n_c,o} S_o\right)^n \right]_l . \tag{6.154}$$

- the water component flow equation ($c=n_c+1$);

$$\sum_{n=\psi_m} \left\{ \left(\mathbb{T}_{n_c+1,g}^{n+1}\right)_{l,n} \begin{bmatrix} (p_{g_n}-p_{g_l})^{n+1}- \\ \rho_{g_n}^{n+1} g(z_{g_n}-z_{g_l}) \end{bmatrix} + \left(\mathbb{T}_{n_c+1,w}^{n+1}\right)_{l,n} \begin{bmatrix} (p_{o_n}-p_{o_l})^{n+1}- \\ \rho_{o_n}^{n+1} g(z_{o_n}-z_{o_l}) \end{bmatrix} \right\} + \sum_{n=\xi_n} \left(q_{n_c+1}^{n+1}\right)_{l,n} - \begin{pmatrix} y_{n_c+1,g}\rho_g Q_g \\ +\rho_w Q_w \end{pmatrix}_l^{n+1}$$

$$= \frac{(U)_l}{\Delta t} \left[\left(\phi\rho_g y_{n_c+1,g} S_g + \phi\rho_w S_w\right)^{n+1} - \left(\phi\rho_g y_{n_c+1,g} S_g + \phi\rho_w S_w\right)^n \right]_l . \tag{6.155}$$

- and the pressure equation.

$$\sum_{n=\psi_n} \left\{ \begin{array}{l} \left(\mathbb{T}_g^{n+1}\right)_{l,n} \rho_g^{n+1} \left[\begin{array}{l} (p_{g_n} - p_{g_l})^{n+1} - \\ \rho_{g_n}^n g(z_{g_n} - z_{g_l}) \end{array} \right] \\ + \left(\mathbb{T}_o^{n+1}\right)_{l,n} \rho_o^{n+1} \left[\begin{array}{l} (p_{o_n} - p_{o_l})^{n+1} - \\ \rho_{o_n}^n g(z_{o_n} - z_{o_l}) \end{array} \right] \\ + \left(\mathbb{T}_w^{n+1}\right)_{l,n} \rho_w^{n+1} \left[\begin{array}{l} (p_{w_n} - p_{w_l})^{n+1} - \\ \rho_{w_n}^n g(z_{o_n} - z_{o_l}) \end{array} \right] \end{array} \right\} + \sum_{n=\xi_n} \left(\begin{array}{l} \rho_g Q_g \\ +\rho_o Q_o \\ +\rho_w Q_w \end{array} \right)_{l,n}^{n+1} - \left(\begin{array}{l} \rho_g Q_g \\ +\rho_o Q_o \\ +\rho_w Q_w \end{array} \right)_l^{n+1}$$

$$= \frac{(U)_l}{\Delta t} \left[\begin{array}{l} \left(\phi \rho_g y_{i,g} S_g + \phi \rho_o y_{i,o} S_o\right)^{n+1} \\ - \left(\phi \rho_g y_{i,g} S_g + \phi \rho_o y_{i,o} S_o\right)^n \end{array} \right]_l \quad (6.156)$$

With this choice of equation ordering, the primary unknowns have the order of S_g, $y_{1,o}$, $y_{2,o}$, $y_{3,o}$, $\cdots$, $y_{n_c-1,o}$, $y_{n_c,o}$, S_w and p_o. Therefore, the vector of primary unknowns for block l is:

$$X_n = \begin{bmatrix} S_g & y_{1,0} & y_{2,0} & y_{3,0} & \cdots & y_{n_c-1,0} & y_{n_c,0} & S_w & p_o \end{bmatrix}_l^T \quad (6.157)$$

6.7.8.1 Well production and injection rate terms

Production wells

The production rates of components i associated with phases $j = g, o, w$ from a single well located in block l are expressed in the following form.

$$q_{i,l}^{n+1} = \left(y_{ig}\rho_g Q_g + y_{io}\rho_o Q_o\right)_l^{n+1} \quad \text{for} \quad i = 1, 2, \cdots, n_c \quad (6.158)$$

$$q_{i,l}^{n+1} = \left(y_{ig}\rho_g Q_g + \rho_w Q_w\right)_l^{n+1} \quad \text{for} \quad i = n_c + 1 \quad (6.159)$$

Fluid production rates in multiphase flow are dependent on each other through at least relative permeabilities. In other words, the specification of the production rate of any phase implicitly dictates the production rates of the other phases. The concern in single-block wells is to estimate the production rate of phase $j = g, o, w$ from well-block l under different well-operating conditions. The equations presented in this section are derived from the recent work of Abou-Kassem (2007a).

(a) Shut-in well

$$Q_{j,l} = 0 \quad \text{for} \quad j = g, o, w \quad (6.160)$$

(b) Specified well flow rate

$$Q_{j,l} = G_{wl} \left(\frac{k_{rj}}{\mu_j}\right)_l (p_o - p_{wf})_l \quad (6.161)$$

where G_{wl} is estimated as suggested by Abou-Kassem (2007a) and p_{wf} is estimated from the well rate specification Q_{sj} using

$$p_{wfl} = p_{ol} + \frac{Q_{sj}}{G_{wl}\sum_{j\in\eta_{prd}} M_{jl}} \qquad (6.162)$$

where η_{prd} and M_j depend on the type of well rate specification as listed in Table 6.14. Substitution Eq. (6.65) into Eq. (6.64) yields

$$Q_{j,l} = \left(\frac{k_{rj}}{\mu_j}\right)_l \frac{Q_{sj}}{\sum_{j\in\eta_{prd}} M_{jl}} \quad \text{for } j = g, o, w. \qquad (6.163)$$

(c) Specified well pressure gradient

For a specified well pressure gradient, the production rate of phase $j = g, o, w$ from well-block l is given by

$$Q_{j,l} = 2\pi r_w k_{Hl} h_l \left(\frac{k_{rj}}{\mu_j}\right)_l \frac{\partial p}{\partial r}\bigg|_{r_w}. \qquad (6.164)$$

(d) Specified well FBHP

If the FBHP of a well p_{wfref} is specified, then the production rate of phase $j = g, o, w$ from well-block l can be estimated using Eq. (6.64) using $p_{wfn} = p_{wfref}$.

Injection wells

For injection wells, one phase (usually water, gas of known composition, or oil of known composition) is injected. The mobility of the injected fluid at reservoir conditions in a well-block is equal to the sum of the mobilities of all phases present in the well-block.

$$M_{inj} = \sum_{j\in\eta_{inj}} M_{jl} \qquad (6.165)$$

where the notations η_{inj} and M_j are:

$$\eta_{inj} = \{o, w, g\}, \quad M_j = k_{rj}/\mu_j \qquad (6.166)$$

TABLE 6.14 Well rate specification and definitions of set η_{prd} and M_j.

Well rate specification, Q_{sj}	Set of specified phases, η_{prd}	Phase relative mobility, M_j
Q_{osj}	$\{o\}$	k_{rj}/μ_j
Q_{Lsj}	$\{o, w\}$	k_{rj}/μ_j
Q_{Tsj}	$\{o, w, g\}$	k_{rj}/μ_j

The concern here is to estimate the injection rate of the injected phase (usually water or gas) into well-block l under different well-operating conditions. The rates of injection of the remaining phases are set to zero.

(a) Shut-in well

$$Q_{j,l} = 0 \text{ for } j = g, o, w \tag{6.167}$$

(b) Specified well flow rate

The injection rate of the injected fluid $j = w$ or g into the well-block l is given by

$$Q_{j,l} = -G_{w_l} M_{inj_l} (p_o - p_{wf})_l \tag{6.168}$$

For a single-block well $Q_{j,\,l} = Q_{sj}$ and Eq. (6.70) is used to estimate p_{wfl}.

$$p_{wf_l} = p_l + \frac{Q_{sj}}{G_{w_l}\left(\frac{M_{inj}}{B_j}\right)_l} \tag{6.169}$$

Then, the injection rate of the injected fluid $j = w$ or g into the well-block $l(Q_{j,\,l})$ is estimated using Eq. (6.70). The use of Eq. (6.70) requires solving for p_{wfl} implicitly along with the reservoir block pressures. An explicit treatment; however, uses Eq. (6.71) at old-time level n to estimate p_{wf}^n, which is subsequently substituted into Eq. (6.70) to estimate the injection rate of the injected phase $j = w$ or g into well-block l.

(c) Specified well pressure gradient

For a specified well pressure gradient, the injection rate of fluid $j = w$ or g into the well-block l is given by

$$Q_{j,l} = -2\pi r_w k_{Hl} h_l M_{injl} \tfrac{\partial p}{\partial r}\big|_{r_w}. \tag{6.170}$$

(d) Specified well FBHP

If the FBHP of a well p_{wfl} is specified, then the injection rate of the injected fluid $j = w$ or g into the well-block l can be estimated using Eq. (6.70).

6.7.9 Fictitious well rate terms (treatment of boundary conditions)

Component i production rates from fictitious wells are

$$\left(q_i^{n+1}\right)_{l,n} = \left(y_{ig}\rho_g Q_g + y_{io}\rho_o Q_o\right)^{n+1}_{l,n} \text{ for } i = 1, 2, \cdots, n_c \tag{6.171}$$

$$\left(q_i^{n+1}\right)_{l,n} = \left(y_{ig}\rho_g Q_g + \rho_w Q_w\right)^{n+1}_{l,n} \text{ for } i = n_c + 1. \tag{6.172}$$

A reservoir boundary can be subject to one of four conditions:

1. a no-flow boundary;
2. a constant flow boundary;
3. a constant pressure gradient boundary;
4. a constant pressure boundary.

The fictitious well rate equations presented are derived from those presented by Abou-Kassem (2007a). The fictitious well rate of phase j ($q_{p,bB}^{n+1}$) reflects the fluid transfer of phase j between the boundary block, (bB), and the reservoir boundary itself, (b), or the block next to the reservoir boundary that falls outside the reservoir. In multiphase flow, a reservoir boundary may (1) separate two segments of one reservoir that has the same fluids, (2) separate an oil reservoir from a water aquifer, or (3) seal off the reservoir from a neighboring reservoir. If the neighboring reservoir segment is an aquifer, then either water invades the reservoir across the reservoir boundary (WOC) or reservoir fluids leave the reservoir block to the aquifer.

(a) Specified Pressure Gradient Boundary Condition

For a specified pressure gradient at the reservoir left (west) boundary,

$$q_{j\ b,bB}^{n+1} = -\left(\frac{k_l k_{rj} A_l}{\mu_j}\right)_{bB}^{n+1}\left[\left.\frac{\partial p_j}{\partial l}\right|_b^{n+1} - (\rho_j g)_{bB}^n \left.\frac{\partial z}{\partial l}\right|_b\right] \quad \text{for } j = g, o, w \quad (6.173)$$

and at the reservoir right (east) boundary,

$$q_{j\ b,bB}^{n+1} = \left(\frac{k_l k_{rj} A_l}{\mu_j}\right)_{bB}^{n+1}\left[\left.\frac{\partial p_j}{\partial l}\right|_b^{n+1} - (\rho_j g)_{bB}^n \left.\frac{\partial z}{\partial l}\right|_b\right] \quad \text{for } j = g, o, w. \quad (6.174)$$

In Eqs. (6.75), (6.76), the specified pressure gradient may replace the phase pressure gradient at the boundary. These two equations were applied to fluid flow across a reservoir boundary that separates two segments of the same reservoir or across a reservoir boundary that represents WOC with fluids being lost to the water aquifer. If the reservoir boundary represents WOC and water invades the reservoir, then

$$q_{j\ b,bB}^{n+1} = -\left(\frac{k_l A_l}{\mu_j}\right)_{bB}^{n+1}(k_{rw})_{aq}^{n+1}\left[\left.\frac{\partial p_j}{\partial l}\right|_b^{n+1} - (\rho_w g)_{bB}^n \left.\frac{\partial z}{\partial l}\right|_b\right] \quad (6.175)$$

for the reservoir left (west) boundary, and

$$q_{w\ b,bB}^{n+1} = \left(\frac{k_l A_l}{\mu_j}\right)_{bB}^{n+1}(k_{rw})_{aq}^{n+1}\left[\left.\frac{\partial p_j}{\partial l}\right|_b^{n+1} - (\rho_w g)_{bB}^n \left.\frac{\partial z}{\partial l}\right|_b\right] \quad (6.176)$$

for the reservoir right (east) boundary. Moreover,

$$q_{o\ b,bB}^{n+1} = q_{g\ b,bB}^{n+1} = 0 \quad (6.177)$$

Note that, in Eqs. (6.77), (6.78), the rock and fluid properties in the aquifer are approximated by those of the boundary block properties because of the lack of geologic control in aquifers and because the effect of oil/water capillary pressure is neglected. In addition, $(k_{rw})_{aq}^{n+1} = 1$ because $S_w = 1$ in the aquifer.

(b) Specified Flow Rate Boundary Condition

If the specified flow rate stands for water influx across a reservoir boundary, then

$$q_w^{n+1}{}_{b,bB} = q_{sp} \tag{6.178}$$

In addition,

$$q_o^{n+1}{}_{b,bB} = q_g^{n+1}{}_{b,bB} = 0. \tag{6.179}$$

If, however, the specified flow rate stands for fluid transfer between two segments of the same reservoir or fluid loss to an aquifer across WOC, then

$$q_j^{n+1}{}_{b,bB} = \frac{\left(\frac{k_{rj}}{\mu_j}\right)_{bB}^{n+1}}{\sum_{l \in \{o,w,fg\}} \left(\frac{k_{rl}}{\mu_l}\right)_{bB}^{n+1}} q_{sp} \quad \text{for } j = g, o, w. \tag{6.180}$$

The effects of gravity forces and capillary pressures are neglected in Eq. (6.82).

(c) No-flow Boundary Condition

This condition results from vanishing permeability at a reservoir boundary or because of symmetry about a reservoir boundary. In either case, for a reservoir no-flow boundary,

$$q_j^{n+1}{}_{b,bB} = 0 \quad \text{for } j = g, o, w. \tag{6.181}$$

(d) Specified Boundary Pressure Condition

This condition arises due to the presence of wells on the other side of a reservoir boundary that operates to maintain voidage replacement and as a result keep the boundary pressure (p_b) constant. The flow rate of phase j across a reservoir boundary that separates two segments of the same reservoir or across a reservoir boundary that represents WOC with fluid loss to an aquifer is estimated using

$$q_j^{n+1}{}_{b,bB} = \left[\frac{k_l k_{rj} A_l}{\mu_j (\Delta l/2)}\right]_{bB}^{n+1} \left[(p_b - p_{bB}^{n+1}) - (\rho_w g)_{bB}^n (z_b - z_{bB})\right] \quad \text{for } j = g, o, w. \tag{6.182}$$

If the reservoir boundary represents WOC with water influx, then

$$q_w^{n+1}{}_{b,bB} = \left[\frac{k_l A_l}{\mu_w (\Delta l/2)}\right]_{bB}^{n+1} (k_{rw})_{bB}^{n+1} \left[(p_b - p_{bB}^{n+1}) - (\rho_w g)_{bB}^n (z_b - z_{bB})\right] \tag{6.183}$$

In addition,

$$q_o^{n+1}{}_{b,bB} = q_g^{n+1}{}_{b,bB} = 0. \tag{6.184}$$

Note that, in Eq. (6.184), the rock and fluid properties in the aquifer are approximated by those of the boundary block properties because of the lack of geologic control in aquifers. It should also take into account that $(k_{rw})_{aq}^{n+1} = 1$ because $S_w = 1$ in the aquifer.

It is worth mentioning that when the reservoir boundary b stands for WOC, the flow rate of phase j across the reservoir boundary is determined from the knowledge of the upstream point between reservoir boundary b and boundary block bB. If b is upstream to bB (i.e., when $\Delta\Phi_w > 0$), the flow is from the aquifer to the reservoir boundary block and Eq. (6.85) applies for water and $q_o^{n+1}{}_{b,\,bB} = q_g^{n+1}{}_{b,\,bB} = 0$. If b is downstream to bB (i.e., when $\Delta\Phi_w < 0$), the flow is from the reservoir boundary block to the aquifer and Eq. (6.84) applies for all phases. The water potential between the reservoir boundary and the reservoir boundary block is defined as $\Delta\Phi_w = (p_b - p_{bB}) - \rho_w g(z_b - z_{bB})$.

6.7.10 Variable temperature reservoir compositional flow equations

6.7.10.1 Energy balance equation

The numerical formulation for the component mass balance equation for a uniform reservoir has been developed in Sections 6.2 and 6.3. When the reservoir temperature is not constant, it is necessary to consider the energy balance equation. The numerical model was introduced in Section 6.1.2.2 for the energy balance equation for two different cases based on the finite control volume method. The formulations should be combined with the Darcy law to obtain a set of equations for practical applications. The relationships for the enthalpy and internal energy of the fluid j using the heat capacity at constant temperature C_p and the heat capacity at constant volume C_v are also applied to express the energy balance equation as a function of reservoir temperature.

$$h_j = C_p(T - T_0) \quad \text{and} \quad \hat{u}_j = C_v(T - T_0) \qquad (6.185)$$

Substituting Eq. (6.87) in Eq. (6.22) and taking into account a 1-D domain as given in Fig. 6.21, the energy balance equation takes the form:

$$\frac{U}{\Delta t}\left\{\begin{bmatrix}(T-T_0)\left(\sum_j \phi\rho_j S_j C_{vj} + (1-\phi)\rho_r C_r\right) - g\phi\sum_j \rho_j S_j Z_j\end{bmatrix}^{n+1} \\ -\begin{bmatrix}(T-T_0)\left(\sum_j \phi\rho_j S_j C_{vj} + (1-\phi)\rho_r C_r\right) - g\phi\sum_j \rho_j S_j Z_j\end{bmatrix}^n\right\}$$

$$+ \begin{bmatrix}\left(\frac{kA x_{i+1/2}}{\Delta x}\right)_{x_{i+1/2}}(T_{i+1} - T_i) \\ + \left(\frac{kA x_{i-1/2}}{\Delta x}\right)_{x_{i-1/2}}(T_{i-1} - T_i)\end{bmatrix}^m$$

$$+ \begin{bmatrix}\sum_j \left[\phi\rho_j S_j C_{pj}(T-T_0)\right]_{x_{i+1/2}} V_{jx_{i+1/2}} A_{x_{i+1/2}} \\ -\sum_j \left[\phi\rho_j S_j C_{pj}(T-T_0)\right]_{x_{i-1/2}} V_{jx_{i-1/2}} A_{x_{i-1/2}}\end{bmatrix}^m + (E_s)^m = 0 \qquad (6.186)$$

Using the discretized form of Darcy's law and substituting in Eq. (6.88), it can be written that

$$\frac{U}{\Delta t}\left\{\left[(T-T_0)\left(\sum_j \phi\rho_j S_j C_{vj} + (1-\phi)\rho_r C_r\right) - g\phi\sum_j \rho_j S_j Z_j\right]^{n+1}\right.$$
$$\left. - \left[(T-T_0)\left(\sum_j \phi\rho_j S_j C_{vj} + (1-\phi)\rho_r C_r\right) - g\phi\sum_j \rho_j S_j Z_j\right]^{n}\right\}$$
$$+ \left[\begin{array}{l}\sum_j\left[\phi\rho_j S_j C_{pj}(T-T_0)\right]_{x_{i+1/2}} A_{x_{i+1/2}} \frac{k_x|_{x_{i+1/2}}}{\Delta x_{i+1/2}}\left(\frac{k_{rj}}{\mu_j}\right)\Big|_{x_{i+1/2}}\left[\begin{array}{l}(p_{j_{i+1}}-p_{j_i})-\\ \rho_{j_{i-1}}g(z_{j_{i+1}}-z_{j_i})\end{array}\right]\\ -\sum_j\left[\phi\rho_j S_j C_{pj}(T-T_0)\right]_{x_{i-1/2}} A_{x_{i-1/2}} \frac{k_x|_{x_{i-1/2}}}{\Delta x_{i-1/2}}\left(\frac{k_{rj}}{\mu_j}\right)\Big|_{x_{i-1/2}}\left[\begin{array}{l}(p_{j_{i-1}}-p_{j_i})-\\ \rho_{j_{i-1}}g(z_{j_{i-1}}-z_{j_i})\end{array}\right]\end{array}\right]^m \quad (6.187)$$
$$+ \left[\begin{array}{l}\left(\frac{kA_{x_{i+1/2}}}{\Delta x}\right)_{x_{i+1/2}}(T_{i+1}-T_i)\\ +\left(\frac{kA_{x_{i-1/2}}}{\Delta x}\right)_{x_{i-1/2}}(T_{i-1}-T_i)\end{array}\right]^m + (E_s)^m = 0$$

Taking into account the definition of phase transmissibility for a grid-block i as:

$$\left(\mathbb{T}_j^m\right)_{i\mp1/2} = \left(\frac{A_x k_x}{\Delta x}\right)_{x_{i\mp1/2}} \left(\frac{k_{ro}}{\mu_o}\right)^m_{x_{i\mp1/2}} \quad (6.188)$$

The energy balance equation may be given as:

$$\frac{U}{\Delta t}\left\{\left[(T-T_0)\left(\sum_j \phi\rho_j S_j C_{vj} + (1-\phi)\rho_r C_r\right) - g\phi\sum_j \rho_j S_j Z_j\right]^{n+1}\right.$$
$$\left. - \left[(T-T_0)\left(\sum_j \phi\rho_j S_j C_{vj} + (1-\phi)\rho_r C_r\right) - g\phi\sum_j \rho_j S_j Z_j\right]^{n}\right\}$$
$$+ \left[\begin{array}{l}\sum_j\left[\phi\rho_j S_j C_{pj}(T-T_0)\right]_{x_{i+1/2}} (\mathbb{T}_j)_{i+1/2}\left[\begin{array}{l}(p_{j_{i+1}}-p_{j_i})-\\ \rho_{j_{i-1}}g(z_{j_{i+1}}-z_{j_i})\end{array}\right]\\ -\sum_j\left[\phi\rho_j S_j C_{pj}(T-T_0)\right]_{x_{i-1/2}} (\mathbb{T}_j)_{i-1/2}\left[\begin{array}{l}(p_{j_{i-1}}-p_{j_i})-\\ \rho_{j_{i-1}}g(z_{j_{i-1}}-z_{j_i})\end{array}\right]\end{array}\right]^m \quad (6.189)$$
$$+ \left[\begin{array}{l}\left(\frac{kA_{x_{i+1/2}}}{\Delta x}\right)_{x_{i+1/2}}(T_{i+1}-T_i)\\ +\left(\frac{kA_{x_{i-1/2}}}{\Delta x}\right)_{x_{i-1/2}}(T_{i-1}-T_i)\end{array}\right]^m + (E_s)^m = 0$$

for $j = g, o, w$. To obtain a generalized form of the equation and provide the ability to extend the formulation for a multidimensional domain, we apply the procedure adopted by Abou-Kassem (2007a) and write the energy balance equation in the following form for a grid block l in a compact form as:

$$\frac{U}{\Delta t}\left\{\left[(T-T_0)\left(\sum_j \phi\rho_j S_j C_{vj} + (1-\phi)\rho_r C_r\right) - g\phi\sum_j \rho_j S_j Z_j\right]^{n+1}\right.$$
$$\left. - \left[(T-T_0)\left(\sum_j \phi\rho_j S_j C_{vj} + (1-\phi)\rho_r C_r\right) - g\phi\sum_j \rho_j S_j Z_j\right]^n\right\}$$
$$\sum_{n=\psi_n}\left\{\sum_j\left[\phi\rho_j S_j C_{pj}(T-T_0)\right]_n (\mathbb{T}_j)_l \begin{bmatrix}(p_{j_n}-p_{j_l})- \\ \rho_{j_n}g(z_{j_n}-z_{j_l})\end{bmatrix}\right.$$
$$\left. + \left(\frac{kA}{\Delta x}\right)_n (T_n - T_l)\right\}^m + (E_s)^m = 0 \qquad (6.190)$$

for $j = g, o, w$. Where $\psi_N = \{l-1, l+1\}$ is the neighboring block to the block l. Eq. (6.94) is for an interior grid block or in other words, is for blocks $l = 2, 3, \cdots, n_x - 1$. To include the boundary blocks, it may be possible to use a fictitious well instead of the boundary conditions and extend a general form of the formulation that can be applied for all gridblocks either interior blocks or boundary blocks. The general form of the formulation is:

$$\frac{U}{\Delta t}\left\{\left[(T-T_0)\left(\sum_j \phi\rho_j S_j C_{vj} + (1-\phi)\rho_r C_r\right) - g\phi\sum_j \rho_j S_j Z_j\right]^{n+1}\right.$$
$$\left. - \left[(T-T_0)\left(\sum_j \phi\rho_j S_j C_{vj} + (1-\phi)\rho_r C_r\right) - g\phi\sum_j \rho_j S_j Z_j\right]^n\right\}$$
$$\sum_{n=\psi_n}\left\{\sum_j\left[\phi\rho_j S_j C_{pj}(T-T_0)\right]_n (\mathbb{T}_j)_l \begin{bmatrix}(p_{j_n}-p_{j_l})- \\ \rho_{j_n}g(z_{j_n}-z_{j_l})\end{bmatrix}\right.$$
$$\left. + \left(\frac{kA}{\Delta x}\right)_n (T_n - T_l)\right\}^m + (E_s)^m + \sum_{n=\xi_n} E_{l,n}^m = 0 \qquad (6.191)$$

The notations ψ_n and ξ_n are the sets showing the neighboring blocks and boundaries, respectively, as described in the derivation of component mass balance equations a previous sections.

Eq. (6.95) is applicable for 1D, 2D, or 3D flow problems. In 1D, ψ_n a block n has a maximum of two elements of $[(i-1), (i+1)]$, and ξ_n is empty or has a maximum of one element of (b_w, b_E). In 2D, ψ_n for n has a maximum of four elements of $[(i-1, j), (i, j-1), (i, j+1), (i+1, j)]$, and ξ_n is empty or has a maximum of two elements of (b_S, b_W, b_E, b_N). In 3D flow, ψ_n for n has a maximum of six elements of $[(i-1, j, k), (i, j-1, k), (i, j, k-1), (i+1, j, k), (i, j+1, k), (i, j, k+1)]$, and ξ_n is empty or has a maximum of three elements of $(b_L, b_S, b_W, b_E, b_N, b_U)$. These are illustrated in Fig. 6.32. In summary, $\psi_n =$ a set whose elements are the existing neighboring blocks to block n in the reservoir, $\xi_n =$ a set whose elements are the reservoir boundaries $(b_L, b_S, b_W, b_E, b_N, b_u)$ that are shared by block n. It is either an empty set for interior blocks or a set that contains one element for boundary blocks that fall on one reservoir boundary, two elements for boundary blocks that fall on two reservoir boundaries, or three elements for blocks that fall on three reservoir boundaries.

6.7.10.2 Implicit formulation of variable temperature reservoir compositional flow equations

A fully implicit formulation for a uniform temperature reservoir was introduced in Section 6.3.1. If the reservoir temperature is not constant in different directions, we should also consider the energy balance equation to obtain the temperature distribution along the reservoir. The implicit form of energy balance equation is in the form:

$$\frac{U}{\Delta t}\left\{\left[(T-T_0)\left(\sum_j \phi\rho_j S_j C_{vj} + (1-\phi)\rho_r C_r\right) - g\phi\sum_j \rho_j S_j Z_j\right]^{n+1} \right.$$
$$\left. - \left[(T-T_0)\left(\sum_j \phi\rho_j S_j C_{vj} + (1-\phi)\rho_r C_r\right) - g\phi\sum_j \rho_j S_j Z_j\right]^{n}\right\} \quad (6.192)$$

$$\sum_{n=\psi_n}\left\{\sum_j\left[\phi\rho_j S_j C_{pj}(T-T_0)\right]_n (\mathbb{T}_j)_l \begin{bmatrix}(p_{j_n}-p_{j_l})- \\ \rho_{j_n}g(z_{j_n}-z_{j_l})\end{bmatrix}^{n+1} + \left(\frac{kA}{\Delta x}\right)_n (T_n - T_l)\right\} + (E_s)^{n+1} + \sum_{n=\xi_n} E_{l,n}^{n+1} = 0$$

This equation with the component mass balance equation and the constraint equations provide a set of algebraic equations to find all primary unknown including the temperature at gridblocks. Following the description in section 6.3.2 and taking into account the energy balance equation, each block contributes $n_c + 4$ equations:

- the oil-phase composition constraint;
- the gas-phase composition constraint;
- $n_c + 1$ oil component flow equations;
- the water component flow equation;
- the pressure equation; and
- the energy balance equation.

The $n_c + 4$ primary unknowns in each block are:

- n_c mass fractions in the oil phase $y_{i,o}$ for $i = 1, 2, \cdots, n_c$;
- the saturation of the gas and water phases S_g, S_w;
- the pressure of the oil-phase p_o; and
- the temperature of the gridblock.

The equations and primary unknowns for each reservoir block are aligned to obtain a diagonally dominant matrix and to alleviate the need for pivoting during forward Gaussian elimination. The model-reduced equations have the following order:

- the gas-phase composition constraint equation;

$$\sum_{i=1}^{n_c+1} y_{i,g}^{n+1} = 1 \quad (6.193)$$

- the oil-phase composition constraint equation;

$$\sum_{i=1}^{n_c} y_{i,o}^{n+1} = 1 \qquad (6.194)$$

- the second lightest oil component flow equation ($i = 2$);

$$\sum_{n=\psi_n} \left\{ \left(\mathbb{T}_{2,g}^{n+1}\right)_{l,n} \begin{bmatrix} (p_{g_n} - p_{g_l})^{n+1} - \\ \rho_{g_n}^{n+1} g(z_{g_n} - z_{g_l}) \end{bmatrix} + \left(\mathbb{T}_{2,o}^{n+1}\right)_{l,n} \begin{bmatrix} (p_{o_n} - p_{o_l})^{n+1} - \\ \rho_{o_n}^{n+1} g(z_{o_n} - z_{o_l}) \end{bmatrix} \right\} + \sum_{n=\xi_n} \left(q_2^{n+1}\right)_{l,n} - \begin{pmatrix} y_{2,g}\rho_g Q_g \\ +y_{2,o}\rho_o Q_o \end{pmatrix}_l^{n+1}$$

$$= \frac{(U)_l}{\Delta t} \begin{bmatrix} \left(\phi\rho_g y_{2,g} S_g + \phi\rho_o y_{2,o} S_o\right)^{n+1} \\ -\left(\phi\rho_g y_{2,g} S_g + \phi\rho_o y_{2,o} S_o\right)^n \end{bmatrix}_l \qquad (6.195)$$

- the third lightest oil component flow equation ($i = 3$);

$$\sum_{n=\psi_n} \left\{ \left(\mathbb{T}_{3,g}^{n+1}\right)_{l,n} \begin{bmatrix} (p_{g_n} - p_{g_l})^{n+1} - \\ \rho_{g_n}^{n+1} g(z_{g_n} - z_{g_l}) \end{bmatrix} + \left(\mathbb{T}_{3,o}^{n+1}\right)_{l,n} \begin{bmatrix} (p_{o_n} - p_{o_l})^{n+1} - \\ \rho_{o_n}^{n+1} g(z_{o_n} - z_{o_l}) \end{bmatrix} \right\} + \sum_{n=\xi_n} \left(q_3^{n+1}\right)_{l,n} - \begin{pmatrix} y_{3,g}\rho_g Q_g \\ +y_{3,o}\rho_o Q_o \end{pmatrix}_l^{n+1}$$

$$= \frac{(U)_l}{\Delta t} \begin{bmatrix} \left(\phi\rho_g y_{3,g} S_g + \phi\rho_o y_{3,o} S_o\right)^{n+1} \\ -\left(\phi\rho_g y_{3,g} S_g + \phi\rho_o y_{3,o} S_o\right)^n \end{bmatrix}_l \qquad (6.196)$$

- …;
- the heaviest oil component flow equation ($i = n_c$);

$$\sum_{n=\psi_n} \left\{ \left(\mathbb{T}_{n_c,g}^{n+1}\right)_{l,n} \begin{bmatrix} (p_{g_n} - p_{g_l})^{n+1} - \\ \rho_{g_n}^{n+1} g(z_{g_n} - z_{g_l}) \end{bmatrix} + \left(\mathbb{T}_{n_c,o}^{n+1}\right)_{l,n} \begin{bmatrix} (p_{o_n} - p_{o_l})^{n+1} - \\ \rho_{o_n}^{n+1} g(z_{o_n} - z_{o_l}) \end{bmatrix} \right\} + \sum_{n=\xi_n} \left(q_{n_c}^{n+1}\right)_{l,n} - \begin{pmatrix} y_{n_c,g}\rho_g Q_g \\ +y_{n_c,o}\rho_o Q_o \end{pmatrix}_l^{n+1}$$

$$= \frac{(U)_l}{\Delta t} \begin{bmatrix} \left(\phi\rho_g y_{n_c,g} S_g + \phi\rho_o y_{n_c,o} S_o\right)^{n+1} \\ -\left(\phi\rho_g y_{n_c,g} S_g + \phi\rho_o y_{n_c,o} S_o\right)^n \end{bmatrix} \qquad (6.197)$$

- the water component flow equation ($i = n_c + 1$);

$$\sum_{n=\psi_m} \left\{ \begin{array}{l} \left(\mathbb{T}^{n+1}_{n_c+1,g}\right)_{l,n} \left[\begin{array}{l} (p_{g_n} - p_{g_l})^{n+1} - \\ \rho^{n+1}_{g_n} g(z_{g_n} - z_{g_l}) \end{array} \right] \\ + \left(\mathbb{T}^{n+1}_{n_c+1,w}\right)_{l,n} \left[\begin{array}{l} (p_{o_n} - p_{o_l})^{n+1} - \\ \rho^{n+1}_{o_n} g(z_{o_n} - z_{o_l}) \end{array} \right] \end{array} \right\} + \sum_{n=\xi_n} \left(q^{n+1}_{n_c+1}\right)_{l,n} - \left(\begin{array}{l} y_{n_c+1,g}\rho_g Q_g \\ +\rho_w Q_w \end{array} \right)^{n+1}_{l}$$

$$= \frac{(U)_l}{\Delta t} \left[\begin{array}{l} \left(\phi\rho_g y_{n_c+1,g} S_g + \phi\rho_w S_w\right)^{n+1} \\ -\left(\phi\rho_g y_{n_c+1,g} S_g + \phi\rho_w S_w\right)^{n} \end{array} \right]_l \qquad (6.198)$$

- the pressure equation; and

$$\sum_{n=\psi_n} \left\{ \begin{array}{l} \left(\mathbb{T}^{n+1}_g\right)_{l,n} \rho^{n+1}_g \left[\begin{array}{l} (p_{g_n} - p_{g_l})^{n+1} - \\ \rho^{n}_{g_n} g(z_{g_n} - z_{g_l}) \end{array} \right] \\ + \left(\mathbb{T}^{n+1}_o\right)_{l,n} \rho^{n+1}_o \left[\begin{array}{l} (p_{o_n} - p_{o_l})^{n+1} - \\ \rho^{n}_{o_n} g(z_{o_n} - z_{o_l}) \end{array} \right] \\ + \left(\mathbb{T}^{n+1}_w\right)_{l,n} \rho^{n+1}_w \left[\begin{array}{l} (p_{w_n} - p_{w_l})^{n+1} - \\ \rho^{n}_{w_n} g(z_{o_n} - z_{o_l}) \end{array} \right] \end{array} \right\} + \sum_{n=\xi_n} \left(\begin{array}{l} \rho_g Q_g \\ +\rho_o Q_o \\ +\rho_w Q_w \end{array} \right)^{n+1}_{l,n} - \left(\begin{array}{l} \rho_g Q_g \\ +\rho_o Q_o \\ +\rho_w Q_w \end{array} \right)^{n+1}_{l}$$

$$= \frac{(U)_l}{\Delta t} \left[\begin{array}{l} \left(\phi\rho_g y_{i,g} S_g + \phi\rho_o y_{i,o} S_o\right)^{n+1} \\ -\left(\phi\rho_g y_{i,g} S_g + \phi\rho_o y_{i,o} S_o\right)^{n} \end{array} \right]_l \qquad (6.199)$$

- the energy balance equation.

$$\frac{U}{\Delta t} \left\{ \left[(T - T_0) \left(\sum_j \phi\rho_j S_j C_{vj} + (1-\phi)\rho_r C_r \right) - g\phi \sum_j \rho_j S_j Z_j \right]^{n+1} - \left[(T - T_0) \left(\sum_j \phi\rho_j S_j C_{vj} + (1-\phi)\rho_r C_r \right) - g\phi \sum_j \rho_j S_j Z_j \right]^{n} \right\}$$

$$+ \sum_{n=\psi_n} \left\{ \sum_j \left[\phi\rho_j S_j C_{pj}(T - T_0)\right]_n (\mathbb{T}_j)_l \left[\begin{array}{l} (p_{j_n} - p_{j_l}) - \\ \rho_{j_n} g(z_{j_n} - z_{j_l}) \end{array} \right]^{n+1} + \left(\frac{kA}{\Delta x}\right)_n (T_n - T_l) \right\} + (E_s)^{n+1} + \sum_{n=\xi_n} E^{n+1}_{l,n} = 0$$

$$(6.200)$$

With this choice of equation ordering, the primary unknowns have the order of S_g, $y_{1,o}$, $y_{2,o}$, $y_{3,o}$, $\ldots$, $y_{n_c-1,o}$, $y_{n_c,o}$, S_w, p_o and T. Therefore, the vector of primary unknowns for the block l is:

$$X_n = \begin{bmatrix} S_g & y_{1,o} & y_{2,o} & y_{3,o} & \cdots & y_{n_c-1,o} & y_{n_c,o} & S_w & p_o & T \end{bmatrix}_l^T \qquad (6.201)$$

6.7.11 Solution method

A complete formulation for a non-uniform temperature reservoir is considered. As described already a fully implicit method is adopted to find the solution for the $n_c + 4$ primary unknowns in each block. In the fully implicit method, transmissibilities, well production rates, and fictitious well rates if present are dated at the current time level (time level $n + 1$). Gravities are dated at the old-time level as mentioned earlier. The model equations (constraint equations and flow equations for all components) for all blocks are solved simultaneously in a fully implicit simulator using Newton's Iteration. However, because the constraint equations do not have inter-block terms, they can be eliminated at the matrix level without influencing the solution or the stability of the fully implicit scheme. This elimination results in considerable savings both in CPU time and memory storage requirements.

6.7.12 Solution of model equations using Newton's iteration

The fully implicit iterative equations for the model are derived using the Coats et al. (1977) procedure. Each equation of the model is written in a residual from at time level $n + 1$, i.e., all terms are placed on one side of an equation and the other side is zero. Each term at the time level $n + 1$ in the resulting equation is approximated by its value at the current iteration level $\nu + 1$, which in turn can be approximated by its value at the last iteration level ν, plus a linear combination of the unknowns arising from partial differentiation with respect to all primary unknown. The unknown quantities in the resulting equation are the changes over an iteration of all the primary unknowns in the original equation. This approach does not use conservative expansions of accumulation terms and, therefore, the resulting equations do not conserve material balance during iterations; but they do preserve it at convergence. The resulting fully implicit iterative equations for the block l are derived as follows.

$$\vec{R}_l^{n+1} = 0 \qquad (6.202)$$

where:

$$\vec{R}_l^{n+1} = \begin{Bmatrix} R_1 \\ R_2 \\ R_3 \\ R_4 \\ R_5 \\ \cdot \\ \cdot \\ \cdot \\ R_{n_c+1} \\ R_{n_c+2} \\ R_{n_c+3} \\ R_{n_c+4} \end{Bmatrix}^{n+1} = \begin{Bmatrix} R^{n+1}_{(y_{i,g})_l} \\ R^{n+1}_{(y_{i,o})_l} \\ R^{n+1}_{(i=2)_l} \\ R^{n+1}_{(i=3)_l} \\ R^{n+1}_{(i=4)_l} \\ \cdot \\ \cdot \\ \cdot \\ R^{n+1}_{(i=n_c)_l} \\ R^{n+1}_{(i=n_c+1)_l} \\ R^{n+1}_{(p)_l} \\ R^{n+1}_{(T)_l} \end{Bmatrix} \qquad (6.203)$$

6.7 The compositional simulator using engineering approach

$$R^{n+1}_{(y_{i,g})_n} = \sum_{i=1}^{n_c+1} y^{n+1}_{i,g} - 1 \qquad (6.204)$$

$$R^{n+1}_{(y_{i,o})_l} = \sum_{i=1}^{n_c} y^{n+1}_{i,o} - 1 \qquad (6.205)$$

$$R^{n+1}_{(i=2,3,\ldots,n_c)_l} = \sum_{n=\psi_n} \left\{ \begin{array}{l} \left(\mathbb{T}^{n+1}_{i,g}\right)_{l,n} \left[\begin{array}{l} (p_{g_n} - p_{g_l})^{n+1} - \\ \rho^{n+1}_{g_n} g(z_{g_n} - z_{g_l}) \end{array} \right] \\ + \left(\mathbb{T}^{n+1}_{i,o}\right)_{l,n} \left[\begin{array}{l} (p_{o_n} - p_{o_l})^{n+1} - \\ \rho^{n+1}_{o_n} g(z_{o_n} - z_{o_l}) \end{array} \right] \end{array} \right\} + $$
$$+ \sum_{n=\xi_n} \left(q^{n+1}_i\right)_{l,n} - \left(\begin{array}{l} y_{i,g}\rho_g Q_g \\ +y_{i,o}\rho_o Q_o \end{array} \right)^{n+1}_l $$
$$-\frac{(U)_l}{\Delta t}\left[\begin{array}{l}\left(\phi\rho_g y_{i,g}S_g + \phi\rho_o y_{i,o}S_o\right)^{n+1}\\ -\left(\phi\rho_g y_{i,g}S_g + \phi\rho_o y_{i,o}S_o\right)^n\end{array}\right]_l \qquad (6.206)$$

$$R^{n+1}_{(i=n_c+1)_l} = \sum_{n=\psi_n} \left\{ \begin{array}{l} \left(\mathbb{T}^{n+1}_{n_c+1,g}\right)_{l,n} \left[\begin{array}{l} (p_{g_n} - p_{g_l})^{n+1} - \\ \rho^{n+1}_{g_n} g(z_{g_n} - z_{g_l}) \end{array} \right] \\ + \left(\mathbb{T}^{n+1}_{n_c+1,w}\right)_{l,n} \left[\begin{array}{l} (p_{o_n} - p_{o_l})^{n+1} - \\ \rho^{n+1}_{o_n} g(z_{o_n} - z_{o_l}) \end{array} \right] \end{array} \right\} + $$
$$+ \sum_{n=\xi_n} \left(q^{n+1}_{n_c+1}\right)_{l,n} - \left(\begin{array}{l} y_{n_c+1g}\rho_g Q_g \\ +\rho_w Q_w \end{array} \right)^{n+1}_l $$
$$-\frac{(U)_l}{\Delta t}\left[\begin{array}{l}\left(\phi\rho_g y_{n_c+1,g}S_g + \phi\rho_w S_w\right)^{n+1}\\ -\left(\phi\rho_g y_{n_c+1,g}S_g + \phi\rho_w S_w\right)^n\end{array}\right]_l \qquad (6.207)$$

$$R_{(p)_l}^{n+1} = \sum_{n=\psi_n} \left\{ \begin{array}{l} \left(\mathbb{T}_g^{n+1}\right)_{l,n} \rho_g^{n+1} \left[\begin{array}{l} \left(p_{g_n} - p_{g_l}\right)^{n+1} - \\ \rho_{g_n}^n g\left(z_{g_n} - z_{g_l}\right) \end{array} \right] \\ + \left(\mathbb{T}_o^{n+1}\right)_{l,n} \rho_o^{n+1} \left[\begin{array}{l} \left(p_{o_n} - p_{o_l}\right)^{n+1} - \\ \rho_{o_n}^n g\left(z_{o_n} - z_{o_l}\right) \end{array} \right] \\ + \left(\mathbb{T}_w^{n+1}\right)_{l,n} \rho_w^{n+1} \left[\begin{array}{l} \left(p_{w_n} - p_{w_l}\right)^{n+1} - \\ \rho_{w_n}^n g\left(z_{o_n} - z_{o_l}\right) \end{array} \right] \end{array} \right\} + \sum_{n=\xi_n} \begin{pmatrix} \rho_g Q_g \\ +\rho_o Q_o \\ +\rho_w Q_w \end{pmatrix}_{l,n}^{n+1}$$

$$- \begin{pmatrix} \rho_g Q_g \\ +\rho_o Q_o \\ +\rho_w Q_w \end{pmatrix}_l^{n+1} - \frac{(U)_l}{\Delta t} \left[\begin{array}{l} \left(\phi\rho_g y_{i,g} S_g + \phi\rho_o y_{i,o} S_o\right)^{n+1} \\ - \left(\phi\rho_g y_{i,g} S_g + \phi\rho_o y_{i,o} S_o\right)^n \end{array} \right]_l \quad (6.208)$$

$$R_{(T)_l}^{n+1} = \frac{U}{\Delta t} \left\{ \begin{array}{l} \left[(T-T_0)\left(\sum_j \phi\rho_j S_j C_{vj} + (1-\phi)\rho_r C_r\right) - g\phi\sum_j \rho_j S_j Z_j \right]^{n+1} \\ - \left[(T-T_0)\left(\sum_j \phi\rho_j S_j C_{vj} + (1-\phi)\rho_r C_r\right) - g\phi\sum_j \rho_j S_j Z_j \right]^n \end{array} \right\}$$

$$+ \sum_{n=\psi_n} \left\{ \sum_j \left[\phi\rho_j S_j C_{pj}(T-T_0)\right]_n \left(\mathbb{T}_j\right)_l \left[\begin{array}{l} \left(p_{j_n} - p_{j_l}\right) - \\ \rho_{j_n} g\left(z_{j_n} - z_{j_l}\right) \end{array} \right] \right\}^{n+1} + (E_s)^{n+1} \quad (6.209)$$

$$+ \left(\tfrac{kA}{\Delta x}\right)_n (T_n - T_l)$$

$$+ \sum_{n=\xi_n} E_{l,n}^{n+1}$$

Let the number of gridblocks is N, it can be written that:

$$\vec{R}^{n+1} = \begin{Bmatrix} \vec{R}_1^{n+1} \\ \vec{R}_2^{n+1} \\ \vec{R}_3^{n+1} \\ \vec{R}_4^{n+1} \\ \cdot \\ \cdot \\ \cdot \\ \vec{R}_{N-2}^{n+1} \\ \vec{R}_{N-1}^{n+1} \\ \vec{R}_N^{n+1} \end{Bmatrix} \quad (6.210)$$

6.7 The compositional simulator using engineering approach

$$\vec{R}^{n+1} \cong \overset{\nu+1}{\vec{R}^{n+1}} \cong \overset{\nu}{\vec{R}^{n+1}} + [J]^{\nu}{}^{n+1} \delta\vec{X}^{\nu+1}{}^{n+1} = 0 \tag{6.211}$$

$$\overset{\nu}{\vec{R}^{n+1}} + [J]^{\nu}{}^{n+1} \delta\vec{X}^{\nu+1}{}^{n+1} = 0 \tag{6.212}$$

$$[J]^{\nu}{}^{n+1} \delta\vec{X}^{\nu+1}{}^{n+1} = -\overset{\nu}{\vec{R}^{n+1}} \tag{6.213}$$

where:

$$\delta\vec{X}^{\nu+1}{}^{n+1} = \left\{ \delta\vec{X}_1^{\nu+1}{}^{n+1} \ \delta\vec{X}_2^{\nu+1}{}^{n+1} \ \delta\vec{X}_3^{\nu+1}{}^{n+1} \cdots\cdots\cdots \delta\vec{X}_{N-2}^{\nu+1}{}^{n+1} \ \delta\vec{X}_{N-1}^{\nu+1}{}^{n+1} \ \delta\vec{X}_N^{\nu+1}{}^{n+1} \right\}^T \tag{6.214}$$

$$\delta\vec{X}_l{}^{\nu+1}_{n+1} = \vec{X}_l{}^{\nu+1}_{n+1} - \vec{X}_l{}^{\nu}_{n+1} \tag{6.215}$$

$$[J] = \{[J_1]\ [J_2]\ [J_3]\ [J_4]\cdots\cdots\cdots[J_{N-2}]\ [J_{N-1}]\ [J_N]\}^T \tag{6.216}$$

$$[J_1] = \left[[J_{1,1}]\ [J_{1,2}]\cdots[J_{1,1+n_x}]\cdots\left[J_{1,1+n_xn_y}\right]\cdots\right] \tag{6.217}$$

$$[J_2] = \left[[J_{2,1}]\ [J_{2,2}]\cdots[J_{2,2+n_x}]\cdots\left[J_{2,2+n_xn_y}\right]\cdots\right] \tag{6.218}$$

$$[J_3] = \left[[J_{1,1}]\ [J_{1,2}]\cdots[J_{1,3+n_x}]\cdots\left[J_{1,3+n_xn_y}\right]\cdots\right] \tag{6.219}$$

$$\cdots\cdots\cdots\cdots$$

$$[J_l] = \left[[J_{1,1}]\ [J_{1,2}]\cdots[J_{1,l+n_x}]\cdots\left[J_{1,l+n_xn_y}\right]\cdots\right] \tag{6.220}$$

$$\cdots\cdots\cdots\cdots$$

$$[J_{N-2}] = \left[[J_{N-2,1}]\ [J_{N-2,2}]\cdots[J_{N-2,N-2+n_x}]\cdots\left[J_{N-2,N-2+n_xn_y}\right]\cdots\right] \tag{6.221}$$

$$[J_{N-1}] = \left[[J_{N-1,1}]\ [J_{N-1,2}]\cdots[J_{N-1,N-1+n_x}]\cdots\left[J_{N-1,N-1+n_xn_y}\right]\cdots\right] \tag{6.222}$$

$$[J_{N-2}] = \left[[J_{N,1}]\ [J_{N,2}]\cdots[J_{N,N+n_x}]\cdots\left[J_{N,N+n_xn_y}\right]\cdots\right] \tag{6.223}$$

In general, $[J_{n,l}]$ for block n, is defined as:

$$[J_{n,l}] = \begin{bmatrix} \frac{\partial R_{(y_{i,g})_n}}{(\partial S_g)_l} & \frac{\partial R_{(y_{i,g})_n}}{(\partial y_{1,o})_l} & \frac{\partial R_{(y_{i,g})_n}}{(\partial y_{1,o})_l} & \cdots & \cdots & \frac{\partial R_{(y_{i,g})_n}}{(\partial S_w)_l} & \frac{\partial R_{(y_{i,g})_n}}{(\partial p_o)_l} & \frac{\partial R_{(y_{i,g})_n}}{(\partial T)_l} \\ \frac{\partial R_{(y_{i,o})_n}}{(\partial S_g)_l} & \frac{\partial R_{(y_{i,o})_n}}{(\partial y_{1,o})_l} & \frac{\partial R_{(y_{i,o})_n}}{(\partial y_{2,o})_l} & \cdots & \cdots & \frac{\partial R_{(y_{i,o})_n}}{(\partial S_w)_l} & \frac{\partial R_{(y_{i,o})_n}}{(\partial p_o)_l} & \frac{\partial R_{(y_{i,o})_n}}{(\partial T)_l} \\ \frac{\partial R_{(i=2)_n}}{(\partial S_g)_l} & \frac{\partial R_{(i=3)_n}}{(\partial y_{1,o})_l} & \frac{\partial R_{(i=2)_n}}{(\partial y_{2,o})_l} & \cdots & \cdots & \frac{\partial R_{(i=2)_n}}{(\partial S_w)_l} & \frac{\partial R_{(i=2)_n}}{(\partial p_o)_l} & \frac{\partial R_{(i=2)_n}}{(\partial T)_l} \\ \frac{\partial R_{(i=3)_n}}{(\partial S_g)_l} & \frac{\partial R_{(i=3)_n}}{(\partial y_{1,o})_l} & \frac{\partial R_{(i=3)_n}}{(\partial y_{2,o})_l} & \cdots & \cdots & \frac{\partial R_{(i=3)_n}}{(\partial S_w)_l} & \frac{\partial R_{(i=3)_n}}{(\partial p_o)_l} & \frac{\partial R_{(i=3)_n}}{(\partial T)_l} \\ & & & \vdots & & & & \\ \frac{\partial R_{(i=n_c+1)_n}}{(\partial S_g)_l} & \frac{\partial R_{(i=n_c+1)_n}}{(\partial y_{1,o})_l} & \frac{\partial R_{(i=n_c+1)_n}}{(\partial y_{2,o})_l} & \cdots & \cdots & \frac{\partial R_{(i=n_c+1)_n}}{(\partial S_w)_l} & \frac{\partial R_{(i=n_c+1)_n}}{(\partial p_o)_l} & \frac{\partial R_{(i=n_c+1)_n}}{(\partial T)_l} \\ \frac{\partial R_{(p)_n}}{(\partial S_g)_l} & \frac{\partial R_{(p)_n}}{(\partial y_{1,o})_l} & \frac{\partial R_{(p)_n}}{(\partial y_{2,o})_l} & \cdots & \cdots & \frac{\partial R_{(p)_n}}{(\partial S_w)_l} & \frac{\partial R_{(p)_n}}{(\partial p_o)_l} & \frac{\partial R_{(p)_n}}{(\partial T)_l} \\ \frac{\partial R_{(T)_n}}{(\partial S_g)_l} & \frac{\partial R_{(T)_n}}{(\partial y_{1,o})_l} & \frac{\partial R_{(T)_n}}{(\partial y_{2,o})_l} & \cdots & \cdots & \frac{\partial R_{(T)_n}}{(\partial S_w)_l} & \frac{\partial R_{(T)_n}}{(\partial p_o)_l} & \frac{\partial R_{(T)_n}}{(\partial T)_l} \end{bmatrix}$$

(6.224)

A complete description of the method and evaluation of the partial differentiation of the elements of $\vec{R}_l^{n+1}$ can be obtained in Abou-Kassem et al. (2006).

6.7.13 The effect of linearization

Nonlinearities include phase compositions, phase transmissibilities, well production (injection) rates, fictitious well rates, and accumulation terms. The flow equation is obtained by combining the mass balance equation and Darcy's law in an integral form for a discrete reservoir. The variation of the integrand is neglected for a selected time interval to recast the flow equation in an algebraic form. This algebraic form of flow equation is nonlinear due to the dependency of involved parameters to each other. The algebraic equations for individual reservoir blocks form a set of nonlinear simultaneous algebraic equations. It is necessary to impose some simplification and linearization to obtain a numerical description of the flow characteristics. There is a need to observe the role of linearizations. The effects of these simplifications and linearizations during the solving process are investigated for two different cases:

- A single-phase flow of natural gas through a reservoir of 20-acre spacing and 30 ft. net thickness. The reservoir is horizontal and described by four gridblocks in the radial direction.
- A multiphase flow of water and oil through a horizontal reservoir of length $L = 1200$ ft., width $W = 350$ ft., and height $H = 40$ ft. The effect of the dimension of the reservoir blocks, the time interval, the pressure-dependent parameters, and the water saturation related parameters are studied

6.7.13.1 Case one: Single-phase flow of a natural gas

An example (Abou-Kassem et al., 2006a) on the natural gas reservoir of 20-acre spacing and 30 ft. net thickness is taken as the test case in this paper. The reservoir is horizontal and described by four gridblocks in the radial direction. It is also assumed that the reservoir has homogeneous and isotropic rock properties with $k = 15$ md and $\varphi = 0.13$. A vertical well ($d = 0.5$ ft) produces from the reservoir at a rate of 1 MMscf/D. The initial reservoir pressure is 4015 psia. The pressure distribution at different time intervals needs to be calculated. The flow is considered to be in the radial direction without any variation in z- and θ-directions.

The governing equations of fluid flow through porous media are obtained by combining two basic engineering concepts including the principle of mass conservation and the constitutive equation. In reservoir simulation, the constitutive property, i.e., the rate of fluid movement into (or out of) the reservoir volume element is described by Darcy's law and is related to the potential gradient. Therefore, the combination of Darcy's law with the conservation of mass results in the flow equation. The resulting differential form of the flow equation is nonlinear. The discrete form of the flow equation may be obtained directly using the finite control volume method described in Section 6.2. The reservoir is divided into gridblocks in different directions and the flow equation is written for each of these gridblocks. Finally, the resulting equations are a system of nonlinear algebraic equations that give the pressure distribution along the reservoir at any time.

The general form of the flow equation for gridblock n can be written as:

$$\sum_{l=\psi_n} \mathbb{T}_{l,n}^{n+1}\left[\left(p_l^m - p_n^m\right) - \gamma_{l,n}^m\left(Z_l^m - Z_n^m\right)\right] + \sum_{l=\xi_n} q_{sc_{l,n}}^m + q_{sc_n}^m = \frac{U}{\alpha_c \Delta t}\left[\left(\frac{\phi}{B}\right)_n^{v+1} - \left(\frac{\phi}{B}\right)_n^v\right] \quad (6.225)$$

The transmissibility along the r-direction in cylindrical coordinates is defined as:

$$\mathbb{T}_{i\mp\frac{1}{2},j,k} = G_{r_{i\mp\frac{1}{2},j,k}} \left(\frac{1}{\mu B}\right)_{i\mp\frac{1}{2},j,k} \quad (6.226)$$

Logarithmic spacing constant, α_{lg} geometric factor (G) in the r-direction and bulk volume (U) is calculated using the recently reported simple and explicit equations (Abou-Kassem, 2007a).

$$\alpha_{lg} = \left(\frac{r_e}{r_w}\right)^{1/n_r}$$

$$G_{r_{i-1/2,j,k}} = \frac{\beta_c \Delta \theta_j}{\left\{\log_e\left[\alpha_{lg} \log_e(\alpha_{lg})/(\alpha_{lg}-1)\right]/\left(\Delta z_{i,j,k} k_{r_{i,j,k}}\right)\right. \\ \left. + \log_e\left[(\alpha_{lg}-1)/\log_e(\alpha_{lg})\right]\left(\Delta z_{i-1,j,k} k_{r_{i-1,j,k}}\right)\right\}}$$

$$G_{r_{i+1/2,j,k}} = \frac{\beta_c \Delta \theta_j}{\left\{\log_e\left[(\alpha_{lg}-1)/\log_e(\alpha_{lg})\right]/\left(\Delta z_{i,j,k} k_{r_{i,j,k}}\right)\right. \\ \left. + \log_e\left[\alpha_{lg} \log_e(\alpha_{lg})/(\alpha_{lg}-1)\right]\left(\Delta z_{i+1,j,k} k_{r_{i+1,j,k}}\right)\right\}}$$

$$V_{b_{i,j,k}} = \left\{\left(\alpha_{lg}^2 - 1\right)^2 / \left[\alpha_{lg}^2 \log_e\left(\alpha_{lg}^2\right)\right]\right\} r_i^2 \left(^1/_2 \Delta \theta_j\right) \Delta z_{i,j,k}$$

for $i = 1, 2, 3, \ldots n_r - 1$; $j = 1, 2, 3, \ldots n_\theta$; and $k = 1, 2, 3, \ldots n_z$;

$$V_{b_{n_r,j,k}} = \left\{1 - \left[\log_e(\alpha_{lg})/(\alpha_{lg} - 1)\right]^2 \left(\alpha_{lg}^2 - 1\right)\left[\alpha_{lg}^2 \log_e\left(\alpha_{lg}^2\right)\right]\right\} r_e^2 \left(^1/_2 \Delta\theta_j\right) \Delta z_{n_r,j,k}$$

for $i = n_r$; $j = 1, 2, 3, \ldots n_\theta$; and $k = 1, 2, 3, \ldots n_z$.

The flow equation can be written as

$$\sum_{l=\psi_n} \mathbb{T}_{l,n}^{n+1}(p_l^m - p_n^m) + q_{sc_n}^m = \frac{U}{\alpha_c \Delta t}\left[\left(\frac{\phi}{B}\right)_n^{v+1} - \left(\frac{\phi}{B}\right)_n^v\right] \quad (6.227)$$

Eq. (6.227) can be simplified to Eq. (6.228) if $\frac{\phi}{B}$ is considered to be a linear function of pressure.

$$\sum_{l=\psi_n} \mathbb{T}_{l,n}^{n+1}(p_l^m - p_n^m) + q_{sc_n}^m = \frac{U}{\alpha_c \Delta t}\left(\frac{\phi}{B}\right)_n' [p_n^{v+1} - p_n^v] \quad (6.228)$$

where, $\left(\frac{\phi}{B}\right)_n'$ is the chord slope of $\left(\frac{\phi}{B}\right)_n$ between p_n^{v+1} and p_n^v.

Although, μ and B are a function of the fluid temperature and pressure, it is assumed that temperature remains constant throughout the reservoir during the production period. The gas formation volume factor, GFVF or B_g, and viscosity, μ as functions of reservoir pressure are shown in Table 6.15.

They are expressed in mathematical form using the polynomial of fourth order for μ and the power function for B as shown in Fig. 6.34A. These two fluid properties are also fitted with the spline functions of different degrees as shown in Fig. 6.34B. It is found that the quadratic and cubic splines give a very good approximation to the variation of μ and B with P.

TABLE 6.15 The variation of the gas formation volume factor and viscosity with reservoir pressure (Abou-Kassem et al., 2006a).

Pressure (psia)	GFVF (RB/scf)	Viscosity (cp)	Pressure (psia)	GFVF (RB/scf)	Viscosity (cp)
215	0.016654	0.0126	2215	0.001318	0.0167
415	0.008141	0.0129	2415	0.001201	0.0173
615	0.005371	0.0132	2615	0.001109	0.0180
815	0.003956	0.0135	2815	0.001032	0.0186
1015	0.003114	0.0138	3015	0.000972	0.0192
1215	0.002544	0.0143	3215	0.000922	0.0198
1415	0.002149	0.0147	3415	0.000878	0.0204
1615	0.001857	0.0152	3615	0.000840	0.0211
1815	0.001630	0.0156	3815	0.000808	0.0217
2015	0.001459	0.0161	4015	0.000779	0.0223

6.7 The compositional simulator using engineering approach

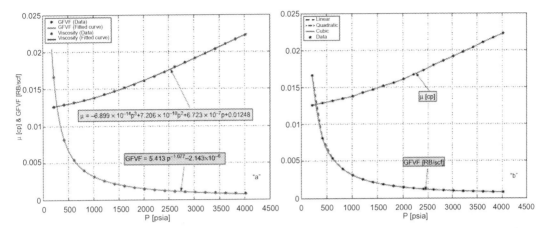

FIG. 6.34 Approximation of variation of μ and $GFVF$ with P (A) using continuous functions (B) using spline functions of different orders.

The flow equation for the gridblocks as specified in Eq. (6.227) can also be expressed in the following form

$$\begin{cases} G\left(\dfrac{1}{\mu B}\right)_j^m (p_2^m - p_1^m) - 10^6 = \beta_1 \left[\left(\dfrac{\phi}{B}\right)_1^{v+1} - \left(\dfrac{\phi}{B}\right)_1^v\right] \\ G\left(\dfrac{1}{\mu B}\right)_j^m (p_1^m - p_2^m) + G\left(\dfrac{1}{\mu B}\right)_j^m (p_3^m - p_2^m) = \beta_2 \left[\left(\dfrac{\phi}{B}\right)_2^{v+1} - \left(\dfrac{\phi}{B}\right)_2^v\right] \\ G\left(\dfrac{1}{\mu B}\right)_j^m (p_2^m - p_3^m) + G\left(\dfrac{1}{\mu B}\right)_j^m (p_4^m - p_3^m) = \beta_3 \left[\left(\dfrac{\phi}{B}\right)_3^{v+1} - \left(\dfrac{\phi}{B}\right)_3^v\right] \\ G\left(\dfrac{1}{\mu B}\right)_j^m (p_3^m - p_4^m) = \beta_4 \left[\left(\dfrac{\phi}{B}\right)_4^{v+1} - \left(\dfrac{\phi}{B}\right)_4^v\right] \end{cases} \quad (6.229)$$

and according to Eq. (6.228) as

$$\begin{cases} G\left(\dfrac{1}{\mu B}\right)_j^m (p_2^m - p_1^m) - 10^6 = \beta_1 \left(\dfrac{\phi}{B}\right)_1' [p_1^{v+1} - p_1^v] \\ G\left(\dfrac{1}{\mu B}\right)_j^m (p_1^m - p_2^m) + G\left(\dfrac{1}{\mu B}\right)_j^m (p_3^m - p_2^m) = \beta_2 \left(\dfrac{\phi}{B}\right)_2' [p_2^{v+1} - p_2^v] \\ G\left(\dfrac{1}{\mu B}\right)_j^m (p_2^m - p_3^m) + G\left(\dfrac{1}{\mu B}\right)_j^m (p_4^m - p_3^m) = \beta_3 \left(\dfrac{\phi}{B}\right)_3' [p_3^{v+1} - p_3^v] \\ G\left(\dfrac{1}{\mu B}\right)_j^m (p_3^m - p_4^m) = \beta_4 \left(\dfrac{\phi}{B}\right)_4' [p_4^{v+1} - p_4^v] \end{cases} \quad (6.230)$$

where $\beta_1 = \frac{U}{a_c \Delta t}$ is constant for each block. The function $\left(\frac{1}{\mu B}\right)^m$ depends on pressure distribution in the reservoir. Two cases regarding subscript j are considered

I. pressure at the upstream, and
II. pressure at the i-th gridblock.

The superscript m is related to the time step, which is taken as $m = \nu + 1$.
Case I: For Case (I), Eq. (6.229) is written as

$$\begin{cases} G\left(\frac{1}{\mu B}\right)_j^{\nu+1} (p_2^{\nu+1} - p_1^{\nu+1}) - 10^6 = \beta_1 \left[\left(\frac{\phi}{B}\right)_1^{\nu+1} - \left(\frac{\phi}{B}\right)_1^{\nu}\right] \\ G\left(\frac{1}{\mu B}\right)_j^{\nu+1} (p_1^{\nu+1} - p_2^{\nu+1}) + G\left(\frac{1}{\mu B}\right)_j^{\nu+1} (p_3^{\nu+1} - p_2^{\nu+1}) = \beta_2 \left[\left(\frac{\phi}{B}\right)_2^{\nu+1} - \left(\frac{\phi}{B}\right)_2^{\nu}\right] \\ G\left(\frac{1}{\mu B}\right)_j^{\nu+1} (p_2^{\nu+1} - p_3^{\nu+1}) + G\left(\frac{1}{\mu B}\right)_j^{\nu+1} (p_4^{\nu+1} - p_3^{\nu+1}) = \beta_3 \left[\left(\frac{\phi}{B}\right)_3^{\nu+1} - \left(\frac{\phi}{B}\right)_3^{\nu}\right] \\ G\left(\frac{1}{\mu B}\right)_j^{\nu+1} (p_3^{\nu+1} - p_4^{\nu+1}) = \beta_4 \left[\left(\frac{\phi}{B}\right)_4^{\nu+1} - \left(\frac{\phi}{B}\right)_4^{\nu}\right] \end{cases} \quad (6.231)$$

and (6.230) is given in the form

$$\begin{cases} G\left(\frac{1}{\mu B}\right)_j^{\nu+1} (p_2^{\nu+1} - p_1^{\nu+1}) - 10^6 = \beta_1 \left(\frac{\phi}{B}\right)_1' \left[p_1^{\nu+1} - p_1^{\nu}\right] \\ G\left(\frac{1}{\mu B}\right)_j^{\nu+1} (p_1^{\nu+1} - p_2^{\nu+1}) + G\left(\frac{1}{\mu B}\right)_j^{\nu+1} (p_3^{\nu+1} - p_2^{\nu+1}) = \beta_2 \left(\frac{\phi}{B}\right)_2' \left[p_2^{\nu+1} - p_2^{\nu}\right] \\ G\left(\frac{1}{\mu B}\right)_j^{\nu+1} (p_2^{\nu+1} - p_3^{\nu+1}) + G\left(\frac{1}{\mu B}\right)_j^{\nu+1} (p_4^{\nu+1} - p_3^{\nu+1}) = \beta_3 \left(\frac{\phi}{B}\right)_3' \left[p_3^{\nu+1} - p_3^{\nu}\right] \\ G\left(\frac{1}{\mu B}\right)_j^{\nu+1} (p_3^{\nu+1} - p_4^{\nu+1}) = \beta_4 \left(\frac{\phi}{B}\right)_4' \left[p_4^{\nu+1} - p_4^{\nu}\right] \end{cases} \quad (6.232)$$

The pressure distribution at the center of gridblocks is computed using Eqs. (6.231), (6.232). Here μ and B vary with pressure, however, k is assumed constant. The time step is chosen as $\Delta t = 1$ month. Fig. 6.28A shows the pressure at gridblock 1 as a function of time when the variation of μ and B with pressure is approximated using continuous functions. It is observed that there is no substantial discrepancy between the pressure values obtained through the original formulation and approximate formulation at different time steps. However, the computational results suggest that, with increasing time, the difference between the pressures from Eqs. (6.231), (6.232) increases. The effect of the simplification of the formulation is more evident when the cubic spline is applied to approximate the variation of μ and B with pressure as shown in Fig. 6.28B.

Effect of interpolation functions and formulation

The pressure distribution at the center of gridblocks is also obtained using different interpolation functions and various formulations. The results for gridblock 1 are shown in

Figs. 6.34A and 6.35B. The non-linear continuous functions as given in Fig. 6.9, linear interpolation, and cubic spline are applied to approximate μ and B in both figures. The transient pressure results using cubic spline and linear interpolation are very close to each other for both linear formulations (Fig. 6.11A) and original formulations (Fig. 6.11B). It is also noticed that the continuous functional interpolation shows higher values for pressure at different time steps. Such observation is found to be true for other gridblocks. The solution with the cubic spline is faster and more accurate.

Effect of time interval

The effect of the time interval on the accuracy of the computation is obtained by using different time steps (Δt), which was varied from 1 day to 4 months. Results are based on formulation following Eqs. (6.231), (6.232) while the interpolation function for μ and B follows the cubic spline. Fig. 6.13, which is the solution with original formulation at different time steps, shows that the accuracy is not affected by the time interval. In most severe cases, the mean relative error is (Figs. 6.35 and 6.36)

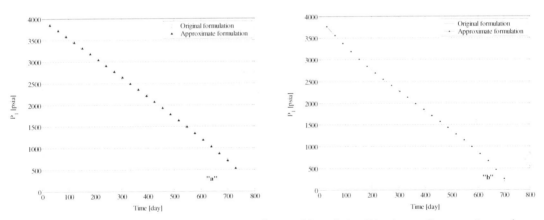

FIG. 6.35 Pressure at gridblock 1 with the linear and original formulation (A) using nonlinear continuous functions (B) using cubic spline.

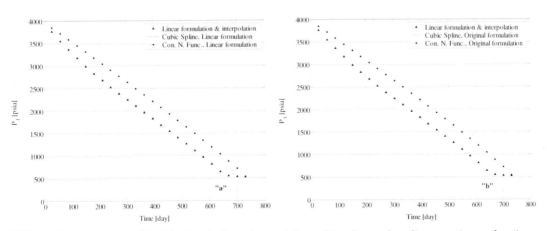

FIG. 6.36 Pressure at gridblock 1 using the linear interpolation, cubic spline, and nonlinear continuous functions (A) with linear formulation (B) with the original formulation.

6. Reserves prediction and deliverability

TABLE 6.16 Relative error at different time steps using linear formulation and cubic spline interpolation.

Time step (Δt)	Relative error
2 days	0.09%
0.5 month	0.68%
1 month	1.46%
2 months	3.19%
4 months	7.82%

$$\text{Relative error} = \sum_{i=1}^{n} \left(\frac{p_{i_{t=4 \text{ months}}} - p_{i_{t=1 \text{ month}}}}{p_{i_{t=4 \text{ months}}}} \right) / n = 0.014\% \tag{6.233}$$

Relative errors are also calculated for the case of linear formulation and the results of it at different time steps are shown in Table 6.16. The table clearly suggests that the linearized formulation is more sensitive to the value of Δt.

Effect of permeability

To investigate the effect of permeability, computation is also carried out with the original formulation. As a case study, a 10% variation in permeability between the boundary blocks is assumed. Such variation can be described by a linear relationship between permeability and pressure through the following equation

$$k = 0.004286 \, p + 15.22 \tag{6.234}$$

During this computation, permeability, k, in each time step and iteration is renewed for the gridblocks. The variation of pressure with time for gridblock 1 with constant and variable permeability is shown in Fig. 6.37A. For the sake of clarity, the results are repeated in a tabular

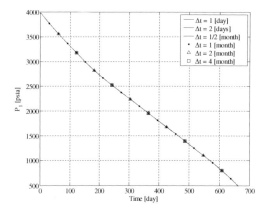 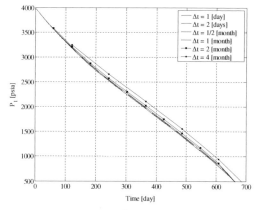

FIG. 6.37 The effect of the time step in computation (A) Cubic spline interpolation using original formulation (B) Cubic spline interpolation using the linear formulation.

Time [months]	P_1 (k=const.) [psia]	P_1(k=$f(p)$) [psia]
1	3765.668915	3762.476196
2	3554.151242	3551.149301
6	2822.768896	2820.363056
12	1959.327022	1957.484815
15	1540.271151	1538.676903
18	1108.891024	1107.635304
20	798.990998	798.110175

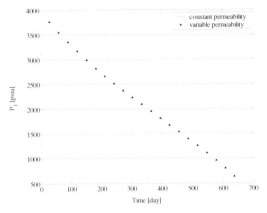

FIG. 6.38 (A) The effect of permeability variation on pressure variation with time (B) tabular presentation.

form 6.14b. It is noticed that there is a small difference between the constant permeability result and the variable permeability result. It is also observed that the margin of difference is more evident in the initial time than later time (Fig. 6.38).

Effect of number of gridblocks

All of the previous computations are carried out with 4 gridblocks are four. To investigate the effect of the number of gridblocks, the number of gridblocks is varied from 4 to 64. For this study, the original formulation as given in (6.227) is applied and the systems of algebraic equations are written for each number of gridblocks. The cubic spline is taken as the interpolation function to include the variation of μ and B; and k is considered to vary according to Eq. (6.234). The Newton method is followed to solve the systems of algebraic equations. Fig. 6.39A–D illustrates the effect of the number of gridblocks at various time steps. It is observed that when the number of gridblocks is increased from 4 to 8, there is a difference in the pressure values predicted. However, increasing from 8 to 16 or more than that provides a smooth curve at all four-time steps.

Spatial and transient pressure distribution using different interpolation functions

Fig. 6.40A–D shows pressure distribution based on the cubic spline and linear interpolation along the reservoir at different times. These figures show that the linear interpolations of μ and B give higher values of pressure than those predicted by the cubic spline technique for all period along the reservoir radius. However, the difference is very small and is only about 5 psia in severe case.

6.7.13.2 Case II

The original formulation is considered and the cubic spline is used to approximate the variation of μ and B. The system of nonlinear algebraic equations is obtained using Eq. (6.229) and is given as

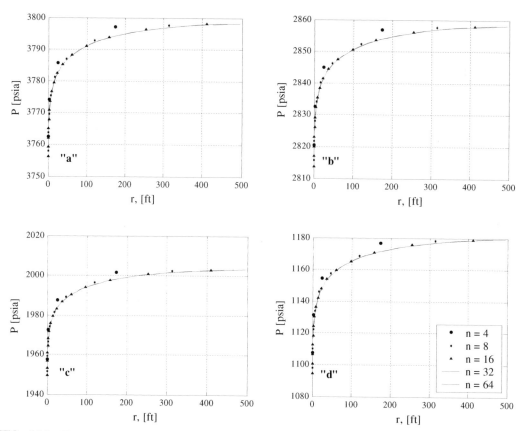

FIG. 6.39 The pressure distribution using different number of gridblocks (A) $t = 1$ month, (B) $t = 6$ months, (C) $t = 12$ months and (D) $t = 18$ months.

$$\begin{cases} G\left(\dfrac{1}{\mu B}\right)_j^{v+1} (p_2^{v+1} - p_1^{v+1}) - 10^6 = \beta_1 \left[\left(\dfrac{\phi}{B}\right)_1^{v+1} - \left(\dfrac{\phi}{B}\right)_1^{v} \right] \\ G\left(\dfrac{1}{\mu B}\right)_j^{v+1} (p_1^{v+1} - p_2^{v+1}) + G\left(\dfrac{1}{\mu B}\right)_j^{v+1} (p_3^{v+1} - p_2^{v+1}) = \beta_2 \left[\left(\dfrac{\phi}{B}\right)_2^{v+1} - \left(\dfrac{\phi}{B}\right)_2^{v} \right] \\ G\left(\dfrac{1}{\mu B}\right)_j^{v+1} (p_2^{v+1} - p_3^{v+1}) + G\left(\dfrac{1}{\mu B}\right)_j^{v+1} (p_4^{v+1} - p_3^{v+1}) = \beta_3 \left[\left(\dfrac{\phi}{B}\right)_3^{v+1} - \left(\dfrac{\phi}{B}\right)_3^{v} \right] \\ G\left(\dfrac{1}{\mu B}\right)_j^{v+1} (p_3^{v+1} - p_4^{v+1}) = \beta_4 \left[\left(\dfrac{\phi}{B}\right)_4^{v+1} - \left(\dfrac{\phi}{B}\right)_4^{v} \right] \end{cases} \quad (6.235)$$

A new method is used to formulate the problem. The first equation of (6.235) is nonlinear, for which the unknown is p_1^{v+1}. As a nonlinear equation, Equation one of (6.235) has the potential to give more than one solution. Using p_1^{v+1} from the solution of The first and the second equations of (6.235) can be solved for p_2^{v+1} that may give more than one solution for each of

6.7 The compositional simulator using engineering approach

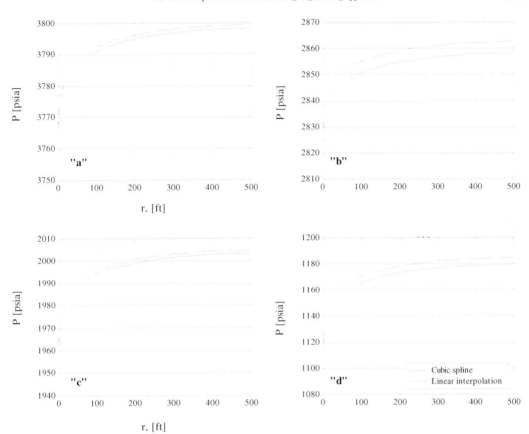

FIG. 6.40 The pressure distribution along the reservoir radius using cubic spline and linear interpolation for variation of μ and B when $n = 64$ (A) $t = 1$ month, (B) $t = 6$ months, (C) $t = 12$ months and (D) $t = 18$ months.

$p_1^{\nu+1}$. The procedure is continued the similar way, i.e., applied to the third and fourth equations of (6.235). Therefore, theoretically, multiple solutions can be expected for $p_3^{\nu+1}$ and consequently, for $p_4^{\nu+1}$.

To examine the feasibility of the technique mentioned earlier, we start with the first equation of (6.235). It is found that there are two solutions for p_1 as obtained in the first iteration. The solutions are: $p_1 = 4004.566$ psia and 9312.639 psia.

The second equation is solved by p_2 using the result of p_1 and similarly, the remaining equations of (6.235) are solved by p_3 and p_4. Table 6.17 shows the results after the first iteration:

TABLE 6.17 Pressure solution at $t = 30.42$ days for four gridblocks.

Time [days]	P_1 [psia]	P_2 [psia]	P_3 [psia]	P_4 [psia]
30.420	9312.639809	−5404.929667	−384.716801	577.929652
30.420	4004.566162	−5147.603731	−310.418879	606.315484
30.420	4004.566162	4009.783158	4012.393337	4012.512134

The first and second sets give results, which are not only unexpected but also unrealistic. During the computation, the second iteration resulted in a breakdown of the solutions. However, the last set provides satisfactory results following several iterations at the desired time interval (the table only shows the pressure at t = 30.42 days).

The pressure values for the gridblocks are also obtained by utilizing (6.235). In this set, the system of algebraic equations is set up with the properties of the gridblock itself. During the computation, μ and B are updated using cubic spline and linear interpolation. Both techniques show almost the same results as evident in Fig. 6.41A. The utilization of gridblock and upstream flow properties are also examined in Fig. 6.41B. The system of algebraic equations (6.231) is based on the upstream flow properties. The figure shows that the results based on these two formulations are also very close to each other. However, the difference increases as time increases (>500 days).

CPU time

In continuing the discussion of the previous section, it is important to note the CPU time required for computation. The main constraint during computation with (6.235) is the computing time. The CPU time required to compute pressure for all four gridblocks for 1.5 years with $\Delta t = 1$ month and using the formulation of (6.235), is approximately 510 s. On the contrary, the same problem when utilized in the formulation of (6.231), takes only 1.156 s, which is significantly lower than the previous ones.

It is important when the time step and the number of gridblocks are increased. The computation of the pressure distribution takes $t = 1.672$[sec] with formulation (6.231) for 64 gridblocks and $\Delta t = 1$ months. The time to compute the pressure with formulation (6) is $t = 31.172$[sec] with $\Delta t = 1$ months for 1.5 years. It indicates that formulation (6) is more efficient in time than formulation (9).

The effects of the nonlinear behavior of some fluid and formation properties and the simplification of the governing equations and the possibility of having multiple solutions are investigated in this example. The pressure distribution along the reservoir is computed with different types of formulations while the viscosity and the fluid formation volume factor are approximated with different types of interpolating functions.

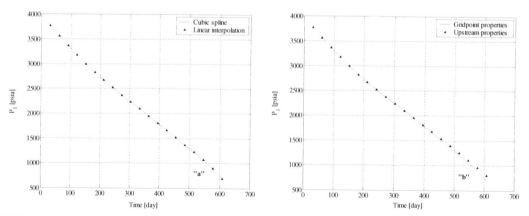

FIG. 6.41 The pressure at gridblock 1 using different sets of the system of algebraic equations and various interpolations (A) Cubic spline and linear interpolation using (6.231) (B) Cubic spline using (6.231) and (6.235).

The effect of the linearization in the governing equations is significant. Linearization of the coefficient of the formulation may lead to the wrong prediction of the pressure distribution along the reservoir. The variation of the fluid and formation properties are obtained more properly if they are approximated with spline functions. The order of the piecewise polynomials has a very minimal effect if the original formulation is applied. However, the computation shows that the pressure-dependent properties have a very weak nonlinearity effect and they may be neglected during the computations.

The problem is also formulated in such a way as to produce multiple solutions for the pressure distribution in the gridblocks. This formulation is promising and shows the mathematical potential of having multiple solutions. More investigation is required to confirm if multiple-valued solutions of physical significance exist.

Case II: An oil/water reservoir

A horizontal reservoir of length $= 1200\,ft$, width $W = 350\,ft$ and height $H = 40\,ft$ is considered, as shown in Fig. 6.42. The flow is in the direction of length to a 7 inches production well located at the end of the reservoir. The pressure at the left side boundary of the reservoir is kept constant the same as the initial reservoir pressure during the production process. The initial pressure of the reservoir is $p_r = 4000$ psia. The gas-oil ratio is $GOR = 400\,SCF/STB$ and the bubble pressure is $p_B = 2000$ psia. The specific gravity of oil and gas are $\gamma_o = 0.876$ and $\gamma_g = 0.75$, respectively. The reservoir temperature is $T_R = 150\,°F$ initially and is assumed to be constant during the production process. It is also assumed that the reservoir has homogeneous rock properties initially with porosity $\varphi = 0.27$ and permeability $k = 270\,md$.

Two cases are considered:

(a) There is a flow of oil into the reservoir in the left hand side boundary to keep this side a constant pressure boundary. Therefore, there is a single-phase flow reservoir.
(b) The pressure is kept constant by the presence of a strong aquifer. The flow of the water into the reservoir caused a multi-phase flow of oil and water.

Single-phase flow: The pressure is kept constant on the left hand side by an oil flow into the reservoir in that boundary. The flow equation can be written as:

$$\sum_{l=\psi_n} T_{ol,n}^{n+1}\left[\left(p_{ol}^{n+1} - p_{on}^{n+1}\right) - \gamma_{ol,n}^n (Z_l - Z_n)\right] + \sum_{l=\xi_n} q_{osc_{l,n}}^{n+1} + q_{osc_n}^{n+1} \qquad (6.236)$$
$$= \frac{U}{\alpha_c \Delta t}\left[\left(\frac{\phi}{B_o}\right)_n^{n+1} - \left(\frac{\phi}{B_o}\right)_n^n\right]$$

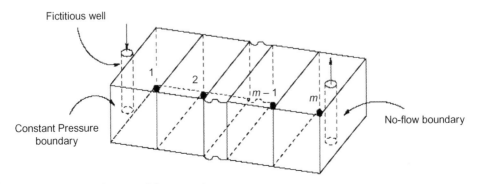

FIG. 6.42 A schematic diagram of the reservoir.

for each grid point. The notation $\mathbb{T}_{ol,n}^{n+1}$ is the oil transmissibility between l and n gridblocks and may be given in the form

$$\mathbb{T}_{ol,n}^{n+1} = G_{l,n} \left(\frac{1}{\mu_o B_o}\right)_{l,n} \tag{6.237}$$

where

$$G_{l,n} = G = \frac{\beta_c A_x k_x}{\Delta x} \tag{6.238}$$

The rate of water flow into the reservoir to keep a constant pressure is obtained by

$$q_{wsc_b,BB}^{n+1} = \left[\frac{\beta_c A_x k_x}{\mu_o B_o \left(\frac{\Delta x}{2}\right)}\right]_{bB}^{n+1} \left[(p_b - p_{bB}^{n+1}) - (\gamma_w)_{bB}^{n+1}(Z_b - Z_{bB})\right] \tag{6.239}$$

where p_b is the boundary pressure (b ≡ boundary) and p_{bB} is the pressure of the boundary block (B ≡ boundary block).

It is considered that the reservoir is divided into m gridblocks of the same length as shown in Fig. 6.43. Since the gridblocks have the same length and same cross-sectional area and the flow is 1 - D, the geometrical factors and bulk volumes are equal for all gridblocks.

$$G_{l,n} = G = \frac{\beta_c A_x k_x}{\Delta x} = \frac{0.001127 \times (350 \times 40) \times 270}{300} = 14.2002$$

$$U = 300 \times 350 \times 40 = 4200000 \, ft^3$$

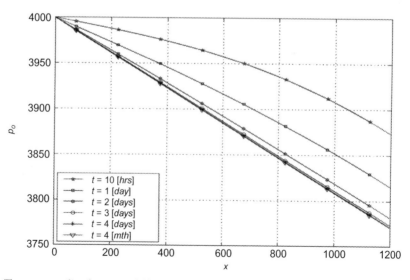

FIG. 6.43 The pressure distribution at different times along the reservoir.

6.7 The compositional simulator using engineering approach

The set of flow equations for the gridblocks is:

$$\begin{cases} G\left(\dfrac{1}{\mu_o B_o}\right)_1^{n+1}\left(p_{o_2}^{n+1}-p_{o_1}^{n+1}\right)+2G\left(\dfrac{1}{\mu_o B_o}\right)_1^{n+1}\left(4000-p_{o_1}^{n+1}\right)=\dfrac{U}{\alpha_c \Delta t}\left[\left(\dfrac{\phi}{B_o}\right)_1^{n+1}-\left(\dfrac{\phi}{B_o}\right)_1^{n}\right] \\ \dots\dots\dots\dots\dots\dots \\ \dots\dots\dots\dots\dots\dots \\ G\left(\dfrac{1}{\mu_o B_o}\right)_1^{n+1}\left(p_{o_{i-1}}^{n+1}-p_{o_i}^{n+1}\right)+G\left(\dfrac{1}{\mu_o B_o}\right)_1^{n+1}\left(p_{o_{i+1}}^{n+1}-p_{o_i}^{n+1}\right)=\dfrac{U}{\alpha_c \Delta t}\left[\left(\dfrac{\phi}{B_o}\right)_i^{n+1}-\left(\dfrac{\phi}{B_o}\right)_i^{n}\right] \\ \dots\dots\dots\dots\dots\dots \\ \dots\dots\dots\dots\dots\dots \\ G\left(\dfrac{1}{\mu_o B_o}\right)_{m-1}^{n+1}\left(p_{o_{m-1}}^{n+1}-p_{o_m}^{n+1}\right)-q_o=\dfrac{U}{\alpha_c \Delta t}\left[\left(\dfrac{\phi}{B_o}\right)_m^{n+1}-\left(\dfrac{\phi}{B_o}\right)_m^{n}\right] \end{cases} \quad (6.240)$$

where, $1 < i < m$. The distribution of the pressure along the reservoir is obtained by solving (6.240). This set of algebraic equations is nonlinear. The nonlinearity is due to the fact that the viscosity and the oil formation volume factor are a function of oil pressure. They change with the variation of pressure. The porosity of the formation is also a function of the fluid pressure inside it.

$$\begin{aligned} &\mu_o=1+c_\mu(p-p_B),\quad c_\mu=6.7\times 10^{-5}\,\dfrac{cp}{psia} \\ &B_o=B_{oB}e^{c_o(p-p_B)},\quad B_{oB}=1.22\,\dfrac{RB}{bbl},\quad c_o=5.666\times 10^{-6}\,psia^{-1} \\ &\phi=\phi^o\left[1+c_\phi(p-p_R)\right],\quad c_\phi=1\times 10^{-6} \end{aligned} \quad (6.241)$$

The Newton method is applied to find the solution for the system of algebraic equations. The solutions are obtained iteratively by updating the coefficients to reach a convergent solution based on an assigned minimum error between two consecutive results. The minimum error is considered to be less than 10^{-6}.

The pressure distribution takes several days to reach a steady-state condition and it does not vary any more as shown in Fig. 6.43. The pressure distributions are shown for different periods of time. The pressure distribution after $t = 4$ [days] is very close to the steady-state condition when the pressure distribution does not change with time anymore. The result is computed with m = 8 and 256. The solution with m = 8 is given with different markers at various periods of time. The solution with m = 256 is given with the solid line at different time. The results are the same with both the number of gridblocks.

The solution with different values of Δt is given in Fig. 6.44. The solution are obtained at different time with $\Delta t = 1$ [min], 1 [hour], 1 [day] and 1 [month]. The time step has no effect on the final steady-state condition. However, it is necessary to use small values of Δt to predict the pressure distribution and the time to reach a steady-state condition. The solution with $\Delta t = 1$ [min] is carried out for only a period of one day with $n = 8$. The results with $\Delta t = 1$ [hour] are obtained with the same number of gridblocks for 1 month. The pressure

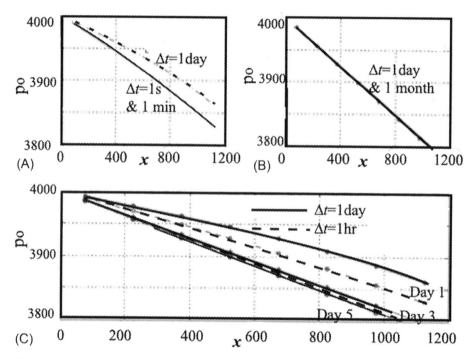

FIG. 6.44 The pressure distribution at different times along the reservoir.

distribution is given for one day with $\Delta t = 1$ [min], 1 [hour], 1 [day] in Fig. 6.44A. It is necessary to have small values of Δt to predict the variation of pressure in the period of the unsteady process. The pressure distribution with $\Delta t = 1$ [day] is more than the ones obtained with smaller values of Δt. The pressure distribution after one day of operation is almost the same with $\Delta t = 1$ [min] and 1 [hour]. This indicates that it is not necessary to adopt very small values of time intervals. The pressure distribution with $\Delta t = 1$ [month] is compared with the solution obtained with $\Delta t = 1$ [hour] in Fig. 6.20B. The solution with $\Delta t = 1$ [month] shows some unsteady behavior after a month of operation that is not correct. It is better to have a smaller time interval to predict more precisely the behavior of fluid flow inside the reservoir. The pressure distributions for the first few days of the production process are given in Fig. 6.44C with $\Delta t = 1$ [hour] and 1 [day]. The solutions are different for the first three days but after five days of operation, the results are almost the same with both time intervals. It seems that a time interval of $\Delta t = 1$ [day] is sufficient to provide precise results for pressure distribution in the unsteady state and the final steady-state conditions.

Multi-phase flow, Oil/Water: The pressure on the left-hand side boundary is kept constant by a strong aquifer that replace the production oil with water. This causes a change in water saturation during the production process. The pressure distribution in the oil phase is also affected by the capillary pressure in the interface of oil and water. The flow equations for the oil/water flow model for each gridblock are:

- for the oil component;

$$\sum_{l=\psi_n} \mathbb{T}^{n+1}_{ol,n}\left[\left(p^{n+1}_{ol}-p^{n+1}_{on}\right)-\gamma^n_{ol,n}(Z_l-Z_n)\right] + \sum_{l=\xi_n} q^{n+1}_{osc_{l,n}} + q^{n+1}_{osc_n}$$
$$= \frac{U}{\alpha_c \Delta t}\left[\left[\frac{\phi(1-S_w)}{B_o}\right]^{n+1}_b - \left[\frac{\phi(1-S_w)}{B_o}\right]^n_n\right] \qquad (6.242)$$

- for the water component,

$$\sum_{l=\psi_n} \mathbb{T}^{n+1}_{wl,n}\left[\left(p^{n+1}_{o_l}-p^n_{o_n}\right)-\left(p^{n+1}_{cow_l}-p^n_{cow_n}\right)-\gamma^n_{w_l,n}(Z_l-Z_n)\right] + \sum_{l=\xi_n} q^{n+1}_{wsc_{l,n}} + q^{n+1}_{wsc_n}$$
$$= \frac{U}{\alpha_c \Delta t}\left[\left[\frac{\phi S_w}{B_w}\right]^{n+1}_b - \left[\frac{\phi S_w}{B_w}\right]^n_n\right] \qquad (6.243)$$

where

$$\mathbb{T}^{n+1}_{ol,n} = G_{l,n}\left(\frac{1}{\mu_p B_p}\right)_{l,n} k_{rp_{l,n}} \qquad (6.244)$$

is the transmissibility of oil ($p \equiv o$) or water ($p \equiv w$). The notation $G_{l,n}$ is the geometric factor between blocks n and l is obtained by

$$G_{l,n} = \left(\frac{\beta_c A_x k_x}{\Delta x}\right)_{l,n} \qquad (6.245)$$

The specific pressure, which is assigned for the boundary of the reservoir, is modeled by the amount of water that is replaced by a strong aquifer. The amount of water to produce a constant pressure at the boundary is obtained by

$$q^{n+1}_{wsc_b,BB} = \left[\frac{\beta_c A_x k_x}{\mu_w B_w\left(\frac{\Delta x}{2}\right)}\right]^{n+1}_{bB} (k_{rw})^{n+1}_{aq}\left[(p_b - p^{n+1}_{bB}) - (\gamma_w)^{n+1}_{bB}(Z_b - Z_{bB})\right] \qquad (6.246)$$

where the relative permeability at the aquifer is an equal one $((k_{rw})^{n+1}_{aq} = 1)$. If the reservoir is assumed to be consist of m gridblocks, Fig. 6.45, the flow equations may be written for the oil phase in the form,

$$\begin{cases} G\left(\frac{k_{ro}}{\mu_o B_o}\right)^{n+1}_1 \left(p^{n+1}_{o_2}-p^{n+1}_{o_1}\right) = \frac{U}{\alpha_c \Delta t}\left\{\left[\frac{\phi(1-S_w)}{B_o}\right]^{n+1}_1 - \left[\frac{\phi(1-S_w)}{B_o}\right]^n_1\right\} \\ \cdots\cdots\cdots\cdots\cdots\cdots \\ \cdots\cdots\cdots\cdots\cdots\cdots \\ G\left(\frac{k_{ro}}{\mu_o B_o}\right)^{n+1}_1 \left(p^{n+1}_{o_{i-1}}-p^{n+1}_{o_i}\right) + G\left(\frac{k_{ro}}{\mu_o B_o}\right)^{n+1}_1 \left(p^{n+1}_{o_{i+1}}-p^{n+1}_{o_i}\right) = \frac{U}{\alpha_c \Delta t}\left\{\left[\frac{\phi(1-S_w)}{B_o}\right]^{n+1}_i - \left[\frac{\phi(1-S_w)}{B_o}\right]^n_i\right\} \\ \cdots\cdots\cdots\cdots\cdots\cdots \\ \cdots\cdots\cdots\cdots\cdots\cdots \\ G\left(\frac{k_{ro}}{\mu_o B_o}\right)^{n+1}_{m-1} \left(p^{n+1}_{o_{m-1}}-p^{n+1}_{o_m}\right) - q_o = \frac{U}{\alpha_c \Delta t}\left\{\left[\frac{\phi(1-S_w)}{B_o}\right]^{n+1}_m - \left[\frac{\phi(1-S_w)}{B_o}\right]^n_m\right\} \end{cases}$$

$$(6.247)$$

S_{wn}	P_c (atm)	S_{wn}	P_c (atm)
0.00	3.9944	0.30	0.3600
0.01	3.5846	0.36	0.2700
0.02	3.1975	0.42	0.1980
0.05	2.2576	0.48	0.1450
0.08	1.6209	0.56	0.0920
0.11	1.2035	0.65	0.0550
0.15	0.8747	0.72	0.0350
0.18	0.7010	0.87	0.0100
0.21	0.5709	0.95	0.0020
0.25	0.4592	1.00	0.0000

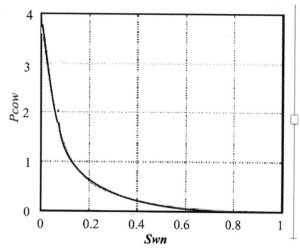

FIG. 6.45 Capillary pressure variation with water saturation.

and for the water phase as

$$\begin{cases}
G\left(\dfrac{k_{rw}}{\mu_w B_w}\right)_1^{n+1}\left[\left(p_{o_2}^{n+1}-p_{o_1}^{n+1}\right)-\left(p_{cow_2}^{n+1}-p_{cow_1}^{n}\right)\right]+2G\left(\dfrac{1}{\mu_w B_w}\right)_1^{n+1}\left[4000-\left(p_{o_1}^{n+1}-p_{cow_1}^{n}\right)\right] \\
=\dfrac{U}{\alpha_c \Delta t}\left[\left(\dfrac{\phi S_w}{B_w}\right)_1^{n+1}-\left(\dfrac{\phi S_w}{B_w}\right)_1^{n}\right] \\
\ldots \ldots \ldots \ldots \ldots \\
\ldots \ldots \ldots \ldots \ldots \\
G\left(\dfrac{k_{rw}}{\mu_w B_w}\right)_{i-1}^{n+1}\left[\left(p_{o_{i-1}}^{n+1}-p_{o_i}^{n+1}\right)-\left(p_{cow_{i-1}}^{n+1}-p_{cow_i}^{n}\right)\right]+G\left(\dfrac{k_{rw}}{\mu_w B_w}\right)_i^{n+1}\left[\left(p_{o_{i+1}}^{n+1}-p_{o_i}^{n+1}\right)-\left(p_{cow_{i+1}}^{n+1}-p_{cow_i}^{n}\right)\right] \\
=\dfrac{U}{\alpha_c \Delta t}\left[\left(\dfrac{\phi S_w}{B_w}\right)_i^{n+1}-\left(\dfrac{\phi S_w}{B_w}\right)_i^{n}\right] \\
\ldots \ldots \ldots \ldots \ldots \\
\ldots \ldots \ldots \ldots \ldots \\
G\left(\dfrac{k_{rw}}{\mu_w B_w}\right)_{m-1}^{n+1}\left[\left(p_{o_{m-1}}^{n+1}-p_{o_m}^{n+1}\right)-\left(p_{cow_{m-1}}^{n+1}-p_{cow_m}^{n}\right)\right]-q_m=\dfrac{U}{\alpha_c \Delta t}\left[\left(\dfrac{\phi S_w}{B_w}\right)_m^{n+1}-\left(\dfrac{\phi S_w}{B_w}\right)_m^{n}\right]
\end{cases}$$

(6.248)

where $1 < i < m$. The unknowns are $p_{o_1}^{n+1}, p_{o_2}^{n+1}, \ldots, p_{o_m}^{n+1}, S_{w_1}^{n+1}, S_{w_2}^{n+1}, \ldots, S_{w_m}^{n+1}$. These two systems of simultaneous algebraic equation (SAE) (6.247), (6.248) are nonlinear due to the dependency of the fluids and formation properties to the variation of the pressure and water saturation along the reservoir.

The water can be considered as an incompressible fluid without loss of accuracy. It is also assumed that the reservoir temperature is constant and therefore, the water phase viscosity is also constant, $\mu_w = 0.52\ cp$, $B_w = 1$. The oil viscosity and the oil formation volume factor are a function of reservoir pressure and temperature (p_R & T_R), bubble point pressure (p_b), and the amount of the gas/oil ratio (GOR), $\mu_o = f_1(T_R, p_R, p_b, GOR)$ and $B_o = f_2(T_R, p_R, p_b, GOR)$. The variation of μ_o, B_o and also the media porosity ϕ are given in (6.241). The relative permeability of oil and water and also the capillary pressure depends on the water saturation of the reservoir, $k_{rw} = f(S_w)$, $k_{ro} = f(S_w)$ & $p_{cow} = f(S_w)$.

Since the pressure is assumed to be kept constant by a high-pressure aquifer, the water saturation is increased, and owing to that both the relative permeabilities of oil and water and the capillary pressure are changed during the production process. It is assumed that the connate water saturation and the minimum possible oil saturation to produce oil are $S_{wc} = 0.18$ and $S_{oc} = 0.1$, respectively. The variation of the water and oil relative permeability is expressed by

$$k_{rw} = 0.59439\, S_{wn}^4, \quad k_{ro} = (1 - S_{wn})^2 \tag{6.249}$$

where $S_{wn} = \frac{S_w - S_{wc}}{1 - S_{wc} - S_{oc}}$ is called the normalized water saturation. The capillary pressure variation is given in Fig. 6.45 in tabular and graphical form as a function of the water saturation.

The oil phase pressures at the grid points may be obtained by combining the equations for the oil phase (6.247) and the water phase (6.248). Eq. (6.247) is multiplied by B_w and (6.248) are multiplied by $B_{o_n}^{n+1}$ and then added together. It gives the following system of simultaneous equations for the oil pressure at

$$\begin{cases} G\left(\dfrac{k_{ro}}{\mu_o B_o B_w}\right)_1^{n+1}\left(p_{o_2}^{n+1} - p_{o_1}^{n+1}\right) + G\left(\dfrac{k_{rw}}{\mu_w B_o B_w}\right)_1^{n+1}\left[\left(p_{o_2}^{n+1} - p_{o_1}^{n+1}\right) - \left(p_{cow_2}^{n+1} - p_{cow_1}^n\right)\right] \\[6pt]
\quad + 2G\left(\dfrac{1}{\mu_w B_o B_w}\right)_1^{n+1}\left[4000 - \left(p_{o_1}^{n+1} - p_{cow_1}^n\right)\right] = \dfrac{U}{\alpha_c \Delta t}\left[\left(\dfrac{\phi}{B_o B_w}\right)_1^{n+1} - \left(\dfrac{\phi}{B_o B_w}\right)_1^n\right] \\[6pt]
\cdots \cdots \cdots \cdots \cdots \\
\cdots \cdots \cdots \cdots \cdots \\
G\left(\dfrac{k_{ro}}{\mu_o B_o B_w}\right)_{i-1}^{n+1}\left(p_{o_{i-1}}^{n+1} - p_{o_i}^{n+1}\right) + G\left(\dfrac{k_{rw}}{\mu_w B_w}\right)_{i-1}^{n+1}\left(\dfrac{1}{B_o}\right)_i^{n+1}\left[\left(p_{o_{i-1}}^{n+1} - p_{o_i}^{n+1}\right) - \left(p_{cow_{i-1}}^{n+1} - p_{cow_i}^n\right)\right] \\[6pt]
G\left(\dfrac{k_{ro}}{\mu_o B_o B_w}\right)_{i-1}^{n+1}\left(p_{o_{i+1}}^{n+1} - p_{o_i}^{n+1}\right) + G\left(\dfrac{k_{rw}}{\mu_w B_o B_w}\right)_i^{n+1}\left[\left(p_{o_{i+1}}^{n+1} - p_{o_i}^{n+1}\right) - \left(p_{cow_{i+1}}^{n+1} - p_{cow_i}^n\right)\right] \\[6pt]
\quad = \dfrac{U}{\alpha_c \Delta t}\left[\left(\dfrac{\phi}{B_o B_w}\right)_i^{n+1} - \left(\dfrac{\phi}{B_o B_w}\right)_i^n\right] \\[6pt]
\cdots \cdots \cdots \cdots \cdots \\
\cdots \cdots \cdots \cdots \cdots \\
G\left(\dfrac{k_{ro}}{\mu_o B_o B_w}\right)_{m-1}^{n+1}\left(p_{o_{m-1}}^{n+1} - p_{o_m}^{n+1}\right) + G\left(\dfrac{k_{rw}}{\mu_w B_w}\right)_{m-1}^{n+1}\left(\dfrac{1}{B_o}\right)_m^{n+1}\left[\left(p_{o_{m-1}}^{n+1} - p_{o_m}^{n+1}\right) - \left(p_{cow_{m-1}}^{n+1} - p_{cow_m}^n\right)\right] \\[6pt]
\quad -\dfrac{q_w}{B_o} - \dfrac{q_o}{B_w} = \dfrac{U}{\alpha_c \Delta t}\left[\left(\dfrac{\phi}{B_o B_w}\right)_m^{n+1} - \left(\dfrac{\phi}{B_o B_w}\right)_m^n\right] \end{cases}$$

$$\tag{6.250}$$

The system of equations (6.250) is solved for oil pressure distribution when the water saturation is assumed to be unvaried. The computed oil pressure is applied to compute the new water saturation by using the set of algebraic equations (6.248). The procedure is to continue for another time step to find the oil pressure and water saturation during the production process.

The systems of algebraic equations (6.248) and (6.250) are nonlinear due to the dependency of the coefficient to the pressure and water saturation distribution. The solution of these systems of algebraic equations is obtained iteratively in each time step by renewing the coefficients to reach a minimum error. The minimum error is considered to be less than 10^{-6}. The Newton method is applied to solve these systems of algebraic equations (6.248) and (6.250). We investigate:

- the effect of the number of gridblocks;
- the effect of the value of time interval Δt; and
- the effect of the variation of the fluid and formation properties.

The effect of gridblocks number: It was shown that the number of gridblocks does not affect the accuracy of the result in the single-phase flow in the previous section. The nonlinearity produces only due to the variation of fluid pressure in single-phase flow. In the multi-phase flow, the nonlinearity is due to the variation of pressure and the saturation of the fluid during the production process. We are going to show the effectiveness of the variation of pressure and fluid saturation in the final solution. In this part, we study the effect of the number of gridblocks on the accuracy of the results. The computation is carried on with a time interval of $\Delta t = 1$[day]. The reservoir is divided into $n = 32, 64, 128$, and 256 the pressure and water saturation distributions are obtained at $t = 1$ [day], 10 [days], 200 [days] and 1 [year].

The results of computations are shown in Fig. 6.46A for $t = 1$[day] the production process. There is not a significant change on S_w and therefore, the fluid flow is almost one phase. The number of gridblocks has no effect on the pressure distribution. However, the water saturation is started to increase along the reservoir and the number of gridblocks has a significant effect on the water saturation distribution. Therefore, it is necessary to have as many as possible gridblocks to predict the water saturation distribution exactly.

The water saturation is increased just near the constant pressure boundary for $t = 10$[days] as shown in Fig. 6.46B. The number of gridblocks is important to predict the distance of the affected area and the amount of water saturation. This is shown that it is necessary to increase the number of gridblocks to get more precise solution. The effect on the pressure distribution is not so significant as the influence on the water saturation. However, the effect of two-phase flows shows that it is also good to have more gridblocks to get more precise pressure distribution.

The pressure distribution for $t = 200$[days] and 1[year] is depicted in Fig. 6.46C and D. The pattern of pressure distributions is different in the area of two-phase flow with the area of single-phase flow. There is a point of discontinuity at the different fluid flow regimes. The number of gridblocks has an influence on the exact prediction of the point of discontinuity in the pressure distribution. Consequently, this affects the pressure distribution in the area of single-phase flow of oil. The water saturation at $t = 200$[days] and 1[year] is shown in Fig. 6.46C with different numbers of gridblocks. The exact distribution of water saturation

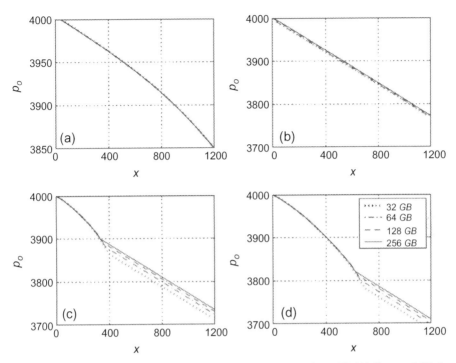

FIG. 6.46 The distribution of oil phase pressure after (A) 1 day, (B) 10 days, (C) 200 days, and (D) 1 year with different numbers of gridblocks (GB).

depends on the number of gridblocks. It is necessary to have as many as possible gridblocks to find the exact distribution of water saturation along the reservoir.

The effect of time interval, Δt: The computation in the single-phase flow showed that the value Δt is important in the unsteady state flow condition and in the prediction of the time that the flow becomes independent of the time variation. It can be realized from the computation in the previous subsection that the multi-phase flow of oil and water does not reach a steady-state condition. The fluid quality and consequently the flow characteristics change with time and a steady-state condition, the flow characteristics are unvaried in respect to time, never reaches. The choice of Δt depends on the computer facilities and the speed and capacity of them but it should be kept as small as possible. We use a PC with an Intel Pentium(R) M processor of 1.73 GHZ and 504 MB of RAM. It is considered that $\Delta t = 1$ [hr], 1 [day] and 15 [days]. The computations are carried out with $n = 256$ gridblocks.

The pressure and water saturation distributions are shown in Figs. 6.47 and 6.48, respectively, at $t = 3, 6, 9$ and 12 [months] of production. The pressure distributions $\Delta t = 1$ [hr], 1 [day] show very small differences. The effect on the water saturation is more pronounced $\Delta t = 1$ [hr], 1 [day]. The results $\Delta t = 15$ [days] show significant differences in both pressure and water saturation distribution with the others. This shows that it is necessary to have a small value of Δt to produce accurate and reliable solutions. The less the Δt more accurate are the results for the pressure and water saturation. However, a time interval $\Delta t = 1$ day gives relatively accurate solutions that can be relied on in engineering problems and can be obtained pretty fast with a normal computer facility.

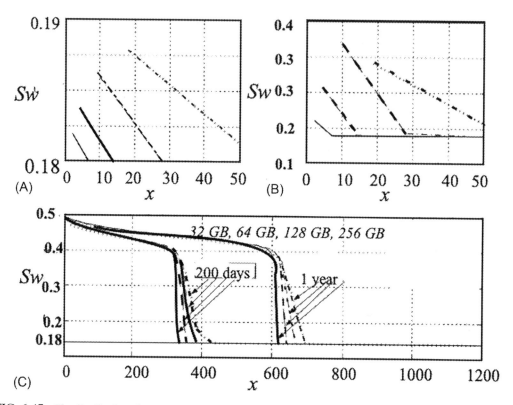

FIG. 6.47 The distribution of water saturation at (A) 1 day, (B) 10 days, and (C) 200 days and 1 year with various numbers of gridblocks (GB).

The effect of variation of parameters: The flow depends on two types of parameters. The first type is those that are changed with the variation of pressure, such as the fluid formation volume factor, the viscosity of the oil, and the porosity of the formation. The second type depends on the variation of water saturation. These are the relative permeability of oil and water and the capillary pressure. The parameters are categorized based on their dependency on the pressure or water saturation and the effect of each group is studied separately.

The effect of variation of parameters: The flow depends on two types of parameters. The first types are those that are changed with the variation of pressure, such as the fluid formation volume factor, the viscosity of the oil, and the porosity of the formation. The second type depends on the variation of water saturation. These are the relative permeability of oil and water and the capillary pressure. The parameters are categorized based on their dependency on the pressure or water saturation and the effect of each group is studied separately.

(a) **The pressure-dependent parameters:** The effect of the variation of the pressure-dependent parameters is first considered separately and then they are considered all together.

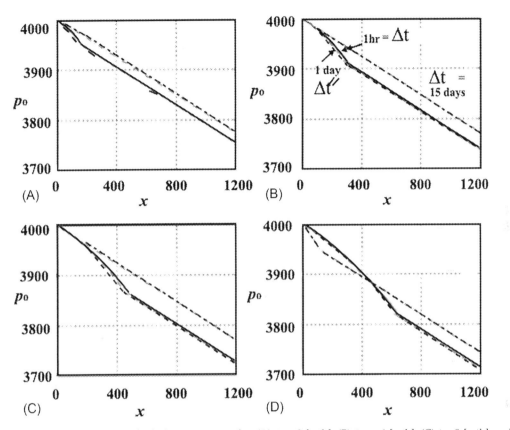

FIG. 6.48 The distribution of oil phase pressure after (A) $t = 3$ [mth], (B) $t = 6$ [mth], (C) $t = 9$ [mth], and (D) $t = 12$ [mth] with different values of Δt.

(i) The oil formation volume factor, B_o

The value of the fluid formation volume factor in the process of oil production is varied 1.2063 to 1.22 from reservoir pressure to bubble point pressure. We neglect the compressibility of the fluid and take that $B_o = 1.22$ during the production process. The pressure and water saturation distribution are computed for a year with $\Delta t = 1$[day] and $n = 256$. The result $t = 3, 6, 9$ and 12 [months] are given in Fig. 6.49. The obtained results are then compared with the solutions when the fluid formation volume factor is changed with a change in pressure. There is not a significant difference in the early months of production. But, the differences are increased with time and with the increase in production time. However, it shows that the variation B_o has a minor effect on the distribution of p_o and S_w. Fig. 6.50 shows the distribution of oil phase pressure and water saturation for constant and variable fluid formation volume factors.

(ii) The oil viscosity, μ_o

The variation of the viscosity with the pressure is neglected and it is assumed that $\mu_o = 1.1340 \, cp$ during the production process. This corresponds to the viscosity at the initial

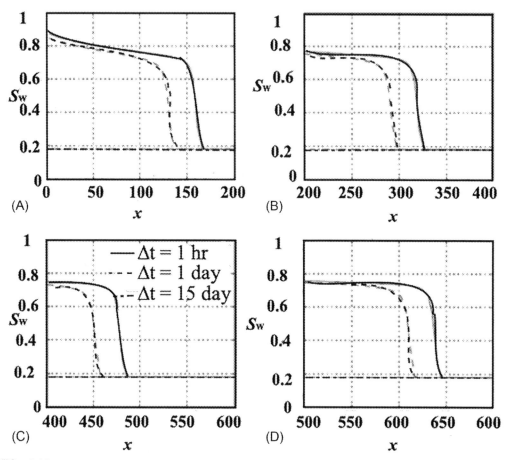

FIG. 6.49 The distribution of water saturation at (A) $t=3$ [mth], (B) $t=6$ [mth], (C) $t=9$ [mth], and (D) $t=12$ [mth] with different values of t.

reservoir pressure, $p_R = 4000$ psia. The computations are carried out for a year with $\Delta t = 1$ [day] and $m = 256$. There was no significant difference between the results when $\mu_o = Const.$ and $\mu_o = f(p_o)$ for various depicted times. It indicates that the variation of viscosity has not a major effect on the pressure and water saturation distributions.

(iii) The porosity of the formation, ϕ

The porosity variation due to pressure change is neglected in this part. It is assumed that $\phi = \phi^o = 0.27$ during the production process. The computations are carried out for a year with $\Delta t = 1$[day] and $m = 256$. The pressure and water saturation distributions do not demonstrate any difference between the two cases of $\phi = Const$ and $\phi = f(p_o)$. There is not any difference between the graphs at a certain time with constant and variable φ, respectively, and the diagrams coincide completely.

6.7 The compositional simulator using engineering approach

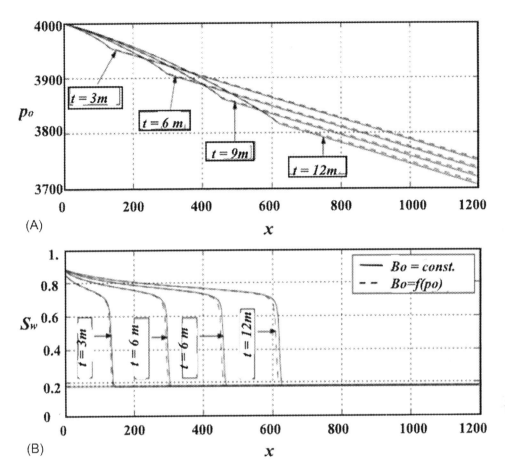

FIG. 6.50 The distribution of oil phase pressure and water saturation for constant and variable fluid formation volume factors.

(iv) The combined effect of variation of B_o, μ_o and ϕ

In this part, the simultaneous effect of all pressure-dependent parameters are considered. It is assumed that $B_o = 1.22$, $\mu_o = 1.1340\ cp$ and $\phi = \phi_o = 0.27$ during the production process. The computations are carried out for a year with $\Delta t = 1[day]$ and $n = 256$. The results are compared with the cases of:

- $B_o = Const.$ when μ_o and ϕ are function of p_o;
- $\mu_o = Const.$ when B_o and ϕ are function of p_o;
- $\phi = Const.$ when B_o and μ_o are function of p_o; and
- B_o, μ_o, and ϕ are function of p_o;

The oil formation volume factor B_o has the major effect while the porosity variation has a minor effect on the oil pressure and water saturation distributions. Neglecting the variation of the pressure-dependent parameters gives maximum relative errors in the pressure distribution as follows with respect to the case when the variation of all pressure-dependent parameters is considered.

- $E_{po} = 0.1623\%$ when B_o, μ_o & $\phi = const.$;
- $E_{po} = 0.1050\%$ when $B_o = const. = $ Const. and μ_o & $\phi = f(p_o)$;
- $E_{po} = 0.0545\%$ when $\mu_o = $ Const. and B_o & $\phi = f(p_o)$; and
- $E_{po} = 0.0014\%$ when $\phi = $ Const. and B_o & $\mu_o = f(p_o)$.

These show that the variation of pressure-dependent parameters do not produce significant effects and they may be assumed constant.

(b) The water saturation dependent parameters

The water permeability of water and oil and the capillary pressure are dependent on the variation of water saturation. If the variation of these parameters with the water saturation is neglected, the problem is reduced to a single-phase flow problem that is addressed in the previous section. The effect of the pressure-dependent parameters was found to be very small compared with the influence of the water saturation variation.

The flow inside a porous media is a nonlinear phenomenon. The derived mathematical models are also nonlinear for such a flow regardless of the fact that many assumptions are made during the formulation of the flow. The nonlinearity of the equations is due to the dependency of the fluid and formation properties on the unknown variables that should be calculated.

In the single-phase flow, the number of grid points creates a little and no loss of accuracy. Increasing the number of gridblocks gives better distribution of pressure but does not affect the accuracy of the results. It is true for both the steady and unsteady state conditions. The time interval value is important to find the time that the regime of the fluid flow change from an unsteady state to a steady-state condition. The smaller the time interval the better description of the unsteady state flow conditions is obtained. A time interval of $\Delta t = 1[day]$ is sufficient to predict the behavior of the flow in unsteady-state conditions for this particular case study.

The quality of the fluid in two phase (oil/water) system changes continuously, as such, no steady-state condition is attainable. The flow properties are a function of the pressure variation and the water saturation. The effect of the variation of the pressure-dependent properties are very small in comparison with the effect of the saturation-dependent parameters variation on the final solution. The computation shows that the effect of the pressure-dependent parameters can be neglected with a little loss of accuracy. However, neglecting the saturation-dependent parameters reduces the two-phase flow problem to a single-phase flow problem.

CHAPTER 7

Field guidelines

Nomenclature

A	area (m^2)
L	reservoir length (m)
W	reservoir width (m)
H	reservoir thickness (m)
c	compressibility (ms^2/kg)
ϕ	porosity fraction
k	permeability (m^2)
k_r	relative permeability (m^2/m^2)
P	pressure (kg/ms^2)
q_i	injection rate (m^3/s)
q_{ia}	injection rate of additive (m^3/s)
q_{prod}	production rate (m^3/s)
D_{Ta}	transverse dispersion of additive (m^2/s)
D_{La}	longitudinal dispersion of additive (m^2/s)
S	saturation
θ	contact angle
t	time (s)
T	reservoir temperature (°C)
E	additive concentration
U	total velocity (m/s)
u	velocity (m/s)
V_r	reservoir volume (m^3)
V_f	fluid volume (m^3)
σ_{go}	interfacial tension between gas and oil phase (kg/s^2)
σ_{ow}	interfacial tension between oil and gas phase (kg/s^2)
υ	dynamic viscosity (s/m^2)
μ	viscosity (kg/ms)
ρ	density (kg/m^3)
r	pore throat radius (m)
g	gravitational acceleration (m/s^2)
τ	tortuosity
K	thermal conductivity (W/m K)
C_p	specific heat capacity (j/kg K)

h enthalpy (j/kg)
L_v latent heat (j/kg)
ξ dummy variable for time (s)

Subscript

f fluid
o oil phase
w water phase
g gaseous phase
i initial
r rock or reservoir
t total

7.1 Introduction

For decades, oil and gas companies have been stepping up their commitment to sustainability. Unfortunately, the term "sustainability" has been politicized. By manipulating the definition of this term, today carbon—the backbone of life—has been designated as the source of unsustainability, and oil and gas—the most abundant naturally available fuel—as inherently unsustainable. With governments across the Middle East and wider world setting targets for lower carbon emissions and greater presence of "renewables" in their national energy mix, the oil and gas industry is looking toward "cleantech innovations" (read non-petroleum) to find the "best ways" to successfully pivot toward this future energy vision. The entire focus is to decarbonize energy resources. Increasingly, we are seeing a lot more oil and gas companies experimenting with such dogmatic practices to gain political support. With so many of the Middle East's OPEC countries pushing the pace of their economic diversification strategies, this is prompting even greater and faster adoption of "sustainability measures" across the industry, so we can expect to see more of the same in upcoming decades. As early as 2006, Ibrahim (2006) demonstrated in his doctoral work that the conventional definition of sustainability is unscientific and creates a false paradigm for industry operations. Later, Islam and his co-workers established in several research monographs that true sustainability can be achieved only with petroleum fuels (e.g., Islam et al., 2010; Khan and Islam, 2007a, 2012). This book deconstructs the current sustainability narrative because that it has no scientific merit (Islam and Khan, 2019). In this chapter, the path to true sustainability is chalked out for the petroleum industry.

7.2 Scaling guidelines

Scaling laws are at the core of engineering designs. Historically, scaled model studies have been used prior to developing numerical and mathematical models (Islam and Ali, 1990). However, over the last few decades, there has been a shift from experimental models to purely mathematical models (Mustafiz et al., 2008). During the late 90s through the early 2000s, the surge of geostatistical models has created the new trend in the scaling up using only mathematical models (e.g., Monteagudo et al., 2001; El-Amin and Subasi, 2019). As we will see in a later section this is not a helpful trend. There is a need to reset the strategy to derive as much information as possible from scaled model studies. Monteagudo et al. (2001) introduced a 3D network model that was used to represent a porous medium and the macroscopic properties of the network (like permeability) were simulated by the Monte Carlo method. It was shown that these macroscopic properties can be related to network parameters (throat-size distribution parameters, network size, and connectivity) through power-law correlations, as can be inferred from percolation theory. In this way, macroscopic properties evaluation during the morphological evolution of a 3D network requires less computational effort, which facilitates the incorporation of this model into oil flow simulators.

Scaled model experiments are extremely important, both from theoretical and engineering application points of view. Irrespective of the actual numbers generated that is used for scaling up the results, the fact remains that each scaled model study generates real data, which then can form the basis for mathematical models. This is particularly important for sustainable oil and gas development operations, for which no rigorous mathematical model exists. Scaled model studies have an important shortcoming. The dimensionless groups are difficult to satisfy simultaneously. This problem has been resolved in early of Farouq Ali and his research group (see, for instance, Kimber et al., 1988; Farouq Ali and Redford, 1977; Farouq Ali et al., 1987). It was suggested that a series of approaches be taken to satisfy different sets of dimensionless groups and then evaluate the results to develop the most realistic plan for the field application. In this process, Islam and Ali (1990, 1992) showed that the scaling group involving time is the most difficult to satisfy, particularly for cases involving chemical and thermal changes. It is of interest that most successful enhanced oil recovery (EOR) schemes involve thermal and chemical changes. Recent work of Hossain (2018) recognized that the conventional approach of using Newtonian flow is not acceptable for many reservoirs. They introduced the concept of memory in fluid flow. As we have seen in Chapter 6, the introduction of memory effects makes governing equations very complex and difficult to solve with numerical solution methods. In their work, they showed that it is more reasonable to approach these problems with scaled physical models.

7.2.1 Approaches to satisfy scaling groups

Rahman et al. (2017) presented a complete set of dimensionless groups for steam flood operations. Following guidelines, provided by Kimber et al. (1988), Rahman et al. derived five sets of scaling criteria, each set consisting of variables for satisfying the scaling criteria by relaxing different scaling phenomena. The different approaches selected different parameters to be relaxed to satisfy the specific requirements. For example, the vertical geometry scale is relaxed to satisfy the viscous and gravitational forces using the concept of the same fluid and same porous medium. Selecting the appropriate approach to use is largely dependent on the specific process being modeled. To choose a proper approach two main factors have to be considered. First, the selection of major mechanisms should be correctly scaled. Second, the minor mechanism should not have a significant effect on the selected approach. The best way to select a suitable approach is the comparison of different approaches. This comparison indicates which mechanism is scaled and which is not with an order of degree. This study will help to select an appropriate steam flooding technique with the minimum number of most influential dimensionless scaling groups.

The similarity groups should be analogous in model and prototype. It is very difficult to satisfy a complete set of scaling criteria, so several groups should be rested to fulfill the scaling criteria. These approaches are applicable for high-pressure reservoir fluids where both reservoir pressure-temperature conditions and different pressure-temperature conditions are used. Approach 2, 3, and 4 is used for porous reservoir medium with reservoir pressure, temperature conditions, and approach 1 and 5 is applicable for other pressure, temperature conditions.

Approach 1 same fluid, different porous media, different pressure drop, and geometric similarity.

Geometric similarity groups can be satisfied by considering pressure drop, gravitational and viscous forces which are different for model and prototype. This condition requires different porous media. Pujol and Boberg (1972) proposed this approach which allows scaling requirements should be satisfied if violates some constitutive relationships, constraints, and boundary conditions. The Saturation pressure and saturation temperature relationship for the steam flooding process cannot be properly scaled by this method, which will ultimately mislead the heat losses from the steam zone. Different steam properties which largely depend on pressure will not be scaled properly. The different porous medium is considered, so the fluid saturations and relative permeability are not scaled accurately. In addition, capillary forces and dispersion effects are not properly scaled (Fig. 7.1 and Table 7.1).

The implementation of these scaling criteria for a model can reduce the length by a scaling factor of a.

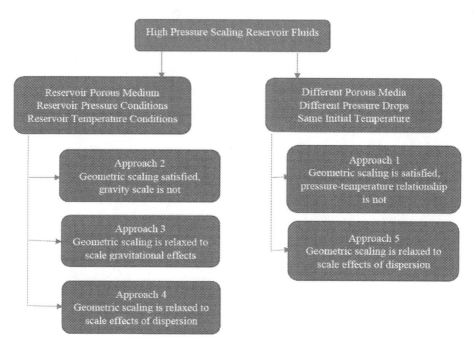

FIG. 7.1 Scaling approaches.

TABLE 7.1 Dimensionless groups from inspectional analysis.

$\pi_1 = \dfrac{t_R u_{xwR}}{\phi x_R s_{wR}}$	$\pi_{45} = \dfrac{\eta_{gR} p_{cowR}}{\Gamma(1-\alpha) t_R^{\alpha-1} \phi_R s_{wR} x_R^2}$	$\pi_{89} = \dfrac{\eta_{wR} p_{wR}}{\Gamma(1-\alpha) t_R^{\alpha-1} \phi_R s_{wR} z_R^2}$	$\pi_{133} = \dfrac{s_{gi}}{s_{gR}}$
$\pi_2 = \dfrac{t_R u_{zwR}}{\phi z_R s_{wR}}$	$\pi_{46} = \dfrac{f_{xoR} \eta_{oR} p_{cowR}}{\Gamma(1-\alpha) t_R^{\alpha-1} \phi_R s_{wR} x_R^2}$	$\pi_{90} = \dfrac{\eta_{gR} p_{wR}}{\Gamma(1-\alpha) t_R^{\alpha-1} \phi_R s_{wR} z_R^2}$	$\pi_{134} = \dfrac{s_{oi}}{s_{oR}}$
$\pi_3 = \dfrac{D_{Lw1R} t_R}{x_R^2 s_{wR}}$	$\pi_{47} = \dfrac{f_{xoR} \eta_{gR} p_{cowR}}{\Gamma(1-\alpha) t_R^{\alpha-1} \phi_R s_{wR} x_R^2}$	$\pi_{91} = \dfrac{f_{xoR} \eta_{oR} p_{wR}}{\Gamma(1-\alpha) t_R^{\alpha-1} \phi_R s_{wR} z_R^2}$	$\pi_{135} = \dfrac{E_{2i}}{E_{o2R}}$
$\pi_4 = \dfrac{D_{Tw1R} t_R}{z_R^2 s_{wR}}$	$\pi_{48} = \dfrac{f_{xwR} \eta_{oR} p_{cowR}}{\Gamma(1-\alpha) t_R^{\alpha-1} \phi_R s_{wR} x_R^2}$	$\pi_{92} = \dfrac{f_{xoR} \eta_{wR} p_{wR}}{\Gamma(1-\alpha) t_R^{\alpha-1} \phi_R s_{wR} z_R^2}$	$\pi_{136} = \dfrac{E_{3i}}{E_{w3R}}$
$\pi_5 = \dfrac{D_{Lw3R} t_R}{x_R^2 s_{wR}}$	$\pi_{49} = \dfrac{f_{xwR} \eta_{oR} p_{cowR}}{\Gamma(1-\alpha) t_R^{\alpha-1} \phi_R s_{wR} x_R^2}$	$\pi_{93} = \dfrac{f_{xoR} \eta_{gR} p_{wR}}{\Gamma(1-\alpha) t_R^{\alpha-1} \phi_R s_{wR} z_R^2}$	$\pi_{137} = \dfrac{E_{1j}}{E_{g1R}}$
$\pi_6 = \dfrac{D_{Tw3R} t_R}{z_R^2 s_{wR}}$	$\pi_{50} = \dfrac{\eta_{gR} p_{cgoR}}{\Gamma(1-\alpha) t_R^{\alpha-1} \phi_R s_{wR} x_R^2}$	$\pi_{94} = \dfrac{f_{xwR} \eta_{oR} p_{wR}}{\Gamma(1-\alpha) t_R^{\alpha-1} \phi_R s_{wR} z_R^2}$	$\pi_{138} = \dfrac{E_{4j}}{E_{w4R}}$
$\pi_7 = \dfrac{D_{Lw4R} t_R}{x_R^2 s_{wR}}$	$\pi_{51} = \dfrac{f_{xoR} \eta_{gR} p_{cgoR}}{\Gamma(1-\alpha) t_R^{\alpha-1} \phi_R s_{wR} x_R^2}$	$\pi_{95} = \dfrac{f_{xwR} \eta_{wR} p_{wR}}{\Gamma(1-\alpha) t_R^{\alpha-1} \phi_R s_{wR} z_R^2}$	$\pi_{139} = \dfrac{\Delta p_{gR}}{p_{gR}}$
$\pi_8 = \dfrac{D_{Tw4R} t_R}{z_R^2 s_{wR}}$	$\pi_{52} = \dfrac{f_{xwR} \eta_{gR} p_{cgoR}}{\Gamma(1-\alpha) t_R^{\alpha-1} \phi_R s_{wR} x_R^2}$	$\pi_{96} = \dfrac{f_{xwR} \eta_{gR} p_{wR}}{\Gamma(1-\alpha) t_R^{\alpha-1} \phi_R s_{wR} z_R^2}$	$\pi_{140} = \dfrac{\rho_{gR} g_R H_R}{p_{gR}}$
$\pi_9 = \dfrac{t_R u_{xoR}}{\phi x_R s_{oR}}$	$\pi_{53} = \dfrac{p_{wR}}{p_{cowR}}$	$\pi_{97} = \dfrac{\eta_{oR} p_{cowR}}{\Gamma(1-\alpha) t_R^{\alpha-1} \phi_R s_{wR} z_R^2}$	$\pi_{141} = \dfrac{\rho_{gR} g_R z_R}{p_{gR}}$
$\pi_{10} = \dfrac{t_R u_{zoR}}{\phi z_R s_{oR}}$	$\pi_{54} = \dfrac{x_R^2}{z_R^2}$	$\pi_{98} = \dfrac{\eta_{gR} p_{cowR}}{\Gamma(1-\alpha) t_R^{\alpha-1} \phi_R s_{wR} z_R^2}$	$\pi_{142} = \dfrac{\Delta p_{oR}}{p_{oR}}$
$\pi_{11} = \dfrac{D_{Lo1R} t_R}{x_R^2 s_{oR}}$	$\pi_{55} = \dfrac{p_{wR} x_R^2}{p_{cowR} z_R^2}$	$\pi_{99} = \dfrac{f_{xoR} \eta_{oR} p_{cowR}}{\Gamma(1-\alpha) t_R^{\alpha-1} \phi_R s_{wR} z_R^2}$	$\pi_{143} = \dfrac{\rho_{oR} g_R H_R}{p_{oR}}$
$\pi_{12} = \dfrac{D_{To1R} t_R}{z_R^2 s_{oR}}$	$\pi_{56} = \dfrac{p_{wR}}{p_{cgoR}}$	$\pi_{100} = \dfrac{f_{xoR} \eta_{gR} p_{cowR}}{\Gamma(1-\alpha) t_R^{\alpha-1} \phi_R s_{wR} z_R^2}$	$\pi_{144} = \dfrac{\rho_{oR} g_R z_R}{p_{oR}}$
$\pi_{13} = \dfrac{D_{Lo2R} t_R}{x_R^2 s_{oR}}$	$\pi_{57} = \dfrac{p_{wR} x_R^2}{p_{cgoR} z_R^2}$	$\pi_{101} = \dfrac{f_{xwR} \eta_{oR} p_{cowR}}{\Gamma(1-\alpha) t_R^{\alpha-1} \phi_R s_{wR} z_R^2}$	$\pi_{145} = \dfrac{\Delta p_{wR}}{p_{wR}}$
$\pi_{14} = \dfrac{D_{To2R} t_R}{z_R^2 s_{oR}}$	$\pi_{58} = \dfrac{p_{cowR}}{p_{cgoR}}$	$\pi_{102} = \dfrac{f_{xwR} \eta_{oR} p_{cowR}}{\Gamma(1-\alpha) t_R^{\alpha-1} \phi_R s_{wR} z_R^2}$	$\pi_{146} = \dfrac{\rho_{wR} g_R H_R}{p_{wR}}$
$\pi_{15} = \dfrac{D_{Lo4R} t_R}{x_R^2 s_{oR}}$	$\pi_{59} = \dfrac{p_{cowR} x_R^2}{p_{cgoR} z_R^2}$	$\pi_{103} = \dfrac{\eta_{gR} p_{cgoR}}{\Gamma(1-\alpha) t_R^{\alpha-1} \phi_R s_{wR} z_R^2}$	$\pi_{147} = \dfrac{\rho_{wR} g_R z_R}{p_{wR}}$
$\pi_{16} = \dfrac{D_{To4R} t_R}{z_R^2 s_{oR}}$	$\pi_{60} = \dfrac{f_{xoR} p_{wR}}{p_{cowR}}$	$\pi_{104} = \dfrac{f_{xoR} \eta_{gR} p_{cgoR}}{\Gamma(1-\alpha) t_R^{\alpha-1} \phi_R s_{wR} z_R^2}$	$\pi_{148} = \dfrac{H U_t}{z_R u_{gxR} z_R}$
$\pi_{17} = \dfrac{u_{xgR} t_R}{\phi x_R s_{gR}}$	$\pi_{61} = \dfrac{f_{xoR} x_R^2}{z_R^2}$	$\pi_{105} = \dfrac{f_{xwR} \eta_{gR} p_{cgoR}}{\Gamma(1-\alpha) t_R^{\alpha-1} \phi_R s_{wR} z_R^2}$	$\pi_{149} = \dfrac{z_R^2 D_{Lo1R}}{x_R^2 D_{To1R}}$
$\pi_{18} = \dfrac{u_{zgR} t_R}{\phi z_R s_{gR}}$	$\pi_{62} = \dfrac{f_{xoR} p_{wR} x_R^2}{p_{cowR} z_R^2}$	$\pi_{106} = \dfrac{t_R D_{Lw1R}}{s_{wR} x_R^2}$	$\pi_{150} = \dfrac{z_R^2 D_{Lo2R}}{x_R^2 D_{To2R}}$
$\pi_{19} = \dfrac{u_{xoR}}{U_{xR}}$	$\pi_{63} = \dfrac{f_{xoR} p_{wR}}{p_{cgoR}}$	$\pi_{107} = \dfrac{t_R D_{Lw3R}}{s_{wR} x_R^2}$	$\pi_{151} = \dfrac{z_R^2 D_{Lo4R}}{x_R^2 D_{To4R}}$
$\pi_{20} = \dfrac{u_{xwR}}{U_{xR}}$	$\pi_{64} = \dfrac{f_{xoR} p_{wR} x_R^2}{p_{cgoR} z_R^2}$	$\pi_{108} = \dfrac{t_R D_{Lw4R}}{s_{wR} x_R^2}$	$\pi_{152} = \dfrac{z_R^2 D_{Lw1R}}{x_R^2 D_{Tw1R}}$

Continued

TABLE 7.1 Dimensionless groups from inspectional analysis—cont'd

$\pi_{21} = \dfrac{u_{xgR}}{U_{xR}}$	$\pi_{65} = \dfrac{f_{xoR} p_{cowR}}{p_{cgoR}}$	$\pi_{109} = \dfrac{t_R D_{Tw1R}}{s_{wR} z_R^2}$	$\pi_{153} = \dfrac{z_R^2 D_{Lw2R}}{x_R^2 D_{Tw2R}}$
$\pi_{22} = \dfrac{u_{zoR}}{U_{zR}}$	$\pi_{66} = \dfrac{f_{xoR} p_{cowR} x_R^2}{p_{cgoR} z_R^2}$	$\pi_{110} = \dfrac{t_R D_{Tw3R}}{s_{wR} z_R^2}$	$\pi_{154} = \dfrac{z_R^2 D_{Lw4R}}{x_R^2 D_{Tw4R}}$
$\pi_{23} = \dfrac{u_{zwR}}{U_{zR}}$	$\pi_{67} = \dfrac{f_{xwR} p_{wR}}{p_{cowR}}$	$\pi_{111} = \dfrac{t_R D_{Tw4R}}{s_{wR} z_R^2}$	$\pi_{155} = \dfrac{H}{z_R}$
$\pi_{24} = \dfrac{u_{zgR}}{U_{zR}}$	$\pi_{68} = \dfrac{f_{xwR} x_R^2}{z_R^2}$	$\pi_{112} = \dfrac{k_{xwR} \lambda_{xwR} p_{wR}}{x_R u_{xwR}}$	$\pi_{156} = \dfrac{L}{x_R}$
$\pi_{25} = \dfrac{u_{xoR}}{f_{xoR} U_{xR}}$	$\pi_{69} = \dfrac{f_{xwR} p_{wR} x_R^2}{f_{xwR} p_{cowR} z_R^2}$	$\pi_{113} = \dfrac{k_{xwR} \lambda_{xwR} \rho_{wR} g z_R \sin\theta_R}{x_R u_{xwR}}$	$\pi_{157} = \dfrac{H U_t}{u_{gxR} z_R}$
$\pi_{26} = \dfrac{u_{zoR}}{f_{zoR} U_{zR}}$	$\pi_{70} = \dfrac{f_{xwR} p_{wR}}{p_{cgoR}}$	$\pi_{114} = \dfrac{k_{xoR} \lambda_{xoR} p_{oR}}{x_R u_{xoR}}$	$\pi_{158} = \dfrac{K_{hf} T A t}{L \Delta T M_1 V}$
$\pi_{27} = \dfrac{u_{xwR}}{f_{xwR} U_{xR}}$	$\pi_{71} = \dfrac{f_{xwR} p_{wR} x_R^2}{p_{cgoR} z_R^2}$	$\pi_{115} = \dfrac{k_{xoR} \lambda_{xoR} \rho_{oR} g z_R \sin\theta_R}{x_R u_{xoR}}$	$\pi_{159} = \dfrac{K_{hf} T A t}{L \Delta T M_1 V}$
$\pi_{28} = \dfrac{u_{zwR}}{f_{zwR} U_{zR}}$	$\pi_{72} = \dfrac{f_{xwR} p_{cowR}}{p_{cgoR}}$	$\pi_{116} = \dfrac{k_{xgR} \lambda_{xgR} p_{gR}}{x_R u_{xgR}}$	$\pi_{160} = \dfrac{t C_{po} \rho_o u_{ox} A}{M_1 V}$
$\pi_{29} = \dfrac{u_{xgR}}{f_{xgR} U_{xR}}$	$\pi_{73} = \dfrac{f_{xwR} p_{cowR} x_R^2}{p_{cgoR} z_R^2}$	$\pi_{117} = \dfrac{k_{xgR} \lambda_{xgR} \rho_{gR} g z_R \sin\theta_R}{x_R u_{xgR}}$	$\pi_{161} = \dfrac{t C_{pw} \rho_w u_{wx} A}{M_1 V}$
$\pi_{30} = \dfrac{f_{xoR}}{f_{xgR}}$	$\pi_{74} = \dfrac{f_{zoR} p_{wR} x_R^2}{p_{cowR} z_R^2}$	$\pi_{118} = \dfrac{k_{zwR} \lambda_{zwR} p_{wR}}{z_R u_{zwR}}$	$\pi_{162} = \dfrac{t C_{po} \rho_o \phi S_o v_x A}{M_1 V}$
$\pi_{31} = \dfrac{f_{xwR}}{f_{xgR}}$	$\pi_{75} = \dfrac{f_{zoR} p_{wR}}{p_{cgoR}}$	$\pi_{119} = \dfrac{k_{zwR} \lambda_{zwR} \rho_{wR} g z_R \sin\theta_R}{z_R u_{xwR}}$	$\pi_{163} = \dfrac{t C_{pw} \rho_w \phi S_w v_x A}{M_1 V}$
$\pi_{32} = \dfrac{u_{zgR}}{f_{zgR} U_{zR}}$	$\pi_{76} = \dfrac{f_{zoR} p_{wR} x_R^2}{p_{cowR} z_R^2}$	$\pi_{120} = \dfrac{k_{zoR} \lambda_{zoR} p_{oR}}{z_R u_{zoR}}$	$\pi_{164} = \dfrac{(1-\phi) C_{pr} t \rho_r v_x A}{\Delta T M_1 V}$
$\pi_{33} = \dfrac{f_{zoR}}{f_{zgR}}$	$\pi_{77} = \dfrac{f_{zoR} p_{wR}}{p_{cgoR}}$	$\pi_{121} = \dfrac{k_{zoR} \lambda_{zoR} \rho_{oR} g z_R \sin\theta_R}{z_R u_{zoR}}$	$\pi_{165} = \dfrac{m_s f_s L_v t}{\Delta T M_1 V}$
$\pi_{34} = \dfrac{f_{zwR}}{f_{zgR}}$	$\pi_{78} = \dfrac{f_{zoR} p_{wR} x_R^2}{p_{cgoR} z_R^2}$	$\pi_{122} = \dfrac{k_{zgR} \lambda_{zgR} p_{wR}}{z_R u_{zgR}}$	$\pi_{166} = \dfrac{m_s C_{pw} t}{M_1 V}$
$\pi_{35} = \dfrac{\eta_{oR} p_{wR}}{\Gamma(1-\alpha) t_R^{\alpha-1} \phi_R s_{wR} x_R^2}$	$\pi_{79} = \dfrac{f_{zoR} p_{cowR}}{p_{cgoR}}$	$\pi_{123} = \dfrac{k_{zgR} \lambda_{zgR} \rho_{gR} g z_R \sin\theta_R}{z_R u_{zgR}}$	$\pi_{167} = \dfrac{(1-\phi) \rho_r C_{pr}}{M_1}$
$\pi_{36} = \dfrac{\eta_{wR} p_{wR}}{\Gamma(1-\alpha) t_R^{\alpha-1} \phi_R s_{wR} x_R^2}$	$\pi_{80} = \dfrac{f_{xoR} p_{cowR} x_R^2}{p_{cgoR} z_R^2}$	$\pi_{124} = \dfrac{\sigma_{owR} J_R(s_w) \cos\theta_R \sqrt{\phi_R/k_R}}{p_{cowR}}$	$\pi_{168} = \dfrac{\phi C_{pw} \rho_w S_w}{M_1}$
$\pi_{37} = \dfrac{\eta_{gR} p_{wR}}{\Gamma(1-\alpha) t_R^{\alpha-1} \phi_R s_{wR} x_R^2}$	$\pi_{81} = \dfrac{f_{xwR} p_{wR}}{p_{cowR}}$	$\pi_{125} = \dfrac{\sigma_{goR} J_R(s_g) \cos\theta_R \sqrt{\phi_R/k_R}}{p_{cgoR}}$	$\pi_{169} = \dfrac{\phi c_{po} \rho_o S_0}{M_1}$
$\pi_{38} = \dfrac{f_{xoR} \eta_{oR} p_{wR}}{\Gamma(1-\alpha) t_R^{\alpha-1} \phi_R s_{wR} x_R^2}$	$\pi_{82} = \dfrac{f_{xwR} x_R^2}{z_R^2}$	$\pi_{126} = \dfrac{s_{oR}}{s_{wR}}$	$\pi_{170} = \dfrac{\phi C_{pg} \rho_g S_g}{M_1}$
$\pi_{39} = \dfrac{f_{xoR} \eta_{wR} p_{wR}}{\Gamma(1-\alpha) t_R^{\alpha-1} \phi_R s_{wR} x_R^2}$	$\pi_{83} = \dfrac{f_{xwR} p_{wR} x_R^2}{f_{xwR} p_{cowR} z_R^2}$	$\pi_{127} = \dfrac{s_{gR}}{s_{wR}}$	$\pi_{171} = \dfrac{\phi L_v \rho_g S_g}{M_1 \Delta T}$
$\pi_{40} = \dfrac{f_{xoR} \eta_{gR} p_{wR}}{\Gamma(1-\alpha) t_R^{\alpha-1} \phi_R s_{wR} x_R^2}$	$\pi_{84} = \dfrac{f_{xwR} p_{wR}}{p_{cgoR}}$	$\pi_{128} = \dfrac{s_o - s_{or}}{1 - s_{wi} - s_{or}}$	$\pi_{172} = \dfrac{K_{hw} S_w}{K_{hf}}$

TABLE 7.1 Dimensionless groups from inspectional analysis—cont'd

$\pi_{41} = \dfrac{f_{xwR}\eta_{oR}p_{wR}}{\Gamma(1-\alpha)t_R^{\alpha-1}\phi_R s_{wR} x_R^2}$		$\pi_{85} = \dfrac{f_{xwR}p_{wR}x_R^2}{p_{cgoR}z_R^2}$		$\pi_{129} = \dfrac{s_w - s_{wi}}{1 - s_{wi} - s_{or}}$		$\pi_{173} = \dfrac{K_{ho}S_o}{K_{hf}}$	
$\pi_{42} = \dfrac{f_{xwR}\eta_{wR}p_{wR}}{\Gamma(1-\alpha)t_R^{\alpha-1}\phi_R s_{wR} x_R^2}$		$\pi_{86} = \dfrac{f_{xwR}p_{cowR}}{p_{cgoR}}$		$\pi_{130} = \dfrac{s_g - s_{gc}}{1 - s_{wi} - s_{or}}$		$\pi_{174} = \dfrac{K_{hg}S_g}{K_{hf}}$	
$\pi_{43} = \dfrac{f_{xwR}\eta_{gR}p_{wR}}{\Gamma(1-\alpha)t_R^{\alpha-1}\phi_R s_{wR} x_R^2}$		$\pi_{87} = \dfrac{f_{xwR}p_{cowR}x_R^2}{p_{cgoR}z_R^2}$		$\pi_{131} = \dfrac{U_t t}{L\phi(1 - s_{wi} - s_{or})}$		$\pi_{175} = \dfrac{\phi K_{hf}}{K_{he}}$	
$\pi_{44} = \dfrac{\eta_{oR}p_{cowR}}{\Gamma(1-\alpha)t_R^{\alpha-1}\phi_R s_{wR} x_R^2}$		$\pi_{88} = \dfrac{\eta_{oR}p_{wR}}{\Gamma(1-\alpha)t_R^{\alpha-1}\phi_R s_{wR} z_R^2}$		$\pi_{132} = \dfrac{s_{wi}}{s_{wD}}$		$\pi_{176} = \dfrac{(1-\phi)K_{hr}}{K_{he}}$	

1. The value of ϕ, s_w, s_o, s_g, E_w, E_g, T, ΔT remain the same.
2. The values of H, Δp should be reduced by a factor of a.
3. The value of k should be increased by a factor of a.
4. The value of t should be reduced by a factor of a^2.

If the gravitational force is important, then this technique is suitable only for the steam flooding process. This approach is unable to scale additive accurately. Relaxed and satisfied scaling groups are given in Table 7.2.

TABLE 7.2 Influence of different dimensionless groups on each approach.

	Approaches						Approaches				
Dimensionless groups	1	2	3	4	5	**Dimensionless groups**	1	2	3	4	5
$\pi_1 = \dfrac{t_R u_{xwR}}{\phi x_R s_{wR}}$	×	↓a	✓	✓	×	$\pi_{89} = \dfrac{\eta_{wR}p_{wR}}{\Gamma(1-\alpha)t_R^{\alpha-1}\phi_R s_{wR} z_R^2}$	×	✓	✓	✓	×
$\pi_2 = \dfrac{t_R u_{ziwR}}{\phi z_R s_{wR}}$	×	↓a	✓		×	$\pi_{90} = \dfrac{\eta_{gR}p_{wR}}{\Gamma(1-\alpha)t_R^{\alpha-1}\phi_R s_{wR} z_R^2}$	×	✓	✓	✓	×
$\pi_3 = \dfrac{D_{Lw1R}t_R}{x_R^2 s_{wR}}$	×	×	×	✓	×	$\pi_{91} = \dfrac{f_{xoR}\eta_{oR}p_{wR}}{\Gamma(1-\alpha)t_R^{\alpha-1}\phi_R s_{wR} z_R^2}$	×	↓$a^{\frac{2}{\alpha-1}}$	↓$a^{\frac{2}{\alpha-1}}$	✓	×
$\pi_4 = \dfrac{D_{Tw1R}t_R}{z_R^2 s_{wR}}$	×	×	↑	✓	×	$\pi_{92} = \dfrac{f_{xoR}\eta_{wR}p_{wR}}{\Gamma(1-\alpha)t_R^{\alpha-1}\phi_R s_{wR} z_R^2}$	×	↓$a^{\frac{2}{\alpha-1}}$	↓$a^{\frac{2}{\alpha-1}}$	✓	×
$\pi_5 = \dfrac{D_{Lw3R}t_R}{x_R^2 s_{wR}}$	×	×	×	✓	×	$\pi_{93} = \dfrac{f_{xoR}\eta_{gR}p_{wR}}{\Gamma(1-\alpha)t_R^{\alpha-1}\phi_R s_{wR} z_R^2}$	×	↓$a^{\frac{2}{\alpha-1}}$	↓$a^{\frac{2}{\alpha-1}}$	✓	×
$\pi_6 = \dfrac{D_{Tw3R}t_R}{z_R^2 s_{wR}}$	×	×	↑	✓	×	$\pi_{94} = \dfrac{f_{xwR}\eta_{oR}p_{wR}}{\Gamma(1-\alpha)t_R^{\alpha-1}\phi_R s_{wR} z_R^2}$	×	↓$a^{\frac{2}{\alpha-1}}$	↓$a^{\frac{2}{\alpha-1}}$	✓	×
$\pi_7 = \dfrac{D_{Lw4R}t_R}{x_R^2 s_{wR}}$	×	×	×	✓	×	$\pi_{95} = \dfrac{f_{xwR}\eta_{wR}p_{wR}}{\Gamma(1-\alpha)t_R^{\alpha-1}\phi_R s_{wR} z_R^2}$	×	↓$a^{\frac{2}{\alpha-1}}$	↓$a^{\frac{2}{\alpha-1}}$	✓	×
$\pi_8 = \dfrac{D_{Tw4R}t_R}{z_R^2 s_{wR}}$	×	×	↑	✓	×	$\pi_{96} = \dfrac{f_{xwR}\eta_{gR}p_{wR}}{\Gamma(1-\alpha)t_R^{\alpha-1}\phi_R s_{wR} z_R^2}$	×	↓$a^{\frac{2}{\alpha-1}}$	↓$a^{\frac{2}{\alpha-1}}$	✓	×
$\pi_9 = \dfrac{t_R u_{xoR}}{\phi x_R s_{oR}}$	×	↓a	✓	✓	×	$\pi_{97} = \dfrac{\eta_{oR}p_{cowR}}{\Gamma(1-\alpha)t_R^{\alpha-1}\phi_R s_{wR} z_R^2}$	×	✓	✓	✓	×

Continued

TABLE 7.2 Influence of different dimensionless groups on each approach—cont'd

Dimensionless groups	1	2	3	4	5	Dimensionless groups	1	2	3	4	5
$\pi_{10} = \dfrac{t_R u_{zoR}}{\phi z_R s_{oR}}$	×	↓a	√	√	×	$\pi_{98} = \dfrac{\eta_{gR} p_{cowR}}{\Gamma(1-\alpha) t_R^{\alpha-1} \phi_R s_{wR} z_R^2}$	×	√	√	√	×
$\pi_{11} = \dfrac{D_{Lo1R} t_R}{x_R^2 s_{oR}}$	×	×	×	√	×	$\pi_{99} = \dfrac{f_{xoR} \eta_{oR} p_{cowR}}{\Gamma(1-\alpha) t_R^{\alpha-1} \phi_R s_{wR} z_R^2}$	×	√	√	√	×
$\pi_{12} = \dfrac{D_{To1R} t_R}{z_R^2 s_{oR}}$	×	×	↑	√	×	$\pi_{100} = \dfrac{f_{xoR} \eta_{gR} p_{cowR}}{\Gamma(1-\alpha) t_R^{\alpha-1} \phi_R s_{wR} z_R^2}$	×	√	√	√	×
$\pi_{13} = \dfrac{D_{Lo2R} t_R}{x_R^2 s_{oR}}$	×	×	×	√	×	$\pi_{101} = \dfrac{f_{xwR} \eta_{oR} p_{cowR}}{\Gamma(1-\alpha) t_R^{\alpha-1} \phi_R s_{wR} z_R^2}$	×	√	√	√	×
$\pi_{14} = \dfrac{D_{To2R} t_R}{z_R^2 s_{oR}}$	×	×	↑	√	×	$\pi_{102} = \dfrac{f_{xwR} \eta_{oR} p_{cowR}}{\Gamma(1-\alpha) t_R^{\alpha-1} \phi_R s_{wR} z_R^2}$	×	√	√	√	×
$\pi_{15} = \dfrac{D_{Lo4R} t_R}{x_R^2 s_{oR}}$	×	×	×	√	×	$\pi_{103} = \dfrac{\eta_{gR} p_{cgoR}}{\Gamma(1-\alpha) t_R^{\alpha-1} \phi_R s_{wR} z_R^2}$	×	√	√	√	×
$\pi_{16} = \dfrac{D_{To4R} t_R}{z_R^2 s_{oR}}$	×	×	↑	√	×	$\pi_{104} = \dfrac{f_{xoR} \eta_{gR} p_{cgoR}}{\Gamma(1-\alpha) t_R^{\alpha-1} \phi_R s_{wR} z_R^2}$	×	√	√	√	×
$\pi_{17} = \dfrac{u_{xgR} t_R}{\phi x_R s_{gR}}$	×	↓a	√	√	×	$\pi_{105} = \dfrac{f_{xwR} \eta_{gR} p_{cgoR}}{\Gamma(1-\alpha) t_R^{\alpha-1} \phi_R s_{wR} z_R^2}$	×	√	√	√	×
$\pi_{18} = \dfrac{u_{zgR} t_R}{\phi z_R s_{gR}}$	×	↓a	√	√	×	$\pi_{106} = \dfrac{t_R D_{Lw1R}}{s_{wR} x_R^2}$	×	×	×	√	×
$\pi_{19} = \dfrac{u_{xoR}}{U_{xR}}$	√	↓a	√	√	√	$\pi_{107} = \dfrac{t_R D_{Lw3R}}{s_{wR} x_R^2}$	×	×	×	√	×
$\pi_{20} = \dfrac{u_{xwR}}{U_{xR}}$	√	↓a	√	√	√	$\pi_{108} = \dfrac{t_R D_{Lw4R}}{s_{wR} x_R^2}$	×	×	×	√	×
$\pi_{21} = \dfrac{u_{xgR}}{U_{xR}}$	√	↓a	√	√	√	$\pi_{109} = \dfrac{t_R D_{Tw1R}}{s_{wR} z_R^2}$	×	×	↑	√	×
$\pi_{22} = \dfrac{u_{zoR}}{U_{zR}}$	√	↓a	√	√	√	$\pi_{110} = \dfrac{t_R D_{Tw3R}}{s_{wR} z_R^2}$	×	×	↑	√	×
$\pi_{23} = \dfrac{u_{zwR}}{U_{zR}}$	√	↓a	√	√	√	$\pi_{111} = \dfrac{t_R D_{Tw4R}}{s_{wR} z_R^2}$	×	×	↑	√	×
$\pi_{24} = \dfrac{u_{zgR}}{U_{zR}}$	√	↓a	√	√	√	$\pi_{112} = \dfrac{k_{xwR} \lambda_{xwR} p_{wR}}{x_R u_{xwR}}$	↑a	√	√	√	↑a
$\pi_{25} = \dfrac{u_{xoR}}{f_{xoR} U_{xR}}$	√	√	√	√	√	$\pi_{113} = \dfrac{k_{xwR} \lambda_{xwR} \rho_{wR} g_R z_R \sin\theta_R}{x_R u_{xwR}}$	↑a	×	√	×	↑a
$\pi_{26} = \dfrac{u_{zoR}}{f_{zoR} U_{zR}}$	√	√	√	√	√	$\pi_{114} = \dfrac{k_{xoR} \lambda_{xoR} p_{oR}}{x_R u_{xoR}}$	↑a	√	√	√	↑a
$\pi_{27} = \dfrac{u_{xwR}}{f_{xwR} U_{xR}}$	√	√	√	√	√	$\pi_{115} = \dfrac{k_{xoR} \lambda_{xoR} \rho_{oR} g_R z_R \sin\theta_R}{x_R u_{xoR}}$	↑a	×	√	×	↑a
$\pi_{28} = \dfrac{u_{zwR}}{f_{zwR} U_{zR}}$	√	√	√	√	√	$\pi_{116} = \dfrac{k_{xgR} \lambda_{xgR} p_{gR}}{x_R u_{xgR}}$	↑a	√	√	√	↑a
$\pi_{29} = \dfrac{u_{xgR}}{f_{xgR} U_{xR}}$	√	√	√	√	√	$\pi_{117} = \dfrac{k_{xgR} \lambda_{xgR} \rho_{gR} g_R z_R \sin\theta_R}{x_R u_{xgR}}$	↑a	×	√	×	↑a
$\pi_{30} = \dfrac{f_{xoR}}{f_{xgR}}$	√	√	√	√	√	$\pi_{118} = \dfrac{k_{zwR} \lambda_{zwR} p_{wR}}{z_R u_{zwR}}$	↑a	√	√	√	↑a

TABLE 7.2 Influence of different dimensionless groups on each approach—cont'd

Dimensionless groups	1	2	3	4	5	Dimensionless groups	1	2	3	4	5
$\pi_{31} = \dfrac{f_{xwR}}{f_{xgR}}$	√	√	√	√	√	$\pi_{119} = \dfrac{k_{zwR}\lambda_{zwR}\rho_{wR}g z_R \sin\theta_R}{z_R u_{xwR}}$	↑a	×	√	×	↑a
$\pi_{32} = \dfrac{u_{zgR}}{f_{zgR}U_{zR}}$	√	√	√	√	√	$\pi_{120} = \dfrac{k_{zoR}\lambda_{zoR}p_{oR}}{z_R u_{zoR}}$	↑a	√	√	√	↑a
$\pi_{33} = \dfrac{f_{zoR}}{f_{zgR}}$	√	√	√	√	√	$\pi_{121} = \dfrac{k_{zoR}\lambda_{zoR}\rho_{oR}g z_R \sin\theta_R}{z_R u_{zoR}}$	↑a	×	√	×	↑a
$\pi_{34} = \dfrac{f_{zwR}}{f_{zgR}}$	√	√	√	√	√	$\pi_{122} = \dfrac{k_{zgR}\lambda_{zgR}p_{wR}}{z_R u_{zgR}}$	↑a	×	√	√	↑a
$\pi_{35} = \dfrac{\eta_{oR}p_{wR}}{\Gamma(1-\alpha)t_R^{\alpha-1}\phi_R s_{wR}x_R^2}$	×	√	√	√	×	$\pi_{123} = \dfrac{k_{zgR}\lambda_{zgR}\rho_{gR}g z_R \sin\theta_R}{z_R u_{zgR}}$	↑a	×	√	×	↑a
$\pi_{36} = \dfrac{\eta_{wR}p_{wR}}{\Gamma(1-\alpha)t_R^{\alpha-1}\phi_R s_{wR}x_R^2}$	×	√	√	√	×	$\pi_{124} = \dfrac{\sigma_{owR} J_R(s_w) \cos\theta_R \sqrt{\dfrac{\phi_R}{k_R}}}{p_{cowR}}$	×	√	×	×	×
$\pi_{37} = \dfrac{\eta_{gR}p_{wR}}{\Gamma(1-\alpha)t_R^{\alpha-1}\phi_R s_{wR}x_R^2}$	×	√	√	√	×	$\pi_{125} = \dfrac{\sigma_{goR} J_R(s_g) \cos\theta_R \sqrt{\dfrac{\phi_R}{k_R}}}{p_{cgoR}}$	×	√	×	×	×
$\pi_{38} = \dfrac{f_{xoR}\eta_{oR}p_{wR}}{\Gamma(1-\alpha)t_R^{\alpha-1}\phi_R s_{wR}x_R^2}$	×	↓$a^{\frac{2}{\alpha-1}}$	↓$a^{\frac{2}{\alpha-1}}$	√	×	$\pi_{126} = \dfrac{s_{oR}}{s_{wR}}$	×	√	√	√	×
$\pi_{39} = \dfrac{f_{xoR}\eta_{wR}p_{wR}}{\Gamma(1-\alpha)t_R^{\alpha-1}\phi_R s_{wR}x_R^2}$	×	↓$a^{\frac{2}{\alpha-1}}$	↓$a^{\frac{2}{\alpha-1}}$	√	×	$\pi_{127} = \dfrac{s_{gR}}{s_{wR}}$	×	√	√	√	×
$\pi_{40} = \dfrac{f_{xoR}\eta_{gR}p_{wR}}{\Gamma(1-\alpha)t_R^{\alpha-1}\phi_R s_{wR}x_R^2}$	×	↓$a^{\frac{2}{\alpha-1}}$	↓$a^{\frac{2}{\alpha-1}}$	√	×	$\pi_{128} = \dfrac{s_o - s_{or}}{1 - s_{wi} - s_{or}}$	×	√	√	√	×
$\pi_{41} = \dfrac{f_{xwR}\eta_{oR}p_{wR}}{\Gamma(1-\alpha)t_R^{\alpha-1}\phi_R s_{wR}x_R^2}$	×	↓$a^{\frac{2}{\alpha-1}}$	↓$a^{\frac{2}{\alpha-1}}$	√	×	$\pi_{129} = \dfrac{s_w - s_{wi}}{1 - s_{wi} - s_{or}}$	×	√	√	√	×
$\pi_{42} = \dfrac{f_{xwR}\eta_{wR}p_{wR}}{\Gamma(1-\alpha)t_R^{\alpha-1}\phi_R s_{wR}x_R^2}$	×	↓$a^{\frac{2}{\alpha-1}}$	↓$a^{\frac{2}{\alpha-1}}$	√	×	$\pi_{130} = \dfrac{s_g - s_{gc}}{1 - s_{wi} - s_{or}}$	×	√	√	√	×
$\pi_{43} = \dfrac{f_{xwR}\eta_{gR}p_{wR}}{\Gamma(1-\alpha)t_R^{\alpha-1}\phi_R s_{wR}x_R^2}$	×	↓$a^{\frac{2}{\alpha-1}}$	↓$a^{\frac{2}{\alpha-1}}$	√	×	$\pi_{131} = \dfrac{U_t t}{L\phi(1 - s_{wi} - s_{or})}$	↓a^2	↓a^2	↓a^2	↓a^2	↓a^2
$\pi_{44} = \dfrac{\eta_{oR}p_{cowR}}{\Gamma(1-\alpha)t_R^{\alpha-1}\phi_R s_{wR}x_R^2}$	×	√	√	√	×	$\pi_{132} = \dfrac{s_{wi}}{s_{wD}}$	×	√	√	√	×
$\pi_{45} = \dfrac{\eta_{gR}p_{cowR}}{\Gamma(1-\alpha)t_R^{\alpha-1}\phi_R s_{wR}x_R^2}$	×	√	√	√	×	$\pi_{133} = \dfrac{s_{gi}}{s_{gR}}$	×	√	√	√	×
$\pi_{46} = \dfrac{f_{xoR}\eta_{oR}p_{cowR}}{\Gamma(1-\alpha)t_R^{\alpha-1}\phi_R s_{wR}x_R^2}$	×	√	√	√	×	$\pi_{134} = \dfrac{s_{oi}}{s_{oR}}$	×	√	√	√	×
$\pi_{47} = \dfrac{f_{xoR}\eta_{gR}p_{cowR}}{\Gamma(1-\alpha)t_R^{\alpha-1}\phi_R s_{wR}x_R^2}$	×	√	√	√	×	$\pi_{135} = \dfrac{E_{2i}}{E_{o2R}}$	×	×	×	√	√
$\pi_{48} = \dfrac{f_{xwR}\eta_{oR}p_{cowR}}{\Gamma(1-\alpha)t_R^{\alpha-1}\phi_R s_{wR}x_R^2}$	×	√	√	√	×	$\pi_{136} = \dfrac{E_{3i}}{E_{w3R}}$	×	×	×	√	√
$\pi_{49} = \dfrac{f_{xwR}\eta_{oR}p_{cowR}}{\Gamma(1-\alpha)t_R^{\alpha-1}\phi_R s_{wR}x_R^2}$	×	√	√	√	×	$\pi_{137} = \dfrac{E_{1j}}{E_{g1R}}$	×	×	×	√	√

Continued

TABLE 7.2 Influence of different dimensionless groups on each approach—cont'd

Dimensionless groups	1	2	3	4	5	Dimensionless groups	1	2	3	4	5
$\pi_{50} = \dfrac{\eta_{gR} p_{cgoR}}{\Gamma(1-\alpha) t_R^{\alpha-1} \phi_R s_{wR} x_R^2}$	×	√	√	√	×	$\pi_{138} = \dfrac{E_{4j}}{E_{w4R}}$	×	×	×	√	√
$\pi_{51} = \dfrac{f_{xoR} \eta_{gR} p_{cgoR}}{\Gamma(1-\alpha) t_R^{\alpha-1} \phi_R s_{wR} x_R^2}$	×	√	√	√	×	$\pi_{139} = \dfrac{\Delta p_{gR}}{p_{gR}}$	×	√	√	√	√
$\pi_{52} = \dfrac{f_{xwR} \eta_{gR} p_{cgoR}}{\Gamma(1-\alpha) t_R^{\alpha-1} \phi_R s_{wR} x_R^2}$	×	√	√	√	×	$\pi_{140} = \dfrac{\rho_{gR} g_R H_R}{p_{gR}}$	√	×	√	×	√
$\pi_{53} = \dfrac{p_{wR}}{p_{cowR}}$	×	√	√	√	√	$\pi_{141} = \dfrac{\rho_{gR} g_R z_R}{p_{gR}}$	√	×	√	×	√
$\pi_{54} = \dfrac{x_R^2}{z_R^2}$	↓a	↓a	↓a	↓a	↓a	$\pi_{142} = \dfrac{\Delta p_{oR}}{p_{oR}}$	×	√	√	√	√
$\pi_{55} = \dfrac{p_{wR} x_R^2}{p_{cowR} z_R^2}$	×	√	√	√	√	$\pi_{143} = \dfrac{\rho_{oR} g_R H_R}{p_{oR}}$	√	×	√	×	√
$\pi_{56} = \dfrac{p_{wR}}{p_{cgoR}}$	×	√	√	√	√	$\pi_{144} = \dfrac{\rho_{oR} g_R z_R}{p_{oR}}$	√	×	√	×	√
$\pi_{57} = \dfrac{p_{wR} x_R^2}{p_{cgoR} z_R^2}$	×	√	√	√	√	$\pi_{145} = \dfrac{\Delta p_{wR}}{p_{wR}}$	√	√	√	√	√
$\pi_{58} = \dfrac{p_{cowR}}{p_{cgoR}}$	×	√	√	√	√	$\pi_{146} = \dfrac{\rho_{wR} g_R H_R}{p_{wR}}$	√	×	√	×	√
$\pi_{59} = \dfrac{p_{cowR} x_R^2}{p_{cgoR} z_R^2}$	×	√	√	√	√	$\pi_{147} = \dfrac{\rho_{wR} g_R z_R}{p_{wR}}$	√	×	√	×	√
$\pi_{60} = \dfrac{f_{xoR} p_{wR}}{p_{cowR}}$	×	√	√	√	√	$\pi_{148} = \dfrac{H U_t}{u_{gxR} x_R}$	√	√	√	√	√
$\pi_{61} = \dfrac{f_{xoR} x_R^2}{z_R^2}$	↓a	√	√	√	√	$\pi_{149} = \dfrac{z_R^2 D_{Lo1R}}{x_R^2 D_{To1R}}$	×	×	×	√	×
$\pi_{62} = \dfrac{f_{xoR} p_{wR} x_R^2}{p_{cowR} z_R^2}$	×	√	√	√	√	$\pi_{150} = \dfrac{z_R^2 D_{Lo2R}}{x_R^2 D_{To2R}}$	×	×	×	√	×
$\pi_{63} = \dfrac{f_{xoR} p_{wR}}{p_{cgoR}}$	×	√	√	√	√	$\pi_{151} = \dfrac{z_R^2 D_{Lo4R}}{x_R^2 D_{To4R}}$	×	×	×	√	×
$\pi_{64} = \dfrac{f_{xoR} p_{wR} x_R^2}{p_{cgoR} z_R^2}$	×	√	√	√	√	$\pi_{152} = \dfrac{z_R^2 D_{Lw1R}}{x_R^2 D_{Tw1R}}$	×	×	×	√	×
$\pi_{65} = \dfrac{f_{xoR} p_{cowR}}{p_{cgoR}}$	×	√	√	√	√	$\pi_{153} = \dfrac{z_R^2 D_{Lw2R}}{x_R^2 D_{Tw2R}}$	×	×	×	√	×
$\pi_{66} = \dfrac{f_{xoR} p_{cowR} x_R^2}{p_{cgoR} z_R^2}$	×	√	√	√	√	$\pi_{154} = \dfrac{z_R^2 D_{Lw4R}}{x_R^2 D_{Tw4R}}$	×	×	×	√	×
$\pi_{67} = \dfrac{f_{xwR} p_{wR}}{p_{cowR}}$	×	√	√	√	√	$\pi_{155} = \dfrac{H}{z_R}$	↓a	↓a	↓a	↓a	↓a
$\pi_{68} = \dfrac{f_{xwR} x_R^2}{z_R^2}$	↓a	√	√	√	√	$\pi_{156} = \dfrac{L}{x_R}$	↓a	↓a	↓a	↓a	↓a
$\pi_{69} = \dfrac{f_{xwR} p_{wR} x_R^2}{f_{xwR} p_{cowR} z_R^2}$	×	√	√	√	√	$\pi_{157} = \dfrac{H U_t}{u_{gxR} z_R}$	√	√	√	√	×

TABLE 7.2 Influence of different dimensionless groups on each approach—cont'd

Dimensionless groups	Approaches 1	2	3	4	5	Dimensionless groups	Approaches 1	2	3	4	5
$\pi_{70} = \dfrac{f_{xwR}p_{wR}}{p_{cgoR}}$	×	√	√	√	√	$\pi_{158} = \dfrac{K_{hr}TAt}{L\Delta T M_1 V}$	×	√	×	√	×
$\pi_{71} = \dfrac{f_{xwR}p_{wR}x_R^2}{p_{cgoR}z_R^2}$	×	√	√	√	√	$\pi_{159} = \dfrac{K_{hf}TAt}{L\Delta T M_1 V}$	×	√	×	√	×
$\pi_{72} = \dfrac{f_{xwR}p_{cowR}}{p_{cgoR}}$	×	√	√	√	√	$\pi_{160} = \dfrac{tC_{po}\rho_o u_{ox}A}{M_1 V}$	×	√	×	√	×
$\pi_{73} = \dfrac{f_{xwR}p_{cowR}x_R^2}{p_{cgoR}z_R^2}$	×	√	√	√	√	$\pi_{161} = \dfrac{tC_{pw}\rho_w u_{wx}A}{M_1 V}$	×	√	×	√	×
$\pi_{74} = \dfrac{f_{zoR}p_{wR}x_R^2}{p_{cowR}z_R^2}$	×	√	√	√	√	$\pi_{162} = \dfrac{tC_{po}\rho_o \phi S_o v_x A}{M_1 V}$	×	√	×	√	×
$\pi_{75} = \dfrac{f_{zoR}p_{wR}}{p_{cgoR}}$	×	√	√	√	√	$\pi_{163} = \dfrac{tC_{pw}\rho_w \phi S_w v_x A}{M_1 V}$	×	√	×	√	×
$\pi_{76} = \dfrac{f_{zoR}p_{wR}x_R^2}{p_{cowR}z_R^2}$	×	√	√	√	√	$\pi_{164} = \dfrac{(1-\phi)C_{pr}t\rho_r v_x A}{\Delta T M_1 V}$	×	√	×	√	×
$\pi_{77} = \dfrac{f_{zoR}p_{wR}}{p_{cgoR}}$	×	√	√	√	√	$\pi_{165} = \dfrac{m_s f_s L_v t}{\Delta T M_1 V}$	×	√	×	√	×
$\pi_{78} = \dfrac{f_{zoR}p_{wR}x_R^2}{p_{cgoR}z_R^2}$	×	√	√	√	√	$\pi_{166} = \dfrac{m_s C_{pw} t}{M_1 V}$	×	√	×	√	×
$\pi_{79} = \dfrac{f_{zoR}p_{cowR}}{p_{cgoR}}$	×	√	√	√	√	$\pi_{167} = \dfrac{(1-\phi)\rho_r C_{pr}}{M_1}$	×	√	×	√	×
$\pi_{80} = \dfrac{f_{xoR}p_{cowR}x_R^2}{p_{cgoR}z_R^2}$	×	√	√	√	√	$\pi_{168} = \dfrac{\phi C_{pw}\rho_w S_w}{M_1}$	×	√	×	√	×
$\pi_{81} = \dfrac{f_{xwR}p_{wR}}{p_{cowR}}$	×	√	√	√	√	$\pi_{169} = \dfrac{\phi c_{po}\rho_o S_o}{M_1}$	×	√	×	√	×
$\pi_{82} = \dfrac{f_{xwR}x_R^2}{z_R^2}$	×	√	√	√	√	$\pi_{170} = \dfrac{\phi C_{pg}\rho_g S_g}{M_1}$	×	√	×	√	×
$\pi_{83} = \dfrac{f_{xwR}p_{wR}x_R^2}{f_{xwR}p_{cowR}z_R^2}$	×	√	√	√	√	$\pi_{171} = \dfrac{\phi L_v \rho_g S_g}{M_1 \Delta T}$	×	√	×	√	×
$\pi_{84} = \dfrac{f_{xwR}p_{wR}}{p_{cgoR}}$	×	√	√	√	√	$\pi_{172} = \dfrac{K_{hw}S_w}{K_{hf}}$	×	√	×	√	×
$\pi_{85} = \dfrac{f_{xwR}p_{wR}x_R^2}{p_{cgoR}z_R^2}$	×	√	√	√	√	$\pi_{173} = \dfrac{K_{ho}S_o}{K_{hf}}$	×	√	×	√	×
$\pi_{86} = \dfrac{f_{xwR}p_{cowR}}{p_{cgoR}}$	×	√	√	√	√	$\pi_{174} = \dfrac{K_{hg}S_g}{K_{hf}}$	×	√	×	√	×
$\pi_{87} = \dfrac{f_{xwR}p_{cowR}x_R^2}{p_{cgoR}z_R^2}$	×	√	√	√	√	$\pi_{175} = \dfrac{\phi K_{hf}}{K_{he}}$	×	√	×	√	×
$\pi_{88} = \dfrac{\eta_{oR}p_{wR}}{\Gamma(1-\alpha)t_R^{\alpha-1}\phi_R s_{wR} z_R^2}$	×	√	√	√	×	$\pi_{176} = \dfrac{(1-\phi)K_{hr}}{K_{he}}$	×	√	×	√	×

√ indicates the group is satisfied; × indicates the group is not satisfied; ↓ indicates the group is reduced; ↑ indicates the group is increased; a indicate the dimension of the scaling factor by which the model is reduced from the prototype.

Approach 2 same fluids, same pressure drop, same porous medium, geometric similarity.

The difficulties raised in approach 1 should be overcome by considering the maximum pressure and temperature difference, and the initial pressure and temperature are the same for the model and prototype. This assumption has allowed the properties to depend on pressure and temperature which are properly scaled. As the same porous medium is used here, so the fluid saturations and relative permeabilities are properly scaled. In addition, viscous forces, diffusion effects, and heat transfers are properly scaled due to these changes. The limitation of this approach is that it cannot accurately scale gravitational forces. Another limitation is that it cannot scale dispersion effects if the flow rate is very high.

The implementation of these scaling criteria for a model can reduce the length by a scaling factor of "a."

1. The value of ϕ, s_w, s_o, s_g, Δp, T, ΔT remain the same.
2. The values of H, U_t should be reduced by a factor of a.
3. The value of t should be reduced by a factor of $a^{\left(\frac{2}{a-1}\right)}$.

This approach is restricted to processes where the gravitational force is not much important such as thin formations with high flow rates. Diffusion effects and PVT properties are scales well for steam flooding with additives. Relaxed and satisfied scaling groups are given in Table 7.2.

Approach 3 same fluids, same pressure drop, same porous media, relaxed geometric similarity.

The advantage of using the same porous medium, same fluid, and similar pressure and temperature conditions help to scale gravitational forces properly but allows the geometric similarity to be relaxed. If the pressure gradient is low due to capillary and viscous forces, then capillary and viscous forces can be scaled for the horizontal well. The vertical direction heat conduction, dispersion effects, and capillary number are not properly scaled.

The implementation of these scaling criteria for a model can reduce the length by a scaling factor of "a." The value of ϕ, s_w, s_o, s_g, Δp, p_{wR}, T, ΔT remain the same.

1. The values of H should be reduced by a factor of "a."
2. The reservoir should be horizontal.
3. The value of t should be reduced by a factor of $a^{\left(\frac{2}{a-1}\right)}$.

This approach is restricted to steam and steam additive processes where a significant reduction of reservoir thickness is considered. Relaxed and satisfied scaling groups are given in Table 7.2.

Approach 4, same fluids, same porous media, same pressure drop, relaxed geometric similarity.

The previous approaches had not attempted to consider the dispersion effect for the case of high flow rate scaling. It is difficult to scale dispersion effects. This approach's objective is to scale transverse dispersion effects. Gravitational and capillary effects are not properly scaled, but viscous and dispersion effects are scaled properly. The merit of this approach is to satisfy all other dimensionless numbers and boundary conditions except capillary and gravity forces. This rigorous method is not very suitable when scaling the steam flooding process is involved.

The implementation of these scaling criteria for a model can reduce the length by a scaling factor of "a."

1. The value of ϕ, s_w, s_o, s_g, Δp, p_{wR}, T, ΔT, k remain the same.
2. The values of H should be reduced by a factor of a.
3. The values of U_t should be reduced by a factor of $a^{(\frac{1}{2})}$.
4. The value of t should be reduced by a factor of a^2.

This approach is restricted to a process where dispersion effects are considered by relaxing gravitational forces. Hot water flooding with a liquid additive is a good option for this approach. In addition, this approach is restricted to thin formations because only a small reduction in thickness is considered. Relaxed and satisfied scaling groups are given in Table 7.2.

Approach 5 same fluid, different pressure drops, different porous media, relaxed geometric similarity.

The main shortcoming of approach 4 is the relaxation of gravitational forces, but this approach tries to satisfy viscous and gravitational forces while still scaling transverse dispersion effects. As a different porous medium and different pressure drop are used, the limitation of approach 1 comes into place. Here, time is scaled by a four-fifth power of the scaling factor rather than squares indicating the longer period of experimental time. Capillary forces and heat conduction are not properly scaled. In addition, saturation pressure and boundary temperature for steam flood are poorly scaled. Relaxed and satisfied scaling groups are given in Table 7.2.

The implementation of these scaling criteria for a model can reduce the length by a scaling factor of "a."

1. The value of ϕ, s_w, s_o, s_g, T, ΔT remain the same.
2. The values of H should be reduced by a factor of a.
3. The values of k should be increased by a factor of a.
4. The value of t should be reduced by a factor of a^2.

7.2.2 Comparison of different scaling approaches

Table 7.2 lists dimensionless numbers and how their effects can change the model for the steam flooding process. Approach 1 can be used by numerous previous researchers. This approach can accurately have scaled the ratio of viscous to gravitation forces, but it had faced difficulty in scaling saturation temperature, saturation pressure, steam injection rate, steam density, the energy stored in the steam phase, and latent heat of vaporization. The different porous medium is used, the relative permeabilities and irreducible saturations can alter also.

Approach 2 would be the most suitable approach for the steam flooding process where gravitational does not play a vital role. When the process is dominated by viscous force, then this approach comes into play a vital role. The effects of gravitational force have been reduced in the model by employing this approach. When this approach creates a significant error under certain conditions which are not studied well, it is restricted to certain conditions for the steam flooding process. In a study of the immiscible isothermal displacement of heavy oil by CO_2 flooding, Rojas (1985) found that the recovery of oil is independent of model prototype ratio when gravitational to viscous forces ratio is less than 5. There should have an upper and lower limit for this approach. The upper limit may represent a point where dispersion effects can be scaled more effectively. On the other hand, the lower limit of gravitational forces can be scaled more rigorously.

Approach 3 can overcome the limitations of approach 1. As the same fluid and same porous medium are used, thus it can ensure the irreducible saturations, and relative permeabilities can be scaled properly. It can also properly scale saturation temperature, saturation pressure, steam injection rate, steam density, the energy stored in the steam phase, and latent heat of vaporization. This approach cannot scale capillary forces along with the vertical conduction of energy. Baker (1973) investigated that the heat losses are a function of time only, it does not depend on injection rate. The effect of transverse dispersion effects will be enhanced in the model.

Approach 4 can easily scale the transverse dispersion effects, but gravitational and capillary forces are poorly scaled. It cannot scale vertical conduction of energy properly. The scaling requirements of irreducible saturations, relative permeabilities, steam density, heat stored in steam, saturation temperature-saturation pressure relationship, and injection temperature are satisfied.

Approach 5 satisfied the requirements for gravitational forces and balanced them with dispersive and viscous forces. However, it has several drawbacks like other approaches. It has required a significant reduction in pressure drop as well as a reduction in time scale factor to satisfy the scale conduction. Therefore, approach 5 poorly scales conduction.

The various aspects of recovery are largely depended on the selection of appropriate approaches. The selection of an appropriate approach is particularly dependent on the properties which are involved within this approach. In approaches 2, 3, and 4 same fluid, same pressure drops, and the same porous medium are used, but the temperature change has a significant effect in simulating these properties even though they have not been properly scaled. There are some important phenomena such as gas solubility, emulsification, distillation, etc., which are not considered here during scaling. The significance of a phenomenon is used as a selection criterion which is not scaled by the selected approach. If capillary force is a prime factor for a process, it is unlikely that approaches 3 and 4 would be satisfied. Similarly, for the case of gravitational forces, approaches 2 and 4 would not be satisfied. Another issue is the selection of the relative significance of a phenomenon. If a phenomenon is considered insignificant for a process, it should remain insignificant in the model also. For selecting a scaling process, the effects of transverse dispersion are considered to be minor in the prototype for the case of approach 3. If it remains insignificant in the model, then it would be considered as a suitable approach for this process.

Hossain and Abu-Khamsin (2012a, b) proposed to use several dimensionless groups in addition to Nusselt and Prandtl numbers. These groups are derived using inspectional and dimensional analysis. Their results show that the proposed numbers are measures of thermal diffusivity and hydraulic diffusivity of a fluid in porous media. This research confirms that the influence of total absolute thermal conductivities of the fluid and rock on the effective thermal conductivity of the fluid-saturated porous medium diminishes after a certain local Nusselt number of the system. Finally, the result confirms that the convective ability of the fluid-saturated porous medium is more pronounced than its conductive ability. This study will help to better understand the modeling of the EOR process thus improving process design and performance prediction.

Dimensionless selling type	Scaling group	Formulation	Comment
Physical effects of flow and fluid properties scaling	Capillary number	$N_C = \dfrac{F_{capillary}}{F_{viscous}}$	Rock fluid interaction, describes set up at the small scale
	Gravity number	$N_g = \dfrac{F_{gravity}}{F_{viscous}}$	Reservoir-fluid shape-dependent, seizures the effect of resistant force
	Mobility ratio	$M = \dfrac{\lambda_{displacedfluid}}{\lambda_{displacingfluid}}$	Fluid-rock-fluid communication effect on the flow performance
Displacement techniques with initial and boundary conditions scaling	Displacement efficiency factor	$E_D = \dfrac{V_{Produced}}{V_{Reference}}$	Dimensionless production response
	Dimensionless time	$t_D = \dfrac{V_{injected}}{V_{pore}}$	Forced injection boundary condition
Reservoir geometry scaling	Aspect ratio	$N_A = \dfrac{L}{H}$	Reservoir shape description scale
	Dip angle	$N_\alpha = \tan \alpha$	Dip angle scaling

Table 7.3 shows the dimensionless groups derived from dimensional analysis (Hossain, 2018).

An extensive literature review is reported by Obembe et al. (2017a,b,c) in the subject area (summarized in Table 7.4). In describing the continuous changes of rock and fluid parameters during thermal EOR, dimensionless heat transfer coefficients play a significant role. Available dimensionless numbers, such as Nusselt, Peclet, and Prandtl numbers, are insufficient in explaining conductive and convective heat transfer, especially when this alteration phenomenon prevails. If the changing variables cannot be explained properly, the mechanics of the EOR process will not be fully understood. Since the progress of any displacement process in porous media is governed by the related variables, it is a challenge to describe the continuous changes of different rock and fluid properties in terms of the existing heat transfer numbers. To solve this difficulty, Obembe et al. (2017a,b,c) have recently proposed five dimensionless heat transfer numbers. Results show that these numbers can explain the above complex phenomena. The proposed numbers are profound to most of the formation properties such as heat capacities, densities, porosity, and permeability. They concluded that dimensionless numbers are helpful to describe the rheological behavior of the fluid and rock system. In addition, it improves the understanding of the effect of heat transfer on changes of effective permeability during thermal recovery operations in a hydrocarbon reservoir. However, none of the above authors did verify these proposed numbers by established techniques such as

TABLE 7.3 Dimensionless groups derived from dimensional analysis (Hossain, 2018).

$\pi_1 = \dfrac{A\rho_o c_f}{T^2}$	$\pi_{10} = \dfrac{k}{A}$	$\pi_{19} = \dfrac{Tt^2 M}{A\rho_o}$	$\pi_{29} = \dfrac{tu_w}{\sqrt{A}}$	$\pi_{39} = \dfrac{t\mu}{A\rho_o}$
$\pi_2 = \dfrac{A\rho_o c_s}{T^2}$	$\pi_{11} = \dfrac{Tt^3 k_t}{A^2 \rho_o}$	$\pi_{20} = \dfrac{t^2 p}{A\rho_o}$	$\pi_{30} = \dfrac{x}{\sqrt{A}}$	$\pi_{40} = \dfrac{\rho_f}{\rho_o}$
$\pi_3 = \dfrac{A\rho_o c_s}{T^2}$	$\pi_{12} = \dfrac{Tt^3 k_f}{A^2 \rho_o}$	$\pi_{21} = \dfrac{t^2 p_i}{A\rho_o}$	$\pi_{31} = \dfrac{t\alpha_{Tf}}{A}$	$\pi_{41} = \dfrac{\rho_g}{\rho_o}$
$\pi_4 = \dfrac{Tt^2 c_{pf}}{A}$	$\pi_{13} = \dfrac{Tt^3 k_g}{A^2 \rho_o}$	$\pi_{22} = \dfrac{tq_i}{A^{3/2}}$ $\pi_{23} = \dfrac{r_{pt}}{\sqrt{A}}$	$\pi_{32} = \dfrac{t\alpha_H}{A}$	$\pi_{42} = \dfrac{\rho_s}{\rho_o}$
$\pi_5 = \dfrac{Tt^2 c_{pg}}{A}$	$\pi_{14} = \dfrac{Tt^3 k_o}{A^2 \rho_o}$	$\pi_{24} = \dfrac{\xi}{t}$	$\pi_{33} = \dfrac{t\alpha_{Ts}}{A}$	$\pi_{43} = \dfrac{\rho_w}{\rho_o}$
$\pi_6 = \dfrac{Tt^2 c_{po}}{A}$	$\pi_{15} = \dfrac{Tt^3 k_s}{A^2 \rho_o}$	$\pi_{25} = \dfrac{T_f}{T}$	$\pi_{34} = \dfrac{t\alpha_{Tb}}{A}$	$\pi_{44} = \dfrac{tq_{prod}}{A^{3/2}}$
$\pi_7 = \dfrac{Tt^2 c_{pw}}{A}$	$\pi_{16} = \dfrac{Tt^3 k_w}{A^2 \rho_o}$	$\pi_{26} = \dfrac{T_i}{T}$	$\pi_{35} = \dfrac{t\alpha_{TB}}{A}$	$\pi_{45} = s_g$
$\pi_8 = \dfrac{Tt^2 c_{ps}}{A}$	$\pi_{17} = \dfrac{L}{\sqrt{A}}$	$\pi_{27} = \dfrac{T_s}{T}$	$\pi_{36} = \dfrac{t\alpha_{Tc}}{A}$	$\pi_{46} = s_o$
$\pi_9 = \dfrac{Tt^3 h_c}{A^{3/2} \rho_o}$	$\pi_{18} = \dfrac{L_c}{\sqrt{A}}$	$\pi_{28} = \dfrac{T_{HW}}{T}$	$\pi_{37} = \dfrac{to}{A}$	$\pi_{47} = s_w$
			$\pi_{38} = \dfrac{\rho_o \eta}{t^{1+\alpha}}$	$\pi_{48} = \phi$

TABLE 7.4 Dimensionless groups arising from inspectional analysis (Hossain, 2018).

$Sg_1 = \dfrac{k_s}{k_c}$	$Sg_5 = \dfrac{h_c L_c}{k_c}$	$Sg_{10} = \dfrac{T_f}{T_i}$	$Sg_{14} = \dfrac{kt}{\phi \mu c_t L^2}$
$Sg_2 = \dfrac{\mu c_{pf}}{k_c}$	$Sg_6 = \dfrac{L}{L_c}$	$Sg_{11} = \dfrac{x}{L}$	$Sg_{15} = \dfrac{k_s + k_f}{k_c}$
$Sg_3 = \dfrac{k\rho_f}{\mu^2 c_t}$	$Sg_7 = \dfrac{L_c \rho_f c_{pf} u_m}{k_c}$	$Sg_{12} = \dfrac{p}{p_i}$	$Sg_{16} = \dfrac{M\alpha_H}{k_c}$
$Sg_4 = \dfrac{(1-\phi)}{\phi} \dfrac{\rho_s c_{ps}}{\rho_f c_{pf}}$	$Sg_8 = \dfrac{T}{T_i}$	$Sg_{13} = \dfrac{q}{q_i}$	$Sg_{17} = \dfrac{v}{\alpha_H}$
	$Sg_9 = \dfrac{T_s}{T_i}$		

inspectional and dimensional analysis. Hossain (2018) addressed the issue and confirmed the viability of the proposed numbers by maintaining the accuracy of the model equations.

No literature or models are available dealing with the continuous alteration phenomenon based on the displacement process with memory other than Obembe et al. (2017a,b,c) and Hossain and Abu-Khamsin (2012a,b). They studied the thermal recovery process using the memory concept for developing dimensionless numbers. They derived five dimensionless groups to explain the heat transfer process (i.e., conduction and convection) when continuous alteration of rock and fluid properties are considered. They explained the significance of those proposed numbers in terms of their influence on permeability, porosity, viscosity, thermal conductivities, heat capacities, etc.

Hossain's (2018) study was aimed at (i) studying the important variables and transform those into dimensionless groups by developing model equation with memory for the displacement processes, (ii) developing links between the numbers proposed by Hossain and Abu-Khamsin and conventional dimensionless numbers such as Nusselt, Peclet, and Prandtl, and finally, (iii) an extensive study is offered on the newly developed dimensionless groups and other established numbers.

7.2.3 Conventional development of various scaling criteria

This review is based on various scaling criteria of fluid flow through porous media for fluid displacement processes within petroleum reservoirs. Scaling groups are very important in describing the influence of parameters on a specific EOR process. Accurate formulation and evaluation of dimensionless scaling groups are important because it can largely affect the physical process. Unscaled physical processes can give erroneous results. Table 7.5 lists the most widely used scaling groups relevant to EOR.

7.2.3.1 Capillary number

The ratio of viscous force to capillary force is termed as capillary number (Fulcher Jr. et al., 1985). Different forms of capillary numbers have been used in the existing literature (Cense and Berg, 2009). Foster (1973) defined the capillary number using the Darcy velocity of displacing fluid, the viscosity, porosity, and the interfacial tension. Sheng (2010) omitted the porosity term and Lake (1989) included the contact angle term. The derivation of capillary

TABLE 7.5 Various scaling dimensionless numbers (Novakovic, 2002).

Dimensionless scaling type	Scaling Group	Formulation	Comment
Physical effects of flow and fluid properties scaling	Capillary number	$N_C = \dfrac{F_{capillary}}{F_{viscous}}$	Rock-fluid interaction describes set-up at the small scale
	Gravity number	$N_g = \dfrac{F_{gravity}}{F_{viscous}}$	Reservoir-fluid shape-dependent, seizures the effect of resistant force
	Mobility ratio	$M = \dfrac{\lambda_{displacedfluid}}{\lambda_{displacingfluid}}$	Fluid-rock-fluid communication effect on the flow performance
Displacement techniques with initial and boundary conditions scaling	Displacement efficiency factor	$E_D = \dfrac{V_{Produced}}{V_{Reference}}$	Dimensionless production response
	Dimensionless Time	$t_D = \dfrac{V_{injected}}{V_{pore}}$	Forced injection boundary condition
Reservoir geometry scaling	Aspect ratio	$N_A = \dfrac{L}{H}$	Reservoir shape description scale
	Dip angle	$N_\alpha = \tan \alpha$	Dip angle scaling

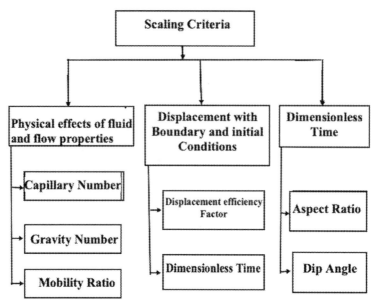

FIG. 7.2   Different scaling groups.

numbers can be found in the literature (Johannesen and Graue, 2007). The capillary number provides satisfactory correlations of mobility of oil with respect to different values of viscosities (Morrow, 1979). The recovery factor is found to be dependent on the capillary factor (Fulcher Jr. et al., 1985) (Fig. 7.2).

7.2.3.2 Bond number

The ratio of the gravity force to the capillary force is called the Bond number. It is of great importance in vertical displacement processes in a reservoir-well system. The Bond number is usually useful for gravity-assisted displacement processes. For the immiscible displacement process, oil recovery improves with increasing Bond number.

7.2.3.3 Gravity number

Gravity number which characterizes the ratio of gravity to viscous force is a dimensionless group. It does not detect the properties of the capillary forces. Gravity number depends on gravity, oil and gas density and viscosity, absolute permeability, and the gravity drainage velocity of the fluid. The gravity number indicates that the gravity effects are larger in thicker reservoirs and the higher the gravity number the better would be the recovery.

7.2.3.4 Mobility ratio

The ratio of effective permeability to phase viscosity of a fluid is expressed as the mobility ratio. The proportion of the mobility of displayed fluid by displacing fluid is termed as the mobility ratio of one fluid by another in terms of the displacement process. The fundamental mechanism behind the displacement of oil by water can be grasped through studying the mobilities of the individual fluids (Dake, 1978).

7.2.3.5 Displacement efficiency

The portion of movable oil that can be extracted from the reservoir using existing technology at any time is defined as displacement efficiency. The microscopic displacement efficiency is dependent on the mobilization or dislocation of oil on a small scale. It can be a criterion for the preliminary oil saturation or remaining oil saturation in the area contacted by the moving fluid. On the other hand, macroscopic or volumetric displacement efficiency depends on the efficiency of the moving fluid in contact with the reservoir in a volumetric sense. The macroscopic displacement efficiency of a fluid can be measured in a way by which the displacing fluid is striking the reservoir volume both areally and vertically.

7.2.3.6 Dimensionless time

The scale-up time of a given prototype field is expressed by the dimensionless time. The expression of dimensionless time (t_d) can be found in the literature (Miguel-H et al., 2004), for the gravity drainage methods and is stated as:

$$t_d = \frac{kk_{ro}\Delta\rho g\left(\frac{K}{\phi}\right)/g_C}{h\phi\rho\mu_o(1-S_{or}-S_{wi})}t \tag{7.1}$$

Eq. (7.1) enables estimation of the time required in the reservoir to reach the same recovery as the scale-up of the run time (in minutes) in the physical model.

7.2.3.7 Aspect ratio

The proportion of a geometric shape magnitude in diverse dimensions is called the aspect ratio of that shape. For illustration, the proportion of width to height or the ratio of the longer side to the shorter side is defined as the aspect ratio of a rectangle. The aspect ratio illustrates the proportion of time for a fluid to flow in vertical and horizontal axes of the reservoir when the equal pressure difference is employed. The determination of the flow regime in the advanced part of the investigation is performed with the help of this explanation. It is important to express the vertical scale along with the horizontal scale on a stratigraphic cross-section to indicate significant details of stratigraphic variation or dip angle of a reservoir. It is imperative to understand the effect that this distortion has on reservoir area or geometry and angular relationships of formation surfaces. The small angular differences among stratigraphic horizons that can consider for thickness variations are strongly exaggerated in such a section.

7.2.4 Scaling classification

Based on the principles on which the scaling criteria are developed, they can be categorized as (a) scaling criteria based on flow and fluid properties, (b) scaling criteria based on displacement techniques, and (c) scaling criteria based on reservoir geometry. This review article will focus on these three types of scaling principles.

7.2.4.1 Flow and fluid properties

This type of scaling principle should be established depending on the fluid/reservoir interaction. It can distinguish wide-ranging flow performance which includes an equilibrium of capillary, gravity, and viscous forces. The flow properties should also be considered for these forces which present the consequence of extrapolation from the initial scale to the scale of concern.

The use of miscible and immiscible multiphase flow scaling has been investigated previously for different EOR techniques. Leverett et al. (1942) studied the dimensionless scaling numbers for immiscible water-induced oil displacement processes. The effect of the water/oil viscosity ratio on immiscible displacement was studied by Croes and Schwarz (1955). The results found for linear displacement of oil by water are presented in the form of a diagram for similar formations. The effect of gravity separation of five-spot models for miscible and immiscible displacement was presented by Craig et al. (1957). It is difficult to build bridges between theoretical multiphase flow behavior and field applications for a hydrocarbon reservoir without simplifying assumptions which result in questionable conclusions (Rapoport, 1955). Rapoport developed the scaling laws for the water–oil displacement process for an incompressible, immiscible, and two-phase flow system. Scaling groups including capillary and gravity numbers have been derived by inspectional analysis. Difficulties have been raised, when reproducing identical capillary pressure and similar relative permeability curves. The different types of model tests performed with different materials can study the water flooding process for a broad range of reservoir settings. In turn, for a specific reservoir, its behavior could be evaluated by interpolating its characteristics into the ranges covered by the scaled model studies. In contrast, Grattoni et al. (2001) described a succession of trials regarding the impacts of water saturation and wettability on multiphase flow. They exceptionally considered the gravity-dominated environments of gas injection. The trials were conducted by instinctive gas injection and dispersion of oil in bead-pack models at very high and low water saturations. Different recovery rates of oil had been found. The procedure seemed to be less effective at irreducible water saturation for the case of the oil-wet condition. Similar recoveries were monitored at residual oil saturation in both cases of water and oil-wet condition, respectively. The authors found a straight-line connection between the derived dimensionless group and all the analyzed conditions of overall recovery. Suzuki and Hewett (2002) demonstrated an innovative technique to scale up the multi-phase flow properties. It ultimately represented the proper boundary conditions in the upscale section. They depicted a technique to scale up an entire finely-gridded model and decide the boundary conditions using injection tubes for two-phase flows. This novel technique can correctly capture the fine-scale two-phase flow behavior, such as saturation distributions, inside each segregated coarse-grid domain. They presented that this method can be pertinent to both viscosity dominated and gravity affected flows for reasonable gravity to viscous ratios. Later, Azoug and Tiab (2004) developed a comprehensive approach using the pseudo function for upscaling three-dimensional anisotropic heterogeneous reservoirs. It was considered for multiphase flow with different capillary and gravity numbers. They compared the performance of several -pseudo-function techniques by considering diverse flow regimes. These are represented by different types of homogeneous small grid models. The researchers became successful to reproduce the oil production level and water cut of the fine grid for equilibrium, viscous dominated, and capillary controlled flows. On the other hand, the authors were unable to match the curves of the fine grid using -pseudo-function techniques for the gravity-controlled flow. Finally, they became successful in upscaling small to large grid simulation for high flow rates using pseudo functions.

Artus and Noetinger (2004) reviewed the main upscaling techniques to derive different capillary and gravity numbers for heterogeneous reservoirs in terms of two-phase flow. They investigated numerical fine to coarse grid methods. Additional physical methods were employed where the statistical arrangement of transport equations was emphasized. They showed a comprehensive logical and numerical study of the dynamic contrast of the oil–waterfront through the study of the viscous connection between the pressure and saturation. There was an extremely effective communication found between the steady or unsteady state character of the fluid flow displacement and heterogeneity of the reservoir. The connection of this type is reliant on a subjective and measurable alteration of the extensive scale conditions. It must be represented by an upscaling procedure. Later Zhang et al. (2005c) demonstrated how conventional upscaling methods may deliver erroneous results and suggested a simple alternative. They reproduced single-phase flow and greatly increased the coarse-scale two-phase flow model using suitable boundary conditions. This method is slower than the local upscaling method and cannot consider the physical disruption caused by heterogeneity in the fine-scaled model. It is not suitable for small-scale heterogeneities where capillary pressure has a significant impact on the fluid flow. Pfister and Chanson (2014) summarized the water-air interfacial properties and the air entrainment rate under a Froude similitude. It represents the physical background of a pore physical model. The smallest values of Weber or Reynolds numbers were considered to limit the scale effects. Based on a literature review, they presented and discussed the existing limit, bringing about a progression of more moderate recommendations in terms of air concentration scaling. As the selection of criteria to examine the scale effects was crucial, it was observed that a couple of factors (e.g., bubble sizes, turbulent scales, etc.) can be influenced by scale consequences, even in comparatively large laboratory models.

7.2.4.2 Capillary to viscous force

Hilfer and Øren (1996) reexamined the multiphase flow equations of small and large scale in porous media through the traditional dimensional analysis. Depending on the category of length scale, porous medium, and saturation history, a macroscopic capillary number was presented that differs from a microscopic capillary number. The macroscopic number could be associated with the Leverett J function. The microscopic number is the ratio of viscous pressure drop to capillary pressure. The sample calculations of desaturation curves are provided when the macroscopic number is equal or close to one for distinctive porous media. Finally, the analytical modification between residual oil saturation of laboratory experiments and field implementations was provided. On the other hand, Wibowo et al. (2004) studied the impact of the correlation of the forces in horizontal good production operation for bottom water drive reservoirs. They successfully constructed a scaled physical model. It can be simulated in the production operation using dimensional analysis and showed that the linking of the reservoir forces increased as the proportion of gravity to viscous forces increases. The significant finding of their work was the good production performance of the reservoir. It will enhance as the capillary pressure is decreasing, and subsequently, the increase of gravity to viscous force ratio will improve the oil recovery. Later, Jonoud and Jackson (2008) showed the capillary or viscous forces flow which validates the steady-state scaling techniques. They found that reservoir flow rates within a reasonable range were valid for viscous limit upscaling techniques. The capillary equilibrium limit technique was limited to exceptionally reduced rates because it overestimates the amount of capillary entrapping. However, the authenticity of capillary equilibrium limit upscaling in a 3D model was not properly captured.

7.2.4.3 *Fluid saturation and relative permeability*

Perkins and Collins (1960) redefined the relative permeabilities and fluid saturations. Their definition permits one to have diverse relative permeability and capillary pressure relationships in the prototype and model. This work proposed a method to authorize a diverse relationship between relative permeability and fluid saturation with capillary pressure. This relationship helps to derive the modified capillary number. They demonstrated one simple example that clarifies how to derive modified scaling criteria. Astarita (1997) discussed the modern viewpoint of dimensional analysis. It is the basis of the theory of scaling to derive gravity and capillary number. The author illustrated several specific examples to show how scaling and dimensional analysis may generate actual important points for the solution of the problem. Finally, the author showed that using scaling, dimensional analysis, and the estimation of the order of magnitude can be used to derive those dimensionless groups. Durlofsky (1998) developed a coarse-scale equation using a volume average saturation calculation of small scale in dissimilar reservoirs. It can be used for two-phase flows to evaluate several approaches for the detailed upscaling method for reservoir characterization. The author discussed the strengths and limitations of each of these techniques. Especially the fundamental assumptions in those calculations using the volume-averaged equations as a framework equations. These equations were rearranged for the unit mobility ratio case and applied to the immediate solution of a coarse-scale model issue. Wang et al. (2009) demonstrated the large error behind the conventional upscaling method. They established a novel approach for the upscaling method of the relative permeability curve. A large model upscaling method was used which best fits with the fine-scaled model. The authors verified this method by constructing a three-dimensional, three-phase, and extremely dissimilar reservoir model. As contrasted with the conventional method the new coarse scaled upscaling method demonstrated a more reasonable result. The outcome can be attained by approximating the consequence of ambiguity through computational time, order, and magnitudes quicker than the earlier methods. Tsakiroglou (2012) developed a model using a network-type multi-scale analysis for immiscible displacement of both wetting and non-wetting phase fluid in dissimilar porous media. The author utilized these methods to decide the transient consequences of the axial dispersion of water saturation, pressure drop. Finally, the functions of relative permeability can be evaluated with its upscaled impact.

None of the above works considers the possibility of having a different set of relative permeability in the field from the set in the laboratory. Islam and Bentsen (1986) introduced the so-called dynamic method to demonstrate that the relative permeability set in an unstable flow regime (with viscous fingering) would be different from stable or pseudostable flow regimes. They suggested that relative permeability measurements be conducted under similar stability conditions. This is contrary to conventional practices in which all tests are-conducted underflow velocity of 1 ft./day, which puts the flow regime under a strictly stable category.

7.2.4.4 Rock and fluid memory

Hossain and Islam (2011) developed new scaling criteria incorporating memory concepts using inspectional and dimensional analysis. They became successful to develop relationships between capillary pressure, saturation, velocities, and fluid pressure for prototype and model. The authors identified a competent tactic for the oil-water displacement process by deriving the sets of similitude groups. Hossain and Abu-Khamsin (2012a) developed new dimensionless groups using mathematical modeling of non-linear energy balance equations. The developed numbers were helpful to demonstrate the rheological behavior of fluid-rock interactions. Their proposed dimensionless numbers described the various types of heat transport mechanisms including convection and conduction in porous media for the processes of thermal recovery. These dimensionless numbers were found to be responsive to a large set of fluid and reservoir rock properties including densities, permeability, heat capacities, porosity, etc. Hossain and Abu-Khamsin (2012b) also developed new dimensionless numbers which can describe convective heat transfer between the fluid and rocks in continuously changing conditions using the memory concept. They employed an energy balance equation to develop the heat transfer coefficient by assuming the rock can attain the temperature of the fluid immediately. The developed new numbers correlate with the Nusselt and Prandtl numbers and the local Peclet number is observed to be responsive to memory.

7.2.4.5 Spontaneous imbibition

Mirzaei-Paiaman and Masihi (2013) developed scaling equations utilizing the countercurrent spontaneous imbibition method for oil and gas recovery from fractured porous media. Earlier scaling equations were defined systematically by linking the primary time squared recovery to squared pore volume. They showed that this settlement does not employ general scaling performances and, if employed, it affects nontrivial sprinkle in the scaling designs. The authors proposed that throughout the expansion of any scaling equations, its reliability with mutual purposes should be a measure that was neglected in the literature. The authors have rewritten scaling equations for two physically expressive numbers, namely, the Darcy number and the Capillary number. It was authenticated by the investigation data from the literature. The author's—efficiently scale scale-up available data and represented different recovery curves by a single master curve.

7.2.5 Compositional flow simulation

Li and Durlofsky (2015) developed an upscaling procedure that is more precise and robust for the simulation of flow composition. They computed the functions related to coarse-scale boundary or block and the prerequisite upscaled factors by using a technique. This technique requires a global fine-scale compositional simulation. The authors introduced near-well behaviors along with a technique for enhancing the α-factors for both production and injection wells. It was combined further to upgrade the coarse-model appropriateness. Finally, they suggested that using their technique the produced upscaled models can be employed to lessen computational difficulties for different purposes including the optimization of good control. The network models fall under this same category.

7.2.6 Immiscible displacements

Rojas (1985) performed scaled model studies for immiscible CO_2 flooding of substantial oil. Lozada and Ali (1987) displayed a group of scaling criteria including six groups of scaling processes. They concluded that a full set of scaling criteria might not be fulfilled at the same time. Thus, few groups had to be excluded to fulfill the major scaling conditions, including the vital factors of a specific method. The authors found that the nature of fluid/rock schemes, flow rate, pressure drop, model geometry, and so on was dissimilar contingent upon the methods exercised. Later, Lozada and Ali (1988) also developed partial differential equations of immiscible carbon dioxide flooding for the moderately heavy oil reservoir. The authors used different sets of scaling criteria to construct scaled models with different operating conditions. A series of similitude numbers were derived for the displacements of moderately heavy oil recovery by dimensional and inspectional analyses. The mass transfer between the phases was considered for immiscible carbon dioxide flooding. So, all the similarity groups were not satisfied in the case of recovery from moderately heavy oil reservoirs. They relaxed some of the groups which had less effect on the physical mechanism and hence found out the dominant scaling groups. Peters et al. (1998) studied the saturation data through a dimensionless self-similitude parameter to develop the dimensionless representative response curve for variable core floods. The authors found that there is a considerable dissimilarity between the response function of oil-wet to water-wet reservoir. Finally, the results showed that the effectiveness of displacement could occur in water-wet reservoirs compared to oil-wet reservoirs. Zhou et al. (1997) defined three dimensionless groups, namely, gravity viscous ratio, shape factor, and viscous capillary ratio. These dimensionless numbers help to detect influential flow regions in numerous situations. They demonstrated the comparative extents of energies in the scheme linked through the reservoir properties. The scaling groups and flow areas governing different kinds of flow performance in the schemes were examined with straightforward heterogeneity formulae. The authors considered three frequently used flow schemes such as immiscible displacement with layered reservoir inhomogeneous media, miscible displacements in the layered reservoir without scattering, and fluid flow in the reservoir with high fracture.

Prosper and Ali (1991) presented a recovery mechanism comprising a two-dimensional and linear scaled model for the water-alternating-gas (WAG) and the low-pressure immiscible carbon dioxide flooding. They compared the results of the Aberfeldy field using the same model at the same pressure and WAG ratios. The authors found the oil recovery of the model involving linear analyses was about one-half at 2.5 MPa pressure. The bottom recovery involving waterflood was 40% and the incremental recovery of 10% was due to the WAG process. On the other hand, the recovery for the two-dimensional model varied from 40% to 50%. Bansal and Islam (1994) performed a study of sequential scaled model by injecting carbon dioxide, propane, and nitrogen gas in the reservoir. Gas injection is a principal method for the recovery of the heavy oil reservoir in Alaska. Nearly 65% of oil initially in place is recovered; the same is indicated by their experimental outcome. For gravity drainage, although the final recovery was the same, it took a long time to recover the same amount of oil. They found the recovery mechanism was different for different gases and the highest recovery was obtained with carbon dioxide. Viscous fingering takes place with different degrees of severity when applying different gas flooding techniques. It is considered harmless as the ultimate recovery is higher by gas injection.

7.2.7 Controlled gravity drainage

Zendehboudi et al. (2011) performed dimensional analysis for scaling the immiscible displacements of the controlled gravity drainage (CGD) method. The authors obtained an empirical model in fracture-dominated porous media by dimensional analysis using Buckingham π theorem to investigate the gravity drainage process. They developed a model to forecast the maximum withdrawal rate, the distance of fluid-gas interface locations, critical pumping rate, and the recovery factor of fluid experiencing the CGD methods. The developed model delivers satisfactory predictions for the oil–gas drainage system.

7.2.8 Immiscible GAGD process

Sharma and Rao (2008) developed a scaled physical model of the gas-assisted gravity drainage (GAGD) technique to describe the enhanced recovery method. They determined the impact of a few dimensionless scaled factors. For example, the Gravity number, Bond number, and Capillary number effect on GAGD technique implementation. Sharma and Rao (2008) found that the Bond number significantly affects GAGD performance than any other number. Finally, they relate the run time of the model to the run time of field development to observe high recoveries. The dimensionless time indicated an augmented rate of recovery when the GAGD method is implemented in field projects. Farahi et al. (2014) developed a few scaling groups by performing inspectional analysis. These groups had analyzed the performance of reservoir fluid displacements by immiscible GAGD technique. They determined five matched scaling groups for homogeneous reservoirs. The authors found a coefficient for the different reservoirs which is called the coefficient of Dykstra-Parson. They determined another new set of dimensionless groups on a large scale that added altogether the prevailing energies. Finally, they evaluated and verified experimental results and found them consistent for rapid forecast of oil recovery for the GAGD technique.

7.2.9 Miscible displacements

Gharbi et al. (1998) studied the miscible displacement scaling in permeable medium utilizing inspectional investigation to produce scaling sets. These sets influence the displacement method in a 2D, similar, different cross-sectional formation. They derived nine groups of dimensionless numbers and from which only one number was found to have no impact on this displacement technique. Babadagli (2008) determined dimensionless scaling groups for miscible displacement utilizing inspectional analysis in a fractured porous and permeable medium. They proposed a new dimensionless number based on the dimensionless group they derived for better characterizing the efficiency of the method. The proposed new group which is called Matrix-Fracture Diffusion Number (N_{M-FD}) was significant in assessing the efficiency of CO_2 sequestration, enhanced oil recovery, and pollutant transportation issues. The authors performed validated laboratory-scale experiments and physically interpreted the Matrix-Fracture Diffusion Number (N_{M-FD}).

7.2.10 Steam flooding

Pujol and Boberg (1972) presented different approaches for scaling the investigation of stream flooding process in viscous oil reservoirs. The scaling of capillary pressure was not considered essential to represent highly viscous oils. On the other hand, for intermediate viscosity oil (<10,000 cP), unscaled capillary pressures can predict the optimistic recovery of oil. They developed a method to convert capillary pressure into the scale, and discovered that it can give qualitative enhancement as the recovery of oil is sensitive to flooding rates. The authors found oil recovery was mainly dependent on per unit volume of heat input to the formation. Kimber et al. (1988) developed novel dimensionless scaling numbers for the recovery of oil by steam or a steam improver and discussed their relative merits. They determined a group of similitude numbers that allow the utilization of similar fluid in prototype and model through inspectional and dimensional analyses. The authors also compared their approach with other approaches which were published in the literature and discussed their relative merits. They outlined a means of developing or selecting a process that best fits the most important characteristics of a specific recovery scheme. Doan et al. (1990) presented mathematical models to derive dimensionless scaling groups of flow inside the horizontal wellbore for performing laboratory investigations. They used variable diameters of a horizontal wellbore and skin factors to conduct the experiments. They carried out a series of steam injection experiments through a development well. Pressure behavior and temperature distribution were controlled to explain the recoveries of oil. They evaluated oil recovery performance for various types of experiments to determine the effectiveness of different horizontal wells and the impact of the perforated casing. Doan et al. (1997) performed steam flood tests utilizing a physical model of the Aberfeldy reservoir (Saskatchewan) to scale up and inspect the recovery of the steam flood technique for horizontal injection and production wells. They analyzed the results from two types of experiments: a base case run steam flooding of a homogeneous reservoir and a reservoir having a 20% net pay bottom water layer. They presented scaling up laboratory outcomes to predict the performance of a prototype. The diagnostic heat loss model demonstrated a 3.1% difference from experimental results. Scaled-up test information data for a base case run showed that approximately 20% of the oil initially in place was recovered after 0.8 PV of steam was added. For a reservoir having 20% net pay, the increase in the oil recovery depends on how the energy contained in the fluid is managed.

7.2.11 Hot fluid injection

Willman et al. (1961) assessed the outcomes of laboratory investigations for steam, cold water, and hot water injection. They studied different cell measurements with various permeabilities. The authors found cold water drive had less recovery than hot water and steam injection drive. Finally, they found the soaked steam with high temperature and pressure is more effective in terms of recovery than steam with low pressure. Moreover, all types of recovery have greatly improved if the temperature of the injected fluid is higher. Cheng and Cheng (2004) provided a fundamental idea of dimensional analysis scaling and reviewed the present research employing these ideas to model the quantities of instrumented indentation. They analyzed the indentation of pyramidal and conical shaping in various viscoelastic materials. They likewise indicated scaling approaches that was best fit for these processes and provide a superior understanding of instrumented indentation measurements. Heron et al. (2005) developed thermally improved remediation techniques that were favorable for the elimination of pollutants at intensely polluted places. They developed methods to incorporate invasion of hot air, high-temperature steam or water using thermal wells or heat blankets; electrical heating with a low frequency; microwave heating; etc.

7.2.12 Solvent/chemical injection

Geertsma et al. (1956) extended the scaling theory to hot water drive and solvent injection by utilizing dimensional and inspectional analyses. They assumed uniform porosity and isotropic permeability. Since not all the scaling groups can be considered in building a model, a comprehensive discussion on which scaling groups are negligible was provided. Nonetheless, experimental studies were performed to verify the feasibility of neglecting some scaling groups. Sundaram and Islam (1994) presented a scaled physical model of petroleum pollutant removal using solutions of surfactants. They developed scaling principles for the decontamination process where viscous forces, aquifer geometry, and the proportion of the viscous to gravitational forces were used. Experiments were conducted to examine the type and concentration of surfactants and injection/production strategies. They found optimum surfactant concentration needed for the removal of a specific contaminant with surface tension. The outcomes of experiments showed that using this decontamination technique more than 90% of the contaminant originally in place may be removed. Basu and Islam (2009) performed a sequence of chemical adsorption experiments to provide the most influential scale-up form. The authors contrasted their outcomes with numerical simulation results. The numerical solutions were offered based on flow rates of the fluid, pore velocity, the amount of adsorbent used, and the adsorption coefficient which were related to field environments. Finally, they developed a guideline for interpreting the investigational outcomes and applied the scaling laws to forecast the field performance. Veedu et al. (2010) presented an upscaling methodology for chemical flooding by comparing results between coarse and fine grid methods. Their technique was quite dissimilar to the other upscaling methods used for the EOR process. They showed that for a heterogeneous reservoir the salinity gradient was not effectively picked up by the coarse grid method. It can lead to slower recovery than the simulations of the fine grid method. Finally, they recommended to use fine grid upscaling for better performance prediction of chemical flooding.

7.2.13 Polymer flooding

Islam and Ali (1989) obtained new dimensionless scaling groups which can incorporate the flow of foams, emulsions, and polymers. They focused on the significance of mass transfer among phases, fractional flow, diffusion, adsorption, trapping, slug size, and interfacial tension. New groups of scaling conditions were derived for co-surfactant improved polymer flooding with a mathematical explanation. The relative permeability and interfacial tension model were also obtained by Islam and Ali (1990). Bai et al. (2008) developed a group of scaling principles by taking into consideration many factors for polymer flooding in the reservoirs. They evaluated the sensitivity analysis of each of the dimensionless numbers. A numerical approach was recommended to enumerate the sensitivity analysis of every dimensionless number. The researcher analyzed the influence of specific physical parameters, such as injection rate, oil viscosity, and permeability, on the predominant level of the dimensionless numbers. Finally, they determined the leading ones for distinctive circumstances. Guo et al. (2012) identified the dimensionless leading scaling groups in heavy oil reservoirs for polymer flooding. They derived 28 dimensionless scaling numbers and build up a mathematical model to authenticate the efficacy of these scaling numbers. The authors performed a numerical sensitivity analysis of individual scaling numbers to find out their consequences on the recovery of oil. They identified nine dominant scaling numbers which were used to design field scale oil recovery experiments.

7.2.14 Micellar flooding

Thomas et al. (1997) discussed the design of micellar flooding experiments using scaling laws. They derived scaling criteria utilizing dimensional and inspectional analyses with six elements for three-phase flow. These criteria were derived in several ways. The partial differential equations, constitutive relations, and initial and boundary conditions are used to form a mathematical model. Finally, the mathematical model was simplified and a group of scaling principles was derived which applied to most laboratory conditions.

7.2.15 In situ combustion

Garon et al. (1982) studied the three-dimensional physical models of tar sand fire flood reservoirs following a pre-heating to explore the reservoir heterogeneity. They used three types of heterogeneity, including communicating and non-communicating bottom water zones and a thin, simulating a fracture heated layer. They chose asymmetrical element pattern of overburden and under burden. It had the same thermal diffusivities as the field was used for the model. They employed actual field crude because its properties affect important features of fire flooding. They increased the characteristic flux in the model in direct proportion to scale for both diffusion and convection of heat and mass transfer. Islam and Ali (1992) provided valuable rules to construct a suitable scaling principles for in situ combustion investigations. They used partial differential equations and imposed initial and boundary conditions to derive a set of scaling criteria. Fire tube tests were employed to investigate the authenticity of the resulting scaling criteria. Their results showed that among the developed scaling groups only a few groups had experimental validation. On the other hand, the outcomes of the research test site fire tubes of wet combustion showed that the measured parameters can mislead the experiments. Kandlikar (2010) developed a local parameter model using scaling analysis of critical heat flux (CHF) in microchannels and insignificant width tubes to estimate the secure working boundaries of refrigeration schemes using flow boiling. The author found a new non-dimensional group K2 with Weber number and capillary number. It represents the proportion of vaporization motion to surface tension forces and rising as the principal sets in enumerating the thin conduit consequences on CHF. The coefficients in the model had found by calculating available experimental data.

Finally, the author evaluated each dataset for individual sets of constants. The outcome showed average inaccuracies of fewer than 10 out of a 100 for entire information groups.

7.2.16 Small scale capillary number

Firstly, the small-scale capillary number was derived by Dombrowski and Brownell (1954) for synthetic media. The modeling of pore-scale is the primary and the smallest scale to consider for the derivation of two-phase flow dimensionless numbers (Moore and Slobod, 1956). The set of connections of wetting and non-wetting phase and the purpose of remaining saturation and scaling numbers that influence these numbers are the basic issue for pore-scale modeling. On the other hand, the medium resolution scale is the second type of scale at which point the subsequent production and flood front performance is detected (Hagoort, 1980). The numerical models of medium-scale and large scale deals with many factors including flow property or barrier distribution (Peters et al., 1993; Willis and White, 2000), geometry, and the parameters which affect the production. The authors with their corresponding scaling numbers are presented in Table 7.6.

7.2.17 Large scale capillary number

Rapoport and Leas (1953) formed the flow regime guide during a large-scale waterflood for scaling the capillary effects. Geertsma et al. (1956) consider the growth of large-scale numbers for both thermal and waterfloods as identical as pore-scale ones. Perkins and Collins (1960), derived gravitational segregation capillary number analogous to a dimensionless number derived by Craig et al. (1957). Shook et al. (1992) developed a scaling number identical to van Daalen and van Domselaar (1972) omitting the conventional capillary number. The authors with their corresponding scaling numbers are presented in Table 7.7.

TABLE 7.6 Small/core/pore-size scale capillary number (N_c).

Reference	Formulation	Comments
Dombrowski and Brownell (1954)	$N_c = \dfrac{k \cdot |\overline{\nabla}\,[Fcy]|}{\sigma \cos\theta}$	Synthetic media, distilled water-pure organics system
Moore and Slobod (1956)	$N_c = \dfrac{v \cdot \mu_1}{\sigma \cos\theta}$	Outcrop sandstone, brine-crude System
Taber (1969)	$N_c = \dfrac{v \cdot \mu_1}{\sigma \cos\theta}$	Berea sandstone, brine-control system
Foster (1973)	$N_c = \dfrac{u \cdot \mu_1}{\sigma}$	Berea sandstone, brine-oil system
Lefebvre du Prey (1973)	$N_c = \dfrac{u \cdot \mu_1}{\sigma}$	Synthetic media, water pure hydrocarbons system
Ehrlich et al. (1974)	$N_c = \dfrac{u \cdot \mu_1}{\sigma}$	Outcrop sandstone, brine crude system
Abrams (1975)	$N_c = \dfrac{v \cdot \mu_1}{\sigma \cdot \Delta S} \cdot \cos\theta \left(\dfrac{\mu_1}{\mu_2}\right)^{0.4}$	Outcrop sandstone, brine crude system

TABLE 7.7 Medium (inter-well)/large (reservoir) scale capillary number.

References	Formulation	Comments
Rapoport and Leas (1953)	$N_{RL} = \sqrt{\dfrac{\phi}{k}} \cdot \dfrac{\mu_1 \cdot u \cdot L_1}{k_{r1}^o \cdot \phi \cdot \sigma_{12} \cdot \cos\theta}$	Capillary dominated regime indicator
Geertsma et al. (1956)	$N_c = \dfrac{\sigma_{12} \cdot \cos\theta \sqrt{k \cdot \phi}}{u \cdot \mu_1 \cdot L}$	Identical to pore-scale N_C
Craig et al. (1957)	$R_c = \dfrac{\mu_1 q_i L}{\sigma_{12} \cdot \cos\theta \sqrt{k_x}}$	Dimensionless scaling number
Perkins and Collins (1960)	$S_c = \dfrac{k_{r1}^o \cdot \sigma}{u \cdot \mu_1 \cdot L_1} \sqrt{\left(\dfrac{\phi}{k}\right)}$	N_c corresponding similarity group
van Daalen and van Domselaar (1972)	$N_{pc} = \dfrac{\lambda_{r2}^o \cdot \sigma}{L \cdot u_T} \sqrt{\phi \cdot k_x}$	Capillary scaling number
Shook et al. (1992)	$N_{pc} = \dfrac{\lambda_{r2}^o \cdot \sigma}{L \cdot u_T} \sqrt{\phi \cdot k_x}$	Scaling dimensionless number of oil-water systems

7.2.18 Gravity number

Gravity number for granular material was first derived by Engelberts and Klinkenberg (1951) for density variation of the system. The two-phase flow gravity number was developed by Rapoport (1955) for the case of petroleum reservoirs. Two different types of gravity numbers were considered by Geertsma et al. (1956) for unconsolidated sand. Gravity numbers that were surveyed in literature also differed from one source to another. Although, the reasonable selection to be considered for gravity number is considerable distinction of density (Craig et al., 1957; Hagoort, 1980), and comprehensive absconding in the structure of the two-liquid scheme. Many researchers (Pozzi and Blackwell, 1963; Peters et al., 1998) have been concerned about the improvement of gravity numbers in a two-liquid scheme. On the other hand, Carpenter et al. (1962) derived a gravity number that was not dimensionless. Using the WAG process Stone (1982) developed the dimensionless group which was different from Wellington and Vinegar (1985) carbon dioxide flooding process. Newley (1989) derived gravity numbers for solvent flooding and Sorbie et al. (1990) developed the number for miscible flooding process. Shook et al. (1992) have proved the consequence of geometric aspect ratio and dip angle that significantly affect the gravity number. The authors with their corresponding scaling numbers are presented in Table 7.8.

7.2.19 Dimensionless scaling groups for GAGD

The most applicable combination of dimensionless scaling groups for the gravity drainage oil recovery process are presented by Edwards et al. (1998), Grattoni et al. (2001), Kulkarni (2005), and Rostami et al. (2005). Grattoni et al. (2001) represented the scaled model as the combination of capillary and Bond number which excluded gravity number. This limitation was eliminated by Kulkarni (2005) with the inclusion of gravity number term and thereby factoring the density ratio in the combination model. Rostami et al. (2005) presented a scaled model with the combination of capillary and Bond number along with the inclusion of viscosity ratio term, but they neglect the gravity number term.

TABLE 7.8 Gravity number (N_g) for petroleum literature.

References	Scaling groups and formulation	Comments
Engelberts and Klinkenberg (1951) and Crocs and Schwartz (1955)	$N_g = \dfrac{\Delta\rho \cdot k_x \cdot \lambda_{T1}}{u_T}$	
Rapoport (1955)	$N_g = \dfrac{\Delta\rho \cdot k_x \cdot \lambda_{T1}}{u_T}$	Two-phase flow
Geertsma et al. (1956)	$N_{g1} = \dfrac{\rho_1 \cdot g \cdot k_x \cdot \lambda_{T1}}{u_T}$ $N_{g2} = \dfrac{\rho_1}{\rho_2}$	
Craig et al. (1957) and Spivak (1974)	$N_g = \dfrac{u_T}{\Delta\rho \cdot g \cdot \sqrt{k_x \cdot k_z} \cdot \lambda_{T2}}$	Zero dip
Perkins and Collins (1960)	$N_g = \dfrac{\Delta\rho \cdot g \cdot k_x \cdot \lambda_{T1}^o}{u_T} \cdot \dfrac{H}{L}$	
Carpenter et al. (1962)	$N_g = \dfrac{q}{\Delta\rho \cdot k_x \cdot \lambda_{T1} \cdot L^2}$	Not dimensionless
Pozzi and Blackwell (1963)	$N_g = \dfrac{u_T}{\Delta\rho \cdot k_x \cdot \lambda_{T2}} \cdot \dfrac{L}{H}$	
Greenkorn (1964)	$N_{g1} = \dfrac{\rho_2 \cdot g \cdot k_x \cdot \lambda_{T2}}{u_T}$ $N_{g2} = \dfrac{\rho_2}{\rho_s}$	Unconsolidated sand
Stone (1982)	$N_g = \dfrac{u_T}{\Delta\rho \cdot g \cdot k_z \cdot (\lambda_1 + \lambda_3)} \cdot \dfrac{L}{H}$	WAG process (injected gas is phase 3)
Wellington and Vinegar (1985)	$\dfrac{\Delta\rho \cdot g \cdot k_z \cdot \lambda_{T1}^o}{u_T} \cdot \dfrac{L}{H}$	CO_2 injection
Newley (1989)	$N_g = \dfrac{u_T}{\Delta\rho g k_x \lambda_{se}} \cdot \sqrt{\dfrac{L}{H}}$	Derived for zero dips and solvent flooding
Lake (1989)	$N_g = \dfrac{\Delta\rho \cdot g \cdot k_x \cdot \lambda_{f2}^o}{u_T}$	Derived using one-dimensional fractional flow theory
Sorbie et al. (1990)	$N_g = \dfrac{\mu \cdot u_T}{\Delta\rho \cdot g \cdot k_x} \cdot \dfrac{L}{H}$	
Vortsos (1991)	$N_g = \dfrac{H k_x \Delta\rho g}{L u_T \mu_2}$	
Shook et al. (1992)	$N_g = \dfrac{\Delta\rho \cdot g \cdot k_x \cdot \lambda_{f2}^o \cdot \cos\alpha}{u_T} \cdot \dfrac{H}{L}$	Buoyancy number
Shook et al. (1992)	$N_g = \dfrac{\Delta\rho g \left(\dfrac{K}{\phi}\right)}{\mu_o v_d}$	Gravity forces viscous forces

7.2.20 Instability number

In 1985, Bentsen used force potential perturbation analysis to introduce a new instability number that lumps together the role of mobility ratio, frontal velocity, size of the displacement, gravity number, and others. For the first time, a single dimensionless number was devised that would dicate the flow regime, namely stable, pseudo stable and unstable. Following is the expression developed by Bentsen (1985).

$$I_{sr} = \frac{\mu_w v(M - 1 - N_g)}{k_{wr}\sigma_e} \times \frac{M^{5/3} + 1}{(M+1)(M^{1/3}+1)^2} \frac{4L_x^2 L_y^2}{L_x^2 + L_y^2} \quad (7.1)$$

where N_g is the gravity number defined as:

$$N_g = \frac{\Delta \rho g k_{or} \cos \alpha}{\mu_o v} \quad (7.2)$$

Note that for a vertical injection N_g assumes the largest value possible. In case N_g is larger than the expression $(M - 1)$ the displacement is unconditionally stable. This one shows the value of a gravity-stabilized displacement process that occurs when gas is injected from the top or water is injected from the bottom. This is the case for both miscible and immiscible displacement processes.

In the above expression, σ_e is the pseudointerfacial tension, which is:

$$\sigma_e = \frac{C_1 \sigma \cos\theta}{\phi} = \frac{2-v}{1-v}\bar{d}A_c. \quad (7.3)$$

A_c is the area under the capillary pressure curve and A_c is a parameter dictated by the curvature of the capillary pressure.

For I_{sr} less than π^2, the flow regime is stable, while it is unstable until the pseudostable state is reached. It turns out that the instability number is much larger than the threshold value unless the expression $(M - 1 - N_g)$ is negative. This analysis highlights the need to conduct relative permeability tests under similar stability conditions (Islam and Bentsen, 1986).

Coskuner and Bentsen (1990) reported a series of instability numbers that deal with miscible displacement. They used linear perturbation in order to obtain the scaling group. The new scaling group differed from those obtained in previous studies because it had taken into account a variable unperturbed concentration profile, both transverse dimensions of the porous medium, and both the longitudinal and the transverse dispersion coefficient.

It has been shown that stability criteria derived in the literature are special cases of the general condition given here. The stability criterion is verified by comparing it with miscible displacement experiments carried out in a Hele-Shaw cell. Moreover, a comparison of the theory with some porous medium experiments from the literature also supports the validity of the theory. The stability criterion is given below.

$$\frac{U\frac{d\mu}{dC} - kg\frac{d\rho}{dC}\sin\gamma}{\bar{\mu}D} \frac{\partial \bar{C} L^2}{\partial x\,\Omega} \cdot \left[\left(\frac{1}{\Omega} + \frac{D'}{D}\right)\left(\frac{1}{\Omega} + 1\right)\right]^{-1} > \pi^2. \quad (7.4)$$

7.2 Scaling guidelines

TABLE 7.9 Dimensionless scaling groups for GAGD EOR process.

References	Scaling groups and formulation	Comment
Edwards et al. (1998)	$N_B = \dfrac{\Delta \rho g l^2}{\sigma}$ and $\dfrac{\Delta \rho g l^2}{\sigma \sqrt{\left(\dfrac{\phi}{k}\right)}}$	Gravity to capillary number
Grattoni et al. (2001)	$N_G = \dfrac{\Delta \rho \cdot g \cdot k}{\Delta \mu \cdot u}$	Gravity to viscous force
Grattoni et al. (2001)	$N_C = \dfrac{v\mu}{\sigma}$ and $\dfrac{v\mu}{P_C R_A} 2\cos\theta$	Capillary forces to viscous forces
Kulkarni (2005)	$N_k = N_G + \left(\dfrac{\rho_G}{\rho_0}(N_C + N_B)\right)$	Improved characterization
Rostami et al. (2005)	$N_{rostami} = \dfrac{N_B(\mu_r)^A}{(N_C)^B}$	Forced gravity drainage

with $\Omega = \dfrac{L^2(B^2 + H^2)}{B^2 H^2}$ (for a two-dimensional system in which $H = 0$, $\Omega = \dfrac{L^2}{B^2}$) and where U, displacement velocity; μ, viscosity of mixture; ρ, density of mixture; C, injectant concentration; k, permeability; g, gravitational acceleration; γ, dip angle; ϕ, porosity.

7.2.21 Other scaling groups

Other dimensionless scaling groups are very important to describe the physical process which affects the model. Dimensionless time is one of the most important scaling groups which was first derived by Rapoport (1955). Mattax and Kyte (1962) were the pioneers who scaled capillary force imbibition under some specific conditions and proposed this number. In this scaling group, the different sauthor-defined viscosity and core length differently (Kazemi et al., 1992; Mattax and Kyte, 1962). Even though the authors applied distinctive equations to identify these factors, every single one of these equations utilized the squared representative length. Kantzas et al. (1988) and Blunt et al. (1995) described the fluid property group and their significance on displacement process. Miguel-H et al. (2004) developed the recent dimensionless time group which was used in different recovery processes. The authors along with their corresponding numbers are presented in Tables 7.9 and 7.10.

TABLE 7.10 Other scaling groups.

References	Scaling groups and formulation	Comments
Rapoport (1955)	$t_{DR} = \dfrac{k \dfrac{d}{ds_w}(p_c)}{u \mu_w L}$	Dimensionless time
Mattax and Kyte (1962)	$t_{DMK} = \dfrac{\sigma \sqrt{\dfrac{k}{\phi}}}{\mu_w L^2} t$	Dimensionless time

Continued

TABLE 7.10 Other scaling groups—cont'd

References	Scaling groups and formulation	Comments
Kazemi et al. (1992)	$t_{KGE} = \dfrac{\sigma \sqrt{\dfrac{k}{\phi}} F_{s,KGE}}{\mu_w} t$	Dimensionless time
Kantzas et al. (1988) and Blunt et al. (1995)	$\alpha = \dfrac{\rho_{ow}(\rho_o - \rho_g)}{\rho_{go}(\rho_w - \rho_o)}$	Fluid property group
Shook et al. (1992)	$R_L = \dfrac{L}{H}\sqrt{\dfrac{k_V}{k_H}}$	Dimensionless geometric group
Edwards et al. (1998)	$N_{DB} = \dfrac{\Delta \rho g k}{\sigma}$	Dombrowski Brownell number
Grattoni et al. (2001)	$N_B = \dfrac{\Delta \rho g \left(\dfrac{K}{\phi}\right)}{\sigma}$	Gravity forces viscous forces
Miguel-H et al. (2004)	$t_D = \dfrac{kk_{ro}\Delta \rho g \left(\dfrac{K}{\phi}\right)/g_C}{h\phi\rho\mu_o(1 - S_{or} - S_{wi})} t$	Dimensionless time

7.3 Planning with reservoir simulators

The simulation of petroleum reservoirs is an essential practice in the development of more efficient techniques to increase hydrocarbon recovery and is considered the main tool for modern reservoir management. For optimal reservoir management, it's critical to determine the reserves, recovery factors, and economic limits as quickly as possible, but that's a difficult job. Using reservoir simulation, engineers can forecast a range of production and depletion scenarios based on different variables. This greatly improves decision-making up-front, before the money is spent to drill new wells, establish infrastructure and surface facilities, and above all damaging the reservoir and losing a great deal of production. Reservoir simulators allow engineers and geoscientists to build dynamic models that predict the movement of oil and gas flowing in reservoirs under in situ conditions.

However, all beneficial aspects of a reservoir simulation are affected by the data that are fed into the simulators, input data, and also a correct recording of the performance of the simulated and real systems, output data. The system consists of a reservoir, wells, and other facilities. The reservoir simulator is normally constructed on very highly uncertain input data regarding the rock and fluid properties. The input data should be delineated based on the output information from the real and simulator systems during the production life of the reservoir. It will help to adopt an optimum strategy to recover the hydrocarbons.

This section intends to provide insight into the related issues beyond the body of the simulator and its construction. The once in-house simulator is built or a commercial simulator is to be used, what are the associated concerns that the user in specific should be wary about. Key features of input and output apprehensions will be discussed.

Reservoir simulators use mathematical expressions, usually in partial differential equations forms, to model the flow behavior of oil, water, and gas inside reservoirs. Abou-Kassem et al. (2006, 2020) have demonstrated that the process of simulator development can be largely simplified by using the so-called engineering approach. The reservoir is divided into discrete gridblocks, and powerful computers are used to compute the changes in conditions over many discrete time intervals. Simulated performance of the field is compared to actual data recorded in the real system. In the oil industry, the predictive capacity of a simulator is adapted such that the predicted results approach measured production data. This process is generally referred to as "*history-matching.*" The history-matching process is carried out due to the uncertainty involve in input data and also the uncertain adopted fluid flow model. A schematic layout of history match of the real and simulated data is given in Fig. 7.3. The production data are used to update the model parameters and therefore, the history-matching process could be considered as a closed-loop control process. A schematic illustration of the geological and geophysical activities is also depicted in Fig. 7.3. The initial geological data are usually obtained by the seismic surveys, well logging, core sampling. These data are delineated with the information obtained during the oil and gas production through the history matching process. The effectiveness of the procedure depends on the duration of updating input data and the history matching process. The traditional history matching process is usually performed on a campaign basis, typically after years. In many cases, the matching

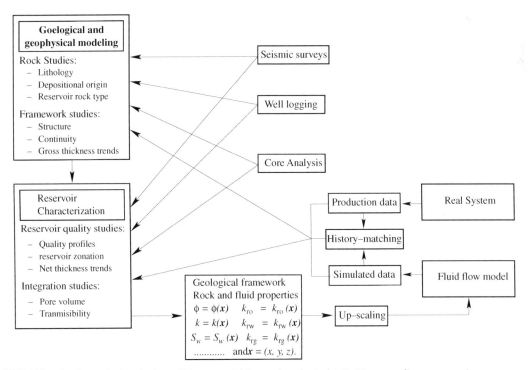

FIG. 7.3 A schematic description of history matching and geological activities regarding a reservoir.

techniques are usually ad hoc and involve manual adjustment of model parameters, instead of systematic parameter updating.

Islam et al. (2016a) showed that history to retool the data is inherently flawed. Instead, the petroleum industry should focus on using scientific filtering techniques and use data of the highest integrity and relevance. The suggestions made in this chapter will bolster accurate description of the reservoir and give further edge if used with the simulators that use the engineering approach with the ability to track multiple solutions (Chapter 6, this book).

Fig. 7.3 may also be considered schematic diagrams to illustrate the input and output data at different levels of reservoir modeling. The first sets of data are generated through geological and geophysical studies. The geological and geophysical model based on the core analysis and well logging with the application of the seismic survey model is a preliminary identification of a hydrocarbon reservoir. The fluid flow model can be used to test the physical model against the production performance of the reservoir. The adjustments are made to the reservoir characterization model until a match is achieved.

Reservoir fluid flow modeling is a mathematical description of the fluid flow through the porous rock. The objective of developing the reservoir flow modeling is to optimize hydrocarbon recovery. The mathematical modeling is transferred into a numerical formulation to find the quantitative description for the fluid flow. The numerical description is usually constructed on a discretized reservoir consists of many units. The size of a unit should allow for using average properties throughout it. However, many factors such as the capabilities of computing facilities and the applied numerical schemes play important role in the selection of the size of a representative computing unit or simply a grid. A detailed analysis indicates that for an accurate reservoir engineering analysis, geostatistical models should have information of 1 m scale. This is the only scale that would satisfy the representative elemental volume (REV) requirement of an enhanced oil recovery (EOR) system. This scale length is orders of magnitude higher than that of core samples and at least an order of magnitude lower than the conventional seismic data. This data gap constitutes the weakest link between geophysical information and reservoir engineering (Islam, 2001). This constitutes to the problem of up- and down-scaling.

Up-scaling is an averaging process from one scale to a larger scale. It is one of the most challenging problems in the modeling of a petroleum reservoir. A medium property is observed at one scale on particular support (volume) of measurement, but the value of that property is needed on a different volume size at a different location. In reservoir modeling, the up-scaling process may be divided into two levels. The first level is scaling up the laboratory measurement on core samples, perhaps a few centimeters in length and diameter, to geophysical cells within several meters. The second level is referring to scale up from the geological fine grid to the fluid flow simulation grid. The upscaling in reservoir simulation is mainly referred to second level of up-scaling process.

7.3.1 Geological and geophysical modeling

The first sets of data are generated through geological and geophysical studies. The reservoirs are normally thousands of feet under the ground level, the size, shape, and constituent of that reservoir are uncertain. The process of finding an oil-bearing rock and estimating the quality and quantity of the rock and fluid is referred to the geological and geophysical activities. Geologists and geophysicists provide information about the reservoir and its contents that will lead to *"geological and geophysical modeling"* and then *"reservoir characterization."*

The geologist and geophysicists usually concentrate on the rock attributes in four stages (Harris, 1975):

a. *rock studies*: to establish lithology and to determine the depositional environment, and to distinguish the reservoir rock from non-reservoir rock,
b. *constructing a geological framework*: to establish the structural style and to determine the three-dimensional continuity character and gross-thickness trends of the reservoir rock,
c. *reservoir-quality studies*: to determine the framework variability of the reservoir rock in terms of porosity, permeability, and capillary properties (the aquifers surrounding the field are similarly studied), and
d. *integration studies*: to develop the hydrocarbon pore volume and fluid transmissibility pattern in three dimensions.

The first two tasks may be categorized under the geological and geophysical modeling of the reservoir. The other two may be classified as reservoir characterization.

The first critical area in the process of constructing a comprehensive reservoir model is building a geological framework of the structure and reservoir architecture from seismic data and well logs (if the oil field has been penetrated by wells). This geological model represents all major geological features that may affect the connectivity of the reservoir. This stage of reservoir studies comes out with a geological framework of the structure and reservoir architecture from the different sources of measurement mainly from the seismic data. However, all major tools and techniques of formation evaluation (mud logging, wireline logging, etc.) should be viewed only as a means to construct a reliable model of the reservoir rock in the subsurface. The geological modeling represents all major geological features (faults, flow barriers, compartments, pinch outs, etc.) that are likely to affect the connectivity of the reservoir.

The distribution of the reservoir- and non-reservoir-rock types and of the reservoir fluids determine the geometry of the model and influence the type of the model to be used. For example, the number and scale of the shale (or dense carbonate) breaks in the physical framework determine the continuity of the reservoir facies and influence the vertical and horizontal dimensions of each cell. Seismic processing is often targeting structure interpretation. For attribute inversion, more high-frequency energy should be captured and relative amplitude should be preserved. In practice, reprocessing is costly; therefore filter and reverse-filter may need to be applied to prepare the seismic data for inversion. Since well logs and geological interpretations are usually available, this preparation should take well logs and geological information into account.

The geological framework is the basic inputs in the reservoir characterization and then simulation of fluid flow inside the porous media. This provides the structural style and reservoir architecture, the connectivity of the reservoir rock, and the gross thickness trends of the pay. Reservoir architecture includes the gross geometrical structure of the reservoir and its physical dimensions. The architecture also belongs to the characteristics of a contiguous water-bearing reservoir and the uniformity or variability of the producing section within the reservoir. The application of 3D seismic interpretation plays a greater role in the development of early-stage and to find out the architecture of the reservoir where limited data are available. The gross geometrical structure of the reservoir represents the shape of the reservoir whether it is in a regular (rectangular, circular, cylindrical, etc.) or any irregular shape. The underground petroleum trap structure is totally dependent on the sedimentation procedure, age of the rock, type of rock, and the geographical location of the reservoir. The use of modern technology is capable to give a general idea of the structure. However, it is still an issue and uncertainty remains to figure out the geometrical structure of the reservoir. Reservoir physical dimension is also an uncertain issue because of the uncertainty behind the prediction of gross geometrical structure. The physical dimensions of a reservoir mean length, width, height or radius, and depth of the reservoir. For any reservoir engineer/geologist, the initial task is to calculate the area or volume of the reservoir.

The geological model that is constructed in such a way is a static model based on initial condition measurements and surveys and bears substantial uncertainty. In other words, this is preliminary reservoir rock type identification that is based on the data obtained in seismic surveys and well loggings. The model should have the ability to convert into a dynamic model to be modified and to capture all possible variations of geological properties using the production data.

7.3.2 Reservoir characterization

The geologists and geophysicists are collaborating to produce a complete map of the reservoir and to characterize it. To portray a reservoir, integration of all data with various qualities and quantities in a consistent manner is required to describe rock and fluid properties through the reservoir. Effective formation evaluation requires the integrated use of every piece of available data.

The basic inputs in the process of reservoir characterization are the geological framework, well log data, core analysis data, seismic amplitude data, acoustic impedance data, well test data, and any other data that can be correlated to rock and fluid properties. The interpretation results from all wells drilled into the reservoir are combined with the seismic data previously acquired to construct a three-dimensional (3D) model of the reservoir. This model not only depicts the shape of the reservoir but details the properties of the rocks and fluids as well. However, the rock and fluid properties should be anticipated in a consistent manner that depends on the quality and quantity of the available data. For the reservoirs with many wells penetrations and consequently considerable well loggings data, the seismic survey data with a high degree of uncertainties may not rely on considerably but in reservoirs with the limited number of penetrating wells such as offshore reservoirs, the seismic data play an important role to characterize the reservoir.

The reservoir characterization should provide the main parameters of the reservoir fluid flow modeling. These parameters are the rock and reservoir fluid properties, i.e., porosity, permeability, density, viscosity, pressure, temperature, etc., and the variation of them through the reservoir. Almost all the reservoir rocks are highly variable in their properties. The heterogeneity of the reservoir matrix may be classified into two main categories. The first is the lithological heterogeneity and the second source of heterogeneity is attributed to inherent spatial rock variability, which is the variation of rock properties from one point to another in space due to the different deposition conditions and different loading histories. The first question arising is how to treat the fluid and the matrix to obtain a complete description of them. It is very complicated, if not impossible, to describe precisely the geometry of the internal solid surfaces that bound the flow domain inside a porous medium. The geometry of the solid cannot be defined by stating the equations that describe the surface bounding the fluid. Therefore, certain macroscopic (or average) geometric properties, such as porosity, are employed as parameters describing or actually reflecting the geometry of a porous matrix. The second question is regarding the size of a representative elementary volume (REV) that reflects the heterogeneity of the solid matrix.

7.3.3 Representative elementary volume (REV)

If we direct our attention to the fluid contained in porous space, we also encounter a lot of difficulties to describe the phenomena associated with the fluid itself, such as motion, mass transport. If we consider the fluid itself, it consists of many molecules. To describe the motion of the fluid, a large number of equations should be provided to solve the problem at the molecular level. Since the final goal is to describe the fluid motion in porous media, it is necessary to have a higher level and treat the fluid as continua. The concept of a particle is essential to the treatment of fluids as continua. A particle is an ensemble of many molecules contained in a small volume. The size of a particle should be larger than the mean free path of a single molecule and, however, be sufficiently small as compared to the considered fluid domain that by averaging fluid and flow properties over the molecules included in it, meaningful values, i.e., values relevant to the description of bulk fluid properties, will be obtained. The mean values are then related to some centroid of the particle and then, at every point in the domain occupied by a fluid, we have a particle possessing definite dynamic and kinematics properties.

The concept of the fluid continuum and the definition of density as an example of the fluid properties are illustrated in Fig. 7.4. It shows that the particle size (elementary volume) should be of the order of magnitude of the average distance λ between the molecules (mean free path

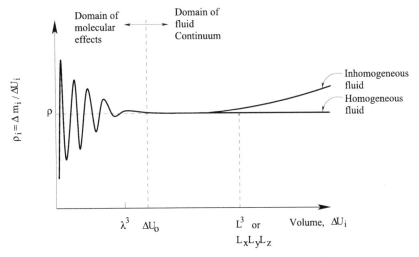

FIG. 7.4 The fluid continuum domain and the variation of fluid density. *Modified from Bear, J., 1975. Dynamics of Flow in Porous Media. American Elsevier Publishing Co., New York., 764 pp.*

of molecules). However, to capture the fluid in-homogeneity, the size of the particle should be less than an upper limit. This may be characterized by the length, L (or L_x, L_y and L_z in the direction of three coordinates). The length L may be considered as characteristic length for the macroscopic changes in fluid properties such as density. The volume L^3 (or $L_x L_y L_z$) may be used as the upper limit for the particle volume, ΔU_i.

If we turn our attention to the solid medium confining the fluid, the size of a representative elementary porous medium volume around a point P should be determined. The size of the representative elementary volume (REV) should be much smaller than the size of the entire fluid flow domain to associate the resulting averages with a point P. On the other hand, it must enough larger than the size of a single pore that it includes a sufficient number of pores to permit the meaningful statistical average required in the continuum concept (Bear, 1975). The definition of REV and representative elementary property (REP) is illustrated in Fig. 7.5. The property of the medium and more importantly permeability which is a property of medium and fluid both, should be averaging around a volume ΔU_0 as REV. The representative elementary volume may be defined as the volume ΔU_0 at which the fluctuation in the REP is not significant beyond it.

The representative elementary volume (REV) may be defined as a critical averaging volume beyond which there is no significant fluctuation in the representative property as the addition of extra voids or solids has a minor effect on the averaged property. It is impossible to carry out actual measurements over enough scales to confirm the behavior shown in this plot. However, it indicates that the variability of a specified property of the rock matrix is varied erratically at a small scale, less than the REV, has a zero (or small) variability at an intermediate scale, in the range of REV. The variation of the specified property such as porosity will be increased with increasing the scale larger than REV. These results are consistent with the general observation. The variability of the average porosity in a set of 1″ diameter core plugs is not zero and this variability is certainly smaller than the variability of porosity on the scale of several micrometers (say 100 µm). The statistical investigation using the variance of the mean of the porosity by Lake and Srinivasan (2004) indicates shows the same trend for

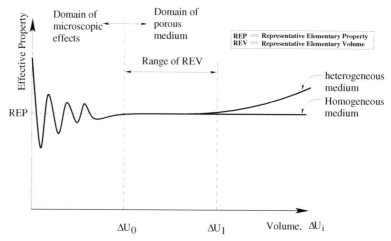

FIG. 7.5 Definition of representative elementary volume (REV) and representative elementary property (REP).

small scale. However, they do not find any region of stable porosity at intermediate even though the parameters were chosen to emphasize such stability. This shows that the only way to obtain a stable average is for the averaging volume to exceed the largest scale of heterogeneity. This is not a satisfactory outcome. It says that the REV is the largest scale of heterogeneity in the field. Since any cell-by-cell computation or any measurement is below the REV scale, there is some uncertainty in the modeled rock properties.

7.3.4 Fluid and rock properties

We discussed the approach in defining the fluid and rock properties. The fluid is considered to be a continuum media and the rock properties are averaging on a representative elementary volume (REV) despite the fact the existence of the REV is tenuous because it has never been identified in real media. The size of the REV is related to how locally correlate the property on the pore (microscopic) scale. It should be large enough for statistically meaningful averaging. If we follow the traditional definition of REV as given in Fig. 7.6, it should be small enough to avoid heterogeneity. As a very rough number, the typical REV size is somewhere around 100–1000 grain diameters. It should be emphasized that despite all ambiguities, the notion of REV is essential to allow us to use continuous mathematics. Our main focus is on all properties needed in the flow computations. These properties formed a central part of the inputs to a numerical simulator.

7.3.4.1 *Fluid properties*

Petroleum deposits vary widely in property. The bulk of the chemical compound present are hydrocarbons that are comprised of hydrogen and carbon. A typical crude oil contains hundreds of different chemical compounds and normally is separated into crude fractions according to the range of boiling points of the compound included in each fraction.

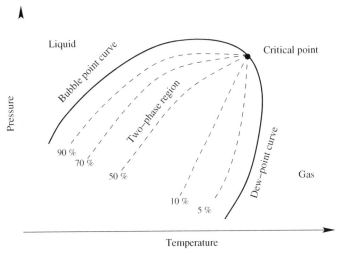

FIG. 7.6 A typical pressure-temperature diagram of an ordinary crude-oil.

Hydrocarbons may be gaseous, liquid, or solid at normal pressure and temperature depending on the number and arrangement of the carbon atoms in the molecules. Hydrocarbons are moved from a gaseous state to a solid-state with increasing the number of carbon atoms. The compound with up to four carbons is gaseous, with 4–20 are in the liquid state and those with more than 20 carbon atoms are solid. Liquid mixtures, such as crude oils, may contain gaseous or solid compounds or both in solution. Several non-hydrocarbons may occur in crude oils and gases such as sulfur (S), nitrogen (N), and oxygen (O).

The petroleum reservoirs may be classified according to the state of the fluid compound and divided into two broad categories of oil and gas reservoirs. These may be subdivided into different groups according to the hydrocarbon compounds, the initial temperature and pressure of the reservoir, and so on. The phase diagrams can be applied to express the behavior of the reservoir fluid in a graphical form. The pressure-temperature diagram is one of them that can be applied to illustrate the reservoir fluid behavior and to classify reservoirs. A typical phase diagram of a multi-components compound is given in Fig. 7.7. In this diagram, the critical point is the point at which all properties of the liquid and gaseous state are identical. If the reservoir temperature T_R is less than the critical point temperature T_C, the reservoir is called an oil reservoir. The reservoir is called a gas reservoir if $T_R > T_C$. A classification of the oil and gas reservoir is given in Fig. 7.7. The parameter denoting by P_R is the initial reservoir pressure and P_{BP} is bubble-point pressure. The gas–oil ratio is denoted by GOR in the diagram and the notation T_{cr} is the maximum temperature that the liquid phase may exist.

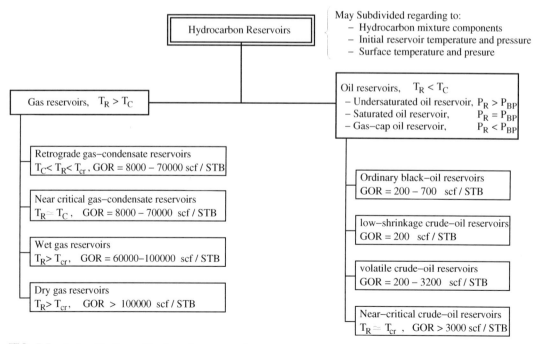

FIG. 7.7 A classifications of hydrocarbon reservoirs.

A complete description of the reservoir fluid properties is necessary to predict the pressure variation inside the reservoir and the other flow parameters. The reservoir rock usually contains some amount of water along with the hydrocarbon compounds. The fluid physical properties of primary interest in reservoir mathematical modeling are:

Specific gravity, γ_o

$$\gamma_o = \frac{\rho_o}{\rho_w} \quad (7.5)$$

ρ_o, ρ_w are both at standard condition (i.e., $P = 14.696$ psi, $T = 520^{oR}$ according to SPE).

- The specific gravity of the solution gas, γ_g
- Gas solubility, R_S $R_S = f(P, T, \gamma_o, \gamma_g)$
- Bubble-point pressure, P_b $P_b = f(R_S, T, \gamma_o, \gamma_g)$
- Oil formation volume factor, $B_o = f(R_S, T, \gamma_o, \gamma_g)$

$$B_o = \frac{(V_o)_{P,T}}{(V_o)_{Std.}} \quad (7.6)$$

- Isothermal compressibility coefficient,

$$c_o = \frac{1}{\rho_o}\left(\frac{\partial \rho_o}{\partial P}\right)_T = -\frac{1}{B_o}\left(\frac{\partial \rho_o}{\partial P}\right)_T \quad (7.7)$$

for the under-saturated crude-oil system,

$$c_o = \frac{1}{B_o}\left(\frac{\partial B_o}{\partial P}\right)_T = -\frac{B_g}{B_o}\frac{\partial R_S}{\partial P} \quad (7.8)$$

for the saturated crude-oil system.

- Oil density, ρ_o

$$\rho_o = \rho_{o0} \exp[c_o(P - P_0)] \quad (7.9)$$

Oil may be considered as slightly compressible fluid. If the series expansion of exponential function is used and consider only the linear part of it, the oil density may express as

$$\rho_o = \rho_{o0}[1 + c_o(P - P_0)] \quad (7.10)$$

where P_0 is a reference pressure.

- Crude oil viscosity, μ_o
- Surface tension, σ

7.3.4.1.1 Natural gas properties

- Apparent molecular weight, M_a

$$M_a = \sum_{i=1} y_i M_i \quad (7.11)$$

where y_i is the mole fraction and M_i is the molecular weight of ith component.
- Specific gravity, γ_g

$$\gamma_g = \frac{\rho_g}{\rho_{air}}, \quad \gamma_g = \frac{M_a}{28.96} \tag{7.12}$$

- Compressibility factor (gas deviation factor), z

$$z = \frac{V_{actual}}{V_{Ideal}}, \quad z = f(y_i, P, T) \tag{7.13}$$

where V_{Ideal} is the gas volume if it is treated as an ideal gas.
- Isothermal gas compressibility coefficient, c_g

$$c_g = -\frac{1}{v}\left(\frac{\partial v}{\partial P}\right)_T, \quad c_g = \frac{1}{P} - \frac{1}{z}\left(\frac{\partial z}{\partial P}\right)_T \tag{7.14}$$

where v is the specific volume of the gas.
- Gas formation volume factor, B_g

$$B_g = \frac{V_{P,T}}{V_{Std}} \tag{7.15}$$

where V_{std} is the volume of the gas at standard condition.
- Gas expansion factor, E_g

$$E_g = \frac{1}{B_g} \tag{7.16}$$

- Viscosity, μ_g $\mu_g = f(y_i, P, T)$

In Chapter 2, fluid characterization rules, derived from the origin and age of petroleum fluids, are presented. In Chapter 4, various applications are suggested for different types of crude oil. This characterization would optimize sustainability and maximize profit without compromising sustainability.

7.3.4.1.2 Water content properties
- Formation volume factor, B_w
- Gas solubility, R_{sw}
- Compressibility coefficient, c_w
- Viscosity, μ_w

The fluid properties and PVT and phase-equilibrium behavior are measured and studied in laboratories to characterize the reservoir fluids and to evaluate volumetric performance at various pressure levels. Different laboratory tests are used to measure the fluid sample properties. The primary tests are involved in the measurement of the specific gravity, gas-oil ratio, viscosity, and so on. These are routine field (on-site) tests. There are several other laboratory tests such as compositional analysis tests, constant-composition expansion test and differential liberation test to obtain:

- the compositional description of the fluid;
- saturation pressure (bubble-point or dew-point pressure);
- isothermal compressibility coefficients;
- compressibility factors of the gas phase;
- total hydrocarbon volume as a function of pressure;
- amount of gas in solution as a function of pressure;
- the shrinkage in the oil volume as a function of pressure;
- properties of the evolved gas including the composition of the liberated gas;
- the gas compressibility factor;
- the gas specific gravity;
- the density of the remaining oil as a function of pressure; and so on.

The sampling pressure may be different from the actual reservoir pressure. In these cases, the PVT measured data should be adjusted to reflect the actual reservoir situations.

7.3.4.2 Rock properties

The basic inputs in the process of reservoir characterization are the geological framework, well log data, core analysis data, seismic amplitude data, acoustic impedance data, well test data, and any other data that can be correlated to rock and fluid properties. The interpretation results from all wells drilled into the reservoir are combined with the seismic data previously acquired to construct a three-dimensional (3D) model of the reservoir, as shown in Fig. 7.8. This model not only depicts the shape of the reservoir but details the properties of the rocks and fluids as well. However, the rock and fluid properties should be anticipated in a consistent manner that depends on the quality and quantity of the available data. For the reservoirs with many wells penetrations and consequently considerable well loggings data, the seismic survey data with a high degree of uncertainties may not rely on considerably but in reservoirs with a limited number of penetrating wells such as offshore reservoirs, the seismic data play an important role to characterize the reservoir.

- *Porosity, ϕ*—Petroleum reservoirs usually have heterogeneous porosity distribution; i.e., porosity changes with location. A reservoir is homogeneous if porosity is constant independent of location. Porosity depends on reservoir pressure because of solid and pore compressibility. It may be defined two types of porosities: absolute porosity ϕ_a and effective porosity ϕ. The effective porosity is used in reservoir engineering calculations.

$$\phi = \frac{\text{interconnected pore volume}}{\text{bulk volume}}$$

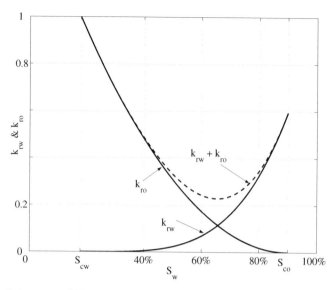

FIG. 7.8 Typical relative permeability curves to water and oil.

- *Saturation, S_o, S_g, S_w*

$$S_o = \frac{V_o}{\phi V_b}, \quad S_g = \frac{V_g}{\phi V_b}, \quad S_w = \frac{V_w}{\phi V_b} \tag{7.17}$$

where V_o is the volume of oil, V_b is the bulk volume of the rock, V_g is the volume of gas, and V_w is the volume of water.

$$S_o + S_g + S_w = 1 \tag{7.18}$$

- *Permeability, k*—Permeability is the capacity of the rock to transmit fluid through its connected pores when the same fluid fills all the interconnected pores. The flux vector, the rate of fluid for the unit area, may be given by using Darcy's law

$$\mathbf{q} = -\left(\frac{1}{\mu}\right) k(\mathbf{x}) \cdot \nabla \Phi, \tag{7.19}$$

where $\mathbf{q}$ is the flux vector; $\mathbf{x} = (x, y, z)$; and $\nabla \Phi = \nabla (P + \rho g z)$. The notation $k(\mathbf{x})$ is permeability tensor that is a directional rock property.

$$k(\mathbf{x}) = \begin{bmatrix} k_{xx} & k_{xy} & k_{xz} \\ k_{yx} & k_{yy} & k_{yz} \\ k_{zx} & k_{zy} & k_{zz} \end{bmatrix} \tag{7.20}$$

The diagonal terms, such as k_{xx}, represent the directional flow, the flow in the same direction as the pressure gradient. The off-diagonal terms, such as k_{xy}, represent a cross-flow, the flow perpendicular to the pressure gradient. If the reservoir coordinates coincide with the principal directions of permeability, then permeability can be represented by k_{xx}, k_{yy}, k_{zz}. The diagonal terms may be simply denoted by k_x, k_y, k_z. The reservoir is described as having isotropic permeability distribution if $k_x = k_y = k_z$.

- *Relative permeability, $k_{ro} = k_{rg} = k_{rw}$*—The relative permeability is defined for the simultaneous flow of two or more immiscible fluids in a porous medium. The relative permeability is the ability of a porous medium to transmit a fluid at a point if part of the porous formation at that point is occupied by other fluids.

$$k_{ro} = \frac{k_o}{k}, \quad k_{rg} = \frac{k_g}{k}, \quad k_{rw} = \frac{k_w}{k} \qquad (7.21)$$

where k_{ro}, k_{rg}, and k_{rw} is the relative permeability of oil, gas, and water, respectively, k_o, k_g, and k_w are the effective permeability of oil, gas, and water, respectively, and k is the absolute permeability of the rock matrix. The relative permeability is a function of the fluids saturation. For the general case of three-phase flow, it may be written that:

$$k_{rw} = f(S_w), \quad k_{rg} = f(S_g), \quad k_{ro} = f(S_w, S_g) \qquad (7.22)$$

A typical diagram for the variation of relative permeability of two-phase flows of oil and water is given in Fig. 7.8. The water as the wetting fluid cease to flow at relatively large saturation that is called the irreducible water saturation S_{iw} or connate water saturation S_{cw}. The oil as the non-wetting fluid also ceases to flow at saturation more than zero, connate oil saturation S_{co}.

- *Capillary pressure, P_c*—The discontinuity of pressure at the interface of two immiscible fluids is the capillary pressure.

$$P_c = P_{nwp} = P_{wp} \qquad (7.23)$$

where P_{nwp} is the non-wetting phase, i.e., oil phase in water/oil flow, and P_{wp} is the wetting phase pressure, i.e., water phase in water/oil flow. The capillary pressure is also a function of the fluids saturation, i.e., $P_c = P_c(S_w)$ in water/oil flow.

The notations and the dimensional units of the quantities that are characterized a reservoir fluid and rock and appeared in flow equations are given in Table 7.11 for different unit systems.

The above properties and the others that are not mentioned are normally obtained through the core analysis test. This information should be combined with the well log data and the other available data to characterize the reservoir rock consistently. Geo-statistical methods may be applied to integrate the sparse data and to propagate reservoir properties in a manner that is statistically coherent and consistent. These algorithms are applied to model spatial autocorrelation, to create continuous maps based on the data for the areal and vertical sections, and to simulate random realizations (dataset) based on a given spatial autocorrelation model. The reservoir is depicted using a very small grid to anticipate the heterogeneity of the reservoir properties. It is the description of the reservoir on the "*geo-cellular model.*" However, the description honors the known and inferred statistics of the reservoir and suffers a considerable degree of uncertainty. The geological description of the cellular grids in many cases is not feasible.

TABLE 7.11 The symbols and units of rock, fluid, and flow properties.

Quantity	Symbol	Customary unit	SI unit	Lab unit
Length	x, y, z, Λ	ft.	m	cm
Area	A, A_x, A_y, Λ	ft.^{2}	m^2	cm^2
Density	ρ_o, ρ_w, ρ_g	lb/ft.^{3}	kg/m^3	g/cm^3
Specific gravity	r_o, r_g	...	...	...
Gas solubility	R_s, R_{sw}	scf/STB	??	??
Bubble point pressure	P_b	psi	Pa	atm
Dew-point pressure	P_{dp}	psi	Pa	atm
Formation volume factor	B_o, B_w, B_g	RB/scf	m^3/std m^3	cm^3/std cm^3
Compressibility coefficient	c_o, c_w, c_g, c_ϕ	...	...	...
Viscosity	μ_o, μ_w, μ_g	Cp	kg/m s	cp
Surface tension	$\sigma_o, \sigma_o, \sigma_g$			
Molecular weight	M_a, M_i, Λ			
Mole fraction	y_i			
Porosity	ϕ	...	...	...
Permeability	k	D, mD	m^2	D, mD
Relative permeability	k_{ro}, k_{rw}, k_{rg}	...	...	...
Saturation	S_o, S_w, S_g	...	...	...
Capillary pressure	CP	psi	Pa	atm
Oil flow rate	q_o	STB/day	std m^3/day	std cm^3/day
gas flow rate	q_g	scf/day	std m^3/day	std cm^3/day
Volumetric velocity	u_o, u_w, u_g	RB/dat ft.^{2}	m/day	cm/s
Time	t	day	day	s

7.3.5 Upscaling

One of the challenging problems in the description of a heterogeneous medium is the problem of averaging and upscaling from one scale to another. A medium property is observed at one scale on particular support (volume) of measurement, but the value of that property is needed on a different volume size at a different location. There are two levels of upscaling. This is the first level of upscaling that a small set of laboratory measurements need to be interpreted at the scale of the reservoir. The laboratory measurement on core samples, perhaps a few centimeters in length and diameter, should be scaled up to geophysical cells within several meters.

Another upscaling problem refers to scale up the geological fine grid to the simulation grid. Geostatistical methods are capable of providing many more values than can be easily accommodated in reservoir fluid flow simulators; hence the geo-statistically derived values are often up-scaled (Paterson et al. 1996). This may be called as the second level of upscaling. It is a process that scales the properties of a fine grid to a coarse grid, such that the fluid flow behavior in the two systems is similar (Qi and Hesketh, 2005). Geological models may include $O(10^7 - 10^8)$ cells for a typical reservoir, whereas practical industrial models typically contain $O(10^5 - 10^6)$ grid blocks (depending on the type of model), with the model size often determined such that the simulation can be run in a reasonable time frame (i.e., overnight) on the available hardware (Gerritsen and Durlofsky, 2005).

The upscaling in reservoir simulation is mainly referred to as scale-up from a geological cell grid to a simulation grid, the second level of upscaling. It is one of the first challenges in reservoir engineering simulations and is subject to intensive research activity (Zhang et al., 2006; Farmer, 2002; Christie, 1996). It should be indicated that upscaling increases the uncertainty of the simulation output. The key point is to minimize the amount of upscaling that must be done. In many cases, the parameters may not be fixed through history-matching when the coarse grids are applied. It is necessary to refine the simulation scale to find a better match between the simulation model and the real system.

An up-scaling algorithm assigns suitable values for porosity, permeability, water saturation, and the other fluid flow properties and functions to a simulation grid based on the reservoir characterization. The major methods in upscaling are power-law averaging methods, arithmetic-, geometric- and harmonic-mean technique; renormalization techniques; pressure-solver method; tensor method; and pseudo-function technique. For the additive parameters such as porosity, fluid saturations, and more generally volume/weight percentages of various phases, the arithmetic averaging method gives a reasonable and fast equivalent number. The main problem in upscaling is related to the non-additive properties like effective permeability. These properties are not intrinsic characters of the heterogeneous medium. They depend on the boundary condition and the distribution of the heterogeneities which depend on the volume being considered (Begg et al. 1989).

7.3.5.1 *Power-law averaging method*

The most obvious approaches of upscaling are the application of mean values. These techniques are very fast but suffer from some limitations in applicability. There are several different kinds of calculations of the mean value.

- Arithmetic mean-value:

$$k_e = \frac{1}{n}\left(\sum_{i=1}^{n} k_i\right) = \frac{1}{n}(k_1 + \cdots + k_n) \tag{7.24}$$

- Geometric mean-value:

$$k_e = \left(\prod_{i=1}^{n} k_i\right)^{1/n} = \sqrt[n]{k_1 \cdot k_2 \cdot \cdots \cdot k_n} \tag{7.25}$$

- Harmonic mean-value:

$$k_e = n \left(\sum_{i=1}^{n} \frac{1}{k_i} \right)^{-1} = \frac{n}{\frac{1}{k_1} + \cdots + \frac{1}{k_n}} \qquad (7.26)$$

Here, k_i is the permeability of a fine gridblock and k_e is the effective permeability of a coarse-grid block (i.e., a simulation grid). The arithmetic averaging method may be used in single phase flow for horizontal flow and the geometric averaging method may be suitable for vertical flow. Journel et al. (1986) derived a model for the effective permeability for shaley reservoirs. They demonstrate by experiments and simulations that flow conditions do not depend on the detail of the permeability spatial distribution, but rather on the spatial connectivity of the extreme permeability values, either low such as impervious shale barriers, or high such as open fractures. They assumed the flow is single-phase and steady-state and approximated the permeability filed by of two extreme modes of k_{sh} and k_{ss}. They established the following model:

$$k_e = \left[r_{sh} k_{sh}^w + (1 - r_{sh}) k_{ss}^w \right]^{1/w} \qquad (7.27)$$

where $r_{sh} \equiv$ volumetric portion of shale; $k_{sh} \equiv$ permeability of shale; $k_{ss} \equiv$ permeability of sandstone; $w \equiv$ power-average.

The power-average w mainly depends on the shale distribution. A power-average with a low power $w = 0.12$ gives a good approximation for the effective permeability in shaley reservoir with $r_{sh} \in [0, 0.5]$.

7.3.5.2 Pressure-solver method

This method is applied for single phase flow. The effective permeability, k_e, of a heterogeneous medium is the permeability of an equivalent homogeneous medium that give the same flux for the same boundary conditions. Thus, it may be derived by equating expressions for the flux through a volume of the real heterogeneous medium with the flux trough an equal volume of the similar homogeneous medium with the same boundary conditions. A steady-state, single-phase incompressible flow can be modeled by using the continuity equation and the Darcy's law,

$$\nabla \cdot \mathbf{q} = 0, \quad \mathbf{q} = -k(\mathbf{x}) \cdot \nabla p \qquad (7.28)$$

where a unit viscosity is assumed, $\mathbf{x} = (x, y, z)$ and $k(\mathbf{x})$ is the permeability tensor. A matrix equation may be set up by combining the continuity equation and the Darcy's law.

$$\nabla \cdot k(\mathbf{x}) \nabla p = 0 \qquad (7.29)$$

The directional effective permeability in each direction is calculated by assuming no flow condition along the sides. This combined equation is solved for p_I at the inlet and p_o at the outlet and sum the flux in a given direction such as x-direction. The effective permeability is given by using the Darcy's law for the equivalent homogeneous medium with of equivalent volume, $k_e^x = -\Delta x^q / A$.

Begg et al. (1989) are obtained the effective permeability in the vertical direction. They modeled the heterogeneous medium by describing the permeability distribution, k_{ijk}, on a fine-scale grid (see Fig. 7.9), and solved Eq. (7.29) for the pressure, p_{ijk}, in each grid

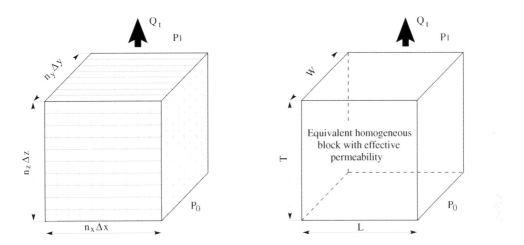

FIG. 7.9 Pressure-solver method illustration. *Redrawn from Begg et al. (1989).*

block. Darcy's law is then used to give the flux out of each block on the outlet face. They found the total flux through the model by summing the flux of each block on the outlet face. The flux through an equivalent homogeneous medium is also given by Darcy's law. Equating the two expressions for the flux gives the effective permeability in vertical direction.

$$k_e^v = \frac{n_z + 1}{n_x n_y (p_I - p_o)} \sum_{i=1}^{n_x} \sum_{j=1}^{n_y} k_{ijn_z} \left(p_{ijn_z} - p_o \right) \quad (7.30)$$

This gives only the directional effective permeability. The full-tensor effective permeability can be obtained by assuming periodic boundary conditions, for a complete description see Pickup et al. (1994).

7.3.5.3 Renormalization technique

The renormalization technique is a stepwise averaging procedure and offers a faster, but less accurate, technique to calculate the effective permeability. The essence of the renormalization procedure is very simple and is to replace a single upscaling step from the fine grid to the coarse grid with a series of steps which pass from the fine grid to the coarse grid through a series of increasingly coarse intermediate grids. In other word, the properties are first distributed on a fine grid, then the effective property is calculated for a group of cells on this fine grid to give a cell value on a coarser grid. This process is continued until the original grid is reduced to a single grid block. This gives an approximation to the effective property value for the original grid. The unit cell can be selected differently to do coarsening procedure. The very simple case that can computed analytically is to adopt a unit cell comprised 2 × 2 (or 2 × 2 × 2 in three dimensions) grid blocks. A schematic illustration of the method is depicted in Fig. 7.10 for 2D. The effective permeability, k_e, of a block of four grid is:

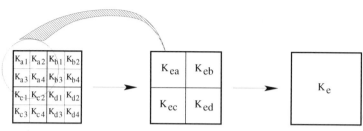

FIG. 7.10   Schematic diagram of upscaling using renormalization technique.

$$k_e = \frac{4(k_1+k_3)(k_2+k_4)[k_2k_4(k_1+k_3)+k_1k_3(k_2+k_4)]}{[k_2k_4(k_1+k_3)+k_1k_3(k_2+k_4)](k_1+k_2+k_3+k_4)+3(k_1+k_2)(k_3+k_4)(k_1+k_3)(k_2+k_4)} \quad (7.31)$$

This gives an approximation to the effective property value for the original grid. This has enabled to carry out calculation of effective permeability for extremely large grids. The effective permeability A complete description of the method and the application of it for single- and two-phase flow can be found in King (1989).

7.3.5.4 Multiphase flow upscaling

The simulation of multiphase flow is more complicated by the fact that it is necessary to consider three forces: viscous, capillary and gravity. The effective permeability tensor will be different for each phase and will depend on phase saturation as well as on the balance between these forces (i.e., the viscous/capillary and viscous/gravity scaling groups) and on the boundary conditions. The pseudo-function techniques are used for multiphase fluid flow upscaling. These methods are referred as pseudoization in conventional reservoir simulation practice. The main objective is to represent the behavior of small-scale multi-phase fluid mechanics and heterogeneity in coarse grid simulation models. The pseudo-functions (e.g., pseudo relative permeability) incorporate the interaction between the fluid mechanics and the heterogeneity as well as correcting for numerical dispersion. One common use of pseudo functions is to reduce the number of gridblocks, and sometimes even reduce the dimension of the problem, such as reducing a 3D field case model to a 2D cross-sectional model. By doing so, we hope to retain fine grid information while carrying out coarse grid simulations. The pseudo-functions may be categorized by the scaling groups under which the fine grid displacement was carried out (i.e., viscous/capillary and viscous/gravity ratio and certain geometrical scaling groups) and are valid for the boundary conditions relevant to the particular flows. If there in non-uniform coarsening to simulate the flow of different region with different rate, it is necessary to use different pseudo curves for different parts of the reservoir.

There are many pseudoization techniques described in the literature. Fig. 7.11 shows two adjacent course gridblocks together with the fine sub-grids. The sub-grids may be labeled by i, j (and k for 3D) and the coarse grid may be denoted by I, J (and K for 3D). The power law averaging methods are adopted for the additive properties such as, saturations, density, viscosity. The pseudo-function can be computed using the upscaling Darcy's Law.

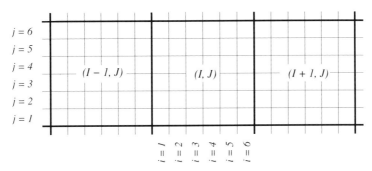

FIG. 7.11 The coarse gridblocks with the fine sub-grids.

Most pseudo-function techniques use a pore-volume weighted average to determine the saturation.

$$S_p^{I,J} = \frac{\sum_i \sum_j (V\phi S_p)_{i,j}}{\sum_i \sum_j (V\phi)_{i,j}} \tag{7.32}$$

The viscosity and the density may be up-scaled with the pore-volume weighted average

$$\mu_p^{I,J} = \frac{\sum_i \sum_j (V\phi \mu_p)_{i,j}}{\sum_i \sum_j (V\phi)_{i,j}}, \quad \rho_p^{I,J} = \frac{\sum_i \sum_j (V\phi \rho_p)_{i,j}}{\sum_i \sum_j (V\phi)_{i,j}} \tag{7.33}$$

or the rate weighted average.

$$\mu_p^{I,J} = \frac{\sum_j (q_p \mu_p)_j}{\sum_j (q_p)_j}, \quad \rho_p^{I,J} = \frac{\sum_j (q_p \rho_p)_j}{\sum_j (q_p)_j} \tag{7.34}$$

It is not clear which averaging should be used to find better results. The up-scaled pressure can be computed using: (a) the pore-volume weighted average or (b) the permeability weighted average (Kyte and Berry, 1975).

$$p_p^{I,J} = \frac{\sum_j \left[V\phi\left(p_p + \rho_p g H\right)\right]_j}{\sum_j (V\phi)_j} \tag{7.35}$$

$$p_p^{I,J} = \frac{\sum_j \left[k_{rp} k h \left(p_p + \rho_p g H\right)\right]_j}{\sum_j (k_{rp} k h)_j} \tag{7.36}$$

Hence, the pseudo relative permeability for phase p can be obtained from Darcy's law.

$$k_{rp}^{I,J} = \frac{-\mu_p \sum_j (q_p)_j}{T^{I,J}\left(\Delta p_p - \rho_p g \Delta H\right)^{I,J}} \qquad (7.37)$$

where T is transmissibility and computed by,

$$T^{I,J} = \left(\frac{kH}{\Delta X}\right)^{I,J} = \sum_j \left(\frac{1}{\sum_i \left(\frac{\Delta x_i}{k_i h_i}\right)}\right) \qquad (7.38)$$

where H and h are the height of the coarse and fine grid, respectively, and X and x are the coarse and fine grid length, respectively.

There are many other methods to do upscaling. For more detailed description, see Qi and Hesketh (2005) and Lohne and Virnovsky (2006). However, it is not clear which approaches should be used and there is no measure of quality for any upscaling routine.

7.3.5.5 Pressure/Production data

Production data is mainly fluids production flow rates of individual wells and eventually the total flow rate of the reservoir. Cumulative production of each fluid produced from the reservoir with production time is also another form of production data. Gas-oil ratio and water cut are part of production data as well. Production data as a function of elapsed production time should be the easiest task and most accurate relatively to other reservoir data. This type of data however, is the most important tool that describes the reservoir behavior and can very well, predict the reservoir production behavior. The economic limits, the need for improving recovery technique and abandoned time can be determined by utilizing decline curve analysis technique.

Surface tanks or separators are where production rates of produced fluids can be measured. The surface pressure can be used to calculate the bottom-hole flowing pressure; this pressure can be measured directly as well using down-hole installed gauges. It is necessary to measure and record well/reservoir productions; we need to calculate the overall hydrocarbon reservoir production capacity to assist the management in planning economic and marketing strategies.

It is necessary to analyze production data on daily basis, such data would help determining if there is any impulsive change in the ratios of produced fluids that may indicate certain production problems therefore; immediate remediation has to be undertaken. Increase in water production may indicate that the water table has reached higher level perforations or leaking water from a top water formation broke to the well. Increase in gas-oil ratio could mean that the gas cap expanded and reached the top perforations. The gradual drop of production sends a clear message to the reservoir management staff that something has to be done to keep production steady before losing reservoir energy. May be pressure maintenance strategy, installing production pumps, gas-lift program are to be implements depending on the reservoir condition and economic feasibility. It may be wise to consider an EOR technique to save the reservoir and maximize its production.

Loading production history data to the simulator, draws bubble maps of production on top of geological prediction maps, and offers cross plot between production and geological model predictions for verification. One has to be fully alerted of the implications of faulty production data and the catastrophic results that might cause to history match process and to the overall simulator's predictions.

7.3.5.6 *Phase saturations distribution*

Initially the reservoir is in a state of equilibrium and reservoir fluids distribution is kept constant. Once the reservoir is disturbed by drilling and consequently production, that state of equilibrium is no longer valid. The common believe is that the reservoir initially saturated by water, after the oil migration, the water is expelled from the reservoir and replaced by oil that fills the pore spaces within the reservoir rock. Practically, water saturation will never get to 0%, some of the water "residual water saturation" will be trapped in small pores and pore throats. Because hydrocarbons are lighter than water, over long geological time the fluids in the reservoir will segregate according to their density. Gas, if any, will be occupying the top portion of the reservoir then the oil at lower level and finally water will occupy the bottom portion.

By calculating water saturation with resistivity log and capillary pressure, water contact and free water level can be determined or verified. Water contact is for the determination of water-free production, and is actually an average number. Between water contact and free water level, which may be as thick as a 1000 ft. or more, a well may produce water, oil, or gas in different proportions depending on the elevation, availability of oil, and porosity. Understanding and modeling of water contact and free water level is important for reserve estimation and production planning when large area is not covered by wells.

Once reservoir fluids are produced or fluids are injected into the reservoir, it is important to understand and track the movement of fluids. Reservoir simulation as a conventional practice is widely used to predict fluid flow in a reservoir. The performance of the simulation model can be improved with production history matching. Fluid saturation within the reservoir is very much affected by the physical properties of these fluids, the reservoir rock properties and the reservoir pressure/stress in situ conditions. Obviously, fractures contained by the reservoir plays a major role in fluid saturations and movements. Rock wettability and capillarity are the main factors that control residual saturations and multiphase flow pattern in the reservoir.

In order to manage a reservoir we need to be able to predict how fluids will flow in response to the production and injection of fluids. Reservoir simulation allows us to predict fluid flow. Before production, a reservoir is characterized using data from geology, seismic exploration, logging, well tests, core and laboratory work. During production, pressure, saturation and temperature change, and the change is predicted by reservoir simulation. Dynamic data collected during production and 4D seismic surveys can be used to improve reservoir simulation. As a first step, reservoir simulation parameters are adjusted so that the model predictions approximate the bottom-hole pressure (BHP), water cut, gas oil ratio (GOR) and other field observations. Reservoir parameters can be further adjusted so that the predicted changes in seismic response approximate the results from 4D seismic surveys. We are currently searching for candidate 4D research sites.

Unfortunately, production data do not tell much about the fluid saturations distribution inside the reservoir. Well logging data (resistivity log, sonic log, SP log, etc.) can be used to determine the fluid saturations surrounding the wellbore. Recently, 4D seismic surveys have been used to observe fluid changes and movement in the reservoir (Sonneland et al., 1997). Four dimensional seismic survey results provide additional constraints that can be used to improve reservoir simulation models.

Both static (initial) and dynamic (post drilling) fluid saturations behaviors has to be understood and well studied before loaded to the reservoir simulator. Such data is very critical and sensitive in determining hydrocarbon in-place, hydrocarbon reserves and flow rates. Two-phase and there-phase models are based on such data, their predictions can be totally misleading if this data is not accurate enough. With the reservoir characterization model ready, the reservoir simulation model can be built to predict fluid changes during the reservoir futuristic production span and useful scenarios can be produced to decide the optimum recovery scheme.

7.3.5.7 *Reservoir simulator output*

Reservoir simulators, either the in-house built or the commercially used ones agree on at least two issues. The first is that a numerical solution approach (finite differences or finite elements in most cases) has to be employed to produce a solution for the model, and second, the output is nothing but a huge data in a digital form—in many cases hundreds of pages. It is impossible to grasp a meaning of the output without a graphical representation of this data. That is why the output of modern simulators always displayed in plotting or graphical, very often in 3D representation.

Input variables for the model include the latest production data, as well as subsurface geologic data gathered from all wells drilled into the reservoir. Surface geology studies and seismic data create the three-dimensional shape of the reservoir—consisting of a big number of grid blocks in the simulation in most cases. Usually, the simulator is including a code written specifically to convert input and output data from simulator raw data to graphical mode.

For every grid block of the reservoir we include data describe each geologic property (rock properties, etc.), fluid properties and a dynamic parameter (commonly called transmissibility). This process is done at an arbitrary initial time (may be 1 min, 1 h or 1 day, etc.), and at the second time segment the same description given to each grid block in the previous time-step is to be updated. The interaction of flow parameters of the grid blocks is also included by imposing suitable boundary conditions at the interface between any adjacent blocks. With this much detail it is difficult to form an image of the reservoir without 3D plots. The images provide an instant picture of the whole reservoir and its features, which is extremely informative. Some simulators equipped with a simulator animation processor capable of transforming data with hundreds of time segments and run them like a video.

Before the reservoir simulation era, our vision to the reservoir in the physical sense is only drawn by those reservoir rock and fluid samples that we retrieve from drilled wells, in addition to the tales of geologists about what happened through millions of years. The introduction of reservoir simulation technology supported by improved robustness and speed solver computers enabled us to see our reservoir in three dimensions. We can imagine how the reservoir has been formed and how the reservoir is behaving as we continue producing. The output of modern reservoir simulators can be a very fancy 3D graphical configuration of the reservoir that enables us to visualize how the reservoir is shaped and how it performs.

With the increased accessibility to reservoir simulators, more oil and gas property evaluations are being substantiated by reservoir simulation studies. Consequently, the need to incorporate reservoir simulation and review industrial requirement for simulation studies have increased in the last 2 decades.

quality production forecast is essential to a realistic reserve evaluation. Production forecasts are usually obtained through conventional reservoir engineering calculations, particularly decline curve analysis and material balance. Reservoirs undergo a series of different depletion mechanisms in their production life that cause the decline trend to shift, often to the extent that the decline trend is no longer predictable. Using a reservoir model to simulate reservoir behavior under different operation strategies becomes necessary for forecasting future performance or devising optimum development scenarios to improve recovery.

Output from seismic simulation is analyzed to investigate the changes in geological characteristics of reservoirs. The output is also processed to guide future oil reservoir simulations. Seismic simulations produce output that represents the traces of sound waves generated by sound sources and recorded by receivers on a three-dimensional grid over many time steps. One analysis of seismic datasets involves mapping and aggregating traces onto a three-dimensional volume through a process called seismic imaging. The resulting three-dimensional volume can be used for visualization or to generate input for reservoir simulators. The seismic imaging process is an example of a larger class of operations referred to as generalized reductions. Processing for generalized reductions consist of three main steps: (1) retrieving data items of interest; (2) applying application-specific transformation operations on the retrieved input items; and (3) mapping the input items to output items and aggregating, in some application specific way, all the input items that map to the same output data item. Most importantly, aggregation operations involve commutative and associative operations, i.e., the correctness of the output data values does not depend on the order input data items are aggregated. There are methods for performing generalized reductions in distributed and heterogeneous environments. These methods include (1) a replicated accumulator strategy, in which the data structures to maintain intermediate results during data aggregation are replicated on all the machines in the environment, (2) a partitioned accumulator strategy, in which the intermediate data structures are partitioned among the processing nodes, and (3) a hybrid strategy, which combines the partitioned and replicated strategies to improve performance in a heterogeneous environment.

The degree of changes in seismic response to production within the reservoir is highly dependent on the physical properties of the reservoir rocks. Accurate modeling of these properties from existing data will allow consideration of their effect on the 4D response. Modeled acoustic/elastic properties include pore volume (Vp), solid volume (Vs), bulk density, bulk Poisson's ratio, and reflectivity, both at normal incidence and non-vertical incidence. The rock physics model will allow variation in fluid saturation, fluid properties, and reservoir pressure, allowing fluid substitution modeling to represent realistic reservoir changes.

The use of dynamic reservoir simulators helps in generating different production scenarios, quantifies oil/gas reserves in the reservoir and predicts future oil, gas and water productions. These important simulation outputs are very crucial in determining production strategies and implementations of improved recovery techniques. It can be used to assess the effect of different reservoir scenarios on the 4D response. Scenarios modeled will determine expectations of reservoir production changes. Examples include, but are not limited to:

a) Saturation change representing water influx from injection well or a natural aquifer.
b) Pressure change in areas where reservoir pressure is not maintained by injection and/or an aquifer.
c) Fluid property changes representing the release of solution gas as reservoir pressure falls below the bubble point.
d) A combination of saturation, pressure, and fluid property changes as predicted by simulation of reservoir production.
e) Changes in the overburden or reservoir physical rock properties brought about by production induced in-situ stress changes.

7.4 Uncertainty analysis

Uncertainty is a measure of the "goodness" of a result. The result can be a measurement quantity, a mathematical or a numerical value. Without such a measure, it is impossible to judge the fitness of the value as a basis for making decisions relating to a scientific and industrial excellence. Evaluation of uncertainty is an ongoing process that can consume time and resources and needs the knowledge of data analysis techniques and particularly statistical analysis. Therefore, it is important for personnel who are approaching uncertainty analysis for the first time to be aware of the resources required and to carefully lay out a plan for data collection and analysis.

The sources of the uncertainty may be classified as:

1. random measurement, systematic (bias) uncertainty and lack of representativeness;
2. upscaling process; and
3. model uncertainty.

A structured approach is needed to estimate uncertainties of different sources. The requirements include:

- A method of determining uncertainties in individual terms in the reservoir simulation.
- A method of aggregating the uncertainties of individual terms.

7.4.1 Measurement uncertainty

There are a number of basic statistical concepts and terms that are central for estimating the uncertainty attributed to random measurement. The process of estimating uncertainties is based on certain characteristics of the variable of interest (input quantity) as estimated from its corresponding dataset. The ideal information includes:

- arithmetic mean (mean) of the dataset;
- standard deviation of the dataset (the square root of the variance);
- standard deviation of the mean (the standard error of the mean);

- probability distribution of the data; and
- covariance of the input quantity with other input quantities.

Measurement systems consist of the instrumentation, the procedures for data acquisition and reduction, and the operational environment, e.g., laboratory, large scale specialized facility, and in situ. Measurements are made of individual variables, x_i to obtain a result, R which is calculated by combining the data for various individual variables through data reduction equations

$$R = R(x_1, x_2, \ldots, x_n) \tag{7.41}$$

Each of the measurement systems used to measure the value of an individual variable x_i is influenced by different elemental error sources. The effects of these elemental errors are manifested as bias errors (estimated by B_i) and precision errors (estimated by P_i) in the measured values of the variable x_i. These errors in the measured values then propagate through the data reduction equation, thereby generating the bias, B_R, and precision, P_R, errors in the experimental result, r.

The total uncertainty in the result R is the root sum-square (RSS) of the bias B_R and precision limits P_R.

$$U_R^2 = B_R^2 + P_R^2 \tag{7.42}$$

where

$$B_R^2 = \sum_{i=1}^{J} \theta_i^2 B_i^2 + 2 \sum_{i=1}^{J} \sum_{k=i+1}^{J-1} \theta_i \theta_k B_{ik} \tag{7.43}$$

$$P_R = \sum_{i=1}^{J} (\theta_i P_i)^2 \tag{7.44}$$

for a single test and

$$P_R \approx P_{\overline{R}} = \sum_{i=1}^{J} (\theta_i P_{\overline{i}})^2, \quad P_{\overline{i}} = w_i S_{\overline{i}}, \quad S_{\overline{i}} = \left[\sum_{k=1}^{M} \frac{(x_k - \overline{x}_i)^2}{M-1} \right]^{1/2} \tag{7.45}$$

for multiple tests measurement. The notations: B_i is the variance of the bias error distribution of variable x_i; B_{ik} is the covariance of the bias distribution of variables x_i and x_k; $\theta_i = \frac{\partial R}{\partial x_i}$ is the sensitivity coefficient; S_i is the standard deviation of M sample results of variable x_i; M is the number of measurement set; and w_i is a weight function that is depend on the number of measurement test.

One complication with reservoir related measurements is that independent repeated measurements that reduce the experimental uncertainty are usually difficult to obtain. It is the case especially for the relative permeability because flooding of the core sample with water and oil may change wetting properties. This implies that the relative uncertainties in the relative permeability curves are determined by the random errors in the measurements of differential pressure and of oil and water rates. However, most of the parameters in reservoir model are not obtained directly from the measurement and inferred through model interpretation and a reduction equation.

The other source of uncertainty is associated with lack of complete correspondence between conditions associated with the available data and the conditions associated with real reservoir condition. Random sampling is usually not feasible since the wells are drilled far from random and core plugs are often taken from the most homogeneous parts of a core. For this reason, classical estimation theory falls short. By geological skill it is still possible to choose core plugs that are representative for a certain geological environment (litho-faces), but it is well known from statistical experience that non-random sampling may have serious pitfalls. In addition, the samples represent only a minor fraction of the total reservoir volume, so the differences between the sample and the population means and variances may be quite substantial.

Another complication is that laboratory measurements are performed on core material that has been contaminated by drilling fluids. Also, the transport to the surface will expose the core to varying temperature and pressure. When the core reaches the laboratory, its state may have changed substantially from its origin. So, even if core floods are performed under reservoir pressure and temperature and with (recombined) reservoir fluids, one should expect deviations from the original reservoir conditions. The representatively of fluid samples is often questionable because it is difficult to obtain a representative in situ sample and bring it to the surface without any leakage of gas. Another favorable situation is when the reservoir has a gas oil contact, and separator samples are available for both reservoir oil and gas. Again, very accurate samples for the oil and the gas zone can be obtained.

Direct reservoir measurements like logs, well tests, and tracer tests are representative for the reservoir properties in the volumes they contact. However, they measure responses on very different scales. While most logs explore volumes comparable to those of core plugs, well and tracer test responses frequently are averages over hundreds of meters. The reservoir simulation model should comply with the rate and pressure data from well tests and tracer production profiles. Traditionally such tests have been applied in the modeling of large-scale "architecture" like reservoir boundaries, major faults, layering, and channeling. Recently, methods have been developed to utilize well test 17 and tracer test 18 information directly in stochastic reservoir description. In such applications, the consequences of the uncertainties in the measured test responses may be less significant than the high degree of non-uniqueness in the interpretations (Bu and Damsleth, 1996).

7.4.2 Risk and uncertainty

Long-term exposure to hazardous material is the paramount risk regarding long-term damage to human health such as asthma and cancer. Safety risk is the acute risk related to short-term damage to the human body such as burns, injuries, and death due to an accident or exposure to explosion.

Risk management is the process that examines the following phases (see Fig. 7.12):

(1) Identification.
(2) Assessment.
(3) Remediation.
(4) Evaluation.
(5) Maintenance.

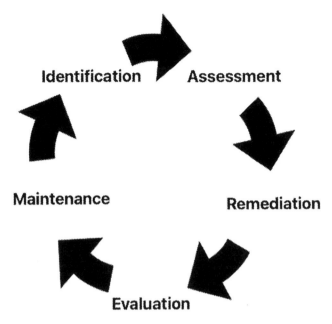

FIG. 7.12   Risk management.

Risk identification deals with:

(1) Site location.
(2) Hazard identification.
(3) Risk analysis.

Risk assessment involves estimating various health and safety risk parameters such as the individual risk. There are two types of risk assessment:

(1) Qualitative.
(2) Quantitative.

The risk remediation stage addresses the following steps:

(1) Strategy proposal.
(2) Strategy implementation.

The components of human health risk assessments are: planning and scoping, exposure assessment, acute hazards, toxicity and risk characterization. The main components of human health risk assessment are shown in Fig. 7.13. There are four different steps in assessing human health risk, which are Planning and Scoping, Exposure Assessment, Acute Hazard Assessment, and Risk Characterization. For efficient risk assessments the "planning and scoping" of the information and data are needed. It should be done before the field investigations and site characterization. The second step of human health risk assessment is

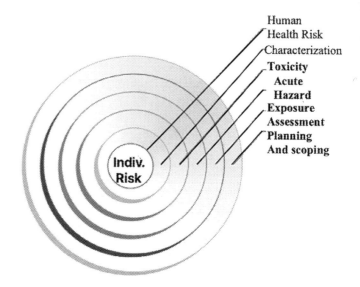

FIG. 7.13 Different components of human health risks assessment.

"exposure assessment" (see Fig. 7.13) that is the contact of leaked petroleum to the human. This process considers how much time, duration and frequencies of the chemical contact with a human in the past, present and future. The "exposure assessment" step should be done following step one. This step should be conducted just once, but if necessary it can be repeated for accuracy of the assessment. In the case of human risk assessment, "acute hazards" mean the conditions that create the potential for injury or damage to occur due to an instantaneous or short duration exposure to the effects of an accidental release. In this study, it is mainly the flammability of natural gas. "Hazard identification" is the process of determining whether exposure to the natural gas can cause an increase in the incidence of a particular adverse health effect. Generally, it is done by the dose responses of particular chemicals. However, this study considered the flammability. The "Risk Characterization" process is the synthesis of results of all other steps and the determination how dangerous the accident is to pipelines. It also considered the major assumptions, and scientific judgments. Finally, the risk characterization estimates the uncertainties embodied in the assessment.

Hossain et al. (2009) have shown the concept of individual risk due to flammability at a locality where dense populations live in. Fig. 7.13 has been redrawn from this reference where detailed analysis has been presented. An accident due to flammability is considered here as the main cause of the incident. In Fig. 7.13, OB is the maximum distance covered by the fire flame within which a fatality or injury can take place. BA and BC are the maximum distances traveled by the flame.

7.4.3 Upscaling uncertainty

The main issue in upscaling is related to the upscaling of non-additive properties such as permeability and relative permeability. There are different methods for upscaling the rock and fluid flow properties. Some of them are mentioned in Section 3.5. The upscaling method may be classified as single-phase and multi-phase upscaling. In the single-phase upscaling, the absolute permeability is only up-scaled from a geo-cellular grid to the simulation grid. The relative permeability is considered identical for both scales. The relative permeability and capillary pressure are also up-scaled along with the absolute permeability from a fine grid into a coarser grid. In many cases, the coarse block properties are obtained by considering only the fine grid scale region corresponding to the target coarse block. It is called as the local upscaling. The global upscaling is referred to the case that the entire fine-scale model is solved and the solution is applied to obtain the coarse-scale behavior. However, there are other categories as extended local upscaling, a border region around the coarse grid is also taken into account, and quasi-global upscaling in which an approximate solution of the entire flow region is adopted to derive the behavior of the coarse block. The local upscaling methods do not consider permeability connectivity.

The most important detrimental effects of the upscaling are homogenization of the medium; and coarsening of the computational grid. The permeability field is made smoother due to the homogenization and the truncation error is increased by using larger computational girds. The error introduced due to the homogenization may be referred to as the "loss of heterogeneity error" and due to the coarsening of the computational grid may be referred to as the "discretization error" (Sablok and Aziz, 2005). The combination of these two gives the total error due to the upscaling process. The total error may be small due to the opposing effect of these two types of errors contributed to the total upscaling error. As mentioned in REV that the representative volume should be as large as the dimension of the field, in the low level upscaling the discretization error may be dominated while the loss of heterogeneity error may dominate for high level upscaling. Li et al. (1995) show that the local upscaling techniques that treat the coarse grid cells as independent, provide more uncertain results than the global upscaling methods.

7.4.4 Model error

The model error includes the approximation in mathematical representation, the solution method that is mostly numeric, and also the model of error estimation. The mathematical formulations in most of the available simulators are based on the material balance equation and Darcy's law. These two fundamental equations do not mimic the reality of the flow inside the porous rock. However, the final solution is nonlinear that is complicated to solve analytically and the solution should be obtained with numerical methods. There is some analytical solution for a few special problems in reservoir engineering such as Buckley-Leveret flow that suffer a quite large number of assumptions and does not show the real situation in the fluid flow inside the reservoir rock. The most applied methods are the finite difference and finite element methods. The finite difference method is based on the Taylor series expansion. The nonlinear parts of the Taylor series expansion are normally neglected to approximate the derivation. If we consider that the normal size of a grid block is in order of 10 m, the neglecting of the nonlinear part of the Taylor series expansion produces a substantial uncertainty. Therefore, if we assume that reservoir model, the geological and characterization model were exactly correct in every detail, we cannot get exact prediction of the reservoir performance due to the fact the reservoir fluid flow model does not mimic reality.

We may have fixed certain aspects of our reservoir model (for example, the reservoir geometry) and only attempted to predict uncertainty with regard to other parameters such as porosity and permeability. Even if we get those parameters exactly correct, we may not get an exact prediction because of errors in the fixed aspects of the model. This second source of model error could in principle be removed by including all possible parameters in the uncertainty analysis but this is never feasible in practice. The true model error is virtually impossible to quantify, so in practice it has to be neglected. The size of the model error depends strongly on the parameterization of the model, and one may hope that if this is well chosen, the model error will be small (Lepine et al., 1999).

7.5 The prediction uncertainty

When the parameters are set according to the production data, the next step is obtaining the uncertainty envelope and the confidence region. This is important in forecasting the reservoir behavior and decision making process. If the set of parameters m^0 are the most likely values after history matching, the uncertainty envelope may be quantify by

$$[C_y] = [A][C_x][A]^T \tag{7.46}$$

where $\lfloor C_y \rfloor$ is the covariance matrix of the predicted quantity, $[A]$ is the sensitivity matrix whose elements are $A_{ij} = \frac{\partial y_i}{\partial x_j}$ and $[C_x]$ is the covariance matrix of the reservoir and model parameters. The first task is to specify the parameters that used for uncertainty analysis. They may be the same as those of the history-matching problem or may be different than them. When the parameters are specified, the most likely history matching values of them using the objective function, Eq. (7.46), is computed and then the sensitivity matrix is obtained. The model and reservoir parameters covariance can be obtained using the Hessian matrix

$$H_{ij} = \frac{\partial^2 E(\mathbf{x})}{\partial x_i \partial x_j}\bigg|_{\mathbf{x}=\mathbf{x}^0} \tag{7.47}$$

where $H_{ij} = \frac{\partial^2 E(\mathbf{x})}{\partial x_i \partial x_j}\big|_{\mathbf{x}=\mathbf{x}^0}$. This provides the uncertainty in the reservoir parameters. The variance of the predicted values is given by the corresponding diagonal element of the covariance matrix. The confident intervals can be obtained from this if its probability distribution is assumed to be Gaussian.

7.6 Recent advances in reservoir simulation

The recent advances in reservoir simulation may be viewed as:

- Speed and accuracy;
- New fluid flow equations;

- Coupled fluid flow and geo-mechanical stress model; and
- Fluid flow modeling under thermal stress.

7.6.1 Speed and accuracy

The need for new equations in oil reservoirs arises mainly for fractured reservoirs as they constitute the largest departure from Darcy's flow behavior. Advances have been made in many fronts. As the speed of computers increased following Moore's law (doubling every 12–18 months), the memory also increased. For reservoir simulation studies, this translated into the use of higher accuracy through inclusion of higher order terms in Taylor series approximation as well as great number of grid blocks, reaching as many as billion blocks.

Large scale reservoir simulation is thought to be essential to understanding various flow processes inside the reservoir (Al-Saadoon et al., 2013). This has prompted the development of high-resolution reservoir modeling using simulation, some (e.g., Saudi Aramco's GigaPOWERS™) capable of simulating multi-billion cell reservoir models.

The implicit assumption is the finer the grid size, the better the description of the reservoir. This notion has motivated researchers to employ high performance computing (HPC) to simulate models larger than even 1 billion cells. As pointed out by Al-Saadoon et al. (2013), Linux clusters have become popular for large-scale reservoir simulation. Many large clusters have been built by connecting processors via high speed networks, such as Infiniband (IB). By connecting multiple computer clusters to build a simulation grid, giant models that are otherwise impossible (due to size limitations) to model with a single cluster can be modeled. A parallel-computing approach would be a suitable technique to tackle these challenges of large simulators. In addition to parallel computing, cloud computing in which a problem is divided into a number of sub-problems and the each sub-problem is solved by a cluster (Armbrust et al., 2010). One such algorithm is MAPREDUCE that has shown positive results in several applications (Dean and Ghemawat, 2008).

Vast majority of efforts in numerical simulation has been in developing faster solution techniques. One such model, called adaptive algebraic multigrid (AMG) solver was reported by Clees and Ganzer (2010). Similarly, other techniques focus on refinement and redistribution of gridblocks, some generating a background gridblock system that is extracted from single phase flow to be used as a base for multiphase flow calculations (Evazi and Mahani, 2010).

The greatest difficulty in this advancement is that the quality of input data did not improve on a par with the advances in speed and memory capacity of computers. As Fig. 7.14 shows, the data gap remains possibly the biggest challenge in describing a reservoir. Note that the inclusion of large number of grid blocks makes the prediction more arbitrary than that predicted by fewer blocks, if the number of input data points is not increased proportionately. The problem is particularly acute when fractured formation is being modeled. The problem of reservoir cores being smaller than the representative elemental volume (REV) is a difficult one, which is more accentuated for fractured formations that have a higher REV. For fractured formations, one is left with a narrow band of grid blocks, beyond which solutions are either meaningless (large grid blocks) or unstable (too small grid blocks). This point is elucidated in Fig. 7.14. Fig. 7.14 also shows the difficulty associated with modeling with both too small or too large grid blocks. The problem is particularly acute when fractured formation is being modeled. The problem of reservoir cores being smaller than the Representative Elemental

802　　　　　　　　　　　　　7. Field guidelines

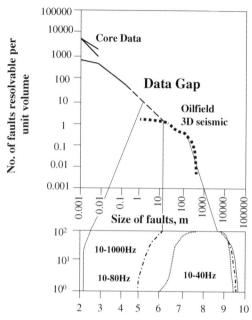

FIG. 7.14　Data gap in geophysical modeling. *After Islam (2001).*

Volume (REV) is a difficult one, which is more accentuated for fractured formations that have a higher REV. For fractured formations, one is left with a narrow band of grid blocks, beyond which solutions are either meaningless (large grid blocks) or unstable (too small grid blocks) (Fig. 7.15).

7.6.2 New fluid-flow equations

A porous medium can be defined as a multiphase material body (solid phase represented by solid grains of rock and void space represented by the pores between solid grains) characterized by two main features: that a representative elementary volume (REV) can be determined for it, such that no matter where it is placed within a domain occupied by the porous medium, it will always contain both a persistent solid phase and a void space. The size of the REV is such that parameters that represent the distributions of the void space and the solid matrix within it are statistically meaningful.

Theoretically, fluid flow in porous media is understood as the flow of liquid or gas or both in a medium filled with small solid grains packed in homogeneous manner. The concept of heterogeneous porous media then introduced to indicate properties change (mainly porosity and permeability) within that same solid-grains-packed system. An average estimation of properties in that system is an obvious solution, and the case is still simple.

Incorporating fluid flow model with a dynamic rock model during the depletion process with a satisfactory degree of accuracy is still difficult to attain from currently used reservoir simulators. Most conventional reservoir simulators, however, do not couple stress changes and rock deformations with reservoir pressure during the course of

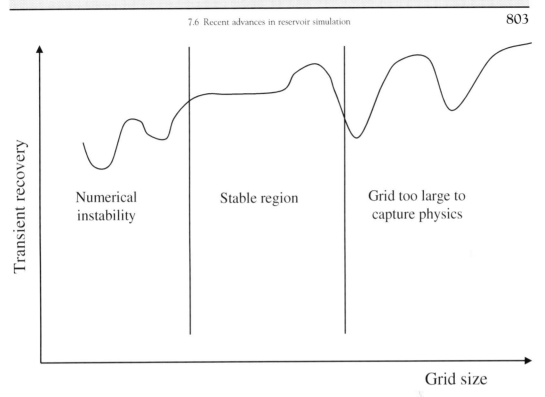

FIG. 7.15   The problem with the finite difference approach has been the dependence on grid size and the loss of information due to scaling up. *From Islam, M.R., 2002. Emerging technologies in subsurface monitoring of petroleum reservoirs. Petroleum Res. J. 13, 33–46.*

production; nor do they include the effect of changes in reservoir temperature during thermal or steam injection recovery. The physical impact of these geo-mechanical aspects of reservoir behavior is neither trivial nor negligible. Pore reduction and-or pore collapse leads to abrupt compaction of reservoir rock, which in turn causes miscalculations of ultimate recoveries, damage to permeability and reduction to flow rates and subsidence at the ground and well casings damage. Using only Darcy's law to describe hydrocarbon fluid behavior in petroleum reservoirs when high gas flow rate is expected or when encountered highly fractured reservoir is totally misleading. Nguyen (1986) has showed that using standard Darcy flow analysis in some circumstances can over-predict the productivity by as much as 100%.

Fracture can be defined as any discontinuity in a solid material. In geological terms, a fracture is any planar or curvy-planar discontinuity that has formed as a result of a process of brittle deformation in the earth's crust. Planes of weakness in rock respond to changing stresses in the earth's crust by fracturing in one or more different ways depending on the direction of the maximum stress and the rock type. A fracture can be said to consist of two rock surfaces, with irregular shapes, which are more or less in contact with each other. The volume between the surfaces is the fracture void. The fracture void geometry is related in various ways to several fracture properties. Fluid movement in a fractured rock depends on discontinuities, at a variety of scales ranging from micro-cracks to faults (in length and width).

Fundamentally, describing flow through fractured rock involves describing physical attributes of the fractures: fracture spacing, fracture area, fracture aperture and fracture orientation and whether these parameters allow percolation of fluid through the rock mass. Fracture parameters also influence the anisotropy and heterogeneity of flow through fractured rock. Thus the conductivity of a rock mass depends on the entire network within the particular rock mass and is thus governed by the connectivity of the network and the conductivity of the single fractures. The total conductivity of a rock mass depends also on the contribution of matrix conductivity at the same time.

A fractured porous medium is defined as a portion of space in which the void space is composed of two parts: an interconnected network of fractures and blocks of porous medium, the entire space within the medium is occupied by one or more fluids. Such a domain can be treated as a single continuum, provided an appropriate REV can be found for it.

For fractured formations, the fundamental premise is that the bulk of the fluid flow takes place through fractures. Such premise is justified based on Darcy's law:

$$\bar{v} = -\frac{k}{\mu}\nabla P \qquad (7.48)$$

Here $\bar{v}$ is the velocity, k the permeability, μ the viscosity, and P is the pressure. Permeability has the dimension of L^2, which means it is exponentially higher in any fracture than the matrix. For a system with very low permeability, fracture flow accounts for 99% of the flow whereas in terms of volume fractures account for 1% of the volume of the void (or total porosity). This is significant, because in classic petroleum engineering, governing equations are always applied without distinction between storage site (where porosity resides) and flow domain (where permeability is conducive to flow). In case fracture network is important for fluid flow whereas the matrix is important for fluid storage volume (e.g. unconventional reservoirs), permeability values are so low in the matrix, typical Darcy's law does not apply to the matric domain. It is recommended that Forchheimer equation be used to describe gas flow. This equation is given by:

$$-\nabla P = \frac{\mu}{k}v + \rho\beta v^2 \qquad (7.49)$$

In the above equation, β is an additional proportionality constant that depends on rock properties. For a system with predominantly fracture flow, β would depend on the fracture density and aspect ratio.

In case, fracture network is insignificant and the reservoir matrix permeability is very low, flow in such system is best described with Brinkman equation, described as:

$$-\frac{\partial P}{\partial x} = u\frac{\mu}{k} - \mu\frac{\partial^2 u}{\partial x^2}. \qquad (7.50)$$

To-date the most commonly used model is that proposed by Warren and Root (1963). This so-called dual porosity model (Fig. 7.16) assumes that two types of porosity are present in the formation, one arising from vug's and fracture system whereas the other from matrix. For unconventional reservoirs, the matrix permeability is negligible compared to fracture permeability (hence depicted with shades). Warren and Root invoked similar assumptions even for a matrix with relatively high permeability. The approach operates on the concept that

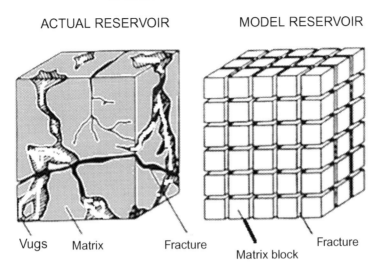

FIG. 7.16 Depiction of Warren and Root model.

fractures have large permeability but low porosity as a fraction of the total pore volume. The matrix rock has the opposite properties: low permeability but relatively high porosity. This approach describes the observation that fluid flow will only occur through the fracture system on a global scale. Locally, fluid may flow between matrix and fractures through interporosity flow, driven by the pressure gradient between matrix and fractures.

Fracture flow is described by Snow's equation (1963), given below:

$$\frac{Q}{\Delta P} = Cw^3 \tag{7.51}$$

Here w is the fracture aperture and C is a proportionality constant that depends on the flow regime that prevails in the formation. Snow's equation emerges from a simple synthesis of parallel plate flow (Poiseuille's law) that assumes permeability to be $b^2/12$, where b is the fracture width.

Fracture geometries are often idealized to simplify modeling efforts. In most cases the width is assumed to be constant, and the fracture is usually considered either a perfect rectangle or a perfect circle. In reality, fracture geometries are very complex (Fig. 7.16), and many different factors could affect the behavior of fluid flow. In Fig. 7.16 that was originally published by Warren and Root (1963), vug's are shown prominently. It is no surprise, they introduced the concept of dual porosity. Indeed, porosity in vugs and in matrix are comparable. For unconventional reservoirs, however, the vugs are non-existent and most fractures have very little storage capacity, making their porosity negligible to that of the matrix. In determining sweet spots within an unconventional reservoir, the consideration of very high fracture to matrix permeability, k_f/k_m is of importance. For application in dynamic reservoir characterization using real-time mud log data, the term "sweet spot" is used to characterize the point at which the drill bit intersects a transverse natural fracture. Such a process is equivalent to numerous passes of history match in the context of reservoir simulation.

The conventional approach involves the use of dual-porosity, dual-permeability models for simulating flow through fractures. Choi et al. (1997) demonstrated that the conventional use of Darcy's law in both fracture and matrix of the fractured system is not adequate. Instead, they proposed the use of the Forchheimer model in the fracture while maintaining Darcy's law in the matrix. Their work, however, was limited to single-phase flow. In future, the present status of this work can be extended to a multiphase system. It is anticipated that gas reservoirs will be suitable candidates for using Forchheimer extension of the momentum balance equation, rather than the conventional Darcy's law. Similar to what was done for the liquid system (Cheema and Islam, 1995); opportunities exist in conducting experiments with gas as well as multiphase fluids in order to validate the numerical models. It may be noted that in recent years several dual-porosity, dual-permeability models have been proposed based on experimental observations (Saghir et al., 2000).

Wu et al. (1997) used Buckley-Leverett-type analytical solution for the displacement in non-Darcy, two-phase immiscible flow in porous media, using Baree and Conway's modification. However, they do not include different equation in different regions, as envisioned by Choi et al. (1997).

He (2015) proposed a single-phase transient flow model that is applicable to naturally fractured carbonate karst reservoirs. It involves Stokes-Brinkman equation and a generalized material balance equation. The model is then generalized to 2D and 3D. They show three examples of field applications with favorable results.

7.6.3 Coupled fluid flow and geo-mechanical stress model

Coupling different flow equations has always been a challenge in reservoir simulators. In this context, Pedrosa and Aziz (1986) introduced the framework of hybrid grid modeling. From this point onward, the focus became coupling local grid systems with the mainframe simulator. For instance, Wasserman (1987) developed and implemented a static local grid refinement technique in a three-dimensional, three-phase reservoir simulator. This one did not allow any transient effect to be incorporated in the fluid transmissibility equations. A more general theoretical formulation on grid refinement using composite grids with variable coefficients was developed and could be incorporated into existing codes without disrupting the basic solution process (Ewing and Lazarov, 1988, 1989). Ewing et al. (1988) further refined the technique to include multi-well, multiphase black oil flow problems. These formulations that related to coupling cylindrical and Cartesian grid blocks, were adapted as a basis for coupling various fluid flow models (Islam, 1990). An adaptive static and dynamic local grid refinement technique was developed for multidimensional, multiphase reservoirs (Biterge and Ertekin, 1992). Nacul and Aziz (1991) recommended using "centered" method yields the most accurate results while tested against five other kinds of gridding. Satisfactory results requiring less computational time were obtained using non-uniform grids and explicit modeling with uniform grid refinement (Wan et al., 1998). Ever since the early work of Pedrosa et al., it has become a common practice to use local grid refinement for various numerical reservoir simulations. The displacement fronts in heterogeneous three dimensional reservoirs with two phase immiscible flow were more accurately tracked using adaptive local grid refinement (Ding and Lemonnier, 1993). The application ranges from steam injection to highly faulted fractures (Haajizadeh and Begg, 1993). It has been found that there is a threshold of grid refinement, beyond which, the results do not improve (Aqeel and Cunha, 2006; Al-Mohannadi et al., 2007). In fact, excessive refinement can lead to spurious results.

Geomechanical stresses are very important in production schemes. However, due to strong seepage flow, disintegration of formation occurs and sand is carried toward the well opening. The most common practice to prevent accumulation as followed by the industry is take filter measures, such as liners and gravel packs. Generally, such measures are very expensive to use and often, due to plugging of the liners, the cost increases to maintain the same level of production. In recent years, there have been studies in various categories of well completion including modeling of coupled fluid flow and mechanical deformation of medium (Vaziri et al., 2001). Vaziri et al. (2001) used a finite element analysis developing a modified form of the Mohr–Coulomb failure envelope to simulate both tensile and shear-induced failure around deep wellbores in oil and gas reservoirs. The coupled model was useful in predicting the onset and quantity of sanding. Nouri et al. (2007) highlighted the experimental part of it in addition to a numerical analysis and measured the severity of sanding in terms of rate and duration. It should be noted that these studies (Nouri et al., 2007; Vaziri et al., 2001) took into account the elasto-plastic stress-strain relationship with strain softening to capture sand production in a more realistic manner. Although, at present these studies lack validation with field data, they offer significant insight into the mechanism of sanding and have potential in smart-designing of well-completions and operational conditions.

Yalamas et al. (2004) developed a poromechanical approach. They identified two zones, namely, the virgin reservoir zone with poroelastic behavior and sand-oil slurry zone that behaves like a Poiseuille fluid. The fluid viscosity varies with sand content in the oil.

Settari et al. (2008) applied numerical techniques to calculate subsidence induced by gas production in the North Adriatic. Due to the complexity of the reservoir and compaction mechanisms, they took a combined approach of reservoir and geo-mechanical simulators in modeling subsidence. As well, an extensive validation of the modeling techniques was undertaken, including the level of coupling between the fluid flow and geo-mechanical solution. The researchers found that a fully coupled solution had an impact only on the aquifer area, and an explicitly coupled technique was good enough to give accurate results. On grid issues, the preferred approach was to use compatible grids in the reservoir domain and to extend that mesh to geo-mechanical modeling. However, it was also noted that the grids generated for reservoir simulation are often not suitable for coupled models and require modification.

In fields, on several instances, subsidence delay has been noticed and related to over consolidation, which is also termed as the threshold effect (Merle et al., 1976). Settari et al. (2008) used the numerical modeling techniques to explore the effects of small levels of over-consolidation in one of their studied fields on the onset of subsidence and the areal extent of the resulting subsidence bowl. The same framework that Settari et al. (2008) used can be introduced in coupling the multiphase, compositional simulator and the geo-mechanical simulator in future.

Zandi (2011) proposed a methodology to simulate the SAGD process, in which she studied the influence of the number of coupling periods in explicit coupling approach. She reported reduction of the computation time through an enhanced coupling method that uses a two-grid system.

In modeling geomechanical systems, accurate results can be obtained with coarser grid resolutions (Guy et al., 2012). In their modeling efforts, they solved a fine Geomechanical Grid (GG) with 6175 elements as well as with 1235 elements. It was observed that coarser model gave rise to quicker approach to optimal time step, in addition to being computationally more efficient. In this regard, the use of Adaptive Mesh Refinement (AMR) can be helpful by dynamically refining the mesh, thus striking a balance between grid refinement and computational efficiency. A number of publications indicate the effectiveness of this technique (Lacroix et al., 2003; Christensen et al., 2004; Wang et al., 2006). Mamaghani et al. (2011) showed on SAGD cases, the reduction of the grid size varies during the simulation and also depends on the criteria used to follow the flow interface. At the beginning of the simulation, when the extent of the steam chamber is small, AMR methods enable to reduce considerably the grid size. The reduction of the cell number decreases afterwards as more oil is heated and flows toward the producer wells but the method still provides substantial savings of computational times.

Guy et al. (2013) combined the approach of Guy et al. (2012) with Lacroix et al. (2003) and Mamaghani et al. (2011) to construct a new coupling scheme of geomechanical and reservoir simulators to obtain precise production forecasts with reduced computational times. They integrated both a field transfer module and an AMR description of the thermal fluid flow. This procedure couples geomechanics and thermal reservoir flow separate grids and Adaptive Mesh Refinements (AMR) to follow the flow interfaces. They showed that it has been shown that the sequential iterative coupling approach combining AMR and non-AMR contexts leads to a comprehensive description of the considered physical phenomena. In addition, the mesh reduction is of about 90% for the Geomechanical Grid (GG) in the coupled area and of more than 70% for the reservoir grid compared to classical coupled analyses without a reduction of accuracy in the computation.

7.6.4 Fluid-flow modeling under thermal stress

The temperature changes in the rock can induce thermo-elastic stresses (Hojka et al., 1993), which can either create new fractures or can alter the shapes of existing fractures, changing the nature of primary mode of production. It can be noted that the thermal stress occurs as a result of the difference in temperature between injected fluids and reservoir fluids or due to the Joule Thompson effect. However, in the study with unconsolidated sand, the thermal stresses are reported to be negligible in comparison to the mechanical stresses (Chalaturnyk and Scott, 1995). Similar trend is noticeable in the work by Chen et al. (1995), which also ignored the effect of thermal stresses, even though a simultaneous modeling of fluid flow and geomechanics is proposed.

Most of the past research has been focused only on thermal recovery of heavy oil. Modeling subsidence under thermal recovery technique (Tortike and Farouq Ali, 1987) was one of the early attempts that considered both thermal and mechanical stresses in their formulation. There are only few investigations that attempted to capture the onset and propagation of fractures under thermal stress. Chaalal et al. (2017) investigated the effects of thermal shock on fractured core permeability of carbonate formations of UAE reservoirs by conducting a series of experiments. Also, the stress-strain relationship due to thermal shocks was noted. Apart from experimental observations, there is also the scope to perform numerical simulations to determine the impact of thermal stress in various categories, such as water injection, gas injection/production, etc. Hossain et al. (2009) showed that new mathematical models must be introduced in order to include thermal effects combined with fluid memory.

Gunnarsson et al. (2011) introduced an interesting case in which thermal changes altered reservoir rock properties in the most tangible fashion. The formation temperature at the originally planned reinjection zone proved to be very hot (>300 °C) when injection wells were drilled there. In order to be able to use that zone for production a new reinjection zone was planned. The wells were drilled into a fault that is active and injection tests have resulted in swarms of small earthquakes. The injectivity of the wells in the new reinjection zone was found to be highly dependent on temperature of the reinjected water. This dependence can be explained by thermo-mechanical effects on fractures in the fracture governed reservoir. The temperature of the reinjected water was varied between 20 and 120 °C. Surprisingly, the injectivity of the wells in the new reinjection zone was much higher for colder water than for warmer water, although the viscosity of the cold water is five times larger than that of the hot water. This effect was observed in a number of wells, and the injectivity ratio between injection with cold and with hot water varied between 3 and 6. This offered an excellent opportunity to model a complex case with all required data available for possible history matching. This was done by Šijačić and Fokker (2015). They implemented a finite element modeling approach for the simulation of enhanced geothermal operations with a full coupling of mechanics, flow and temperature. They observed that the thermal effects play a crucial role in the development of injectivity, and thus should be taken into account in the operational design. The thermal effects are not limited to the time of injection but persist during the entire shut-in period. In all measures, they concluded that neglecting the thermal effects is not justified, since they do change the physics of the system. The von Mises stress was found to be more than three times higher for the system where thermal effects are included, and so is the permeability enhancement. Also, stress effects are localized at the fracture walls for the fully coupled system. This is commensurate with physical observations by many. In addition, authors included the effect of seismicity. Starting from first principles, they derived a seismicity model that was included in the fully coupled thermo-hydro-mechanical. This seismic model included frictional weakening and healing. Those processes ensure that the estimated seismicity is realistic and that the events of larger magnitudes are included. The fully coupled model triggered more seismicity during the stimulation period and the largest seismic event had larger magnitude compared to that of the model where thermal effects were neglected. This work lays the foundation for modeling thermally active reservoirs, both in heavy and light oil reservoirs.

Fig. 7.17 shows to what extent coupling of various governing equations has to be made. In case, non-Darcy flow is prevalent, the momentum balance equation will have to be non-Darcian flow equations (e.g., Forchheimer or Brinkman equations). Poroelasticity describes the influence of pore pressure on stress and strain. The changes in stress and strain change permeability and porosity that affect fluid flow. Heat-convecting fluid flow (cold water injected in hot rock or hot water/steam in a cold formation) influences the temperature distribution, which in turn influences the fluid viscosity (and density), altering again the flow itself. The temperature change will also create thermal stresses, effecting the geomechanics. In the geomechanical model we used Mohr-Coulomb failure and associated shear dilation for fault reactivation.

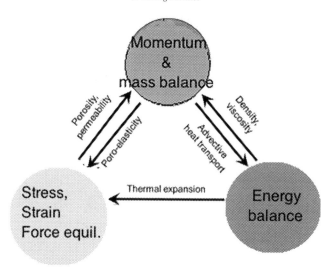

FIG. 7.17 Coupling of geomechanics, fluid flow and heat transport in EGS.

7.6.5 Memory models

Continuous time function is of utmost importance in turning simulation into emulation. In scientific term, the time function stands for "memory." Zhang (2003) defined memory as a function of time and space, and forward time events depend on previous time events; Zavala-Sanchez et al. (2009) showed that the system "remembers" its initial state, which was defined as memory effects for the effective transport coefficients; and Hossain and Abu-Khamsin (2012a, b) defined memory as the effect of past events on the present and future course of developments. When a complex fluid flows through porous media, there is a change in rock/fluid properties due to thermal alterations, chemical reaction, mineral precipitation, etc. and therefore, they diminish over time. As a result, solving equation models governing the diffusion brings huge analytical and/or numerical challenges due to the variation of rock/fluid properties.

7.6.6 Thermal hysteresis

In oil–water systems, hysteresis is due to several factors, such as the contact angle hysteresis, oil trapping, capillary hysteresis and an asymmetric role played by various types of pores in drainage and imbibition (Caputo, 1997, 2000). Hysteresis properties may be altered by the elevated heat and fluid flows (Hossain et al., 2011a, 2011b). The dynamic theories of multi-phase porous media developed for hydrocarbon reservoirs are tailored to the requirements of wave propagation problems and, as a result, drag forces have attracted more attention than capillary hysteresis. Effects of temperature on capillary properties and the relaxation time of the liquid flux by capillary effects are also taken into account and studied. Neglecting hysteresis effects in simulating a dynamic transport process can lead to errors in the instantaneous heat flux of up to 51% for an initially dry fiberglass slab and 23% for an initially moist slab (Zhang, 2003). Therefore modeling the dynamic process (e.g., oil-water displacement process) with thermal hysteresis and capturing the phenomena through experiment is very important.

Previous literature shows that the most important knowledge gap occurs in including thermal hysteresis during the study of thermal alterations of rock and fluid. This phenomenon, even though recognized in mid-80s in both Darcian and non-Darcian flow (Islam and Nandakumar, 1986, 1990; Islam, 1993) has eluded most researchers of petroleum reservoir modeling.

7.6.7 Mathematical and numerical models

The literature on mathematical modeling of rock/fluid interactions in porous media shows that majority of researchers aimed at developing models with memory (Hossain and Islam, 2009). Closing the gap of developing a model which combines rock and fluid interactions incorporating thermal hysteresis becomes an important future task in modeling. Associated with numerical modeling, Hossain et al. (2009) showed that the main sources of nonlinearity are the inherent and continuous alterations of rock/fluid properties during the flow process within a reservoir. This notion leads to multiple solutions, which have not been considered during the development of existing fluid flow models through porous media. Multiple solutions were observed decades ago, and were correctly dubbed as "spurious" by the ground-breaking work of Buckley and Leverett (Caputo, 1997).

The current literature shows that memory-based models have proven very effective in tackling the complex nature of the actual problem (Hossain et al., 2009; Christensen, 2003; Zavala-Sanchez et al., 2009). However, these models do not include thermal hysteresis. Therefore, new sets of mathematical models need to be developed in terms of a group of heat transfer dimensionless numbers that contain terms related to varying rock and fluid properties using memory and thermal hysteresis. A challenge with this approach is that more complex constitutive equations will be used for fluid dynamics calculations, because of improved numerical methods/schemes and the rapid developments in computer memory. An underlying challenge is to identify representative equations, pertaining to thermal hysteresis effects. It may not be possible to use constitutive equations that contain a full spectrum of time constraints or spurious solutions may occur. However, the attempts should be made to track at least a few hysteresis maps within reasonable time frame.

7.6.8 Future challenges in reservoir simulation

The future development in reservoir modeling may be look at different aspects. These are may be classified as:

- Experimental challenges;
- Numerical Challenges; and
- Remote sensing and real-time monitoring.

7.5.9.1. Experimental challenges

The need for well-designed experimental work to improve the quality of reservoir simulators cannot be over-emphasized. Most significant challenge in experimental design arise from determining rock and fluid properties. Even though, progress has been made in terms of specialized core analysis and PVT measurements, numerous problems persist due to difficulties associated with sampling techniques and core integrity. Recently, Belhaj et al. (2009) used a 3D spot gas permeameter to measure permeability at any spot on the surface of the sample, regardless of the shape and size. Moreover, a mathematical model was derived to describe the flow pattern associated with measuring permeability using the novel device.

In a reservoir simulation study, all relevant thermal properties including coefficient of thermal expansion, porosity variation with temperature, and thermal conductivity need to be measured in case such information are not available. Experimental facilities e.g., double diffusive measurements, transient rock properties; point permeability measurements can be very important in fulfilling the task. In this regard, the work of Belhaj et al. (2009) is noteworthy.

In order to measure the extent of 3D thermal stress, a model experiment is useful to obtain temperature distribution in carbonate rock formation in the presence of a heat source. Examples include microwave heating water-saturated carbonate slabs with in order to model only conduction and radiation. An extension to the tests can be carried out to model thermal stress induced by cold fluid injection for which convection is activated. The extent of fracture initiation and propagation can be measured in terms of so-called damage parameter. Time-dependent crack growth still is an elusive topic in petroleum applications (Kim and van Stone, 1997). The methodology outlined by Yin and Liu (1994) can be considered to measure fracture growth. The mathematical model can be developed following the numerical method developed by Wang and Meguid (1995). Young's modulus, compressive strength, and fracture toughness are important for modeling the onset and propagation of induced fracture for the selected reservoir.

The most relevant application of double diffusive phenomena, involving thermal and solutal transfer is in the area of vapor-extraction (VAPEX) of heavy oil and tar sands. From the early work of Roger Butler, numerous experimental studies have been reported.

In modeling memory, there has been significant progress made, albeit in non-petroleum engineering applications. For instance, Gaede-Koehler et al. (2012) used ultrasound to measure hysteresis indirectly. More recently, magnetic properties have been instrumental in mapping thermal hysteresis (Lochenie et al., 2014). Chen et al. (2014) provided an eloquent discussion on the phenomenon of thermal hysteresis and outlined a procedure for comprehensive mapping of thermal hysteresis. To capture the memory and other continuous alteration parameters, the experiments should involve flooding a real core sample with hot water/steam and monitoring both the temperature and pressure along the length of the core. Because the concept of memory is not clearly defined in terms of tangible properties of natural elements, such as crude oil and reservoir formation, they cannot be visualized and tracked easily with conventional experimental practices. However, with currently available advanced and more sensitive technologies, it is possible to capture the memory effect existing in natural phenomena.

Upscaling becomes a formidable challenge in situations involving complex solid-fluid and fluid-fluid interactions, which are predominant in a complex reservoir system. One of the most important conditions of scaling is that dimensionless properties must be the same functions of dimensionless variables in the model and prototype. In reality, this may not be the case. In addition, the complete set of scaling criteria is very difficult to satisfy, leaving room for errors in prioritizing more relevant scaling groups. Hossain and Islam (2009) have developed the complete set of scaling groups involving fluid with memory. However, scaling criteria for which both fluid and rock memories are accounted for are still to be developed.

7.6.8.1 Numerical challenges

7.6.8.1.1 Theory of onset and propagation of fractures due to thermal stress

Fundamental work needs to be performed in order to develop relevant equations for thermal stresses. Similar work has been initiated by Wilkinson et al. (1997), who used finite element modeling to solve the problem. There has been some progress in the design of material manufacturing for which in situ fractures and cracks are considered to be fatal flaws. Similarly, there have been numerical models using finite difference modeling to predict onset of fractures in medical applications (Graeff et al., 2013). Therefore, formulation complete equations are required in order to model thermal stress and its effect in petroleum reservoirs. It is to be noted that this theory deals with only transient state of the reservoir rock.

Similar progress has been made in modeling fracture propagation using finite element methods. Tsai et al. (2013) published a series of papers on experimental and numerical modeling of fracture onset and propagation. They use symmetric smoothed particle hydrodynamics (SSPH) and moving least squares (MLS) basis functions to analyze six linear elastostatics problems by first deriving their Petrov-Galerkin approximations. With SSPH basis functions one can approximate the trial solution and its derivatives by using different basis functions whereas with MLS basis functions the derivatives of the trial solution involve derivatives of the basis functions used to approximate the trial solution. The two basis functions are also used to analyze crack initiation and propagation in plane stress mode-I deformations of a plate made of a linear elastic isotropic and homogeneous material with particular emphasis on the computation of the T-stress. The crack trajectories predicted by using the two basis functions agree well with those found experimentally.

7.6.8.1.2 Viscous fingering during miscible displacement

The term "fingering" originally refers to the phenomenon that the water/oil dgfs" during water flooding. It also refers to non-uniformity along the interface, often associated with instability. In essence, fingering is the performance of hydrodynamic instability and chaotic phenomenon which is often encountered in nonlinear dynamics (Sharma et al., 2012). Such fingering can be onset through any of the following: viscous fingering (Aubcrtin et al., 2009), capillary fingering (Ferer et al., 2007), density fingering (Rica et al., 2005) and gravity fingering (Di Giuseppe et al., 2010). In multiphase flow, fingering is one of the most important features that prevail in majority of the displacement cases, even though to-date no reservoir simulator can account for this phenomenon. It is because all governing equations are derived for stable and stabilized displacement fronts that are practically absent in the field (Yadali Jamaloei et al., 2015). However, there has not any lack of in-depth research on the phenomenon for decades (Hill, 1952; Saffman and Taylor, 1958; Bentsen, 1985; Islam and Bentsen, 1986; Bokhari and Islam, 2005). The interest in this topic emerges from its wide range of applications for energy, environment, industrial and engineering, such as enhanced oil recovery, underground storage of carbon dioxide, pollution of groundwater, biological treatment, chemical industrial processes and others.

Even though the academic significance has been recognized through numerous research studies, reservoir scale modeling of fingering phenomena eluded the petroleum industry. At present the research projects are mostly concentrated on viscous fingering and capillary fingering such as water flooding, polymer flooding and alkaline flooding. However, viscous fingering phenomenon occurs even in the displacement between a same fluid, for example, a low viscosity and miscibility fluid with crude oil displacing the crude oil in the stratum. In some applications, such as acid fracturing, the presence of finger can play a dual role, it increases access to rock and also decreases the frontal surface area. In water flooding, the fingering is undesirable since it reduces the sweep efficiency and leads to early breakthrough which will consequently reduce oil displacement efficiency, but it is favorable in acid fracturing because it could extend the effective distance acid, which gains the acid fracture conductivity.

Limited success has been achieved by researchers from Shell that attempted to model viscous fingering with the chaos theory. In principle, the problem of modeling viscous fingering can be handled by tracking periodic and even chaotic flow in a displacement process (Islam, 1993). This tracking is possible by solving the governing partial differential equations with improved accuracy (Δx^4, Δt^2). The tracking of chaos (and hence viscous fingering) in a displacement system can be further enhanced by studying phenomena that onset fingering in a reservoir. It eventually will lead to developing operating conditions that would avoid or minimize viscous fingering. However, to date, no reservoir simulation has adopted this solution scheme.

7.7 Real-time monitoring

The history-matching process can be considered as a closed-loop process. The production data from the real system are adopted to modify the reservoir and model parameters and to use for future prediction. The history matching process is usually carried out after period of years. The traditional history-matching involve manual adjustment of model parameters and is usually ad hoc. According to Brouwer et al. (2004), the draws backs of a traditional history-matching are:

a. It is usually only performed on a campaign basis, typically after periods of years.
b. The matching techniques are usually ad hoc and involve manual adjustment of model parameters, instead of systematic parameter updating.
c. Uncertainties in the state variables, model parameters and measured data are usually not explicitly taken into account.
d. The resulting history-matched models often violate essential geological constraints.
e. The updated model may reproduce the production data perfectly, but have no predictive capacity, because it has been over-fitted by adjusting a large number of unknown parameters using a too small number of measurements.

To adopt an optimum production strategy and to produce oil and gas in challenging physical environments such as deepwater reservoirs and oil-bearing formation in the

Arctic, it is required to update the model more frequently and systematically. If the model is also smart, it will produce real-time data. Smart field technologies are currently generating significant interest in the petroleum industry, primarily because it is estimated that their implementation could increase oil and gas reserves by 10%–15% (Severns, 2006). This help to optimize the reservoir performance under geological uncertainty and also incorporate dynamic information in real-time and reduce uncertainty. A schematic layout of a smart reservoir modeling is depicted in Fig. 7.18 (redraw from Brouwer et al., 2004). It is a real-time closed-loop to control the reservoir behavior and to attain an optimum production. This figure illustrate a true closed-loop optimal control approach that shifts from a campaign-based ad hoc history matching to a more frequent systematic updating of system models, based on data from different sources, while honoring geological constraints and the various sources of uncertainty.

This simulation model relates the simulated output data with a cost function. The cost function, designated by $J(u)$, might be net present value (NPV) or cumulative oil produced. The notation u is the vector of control variables, such as well rates, bottom hole pressures (BHP). The closed-loop process initiates with an optimization loop (marked in blue in Fig. 7.20). The control variables are set by minimizing or maximizing the cost function. The optimizing process must be performed on an uncertain simulation model. The control variables, i.e. well rates, BHP, are then applied to the real system, i.e. reservoir, wells and facilities, as input over the control step. These inputs impact the outputs from the real system. The new measured output data (production information) are applied to updated the model parameters to reduced the uncertainty. This is called the history-matching process that here called as the model-updating loop, marked in red in Fig. 7.18. The process repeated over the life of the reservoir to obtain an optimized performance of the system. The closed-loop approach for efficient real-time optimization consists of three key components: efficient

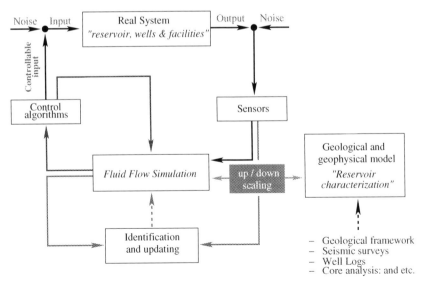

FIG. 7.18　Schematic layout of a comprehensive smart reservoir modeling. *Redrawn from Brouwer et al. (2004).*

optimization algorithms, efficient model updating algorithms, and techniques for uncertainty propagation.

The optimization process can be described in general form according to Sarma et al. (2005) as:

$$\max_{u^n} \left[J(u^0, u^1, \ldots, u^{N-1}) = \phi(x^N) + \sum_{n=0}^{N-1} L^n(x^{n+1}, u^n, m) \right] \text{ and } \forall n \in (0, 1, \ldots, N-1)$$

Subject to:

$g^n(x^{n+1}, x^n, u^n, m) = 0$ and $\forall n \in (0, 1, \ldots, N-1)$

$x^0 = x_0$ Initial condition of the system (reservoir, wells and the facilities)

$c^n(x^{n+1}, x^n, u^n, m) \leq 0$ and $\forall n \in (0, 1, \ldots, N-1)$

$Au^n \leq b$ and $\forall n \in (0, 1, \ldots, N-1)$

$LB \leq u^n \leq UB$ and $\forall n \in (0, 1, \ldots, N-1)$

(7.52)

Here, n is the control step index and N is the total number of control steps, x^n refers to the dynamic states of the system, such as pressures, saturations, compositions, etc., and m is the model parameters (permeability, porosity, etc.). The first equation is the cost function that should be maximize by controlling the variable vector u^n. The cost function subjected to a set of constraints consists of: the set of reservoir simulation equations g^n for each grid block at each time step, initial system condition and additional constraints for the controls including—nonlinear inequality path constraints, linear path constraints, and bounds on the controls. These path constraints must be satisfied at every time step. These last constraint can be maximum injection rate constraint, maximum water cut constraint, maximum liquid production rate constraint, etc. The function $\phi(x^N)$ is the dynamic states of the last control step (e.g., abandonment cost). The function L^n involves the quantity related to the well parameters.

This problem that same as history-matching problem may be solved using stochastic algorithms, such as genetic algorithm and simulated annealing algorithm, or deterministic algorithm, i.e., gradient-based methods. The deterministic methods are generally very efficient, requires few forward model evaluations and also guarantees reduction of the objective function at each iteration, but only assures local optima for non-convex problems (Gill et al., 1982). However, the computation of the gradients of the objective function respect to the control variables are complicated to obtain analytically and should be used numerical methods to compute them. The number of required gradient depends on the well number and well variables, i.e., BHP, that is taken into account for the optimization process. The total number is the product of the well number and the number of well variables.

The most efficient algorithms for calculating gradients are the adjoint techniques especially for a large number of controls, as the algorithm is independent of the number of controls. The adjoint model equations are obtained from the necessary conditions of optimality of the optimization problem defined by Eqs. (3.44). The essence of the theory is that the cost function along with all the constraints can be written equivalently in the form of an augmented cost function using the a set of Lagrange multipliers. The result is a modified objective function. The solution procedure and more detail description may be found in Brouwer (2004) and Sarma et al. (2006).

7.8 Sustainability analysis of a zero-waste design

The concept of "zero waste living" has been generated from the undisturbed activities of nature. Nature is operating in zero-waste modes, generating tangible, intangible and long term benefits for the whole world. Natural activities increase their orderliness on paths that converge at infinity after providing maximum benefits over time. However, any product that resembles a natural product does not guarantee its sustainability unless the natural pathway is followed (Ibrahim, 2006). The term "sustainable" is a growing concept in today's technology development. Simply, it can be inferred that the word "sustainable" implies the benefits from the immediate to the long-term for all living beings. That is why any natural product or process is said to be inherently sustainable. Immediate benefits are termed as tangibles and long-term benefits are termed as intangibles. However, focus only on tangible benefits might mislead technological development, unless intangible benefits are also considered.

Even after extensive development of different technologies from decade to decade, it is found that the world is becoming a container of toxic materials and losing its healthy atmosphere continuously. That is why it is necessary to test the sustainability of any process and the pathway of the process. To date, a number of definitions of sustainability are found in the literature: the most common of all definitions of sustainability is the concern about the well-being of future generations. However, most of the assessment processes of sustainability are incomplete and because of the lack of appropriate assessment processes, there exists a tendency to claim a product or technology sustainable without proper analysis. Recently, Khan and Islam (2007a) developed a sustainability test model distinct from others, as they emphasized the time factor and showed the direction of both new and existing technologies. According to this model, if and only if, a process travels a path that is beneficial for an infinite span of time, is it sustainable; otherwise the process must fall in a direction that is not beneficial in the long run. The pro-nature technology is the long-term solution, while the anti-nature one is the result of $\triangle t$ approaching 0.

The most commonly used schema would select technologies that are good for $t =$ "right now," or $\triangle t = 0$. In reality, such models are non-existent and, thus, aphenomenal and cannot be placed on the graph (Fig. 7.19). However, "good" technologies, can be developed following the principles of nature. In nature, all functions or techniques are inherently sustainable, efficient and functional for unlimited time periods. In other words, as far as natural processes are concerned, "*time tends to Infinity,*" this can be expressed as t or, for that matter, $\triangle t \to \infty$ (Khan and Islam, 2007a).

It is found from the figure that perception does not influence the direction and intensity of the process. Only the base is found shifted toward top or bottom, depending on whether the direction of the process is explicit as the process advances with time.

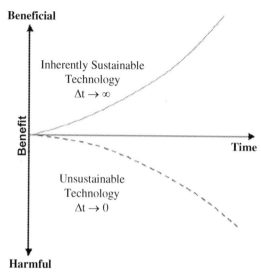

FIG. 7.19 Direction of sustainability/green technology.

The model has been developed observing nature and by the analysis of human intervention in the past years. Today's regulatory organizations are banning lots of chemicals and synthetic materials from production and regular use. However, the goal of banning remains controversial (Thompson et al., 2009). The vulnerable effects of products or processes that are tangible in nature are selected for bans. However, the products or processes that have intangible, long-term adverse effects are allowed to continue. Islam et al. (2010) has shown that intangible and long-term harm is more detrimental than tangible, short-term harm. Any process or product should be analyzed carefully to understand both tangible and intangible effects.

Khan and Islam (2012, 2016) proposed a "zero-waste model" that has been tested for sustainability according to the above definition and a detailed analysis has been presented here. The technology that exists in nature must be sustainable because natural processes have already proved to be beneficial for the long term. The bio-digestion process is an adequate example of natural technology and that is why the technology is sustainable. The burning of marsh gas is seen in nature. So the burning of biogas would not be anti-nature. It is assumed that the carbon dioxide production from the petroleum fuel (the old food) and the carbon dioxide from the renewable biomaterials (new food) are not the same. The same chemical with different isotope numbers must not be identical. The exact difference between a chemical that existed for millions of years and the same chemical that existed for only a few years, is not clearly revealed. According to nature, it can be said that the waste/exhaust from the biomaterial is not harmful.

Khan and Islam (2012) introduced a mathematical model to this concept and explained the role of intention in the sustainable development process. The intention is found as the key tool for any process to be sustainable. In this chapter, this model has been further developed by analyzing the term "perception" which was found important at the beginning of

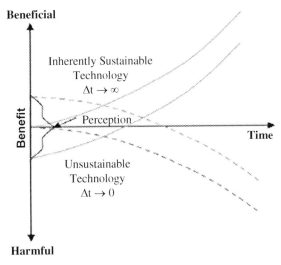

FIG. 7.20 Sustainability model with perception.

any process, as explained by Khan and Islam (2012). Perception varies from person to person. It is very subjective and there is no way to prove if perception is true or false. Perception is completely dependent on one's personal opinion and developed from one's experience without appropriate knowledge. That is why perception cannot be used as the base of the model. However, if perception is used in the model, the model would look as follows (Fig. 7.20):

The problem of all renewable and nonrenewable is the use of highly pure fossil fuels, synthetic and toxic electrolytes, and ultra-pure metallic electrodes at various stages of the processing. However, at each stage, the use of natural energy sources or materials can be introduced at its natural state can render a technology sustainable.

As an example of 100% sustainable technology, Khan and Islam (2012) presented the use of the Einstein refrigeration cycle, coupled with direct solar heating. This system uses ammonia, butane and water exist in nature. So the uses of these materials make the refrigeration process sustainable.

Solar energy is non-toxic and free. The utilization of solar energy must be sustainable. The replacement of thermal oil with waste oil makes this process sustainable. The absence of processing chemicals keeps the equipment corrosion-free (Al-Darbi et al., 2002a, b). Even any use of solar energy, in turn, reduces the need for refined fuel in a remote location.

Two types of refrigeration systems are in common use, viz. (a) a vapor compression type and (b) an absorption type. Vapor compression refrigeration cycles are energy-inefficient technology, in which electricity is the source of energy. The energy consumption, as well as energy loss of vapor compression cycle, is very high. On the other hand, an absorption-type refrigeration/cooling system is operated by direct firing. Among different absorption refrigeration cycles, we are particularly interested in the single-pressure thermally-driven refrigeration cycle derived from a patent (US Patent 1,781,541) issued to Einstein and Szilard. Most absorption cooling systems have at least one pump to lift the fluid, but Einstein's refrigeration system is solely heat-dependent. Delano (1998) in his research work created a thermodynamic model of the Einstein refrigeration cycle to calculate the cycle performance.

Around 1930, General Motors and Du Pont developed the first synthetic refrigerant (trade name: Freon) which was claimed to be nontoxic during that time. These chemical compounds are extremely stable, non-flammable, non-corrosive, and cheap to produce, which is why they have been widely used as refrigerants, especially in vapor compressor cooling systems (Auffhammer et al., 2005). However, scientists recognized as early as 1974 that the extensive use of Freon (CFC or HCFC) is depleting the Ozone Layer. To reduce the destruction of stratospheric ozone, a treaty known as The Montreal Protocol in 1987, was signed by 24 nations and the European Economic Community to regulate the production and trade of ozone-depleting substances. It was thereafter revised several times (Auffhammer et al., 2005). The Montreal Protocol on Substances that Deplete the Ozone Layer (1987) is considered one of the most successful and important pieces of international environmental legislation in history (Auffhammer et al., 2005).

It took nearly 60 years to ban Freon, after realizing its detrimental effects on the earth. On the contrary, Albert Einstein and Leo Szilard patented the adsorption refrigeration system in 1930. Almost 65 years later, Delano (1998) first validated it and found the process sustainable. In fact, to date, this process is one of the very few sustainable processes, which uses environmentally benign fluids. It is one of the reasons to choose the Einstein cooling system in addition to the use of direct solar heat.

Solar energy is the only ultimate source of energy available to planet Earth. This source is clean, abundant, and free of cost. The solar constant of solar energy is 1367.7 W/m^2, which is defined as the quantity of solar energy (W/m^2) at normal incidence outside the atmosphere (extraterrestrial) at the mean sun-earth distance. In space, solar radiation is practically constant; on earth, it varies with the time of the day and year as well as with the latitude and weather. The maximum value on earth is between 0.8 and 1.0 kW/m^2 (Khan and Islam, 2016). But the method of utilizing solar energy is different from one application to another. Even when solar energy is utilized, the mere fact that the most common usage is the use of photovoltaic, the maximum efficiency can be only 15%–20%.

Solar energy is radiant light and heat from the sun that is harnessed using a range of technologies such as solar heating, solar photovoltaics, solar thermal electricity, solar architecture, and artificial photosynthesis. Solar technologies are broadly characterized as either (1) passive or (2) active, depending on the method used to capture, convert and distribute solar energy. Active solar techniques include the use of photovoltaic panels and solar thermal collectors to harness the energy. Passive solar techniques include (1) orienting the solar panels on a building toward the sun, (2) selecting materials with favorable thermal mass or light dispersing properties, and (3) designing spaces that naturally circulate air.

The development of affordable, inexhaustible, and clean solar energy technologies will have huge longer-term benefits. The use of solar energy technologies will increase energy security for energy-importing countries through (1) reliance on an indigenous, inexhaustible, and mostly import-independent resource, (2) enhancement of energy sustainability, and (3) reduction in pollution.

The design of any system that operates under the direct use of solar energy has maximum energy conversion efficiency. Solar energy is an excellent source for thermally driven absorption refrigeration cycles. They do not require a costly electric power plant and they use environmentally benign natural fluids. Current absorption systems are dominated by dual-pressure cycles, using a solution pump (which still requires a small electrical power source). Single-pressure cycles remove the need for a pump and any electrical power. Heat is the driving source for this heat pump.

The direct thermally-driven refrigeration system offers the following advantages over vapor compression cycles (Cui et al., 2005):

1) *Silent operation*: Thermally driven absorption refrigeration does not have any moving parts (compressor), unlike the vapor compression refrigeration systems.
2) *Simple structure*: It can operate at fairly low pressure compared to vapor compression refrigeration cycles, which is why it does not need any high-pressure equipment and piping systems.
3) *No need for electricity*: Upon efficient design, an absorption refrigeration system can solely be operated by heat. So no electricity is required.
4) *Higher heating efficiency*: Heat is directly used in the absorption system. So it eliminates the loss of heat during the conversion of heat to electricity. Due to direct use, higher heating efficiency is obtained.
5) *Inexpensive equipment*: No high-pressure equipment, piping systems, and moving parts are required. Only some low-pressure reservoirs and piping systems are required, which are inexpensive compared to the vapor cycle refrigerator.
6) *No moving parts*: As there is no compressor and moving parts, it is safer than other systems.
7) *High reliability*: The absorption system is highly reliable. It is not subjected to any electrical disturbance or any moving parts" problems.
8) *Portability*: The independence of electricity has made it portable. It can be installed in any place where there is a heat source. As solar energy is everywhere, it can be operated anywhere.

Most of the technologies in use today are not beneficial to living beings. Even after extensive development of different technologies from decade to decade, it is found that the world is becoming a container of toxic materials and losing its healthy atmosphere continuously. That is why it is necessary to test the sustainability of any process and the pathways of such processes. In this study, we have used the sustainability test model proposed by Khan and Islam (2007a) to analyze the two different systems of cooling. According to this model, if and only if, a process travels a path that is beneficial for an infinite span of time, is it sustainable; otherwise, the process must fall in a direction that is not beneficial in the long run. The pro-nature technology is the long-term solution, while the anti-nature one is the result of $\triangle t$ approaching 0. The most commonly used theme is to select technologies that are good for $t =$ "right now," or $\triangle t = 0$. In reality, such models are non-existent and, thus, aphenomenal and cannot be placed on the graph (Fig. 7.21). However, "good" technologies can be developed following the principles of nature. In nature, all functions or techniques are inherently sustainable, efficient, and functional for an unlimited time. In other words, as far as natural processes are concerned, "*time tends to Infinity.*" This can be expressed as t or, for that matter, $\triangle t \to \infty$. For example, Freon was extremely stable, non-flammable, non-corrosive, and cheap to produce at the beginning, but its detrimental effects were understood after a long time and thus proved as unsustainable products.

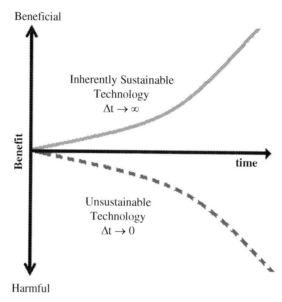

FIG. 7.21 Direction of sustainability. *Redrawn from Khan, M.I., Islam, M.R., 2007. The Petroleum Engineering Handbook: Sustainable Operations. Gulf Publishing Company, Houston, TX, 461 pp.*

Mathematically, the expression can be written as:

$$B = b + (\pm a)^i t e^{kt} \tag{7.53}$$

where B indicates the benefit of the process with time, b is the initial condition of the process or the product to be analyzed. The value of b can be positive, negative, or zero depending on the initial perception of the process or product. Symbols a and k are constants, depending on various factors such as social, environmental, and ecological factors of the products or processes to be analyzed. The index "i," is considered to be the directional parameter, which depends only on intention and indeed certifies whether the technology will be sustainable or not.

The positive and negative values indicate sustainable and unsustainable technology. The slope of the above equation can be written as:

$$\frac{dB}{dt} = (\pm a)^i k e^{kt} \tag{7.54}$$

The positive slope indicates the rate at which benefits are being achieved. Similarly, a negative slope demonstrates how fast a process/technology will lead to un-sustainability. It must be added here that if the intention is intended for the long-term or even meant for self-interest, it can be described by the positive slope. However, any action serving self-interest for the short-term is bound to collapse and must be described by the negative slope in the figure. For any new technology to be tested, mostly depends on the value of "i." But any existing technology that is already in operation can be tested from the field data.

Fig. 7.21 has the inherent assumption that nature is perfect. Even though this concept is as old as the universe, only recently is the science of this concept being discussed (TIME, 2005). Because no mass can be created or destroyed, and no mass can be isolated from the rest of the universe, any entity must play a role in creating overall balance (Khan and Islam, 2012). This is where intention becomes important for human beings. An intention to benefit others at present time translates into multifold benefits for the subject of the intention. Therefore, benefits for infinite numbers of entities in space, at any given time, are equivalent to multifold benefits for the individual in the long run.

The proposed model of this study has been tested for sustainability according to the above definition and the detailed analysis has been presented here.

The technology that exists in nature must be sustainable because natural processes have already proved to be beneficial for the long term. Ammonia, butane, and water exist in nature. So the uses of these materials make the refrigeration process sustainable. Solar energy is non-toxic and is a free source of energy. The utilization of solar energy must be a sustainable way of thinking. In this experimental solar collector, it has been recommended to use waste vegetable oil. The utilization of vegetable oil, keeps all of the equipment used, free of corrosion. Even any use of solar energy, in turn, reduces the dependency on fossil fuels and thereby saves the world from more air pollution. So, this is also sustainable technology.

On the other hand, alternate electricity is not natural. The process of producing electricity poses a vulnerable effect at every stage. The use of synthetic refrigeration is also anti-nature. The entirety of this makes the vapor refrigeration system aphenomenal.

7.8.1 Einstein refrigeration cycle

All thermally-driven heat pump cycles exchanging heat with only three temperature reservoirs are shown in Fig. 7.22. The Einstein cooling system is not an exception to this, but the use of three fluids that circulate in different reservoirs makes it solely heat-dependent and unique. Einstein and Szilard proposed the use of butane, ammonia, and water as the working fluids in their suggested absorption refrigeration systems (Einstein and Szilard, 1930).

This cycle is a completely thermally driven cycle employing environmentally benign fluids (Delano, 1998). In the cycle, butane acts as the refrigerant; ammonia as an inert gas; and water as an absorbent. It has three main components: (a) evaporator, (b) condenser/absorber, and (c) generator. The fluid cycles in different reservoirs are depicted in Fig. 7.23.

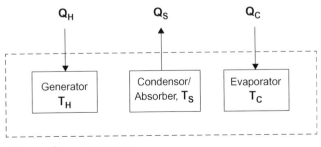

FIG. 7.22 Thermally driven refrigeration system.

824   7. Field guidelines

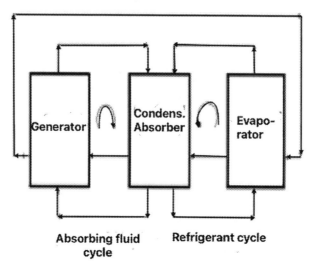

FIG. 7.23   Fluid cycles in the Einstein refrigeration cycle.

In the Einstein cycle, the evaporator is fed with liquid butane from the absorber and gaseous ammonia from the generator (Fig. 7.24). The presence of ammonia in the evaporator reduces the partial pressure of butane that results in a decrease in saturation temperature. This causes the butane to evaporate and cools the system and its surroundings. As soon as the gaseous mixture (butane ammonia), reaches the condenser/absorber through the pre-cooler, ammonia is readily absorbed by the spraying water coming from the generator. As a result, the partial pressure of butane increases nearly to the total pressure and it condenses in high saturation temperatures at that total pressure. The immiscibility and lighter density cause butane to float atop the liquid ammonia-water. Butane is siphoned back to the evaporator. The liquid ammonia-water leaves the absorber/condenser and reaches the generator through the heat exchanger. The application of heat inside the generator drives off ammonia vapor to the evaporator. The remaining weak ammonia solution is pumped up to a reservoir via a thermally driven bubble pump. The ammonia from the reservoir is separated and sent to the absorber/condenser and the liquids from the reservoir—mainly, water is passed through a heat exchanger and sprayed over a superheated gaseous mixture and thus, the whole cycle is completed.

7.8.2 Thermodynamic model and its Cycle's energy requirement

Delano (1998), in his research work, created a thermodynamic model of the Einstein refrigeration cycle to calculate the cycle performance. The cycle, shown in Fig. 7.25, has been modified from Einstein's original refrigeration cycle configuration. Various heat exchangers, such as the internal generator solution heat exchanger and the evaporator precooler have been added in order to create a cycle with higher efficiency. After an extensive

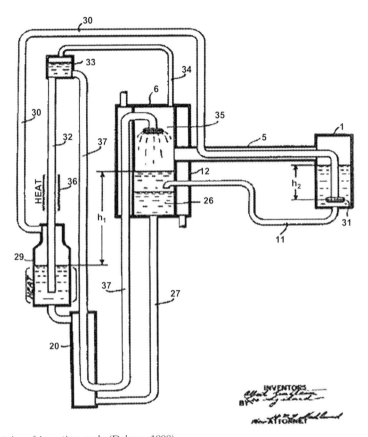

FIG. 7.24 Einstein refrigeration cycle (Delano, 1998).

analysis, Delano (1998) fixed the system pressure at 4 bar to obtain realistic operating conditions. Delano (1998) used the Patel-Teja equation of state to get the behavior of an ammonia-butane mixture at the evaporator and the behavior of the ammonia-water mixture at the generator.

A temperature-concentration plot of the ammonia-butane mixture obtained by Delano (1998) is shown in Fig. 7.26. From this plot, it is clear that for a given refrigerant (butane) at a fixed system pressure (4 bars), there is a maximum temperature (315 K) and a minimum temperature (266 K) at which the system can operate. These two extreme temperatures were chosen as the condenser/absorber temperature and as the evaporator temperature respectively for the thermodynamic model of Delano (1998).

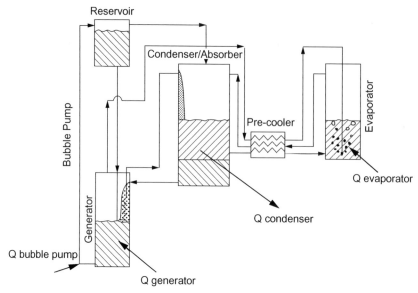

FIG. 7.25 Einstein refrigeration cycle schematic. *Redrawn from Shelton, S.V., Delano, A., Schaefer, L., 1999. Second law study of the Einstein refrigeration cycle. In: Proceedings of the Renewable and Advanced Energy Systems for the 21st Century, April 11–15, 1999, Lahaina, Maui, Hawaii.*

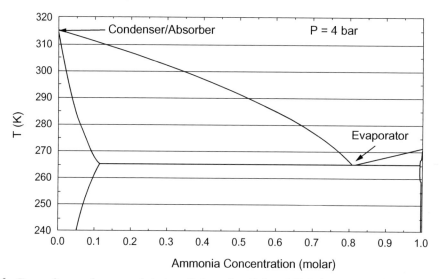

FIG. 7.26 *T-x-x-y* diagram for ammonia-butane. *Redrawn from Shelton, S.V., Delano, A., Schaefer, L., 1999. Second law study of the Einstein refrigeration cycle. In: Proceedings of the Renewable and Advanced Energy Systems for the 21st Century, April 11–15, 1999, Lahaina, Maui, Hawaii.*

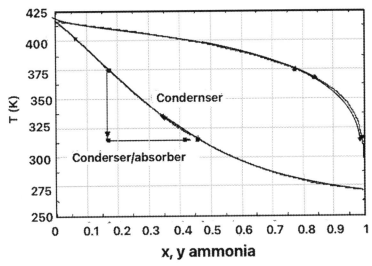

FIG. 7.27 *T-x-y* diagram for ammonia-water, $p = 4$ bar. *Redrawn from Shelton, S.V., Delano, A., Schaefer, L., 1999. Second law study of the Einstein refrigeration cycle. In: Proceedings of the Renewable and Advanced Energy Systems for the 21st Century, April 11–15, 1999, Lahaina, Maui, Hawaii.*

The selection of a suitable generator temperature needs to establish a temperature-concentration diagram for the ammonia-water mixture. A *T-x-y* diagram of the ammonia-water mixture was obtained by Delano (1998), as shown in Fig. 7.27. Lower temperatures would reduce the amount of the desorbed ammonia vapor, and higher temperatures would boil unwanted water vapor (Fig. 7.27). After a detailed analysis, Delano (1998) selected 375 K as the maximum generator temperature to get higher efficiency.

Energy and mass balance over each component provides sufficient information for the total cycle. The detailed calculations are given in the research works of Delano (1998). Delano (1998) also carried out a comprehensive first and second law of thermodynamics analysis of the cycle on each process to identify the thermodynamic sources of irreversibility and, therefore, the sources of the low efficiency. They identified the generator and the evaporator as the largest source of irreversibility.

The key component of this system is a bubble pump that makes the whole system solely thermally driven without any electrical interference. Bubble pumps are thermally driven pumps that use buoyancy lift arising from the thermal differences in the liquid reservoir relative to the liquid in the generator. The details of this pump have been discussed in the research works of Delano (1998). Delano (1998) demonstrated a prototype of the Einstein cooling system and proved the practical viability of Einstein's cooling system (Fig. 7.28).

The thermal energy needed to supply the Einstein refrigeration system can be supplied directly from solar energy and makes the Einstein refrigeration system more economical and environmentally friendly.

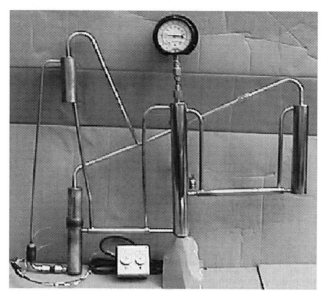

FIG. 7.28 Conceptual demonstration prototype (Delano, 1998).

7.8.3 Solar cooler and heat engine

Besides solar refrigerators, solar air coolers have received much attention recently. Most of the solar absorption air cooler systems found in the literature are single-stage, two-fluid systems using either LiBr/H_2O or H_2O/NH_3 as working fluids (Sayed et al., 2005). The detailed analysis of those two fluid systems shows that those systems require a cooling water supply and several pumps. Those pumps create dependency on electricity if they are not solar pumps. Cooling water supplies also need an installation of a cooling tower, which is recognized as an impediment to smooth maintenance (Li and Sumathy, 2001).

Using the same Einstein principle with some modification, the Einstein absorption refrigerator can be extended to be used as a solar air cooler for space-cooling a home. The additional parts of the air-cooling system, compared to the refrigeration system are the chilled water storage and the chilled water circulation systems that circulate chilled water from the evaporator to the cooling spaces. Air-conditioning systems generally require a higher cooling capacity with 5–7 °C cooling temperature. In the solar air cooler system, the condenser/absorber unit can be placed outside the room space, so that better heat exchange is obtained due to the temperature gradient and outside natural air circulation. Even though the evaporator temperature can reach down to −7 °C, sufficient water circulation through the evaporator gives a chilled water storage temperature of 5–7 °C. The water circulation in the room to be cooled can be facilitated with a solar pump. The extended surface of the water tube will enhance the heat absorption. Fig. 7.29 shows an air conditioning unit. The same unit can be used as a heat pump during winter if the absorption/condenser unit is kept in the room and the evaporator is kept outside of the room. The absorption/condenser unit is considered as the heat transferring unit during winter.

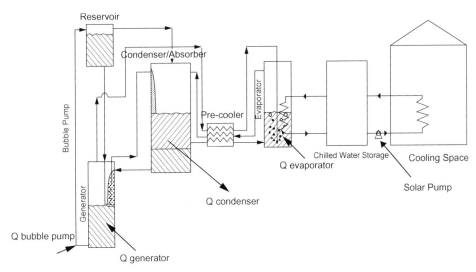

FIG. 7.29 Air cooling system.

7.8.4 Actual coefficient of performance (COP) calculation

In this chapter, the absorption refrigeration system and the vapor compression cycle refrigeration system have been compared to identify the true efficient system. To date, the efficient cooling system is identified by calculating its coefficient of performance (COP) (Smith et al., 2001).

The general definition for COP of a refrigerator can be defined as the ratio of the energy removed from the desired space to the energy required to drive the process.

For the vapor compression cycle, COP is defined as follows (Smith et al., 2001):

$$COP_v = \frac{\text{Heat(removed)}}{\text{Net work}} \tag{7.55}$$

Whereas the COP of absorption air-conditioner is defined as the ratio of the heat transfer rate into the evaporator to the heat transfer rate into the generator (Li and Sumathy, 2001):

$$COP_a = \frac{\text{Heat transfer rate into the evaporator}}{\text{Heat transfer rate into the generator}} \tag{7.56}$$

The energy balance from the primary energy source to the refrigeration system shows the definition of COP since the vapor compression system does not indicate the true COP. It is found that in the vapor compression refrigeration system, the energy input is the work done by a compressor which is driven by electricity. Generally, this electricity comes from a power plant which is driven primarily by heat. On the other hand, the COP calculation of the absorption system includes heat as the primary source of energy. So, the vapor compression cycle is found as a process of converting heat into work and then using work to pump heat through a temperature lift, whereas the absorption system uses heat to pump heat directly. The concern is that the absorption system eliminates the requirement of a power plant and thus becomes a better choice for the refrigeration system.

7.8.5 Vapor compression cycle refrigeration system

Any vapor compression cycle that follows a Carnot cycle operating within two adiabatic steps and two isothermal steps in which heat $|Q_C|$ is absorbed at in the lower temperature T_C, requiring a net amount of work $|W|$, and heat $|Q_S|$ is rejected at the higher temperature T_S; the first law of thermodynamics, therefore, reduces to (Fig. 7.30):

$$|w| = |Q_S| - |Q_C| \tag{7.57}$$

According to Eq. (7.1) the COP can be written as:

$$COP_v = \frac{|Q_C|}{W} \tag{7.58}$$

From the Carnot cycle it is found that (Smith et al., 2001):

$$\frac{|Q_S|}{|Q_C|} = \frac{|T_S|}{|T_C|} \tag{7.59}$$

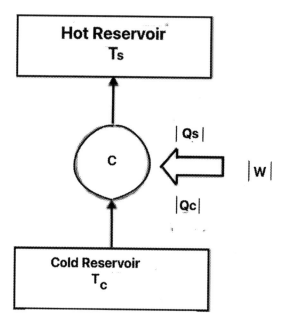

FIG. 7.30 A schematic diagram Carnot refrigerator.

Using Eqs. (7.3)–(7.5), COP can be further defined as:

$$COP_v = \frac{T_C}{T_S - T_C} \tag{7.60}$$

This is the COP equation for the vapor compression refrigeration cycle system that follows the Carnot cycle.

7.9 Global efficiency calculations

To demonstrate the essential features of global efficiency, first introduced by Chhetri and Islam (2008), Khan and Islam (2012) used the example of the absorption refrigeration system. The absorption refrigeration systems are operated at the three temperature levels (T_H, T_S, T_C), as shown in Fig. 7.19. For the COP calculation, two pairs of temperature levels (T_H, T_S, and T_S, T_C) are chosen in such a way as if two Carnot cycles are operating in those pairs.

The Carnot cycle that operated between temperature T_H (generator's temperature) and T_S (surrounding temperature) is a heat engine. The thermal efficiency of a Carnot heat engine that operated within two temperature levels (T_H and T_S) can be defined as (Smith et al., 2001):

$$\eta = \frac{\text{Net work output}}{\text{Heat absorbed}} \tag{7.61}$$

$$= \frac{|W|}{|Q_H|} = \frac{|Q_H| - |Q_S|}{|Q_H|} \tag{7.62}$$

$$= 1 - \frac{Q_S}{Q_H}$$

$$= 1 - \frac{T_S}{T_H} \tag{7.63}$$

From Eqs. (7.8) and (7.9) the following equation can be written:

$$|Q_H| = |W| \frac{T_H}{T_H - T_S} \tag{7.64}$$

Another Carnot cycle that is operated between temperature T_S (surrounding temperature) and T_C (evaporator's temperature) is a Carnot refrigerator.

From Eqs. (7.4) and (7.6) the following equation can be obtained:

$$W = \frac{T_S - T_C}{T_C} |Q_C| \tag{7.65}$$

Substituting W in the Eq. (7.11), the following equation can be obtained:

$$|Q_H| = |Q_C| \frac{T_H}{T_H - T_S} \frac{T_S - T_C}{T_C} \tag{7.66}$$

According to the definition of COP for absorption refrigeration:

$$COP_a = \frac{|Q_C|}{|Q_H|} \frac{T_H - T_S}{T_H} \frac{T_C}{T_S - T_C} \qquad (7.67)$$

From Eqs. (7.6), (7.9), and (7.13), the following relation can be found (Smith et al., 2001):

$$COP_a = \eta_{\text{Carnot heat engine}} \times COP_{\text{Carnot vapor compression cycle}} \qquad (7.68)$$

It is found that for the absorption system, the COP is equal to the multiplication of the efficiency of the Carnot heat engine and the COP of the Carnot vapor compression refrigeration system.

The COP calculation for the absorption system includes primary heating to the final cooling load. However, the COP equation of the vapor compression cycle is not complete as it does not include the primary heating load. This exclusion makes the calculated COP of the vapor cycle system higher than that of the absorption system for the same cooling and surrounding temperatures. That is why conventional COP is not a true indication of cooling efficiency.

Identifying the above anomaly, this study has suggested a new way to calculate the COP for vapor cycle refrigeration systems. According to this study, for an actual and complete COP calculation for the vapor cycle system, the global efficiency (η_{Global}) from primary heat input to the compressor output should be multiplied with the conventional COP so that the conversion of heat to electricity is counted. So, the proposed COP of the vapor compression system will be:

$$COP_{vp} = \eta_{\text{Global}} \times COP_v \qquad (7.69)$$

The calculation of global efficiency is necessary to obtain the theoretically true COP (proposed) of the vapor compression system.

A detailed analysis has been performed to calculate the efficiency from heating at the boiler to the output of the compressor. Fig. 7.31 shows a widely used steam power plant which transforms heat energy into electrical energy.

This electricity is the input of the compressor of the vapor compression cycle refrigeration system as shown in Fig. 7.32.

To obtain the global efficiency of energy conversion, the efficiency of each component of the process should be calculated.

7.9.1 Heat transfer efficiency

Heat transfer efficiency indicates what fraction of heat is actually transferred to the vessel and water (Gupta et al., 1998). It measures the ability of the exchanger to transfer heat from the combustion process to water or steam in the boiler. It includes the radiation and convection loss of the boiler and the loss of heat in the stack gas. Instead of heat transfer efficiency, it is better to use the term boiler efficiency or fuel-to-steam efficiency. Boiler efficiency is a measurement of how much combustion energy is converted into steam energy (Williams, 2003). With recent improvements, the boiler efficiency is found to be up to 70% (Barroso et al., 2003).

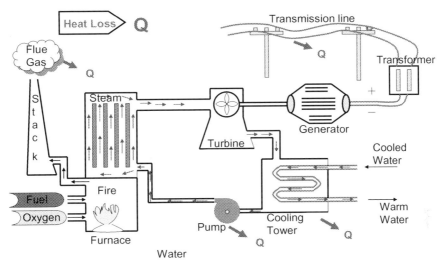

FIG. 7.31 Typical steam power plant.

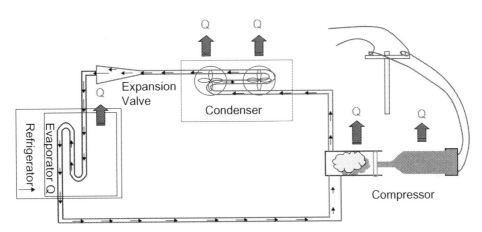

FIG. 7.32 Vapor compression refrigeration system.

The increase of access air up to a limit increases the combustion efficiency, but it decreases the heat transfer efficiency as the access air takes more heat and losses through the stack gas.

7.9.2 Turbine efficiency

A team turbine is a mechanical device that extracts thermal energy from pressurized steam and converts it into useful mechanical work. Turbine efficiency is dependent on thermal efficiency. The thermal efficiency of the engine is defined by Eqs. (7.7) and (7.9). According to the second law of thermodynamics, Carnot's equation gives the theoretical maximum efficiency of any heat engine operating between two temperature levels.

The Carnot cycle comprises some systems and processes which have no existence in reality. A very simple analysis of the Carnot cycle shows the following impossibilities:

(1) Ideal gas
(2) Reversible process
(3) Adiabatic process
(4) Isothermal process (without phase change)

The definition of ideal gas shows that it is a hypothetical gas consisting of identical particles of negligible volume, with no intermolecular forces. There is no practical existence of an ideal gas. Similarly, no process can be perfectly reversible, adiabatic or isothermal, unless the process is extended infinitely in space—a far cry from other conventionally "closed" systems. Thus, the Carnot equation based on the above assumptions is a theoretical equation, lacking a real basis.

Any heat engine that operates between 300 and 30 °C, has a Carnot efficiency of 53%. But the actual efficiency (37%) is less than 70% of the maximum possible value when steam quality, friction loss, and other non-ideal conditions are considered.

7.9.3 Generator efficiency

The generator converts mechanical energy into electrical energy by the Faraday Effect. Most of the power plants produce AC currents by means of alternators (synchronous generators). The generator efficiency of new large plants has been reported in excess of 90% (Weisman and Eckart, 1985). The generator efficiency of the average type of power plant can be taken as 80%.

7.9.4 Transmission efficiency

After the generation of electricity, it is necessary to deliver it to the consumer. That is why a network is established to transmit and distribute electricity to the user's point. The network losses can represent 5%–10% of the total generation (Unsihuay-Vila and Saavedra, 2006). This number, however, is grossly conservative. One significant loss of transmission is the heat loss due to the resistance of the transmission wire. The more the transmission line length is from the origin, the more the energy loss.

7.9.5 Compressor efficiency

For vapor cycle refrigeration of cooling systems, compressors are considered to be the heart of those systems. For reciprocating compressors, isentropic efficiencies are generally in the range of 70%–90% (Peters et al., 2002). In this study, 80% is chosen as the practical compressor efficiency.

7.9.6 Global efficiency

The global efficiency from primary heating to the cooling input includes the efficiencies of several units (Fig. 7.33).

Global heat transfer efficiency (η_{Global}) = Heat − to − Steam efficiency (70%) ×
Turbine eficiency or thermal efficiency (η_t) × Generator efficiency (80%) × (7.70)
Transmission efficiency (90%) × Compressor's rotor efficiency (80%).

$$\text{Global heat transfer efficiency } (\eta_{Global}) = 40\% \times (\eta_t) \tag{7.71}$$

From Eqs. (7.15) and (7.17), we get the new COP equation for the vapor compression cycle:

$$COP_{vp} = 40\% \times (\eta_t) \times COP_v \tag{7.72}$$

On the other hand, if the turbine efficiency is taken η_t, the COP for the absorption system will be (from Eq. 7.14):

$$COP_a = \eta_t \times COP_v \tag{7.73}$$

Comparing Eqs. (7.18) and (7.19), it is found:

$$\frac{COP_a}{COP_{vp}} = \frac{\eta_t}{40\% \times \eta_t} = 2.5 \tag{7.74}$$

So, it is found that if a true pathway is analyzed and included, the cycle for the same surroundings and cooling temperature levels of the COP of an absorption system is almost 2.5 times greater than that of a vapor compression. This result indicates that direct use of heat has the maximum efficiency.

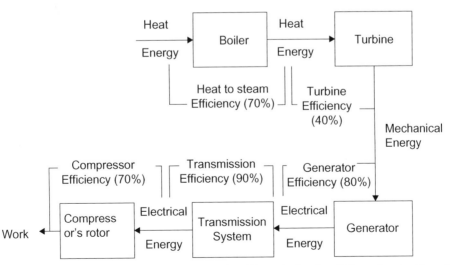

FIG. 7.33 The efficiency calculation for a steam power plant. If we calculate the efficiency from heat transfer to the boiler water to the compressor's output of the cooling/refrigeration system, the global efficiency for a typical power plant will be:

According to Eq. (7.13), any absorption refrigeration system, which operates with a generator temperature of 100 °C, condenser/absorption temperature at 42 °C, and cooling temperature −7 °C, has a COP 0.84, which is a theoretical maximum. Using Eq. (7.18) for the same operating conditions of the vapor compression refrigeration system, the proposed COP becomes 0.336. But the actual COP is found 30%–40% of Carnot COP. So, the actual COP of the absorption refrigeration system and the vapor compression refrigeration system will be 0.336 and 0.134, respectively. The estimated COP of the absorption refrigeration system is higher than that of the vapor refrigeration system when the true path is followed. According to this calculated value, for any absorption refrigeration systems having a cooling load of 3.517 kW (1 ton), the heating load will be 10.47 kW (according to Eq. 7.2). The choice of heating source is important.

Many sources of energy are present on the earth. Fossil fuels are considered non-renewable energy sources, even though they are renewable over very long periods. The increasing rate of fossil fuel utilization is alarming, as fossil fuel is not readily renewed. So, only the dependency of the constant source of energy can eliminate the depletion of energy sources. Solar energy is considered a constant source of energy. The energy extraction process from the source is different for different energy sources. The extraction efficiency of energy from its original source to heating is also an important factor to understand the efficiency of the extraction process.

7.9.7 Fossil fuel combustion efficiency

Energy is released from the fossil fuel by the combustion process. So, the efficiency of combustion is important to understand the efficiency of energy transfer. Combustion efficiency measures the extent to which the chemical energy of a fuel is converted into heat (Gupta et al., 1998). The amount of unburned fuel and excess air in the exhaust is used to assess the combustion efficiency of a furnace. A quality furnace design will allow firing at minimum excess air levels of 15% over the stoichiometric amount. Combustion efficiency varies from 50% to 90%, depending upon fuel specifications, and access to air levels, and the furnace's design.

7.9.8 Solar energy

For any solar high-temperature system, a solar contractor is necessary. For the Einstein cooling system, a parabolic solar trough is sufficient to meet the operating temperatures and heating requirements. This process works in only to few processes such as solar collector and transmission.

7.9.9 Solar collector efficiency

The solar collector efficiency indicates the fraction of solar energy that can be transferred to the thermal fluid in the receiver. The parabolic solar collector efficiency varies much on the fluid temperature. Eck and Steinmann (2005) reported that the collector efficiency shows higher at low-temperature ranges (Fig. 7.34).

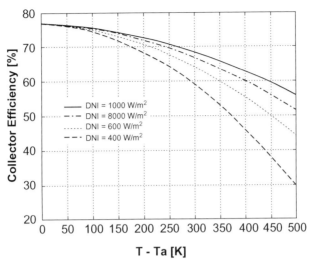

FIG. 7.34 Collector efficiency at different direct normal irradiance (DNI) as a function of fluid temperatures above the ambient temperatures. *Redrawn from Eck, M., Steinmann, W.D., 2005. Modelling and design of direct solar steam generating collector fields. J. Sol. Energy Eng. Trans. ASME 127, 371–380.*

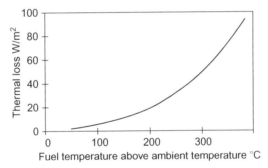

FIG. 7.35 The thermal loss of the collector with respect to a fluid temperature above the ambient temperature. *Redrawn from Odeh et al. (1998).*

At low fluid temperatures, the thermal loss is minimal, as shown in Fig. 7.35. From the figure, it is found that at a fluid temperature of 100 °C (78 °C above ambient temperature), the efficiency of the solar collector is 75%. The Einstein cooling system, described in the paper, has a maximum generator temperature of nearly 100 °C. So the solar collector can be operated at a lower fluid temperature which thus increases the collector efficiency.

7.9.10 Transmission efficiency

The solar transmission efficiency is dependent on the heat transfer loss from the thermal fluid to the fluid in the generator and the bubble pump. An efficient system will have more than 90% efficiency of transmission.

If the efficiency of the solar system is calculated from the solar energy on the parabolic surface to the heat transfers to the refrigerator's generator fluid, the overall efficiency will be:

$$\text{Overall energy transfer efficiency} = \text{Collector efficiency (75\%)} \\ \times \text{Transmission efficiency (90\%)}.$$

$$\text{Overall energy transfer efficiency} = 67.5\% \tag{7.75}$$

It can be speculated that the extraction process of energy from different processes does not differ much. So the consideration of a solar system is beneficial as it has other benefits as discussed earlier.

7.9.11 Solar energy utilization in the refrigeration cycle

There are some existing efficient methods to concentrate the diluted solar energy and transfer it to the desired places. The most common method is the use of a parabolic trough (Fig. 7.36) for the concentration of solar energy to obtain high temperatures without any serious degradations in the collector's efficiency (Bakos et al., 2001). The solar refrigerator was first proposed by DeSa (1965), but that was not very successful because of a poor solar collector. Later on, many improvements on solar collectors have been noticed. The reflector, which concentrates the sunlight to a focal line or focal point, has a parabolic shape. The parabolic trough collector consists of a large curved mirror, which can concentrate the sunlight by a factor of 80 or more to a focal line depending upon the surface area of the trough. In the focal line of these is a metal absorber tube, which is usually embedded into an evacuated glass tube that reduces heat losses (Fig. 7.37). A special high-temperature, resistive selective coating additionally reduces radiation heat losses.

California power plants, known as solar electric generating systems have a total installed capacity of 354 MW (Kalogirou et al., 2004). These systems use thermo-oil as a heat transfer

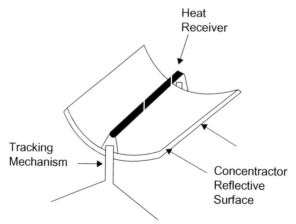

FIG. 7.36 Parabolic trough.

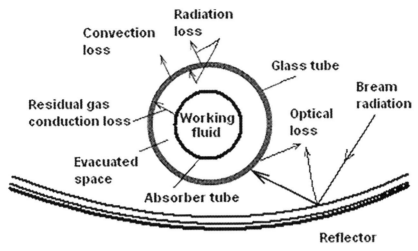

FIG. 7.37 Cross-section of collector assembly. *Redrawn from Odeh et al. (1998).*

fluid, which can reach up to 400 °C (Herrmann et al., 2004). The parabolic collector effectively produces heat at a temperature between 50 and 400 °C (Kalogirou et al., 2004).

7.9.12 The new system

A parabolic trough was constructed that is adjustable and moves along the direction of the sun so that maximum solar energy can be achieved at any time of the day. Each parabolic trough has a surface area of 4 m^2 (2.25 m × 1.8 m) that can radiate almost 1.6–4 kW to the absorber, depending on the direct normal irradiance, which is again dependent on the geographical area. Taking 600 w/m^2 as DNI (direct normal irradiance) and considering the energy transfer efficiency (Eq. 7.21) from the solar surface to the heating point, it is found that one surface (4 m^2) can supply 1.62 kW. A heating load of 10.47 kW will require 7 such parabolic collectors, which can supply the necessary energy to run a refrigerator or an air cooler having a 1-ton cooling load. The number of collectors will vary from place to place, depending on the DNI of any place and the climate of that place. The experimental data show that the parabolic collector can absorb 0.80 kW during early summer in a cold country when the environmental temperature is nearly 21 °C. This parabolic trough can be placed on the roof of a house or can be mounted onto an outside wall of a house where the availability of sunlight is the highest. Proper insulation of the carrier tube can reduce the heat loss of the transferring process. The thermal fuel (vegetable oil) is circulated by the solar pump and that is why no electricity is needed (Fig. 7.38).

A household kitchen-based biogas production system can enhance the operability of this refrigeration in the absence of sun. Biogas can be burnt to get the necessary heat to operate the cooling system, especially at night.

In that study, thermal oil was replaced with waste vegetable oil. This study claims the first reporting of using vegetable oil as thermal oil.

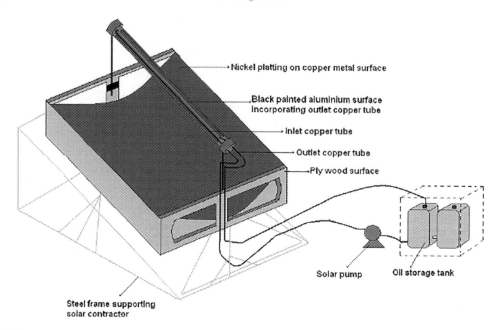

FIG. 7.38 Experimental solar trough. *Redrawn from Khan, M.I., Islam, M.R., 2007. The Petroleum Engineering Handbook: Sustainable Operations. Gulf Publishing Company, Houston, TX, 461 pp.*

7.9.13 Pathway analysis

Most existing processes are energy inefficient. That is why much attention is needed to increase energy efficiency. To obtain true efficiency, it is needed to analyze the pathway of any process. Recently, Islam et al. (2010) identified three basic factors of energy that need to be analyzed before addressing a process as an efficient process. These three factors are:

a. Global economics
b. Environmental impact
c. Quality

Starting from the extraction of energy from the source to its application and its affect on the products and the consumers is something that should be carefully analyzed to identify the efficiency of the processes.

7.9.14 Environmental pollution observation

Energy and the environment are two of the most concerning issues in the current world. Today, most electricity is generated by consuming fossil fuels such as coal, oil, and natural gas. Not only do fossil fuels have a limited life, but their combustion emissions have serious negative impacts on our environment, such as adding to the greenhouse effect and causing acid rain. That is why, for any process, pollution should be considered as an important factor which is eluded to in most of these cases. For some processes, only the consideration of pollution can be the decisive factor. For the steam power plant process, the pollution and loss of heat can be analyzed by considering the following stages:

- Fuel processing stage
- Combustion stage
- Transmission stage
- Refrigeration stage

The drilling of toxic fluids, synthetic drilling mud, toxic surfactant additions or chemical treatments to enhance oil recovery (EOR) projects, etc., cause great environmental pollution during the oil production from the reservoirs. This oil needs to be refined before being used in steam power plants. During this refining process, lots of toxic materials are added, which makes the fuel environmentally vulnerable. Chhetri and Islam (2008) described the toxicity addition of fuel during the processing stage (Fig. 7.39).

Additionally, producing oil and gas from the reservoir reduces the necessary reservoir pressure, leading to significant subsidence. Recent investigation shows that this subsidence is not negligible.

In the absence of access to air, the fuel can be burnt incompletely and produce carbon monoxide. All the toxic gases realized through the stack gas, greatly pollute the environment. This vulnerable effect increases with the treated fuel. The more the treatment of fuel, the more it is likely to cause pollution during the combustion stage. At this stage, the toxic constituents can escape from the fuel and mix with the air through the stack gas. Chhetri and Islam (2008) have reported that the burning of untreated fuelwood causes less damage to the environment than treated fuelwood. The addition of different toxic materials during the fuel processing unit increases this tendency.

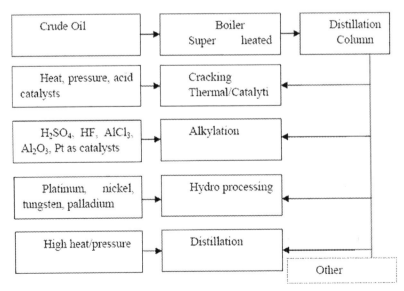

FIG. 7.39 Pathway of an oil refining process. *Redrawn from Chhetri, A.B., Islam, M.R., 2008. Inherently Sustainable Technologies. Nova Science Publishers, 452 pp.*

It is now a concern whether human exposure to power-frequency electromagnetic fields has significant health consequences (Morgan et al., 1985). Hamza et al. (2002), speculate that one source of environmental pollution is the magnetic field produced near high and extra-high voltage (EHV) transmission lines. It has become a controversial issue since no detailed research work has been found to address this problem. Qualitatively, some researchers found it as a source of causing certain types of cancers (Islam et al., 2016b).

Vapor compression cooling systems generally use synthetic refrigerant fluids. The negative impacts of chlorofluorocarbons (CFC) and hydrochlorofluorocarbons (HCFC) are well documented. Besides participating in the destruction of stratospheric ozone, the release of CFCs may also contribute to global warming, which means that CFCs influence the reflection of infrared radiation from the surface of the earth and thus cause global climate change (Hayman and Derwent, 1997). Because of similar physiochemical properties, HCFCs have been used as the replacement for CFCs. The release of chlorine from the HCFCs to the stratosphere has been identified to be detrimental to the earth for the long term. According to the updated Montreal Protocol, a virtual phase-out of HCFCs is scheduled by 2020.

Research on the replacement of one synthetic refrigerant by another synthetic refrigerant is continuing. Replacement is taking place when it is assumed safe for the environment. But previous experience indicated that no synthetic refrigerants have been found to be ultimately environmentally safe. There were, at first claimed safe during initial replacement, but ultimately, they were found detrimental to the environment. Sometimes there is a delay in understanding the vulnerable effects due to a lack of appropriate knowledge. That is why it is safe to use natural refrigerant fluids instead of any other synthetic refrigerants.

Thus, if all these pollution stages are considered, it can be shown that the vapor compression cycle is a source of environmental pollution. The pollution can be in different phases, depending upon the fuel system and the refrigerant fluids. Radiation, greenhouse pollution, air/water pollution, disruption of the ecosystem, and the sickness of human beings are the direct and indirect effects of this system.

Furthermore, there are heat losses in every step of this process, which suffers from a huge overall loss of heat. This is truly an indication of generating more heat by removing heat as shown in Figs. 7.41 and 7.42.

7.9.15 environmentally friendly system

Solar absorption in the Einstein cooling system is found to be an environmentally friendly system. The solar collection system does not produce any negative impacts on the environment because solar energy is clean. Again, the use of vegetable or waste vegetable oil makes it more environmentally acceptable.

Due to ozone depletion and global warming effects, environmentally friendly refrigerants with zero ozone-depletion potential should be used in refrigerators and heat pumps (Fernandez et al., 2001; Hwang et al., 2002; Zhang et al., 2002). The cooling system is found to use naturally benign fluids which spare any significant long-term effects upon the environment.

7.9.16 Global economics of the systems

It is important to find a means to calculate the global economics of any process. It has already been shown that the vapor compressor cooling systems are involved in the pollution of the environment in different directions. Again, this is a process that depletes necessary but concentrated energy sources. If the costs of the remedial processes of all the vulnerable effects along with the plant costs are considered, the actual cost of the total process can be obtained. The cost involves the remedial cost of soil contamination and air/water pollution, ecological restoration. The cost of medical intervention and medicines, etc. (Zatzman and Islam, 2007). There are some more costs that are difficult to understand as the direct effect of the processes cannot be readily identified.

On the other hand, the pathway of the absorption cooling system indicates that it is not associated with the above vulnerable effects and that is why no additional costs are required.

7.9.17 Quality of energy

The quality of energy in the environment in the environment in the environment includes an important phenomenon that is little understood. For example, heating a home by wood is better than heating by electricity. The radiation due to the electro-magnetic rays might cause interference with the human's radiation frequency which can cause acute long-term damage to human beings. Energy with natural frequency is the most desirable. Alternating current is not natural; hence there are bound to be some vulnerable effects of this frequency to the environment and human beings (Islam et al., 2015).

Even cooling by using synthetic fluid, the leakiness of this fluid might destroy the quality of the food subject to cooling.

Considering all of the effects, the vapor refrigeration system seems to be an inefficient system, compared to the solar absorption system.

Conclusions

8.1 Facts and fictions

Oil and gas resources are unique in many ways (Chapter 2), but perhaps the most unique feature of them is in politicization. It is an incontrovertible truth that nearly 40% of today's energy needs are met by the petroleum industry. It is also widely accepted that carbon is the source of unsustainability, thus making the term, "sustainable petroleum" an oxymoron. This is based on the fictional premise that carbon—the essence of life—is the enemy. This adherence to a ludicrous premise has become the hallmark of modern-day climate alarmism. The politicization of this dogma-based "science" is perhaps the most toxic narrative that has set the petroleum industry up for failure. Petroleum products have been used by every ancient civilization, yet they never created any negative impact. In fact, direct use of crude oil for medicinal purposes was ubiquitous as late as the medieval era. There are other fictional theories and assertions that have been propelled into the status of "settled science." No scientific discourse should be founded on such premises. This book series of sustainable petroleum development moves away from such flawed premises, thus establishing a scientifically sound sustainability track.

New Science has introduced a series of false premises, then doubled down on promoting them as "settled science," erecting absurd notions such as dimming the sun, blaming the carbon, and many others while dismissing scientific solutions, like the zero-waste approach. In this book series, the prevalent myths are deconstructed and replaced with knowledge-based fundamentally sound phenomenal premises.

Myth	Consequence	Correct premise
Chemicals are chemicals	Material products can be detached from their origins and be reduced to their major ingredients. The role of artificial chemicals in creating long-term harm is ignored	Each chemical is sustainable in its natural state. Toxicity comes from artificial chemicals
Energy is "energy"	Energy can be detached from its origin and be reduced to its tangible effects, such as temperature, brightness, electric charge. The false characterization of energy	Sustainable energy comes from the natural source
If you cannot "see," it doesn't exist	All environmental impacts can be made to disappear if they are covered up for a certain period, therefrom emerges the "dilution is the solution to pollution" mantra	Contamination is independent of the detectability of a chemical at the outlet
Doing good to the environment is costly and cannot be profitable	Short-term approaches govern the operating principles and long-term sustainability is shunned. The original slogan: In the long run, we are all dead	Sustainable processes are inherently profitable and good for the environment
We are progressing as the human race and solutions lie within incremental advancements over current practices	The status quo is the best we can do, triggering the TINA (there is no alternative) syndrome	Our progress depends on sustainable developments. If not sustainable, technologies are regressive
Crude oil is uniquely formed from organic matter	A significant amount, belonging to non-organic origins, remains unexplored	New opportunities open up when non-organic sources are explored
Current usage of crude oil is sufficient and necessary	Promotes status quo, making it impossible to alter unsustainable refining practices	Unsustainable practices must be identified and replaced at the onset
Current reservoir modeling approaches are adequate	Promotes flawed simulation approaches or linear solutions and linearization techniques	Multiple solutions of non-linear governing equations are necessary as a starting point
Scaled model studies are replaceable	A surge of numerical models, which are big on computing but devoid of physics	Observation of physical phenomena should come before the development of mathematical models

8.2 Conclusions

8.2.1 Chapter 2

Based on the discussion in Chapter 2, the following conclusions can be made.

1. Unique features of oil and gas are not commensurate with conventional processing and utilization of crude oil. By custom designing both processing and utilization in line with natural traits of petroleum fluids.

2. Conventional rock fluid characterization leads to unsustainable practices.
3. Fluid characterization based on origin makes way for diversifying applications of the fluid, thus increasing economic benefits tremendously.
4. Scientific rock characterization unlocks significant petroleum reserves, previously unrecognized.
5. As much as 10% of new reserves worldwide can be unlocked and produced with currently available technologies.
6. As much as 20% gain in returns can be achieved with sustainable technologies.

8.2.2 Chapter 3

Based on the discussion carried out in Chapter 3, the following conclusions can be made.

7. Natural systems are extremely complex and have nothing in common with the simplified version used by New Science, which has implemented false premises that are paradoxical and in conformance with the myths, discussed in the previous section.
8. Current reserve calculation and reservoir characterization methods are not scientific, leading to confusion, when it comes to comparing the true reserves of different countries.
9. The confusion in calculating reserve comes from the inadequate definition of "recoverable reserve," the dynamic nature of technology development, and over-reliance on economic models. In this exercise, there is no provision for sustainable technologies, making the entire process incapable of determining true reserve potentials.
10. Complex reservoirs must be treated with unconventional tools, especially suited for such reservoirs.
11. Characterization with non-linear tools creates an optimistic picture of the complex reservoirs.
12. Thermal changes affect complex reservoirs more intensely than in less heterogeneous reservoirs. Such changes can occur even during so-called isothermal primary recovery.
13. Recovery techniques should be custom designed for the type of complex reservoirs.

8.2.3 Chapters 4 and 5

Based on the discussion presented in Chapters 4 and 5, the following conclusions can be made.

14. Unconventional reservoirs hold great potential for the sustainable development of oil and gas.
15. Scientifically, the boundary between non-renewable and renewable energy sources is fictitious and is mainly a matter of sematic. In reality, the transition between various energy sources is a continuous process, in which sustainable development plays a complementary role.
16. Sustainable development of oil and gas can make the petroleum exploitation process perpetual, meaning the world will not run out of this resource.
17. Scientific characterization of petroleum products can add value to petroleum products, depending on the applications. This also leads to long-term environmental and economic sustainability of petroleum products.

18. Reassessment of current reserve reveals an optimistic picture of marginal and unconventional reservoirs.
19. Each stage of recovery can be rendered sustainable by selecting naturally available chemicals.
20. Scientific characterization also helps with the planning of future EOR projects.
21. The reserve can be bolstered significantly just by introducing updated technologies. Many of these reservoirs can become excellent candidate floor EOR operations.

8.2.4 Chapter 6

Based on the discussion presented in Chapter 6 the following conclusions can be made.

22. Sustainable development involves the treatment of material balance equations with fluids with memory functions.
23. Such treatment improves accuracy and makes a great impact on the proper characterization of the reservoirs.
24. Currently available well test programs are deficient.
25. Well test interpretation can be improved with numerical simulation, with the possibility of solving non-linear equations with cloud points, rather than unique solutions, corresponding to the linearized formulation.
26. The engineering approach is found to be simpler and more amenable to solving with non-linear solvers.

8.3 Recommendations

Based on Chapter 7, the following recommendations can be made as a field guideline.

8.3.1 Better use of data

Every operation in an oil/gas field generates signals that can help improve the picture of the reservoir. Data processing technologies, presented in this book, can become a powerful tool for creating a knowledge base. Correctly implemented data analytics systems and tools can overcome the operational complexity of oil and gas operations, yielding returns of as much as 30–50 times the original investment and reducing ecological impact by reducing wastage, accidents, and bottlenecks.

8.3.2 Decreasing freshwater usage

Water is an essential element in various oil production processes, from fracking to separating oil from other elements present in oil sands. Hundreds of millions of barrels of water are utilized every single day, and while the global petroleum industry currently manages to recycle the vast majority of this water (between 80% and 95%), the use of freshwater usage can be eliminated altogether. The sustainable practice involves the use of seawater and produced water exclusively within a zero-waste cycle. At present, companies are aiming to use 100% non-potable water by improving filtration oxidization methods, as well as advanced chemical-free water treatment solutions to neutralize bacterial contaminants such as sulfate-reducing and iron-oxidizing bacteria. In this book, a sustainable, 100% zero-waste water purification technique is presented.

8.3.3 Reducing methane leaks

Finding ways to reduce methane leaks is a cost-effective opportunity for the industry. Recent figures from the International Energy Agency have outlined that it is financially possible to reduce oil and gas methane emissions utilizing currently available and emerging technologies. However, none introduces sustainable technologies. True sustainability can be achieved through instant recycle and alternate usage of the escaped gas. This aspect will be discussed in upcoming books of this book series.

8.3.4 Used oil recycling

More companies are utilizing small-scale waste-oil micro-refinery units that transform used oil into diesel fuel. Not only does this approach yield fuel for ongoing operations, but it's also a relatively inexpensive alternative to more traditional oil disposal methods. However, these technologies are not sustainable. True sustainability is in reusing the waste oil without refining it. Various options are presented in this book.

8.3.5 Streamlining/improving processes

Innovations that don't specifically make oil and gas processing greener and cleaner can still help improve the industry's overall sustainability by allowing for more cost-efficient processes. For example, new ultrasound technology allows companies to create 3D images of the inside of oil wells, enabling them to make more informed and cost-effective production decisions. Similarly, IIOT, analytics, automation, reserve replacement, and enhancement capabilities, and emerging artificial intelligence programs can all help find and eliminate operational inefficiencies. Each of these innovations can be retooled to align with sustainable developments. For instance, artificial intelligence can be used to implement the zero-waste concept as well as select the correct solution from among multiple solutions of the non-linear governing equations.

8.3.6 Creating real-time data acquisition and adaptive control

Going beyond incremental operational improvements, the quickening pace of digitalization of the petroleum industry has allowed for the creation of the "digital oilfield," a process that is starting to come to prominence. Through the use of cloud technologies and big data, the digital oilfield allows for all operational data to be monitored, analyzed, and utilized in real-time. These data can become an excellent feed for the simulators, which then can serve as the adaptive control engine for the entire operation.

8.3.7 Scaled model studies to develop new mathematical models

Scaled model studies open up opportunities to develop new models for both fluid flow and heat transfer within the porous medium. The conventional practice has been to conduct scaled model studies, the use the results directly to design a pilot project. Of late, that tradition has been replaced with direct modeling, often with network models. This approach is not sustainable and has obstructed progress in terms of the greening of petroleum operations. It is suggested that scaled models, in which certain sets of dimensionless groups are scaled properly, are used to generate data, which are then used to develop non-linear models. These models then can be solved directly using the engineering approach, which is more amenable to possible multiple solutions. Thus, cloud points, instead of a single point, are created. Fluid selection for enhanced oil recovery must focus on using waste gas, produced water, combined with the natural source of energy (direct solar). Any EOR must consider the flow regime for designing the operating conditions, instead of the conventional velocity of 1 ft/day. If the mobility ratio is unfavorable, flow regime would likely be unstable, making way to early breakthrough. This scenario is averted if the gravity number is negative.

8.3.8 Revisiting petrophysical data

Conventionally, core permeabilities have been used for both porosity and permeability data. It is recommended that permeability data, as derived from well tests, is then filtered with shale breaks, fracture frequency, and other properties. For porosity data, it is recommended that cut-off points be eliminated, particularly for basement, unconventional reservoirs. The relative permeability data should be focused on conducting based on instability number, which dictates the flow regime, unless the mobility ratio is favorable (smaller than 1).

References

AAPG, 1970. Geology of giant petroleum fields. Memoir 14. 575 pp.

AAPG, 1994. The petroleum system—from source to trap. AAPG Mem. 60. 655 pp.

AAPG, 2015. Unconventional energy resources: 2015 review. Nat. Resour. Res. 24, 443–508. https://doi.org/10.1007/s11053-015-9288-6 (American Association of Petroleum Geologists, Energy Minerals Division).

Aase, N.E., Bjørkum, P.A., Nadeau, P.H., 1996. The effect of grain coating micro quartz on preservation of reservoir porosity. AAPG Bull. 80 (10), 1654–1673.

Abdelazim, R., Rahman, S.S., 2016. Estimation of permeability of naturally fractured reservoirs by pressure transient analysis: an innovative reservoir characterization and flow simulation. J. Pet. Sci. Eng. 145, 404–422.

Abdelazim, R., et al., 2014. 3D poro-thermo-elastic numerical model for analysing pressure-transient response to improve the characterization of naturally fractured geothermal reservoirs. Trans.-Geotherm. Resour. Counc. 38, 907–915.

Abe, H., Sekine, H., Shibuya, Y., 1983. Thermoelastic analysis of a crack-like reservoir in a hot dry rock during extraction of geothermal energy. J. Energy Resour. Technol. 105, 503–508.

Abe, H., Hayashi, K., Yamamoto, T., 1995. Process of crack growth from a reservoir formed along a joint by injection of fluid for extraction of geothermal heat. In: Mechanics of Jointed and Faulted Rock. MJFR-2, Vienna, pp. 423–428.

Abou-Kassem, J.H., 2007a. Reconstruction of the compositional reservoir simulator using the engineering approach. J. Nat. Sci. Sustain. Technol. 2 (1–2), 37–110.

Abou-Kassem, J.H., 2007b. Engineering approach vs the mathematical approach in developing reservoir simulators. J. Nat. Sci. Sustain. Technol. 1 (1), 35–68.

Abou-Kassem, J.H., Farouq Ali, S.M., Islam, M.R., 2006a. Petroleum Reservoir Simulation—A Basic Approach. Gulf Publications. 445 pp.

Abou-Kassem, J.H., Ali, S.M.F., Islam, M.R., 2006b. Petroleum Reservoir Simulation: A Basic Approach, second ed. Gulf Publishing Company, Houston, TX. 480 pp.

Abou-Kassem, J.H., Islam, M.R., Ali, S.M.F., 2020. Petroleum Reservoir Simulation: The Engineering Approach. Elsevier. 491 pp.

Abrams, A., 1975. The influence of fluid viscosity, interfacial tension, and flow velocity on residual oil saturation left by waterflood. Soc. Pet. Eng. J. 15, 437.

Accenture, n.d. Accenture Fact Sheet. https://newsroom.accenture.com/fact-sheet/.

Addis, M.A., Hanssen, T.H., Yassir, N., Willoughby, D.R., Enever, J., 1998. A comparison of leak-off test and extended leak-off. In: Proceedings of Eurock'98, SPE/ISRM Symposium: Rock Mechanics in Petroleum Engineering, 8–10 July 1998.

Adil, M., et al., 2018. Experimental study on electromagnetic-assisted ZnO nanofluid flooding for enhanced oil recovery (EOR). PLoS ONE 13 (2). https://doi.org/10.1371/journal.pone.0193518, e0193518.

Adjedj, B., 1999. Application of the decomposition method to the understanding of HIV immune dynamics. Kybernetes 28 (3), 271–283.

Adomian, G., 1986. Nonlinear Stochastic Operator Equations. Academic, Orlando, FL.

Adomian, G., 1994. Solving Frontier Problems of Physics: The Decomposition Method. Kluwer Academic, Dordrecht.

AEO, 2021. Available from: https://www.eia.gov/outlooks/aeo/pdf/03%20AEO2021%20Natural%20gas.pdf.

AER (Alberta Energy Regulator), 2015. Alberta's Energy Reserves 2014 and Supply/Demand Outlook 2015-2024. Alberta Energy Regulator (AER), Statistical Series, ST98-2015. Available from: http://www.aer.ca/data-and-publications/statistical-reports/st98.

Ageev, P.G., Molchanov, A.A., 2015. Plasma source for generating nonlinear, wide-band, periodic, directed elastic oscillations and a system and method for stimulating wells, deposits and boreholes using the plasma source. US Patent No. US 9,181,788 B2, 10.

Aguilera, R., 1974. Analysis of naturally fractured reservoirs from sonic and resistivity logs. J. Pet. Technol. 26 (11), 1233–1238.
Aguilera, R., 1976. Analysis of naturally fractured reservoirs from conventional well logs. J. Pet. Technol. 28 (7), 764–772.
Aguilera, R., 1999. Recovery factors and reserves in naturally fractured reservoirs. J. Can. Pet. Technol. 38, 15–18.
Aguilera, M.S., Aguilera, R., 2003. Improved models for petrophysical analysis of dual porosity reservoirs. Petrophysics 44 (1).
Ahmed, T.H., Centilmen, A., Roux, B.P., 2006. A generalized material balance equation for coalbed methane reservoirs. In: SPE Annual Technical Conference and Exhibition, January. Society of Petroleum Engineers.
Ahn, J.S., Lee, K.H., 1986. Studies on the volatile aroma components of edible mushroom (*Tricholoma matsutake*) of Korea. J. Korean Soc. Food Nutr. 15, 253–257.
Ahr, W.M., 2011. Geology of Carbonate Reservoirs: The Identification, Description and Characterization of Hydrocarbon Reservoirs in Carbonate Rocks. Wiley, 296 pp.
Airey, E.M., 1968. Gas emission from broken coal: an experimental and theoretical investigation. Int. J. Rock Mech. Min. Sci. 5, 475–494.
Aboudwarej, H., et al., 2006. Highlighting heavy oil. Oilfield Rev. 18 (2), 34–53.
Alexander, R., Kagi, R., Noble, R.J., 1983. Identification of the bicyclic sesquiterpenes drimane and eudesmane in petroleum. J. Chem. Soc. Chem. Commun. 1983, 226–228.
Alexandrino, G.L., Augusto, F., 2018. Comprehensive two-dimensional gas chromatography-mass spectrometry/selected ion monitoring (GC x GC-MS/SIM) and chemometrics to enhance inter-reservoir geochemical features of crude oils. Energy Fuel 32 (8), 8017–8023. https://doi.org/10.1021/acs.energyfuels.8b00230.
Al-Darbi, M.M., Jaralla, M., Islam, M.R., 2002a. The inhibitive characteristics of mixed inhibitor combinations under heat and mass transfer conditions. Energy Sources 24 (11), 1019–1030.
Al-Darbi, M.M., Muntasser, Z., Tango, M., Islam, M.R., 2002b. Control of microbial corrosion using coatings and natural additives. Energy Sources 24 (11), 1009–1018.
Al-Maghrabi, I., Bin-Aqil, A.O., Chaalal, O., Islam, M.R., 1999. Use of thermophilic bacteria for bioremediation of petroleum contaminants. Energy Sources 21 (1/2), 17–30.
Al-Mohannadi, N.S., Ozkan, E., Kazemi, H., 2007. Grid-system requirements in numerical modeling of pressure-transient tests in horizontal wells. SPE Reserv. Eval. Eng. 10 (2), 122–131.
ALS, 2019a. Fluid Inclusion Studies of 5 samples from Lancaster 205/21a-7 and 5 samples from Lincoln 205/26b-12. Confidential report prepared for Hurricane Energy, No. GB-18-J513.
ALS, 2019b. Thin Section Petrography of 64 Rotary Sidewall Core Samples, Lancaster Appraisal Well 205/21a-7, UKCS. Confidential report prepared for Hurricane Energy No. ALSPR017/029-4.
Al-Saadoon, O., Hayder, M.E., Baddourah, M., 2013. Building a simulation grid for large scale reservoir simulation. In: Paper Presented at the EAGE Annual Conference & Exhibition incorporating SPE Europec, London, UK, June 2013. Paper Number: SPE-164848-MS.
Altowilib, A., et al., 2020. Reserves estimation for coalbed methane reservoirs: a review. Sustainability 12 (24), 10621. https://doi.org/10.3390/su122410621.
Alvarado, V., Manrique, E., 2010. Enhanced oil recovery: an update review. Energies 3 (9), 1529–1575. https://doi.org/10.3390/en3091529.
Amadei, B., Stephansson, O., 1997. Rock Stress and its Measurement. Chapman and Hall, London.
Amthor, J.E., Okkerman, J., 1998. Influence of early diagenesis on reservoir quality of Rotliegende sandstones, northern Netherlands. AAPG Bull. 82 (12), 2246–2265.
Anderson, S. (Ed.), 2005. Making Medicines—A Brief History of Pharmacy and Pharmaceuticals. Pharmaceutical Press, Grayslake, IL, ISBN: 0-85369-597-0. 318 pages, Hardcover.
Anderson, R.B., Bayer, J., Hofer, L.J.E., 1966. Equilibrium sorption studies of methane on pittsburgh seam and pocahontas no. 3 scam coal. In: Gould, R.F. (Ed.), Coal Science. Advances in Chemistry Series No. 55, American Chemical Society, pp. 386–399.
Antman, K.H., 2001. Introduction: the history of arsenic trioxide in cancer therapy. Oncologist 6, 1–2.
APPEA, 2009. Australian Petroleum Production & Exploration Association Limited. Annual Production Statistics. www.appea.com.au/content/pdfs_docs_xls/annualproduction_statistics.xls.
Aqeel, N., Cunha, N.B., 2006. An investigation of grid design characteristics in numerical modelling of horizontal well pressure transient tests. In: Paper Presented at the Canadian International Petroleum Conference, Calgary, Alberta, June 2006., https://doi.org/10.2118/2006-161. Paper Number: PETSOC-2006-161.

References

Arenzon, J.J., Levin, Y., Sellitto, M., 2003. Slow dynamics under gravity: a nonlinear diffusion model. Phys. A Stat. Mech. Appl. 325 (3), 371–395.

Armbrus, M., et al., 2010. A view of cloud computing. Commun. ACM 53 (4), 50–58.

Arps, J.J., 1945. Analysis of decline curves. Trans. Am. Inst. Min. Metall. Eng. 160, 228–247.

Artus, V., Noetinger, B., 2004. Up-scaling two-phase flow in heterogeneous reservoirs: current trends. Oil Gas Sci. Technol. 59 (2), 185–195.

Asanuma, H., Soma, N., Kaieda, H., Kumano, Y., Izumi, T., Tezuka, K., Niitsuma, H., Wyborn, D., 2005. Microseismic monitoring of hydraulic stimulation at the Australian HDR project in Cooper Basin. In: Proceedings of the World Geothermal Congress, Antalya. 5 pp.

Ash, N., Jenkins, M., 2006. Biodiversity and poverty reduction: the importance of biodiversity for ecosystem services. In: Final Report Prepared by the United Nations Environment Programme World Conservation Monitoring Centre (UNEP-WCMC) for the Department for International Development (DFID). http://www.unep-wcmc.org.

Asquith, G., Krygowski, D., 2004. Basic Well Log Analysis. AAPG. 247 pp.

Astarita, G., 1997. Dimensional analysis, scaling, and orders of magnitude. Chem. Eng. Sci. 52 (24), 4681–4698.

Attanasi, E.D., Meyer, R.F., 2010. In: Clarke, A., Trinnaman, J.A. (Eds.), Natural Bitumen and Extra-Heavy Oil. World Energy Council, ISBN: 978 0 946121 021, pp. 137–150.

Attanasi, E.D., Mast, R.F., Root, D.H., 1999. Oil, gas field growth projections—wishful thinking or reality? Oil Gas J. 97 (14), 79–81.

Aubcrtin, A., et al., 2009. Miscible viscous fingering in microgravity. Phys. Fluids 21, 054107.

Auffhammer, M., Morzuch, B., Stranlund, J.K., 2005. Production of chlorofluorocarbons in anticipation of the Montreal protocol. Environ. Resour. Econ. 30 (4), 377–391.

Axelsson, G., 2010. Sustainable geothermal utilization—case histories; definitions; research issues and modeling. Geothermics 39 (4), 283–291.

Ayres, N.A., et al., 2013. Geotechnical influence on the acoustic properties of marine sediments of the Santos Basin, Brazil. Mar. Georesour. Geotechnol. 31, 125–136.

Azoug, Y., Tiab, D., 2004. Scaling-up fine grid models using pseudo functions in heterogeneous porous media. In: Paper 2004-022, Canadian International Petroleum Conference. Petroleum Society of Canada, https://doi.org/10.2118/2004-022.

Bažant, Z.P., Ohtsubo, H., 1977. Stability conditions for propagation of a system of cracks in a brittle solid. Mech. Res. Commun. 4, 353–366.

Bažant, Z.P., Ohtsubo, H., Aoh, K., 1979. Stability and post-critical growth of a system of cooling or shrinkage cracks. Int. J. Fract. 15, 443–456.

Babadagli, T., 2008. Scaling miscible displacement in fractured porous media using dimensionless groups. J. Pet. Sci. Eng. 61 (2), 58–66.

Bagrintseva, K.I., Dmitrievsky, A.N., Bochko, R.A., 1989. Atlas of Carbonate Reservoir Rocks of the Oil and Gas Fields of the East European and Siberian Platforms. VNIGNI, Moscow, Russia.

Bai, Y., Li, J., Zhou, J., Li, Q., 2008. Sensitivity analysis of the dimensionless parameters in scaling a polymer flooding reservoir. Transp. Porous Media 73 (1), 21–37.

Baisch, S., Weidler, R., Voros, R., Wyborn, D., de Graaf, L., 2006. Induced seismicity during the stimulation of a geothermal HFR reservoir in the Cooper Basin, Australia. Bull. Seismol. Soc. Am. 96, 2242–2256.

Bajaj, S., Vohora, S.B., 1998. Analgesic effects of gold preparations used in Ayurveda and Unan-Tibb. Indian J. Med. Res. 108, 10–11.

Bajaj, S., Vohora, S.B., 2000. Ant-cataleptic, ant-anxiety and ant-depressant activity of gold preparations used in Indian systems of medicine. Indian J. Pharm. 32, 339–346.

Baker, P.E., 1973. Effect of pressure and rate on steam zone development in steam flooding. Soc. Pet. Eng. J. 13 (5), 274–284.

Baker, E.T., Massoth, G.J., 1987. Characteristics of hydrothermal plumes from two vent fields on the Juan de Fuca Ridge, northeast Pacific Ocean. Earth Planet. Sci. Lett. 85, 59–73.

Bakos, G.C., et al., 2001. Design, optimisation and conversion-efficiency determination of a line-focus parabolic-trough solar-collector (PTC). Appl. Energy 68 (1), 43–50.

Balhoff, M., et al., 2012. Numerical algorithms for network modeling of yield stress and other non-Newtonian fluids in porous media. Transp. Porous Media 93 (3), 363–379.

Bandis, S.C., Lumsden, A.C., Barton, N., 1983. Mechanical properties of rock joints. Int. J. Rock Mech. Min. Sci. Geomech. Abstr. 20, 249–263.

Bandopadhyay, A.K., Mohanty, D., 2014. Variation in hydrogen content of vitrinite concentrates with rank advance. Fuel 134, 220–225.

Bansal, A., Islam, M.R., 1994. Scaled model studies of heavy oil recovery from Alaskan reservoir using gas injection with horizontal wells. J. Can. Pet. Technol. 33 (6), 52–62.

Baqués, V., et al., 2020. Fracture, dissolution, and cementation events in middle Ordovician carbonate reservoirs, Tarim Basin, NW, China. Geofluids, 9037429. https://doi.org/10.1155/2020/9037429.

Barenblatt, G.E., Zheltov, I.P., Kochina, I.N., 1960. Basic concepts in the theory of homogeneous liquids in fissured rocks. J. Appl. Math. Mech. 24, 1286–1303.

Baria, R., Baumgärtner, J., Gerard, A., Jung, R., Garnish, J., 1999. European HDR research programme at Soultz-sous-Forets (France) 1987-1996. Geothermics 28, 655–669.

Barker, C., 1980. Primary Migration: The Importance of Water Organic Mineral Matter Interaction in the Source Rock. AAPG Studies in Geology, Tulsa.

Barker, C., 1990. Calculated volume and pressure changes during the thermal cracking of oil to gas in reservoirs. AAPG Bull. 74, 1254–1261.

Barker, C., Takach, N.E., 1992. Prediction of natural gas composition in ultradeep sandstone reservoirs. AAPG Bull. 76 (12), 1859–1873.

Barnum, R.S., et al., 1995. Gas condensate reservoir behaviour: productivity and recovery reduction due to condensation. In: SPE 30767, Paper Presented at the SPE Annual Technical Conference and Exhibition, Dallas, Texas, October 1995.

Barr, D.T., Cleary, M.P., 1983. Thermoelastic fracture solutions using distributions of singular influence functions—II: numerical modelling of thermally self-driven cracks. Int. J. Solids Struct. 19 (1), 83–91.

Barroso, J., et al., 2003. On the optimization of boiler efficiency using bagasse as fuel. Fuel 82, 1451–1463.

Bartley, J.T., Ruth, D.W., 2001. Relative permeability analysis of tube bundle models, including capillary pressure. Transp. Porous Media 45 (3), 447–480.

Barton, N.R., Bakhtar, K., 1982. Rock joint description and modeling for the hydrothermomechanical design of nuclear waste repositories. TerraTek Engineering, Salt Lake City, Utah. Tech Rep 83–10.

Barton, N.R., Choubey, V., 1977. The shear strength of rock joints in theory and practice. Rock Mech. 10, 1–54.

Barton, N.R., Bandis, S., Bakhtar, K., 1985. Strength, deformation and conductivity coupling of rock joints. Int. J. Rock Mech. Min. Sci. Geomech. Abstr. 22, 121–140.

Baskin, D.K., 1997. Atomic ratio H/C of kerogen as an estimate of thermal maturity and organic matter conversion. AAPG Bull. 81 (9), 1437–1450.

Basu, A., Islam, M.R., 2009. Scaling up of chemical injection experiments. Pet. Sci. Technol. 27 (7), 654–665.

Bateman, R.M., Konen, C.E., 1977. The log analyst and the programmable pocket calculator. Log. Anal. 18 (5), 3–10.

Batini, A.F., et al., 2002. Geophysical well logging—A contribution to the fractures characterization. In: Proceedings, Twenty-Seventh Workshop on Geothermal Reservoir Engineering, January 28–30. SGP-TR-171. Stanford University, Stanford, California.

Baumgartner, J., Rummel, F., 1989. Experience with "fracture pressurization tests" as a stress measuring technique in a jointed rock mass. Int. J. Rock Mech. Min. Sci. Geomech. Abstr. 26, 661–671.

Bawazer, W., Lashin, A., Kinawy, M.M., 2018. Characterization of a fractured basement reservoir using highresolution 3D seismic and logging datasets: a case study of the Sab'atayn Basin, Yemen. PLoS ONE 13 (10). https://doi.org/10.1371/journal.pone.0206079, e0206079.

Beacom, L.E., Holdsworth, R.E., McCaffrey, K.J.W., Anderson, T.B., 2001. A quantitative study of the influence of preexisting compositional and fabric heterogeneities upon fracture-zone development during basement reactivation. In: Holdsworth, R.E., Strachan, R.A., Magloughlin, J.F., Knipe, R.J. (Eds.), The Nature and Tectonic Significance of Fault Zone Weakening. Geological Society, London, Special Publications, vol. 186, pp. 195–211, https://doi.org/10.1144/GSL.SP.2001.186.01.12.

Beamish, B.B., Crosdale, P.J., 1995. The influence of maceral content on the sorption of gases by coal and the association with outbursting. In: Management and Control of High Gas Emission and Outbursts. Wollongong, 20-24 March 1995, pp. 353–361.

Bear, J., 1972. Dynamics of Fluids in Porous Media. Elsevier, Amsterdam.

Bear, J., 1975. Dynamics of Flow in Porous Media. American Elsevier Publishing Co., New York. 764 pp.

Beeson, D., Hoffman, K., Laure, D., McNaboe, J., Singer, J., 2014. Creation and utility of a large fit-for-purpose earth model in a giant mature field: Kern River field, California. Am. Assoc. Geol. Bull. 98, 1305–1324.

Begg, S.H., Carter, R.R., Dranfield, P., 1989. Assigning effective values to simulator grigblock parameters for heterogeneous reservoir. In: SPE Reserv. Eng. SPE'Paper 16754.

Behrens, C., et al., 2011. U.S. Fossil Fuel Resources: Terminology, Reporting, and Summary.

Belaidi, A., Bonter, D.A., Slightam, C., Trice, R.C., 2016. The lancaster field: progress in opening the UK's fractured basement play. In: Bowman, M., Levell, B. (Eds.), Petroleum Geology of NW Europe: 50 Years of Learning—Proceedings of the 8th Petroleum Geology Conference. Geological Society, London, pp. 385–398, https://doi.org/10.1144/PGC8.20.

Belhaj, H.A., Nouri, A., Vaziri, H.H., Islam, M.R., 2009. Laboratory investigation of effective stresses influence on petrophysical properties of sandstone reservoirs during depletion. J. Can. Pet. Technol., 47–53.

Bell, J.S., 2003. Practical methods for estimating in situ stresses for borehole stability applications in sedimentary basins. J. Pet. Sci. Eng. 38, 111–119.

Bell, J.S., Gough, D.I., 1979. Northeast-southwest compressive stress in Alberta: evidence from oil wells. Earth Planet. Sci. Lett. 45, 475–482.

Bennion, D.B., et al., 2006. A correlation of the low and high temperature water-oil relative permeability characteristics of typical western Canadian unconsolidated bitumen producing formations. In: Paper Presented at the Canadian International Petroleum Conference, Calgary, Alberta, June. Paper Number: PETSOC-2006-136.

Bentley, R., Boyle, G., 2007. Global oil production: forecasts and methodologies. Environ. Plann. B Plann. Des. 34, 609–626.

Bentley, R., Mannan, S.A., Wheeler, S.J., 2007. Assessing the date of the global oil peak: the need to use 2P reserves. Energy Policy 35 (12), 6364–6382.

Bentsen, R., 1978. Conditions under which the capillary term may be neglected. J. Can. Pet. Technol. 17 (4), 25–30. https://doi.org/10.2118/78-04-01.

Bentsen, R., 1985. A new approach to instability theory in porous media. Soc. Pet. Eng. J. 25, 765–779. https://doi.org/10.2118/12725-PA.

Berard, T., Cornet, F.H., 2003. Evidence of thermally induced borehole elongation: a case study at Soultz, France. Int. J. Rock Mech. Min. Sci. 40, 1121–1140.

Berkenpas, P.G., 1991. The Milk River shallow gas pool: role of the up dip water trap and connate water in gas production from the pool. SPE J. 229 (22), 371–380.

Bernabe, Y., Evans, B., 2007. Numerical modelling of pressure solution deformation at axisymmetric asperities under normal load. Geol. Soc. London Spec. Publ. 284, 1185–1205.

Bertani, R., 2012. Geothermal power generation in the world 2005-2010 update report. Geothermics 41, 1–19.

Bertrand, G., Rangin, C., 2003. Tectonics of the western margin of the Shan plateau (Central Myanmar): implication for the India-Indochina oblique convergence since the Oligocene. J. Asian Earth Sci. 21 (10), 1139–1157. https://doi.org/10.1016/S1367-9120(02)00183-9.

Bethel, F.T., Calhoun, J.C., 1953. Capillary desaturation in unconsolidated beads. J. Pet. Technol. 5 (8), 197–202.

Beyer Jr., K.H., et al., 1983. Final report on the safety assessment of triethanolamine, diethanolamine and monoethanolamine. J. Am. Coll. Toxicol. 2, 183–235.

Biazar, J., Ebrahimi, H., 2011. A strong method for solving systems of integro-differential equations. Appl. Math. 2 (9), 1105–1113. https://doi.org/10.4236/am.2011.29152.

Biazar, J., Rahman, M.A., Islam, M.R., 2005. Solution of partial diffusivity equation through perforation tunnel created by PS and PD technique including non-linear pressure gradient contributions. Int. Math. J. 6 (1), 1–6.

Bickle, M., et al., 2012. Shale Gas Extraction in the UK: A Review of Hydraulic Fracturing. Royal Society & Royal Academy of Engineering, p. 105.

Biddle, K.T., Wielchowsky, C.C., 1994. Hydrocarbon traps: Chapter 13: Part III. Processes. In: Magoon, L., Dow, W. (Eds.), The Petroleum System—From Source to Trap. AAPG, Memoirs, vol. 60. American Association of Petroleum Geologists, pp. 219–235.

Biello, D., 2009. The origin of oxygen in Earth's atmosphere. Am. Sci. (August).

Biot, M.A., 1941. General theory of three-dimensional consolidation. J. Appl. Phys. 12 (2), 155–164.

Biot, M.A., 1956. General solutions of the equations of elasticity and consolidation for a porous material. J. Appl. Mech. 78, 91.

Biterge, M.B., Ertekin, T., 1992. Development and testing of a static/dynamic local grid-refinement technique. SPE J. Pet. Technol. 44 (4), 487–495.

Bjørkum, P.A., et al., 1998. Porosity prediction in quartzose sandstones as a function of time, temperature, depth, stylolite frequency, and hydrocarbon saturation. AAPG Bull. 82 (4), 637–647.

Bjorndalen, N., Islam, M.R., 2004. The effect of microwave and ultrasonic irradiation on crude oil during production with a horizontal well. J. Pet. Sci. Eng. 43 (3–4), 139–150.

Bjorndalen, N., Mustafiz, S., Islam, M.R., 2005. No-flare design: converting waste to value addition. Energy Sources 27 (3), 371–380.

Blackmon, D., 2017. The news media finally notices trump's energy policy sea change. Forbes. August.

Blanc, P., Connan, J., 1994. Preservation, Degradation, and Destruction of Trapped Oil. Part III. Processes. (Chapter 14).

Blas, J., 2018. The U.S. just became a net oil exporter for the first time in 75 years. Bloomberg (December 6).

Blevins, R.D., 1984. Applied Fluid Dynamics Handbook. Van Nostrand Reinhold Co., New York. 568 pp.

Blomstrom, D.C., Beyer Jr., E.M., 1980. Plants metabolise ethylene to ethylene glycol. Nature 283, 66–68.

Bloom, J.C., et al., 1988. The effect of long term treatment with auranofin and gold sodium thiomalate on immune function of dog. J. Rheumatol. 15, 409–415.

Bloomberg, 2013. https://www.bloomberg.com/news/articles/2013-06-05/crude-output-exceeds-imports-for-first-time-in-16-years.

Blunt, M., Zhou, D., Fenwick, D., 1995. Three-phase flow and gravity drainage in porous media. Transp. Porous Media 20 (1–2), 77–103.

Bluokeny, T.B., 2001. Development properties under deep abnormally high pressure (D. Shi, Trans.). Nat. Gas Geosci. 12 (4–5), 61–64.

Bobet, A., Einstein, H.H., 1998. Numerical modeling of fracture coalescence in a model rock material. Int. J. Fract. 92, 221–252.

Bockmeulen, H., Barker, C., Dickey, P.A., 1983. Geology and geochemistry of crude oil, Bolivar Coastal fields, Venezuela. Am. Assoc. Pet. Geol. Bull. 67, 242–270.

Bodvarsson, G., 1976. Thermoelastic phenomena in geothermal systems. In: Proceedings of the Second UN Symposium on the Development and Use of Geothermal Resources, San Francisco, pp. 903–907.

Boghi, A., et al., 2015. Stochastic fracture generation accounting for the stratification orientation in a folded environment based on an implicit geological model. Eng. Geol. 187, 135–142.

Boley, B.A., Weiner, J.H., 1960. Theory of Thermal Stresses. John Wiley & Sons, New York.

Bombardelli, F.A., González, A.E., Niño, Y.I., 2008. Computation of the particle Basset force with a fractional-derivative approach. J. Hydraul. Eng. 134 (10), 1513–1520.

Bonnet, E., Bour, O., Odling, N.E., Davy, P., Main, I., Cowie, P., Berkowitz, B., 2001. Scaling of fracture systems in geological media. Rev. Geophys. 39, 347–383. https://doi.org/10.1029/1999RG000074.

Bonter, D.A., Trice, R., 2019. An integrated approach for fractured basement characterization: the Lancaster field, a case study in the UK. Pet. Geosci. 25 (4), 400–414.

Bonter, D., Trice, R., Cavalleri, C., Delius, H., Singh, K., 2018. Giant oil discovery West of Shetland—challenges for fractured basement formation evaluation. In: SPWLA 59th Annual Logging Symposium, 2–6 June. Society of Petrophysicists and Well-Log Analysts, London. https://www.onepetro.org/conference-paper/SPWLA-2018-M.

Booker, J.R., Savvidou, C., 1984. Consolidation around a spherical heat source. Int. J. Solids Struct. 20, 1079.

Boone, T.J., Ingraffea, A.R., Roegiers, J.-C., 1991. Simulation of hydraulic fracture propagation in poroelastic rock with application to stress measurement technique. Int. J. Numer. Anal. Methods Geomech. 28, 1–14.

Boucher, J., 2012. In: Hein, F. (Ed.), Canadian Heavy Oil Association: The First Quarter Century: 1985–2011, 2012. Canadian Heavy Oil Association, Calgary, AB, pp. 46–48.

Bourbiaux, B., Granet, S., Landereau, P., Noetinger, B., Sarda, S., Sabathier, J.C., 1999. Scaling up matrix-fracture transfers in dual-porosity models: theory and application. In: Proceedings of Paper SPE 56557 Presented at the SPE Annual Technical Conference and Exhibition, 3–6 October, Houston.

Bourdet, D., 2002. Well Test Analysis the Use of Advanced Interpretation Model: Handbook of Petroleum Exploration & Production (HPEP). vol. 3 Elsevier.

Bourdin, B., Francfort, G.A., Marigo, J.-J., 2008. The variational approach to fracture. J. Elast. 91 (1–3), 5–148.

BP, 2017. BP annual report. Available from: https://www.bp.com/content/dam/bp/business-sites/en/global/corporate/pdfs/investors/bp-annual-report-and-form-20f-2017.pdf.

BP, 2018. BP annual report. https://www.bp.com/content/dam/bp/business-sites/en/global/corporate/pdfs/investors/bp-annual-report-and-form-20f-2018.pdf.

BP, 2019. Statistical Review of World Energy, sixty-eighth ed. https://www.bp.com/content/dam/bp/business-sites/en/global/corporate/pdfs/energy-economics/statistical-review/bp-stats-review-2019-full-report.pdf.

BP, 2020. BP annual report. https://www.bp.com/content/dam/bp/business-sites/en/global/corporate/pdfs/investors/bp-annual-report-and-form-20f-2020.pdf.

BP, 2021. BP annual report. Available from: https://www.bp.com/content/dam/bp/business-sites/en/global/corporate/pdfs/energy-economics/statistical-review/bp-stats-review-2021-full-report.pdf.

Brinkman, H.C., 1949. A calculation of the viscous force exerted by a flowing fluid on a dense swarm of particles. Appl. Sci. Res. 1 (1), 27–34.

Brons, F., Marting, V., 1961. The effect of restricted fluid entry on well productivity. J. Pet. Technol. 13 (2), 172–174.

Brooks, J.M., 1979. Deep methane maxima in the Northwest Caribbean Sea: possible seepage along the Jamaica ridge. Science 206 (4422), 1069–1071. https://doi.org/10.1126/science.206.4422.1069. 17787481.

Broszeit, J., 1997. Finite-element simulation of circulating steady flow for fluids of the memory-integral type: flow in a single-screw extruder. J. Non-Newton. Fluid Mech. 70 (1–2), 35–58.

Brouwer, D.R., 2004. Dynamic Water Flood Optimization with Smart Wells Using Optimal Control Theory (PhD thesis). Delft University of Technology, Delft.

Brouwer, D.R., Naevdal, G., Jansen, J.D., Vefring, E.H., van Kruijsdiik, C.P.J.W., 2004. Improved reservoir management through optimal control and continuous model updating. In: SPE Paper 90149.

Brudy, M., Zoback, M.D., Fuchs, K., Baumgartner, J., 1997. Estimation of the complete stress tensor to 8 km depth in the KTB scientific drill holes: implications for crustal strength. J. Geophys. Res. 102, 18453–18475.

Bruel, D., 2002. Impact of induced thermal stresses during circulation tests in an engineered fractured geothermal reservoir; example of the Soultzsus-Forets European Hot Fractured rock Geothermal Project, Rhine Graben, France. Oil Gas Sci. Technol. 57, 459–470.

Bu, T., Damsleth, E., 1996. Errors and uncertainties in reservoir performance predictions. In: SPE Paper 30604.

Buckley, S.E., Leverett, M.E., 1942. Mechanism of fluid displacements in sands. Trans. AIME 146, 107–116.

Burland, J.B., 1990. On the compressibility and shear strength of natural clays. Geotechnique 40, 329–378.

California Department of Conservation, 2007. Annual report of the state oil and gas supervisor. Division of Oil, Gas and Geothermal Resources. ftp://ftp.consrv.ca.gov/pub/oil/annual_reports/2007/PR06_2007.pdf.

Camacho, G., Ortiz, M., 1996. Computational modeling of impact damage in brittle materials. Int. J. Solids Struct. 33, 2899–2938.

Cant, D.J., 1986. Diagenetic traps in sandstones. AAPG Bull. 70 (2), 155–160.

Cao, H., Aziz, K., 1999. Evaluation of pseudo functions. In: SPE Paper 54589.

Cao, Y., et al., 2013. Fermentative succinate production: an emerging technology to replace the traditional petrochemical processes. Biomed. Res. Int. 2013. https://doi.org/10.1155/2013/723412, 723412.

Caputo, M., 1997. Rigorous time domain responses of polarizable media. Ann. Geophys. 40 (2), 399–407.

Caputo, M., 1998. Diffusion of fluids in porous media with memory. Geothermics 28, 2–19.

Caputo, M., 1999. Diffusion of fluids in porous media with memory. Geothermics 28 (1), 113–130.

Caputo, M., 2000. Models of flux in porous media with memory. Water Resour. Res. 36, 693–705.

Caputo, M., Cametti, C., 2008. Diffusion with memory in two cases of biological interest. J. Theor. Biol. 254, 697–703.

Caputo, M., Carcione, J.M., 2013. A memory model of sedimentation in water reservoirs. J. Hydrol. 476, 426–432.

Caputo, M., Kolari, J., 2001. An analytical model of the fisher equation with memory function. Altern. Persp. Financ. Account. 1, 1–16.

Caputo, M., Mainardi, F., 1971. A new dissipation model based on memory mechanisms. Pure Appl. Geophys. 8 (91), 134–147.

Caputo, M., Plastino, W., 1998. Rigorous time domain responses of polarizable media. Ann. Geofis. 41 (3), 399–407.

Caputo, M., Plastino, W., 2003. Diffusion with space memory. In: Grafarend, E.W., Krumm, F.W., Schwarze, V.S. (Eds.), Geodesy—The Challenge of the 3rd Millennium. Springer, Berlin, Heidelberg, https://doi.org/10.1007/978-3-662-05296-9_45.

Caputo, M., Plastino, W., 2004. Diffusion in porous layers with memory. Geophys. J. Int. 158, 385–396.

Carey, G., 2018. Why hunger rivals bombs as the biggest danger in Yemen. Bloomberg (March 18).

Carillo, S., 2005. Some remarks on materials with memory: heat conduction and visco-elasticity. J. Nonlinear Math. Phys. 12 (Suppl. 1), 163–178.

Carillo, S., Valente, V., Vergara Caffarelli, G., 2014. Heat conduction with memory: a singular kernel problem. Evol. Equ. Control Theory 3 (3), 399–410.

Carls, M.G., et al., 2016. Petroleum biomarkers as tracers of *Exxon Valdez* oil. Environ. Chem. (11 April). https://doi-org.ezproxy.library.dal.ca/10.1002/etc.3454.

Carlson, E.S., Mercer, J.C., 1991. Devonian shale gas production: mechanisms and simple models. J. Pet. Technol. 43 (04), 476–482.

Carpenter Jr., C.W., Bail, P.T., Bobek, J.E., 1962. A verification of waterflood scaling in heterogeneous communicating flow models. Soc. Pet. Eng. J. 2 (1), 9–12. https://doi.org/10.2118/171-PA.

Carter, J.P., Booker, R.J., 1982. The analysis of consolidation and creep around a deep circular tunnel in clay. In: 4th International Conference on Numerical Methods in Geomechanics in Edmonton, Canada, January.

Carter, B.J., Desroches, J., Ingraffea, A.R., Wawrzynek, P.A., 2000. Simulating fully 3D hydraulic fracturing. In: Zaman, M., Booker, J., Gioda, G. (Eds.), Modeling in Geomechanics. Wiley Publishers, New York.

Carver, P.L., 2019. Essential metals in medicine: the therapeutic use of metal ions in the clinic. In: Sigel, A., et al. (Eds.), Essential Metals in Medicine: Therapeutic Use and Toxicity of Metal Ions in the Clinic. Metal Ions in Life Sciences. vol. 19. de Gruyter GmbH, Berlin, ISBN: 978-3-11-052691-2, pp. 1–16, https://doi.org/10.1515/9783110527872-007.

Cattaneo, C., 1958. A form of heat conduction equation which eliminates the paradox of instantaneous propagation. Compt. Rendus 247 (4), 431–433.

Cense, A.W., Berg, S., 2009. The viscous-capillary paradox in 2-phase flow in porous media. In: International Symposium of the Society of Core Analysts Held in Noordwijk, The Netherlands, pp. 27–30.

Chaalal, O., Islam, M.R., 2001. Integrated management of radioactive strontium contamination in aqueous stream systems. J. Environ. Manag. 61, 51–59.

Chaalal, O., Islam, M.R., Zekri, Y.A., 2017. A comprehensive study of thermal stress on limestone rocks. Int. J. Pet. Res. 1 (1), 19–25.

Chalaturnyk, R., Scot, J.D., 1995. Geomechanics issues of steam assisted gravity drainage. In: Paper SPE 30280 Presented at the International Heavy Oil Symposium held in Calgary, Alberta, Canada, 19–21 June.

Chambers, L., Darbyshire, F., Noble, S., Ritchie, D., 2005. NW UK continental margin: chronology and isotope geochemistry. British Geological Survey. Commissioned Report, CR/05/095.

Chan, Y.T., Banerjee, S., 1981. Analysis of transient three-dimensional natural convection in porous media. J. Heat Transf. 103 (2), 242–248. https://doi.org/10.1115/1.3244448.

Chandra, D., Mandal, A.K., 2000. Toxicological and pharmacological study of Navbal Rasayan—a metal based formulation. Indian J. Pharm. 32, 36–71.

Chang, X., et al., 2013. Ordovician gas exploration breakthrough in the Gucheng lower uplift of the Tarim Basin and its enlightenment. J. Pet. Geol. 36, 383–398.

Charlou, J.L., Donval, J.P., 1993. Hydrothermal methane venting between 12°N and 26°N along the Mid-Atlantic Ridge. J. Geophys. Res. 98, 9625–9642.

Charlou, J.L., Donval, J.P., Fouquet, Y., Jean-Baptist, P., Holm, N., 2002. Geochemistry of high H2 and CH4 vent fluids issuing from ultramafic rocks at the rainbow hydrothermal field (36°14′N, MAR). Chem. Geol. 184 (1–2), 37–48.

Chattaraj, P., et al., 2016. Thermodynamics, kinetics and modeling of sorption behaviour of coalbed methane—a review. J. Unconv. Oil Gas Resour. 16, 14–33.

Chatzis, I., et al., 1983. Magnitude and detailed structure of residual oil saturation. SPE J. 23, 311–326.

Cheema, T.J., Islam, M.R., 1995. A new modeling approach for predicting flow in fractured formations. In: El-Kady, A.I. (Ed.), Groundwater Models for Resources Analysis and Management. Lewis Publishers, Boca Raton, FL, pp. 327–338.

Chen, P.Y.P., 1988. Transient thermal stresses in a rectangular plate due to nonuniform heat transfer coefficients. J. Therm. Stresses 11, 115–125. https://doi.org/10.1080/01495738808961924.

Chen, C.-C., et al., 1985. Pressure transient analysis methods for bounded naturally fractured reservoirs. SPE J. 25 (03), 451–464.

Chen, H.Y., Teufel, L.W., Lee, R.L., 1995. Couple fluid flow and Geomechanics in reservoir study-I. Theory and governing equations. In: Paper SPE 30752, Paper Presented at the SPE Annual Technical Conference and Exhibition, Held in Dallas, USA.

Chen, H.-B., et al., 1999. Influence of trivalent metal ions on the surface structure of a copper-based catalyst for methanol synthesis. Appl. Surf. Sci. 147, 85–93.

Chen, J.P., Liang, D.G., Wang, X.L., et al., 2003. Mixed oils derived from multiple source rocks in the Cainan oilfield, Junggar Basin, Northwest China. Part II: artificial mixing experiments on typical crude oils and quantitative oil-source correlation. Org. Geochem. 34 (7), 911–930.

Chen, Z., Pan, Z., Liu, J., Connell, L.D., Elsworth, D., 2011. Effect of the effective stress coefficient and sorption-induced strain on the evolution of coal permeability: experimental observations. Int. J. Greenhouse Gas Control 5, 1284–1293.

Chen, S., et al., 2014. Unraveling mechanism on reducing thermal hysteresis width of VO2 by Ti doping: a joint experimental and theoretical study. J. Phys. Chem. C 118 (33), 18938–18944.

Cheng, Y.T., Cheng, C.M., 2004. Scaling, dimensional analysis, and indentation measurements. Mater. Sci. Eng. R. Rep. 44 (4), 91–149.

Cherizol, R., Mohini, I., Tjong, S., 2015. Review of non-Newtonian mathematical models for rheological characteristics of viscoelastic composites. Green Sustain. Chem. 05 (01), 6–14.

Chhetri, A.B., Islam, M.R., 2008. Inherently Sustainable Technologies. Nova Science Publishers. 452 pp.

Chmielowski, J., 2013. BP Alaska heavy oil production from the Ugnu fluvial-deltaic reservoir. In: 2013 SPE Western Region/Pacific Section AAPG Joint Technical Conference, Monterey, California, April 20–25, Oral Presentation. http://www.searchanddiscovery.com/documents/2013/80289chmielowski/ndx_chmielowski.pdf.

Choi, E., Cheema, T., Islam, M., 1997. A new dual-porosity/dual-permeability model with non-Darcian flow through fractures. J. Pet. Sci. Eng. 17 (3), 331–344.

Chopra, A., Doiphode, W., 2002. Ayurvedic medicine: core concept, therapeutic principles, and current relevance. Med. Clin. North Am. 86, 75–89.

Chopra, R.N., Chopra, I.C., Handa, K.L., Kapur, L.D., 1958. Chopra's Indigenous Drugs of India, second ed. vol. 198 Academic Publisher, Calcutta, pp. 46–64.

Chopra, S., Castagna, J.P., Portniaguine, O., 2006a. Seismic resolution and thin-bed reflectivity inversion. CSEG Rec. 31, 19–25.

Chopra, R.N., et al., 2006b. Chopra's Indigenous Drugs of India. vol. 198 Academic Publishers, Kolkata, pp. 46–64 (1933, original).

Chopra, S., et al., 2012. Shale gas reservoir characterization workflows. In: Seg Technical Program Expanded Abstracts.

Chow, T.S., 2005. Fractional dynamics of interfaces between soft-nanoparticles and rough substrates. Phys. Lett. A 342, 148–155. https://doi.org/10.1016/j.physleta.2005.05.045.

Christensen, J.R., Darche, G., Déchelette, B., Ma, H., Sammon, P.H., 2004. Applications of dynamic gridding to thermal simulations. In: Proceedings of the SPE International Thermal Operations and Heavy Oil Symposium and Western Regional Meeting, Bakersfield, California, 16–18 March. SPE 86969.

Christie, M.A., 1996. Upscaling for reservoir simulation. JPT, 1004–1009.

Christie, M.A., Blunt, M.J., 2001. Tenth SPE Comparative Solution Project: A Comparison of Upscaling Techniques, SPE Paper 66599.

Chronopoulos, J., Haidouti, C., Chronopoulou, A., Massas, I., 1997. Variations in plant and soil lead and cadmium content in urban parks in Athens, Greece. Sci. Total Environ. 196, 91–98.

Chuoke, R.L., van Meurs, P., van der Poel, C., 1959. The instability of slow, immiscible viscous liquid-liquid displacements in permeable media. Trans. AIME 216, 188–194.

CIA Factbook, 2020. https://www.cia.gov/the-world-factbook/. (Accessed December 2020).

Ciarletta, M., Scarpetta, E., 1989. Minimum problems in the dynamics of viscous fluids with memory. Int. J. Eng. Sci. 27 (12), 1563–1567.

Cinco-Ley, H., Fernando, S.V., 1982. Pressure transient analysis for naturally fractured reservoirs. In: Paper Presented at the SPE Annual Technical Conference and Exhibition, New Orleans, Louisiana, September. Paper Number: SPE-11026-MS.

Clark, G., Jacks, D., 2007. Coal and the industrial revolution, 1700-1869. Eur. Rev. Econ. Hist. 11, 39–72.

Clees, T., Ganzer, L., 2010. An efficient algebraic multigrid solver strategy for adaptive implicit methods in oil reservoir simulation. SPE J. 15 (3), 670–681.

Cluff, R.M., Cluff, S.G., 2004. The origin of Jonah field, Northern Green River Basin, Wyoming. In: Robinson, J.W., Shanley, K.W. (Eds.), Jonah Field: Case Study of a Tight-Gas Fluvial Reservoir. AAPG Stud Geol, vol. 52, pp. 215–241.

Coleman, B.D., Dill, E.H., 1973. On thermodynamics and the stability of motions of materials with memory. Arch. Ration. Mech. Anal. 51 (1), 1–53.

Coney, D., Fyfe, T.B., Retail, P., Smith, P.J., 1993. Clair appraisal—the benefits of a co-operative approach. In: Parker, J.R. (Ed.), Petroleum Geology of Northwest Europe: Proceedings of the 4th Conference. Geological Society, London, pp. 1409–1420, https://doi.org/10.1144/0041409.

Cook, A.M., Myer, L.R., Cook, N.G.W., Doyle, F.M., 1990. The effect of tortuosity on flow through a natural fracture. In: Hustrulid, W.A., Johnson, W.A. (Eds.), Rock Mechanics Contributions and Challenges, Proceedings of the 31st US Symposium on Rock Mechanics. Golden, CO.

Cook, C.C., Andersen, M.A., Halle, G., Gislefoss, E., Bowen, G.R., 1997. Simulating the effects of water-induced compaction in a North Sea reservoir. In: SPE Reservoir Simulation Symposium, 8–11 June, Dallas, Texas. SPE Paper 52891.
Cornelius, 1987. Available from: https://collections.lib.utah.edu/ark:/87278/s67h4hn3.
Cornet, F.H., 1987. Results from Le Mayet de Montagne project. Geothermics 16 (4), 355–374.
Cornet, F.H., 2000. Comment on large-scale in situ permeability tensor of rocks from induced microseismicity by S.A. Shapiro, P. Audigane and J.-J. Royer. Geophys. J. Int. 140, 465–469.
Cornet, F.H., Jianmin, Y., 1995. Analysis of induced seismicity for stress field determination and pore pressure mapping. Pure Appl. Geophys. 145 (3/4), 677–700.
Cornet, F.H., Valette, B., 1984. In situ stress determination from hydraulic injection test data. J. Geophys. Res. 89, 11527–11537.
Cornet, F.H., Helm, J., Poitrenaud, H., Etchecopar, A., 1997. Seismic and aseismic slips induced by large-scale fluid injection. Pure Appl. Geophys. 150, 563–583.
Cortis, A., Birkholzer, J., 2008. Continuous time random walk analysis of solute transport in fractured porous media. Water Resour. Res. 44, W06414.
Coskuner, G., Bentsen, R.G., 1990. An extended theory to predict the onset of viscous instabilities for miscible displacements in porous media. Transp. Porous Media 5 (5), 473–490.
Craig, H., 1965. The measurement of oxygen isotope paleotemperatures. In: Tongiorgi, E. (Ed.), Proceedings of the Spoleto Conference on Stable Isotopes in Oceanographic Studies and Paleotemperatures, pp. 161–182.
Craig, F.F., Sanderlin, J.L., Moore, D.W., Geffen, T.M., 1957. A laboratory study of gravity segregation in frontal drives. Trans. AIME 210 (275), 1957.
CRC Handbook of Chemistry and Physics, 1983. Handbook of Chemistry and Physics, sixty-second ed. Hardcover. ISBN-10: 0849304628.
Croes, G.A., Schwarz, N., 1955. Dimensionally scaled experiments and the theories on the water-drive process. Trans. AIME 204 (35), 8.
Crovelli, R.S., Schmoker, J.W., 2001. Probabilistic method for estimating future growth of oil and gas reserves. In: Dyman, T.S., Schmoker, J.W., Verma, M.K. (Eds.), Geologic, Engineering, and Assessment Studies of Reserve Growth: US Geological Survey Bulletin 2172-C. 12 p.
Crowe, C.T., et al., 2011. Multiphase Flows with Droplets and Particles. CRC Press.
Crowley, Q.C., Key, R., Noble, S.R., 2015. High-precision U-Pb dating of complex zircon from the Lewisian Gneiss Complex of Scotland using an incremental CA-ID-TIMS approach. Gondwana Res. 27, 1381–1391. https://doi.org/10.1016/j.gr.2014.04.001.
CSS (Center for Sustainable Systems), 2020. Unconventional Fossil Fuels Factsheet. University of Michigan. Pub. No. CSS13-19.
Cui, Q., et al., 2005. Environmentally benign working pairs for adsorption refrigeration. Energy 30 (2), 261–271.
Cuong, T.X., Warren, J.K., 2009. Bach Ho field, a fractured granitic basement reservoir, Cuu Long Basin, offshore SE Vietnam—a 'buried-hill' play. J. Pet. Geol. 32, 129–156. https://doi.org/10.1111/j.1747-5457.2009.00440.x.
Curran, R., 2014. Low oil prices taking toll on Canadian E&P. World Oil 336 (2), 79–83.
Dabrowiak, J.C., 2012. Metals in medicine. Inorg. Chim. Acta. Preface. Wiley, 344 pp.
Dai, J.X., et al., 2004. Origins of partially reversed alkane δ13C values for biogenic gases in China. Org. Geochem. 35, 405–411.
Dake, L., 1978. Fundamentals of Reservoir Engineering. Elsevier Science Publishing Company Inc., New York, NY.
Dandekar, A.Y., 2006. Petroleum Reservoir Rock and Fluid Properties, first ed. CRC. 488 pp.
Darcy, H., 1856. Les Fountaines publiques de la ville de Dijon. Exposition et application á suivre et des forules á employer dans les questions de distribution d'eau.
Das, S., 2008. Electro-magnetic heating in viscous oil reservoir. In: Proceedings of International Thermal Operations and Heavy Oil Symposium, Calgary, AB, Canada, 20–23 October. SPE-117693.
Das, Y.C., Navaratna, D.R., 1962. Thermal bending of rectangular plates. J. Aerosp. Sci. 29, 1397–1399.
Davies, J.H.F.L., Heaman, L.M., 2014. New U-Pb baddeleyite and zircon ages for the Scourie dyke swarm: a long-lived large igneous province with implications for the Palaeoproterozoic evolution of NW Scotland. Precambrian Res. 249, 180–198. https://doi.org/10.1016/j.precamres.2014.05.007.
Davis, T.J., et al., 1995. Phase-contrast imaging of weakly absorbing materials using hard X-rays. Nature (London) 373, 595–598.

De Bree, P., Walters, J.V., 1989. Micro/Minifrac test procedures and interpretation for in situ stress determination. Int. J. Rock Mech. Min. Sci. Abstr. 26, 515–521.
de Vries, A.J., 1958. Foam stability: part. II. Gas diffusion in foams. Recl. Trav. Chim. 77, 209–223.
Dean, J., Ghemawat, S., 2008. MapReduce: simplified data processing on large clusters. In: USENIX Association OSDI'04: 6th Symposium on Operating Systems Design and Implementation, pp. 137–149.
Deines, P., Harris, J.W., Spear, P.M., Gurney, J.J., 1989. Nitrogen and ^{13}C content of Finch and Premier diamonds and their implication. Geochim. Cosmochim. Acta 53 (6), 1367–1378.
Delamaide, E., Corlay, P., Wang, D., 1994. Daqing oil field: the success of two pilots initiates first extension of polymer injection in a giant oil field. In: Proceedings of SPE/DOE Improved Oil Recovery Symposium, Tulsa, OK, USA, 17–20 April. SPE-27819.
Delano, A., 1998. Design Analysis of the Einstein Refrigeration Cycle (Ph.D. thesis). Georgia Institute of Technology, Atlanta, GA.
Dempsey, E.D., Holdsworth, R.E., Imber, J., Bistacchi, A., Di Toro, G., 2014. A geological explanation for intraplate earthquake clustering complexity: the zeolite-bearing fault/fracture networks in the Adamello massif (Southern Italian Alps). J. Struct. Geol. 66, 58–74. https://doi.org/10.1016/j.jsg.2014.04.009.
Dennis, J., 1967. International Tectonic Dictionary. AAPG Memoir No. 7, p. 196.
DeSa, V.G., 1965. Experiments with solar-energy utilization at Dacca. Sol. Energy 8 (3), 83–90.
Detwiler, R.L., 2010. Permeability alteration due to mineral dissolution in partially saturated fractures. J. Geophys. Res. 115, B09210–B09220.
Detwiler, R.L., Glass, R.J., Bourcier, W.L., 2003. Experimental observations of fracture dissolution: the role of Peclet number on evolving aperture variability. Geophys. Res. Lett. 30 (12), 1648–1652.
Di Giuseppe, E., Moroni, M., Caputo, M., 2010. Flux in porous media with memory: models and experiments. Transp. Porous Media 83 (3), 479–500.
Dickson, M.H., Fanelli, M., 2003. Geothermal Energy: Utilization and Technology. UNESCO Renewable Energy Series, Earthscan Publications Ltd, London.
Dillon, W.P., et al., 1993. Gas Hydrates on the Atlantic Continental Margin of the United States—Controls on Concentration. USGS Paper No. 1570.
Dincer, I., Rosen, M.A., 2005. Thermodynamic aspects of renewable and sustainable development. Renew. Sust. Energ. Rev. 9, 169–189.
Dincer, I., Rosen, M.A., 2007. Exergy: Energy, Environment and Sustainable Development. Elsevier, Oxford, UK.
Dincer, I., Rosen, M.A., 2011. Thermal Energy Storage Systems and Applications, second ed. Wiley, London, UK.
Ding, Y., Lemonnier, M., 1993. Development of dynamic local grid refinement in reservoir simulation. In: Paper Presented at the SPE Symposium on Reservoir Simulation, New Orleans, Louisiana, February 1993., https://doi.org/10.2118/25279-MS. Paper Number: SPE-25279-MS.
Djurhuus, J., Aadnoy, B.S., 2003. In situ stress state from inversion of fracturing data from oil wells and borehole image logs. J. Pet. Sci. Eng. 38 (3–4), 121–130.
Doan, Q., Ali, S.M.F., George, A.E., 1990. Scaling criteria and model experiments for horizontal wells. In: Proc. CIM/SPE Int. Tech. Meeting, Calgary, Alta., Pap. CIM/SPE. 90–128, pp. 128-1–128-13, https://doi.org/10.2118/92-09-07.
Doan, L.T., Doan, Q.T., Ali, S.M.F., George, A.E., 1997. Analysis of scaled steam flood experiments. In: SPE 37522, International Thermal Operations.
Dobroskok, A.A., Ghassemi, A., Linkov, A.M., 2005. Numerical simulation of crack propagation influenced by thermal and porous liquid stresses. Int. J. Fract. 134, L29–L34.
Dobson, P.F., Kneafsey, T.J., Sonnenthal, E.I., Spycher, N., Apps, J.A., 2003. Experimental and numerical simulation of dissolution and precipitation: implications for fracture sealing at Yucca Mountain, Nevada. J. Contam. Hydrol. 62–63, 459–476.
DoE/EIA, 2013. https://www.eia.gov/outlooks/ieo/pdf/0484(2013).pdf.
Dombrowski, H.S., Brownell, L.E., 1954. Residual equilibrium saturation of porous media. Ind. Eng. Chem. 46 (6), 1207–1219.
Donaldson, A.B., 1997. Reflections on a downhole steam generator program. In: Proceedings of SPE Western Regional Meeting, Long Beach, CA, USA, 25–27 June. SPE-38276.
Doscher, T.M., Hammershaimb, E.C., 1981. Field demonstration of steam drive with ancillary materials. J. Pet. Technol. 34 (07), 1535–1542.

Dott, R.H., 1969. Hypotheses for an organic origin. In: Dott, R.H., Reynolds, M.J. (Eds.), Sourcebook for Petroleum Geology, Part 1—Genesis of Petroleum. American Association of Petroleum Geologists, Tulsa, pp. 1–244.

Doublet, L.E., Pande, P.K., McCollom, T.J., Blasingame, T.A., 1994. Decline curve analysis using type curves—analysis of oil well production data using material balance time: application to field cases. In: Society of Petroleum Engineers Paper Presented at the International Petroleum Conference and Exhibition of Mexico, 10–13 October, Veracruz, Mexico. 24 pp. SPE Paper 28688-MS.

Down Stream Strategies Website 4. http://www.downstreamstrategies.com/documents/reports_publication/marcellus_wv_pa.pdf.

Drab, D., et al., 2013. Hydrocarbon synthesis from carbon dioxide and hydrogen: a two-step process. Energy Fuel 27 (11). https://doi.org/10.1021/ef4011115.

Duguid, J.O., Lee, P.C.Y., 1977. Flow in fractured porous media. Water Resour. Res. 26, 351–356.

Dunnington, H.V., 1967. Stratigraphic distribution of oil fields in the Iraq-Iran-Arabia Basin. Inst. Pet. J. 53 (520), 129–161.

Durlofsky, L.J., 1998. Coarse scale models of two phase flow in heterogeneous reservoirs: volume averaged equations and their relationship to existing upscaling techniques. Comput. Geosci. 2 (2), 73–92.

Dusseault, M.B., Simmons, J.V., 1982. Injection induced stress and fracture orientation changes. Can. Geotech. J. 19 (4), 483–493.

Dyke, C.G., 1996. How sensitive is natural fracture permeability at depth to variation in effective stress? In: Fractured and Jointed Rock Masses, Proceedings of the International ISRM Symposium on Fractured and Jointed Rock Masses. A. A. Balkema, Rotterdam.

Dyman, T.S., Schmoker, J.W., 1998. The use of well-production data in quantifying gas-reservoir heterogeneity. U.S. Geological Survey Open-File Report 98-778, 7 pp.

Dyman, T.S., Schmoker, J.W., Quinn, J.C., 2000. Heterogeneity of Morrow Sandstone reservoirs measured by production characteristics. In: Johnson, K.S. (Ed.), Marine Clastics in the Southern Midcontinent, 1997 Symposium: Oklahoma Geological Survey Circular 103, pp. 119–127.

Dyni, J.R., 2005. Geology and resources of some world oil-shale deposits. USGS Scientific Investigations Report 2005-5294, 42 pp. http://pubs.usgs.gov/sir/2005/5294/pdf/sir5294_508.pdf.

EASAC, 2007. EASAC report. https://easac.eu/publications/details/study-on-the-eu-oil-shale-industry/.

Eck, M., Steinmann, W.D., 2005. Modelling and design of direct solar steam generating collector fields. J. Sol. Energy Eng. Trans. ASME 127, 371–380.

Edwards, J.T., et al., 1998. Validation of gravity-dominated relative permeability and residual oil saturation in a giant oil reservoir. In: Paper SCA-9903, Presented at the 1998 SPE Annual Technical Conference and Exhibition, pp. 27–30, https://doi.org/10.2118/49316-MS.

Eggleston, W.S., 1948. Summary of oil production from fractured rocks reservoirs in California. AAPG 32, 1352–1355.

Eglinton, G., Calvin, M., 1967. Chemical fossils. Sci. Am. 216 (1), 32–43.

Ehrenberg, S.N., 2006. Porosity destruction in carbonate platforms. J. Pet. Geol. 29 (1), 41–52.

Ehrenberg, S., Nadeau, P.H., Steen, Ø., 2009. Petroleum reservoir porosity versus depth: influence of geological age. AAPG Bull. 93 (10), 1281–1296.

Ehrenberg, S.N., Walderhaug, O., Bjørlykke, K., 2012. Carbonate porosity creation by mesogenetic dissolution: reality or illusion? AAPG Bull. 96 (2), 217–233.

Ehrlich, R., Hasiba, H.H., Raimondi, P., 1974. Alkaline waterflooding for wettability alteration-evaluating a potential field application. J. Pet. Technol. 26 (12), 1335–1343.

EIA, 2006. Technology-Based Oil and Natural Gas Plays: Shale Shock! Could There Be Billions in the Bakken? Office of Oil and Gas, Reserves and Production Division. Available from: http://tonto.eia.doe.gov/ftproot/features/ngshock.pdf.

EIA, 2018a. Report on monthly supply. https://www.eia.gov/petroleum/.

EIA, 2018b. Drilling productivity report. October 15 https://www.eia.gov/petroleum/drilling/.

EIA, 2020a. EIA report. Available from: https://www.eia.gov/energyexplained/coal/how-much-coal-is-left.php.

EIA, 2020b. https://www.eia.gov/energyexplained/natural-gas/where-our-natural-gas-comes-from.php.

EIA, 2020c. World Energy Outlook. https://www.eia.gov/outlooks/steo/report/global_oil.php.

EIA, 2020d. https://www.eia.gov/todayinenergy/detail.php7id=38672.

EIA, 2020e. World Shale Resource Assessment. https://www.eia.gov/analysis/studies/worldshalegas/.

EIA, 2021. Available from: https://www.eia.gov/outlooks/aeo/pdf/03%20AEO2021%20Natural%20gas.pdf.

EIA Website 5, EIA report. https://www.eia.gov/energyexplained/oil-and-petroleum-products/.

Einstein, A., Szilard, L., 1930. U.S. Patent 1,781,541.

Eisenack, K., et al., 2007. Qualitative modelling techniques to assess patterns of global change. In: Kropp, J., Scheffran, J. (Eds.), Advanced Methods for Decision Making and Risk Management in Sustainability Science. Nova Science Publishers, pp. 83–127.

Ekstrom, M.P., et al., 1987. Formation imaging with microelectrical scanning arrays. Log. Anal. 28, 294–306.

El-Amin, M.F., Subasi, A., 2019. Developing a generalized scaling-law for oil recovery using machine learning techniques. Proc. Comput. Sci. 163, 237–247.

El-Banbi, A.H., 1998. Analysis of Tight Gas Wells (Ph.D. dissertation). Texas A&M University.

Elias, B.P., Hajash, A.J., 1992. Changes in quartz solubility and porosity due to effective stress: an experimental investigation of pressure solution. Geology 20, 451–454.

El-Sayed, S.M., Abdel-Aziz, M.R., 2003. A comparison of Adomian's decomposition method and Wavelet-Galerkin method for solving integro-differential equations. Appl. Math. Comput. 136, 151–159. https://doi.org/10.1016/S0096-3003(02)00024-3.

El-Shahed, M., 2003. Fractional calculus model of the semilunar heart valve vibrations. In: 19th Biennial Conference on Mechanical Vibration and Noise, Parts A, B, and C. ASME, pp. 711–714.

Elliott, L., 2019. Ending the Iranian sanctions waiver could be own goal for trump. The Guardian (April 23).

Elsworth, D., 1989. Thermal permeability enhancement of blocky rocks: one-dimensional flows. Int. J. Rock Mech. Min. Sci. 26 (3/4), 329–339.

Elsworth, D., Yasuhara, H., 2009. Mechanical and transport constitutive models for fractures subject to dissolution and precipitation. Int. J. Numer. Anal. Methods Geomech. 34 (5), 533–549.

Emmermann, R., Lauterjung, J., 1997. The German continental deep drilling program KTB: overview and major results. J. Geophys. Res. 102 (B8), 18179–18201. https://doi.org/10.1029/96JB03945.

Encyclopaedia Britannica, 2008. Available from: https://www.britannica.com.

Enerdata, 2018. Enerdata annual report. Available from: https://www.enerdata.net/publications/reports-presentations/2018-world-energy-trends-projections.html.

Engelberts, W.F., Klinkenberg, L.J., 1951. Laboratory experiments on the displacement of oil by water from packs of granular material. In: Paper Presented at the 3rd World Petroleum Congress, The Hague, The Netherlands, May 1951. Paper Number: WPC-4138.

Engineering ToolBox Website 3. https://www.engineeringtoolbox.com/heating-values-fuel-gases-d_823.html.

EPA, 2001. EPA report. https://www.api.org/~/media/Files/Policy/Exploration/cbmstudy_attach_uic_exec_summ.pdf.

Ernst, E., 2002. Heavy metals in traditional Indian remedies. Eur. J. Clin. Pharmacol. 57, 891–898.

Escojido, D., 1981. Subsidence in the Bolivar Coast—The Future of Heavy Crude and Tar Sands. McGraw-Hill, New York, pp. 761–767.

Eson, R.L., 1982. Downhole steam generator—field tests. In: Proceedings of SPE California Regional Meeting, San Francisco, CA, USA, 24–26 March. SPE-10745.

Esteban, M., 1989. Paleokarst Reservoirs in Unconformity Plays; Strategies and Applications to Miocene Reefs. Erico Petroleum Information, Jakarta.

Etiope, G., Lollar, B.S., 2013. Abiotic methane on earth. Rev. Geophys. 51, 276–299.

EurekAlert, 2020. News Release 10-Dec-2020. COVID-19 lockdown causes unprecedented drop in global CO2 emissions in 2020. EurekAlert (December 10).

Evans, K.F., Cornet, F.H., Hashida, T., Hayashi, K., Ito, T., Matsuki, K., Wallroth, T., 1999. Stress and rock mechanics issues of relevance to HDR/HWR engineered geothermal systems: review of developments during the past 15 years. Geothermics 28, 455–474.

Evans, K.F., Zappone, A., Kraft, T., Deichmann, N., Moia, F., 2012. A survey of the induced seismic responses to fluid injection in geothermal and CO_2 reservoirs in Europe. Geothermics 41, 30–54.

Evazi, M., Mahani, H., 2010. Unstructured-coarse-grid generation using background-grid approach. SPE J. 15 (2), 326–340.

Ewing, R.E., Lazarov, R.D., 1988. Adaptive local grid refinement. In: SPE Rocky Mountain Regional Meeting, Casper, Wyoming. Paper no. SPE 17806-MS.

Ewing, R.E., Lazarov, R.D., 1989. Local refinement techniques in the finite element and the finite difference methods. In: Proc. Conf. on Numerical Methods and Appl.'88, Sofia, pp. 148–159. ISC Report no. 1988-14.

Ewing, R.E., Lazarov, R.D., Vassilevski, P.S., 1988. Local Refinement Techniques for Elliptic Problems on Cell-Centered Grids. ISC, University of Wyoming, pp. 1–73. Preprint 1988-16.

Ewing, R.E., et al., 1989. Efficient use of locally refined grids for multiphase reservoir simulation. In: SPE Symposium on Reservoir Simulation. Society of Petroleum Engineers, Houston, Texas.

Falls, A.H., et al., 1988. Development of a mechanistic foam simulator: the population balance and generation by snap-off. SPE Reserv. Eng. 3 (03), 884–892.

Fan, L.-S., Zhu, C., 2005. Principles of Gas-Solid Flows. Cambridge University Press, Cambridge.

Farahi, M.M.M., et al., 2014. Scaling analysis and modeling of immiscible forced gravity drainage process. J. Energy Resour. Technol. 136 (2), 022901–022908.

Fardin, N., Jing, L., Stephansson, O., 2001a. Heterogeneity and nisotropy of roughness of rock joints. In: Proceedings of the ISRM Regional Symposium EUROCK. Espoo, Finland, pp. 223–227.

Fardin, N., Stephansson, O., Jing, L., 2001c. The scale dependence of rock joint surface roughness. Int. J. Rock Mech. Min. Sci. 38, 659–669.

Farmer, C.L., 2002. Upscaling: a review. Int. J. Numer. Methods Fluids 40, 63–78.

Farnetti, E., Di Monte, R., Kašpar, J., 2000. Inorganic and Bio-Inorganic Chemistry—Vol. II—Homogeneous and Heterogeneous Catalysis. Encyclopedia of Life Support Systems (EOLSS).

Farooqui, M.Y., et al., 2009. Evaluating volcanic reservoirs. Oilfield Rev. 21 (1), 36–47.

Farouq Ali, S.M., Redford, D.A., 1977. Physical modeling of in situ recovery methods for oilsands. In: Oil Sands-Canada-Venezuela. CIMM, Montreal.

Farouq Ali, S.M., Redford, D.A., Islam, M.R., 1987. Scaling laws for enhanced oil recovery experiments. In: Proceedings of the China-Canada Heavy Oil Symposium, Zhu Zhu City, China.

Farquhar, J., Bao, H., Thiemens, M.H., 2000. Atmospheric influence of Earth's earliest sulfur cycle. Science 289 (5480), 756–759. https://doi.org/10.1126/science.289.5480.756.

Fatti, J.L., et al., 1994. Detection of gas in sandstone reservoirs using AVO analysis: a 3D seismic case history using the Geostack technique. Geophysics 59, 1362–1376.

Fayers, F.J., Sheldon, J.W., 1959. The effect of capillary pressure and gravity on two- phase fluid flow in a porous medium. Pet. Trans. 216 (01), 147–155.

Fekete, 2021. http://www.fekete.com/. (Accessed 23 February 2021).

Ferer, M., Bromhal, G.S., Smith, D.H., 2007. Crossover from capillary fingering to compact invasion for two-phase drainage with stable viscosity ratios. Adv. Water Resour. 30 (2), 284–299.

Fernandes, M., Brooks, P., 2003. Characterization of carbonaceous combustion residues: II. Nonpolar organic compounds. Chemosphere 53, 447–458.

Fianu, J., Gholinezhad, J., Hassan, M., 2018. Comparison of temperature-dependent gas adsorption models and their application to shale gas reservoirs. Energy Fuel 32 (4), 4763–4771.

Fick, A., 1855. Ueber diffusion. Ann. Phys. Chem. 170, 59–86.

Fick, A., 1995. On liquid diffusion. J. Membr. Sci., 33–38.

Finlay, A., Selby, D., Osborne, M., 2011. Re-Os geochronology and fingerprinting of United Kingdom Atlantic margin oil: temporal implications for regional petroleum systems. Geology 39, 475–478. https://doi.org/10.1130/G31781.1.

Finnie, I., Cooper, G.A., Berlie, J., 1979. Fracture propagation in rock by transient cooling. Int. J. Rock Mech. Min. Sci. 16, 11–21.

Finol, A.S., Sancevic, Z.A., 1995. Subsidence in Venezuela. In: Chilingian, G.V. (Ed.), Subsidence due to Fluid Withdrawal. Elsevier Science. 516 pp.

Firanda, E., 2011. The development of material balance equations for coalbed methane reservoirs. In: SPE Asia Pacific Oil and Gas Conference and Exhibition, January. Society of Petroleum Engineers.

Fischer, A., Hahn, C., 2005. Biotic and abiotic degradation behaviour of ethylene glycol monomethyl ether (EGME). Water Res. 39 (10), 2002–2007. https://doi.org/10.1016/j.watres.2005.03.032. 15878604.

Fishman, N.S., Turner, C.E., Peterson, F., Dyman, T.S., Cook, T., 2008. Geologic controls on the growth of petroleum reserves. In: U.S. Geological Survey Bulletin 2172–I. 53 pp.

Flores, R.M., 2013. Coal and Coalbed Gas: Fueling the Future. Elsevier Science & Technology, Saint Louis.

Ford, D., Williams, P., 2007. Karst Hydrogeology and Geomorphology. Wiley, Chichester.

Fortin, M., Archambault, G., Aubertin, M., Gill, D.E., 1988. An algorithm for predicting the effect of a variable normal stiffness on shear strength of discontinuities. In: Proceedings of the 15th Canadian rock Mechanics Symposium, Toronto, pp. 109–117.

Foster, W.R., 1973. A low-tension waterflooding process. J. Pet. Technol. 25 (2), 205–210. https://doi.org/10.2118/3803-PA.

Foulger, G.R., Julian, B.R., Hill, D.P., Pitt, A.M., Malin, P.E., Shalev, E., 2004. Non-double-couple microearthquakes at Long Valley caldera, California, provide evidence for hydraulic fracturing. J. Volcanol. Geotherm. Res. 132, 45–71.

Franks, K.A., Lambert, P.F., 1985. Early California Oil: A Photographic History, 1865–1940. Montague History of Oil Series, Number Four, Hardcover.

Fredrich, J.T., Wong, T.-F., 1986. Mechanics of thermally induced cracking in three crustal rocks. J. Geophys. Res. 91 (B12), 12743–12754.

Freedman, R., et al., 1998. Combining NMR and density logs for petrophysical analysis in gas-bearing formations. In: Paper II, SPAWLA 39th Annual Meeting, Colorado, 26–29 May.

Frenzel, M., Woodcock, N.H., 2014. Cockade breccia: product of mineralisation along dilational faults. J. Struct. Geol. 68, 194–206. https://doi.org/10.1016/j.jsg.2014.09.001.

Frost, B.R., Frost, C.D., 2008. On charnockites. Gondwana Res. 13, 30–44. https://doi.org/10.1016/j.gr.2007.07.006.

Fu, C., Liu, N., 2019. Waterless fluids in hydraulic fracturing—a review. J. Nat. Gas Sci. Eng. 67, 214–224.

Fuchs, K., Müller, B., 2001. World Stress Map of the Earth: a key to tectonic processes and technological applications. Naturwissenschaften 88, 357–371.

Fulcher Jr., R.A., Ertekin, T., Stahl, C.D., 1985. Effect of capillary number and its constituents on two-phase relative permeability curves. J. Pet. Technol. 37 (2), 249–260. https://doi.org/10.2118/12170-PA.

Gaede-Koehler, A., et al., 2012. Direct measurement of the thermal hysteresis of antifreeze proteins (AFPs) using sonocrystallization. Anal. Chem. 84 (23), 10229–10235.

Gai, X., 2004. A Coupled Geomechanics and Reservoir Flow Model on Parallel Computers (Ph.D. thesis). University of Texas at Austin, Texas.

Gale, J.E., 1982. The effects of fracture type (induced vs natural) on the stress—fracture closure—fracture permeability relationship. In: Proceedings of the 23rd US Symposium on Rock Mechanic, Berkeley, CA, pp. 290–296.

Gale, J., Reed, R., Holder, J., 2007. Natural fractures in the Barnett shale and their importance for hydraulic fracture treatments. AAPG Bull. 91, 603–622.

Galloway, W.E., Hobday, D.K., Magara, K., 1982. Frio formation of Texas Gulf Coastal Plain: depositional systems, structural framework, and hydrocarbon distribution. Am. Assoc. Pet. Geol. Bull. 66 (1). OSTI No. 664663.

Gao, Z.Q., Fan, T.L., 2013. Ordovician intra-platform shoal reservoirs in the Tarim Basin, NW China: characteristics and depositional controls. Bull. Can. Petrol. Geol. 61, 83–100.

Gao, H., Klein, P., 1997. Numerical simulation of crack growth in an isotropic solid with randomized internal cohesive bonds. J. Mech. Phys. Solids 46, 187–218.

Gao, G., Titi, A., Yang, S., Tang, Y., Kong, Y., He, W., 2017a. Geochemistry and depositional environment of fresh lacustrine source rock: a case study from the Triassic Baijiantan Formation shales in Junggar Basin, Northwest China. Org. Geochem. 113, 75–89. https://doi.org/10.1016/j.orggeochem.2017.08.002.

Gao, G., Ren, J., Yang, S., Xiang, B., Zhang, W., 2017b. Characteristics and origin of solid bitumen in Glutenites: a case study from the Baikouquan Formation Reservoirs of the Mahu Sag in the Junggar Basin. China. Energy Fuel 31 (12), 13179–13189. https://doi.org/10.1021/acs.energyfuels.7b01912.

Garland, J., et al., 2012. Advances in carbonate exploration and reservoir analysis. Geol. Soc. Spec. Publ. 370, 17–37.

Garon, A.M., Geisbrecht, R.A., Lowry Jr., W.E., 1982. Scaled model experiments of fire flooding in tar sands. J. Pet. Technol. 34 (9), 2–158. https://doi.org/10.2118/9449-PA.

Gaspar, A.T.F.S., et al., 2005. Enhanced oil recovery with CO_2 sequestration: a feasibility study of a Brazilian mature oil field. In: Paper Presented at the SPE/EPA/DOE Exploration and Production Environmental Conference, Galveston, Texas, March., https://doi.org/10.2118/94939-STU. Paper Number: SPE-94939-STU.

Gasparik, M., Gensterblum, Y., Ghanizadeh, A., Weniger, P., Krooss, B.M., 2015. High-pressure/high-temperature methane-sorption measurements on carbonaceous shales by the manometric method: experimental and data-evaluation considerations for improved accuracy. SPE J. 20 (04), 790–809.

Gatens, J.M., et al., 1987. Analysis of eastern Devonian gas shales production data. J. Pet. Technol. 41 (05), 519–525.

Gatti, S., Vuk, E., 2005. Singular limit of equations for linear viscoelastic fluids with periodic boundary conditions. Int. J. Non-Linear Mech. 41 (4), 518–526.

Geertsma, J., de Klerk, F.A., 1969. A rapid method of predicting width and extent of hydraulically induced fractures. J. Pet. Technol., 1571–1581.

Geertsma, J., Croes, G.A., Schwarz, N., 1956. Theory of dimensionally scaled models of petroleum reservoirs. Trans. AIME 207, 118–127.

George, S.C., 1992. Effect of igneous intrusion on the organic geochemistry of a siltstone and an oil shale horizon in the Midland Valley of Scotland. Org. Geochem. 18 (5), 705–723.

George, D.S., Islam, M.R., 1998. A micromodel study of acidization foam. In: Proc. SPE ADIPEC 98, Abu Dhabi, UAE. Paper No. 49489.

Gephart, J.W., Forsyth, D.W., 1984. An improved method for determining the regional stress tensor using earthquake focal mechanism data: application to the San Fernando earthquake sequence. J. Geophys. Res. 89, 9305–9320.

Gerritsen, M.G., Durlofsky, L.J., 2005. Modelling fluid flow in oil reservoirs. Annu. Rev. Fluid Mech. 37, 211–238.

Gharbi, R., Peters, E., Elkamel, A., 1998. Scaling miscible fluid displacements in porous media. Energy Fuel 12 (4), 801–811.

Ghassemi, A., 2012. A review of some rock mechanics issues in geothermal reservoir development. Geotech. Geol. Eng. 30 (3), 647–664. https://doi.org/10.1007/s10706-012-9508-3.

Ghassemi, A., Kumar, S., 2007. Changes in fracture aperture and fluid pressure due to thermal stress and silica dissolution/precipitation induced by heat extraction from substrate rocks. Geothermics 36, 115–140.

Ghassemi, A., Roegiers, J.-C., 1996. A three-dimensional poroelastic hydraulic fracture simulator using the displacement discontinuity method. In: Proceedings of the 2nd North American Rock Mechanics Symposium, Montreal, pp. 982–987.

Ghassemi, A., Suresh Kumar, G., 2007. Variation of fracture aperture and pressure due to combined heat extraction-induced thermal stress and silica dissolution/precipitation. Geothermics 36, 115–140.

Ghassemi, A., Tarasovs, S., 2015. Analysis of fracture propagation under thermal stress in geothermal reservoirs. In: Proceedings World Geothermal Congress 2015 Melbourne, Australia, 19–25 April 2015.

Ghassemi, A., Zhang, Q., 2004a. Poro-thermoelastic mechanisms in wellbore stability and reservoir stimulation. In: Proceedings of the 29th Workshop on Geothermal Reservoir Engineering. Stanford University, Stanford, CA.

Ghassemi, A., Zhang, Q., 2004b. A transient fictitious stress boundary element method for poro-thermoelastic media. J. Eng. Anal. Bound. Elem. 28 (11), 1363–1373.

Ghassemi, A., Zhang, Q., 2006. Poro-thermoelastic response of a stationary crack using the displacement discontinuity method. ASCE J. Eng. Mech. 132 (1), 26–33.

Ghassemi, A., Zhou, X.X., 2011. A three-dimensional poro- and thermoelastic model for analysis of fracture response in EGS. Geothermics 40, 39–49.

Ghassemi, A., Tarasovs, S., Cheng, A.D.-H., 2003. An integral equation method for modeling three-dimensional heat extraction from a fracture in hot dry rock. Int. J. Numer. Anal. Methods Geomech. 27 (12), 989–1004.

Ghassemi, A., Tarasovs, S., Cheng, A.H.-D., 2005. Integral equation solution of heat extraction-induced thermal stress in enhanced geothermal reservoir. Int. J. Numer. Anal. Methods Geomech. 29, 829–844.

Ghassemi, A., Tarasovs, S., Cheng, A.H.-D., 2007. A three-dimensional study of the effects of thermo-mechanical loads on fracture slip in enhanced geothermal reservoir. Int. J. Rock Mech. Min. Sci. 44, 1132–1148.

Ghassemi, A., Nygren, A., Cheng, A.D.-H., 2008. Effects of heat extraction on fracture aperture: a poro-thermoelastic analysis. Geothermics 37 (5), 525–539.

Giardini, A.A., Melton, C.E., Mitchel, R.S., 1982. The nature of the upper 400 km of the Earth and its potential as a source for nonbiogenic petroleum. J. Pet. Geol. 5 (2), 130–137.

Gieskes, J.M., et al., 1988. Hydrothermal fluids and petroleum in surface sediments of Guaymas Basin, Gulf of California: a case study. Can. Mineral. 26 (3), 589–602.

Gill, P.E., Murray, W., Wright, M., 1982. Practical Optimization. Academic Press, New York.

Gilman, J., 1986. An efficient finite-difference method for simulating phase segregation in the matrix blocks in double-porosity reservoirs. SPE Reserv. Eng. 1 (4), 403–413.

Glass, R.J., Rajaram, H., Detwiler, R.L., 2003. Immiscible displacements in rough-walled fractures: competition between roughening by random aperture variations and smoothing by in-plane curvature. Phys. Rev. E 68, 061110–061116.

Glossary, G.L., 2000. U.S. Geological Survey. U.S. Department of the Interior, Reston, VA.

Gochfeld, M., 2003. Cases of mercury exposure, bioavailability, and absorption. Ecotoxicol. Environ. Saf. 56, 17–19.

Gogtay, N.J., Bhatt, H.A., Dalvi, S.S., Kshirsagar, N.A., 2002. The use and safety of non-allopathic Indian medicines. Drug Saf. 25, 1005–1019.

Gold, T., 1999. The Deep Hot Biosphere. Copernicus, New York.

Gold.org, 2021. https://www.gold.org/goldhub/data/gold-prices?utm_source=google&utm_medium=cpc&utm_campaign=rwm-goldhub&utm_content=466092663098&utm_term=gold%20price%20history&gclid=CjwKCAiAgJWABhArEiwAmNVTBxJWHZxBdKWr7Oq7x8oLHxOWK-rkOx2JkLgLUgu-5td8taGqMUvTbhoCoLIQAvD_BwE.

Gollapudi, U.K., Knutson, C.L., Bang, S.S., Islam, M.R., 1995. A new technique for plugging permeable formations. Chemosphere 30 (4), 695–705.

Gong, B., Karimi-Fard, M., Durlofsky, L.J., 2008. Upscaling discrete fracture characterizations to dual-porosity, dual-permeability models for efficient simulation of flow with strong gravitational effects. SPE J. 13 (1), 58.

González-Gomez, W.S., Alvarado-Gil, J.J., 2015. Thermal effects on the physical properties of limestones from the Yucatan Peninsula. Int. J. Rock Mech. Min. Sci. 75, 182–189. https://doi.org/10.1016/j.ijrmms.2014.12.010.

Goodarzi, S., Settari, A., Zoback, M., Keith, D.W., 2010. Thermal aspects of geomechanics and induced fracturing in CO_2 injection with application to CO_2 sequestration in Ohio River Valley. In: SPE International Conference on CO_2 Capture, Storage, and Utilization, New Orleans, Louisiana, USA.

Goodman, R.E., 1980. Introduction to Rock Mechanics. Wiley, New York.

Goodman, R.E., St John, C., 1977. Finite element analysis for discontinuous rocks. In: Numerical Methods in Geotechnical Engineering, pp. 148–175.

Gordon, D., 2012. Understanding Unconventional Oil. Carnegie Endowment for International Peace, Washington, DC.

Gough, D.I., Bell, J.S., 1981. Stress orientations from oil well fractures in Alberta and Texas. Can. J. Earth Sci. 18, 1358–1370.

Graeff, C., et al., 2013. High resolution quantitative computed tomography-based assessment of trabecular microstructure and strength estimates by finite-element analysis of the spine, but not DXA, reflects vertebral fracture status in men with glucocorticoid-induced osteoporosis. Bone 52 (2), 568–577.

Grasselli, G., 2006. Shear strength of rock joints based on quantified surface description. Rock Mech. Rock. Eng. 39, 295–314.

Grasselli, G., Egger, P., 2003. Constitutive law for the shear strength of rock joints based on three-dimensional surface parameters. Int. J. Rock Mech. Min. Sci. 40, 25–40.

Grasselli, G., Wirth, J., Egger, P., 2002. Quantitative three-dimensional description of rough surface and parameter evolution with shearing. Int. J. Rock Mech. Min. Sci. 39, 789–800.

Grattoni, C.A., Jing, X.D., Dawe, R.A., 2001. Dimensionless groups for three-phase gravity drainage flow in porous media. J. Pet. Sci. Eng. 29 (1), 53–65.

Greaves, M., Xia, T., 2004. Downhole upgrading of Wolf Lake oil using THAI/CAPRI processes-tracer tests. Am. Chem. Soc. Div. Pet. Chem. Prepr. 49 (1), 69–72.

Green, D., et al., 2010. Water and its influence on the lithosphere–asthenosphere boundary. Nature 467, 448–451. https://doi.org/10.1038/nature09369.

Greenkorn, R.A., 1964. Flow models and scaling laws for flow through porous media. Ind. Eng. Chem. 56 (3), 32–37.

Guellal, S., Grimalt, P., Cherruault, Y., 1997. Numerical study of Lorenz's equation by the Adomian method. Comput. Math. Appl. 33 (3), 25–29.

Gui, B., Yang, Q.Y., Wu, H.J., Zhang, X., Yao, L., 2010. Study of the effects of low-temperature oxidation on the chemical composition of a light crude oil. Energy Fuel 24 (2), 1139–1145. https://doi.org/10.1021/ef901056s.

Guiducci, C., Pellegrino, A., Radu, J.P., Collin, F., Charlier, R., 2002. Numerical modeling of hydro-mechanical fracture behavior. In: Numerical Models in Geomechanics (NUMOG VIII). Balkema, Rotterdam, pp. 293–299.

Gunal, G.O., Islam, M.R., 2000. Alteration of asphaltic crude rheology with electromagnetic and ultrasonic irradiation. J. Pet. Sci. Eng. 26 (1–4), 263–272.

Gunnarsson, G., Arnaldsson, A., Oddsdóttir, A.L., 2011. Model simulations of the Hengill Area, Southwestern Iceland. Transp. Porous Media 90 (1), 3–22.

Gunnell, Y., Louchet, A., 2000. The influence of rock hardness and divergent weathering on the interpretation of apatite fission-track denudation rates. Evidence from charnockites in South India and Sri Lanka. Z. Geomorphol. 44, 33–57.

Guo, F., Morgenstern, N.R., Scott, J.D., 1993. Interpretation of hydraulic fracturing breakdown pressure. Int. J. Rock Mech. Min. Sci. Geomech. Abstr. 30 (6), 617–626.

Guo, Z., et al., 2012. Dominant scaling groups of polymer flooding for enhanced heavy oil recovery. Ind. Eng. Chem. Res. 52 (2), 911–921.

Gupta, S., Saksena, S., Shankar, V.R., Joshi, V.U., 1998. Emission factors and thermal efficiencies of cooking biofuels from five countries. Biomass Bioenergy 14, 547–559.

Gutierrez-Negrin, L.C.A., Quijano-Leon, J.L., 2003. Analysis of seismicity in the Los Humeros, Mexico, geothermal field. Geotherm. Resour. Counc. Trans. 28, 467–472.

Gutmanis, J.C., 2009. Basement reservoirs—a review of their geological and production histories. In: International Petroleum Technology Conference held in Doha, Qatar, 7–9 December. IPTC, p. 13156, https://doi.org/10.2523/IPTC-13156-MS.

Gutmanis, J.C., 2010. Basement reservoirs—a review of their geological and production characteristics. In: Proc. of the International Petroleum Technology Conference., https://doi.org/10.2523/13156-MS.

Guy, N., Enchery, G., Renard, G., 2012. Numerical modelling of thermo-hydro-mechanics involving AMR-based thermal fluid flow and geomechanics: application to thermal EOR. In: Paper Presented at the 46th U.S. Rock Mechanics/Geomechanics Symposium, Chicago, Illinois, June 2012. Paper Number: ARMA-2012-530.

Guy, N., Enchéry, G., Renard, G., 2013. Numerical modeling of thermal EOR: comprehensive coupling of an AMR-based model of thermal fluid flow and geomechanics. Oil Gas Sci. Technol. Rev. 67 (6), 1019–1027.

Haajizadeh, M., Begg, S.H., 1993. Sensitivity of oil recovery to grid size and reservoir description in fluvially dominated deltaic facies. In: Paper Presented at the SPE Western Regional Meeting, Anchorage, Alaska, May 1993. Paper Number: SPE-26080-MS.

Haavardsson, N.F., Huseby, A.B., 2007. Multisegment production profile models—a tool for enhanced total value chain analysis. J. Pet. Sci. Eng. 58, 325–338.

Hadamard, J., 1923. Lectures on the Cauchy Problem in Linear Partial Differential Equations. Yale Univ. Press, New Haven, London.

Hagoort, J., 1980. Oil recovery by gravity drainage. Soc. Pet. Eng. J. 20 (3), 139–150. https://doi.org/10.2118/7424-PA.

Hamada, G.M., 2009. Petrophysical properties evaluation of tight gas sand reservoirs using NMR and conventional openhole logs. Open Renew. Energ. J. 2, 6–18.

Han, D.K., Yang, C.Z., Zhang, Z.Q., Lou, Z.H., Chang, Y.I., 1999. Recent development of enhanced oil recovery in China. J. Pet. Sci. Eng. 22, 181–188.

Handy, L.L., et al., 1982. Thermal stability of surfactants for reservoir application. SPE J. 22 (05), 722–730.

Haring, M., Ulich, S., Ladner, F., Dyer, B., 2008. Characterization of the Basel 1 enhanced geothermal system. Geothermics 37 (5), 469–495.

Hariri, M.M., Lisenbee, A.L., Paterson, C.J., 1995. Fracture control on the tertiary epithermal-mesothermal gold deposits northern Black Hills, South Dakota. Explor. Min. Geol. 3 (4), 205–214.

Harpalani, S., Schraufnagel, R.A., 1990. Shrinkage of coal matrix with release of gas and its impact on permeability of coal. Fuel 69, 551–556.

Harpalani, S., Prusty, B.K., Dutta, P., 2006. Methane/CO_2 sorption modeling for coalbed methane production and CO2 sequestration. Energy Fuel 20, 1591–1599.

Harris, D.G., 1975. The role of geology in reservoir simulation studies. J. Pet. Eng., 625–632. SPE Paper 5022.

Hascakir, B., Babadagli, T., Akin, S., 2008. Experimental and numerical modeling of heavy-oil recovery by electrical heating. In: Proceedings of International Thermal Operations and Heavy Oil Symposium, Calgary, AB, Canada, 20–23 October. SPE-117669.

Haven, H.L., et al., 1988. Application of biological markers in the recognition of palaeohypersaline environments. In: Fleet, A.J., Kelts, K., Talbot, M.R. (Eds.), Lacustrine Petroleum Source Rocks. Geological Society of London Special Publication 40, Blackwell, Oxford, pp. 123–130.

Hayashi, K., Sayama, T., Abe, H., 1990. Stability of a geothermal reservoir crack connecting with a horizontal natural fracture in the earth's crust. In: Rossmanith, H.-P. (Ed.), Mechanics of Jointed and Faulted Rock. MJFR. A.A. Balkema, Rotterdam, pp. 659–666.

Hayat, T., et al., 2011. Steady flow of Maxwell fluid with convective boundary conditions. Z. Naturforsch. A 66 (6–7), 417–422.

Hayman, G.D., Derwent, R.G., 1997. Atmospheric chemical reactivity and ozone-forming potentials of potential CFC replacements. Environ. Sci. Technol. 31 (2), 327–336. https://doi.org/10.1021/es9507751.

Haynes and Boone, 2020. Oil Patch Bankruptcy Monitor.

Hazen, R.M., Hemley, R.J., Mangum, A.J., 2012. Carbon in Earth's interior: storage, cycling, and life. EOS Trans. Am. Geophys. Union 93, 17–28.

He, J., 2015. A unified finite difference model for the simulation of transient flow in naturally fractured carbonate karst reservoirs. In: SPE Reservoir Simulation Symposium, Houston, Texas, February 2015. SPE-173262-MS.

Healy, D., Jones, R.R., Holdsworth, R.D., 2006. Three-dimensional brittle shear fracturing by tensile crack interaction. Nature 439. https://doi.org/10.1038/nature04346.

Heidbach, O., et al., 2004. Stress maps in a minute: the 2004 world stress map release. Eos Trans. 85 (49), 521–529.

Heidbach, O., et al., 2008. The World Stress Map Database Release., https://doi.org/10.1594/GFZ.WSM.Rel2008.

Henda, R., Hermas, A., Gedye, I., M.R., 2005. Microwave enhanced recovery of nickel-copper ore: communition floatability aspects. J. Microw. Power Electromagn. Energy. 40 (1), 7–16.

Heron, G., Carroll, S., Nielsen, S.G., 2005. Full-scale removal of DNAPL constituents using steam-enhanced extraction and electrical resistance heating. Groundw. Monit. Remediat. 25 (4), 92–107.

Herrmann, U., Kelly, B., Price, H., 2004. Two-tank molten salt storage for parabolic trough solar power plants. Energy 29 (5), 883–893. https://doi.org/10.1016/S0360-5442(03)00193-2.

Hettema, M., Papamichos, E., Schutjens, P., 2002. Subsidence delay: field observations and analysis. Oil Gas Sci. Technol. – Rev. IFP 57 (5), 443–458.

Hickman, S.H., Davatzes, N.C., 2010. In situ stress and fracture characterization for planning of an EGS stimulation in the Desert Peak geothermal field, Nevada. In: Proceedings 35th Workshop on Geothermal Reservoir Engineering. Stanford University, Stanford, CA.

Hilfer, R., Øren, P.E., 1996. Dimensional analysis of pore scale and field scale immiscible displacement. Transp. Porous Media 22 (1), 53–72.

Hill, S., 1952. Channelling in packed columns. Chem. Eng. Sci. 1, 247–253.

Hillis, R.R., Reynolds, S.D., 2000. The Australian stress map. J. Geol. Soc. 157, 915–921.

Hoffmann, C.F., Strausz, O.P., 1986. Bitumen accumulation in Grosmont platform complex, Upper Devonian, Alberta, Canada. Am. Assoc. Pet. Geol. Bull. 70 (9), 1113–1128.

Hojka, K., Dusseault, M.B., Bogobowicz, A., 1993. Analytical solutions for transient thermoelastic stress fields around a borehole during fluid injection into permeable media. J. Can. Pet. Technol. 32 (4), 49–57. https://doi.org/10.2118/93-04-03.

Holdsworth, R.E., Morton, A., et al., 2018. The nature and significance of the Faroe-Shetland terrane: linking Archaean basement blocks across the North Atlantic. Precambrian Res. 321, 154–171. https://doi.org/10.1016/j.precamres.2018.12.004.

Holdsworth, R.E., McCaffrey, K.J.W., et al., 2019. Natural fracture propping and earthquake-induced oil migration in fractured basement reservoirs. Geology 47, 700–704. https://doi.org/10.1130/G46280.1.

Holdsworth, R.E., et al., 2020. The nature and age of basement host rocks and fissure fills in the Lancaster field fractured reservoir, West of Shetland. J. Geol. Soc. https://doi.org/10.1144/jgs2019-142.

Holdway, D.A., 2002. The acute and chronic effects of wastes associated with offshore oil and gas production on temperate and tropical marine ecological processes. Mar. Pollut. Bull. 44 (3), 185–203. https://doi.org/10.1016/s0025-326x(01)00197-7.

Holland, M., Van Gent, H.W., Bazalgette, L., Yassir, N., Hoogerduijn-Strating, E.H., Urai, J.L., 2011. Evolution of dilatant fracture networks in normal faults—evidence from 4D model experiments. Earth Planet. Sci. Lett. 304, 399–406. https://doi.org/10.1016/j.epsl.2011.02.017.

Holmes, A.J., Griffith, C.E., Scotchman, I.C., 1999. The Jurassic petroleum system of the West of Britain Atlantic margin—An integration of tectonics, geochemistry, and basin modelling. In: Fleet, A.J., Boldy, S.A.R. (Eds.), Petroleum Geology of Northwest Europe: Proceedings of the 5th Conference. Geological Society, London, pp. 1351–1365, https://doi.org/10.1144/0051351.

Holmgren, C.R., Morse, R.A., 1951. Effect of free gas saturation on oil recovery by water flooding. Petrol. Trans. AIME 192, 135–140.

Holtz, M.H., 2008. Summary of sandstone Gulf Coast CO_2 EOR flooding application and response. In: Proceedings of SPE/DOE Symposium on Improved Oil Recovery, Tulsa, OK, USA, 20–23 April. SPE-113368.

Honglin, L., Guizhong, L., Yanxiang, L., 2007. High coal rank potential of coalbed methane and its distribution in China. In: International Coalbed Methane and Shale Gas Symp., Tuscaloosa, Alabama, May 23–24. Paper 0705.

Hopkirk, R.J., Sharma, D., Pralong, P.J., 1981. Coupled convective and conductive heat transfer in the analysis of HDR geothermal sources. In: Lewis, R.W., et al. (Eds.), Numerical Methods in Heat Transfer. Wiley, New York, pp. 261–307.

Hossain, M.E., 2018. Dimensionless scaling parameters during thermal flooding process in porous media. J. Energy Resour. Technol. 140, 072004-3. ASME.

Hossain, M.E., Abu-Khamsin, S.A., 2012a. Development of dimensionless numbers for heat transfer in porous media using memory concept. J. Porous Media 15 (10), 957–971.

Hossain, M.E., Abu-Khamsin, S.A., 2012b. Utilization of memory concept to develop heat transfer dimensionless numbers for porous media undergoing thermal flooding with equal rock and fluid temperatures. J. Porous Media 15 (10), 937–953.

Hossain, M.E., Islam, M.R., 2009a. An Advanced Analysis Technique for Sustainable Petroleum Operations. VDM Publishing Ltd., Germany. 750 pp.

Hossain, M.E., Islam, M.R., 2009b. A comprehensive material balance equation with the inclusion of memory during rock-fluid deformation. Adv. Sustain. Pet. Eng. Sci. 1 (2), 141–162.

Hossain, M.E., Islam, M.R., 2011. Development of new scaling criteria for a fluid flow model with memory. Adv. Sustain. Pet. Eng. Sci. 2 (3), 239–261.

Hossain, M.E., Mousavizadegan, S.H., Ketata, C., Islam, M.R., 2007. A novel memory based stress-strain model for reservoir characterization. J. Nat. Sci. Sustain. Technol. 1 (4), 653–678.

Hossain, M.E., Ketata, C., Khan, M.I., Islam, M.R., 2009a. Flammability and individual risk assessment for natural gas pipelines. Adv. Sustain. Pet. Eng. Sci. 1 (1), 33–44.

Hossain, M.E., Liu, L., Islam, M.R., 2009b. Inclusion of the memory function in describing the flow of shear-thinning fluids in porous media. Int. J. Eng. 3 (5).

Hovanessian, S.A., Fayers, F.J., 1961. Linear water flood with gravity and capillary effects. Soc. Pet. Eng. J. 1, 32–36.

Hristov, J., 2011. Transient flow of a generalized second grade fluid due to a constant surface shear stress: an approximate integral-balance solution. Int. Rev. Chem. Eng. 3, 802–809.

Hristov, J., 2013. A note on the integral approach to non-linear heat conduction with Jeffrey's fading memory. Therm. Sci. 17, 733–737.

Hristov, J., 2014. Diffusion models with weakly singular kernels in the fading memories: how the integral-balance method can be applied? Therm. Sci. 73.

Hristov, J., 2017a. Sub diffusion model with time dependent diffusion concept. Therm. Sci. 21 (1A), 69–80.

Hristov, J., 2017b. Space fractional diffusion with a potential power-law coefficient: transient approximate solution. Progr. Fract. Differ. Appl. 3 (1), 19–39.

Huang, J., Griffiths, D.V., Wong, S.W., 2011. Characterizing natural-fracture permeability from mud-loss data. SPE J. 16 (1), 111–114. https://doi.org/10.2118/139592-PA.

Huenges, E., Holl, H.-G., Legarth, B., Zimmermann, G., Saadat, A., Tischner, T., 2004. The stimulation of a sedimentary geothermal reservoir in the North German basin: case study Groß Schönebeck. Z. Angew. Geol. 2, 24–27.

Hughes, B., Sarma, H.K., 2006. Burning reserves for greater recovery? Air injection potential in Australian light oil reservoirs. Environ. Sci. https://doi.org/10.2118/101099-MS.

Hull, T.R., et al., 2002. Combustion toxicity of fire retarded EVA. Polym. Degrad. Stab. 77, 235–242.

Hulm, E., et al., 2013. Integrated reservoir description of the Ugnu heavy-oil accumulation, North Slope, Alaska. In: Hein, F.J., Leckie, D., Larter, S., Suter, J.R. (Eds.), Heavy-Oil and Oil-Sand Petroleum Systems in Alberta and Beyond. AAPG Studies in Geology, vol. 64. AAPG, Tulsa, OK, pp. 481–508.

Hunt, J.M., 1995. Petroleum Geochemistry and Geology, second ed. W. H. Freeman and Company, New York.

Hurst, R., 1934. Unsteady flow of fluids in oil reservoirs. J. Appl. Phys. 5 (20), 20–30.

Hutchinson, G.E., 1957. A Treatise on Limnology, Geography, Physics and Chemistry. John Wiley and Sons, New York.

Hyne, N.J., 2001. Nontechnical Guide to Petroleum Geology, Exploration and Production, second ed. Pennwell Corporation. 575 pp.

Iaffaldano, G., Caputo, M., Martino, S., 2006. Experimental and theoretical memory diffusion of water in sand. Hydrol. Earth Syst. Sci. 10, 93–100. https://doi.org/10.5194/hess-10-93-2006.

IEA, 2008. World Energy Outlook 2008. http://www.worldenergyoutlook.org/.

IEA, 2012. Golden Rules for a Golden Age of Gas: World Energy Outlook Special Report.

IEA, 2013. World Energy Outlook. https://www.iea.org/reports/world-energy-outlook-2013.

IEA, 2014. https://www.iea.org/reports/world-energy-outlook-2014.

IEA, 2016. International Energy Outlook Report. Available from: https://www.eia.gov/outlooks/ieo/pdf/0484(2016).pdf.

IEA, 2017. World Energy Outlook: Poverty and Prosperity. https://www.iea.org/publications/freepublications/publication/WEO2017SpecialReport_EnergyAccessOutlook.pdf.

IEA, 2018a. World Energy Outlook 2017, China. https://www.iea.org/weo/china/.

IEA, 2018b. https://www.iea.org/weo/.

IEA, 2020a. https://www.iea.org/reports/oil-2020.

IEA, 2020b. World energy balances, report. Available from: https://iea.blob.core.windows.net/assets/4f314df4-8c60-4e48-9f36-bfea3d2b7fd5ZWorldBAL_2020_Documentation.pdf.

IEA, 2020c. World Energy Outlook. https://www.iea.org/reports/world-energy-outlook-2020.

IEA, 2021. Oil Demand Forecast, 2010–2026. Pre-Pandemic and in Oil 2021. IEA, Paris. https://www.iea.org/data-and-statistics/charts/oil-demand-forecast-2010-2026-pre-pandemic-and-in-oil-2021.

Ibrahim, M.I., 2006. Towards Sustainability in Offshore Oil and Gas Operations (PhD dissertation). Dalhousie University.

IEA Energy Technology Network, 2010. Unconventional Oil & Gas Production.

IHS, 2007. Growth of world oil fields. In: Presentation held by Keith King and IHS Energy at IHS London Symposium, 17–18 April. http://energy.ihs.com/Events/london-symposium-2007/Presentations.htm.

Imber, J., et al., 2001. A reappraisal of the Sibson-Scholz fault zone model: the nature of the frictional to viscous ('brittle-ductile') transition along a long-lived, crustal-scale fault, Outer Hebrides, Scotland. Tectonics 20, 601–624. https://doi.org/10.1029/2000TC001250.

International Energy Agency (IEA), 2019. Key World Energy Statistics 2019.

Ipsos, 2011. Public attitudes to science. Main report, May. 122 pp.

Irvine, T.N., 1989. A global convection framework: concepts of symmetry, stratification, and system in the Earth's dynamic structure? Econ. Geol. 84 (8), 2059–2114.

Islam, M.R., 1990. Comprehensive mathematical modelling of horizontal wells. In: Proc. of the 2nd European Conf. on the Mathematics of Oil Recovery, Latitudes Camargue, Arles, France, September 11–14. Paper no. 69.

Islam, M.R., 1993. Route to chaos in chemically enhanced thermal convection in porous media. Chem. Eng. Commun. 124, 77–95.

Islam, M.R., 2001. Advances in petroleum reservoir monitoring technologies. In: SPE Paper 68804.

Islam, M.R., 2002. Emerging technologies in subsurface monitoring of petroleum reservoirs. Pet. Res. J. 13, 33–46.

Islam, M.R., 2014. Unconventional Gas Reservoirs. Elsevier.

Islam, M.R., 2015. Unconventional Gas Reservoirs. Elsevier.

Islam, M.R., 2020. Environmentally and Economically Sustainable Enhanced Oil Recovery. Wiley-Scrivener. 816 pp.

Islam, M.R., Ali, S.M.F., 1989. New scaling criteria for polymer, emulsion and foam flooding experiments. J. Can. Pet. Technol. 28 (4), 79–87. https://doi.org/10.2118/89-04-05.

Islam, M.R., Ali, S.M.F., 1992. New scaling criteria for in-situ combustion experiments. J. Pet. Sci. Eng. 6 (4), 367–379. https://doi.org/10.2118/SS-89-01.

Islam, M.R., Bentsen, R.G., 1986. A dynamic method for measuring relative permeability. J. Can. Pet. Technol. 25 (1), 39–50.

Islam, M.R., Chakma, A., 1992. A new recovery technique for heavy oil reservoirs with bottomwater. SPE Reserv. Eng. 7 (2), 180–186.

Islam, M.R., Chakma, A., 1993. Storage and utilization of CO_2 in petroleum reservoirs—a simulation study. In: IEA CO_2 Disposal Symposium, Oxford, England.

Islam, M.R., Chilingar, G.V., 1995. A new technique for recovering heavy oil and tar sands. Int. J. Sci. Technol. 2 (1), 15–28.

Islam, M.R., Farouq Ali, S.M., 1990. New scaling criteria for chemical flooding experiments. J. Can. Pet. Technol. 29 (1), 29–36.

Islam, M.R., Farouq Ali, S.M., 1991. Scaling of in-situ combustion experiments. J. Pet. Sci. Eng. 6, 367–379.

Islam, M.R., George, A.E., 1991. Sand control in horizontal wells in heavy oil reservoirs. J. Pet. Technol. 43 (7), 844–853. https://doi.org/10.2118/18789-PA.

Islam, M.R., Hossain, M.E., 2020. Drilling Engineering: Towards Achieving Total Sustainability. Elsevier. 800 pp.

Islam, M.R., Khan, M.M., 2019. The Science of Climate Change. Scrivener Wiley.

Islam, M.R., Nandakumar, K., 1986. Multiple solution for buoyancy-induced flow in saturated porous media for large Peclet numbers. Trans. ASME J. Heat Transf. 108 (4), 866–871.

Islam, M.R., Nandakumar, K., 1990. Transient convection in saturated porous layers with internal heat sources. Int. J. Heat Mass Transf. 33 (1), 151–161.

Islam, M.R., Wadadar, S.S., 1991. Enhanced oil recovery of Ugnu tar sands of alaska using electromagnetic heating with horizontal wells. In: Paper Presented at the International Arctic Technology Conference, Anchorage, Alaska, May. Paper Number: SPE-22177-MS.

Islam, M.R., Chakma, A., Farouq Ali, S.M., 1989. State-of-the-art of in situ combustion modeling and operations. In: Paper SPE-18755, Proc. Of the SPE California Regional Meeting, Bakersfield, CA, (April 1989).

Islam, M.R., et al., 2000. Advanced Petroleum Reservoir Simulation. Willey-Scrivener. 572 pp.

Islam, M.R., Chhetri, A.B., Khan, M.M., 2010a. Greening of Petroleum Operations. Wiley-Scrivener. 852 pp.

Islam, M.R., Mousavizadeghan, H., Mustafiz, S., Abou-kassem, J.H., 2010b. Reservoir Simulation: Advanced Approach. Scrivener-Wiley. 468 pp.

Islam, M.R., Islam, J.S., Mughal, M.A.H., Rahman, M.S., Zatzman, G.M., 2015. The Greening of Pharmaceutical Engineering, Volume 1: Practice, Analysis, and Methodology. Scrivener-Wiley, New York. 778 pp.

Islam, M.R., et al., 2016a. Advanced Reservoir Simulation: Towards Developing Reservoir Emulators. Scrivener-Wiley. 520 pp.

Islam, M.R., et al., 2016b. The Greening of Pharmaceutical Engineering. Theories and Solutions, vol. 2 Scrivener-Wiley. 482 pp.

Islam, J.S., et al., 2018b. Economics of Sustainable Energy. Scrivener-Wiley. 628 pp.

Islam, M.R., Islam, A.O., Hossain, M.E., 2018a. Hydrocarbons in Basement Formation. Wiley-Scrivener.

Ito, T., Hayashi, K., 1991. Physical background to the breakdown pressure in hydraulic fracturing tectonic stress measurements. Int. J. Rock Mech. Min. Sci. Geomech. Abstr. 28, 285–293.

Ivanov, K.S., et al., 2014. Composition and age of the crystalline basement in the Northwestern part of the West Siberian oil-and-gas megabasin. Dokl. Earth Sci. 459, 1582–1586.

Ivanov, K.S., Erokhin, Y.V., Ponomarev, V.S., 2018. Age and composition of granitoids from the basement of Krasnoleninsky oil and gas region (Western Siberia). Izv. UGGU 2 (50), 7–14.

Iwai, K., 1976. Fundamental Studies of Fluid Flow through a Single Fracture (Ph.D. dissertation). University of California, Berkeley. 208 pp.

Jack, T.R., Stehmeier, L.G., Ferris, F.G., Islam, M.R., 1991. Microbial selective plugging to control water channeling. In: Donaldson, E.C. (Ed.), Microbial Enhancement of Oil Recovery—Recent Advances. Elsevier Science Publishing Co., New York, pp. 433–440.

Jadhunandan, P.P., Morrow, N.R., 1995. Effect of wettability on waterflood recovery for crude-oil/brine/rock systems. SPE Reserv. Eng. 10 (1), 40–46.

James, N.P., Choquette, P.W., 1983. Diagenesis 6. Limestones—the sea floor diagenetic environment. Geosci. Can. 10 (4).

Jarvie, D.M., 2014. Components and processes affecting producibility and commerciality of shale resource systems. Geol. Acta 12 (4), 307–325.

Jeffrey, S.W., Mantoura, R.F.C., Wright, S.W. (Eds.), 1997. Phytoplankton Pigments in Oceano-Graphy: Guidelines to Modern Methods. UNESCO, Paris.

Jeffrey, R.G., Zhang, X., Bunger, A.P., 2010. Hydraulic fracturing of naturally fractured reservoirs. In: Proceedings of the 35th Workshop on Geothermal Reservoir Engineering. Stanford University, Stanford, CA.

Jensen, D., Smith, L.K., 1997. A practical approach to coalbed methane reserve prediction using a modified material balance technique. In: Proceedings, International Coalbed Methane Symposium, May, Tusaloosa, Alabama, USA.

Jia, C., 2017. Breakthrough and significance of unconventional oil and gas to classical petroleum geology theory. Pet. Explor. Dev. 44 (1).

Jing, L., Nordlund, E., Stephansson, O., 1992. An experimental study on the anisotropy and stress-dependency of the strength and deformability of rock joints. Int. J. Rock Mech. Min. Sci. Geomech. Abstr. 29 (6), 535–542.

Jiu, B., Wenhui, H., Yuan, L., 2020. The effect of hydrothermal fluids on Ordovician carbonate rocks, southern Ordos Basin, China. Ore Geol. Rev. 126 (3). https://doi.org/10.1016/j.oregeorev.2020.103803, 103803.

Johannesen, E.B., Graue, A., 2007. Mobilization of remaining oil-emphasis on capillary number and wettability. Soc. Pet. Eng. https://doi.org/10.2118/108724-MS.

Johnson, R.W., 1998. Handbook of Fluid Johnson. CRC Press.

Jolley, S.J., Barr, D., Walsh, J.J., Knipe, R.J., 2007. Structurally complex reservoirs: an introduction. Geol. Soc. Lond., Spec. Publ. 292, 1–24.

Jones, F.O., 1975. A laboratory study of the effects of confining pressure on fracture flow and storage capacity in carbonate rocks. J. Pet. Technol. 27, 21–27.

Jonoud, S., Jackson, M.D., 2008. New criteria for the validity of steady-state upscaling. Transp. Porous Media 71 (1), 53–73.

Journel, A.G., Deutsch, C., Desbarats, A.J., 1986. Power averaging for block effective permeability. In: SPE Paper 15128.

Julian, B.R., Miller, A.D., Foulger, G.R., 1998. Non-double-couple earthquakes, 1, theory. Rev. Geophys. 36, 525–549.

Julian, B.R., Foulger, G.R., Richards-Dinger, K., Monastero, F., 2006. Time dependent seismic tomography of the Coso geothermal area, 1996-2004. In: Proceedings of the 31st Workshop on Geothermal Reservoir Engineering. Stanford University, Stanford, CA.

Jung, R., 1989. Hydraulic in situ investigations of an artificial fracture in Falkenberg granite. Int. J. Rock Mech. Min. Sci. Geomech. Abstr. 26 (3–4), 301–308.

Kahrilas, G.A., et al., 2015. Biocides in hydraulic fracturing fluids: a critical review of their usage, mobility, degradation, and toxicity. Environ. Sci. Technol. 49 (1), 16–32.

Kalathingal, P., Kuchinski, R., 2010. The role of resistivity image logs in deep natural gas reservoirs. In: Paper Presented at the SPE Deep Gas Conference and Exhibition, Manama, Bahrain, January. https://doi.org/10.2118/131721-MS. Paper Number: SPE-131721-MS.

Kalogirou, S., 1997. Survey of solar desalination systems and system selection. Energy 22 (1), 69–81.

Kalogirou, S.A., et al., 2004. Solar thermal collectors and applications. Prog. Energy Combust. Sci. 30 (3), 231–295.

Kandlikar, S.G., 2010. A scale analysis based theoretical force balance model for critical heat flux (CHF) during saturated flow boiling in microchannels and mini channels. J. Heat Transf. 132 (8), 081501.

Kang, Z., 2003. Discovery and exploration of like-layered reservoir in Tahe Oilfield of Tarim Basin. Acta Petrol. Sin. 24, 4–9.

Kann, D., et al., 2019. The most effective ways to curb climate change might surprise you. CNN (April 19).

Kantzas, A., Chatzis, I., Dullien, F.A.L., 1988. Mechanisms of capillary displacement of residual oil by gravity-assisted inert gas injection. In: SPE 17506. Proc. SPE Rocky Mountain Regional Meeting, Casper, Wyoming., https://doi.org/10.2118/17506-MS.

Kappelmeyer, O., Gerard, A., 1987. Production of heat from impervious hot crystalline rock sections (hot-dry-rock concept). In: Terrestrial Heat From Impervious Rocks-Investigation in the Falkenberg Granite Massif. Geologisches Jahrbuc, Reihr E, Heft. 39, pp. 7–22.

Kar, M., et al., 2014. Heat and mass transfer effects on a dissipative and radiative visco-elastic MHD flow over a stretching porous sheet. Arab. J. Sci. Eng. 39 (5), 3393–3401.

Karacan, C.O., Okandan, E., 2000. Assessment of energetic heterogeneity of coals for gas adsorption and its effect on mixture predictions for coalbed methane studies. Fuel 79, 1963–1974.

Karge, H., Weitkamp, J., Ruthven, D., 2008. Fundamentals of adsorption equilibrium and kinetics in micro porous solids, adsorption and diffusion. In: Molecular Sieves—Science and Technology. Springer, Berlin/Heidelberg, pp. 1–43.

Kavald, P.A., Guven, O., 2004. Removal of concentrated heavy metal ions from aqueous solutions using polymers with enriched amidoxime groups. J. Appl. Polym. Sci. 93, 1705–1710.

Kazemi, H., et al., 1976. Numerical simulation of water-oil flow in naturally fractured reservoirs. Soc. Petrol. Eng. J. 16 (6), 317–326.

Kazemi, H., Gilman, J.R., Elsharkawy, A.M., 1992. Analytical and numerical solution of oil recovery from fractured reservoirs with empirical transfer functions (includes associated papers 25528 and 25818). SPE Reserv. Eng. 7 (2), 219–227. https://doi.org/10.2118/19849-PA.

Kelafant, J., 2016. International coal seam gas activities. In: North American Coalbed Methane North American Coalbed Methane Forum Coal Seam Gas. Quarterly Newsletter, vol. 5(3).

Kelley, S.P., Reddy, S.M., Maddock, R., 1994. Laser-probe 40Ar/39Ar investigation of a pseudotachylyte and its host rock from the Outer Isles thrust, Scotland. Geology 22, 443–446. https://doi.org/10.1130/0091-7613(1994)022<0443:LPAAIO>2.3.CO;2.

Kemp, J., 2018. U.S. oil output surges but growth likely to moderate in 2019. Reuters (November 1).

Kenney, J.F., 1996. Considerations about recent predictions of impending short-ages of petroleum evaluated from the perspective of modern petroleum science. In: "The Future of Petroleum" in Energy World, Special ed. British Institute of Petroleum, London.

Kenney, J.F., et al., 2002. The evolution of multicomponent systems at high pressures: VI. The thermodynamic stability of the hydrogen-carbon system: the genesis of hydrocarbons and the origin of petroleum. Proc. Natl. Acad. Sci. U. S. A. 99 (17), 10976–10981.

Kenney, J.F., Sozanksy, V.I., Chepil, P.M., 2009. On the spontaneous renewal of oil and gas fields. Energy Polit. XVII.

Khan, M.I., Islam, M.R., 2007a. The Petroleum Engineering Handbook: Sustainable Operations. Gulf Publishing Company, Houston, TX. 461 pp.

Khan, M.I., Islam, M.R., 2007b. True Sustainability in Technological Development and Natural Resource Management. Nova Science Publishers, New York. 381 pp.

Khan, M.M., Islam, M.R., 2012. Zero-Waste Engineering. Scrivener-Wiley. 465 pp.

Khan, M.M., Islam, M.R., 2016. Zero Waste Engineering: A New Era of Sustainable Technology Development, second ed. Wiley, Scrivener.

Kharabaf, H., Yortsos, Y.C., 1997. Invasion percolation with memory. Phys. Rev. E 55 (6), 7177.
Kholpanov, L.P., Ibyatov, R.I., 2005. Mathematical modeling of the dispersed phase dynamics. Theor. Found. Chem. Eng. 39 (2), 190–199.
Kim, A.G., 1977. Estimating Methane Content of Bituminous Coalbeds from Adsorption Data. 8245 US Bur. Mines, Rep. Invest, p. 22.
Kim, K.S., van Stone, R.H., 1997. Crack growth under thermo-mechanical and temperature gradient loads. Eng. Fract. Mech. 58, 133–147.
Kimber, K.D., Ali, S.M.F., Puttagunta, V.R., 1988. New scaling criteria and their relative merits for steam recovery experiments. J. Can. Pet. Technol. 27 (4), 86–94. https://doi.org/10.2118/88-04-07.
King, P.R., 1989. The use of renormalization for calculating effective permeability. Transp. Porous Media 4, 37–58.
King, G.R., 1993. Material-balance techniques for coal-seam and devonian shale gas reservoirs with limited water influx. SPE Reserv. Eng. 8 (01), 67–72.
King, P.R., Muggeridge, A.H., Price, W.G., 1993. Renormalization calculations of immiscible flow. Transp. Porous Media 12, 237–260.
Kinny, P.D., Strachan, R.A., et al., 2019. The Neoarchean Uyea Gneiss Complex, Shetland: an onshore fragment of the Rae Craton on the European Plate. J. Geol. Soc. Lond. https://doi.org/10.1144/jgs2019-017.
Kirk Petrophysics, 2011. Petrographic and Fluid Inclusion Study of the Lewisian Basement of Lancaster Prospect Well 205/21a-4z. Confidential report prepared for Hurricane Exploration, No. KP10004-3.
Kjärstad, J., Johnsson, F., 2009. Resources and future supply of oil. Energy Policy 37 (2), 441–464.
Klein, P., Gao, H., 1998. Crack nucleation and growth as strain localization in a virtual-bond continuum. Eng. Fract. Mech. 61 (1), 21–48.
Klett, T.R., 2003. Graphic comparison of reserve-growth models for conventional oil and gas accumulation. In: Dyman, T.S., Schmoker, J.W., Verma, M.K. (Eds.), Geologic, Engineering, and Assessment Studies of Reserve Growth. U.S. Geological Survey Bulletin 2172–F.
Koenders, M.A., Petford, N., 2003. Thermally induced primary fracture development in tabular granitic plutons: a preliminary analysis. In: Petford, N., McCaffrey, K.J.W. (Eds.), Hydrocarbons in Crystalline Rocks. Geological Society, London, Special Publications, 214, pp. 143–150, https://doi.org/10.1144/GSL.SP.2003.214.01.09.
Kohl, T., Evans, K.F., Hopkirk, R.J., Ryback, L., 1995. Coupled hydraulic, thermal, and mechanical considerations for the simulation of hot dry rock reservoirs. Geothermics 24, 345–359.
Kolesnikov, A.V., Kutcherov, G., Goncharov, A.F., 2009. Methane-derived hydrocarbons produced under upper-mantle conditions. Nat. Geosci. 2, 566–570.
Kolomietz, V.M., 2014. Memory effects in nuclear Fermi-liquid. Phys. Part. Nucl. 45, 609–627.
Koning, T., 2000. Oil production from basement reservoirs—examples from Indonesia, USA and Venezuela. In: Proceedings of the 16th World Petroleum Congress, Calgary.
Koning, T., 2003a. Oil production from basement reservoirs: examples from Indonesia, USA and Venezuela. In: Petford, N., McCaffrey, K.J.W. (Eds.), Hydrocarbons in Crystalline Rocks. Geological Society, London, Special Publications, vol. 214, pp. 83–92, https://doi.org/10.1144/GSL.SP.2003.214.01.05.
Koning, T., 2003b. Oil and gas production from basement reservoirs. In: Landes, K.K., Amoruso, J.J., Charlesworth, L.J., Heany, F., Lesperan (Eds.), Hydrocarbons in Crystalline Rocks. Geological Society of London, Special Publication, vol. 214.
Koning, T., 2007. Remember basement in your oil and exploration: examples of producing basement reservoirs in Indonesia, Venezuela and USA. In: Let it Flow—2007 CSPG CSEG Convention.
Koottungal, L., 2012. 2012 worldwide EOR survey. Oil Gas J. 110 (2). Available from: www.ogj.com/articles/print/vol-110/issue4/general-interest/special-report-eor-heavy-oil-survey/2012-worldwide-eor-survey.html.
Koshelev, V.F., Ghassemi, A., 2003. Hydraulic fracture propagation near a natural discontinuity. In: Proceedings of the 28th Workshop on Geothermal Reservoir Engineering. Stanford University, Stanford, CA.
Koski, R.A., Shanks-III, W.C., Bohrson, W.A., Oscarsen, R.L., 1988. The composition of massive sulphide deposits from the sediment-covered floor of Escanaba Trough, Gorda Ridge: implication for depositional processes. Can. Mineral. 26 (3), 655–673.
Kovscek, A.R., et al., 2005. https://gcep.stanford.edu/pdfs/QeJ5maLQQrugiSYMF3ATDA/3.3.1.kovscek_06.pdf.
Krayushkin, V.A., 2002. The basement petroleum presence in the Dnieper-Donets Basin. Geol. Geophys. Dev. Oil Gas Fields 1, 9–16.
Kreipl, M.P., Kreipl, A.T., 2017. Hydraulic fracturing fluids and their environmental impact: then, today, and tomorrow. Environ. Earth Sci. 76, 160.

References

Kucuk, F., Sawyer, W.K., 1988. Transient flow in naturally fractured reservoirs and its application to Devonian gas shales. In: Paper Presented at the SPE Annual Technical Conference and Exhibition, Dallas, Texas, September 1980. Paper Number: SPE-9397-MS.

Kudryaavtsev, N., 1951. Against the organic hypothesis of the origin of petroleum (in Russian). Pet. Econ. (Neftianoye Khozyaistvo) 9, 17–29.

Kudryavtsev, N.A., 1951. Against the organic hypothesis of oil origin. Oil Econ. J. 9, 17–29.

Kulatilake, P.H.S.W., Um, J., 1999. Requirements for accurate quantification of self-affine roughness using the roughness-length method. Int. J. Rock Mech. Min. Sci. Geomech. Abstr. 36 (1), 5–18.

Kulatilake, P.H.S.W., Shou, G., Huang, T.H., Morgan, R.M., 1995. New peak shear strength criteria for anisotropic rock joints. Int. J. Rock Mech. Min. Sci. Geomech. Abstr. 32 (7), 673–697.

Kulkarni, M.M., 2005. Multiphase Mechanisms and Fluid Dynamics in Gas Injection Enhanced Oil Recovery Processes (Doctoral dissertation). Louisiana State University. 267 pp.

Kumar, S., Mandal, A., 2017. A comprehensive review on chemically enhanced water alternating gas/CO_2 (CEWAG) injection for enhanced oil recovery. J. Pet. Sci. Eng. 157. https://doi.org/10.1016/j.petrol.2017.07.066 S.

Kundu, P.K., Cohen, I.M., Dowling, D.R., 2015. Fluid Mechanics. Elsevier.

Kurashige, M.A., 1989. Thermoplastic theory of fluid-filled porous materials. Int. J. Solids Struct. 25 (9), 1039–1052. https://doi.org/10.1016/0020-7683(89)90020-6.

Kutcherov, V.G., 2013. Abiogenic deep origin of hydrocarbons and oil and gas deposits formation. In: Kutcherov, V. (Ed.), Hydrocarbon., https://doi.org/10.5772/51549.

Kutcherov, V.G., et al., 2002. Synthesis of hydrocarbons from minerals at pressure up to 5 GPa. Proc. Russ. Acad. Sci., 3876789792.

Kutcherov, V., et al., 2008. Theory of abyssal abiotic petroleum origin: challenge for petroleum industry. AAPG Eur. Reg. Newslett. 3, 2–4.

Kutcherov, V.G., et al., 2010. Synthesis of complex hydrocarbon systems at temperatures and pressures corresponding to the Earth's upper mantle conditions. Dokl. Phys. Chem. 433 (1), 132. https://doi.org/10.1134/S0012501610070079.

Kuuskraa, V.A., Hammershaimb, E.C., Pague, M., 1987. Major tar sand and heavy-oil deposits in the United States. In: Meyer, R.F. (Ed.), Exploration for Heavy Crude Oil and Natural Bitumen. AAPG Studies in Geology, vol. 64. AAPG, Tulsa, OK, pp. 123–135.

Kvenvolden, K.A., 1993a. Gas hydrates – geological perspective and global change. Rev. Geophys. 31 (2), 173–187.

Kvenvolden, K.A., 1993b. A primer on gas hydrates. In: The Future of Energy Gases. U.S. Geological Survey Professional Paper 1570, pp. 279–1008.

Kvenvolden, K.A., 2006. Organic geochemistry—a retrospective of its first 70 years. Org. Geochem. 37 (1), 1–11.

Kvenvolden, K.A., Kastner, M., 1990. Gas hydrates of the Peruvian outer continental margin. In: Suess, E., von Huene, R., et al. (Eds.), Proceedings of the Ocean Drilling Program, Scientific Results. vol. 112. Ocean Drilling Program, College Station, TX, pp. 517–526.

Kvenvolden, K.A., Rogers, B.W., 2005. Gaia's breath—global methane exhalations. Mar. Pet. Geol. 22, 579–590.

Kvenvolden, K.A., Simoneit, B.R.T., 1987. Petroleum from Northeast Pacific Ocean hydrothermal systems in Escanaba Trough and Guaymas Basin. Am. Assoc. Pet. Geol. Bull. 71 (5), 580.

Kyte, J.R., Berry, D.W., 1975. New pseudo functions to control numerical dispersion. SPE J., 269–276.

Lacroix, S., Lemonnier, P., Renard, G., Taieb, C., 2003. Enhanced numerical simulations of IOR processes through dynamic subgridding. In: Proceedings of the Canadian International Petroleum Conference, Calgary, Alberta, 10–12 June.

Ladanyi, B., Archambault, G., 1970. Simulation of shear behavior of a jointed rock mass. In: Proceedings of 11th US Symposium on Rock Mechanics, pp. 105–125.

Laffez, P., Abbaoui, K., 1996. Modelling of the thermic exchanges during a drilling. Resolution with Adomian's decomposition method. Math. Comput. Model. 23 (10), 11–14.

Laherrere, J., 2003. Future of oil supplies. Energy Explor. Exploit. 21 (3), 227–267.

Lai, C.H., et al., 1983. A new model for well test data analysis for naturally fractured reservoirs. In: Paper Presented at the SPE California Regional Meeting, Ventura, California, March.

Lake, L.W., 1989. Enhanced Oil Recovery. Prentice-Hall. 550 pp.

Lake, L.W., Srinivasan, S., 2004. Statistical scale-up of reservoir properties: concepts and applications. J. Pet. Sci. Eng. 44, 27–39.

Lakhal, S., H'mida, S., Islam, M.R., 2007. Green supply chain parameters for a Canadian Petroleum Refinery Company. Int. J. Environ. Technol. Manag. 7 (1/2), 56–67.

Landau, L.D., 1959. On the theory of the Fermi liquid. Sov. Phys. Jetp-Ussr. 8, 70–74.

Lander, R.H., Laubach, S.E., 2015. Insights into rates of fracture growth and sealing from a model for quartz cementation in fractured sandstones. Geol. Soc. Am. Bull. 127, 516–538. https://doi.org/10.1130/B31092.1.

Landereau, P., Noetinger, B., Quintard, M., 2001. Quasi-steady two-equation models for diffusive transport in fractured porous media: large-scale properties for densely fractured systems. Adv. Water Resour. 24 (8), 863–876.

Landes, K.K., Amoruso, J.J., Charlesworth Jr., L., Heany, F., Lesperance, P., 1960. Petroleum resources in basement rocks. AAPG Bull. 44, 1682–1691.

Laubach, T., 2009. Laurentian palaeostress trajectories and ephemeral fracture permeability, Cambrian Eriboll Formation sandstones west of the Moine Thrust Zone, NW Scotland. J. Geol. Soc. 166 (2), 349–362.

Lauriat, G., Prasad, V., 1989. Non-Darcian effects on natural convection in a vertical porous enclosure. Int. J. Heat Mass Tranf. 32, 2135–2148.

Lee, J., 2019. Any damage to exports will tighten global supplies already constrained by chaos Venezuela and sanctions on Iran. Bloomberg (April 8).

Lee, H.-Y., et al., 2006. On three-dimensional continuous saltating process of sediment particles near the channel bed. J. Hydraul. Res. 44 (3), 374–389.

Lefebvre du Prey, E.J., 1973. Factors affecting liquid-liquid relative permeabilities of a consolidated porous medium. Soc. Pet. Eng. J. 13, 39–47.

Lehman-McKeeman, L.D., Gamsky, E.A., 1999. Diethanolamine inhibits choline uptake and phosphatidylcholine synthesis in Chinese hamster ovary cells. Biochem. Biophys. Res. Commun. 262 (3), 600–604. https://doi.org/10.1006/bbrc.1999.1253. 10471370.

Lehner, F.K., 1995. A model for intergranular pressure solution in open systems. Tectonophysics 245, 153–170.

Lepine, O.J., et al., 1999. Uncertainty analysis in predictive reservoir simulation using gradient information. SPE J. 4 (3), 251–259.

Leverett, M.C., 1941. Capillary behavior in porous solids. Trans. AIME 142, 152.

Leverett, M.C., Lewis, W.B., True, M.E., 1942. Dimensional-model studies of oil-field behavior. Trans. AIME 146 (1), 175–193. https://doi.org/10.2118/942175-G.

Levine, J.R., 1987. Influence of coal composition on the generation and retention of coalbed natural gas. In: Proceedings of 1987 Coalbed Methane Symposium, pp. 15–18.

Levy, J.H., Day, S.J., Killingley, J.S., 1997. Methane capacities of Bowen Basin coals related to coal properties. Fuel 76 (9), 813–819.

Li, H., Durlofsky, L.J., 2015. Upscaling for compositional reservoir simulation. SPE J. 873–887. https://doi.org/10.2118/173212-MS, SPE-173212-PA.

Li, Z.F., Sumathy, K., 2001. Experimental studies on a solar powered air conditioning system with portioned hot water storage tank. Sol. Energy 71 (5), 285–297.

Li, J., Yin, L., 2019. Rhenium-osmium isotope measurements in marine shale reference material SBC-1: implications [Q2] for method validation and quality control. Geostand. Geoanal. Res. 43 (3), 497–507.

Li, D., Cullick, A.S., Lake, L.W., 1995. Global scale-up of reservoir model permeability with local grid refinement. J. Pet. Sci. Eng. 14, 1–13.

Li, X., Cui, L., Roegiers, J.-C., 1998. Thermoporoelastic modeling of wellbore stability in non-hydrostatic stress field. Int. J. Rock Mech. Min. Sci. 35 (4–5). Paper No. 063.

Li, R., Reynolds, A.C., Oliver, D.S., 2001. History matching of three-phase flow production data. In: SPE Paper 66351.

Li, D., Liu, Q., Weniger, P., Gensterblum, Y., Busch, A., Krooss, B.M., 2010. Highpressure sorption isotherms and sorption kinetics of CH4 and CO2 on coals. Fuel 89 (3), 569–580.

Liang, L., Michaelides, E.E., 1992. The magnitude of Basset forces in unsteady multiphase flow computations. J. Fluids Eng. 114 (3), 417–419.

Liétard, O., 1999. Permeabilities and skins in naturally fractured reservoirs: an overview and an update for wells at any deviation. In: Paper SPE 54725 Presented at the SPE European Formation Damage Conference, The Hague, 31 May–1 June., https://doi.org/10.2118/54725-MS.

Liétard, O., Unwin, T., Guillot, D., Hodder, M.H., 1999. Fracture width logging while drilling and drilling mud/loss-circulation-material selection guidelines in naturally fractured reservoirs. SPE Drill. Complet. 14 (3), 168–177. SPE-57713-PA https://doi.org/10.2118/57713-PA.

Lin, W., et al., 2010. Localized rotation of principal stress around faults and fractures determined from borehole breakouts in hole B of the Taiwan Chelungpu-fault Drilling Project (TCDP). Tectonophysics 482 (1), 82–91.

Lin, C.S., et al., 2012. Sequence architecture and depositional evolution of the Ordovician carbonate platform margins in the Tarim Basin and its response to tectonism and sea-level change. Basin Res. 24, 559–582.

Lippard, S.J., 1994. Metals in medicine. In: Bioinorganic Chemistry. University Science Books, Mill City, pp. 505–583.

Litasov, K.D., Goncharov, A.F., Hemley, R.J., 2011. Crossover from melting to dissociation of CO_2 under pressure: implications for the lower mantle. Earth Planet. Sci. Lett. 309, 318–323.

Liu, Z., Hamid Emami-Meybodi, H., 2021. A unified approach to the nonlinearity of the diffusivity equation and assessment of pseudotime. SPE J. 26 (01), 241–261.

Liu, S., Valkó, P.P., 2019. Production-decline models using anomalous diffusion stemming from a complex fracture network. SPE J. 24 (06), 2609–2634.

Liu, C.A., Xiong, J.F., 1991. New understanding of Carboniferous and Permian stratigraphy in the northern region of Tarim basin. In: Jia, R. (Ed.), Research of Petroleum Geology of the Northern Tarim Basin in China. vol. 1. University of Geoscience Press, Wuhan, pp. 64–73.

Liu, J., Sheng, J., Polak, A., Elsworth, D., Yasuhara, H., Grader, A., 2006. A fully-coupled hydrological-mechanical-chemical model for fracture sealing and preferential opening. Int. J. Rock Mech. Min. Sci. 43, 23–36.

Liu, J., Chen, Z., Elsworth, D., Miao, X., Mao, X., 2011a. Evolution of coal permeability from stress-controlled to displacement-controlled swelling conditions. Fuel 90 (10), 2987–2997.

Liu, J., Chen, Z., Elsworth, D., Qu, H., Chen, D., 2011b. Interactions of multiple processes during CBM extraction: a critical review. Int. J. Coal Geol. 87 (3–4), 175–189.

Liu, J., Wang, J., Chen, Z., Wang, S., Elsworth, D., Jiang, Y., 2011c. Impact of transition from local swelling to macro swelling on the evolution of coal permeability. Int. J. Coal Geol. 88 (1), 31–40.

Liu, J., et al., 2012. Volcanic Rock-Hosted Natural Hydrocarbon Resources: A Review. Openaccess. Available from: https://www.intechopen.com/chapters/41663.

Liu, Q., et al., 2014. Origin of marine sour natural gas and gas-filling model in the Puguang giant gas field, Sichuan Basin, China. Energy Explor. Exploit. 32 (1), 113–138. http://www.jstor.org/stable/90006072.

Liu, J., Li, Z., Cheng, L., Li, J., 2017a. Multiphase calcite cementation and fluids evolution of a deeply buried carbonate reservoir in the Upper Ordovician Lianglitage Formation, Tahe Oilfield, Tarim Basin, NW China. Geofluids, 1–19. https://doi.org/10.1155/2017/4813235.

Liu, L.H., Ma, Y.S., Liu, B., Wang, C.L., 2017b. Hydrothermal dissolution of Ordovician carbonates rocks and its dissolution mechanism in Tarim Basin, China. Carbonates Evaporites 32, 525–537. https://doi.org/10.1007/s13146-016-0309-2.

Liu, S., Gao, G., Gang, W., Tong, Q., Dang, W., Zhang, W., Yang, S., Zhu, K., 2020. Implications of organic matter source and fluid migration from geochemical characteristics of stylolites and matrix in carbonate rocks: a case study from the Carboniferous and the Ordovician in the Sichuan Basin, SW China. J. Pet. Sci. Eng. 186. https://doi.org/10.1016/j.petrol.2019.106606, 106606.

Lochenie, C., et al., 2014. Large thermal hysteresis for iron(II) spin crossover complexes with N-(Pyrid-4-yl) isonicotinamide. Inorg. Chem. 53 (21), 11563–11572.

Lohne, A., Virnovsky, G., 2006. Three-phase upscaling in capillary and viscous limit. In: SPE Paper 99567.

Lollar, B.S., Westgate, T.D., Ward, J.A., Slater, G.F., Lacrampe-Couloume, G., 2002. Abiogenic formation of alkanes in the Earth's crust as a minor source for global hydrocarbon reservoirs. Nature 416 (6880), 522–524.

Loucks, R., 1999. Modern analogs for paleocave sediment fills and their importance in identifying paleocave reservoirs. AAPG Bull. 46, 195–206.

Long, X.X., Yang, X.E., Ni, W.Z., 2002. Current status and prospective on phytoremediation of heavy metal polluted soils. J. Appl. Ecol. 13, 757–762.

Lough, M.F., Lee, S.H., Kamath, J., 1998. A new method to calculate the effective permeability of grid blocks used in the simulation of naturally fractured reservoirs. In: Proceedings of Paper Presented at the SPE Annual Technical Conference.

Love, G.J.L., Kinny, P.D., Friend, C.R.L., 2004. Timing of magmatism and metamorphism in the Gruinard Bay area of the Lewisian Gneiss Complex: comparisons with the Assynt terrane and implications for terrane accretion. Contrib. Mineral. Petrol. 146, 620–636. https://doi.org/10.1007/s00410-003-0519-1.

Lovell, M.A., Williamson, G., Harvey, P.K. (Eds.), 1999. Borehole Imaging: Applications and Case Histories. Geological Society, London, pp. 1–43.

Lowell, R.P., 1990. Thermoelasticity and the formation of black smokers. Geophys. Res. Lett. 17, 709–712.

Lowell, R.P., Germanovich, L.N., 1995. Dike injection and the formation of megaplumes at ocean ridges. Science 267, 1804–1807.

Lowell, R.P., Van Cappellen, P., Germanovich, L.N., 1993. Silica precipitation in fractures and the evolution of permeability in hydrothermal upflow zones. Science 260, 192–194.

Lozada, D., Ali, S.M.F., 1987. New scaling criteria for partial equilibrium immiscible carbon dioxide drive. In: Paper No. 87-38-23, Presented in 38th Annual Technical Meeting of the Petroleum Society of CIM, Calgary, Canada, June 7–10, pp. 393–410, https://doi.org/10.2118/87-38-23.

Lozada, D., Ali, S.M.F., 1988. Experimental design for non-equilibrium immiscible carbon dioxide flood. In: Paper No. 159, Presented in 4th UNITAR/UNDP International Conference on Heavy Crude and Tar Sands, August 7–12, Edmonton, Alberta, Canada.

Lu, J.-F., Hanyga, A., 2005. Wave field simulation for heterogeneous porous media with singular memory drag force. J. Comput. Phys. 208, 651–674.

Lu, X.C., Li, F.C., Watson, A.T., 1995. Adsorption studies of natural gas storage in Devonian shales. SPE Form. Eval. 10 (02), 109–113.

Lucia, F.J., 2007. Carbonate Reservoir Characterization. Springer-Verlag, Berlin Heidelberg. 336 pp.

Lucido, G., 1983. A new hypothesis on the origin of water from magma. J. Geol. 91 (4), 456–461.

Lukerchenko, N., 2010. Basset history force for the bed load sediment transport. In: First IAHR European Division Congress Edinburgh FMIId.

Lund, J.E., Freeston, D.H., Tonya, L.B., 2011. Direct utilization of geothermal energy 2010 worldwide review. Geothermics 40 (3), 159–180.

Luo, Z., et al., 2004. Research on biomass fast pyrolysis for liquid fuel. Biomass Bioenergy 26, 455–462.

Lynch, E., Braithwaite, R., 2005. A review of the clinical and toxicological aspects of 'traditional' (herbal) medicines adulterated with heavy metals. Expert Opin. Drug Saf. 4, 76–78.

Maceka, R.W., Silling, A.A., 2007. Peridynamics via finite element analysis. Finite Elem. Anal. Des. 43, 1169–1178.

Macuda, J., Nodzenski, A., Wagner, M., Zawisza, L., 2011. Sorption of methane on lignite from Polish deposits. Int. J. Coal Geol. 87 (1), 41–48.

Majer, E.L., 2007. White Paper: Induced Seismicity and Enhanced Geothermal Systems. Center for Computational Seismology, Ernest Orlando Lawrence Berkeley Laboratory, Berkeley. 32 pp.

Majer, E.L., Baria, R., Stark, M., Oates, S., Bommer, J., Smith, B., Asanuma, H., 2007. Induced seismicity associated with enhanced geothermal systems. Geothermics 36, 185–227.

Makeev, A.B., Ivanukh, V., 2004. Morphology of crystals, films, and selvages on a surface of the Timanian and Brazil diamonds. In: Problems of Mineralogy, Petrography and Metal-logeny. Perm Univ. Press, Perm, Russia, pp. 193–216.

Malashetty, M.S., Begum, I., 2011. Effect of thermal/gravity modulation on the onset of convection in a Maxwell fluid saturated porous layer. Transp. Porous Media 90 (3), 889–909.

Mamaghani, M., Enchéry, G., Chainais-Hillairet, C., 2011. Development of a refinement criterion for adaptive mesh refinement in steam-assisted gravity drainage simulation. Comput. Geosci. 15 (1), 17–34.

Manhadieph, N.R., 2001. Geothermal conditions of deep hydrocarbon-bearing layers (D. Shi, Trans.). Nat. Gas Geosci. 12 (2), 56–60.

Mao, Z., et al., 2011. Dolomite III: a new candidate lower mantle carbonate. Geophys. Res. Lett. 38 (22), L22303. https://doi.org/10.1029/2011GL049519.

Mardia, K.V., 1972. Statistics of Directional Data: Probability and Mathematical Statistics. Academic Press, London. 357 pp.

Marecka, A., Mianowski, A., 1998. Kinetics of CO_2 and CH_4 sorption on high rank coal at ambient temperatures. Fuel 77 (14), 1691–1696.

Markus Eck, M., Steinmann, W.D., 2005. Modelling and design of direct solar steam generating collector fields. J. Sol. Energy Eng. 127 (3), 371–380.

Márquez, G., et al., 2016. Intra- and inter-field compositional changes of oils from the Misoa B4 reservoir in the Ceuta Southeast Area (Lake Maracaibo, Venezuela). Fuel 2016 (167), 118–134. https://doi.org/10.1016/j.fuel.2015.11.046.

Martin, J.T., Lowell, R.P., 1997. On thermoelasticity and silica precipitation in hydrothermal systems: numerical modeling of laboratory experiments. J. Geophys. Res. 102 (B6), 12095–12107.

Martinius, A.W., Hegner, J., Kaas, I., Mjos, R., Bejarano, C., Mathieu, X., 2013. Geologic reservoir characterization and evaluation of the Petrocedeño field, early Miocene Oficina Formation, Oronoco Heavy Oil Belt, Venezuela. In:

Hein, F.J., Leckie, D., Larter, S., Suter, J.R. (Eds.), Heavy-Oil and Oil-Sand Petroleum Systems in Alberta and Beyond. AAPG Studies in Geology, 64, pp. 103–131.

Mastalerz, M., 2014. Coalbed methane: reserves, production, and future outlook. In: Letcher, T. (Ed.), Future Energy, second ed, pp. 145–158.

Masters, J., 1979. Deep Basin gas trap, West Canada. Am. Assoc. Pet. Geol. Bull. 63 (2), 152–181.

Mavor, M., Nelson, C.R., 1997. Coalbed reservoir gas-in-place analysis. Gas Research Institute Report 97/0263, p. 134.

Mavor, M.J., Close, J.C., McBane, R.A., 1990a. Formation evaluation of exploration coalbed methane wells. In: CIM/SPE 90-101, SPE/CIM International Technical Meeting, Calgary, Alberta Canada, June 10–13.

Mavor, M.J., Owen, I.B., Pratt, T.J., 1990b. Measurement and evaluation of coal sorption isotherms data. In: SPE 65th Annual Technical Conference and Exhibition, New Orleans, Louisiana, September 23–26.

Mavor, M.J., Pratt, T.J., Nelson, C.R., Casey, T.A., 1996. Improved gas-in-place determination for coal gas reservoirs. In: SPE Gas Technology Symposium. SPE 35623.

Mazumder, S., van Hemert, P., Busch, A., Wolf, K.H.A., Tejera-Cuesta, P., 2006. Flue gas and pure CO_2 sorption properties of coal: a comparative study. Int. J. Coal Geol. 67, 267–279.

McCabe, P.J., 1998. Energy resources—cornucopia or empty barrel? Am. Assoc. Pet. Geol. Bull. 82 (11), 2110–2134.

McCain Jr., W.D., 1993. Chemical composition determines behavior of reservoir fluids. Pet. Eng. Int. 65 (10), 0164-8322.

McFall, K.S., Wicks, D.E., Kuuskraa, V.A., 1986. A geologic assessment of natural gas from coal seams in the warrior basin, Alabama. Gas Research Inst., Topical Rep. GRI-86/0272.

McGuire, A., 2015. 25 Important events in crude oil history since 1862. Money Morning (July 22).

Mcintyre, T., 2003. Phytoremediation of heavy metals from soils. Adv. Biochem. Eng. Biotechnol. 78, 97–123.

McLean, K., McNamara, D.D., 2011. Fractures interpreted from acoustic formation imaging technology: correlation to permeability. In: Thirty-Sixth Conf. Geothermal Reservoir Eng., Stanford, California.

McLennan, J.D., Schafer, P.S., Pratt, T.J., 1995. A guide to determining coalbed gas content, gas research institute report no. GRI-94/0396, Chicago, Illinois, models and testing data. Int. J. Coal Geol. 92, 1–44.

McNamara, K., et al., 2018. Using lake sediment magmatic and volcanic processes in continental rifts. Geochem. Geophys. Geosyst. 19 (9), 3164–3188.

Megel, T., Kohl, T., Rose, P., 2005. Reservoir modeling for stimulation planning at the Coso EGS project. Geotherm. Res. Counc. Trans. 29, 173–176.

Melienvski, B.N., 2001. Discussion on the deep zonation of the oil-gas formation (D. Shi, Trans.). Nat. Gas Geosci. 12 (4–5), 52–55.

Meling, L.M., 2005. Filling the gap. In: Paper Presented at the Seminar: Running Out of Oil—Scientific Perspectives on Fossil Fuels, 26 May. Royal Swedish Academy of Sciences and the Royal Swedish Academy of Engineering Sciences.

Melton, C.E., Giardini, A.A., 1974. The composition and significance of gas released from natural diamonds from Africa and Brazil. Am. Mineral. 59 (7–8), 775–782.

Melton, C.E., Giardini, A.A., 1975. Experimental results and theoretical interpretation of gaseous inclusions found in Arkansas natural diamonds. Geochim. Cosmochim. Acta 60 (56), 413–417.

Mendeleev, D., 1877. L'Origine du Petrole. Rev. Sci. 2 (8), 409–416.

Merle, H.A., et al., 1976. The Bachaquero study—a composite analysis of the behavior of a compaction drive/solution gas drive reservoir. J. Pet. Technol., 1107–1115.

Metzler, R., Barkai, E., Klafter, J., 1999. Anomalous diffusion and relaxation close to thermal equilibrium: a fractional Fokker-Planck equation approach. Phys. Rev. Lett. 82, 3563–3567. https://doi.org/10.1103/PhysRevLett.82.3563.

Meyer, R., Attanasi, E., Freeman, P., 2007. Heavy Oil and Natural Bitumen Resources in Geological Basins of the World Open File Report.

Michael, A.J., 1987. Use of focal mechanisms to determine stress: a control study. J. Geophys. Res. 92, 357–368.

Michaelides, E.E., 1992. A novel way of computing the basset term in unsteady multiphase flow computations. Phys. Fluids A 4 (7), 1579–1582.

Mifflin, R.T., Schowalter, W.R., 1986. A numerical technique for three-dimensional steady flows of fluids of the memory-integral type. J. Non-Newton. Fluid Mech. 20, 323–337.

Miguel-H, N., Miller, M.A., Sepehrnoori, K., 2004. Scaling parameters for characterizing gravity drainage in naturally fractured reservoirs. In: SPE-89990-MS, SPE Annual Technical Conference and Exhibition., https://doi.org/10.2118/89990-MS.

Miller, R.G., 1995. A future for exploration geochemistry. In: Grimalt, J.O., Dorronsoro, C. (Eds.), Organic Chemistry: Developments and Applications to Energy, Climate Environment and Human History. AIGOA, Donostia. 1150 pp.

Miller, K.S., Ross, B., 1993. An Introduction to the Fractional Calculus and Fractional Differential Equations. Wiley, New York.

Min, K.S., Ghassemi, A., 2011. Three-dimensional numerical analysis of thermal fracturing in rock. In: Proceedings of the 45th ARMA Conference, San Francisco, CA.

Miralai, S., Khan, M.M., Islam, M.R., 2007. Replacing artificial additives with natural alternatives. J. Nat. Sci. Sustain. Technol. 1 (3), 403–434.

Mirzaei-Paiaman, A., Masihi, M., 2013. Scaling equations for oil/gas recovery from fractured porous media by counter-current spontaneous imbibition: from development to application. Energy Fuel 27 (8), 4662–4676.

Moffat, D.H., Weale, K.E., 1955. Sorption by coal of methane at high pressures. Fuel 34, 449–462.

Moghadam, S., Jeje, O., Mattar, L., 2011. Advanced gas material balance in simplified format. J. Can. Pet. Technol. 50 (1), 90–98.

Mohanty, D., 2011. Geologic and genetic aspects of coal seam methane. In: Singh, A.K., Mohanty, D. (Eds.), First Indo-US Workshop on Coal Mine Methane, 17–20 October. CIMFR, Dhanbad, pp. 45–60 (Chapter 4).

Mohanty, D., et al., 2013. A comparative account of shale gas potentiality of Cambey Shale, Cambey Basin and Raghavapuram Shale, Krishna-Godavari Basin of India. In: International Conference on Future Challenges in Earth Sciences for Energy & Mineral Resources (ESEMR 2013). Indian School of Mines, Dhanbad. 149 pp.

Mokhatab, S., Fresky, M.A., Islam, M.R., 2006. Applications of nanotechnology in oil and gas E&P. J. Pet. Technol. 18.

Montana and Wyoming U.S. Geological Survey, 2006. Coalbed Methane Extraction and Soil Suitability Concerns in the Powder River Basin.

Monteagudo, J.E.P., Rajagopal, K., Lage, P.L.C., 2001. Scaling laws in network models: porous medium property prediction during morphological evolution: petroleum production research in Brazil. J. Pet. Sci. Eng. 32 (2–4), 179–190.

Montenat, C., Barrier, P., Ott d'Estevou, P., 1991. Some aspects of the recent tectonics in the Strait of Messina, Italy. Tectonophysics 194, 203–215. https://doi.org/10.1016/0040-1951(91)90261-P.

Moore, T.A., 2012. Coalbed methane: a review. Int. J. Coal Geol. 101, 36–81.

Moore, T.F., Slobod, R.L., 1956. The effect of viscosity and capillarity on the displacement of oil by water. Prod. Mon. 20 (10), 20–30.

Moos, D., Zoback, M.D., 1990. Utilization of observations of well bore failure to constrain the orientation and magnitude of crustal stresses: application to continental, deep sea drilling project, and ocean drilling program boreholes. J. Geophys. Res. 95 (B6), 9305–9325. https://doi.org/10.1029/JB095iB06p09305.

Morgan, M.G., et al., 1985. Powerline frequency electric and magnetic fields: a pilot study of risk perception. Risk Anal. 5 (2), 139–149.

Moridis, G.J., Blasingame, T.A., Freeman, C.M., 2010. Analysis of mechanisms of flow in fractured tight-gas and shale-gas reservoirs. In: Paper Presented at the SPE Latin American and Caribbean Petroleum Engineering Conference, Lima, Peru, December 2010., https://doi.org/10.2118/139250-MS. Paper Number: SPE-139250-MS.

Moritis, G., 2008. Worldwide EOR survey. Oil Gas J. 106 (41–42), 44–59.

Morrow, N.R., 1979. Interplay of capillary, viscous and buoyancy forces in the mobilization of residual oil. J. Can. Pet. Technol. 18 (3), 35–46. https://doi.org/10.2118/79-03-03.

Moshier, S.O., Waples, D.O., 1985. Quantitative evaluation of Lower Cretaceous Mannville Group as source rock of Alberta's oil sands. Am. Assoc. Pet. Geol. Bull. 69, 161–172.

Mossop, A., 2001. Seismicity, Subsidence and Strain at the Geysers Geothermal Field (Ph.D. dissertation). Stanford University.

MSDS, 2005. Material Safety Data Sheet. http://www.btps.ca/documents/general/glucose.pdf.

Murray, D., 1996. Coalbed methane in the U.S.A.: analogues for worldwide development. In: Gayer, R., Harris, I. (Eds.), Coalbed Methane and Coal Geology. vol. 109. Geological Society Special Publication, London, pp. 1–12.

Mustafiz, S., Islam, M.R., 2005. Adomian decomposition of two-phase, two-dimensional non-linear PDEs as applied in well testing. In: Bennacer, R. (Ed.), Proceedings of 4th International Conference on Computational Heat and Mass Transfer, Paris-Cachan, May 17–20, pp. 1353–1356.

Mustafiz, S., Biazar, J., Islam, M.R., 2005. Adomian decomposition of two-phase, two-dimensional non-linear PDEs as applied in well testing. In: Bennacer, R. (Ed.), Proceedings of 4th International Conference on Computational Heat and Mass Transfer. Paris-Cachan, May 17–20, pp. 1353–1356.

Mustafiz, S., Mousavizadegan, H., Islam, M.R., 2008a. Adomian decomposition of Buckley Leverett equation with capillary terms. Pet. Sci. Technol. 26 (15), 1796–1810.

Mustafiz, S., Mousavizadegan, S.H., Islam, M.R., 2008b. The effects of linearization on solutions of reservoir engineering problems. Pet. Sci. Technol. 26 (10–11), 1224–1246.

Myal, F.R., Frohne, K.-H., 1992. Drilling and early testing of a sidetrack to the slant-hole completion test well: a case study of gas recovery research in Colorado's Piceance Basin. In: Paper SPE 24382 Presented at the SPE Rocky Mountain Regional Meeting, Casper, Wyoming, USA, 18–21 May., https://doi.org/10.2118/24382-MS.

Myers, A.L., 2002. Thermodynamics of adsorption in porous materials. AICHE J. 48 (1), 145–160.

Nacul, E.C., Aziz, K., 1991. Use of irregular grid in reservoir simulation. In: Paper Presented at the SPE Annual Technical Conference and Exhibition, Dallas, Texas, October., https://doi.org/10.2118/22886-MS.

Nagel, N., 2001. Ekofisk geomechanics monitoring. In: International Workshop on Geomechanics in Reservoir Simulation. IFP, Reuil-Malmaison, France.

Naím, M., 2019. https://carnegieendowment.org/2016/08/17/six-political-events-that-have-distorted-world-of-oil-pub-64514.

Nandi, S.P., Walker, P.L., 1970. Activated diffusion of methane in coal. Fuel 49, 309–323.

Nash, R.A., 2005. Metals in medicine. Altern. Ther. II (4), 18–25.

Natural Resources Canada, 2019. Crude Oil Facts.

Nelson, R.A., 1985. Geologic Analysis of Naturally Fractured Reservoirs. Contributions in Petroleum Geology & Engineering. vol. 1 Gulf Publishing, Houston, TX.

Nelson, R.A., 2001. Geologic Analysis of Naturally Fractured Reservoirs, second ed. Gulf Professional Publishing.

Nelson, C.R., Hill, D.G., Pratt, T.J., 2000. Properties of paleocene fort union formation canyon seam coal at the triton federal coalbed methane well. In: SPE/CERI Gas Technology Symposium, Calgary, Alberta Canada 3-5 April, pp. 639–649.

Nemat-Nasser, S., 1983. Numerical modelling of thermally self-driven cracks: discussion of David T. Barr and Michael P. Cleary: thermoelastic fracture solutions using distributions of singular influence functions—II. Int. J. Solids Struct. 22 (11), 1369–1373.

Nemat-Nasser, S., Keer, L.M., Parihar, K.S., 1978. Unstable growth of thermally induced interacting cracks in brittle solids. Int. J. Solids Struct. 14, 409–430.

Nessa, F., Khan, S.A., Abu Shawish, K.Y.I., 2016. Lead, cadmium and nickel contents of some medicinal agents. Indian J. Pharm. Sci. 78 (1), 111–119.

Neuman, S.P., 1977. Theoretical derivation of Darcy's law. Acta Mech. 25 (3–4), 153–170.

Neuzil, C.E., Tracy, J.V., 1981. Flow through fractures. Water Resour. Res. 17, 191–199.

Newley, T.M.J., 1989. Comparisons of empirical models for unstable miscible displacement. In Situ 13, 4.

Nguyen, T.V., 1986. Experimental study of non-Darcy flow through perforations. In: SPE Annual Technical Conference and Exhibition, 5–8 October, New Orleans, Louisiana. SPE 15473.

Nguyen, V.U., 1989. A fortran program for modeling methane gas desorption from coal. Comput. Geosci. 15 (5), 695–707.

Nibbi, R., 1994. Some properties for viscous fluids with memory. Int. J. Eng. Sci. 32 (6), 1029–1036.

Niitsuma, H., Fehler, M., Jones, R., et al., 1999. Current status of seismic and borehole measurements for HDR/HWR development. Geothermics 28, 475–490.

Nishino, J., 2001. Adsorption of water vapour and carbon dioxide at carboxylic functional groups on the surface of coal. Fuel 80, 757–764.

Noetinger, B., Estebenet, T., 2000. Up-scaling of double porosity fractured media using continuous-time random walks methods. Transp. Porous Media 39, 315–337.

Norbeck, J.H., 2010. Identification and Characterization of Natural Fractures while Drilling Underbalanced (MS thesis). Stanford University.

Nouri, A., et al., 2002. Evaluation of hydraulic fracturing pressure in a porous medium by using the finite element method. Energy Sources 24 (8), 715–725.

Nouri, A., Vaziri, H., Belhaj, H., Islam, M.R., 2006. Sand-production prediction: a new set of criteria for modeling based on large-scale transient experiments and numerical investigation. SPE J. 11 (2), 227–237.

Nouri, A., Vaziri, H., Belhaj, H., Kuru, E., Islam, M.R., 2007. Physical and analytical studies of sand production from a supported wellbore in unconsolidated sand media with single- and two-phase flow. J. Can. Pet. Technol. 46 (6), 41–48.

Novakovic, D., 2002. Numerical Reservoir Characterization Using Dimensionless Scale Numbers With Application in Upscaling. (LSU Doctoral Dissertations). No. 443 https://digitalcommons.lsu.edu/gradschool_dissertations/443.

NRCan, 2021. Report. Available from: https://www.nrcan.gc.ca/science-and-data/data-and-analysis/energy-data-and-analysis/energy-facts/crude-oil-facts/20064.

Nygren, E., 2008. Aviation Fuels and Peak Oil (Diploma thesis). Uppsala University. UPTEC ES08 033 http://www.tsl.uu.se/uhdsg/Publications/Aviationfuels.pdf.

Nygren, A., Ghassemi, A., 2004. An estimate of S Hmax from drilling-induced tensile fractures in Coso geothermal reservoir, CA. In: Poster Presentation 2004 GRC Annual Meeting, Palm Springs.

Nygren, A., Ghassemi, A., 2005. Influence of cold water injection on critically stressed fractures in Coso geothermal field, CA. In: Proceedings of the 39th US Rock Mechanics Symposium, Anchorage, Alaska.

Obembe, A.D., Abu-Khamsin, S.A., Hossain, M.E., 2016. A review of modeling thermal displacement processes in porous media. Arab. J. Sci. Eng. 41 (12), 4719–4741. https://doi.org/10.1007/s13369-016-2265-5.

Obembe, A.D., Al-Yousef, H.Y., Hossain, M.E., Abu-Khamsin, S.A., 2017a. Fractional derivatives and its application in reservoir engineering: a review. J. Pet. Sci. Eng. 157, 312–327.

Obembe, A.D., Hossain, M.E., Abu-Khamsin, S.A., 2017b. Variable-order derivative time fractional diffusion model for heterogeneous porous media. J. Pet. Sci. Eng. 152, 391–405.

Obembe, A.D., Hossain, M.E., Mustapha, K.A., Abu-Khamsin, S.A., 2017c. A modified memory-based mathematical model describing fluid flow in porous media. Comput. Math. Appl. 73 (6), 1385–1402.

Odeh, A.S., Babu, D., 1990. Transient flow behavior of horizontal wells pressure drawdown and buildup analysis. SPE Form. Eval. 5 (1), 7–15.

Odeh, S., Morrison, G.L., 2006. Optimization of parabolic solar collector system. Int. J. Energy Res. 30 (4), 259–271. https://doi.org/10.1002/er.1153.

Odell, P., 2004. Why Carbon Fuels will Dominate the 21st Century's Global Energy Economy. Multi-Science Publishing Co. Ltd, Brentwood, UK.

Okiongbo, K.S., 2011. Effective stress-porosity relationship above and within the oil window in the North Sea Basin. Res. J. Appl. Sci. Eng. Technol. 3 (1), 32–38.

Oldenburg, C.M., Pruess, K., Benson, S.M., 2001. Process modeling of CO_2 injection into natural gas reservoirs for carbon sequestration and enhanced gas recovery. Energy Fuel 15, 293–298.

Oldenburg, C.M., et al., 2003. Carbon sequestration in natural gas reservoirs: enhanced gas recovery and natural gas storage. In: Proceedings, Tough Symposium, Lawrence Berkeley National Laboratory, Berkeley, California, May 12–14.

Olorode, O.M., 2011. Numerical Modeling of Fractured Shale-Gas and Tight-Gas Reservoirs Using Unstructured Grids (MSc thesis). Texas A&M.

Olson, J.E., 2008. Multi-fracture propagation modeling: application to hydraulic fracturing in shale and tight gas sands. In: Proceedings of the 42nd American Rock Mechanics Symposium, San Francisco.

Onishi, T., et al., 2007. High pressure air injection into light oil reservoirs: experimental study on artificial ignition. In: Proceedings of IEA Collaborative Project on Enhanced Oil Recovery 28th Annual Workshop and Symposium, Vedbæk, Denmark, 4–7 September.

Ozkan, E., Ohaeri, U., Raghavan, R., 1987. Unsteady flow to a well produced at a constant pressure in a fractured reservoir. SPE Form. Eval. 2 (02), 186–200.

Paillet, F.L., et al., 1990. Borehole Imaging. SPWLA Reprint Series, Society of Professional Well Log Analysts, Houston, TX.

Palmer, E., 2015. Audi creates green 'e-diesel' fuel of the future using joint carbon dioxide and water. Int. Bus. Times (April 26).

P'an, C.H., 1982. Petroleum in basement rocks. AAPG Bull. 66, 1597–1643.

Pandey, B.L., 1983. A study of the effect of Tamra bhasma on experimental gastric ulcers and secretions. Indian J. Exp. Biol. 21, 25–64.

Pang, X.-Q., Jia, C.-Z., Wang, W.-Y., 2015. Petroleum geology features and research developments of hydrocarbon accumulation in deep petroliferous basins. Pet. Sci. 12, 1–53.

Papay, J., 2003. Development of Petroleum Reservoirs: Theory and Practice. Akademiai Kiads, Hungary. 940 pp.

Park, R.G., 2005. The Lewisian terrane model: a review. Scott. J. Geol. 41, 105–118. https://doi.org/10.1144/sjg41020105.

Parotidis, M., Shapiro, S.A., Rothert, E., 2004. Backfront of seismicity induced after termination of borehole fluid injection. Geophys. Res. Lett. 31. https://doi.org/10.1029/2003GL018987, L02612.

Pashin, J.C., McIntyre, M.R., 2003. Temperature–pressure conditions in coalbed methane reservoirs of the Black Warrior basin: implications for carbon sequestration and enhanced coalbed methane recovery. Int. J. Coal Geol. 54 (3–4), 167–183.

Pashin, J.C., et al., 2011. Geological Foundation for Production of Natural Gas from Diverse Shale Formations. Geological Survey of Alabama, Tuscaloosa, AL.

Passchier, C.W., Trouw, R.A.J., 2005. Microtectonics, second ed. Springer, Berlin.

Passey, Q.R., et al., 1990. A practical model for organic richness from porosity and resistivity logs. AAPG Bull. 74 (12), 1777–1794.

Patel, K., Shah, M., Sircar, A., 2018. Plasma pulse technology: an uprising EOR technique. Pet. Res. 3 (2), 180–188. https://doi.org/10.1016/j.ptlrs.2018.05.001.

Paterson, L., Painter, S., Zhang, X., Pinczewski, V., 1996. Simulating residual saturation and relative permeability in heterogeneous formations. In: SPE Paper 36523.

Pattanaik, N., 2003. Toxicology and free radicals scavenging property of Tamra Bhasma. Indian J. Clin. Biochem. 18, 18–89.

Pearson, C., 1981. The relationship between micro-seismicity and high pore pressure during hydraulic stimulation experiments in low permeability rocks. J. Geophys. Res. 86 (B9), 7855.

Pearson, I., Zeniewski, P., Pavel, Z., Gracceva, F., 2012. Unconventional gas: potential energy market impacts in the European Union. JRC Sci. Policy Rep., 328.

Pearson, D., et al., 2014. Hydrous mantle transition zone indicated by ringwoodite included within diamond. Nature 507, 221–224.

Pedrosa, O.A., Aziz, K., 1986. Use of a hybrid grid in reservoir simulation. SPE Reserv. Eng. 1, 611–621.

Peraccini, M., et al., 2009. Propagation of large bandwidth microwave signals in water. IEEE Trans. Antennas Propag. 57 (11), 3612–3618. https://doi.org/10.1109/TAP.2009.2025674.

Perkins Jr., F.M., Collins, R.E., 1960. Scaling laws for laboratory flow models of oil reservoirs. J. Pet. Technol. 12 (8), 69–71. https://doi.org/10.2118/1487-G.

Perkins, T.K., Gonzalez, J.A., 1985. The effect of thermoelastic stress on injection well fracturing. SPE J. 2, 78–88.

Perkins, T.K., Kern, L.R., 1961. Widths of hydraulic fractures. J. Pet. Technol., 937–949.

Peska, P., Zoback, M.D., 1995. Compressive and tensile failure of inclined well bores and determination of in situ stress and rock strength. J. Geophys. Res. 100 (12791-12), 811.

Peters, E.J., Flock, D.L., 1981. The onset of instability during two-phase immiscible displacement in porous media. SPE J. 21 (2), 249–258.

Peters, K.E., Moldowan, J.M., 1992. The Biomarker Guide: Interpreting Molecular Fossils in Petroleum and Ancient Sediments. Prentice Hall, Englewood Cliffs, NJ.

Peters, E.J., Afzal, N., Gharbi, R., 1993. On scaling immiscible displacements in permeable media. J. Pet. Sci. Eng. 9 (3), 183–205.

Peters, B.M., Denzen, Z., Blunt, M.J., 1998. Experimental investigation of scaling factors that describe miscible floods in layered systems. In: Symposium on Improved Oil Recovery, pp. 211–218, https://doi.org/10.2118/39624-MS.

Peters, M., Timmerhaus, K., Ronald West, R., 2002. Plant Design and Economics for Chemical Engineers. McGraw Hills, Inc., New York.

Peters, K.E., Walters, C.C., Moldowan, J.M., 2005. The Biomarker Guide, second ed. Cambridge University Press, New York; Cambridge, UK.

Petford, N., McCaffrey, K., 2003. Hydrocarbons in crystalline rocks: an introduction. Geol. Soc. Spec. Publ. 214, 1–5.

Petty, S., 2002. Thermal stimulation of well 83–16. In: Rose, P. (Ed.), EGS Quarterly Report. University of Utah Energy and Geoscience Institute, Utah.

Pfister, M., Chanson, H., 2014. Two-phase air-water flows: scale effects in physical modeling. J. Hydrodyn. Ser. B 26 (2), 291–298.

Pham, C., et al., 2020. Effect of faults and rock physical properties on in situ stress within highly heterogeneous carbonate reservoirs. J. Pet. Sci. Eng. 185, 106601.

Phillip, P., Lewis, C.A., 1987. Organic geochemistry of biomarkers. Annu. Rev. Earth Planet. Sci. 15 (1), 363–395.

Philp, R.P., 2007. The emergence of stable isotopes in environmental and forensic geochemistry studies: a review. Environ. Chem. Lett. 5, 57–66.

Philp, R.P., Gilbert, T.D., 1986. Biomarkers distributions in oils and predominantly derived from terrigenous source material. In: Leythauser, D., Rullkötter, J. (Eds.), Advances in Organic Geochemistry 1985. Pergamon Press, London, England.

Philp, R.P., Mansuy, L., 1997. Petroleum geochemistry: concepts, applications, and results. Energy Fuel 11 (4), 749–760.

Philp, R.P., Gilbert, T.D., Friedrich, J., 1981. Methylated hopanes in crude oils and their applications in petroleum geochemistry. J. Geochim. Cosmochim. Acta 45, 1173–1180.

Pickup, G.E., Ringrose, P.S., Jensen, J.L., Sorbie, K.S., 1994. Permeability tensors for sedimentary structures. Math. Geol. 26 (2), 227–250.

Pidwirny, M., 2008. Global distribution of precipitation. In: Fundamentals of Physical Geography, second ed. 17 April.

Pierce, R.L., 1970. Reducing land subsidence in the Wilmington oil field by use of saline waters. Water Resour. Res. 6 (5), 1505–1514.

Piipari, R., Tuppurainen, M., Tuomi, T., Mäntylä, L., Henriks-Eckerman, M.L., Keskinen, H., Nordman, H., 1998. Diethanolamine-induced occupational asthma, a case report. Clin. Exp. Allergy 28 (3), 358–362. https://doi.org/10.1046/j.1365-2222.1998.00232.x. 9543086.

Pine, R.J., Batchelor, A.S., 1984. Downward migration of shearing in jointed rock during hydraulic injections. Int. J. Rock Mech. Min. Sci. Geomech. Abstr. 21 (5), 249–263.

Pizarro, J.O., Trevisan, O.V., 1990. Electrical heating of oil reservoirs: numerical simulation and field test results. J. Pet. Technol. 42, 1320–1326.

Plumb, R., Hickman, S.H., 1985. Stress-induced borehole elongation: a comparison between the four-arm dipmeter and the borehole televiewer in the Auburn geothermal well. J. Geophys. Res. 90, 5513–5521.

Podlubny, I., 1998. Fractional differential equations: an introduction to fractional derivatives, fractional differential equations, to methods of their solution and some of their applications. Math. Sci. Eng., 198.

Podlubny, I., 2001. Geometric and physical interpretation of fractional integration and fractional differentiation. arXiv. preprint math/0110241.

Portella, R.C., Prais, F., 1999. Use of automatic history matching and geostatistical simulation to improve. In: SPE Paper 53976.

Pospisil, G., 2011. BP Exploration (Alaska) Inc. Presentation on January 6. Available from: www.aoga.org/wp-content/uploads/2011/01/8-Pospisil-Heavy-Viscous-Oil.pdf.

Postelnicu, A., 2007. Thermal hydrodynamic instability of a Walters B viscoelastic fluid in a fluid-saturated anisotropic porous medium with fast chemical reaction. In: Reactive Heat Transfer in Porous Media, Ecole des Mines d'Albi, France. vols. ET81-XXXX, pp. 1–8.

Pozzi, A.L., Blackwell, R.J., 1963. Design of laboratory models for study of miscible displacement. Soc. Pet. Eng. J. 3 (1), 28–40. https://doi.org/10.2118/445-PA.

Pride, S.R., Berryman, J.G., 2003. Linear dynamics of double-porosity dual-permeability materials. I. Governing equations and acoustic attenuation. Phys. Rev. E 68 (3), 036603.

Prosper, G.W., Ali, S.M.F., 1991. Scaled model studies of the immiscible carbon dioxide flooding process at low pressures. In: PETSOC-91-92, Annual Technical Meeting. Pet. Soc. Can., https://doi.org/10.2118/91-2.

Pruess, K., 1985. A practical method for modeling fluid and heat flow in fractured porous media. Soc. Pet. Eng. J. 25 (1), 14–26.

Pruess, K., 1991. TOUGH2: a general numerical simulator for multiphase fluid and heat flow. Lawrence Berkeley Laboratory, Berkeley, CA. Report LBL-29400.

Psyrillos, A., Burley, S.D., Manning, D.A.C., Fallick, A.E., 2003. Coupled mineral-fluid evolution of a basin and high: kaolinization in the SW England granites in relation to the development of the Plymouth Basin. In: Petford, N., McCaffrey, K.J.W. (Eds.), Hydrocarbons in Crystalline Rocks. Geological Society, London, Special Publications, 214, pp. 175–195, https://doi.org/10.1144/GSL.SP.2003.214.01.11.

Pujol, L., Boberg, T.C., 1972. Scaling accuracy of laboratory steam flooding models. In: SPE-4191-MS, SPE California Regional Meeting., https://doi.org/10.2118/4191-MS.

Purdy, J., 1995. Study Guide-Environmental Studies 200: Impacts of Human Activities on Marine Ecosystems. Bowdoin College, Brunswick, Maine, pp. 190–200.

Puryear, C.I., Castagna, J.P., 2008. Layer-thickness determination and stratigraphic interpretation using spectral inversion: theory and application. Geophysics 73, R37–R48.

Qi, D., Hesketh, T., 2005. An analysis of upscaling techniques for reservoir simulation. Pet. Sci. Technol. 23, 827–842.

Qi, Y., Ju, Y., Tan, J., Bowen, L., Cai, C., Yu, K., Zhu, H., Huang, C., Zhang, W., 2020. Organic matter provenance and depositional environment of marine-to-continental mudstones and coals in eastern Ordos Basin, China—evidence from molecular geochemistry and petrology. Int. J. Coal Geol. 217. https://doi.org/10.1016/j.coal.2019.103345, 103345.

Qian, W., Pedersen, L.B., 1991. Inversion of borehole breakout orientation data. J. Geophys. Res. 96, 20093–20107.

Rabemananaa, V., Durstb, P., Bachlerc, D., Vuataza, F.D., Kohl, T., 2003. Geochemical modelling of the Soultz-sous-Forets hot fractured rock system: comparison of two reservoirs at 3.8 and 5 km depth. Geothermics 32 (4–6), 645–653.

Rahman, M.A., Mustafiz, S., Biazar, J., Islam, M.R., 2007. Experimental and numerical modeling of a novel rock perforation technique. J. Frankl. Inst. 344 (5), 777–789.

Rahman, A., Happy, F.A., Ahmed, S., Hossain, M.E., 2017. Development of scaling criteria for enhanced oil recovery: a review. J. Pet. Sci. Eng. 158, 66–79.

Raimi, D., 2018. The Health Impacts of the Shale Revolution. Resources for the Future.

Ramírez, A., et al., 2007. Mathematical simulation of oil reservoir properties. Chaos, Solitons Fractals 38 (3), 778–788.

Ramseur, J., et al., 2014. Oil Sands and the Keystone XL Pipeline. Congressional Research Service.

Ranalli, G., Gale, A., 1976. Lectures on the Rheology of the Earth, Part I: Basic Concepts. vol. 76-1 Carleton University, p. 157.

Rapoport, L.A., 1955. Scaling laws for use in design and operation of water-oil flow models. Trans. AIME 204, 143.

Rapoport, L.A., Leas, W.J., 1953. Properties of linear waterfloods. J. Pet. Technol. 5, 139–148. https://doi.org/10.2118/213-G.

Raven, K.G., Gale, J.E., 1985. Water flow in a natural rock fracture as a function of stress and sample size. Int. J. Rock Mech. Min. Sci. Geomech. Abstr. 22, 251–261.

Rawal, C., Ghassemi, A., 2014. A reactive thermo-poroelastic analysis of cold water injection into an enhanced geothermal reservoir. Geothermics 50, 10–23.

Raymond, A.C., Murchison, D.G., 1988. Effect of volcanic activity on level of organic maturation in Carboniferous rocks of East Fife, Midland Valley of Scotland. Fuel 67 (8), 1164–1166.

Read, R.S., 2004. 20 years of excavation response studies at AECL's underground research laboratory. Int. J. Rock Mech. Min. Sci. 41, 1251–1275.

Reilly, C., 1991. Metal Contamination of Food, second ed. Elsevier Science Publishers Ltd, London and New York.

Reinecker, J., Tingay, M., Müller, B., 2003. Borehole Breakout Analysis from Four-arm Caliper Logs. World Stress Map Project (WSM). http://dc-app3-14.gfz-potsdam.de/pub/guidelines/WSM_analysis_guideline_breakout_caliper.pdf.

Revil, A., 1999. Pervasive pressure-solution transfer: a poro-visco-plastic model. Geophys. Res. Lett. 26, 255–258.

Riber, L., Dypvik, H., Sørlie, R., 2015. Altered basement rocks on the Utsira High and its surroundings, Norwegian North Sea. Nor. J. Geol. 93, 57–89.

Riber, L., Dypvik, H., Sørlie, R., Naqvis, A.M., Stangvik, K., Oberhardt, N., Schroeder, P.A., 2017. Comparison of deeply buried palaeoregolith profiles, Norwegian North Sea, with outcrops from southern Sweden and Georgia, USA—implications for petroleum exploration. Palaeogeogr. Palaeoclimatol. Palaeoecol. 471, 82–95. https://doi.org/10.1016/j.palaeo.2017.01.043.

Rica, T., Horváth, D., Toth, A., 2005. Density fingering in acidity fronts: effect of viscosity. Chem. Phys. Lett. 408 (4), 422–425.

Rice, J., Cleary, M.P., 1976. Some basic diffusion solutions for fluid-saturated elastic porous M constituents. Rev. Geophys. Space Phys. 14 (2), 227–241.

Ricklefs, R.E., Miller, G.L., 2000. Ecology, fourth ed. Macmillan, ISBN: 978-0-7167-2829-0.

Rickman, R., et al., 2008. A practical use of shale petrophysics for stimulation design optimization: all shale plays are not clones of the Barnett Shale. In: SPE 11528., https://doi.org/10.2118/115258-MS.

Rintoul, W., 1990. Drilling Through Time: 75 years with California's Division of Oil & Gas. California Division of Oil.

Ritchie, J.D., Ziska, H., Johnson, H., Evans, D., 2011. Geology of the Faroe-Shetland Basin and adjacent areas. British Geological Survey Research Report, RR/11/01.

Robelius, F., 2007. Giant Oil Fields—The Highway to Oil: Giant Oil Fields and their Importance for Future Oil Production (Doctoral thesis). Uppsala University. 156 pp http://publications.uu.se/abstract.xsql?dbid=7625.

Rodrenvskaya, М.И., 2001. Hydrocarbon potential in deep layers (D. Shi, Trans.). Nat. Gas Geosci. 12 (4–5), 49–51.

Rogers, J.D., Grigg, R.B., 2000. A literature analysis of the WAG injectivity abnormalities in the CO_2 process. In: Paper SPE 59329 Presented at the 2000 Review SPE/DOE Improved Oil Recovery Symposium, Tulsa, OK.

Rogner, H.-H. (Ed.), 2012. Energy Resources and Potentials. IEA Publication, pp. 458–459. Available from: https://iiasa.ac.at/web/home/research/Flagship-Projects/Global-Energy-Assessment/GEA_Chapter7_resources_lowres.pdf.

Rojas, G.A., 1985. Scaled Model Studies of Immiscible Carbon Dioxide Displacement of Heavy Oil (PhD dissertation). University of Alberta, Edmonton, AB, Canada.

Rona, P.A., 1988. Hydrothermal mineralization at oceanic ridges. Can. Mineral. 26 (3), 431–465.

Root, D.H., Attanasi, E.D., 1993. A primer in field-growth estimation. In: Howell, D.G. (Ed.), The Future of Energy Gases: U.S. Geological Survey Professional Paper 1570, pp. 547–554.

Rosa, M.B., et al., 2018. The Giant Lula Field: world's largest oil production in ultra-deep water under a fast-track development. In: Paper Presented at the Offshore Technology Conference, Houston, Texas, USA, April., https://doi.org/10.4043/29043-MS.

Rose, W., 2000. Myths about later day extensions of Darcy's law. J. Pet. Sci. Eng. 26 (1–4), 187–198.

Rostami, B., et al., 2005. Identification of fluid dynamics in forced gravity drainage using dimensionless groups. Transp. Porous Media 83 (3), 725–740.

Rothert, E., Shapiro, S.A., 2003. Microseismic monitoring of borehole fluid injections: data modeling and inversion for hydraulic properties of rocks. Geophysics 68, 685–689. https://doi.org/10.1190/1.1567239.

Rottenfusser, B., Ranger, M., 2004. A geological comparison of six projects in the Athabasca oil sands. In: Presentation Given at CSPG/CSEG 2004 GeoConvention, Calgary, AB, Canada, May 31–June 4.

RPS, 2017. Evaluation of Lancaster Field within Licence P. 1368 on behalf of Hurricane Energy plc. Competent persons report, ECV2210.

Rudnicki, J.W., 1999. Alteration of regional stress by reservoirs and other inhomogeneities: stabilizing or destabilizing? In: Proceedings of the Ninth International Congress on Rock Mechanics, Paris, France, pp. 1629–1637.

Rudraiah, N., et al., 1990. Effect of modulation on the onset of thermal convection in a viscoelastic fluid-saturated sparsely packed porous layer. Can. J. Phys. 68 (2), 214–221.

Ruthven, D.M., 1984. Principles of Adsorption and Adsorption Processes. John Wiley, New York.

Rutqvist, J., Stephansson, O., 2003. The role of hydromechanical coupling in fractured rock engineering. Hydrogeol. J. 11, 7–40.

Rutqvist, J., Tsang, C.-F., Stephansson, O., 1998. Determination of fracture storativity in hard rocks using high pressure testing. Water Res. 34, 2551–2560.

Rutqvist, J., Majer, E., Oldenburg, C., Peterson, J., Vasco, D., 2006. Integrated modeling and field study of potential mechanisms for induced seismicity at the geysers geothermal field, California. Geotherm. Resour. Counc. Trans. 30, 629–633.

Ryan, B.D., Dawson, M.F., 1993. British Columbia Geological Survey Report. http://cmscontent.nrs.gov.bc.ca/geoscience/publicationcatalogue/Paper/BCGS_P1994-01-17_Ryan.pdf.

Sablok, R., Aziz, K., 2005. Upscaling and discretization errors in reservoir simulation. In: SPE Paper 93372.

Saeb, S., Amadei, B., 1990. Modelling joint response under constant or variable normal stiffness boundary conditions. Int. J. Rock Mech. Min. Sci. Geomech. Abstr. 27 (3), 213–217.

Saeed, N.O., Ajijolaiya, L.O., Al-Darbi, M.M., Islam, M.R., 2003. Mechanical properties of mortar reinforced with hair fibre. In: Proc. Oil and Gas Symposium, CSCE Annual Conference, Refereed Proceeding, Moncton, June.

Safari, M.R., Ghassemi, A., 2011. A 3D analysis of enhanced geothermal reservoir: shear slip and micro-seismicity. In: Proceedings of the 45th US Rock Mechanics/Geomechanics Symposium, June 26–29, San Francisco, CA.

Saffman, P.G., Taylor, G.I., 1958. The penetration of a fluid into a porous medium or Hele-Shaw cell containing a more viscous liquid. Proc. R. Soc. Lond. A 245, 312–329.

Saghir, M.Z., Islam, M.R., 1999. Viscous fingering during miscible liquid-liquid displacement in porous media. Int. J. Fluid Mech. Res. 26 (4), 215–226.

Saghir, Z., Chaalal, O., Islam, M.R., 2000. Experimental and numerical modeling of viscous fingering. J. Pet. Sci. Eng. 26 (1–4), 253–262.

Saha, S., Chakma, A., 1992. Separation of CO_2 from gas mixtures with liquid membranes. Energy Convers. Manag. 33 (5–8), 413–420.

Samvelov, P.Г., 1997. Deep oil-gas reservoir forming characteristics and distribution (F.X. Guan, Trans.). Northwest Oil Gas Explor. 9 (1), 52–57.

Sánchez-Lemus, M.C., et al., 2016. Physical properties of heavy oil distillation cuts. Fuel 180, 457–472.

Sanderson, D., Nixon, C., 2015. The use of topology in fracture network characterization. J. Struct. Geol. 72, 55–66. https://doi.org/10.1016/j.jsg.2015.01.005.

Sandler, S.R., Karo, W., Bonesteel, J., Pearce, E.M., 1998. Polymer Synthesis and Characterization: A Laboratory Manual. Academic Press.

Sarkar, S., Toksoz, M.N., Burns, D.R., 2002. Fluid Flow Simulation in Fractured Reservoirs. Massachusetts Institute of Technology, Earth Resources Laboratory.

Sarma, H., Das, S., 2009. Air Injection Potential in Kenmore Oilfield in Eromanga Basin, Australia: A Screening Study Through Thermogravimetric and Calorimetric Analyses., https://doi.org/10.2118/120595-MS.

Sarma, P., Aziz, K., Durlofsky, L.J., 2005. Implementation of adjoint solution for optimal control of smart wells. In: SPE Paper 92864.

Sarma, L., Durlofsky, J., Aziz, K., Chenb, W.H., 2006. Efficient real-time reservoir management using adjoint-based optimal control and model updating. Comput. Geosci. 10, 3–36.

Sarsekeyeva, F., et al., 2015. Cyanofuels: biofuels from cyanobacteria. Reality and perspectives. Photosynth. Res. 125, 1–2. https://doi.org/10.1007/s11120-015-0103-3.

Satter, A., Iqbal, G., Buchwalter, J., 2008. Practical Enhanced Reservoir Engineering. Pennwell Corp. 688 pp.

Saulsberry, J.L., Schafer, P.S., Schraufnagel, R.A. (Eds.), 1996. A guide to coalbed methane reservoir engineering. In: Gas Research Institute Report GRI-94/0397, Chicago, IL.

Sayed, A., et al., 2005. A novel experimental investigation of a solar cooling system in Madrid. Int. J. Refrig. 28 (6), 859–871.

Schamel, S., 2013a. Unconventional oil resources of the Uinta Basin, Utah. In: Hein, F.J., Leckie, D., Larter, S., Suter, J.R. (Eds.), Heavy-Oil and Oil-Sand Petroleum Systems in Alberta and Beyond. AAPG Studies in Geology, vol. 64. AAPG, Tulsa, OK, pp. 437–480.

Schamel, S., 2013b. Tar sand triangle bitumen deposit, Garfield and Wayne counties, Utah. In: Morris, T.H., Ressetar, R. (Eds.), The San Rafael Swell and Henry Mountains Basin—Geologic Centerpiece of Utah. Utah Geological Association Publication 42, Utah, pp. 497–522.

Schenk, C.J., Pollastro, R.M., Hill, R.J., 2006. Natural bitumen resources of the United AAPG EMD Bitumen and Heavy Oil Committee Commodity Report—May 2019. U. S. Geological Survey Fact Sheet 2006-3133, 2 pp. http://pubs.usgs.gov/fs/2006/3133/pdf/FS2006-3133_508.pdf.

Scher, H., Lax, M., 1973. Stochastic transport in a disordered solid. I. Theory. Phys. Rev. B 7, 4491–4502.

Schiffman, R.L., 1971. A thermoelastic theory of consolidation. In: Environmental Geophysics and Heat Transfer. vol. 4, p. 78.

Schimmelmann, M., et al., 2009. Dike intrusions into bituminous coal, Illinois Basin: H, C, N, O isotopic responses to rapid and brief heating. Geochim. Cosmochim. Acta 73 (20), 6264–6281. https://doi.org/10.1016/j.gca.2009.07.027.

Schlumberger Information Solutions, 2009. ECLIPSE Black-Oil Simulation. http://www.slb.com/content/services/software/reseng/eclipse_simulators/blackoil.asp.

Schmeeckle, M.W., Nelson, J.M., 2003. Direct numerical simulation of bedload transport using a local, dynamic boundary condition. Sedimentology 50 (2), 279–301.

Schutter, S.R., 2003a. Hydrocarbon occurrence and exploration in and around igneous rocks. Geol. Soc. Spec. Publ. 214, 7–33.

Schutter, S.R., 2003b. Occurrences of hydrocarbons in and around igneous rocks. Geol. Soc. Spec. Publ. 214 (1), 35–68. https://doi.org/10.1144/GSL.SP.2003.214.01.03.

Science News, 2018. https://www.sciencedaily.com/releases/2018/01/180124113951.htm.

Seeburger, D.A., Zoback, M.D., 1982. The distribution of natural fractures and joints at depth in crystalline rock. J. Geophys. Res. 87, 5517–5534. https://doi.org/10.1029/JB087iB07p05517.

Segall, P., 1989. Earthquakes triggered by fluid extraction. Geology 17, 942–946.

Segall, P., Fitzgerald, S.D., 1998. A note on induced stress changes in hydrocarbon and geothermal reservoirs. Tectonophysics 289, 117–128.

Seidle, J.P., 1999. A modified p/Z method for coal wells. In: Paper Presented at the 1999 SPE Rocky Mountain Regional Meeting, Gillette, Wyoming. 15–18 May.

Seifert, S.R., Lennox, T.R., 1985. Developments in tar sands in 1984. Am. Assoc. Pet. Geol. Bull. 69 (10), 1890–1897.

Selby, D., Creaser, R.A., 2005. Direct radiometric dating of hydrocarbon deposits using Rhenium-Osmium isotopes. Science 308 (5726), 1293–1295. https://doi.org/10.1126/science.1111081. 15919988.

Selley, R., 1998. Elements of Petroleum Geology, second ed. Academic Press, USA. 470 pp.

Sesetty, V., Ghassemi, A., 2012. Modeling and analysis of stimulation for fracture network generation. In: Proceedings of the 37th Workshop on Geothermal Reservoir Engineering. Stanford University, Stanford, CA.

Seto, Y., et al., 2008. Fate of carbonates within oceanic plates subducted to the lower mantle, and a possible mechanism of diamond formation. Phys. Chem. Miner. 35, 223–229.

Settari, A., Sullivan, R.B., 2009. A novel hydraulic fracturing model fully coupled with geomechanics and reservoir simulation. SPE J. 14 (3), 423–430.

Settari, A., et al., 2008. Numerical techniques used for predicting subsidence due to gas extraction in the North Adriatic Sea. Pet. Sci. Technol. 26 (10–11), 1205–1223.

Severns, W., 2006. Can technology turn the tide on decline? In: Keynote Presentation at the SPE Intelligent Energy Conference and Exhibition, Amsterdam, Netherlands.

Shanley, K.W., Cluff, R.M., Robinson, J.W., 2004. Factors controlling prolific gas production from low-permeability sandstone reservoirs: implications for resource assessment, prospect development, and risk analysis. AAPG Bull. 88 (8), 1083–1121.

Shapiro, S., Huenges, E., Borm, G., 1997. Estimating the crust permeability from fluid-injection-induced seismic emission at the KTB site. Geophys. J. Int. 131, F15–F18.

Shapiro, S., Audigane, P., Royer, J., 1999. Large-scale in situ permeability tensor of rocks from induced microseismicity. Geophys. J. Int. 137, 207–213.

Shapiro, S.A., Rothert, E., Rath, V., Rindschwentner, J., 2002. Characterization of fluid transport properties of reservoirs using induced microseismicity. Geophysics 67, 212–220.

Sharma, A., Rao, D.N., 2008. Scaled physical model experiments to characterize the gas-assisted gravity drainage EOR process. In: SPE-113424-MS, SPE Symposium on Improved Oil Recovery., https://doi.org/10.2118/113424-MS.

Sharma, M.K., Kumar, M., Kumar, A., 2002. *Ocimum sanctum* aqueous leaf extract provides protection against mercury induced toxicity in Swiss albino mice. Indian J. Exp. Biol. 40, 107–182.

Sharma, R.K., Agrawal, M., Marshall, F.M., 2009. Heavy metals in vegetables collected from production and market sites of a tropical urban area of Indi. Food Chem. Toxicol. 47, 58–91.

Sharma, J., Inwood, S., Kovscek, A., 2012. Experiments and analysis of multiscale viscous fingering during forced imbibition. In: SPE Annual Technical Conference and Exhibition, SPE-143946-MS. vol. 17(4), pp. 1142–1159.

Shelton, S.V., Delano, A., Schaefer, L., 1999. Second law study of the Einstein refrigeration cycle. In: Proceedings of the Renewable and Advanced Energy Systems for the 21st Century, April 11–15, 1999, Lahaina, Maui, Hawaii.

Shen, F., 1998. Seismic Characterization of Fractured Reservoirs (Part I) Crustal Deformation in Tibetan Plateau (Part II) (Ph.D. dissertation). MIT.

Sheng, J., 2010. Modern Chemical Enhanced Oil Recovery: Theory and Practice. Gulf Professional Publishing, Burlington, USA, p. 617.

Sheridan, J.M., Hickman, S.H., 2004. In situ stress, fracture, and fluid flow analysis in Well 38C-9: an enhanced geothermal system in the Coso geothermal field. In: Proceedings of the 29th Workshop on Geothermal Reservoir Engineering, January 26–28, 2004. Stanford University, Stanford, CA.

Sheridan, J., Kovac, K., Rose, P., Barton, C., McCulloch, J., Berard, B., Moore, J., Petty, S., Spielman, P., 2003. In situ stress, fracture and fluid flow analysis-East Flank of the Coso geothermal field. In: Proceedings of the Twenty-Eighth Workshop on Geothermal Reservoir Engineering. Stanford University, Stanford, CA, pp. 34–49.

Sherlock, S.C., Strachan, R.A., Jones, K.A., 2009. High spatial resolution 40Ar/39Ar dating of pseudotachylites: geochronological evidence for multiple phases of faulting within basement gneisses of the Outer Hebrides. J. Geol. Soc. Lond. 166, 1049–1059. https://doi.org/10.1144/0016-76492008-125.

Shi, J.Q., Durucan, Q.S., 2003. A bidisperse pore diffusion model for methane displacement desorption in coal by CO2 injection. Fuel 82, 1219–1229.

Shi, J., et al., 2015. Study of Longkou oil shale pyrolysis behavior with bitumen as the intermediate. In: 4th International Conference on Mechatronics, Materials, Chemistry and Computer Engineering., https://doi.org/10.2991/icmmcce-15.2015.539.

Shibulal, B., et al., 2014. Microbial enhanced heavy oil recovery by the aid of inhabitant spore-forming bacteria: an insight review. Sci. World J. 12. https://doi.org/10.1155/2014/309159, 309159.

Shirey, S.B., Richardson, S.H., 2011. Start of the Wilson cycle at 3 Ga shown by diamonds from the subcontinental mantle. Science 333, 434–436. https://doi.org/10.1126/science.1206275.

Shivakumara, I.S., Lee, J., et al., 2011. Effect of thermal modulation on the onset of convection in Walters B viscoelastic fluid-saturated porous medium. Transp. Porous Media 87 (1), 291–307.

Shook, M., Li, D., Lake, L.W., 1992. Scaling immiscible flow through permeable media by inspectional analysis. In Situ-NY 16, 311.

Sibley, D.N., 2010. Viscoelastic Flows of PTT Fluids (Ph.D. dissertation). University of Bath.

Sibson, R.H., 2002. 29 Geology of the crustal earthquake source. Int. Geophys. 81, 455–473. https://doi.org/10.1016/S0074-6142(02)80232-7.

Sierra, R., et al., 2001. Promising progress in field application of reservoir electrical heating methods. In: Proceedings of SPE International Thermal Operations and Heavy Oil Symposium, Margarita Island, Venezuela, 12–14 March. SPE-69709.

Šijačić, D., Fokker, P.A., 2015. Thermo-hydro-mechanical modeling of EGS using COMSOL multiphysics. In: 40th Workshop on Geothermal Reservoir Engineering, p. 52.

Simmons, M.R., 2002. The world's giant oilfields. Hubbert Center Newslett. 1, 1–62.

Simoneit, B.R.T., 1988. Petroleum generation in submarine hydrothermal systems: an update. Can. Mineral. 26 (3), 827–840.

Simoneit, B.R.T., Lonsdale, P.F., 1982. Hydrothermal petroleum in mineralized mounds at the seabed of Guaymas Basin. Nature 295 (5846), 118–202.

Singh, R., et al., 2011a. Heavy metals and living systems: an overview. Indian J. Pharm. 43 (3), 246–253.

Singh, K., et al., 2011b. Reservoir fluid characterization and application for simulation study. In: SPE 143612, SPE EUROPEC/EAGE Annual Conference and Exhibition held in Vienna, Austria, 23–26 May.

Singh, J.S., et al., 2016. Cyanobacteria: a precious bio-resource in agriculture, ecosystem, and environmental sustainability. Front. Microbiol. 7, 529. https://doi.org/10.3389/fmicb.2016.00529.

Singurindy, O., Berkowitz, B., 2003. Evolution of hydraulic conductivity by precipitation and dissolution in carbonate rock. Water Resour. Res. 39, 1016–1030.

Sircar, A., 2004. Hydrocarbon production from fractured basement formations. Curr. Sci. 87 (2).

Skempton, A.W., 1970. The consolidation of clays by gravitational compaction. Q. J. Geol. Soc. Lond. 125, 373–411.

Skrobot, V.L., Castro, E.V.R., Pereira, R.C.C., Pasa, V.M.D., Fortes, I.C.P., 2005. Identification of adulteration of gasoline applying multivariate data analysis techniques HCA and KNN in chromatographic data. Energy Fuel 19 (6), 2350–2356. https://doi.org/10.1021/ef0500311.

Skrobot, V.L., Castro, E.V.R., Pereira, R.C.C., Pasa, V.M.D., Fortes, I.C.P., 2007. Use of principal component analysis (PCA) and linear discriminant analysis (LDA) in gas chromatographic (GC) data in the investigation of gasoline adulteration. Energy Fuel 21 (6), 3394–3400. https://doi.org/10.1021/ef0701337.

Slattery, J.C., 1967. Flow of viscoelastic fluids through porous media. AICHE J., 1066–1077.

Slightam, C., 2012. Characterizing seismic-scale faults pre- and post-drilling; Lewisian Basement, West of Shetlands, UK. In: Spence, G.H., Redfern, J., Aguilera, R., Bevan, T.G., Cosgrove, J.W., Couples, G.D., Daniel, J.M. (Eds.), Advances in the Study of Fractured Reservoirs. Geological Society, London, Special Publications, vol. 374, pp. 311–332, https://doi.org/10.1144/SP374.6.

Smayda, T.J., 1978. From phytoplankters to biomass. In: Sournia, A. (Ed.), Phytoplankton Manual. UNESCO, Paris, pp. 273–279.

Smith, G., 2019. Iraq will deliver third-biggest chunk of New Oil over next decade, IEA says. Bloomberg (April 24).

Smith, J.M., van Ness, J.N., Abott, M.M., 2001. Introduction to Chemical Engineering Thermodynamics. McGraw-Hill Education.

Sochi, T., 2010. Modelling the flow of yield-stress fluids in porous media. Transp. Porous Media 85, 489–503.

Sochi, T., Blunt, M.J., 2008. Pore-scale network modeling of Ellis and Herschel–Bulkley fluids. J. Pet. Sci. Eng. 60 (2), 105–124.

Society of Petroleum Engineers (SPE), 2009. Glossary of Industry Terminology. http://www.spe.org/spe-app/spe/industry/reference/glossary.htm.

Sonneland, L., Veire, H.H., Raymond, B., Signer, C., Pedersen, L., Ryan, S., Sayers, C., 1997. Seismic reservoir monitoring on Gullfaks. Lead. Edge 16 (9), 1247–1252.

Sorbie, K.S., et al., 1990. Scaled miscible floods in layered bead packs investigating viscous crossflow, the effects of gravity, and the dynamics of viscous slug breakdown. In: SPE-20520-MS, SPE Annual Technical Conference and Exhibition., https://doi.org/10.2118/20520-MS.

Sorrell, S., et al., 2009. An assessment of the evidence for a near-term peak in global oil production. A report produced by the Technology and Policy Assessment function of the UK Energy Research, ISBN: 1-903144-0-35.

Speight, J.G., Islam, M.R., 2016. Peak Energy: Myth or Reality? Wiley-Scrivener. 400 pp.

Speight, J.G., Moschopedis, S.E., 1989. In: Bunger, J.W., Li, N.C. (Eds.), Chemistry of Asphaltenes. Advances in Chemistry Series. vol. 195. American Chemical Society, Washington, DC, pp. 1–15.

Sperner, B., et al., 2003. Tectonic stress in the Earth's crust: advances in the World Stress Map Project. In: Nieuwland, D. (Ed.), New Insights into Structural Interpretation and Modeling. Geol. Soc. Spec. Publ, vol. 212, pp. 101–116.

Spivak, A., 1974. Gravity segregation in two-phase displacement processes. Soc. Pet. Eng. J. 14 (6), 619–632. https://doi.org/10.2118/4630-PA.

Spivey, J.P., Semmelbeck, M.E., 1995. Forecasting long-term gas production of dewatered coal seams and fractured gas shales. In: Paper Presented at the Low Permeability Reservoirs Symposium, Denver, Colorado, March., https://doi.org/10.2118/29580-MS. Paper Number: SPE-29580-MS.

Sposito, G., 1980. General criteria for the validity of the Buckingham-Darcy flow law. Soil Sci. Soc. Am. J. 44 (6), 1159.

Sprouse, B.P., 2010. Computational efficiency of fractional diffusion using adaptive time step memory and the potential application to neural glial networks. ArXiv. 1004.5128.

Stark, M.A., 1990. Imaging injected water in the Geysers reservoir using microearthquakes data. Geotherm. Resour. Counc. Trans. 14 (Part II), 1697–1704.

Statista, 2021a. https://www.statista.com/statistics/1088739/global-oil-discovery-volume/#:~:text=Global%20oil%20discoveries%202011%2D2018&text=The%20total%20volume%20of%20crude,5.45%20billion%20barrels%20of%20oil.

Statista, 2021b. https://www.statista.com/statistics/183943/us-carbon-dioxide-emissions-from-1999/#:~:text=The%20statistic%20shows%20the%20total,carbon%20dioxide%20was%20emitted%20globally.

Stavsky, Y., 1963. Thermoelasticity of heterogeneous aeolotropic plates. ASCE J. Eng. Mech. Div. 89, 89–105.

Steefel, C., Lasaga, A.C., 1994. A coupled model for transport of multiple chemical species and kinetic precipitation/dissolution reactions with application to reactive flow in single phase hydrothermal systems. Am. J. Sci. 294, 529–592.

Steefel, C.I., Litchner, P.C., 1998. Multicomponent reactive transport in discrete fractures: I. Controls on reaction front geometry. J. Hydrol. 209, 186–199.

Stewart, G., Asharsobbi, F., 1988. Well test intepretation for naturally fractured reservoirs. In: Paper Presented at the SPE Annual Technical Conference and Exhibition, Houston, Texas, October., https://doi.org/10.2118/18173-MS. Paper Number: SPE-18173-MS.

Stoker, M.S., Holford, S.P., Hillis, R.R., 2018. A rift-to-drift record of vertical crustal motions in the Faroe-Shetland Basin, NW European margin: establishing constraints on NE Atlantic evolution. J. Geol. Soc. Lond. 175, 263–274. https://doi.org/10.1144/jgs2017-076.

Stone, H.L., 1982. Vertical, conformance in an alternating water-miscible gas flood. In: SPE-11130-MS, SPE Annual Technical Conference and Exhibition., https://doi.org/10.2118/11130-MS.

Streltsova, T.D., 1983. Well pressure behavior of a naturally fractured reservoir. SPE J. 23 (05), 769–780.

Stüben, K., Clees, T., Klie, H., Lou, B., Wheeler, M.F., 2007. Algebraic multigrid methods (AMG) for the efficient solution of fully implicit formulations in reservoir simulation. In: Paper Presented at the 2007 SPE Reservoir Simulation Symposium, Houston, TX, Feb. 28–30., https://doi.org/10.2118/105832-MS. Paper Number: SPE-105832-MS.

Sugie, H., 2004. Three cases of sudden death due to butane or propane gas inhalation: analysis of tissues for gas components. Forensic Sci. Int. 143 (2–3), 211–214. https://doi.org/10.1016/j.forsciint.2004.02.038.

Sugisaki, R., 1981. Deep-seated gas emission induced by the earth tide: a basic observation for geochemical earthquake prediction. Science 212 (4500), 1264–1266.

Sugisaki, R., Mimura, K., 1994. Mantle hydrocarbons: abiotic or biotic? Geochim. Cosmochim. Acta 58 (11), 2527–2542. https://doi.org/10.1016/0016-7037(94)90029-9. 11541663.

Sun, H. (Ed.), 2015. Fundamentals of advanced production decline analysis. In: Advanced Production Decline Analysis and Application. Elsevier.

Sundaram, N.S., Islam, M.R., 1994. Scaled model studies of petroleum contaminant removal from soils using surfactant solutions. J. Hazard. Mater. 38 (1), 89–103.

Suresh Kumar, G.S., Ghassemi, A., 2005. Numerical modeling of non-isothermal quartz dissolution/precipitation in a coupled fracture-matrix system. Geothermics 34 (4), 411–439.

Sutton, J., Watson, J., 1951. The pre-Torridonian metamorphic history of the Loch Torridon and Scourie areas in the north-west Highlands, and its bearing on the chronological classification of the Lewisian. Q. J. Geol. Soc. 106, 241–307. https://doi.org/10.1144/GSL.JGS.1950.106.01-04.16.

Suzuki, K., Hewett, T.A., 2002. Sequential upscaling method. Transp. Porous Media 46 (2–3), 179–212.

Svenson, E., Schweisinger, T., Murdoch, L.C., 2007. Analysis of the hydromechanical behavior of a flat-lying fracture during a slug test. J. Hydrol. 347, 35–47.

Swaan, D., 2013. Influence of shape and skin of matrix-rock blocks on pressure transients in fractured reservoirs. SPE Form. Eval. 5 (04). https://doi.org/10.2118/15637-PA.

Swart, P.K., Cantrell, D.L., Arienzo, M.M., Murray, S.T., 2016. Evidence for high temperature and ^{18}O-enriched fluids in the Arab-D of the Ghawar Field, Saudi Arabia. Sedimentology 63 (6), 1739–1752. https://doi.org/10.1111/sed.12286.

Swenson, D., Hardemana, B., 1997. The effects of thermal deformation on flow in a jointed geothermal reservoir. Int. J. Rock Mech. Min. Sci. 34 (3–4), 308.E1–308.E19.

Swenson, D., et al., 1997. GEOCRACK: A Coupled Fluid Flow/Heat Transfer/Rock Deformation Program for Analysis of Fluid Flow in Jointed Rock. Mechanical Engineering Department, Kansas State University, Manhattan.

Sylte, J.E., Thomas, L.K., Rhett, D.W., Bruning, D.D., Nagel, N.B., 1999. Water induced compaction in the Ekofisk field. In: SPE Annual Technical Conference and Exhibition, 3–6 October, Houston, Texas. SPE Paper 56426-MS.

Taber, J., 1969. Dynamic and static forces required to remove a discontinuous oil phase from porous media containing oil and water. SPEJ 9, 3–12.

Tang, C.A., Tham, L.G., Lee, P.K.K., Yang, T.H., Li, L.C., 2002. Coupled analysis of flow, stress and damage (FSD) in rock failure. Int. J. Rock Mech. Min. Sci. 39, 477–489.

Tao, Q., Ghassemi, A., 2010. Poro-thermoelastic borehole stress analysis for determination of the in situ stress and rock strength. Geothermics 39 (3), 250–259.

Tarasovs, S., Ghassemi, A., 2010. A study of propagation of cooled cracks in a geothermal reservoir. In: Geothermal Resources Council Annual Meeting. 2010: Sacramento, CA.

Taylor, K.C., Hawkins, B.F., Islam, M.R., 1990. Dynamic interfacial tension in surfactant-enhanced alkaline flooding. J. Can. Pet. Technol. 29 (1), 50–55.

Teimoori, A., et al., 2005. Effective permeability calculation using boundary element method in naturally fractured reservoirs. Pet. Sci. Technol. 23 (5–6), 693–709.

Teng, L., Mavko, G., 1996. Fracture signatures on P-wave AVOZ. In: 66th Ann. Intern. Mtg. Soc. Expl. Geophys, pp. 1818–1821. Expanded Abstract.

Tennyson, M.E., 2005. Growth history of oil reserves in major California oil fields during the Twentieth Century. In: Dyman, T.S., Schmoker, J.W., Verma, M. (Eds.), Geologic, Engineering, and Assessment Studies of Reserve Growth. U.S. Geological Survey Bulletin 2172-H, pp. 1–15.

Teramoto, T., et al., 2005. Air injection EOR in highly water saturated oil reservoir. In: Proceedings of IEA Collaborative Project on Enhanced Oil Recovery 25th Annual Workshop and Symposium, Stavanger, Norway, 5–8 September.

Terwilliger, P.L., et al., 1951. An experimental and theoretical investigation of gravity drainage performance. Pet. Trans. AIME 192, 285.

Thakur, P., et al. (Eds.), 2020. Coal Bed Methane: Theory and Applications, second ed. Elsevier. 426 pp.

Thiéry, R.T., et al., 2002. Individual characterization of petroleum fluid inclusions (composition and P-T trapping conditions) by microthermometry and confocal laser scanning microscopy: inferences from applied thermodynamics of oils. Mar. Pet. Geol. 19 (7), 847–859.

Thomas, S., Ali, S.M.F., Thomas, N.H., 1997. Scale-up methods for micellar flooding and their verification. In: Paper Presented at the Annual Technical Meeting, Calgary, Alberta, June 1997., https://doi.org/10.2118/97-5. Paper Number: PETSOC-97-50.

Thompson, G., Humphris, S.E., Shroeder, B., 1988. Active vents and massive sulphides at 26°N (TAG) and 37°N (Snakepit) on the Mid-Atlantic Ridge. Can. Mineral. 26 (3), 697–711.

Thompson, R.C., et al., 2009. Plastics, the environment and human health: current consensus and future trends. Philos. Trans. R. Soc. Lond. Ser. B Biol. Sci. 364 (1526), 2153–2166.

Thomsen, L., 1986. Weak elastic anisotropy. Geophysics 51, 1954–1966.

Thorat, S., Dahanukar, S., 1991. Can we dispense with ayurvedic samskaras? J. Postgrad. Med. 37, 15–59.

TIME, 2005. The science of happiness. Cover story (January 17).

Timoshenko, S., Woinowsky-Krieger, S., 1959. Theory of Plates and Shells, second ed. McGraw-Hill, New York.

Tingay, M., Reinecker, J., Müller, B., 2008. Borehole breakout and drilling-induced fracture analysis from image logs. In: World Stress Map Project—Guidelines: Image Logs, Helmholtz Cent. GFZ German Research Centre for Geosciences, Potsdam, Germany. http://dc-app3-14.gfz-potsdam.de/pub/guidelines/WSM_analysis_guideline_breakout_image.pdf.

Tiratsoo, E.N., 1984. Oilfields of the World, third ed. Scientific Press Ltd., England. 392 pp.

Tissot, B.P., Welte, D.H., 1978. Petroleum Formation and Occurrence—A New Approach to Oil and Gas Exploration. Springer, Berlin. 538 pp.

Tissot, B.P., Welte, D.H., 1984. Petroleum Formation and Occurrence, Second Revised and Enlarged Version. Springer-Verlag. 699 pp.

Toninello, A., Pietrangeli, P., De Marchi, U., Salvi, M., Mondovì, B., 2006. Amine oxidases in apoptosis and cancer. Biochim. Biophys. Acta 1765 (1), 1–13. https://doi.org/10.1016/j.bbcan.2005.09.001. 16225993. Epub 2005 Sep 29.

Tortike, W.A., Farouq Ali, S.M., 1987. A framework for multiphase nonisothermal fluid flow in a deforming heavy oil reservoir. In: Paper SPE 16030 Presented at the Ninth SPE Symposium on Reservoir Simulation, Held in San 1991. Antonio, TX, Feb.

Trice, R., 2005. Challenges and insights of optimising oil production from Middle East mega karst reservoirs. Soc. Pet. Eng. https://doi.org/10.2118/93679-MS. SPE-93679-MS.

Trice, R., 2013. Strategies for fractured basement exploration: a case study from the West of Shetland. In: NGF Abstracts and Proceedings, No 2.

Trice, R., 2014. Basement exploration West of Shetlands, progress in opening a new play on the UKCS. In: Cannon, S.J.C., Ellis, D. (Eds.), Hydrocarbon Exploration to Exploitation West of Shetlands. Geological Society, London, Special Publications, vol. 397, pp. 81–105, https://doi.org/10.1144/SP397.3.

Trice, R., Holdsworth, R.E., Rogers, S., McCaffrey, K.J., 2018. Characterising the fracture properties of Lewisian Gneiss basement reservoirs, Rona Ridge, West of Shetland. In: The Geology of Fractured Reservoirs, 24–25 October, London. Programme and Abstract Volume. Geological Society Petroleum Group, London, p. 55.

Trice, R., Hiorth, C., Holdsworth, R.E., 2019. Fractured basement play development in the UK and Norwegian rifted margins: a discussion and implications for exploitation. In: Jackson, C.A.-L. (Ed.), Cross-Border Petroleum Geology and Exploration: The British and Norwegian Continental Margins. Geological Society, London, Special Publications, https://doi.org/10.1144/SP495-2018-174.

Tsai, C.L., et al., 2013. Comparison of the performance of SSPH and MLS basis functions for two-dimensional linear elastostatics problems including quasistatic crack propagation. Comput. Mech. 51 (1), 19–34.

Tsakiroglou, C.D., 2012. A multi-scale approach to model two-phase flow in heterogeneous porous media. Transp. Porous Media 94 (2), 525–536.

Tuo, J.C., 2002. Research status and advances in deep oil and gas exploration. Adv. Earth Science 17 (4), 565–571 (in Chinese).

Turek, A., Robinson, R.N., 1984. Geology of the Precambrian basement in southern Ontario. Ontario Geological Survey Open File Report 5496, 135 pp.

Turek, M., Skryzynksi, K., Smolinksi, A., 2008. Structure and Changes of Production Costs in 1998–2005 in the Polish Hard Coal Industry. Glückauf Essen.

Turkevich, P., Ono, Y., 1969. Catalytic research on zeolites. Adv. Catal. 20, 135–152.

Turner, S., et al., 2003. Case studies of plagioclase growth and residence times in island arc lavas from Tonga and the Lesser Antilles, and a model to reconcile discordant age information. Earth Planet. Sci. Lett. 214, 279–294.

Turta, A.T., Singhal, A.K., 2007. Reservoir engineering aspects of light-oil recovery by air injection. SPE Reserv. Eval. Eng. 4 (04), 336–344.

U.S. Congress, 2005. Energy Policy Act of 2005. 109th Congress.

U.S. EIA, 2013. Technically Recoverable Shale Oil and Shale Gas Resources: An Assessment of 137 Shale Formations in 41 Countries Outside the United States.

U.S. EIA, 2017. Annual Energy Outlook 2017.

U.S. EIA, 2018. Annual Energy Outlook 2018.

U.S. EIA, 2019. U.S. Crude Oil Production Grew 17% in 2018, Surpassing the Previous Record in 1970.

U.S. EIA, 2020a. How Much Shale (Tight) Oil is Produced in the United States?.

U.S. EIA, 2020b. Spot Prices for Crude Oil and Petroleum Products.

U.S. EIA, 2020c. Annual Energy Outlook 2020.

U.S. EIA, 2020d. Assumptions to the Annual Energy Outlook 2020: Oil and Gas Supply Module.

U.S. EIA, 2020e. Crude Oil Production.

U.S. EIA, 2020f. Tight Oil Production Estimates by Play.

U.S. EIA, 2020g. U.S. Crude Oil Imports by Country of Origin.

U.S. Energy Information Administration (EIA), 2020a. Monthly Energy Review June 2020.

U.S. Energy Information Administration (EIA), 2020b. U.S. Coal Reserves. Table 15, October.

U.S. EPA, 2020. Inventory of U.S. Greenhouse Gas Emissions and Sinks: 1990-2018.

References

Ukar, E., Laubach, S.E., 2016. Syn- and postkinematic cement textures in fractured carbonate rocks: insights from advanced cathodoluminescence imaging. Tectonophysics 690, 190–205.

Ukar, E., Baques, V., Laubach, S.E., Marrett, R., 2020. The nature and origins of decametre-scale porosity in Ordovician carbonate rocks, Halahatang oilfield, Tarim Basin. China. J. Geol. Soc. 177 (5), 1074–1091.

Union of Concerned Scientists, 2016. What Is Tight Oil?.

Unsihuay-Vila, C., Saavedra, O.R., 2006. Transmission loss unbundling and allocation under pool electricity markets. IEEE Trans. Power Syst. 21 (1), 77–84.

US Geological Survey, 2000. World Petroleum Assessment 2000: New Estimates of Undiscovered Oil and Natural Gas, Including Reserve Growth, Outside the United States—Chapter.

USGS, 2006. https://www.usgs.gov/centers/nmic.

USGS, 2014. Map of Assessed Coalbed-Gas Resources in the United States. https://pubs.usgs.gov/dds/dds-069/dds-069-ii/pdf/dds69ii.pdf.

Vafai, K., Sozan, M., 1990. Analysis of energy and momentum transport for fluid flow through a porous bed. J. Heat Transf. 112 (3), 690–699. https://doi.org/10.1115/1.2910442.

Van Daalen, F., Van Domselaar, H.R., 1972. Scaled fluid-flow models with geometry differing from that of prototype. Soc. Pet. Eng. J. 12 (3), 220–228. https://doi.org/10.2118/3359-PA.

van Everdingen, A.F., Hurst, W., 1949. The application of the Laplace transformation to flow problems in reservoirs. Trans. Am. Inst. Min. Metall. Eng. 186, 305–324.

van Gent, H.W., Holland, M., Urai, J.L., Loosveld, R., 2010. Evolution of fault zones in carbonates with mechanical stratigraphy—insights from scale models using layered cohesive powder. J. Struct. Geol. 32, 1375–1391. https://doi.org/10.1016/j.jsg.2009.05.006.

Vaziri, H., Jalali, J.S., Islam, M.R., 2001. An analytical model for stability analysis of rock layers over a circular opening. Int. J. Solids Struct. 38, 3735–3757.

Veedu, F.K., Delshad, M., Pope, G.A., 2010. Scaleup methodology for chemical flooding. In: SPE-135543-MS, SPE Annual Technical Conference and Exhibition., https://doi.org/10.2118/135543-MS.

Verma, M.K., 2003. Modified Arrington method for calculating reserve growth—a new model for United States oil and gas fields. In: Dyman, T.S., Schmoker, J.W., Verma, M.K. (Eds.), Geologic, Engineering, and Assessment Studies of Reserve Growth. U.S. Geological Survey Bulletin 2172–D.

Vernik, L., Zoback, M.D., 1990. Strength anisotropy in crystalline rock-implications for assessment of in situ stresses from wellbore breakouts. In: Balkema, A.A. (Ed.), Proceedings 31st US Symposium on Rock Mechanics. Rotterdam, pp. 841–848.

Vigrass, L.W., 1968. Geology of Canadian heavy oil sands. Am. Assoc. Pet. Geol. Bull. 52 (10), 1984–1999.

Villarroel, T., Mambrano, A., Garcia, R., 2013. New progress and technological challenges in the integral development of the Faja Petrolifera del Oronoco, Venezuela. In: Hein, F.J., Leckie, D., Larter, S., Suter, J.R. (Eds.), Heavy-Oil and Oil-Sand Petroleum Systems in Alberta and Beyond. AAPG Studies in Geology, vol. 64. AAPG, Tulsa, OK, pp. 669–688.

Vojir, D.J., Michaelides, E.E., 1994. Effect of the history term on the motion of rigid spheres in a viscous fluid. Int. J. Multiphase Flow 20 (3), 547–556.

von Hagke, C., Kettermann, M., Bitsch, N., Bucken, D., Weismuller, C., Urai, J.L., 2019. The effect of obliquity of slip in normal faults on distribution of open fractures. Front. Earth Sci. 7, 18. https://doi.org/10.3389/feart.2019.00018.

Vortsos, Y.C., 1991. A theoretical analysis of vertical flow equilibrium. Soc. Pet. Eng. https://doi.org/10.2118/22612-MS.

Wadadar, S.S., Islam, M.R., 1994. Numerical simulation of electromagnetic heating of Alaskan tar sands using horizontal wells. J. Can. Pet. Technol. 33 (7), 37–43.

Wagner, S., et al., 2014. Spot the difference: engineered and natural nanoparticles in the environment—release, behavior, and fate. Angew. Chem. Int. Ed. 53 (46), 12398–12419.

Walker, R.J., Holdsworth, R.E., Imber, J., Ellis, D., 2011. The development of cavities and clastic infills along fault-related fractures in tertiary basalts on the NE Atlantic margin. J. Struct. Geol. 33, 92–106. https://doi.org/10.1016/j.jsg.2010.12.001.

Walters, C.C., 2006. The origin of petroleum. In: Hsu, C.S., Robinson, P.R. (Eds.), Practical Advances in Petroleum Processing. Springer, pp. 79–101.

Wan, J., et al., 1998. Effects of grid systems on predicting horizontal well productivity. SPE J. 5 (3).

Wang, X.D., Meguid, S.A., 1995. On the dynamic crack propagation in an interface with spatially varying elastic properties. Int. J. Fract. 69, 87–99.

Wang, S., Tan, W., 2008. Stability analysis of double-diffusive convection of Maxwell fluid in a porous medium heated from below. Phys. Lett. A 372 (17), 3046–3050.

Wang, S., Tan, W., 2011. Stability analysis of soret-driven double-diffusive convection of Maxwell fluid in a porous medium. Int. J. Heat Fluid Flow 32 (1), 88–94.

Wang, D., Cheng, J., Wu, J., Wang, G., 2002. Experiences learned after production of more than 300 million barrels of oil by polymer flooding in Daqing Oil Field. In: Proceedings of SPE Annual Technical Conference and Exhibition, San Antonio, TX, USA, 29 September–2 October. SPE-77693.

Wang, X.H., Quintard, M., Darche, G., 2006. Adaptive mesh refinement for one dimensional three-phase flow with phase change in porous media. Numer. Heat Transf. B Fundam. 50, 231–268.

Wang, K., Killough, J.E., Sepehrnoori, K., 2009. A new upscaling method of relative permeability curves for reservoir simulation. In: SPE-124819-MS, SPE Annual Technical Conference and Exhibition., https://doi.org/10.2118/124819-MS.

Wang, Z., et al., 2014. Ordovician gas exploration breakthrough in the Gucheng lower uplift of the Tarim Basin and its enlightenment. Nat. Gas Ind. B 1, 32–40.

Wang, J., et al., 2016. Analysis of resource potential for China's unconventional gas and forecast for its long-term production growth. Energy Policy 88, 389–401.

Warpinski, N.R., Teufel, L.W., 1987. Influence of geological discontinuities on hydraulic fracture propagation. JPT 39, 209–220.

Warpinski, N.R., Wright, T.B., Uhl, J.E., Engler, B.P., Drozda, P.M., Pearson, R.E., 1996. Micro-seismic monitoring of the B-sand hydraulic fracture experiment at DOE/GRI multi-site projects. In: SPE 36450.

Warpinski, N.R., Wolhart, S.L., Wright, C.A., 2001. Analysis and prediction of microseismicity induced by hydraulic fracturing. In: SPE Paper # 71649.

Warren, J., Root, P.J., 1963. The behaviour of naturally fractured reservoirs. SPE J. 3 (3), 245–255.

Wasiuddin, N.M., Ali, N., Islam, M.R., 2002. Use of offshore drilling waste in Hot Mix Asphalt (HMA) concrete as aggregate replacement. In: ETCE'02, Feb. 4–6, Houston, Texas. Paper No. EE 29168.

Wasserman, M.L., 1987. Local grid refinement for three-dimensional simulators. In: Paper Presented at the SPE Symposium on Reservoir Simulation, San Antonio, Texas, February., https://doi.org/10.2118/16013-MS. Paper Number: SPE-16013-MS.

Watson, A.T., 1989. An analytical model for history matching naturally fractured reservoir production data. In: Paper Presented at the SPE Production Operations Symposium, Oklahoma City, Oklahoma, March. Paper Number: SPE-18856-MS.

Wazwaz, A.M., El-Sayed, S.M., 2001. A new modification of the Adomian decomposition method for linear and nonlinear operators. Appl. Math. Comput. 122, 393–405.

Weber, W.J., Morris, J.C., 1963. Kinetics of adsorption on carbon from solutions. J. Sanit. Eng. Div. 89, 31–60.

Weisman, J., Eckart, R., 1985. Modern Power Plant Engineering. Prentice-Hall. 401 pp.

Welge, H.J., 1952. Simplified method for computing oil recovery by gas or water drive. J. Pet. Technol. 4 (04), 91–98.

Welhan, J.A., Craig, H., 1979. Methane and hydrogen in East Pacific rise hydrothermal fluids. Geophys. Res. Lett. 6 (11), 829–831.

Wellington, S.L., Vinegar, H.J., 1985. CT studies of surfactant-induced CO_2 mobility control. In: SPE-14393-MS, SPE Annual Technical Conference and Exhibition., https://doi.org/10.2118/14393-MS.

Wells, J.T., Ghiorso, M.S., 1991. Coupled fluid flow and reaction in mid-ocean ridge hydrothermal systems I: the behavior of silica. Geochim. Cosmochim. Acta 55, 2467–2482.

Weng, X., Kresse, O., Cohen, C., Wu, R., Gu, H., 2011. Modeling of hydraulic fracture network propagation in a naturally fractured formation. In: SPE 140253, Presented at the SPE Hydraulic Fracturing Technology Conference and Exhibition Held in the Woodlands. Texas, USA.

Wennekers, J.H.M., 1981. Tar sands. Am. Assoc. Pet. Geol. Bull. 66 (10), 2290–2293.

Werner, M.R., 1987. Tertiary and Upper Cretaceous heavy-oil sands, Kuparuk River Unit Area, Alaskan North Slope. In: Meyer, R.F. (Ed.), Exploration for Heavy Crude Oil and Natural Bitume. AAPG Studies in Geology, vol. 25. AAPG, Tulsa, OK, pp. 537–547.

Wessling, S., Junker, R., Rutqvist, J., Silin, D., Sulzbacher, H., Tischner, T., Tsang, F.-F., 2009. Pressure analysis of the hydromechanical fracture behaviour in stimulated tight sedimentary geothermal reservoirs. Geothermics 38, 211–226.

Whitaker, S., 1986. Flow in porous media I: a theoretical derivation of Darcy's law. Transp. Porous Media 1 (1), 3–25.

White, F.M., Corfield, I., 2006. Viscous Fluid Flow. McGraw-Hill, New York.

Whitman, W.B., Coleman, D.C., Wiebe, W.J., 1998. Prokaryotes: the unseen majority. Proc. Natl. Acad. Sci. U. S. A. 95 (12), 6578–6583. https://doi.org/10.1073/pnas.95.12.6578.

Wibowo, W., et al., 2004. Behavior of water cresting and production performance of horizontal well in bottom water drive reservoir: a scaled model study. In: SPE-87046-MS, Presented at the SPE Asia Pacific Conference on Integrated Modelling for Asset Management., https://doi.org/10.2118/87046-MS.

Wilkinson, D.S., Maire, E., Embury, J.D., 1997. The role of heterogeneity on the flow and fracture of two-phase materials. Mater. Sci. Eng. A 233 (1/2), 145–154.

Williams, C., 2003. Boiler efficiency vs. steam quality. In: Heating, Piping, and Air Conditioning, pp. 40–45.

Williams, H., McBirney, A.R., 1979. Volcanology. Freeman, Cooper & Co, San Francisco, CA. 319 pp.

Willis, B.J., White, C.D., 2000. Quantitative outcrop data for flow simulation. J. Sediment. Res. 70, 788–802.

Willis-Richards, J., Watanabe, K., Takahashi, H., 1996. Progress toward a stochastic rock mechanics model of engineered geothermal systems. J. Geophys. Res. 101 (B8), 17481–17496.

Willman, B.T., et al., 1961. Laboratory studies of oil recovery by steam injection. J. Pet. Technol. 13 (7), 681–690. https://doi.org/10.2118/1537-G-PA.

Witt, A.J., Fowler, S.R., Kjelstadli, R.M., Draper, L.F., Barr, D., McGarrity, J.P., 2011. Managing the Start-Up of a Fractured Oil Reservoir: Development of the Clair Field, West of Shetland.

Woodrow, C.K., et al., 2008. One company's first exploration UBD well for characterizing low permeability reservoirs. In: Paper Presented at the SPE North Africa Technical Conference & Exhibition, Marrakech, Morocco, March., https://doi.org/10.2118/112907-MS. Paper Number: SPE-112907-MS.

World Bank, 2019. https://www.worldbank.org/en/country/china/overview.

World Energy Council, 2010. 2010 Survey of Global Resources.

World Energy Council, 2016. World Energy Resources 2016.

Wright, et al., 2010. Techno-economic analysis of biomass fast pyrolysis to transportation fuels. Technical report, NREL/TP-6A20-46586, November. Contract No. DE-AC36-08GO28308.

Wu, S., 1997. Analysis on transient thermal stresses in an annular fin. J. Therm. Stresses 20, 591–615. https://doi.org/10.1080/01495739708956120.

Wu, Y.-S., Pruess, K., Chen, Z.X., 1997. Buckley-Leverett flow in composite porous media. SPE Adv. Technol. Ser. 1 (2). https://doi.org/10.2118/22329-PA.

Xu, X.-P., Needleman, A., 1994. Numerical simulations of fast crack growth in brittle solids. J. Mech. Phys. Solids 42, 1397–1434.

Xu, T., Pruess, K., 2001. Modeling multiphase non-isothermal fluid flow and reactive geochemical transport in variably fractured rocks: 1. Methodology. Am. J. Sci. 301, 16–33.

Xu, Y.G., et al., 2014. The Early Permian Tarim Large Igneous Province: main characteristics and a plume incubation model. Lithos 204, 20–35.

Xue, Z., et al., 2016. Viscosity and stability of ultra-high internal phase CO_2-in-water foams stabilized with surfactants and nanoparticles with or without polyelectrolytes. J. Colloid Interface Sci. 461, 383–395.

Xue, L., Zhou, N., Xie, X., 2018. Reservoir characteristics and three-dimensional architectural structure of a complex fault-block reservoir, beach area, China. J. Pet. Explor. Prod. Technol. 8, 1535–1545.

Yadali Jamaloei, B., Kharrat, R., Asghari, K., Ahmadloo, F., 2015. Phase trapping effects in viscous-modified low interfacial tension flow during surfactant-polymer flooding in heavy oil reservoirs. Energy Sources A Recovery Util. Environ. Eff. 37 (2), 139–148. https://doi.org/10.1080/15567036.2011.584115.

Yadav, S.K., 2010. Heavy metals toxicity in plants: an overview on the role of glutathione and phytochelatins in heavy metal stress tolerance of plants. S. Afr. J. Bot. 76, 16–179.

Yalamas, T., et al., 2004. Sand erosion in cold heavy-oil production. In: Paper Presented at the SPE International Thermal Operations and Heavy Oil Symposium and Western Regional Meeting, Bakersfield, California, March., https://doi.org/10.2118/86949-MS. Paper Number: SPE-86949-MS.

Yang, Y., Aplin, A.C., 2004. Definition and practical application of mudstone porosity-effective stress relationships. Pet. Geosci. 10, 153–162.

Yang, X.E., Long, X.X., Ni, W.Z., Fu, C.X., 2002. *Sedum alfredii* H. A new Zn hyperaccumulating plant first found in China. Chin. Sci. Bull. 47, 1634–1637.

Yasuhara, H., Elsworth, D., 2006. A numerical model simulating reactive transport and evolution of fracture permeability. Int. J. Numer. Anal. Methods Geomech. 30, 1039–1062.

Yasuhara, H., Elsworth, D., Polak, A., 2003. A mechanistic model for compaction of granular aggregates moderated by pressure solution. J. Geophys. Res. 108 (B11), 2530.

Yazid, A., Abdelkader, N., Abdelmadjid, H., 2009. State-of-the-art review of the X-FEM for computational fracture mechanics. Appl. Math. Model. 33, 4269–4282.

Ye, Z., Chen, D., Pan, Z., Zhang, G., Xia, Y., Ding, X., 2016. An improved Langmuir model for evaluating methane adsorption capacity in shale under various pressures and temperatures. J. Nat. Gas Sci. Eng. 31, 658–680.

Yglesias, M., 2019. What trump has actually done in his first 3 years. Vox (Dec 2, 2019).

Yilmaz, B., Müller, U., 2009. Catalytic applications of zeolites in chemical industry. Top. Catal. 52, 888–895. https://doi.org/10.1007/s11244-009-9226-0.

Yin, X.C., Liu, X.H., 1994. Investigations of fracture instability in crack growth for several metals—part I: experimental results. Int. J. Fract. 69 (2), 123–143.

Yin, G., et al., 2018. Glia maturation factor beta is required for reactive gliosis after traumatic brain injury in zebrafish. Exp. Neurol. 305, 129–138.

Young, J.P., Mathews, W.L., Hulm, E.J., 2010. Alaskan heavy oil: first CHOPS at a vast, untapped arctic resource. In: SPE Conference Paper 133592-MS, SPE Western Regional Meeting, 27–29 May, Anaheim, California.

Yu, H., et al., 2008. Air foam injection for IOR: from Laboratory to Field implementation in Zhongyuan Oilfield China. In: Paper Presented at the SPE Symposium on Improved Oil Recovery, Tulsa, Oklahoma, USA, April. Paper Number: SPE-113913-MS.

Yuan, S.C., Harrison, J.P., 2006. A review of the state of the art in modeling progressive mechanical breakdown and associated fluid flow in intact heterogeneous rock. Int. J. Rock Mech. Min. Sci. 43, 1001–1022.

Zandi, S., 2011. Numerical modeling of geomechanical effects of steam injection in SAGD heavy oil recovery. In: Applied Geology. École Nationale Supérieure des Mines de Paris. English. NNT: 2011ENMP0058. pastel-00671450.

Zatzman, G.M., 2012a. Sustainable Resource Development. Wiley-Scrivener. 544 pp.

Zatzman, G.M., 2012b. Sustainable Energy Pricing: Nature, Sustainable Engineering, and the Science of Energy Pricing. Scrivener-Wiley, ISBN: 978-0-470-90163-2. 608 pp.

Zatzman, G., Islam, M.R., 2007. Economics of Tangibles. Nova Science Publishers. 407 pp.

Zavala-Sanchez, V., Dentz, M., Sanchez-Vila, X., 2009. Characterization of mixing and spreading in a bounded stratified medium. Adv. Water Resour. 32 (5), 635–648.

Zdravkov, A., et al., 2017. Palaeoenvironmental implications of coal formation in Dobrudzha Basin, Bulgaria: insights from organic petrological and geochemical properties. Int. J. Coal Geol. 2017 (180), 1–17. https://doi.org/10.1016/j.coal.2017.07.004.

Zellou, A.M., Ouenes, A., 2007. Integrated fractured reservoir characterization using neural networks and fuzzy logic: three case studies. J. Pet. Geol. 24 (4), 459–476. https://doi.org/10.1111/j.1747-5457.2001.tb00686.x.

Zendehboudi, S., Chatzis, I., Mohsenipour, A.A., Elkamel, A., 2011. Dimensional analysis and scale-up of immiscible two-phase flow displacement in fractured porous media under controlled gravity drainage. Energy Fuel 25 (4), 1731–1750.

Zhang, K., 1990. The "in-place" resource evaluation and directions of exploration for the major natural gas accumulations of the non-biogenic origin in Xinjang. Pet. Explor. Dev. 17 (1), 14–21.

Zhang, H.M., 2003. Driver memory, traffic viscosity and a viscous vehicular traffic flow model. Transp. Res. B Methodol. 37, 27–41. https://doi.org/10.1016/S0191-2615(01)00043-1.

Zhang, Z.N., Ge, X.R., 2005a. Micromechanical consideration of tensile crack behavior based on virtual internal bond in contrast to cohesive stress. Theor. Appl. Fract. Mech. 43 (3), 342–359.

Zhang, Z.N., Ge, X.R., 2005b. A new quasi-continuum constitutive model for crack growth in an isotropic solid. Eur. J. Mech. A. Solids 24 (2), 243–252.

Zhang, Z.N., Ge, X.R., 2006. Micromechanical modelling of elastic continuum with virtual multi-dimensional internal bonds. Int. J. Numer. Methods Eng. 65, 135–146.

Zhang, Z., Ghassemi, A., 2011. Simulation of hydraulic fracture propagation near a natural fracture using virtual multidimensional internal bonds. Int. J. Numer. Anal. Methods Geomech. 35, 480–495.

Zhang, X., Jeffrey, R.J.G., 2006. The role of friction and secondary flaws on deflection and re-initiation of hydraulic fractures at orthogonal pre-existing fractures. Geophys. J. Int. 166, 1454–1465.

Zhang, Z., Wu, S., Gao, Z., Ziao, S.B., 1983. Research on sedimentary model from Late Carboniferous to Early Permian Epoch in Kalpin region, Xinjiang. Xinjiang Geol. 1, 9–20.

Zhang, P., Pickup, G.E., Christie, M.A., 2005. A new upscaling approach for highly heterogeneous reservoirs. In: Paper SPE 93339 Presented at the SPE Symposium on Reservoir Simulation. Houston, Texas, January., https://doi.org/10.2118/93339-MS.

Zhang, P., Pickup, G., Christie, M., 2006. A new method for accurate and practical upscaling in highly heterogeneous reservoir models. In: SPE Paper 103760.

Zhang, Q., et al., 2010. Heavy metals chromium and neodymium reduced phosphorylation level of heat shock protein 27 in human keratinocytes. Toxicol. in Vitro 24, 1098–1104.

Zhong, L., Siddiqui, S., El-Hordalo, S., Islam, M.R., 1999. New screening criteria for selection of acid-foam surfactants. In: CIM Conference, Regina, Oct. 5. CIM Paper No. 99-92.

Zhou, X., Ghassemi, A., 2011. Three-dimensional poroelastic analysis of a pressurized natural fracture. Int. J. Rock Mech. Min. Sci. https://doi.org/10.1016/j.ijrmms.2011.02.002.

Zhou, D., Fayers, F.J., Orr Jr., F.M., 1997. Scaling of multiphase flow in simple heterogeneous porous media. SPE Reserv. Eng. 12 (3), 173–178. https://doi.org/10.2118/27833-PA.

Zhu, G., Zhang, S., et al., 2012. The occurrence of ultra-deep heavy oils in the Tabei Uplift of the Tarim Basin, NW China. Org. Geochem. 52, 88–102.

Zhu, G., Su, J., et al., 2013. Formation mechanisms of secondary hydrocarbon pools in the Triassic reservoirs in the northern Tarim Basin. Mar. Pet. Geol. 46, 51–66.

Zhu, G., Wang, H., Weng, N., 2016. TSR-altered oil with high-abundance thiaadamantanes of a deep-buried Cambrian gas condensate reservoir in Tarim Basin. Mar. Pet. Geol. 69, 1–12. https://doi.org/10.1016/j.marpetgeo.2015.10.007.

Zhu, G., Liu, X., Yang, H., Su, J., Zhu, Y., Wang, Y., Sun, C., 2017. Genesis and distribution of hydrogen sulfide in deep heavy oil of the Halahatang area in the Tarim Basin, China. J. Nat. Gas Geosci. 2, 57–71. https://doi.org/10.1016/j.jnggs.2017.03.00.

Zhu, G., et al., 2019. Formation and preservation of a giant petroleum accumulation in superdeep carbonate reservoirs in the southern Halahatang oil field area, Tarim Basin, China. AAPG Bull. 103, 1703–1743. https://doi.org/10.1306/11211817132.

Zoback, M.D., Healy, J.H., 1992. In situ stress measurements to 3:5 km depth in the Cajon Pass scientific research borehole: implications for the mechanics of crustal faulting. J. Geophys. Res. 97, 5039–5057.

Zoback, M.D., Moos, D., Mastin, L., Anderson, R.N., 1985. Wellbore breakouts and in situ stress. J. Geophys. Res. 90, 5523–5530.

Zou, C., et al., 2013. Formation mechanism, geological characteristics and development strategy of nonmarine shale oil in China. Pet. Explor. Dev. 40 (1), 15–27. https://doi.org/10.1016/S1876-3804(13)60002-6.

Index

Note: Page numbers followed by *f* indicate figures and *t* indicate tables.

A

AARD. *See* Average absolute relative deviations (AARD)
Abiogenic petroleum origin theory
  crystal structure, 81
  diamond, 86–88
  fluid characterization, 101–106
  gas hydrates, 97–101
  hydrocarbons
    polymerization, 79
    synthesis, 79, 81
  hydrothermal activity, 85
  methane concentration, 82
  oil and gas deposit, 82–83, 89–94
  pure methane, 81–82
  redox conditions, 82
  seafloor spreading, 85
  smokers, 84
  sub-bottom convectional hydrothermal systems, 84–85
  supergiant oil and gas accumulations, 94–97
  synchrotron X-ray diffraction, 80
  thermobaric conditions, 82
  water-storage capacity, 81
Abiogenic sources, 595–597
Accelerated over-relaxation (AOR), 653
Acid catalysts, 494–495
Acid gases, 503
Adaptive algebraic multigrid (AMG) solver, 801
Adomian decomposition method (ADM), 483–485, 487–491, 493*f*
Adomian polynomials, 488
Adsorbed gas in-place (GIP)
  combining free, 290*f*
  gas content (GC), 289–291
  Langmuir isotherm, 289
  Marcellus shale content, 290*f*
Advection–dispersion models, 649–650
Alberta energy regulator (AER), 324
Alberta government strategy, 390–392, 392*f*
Alkylation, 517
American Petroleum Institute (API), 610
Amines, 504–507, 513
Anisotropy, 203–204

Annual Energy Outlook 2016, 383–384
Anomalous diffusion
  Brownian motion models, 649
  fractional calculus, 649
  fractional-order transport equations and numerical schemes, 650–653
  implicit formulation, 660–662
  mass conservation, 658–660
  methods, 649–650
  numerical simulation, 662, 662*t*
  porous media considerations, 653–656
  theoretical development, 656–658
AOR. *See* Accelerated over-relaxation (AOR)
API. *See* American Petroleum Institute (API)
Archie's law, 193
Arctic National Wildlife Refuge (ANWR), 381
Arctic Technology Conference, 316
Aspect ratio, 755
Average absolute relative deviations (AARD), 102

B

Bakken formation, 452–454, 454*t*
Barnett shale, 337*f*, 452, 453*t*
Basement reservoirs, 51
  case studies
    Bateman-Konen technique, 560–562
    bulk porosity logs, 562
    challenges, 557
    conventional logs, 559–560
    deep dissolution, 553–554
    defined, 533–534
    discrete fracture network model building, 566
    dissolution features, 572
    dolomitic sequences, 568
    dual-porosity model porosity, 559*t*
    early production system, 557
    fracture characterization technique, 560
    fracture identification, 559
    fracture-cavity reservoirs, 571–572
    full-bore formation microimages, 568–571
    full-field simulation model, 566–567
    geological model, 556

Basement reservoirs *(Continued)*
    Halahatang oilfield, 568
    high-resolution gas chromatography, 557
    hydrocarbon reservoirs, 568
    isochron age, 556
    Lancaster reservoir, 562, 566–567
    logging while drilling, 557
    mixed metal oxide, 562–563
    mudlogging data, 557
    nuclear magnetic resonance, 560–562
    oil-bearing fissure networks, 554–555
    Paleozoic-Cenozoic sediments, 569–570f
    permeability values, 559t
    plane-light optical photomicrographs, 571f
    productivity index, 557
    representative elementary volume, 565–566
    sediment-filled fissure systems, 555f, 556
    seismic-scale faults, 562
    Sm-Nd fluorite, 573–575
    3D seismic data coverage, 554
  characterization
    basement formation, 551
    convective system, 550
    countries, 549
    daunting task, 550
    distribution, 548–549, 548f
    fault-valving mechanisms, 551
    formation evaluation, 552
    fractured basement reservoirs, 548
    hydrocarbons, 549–552
    interference, 551
    logging data, 551
    new igneous rock, 550
    open fracture systems, 551
    pressure transient analysis, 551
    ranking of petroleum (*see* Ranking of petroleum)
    reserve growth potential, 552–553
    rock reservoirs, 548
    seismic pumping mechanisms, 551
  Hurricane
    interpretation, 556
    well data acquisition, 558t
    well tests, 559t
  petroleum
    nonconventional sources (*see* Nonconventional sources, petroleum)
    ranking (*see* Ranking of petroleum)
  unconventional reservoirs, 534
Basset force, 646–649
Basset-Boussinesq-Oseen (BBO) equation, 648
BC. *See* Black carbon (BC)
Bi-Langmuir model, 618–619
Biomass, 60–61
Bio-membranes, 515
Biot's theory, 245, 247
Bitumen/heavy oil
  Alberta's energy reserves, resources and production, 325t
  Alberta's Peace River, 325t
  Canada, 324–325
  crude oil resources, 323
  estimated resource and richness, Utah, 332t
  global in-place resources, 323
  immature oil generation, 323
  initial in-place volumes, 326t
  oil density *vs.* viscosity, 322f
  porous sandstone and carbonates, 322–323
  Russia, 333
  shallow accumulations, 331f
  United States, 328–333, 329t
  Venezuela, 325–328
Black carbon (BC), 518–519
Blocked-based permeability tensors, 190
Blue-green algae, 591
Bottom simulating reflectors (BSRs), 317
Boyle's law, 625–626
BrightWater, 414
Brownian motion models, 649
Buckley-Leverett equation, 226, 483–491
Bulk rock-fracture system, 193
Burial-thermal diagenesis, 427–428, 428f

C

CAPRI. *See* Controlled atmospheric pressure resin infusion (CAPRI)
Carbon capture utilization and storage (CCUS), 346–347
Carbon dioxide (CO_2) injection, 357
  backbone, 394
  considerations, 389
  culture, 394
  drilled but uncompleted wells (DUCs), 399–400f, 400
  drilling activities, 399, 399f
  economics, 394
  enhanced oil recovery (EOR), 391f
  environmental integrity, 396
  evolution, 391f
  gas recovery, 392–393
  geological formations, 394
  global warming, 389
  greenhouse gas mitigation, 394
  natural gas production, 393f
  natural sources, 390
  oil production rate history, 396f
  oil recovery, 396
  oil wars, 397–399
  Permian Basin, 400

Index

pipeline system, 390
proven reserve, 397, 397f
scaled model, 390–392
screening criteria, 392t
sequestration, 393, 395f
sustainability, energy management, 397f
usage, 389
viscosity change, 398f
Carbon module, 262
Carbon sequestration enhanced gas recovery (CSEGR)
    air injection, 435f
    biomass, 426
    burial-thermal diagenesis, 427–428, 428f
    carbon dioxide sequestration, 420
    CH_4, 421, 422f
    challenges, 425
    chemically active water/gas, 435f
    coal burning, 429–430
    components, 420, 420f
    cross section, 424f
    cycles, tectonic events, 427, 428f
    cyclic thermal fluid injection, 429
    depressurization, 430–432
    diamond formation, 425–426
    displacement-type recovery scheme, 432
    factors, 421
    feasibility, 423
    flow instability, 423
    flue gas vs. CO_2 efficiency, 424–425, 424t
    fossil-fueled power plants, 420
    fracture-free hydrate reservoirs, 432
    fractured volcanic reservoir properties vs. fractured conventional, 429f
    gas qualities, 425f
    gas traps, 430–432, 430f
    heavy metals/toxic additives, 423
    heterogeneity, 423
    hydrate instability, 430–432, 431f
    hydraulic fractures, 432–433, 436f
    in situ combustion, 430f
    isotopic compositions, 433f
    mineralogy, 426
    multilateral conditions, 433–436
    natural gas, 432
    oil-bearing volcanic reservoirs, 430
    permeability jail, 429f
    permeability-porosity characteristics, 428
    petroleum fluids, 425
    power plant, 420–421
    pressure gradient and gravitational effects, 421
    productivity improvement factor (PIF), 436
    reservoirs types, 436f
    steam assisted gravity drainage (SAGD)-like recovery scheme, 435f
    supercritical conditions, 421
    thermal cracking equivalent, 432
    transport of gas, 432
    types, unconventional reservoirs, 437t
    unconventional gas production schemes, 434f
    volcanic rocks, 426–427, 427f
    water coning, 421
Carnot refrigerator, 830f, 833–834
Cartesian coordinate system, 630
Catalytic cracking, 498
CBIL. See Circumferential borehole imaging log (CBIL)
CBM. See Coal bed methane (CBM)
CCE. See Constant composition expansion (CCE)
CCUS. See Carbon capture utilization and storage (CCUS)
CDGs. See Colloidal dispersion gels (CDGs)
Chemical injection, 352, 356
Chemically enhanced water alternating gas injection (CEWAG), 358
Chlorofluorocarbons (CFC), 842
CHOPS. See Cold heavy oil production with sand (CHOPS)
Circumferential borehole imaging log (CBIL), 156–169
Cleaner crude oil, 517–519
Cleantech innovations, 738
Climate change hysteria, 1
CMG. See Computer Modeling Group (CMG)
CMM. See Coal mine methane (CMM)
Coal, 58
Coal bed methane (CBM), 274
    1989–2010 production, 315t
    1990–2010 production, 313f
    2014 and 2018 estimation, 303t
    annual production rate, 304t
    basins, 306
    benefits, 306
    Canada, 311
    cleats, 301
    depressurization, 309
    dual porosity, 309–310
    estimation, 308t
    forms, 309
    gas hydrate, 316–322
    gas transport, 310–311
    gas-in-place, 300
    geographic distribution, 303f
    gross and net heating values, 302t
    hydrocarbons, 301, 306–307
    infra-red radiations, 299
    natural gas, 300–301, 304f
    production, 301, 301f

Coal bed methane (CBM) (Continued)
  rank enhancement, 309f
  research and developments, 300
  shale/tight gas reservoirs, 616
  thermal maturity, 307–308
  transportation stages, 309–310
  United Kingdom, 311
  United States, 311–316
    coal basins, 308t
    production, 304–305, 305f
    reserves, 306f
  world natural gas consumption, 307f
Coalbed natural gas. See Coal bed methane (CBM)
Coal mine methane (CMM), 300
Coal seam gas. See Coal bed methane (CBM)
Coal-to-liquid (CTL), 267–268
Coefficient of Dykstra-Parson, 761
Cold heavy oil production with sand (CHOPS), 330–331
Collector assembly, 509f
Colloidal dispersion gels (CDGs), 414
Complex reservoirs, 51
  avoiding spurious solutions, 265–266
  core analysis, 214–221
  decision-making process, 185
  deep basins, 186, 187t
  deep formations, 186, 188
  deep hydrocarbon reservoir formation, 188
  essence of reservoir simulation (see Reservoir simulation)
  fluid and rock properties, 264–265
  geochemistry, 251–264
  geological scale, fracture mechanics (see Fracture mechanics)
  linear models, 186f
  material balance equation, 233–235
  modeling unstable flow, 221–232
  oil and gas depths, 189, 189f
  petroleum formation, 185
  pressure-temperature diagram, crude oil, 265f
  representative elemental volume (REV), 236–238, 237f
  research tasks, 188–189
  structural complexity, 186
  technical challenges, 186
  thermal stress, 238–250
Compositional simulator
  block vs. neighboring blocks, 675–676f
  directional geometric factors, 689t
  discretization of motion equation, 685–686
  discretized equations, 673–674
  engineering notation, 691f
  fictitious well rate terms, 699–702
  finite control volume method, 674
  formulations, 673
  governing equations, 677–685
  gridblocks, 673–674
  heterogeneity and anisotropy, 674
  linearization effect, 712–736
  multidimensional domain
    implicit formulation, 692–694
    mass balance equation, 690–692
    reduced equations, implicit model, 694–699
  natural ordering, 692f
  1-D domain, uniform temperature reservoir flow equations, 686–690
  partial differential equations, 673
  reservoir discretization, 674–676, 675f
  solution method, 708
  variable temperature reservoir
    energy balance equation, 702–704
    implicit formulation, 705–708
Comprehensive carbon management scheme, 395f
Comprehensive modeling
  adomian decomposition method (ADM), 484–485
  Darcian and non-Darcian domains, 483
  fluid flow, 481, 482–483f
  Forchheimer's model, 476–479
  governing equations, 475–476
  grain particles, 481
  modified Brinkman's model (MBM), 479–480
  Navier-Stokes equations, 481, 481f
  nonlinear equations, 483–484
  porosity, 481–482
  porous material, 481–482
  pressure gradient, 482–483
Computer Modeling Group (CMG), 266
Connate water saturation, 783
Constant composition expansion (CCE), 102
Constant pressure boundary, 631
Constructed parabolic trough, 510f
Continuous catalytic reforming, 498
Continuous depositional mine, 270
Continuous time random walk (CTRW), 190
Continuous-type unconventional sources, 461
Controlled atmospheric pressure resin infusion (CAPRI), 412–413, 413f
Conventional oil, 267
Copenhagen agreement, 340
Core analysis
  burial and basin evolution, 216–218
  capillary pressure vs. position, 217f
  carbonate formations, 217f
  conventional reservoirs, 215
  effective stress, 218f
  factors, 219
  fracture density and spacing, 214–215
  geological age effect, 220f

issues, 214–215
matrix and fracture permeability, 221f
net overburden conditions, 220f
organic matter, 218–219
permeability jail, 215
porosity, 216
stress sensitivity, 219–221
unconventional basement, 215
unconventional gas reservoir, 216f
wetting-phase saturations, 215
Cost-effective opportunity, 849
Coupled fluid flow/geomechanical stress model
　challenges, unconventional gas reservoirs, 469–475
　comprehensive modeling, 475–485
　governing equations, 485–491
　grid issues, 468
　hybrid grid modeling, 465
　multiphase flow, 465
　production schemes, 466–468
　reservoir and compaction mechanisms, 468
　thermal stress, 468
　threshold effect, 468
Covid-19 pandemic, 53
　CO_2 emission, 9
　effect, 9–11
　global energy outlook, 9
　restrictions, 9–11
Cracking, 270
Crank-Nicholson method, 652
Crude oil formation, 492–493, 494f
CTL. See Coal-to-liquid (CTL)
CTRW. See Continuous time random walk (CTRW)
Cumulative production, 314, 461–464
Cyanobacteria, 591
Cyanophyta, 591
Cyclic steam injection, 407–408
Cyclic steam stimulation (CSS), 324–325

D

Darcy's diffusivity equation, 475
Darcy's law, 233, 609, 613–614, 626, 637, 645–646
Darcy's model, 476
Darcy velocity, 223–224
Data acquisition technique, 195
DDP. See Double displacement (DDP)
Decline curves, 626–631, 628t
Decline rate analysis, 625–626
Deep basins, 186, 187t
Deep formations, 186, 188
Deep hydrocarbon reservoir formation, 188
Deep vacuum fractionation apparatus (DVFA), 102
Deeper marine shales
　Bakken formation, 452–454, 454t
　Barnett shale, 452, 453t
Demonstrated reserve base (DRB), 3–5
Depletion-driven decline, 625–626
Depositional environment, 283
Depth criterion, 283
Desalination technique, 515–516
De Swaan's theory, 472
Diethanolamines (DEA), 503–505, 513
Diethylene glycol (DEG), 504
Differential liberation expansion (DLE), 102
Digital oilfield, 849
DIM. See Double-integration method (DIM)
Dimensional/dimensionless variables, 634t
Direct normal irradiance (DNI), 508f, 510
Direction of sustainability/green technology, 818f, 822f
Discretization error, 799
Discretized equations, 673–674
Displacement efficiency, 755
Diterpanes, 258
DNI. See Direct normal irradiance (DNI)
Double displacement (DDP), 410
Double-integration method (DIM), 642
Downhole refinery, 516
Drilled but uncompleted wells (DUCs), 399–400f, 400
Drilling Productivity Report (DPR), 399
Dry gas, 533
Dual continuum models, 190
Dual permeability models, 190
Dual porosity model, 190, 804–805
DVFA. See Deep vacuum fractionation apparatus (DVFA)

E

Economic developments, 267
Economics-driven decline, 625–626
Efficiency calculations, refrigeration cycle
　compressor efficiency, 834
　environmental pollution observation, 840–842
　fossil fuel combustion efficiency, 836
　generator efficiency, 834
　global economics, 843
　global efficiency, 835–836, 835f
　parabolic trough, 838–840f, 839
　pathway analysis, 840, 841f
　quality, energy, 843
　solar collector efficiency, 836–837
　solar energy, 836, 838–839
　transmission efficiency, 834
　turbine efficiency, 833–834
EIA. See Energy Information Administration (EIA)

Einstein refrigeration cycle, 823–824, 824–825f, 828f
   coefficient of performance (COP) calculation, 829–830
   redrawn, Shelton et al., 826f
   solar cooler and heat engine, 828, 829f
   thermodynamic model, 824–827, 826–827f
   vapor compression cycle refrigeration system, 830–831
Electric submersible pumps (ESPs), 382–383
Electrical formation microscanner (EMS), 205
Ellenburger group, 455, 456–457t, 467f
Energy balance equation, 702–704
   different temperature, 684
   equal temperature, 683
Energy consumption, 1
Energy Information Administration (EIA), 300, 312–313, 383, 390, 610
Engineered nanomaterials (ENMs), 527
Engineering approach. *See* Compositional simulator
Enhanced oil recovery (EOR), 545–546, 671–673, 739, 850
   during 1971–2006, 341f
   costs, 345
   definition, 339
   environmental concerns, 339
   global oil production, 343f
   government incentive, 344–345
   need for
     energy source utilization techniques, 351
     natural processing time, 350, 351f
     proven reserve, 350, 352f
     reasons, 349–350
     recoverable oil, 351–352
     recoverable reserve, 351–352
     recovery factor, 351–352
     reserve production ratio, 351–352
     techniques, 352
     tertiary recovery, 352f
   nonconventional petroleum extraction, 348–349
   oil sands, Canada, 349f
   percentage incremental recovery, 353f
   petroleum engineering, 353
   pivotal criterion
     declared reserve for countries, 361, 364f
     distribution, proved reserve, 364, 368f
     global reserve shares, 364, 366–367f
     oil, natural gas and coal, 362–363, 365t
     proven reserve, 358, 359–360t, 361
     reserve numbers, 361
     reserve recovery ratios, 360, 362t, 364, 369t
     reservoir/production ratio (R/P), 358, 361, 365t, 370f
     reservoir/production ratio (R/P) *vs.* proven reserve, 360, 361f
     top oil producing countries, 361, 363t
     waterflood schemes, 363
   polymer-viscosified water, 328
   post Cold war era, 342f
   recovery rates, 345
   refining efficiency, 349, 350f
   reserve production ratio, 347f
   solar, Oman, 344f
   subcategories, 359f
   sustainable recovery techniques, 348
   sustainable technologies, 343–344
   unconventional formations, 380–389
   victim, 340–341
   water alternating gas, 342
Environmental sustainability, 357
Eolian reservoirs
   Minnelusa formation, 445t, 446
   Norphlet formation, 439–446, 443t
   phases, conventional reserve, 444f
   unconventional reserve growth, 444f
EOR. *See* Enhanced oil recovery (EOR)
Equation-of-state (EOS), 101
ESPs. *See* Electric submersible pumps (ESPs)
Ethylene-vinyl acetate copolymer (EVA), 504
European Cooperation for Science and Technology (COST), 320
Experimental solar trough, 511f
Exploration geochemistry
   conventional and unconventional reservoirs, 257–262
   crude oils, 259f
   diterpanes, 258
   isoprenoids, 258
   m/z 191 chromatogram, 260f
   n-alkanes, 257
   natural and artificial evolution paths, 261f
   nonhopanoid terpanes, 260
   pentacyclic terpanes, 259
   sesquiterpanes, 258
   sesterterpanes, 258
   steranes, 261
Extended finite element method (XFEM), 242
Extra high voltage (EHV), 842
Extra-heavy oil, 610

F

Fatti's equation, 211–212
FDEs. *See* Fractional differential equations (FDEs)
FEM. *See* Finite element method (FEM)
FFA. *See* Frozen front approach (FFA)
Fick's diffusion equation, 646
Fick's law, 474
Fictitious well rate
   conditions, 700
   multiphase flow, 700
   no-flow boundary condition, 701

Index

specified boundary pressure condition, 701
specified flow rate boundary condition, 701
specified pressure gradient boundary condition, 700
Field guideline, recommendations
    data usage, 848
    oil recycling, 849
    petrophysical data, 850
    real-time data acquisition and adaptive control, 849
    reduce freshwater usage, 848–849
    scaled model studies, 850
    streamlining/improving processes, 849
Finite control volume method, 674, 686
Finite element method (FEM), 242
Fire flood, 355
Fischer-Tropsch process, 432
Flow equations, 647–648t
Fluid characterization, 101–106
Fluid classification, 610
Fluid continuum, 236–237, 236f
Fluid flows, 624–625
Fluid memory
    Basset force, 646–649
    constitutive equations, 639
    definition, 641–646
    Navier-Stokes equation, 638
    Newtonian fluid mechanics, 638
    porous media, 637–641
    representations, 639f
    rheology, 638
    stress tensor, 638–639
    viscoelastic models, 640–641t
Fluid/fluidized catalytic cracking, 498
Fluid/rock properties, 264–265
Fluvial reservoir, Wasatch formation, 461
Forchheimer's model, 476–479
Forward in time and centered in space (FTCS), 651–652
Fracking, 348–349, 373, 383
Fractional differential equations (FDEs), 650
Fractional-order transport equations
    applications, 650, 650f
    classical Crank-Nicholson method, 651
    diffusion equations, 650–651
    explicit- and semi-explicit scheme, 651
    Fourier analysis, 652
    fourth-order approximation, 651
    hybrid implicit scheme, 652–653
    modified Grünwald approximation, 651
    net killing rate, 653
    ordinary diffusion equations, 652
    partial differential equation models, 653, 654–655t
    second-order approximation, 651
    stability analysis, 652
Fractured porous medium, 804

Fracture intensity *vs.* mean square permeability, 214f
Fracture mechanics
    discrete representation, 192f
    dual continuum approach, 190
    electron hopping, heterogeneous physical systems, 190
    estimation, fracture properties, 210
    HB resistivity equation, 193
    and matrix flow, 191f
    meaningful modeling, 195
    nonlinear filtering permeability data, 205–209
    original oil in place, 196–197
    phases, conventional reserves, 198f
    pressure change and derivatives, 214f
    probability distributions, 200f
    quantitative measures, well production variability, 199–201
    reservoir heterogeneity, 201–205
    shale considerations, 210–213
    single continuum approach, 190
    subsurface fracture map, 213f
    total volume estimate, 198–199
    transfer functions, 190
    triple porosity, 194
    unconventional reserve growth, 198f
    unconventional reservoirs, 211f
    vuggy reservoir, 192–193
    vugs and oomoldic porosity, 194
Fractures
    characterization, 127
    condensation shrinkage, 126
    diagenetic, 126
    dissolving, 127
    effects
        anisotropic parameters properties, 132–134
        normal moveout velocities, 132
        P-wave azimuthal amplitude *vs* offset (AVO) response, 132
        P-wave normal moveout velocities (NMO) properties, 132–134
    families, 128–129
    gas hydrates, 129
    genetic types, 129
    high-angle fractures, 127
    identification, 469f
    low-angle fractures, 127
    natural, 470
    nature, 125
    oblique fractures, 127
    orientation, 123–125, 142f
    primary/hydraulic, 469
    radial, 470
    reconstructing history, 141f

Fractures *(Continued)*
  secondary, 469
  steps, 129
  stochastic discrete fracture network, 128–129
  surface, 125*f*
  tectonic micro-fractures, 126–127
  total organic content, 129
  types, 143*f*
Free gas in-place (GIP)
  gas-filled porosity, 288
  net organically rich shale thickness, 287
  pressure, 288
  standard reservoir engineering equation, 288–289
  temperature, 288
Frio formation, 446, 447–449*t*
Frozen front approach (FFA), 642
FTCS. *See* Forward in time and centered in space (FTCS)

G

Gamma ray *vs.* total organic carbon, 284*f*
Garbage in and garbage out (GIGO), 185
Gas-assisted gravity drainage (GAGD), 761
Gas diffusion, 310
Gas hydrate, 533
  China, 317
  European Union, 320
  Ignik Sikumi gas hydrate exchange trial, 316–317
  India, 317, 319
  Japan, 318
  Nankai test, 316
  New Zealand, 320–322, 321*f*
  South Korea, 320
  sustained natural gas production, 316
  Turkey, 320
  United States, 317–318
  United States gas hydrate program, 316
Gas injection, 352–354
  chemical flooding, 418
  CO_2 flooding, 417
  CO_2 injection, 417–418, 417*t*
  economics, 418
  emerging technologies, 418–419
  environmental integrity, 418
  flow instability, 418
  gravitational forces, 415
  gravity, 419
  greenhouse gas sequestration, 417–418
  hydrocarbon, 416–417
  miscible and immiscible schemes, 415
  nitrogen (N_2) injection, 416
  pressure maintenance program, 416*f*
  sequential screening, 419
  steps, 419
  transition zone, 419
  viscous fingering, 415–416
  Wyoming sandstone reservoirs, 417
Gas processing, 500–503, 504*f*
Gas reservoir, 778
Gas-to-liquid (GTL), 267–268
Gauss-Seidel (GS) iterative method, 653
Gaussian stochastic simulation, 212
Generative organic carbon (GOC), 270
Geographic location, 285
Geological sequestration, 394
Geomechanics modeling, 239, 246
Geothermal system, 240*f*, 242
GIGO. *See* Garbage in and garbage out (GIGO)
GlassPoint's technology, 343–344, 355
Global energy security, 276
Global petroleum industry, 848–849
Glycol, 504–507, 505*f*, 513
  dehydration, 503
GOC. *See* Generative organic carbon (GOC)
Governing equations
  Adomian decomposition, 487–491, 493*f*
  assumptions, 486
  Buckley-Leverett equation, 485–491, 493*f*
  capillary pressure, 486–487, 490*t*, 492*f*
  Darcy's diffusivity equation, 475
  Darcy's model, 476
  energy balance equation, 683–685
  hydraulic fracturing, 476
  injection/flooding recovery techniques, 476
  linear differential equation, 486
  mass conservation equation, 677–683
  mobility ratio, 486
  modified Brinkman's model (MBM), 475
  non-isothermal reservoir, 677
  water and oil relative permeability, 490*f*
  water saturation distribution, 486–487
Gravity number, 610
Gravity separation systems, 519
Green River Formation, 331–332
Greenhouse gas sequestration, 353–354
Gridblocks, 673–674, 713, 715–716, 722
Grunwald-Letnikov approximation, 609–610
GTL. *See* Gas-to-liquid (GTL)

H

Heat-balance integral method (HBIM), 642
High pressure air injection (HPAI), 388–389, 410, 411*f*, 418–419
History-matching process/model-updating loop, 814–816
Holistic approach, 2
Honey-sugar-saccharine-aspartame (HSSA), 2, 506*t*

Hopanes, 259
Horizontal drilling technology, 306
HTPF. *See* Hydraulic testing of preexisting fractures (HTPF)
Human civilization, 1
Humanity, 1
Hydraulic fracturing, 432–433, 436f, 476
   apparent viscosity, 527t
   aqueous fluids, 524–525
   chemicals, 530, 530–531t
   eco-toxicity, 524–525
   engineered nanomaterials (ENMs), 527
   fate and transport, 525f
   foam properties, 525–526
   gelling agents, 524–525
   injecting gases, 523–524
   iron oxide nanoparticles, 529f
   mechanisms, 524
   mobile natural colloids, 528
   nano iron oxide particle (NIOP), 527–528, 528f
   nanoparticles, 526–527
   natural surfactants and nanomaterials, 527
   polymers, 526–527
   risk assessment strategies, 527
   saturated porous media, 529f
   waterless fracturing, 525–526
Hydraulic testing of preexisting fractures (HTPF), 239–240
Hydrocarbon
   form, 533
   fracture system, 533–534
   organic matters, 533
   organic sources
      algal carbon content, 584
      burial, 587–589, 587–588f
      carbon, 576
      chlorophyll content, 585t
      composition, 581t
      composition of biomass, 580t
      effect of temperature, 578, 581t
      extraction of liquid hydrocarbon, 589–590
      global biomass production, 586t
      inorganic, 589–590
      marine phytoplankton, 577
      microbial natural gas, 580
      natural products, 578
      oil shale, 585
      organic-rich sediment layers, 579, 582f
      pathways, 576–577
      phytobentos, 577
      phytoplankton, 583–585, 584t
      sedimentologic settings, 580–583
      stagnant water conditions, 578
      steps in the transformation, 577
      thermal alteration, 577
      thermal maturation, 583
      wood-derived crude bio-oil, 581t
   types, 533
Hydrofluoric acid (HF), 517, 522
Hydrogen fluoride, 522
Hydrotreating, 498

I

Ideal gas, 834
IEA. *See* International Energy Agency (IEA)
IFT. *See* Interfacial tension (IFT)
Ignik Sikumi gas hydrate exchange trial, 316–317
Implicit-Euler scheme, 651
Implicit formulation, 660–662
Improved oil recovery, 352
In-situ combustion (ISC), 355–356, 355t, 410–411, 411f, 430f
In situ stress orientations
   borehole imaging tools, 123
   dip angle, 135f
   dual-caliper tools, 123, 130f
   fracture
      data analysis, 123
      frequency *vs.* radius, 123
      intensity *vs.* mean square permeability, 139f
   reservoir pressure, 136f
Indirect liquefaction, 267–268
Industrial catalytic processes, 497f
Infinite acting radial flow, 631
Injection wells
   shut-in well, 699
   specified well FBHP, 699
   specified well flow rate, 699
   specified well pressure gradient, 699
Injector/producer distribution, 664, 665t, 666–667f
Inorganic membranes, 514–515
Integral-balance method, 642
Interfacial tension (IFT), 358, 405–407
International Energy Agency (IEA), 272, 275, 306, 307f, 323, 849
Invasion percolation with memory (IPM), 642–643
Inverse aqua regia, 104
Irreducible water saturation, 783
ISC. *See* In-situ combustion (ISC)
Isomerization, 498–499
Isoprenoids, 258
Isothermal viscoelasticity problems, 643
Isotropic elastic stress analysis, 241

J

Jacobian matrix, 266
Japan Methane Hydrate Operating Company (JMH), 318
Japan's Agency for Natural Resources and Energy (ANRE), 318, 319f
Joint Industry Project (JIP), 317
Joule Thompson effect, 468

K

Kelvin-Voigt linear model, 638–639
Kimmeridge Clay Formation (KCF), 216–218
Knowledge model, 196f
Knudsen diffusion, 310
Korean National Gas Hydrate Program, 320
Kozeny-Carman equation, 655–656
Kyoto Agreement, 340

L

Lagrangian approach, 485
Langmuir isotherm, 617, 619
Langmuir pressure (PL), 289–291
Langmuir volume (VL), 289–291, 619
Laplace space equation, 474
LASER. See Liquid addition to steam for enhancing recovery (LASER)
Leo Formation, 446
Leo sandstone, 446
  of Minnelusa Formation, 446
Leo section, 446
Linear isotropic poroelastic system, 247
Linear rock mechanics model, 188
Linearization effect
  multiphase flow of water and oil, 719–736
  nonlinearities, 712–713
  single-phase flow of natural gas, 713–719
Liquefied natural gas (LNG), 314–315
Liquid addition to steam for enhancing recovery (LASER), 408
Liquid hydrocarbons, 533
Liquid membranes, 513–514
Local upscaling, 798–799
Logging-while-drilling (LWD), 317, 557
Loss of heterogeneity error, 799

M

Mackover formation, 458–460t
Macroscopic sweep efficiency, 226
Marine carbonate reservoirs
  Ellenburger group, 455, 456–457t, 467f
  Smackover formation, 455
Mass conservation
  composite variable, 659–660
  control volume, 658f
  porous medium, 658

Mass conservation equation
  accumulation term, 677
  divergence theorem, 677
  fluid production rates, 682
  flux through boundaries, 678
  grid block, 682f
  numerical integration, 680–681
  source term, 682
  time integral, 680f
  total net mass flux, 681
Material balance equation (MBE), 233–235
  algorithm, 620f
  coal bed methane (CBM), 616
  coal/shale reservoirs, 617
  compressibility factor (Z), 615
  gas adsorption, 618–619
  methods, 617–618t
  non-homogeneous adsorbents, 619
  overpressured reservoir, 621
  principle, 615
  real gas law, 615
  reservoir types, 617
  steps, 619–621
  stress-strain formulation, 644
  terms, 616t
  volumetric gas reservoir, 616
  water-drive reservoir, 622
Matrix-fracture diffusion number, 761
Matrix transient linear, 473
Maturation studies, 263–264
Maxwell's one-dimensional linear model, 638–639
MBE. See Material balance equation (MBE)
MBM. See Modified Brinkman's model (MBM)
Meaningful modeling, 195
Memory, 810
  diffusion model, 645
  formalism, 612, 646
  model validation, 662, 663f
MEOR. See Microbial enhanced oil recovery (MEOR)
Methane Hydrate Advisory Committee, 318
Methane hydrate production methodology, 316
Methane-rich gases, 267–268
Methanolamines (MEA), 503
Methoxy acetaldehyde (MALD), 505
Methylhopanes, 259
Microbial enhanced oil recovery (MEOR), 357
Microbial injection, 356–357
Middle East's OPEC countries, 738
MINC. See Multiple Interacting Continua (MINC)
Mineral matrix effect, 262
Mineralogy, 296, 426
Minimum threshold path (MTP), 642–643
Minnelusa formation, 445t, 446
Miocene Oficina Formation, 327

Index

Mittag-Leffler function, 645
Modified Brinkman's model (MBM), 475, 479–480
Molecular/bulk diffusion, 310
Monoethanolamines (MEA), 504, 513
Monte Carlo method, 738–739
The Montreal Protocol, 820
Morrow formation, 446–452, 450–451t
MTP. *See* Minimum threshold path (MTP)
Multiphase flow of water/oil
    algebraic equations, 722f, 730
    capillary pressure variation, 728f
    CPU time, 722
    equations, 726–727
    gridblocks, 722, 730
    interpolations, 722f
    nonlinear algebraic equations, 719–720
    pressure distribution, 724f, 726f
    pressure solution, 721t
    pressure-dependent parameters, 732
    relative permeability, 729
    reservoir, 723, 723f
    single-phase flow, 723–724
    time interval, 731
    variation of parameters, 732–733
    water saturation dependent parameters, 736
Multiple Interacting Continua (MINC), 473

N

n-alkanes, 257
Nano iron oxide particle (NIOP), 527–528, 528f
National Centre of Statistics and Information (NCSI), 343–344
National Institute of Water and Atmospheric Research (NIWA), 321
National Petroleum Reserve-Alaska (NPR-A), 381
Natural and artificial fractures
    borehole breakout, 120, 121f
    borehole imaging tools, 121–122
    compressive stress orientation, 120
    drilling-induced fractures, 120, 124f
    dual-caliper logging, 121
    4D features, 120
    high-resolution borehole image, 122–123
    hydraulic fracturing operations, 120–121
    in situ stress, 119, 121
    orientation of elongations, 120
    stages, fracture characterization, 123
Natural bitumen, 610
Natural chemicals, 500f
Natural fracture distribution, 183
Natural gas, 54–56
Natural gas "well to wheel" pathway, 503f
Natural gas hydrates, 56
Natural gas plant liquids (NGPL), 272

Navier-Stokes equations, 481, 481f, 638
NBP. *See* Normal boiling point (NBP)
NCSI. *See* National Centre of Statistics and Information (NCSI)
Neuquen Basin, 282, 282f, 286f
New York State's Draft Supplemental Generic Environmental Impact Statement (SGEIS), 530
Newton's iteration, 708–712
Newtonian flow, 739
Newtonian fluid mechanics, 638
NGPL. *See* Natural gas plant liquids (NGPL)
NIOP. *See* Nano iron oxide particle (NIOP)
NIWA. *See* National Institute of Water and Atmospheric Research (NIWA)
No-flaring method, 506f
No-flow boundary condition, 701
Nonconventional sources, petroleum
    abiogenic sources, 595–597
    basement oil, 594–595
    cosmic theories, 590–591
    cynobacteria, 591–594
    exploring basement reservoirs, 595
    microbes, 591
    organisms, 591
Nonhopanoid terpanes, 260
Nonlinear equations, 483–484
Nonlinear filtering permeability data
    enhanced oil recovery (EOR) design, 205–207
    fractured formation, 209
    hk data, 209, 209f
    overburden pressure, 208f
    representative elemental volume (REV), 205–207, 207f
    scaling laws, 207
    steady state, conventional terminology, 208
Nonlinear version of linear equations, 185
Non-Markovian Navier-Stokes equation, 642
Normal boiling point (NBP), 102
Normalized water saturation, 729
Norphlet formation, 439–446, 443t
Novel desalination technique, 47–50
Nuclear Fermi liquid, 642

O

Offshore natural gas production, 272
OGIP. *See* Original gas-in-place (OGIP)
OGJ. *See* Oil and Gas Journal (OGJ)
OI. *See* Oxygen index (OI)
OIIP. *See* Oil initially in place (OIIP)
Oil and Gas Journal (OGJ), 341
Oil and gas production
    in Brazil, 32–37
    in China, 29–31
    coronavirus outbreak, 24–25
    energy data, 14–15

Oil and gas production (Continued)
　environmental impact, 25–27
　fast-reacting fossil fuel, 18
　financial collapse, 21
　fossil fuel consumption, 14, 18
　global environmental challenges, 17–18
　the Middle East crisis, 31
　natural gas consumption, 19
　oil consumption, 18–19
　prices, 20–24
　in Russia, 31–32
　transformation, 15–17t
　in USA, 27–28
　in Venezuela, 32–37
　wind and solar facilities, 18
　world energy balances, 15–17
Oil and gas resources, myth and consequences, 846t
Oil and petroleum
　carbon cycle, 72–79
　feature, 63–65
　formation, 61, 61f
　geochemical research, 62
　hydrocarbon, 72–79
　hydrocarbon formation theory, 62
　medicinal benefits, 65
　medicines, 66
　origin, 61–63
　petrochemicals, 66
　petroleum jelly, 66
　pharmaceutical drugs, 66
　pill capsules, 66
　role of magma, 79
　Seneca Oil, 66
　steps, formation, 64
　stratigraphic formations, 65
　sustainability criteria, 57
　use, 66
　vaseline, 66
Oil demand, 277, 277f
Oil discoveries, 276–277, 276f
Oil field formation, 612–615
Oil/gas recovery improvement
　carbon dioxide (CO_2) injection, 357, 389–400
　carbon sequestration enhanced gas recovery (CSEGR), 420–436
　chemical injection, 356
　chemical methods, 414–415
　enhanced oil recovery (see Enhanced oil recovery (EOR))
　existing projects
　　climate change, 340
　　enhanced oil recovery (see Enhanced oil recovery (EOR))
　　greenhouse gas emissions, 340
　　polymers and surfactants, 341
　　primary, secondary and tertiary recovery, 339
　　thermal recovery, 340–341
　　toxic chemicals and catalysts, 340
　　waterflood schemes, 341
　gas injection, 353–354, 415–419
　microbial injection, 356–357
　mobility control issues, 358
　thermal injection, 354–356
　thermal methods, 401–413
　ultrasonic irradiation, 358
　United States leadership, 364–380
　water-alternating-gas (WAG), 357–358
Oil initially in place (OIIP), 611
Oil in-place (OIP), 611
　net organically rich shale thickness, 287
　oil- and gas-filled porosity, 287
　pressure, 287
　temperature, 287
Oil originally in place (OOIP), 611
Oil production, 623–625, 624f, 629, 630f
Oil refining, 493–498, 494–495f, 496t
Oil reservoir, 778
Oil sand, 336f
Oil shale, 58–59
OIP. See Oil in-place (OIP)
Onshore tight oil development, 271
OOIP. See Oil originally in place (OOIP)
Organic geochemistry
　analytical techniques, 254
　biomarkers, 251–254, 252–253t
　depositional environment, 251–254
　exploration, 257–262
　fluid sampling, 251
　fluid's phase diagram, 251
　fossil fuels, 251
　natural vs. artificial evolution paths, 264f
　organic matter and artificial maturation
　　maturation studies, 263–264
　　source rock evaluation, 262–263
　pattern-matching procedures, 251
　petroleum exploration, 255
　phases, 255–256
　porphyrins, 251–254
　pyrolysis techniques, 255
　secondary and tertiary recovery techniques, 254–255
　timeline, 256–257, 256f
　wax deposition, 251, 256
Organic waste products, 510–511
Organoporosity, 270

Original gas-in-place (OGIP), 616
Overpressured reservoir, 621
Oxygen index (OI), 262

P
Paleozoic Exshaw Formation, 104
Parabolic trough, 509f
Paradigm shift, 499
Parallel resistance, 190–192
Path of minimum pressure (PMP), 642–643
Peng-Robinson (PR) equation, 102
Pentacyclic terpanes, 259
Permeability correlations, 656, 657t
Petroleos de Venezuela (PDVSA), 327–328, 327f
Petroleum, 53–54
Petroleum-bearing formations, 609
Petroleum Development Oman (PDO), 355
Petroleum refining processes, 501–502t
Phenomenal model, 196f
Photovoltaic cells, 17–18
PIF. See Productivity improvement factor (PIF)
Pipe line quality gas, 500–502, 512
Plasma-pulse technology (PPT), 358
Plastic hydrocarbons, 533
PLT. See Production logging tool (PLT)
PMP. See Path of minimum pressure (PMP)
Poisson's ratio, 210–212
Politics-driven decline, 625–626
Polymer flooding, 414–415
Polymeric membranes, 514
Pore throat, 613–614
Poroelastic/thermoelastic displacement discontinuity methods, 242
Porous medium, 653–656, 802
Porous system, 477f
PPT. See Plasma-pulse technology (PPT)
Pressure-dependent constants, 635t
Pressure-dependent parameters
  combined variation effect, 735
  oil formation volume factor, 733
  oil viscosity, 733
  porosity formation, 734
Pressure profile across reservoir
  classic and memory model, 671f
  porosity change
    with distance, 669
    with time, 670f, 671
  porosity history, 672f
  variation of pressure with distance, 667
  variation of wellbore pressure with time, 667–669, 668f
  wellbore pressure history, 673f
Pressure profile, reservoir length, 663, 664–665f
Primary recovery method, 611–612

Production logging tool (PLT), 557
Production wells
  shut-in well, 697
  specified well FBHP, 698
  specified well flow rate, 697, 698t
  specified well pressure gradient, 698
Productivity improvement factor (PIF), 436
Productivity index (PI), 262, 557
Proppant movement, 335f
Pseudocapillary pressure, 228
Pseudogas method, 266
Pseudoization, conventional reservoir, 788
Pseudo-pressure-based diffusivity equation, 634, 637
Pseudo-steady state flow, 631–637, 632f
Py-GC. See Pyrolysis-gas chromatography (Py-GC)
Py-GC-MS. See Pyrolysis-gas chromatography-mass spectrometry system (Py-GC-MS)
Pyrolysis techniques, 255
Pyrolysis-gas chromatography (Py-GC), 262
Pyrolysis-gas chromatography-mass spectrometry system (Py-GC-MS), 262

Q
q-homotopy analysis method (q-HAM), 653
Quantity of solar energy, 820

R
Radioactive elements, 519–520
Ranking of petroleum
  advantage, 601–602
  API gravity, 602–604
  basement rocks, 107, 598
  bismuth, 111
  breeds sustainability, 598–600
  coal bed methane, 604
  copper, 64, 110, 112–113
  crude oil, 117
  efficiency and environmental impact, 600–601, 602f
  energy source utilization techniques, 606
  essential oils, 600
  fluid-metasomatic processes, 597–598
  gold, 111
  heavy metals, 109–113, 113t
  hydrocarbon age, 106, 597
  iron, 110
  isotopic mineral ages, 598t
  lanthanum carbonate, 111
  lithium, 111
  mercury, 112
  metal ions, 110–111
  natural gas, 605
  natural processing time, 107–108, 108f, 598–600, 599f, 606f

Ranking of petroleum *(Continued)*
  plagiogneis, 106–107, 106t
  platinum, 111
  refining techniques, 602
  silver, 111
  Tamra Bhasma, 112, 114t
  therapeutic value of metals, 111
  titanium, 111
  toxic elements, 602
  types of oils, 107–108, 598–600
  unconventional resources, 601, 601f
  vanadium, 111
  volume, 115, 115f
  volume of resources, 604, 605f, 607f
  whole-rock Rb-Sr isochron, 599f
  zinc, 110–111
Real-time monitoring, reservoir modeling, 814–817
Recoverable reserve, 847
Recovery factor (RF), 611
Recovery techniques, 847
Refinery emission, 495t
Refining technique, 516
Reforming, 499
Relative permeability, 783
Renewable technologies, 1–2, 738
Representative elemental volume (REV), 205–207, 207f, 236–238, 237f
Representative elementary property (REP), 237, 237f
Reserve estimates, 611–612
Reserve growth potential
  fluid-flow pathways, 438
  geologic and nongeologic factors, 437
  observational techniques, 438
  proven reserves, 437
  reserve growth, 437–438
  United States
    deeper marine shales, 452–454
    depositional environments, 438, 438–439t, 442f
    environments, 439
    Eolian reservoirs, 439–446
    fluvial reservoir, 461
    Gulf of Mexico Basin region, 440f
    interconnected fluvial, deltaic and shallow marine reservoirs, 446–452
    marine carbonate reservoirs, 455
    submarine fan reservoir, 455–461
Reservoir characterization
  biomass, 60–61
  circumferential borehole imaging log, 156–169
  coal, 58
  core analysis, 148–151, 169–171
  data collection, 118–119
  drill stem test, 119
  during drilling, 135–139
  forces of oil and gas reservoirs, 171–178
  fractured reservoirs, 117
  fractures (*see* Fractures)
  geophysical logs, 152–155
  identify breakouts, borehole images, 123
  in situ stress orientations (*see* In situ stress orientations)
  with image log, 148–151
  mechanism of fluid flow, 117
  multi-source information, 118
  natural and artificial fractures (*see* Natural and artificial fractures)
  natural gas, 54–56
  natural gas hydrates, 56
  oil shale, 58–59
  overbalanced drilling approaches, 140–141
  petroleum, 53–54
  production cycle, 119
  seismic fracture characterization, 129–132
  tar sand bitumen, 56–58
  underbalanced drilling, 141–148
  wax, 59
  WSM quality ranking scheme, 118
Reservoir discretization, 674–676, 675f
Reservoir flow relations, 622–623
Reservoir heterogeneity
  anisotropy, 180, 203
  carbonate-cemented samples, 201
  consolidated formations, 178, 201
  degree of interconnection, 178
  enhanced oil recovery (EOR) process, 181
  filtering permeability data, 181–183
  fractures, 178
    effect, 181f
    frequency, 203
    open, 180–181
    properties estimation, 183
    role, 181
  function of fracture frequency, 180–181
  improvement factor, open fractures, 204f
  macroscopic and microscopic scales, 202
  matrix permeability, 179, 202–203
  parameters, 203–204
  phenomenon of heterogeneous permeability, 201
  porosity-permeability correlation, 179–180, 180f
  porosity *vs.* log (permeability), 202–203, 202–203f
  representative elemental volume, 181–182
  residual oil mobilization, 204, 205f
  Rose diagram, 205, 206f
  sandstones, 201–202
  shale, 183–184
  stratification, 203–204

Index

total volume estimate, 183
water/gas injection, 204
Reservoir simulation
  assumptions, 232–233
  coupled fluid flow and geo-mechanical stress model, 806–808, 810f
  custom-designed simulator, 232
  development, reservoir modeling, 811–814
  engineering approach, 232
  fluid-flow equations, 802–806, 805f
  fluid-flow modeling, 808–809
  formulation step, 232
  mathematical and numerical models, 811
  memory model, 810
  sources of errors, 233f
  speed and accuracy, 801–802, 802–803f
  thermal hysteresis, 810–811
Reservoir simulators, 770–794
  characterization, 771f, 774–775
  geological and geophysical modeling, 772–774
  history matching and geological activities, 771f
  power-law averaging method, 785–786
    fluid properties, 777–781, 778f
    rock properties, 781–783, 782f
  representative elementary volume (REV), 775–777, 775–776f
  upscaling, 784–794
    multiphase flow, 788–790, 789f
    phase saturations distribution, 791–792
    power-law averaging method, 785–786
    pressure/production data, 790–791
    pressure-solver method, 786–787, 787f
    renormalization technique, 787–788, 788f
    reservoir simulator output, 792–794
Resource-saving methods, 649
REV. *See* Representative elemental volume (REV)
RF. *See* Recovery factor (RF)
Rheology, 638
Rose diagram, 205, 206f

S

SAE. *See* Simultaneous algebraic equation (SAE)
SAGD. *See* Steam-assisted gravity drainage (SAGD)
Saudi Aramco's Safaniya Master Development Plan, 382–383
Sawdust fueled electricity generator, 518f
Scaled model experiments, 739
Scaling approaches
  comparison, 749–753
  dimensionless groups derived from dimensional analysis, 752t
  dimensionless groups from inspectional analysis, 741–743t, 752t
  influence of different dimensionless groups, 743–747t
Scaling classification, 755–759
  capillary to viscous force, 757–758
  flow and fluid properties, 755–757
  fluid saturation and relative permeability, 758
  rock and fluid memory, 759
  spontaneous imbibition, 759
Scaling criteria, conventional development
  aspect ratio, 755
  bond number, 754
  capillary number, 753–754
  different scaling groups, 754f
  dimensionless numbers, 753t
  dimensionless time, 755
  displacement efficiency, 755
  gravity number, 754
  mobility ratio, 754
Scaling laws, 738–739
Scientific characterization, petroleum, 847–848
SCO. *See* Synthetic crude oil (SCO)
SDS. *See* Sustainable Development Scenario (SDS)
Second level of upscaling, 785
Secondary recovery method, 611–612
Seismic imaging, 793
Selective surface flow (SSF), 514–515
Separation technique, 519–520, 520f
Sesquiterpanes, 258
Sesterterpanes, 258
Settled science, 845
Shale gas/tight oil reserves
  adsorbed gas in-place (GIP), 289
  formations, 278
  free gas in-place (GIP), 288
  global map, 281f
  global reserve, 279–280t
  methodology, 281
  oil in-place, 286
  resource assessment
    areal extent, 282
    oil and gas in-place (OIP/GIP), 286
    preliminary geologic and reservoir characterization, 281
    prospective area, 283
    success/risk factors, 290
    technically recoverable resource, 291
Shale mineralogy, 283f
Shale oil, 267–268
Short memory approach, 644
Simulation of flow composition
  controlled gravity drainage (CGD) method, 761
  dimensionless scaling group, 766–767, 769, 769–770t

Simulation of flow composition (Continued)
  gravity number, 766, 767t
  hot fluid injection, 762–763
  immiscible displacements, 760
  immiscible GAGD process, 761
  in situ combustion, 764–765
  instability number, 768–769
  large scale capillary number, 765, 766t
  micellar flooding, 764
  miscible displacements, 761
  polymer flooding, 763–764
  small scale capillary number, 765, 765t
  solvent/chemical injection, 763
  steam flooding, 762
Simultaneous algebraic equation (SAE), 727–728
Simultaneous water-alternating-gas (SWAG), 357–358
Single carbon number (SCN) fractions, 102
Single-phase flow of natural gas
  case I, 716
  cubic spline and linear interpolation, 721f
  gas formation volume factor and viscosity, 714t
  governing equations, 713
  gridblocks, 715–716, 717f
  homogeneous and isotropic rock properties, 713
  interpolation functions and formulation, 716
  logarithmic spacing constant, 713–714
  number of gridblocks, 719, 720f
  permeability effect, 718, 719f
  relative error, 718t
  spatial and transient pressure distribution, 719
  time interval, 717
Smackover formation, 455
Small-scale waste-oil micro-refinery units, 849
Smart reservoir modeling, 815f
Society of Petroleum Engineers (SPE), 610
Solar-based setups, 355
Solar electric generating systems, 510, 838–839
Solar energy, 507–510
Solid hydrocarbons, 533
SOR. See Successive over-relaxation (SOR)
Source rock evaluation, 262–263
SPE. See Society of Petroleum Engineers (SPE)
Specific gravity, 610
Specified boundary pressure condition, 701
Specified flow rate boundary condition, 701
Specified pressure gradient boundary condition, 700
Spraberry formation, 455–461, 462t
SRB. See Sulfate reducing bacteria (SRB)
SSF. See Selective surface flow (SSF)
Steady-state flow, 631, 632f
Steam-assisted gravity drainage (SAGD), 324–325, 356, 407–408
Steamflooding, 354, 407–408

Steam power plant, 507f, 833f, 835f
Steranes, 261
Stress
  data, 183
  evaluation, 210f
  sensitivity, 219–221
Stress-induced wellbore breakouts, 240–241
Stress-strain law, 246
Structural complexity, 186
Submarine fan reservoir, 455–461
Successive over-relaxation (SOR), 653
Suitable approach, 739
Sulfate reducing bacteria (SRB), 357
Surface diffusion, 310
Sustainability, 738
Sustainability model with perception, 819f
Sustainability pathways
  breakdown heavier hydrocarbons, 517
  catalytic cracking, 498
  cleaner crude oil, 517–519
  crude oil formation, 492–493, 494f
  effective separation
    gas from gas, 512
    liquid from liquid, 511–512
    solid from liquid, 510–511
  fossil fuel utilization, 492
  gas processing, 500–503, 504f
  gas processing chemicals (glycol and amines), 513
  glycol and amines, 504–507, 505f
  gravity separation systems, 519
  isomerization, 499
  membranes and absorbents, 513–515
  novel desalination technique, 515–516
  novel refining technique, 516
  novel separation technique, 519–520, 520f
  oil refining, 493–498, 494–495f, 496t
  pH, de-ionized water, 514f
  reforming, 499
  solar energy, 507–510
  solid acid catalyst, alkylation, 517
  toxic chemicals and catalysts, 492
Sustainability status, current technologies
  crude oil, 38
  drilling, 39, 41–42
  human knowledge, 43
  lesser-cited categories, 45
  modern civilization, 43
  novel desalination technique, 47–50
  peripheral technologies, 43–45
  petroleum engineering, 38
  petroleum operations, 38–39
  technology selection, 39–41
  waste management, 45–46

work of planning, 43
Zipper fracturing, 45
Sustainable development, 268, 337–339
Sustainable Development Scenario (SDS), 345
Sustainable petroleum development, 1, 53
Sustainable technologies, 849
SWAG. *See* Simultaneous water-alternating-gas (SWAG)
Synthetic crude oil (SCO), 324

T

Tangent construction method, 485
Tar sand bitumen, 56–58
Technically recoverable resource
  average gas recovery, 295
  average oil recovery, 292
  efficiency factor, 292
  extended *vs.* limited length lateral wells, 298*f*
  favorable gas recovery, 295
  favorable oil recovery, 291
  geologic complexity
    deep seated fault system, 297
    extensive fault systems, 297
    thrust faults and high stress geological features, 297
  less favorable gas recovery, 292, 295
  mineralogy, 296
  oil recovery efficiency, 294–295*t*
  oil recovery technologies
    intensive well stimulation, 296
    long horizontal wells, 295
    well drilling and completion techniques, 295, 296*f*
  shale gas reservoir properties and Argentina resources, 298*t*
  shale oil reservoir properties and Argentina resources, 299*t*
  shale well productivity, 291
  tight oil data base, 292–293*t*
Technological disaster, 2
THAI. *See* Toe-to-heel air injection (THAI)
Thermal cracking equivalent, 432
Thermal enhanced oil recovery (TEOR), 354
Thermal injection, 352, 354–356
Thermal loss, 508*f*
Thermal maturity, 235, 285
Thermal methods
  air and water injection, 411
  average permeability, 408–409, 409*f*
  controlled atmospheric pressure resin infusion (CAPRI), 412–413, 413*f*
  conventional light oil recovery processes, 407–408
  electromagnetic heating, 413
  factors, 403
  geology and reservoir characteristics, 409–410
  improvements, 410
  in-situ combustion (ISC), 410–411, 411*f*
  light oil reservoirs, 408
  microscopic behavior, surfactant-steamflooding, 404
  oil viscosity spectrum, 408
  pre and post steamflood residual oil saturation, 407, 407*f*
  residual oil on foam, 404
  residual oil saturation, 405, 406*f*
  steam injection, 401–404
  steam-related recovery techniques, 408
  surface tension, 405, 406*f*
  surfactant selection criteria, steam flooding, 404
  synthetic crude and bitumen production, 403, 403*f*
  toe-to-heel air injection (THAI), 411–413, 412–413*f*
  viscosity *vs.* temperature, 405, 405*f*
  water and petroleum features, 401, 401–402*t*
Thermal recovery, 468
Thermal stress, 468
  boundary integral equation, 244
  breakout inversion techniques, 241
  crack length, 243–244, 244*f*
  directions and resulting fracture, 249*f*
  edge cracks, 243*f*
  elastic theory, 238–239
  energy balance equation, 245
  experimental analysis, 242–243
  fluid-infiltrated porous solids, 247
  formation and propagation, 243–244
  geothermal systems, 242
  hydraulic fracturing, 238–239
  in-situ stress, 238–241
  inversion techniques, 242
  Joule Thompson effect, 238–239
  leak-off tests and mini-fracs, 239–240
  linear poroelastic processes, 247
  load decomposition, 248
  modes of fracture, 248, 248*f*
  nonlinear governing equations, 238–239
  nonlinear problem, 245–246
  perturbation approach, 238–239
  petroleum reservoir development, 242
  plane strain borehole, 246
  pore-elastic theory, 245
  rock mechanics research, 239
  secondary thermal cracks, 243*f*
  semianalytical methods, 239
  steady-state pressure distribution, 245
  stress-induced wellbore breakouts, 240–241
  temperature changes, 244, 246
  thermo-poroelastic effects, 241–242
  3D mesh, 249
  transient fracture opening profiles, 249*f*
  transient temperature and stress fields, 245
  unsymmetrical distributions, 250*f*
Thermally-driven refrigeration system, 821, 823*f*

Thistle field in North Sea, 627–628, 629f
Three-dimensional finite-difference approach, 624–625
3D network model, 738–739
Threshold effect, 468
Through Heat Recovery Steam Generators (OTSGs), 344
Tight oil, 267–268
Toe-to-heel air injection (THAI), 411–413, 412–413f
Total organic carbon (TOC), 262, 270
Total organic content (TOC), 183–184, 210, 283
TPAO. *See* Turkish National Oil Company (TPAO)
Transition stage, 631
Transmissibility, 792
Triethanolamine (TEA), 504, 513
Triethylene glycol (TEG), 504
Turkish National Oil Company (TPAO), 320

U

Uncertainty analysis, 794–800
  measurement, 794–796
  model of error estimation, 799–800
  prediction, 800
  risk management, 796–798, 798f
  upscaling, 798–799
Unconventional formations
  Algeria, 384
  Argentina, 384
  Canada, 384
  carbon dioxide projects, 388–389
  chemical flooding schemes, 389
  chemical/polymer processes, 388–389
  China, 384
  climate change, 386
  completion techniques, 383
  contribution, 388f
  ease of implementation, 385
  gas hydrate, 386
  market attractiveness, 385
  Mexico, 384
  natural cleaning technique, 380
  nitrogen injection project, 382
  oil and gas, 383
  oil sands, 387f
  past and projected emissions, 387f
  pilot projects, 382
  proven reserve, 386f
  robust reservoir-oriented strategy, 382
  scenarios, 380–381
  shale production volumes, 383
  steam flooding, 381–382
  thermal and chemical recovery processes, 380
  tight oil, 381f
  ultra-deep offshore environment, 382
  United States, 384, 388–389
Unconventional gas reservoirs, 51
  challenges, 470–471
  conventional simulation techniques, 475f
  dual porosity, 473
  dynamic reservoir characterization tool, 474–475
  flow equations, 469
  flow regimes, 472–473
  fracture geometries/orientations
    irregular/nonideal, 470
    regular/ideal, 470
    simplification, 471f
  fracture systems, 469–470
  gradient model, 473
  matrix transient linear, 473
  nonplanar and nonorthogonal fractures, 469–470, 472f
  pressure distribution, 472
  production data methods, 474
  pseudosteady state, 472
  slab matrix model, 470–471, 474
  transient model, 473
  Warren Root model, 470–471
Unconventional oils, 51
  definition, 267–268
  and gas production
    aging fields, 273
    average break-even price, 274
    crude oil and natural gas, 271
    electric power consumption, 272
    fossil fuels, 275
    global demand, 274
    inflection point and peak demand, 274
    liquid petroleum and biofuel, 275f
    natural gas plant liquids, 272
    offshore natural gas production, 272
    offshore sector, 274
    oil production, United States, 271f
    onshore tight oil development, 271
    shale and light, tight oil (LTO), 272–273
    two-speed oil market, 274
  types, 267–268
Unconventional reservoirs
  category, 466t
  characteristics, 269–270
  coupled fluid flow and geomechanical stress model, 465–491
  current potentials, 275–337
  custom-designed schemes, 267
  depositional basins, 270
  energy resource, 268

geology and reservoir features, 268–269
global oil demand, 269f
hybrid systems, 270
hydraulic fracturing, 523–530
hydrodynamic effect, 270
improving oil and gas recovery, 339–436
International Energy Agency (IEA) vision, 268
oil and gas production, 271–275
open fracture-related porosity, 270
organic richness and hydrogen content, 270
probability distributions, 464f
quantitative measures, well production variability, 461–465
reserve growth potential, 437–461
shale resource systems, 269–270
sustainability pathways, 492–520
sustainable development, 337–339
total supply, United States, 278f
zero-waste concept, 268
zero-waste operations, 520–523
Underbalanced drilling (UBD), 141–148
Underground refining, 333–337
United States gas hydrate program, 316
United States leadership
  2008 financial crisis, 377
  crude oil and lease condensate proved reserve, 373f
  crude oil production, 372f
  declared reserve, 376f
  enhanced oil recovery (EOR), 370, 371f, 374–376, 378–380, 379f
  gas production history, 373, 374f
  gas reserve-production history, 375f
  heavy oil reservoirs, 366–369
  hydraulic fracturing techniques, 377
  local fluids, 378–380
  natural gas, 373
  offshore oilfields, 374–376
  oil production, 370f
  peak recovery rate, 374f
  proved reserve, 373
  recovery rates decline, 375f
  reserve/production ratio, 366–369
  reserve variation, 374f, 377f
  reservoir/production (R/P) ratio variation, 371–372, 371–372f
  sequential development, 366
  technical recoverability, 378, 378f
  technology development, 364–365
  uneconomical formations, 373
  waterflood schemes, 376–377
Unstable flow modeling
  breakthrough recovery and instability number, 223f
  breakthrough recovery and Peters-Flock stability number, 225f
  capillary number, 222
    vs. residual oil saturation, 223f
    vs. water breakthrough, 228f
  critical gas saturation, 231, 231f
  Darcy velocity, 223–224
  displacement process, 229–230
  forces, petroleum reservoir, 221
  instability number *vs.* breakthrough recovery, 224f
  interfacial tension (IFT), 224, 229f
  microscopic and macroscopic recoveries, 226
  mobility ratio, 221, 225–226, 226f
  N_c *vs.* residual saturation, 222f
  permeability jail, 230f
  pseudocapillary pressure, 228
  pseudostable region, 224
  residual hydrocarbon saturation, 222
  residual saturations, 230
  saturation endpoints, 229
  shock front, 225–226
  stability analysis, 227
  viscous fingering, 227
  water alternating gas (WAG) ratio, 231
  water/gas injection, 221
  wettability, 230
US Department of Energy (DOE), 267–268
US Energy Information Administration's International Energy Outlook 2016, 383–384
US Geological Survey (USGS), 312
US natural gas production, 312f

V
Van der Waal forces, 309
Van Krevelen diagram, 235
Vapor compression cycles, 821
Vapor-extraction (VAPEX), 356, 812
Variation coefficient (VC), 199, 461–464
Vitrinite reflectance, 235
Volcanic rocks, 426–427, 427f
Volterra-type integral, 643

W
Warren Root model, 470–471
Wasatch formation, 461
Waste management, 45–46
Water-alternating gas (WAG), 342, 357–358, 382, 416–417
Water disposal, 312–313
Water-drive reservoir, 622
Water saturation, 235
  dependent parameters, 736
Water vapor absorption, 513f
Wax, 59

Weighted and shifted Grunwald difference (WSGD), 651
Weighted and shifted Lubich difference (WSLD), 651
West Sak Formation, 330
Wet gas, 533
Weyburn CO_2 project, 205, 206f
World energy
  coal, 13
  Covid-19 pandemic, 9–11
  decarbonization scheme, 8–9
  deforestation fires, 11
  demonstrated reserve base, 4–5
  energy information administration, 3–4
  evolving transition, 8
  fossil fuel consumption, 12
  fractions, 12–13, 12f
  fuel consumption, 5
  greenhouse gas, 8–9
  growth, 5–6, 6t, 9
  human civilization, 2–3
  hydro energy, 13–14
  internal combustion engine, 8–9
  nuclear energy, 13–14
  peak oil theory, 6–7
  renewable energy, 7–8, 13–14
  shares, 13, 13t
  surface transport, 9
  U.S. coal resources, 3–4
World reserve
  CIA factbook, 540–542t
  crude oils, 546–548, 546f
    American Petroleum Institute (API) variation, 547f
    quality, 547f
    sulfur content, 546f
  Energy Information Administration (EIA) estimates, 544–545
  enhanced oil recovery, 545–546
  pre-COVID-19 peak production, 545
  proven reserve, 534
  recovery factors, 535–536t
  reservoir/production (R/P) ratios, 543
  Russia, 545
  sustainable recovery techniques, 534
  United States, 546f
WSGD. *See* Weighted and shifted Grunwald difference (WSGD)
WSLD. *See* Weighted and shifted Lubich difference (WSLD)

X

XFEM. *See* Extended finite element method (XFEM)

Y

Young's modulus, 210–212

Z

Zero emissions, 521
Zero use of toxics, 522
Zero-waste
  concept, 268, 818, 849
  design, sustainability analysis, 817–831
  engineering, 1–2
  in administration activities, 521–522
  in product life cycle, 522
  in reservoir management, 523
  of resources, 521
  oil recovery scheme, 524f
Zero-waste operations
  administration activities, 521–522
  aspects, 520
  emissions (air, soil, water, solid waste, hazardous waste), 521
  green inputs and outputs, 520–521
  product life cycle (transportation, use and end of life), 522
  reservoir management, 523
  of resources (energy, material and human), 521
  unconventional reserve growth, 523f
  use of toxics (processes and products), 522
Zipper fracturing, 45
Zuata Sweet, 328

Printed in the United States
by Baker & Taylor Publisher Services